Inquiry into
CHEMISTRY

Authors

Dr. Audrey Chastko
Springbank Community High School
Calgary, Alberta

Jeff Goldie
Strathcona High School
Edmonton, Alberta

Dr. Frank Mustoe
University of Toronto Schools (retired)
Toronto, Ontario

Dr. Ian Phillips
McNally High School
Edmonton, Alberta

Sandy Searle
Western Canada High School
Calgary, Alberta

Toronto   Montréal   Boston   Burr Ridge, IL   Dubuque, IA   Madison, WI   New York   San Francisco
St. Louis   Bangkok   Bogotá   Caracas   Kuala Lumpur   Lisbon   London   Madrid   Mexico City
Milan   New Delhi   Santiago   Seoul   Singapore   Sydney   Taipei

The McGraw-Hill Companies

COPIES OF THIS BOOK
MAY BE OBTAINED BY
CONTACTING:
McGraw-Hill Ryerson Ltd.

WEB SITE:
http://www.mcgrawhill.ca

E-MAIL:
orders@mcgrawhill.ca

TOLL-FREE FAX:
1-800-463-5885

TOLL-FREE CALL:
1-800-565-5758

OR BY MAILING
YOUR ORDER TO:
McGraw-Hill Ryerson
Order Department
300 Water Street
Whitby, ON L1N 9B6

Please quote the ISBN and
title when placing your order.

McGraw-Hill Ryerson
Inquiry Into Chemistry

The information and activities in this textbook have been carefully developed and reviewed by professionals to ensure safety and accuracy. However, the publisher shall not be liable for any damages resulting, in whole or in part, from the reader's use of the material. Although appropriate safety procedures are discussed and highlighted throughout the textbook, the safety of students remains the responsibility of the classroom teacher, the principal, and the school board district.

13-Digit ISBN 978-0-07-096051-0
10-Digit ISBN 0-07-096051-8

www.mcgrawhill.ca

6 7 8 9 10 TCP 18 17 16 15

Printed and bound in Canada

Care has been taken to trace ownership of copyright material contained in this text. The publishers will gladly accept any information that will enable them to rectify any reference or credit in subsequent printings.

SCIENCE PUBLISHER: Keith Owen Richards
PROJECT MANAGER: Susan Girvan
SENIOR DEVELOPMENTAL EDITOR: Lois Edwards
DEVELOPMENTAL EDITORS: Katherine Hamilton, Christy Hayhoe, Sara Goodchild, Leslie Macumber, Natasha Marko, Christine Weber
MANAGER, EDITORIAL SERVICES: Crystal Shortt
SUPERVISING EDITOR: Kristi Clark, Shannon Martin
PHOTO RESEARCH & PERMISSIONS: Pronk&Associates
EDITORIAL ASSISTANTS: Erin Hartley, Michelle Malda
MANAGER, PRODUCTION SERVICES: Yolanda Pigden
PRODUCTION COORDINATOR: Andree Davis
SET-UP PHOTOGRAPHY: Dave Starrett/Pronk&Associates
COVER DESIGN: Pronk&Associates
ART DIRECTION: Pronk&Associates
ELECTRONIC PAGE MAKE-UP: Pronk&Associates
COVER IMAGES: (background), © Jim Richardson/CORBIS; (large inset), © FRED HABEGGER/Grant Heilman Photography; (medium inset), © PETER APRAHAMIAN/SPL/PUBLIPHOTO; (small inset), © LAGUNA DESIGN/SPL/PUBLIPHOTO

Acknowledgements

Producing a textbook of high quality is a true team effort, requiring the input and expertise of a very large number of people. The authors, editorial team, and publishers of this book would like to convey our sincere thanks to the reviewers listed below who provided crucial analyses of our draft manuscript, and often provided reviews of designed pages. Their assistance was invaluable in helping us develop a text that we hope you will find completely appropriate for your teaching and your students' learning. In addition, the comments from those teachers (and their students!) who took part in the field testing of this material were very helpful and much appreciated. We realize that tackling a new curriculum with a text in-progress is a big challenge, and we thank you for being up to the task.

We also thank the following writers who researched and prepared the Special Features in *Inquiry into Chemistry*: Kirsten Craven, Jenna Dunlop, Eric Grace, Ann Heide, and Alexandra Venter. Finally, we thank the talented and dedicated members of the team at Pronk&Associates who did their best no matter what challenges they faced.

Reviewers

Monika Amies
Western Canada High School
Calgary, Alberta

John Callegari
M.E. LaZerte High School
Edmonton, Alberta

Brock Campbell
W.R. Myers High School
Taber, Alberta

Genelee Chiong
St. Mary's High School
Calgary, Alberta

Roger Cowan
Brooks Composite High School
Brooks, Alberta

Leno Delcioppo
Harry Ainlay High School
Sherwood Park, Alberta

Michelle Duke
University of Lethbridge
Lethbridge, Alberta

Ken Ealey
St. Jerome, A Science Academy
Edmonton, Alberta

Sandra Feigel
Henry Wise Wood Senior High School
Calgary, Alberta

Kelty Findlay
Eagle Butte High School
Dunmore, Alberta

Evelyn Fray
Ross Sheppard High School
Edmonton, Alberta

Shaun Grainger
Fort Saskatchewan High School
Fort Saskatchewan, Alberta

Patricia Henderson
Henry Wise Wood Senior High School
Calgary, Alberta

Caroline Heppell
Strathcona Senior High School
Edmonton, Alberta

Dania Hill
Peace River High School
Peace River, Alberta

Jennifer Hopkins
Louise Dean Centre
Calgary, Alberta

Stephen Jeans
University of Calgary
Calgary, Alberta

Sharon LaCour
St. Mary's High School
Calgary, Alberta

Jeff Lailey
Western Canada High School
Calgary, Alberta

Mark Lewis
Centennial High School
Calgary, Alberta

Contents

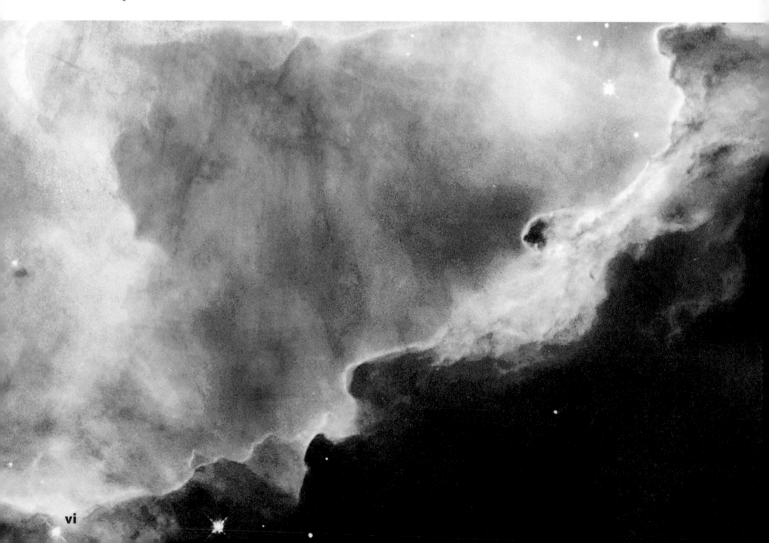

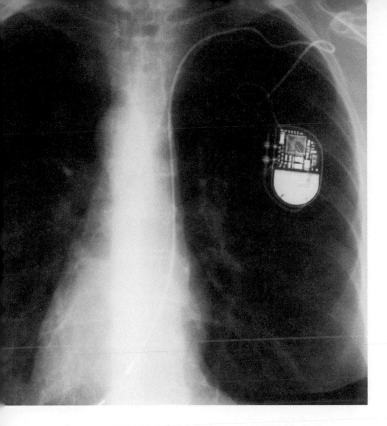

UNIT 6

UNIT 7

CHEMICAL CHANGES OF ORGANIC COMPOUNDS

Safety in Your Chemistry Lab and Classroom

Actively engaging in laboratory investigations is essential to gaining a hands-on understanding of chemistry. Following safe laboratory procedures should not be seen as an inconvenience in your investigations. Instead, it should be seen as a positive way to ensure your safety and the safety of others who share a common working environment. Familiarize yourself with the following general safety rules and procedures. It is your responsibility to follow them when completing any of the investigations in this textbook, or when performing other laboratory procedures.

General Precautions

- Always wear safety glasses and a lab coat or apron in the laboratory. Wear other protective equipment, such as gloves, as directed by your teacher or by the Safety Precautions at the beginning of each investigation.

- If you wear contact lenses, always wear safety goggles or a face shield in the laboratory. Inform your teacher that you wear contact lenses. Generally, contact lenses should not be worn in the laboratory. If possible, wear eyeglasses instead of contact lenses, but remember that eyeglasses are not a substitute for proper eye protection.

- Know the location and proper use of the nearest fire extinguisher, fire blanket, fire alarm, first aid kit, and eyewash station (if available). Read "Fire Safety" on the next page, and discuss with your teacher what type of fire-fighting equipment should be used on particular types of fires.

- Do not wear loose clothing in the laboratory. Do not wear open-toed shoes or sandals. Accessories may get caught on equipment or present a hazard when working with a Bunsen burner. Ties, scarves, long necklaces, and dangling earrings should be removed before starting an investigation.

- Tie back long hair and any loose clothing before starting an investigation.

- Lighters and matches must not be brought into the laboratory.

- Food, drinks, and gum must not be brought into the laboratory.

- Inform your teacher if you have any allergies, medical conditions, or physical problems (including hearing impairment) that could affect your work in the laboratory.

Before Beginning Laboratory Investigations

- Listen carefully to the instructions that your teacher gives you. Do not begin work until your teacher has finished giving instructions.

- Obtain your teacher's approval before beginning any investigation that you have designed yourself.

- Read through all of the steps in the investigation before beginning. If there are any steps that you do not understand, ask your teacher for help.

- Be sure to read and understand the Safety Precautions at the start of each investigation.

- Always wear appropriate protective clothing and equipment, as directed by your teacher and the Safety Precautions.

- Be sure that you understand all safety labels on materials and equipment. Familiarize yourself with the WHMIS symbols in this section.

- Make sure that your work area is clean and dry.

During Laboratory Investigations

- Make sure that you understand and follow the safety procedures for different types of laboratory equipment. Do not hesitate to ask your teacher for clarification if necessary.

- Never work alone in the laboratory.

- Remember that gestures or movements that may seem harmless could have dangerous consequences in the laboratory. For example, tapping people lightly on the shoulders to get their attention could startle them. If they are holding a beaker that contains an acid, for example, the results could be very serious.

- Make an effort to work slowly and steadily in the laboratory. Be sure to make room for other students.

- Organize materials and equipment neatly and logically. For example, do not place materials that you will need during an investigation on the other side of a Bunsen burner from you. Keep your bags and books off your work surface and out of the way.

- Never taste any substances in the laboratory.

- Never touch a chemical with your bare hands.

- Never draw liquids or any other substances into a pipette or a tube with your mouth.

- If you are asked to smell a substance, do not hold it directly under your nose. Keep the object at least 20 cm away, and waft the fumes toward your nostrils with your hand.

- Label all containers holding chemicals. Do not use chemicals from unlabelled containers.

- Hold containers away from your face when pouring liquids or mixing reactants.

- If any part of your body comes in contact with a potentially dangerous substance, wash the area immediately and thoroughly with water.

- If you get any material in your eyes, do not touch them. Wash your eyes immediately and continuously for 15 minutes, and make sure that your teacher is informed. A doctor should examine any eye injury. If you wear contact lenses, take them out immediately. Failing to do so may result in material becoming trapped behind the contact lenses. Flush your eyes with water for 15 minutes, as above.

- Do not touch your face or eyes while in the laboratory unless you have first washed your hands.

- Do not look directly into a test tube, flask, or the barrel of a Bunsen burner.

- If your clothing catches fire, smother it with the fire blanket or with a coat, or get under the safety shower.

- If you see any of your classmates jeopardizing their safety or the safety of others, let your teacher know.

Heat Source Safety

- When heating any item, wear safety glasses, heat-resistant safety gloves, and any other safety equipment that your teacher or the Safety Precautions suggests.

- Always use heat-proof, intact containers. Check that there are no large or small cracks in beakers or flasks.

- Never point the open end of a container that is being heated at yourself or others.

- Do not allow a container to boil dry unless specifically instructed to do so.

- Handle hot objects carefully. Be especially careful with a hot plate that may look as though it has cooled down, or glassware that has recently been heated.

- Before using a Bunsen burner, make sure that you understand how to light and operate it safely. Always pick it up by the base. Never leave a Bunsen burner unattended.

- Before lighting a Bunsen burner, make sure there are no flammable solvents nearby.

- If you do receive a burn, run cold water over the burned area immediately. Make sure that your teacher is notified.

- When you are heating a test tube, always slant it. The mouth of the test tube should point away from you and from others.

- Remember that cold objects can also harm you. Wear appropriate gloves when handling an extremely cold object.

Electrical Equipment Safety

- Ensure that the work area, and the area of the socket, is dry.

- Make sure that your hands are dry when touching electrical cords, plugs, sockets, or equipment.

- When unplugging electrical equipment, do not pull the cord. Grasp the plug firmly at the socket and pull gently.

- Place electrical cords in places where people will not trip over them.

- Use an appropriate length of cord for your needs. Cords that are too short may be stretched in unsafe ways. Cords that are too long may tangle or trip people.

- Never use water to fight an electrical equipment fire. Severe electrical shock may result. Use a carbon dioxide or dry chemical fire extinguisher. (See "Fire Safety" on the next page.)

- Report any damaged equipment or frayed cords to your teacher.

Glassware and Sharp Objects Safety

- Cuts or scratches in the chemistry laboratory should receive immediate medical attention, no matter how minor they seem. Alert your teacher immediately.

- Never use your hands to pick up broken glass. Use a broom and dustpan. Dispose of broken glass as directed by your teacher. Do not put broken glassware into the garbage can.

- Cut away from yourself and others when using a knife or another sharp object.

- Always keep the pointed end of scissors and other sharp objects pointed away from yourself and others when walking.

- Do not use broken or chipped glassware. Report damaged equipment to your teacher.

Fire Safety

- Know the location and proper use of the nearest fire extinguisher, fire blanket, and fire alarm.

- Understand what type of fire extinguisher you have in the laboratory, and what type of fires it can be used on. (See below.) Most fire extinguishers are the ABC type.

- Notify your teacher immediately about any fires or combustible hazards.

- Water should only be used on Class A fires. Class A fires involve ordinary flammable materials, such as paper and clothing. *Never use water* to fight an electrical fire, a fire that involves flammable liquids (such as gasoline), or a fire that involves burning metals (such as potassium or magnesium).

- Fires that involve a flammable liquid, such as gasoline or alcohol (Class B fires) must be extinguished with a dry chemical or carbon dioxide fire extinguisher.

- Live electrical equipment fires (Class C) must be extinguished with a dry chemical or carbon dioxide fire extinguisher. Fighting electrical equipment fires with water can cause severe electric shock.

- Class D fires involve burning metals, such as potassium and magnesium. A Class D fire should be extinguished by smothering it with sand or salt. Adding water to a metal fire can cause a violent chemical reaction.

- If someone's hair or clothes catch on fire, smother the flames with a fire blanket. Do not discharge a fire extinguisher at someone's head.

Clean-Up and Disposal in the Laboratory

- Clean up all spills immediately. Always inform your teacher about spills.

- If you spill acid or base on your skin or clothing, wash the area immediately with a lot of cool water.

- You can neutralize small spills of acid solutions with sodium hydrogen carbonate (baking soda). You can neutralize small spills of basic solutions with sodium hydrogen sulfate or citric acid.

- Clean equipment before putting it away, as directed by your teacher.

- Dispose of materials as directed by your teacher, in accordance with your local School Board's policies. Do not dispose of materials in a sink or a drain unless your teacher directs you to do so.

- Wash your hands thoroughly after all laboratory investigations.

The following Safety Precautions symbols appear throughout *Inquiry into Chemistry*, whenever an investigation presents possible hazards.

Safety Symbols

	appears when there is a danger to the eyes, and safety goggles, safety glasses, or a face shield should be worn
	appears when substances that could burn or stain clothing are used
	appears when objects that are hot or cold must be handled
	appears when sharp objects are used, to warn of the danger of cuts and punctures
	appears when toxic substances that can cause harm through ingestion, inhalation, or skin absorption are used
	appears when corrosive substances, such as acids and bases, that can damage tissue are used
	warns of caustic substances that could irritate the skin
	appears when chemicals or chemical reactions that could cause dangerous fumes are used and ventilation is required
	appears as a reminder to be careful when you are around open flames and when you are using easily flammable or combustible materials
	warns of danger of electrical shock or burns from live electrical equipment

WHMIS Symbols

Symbol	Description
	Poisonous and Infectious Material Causing Immediate and Serious Toxic Effects
	Poisonous and Infectious Material Causing Other Toxic Effects
	Flammable and Combustible Material
	Compressed Gas
	Corrosive Material
	Oxidizing Material
	Dangerously Reactive Material
	Biohazardous Infectious Material

Safety in Your Online Activities

The Internet is like any other resource you use for research—you should confirm the source of the information and the credentials of those supplying it to make sure the information is credible before you use it in your work.

Unlike other resources, however, the Internet has some unique pitfalls you should be aware of, and practices you should follow.

- It's easy to waste a lot of time following links that "look interesting" long after you've found the information you need. Take advantage of the online links provided at *www.albertachemistry.ca* to use your Internet research time efficiently. Develop your Internet discipline early: focus on what you need to know, find it, and log off.

- Online content is constantly changing. If you find some useful information once, there's no guarantee that it will be there when you go back to look for it. You may want to print it in order to have a permanent record. Always include the source and date of the information you're saving.

- When you copy or save something from the Internet, you could be saving more than information. Be aware that information you pick up could also include hidden, malicious software code (known as "worms" or "Trojans") that could damage your system or destroy data.

- It's easy to find your way into sites that are considered to be "off limits" by teachers and parents. Why are they judged this way? They are off-limits because they contain material that is disturbing, illegal, harmful, and/or was created by exploiting others. There are rules about what is acceptable in print and on the airwaves; they apply to Internet material as well. Also be aware that these site visits can come back to "haunt" you if you pick up "cookies" (electronic tags), that identify your computer as a target for more of the same.

- *Never, ever* give out personal information online. This includes your name, your age, your gender, your email address, street address, phone number, or your picture. Protect your privacy, even if it means not registering to use a site that looks helpful. Discuss ways to use the site while protecting your privacy with your teacher. There may be a way to access it through the school or the school library.

- Report any online content or activity that you suspect is illegal to your teacher. This can include online hate, harassment, cyberstalking, cyberbullying, or attempts to lure you into a face-to-face meeting with a stranger; dangerous activities concerning terrorism or illegal weapons; or physical threats. Discuss ways to deal with such material with your teacher; report it to the Internet Supervisor at your school, and find out what the school policy is for dealing with such material.

With your teacher and fellow students, discuss ways to apply critical thinking to online research and develop safe Internet practices.

Introducing *Inquiry into Chemistry*

Chemistry—the study of the physical and chemical properties of substances— has long been a part of your science education. *Inquiry into Chemistry* will take you to a deeper level of understanding, helping you build on your knowledge of these properties, in particular, how these properties are determined by the chemical bonds between atoms in molecules, and by the forces of attraction between molecules.

Introducing Your Inquiry

Start with the **Unit Opener** to get a sense of the topics you are about to investigate. Try to answer the **Focussing Questions** based on what you know now. Before you begin a unit, log on to *www.albertachemistry.ca* to take the **Unit Prequiz.** You can use the quiz to find out what you recall from earlier science studies—results are available at the click of a button. If you need to refresh your memory, turn to the **Unit Preparation** feature.

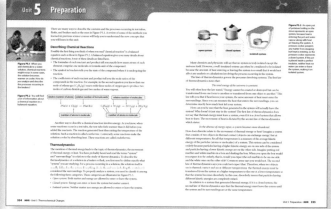

Using *Inquiry into Chemistry*

Each chapter opens with a list of **Chapter Concepts,** giving you an overview of key points and how they relate to one another. The opening text and photograph will give you the context, and you will use the **Launch Lab** to do or discuss an idea that will be explored further in the chapter.

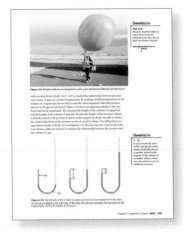

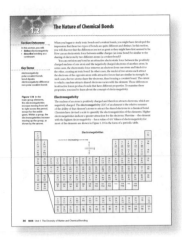

Each section opens with **Section Outcomes** that outline the knowledge and skills you will develop in each section. The list of **Key Terms** highlights chemistry vocabulary you will be using. Each key term appears in boldface type in the section, where it is explained in context, and it is defined again in the **Glossary** at the back of the book.

Watch for the **ChemistryFile** feature in the margins of many pages. **FYI** gives you instant facts and fascinating tidbits. **Try This** challenges you to test an idea. **Web Link** connects you to the *Inquiry into Chemistry* web site, where you can find out more about a topic of interest.

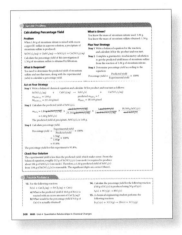

Sample Problems demonstrate techniques and procedures for solving equations and math problems. **Practice Problems** are supplied so you can try your hand, with **Problem Tips** and **Math Tips** available to help you get the job done.

The variety of **Investigations** presented may be directing you to conduct an inquiry to confirm results of experiments done by others, design a procedure to conduct your own investigation, or hone your skills in decision-making by gathering data and evaluating evidence in order to make decisions and solve practical problems. **Thought Labs** guide you in analyzing data or researching information to look for patterns, form opinions, and evaluate points of view.

Check out the back of the book for **Appendices** that detail basic scientific practices and procedures—from how to design and report on investigations, how to titrate, and how to prepare and dilute solutions to a quick review of math related to chemistry and tips for writing Diploma Exams. You will also find answers to practice problems and supplementary practice problems in the appendices. The **Glossary** and the **Index** can also aid your studies.

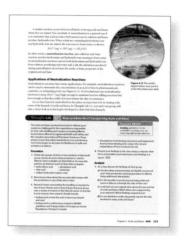

Real-world Applications

The **Connections** feature in each chapter spotlights a specific idea, technology, or issue that touches your life. Find out, for example, how drain cleaners, hair care products, and barometers work or how food fuels you and how trans fats can harm you. Or you can tackle the challenges of car pollution and home heating, or get oil out of the tar sands.

The **Career Focus** at the end of each unit features an interview with someone whose work in a particular field of chemistry may inspire your own career aspirations. You will also find a sampling of related careers that draw upon other fields within and beyond those of chemistry.

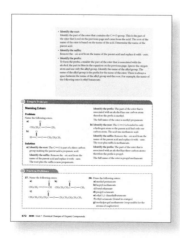

Assessing Your Learning

Use the following opportunities to pause and reflect on your learning.

- **Questions for Comprehension** ("Q questions") check basic understanding of concepts.

- The **Section Summary** lists key points.
- The **Section Review** helps you gauge your understanding of essential knowledge and applications.

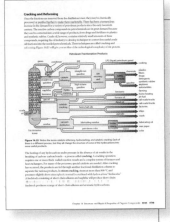

- The **Chapter Summary** and **Chapter Review** broaden your focus and encourage you to apply your knowledge to different situations and contexts.

- The **Unit Review** enables you to reflect on, consolidate, and apply your learning of the entire unit.

Inquiring Further at *Inquiry into Chemistry* Online

In addition to the Unit Prequizzes and Web Links, *Inquiry into Chemistry* Online at *www.albertachemistry.ca* highlights and reinforces key points from each chapter. The Electronic Study Partner aids and reinforces your understanding of key concepts and skills. Your web resource is also packed with study tips, strategies, research tools, and opportunities to extend your learning in many new directions.

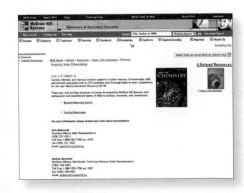

UNIT 1

The Diversity of Matter and Chemical Bonding

General Outcomes

- describe the role of modelling, evidence, and theory used in explaining and understanding the structure, chemical bonding, and properties of ionic compounds

- describe the role of modelling, evidence, and theory used in explaining and understanding the structure, chemical bonding, and properties of molecular substances

Unit Contents

Focussing Questions

1. Why do some substances dissolve easily, while others do not?

2. Why do different substances have different boiling and melting points and different enthalpies of fusion and vaporization?

3. How can models increase understanding of bonding?

Unit PreQuiz ⑦
www.albertachemistry.ca

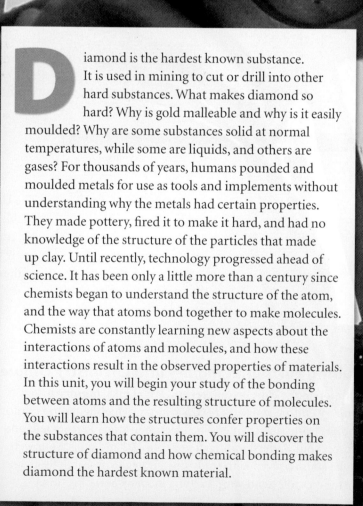

Diamond is the hardest known substance. It is used in mining to cut or drill into other hard substances. What makes diamond so hard? Why is gold malleable and why is it easily moulded? Why are some substances solid at normal temperatures, while some are liquids, and others are gases? For thousands of years, humans pounded and moulded metals for use as tools and implements without understanding why the metals had certain properties. They made pottery, fired it to make it hard, and had no knowledge of the structure of the particles that made up clay. Until recently, technology progressed ahead of science. It has been only a little more than a century since chemists began to understand the structure of the atom, and the way that atoms bond together to make molecules. Chemists are constantly learning new aspects about the interactions of atoms and molecules, and how these interactions result in the observed properties of materials. In this unit, you will begin your study of the bonding between atoms and the resulting structure of molecules. You will learn how the structures confer properties on the substances that contain them. You will discover the structure of diamond and how chemical bonding makes diamond the hardest known material.

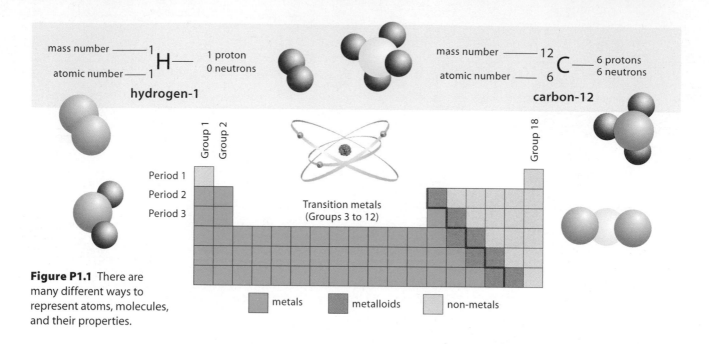

Figure P1.1 There are many different ways to represent atoms, molecules, and their properties.

How much do you remember about the topics in chemistry that you studied in previous science courses? Does Figure P1.1 give you any hints? Chemical bonding, the subject presented in Unit 1, builds on several of the concepts that you encountered previously. The following pages will help refresh your memory of the concepts on which Unit 1 relies.

Atoms, Elements, and Isotopes

According to modern theory, atoms consist of a tiny, positively charged nucleus surrounded by a "cloud" of negatively charged electrons (see Figure P1.2). The nucleus contains nearly all of the mass of the atom in the form of protons and neutrons. The protons carry the positive charge while the neutrons have no charge. Although the total mass of the electrons is much smaller than the mass of the nucleus, the electron cloud is much larger than the nucleus. Electrons are negatively charged, thus, in a neutral atom, the number of electrons is equal to the number of protons.

The chemical nature of a given element is determined by the number of protons in the nucleus. Also called the *atomic number*, the number of protons is symbolized by Z. The number of neutrons (N) in the nucleus of a given element can vary, but it does not affect the chemical properties of the atom. Atoms of the same element that have different numbers of neutrons are called *isotopes*. Isotopes are represented by the chemical symbol for the element, the atomic number (Z), and the *mass number* (A) which is the sum of the numbers of protons and neutrons (A = Z + N). Each isotope can be uniquely represented as shown in Figure P1.3 where X represents the chemical symbol for the element.

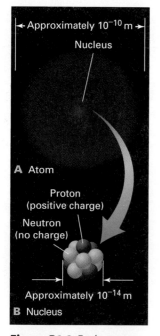

Figure P1.2 Each proton and neutron in the nucleus has nearly 2000 times the mass of an electron, yet they occupy a very small part of the atom.

Consider, for example, the isotope of potassium that has 21 neutrons. The chemical symbol for potassium is K. Potassium has 19 protons so the atomic number is 19. The mass number is the sum of the protons and neutrons or A = Z + N. Thus, for potassium, A = 19 + 21 = 40. Therefore, the complete symbol for the isotope is $^{40}_{19}K$. You would refer to this isotope as "potassium-40."

Periodic Table

As chemists began to discover new elements and observe their properties, a pattern emerged. They found elements of widely differing atomic numbers that had similar chemical properties. In 1869, Dmitri Mendeleev (1834–1907) published his organization of all of the elements that had been discovered at that time. He placed them on a chart according to increasing atomic number in one direction and similar chemical characteristics in the perpendicular direction. Where gaps existed in his chart, Mendeleev correctly predicted the existence of elements yet to be discovered. Mendeleev's chart was the predecessor of the modern periodic table (shown in Appendix C). The horizontal rows contain elements that increase in atomic number by one. The rows start over when the chemical properties of the next element are similar to those of the elements in the first column. As shown in the outline form of a periodic table in Figure P1.4, the rows are called *periods* and the columns are called *groups*. The periods (horizontal) contain the elements in sequential order of atomic number or number of protons. The elements within a group (vertical) have similar chemical properties.

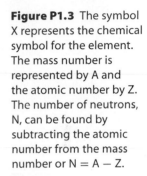

Figure P1.3 The symbol X represents the chemical symbol for the element. The mass number is represented by A and the atomic number by Z. The number of neutrons, N, can be found by subtracting the atomic number from the mass number or N = A − Z.

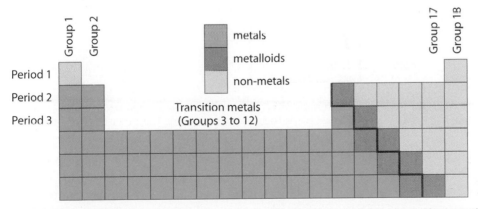

Figure P1.4 To identify the element that fits in each square, turn to Appendix C. Also note that another set of elements, called the inner transition elements, are included on the periodic table in Appendix C, but are not included in this simplified diagram.

The elements in Group 1 are called the *alkali metals*. These metals are highly reactive. In fact, they react explosively with water. The metals in Group 2 are called the *alkaline earth metals*. They are not as reactive as the alkali metals, but are nevertheless quite reactive. The non-metals in Group 17 are called the *halogens* and are quite reactive. The non-metals in Group 18 are the *noble gases*. They are almost completely unreactive.

Chemistry File

Try This

Use the periodic table to answer the following.

- Radioactive strontium is harmful because it tends to be incorporated into bones and remains there for long periods of time. Explain why.

- Metallic sodium reacts violently with water and forms sodium hydroxide and hydrogen gas. What happens when metallic potassium is placed in water?

Table P1.1 lists some of the characteristics of three major classes of elements, the metals, the metalloids, and the non-metals. The colours in the table designate these major classes.

Table P1.1 Properties of Three Classes of Elements

Class	State (at room temperature)	Appearance	Conductivity	Malleability and ductility
Metals	• solid except mercury, which is liquid	• shiny lustre	• good conductors of heat and electric current	• malleable • ductile
Metalloids	• solids	• some are shiny while others are dull	• some conduct electric current somewhat • poor conductors of heat	• brittle • not ductile
Non-metals	• some are gases • some are solids • one liquid (bromine)	• not shiny	• poorer conductors of heat and electric current	• brittle • not ductile

Each square in the periodic table in Appendix C contains a large amount of information about each element. The example in Figure P1.5 details the type of information you will find.

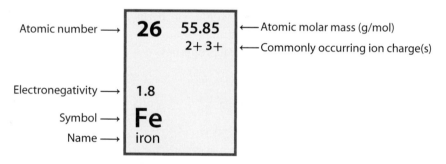

Figure P1.5 label pointers: Atomic number → 26; Atomic molar mass (g/mol) → 55.85; Commonly occurring ion charge(s) → 2+ 3+; Electronegativity → 1.8; Symbol → Fe; Name → iron

Figure P1.5 The periodic table in Appendix C provides this type of information about each element.

Most of the labels in Figure P1.5 are self-explanatory, except for electronegativity and possibly atomic molar mass. You will learn about electronegativity in Chapter 1. The *atomic molar mass* (mass of one mole of atoms of the element) is a weighted average of the molar masses of the naturally occurring isotopes of the element. The molar masses of individual isotopes can be very accurately determined with an instrument called a mass spectrometer. These masses are then averaged. For example, the two naturally occurring isotopes of chlorine are chlorine-35 and chlorine-37. The data for these isotopes are shown in Table P1.2 and are used to calculate the weighted average by the method shown on the next page. You will find this average value on your periodic table.

Table P1.2 Data for Molar Mass of Chlorine

Isotope	Molar mass	Abundance
$^{35}_{17}\text{Cl}$	34.96885 g/mol	75.77%
$^{37}_{17}\text{Cl}$	36.96590 g/mol	24.23%

$$\frac{\text{percent abundance}}{100\%} \times \text{molar mass of isotope} = \text{contribution}$$

$$(^{35}_{17}\text{Cl}) \ \frac{75.77\%}{100\%} \times 34.96885 \ \frac{\text{g}}{\text{mol}} = 26.49590 \ \frac{\text{g}}{\text{mol}}$$

$$(^{37}_{17}\text{Cl}) \ \frac{24.23\%}{100\%} \times 36.96590 \ \frac{\text{g}}{\text{mol}} = 8.956838 \ \frac{\text{g}}{\text{mol}}$$

$$\text{contribution by } (^{35}_{17}\text{Cl}) + \text{contribution by } (^{37}_{17}\text{Cl}) = \text{molar mass of chlorine}$$

$$8.956848 \ \frac{\text{g}}{\text{mol}} + 26.49590 \ \frac{\text{g}}{\text{mol}} = 35.452738 \ \frac{\text{g}}{\text{mol}}$$

$$\approx 35.45 \ \frac{\text{g}}{\text{mol}}$$

• • •

1 Copy the following table into your notebook. Fill in all of the missing information. Use a periodic table to find the missing data.

Chemical notation	Element	Atomic number (Z)	Mass number (A)	Number of protons	Number of neutrons
$^{11}_{5}\text{B}$					
$^{208}_{82}\text{Pb}$					
	tungsten				110
	helium				2
$^{239}_{94}\text{Pu}$					
$^{56}_{26}\square$					
	bismuth				126
				47	60
$^{20}_{10}\square$					
			108	47	
				33	42
		35			45
		79	197		
				50	69

2 List the following elements with the help of a periodic table:

a) noble gases

b) alkaline earth metals

c) halogens

d) alkali metals

• • •

Electron Energy Levels

According to the modern (quantum mechanical) theory of the atom, electrons exist in a cloud around the nucleus. Nevertheless, the motion and energy of the electrons are not random. Electrons can have only specific amounts of energy. They are said to exist in allowed energy levels. Only two electrons can occupy the first energy level.

Eight electrons are allowed in the second energy level. No more than eight electrons exist in the outer (highest) energy level of any atom. The condensed periodic table in Figure P1.6 shows the number of electrons occupying each energy level for the first 20 elements.

	Energy Level	Group 1	2	13	14	15	16	17	18
Period 1	1st	1 H							2 He
Period 2	2nd 1st	1 2 Li	2 2 Be	3 2 B	4 2 C	5 2 N	6 2 O	7 2 F	8 2 Ne
Period 3	3rd 2nd 1st	1 8 2 Na	2 8 2 Mg	3 8 2 Al	4 8 2 Si	5 8 2 P	6 8 2 S	7 8 2 Cl	8 8 2 Ar
Period 4	4th 3rd 2nd 1st	1 8 8 2 K	2 8 8 2 Ca						

Figure P1.6 For the first 20 elements, the number of electrons in the outer energy level is the same as the number of the column from one to eight. This is not the same as the group number because the elements above number 20 create more groups.

Examination of Figure P1.6 reveals several significant patterns. The number of the period (row) is the same number as the highest occupied energy level. Because the first energy level can have only two electrons, there are only two elements in the first period. Also, with the exception of helium in Period 1, the elements in a given group (column) have the same number of electrons in their outer energy level. It is these electrons that participate in chemical reactions and, therefore, determine the chemical properties of the element. For this reason, the electrons in the outer energy level are given a special name—*valence electrons*.

Electron Dot Diagrams

Electron dot diagrams are excellent models for visualizing valence electrons in atoms as they interact in chemical reactions. An electron dot diagram consists of the chemical symbol for the element with dots representing the valence electrons. The electrons in the lower, filled energy levels are not shown in these diagrams. When drawing the dots (electrons), you usually place the first dot at the top of the symbol and then add dots by moving clockwise, spacing the dots equal distances apart. For elements that have more than four valence electrons, you place the fifth dot beside the first at the top of the symbol, as shown in Figure P1.7.

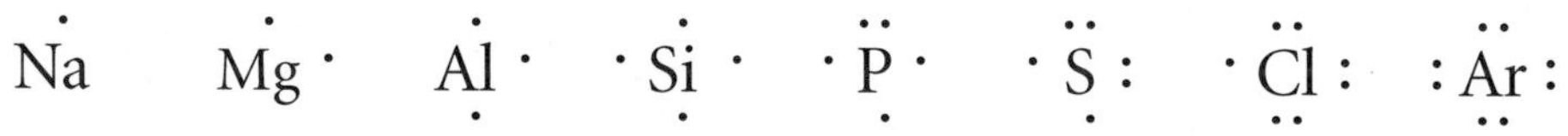

Figure P1.7 These electron dot diagrams of the Period 3 elements show you how dots are added as the number of valence electrons increases.

Forming Bonds

As you read above, no more than eight electrons exist in the outer energy level of any atom. The outer energy level is said to be "filled" when it has eight electrons. In addition, when bonds form between atoms, the outer energy level usually becomes filled. The *octet rule* states that when bonds form between atoms, the atoms gain, lose, or share electrons in such a way that they create outer energy levels with eight electrons. Although the octet rule is helpful for understanding bond formation and making predictions, it should not be considered to be a law of nature. It is possible for bonds to form that result in one atom having more than eight electrons in its outer energy level. You will not encounter such bonds in this textbook.

An *ionic bond* is the electrostatic attraction between two oppositely charged ions. Thus, to form an ionic bond, one atom must lose one or more electrons and another atom must gain one or more electrons. Metal atoms tend to loose their valence electrons and become positively charged ions. When the valance electrons are gone, the filled energy level below the valence level becomes the filled outer level. Non-metals gain electrons and become negatively charged ions. They gain just enough electrons to fill their outer (valence) energy level. Ionic compounds must be neutrally charged so the total number of electrons lost by the metal atoms must be equal to the total number of electrons gained by the non-metal atoms. Figure P1.8 shows two examples of the formation of ionic bonds. To help you visualize the exchange of electrons, the electrons on the metal atoms are drawn as open circles instead of dots.

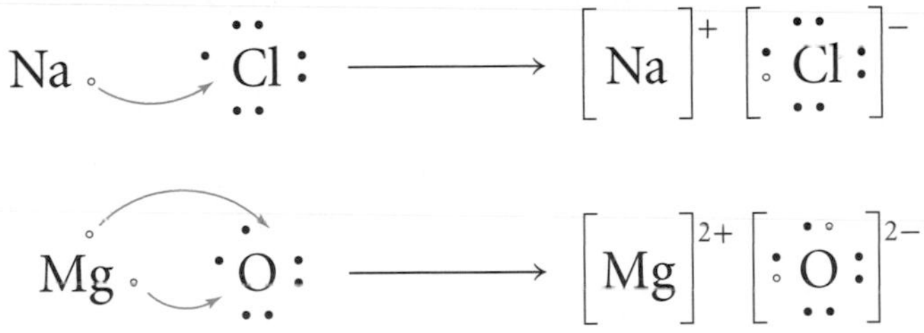

$$Na\,\overset{\circ}{\underset{\frown}{\quad}}\,{:}\overset{..}{\underset{..}{Cl}}{:} \longrightarrow \left[Na\right]^{+}\left[{:}\overset{..}{\underset{..}{Cl}}{:}\right]^{-}$$

$$Mg\,\overset{\circ}{\underset{\frown}{\quad}}\,{\cdot}\overset{..}{O}{:} \longrightarrow \left[Mg\right]^{2+}\left[{:}\overset{.\,\circ}{\underset{..}{O}}{:}\right]^{2-}$$

Figure P1.8 When metal atoms such as sodium and magnesium lose electrons, they have no valence electrons remaining. Therefore, there are no dots around the symbol for the metal ion.

A *covalent bond* forms when two non-metal atoms share one or more pairs of electrons. Including the shared electrons, both atoms achieve a filled outer energy level. The electron dot diagrams in Figure P1.9 show the individual atoms without filled outer energy levels and then the sharing of electrons. The third row in each diagram has circles around the electrons that "belong" to each atom. The electrons in the overlapping parts of the circles are the shared electrons that constitute a covalent bond.

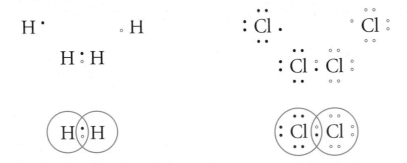

Figure P1.9 The shared electrons in a covalent bond "belong" to both atoms when you are counting electrons in the outer energy level.

At first glance, you might think that the hydrogen molecule is not obeying the "octet" rule because there are only two electrons in the outer energy level of the atoms. However, only the first energy level in hydrogen contains electrons and two electrons fill this energy level. So the principle of completing a filled outer energy level is still followed.

Table P1.3 Names of Some Common Non-metal Ions

Formula	Name
F^-	fluoride
Cl^-	chloride
Br^-	bromide
I^-	iodide
O^{2-}	oxide
S^{2-}	sulfide
N^{3-}	nitride

Names and Formulas for Binary Compounds

Ionic Compounds

A *binary ionic compound* is an ionic compound that consists of two different elements. The rules for naming binary ionic compounds are as follows.

- The name of the metal ion is first, followed by the name of the non-metal (negative) ion.
- The name of the metal ion is the same as the name of the metal atom.
- For transition metals that can have more than one possible charge, the magnitude of the charge is indicated by roman numerals in round brackets after the name.
- To name the non-metal ion, change the ending of the name to "*-ide*."

The names of several common non-metal ions are listed in Table P1.3

For example, when sodium atoms and chlorine atoms ionize and form an ionic compound, the name is sodium chloride. When magnesium atoms and oxygen atoms ionize and form an ionic compound, the name is magnesium oxide. When iron atoms form ions with two positive charges and oxygen atoms ionize, they form an ionic compound called iron(II) oxide. The name does not need to indicate the number of each of the ions involved, because when you know which ions are involved, the charge tells you how many of each ion are necessary to create a net neutral charge for the compound.

The chemical formulas for ionic compounds indicate the number of metal and non-metal ions in the compound by using subscripts on the chemical symbols. When a symbol has no subscript, it is understood that the number is one. Table P1.4 shows the electron dot diagrams, names, and chemical formulas for several compounds. Note that the subscripts on the chemical formulas show the number of ions of each charge that are necessary to make the compound neutral.

Table P1.4 Sample Structures, Names, and Formulas

Electron dot diagram	$\left[Na \right]^+ \left[\ddot{\underset{\cdot\cdot}{Cl}} \right]^-$	$\left[Mg \right]^{2+} \left[\ddot{\underset{\cdot\cdot}{O}} \right]^{2-}$	$\left[Ca \right]^{2+} \begin{matrix} \left[\ddot{\underset{\cdot\cdot}{F}} \right]^- \\ \left[\ddot{\underset{\cdot\cdot}{F}} \right]^- \end{matrix}$	$\begin{matrix} \left[K \right]^+ \\ \left[K \right]^+ \end{matrix} \left[\ddot{\underset{\cdot\cdot}{S}} \right]^{2-}$	$\begin{matrix} \left[Fe \right]^{3+} \\ \left[Fe \right]^{3+} \end{matrix} \begin{matrix} \left[\ddot{\underset{\cdot\cdot}{O}} \right]^{2-} \\ \left[\ddot{\underset{\cdot\cdot}{O}} \right]^{2-} \\ \left[\ddot{\underset{\cdot\cdot}{O}} \right]^{2-} \end{matrix}$
Name	sodium chloride	magnesium oxide	calcium fluoride	potassium sulfide	iron(III) oxide
Chemical formula	NaCl	MgO	CaF_2	K_2S	Fe_2O_3

Covalently Bonded Compounds

Naming binary, covalently bonded compounds requires more detail because non-metals can form a greater variety of compounds. For example, nitrogen and oxygen can form several different combinations, such as one nitrogen atom and one oxygen atom, $NO(g)$; two nitrogen atoms and one oxygen atom, $N_2O(g)$; and one nitrogen atom and two oxygen atoms, $NO_2(g)$. If you heard the name "nitrogen oxide," you would have no way of knowing which compound was being named. The formula for the compound will be clearly described by the name if you apply the following rules.

- The non-metal with the lowest group number (to the left on the periodic table) is named first.
- The element with the highest group number (to the right on the periodic table) is named second.
- The name of the first non-metal is unchanged.
- The name of the second element is formed by using the root name of the element and adding the suffix "-ide."
- An exception to the above rules occurs when oxygen is combined with a halogen. In this case, the halogen with the higher group number is named first.
- If the number of atoms of the first name is greater than one, it has a prefix that indicates the number.
- The second name always has a prefix indicating the number of atoms of that element in the compound.
- Compounds that contain carbon and hydrogen are in a class of their own (hydrocarbons) and are named according to the organic naming system.

Prefixes for binary molecular compounds are listed in Table P1.5.

For example, the names of the combinations of nitrogen and oxygen mentioned on the previous page are nitrogen monoxide, $NO(g)$; dinitrogen monoxide, $N_2O(g)$; and nitrogen dioxide $NO_2(g)$. Nitrogen monoxide and nitrogen dioxide are produced in automobile engines and are components of photochemical smog. Dinitrogen monoxide is also called laughing gas because it has an intoxicating effect. It is sometimes used as an anaesthetic in dentistry.

Formulas for covalently bonded compounds specify the number of atoms of each element in a unit or molecule. Once again, molecules consisting of nitrogen and oxygen provide excellent examples. The molecules of nitrogen dioxide, $NO_2(g)$, consist of one nitrogen atom and two oxygen atoms. Molecules of dinitrogen tetroxide, $N_2O_4(\ell)$, consist of two nitrogen atoms and four oxygen atoms. The ratio of atoms in both compounds is one nitrogen atom to two oxygen atoms, yet they are different compounds. Under standard conditions, nitrogen dioxide is a gas and dinitrogen tetroxide is a liquid.

The rules for writing chemical formulas for binary covalent compounds (compounds that contain atoms of two different non-metal elements) are essentially the same as those for writing the names.

- The symbol for the non-metal with the lowest group number (to the left on the periodic table) is written first.
- If the two non-metals are in the same group, the one in the larger period or with the larger atomic number (below on the periodic table) is named first.
- The symbol for the non-metals with the highest group number (to the right on the periodic table) is written second.
- An exception to the above rules occurs when oxygen is combined with a halogen. In this case, the halogen is written first.
- The number of atoms of each element is written as a subscript on the symbol. If the number is one, it is understood and is not written.

Table P1.5 Prefixes for Binary Molecular Compounds

Number	Prefix
1	mono-
2	di-
3	tri-
4	tetra-
5	penta-
6	hexa-
7	hepta-
8	octa-
9	nona-
10	deca-

Ionic Compounds with Polyatomic Ions

In many ionic compounds, one or both of the ions is a *polyatomic ion*—an ion consisting of several atoms that are covalently bound together and are charged. The polyatomic ion functions as a unit in these compounds. The names and structures of some common polyatomic ions are listed in Table P1.6.

Table P1.6 Some Common Polyatomic Ions

Name	Structure	Name	Structure
ethanoate (acetate)	CH_3COO^-	nitrate	NO_3^-
ammonium	NH_4^+	nitrite	NO_2^-
benzoate	$C_6H_5COO^-$	oxalate	$OOCCOO^{2-}$
borate	BO_3^{3-}	hydrogen oxalate	$HOOCCOO^-$
carbonate	CO_3^{2-}	permanganate	MnO_4^-
hydrogen carbonate (bicarbonate)	HCO_3^-	phosphate	PO_4^{3-}
perchlorate	ClO_4^-	hydrogen phosphate	HPO_4^{2-}
chlorate	ClO_3^-	dihydrogen phosphate	$H_2PO_4^-$
chlorite	ClO_2^-	silicate	SiO_3^{2-}
hypochlorite	ClO^-	sulfate	SO_4^{2-}
chromate	CrO_4^{2-}	hydrogen sulfate	HSO_4^-
dichromate	$Cr_2O_7^{2-}$	sulfite	SO_3^{2-}
cyanide	CN^-	hydrogen sulfite	HSO_3^-
hydroxide	OH^-	hydrogen sulfide	HS^-
iodate	IO_3^-	thiocyanate	SCN^-
		thiosulfate	$S_2O_3^{2-}$

Determining the names and chemical formulas for ionic compounds that contain polyatomic negative ions is very similar to naming the binary ionic compounds, with one addition.

- The name of the positive ion is first, followed by the name of negative ion.
- The net charge on the compound must be zero.
- If more than one polyatomic ion is needed to make the net charge equal to zero, the symbol for the ion is placed inside the round brackets, and the subscript outside the round brackets.
- There are no brackets around the monatomic ion.

For example, the compound $Na_2CO_3(s)$ is sodium carbonate. Because there is only one carbonate ion, CO_3^{2-}, there is no need for round brackets. The chemical symbol for calcium nitrate is $Ca(NO_3)_2(s)$. Since there are two nitrate ions, NO_3^-, round brackets must be used.

There are no comprehensive rules for naming polyatomic ions; thus, it is best to learn the names. The following generalizations will, nevertheless, help you remember some of the names.

1. Most of the polyatomic ions are negatively charged and have a non-metal other than oxygen combined with one or more oxygen atoms. Many of these exist in families with varying numbers of oxygen atoms. For families of two,

   • the one with more oxygen atoms ends with "-*ate*."
   • the one with fewer oxygen atoms ends with "-*ite*."

   For families of four, Table P1.7 lists prefixes and suffixes.

Table P1.7 Prefixes and Suffixes for Families with Four Oxygen Atoms

Relative number of oxygen atoms	Prefix	Suffix	Example	
most	per-	-ate	ClO_4^-	perchlorate
second most	(none)	-ate	ClO_3^-	chlorate
second fewest	(none)	-ite	ClO_2^-	chlorite
fewest	hypo-	-ite	ClO^-	hypochlorite

2. When one hydrogen atom is included with the negatively charged polyatomic ion, the word *hydrogen* precedes the name of the polyatomic ion. When two hydrogen atoms are included, the word *dihydrogen* precedes the name of the negatively charged polyatomic ion. For example, the anion $H_2PO_4^-$ is dihydrogen phosphate.

3. The prefix "*thio-*" indicates that a sulfur atom has taken the place of an oxgen atom. For example, cyanide is CN^- and thiocyanate is SCN^-.

• • •

3 Name the following binary ionic compounds:

   a) $Al_2O_3(s)$ **b)** $HgI_2(s)$ **c)** $Na_3P(s)$ **d)** $CaBr_2(s)$

4 Write the chemical formula for the following binary ionic compounds:

   a) zinc oxide **c)** magnesium iodide
   b) iron(II) sulfide **d)** cobalt(III) chloride

5 Name the following binary covalently bonded compounds:

   a) $SF_6(g)$ **b)** $N_2O_5(s)$ **c)** $PCl_5(s)$ **d)** $CF_4(g)$

6 Write the chemical formula for the following covalently bonded compounds:

   a) dihydrogen monoxide **c)** dinitrogen tetroxide
   b) sulfur trioxide **d)** dinitrogen monoxide

7 Name the following polyatomic ionic compounds:

   a) $K_3PO_4(s)$ **b)** $NH_4Cl(s)$ **c)** $LiClO_4(s)$ **d)** $NaHCO_3(s)$

8 Write the chemical formula for the following polyatomic ionic compounds:

   a) potassium hypochlorite **c)** sodium cyanide
   b) magnesium oxalate **d)** ammonium sulfate

• • •

Prerequisite Skills

• **use** model building kits or software to model compounds and molecules

• **demonstrate** procedures for safe handling, storing, and disposal of materials used in the laboratory

In Unit 1, you will build on the concepts you have reviewed here. You will also use some skills that are listed here.

Chemical Bonding

Chapter Concepts

1.1 Forming and Representing Compounds

- When atoms form bonds, they tend to gain, lose, or share electrons to achieve a filled outer energy level.

- Ionic bonds form when metal atoms lose electrons and non-metal atoms gain electrons. The resulting positive and negative ions attract each other by an electrostatic force.

- Covalent bonds form when non-metal atoms achieve a filled outer energy level by sharing electrons.

- Names of ionic compounds indicate which elements make up the compound.

- Names of some simple molecular compounds indicate the number of each type of atom in the molecule.

- Structural formulas show the bonds between atoms in a molecule.

1.2 The Nature of Chemical Bonds

- Electronegativity is a measure of the relative ability of an element's atoms to attract electrons.

- Smaller atoms tend to have higher electronegativities.

- The difference in the electronegativities between two elements determines the type of bond that will form between atoms of those elements.

Chemist Dr. Harry Coover was attempting to develop a clear plastic with optical properties suitable for lenses when he synthesized a compound called cyanoacrylate. It had some of the properties Coover desired, but it stuck to everything. Dr. Coover had invented super glue. Today, super glue is used for everything from toys to forensic science. Super glue is also used to prepare and maintain fossils for museum displays, such as this dinosaur skeleton.

In another laboratory, Dr. Roy Plunkett was attempting to develop a refrigerant. He decided to try to react a compound called tetrafluoroethylene with hydrochloric acid. He first synthesized tetrafluoroethylene and then stored it in cylinders that could hold high-pressure gas. When he opened the valve on one of the cylinders, no gas came out. He cut open the cylinder to see what had happened and found a white, slippery powder. Dr. Plunkett had just discovered PTFE (polytetrafluoroethylene), now known as Teflon™—a compound that will stick to very few substances. Today, Teflon™ is used on cookware, in artificial heart valves, and for many other purposes.

Two chemists conducted experiments that yielded results that they had not predicted. If these two chemists had not had a thorough understanding of the bonding between atoms, they would not have understood what had occurred in these "failed" experiments. Moreover, society would not have super glue and Teflon™. In science, discoveries often arise from "failed" experiments.

In this chapter, you will learn about the nature of bonding between atoms. You will discover why bonds form between certain combinations of atoms and not between others. You will draw and model many types of bonds.

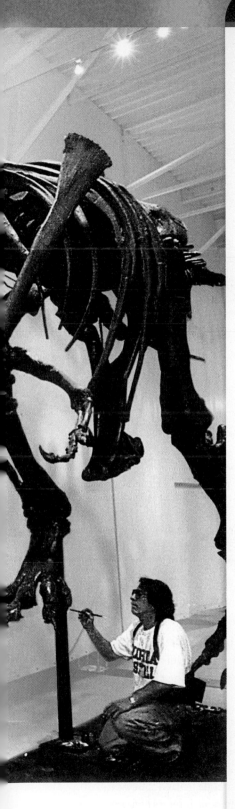

Chemistry Recall

As you begin your study of chemistry, you will need to recall concepts you learned in previous science courses. This activity will give you an opportunity to find out how much you remember and how much you need to review. A review is provided for you in the Unit Preparation on page 4. Work with a partner.

Materials

- pencil
- paper

Procedure

H	Li	Mg	Na	K
Ca	Ba	C	N	O
P	S	Cl	F	Br
Se	I	Fe	Ag	Au
Pb	Zn	Si	B	Al

1. On small pieces of paper (each about 2 cm × 2 cm), you and your partner should each write the chemical symbols shown here.

2. Turn the pieces of paper upside down on a desk or table. Mix them thoroughly.

3. Pick up two pieces of paper and read the symbols.

4. Without letting your partner see the symbols, write them down on another piece of paper.

5. Below the symbols, write the names of the two elements.

6. Imagine that you try to form a compound with these two elements. Below the names of the elements, indicate whether you think you could successfully create a compound by writing "yes" or "no." If you are not sure, write "Do not know."

7. If you think a compound would form, write the name and formula of the compound beside your answer to step 6. Indicate whether the bond would be ionic, covalent, or neither. If you are not sure, write "Do not know."

8. Repeat Procedure steps 3 to 7 until you have at least four different compounds.

9. On separate small pieces of paper, write the names of the compounds and exchange them with your partner.

10. Try to identify the elements written on the papers your partner picked up and have your partner identify the elements you picked up. After determining whether you communicated the elements correctly, compare the information that you and your partner wrote and discuss the results.

Analysis

1. How well did you recall the names of the symbols? How well do you think you named the compounds and identified the type of bonding?

2. How much review do you need to name the elements, compounds, and bonding?

3. If you did not recall the names of the elements or could not answer all of the questions, study the appropriate sections in the Unit 1 Preparation.

4. Design a game that you could play with one or more partners that would help everyone develop a good chemistry vocabulary.

Forming and Representing Compounds

Section Outcomes

In this section, you will:
- **explain** how an ionic bond results from the simultaneous attraction of oppositely charged ions
- **relate** electron pairing to multiple and covalent bonds
- **draw** electron dot diagrams of atoms and molecules
- **recall** principles for naming ionic and molecular compounds

Key Terms

electron pairs
unpaired electrons
bonding pair
lone pairs
main group elements
molecular compounds
Lewis structures
single bond
double bond
triple bond
structural formula
metallic bonding
delocalized
nomenclature

Figure 1.1 The helium used in these balloons is a noble gas. Noble gases are some of the very few elements that are found in nature in their pure form.

Ninety-two naturally occurring elements combine to form the millions of different compounds in nature. In fact, very few elements are found naturally in their elemental form (see Figure 1.1). Atoms of most elements are combined with atoms of other elements in a wide variety of chemical compounds and molecules. What property of the elements causes them to combine with other elements? Why are some combinations of elements much more common than others? How can you predict the type of combinations that will occur? Answers to these questions are based on the types of bonds that form between atoms of elements. To find clues to the nature of chemical bonds, examine some naturally occurring compounds. Look for patterns in these compounds.

Clues in Naturally Occurring Substances

Ores are metal compounds that are mined to produce metals, such as iron and copper. Most ores consist of a metal combined with a non-metal, such as oxygen, sulfur, a halogen, or carbonate ions. Very few metals are found in their pure form in nature. Gold, silver, and platinum are called precious metals because they can be found in their pure, metallic form. In both the combined and the pure form, all metals except mercury are solid at room temperature. (Metallic mercury is a liquid.) For a compound to be solid, some type of attractive force must be holding the individual atoms together. Is the attractive force between metal atoms and non-metal atoms the same as the force that attracts metal atoms to other metal atoms?

Figure 1.2 The ore from this open pit mine consists of metal ions combined with non-metal ions. The ore must be refined to extract the pure metals.

You can gain more insight by considering the atmosphere, which contains the oxygen that you inhale and the carbon dioxide that you exhale. The atmosphere also contains the water vapour that forms clouds, which can then turn into rain or snow. Additionally, the atmosphere has gases, such as nitrogen, argon, and traces of methane, ozone, hydrogen, and other noble gases. Carbon dioxide and methane gases consist of non-metal elements combined to form molecules. Water consists of the non-metals hydrogen and oxygen. Water is the only compound found naturally in all three states: liquid, gas, and solid. Oxygen, nitrogen, and hydrogen are found as elements, and they are all diatomic molecules. This means they are made up of two identical bonded atoms. Only the noble gases, helium, argon, neon, krypton, xenon, and radon are found in an uncombined form. In nature, the noble gases are found as monatomic gases because they are uncombined.

The following patterns have emerged from these observations.

- Metals are usually found in combination with non-metals, and these compounds are solid.
- Metals (other than mercury) in pure form are solids.
- Non-metals can combine with one another to form gases, liquids, or solids.
- The only elements that are *never* found in nature in a combined form are the noble gases.

Because atoms of the noble gases are always found as monatomic gases and because atoms of all other elements are usually found chemically bonded to other atoms, you can infer that there is something very unique about the chemistry of noble gases.

Chemistry File

Web Link
Because the noble gases rarely undergo chemical reactions, you might think that they have no practical applications. Research using the Internet to find out why helium is used in breathing mixtures for deep-sea diving instead of nitrogen. How is argon used in welding?

www.albertachemistry.ca
WWW

Examine Figure 1.3, which shows the electron dot diagrams of the noble gases. What is unique about these elements? They are the only elements that have filled outer (valence) energy levels. Because helium has only two electrons, and the lowest energy level is limited to two electrons, its outer energy level is filled. All other noble gases have an octet of electrons, that fills their outer energy level.

$$\text{He} \quad :\text{Ne}: \quad :\text{Ar}: \quad :\text{Kr}: \quad :\text{Xe}: \quad :\text{Rn}:$$

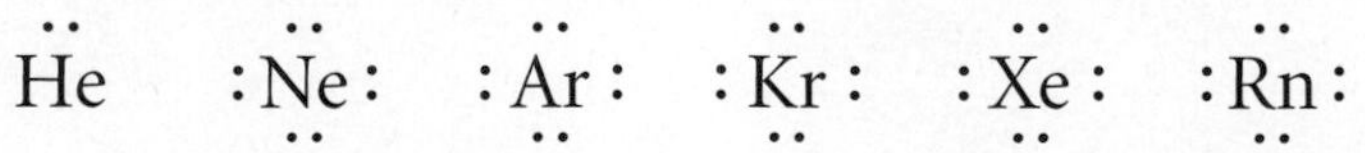

Figure 1.3 Although the different noble gases have differing numbers of electrons, the outer energy level, in each case, is filled.

Chemical Bonds

In nature, atoms of elements that have less than eight electrons in their outer energy levels nearly always form bonds with other atoms. Atoms of elements that have eight electrons in their outer energy level never form bonds with other atoms. Thus, having filled valence energy levels makes atoms, ions, or molecules stable. The stability of a filled outer energy level is the basis of the octet rule (see page 8) for forming ionic and covalent bonds. The octet rule allows you to predict whether bonds will form between atoms of specific elements and what type of bond will form.

Electrons dot diagrams (see page 8) are also very helpful when analyzing bond formation. The pairing of dots for atoms with five or more valence electrons has more significance than simply being a convenient way to draw diagrams. Electrons, being negatively charged, repel one another and remain as far apart as possible in the atom. The modern model of the atom—called the quantum mechanical model—shows that when more than four electrons occupy one energy level, electrons form pairs that interact in a unique way that allows them to be closer together. These **electron pairs**, as they are called, are less likely to participate in bond formation than are the **unpaired electrons** (see Figure 1.4).

When bonds form, the unpaired electrons are the ones that become involved in the bonds. For example, when a metal loses an electron and forms an ion, that electron is accepted by a non-metal atom and it forms a pair with one of the unpaired electrons in the non-metal atom. When two non-metal atoms share electrons, the electrons that are involved in the covalent bond are those that were previously unpaired. When the atoms share the electrons and form a covalent bond, the electrons are called a **bonding pair**. The electrons that were originally paired in an atom and do not form a bonding pair in the molecule are called a **lone pair**. Figure 1.4 illustrates all of these classes of electrons.

unpaired electrons   electron pairs   lone pairs   bonding pairs

Figure 1.4 Electrons in a free atom are either unpaired or part of an electron pair. In a molecule, the shared electrons are called bonding electrons. Electrons that are not part of a bonding pair are called a lone pair.

Use the octet rule and electron dot diagrams to answer the following questions. If you have any trouble answering the questions, review pages 8 and 9.

Q1 Which of the following pairs of atoms would be capable of forming an ionic bond? For those that would, use electron dot diagrams and the octet rule to show how they would form a bond. For those that could not form an ionic bond, explain why.

a) bromine and potassium   **c)** magnesium and lithium

b) calcium and sulfur   **d)** phosphorous and oxygen

Q2 Which of the following pairs of atoms would be capable of forming a covalent bond? For those that would, use electron dot diagrams and the octet rule to show how they would form a bond. For those that could not form a covalent bond, explain why.

a) neon and sodium   **c)** lithium and fluorine

b) chlorine and iodine   **d)** hydrogen and oxygen

Q3 State the octet rule.

Q4 Describe the steps you would use to draw an electron dot diagram of nitrogen.

Q5 Define valence electron.

Forming Ionic Bonds

As you know, the electrostatic attraction between oppositely charged ions is an ionic bond. Thus, to form ionic bonds, atoms must first become ionized (see page 8). When the valence level is less than half filled as it is in metal atoms, they can achieve a filled outer level (the energy level below the valence level) by losing all of their valence electrons. So when you know the number of valence electrons, you can predict the charge that the atom will have. For example, calcium has two valence electrons and, when it loses them, it has a net of two positive charges or Ca^{2+}. When the valence level is more than half filled as it is in most non-metal atoms, the atoms can achieve a filled outer energy by gaining enough electrons to fill the valence level. Once again, when you know the number of electrons in the valence level, you can predict the charge that the ion will have.

Until now, you have been considering only the first 20 elements on the periodic table. These elements belong to the **main group elements**—Groups 1, 2, and 13 through 18. For most of the main group elements, you can predict the number of valence electrons from their position on the periodic table. Group 1 elements have one valence electron and Group 2 elements have two valence electrons. Group 13 elements have three valence electrons and the numbers increase sequentially up to Group 18 elements, which have eight valence electrons.

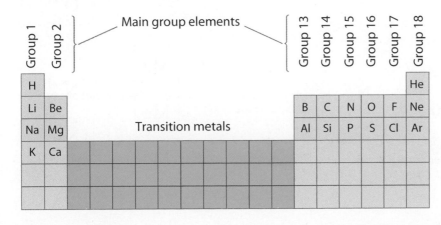

Figure 1.5 Elements in Groups 1, 2, and 13 through 18 are called the main group elements.

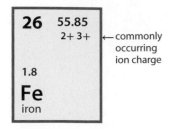

26	55.85
	2+ 3+ ← commonly occurring ion charge
1.8	
Fe	
iron	

Figure 1.6 The numbers 2+ and 3+ indicate that iron can lose two electrons and form iron(II) (Fe^{2+}) or lose three electrons and form iron(III) (Fe^{3+}).

For transition metals and a few of the larger metals in the main group, the number of valence electrons is not as predictable as it is for most of the main group elements. Thus you cannot always predict the charge that the atoms of those elements will have. In fact, due to differences in the environment around the ion, some metal atoms can form ions in more than one way. For example, copper can lose either one or two electrons. You can use data from the periodic table to determine the nature of the charge that an ion will usually have. For elements that typically become ionized, the square in the periodic table has one or more numbers labelled "commonly occurring ion charge." These are the charges that an ion of that element is likely to have. For example, Figure 1.6 shows you that iron can have a charge of 2+ or 3+.

When ionic compounds form, the number of electrons lost by the metal atoms must equal the number of electrons gained by the non-metal atoms. The compounds must have a net neutral charge. Nevertheless, ionic bonds can form between ions that have an unequal number of charges. For example, a calcium atom can donate its two electrons to two different fluorine atoms. As well, two atoms of potassium can both donate their individual electrons to one atom of sulfur. Figure 1.7 shows you how to represent these reactions using electron dot diagrams.

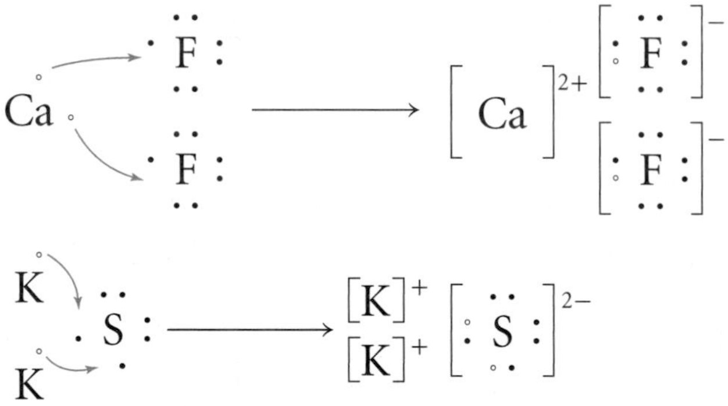

Figure 1.7 Calcium fluoride has a ratio of two fluoride ions to one calcium ion. Potassium sulfide has a ratio of two potassium ions to one sulfide ion.

Chemistry File

Try This
When the transition elements are placed in the periodic table, the table has 18 columns. What can you infer about the number of electrons that are allowed in the third inner filled energy level of the transition elements?

Figure 1.8 shows how the distribution of electrons from metals to non-metals can be complex when six electrons are involved in the transfer. The only criterion is that the net charge of the ions in the compound is zero.

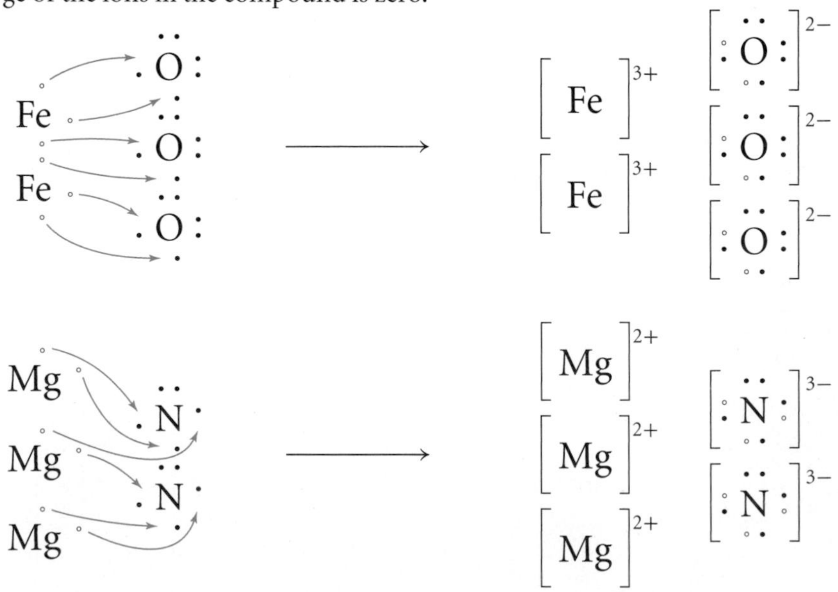

Figure 1.8 In the formation of bonds between magnesium and nitrogen, and between iron and oxygen, notice that the metal atoms donate six electrons and the non-metal atoms accept six electrons. The numbers of positive and negative charges are balanced.

6 Use electron dot diagrams to show the formation of the following compounds:

a) potassium chloride   **c)** magnesium fluoride   **e)** aluminium oxide

b) barium oxide   **d)** potassium sulfide

Chemistry File

Try This
Read the following statement and then decide whether you agree with it. Explain why or why not. "In general, the greater the horizontal distance between two elements on the periodic table, the more likely they are to be involved in ionic bonding."

Using the Charge to Determine the Chemical Formula of an Ionic Compound

How can you predict the chemical formula of an ionic compound consisting of oppositely charged ions that do not have the same magnitude of charge without using electron dot diagrams? You could "guess and check." However, there is a more direct method for determining the number of each type of ion that will result in a net charge of zero. You can always find a combination giving a net charge of zero if the subscript on the positively charged ion is equal to the charge on the negatively charged ion and the subscript on the negatively charged ion is equal to the charge on the positively charged ion. Then examine the subscripts to find the simplest whole number ratio. Some examples are shown in Figure 1.9.

$Fe^{3+}\quad S^{2-}$

Fe_2S_3

$2(+3) + 3(-2) = 6 - 6 = 0$

$Ca^{2+}\quad N^{3-}$

Ca_3N_2

$3(+2) + 2(-3) = 6 - 6 = 0$

$Co^{3+}\quad Cl^{-}$

$CoCl_3$

$1(+3) + 3(-1) = 3 - 3 = 0$

$Ca^{2+}\quad O^{2-}$

Ca_2O_2

Lowest whole number ratio is 1:1

CaO

$1(+2) + 1(-2) = 2 - 2 = 0$

$Al^{3+}\quad N^{3-}$

Al_3N_3

Lowest whole number ratio is 1:1

AlN

$1(+3) + 1(-3) = 3 - 3 = 0$

Figure 1.9 When you make the subscript on each ion equal to the magnitude of the charge on the opposite ion, the compound is neutral. If the subscripts you obtain are not the simplest whole number ratio, reduce them.

7 Write the chemical formula for the following compounds without using electron dot diagrams:

a) zinc oxide   **c)** magnesium iodide

b) iron(II) sulfide   **d)** cobalt(III) chloride

Chemistry File

FYI
The technique of using Roman numerals to indicate the charge on transition metal ions is called the Stock system, in honour of Alfred E. Stock (1876–1946), who devised the system. An older nomenclature system uses the suffixes "-ic" for the larger charge and "-ous" for the smaller charge. This system also uses many of the older names of the metals—those names for which the symbols were formulated. For example, iron(III) was called the ferric ion and iron(II) was the ferrous ion. Copper(II) was the cupric ion and copper(I) was the cuprous ion.

Ionic Liquids

Society is shaped by molecular bonding and chemical reactions. The food people eat, the products people use to clean and paint their homes, and even the clothes people wear—all these products are based on chemical processes. Unfortunately, many of these processes use solvents that are dangerous to people and to the environment. Green chemistry is a movement to create environmentally friendly chemical processes. Scientists have been conducting research on ionic liquids and have found that they are excellent "green" solvents that help reduce the damage caused by standard industrial processes.

Ionic liquids could be used to purify the air in submarines.

Background

Room temperature ionic liquids, or RTILs, are salts that are liquid at room temperature. The lower melting points of RTILs result from their chemical composition. Ionic liquids, particularly RTILs, are made up of large bulky organic cations and inorganic anions. The resulting asymmetry decreases the lattice energy—the energy needed to completely separate the ions in an ionic solid—which in turn reduces the melting point. The composition and properties of ionic liquids depend on the cation and anion combinations, and there are billions of possibilities. This means that ionic liquids can be designed for specific purposes, with chemical and physical properties that achieve very specific tasks.

Most solvents used in industrial processes are called volatile organic compounds, or VOCs. The interest in ionic liquids for green chemistry is based on their non-measurable vapour pressure. Ionic liquids do not evaporate into the air. Therefore, unlike VOCs, ionic liquids can be easily contained and will not pollute the environment. The ability to contain ionic liquids also means that they can be recycled and re-used.

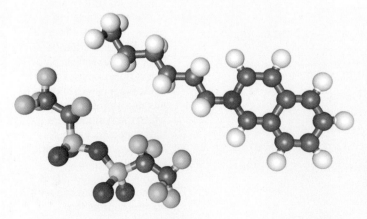

Testing an Idea

Scientists are also hypothesizing about other ways to use and apply ionic liquids. For example, ionizing radiation does not seem to affect ionic liquids, so they might be useful in treating high-level nuclear waste. The ability of ionic liquids to dissolve specific gases means they could be used to purify the air in small vessels, such as spacecraft and submarines. One newly designed ionic liquid removes mercury from contaminated water. The ionic liquid is water-insoluble, so when it comes in contact with contaminated water, the mercury ions are removed from the water and bond in the ionic liquid.

• • •

1. Most ionic salts have very high melting points, such as sodium chloride, which melts at 801 °C, and, in solid form, are unable to conduct electric current. Explain why ionic liquids are able to conduct electric current even though they are ionic salts. Conduct research to find one possible application for this conductivity.

2. What does green chemistry mean to you? Survey your friends and family to find out what they know about green chemistry. Do they feel it is important?

3. Suppose that you could design your own ionic liquid. Identify one problem that may be solved by using your ionic liquid, and explain how it would work.

• • •

This polycyclic N-alkylisoquinolinium cation ([Cnisoq]⁺) combined with the bis(perfluoroethyl sulfonyl)imide anion ([BETI]⁻) creates an ionic liquid that can separate an aromatic solute, such as a chlorobenzene, from contaminated water.

Forming Covalent Bonds

A covalent bond is the sharing of electrons by two atoms, which are nearly always non-metal atoms (see page 9). Compounds in which the atoms are covalently bonded together are correctly called **molecular compounds**. Electron dot diagrams of molecules are called **Lewis structures**, in honour of Gilbert N. Lewis (1875-1946), the chemist who developed the technique and furthered the understanding of bond formation.

Two examples of unlike atoms sharing electrons are shown in Figure 1.10. You may recognize the first structure as methane, $CH_4(g)$, the principle component of natural gas. Notice that there are no lone pairs in the methane molecule. Hydrogen, of course, can have only two electrons, therefore, it has no lone pairs. Carbon has an octet of electrons, but they are all involved in bonding pairs. Carbon is found in thousands of stable compounds, but in only one compound does the carbon atom have a lone pair. That compound is carbon monoxide. The second example in Figure 1.10 is water. Once again, the hydrogen atoms have no lone pairs. The oxygen has an octet of electrons and shares two electrons with each hydrogen atom. Therefore, the oxygen atom has two lone pairs.

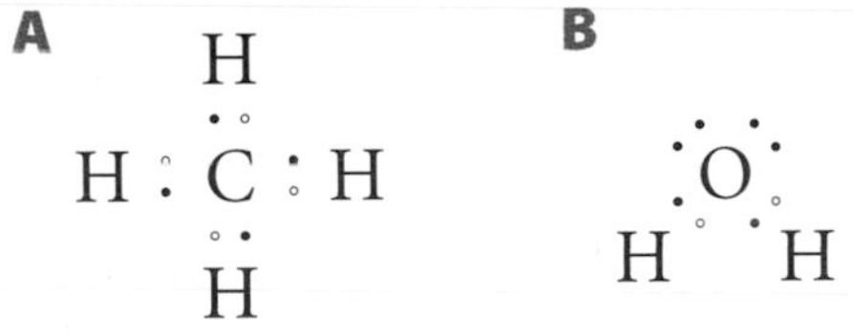

Figure 1.10 (A) In methane, $CH_4(g)$, carbon shares each of its electrons with a different hydrogen atom. **(B)** The oxygen atom in water, $H_2O(\ell)$, shares an electron with each of the two hydrogen atoms and also has two lone pairs.

In some molecules, there are not enough valence electrons for two atoms to share one pair of electrons and then fill an octet with lone pairs. For example, in carbon dioxide, the carbon atom has four valence electrons and the two oxygen atoms each have six valence electrons (Figure 1.11A). If the carbon atom shared one pair of electrons with each oxygen atom, each oxygen atom would have only seven electrons in the outer energy level (Figure 1.11B). This configuration has not been found to occur. To complete an octet for each atom, the atoms share four electrons (Figure 1.11C). Because one pair of shared electrons forms one bond, called a **single bond**, the sharing of two pairs of electrons is called a **double bond**.

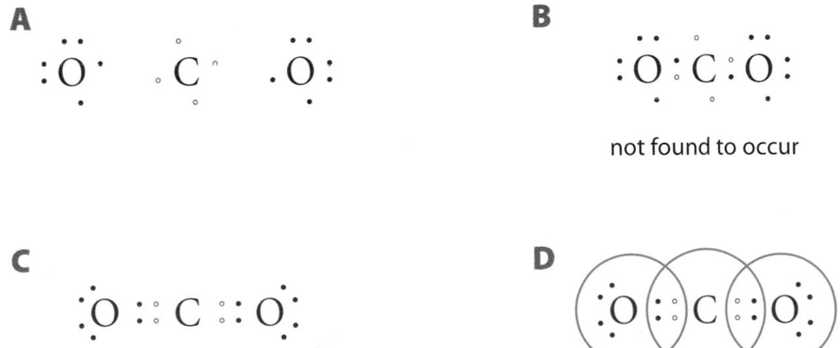

B

not found to occur

Figure 1.11 By sharing two pairs of electrons, each oxygen atom and the carbon atom in a carbon dioxide molecule can acquire an octet of electrons.

A similar situation exists for the nitrogen molecule, $N_2(g)$. Each nitrogen atom in a nitrogen molecule has five valence electrons. To form an octet of electrons around each nitrogen atom in the molecule, the two atoms must share three pairs of electrons, as shown in Figure 1.12. Because each pair of shared electrons constitutes one bond, three pairs of shared electrons makes a **triple bond**.

Figure 1.12 The arrows indicate the placement of the electrons in nitrogen that create a triple bond between the atoms.

Carbon is a versatile element. As shown in Figure 1.13, carbon can have four single bonds, two single bonds and one double bond, two double bonds, or one single bond and one triple bond. Circles are added to the Lewis structures to indicate the electrons that are associated with each atom.

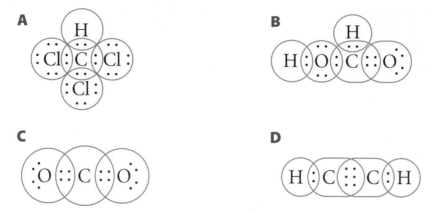

Figure 1.13 Compound **(A)** is trichloromethane ($CHCl_3(\ell)$), which is also called chloroform. It used to be used as an anaesthetic. Compound **(B)** is methanoic acid ($HCOOH(\ell)$), which is also called formic acid. It causes the sting left by some insect bites. Compound **(C)** is carbon dioxide ($CO_2(g)$), the gas that animals exhale. Compound **(D)** is ethyne ($C_2H_2(g)$), which is also called acetylene. It is commonly used in welders' torches.

For simple molecules, given good information, Lewis structures can be drawn quite easily. Examine the Sample Problem and then complete the Practice Problems.

Sample Problem

Drawing Lewis Structures

Problem
One carbon atom is bonded to one oxygen atom and two hydrogen atoms. Draw a Lewis structure to represent the bonds in this molecule.

What Is Required?
A Lewis structure for one carbon atom bonded to one oxygen atom and two hydrogen atoms.

What Is Given?
You know that a carbon atom has four valence electrons and can share four electrons for a filled outer energy level. A carbon atom, therefore, can form four bonds. These bonds can all be single bonds or combinations of single, double, or triple bonds. Oxygen has six valence electrons and can form two single bonds or one double bond. Hydrogen has one valence electron and can form only one single bond.

The bonds must be arranged in a way that will create two electrons in the outer energy level for hydrogen and eight electrons in the outer energy levels for carbon and oxygen. The problem states that the oxygen atom and the two hydrogen atoms are all bonded to the carbon atom.

Plan Your Strategy
Draw each of the atoms alone and find a way to fit them together to satisfy all the criteria.

Act on Your Strategy
Each hydrogen atom shares two electrons with the carbon atom. The carbon and oxygen atoms share four electrons, making a double bond.

$$H \cdot \qquad H \cdot \qquad \cdot \overset{\cdot}{C} \cdot \qquad \overset{\cdot \cdot}{\underset{\cdot \cdot}{O}} \cdot$$

$$\overset{H}{\underset{H}{}} \cdot \overset{\cdot \cdot}{C} :: \overset{\cdot \cdot}{\underset{\cdot \cdot}{O}}$$

Check Your Solution
The outer energy level of each hydrogen atom is filled with two electrons. The outer energy levels on the carbon atom and the oxygen atom are both filled with an octet of electrons. The name of this compound is methanal, but it is also called formaldehyde.

Draw Lewis structures to represent each of the following molecules:

1. One carbon atom is bonded to two hydrogen atoms and two chlorine atoms. This compound is dichloromethane. It is also called methylene chloride and is sometimes used as a paint stripper.

2. One hydrogen atom is bonded to a fluorine atom. The compound, hydrogen fluoride, is sometimes used to etch glass.

3. Three hydrogen atoms are bonded to one nitrogen atom in this compound that is found in some cleaning solutions.

4. One carbon atom is bonded to two sulfur atoms. This compound, carbon disulfide, is sometimes found in gases from volcanic eruptions and in marsh gas.

5. A central carbon atom is bonded to a hydrogen atom and to a nitrogen atom. This is a toxic compound called hydrogen cyanide.

6. Two carbon atoms are bonded to each other and two hydrogen atoms are bonded to each of the two carbon atoms. This compound, called ethene, or, more commonly, ethylene, is used in the plastics industry.

When you are told which atoms are bonded, you can draw Lewis structures readily. However, you are usually given a chemical formula and asked to draw a Lewis structure. A chemical formula does not identify the atoms between which the bonds are located. You need a method to make these predictions. The steps listed below give you enough information to draw Lewis structures for simple compounds that have a single central atom. You can extend these rules to some slightly more complex molecules by making a few common observations. For example, when a molecule has more than one carbon atom, the carbon atoms are quite often bonded to each other. As well, chemical formulas are often written in ways that provides some hints about their structure. For example, the formula $C_2H_5SCH_3$ tells you that there are two carbons bonded together and then bonded to a sulfur atom. The final carbon atom is also bonded to the sulfur atom.

Read through the steps for drawing Lewis structures and then examine the Sample Problem that follows. Finally, develop your skills in drawing Lewis structures by completing the Pratice Problems.

Drawing Lewis Structures for Simple Molecules and Polyatomic Ions

Step 1 Determine the total number of valence electrons in all the atoms in the molecule. If the molecule is charged, add or subtract electrons to account for the charge.

Step 2 Choose the atom with the most unpaired electrons as the central atom. Draw a skeleton structure for the molecule by placing the other atoms around the central atom. Draw one pair of electrons between each pair of atoms to represent a single bond.

Step 3 Place lone pairs of electrons around all the atoms *except* the central atom to form an octet of electrons. Hydrogen, of course, can only have two electrons in its outer shell.

Step 4 a) If all the valence electrons determined in step 1 have not been accounted for, add one or more lone pairs around the central atom to complete an octet of electrons.

b) If all the valence electrons have been used up but the central atom still does not have an octet of electrons, move one or more of the lone pairs to form double or triple bonds between the central atom and an adjacent atom.

Drawing the Lewis Structure of a Molecule

Problem
Draw the Lewis structure for a methanal (formaldehyde), $CH_2O(\ell)$, molecule.

What Is Required?
Lewis structure of $CH_2O(\ell)$ molecule

What Is Given?
The molecular formula, $CH_2O(\ell)$, tells you that there are two hydrogen atoms, one carbon atom, and one oxygen atom in a molecule of methanal.

Plan Your Strategy
Apply the steps for drawing Lewis structures.

• • •

Act on Your Strategy

Step 1 Determine the total number of valence electrons in all the atoms in the molecule:

$$\left(1 \,\overline{\text{C atom}} \times \frac{4e^-}{\overline{\text{C atom}}}\right) + \left(1 \,\overline{\text{O atom}} \times \frac{6e^-}{\overline{\text{O atom}}}\right) + \left(2 \,\overline{\text{H atoms}} \times \frac{1e^-}{\overline{\text{H atom}}}\right) = 4e^- + 6e^- + 2e^-$$
$$= 12e^-$$

Step 2 Select the atom with the most unpaired electrons. Carbon has four unpaired electrons, oxygen has two unpaired electrons, and hydrogen has one unpaired electron. Therefore, carbon is the central atom. Draw a skeleton structure around carbon with one pair of electrons—a covalent bond—between each pair of atoms.

$$\begin{array}{c} \text{O} \\ \text{H} \! : \! \overset{\displaystyle ..}{\text{C}} \! : \! \text{H} \end{array}$$

Step 3 Place lone pairs of electrons around the oxygen atoms to form an octet.

$$\begin{array}{c} : \! \overset{\displaystyle ..}{\underset{\displaystyle ..}{\text{O}}} \! : \\ \text{H} \! : \! \text{C} \! : \! \text{H} \end{array}$$

Step 4 Compare the number of electrons in the structure, as written, with the number determined in step 1. There are 12 electrons in the structure, which is the same as the number determined in step 1. The number of electrons is correct. However, the carbon atom has only six electrons. Therefore, move one of the lone pairs on the oxygen atom to share with the carbon, thus forming a double bond.

$$\begin{array}{c} : \! \overset{\displaystyle ..}{\underset{\displaystyle ..}{\text{O}}} \! : \\ \text{H} \! : \! \text{C} \! : \! \text{H} \end{array} \quad \longrightarrow \quad \begin{array}{c} \overset{\displaystyle ..}{\underset{\displaystyle ..}{\text{O}}} \\ \text{H} \! : \! \text{C} \! : \! \text{H} \end{array}$$

Check Your Solution
Each atom has achieved a noble gas configuration. The oxygen and carbon atoms each have eight electrons and the hydrogen atoms have two electrons. Thus, you can be confident that this is a reasonable Lewis structure.

Notice that the molecule in this Sample Problem is the same as the molecule in the Sample Problem on page 24. Did you find it easier to develop the Lewis structure by following the steps rather than guessing and checking? Use the steps to complete the following Practice Problems.

Draw Lewis structures for each of the following molecules:

7. $NH_3(g)$ **9.** $CF_4(g)$ **11.** $H_2S(g)$ **13.** $ClNO(g)$ **15.** $CS_2(\ell)$ **17.** $C_2H_2(g)$

8. $CH_4(g)$ **10.** $AsH_3(g)$ **12.** $H_2O_2(\ell)$ **14.** $C_2H_4(g)$ **16.** $HOCl(g)$

Thought Lab 1.1 | Lewis Structures

The rules that you learned for drawing Lewis structures of simple molecules were based on molecules having a central atom. Many molecules do not have a central atom but are often written in a way that gives you clues about their structures. In this lab, you will attempt to draw Lewis structures of some of these molecules. You have already attempted two such molecules, C_2H_2 and C_2H_4. The symmetry of these molecules helps you determine their structure. In many cases, chemists and biochemists write molecular formulas in groups of atoms. For example, a certain amino acid (glycine) can be written, $NH_2CH_2COOH(s)$. You can try to isolate the groups, draw their structures, then attach them together.

Work with a partner. First, attempt to draw the structures individually. When each of you has found what you believe to be the correct structure, compare your structures. If you do not have the same structure, discuss them and try to come to agreement.

Draw Lewis structures of the following compounds:
a) $NH_2CH_2COOH(s)$
b) $CH_3OCH_3(s)$
c) $CH_3COOH(\ell)$
d) $CF_3CHF_2(g)$
e) $ClCH_2CH_2SCH_2CH_2Cl(\ell)$
f) $(NH_2)_2CO(s)$
g) Challenge! Try to draw the Lewis structure of glucose, $C_6H_{12}O_6(s)$. (**Hint:** The carbon atoms are all bonded to each other in a straight chain. Each carbon atom is bonded to at least one hydrogen atom and one oxygen atom.)

More Than One Possible Lewis Structure

○ **Begin extension material**

When applying the rules for drawing Lewis structures for certain molecules, you might find a reasonable structure that does not agree with experimental data. For example, the structure below for a *sulfur dioxide*, $SO_2(g)$, molecule fits all the criteria for an acceptable structure:

$$:\ddot{O}:\ddot{S}::\ddot{O}:$$

Experimental data, however, indicate that the two bonds in $SO_2(g)$ are identical, so there cannot be one single and one double bond. The data indicate that the two identical bonds have properties that could be described as half way between a single bond and a double bond. Chemists call such structures as this resonance structures.

Resonance structures are sometimes written as shown here, indicating that the molecule is changing back and forth between two structures.

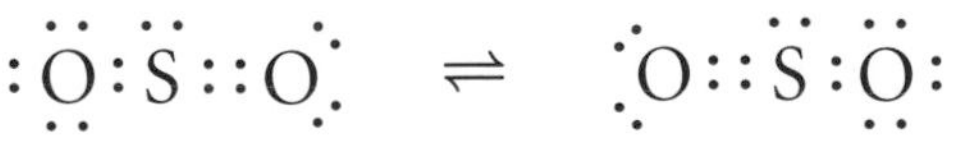

$$:\ddot{O}:\ddot{S}::\ddot{O}: \quad \rightleftharpoons \quad :\ddot{O}::\ddot{S}:\ddot{O}:$$

Neither structure actually exists but instead the molecule is a "hybrid" of these two structures. You can envision some of the electrons as being shared by all three atoms. Sulfur dioxide is not unique in having a resonance structure. Many molecules have resonance structures. Two common examples are ozone, $(O_3(g))$, and nitrogen dioxide, $(NO_2(g))$.

Coordinate Covalent Bonds

In all the examples of covalent bonds that you have seen thus far, each of the two atoms involved have contributed one of the two electrons in the bond. In some cases, however, one of the atoms can contribute both the electrons. The bond, in these cases, is called a coordinate covalent bond. Once formed, it is not possible to distinguish a coordinate covalent bond from any other covalent bond. The following Sample Problem involves a coordinate covalent bond.

Sample Problem

Drawing a Lewis Structure That Includes a Coordinate Covalent Bond

Problem
Draw the Lewis structure for a hydronium ion, H_3O^+(aq).
(**Note:** Hydrogen ions (H^+) do not exist unbonded in water solutions. Instead, the hydrogen ions bond to water molecules, producing hydronium ions.)

What Is Required?
Lewis structure of H_3O^+(aq)

What Is Given?
The chemical formula tells you that the hydronium ion has three hydrogen atoms, one oxygen atom, and a charge of 1+.

Plan Your Strategy
Follow the four-step procedure for drawing Lewis structures.

• • •

Act on Your Strategy

Step 1 Determine the total number of valence electrons in all the atoms in the molecule. The oxygen atom has six valence electrons and hydrogen atoms have one valence electron each. Subtract one electron to give the molecule its one positive charge.

$$\left(1 \; \cancel{O \; atom} \times \frac{6e^-}{\cancel{O \; atom}}\right) + \left(3 \; \cancel{H \; atoms} \times \frac{1e^-}{\cancel{H \; atom}}\right) - 1e^- = 6e^- + 3e^- - 1e^-$$
$$= 8e^-$$

Step 2 Select the atom with the most unpaired electrons. Oxygen atoms have two unpaired electrons and hydrogen atoms have one unpaired electron. Therefore, oxygen is the central atom. Draw a skeleton structure of the atom with one pair of bonding electrons between each of the atoms.

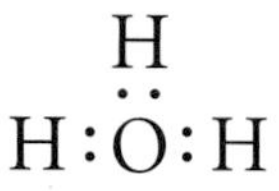

$$\begin{array}{c} H \\ H \!:\! \overset{\cdot\cdot}{O} \!:\! H \end{array}$$

Step 3 There are no lone pairs in hydrogen atoms, so no electrons should be added in this step.

Step 4 The structure has six electrons, and a total of eight valence electrons were determined in step 1. Therefore, add one more lone pair to the central atom, oxygen.

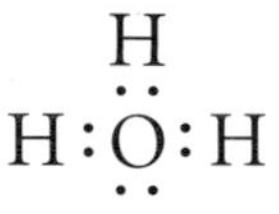

$$\begin{array}{c} H \\ H \!:\! \overset{\cdot\cdot}{\underset{\cdot\cdot}{O}} \!:\! H \end{array}$$

Because the molecule has a charge of 1+, you must indicate this with square brackets and a plus sign as a superscript.

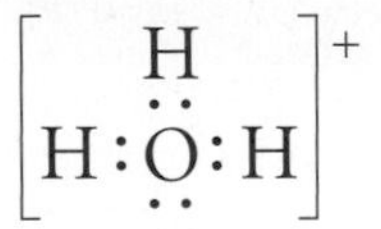

$$\left[\begin{array}{c} H \\ H\!:\!\overset{\displaystyle\cdot\cdot}{\underset{\displaystyle\cdot\cdot}{O}}\!:\!H \end{array} \right]^+$$

Check Your Solution

Each atom has achieved a noble gas configuration. The number of valence electrons is correct. The positive charge is included. This is a reasonable Lewis structure for $H_3O^+(aq)$.

Practice Problems

Draw Lewis structures for each of the following ions:

18. $BrO^-(aq)$

19. $NO^+(aq)$

20. $ClO_3^-(aq)$

21. $SO_3^{2-}(aq)$

22. Dichlorofluoroethane, $CH_3CFCl_2(g)$, has been proposed as a replacement for chlorofluorocarbons (CFCs). The presence of the hydrogen atoms in $CH_3CFCl_2(g)$ markedly reduces the ozone-depleting ability of the compound. Draw a Lewis structure for this molecule.

23. Draw Lewis structures for the following molecules, which do not have a single central atom:
   a) $N_2H_4(g)$
   b) $N_2F_2(g)$

24. Although Group 18 elements are never found in compounds in nature, chemists are able to synthesize compounds of several noble gases, including xenon, $Xe(g)$. Draw a Lewis structure for the xenon tetroxide, $XeO_4(aq)$, molecule.

The hydronium ion in the Sample Problem is similar to the ammonium ion. In both cases, you can visualize a neutral molecule with a lone pair. A hydrogen ion is attracted to the lone pair, which becomes a bonding pair, as shown in Figure 1.14. In the final structures of the ammonium ion and the hydronium ion, it is not possible to determine which bonding pair was the lone pair or which hydrogen atom was the free hydrogen ion. The positive charge is not found or localized at any one point in the ion but is a property of the entire ion.

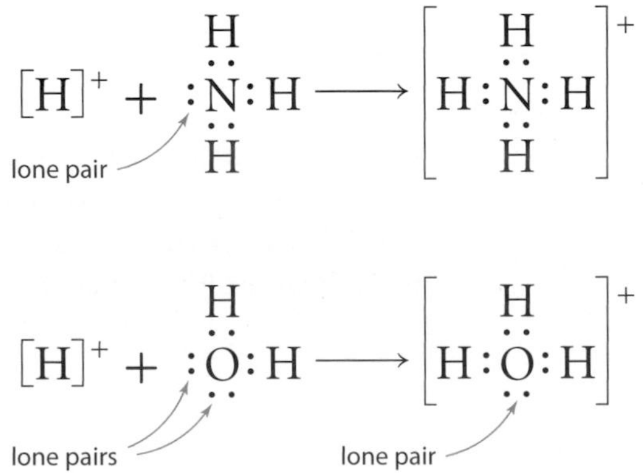

Figure 1.14 Ammonium and hydronium ions form in much the same way. A hydrogen ion, H^+, is attracted to a lone pair and then becomes covalently bonded to the molecule.

● **End extension material**

INVESTIGATION 1.A

Target Skills

Drawing electron dot diagrams and **building** models

Analyzing and **evaluating** models constructed by others

Modelling Molecules

Models are very important tools for chemists. You cannot see detailed features of molecules, even with a microscope. However, you can build models that fit some of the properties that chemists have determined through experimentation. In this investigation, you will use a molecular model kit to assemble models of a few molecules.

Question

What can you predict about the structure of molecules by building models?

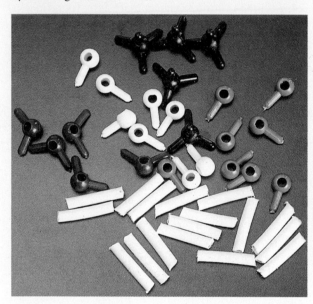

Materials

- molecular model kit
- pen
- paper

Procedure

1. Obtain a model kit from your teacher.
2. In your notebook, make a table with the headings shown in the next column. Give your table a title. Use this table for the drawings that you will make in Procedure steps 3, 4, and 6.

Name	Formula	Lewis structure and compound	Sketch of shape of molecule

3. In your table, draw Lewis structures of each molecule in the list below:
   a) hydrogen bonded to hydrogen: $H_2(g)$
   b) chlorine bonded to chlorine: $Cl_2(g)$
   c) oxygen bonded to two hydrogens: $H_2O(\ell)$
   d) carbon bonded to two oxygens: $CO_2(g)$
   e) nitrogen bonded to three hydrogens: $NH_3(g)$
   f) carbon bonded to four chlorines: $CCl_4(\ell)$
   g) nitrogen bonded to three fluorines: $NF_3(g)$
4. Look through the chapter and choose three other molecules that are not in the list above. Draw Lewis structures of your selected molecules in your table.
5. Based on your Lewis structures, build models of all the molecules. Make a sketch of your model.

Analysis

1. Compare your models and sketches with those of your classmates. Discuss any differences.
2. What can you learn from models that you cannot learn from Lewis structures?

Conclusion

3. Summarize the strengths and limitations of creating molecular models using kits. What can you deduce from the models? What features of molecules cannot be deduced from models?

Structural Formulas

Lewis structures are very helpful for understanding bonding and making predictions about structures. They are, however, very tedious to draw. To provide a clear image of a structure in a simpler form, you can draw a structural formula. A **structural formula** shows every individual atom in the molecule and shows the bonds between atoms. A structural formula uses a line to represent a bond, or pair of shared electrons, but it does not show the lone pairs. Two lines represent a double bond and three lines represent a triple bond. Compare the Lewis structures and the structural formulas in Figure 1.15. A Lewis structure is helpful for predicting the structure of a molecule, but a structural formula is clearer and simpler to interpret.

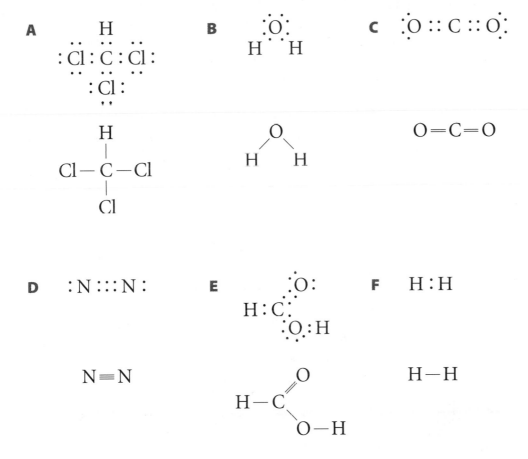

Figure 1.15 Structural formulas provide almost as much information as Lewis structures but are clearer and somewhat easier to read.

• • •

8 Draw structural formulas for each of the following compounds:

a) $NH_3(g)$ **d)** $AsH_3(g)$ **g)** $ClNO(g)$ **i)** $HOCl(g)$

b) $CH_4(g)$ **e)** $H_2S(g)$ **h)** $C_2H_4(g)$ **j)** $C_2H_2(g)$

c) $CF_4(g)$ **f)** $H_2O_2(\ell)$

• • •

Begin extension material

Metallic Bonding

You have learned about non-metals forming ionic bonds with metals and about non-metals forming covalent bonds with non-metals. Metals also bond with metals. How do they interact to bond with each other? Metallic atoms can lose electrons to other atoms, but because they cannot accept enough electrons to fill their outer energy level, they cannot form typical ionic bonds. Can metal atoms share electrons to form an octet of electrons around each atom? Consider sodium atoms that have only one valence electron. Try to imagine eight sodium atoms attempting to share the same eight electrons. Could you draw a Lewis structure for such a compound? Metallic atoms do not form covalent bonds with other metal atoms. Yet all metals except mercury are solid at standard temperatures. (Mercury is a liquid at standard temperature.) No substance could exist as a solid unless relatively strong attractive forces were holding the atoms together. If there were no attractive forces between the atoms, metals would be gases. The bonds that hold metal atoms together are neither ionic bonds nor covalent bonds but instead comprise a separate category of bonding.

In **metallic bonding**, all the atoms share all the valence electrons. The valence electrons are **delocalized,** so they are free to move from one atom to the next. You can visualize metallic bonding as a network of positive ions in a "sea" of electrons. The electrostatic attractive force between positively charged metal ions and the sea of negative electrons constitutes a metallic bond. The model, shown in Figure 1.16, is called the free-electron model.

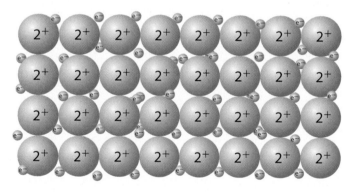

Figure 1.16 The diagram represents magnesium atoms that have released their electrons and are embedded in a sea of electrons.

End extension material

ChemistryFile

Web Link
Metal alloys such as bronze and pewter often have more desirable properties than pure metals. What is an alloy? What type of bonding holds alloys together.

www.albertachemistry.ca
WWW

Naming Compouds

"Please pass the sodium chloride." "Do you have enough sucrose for your iced tea"? "Oh, these biscuits are so hard and flat. I must have forgotten to put the sodium hydrogen carbonate in the dough"! "I'm so thirsty, I could drink a whole pitcher of dihydrogen monoxide"! It is unlikely that you have ever heard these statements around the dinner table. However, you have probably heard similar statements, but the terminology was different (see Figure 1.17). Some of the terms in these sentences are chemical terms for common substances. Do you know what they are?

Figure 1.17 Do you know the chemical names for any of the foods that you eat?

Briefly review the naming of compounds by analyzing two of the compounds named in the discussion at the dinner table. First, sodium chloride (NaCl(s)) is common table salt. Sodium is a metal and chlorine is a non-metal making sodium chloride an ionic compound. According to the periodic table, when sodium is ionized, it has one positive charge and when a chlorine atom is ionized it has one negative charge. Because they must combine in a one to one ratio, there is no need to identify the relative numbers of ions in a formula unit. The name of the metal is first and the name of the non-metal, with the suffix changed to "-ide", is last. The name, sodium chloride, completely describes the compound.

Sucrose ($C_{12}H_{22}O_{11}$(s)) is common table sugar. Carbon, hydrogen, and oxygen are all non-metal elements. Therefore, sucrose is a covalently bonded or molecular compound. Because sucrose is a carbon-based compound (see page 11), it is named according to the rules for naming organic compounds (Chapter 14).

Sodium hydrogen carbonate (NaHCO$_3$(s)) contains sodium Na$^+$(s), a metal, and hydrogen carbonate HCO$_3{}^-$(s), a negatively charged polyatomic ion. Therefore, it is an ionic compound. The name of the metal comes first and the name of the negatively charged polyatomic ion comes last. To find the name of any of the polyatomic ions that you will encounter in this text, you can consult your table of polyatomic ions.

The common name for dihydrogen monoxide ($H_2O(\ell)$) is simply water. Although hydrogen is in Group 1, it does not have the properties of a metal under standard conditions but behaves more like a non-metal. In water, the oxygen and hydrogen atoms are covalently bonded. Therefore, it is named according to the rules for naming molecular compounds. The name of the element to the left of the periodic table comes first and the element to the right is named last. Because water has two hydrogen atoms, the prefix "di-" must be used. Because the oxygen atom is named last, the number or oxygen atoms must be indicated by the prefix "mono-." Thus the IUPAC name of water is dihydrogen monoxide. The following concept map will help you recall the method for naming some simple chemical compounds.

Naming Simple Compounds

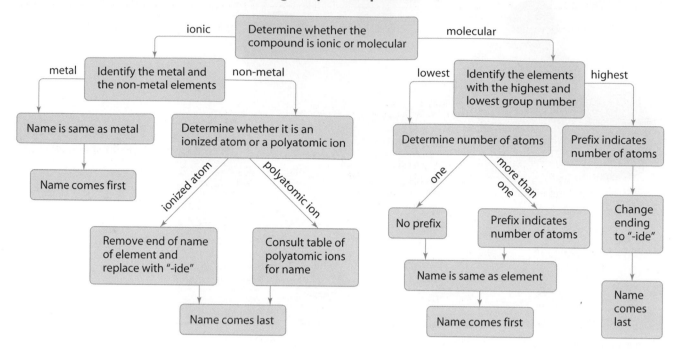

Before proceeding to the next section, check your skills of chemical **nomenclature**—the naming of chemical compounds—by answering the questions below. If you do not feel confident about your naming skills, review pages 10 through 12.

9 Name the following compounds:

a) $SiO_2(s)$ d) $N_2O_5(s)$ g) $(NH_4)_2HPO_4(s)$ j) $Mg(NO_3)_2(s)$

b) $CsI(s)$ e) $Ba(OH)_2(s)$ h) $HClO_3(\ell)$ k) $KClO(s)$

c) $Na_2HPO_4(s)$ f) $NBr_3(\ell)$ i) $Al_2O_3(s)$ l) $KMnO_4(s)$

Section 1.1 Summary

- A filled outer energy level makes an atom stable, or unreactive.
- When chemical bonds form, the atoms gain, lose, or share electrons in such a way that they create a filled outer energy level.
- Unpaired electrons participate in bonding.
- You can determine the valence of main group elements by the position of their column in the periodic table.
- The periodic table gives you the charge of the commonly occurring ions of the transition metals.
- Metals lose electrons to form positively charged ions. Non-metals gain electrons to form negatively charged ions. Oppositely charged ions attract each other and form ionic bonds.
- Non-metal atoms share electrons to form covalent bonds.
- Electron dot diagrams of molecules are called Lewis structures.
- In structural formulas, a covalent bond is represented by a straight line. Lone pairs are not included.
- In metallic bonding, the valence electrons of all of the atoms are delocalized and can move from one atom to the next. The metal ions are attracted to all of the nearby electrons.

1. Explain how observations of the noble gases lead, in part, to the octet rule.

2. Define the following terms:
   a) valence electrons   c) unpaired electron
   b) lone pair   d) ionic bond

3. Use electron dot diagrams to predict the ratio of metal and non-metal ions in ionic compounds formed by the following pairs of elements:
   a) magnesium and fluorine
   b) potassium and bromine
   c) rubidium and chlorine
   d) calcium and oxygen

4. In some electron dot diagrams, a chemical symbol may have no dots. Explain the meaning of such a symbol.

5. Draw Lewis structures of molecules formed by the following combinations of atoms:
   a) one carbon atom bonded to two sulfur atoms
   b) one carbon atom, three hydrogen atoms, and one chlorine atom
   c) two iodine atoms
   d) two bonded carbon atoms, with three hydrogen atoms bonded to one of the carbon atoms, and one hydrogen atom and one oxygen atom bonded to the other carbon atom

6. Examine the Lewis structures shown below. Identify one error in each diagram.

   a) 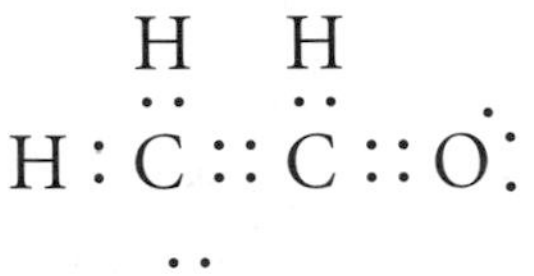 $H : C :: C :: O$ (with H, H above and dots)

   b) $: F : C : F :$ with $: F :$ above and $: H :$ below

   c) $H : C :: O :$ with H above and H below

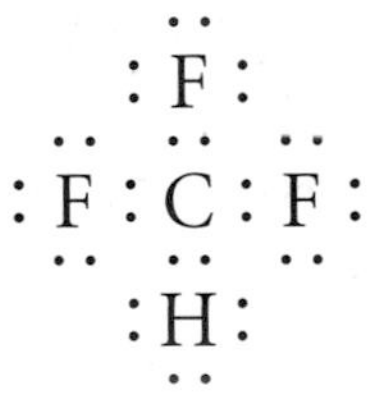

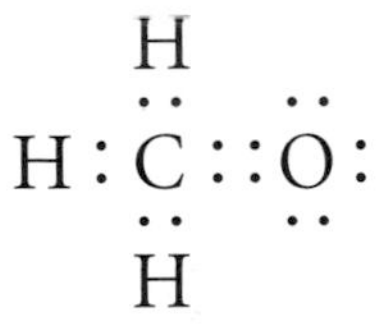

7. Name the following binary ionic compounds:
   a) $MgCl_2(s)$ c) $FeCl_3(s)$ e) $ZnS(s)$
   b) $Na_2O(s)$ d) $CuO(s)$ f) $AlBr_3(s)$

8. Write formulas for the following compounds:
   a) iron(III) chloride   d) calcium phosphide
   b) magnesium oxide   e) manganese(II) sulfide
   c) potassium bromide   f) barium iodide

9. Name the following ionic compounds:
   a) $K_2CrO_4(s)$ d) $Sr_3(PO_4)_2(s)$
   b) $NH_4NO_3(s)$ e) $KNO_2(s)$
   c) $Na_2SO_4(s)$ f) $Ba(ClO)_2(s)$

10. Write formulas for the following compounds:
    a) sodium hydrogen carbonate
    b) sodium thiosulfate
    c) sodium hypochlorite
    d) lithium nitrite
    e) potassium permanganate
    f) ammonium chloride

11. Name the following binary molecular compounds:
    a) $SO_2(g)$ c) $N_2O_5(s)$
    b) $CO(g)$ d) $Cl_2O(g)$

12. Write formulas for the following compounds:
    a) dihydrogen monoxide   c) silicon tetrachloride
    b) sulfur trioxide

13. In the following table, the letters A through G represent an ion or atom of some element. Answer the following questions:
    a) Which letters represent atoms and which represent ions?
    b) For a letter that represents an ion, what is the charge?
    c) Write the chemical symbol for each atom or ion.

	Atom or ion						
	A	B	C	D	E	F	G
Number of electrons	5	10	18	28	36	5	9
Number of protons	5	7	19	30	35	5	9
Number of neutrons	5	7	20	36	46	6	10

14. Identify the error in each of the following statements or phrases and correct it:
    a) four molecules of potassium bromide
    b) the compound $NaHSO_4(s)$ is sodium sulfate
    c) the compound $KNO_2(s)$ is potassium nitrate

15. Write structural formulas for the following Lewis structures:

    a) 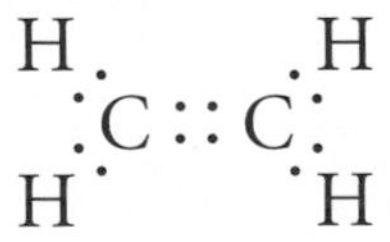 $: C :: C :$ with H, H on left and H, H on right

    b) $: F : N : F :$ with $: F :$ below

    c) $H : C :: N :$

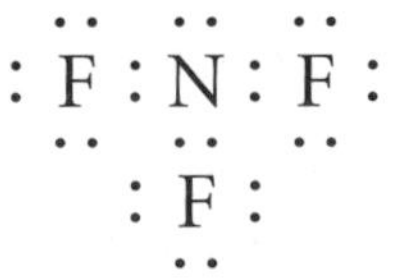

The Nature of Chemical Bonds

Section Outcomes

In this section, you will:
- **define** electronegativity
- **describe** bonding as a continuum

Key Terms

electronegativity
polar covalent bonds
bond dipoles
electronegativity difference
non-polar covalent bonds

Figure 1.18 In the main group elements, the electronegativities increase moving from left to right across the period, except for the noble gases. Within a group, the electronegativities increase moving up the group, as shown by the arrows.

When you began to study ionic bonds and covalent bonds, you might have developed the impression that these two types of bonds are quite different and distinct. In this section, you will discover that the differences are not as great as they might have first seemed to be. How can an electrostatic force between unlike charges (an ionic bond) be similar to the sharing of electrons by two different atoms (a covalent bond)?

You can envision any bond as an attractive electrostatic force between the positively charged nucleus of one atom and the negatively charged electrons of another atom. In some cases, the electrostatic force removes an electron from one atom and binds it to the other, creating an ionic bond. In other cases, the nuclei of two atoms each attract the electrons of the opposite atom with attractive forces that are similar in strength. In such cases, the two atoms share the electrons, thus forming a covalent bond. The extent to which a nucleus attracts shared electrons varies with the element. These differences in attractive forces produce bonds that have different properties. To examine these properties, you need to learn about the concept of electronegativity.

Electronegativity

The nucleus of an atom is positively charged and therefore attracts electrons, which are negatively charged. The **electronegativity** (*EN*) of an element is the relative measure of the ability of that element's atoms to attract the shared electrons in a chemical bond. Chemists have devised a scale to quantify the electronegativities of the elements. Higher electronegativities indicate a greater attraction for the electrons. Fluorine—the element with the highest electronegativity—has a value of 4.0. Values of electronegativity for most of the elements are shown in Figure 1.18 in the form of a periodic table.

Electronegativities

increasing →

increasing ↑

1 H 2.2																	2 He -
3 Li 1.0	4 Be 1.6											5 B 2.0	6 C 2.6	7 N 3.0	8 O 3.4	9 F 4.0	10 Ne -
11 Na 1.0	12 Mg 1.3											13 Al 1.6	14 Si 1.9	15 P 2.2	16 S 2.6	17 Cl 3.2	18 Ar -
19 K 0.8	20 Ca 1.0	21 Sc 1.4	22 Ti 1.5	23 V 1.6	24 Cr 1.7	25 Mn 1.6	26 Fe 1.8	27 Co 1.9	28 Ni 1.9	29 Cu 1.9	30 Zn 1.7	31 Ga 1.8	32 Ge 2.0	33 As 2.2	34 Se 2.6	35 Br 3.0	36 Kr -
37 Rb 0.8	38 Sr 1.0	39 Y 1.2	40 Zr 1.3	41 Nb 1.6	42 Mo 2.2	43 Tc 2.1	44 Ru 2.2	45 Rh 2.3	46 Pd 2.2	47 Ag 1.9	48 Cd 1.7	49 In 1.8	50 Sn 2.0	51 Sb 2.1	52 Te 2.1	53 I 2.7	54 Xe -
55 Cs 0.8	56 Ba 0.9		72 Hf 1.3	73 Ta 1.5	74 W 1.7	75 Re 1.9	76 Os 2.2	77 Ir 2.2	78 Pt 2.2	79 Au 2.4	80 Hg 1.9	81 Tl 1.8	82 Pb 1.8	83 Bi 1.9	84 Po 2.0	85 At 2.2	86 Rn -
87 Fr 0.7	88 Ra 0.9		104 Rf -	105 Db -	106 Sg -	107 Bh -	108 Hs -	109 Mt -	110 Uun -	111 Uuu -	112 Uub -	113 -	114 Uuq -	115 -	116 Uuh -		

57 La 1.1	58 Ce 1.1	59 Pr 1.1	60 Nd 1.1	61 Pm -	62 Sm 1.2	63 Eu -	64 Gd 1.2	65 Tb -	66 Dy 1.2	67 Ho 1.2	68 Er 1.2	69 Tm 1.3	70 Yb -	71 Lu 1.0
89 Ac 1.1	90 Th 1.3	91 Pa 1.5	92 U 1.7	93 Np 1.3	94 Pu 1.3	95 Am -	96 Cm -	97 BK -	98 Cf -	99 Es -	100 Fm -	101 Md -	102 No -	103 Lr -

Placing the electronegativity values in a periodic table allows you to observe trends. Choose any period and look only at the main group elements. You will see that the electronegativities increase as you go from the metals on the left to the non-metals on the right. When you look to the far right on the table, you will see that the noble gases do not have values for electronegativities. Because the noble gases do not naturally form bonds, and very few artificial compounds containing the noble gases have been synthesized, electronegativities have not been assigned for these elements.

Choose any one group and compare the electronegativities of the elements in that group to see another trend. As you look down the column, the values for electronegativity decrease. Francium, the element in the lower left corner of the periodic table, has the lowest electronegativity of any element. Interestingly, francium and fluorine, with the highest and lowest electronegativities, are two of the most chemically reactive elements.

Size and Electronegativity

Because the order of the elements in the periodic table is based on atomic number, it would seem logical that there should be trends in the size of the atoms of each element within the periodic table. You may, however, be surprised at the nature of those trends. The diagram in Figure 1.19 shows both the size and the electronegativity of the main group elements. The height of each bar represents the electronegativity, and the size of the sphere shows the relative size of a neutral atom of the element. Notice that, for any given period, as you scan from left to right, the electronegativity and the atomic number increase, while the size of the neutral atoms decreases. How can the size of atoms decrease when the mass is getting larger? How might this phenomenon affect electronegativity?

Chemistry File

FYI
Until the early 1960s, chemists believed that it was not possible for the noble gases to form bonds with any atoms. In 1962, Neil Bartlett of the University of British Columbia synthesized the first compound that contained a noble gas.

Chemistry File

Web Link
Periodic tables provide a wealth of information. What did the first periodic table look like? How many different trends have been represented on periodic tables?

www.albertachemistry.ca
WWW

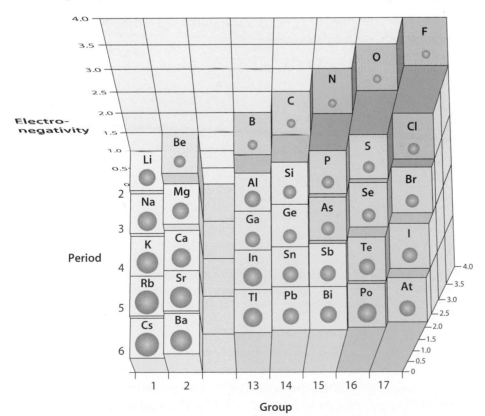

Figure 1.19 In this three-dimensional image of the main group elements, the electronegativity scale is "vertical." The period and group numbers are on the left and bottom, respectively.

As you learned when you studied models of the atom, most of an atom is empty space. The size of an atom depends on the radius of the energy level containing the valence electrons. In any specific period, the valence electrons of each atom occupy the same energy level. Based on this information, you may think that the atoms in one period should be the same size. However, the charge of the nucleus affects the radius of the energy levels. As the number of protons in the nucleus (atomic number) increases, the net attractive force on the electrons increases, pulling them closer to the nucleus. Therefore, as the atomic number increases within a period, the radius of every energy level decreases and the atom is smaller. Scanning across any period in Figure 1.19, you can see that the atoms become smaller. For example, examine Period 2, starting with lithium and ending with fluorine. Notice that fluorine has nearly three times the mass of lithium, yet its radius is less than half that of lithium. As you scan down through any individual group in Figure 1.19, you will see that the size of the atoms becomes larger. This observation agrees with the concept that the valence electrons are in sequentially higher energy levels.

To find explanations for the trends in electronegativity, consider the factors that affect the strength of the attractive electrostatic force between a positive charge and a negative charge.

- The attractive force between opposite charges decreases with the square of the distance between the charges. For example, if the distance between them doubles, the force is four times smaller.

- The attractive force between opposite charges is directly proportional to the magnitude of the charges.

Now apply these properties of electrostatic forces to the negatively charged electrons in the valence level of one atom and the positively charged nucleus of an adjacent atom. As shown in Figure 1.20, the outer electrons from one atom can get as close to the nucleus of another atom as the radius of that atom.

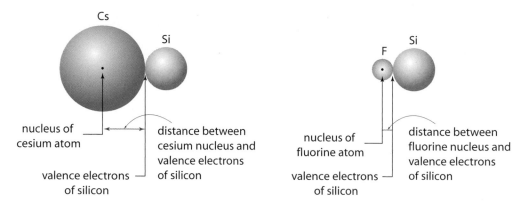

Figure 1.20 Because the valence electrons of an atom, for example, silicon, can get much closer to the nucleus of fluorine than they can to cesium, the attractive force of the fluorine nucleus on those valence electrons is much greater than the attractive force of the cesium nucleus on the valence electrons.

Because fluorine has the smallest radius of all the elements, electrons from other atoms can get closer to its nucleus than to that of any other atom. As well, the nucleus of a fluorine atom has a larger positive charge than an atom of any other element in the second period. Consequently, the positively charged fluorine nucleus can attract valence electrons of other atoms more strongly than can any other nucleus.

Because the strength of the attractive electrostatic force increases with the charge of the nucleus, you might expect that atoms of elements with very large atomic numbers would have larger electronegativities than they have. However, you must also consider the negatively charged electrons in the atom. The electrons in lower, filled energy levels reduce the effective charge of the nucleus by creating a "cloud" of negative charge around the positive nucleus. This "cloud" of electrons shields the positive charge of the nucleus so nearby electrons "see" a less positive nucleus.

Bond Type and Electronegativity

How do these trends in electronegativity relate to the properties of ionic and covalent bonds you learned about? First, consider ionic bonds that form between metal and non-metal atoms. When you examine the electronegativities of metals in Figure 1.18, you can see that they are relatively small. Metals do not strongly attract electrons in a bond. As well, you can see that non-metals have large electronegativities and, thus, strongly attract electrons involved in a bond. These observations correlate very well with the concept that metals lose electrons, non-metals gain electrons, and the resulting ions are attracted to each other, thus forming ionic bonds.

Next, consider the opposite extreme—bonds between atoms of elements that have the same electronegativity. For example, all of the elements that exist as diatomic molecules, such as hydrogen, nitrogen, and oxygen gases, have bonds between two identical atoms. Because the electronegativities are identical, the two nuclei attract the electrons of the bond with exactly the same strength. Neither atom loses nor gains electrons. They share electrons in covalent bonds and the molecules are symmetrical.

Thus far, the concept of electronegativity has given you the same information as you can learn by applying the octet rule. Some variations in bond structure begin to emerge when you consider covalent bonds between atoms of unlike elements, such as the bond between carbon and chlorine atoms in $CHCl_3(g)$ (see Figure 1.13 for trichloromethane or chloroform). Notice that the electronegativity of carbon is 2.6, while that of chlorine is 3.2. The electronegativity of chlorine is significantly higher than that of carbon, indicating that chlorine attracts electrons more strongly than does carbon. As a result, the atoms do not share electrons equally and those that are shared spend more time near the chlorine atom than near the carbon atom. Thus, the end of the bond near the carbon atom will be slightly positive. The other end of the bond—the end near the chlorine atom—will be slightly negative.

Chemists call these partial charges and designate them with the lowercase Greek letter delta (δ). The symbol δ^+ means that the end of the molecule has a partial positive charge that is less than 1. The symbol δ^- means that the end of the molecule has a partial negative charge that is less than 1. Bonds that have this separation of charge or unequal sharing of electrons are called **polar covalent bonds**. An arrow is often used to indicate the end at which electrons tend to spend more time (see Figure 1.21). Because these bonds have a negative "pole" and a positive "pole," they are sometimes called **bond dipoles**.

The difference in the electronegativities of the elements involved in a bond determines some of the characteristics of the bond. Chemists have devised a scheme by which you can use the **electronegativity difference** (ΔEN) of two elements to predict the nature of the bond between atoms of those two elements. You can calculate the electronegativity difference between any two elements—call them A and B—by finding the electronegativity of each element on a table or chart, and then subtracting the smaller electronegativity from the larger electronegativity.

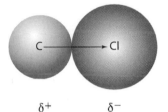

δ^+ δ^-

Figure 1.21 The chlorine nucleus attracts the electrons more strongly than the carbon nucleus does. Therefore, the chlorine end of the bond is slightly negative and the carbon end of the bond is slightly positive.

Begin extension material

As shown in Figure 1.22, if the electronegativity difference is greater than 1.7, the bond can be considered ionic. The term "mostly" is included in the figure because there is always some attraction between the nucleus of one atom and the electrons of the other. If the electronegativity difference for two atoms is less than 0.5, the bond will be only slightly polar. If the magnitude of the electronegativity difference for two elements is between 0.5 and 1.7, the bond will be a polar covalent bond. If the electronegativity difference is zero, the atoms are identical and the bond will not be polar. Such bonds are called **non-polar covalent bonds**.

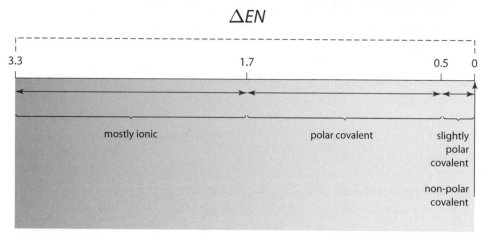

Figure 1.22 The nature of chemical bonds changes in a continuous way, creating a broad range of characteristics.

Some examples will clarify the meaning of the diagram in Figure 1.22. Analyze the bond between potassium and fluorine. The electronegativity of fluorine is 4.0 and the electronegativity of potassium is 0.8. The electronegativity difference is calculated by subtracting the smaller number from the larger number, as shown below:

$$\Delta EN = EN_F - EN_K$$
$$\Delta EN = 4.0 - 0.8$$
$$\Delta EN = 3.2$$

Because 3.2 is much greater than 1.7, the bond between potassium and fluorine is mostly ionic. Next, look at the bond between the two oxygen atoms in an oxygen molecule. The electronegativity of oxygen is 3.4. The electronegativity difference shown below is zero, indicating that the bond is non-polar covalent.

$$\Delta EN = EN_O - EN_O$$
$$\Delta EN = 3.4 - 3.4$$
$$\Delta EN = 0$$

Finally, calculate the electronegativity difference of a carbon to chlorine bond that was used in the discussion above. The electronegativity of carbon is 2.6 and the electronegativity of chlorine is 3.2. The calculation below shows that the electronegativity difference is 0.6. This magnitude of the value is between 1.7 and 0.5, which indicates that the bond is a polar covalent bond.

$$\Delta EN = EN_{Cl} - EN_C$$
$$\Delta EN = 3.2 - 2.6$$
$$\Delta EN = 0.6$$

10. For each of the following pairs of atoms, predict whether a bond between them will be non-polar covalent, slightly polar covalent, polar covalent, or mostly ionic:

a) carbon and fluorine

b) oxygen and nitrogen

c) chlorine and chlorine

d) copper and oxygen

e) silicon and hydrogen

f) sodium and fluorine

g) iron and oxygen

h) manganese and oxygen

11. For each polar bond in Q10, indicate the locations of the partial positive and partial negative charges.

12. Arrange the bonds in sets (a) and (b) below in order of increasing polarity:

a) hydrogen bonded to chlorine, oxygen bonded to nitrogen, carbon bonded to sulfur, sodium bonded to chlorine

b) carbon bonded to chlorine, magnesium bonded to chlorine, phosphorous bonded to oxygen, nitrogen bonded to nitrogen

Because there is no clear distinction between ionic and covalent bonds, and because bond categories can be viewed as a continuum, chemists sometimes assign a percent ionic or covalent character to bonds. Table 1.1 lists some electronegativity differences with the assigned percent character of the two types of bonds.

Table 1.1 Character of Bonds

Electronegativity difference	0.00	0.65	0.94	1.19	1.43	1.67	1.91	2.19	2.54	3.03
Percent ionic character (%)	0	10	20	30	40	50	60	70	80	90
Percent covalent character (%)	100	90	80	70	60	50	40	30	20	10

A good example of the overlap in ionic and covalent bond character is found in hydrogen chloride (see Figure 1.23). The electronegativity difference between hydrogen and chlorine is 1.0, which lies in the polar covalent region. When hydrogen chloride is a gas, it behaves as a covalent compound. However, when hydrogen chloride dissolves in water, it ionizes and becomes hydrochloric acid. As you can see, the classification of bond type is not always simple.

gas

in solution in water

δ^+ δ^- $\longleftrightarrow$

H : Cl :   HCl $\left[\text{H} \right]^+$ $\left[: \text{Cl} : \right]^-$

Figure 1.23 The bond between the hydrogen and chlorine atoms appears to have different properties when it is in air and when it is in water. How might water affect the bonding?

End extension material

Note, also, in Figure 1.23 a symbol that chemists sometimes use to indicate the polarity of a compound in conjunction with the chemical formula. The tail of the arrow (above the H) looks like a plus sign and indicates that this end of the molecule is slightly positive. The arrowhead pointing to the right indicates that the electrons spend more time around the chlorine atom, making it slightly negative.

In summary, bonds seem to form a continuum. There are no distinct differences between slightly polar covalent bonds and polar covalent bonds. There is not even a clear distinction between covalent and ionic bonds. This continuum of polarity is one of the factors that gives molecules and compounds such a great variety of characteristics.

Section 1.2 Summary

- The electronegativity of an element is the ability of that element's atoms to attract the electrons that form a bond with another atom.
- For the main group elements, the electronegativity tends to increase as the group number increases (moving left to right on the periodic table).
- The electronegativity of an element tends to increase as the size of the neutral atom of that element decreases. In other words, the electronegativity of an element tends to decrease as you move down a group.
- Within each period, as the charge of a nucleus increases, the radius of the neutral atom decreases.
- Non-metals have high electronegativities and metals have low electronegativities, allowing the non-metals to remove electrons from metals and thus form ionic bonds.
- When their electronegativities are similar in magnitude, atoms of the two elements tend to share electrons in a covalent bond.
- When the electronegativities of two bonded non-metals are quite different, the atom with the higher electronegativity attracts the electrons in the bond more strongly, resulting in a polar bond.

SECTION 1.2 REVIEW

1. Define electronegativity.

2. Explain how an electrostatic attractive force can be responsible for both ionic and covalent bonds.

3. Describe the trend in electronegativities within one period on the periodic table as the atomic number increases.

4. As you scan down one group on the periodic table, what is the trend in the electronegativities of the elements?

5. What is the relationship between the size of an atom and the electronegativity of the element?

6. Explain the meaing of "bonding continuum."

7. Arrange the elements in sets (a) through (d) in order of increasing attraction for the electrons involved in a bond:
   a) N and O   c) H and Cl
   b) Mn and O   d) Ca and Cl

8. What types of atoms form bonds that are:
   a) mostly ionic
   b) polar covalent
   c) non-polar covalent

9. Using Figure 1.18 on page 36, find atoms of two different elements that, when bonded together, would form a non-polar bond.

10. Describe the nature of the chemical bond that would form between the following pairs of elements:
   a) Na and Cl
   b) N and O
   c) P and H
   d) H and F

Nature provides many clues about the fundamentals of chemical bonding. Metals combined with non-metals (such as ores) are solid, indicating strong bonds. Non-metals bonded to other non-metals can be solid, liquid, or gaseous under normal conditions, indicating a variety of possible bond strengths. Atoms of the noble gases are the only elements that are never found bonded to other atoms in nature. These observations, along with the experimental discovery that all noble gases have eight electrons in their outer energy level, leads to the octet rule. When bonds form between atoms, the atoms gain, lose, or share electrons in such a way that they create outer energy levels with eight electrons.

To form ionic compounds, metal atoms lose electrons and become positively charged while non-metal atoms gain electrons and become negatively charged. The oppositely charged ions are then attracted to one another and form ionic bonds. The octet rule allows you to predict the number of electrons that each element must gain or lose to become a stable ion. The fact that ionic compounds must have a neutral net charge leads you to the ratios in which specific metal and non-metal atoms must combine.

To form molecular compounds, non-metals share electrons according to the octet rule. The octet rule thus allows you to use electron dot diagrams of non-metal atoms to predict the bonding patterns, in the form of Lewis structures, for molecules. A set of steps makes it possible to draw Lewis structures for many simple molecules.

It is critical to note that the octet rule is not a rigorous law of nature but instead a guideline for many compounds. Although you will not encounter any such compounds in this text, compounds that do not follow the octet rule do exist.

Although Lewis structures provide a detailed description of the bonding in a molecule, they are tedious to draw and can be difficult to analyze. Structural formulas are easier to draw and analyze. To convert a Lewis structure to a structural formula, draw a single line between atoms for every pair of shared electrons. Lone pairs are not represented in structural formulas.

Chemical formulas for compounds include the symbol for the compound with subscripts to indicate the ratio of atoms of the given element in a formula unit (ionic compounds) or molecule (covalently bonded compounds). The names of ionic compounds include the name of the metal followed by the name of the non-metal with the ending changed to "-ide." The names of molecular (covalently bonded) compounds include prefixes to indicate the number of atoms of each element in a molecule. The only exception is the case in which the molecule contains only one atom of the element named first, and no prefix is used. The element name that comes second always has a prefix. The element with the lower group number is named first. The element with the largest group number is named second and its ending is replaced with "-ide."

Concept Organizer Bonding

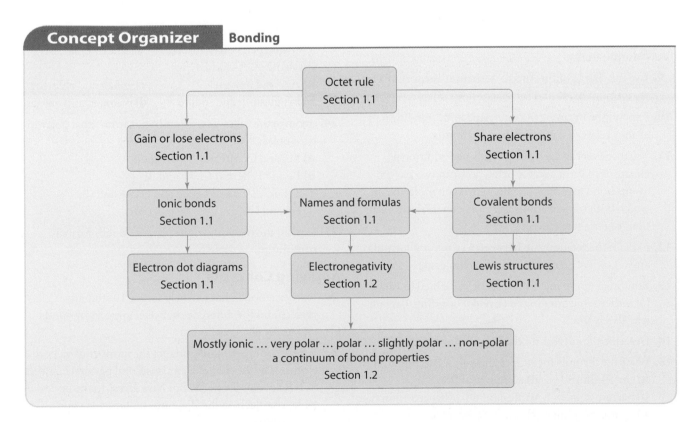

Understanding Concepts

1. Explain how a property of the noble gases leads to the octet rule.

2. When hydrogen is involved in a bond, why does it not conform to the octet rule? What other element does not conform to the octet rule?

3. Describe an electron dot diagram.

4. Draw electron dot diagrams for the following atoms:
   a) H   **b)** N   **c)** Ba   **d)** I

5. Use electron dot diagrams to predict the formulas of the following compounds:
   a) potassium bromide   **c)** magnesium oxide
   b) calcium fluoride   **d)** lithium oxide

6. When you examine the electron dot diagram of a non-metal, such as nitrogen or sulfur, how can you predict the number of electrons that will most likely be involved in bonding? How can you predict the number of electrons that will *not* be involved in a bond?

7. Draw Lewis structures of each of the following compounds:
   a) $CS_2(s)$
   b) $Br_2O(g)$ (unstable at standard ambient temperature and pressure, or SATP)
   c) $ClF(g)$
   d) $NF_3(g)$

8. Use electron dot diagrams to illustrate, in detail, the formation of ionic bonds between calcium atoms and chlorine atoms.

9. Use the charge of the ions to determine the formula for strontinum oxide and for magnesium nitride.

10. Explain the meaning of "lone pair" and "bonding pair" as they relate to covalently bonded atoms.

11. Because every oxygen atom needs two additional electrons to form a stable octet, how can two oxygen atoms bond together to make an oxygen molecule in which both oxygen atoms have an octet of electrons in their outer energy levels?

12. Explain the relationship between a structural formula and a Lewis structure for the same molecule.

13. Bonding in pure metals cannot be explained by models for either ionic bonding or covalent bonding. Explain why this is true.

14. Describe the current model for metallic bonding.

15. Write the formula for each of the following compounds:
   a) tin(II) fluoride   **d)** cesium bromide
   b) barium sulfate   **e)** ammonium hydrogen phosphate
   c) hydrogen cyanide   **f)** sodium periodate

   g) potassium bromate   **i)** calcium hypochlorite
   h) sodium cyanate   **j)** lead (IV) phosphate

16. Name the following compounds:
   a) $HIO_2(s)$ **d)** $K_2Cr_2O_7(s)$ **g)** $Al_2(SO_4)_3(s)$
   b) $CsF(s)$ **e)** $KClO_4(s)$ **h)** $Fe(IO_4)_3(s)$
   c) $NaHSO_4(s)$ **f)** $N_2Cl_2(g)$

17. Write formulas for the following compounds:
   a) chlorine monofluoride
   b) phosphorous pentoxide
   c) silicon disulfide
   d) diphosphorus triselenide

18. Name the following compounds:
   a) $CS_2(\ell)$ **c)** $SO_2(g)$ **e)** $PCl_3(\ell)$
   b) $ICl(s)$ **d)** $N_2O_3(s)$ **f)** $SeBr_2(g)$

19. The concept of electronegativity applies to an affinity for which electrons in an atom?

20. Summarize the trends in electronegativity within a period and within a group on the periodic table.

21. Based only on their position in the periodic table, arrange the elements in each set in order of increasing attraction for electrons in a bond.
   a) Li, Br, Zn, La, Si
   b) P, Ga, Cl, Y, Cs

22. How does the sharing of bonding electrons differ between a non-polar covalent bond and a polar covalent bond?

23. State whether a bond that would form between the following pairs of atoms would be ionic or polar covalent:
   a) zinc and oxygen   **c)** cobalt and chlorine
   b) magnesium and iodine   **d)** nitrogen and oxygen

24. Arrange the following sets of bonds from most polar to least polar:
   a) Mn and O, Mn and N, Mn and F
   b) Be and F, Be and Cl, Be and Br
   c) Ti and Cl, Fe and Cl, Cu and Cl, Ag and Cl, Hg and Cl

25. Explain the meaning of the phrase, "bonds form a continuum."

Applying Concepts

26. Design a model that will help Grade 10 students understand the difference between ionic compounds and molecular compounds.

27. Explain how you would predict the number of valence electrons in the elements of the second period (Li, Be, B, C, N, O, F, Ne) if you did not have access to the periodic table. Use electron dot diagrams to illustrate your explanation.

28. a) Complete the following table:

Group #	Example	Electron dot diagram	# valence electrons	# lone pairs	# unpaired electrons (# bonds normally formed)	Nature of bonds formed (ionic, metallic, covalent)
1						
2						
13						
14						
15						
16						
17						
18						

b) Describe and explain the trends that you see.

29. Create a concept map to summarize the nature of chemical bonds.

30. Given the following information, identify element X.
   a) XF_4, where X is in Period 2
   b) CaX where X is in Period 2
   c) XH_2O, where X is in Period 2, and the molecule contains two single bonds and one double bond
   d) XH_3, where X is in Period 2 and the molecule contains only single bonds
   e) X_2H_2, where X is in Period 2, and the molecule contains one triple bond and two single bonds

31. Sparklers contain iron. When ignited, the iron combines with oxygen from the air.

Describe the nature of the bond:
   a) between the atoms of iron in the sparkler
   b) between the atoms of oxygen in the air
   c) between the iron and oxygen in the compound that is formed during combustion

32. Ammonia, NH_3, is a common ingredient found in household cleaners.

   a) Draw an electron dot diagram for both hydrogen and nitrogen.
   b) Draw a Lewis structure for the ammonia molecule.
   c) Indicate the electron pairs and unpaired electrons.
   d) Describe the nature of the bonds formed between the nitrogen and hydrogen atoms.

Making Connections

33. At the beginning of the chapter, you read about a few naturally occurring substances that provided some clues about the nature of chemical bonding. You read about ores that consist of metals and non-metals. You also read about properties of gases in the atmosphere and properties of water. Think of at least two other naturally occurring substances, their properties, and their chemical constituents, and relate these properties to the nature of bonding. Prepare a multimedia presentation that communicates the information about the substances you chose. **ICT**

34. Locate three cleaning products in your home and read the labels. For each product, predict whether it is an ionic compound, a molecular compound, or a combination of both.

35. The bonds in water are very polar. Table salt, NaCl(s), dissolves in water. Vegetable oil has only very slightly polar bonds and does not dissolve in water. Make a prediction about the way that polar bonds influence the solubility of a compound in water.

CHAPTER 2

Diversity of Matter

Chapter Concepts

2.1 Three-Dimensional Structures

- The shape of the smallest repeating unit in a crystal determines the shape of the entire crystal.

- The Valence-Shell Electron-Pair Repulsion (VSEPR) theory allows you to predict the three-dimensional structure of simple molecules.

2.2 Intermolecular Forces

- Some molecules form permanent dipoles, where the positive end of one molecule is attracted to the negative end of another.

- Hydrogen bonds can occur between molecules when, in each molecule, a hydrogen atom is covalently bonded to an atom with a high electronegativity.

- All molecules are attracted to one another by London (dispersion) forces. In non-polar molecules, these forces are the only attractive forces.

2.3 Relating Structures and Properties

- The state of a substance depends on the strength of the forces between molecules and the average kinetic energy (temperature) of the particles.

- Melting and boiling points of pure substances are an indication of the nature of the bonding in the substance.

- The conductivity of a substance depends on the freedom of electrons or ions to move past one another.

Water is a remarkable substance. It exists in nature in all three phases: solid, liquid, and gas. As a solid, water can be as delicate as a snowflake or as strong as an iceberg. As a liquid, it can form wispy clouds or support gigantic ocean liners. As a gas, water permeates the air you breathe.

In contrast to the complex properties of water, the water molecule is simple. It consists of two hydrogen atoms covalently bonded to one oxygen atom. What are the properties of these three atoms? What is the nature of the two identical bonds that hold the atoms together? The bonds holding water molecules together give it most of its important properties.

In this chapter, you will use the information about bonds that you learned in Chapter 1 and find out how these bonds lead to the three-dimensional structure of water and other molecules. You will learn how the structures influence the interactions among the molecules. Finally, you will discover how the shape of the molecules and their interactions are linked to the remarkable properties of water.

Crystalline Columns

How do stalagmites and stalactites form in caves? Why do crystals of different substances have different shapes? In this activity, you will make some observations and attempt to answer these questions.

Safety Precautions

- Wash your hands when you have completed the investigation.

Materials

- water
- sodium acetate trihydrate crystals* (T) (☠)
- balance
- 10 mL graduated cylinder
- 100 mL Erlenmeyer flask
- hot plate
- water bottle
- burette and stand
- glass plate
- forceps

Procedure

1. Measure 50 g of sodium acetate trihydrate (sodium ethanoate) and place it in a clean 100 mL Erlenmeyer flask.

2. Add 5 mL of water and heat slowly using a hot plate.

3. Swirl the flask until the solid dissolves. Wash any crystals on the inside of the flask down with a small quantity of water.

4. Pour the solution into a clean, dry burette.

5. Raise the burette as high as it will safely go on the burette stand. Place it on the lab bench with a glass plate directly below the burette.

6. Obtain a relatively large sodium acetate trihydrate crystal. With clean, dry forceps, place the crystal on the glass plate directly under the burette.

7. Gradually turn the burette stopcock until the solution drips out very slowly. There should be from one to three seconds between drops. Adjust the position of the glass plate so the drops fall directly on the crystal.

8. Observe the crystal for 10 min and record your observations. Continue to observe and record your observations every 10 min until the end of the class period.

Analysis

1. Describe your observations.

2. Is sodium acetate trihydrate an ionic or molecular compound?

3. Based on your knowledge of bonding for this type of compound, describe what you think might have been happening as the solution dripped onto the crystal.

4. Why do you think that the crystal on the plate was necessary?

5. Stalagmites and stalactites are formed from calcium carbonate. Based on your observations of the sodium acetate trihydrate, explain how the stalagmites and stalactites could form in caves.

6. Suggest possible reasons why solids made of different compounds might make different-shaped crystals or might not make crystals at all.

water molecule

*Note: The term "trihydrate" in sodium acetate trihydrate means there are three water molecules associated with every pair of sodium and acetate ions in the compound.

Three-Dimensional Structures

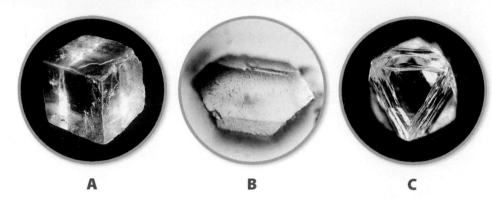

Figure 2.1 Photograph **(A)** is a crystal of sodium chloride or common table salt. **(B)** is a crystal of sucrose or common table sugar. **(C)** is a crystalline form of pure carbon or an uncut diamond. How are these crystals similar? How are they different?

What was the shape of the crystals you observed in the Launch Lab? Did they look like any of the crystals in Figure 2.1? Do all crystals consist of ionic compounds? Why do crystals have such different shapes? What characteristics of the chemical bonds that hold crystals together determine the structure of the crystal? You will find answers to these questions as you read this section.

Ionic Crystals

In Chapter 1, you focussed on the ionization of individual atoms and the combinations of the ions that resulted in a net neutral charge. For example, one sodium ion and one chloride ion make a neutral unit of sodium chloride (Figure 2.2A). In a real crystal of solid sodium chloride with a mass of approximately 1 mg, however, there are approximately 1×10^{21} sodium ions and the same number of chloride ions. Because an attractive force exists between positive and negative charges, the ions pack together very tightly, similar to the structure shown in Figure 2.2B. This three-dimensional pattern of alternating positive and negative ions is called a crystal lattice. To visualize the attractive forces, examine the "ball and stick" model in Figure 2.2C. The sticks represent the attractive forces between oppositely charged ions. Notice that each chloride ion is attracted to six adjacent sodium ions and that each sodium ion is attracted to the six chloride ions adjacent to it. Because all of the attractive forces are the same, there are no specific pairs of sodium and chloride ions that you could identify as "molecules." Each ion is attracted to *all* of the adjacent ions of the opposite charge. Therefore, the formula NaCl(s) simply means that the ratio of sodium to chloride ions is one to one in the entire crystal. Similarly, the formula $CaF_2(s)$ means that there is a ratio of one calcium ion to two fluoride ions in a crystal of calcium fluoride. The smallest ratio of ions in a crystal such as NaCl(s) is called a **formula unit** and not a molecule.

Figure 2.2 Sodium and chloride ions **(A)** pack together in crystals, as shown in the space-filling model **(B)**. However, it is easier to visualize the interactions among the ions when you examine a ball and stick model **(C)**. The relative sizes and packing of ions in a ball and stick model are not to scale.

A   **B**   **C**

⬤ Cl⁻   ● Na⁺

Sodium chloride forms cubic crystals, as you can see in Figure 2.1. Crystals of some compounds appear needle-like, while others form hexagonal columns. The shape of macroscopic crystals is influenced by the way the individual ions pack together. Because oppositely charged particles attract one another, ions will pack together so that oppositely charged ions are as close together as possible. The relative sizes and the charges of the ions also affect the pattern of the packing. If the crystals form very slowly, the packing of the ions will be very uniform, creating the repeating pattern that gives the crystal its symmetrical shape. Some examples will help you visualize the formation of crystals.

The smallest set of ions in a crystal for which the pattern is repeated over and over is called the *unit cell*. For example, the structure in Figure 2.2C represents a unit cell of sodium chloride. Since the ions on the sides of each unit cell are shared by the next unit cell, the depiction in Figure 2.3 is a more realistic, space-filling diagram of a sodium chloride unit cell. Notice that the ions on the sides of the unit cell appear to be sliced into pieces. The other parts of the ions belong to the adjacent unit cell.

When examining the sodium chloride models in Figures 2.2 and 2.3, you might have noticed that the chloride ion is roughly twice the size of the sodium ion. These relative sizes might appear to contradict the information about the sizes of atoms portrayed in Figure 1.19 on page 37. Figure 1.19 showed that sodium atoms are larger than chlorine atoms. Both pieces of information are, in fact, correct. The reason for the apparent contradiction is that the size of an ion is different than the size of the original atom. When a sodium atom becomes ionized, it loses the only electron that occupied the third energy level. The ion, therefore, has electrons in only the first and second energy level which have smaller radii than the third energy level. Therefore, the sodium ion is much smaller than the sodium atom. Conversely, when a chlorine atom becomes ionized to a chloride ion, it gains an electron, filling the third energy level. The chloride ion is thus larger than a sodium ion. The end result is a chloride ion that is nearly twice the size of a sodium ion.

When the relative size of ions in a crystal is different than it is in sodium chloride, the pattern of packing of ions is different. For example, the zinc ion ($Zn^{2+}(s)$) and the sulfide ion ($S^{2-}(s)$) in zinc sulfide are nearly the same. You can see the pattern of the packing of zinc and sulfide ions in Figure 2.4A. Another factor that affects the pattern of ions in a crystal is the relative charge on the ions. In Figure 2.4B, you see the pattern of ions in a unit cell of calcium fluoride ($CaF_2(s)$). Since calcium ($Ca^{2+}(s)$) has a charge of 2+ and the fluoride ion ($F^-(s)$) has only one negative charge, there must be two fluoride ions for each calcium ion. Examine Figure 2.4 until you can visualize the difference in the patterns.

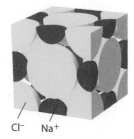

Figure 2.3 This is a space-filling model of a sodium chloride unit cell. There are about 10^{20} of these unit cells in a milligram of sodium chloride.

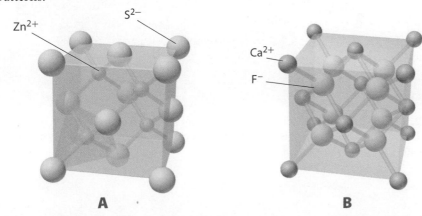

Figure 2.4 (A) The ions in this unit cell of ZnS(s) are not drawn to scale so that you can see the positions of the ions relative to on another. **(B)** This unit cell of $CaF_2(s)$ is an example of a crystal that has twice as many negative ions ($F^-(s)$) as positive ions ($Ca^{2+}(s)$). Remember that all of the calcium ions on the surface of the unit cell are shared with other unit cells.

• • •

(1) Name two characteristics of ions that help determine the way they will pack together in crystals.

(2) Give one reason why different combinations of ions pack together with quite different types of symmetry.

• • •

These examples are just a few of the wide variety of crystalline structures that exist. Nevertheless, these examples should give you a sense of the way in which ions align to form crystals. In case you think that crystals form only under carefully controlled laboratory conditions, study the photographs in Figure 2.5. The first photograph (A) is a geological feature named Devils Tower. Although geologists are not certain of the details of the formation of Devils Tower, they agree that it was created when molten rock pushed upward through sedimentary rock. When the molten rock began to cool, it was still below the surface of the ground. Some geologists propose that the lower part of the tower cooled so slowly that it took on a crystalline structure. Because the top part of the molten rock was closer to the surface, it cooled more rapidly and has no symmetrical shape. Over many thousands of years, the surrounding rock and soil eroded away, leaving the tower standing 264 m above the ground at its base.

Figure 2.5B is a photograph of possibly the largest geode ever discovered. A geode forms when something, such as a gas bubble, is trapped inside molten rock as it solidifies. As water with dissolved salts seeps into and through a hollow rock, the salt crystallizes, leaving a pocket of crystals inside a solid rock. The geode in the photograph is 8 m long, 1.8 m wide, and 1.7 m high. Some of the crystals are half a metre long.

Figure 2.5 (A) Each hexagonal column in Devils Tower is about 6 m across. **(B)** A geode is a rounded hollow rock lined with crystals. Most geodes can be held in one hand. As many as ten people can sit inside this giant geode found in Spain.

Building Ionic Crystals

Seeing a two-dimensional image of a three-dimensional structure often does not give you a clear picture of the object. Building models of crystal structures will provide a much clearer understanding of the structure of ionic compounds.

Materials

- polystyrene balls of two different sizes or gumdrops in two different sizes or colours
- toothpicks

Procedure

1. Carefully study the arrangement of the ions in sodium chloride in Figure 2.2 and zinc sulfide and calcium fluoride in Figure 2.4. Discuss with your partner how the ions are arranged in the three different types of crystals.

2. If you are using polystyrene balls, decide which size should represent sodium and which should represent chloride in a sodium chloride crystal. If you are using gumdrops, assign one colour to sodium and another colour to chloride.

3. Build a model of at least two unit cells of a sodium chloride crystal.

4. Repeat Procedure steps 2 and 3 for a zinc sulfide crystal.

5. Repeat Procedure steps 2 and 3 for a calcium fluoride crystal.

6. After you and your partner are satisfied that you have built your models correctly, compare your models with the models of another team. If your models are not the same, discuss the differences in the models with the other set of partners.

7. As a class, agree on the correct model for each type of crystals.

Analysis

1. What aspect did you find most difficult about building your models?

2. In general, were the class models nearly all the same? If not, what do you think caused different partners to build their models differently?

3. What do you think you can learn about crystals by building models that you cannot learn by looking at two-dimensional pictures?

Conclusion

4. How well do you think your models represent real crystals? Describe ways in which your models are similar to real crystals and ways in which they are different from real crystals.

Structures of Molecules

The structures of molecular compounds are much more varied than are those of ionic compounds, in part because of the nature of covalent bonds. As you read earlier, molecular compounds can be gases, liquids, or solids under standard conditions of temperature and pressure. To understand the structure and thus the properties of molecular compounds on the macroscopic level, you need to focus first on the structure of individual molecules.

In contrast to ionic compounds, in which ions are attracted to all the oppositely charged ions adjacent to them, molecular compounds have covalent bonds that exist between specific pairs of atoms. You can identify individual units—molecules—in which all atoms are covalently bonded together. A technique developed in 1957 makes it possible to predict the three-dimentional shape of many small molecules based on their Lewis structures.

Chemistry File

FYI

English chemist Ronald Gillespie and colleague Ronald Nyholm developed the method, now called VSEPR, for predicting the shape of molecules. In 1957, Dr. Gillespie joined the faculty of McMaster University in Hamilton, Ontario, where he had a long and distinguished career as a professor of chemistry.

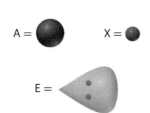

Figure 2.6 In the figures that show the VSEPR shapes, the red sphere (A) will represent the central atom, a black sphere (X) will represent any atom bonded to the central atom, and the teardrop shape with two black dots (E) will represent the electron cloud formed by a lone pair of electrons.

VSEPR and the Structures of Molecular Compounds

Valence-Shell Electron-Pair Repulsion theory allows you to predict the three-dimensional shape of a molecule. The name is usually abbreviated to VSEPR and pronounced "vesper." VSEPR theory is based on the electrostatic repulsion of electron pairs that causes them to take positions as far apart as possible in the molecule. A lone pair (LP) spreads out more than a bonding pair does. Therefore, the repulsion is greatest between two lone pairs (LP—LP). Bonding pairs are more localized between the atomic nuclei and, therefore, spread out less than do lone pairs. As a result, the repulsion between bonding pairs (BP—BP) is less than that between lone pairs. The repulsion between a bonding pair and a lone pair (BP—LP) is intermediate between the other two. Thus, the order of repulsion can be expressed as follows:

$$LP—LP > LP—BP > BP—BP$$

Another feature of the molecule that you must consider when you apply the VSEPR theory is multiple bonds: double bonds and triple bonds. Any bond—single, double, or triple—functions as one group when repelling other electrons around a central atom. In a molecule with several atoms, you can choose more than one central atom and analyze them one at a time. You could then predict the shape around two or even three atoms in the molecule.

When applying VSEPR theory to a central atom that has an octet of electrons in its valence level, you can classify the shapes into three general categories—linear, trigonal planar, and tetrahedral. Both trigonal planar and tetrahedral electron arrangements can then be divided into subcategories. As you examine the figures that show all of the categories, use the symbols and images in Figure 2.6 as a key to understand what the shapes represent. Notice that the letter "A" represents the central atom, "X" represents any atom bonded to the central atom, and "E" represents a lone pair. Each class of shapes in VSEPR can be expressed by AX_mE_n where the subscript "m" indicates the number of bonded atoms and the subscript "n" indicates the number of lone pairs on the central atom.

Linear

A central atom with two electron groups forms a **linear** shape because the electron groups exert repulsive forces on each other that place them as far apart as possible. Another way to describe the shape is to state that the angle between the two bonds—the **bond angle**—is 180° as shown in Figure 2.7A. These two electron groups could consist of two double bonds or a combination of one single bond and one triple bond. In any case, the central atom is bonded to two other atoms as shown in Figure 2.7B. The symbol for the linear shape is AX_2 because there are two atoms bonded to the central atom and there are no lone pairs. Examples of the two possible combinations of bonds are shown in Figure 2.7C. Carbon dioxide has two double bonds and hydrogen cyanide has one single bond and one triple bond yet both molecules are linear in shape.

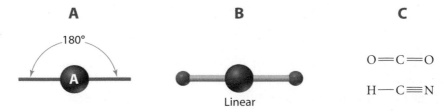

Figure 2.7 Carbon dioxide and hydrogen cyanide are both linear molecules of class AX_2.

Trigonal Planar

A central atom with three electron groups forms a **trigonal planar** shape as shown in Figure 2.8. As you can see in the figure, in the ideal trigonal planar shape, the bond angles are all 120°. The measured angles in real molecules are very close to 120°, as you will see in the examples.

Figure 2.9 shows an example of a trigonal planar shape in which all three electron groups consist of bonding electrons—group AX_3. The example in Figure 2.9B is methanal, $HCHO(\ell)$, which is also known as formaldehyde. Formaldehyde was previously used to preserve once-living specimens.

Trigonal Planar

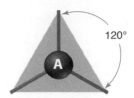

Figure 2.8 Trigonal planar shapes have one central atom with three electron groups that lie in a flat plane. All of the electron groups may consist of bonding electrons or one of the electron groups may be a lone pair.

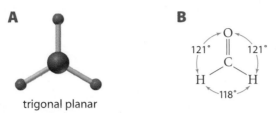

trigonal planar

Figure 2.9 The bond angle between the hydrogen–carbon bonds in methanal is 118° and the bond angles between each hydrogen–carbon bond and the oxygen–carbon bond are 121°.

Figure 2.10 shows an example of a trigonal planar shape in which one of the electron groups is a lone pair—group AX_2E. You probably recall from Chapter 1 that sulfur dioxide $(SO_2(g))$ is really a resonance structure and the structural formula in the figure, with one single bond and one double bond, is not totally accurate. Nevertheless, it is a good example of a group AX_2E shape. The bond angle between the sulfur–oxygen bonds is 119.5°. Molecules with the AX_2E shape are often called **bent**. The reason is clear when you look at the structural formula that shows only the atoms and bonds and not the lone pairs.

Bent (V shaped)

Figure 2.10 Sulfur dioxide is among the AX_2E molecules that are also called bent. Some other examples are ozone, $O_3(g)$, and nitrogen dioxide, $NO_2(g)$. Both of these molecules, like sulfur dioxide, are resonance structures with bonds that are partially single and partially double.

Tetrahedral

A central atom with four electron groups forms a **tetrahedral** shape as shown in Figure 2.11. As you can see in the figure, in the ideal tetrahedral shape, the bond angles are all 109.5°. Tetrahedral shapes can consist of four bonding groups (AX_4), three bonding groups and one lone pair (AX_3E), or two bonding groups and two lone pairs (AX_2E_2).

Figure 2.12 shows the AX_4 shape with four bonding groups. Methane, $CH_4(g)$, is an excellent example because it is a perfect tetrahedron. Because methane and other tetrahedral molecules are three dimensional, typical structural formulas cannot depict them correctly. The wedge-shaped bonds that you see in Figure 2.12 indicate whether a bond is directed out of the plane of the page or into the place of the page. In each case, the wide part of the wedge is closer to you, the observer, and the narrow end of the wedge is directed away from the observer. In the methane molecule in Figure 2.12B, the carbon–hydrogen bond that is a uniform line is in the plane of the page. The carbon–hydrogen bonds on either side have the wide part of the wedge by the carbon atom meaning that

Tetrahedral

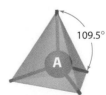

Figure 2.11 Tetrahedral shapes are three dimensional. You need to analyze the perspective until you can see that, in this image, lines going up to the left are in the plane of the page. The lower line on the right is coming out from the page and the upper line on the right is going into the page.

from the carbon atom, the bonds go into the plane of the page and the hydrogen atoms are behind the plane of the page. Conversely, the lower carbon–hydrogen bond has the wide part at the hydrogen atom and the point at the carbon atom. This hydrogen atom comes out from the plane of the page.

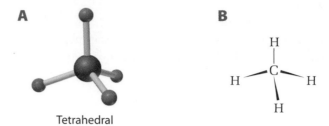

Figure 2.12 Bond angles are not shown for methane because they are the same as the ideal tetrahedral shape or 109.5°.

When a tetrahedral molecule has three bonding groups and one lone pair (AX_3E), it is also called **pyramidal**. The lone pair exerts a repulsive force on the three bonding groups and pushes them away causing the central atom to be the peak of the pyramid and the three other atoms to be the triangular base of the pyramid. Examine Figure 2.13 to visualize the pyramidal shape of the ammonia molecule. Because the lone pair exerts a greater force on the bonding groups than they exert on one another, the bond angles are smaller than the ideal 109.5°.

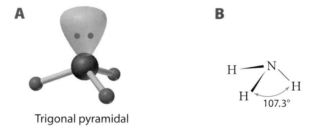

Figure 2.13 In AX_3E molecules such as ammonia, the shape is sometimes called trigonal pyramidal to indicate that the base of the pyramid has three corners.

When a tetrahedral molecule has two bonding groups and two lone pairs (AX_2E_2), it is also called **bent**. The two lone pairs exert a repulsive force on each other and on the two bonding groups than the bonding groups do on each other. Therefore, the bond angle between is much smaller than the ideal 109.5°. The bond angle in water, an excellent example of a bent molecule, is 104.5° as shown in Figure 2.14.

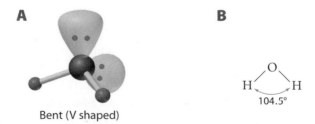

Figure 2.14 Water is a tetrahedral molecule because it has four electron groups. It is also a bent molecule because it has only three atoms but they are not linear.

You probably noticed that two different classes of shapes produce the bent shape. Sulfur dioxide is bent because it is trigonal planar with one lone pair (AX_2E). Water is bent because it is tetrahedral with two lone pairs (AX_2E_2).

Using VSEPR Theory to Predict Molecular Shapes

Table 2.1 summarizes the five molecular shapes that were described. Following the table, you will find the steps that will allow you to predict the shape of many small molecules. Refer to these steps and Table 2.1 as you work through the Sample and Practice Problems.

Table 2.1 Summary of Molecular Shapes

VSEPR class	Name of molecular shape	Type of electron pairs	Shape	Example
AX_2	linear	all BP		CO_2
AX_3	trigonal planar	all BP	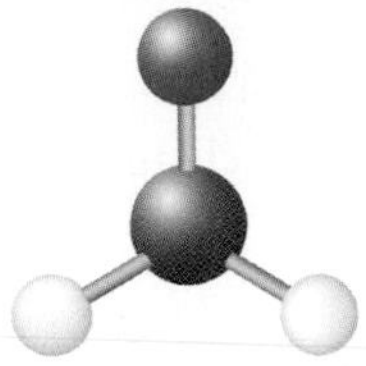	CH_2O
AX_2E	bent (trigonal planar electron groups)	2 BP, 1 LP	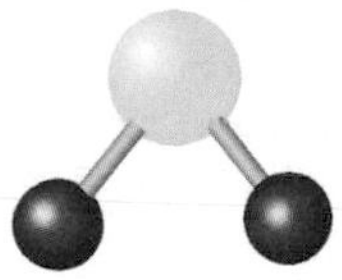	SO_2
AX_4	tetrahedral	all BP	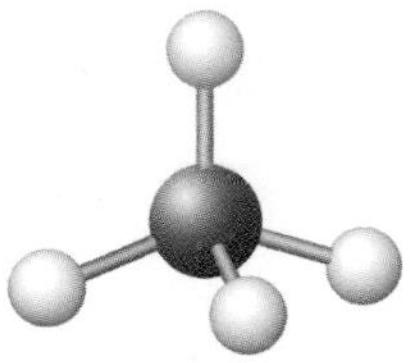	CH_4
AX_3E	trigonal pyramidal (tetrahedral electron groups)	3 BP, 1 LP	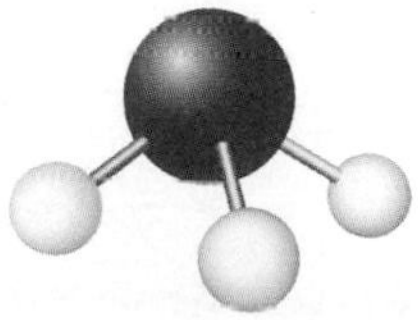	NH_3
AX_2E_2	bent (tetrahedral electron groups)	2 BP, 2 LP	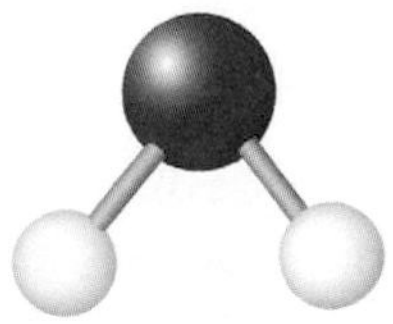	H_2O

Steps for Predicting Molecular Shapes

Step 1 Draw a preliminary Lewis structure of the molecule based on the chemical formula.

Step 2 Determine the total number of electron groups around a central atom. A double bond or a triple bond is counted as one electron group.

Step 3 Determine the types of electron groups (bonding pairs or lone pairs).

Step 4 Determine which one of the shapes in Table 2.1 will accommodate this combination of electron groups.

Predicting Molecular Shape for a Simple Compound

Problem
Determine the molecular shape of the hydronium ion, $H_3O^+(aq)$.

What Is Required?
molecular shape of $H_3O^+(aq)$

What Is Given?
The chemical formula tells you that the hydronium ion has three hydrogen atoms, one oxygen atom, and a charge of $1+$.

Plan Your Strategy
Follow the four-step procedure to help you predict molecular shapes.

Act on Your Strategy
Step 1 A possible Lewis Structure for $H_3O^+(aq)$ is:

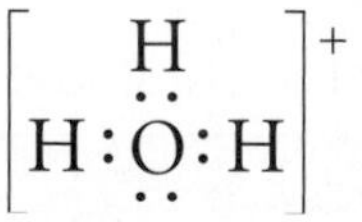

$$\left[\begin{array}{c} \text{H} \\ \text{H} : \overset{\cdot\cdot}{\underset{\cdot\cdot}{\text{O}}} : \text{H} \end{array}\right]^+$$

Step 2 The Lewis structure shows four electron groupings around the central oxygen atom.

Step 3 The Lewis structure shows three bonding pairs (BP) and one lone pair (LP) around the central oxygen atom.

Step 4 For three bonding groups and one lone pair, the molecular shape is trigonal pyramidal. The class is AX_3E.

Check Your Solution
The answer is in agreement with the shapes in Table 2.1.

Use VSEPR theory to predict the molecular shape for each of the following molecules:

1. $HCN(g)$ **3.** $SO_3(g)$ **5.** $AsCl_3(\ell)$ **7.** $CH_3F(g)$ **9.** $CS_2(\ell)$

2. $PF_3(g)$ **4.** $COCl_2(aq)$ **6.** $SI_2(g)$ **8.** $CH_2F_2(g)$ **10.** $CH_3COCH_3(\ell)$ (**Hint:** Treat the two CH_3 groups as single units)

INVESTIGATION 2.B

Target Skills
Building models of molecular substances

Evaluating and **analyzing** models constructed by others

Soap Bubble Molecules

Using molecular modelling kits helped you visualize the bonds between atoms and clearly identify which atoms are bonded. However, you do not get a good sense of the more realistic shape that is shown by space-filling molecules. Soap bubbles produce good models of molecular shapes. In this investigation, you will model some three-dimensional features of simple molecules.

Materials
- soap solution (mixture of 80 mL distilled water, 15 mL dish soap, and 5 mL glycerin)
- 100 mL beaker
- hard, flat surface
- straw
- 2 transparent 15 cm rulers
- protractor
- paper towels

Safety Precautions

- Ensure that each person uses a clean straw.
- Do not get any of the detergent solution in your mouth.
- Clean up all spills immediately.

Procedure

1. Obtain approximately 25 mL of the prepared soap solution in a 100 mL beaker.

2. Use the soap solution to wet an area of about 10 cm × 10 cm on a hard, flat surface.

3. Dip a straw into the soap solution in the beaker and blow a bubble on the wet surface. Then blow a second bubble of the same size so that it touches the first bubble. Record the shape of this simulated molecule and measure the bond length between the centres of the two bubbles where the nuclei of the atoms would be located. You could stand above the bubbles and hold a transparent ruler over, but not touching them, to roughly measure the centre to centre distance.

4. Repeat Procedure step 3 with three bubbles of the same size, to simulate three bonding pairs. Record the shape and the bond angles (the angles of the planes of contact of the soap bubbles). To measure angles, hold two rulers over the bubbles. Hold the ends of the rulers together with one hand, as shown, and adjust the angle between the rulers to match the angles formed by the bubbles. Then measure the angle between the rulers.

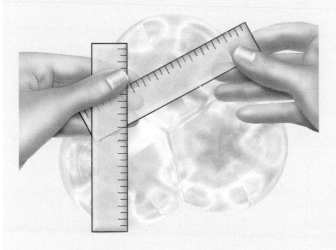

5. On top of the group of three equal-sized bubbles, blow a fourth bubble of the same size that touches the centre. Estimate the angle that is formed where bubbles meet.

6. Repeat Procedure step 3, but this time, to simulate a lone pair, make one bubble slightly larger than the other two.

7. To simulate two lone pairs, make a fourth bubble that is the same size as the large bubble and rest it on top of the three bubbles you made in Procedure step 6. Record the resulting bond angles. Record any changes in the original bond angles when the fourth bubble was added.

8. Clean your work area using dry paper towels first. Then wipe the area with wet paper towels.

Analysis

1. What shapes and bond angles were associated with two, three, and four same-sized bubbles? Give an example of a molecule that matches each of these shapes.

2. Give an example of a molecule that matches each of the shapes in Procedure steps 6 and 7.

Conclusion

3. What property of soap bubbles causes them to assume shapes similar to atoms in a molecule?

• • •

3 What features of a molecule must you obtain from Lewis structures in order to apply VSEPR theory to determine the shape of a molecule?

4 Define an "electron group" as it applies to VSEPR theory.

• • •

Polar Bonds and Polar Molecules

If a molecule includes one or more polar bonds, is it necessarily a polar molecule? To answer that question, analyze the structures of carbon dioxide and water—two common small molecules that both have two polar bonds. Combine the concepts that you learned about molecular shape and polar bonds. You know from Chapter 1 that the O—H bond and the C═O bonds are both polar. You would represent them as shown here:

 O—H   C═O

You have also learned that carbon dioxide is a linear molecule and water is a bent molecule. To determine whether water or carbon dioxide is a polar molecule, draw the structures and add the arrows that represent polarity, as shown in Figure 2.15. The arrows can be considered to be vectors or lines that show the direction of the polarity of the bonds. You can determine the polarity of the molecules as a whole by adding the vectors, as shown below the structures in Figure 2.15. In the water molecule, the horizontal parts (components) of the vectors point toward each other and, therefore, cancel each other. The vertical parts (components) of the vectors are added to give the final vector, labelled F_{total}. Thus, water is a polar molecule having an oxygen atom with a partial negative charge and hydrogen atoms with a partial positive charge. The carbon dioxide molecule is linear, and the vectors that represent the polarity of the bonds point directly away from each other. Because the bonds are identical, the polarities of the bonds are the same, and the polarity vectors cancel each other. The carbon dioxide molecule is *not* polar.

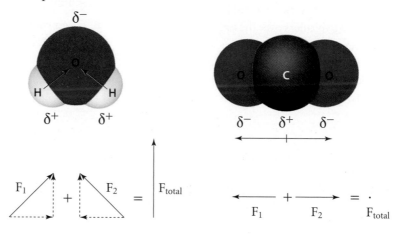

Figure 2.15 The presence of polar bonds does not ensure that the molecule is polar.

You can apply the method used above for carbon dioxide and water to many small molecules. Some molecules are more complicated, but one feature, symmetry, often helps in such cases. Consider the two molecules in Figure 2.16. Carbon tetrachloride, $CCl_4(\ell)$, is a symmetrical molecule. All the polar C—Cl bonds are identical. Therefore, the polarities of the four bonds cancel one another and the molecule is non-polar. If you replace one of the chlorine atoms in $CCl_4(\ell)$ with a hydrogen atom, you have trichloromethane, $CHCl_3(\ell)$, which is polar. Examine the diagram in Figure 2.16 to see why. The C—Cl bond polarities cancel each other horizontally, but the polarities of all the vertical bonds point downward. The total polarity of trichloromethane is represented by an arrow from the hydrogen atom downward through the central carbon atom.

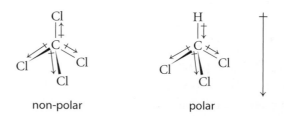

non-polar   polar

Figure 2.16 In carbon tetrachloride, the vectors representing the polarity of the bonds are identical in magnitude and combine to give a net zero polarity for the molecule. In trichloromethane, the net polarity is straight down.

Table 2.2 summarizes the combination of molecular shape and bond polarity for many simple molecules. In the table, the symbol "A" represents the central atom, while symbols "X" and "Y" represent atoms that are bonded to the central atom. Two different symbols, X and Y, are used for atoms bonded to the central atom to represent different magnitudes and directions of polarity of bonds. Refer to the table when completing the Practice Problems that follow.

Table 2.2 Summary of Molecular Shapes and Molecular Polarity

Molecular shape	Bond polarity	Molecular polarity
linear	X—A—X	non-polar
linear	X—A—Y	polar
bent		polar
trigonal planar		non polar
trigonal planar		polar
tetrahedral		non-polar
tetrahedral		polar
trigonal pyramidal		polar

Practice Problems

Use VSEPR theory to predict the shape of each of the following molecules. From the molecular shape and the polarity of the bonds, determine whether the molecule is polar. Justify your conclusions.

11. $CH_3F(g)$ **13.** $AsI_3(s)$

12. $CH_2O(\ell)$ **14.** $H_2O_2(\ell)$

15. Freon-12, $CCl_2F_2(g)$, was used as a coolant in refrigerators until it was suspected to be a cause of ozone depletion. Determine the molecular shape of CCl_2F_2 and discuss whether the molecule is polar or non-polar.

16. Which is more polar, $NF_3(g)$ or $NCl_3(\ell)$? Justify your answer.

Begin extension material ○

Network Solids

Pure carbon can be found in the form of the hardest substance on Earth—diamond. Pure carbon can also be found in the form of a slippery, black substance used in pencils and as a dry lubricant—graphite. How can the same pure element exist in such different forms? The answer to that question lies in another example of substances stabilized by covalent bonding. In these compounds, called **network solids**, dozens or even millions of atoms of the same element or two different elements are covalently bonded in a variety of patterns. Four examples of network solids consisting only of carbon atoms are shown in Figure 2.17. These different network solids, which consist of the same element but have different physical properties, are called allotropes.

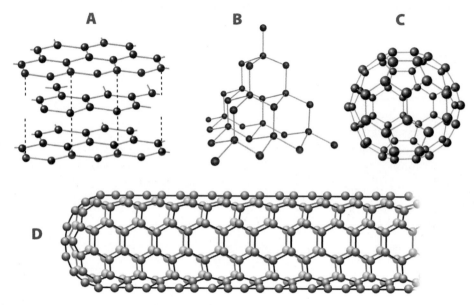

Figure 2.17 In graphite **(A)**, the dotted lines represent bonds that are weak and can be broken and reformed easily. In diamond **(B)**, each carbon atom is perfectly tetrahedral, creating the strength and hardness of the gem. In "buckyballs" **(C)** and nanotubes **(D)**, alternate bonds are double bonds, giving each carbon atom the requisite four bonds.

Chemistry File

FYI

At temperatures around 600 °C and extremely low pressures, methane will decompose and deposit a thin film of carbon, in the form of diamond, on a surface. The process, known as chemical vapour decomposition (CVD), can be used to coat a wide variety of surfaces. The hardness of this diamond film can be used for razor blades and applications that require non-scratch surfaces, such as eyeglasses.

The carbon atoms in graphite, shown in Figure 2.17A, have three short bonds with other carbon atoms. This array forms layers. The fourth bond on each carbon atom is longer and not as strong. As a result, this bond can be easily broken and reformed with another atom, allowing the layers to slide past each other. This property gives graphite its slippery feel and its ability to act as a solid lubricant. The same property makes graphite a good substance to use in pencils. As you push the pencil across paper, the layers slide off the pencil and onto the paper.

In diamond, each carbon atom is perfectly tetrahedral as shown in Figure 2.17B. Carbon–carbon bonds have their greatest strength in this configuration. This three-dimensional array of covalently bound carbon atoms makes diamond the hardest naturally occurring substance known. This structure also creates many planes of carbon atoms within diamonds, which reflect light and give diamonds their brilliance and sparkle.

In 1985, chemists Robert F. Curl Jr. (1933–), Richard E. Smalley (1943–2005), and Sir Harold W. Kroto (1939–) discovered a class of spherical allotropes of carbon that they called *fullerenes*. Figure 2.17C shows C_{60} or buckminsterfullerene, named after the architect R. Buckminster Fuller who designed geodesic dome structures that had an appearance similar to the fullerenes. The bonds in C_{60}, also called "buckyballs," form a pattern that is nearly the same as the stitching on a soccer ball (see Figure 2.18). If you examine Figure 2.17C, you will see five-membered rings and six-membered rings of carbon atoms. Notice that each carbon atom appears to have only three bonds.

However, alternate bonds are double bonds, giving each carbon atom four bonds. Chemists have also identified and studied C_{70}, C_{74}, and C_{82}, which have properties very similar to C_{60}. Although there is currently no practical application for buckyballs, chemists are studying their properties with an eye toward practical applications. Some chemists have speculated that they might be used as dry lubricants similar to graphite.

The allotrope of carbon shown in Figure 2.17D is called a nanotube. You probably recognize the prefix "nano-" as representing the 10^{-9}. Nanotubes are one of the first products of the new field of nanotechnology, which includes research and development of devices that have sizes ranging from 1 to 100 nanometres. Nanotubes with diameters of just a few nanometres and lengths in the range of micrometres have been assembled. Such nanotubes have great promise in the fields of microelectronics and medicine.

Some network solids consist of two different elements instead of one. The most common compound in Earth's crust, silicon dioxide, $SiO_2(s)$, exists in the form of a network solid. Sand and quartz consist almost entirely of silicon dioxide (or silica). If you were asked to write the Lewis structure for silicon dioxide, you might draw a structure similar to carbon dioxide. Silicon, however, does not typically form double bonds with oxygen. Instead, silicon bonds to four oxygen atoms to form a tetrahedral shape around the silicon atom. Each oxygen atom is then bonded to another silicon atom to form a network, as shown in Figure 2.19. The network can be represented by the formula $(SiO_2)_n$, indicating that SiO_2 is the simplest repeating unit.

Figure 2.18 On this soccer ball, you can see that the stitching forms five six-membered rings surrounding each five-membered ring. This pattern formed by the stitching is the same as the pattern of bonds in C_{60} or buckminsterfullerene (buckyballs).

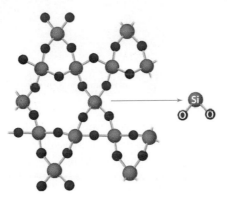

Figure 2.19 The repeating unit, SiO_2, is shown on the right. In the network solid, each oxygen atom in the SiO_2 unit is bonded to another silicon atom and each silicon atom is bonded to another oxygen atom.

End extension material

Section 2.1 Summary

- The packing of ions in a crystal is influenced by the relative sizes and charges of the positively and negatively charged ions.

- Since positively charged ions are attracted to all negatively charged ions, there are no combinations of ions that can be identified as a "molecule."

- Valence-shell electron-pair repulsion (VSEPR) theory allows you to predict the shape around a central atom from the Lewis structure.

- VSEPR is based on the repulsion of electron groups: LP—LP > LP—BP > BP—BP.

- Classes within VSEPR depend on the total number of electron groups (two, three, or four) and the combinations of bonding groups and lone pairs.

- The polarity of a molecule can be found by addition of vectors that represent the polarity of individual bonds.

1. What is a *unit cell*? Provide an example.

2. How do the sizes of cations (positive ions) and anions (negative ions) compare with the size of neutral atoms? How does the size of a cation compare with the size of an anion in the same period? Explain.

3. Ionic crystals come in many shapes and sizes, such as those in the photograph. What are the two factors that affect the shape of an ionic crystal?

4. Why do ionic compounds exist in crystals while molecular compounds exist in molecules?

5. Explain how VSEPR theory is used to predict molecular shape.

6. VSEPR theory uses "order of repulsion" to refine the shapes that it predicts. List the order of repulsion as described by VSEPR theory and explain the implications.

7. How do single, double, and triple bonds affect the overall shape of a molecule, if at all? Explain.

8. List and describe the five basic shapes predicted by VSEPR theory. In your own words, provide the criteria for determining the bonding pattern that results in each of the five shapes. Organize your thoughts in a table.

9. The predicted bond angles in a trigonal planar molecule are all 120°. Explain. The observed bond angles in a molecule of $COCl_2$ are 124.5° between the carbon–chlorine and carbon–oxygen bonds, and only 111° between the two carbon–chlorine bonds.

   a) Draw a structural formula for this molecule, including bond angles.

   b) Provide a possible explanation for the deviation from predicted bond angles.

   c) How do these bond angles compare with those found in a molecule of methanal, CH_2O (see Figure 2.9 on page 53)? Explain the differences.

10. For each of the following molecules, use VSEPR theory to predict molecular shape. Draw a three-dimensional diagram for each of the molecules and determine whether the molecule is polar or non-polar:

    a) $AsCl_3(\ell)$

    b) $CH_3CN(g)$

    c) $Cl_2O(g)$

    d) $SiCl_4(g)$

    e) $CH_3COOH(\ell)$

    f) $N_2H_4(g)$

    g) $SiH_3Cl(g)$

    h) $H_2S(g)$

    i) $COS(g)$

    j) $PICl_2(g)$

11. Candles are often made from paraffin wax, $C_{20}H_{42}(s)$. Describe the shape of this molecule and the polarity of the bonds. Is this molecule polar or non-polar? Explain.

12. Discuss the validity of the following statement: "All polar molecules must have polar bonds, and all non-polar molecules must have non-polar bonds."

13. Identify and explain the factors that determine the structure and polarity of molecules.

14. Use VSEPR theory to predict the molecular shape and polarity of $SF_2(g)$. Explain your answer.

15. What similarities and differences would you expect in the molecular shape and the polarity of $CH_4(g)$ and $CH_3OH(\ell)$? Explain your answer.

16. Draw a Lewis structure for $PCl_4^+(aq)$. Use VSEPR theory to predict the molecular shape of this ion.

17. The molecules $BF_3(g)$ and $NH_3(g)$ are known to undergo a combination reaction in which the boron and nitrogen atoms of the respective molecules join together. Draw a Lewis structure of the molecule that you would expect to form from this reaction. Give a reason for your answer.

Intermolecular Forces

Section Outcomes

In this section, you will:
- **explain** dipole-dipole attractions, hydrogen bonding, and London (dispersion) forces
- **explain** how research and technology interact in the production of beneficial materials

Key Terms

intramolecular forces
intermolecular forces
dipoles
dipole-dipole attraction
hydrogen bond
dispersion force
London force

Figure 2.20 How can salt, an ionic compound, and sugar, a molecular compound, look so much alike? What properties distinguish them from each other?

Common table salt and sugar are both white, crystalline solids (see Figure 2.20). They look so much alike that you probably could not distinguish one from the other without tasting them. As well, salt and sugar are both highly soluble in water. Chemically, however, the compounds are very different. Table salt (sodium chloride) is an ionic compound, and table sugar (sucrose) is a molecular compound. You have learned how the atoms in ionic compounds are held together. Each ion is attracted to several oppositely charged ions surrounding it. A continuous set of bonds exist from one ion to the next and to the next. In molecular compounds, each atom is covalently bonded to the other atoms in one molecule. The forces involved in these bonds act only between atoms within the same molecule and are, therefore, called **intramolecular forces**. Covalent bonds, however, do not act between atoms of different molecules. What then, holds molecules together in a solid, such as sugar, or even in a liquid, such as water? In this section, you will learn about **intermolecular forces**—the forces that are involved in the interactions *between* molecules. These forces are also called van der Waals forces in honour of the Dutch physicist Johannes van der Waals (1837–1923), who studied them extensively.

Dipole-Dipole Forces

In Section 2.1, you read that molecules can be polar, even if their net charge is zero. Water is an excellent example. In a polar molecule, one end is slightly positive and the other end is slightly negative. Because these molecules have two different "poles," they are called **dipoles**. When any two dipoles are near each other, the positive end of one attracts the negative end of the other. Figure 2.21 shows how one polar molecule can be attracted to two, three, four, or more other polar molecules at the same time. This attractive force is called a **dipole-dipole attraction**. Although the dipole-dipole attraction is not nearly as strong as the force between ions, it can still cause polar molecules to be attracted to each other enough to stabilize a solid crystal, such as table sugar.

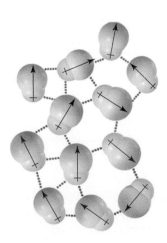

Figure 2.21 The dotted lines represent dipole-dipole attractions among the polar molecules. These attractive forces are electrostatic forces acting between positive and negative regions of the molecules.

Dipole Balloons

You cannot observe the motion of individual dipoles, but you can observe models. In this activity, you will model dipoles by using charged balloons. You will consider them as dipoles because they will have one charged end and one uncharged end.

Question

How do charged balloons model polar molecules?

Prediction

Read the entire procedure and make a prediction about the response of the balloons to each situation described in the procedure.

Materials

- 2 round balloons
- string
- marker

Procedure

1. Blow up the balloons so they are firm but not over-inflated.

2. Tie a 40 cm to 50 cm string to each.

3. Vigorously rub one side of the first balloon on wool or your hair to give it a charge. Gently mark the area on the balloon that you rubbed.

4. Suspend the balloon from the string and hold it or tie it away from other objects.

5. Very slowly, move your hand near the charged side of the balloon and observe the response of the balloon.

6. Charge the second balloon in the same way that you charged the first one. Mark the charged area.

7. Suspend the second balloon by the string and slowly bring it near the first balloon with their charged areas toward each other. Observe how the two balloons respond.

8. With the two balloons near each other, slowly move your hand into a location between the balloons. Observe the response of the balloons.

9. Work with another group and place the four balloons in different orientations relative to one another. In some of the orientations, place your hand between two of the balloons and observe any changes in the positions of the balloons.

Analysis

1. Explain why the balloon moved as you observed in Procedure step 5.

2. Why did the balloons move as they did when you brought them together?

3. Why did the balloons respond as they did when you put your hand between the two charged balloons?

Conclusion

4. Discuss how the balloons are models of polar molecules.

Hydrogen Bonding

One specific dipole-dipole interaction, called hydrogen bonding, is so strong and so important in both water and biological molecules that it is given a category of its own. When a hydrogen atom is covalently bonded to a highly electronegative atom, such as oxygen or nitrogen, the electronegative atom draws the electrons away from the hydrogen. Because hydrogen has no electrons other than the bonding electrons, the positive proton that makes up the hydrogen nucleus is nearly bare. A **hydrogen bond** is the electrostatic attraction between the exposed proton—or nucleus—of the hydrogen atom and the partial negative charge on the highly electronegative atom (usually oxygen or nitrogen) on an adjacent molecule. Lone pairs on the electronegative atom enhance the attractive force. This attractive force is often much stronger than other types of dipole-dipole interactions.

Because water consists entirely of oxygen and hydrogen atoms and has two O—H bonds, hydrogen bonding is an important factor in the structure and properties of water. All hydrogen atoms in water have a partial positive charge, and all oxygen atoms have a partial negative charge. A single oxygen atom of one water molecule can be hydrogen bonded to as many as six hydrogen atoms in different water molecules at the same time, as shown in Figure 2.22. The hydrogen bonding of water molecules, together with the bond angle between the oxygen and hydrogen atoms, gives snowflakes their characteristic six-sided shape.

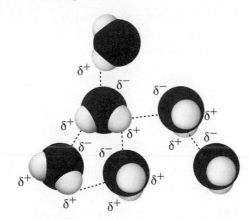

Figure 2.22 Individual hydrogen bonds are much weaker than covalent bonds. However, when many hydrogen bonds are acting on each water molecule at the same time, they have a significant effect on the properties of water.

In solid water (ice), each water molecule is hydrogen bonded to four other water molecules, as shown in Figure 2.23A. The molecules are farther apart in ice than in liquid water and thus the hydrogen bonds are longer. Nevertheless, the hydrogen bonds are stronger because of the orientation of the atoms and the bonds. Hydrogen bonds are strongest when the molecules are oriented as shown in Figure 2.23B. Notice that the line from the oxygen atom through the hydrogen atom in that molecule and to the oxygen atom of the next water molecule is straight. In ice, all the hydrogen bonds have this orientation.

Figure 2.23 The hydrogen bonds in ice are longer than they are in water, making ice less dense than water. It is the orientation of the hydrogen bonds in ice that gives them their maximum strength.

Investigating the Properties of Water

In this investigation, you will infer a property of water from an observation. You will then use that property to explain other observations.

Question

How can you explain the properties of water based on its shape and charge properties?

Materials

- water
- vegetable oil
- ethanol (🔥) (T) (☠)
- pepper
- liquid dish detergent
- acetate strip
- cotton cloth
- vinyl strip
- wool cloth
- 500 mL beaker
- 2 shallow dishes
- sewing needles
- 2 small glasses
- 150 mL beaker

Safety Precautions

- Use care in handling sharp objects.

Procedure

1. Turn on a tap so that a very small steady stream of water is flowing.

2. Vigorously rub the end of the acetate strip with the cotton cloth.

3. Slowly move the acetate strip near the stream of water without touching the water. Record your observation.

4. Repeat Procedure steps 2 and 3 using the vinyl strip and wool cloth.

5. Have one lab partner slowly pour vegetable oil into the 500 mL beaker, forming a small steady stream of vegetable oil similar to the water in Procedure step 1.

6. Repeat Procedure steps 2 through 4 using the vegetable oil. Record any differences between the behaviour of the oil and the water.

7. Place water in one shallow dish. Very carefully place a pin or needle onto the surface of the water in a horizontal position. (Your teacher may choose to do this step.) Record the appearance of the needle.

8. Gently touch the needle. Record your observations.

9. Place ethanol in the second shallow dish. Try to repeat Procedure steps 7 and 8 with the ethanol instead of the water. Record your observations.

10. Place one of the glasses in an empty shallow dish. Pour water into the glass from a beaker. When the glass is nearly full, pour very slowly and carefully. Observe and record the shape of the top of the water just before it starts to spill over into the dish.

11. Repeat Procedure step 10 with ethanol. Record your observations.

12. Fill the 150 mL beaker about two thirds full of water.

13. Sprinkle pepper onto the water until a thin film of pepper forms on the surface.

14. Add a few drops of liquid dish detergent to the water. Observe the response of the film of pepper. Record your observations.

Analysis

1. The cotton-rubbed acetate strip was positively charged and the wool-rubbed vinyl strip was negatively charged. Was the stream of water attracted or repelled by the positive charge? Was the stream of water attracted or repelled by the negative charge? Was the stream of vegetable oil attracted or repelled by the positive charge? Was the stream of vegetable oil attracted or repelled by the negative charge? What do these observations tell you about the properties of water and of vegetable oil?

2. What happened when you touched the needle that was floating on the water? Is the needle denser than water? Use the property of water that you described in your answer to Analysis question 1 to explain your observations of the needle.

3. What happened when you attempted the same procedure with ethanol? Explain your observations.

4. Describe the shape of the water in the glass just before it started to run over? How does the property of water that you described in Analysis question 1 account for this shape?

5. How did filling the glass with ethanol differ from filling it with water? Explain the difference in the two liquids.

6. How did the dish detergent affect the film of pepper on the water? The molecules in the dish detergent have a long non-polar end and a charged head. Use this information to explain the effect on the pepper film that you described.

Conclusion

7. Describe the differences in the physical properties of ethanol and water. What is the significance of these properties of water to living systems, considering the fact the living cells contain a large percentage of water?

London (Dispersion) Forces

The intermolecular forces discussed thus far are all due to the partial charges at different points on polar molecules. Do non-polar molecules interact with each other? If so, what is the nature of the interaction? The photographs in Figure 2.24 should give you some hints. The three photographs were taken under standard conditions of temperature and pressure. Photograph (A) is chlorine, $Cl_2(g)$; (B) is bromine, $Br_2(\ell)$; and (C) is iodine, $I_2(s)$—three of the halogens in increasing order of molar mass. The three halogens—chlorine, bromine, and iodine—are all diatomic molecules that are completely symmetrical and non-polar. Nevertheless, as the molar mass increases, the state of the substances goes from a gas to a liquid and then to a solid. In the gaseous state, there is very little interaction among the molecules. To form a liquid, the molecules must have some attraction for one another to remain close together. In a solid, such as iodine, the intermolecular forces must be relatively strong because the molecules remain in the same positions relative to one another.

Figure 2.24 Under standard conditions of temperature and pressure, chlorine is a yellowish-orange gas **(A)**, bromine is a reddish-brown liquid **(B)**, and iodine is a violet-dark grey solid **(C)**.

The attractive force that acts between non-polar molecules is called the **dispersion force**. Dispersion forces act between all molecules, but in non-polar molecules they are the only force. In honour of the German physicist Fritz London (1900–1954), who explained the force mathematically, the dispersion force is often called the **London force**. An analogy will help you understand the basis of the London (dispersion) force.

At some point, you might have rubbed a balloon against some cloth or your hair and then placed it against a wall, where it stuck. Figure 2.25 shows you why the balloon sticks to the wall. When you rub a balloon, friction causes it to lose electrons and become positively charged. When you bring the balloon near the wall, the charge on the balloon attracts negative charges and repels positive charges in the molecules in the wall. This attractive force distorts the molecules, causing them to become dipoles. Their negative ends are then directed outward and their positive ends point inward, as shown in Figure 2.25. Now the positive charges on the balloon are attracted to the negatively charged ends of the dipoles in the surface of the wall and the balloon sticks to the wall. The charges on the balloon induced (caused) the molecules in the wall to become dipoles.

ChemistryFile

FYI

London (dispersion) forces are responsible for the "stickiness" of geckos' feet. A gecko can hang from a piece of glass with only one toe. Recently, a new type of tape has been manufactured that is based on the structure of geckos' feet.

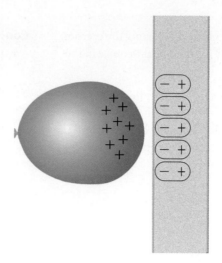

Figure 2.25 The force that holds the balloon to the wall is an electrostatic force between the positively charged spot on the balloon and the negative end of the induced dipoles in the wall. The positively charged balloon induces the dipoles to form and then sticks to them.

You are probably wondering how a charged balloon that induces the formation of dipoles in a wall is related to interactions between non-polar molecules. There are no net charges or permanent dipoles in non-polar substances. However, it is possible to induce the formation of dipoles in molecules that are classified as non-polar. As well, non-polar molecules spontaneously form temporary dipoles. Electrons in atoms and molecules are in constant, rapid motion. For a brief instant, the distribution of electrons can become distorted so that one point in a molecule is very slightly positive and another point is slightly negative. This momentary separation of charge (temporary dipole) induces a temporary dipole in the molecule beside it. This induced dipole can induce another instantaneous dipole in another adjacent molecule. The process "disperses" through the substance, creating flickering dipoles that attract one another. Each individual attractive force is extremely weak and lasts only an instant, but when many interactions occur at the same time, the overall effect is significant. These interactions are the dispersion forces.

Two factors affect the magnitude of London (dispersion) forces. First, as you read on the previous page, in relation to the halogen molecules, the attraction becomes larger as the mass of the molecules becomes larger. The basis for this increase in London forces is the greater number of electrons. The probability that a temporary dipole will form increases as the number of electrons increases. Second, the shape of the molecule affects the strength of the London forces. As you can see in Figure 2.26, the area of contact between two spherically shaped molecules is very small, while the area of contact between linear molecules is larger. When two molecules have about the same number of electrons, London forces are greater for the molecule with a linear shape.

Figure 2.26 Both types of molecules have five carbon atoms and 12 hydrogen atoms, but their shapes are quite different. London (dispersion) forces are greater between the linear molecules than between the spherical molecules.

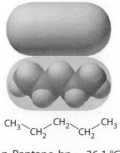

n-Pentane, bp = 36.1 °C

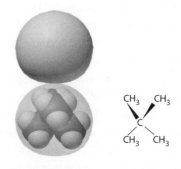

2,2-dimethylpropane, bp = 9.5 °C

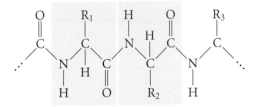

5 Explain the difference between intramolecular forces and intermolecular forces.

6 How are hydrogen bonds similar to and distinct from other types of dipole-dipole interactions?

7 Why do London (dispersion) forces have such a significant effect on the structures of many substances?

Bonding in Biological Molecules

Van der Waals forces play a major role in the structure, and therefore the function, of several of the most important classes of biological molecules. For example, examine the structure of DNA (deoxyribonucleic acid) molecules shown in Figure 2.27. DNA molecules consist of long chains of units called nucleotides. Each nucleotide contains a sugar, a phosphate group, and one of four bases represented by the letters A, C, G, and T. The sugar and phosphate groups form a "backbone," and the bases are bonded to each sugar. Because of the structure of the bases, hydrogen bonds form between bases C and G and between bases A and T. Individual hydrogen bonds are only about five percent as strong as covalent bonds but, because DNA consists of thousands of bases, the large number of hydrogen bonds creates a very strong force. As you can see in the Figure 2.27, the DNA strands twist around each other to form the famous "double helix." Although it is difficult to show in this diagram, the flat planes of the bases lie perpendicular to the direction of the helix. London forces act between the bases and draw them close together, stabilizing the helical structure.

The sequence of the bases in DNA is the information that directs the synthesis (production) of proteins—the molecules that carry out the functions of the cells. Proteins are long chains of units called amino acids. When incorporated into protein molecules, each amino acid has an amino group (N—H) and a carbonyl group (C=O), as shown in Figure 2.28. The 20 amino acids found in proteins are distinguished only by differences in chemical groups called R groups.

Begin extension material

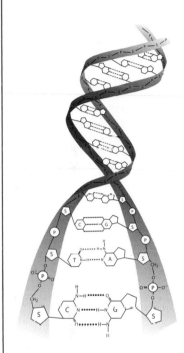

Figure 2.27 In DNA, the C and G bases and the A and T bases fit together like pieces of a puzzle. Three hydrogen bonds hold C and G together and two hydrogen bonds hold A and T together.

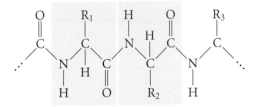

Figure 2.28 The hydrogen atom on the N—H group in each amino acid has a partial positive charge and the oxygen atom on the C=O group has a partial negative charge. The subscripts, 1, 2, or 3, on the Rs indicate that these are different amino acids.

Proteins vary tremendously in size. The smallest protein—insulin—has 51 amino acids, but some proteins have more than 1000 amino acids. Each protein must have the correct sequence of amino acids and must be folded in a precise three-dimensional structure to carry out its function. Hydrogen bonds stabilize two common structures within many proteins, the alpha helix and the beta structure, shown in Figure 2.29. In the alpha helix, the chain coils back onto itself and the N—H group of one amino acid hydrogen bonds to the C=O group of another amino acid three units down the chain. In the beta structure, also called the pleated sheet, separate portions of the amino acid chain fold back and lie beside each other. The two strands are hydrogen bonded to each other, as shown in the figure.

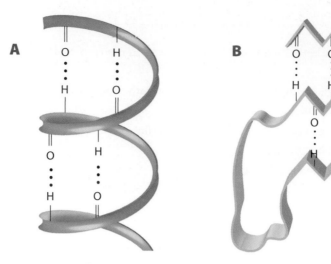

Figure 2.29 Portions of alpha helix **(A)** are found in most proteins. The keratin in hair and skin has large amounts of alpha helix. The beta structure **(B)** is found in a few proteins. The beta structure predominates in wool.

Figure 2.30 The black line within the blue tubes represents the chain of amino acids of myoglobin. Each dot indicates the connection between amino acids. You can see that the chain forms a helix in the straight sections of the tubular shape.

End extension material

To complete the correct three-dimensional structure of a protein, the segments that have formed an alpha helix or a beta structure fold back onto each other. London forces then hold the non-polar R groups in place.

The three-dimensional structure of the protein myoglobin is shown in Figure 2.30. Myoglobin is the protein that binds oxygen and stores it in muscle cells. About 70 percent of the amino acids in myoglobin are found in segments of alpha helix—the tubular sections in the figure. Myoglobin contains no beta structure.

Section 2.2 Summary

- Polar molecules, possessing dipoles, are attracted to one another by electrostatic attractions between their oppositely charged ends.

- Dipole-dipole interactions are much weaker than covalent bonds, but they have a significant effect on the structure and function of a compound.

- A very strong dipole-dipole interaction—the hydrogen bond—is critical to the structure of water and many biological molecules.

- Molecules also interact through London (dispersion) forces, which are attractions between temporary, induced dipoles. In non-polar molecules, these are the only attractive forces.

SECTION 2.2 REVIEW

1. What is a dipole? Explain how polar molecules interact.

2. What characteristics of molecules are necessary for hydrogen bonds to form between them?

3. Water molecules in the liquid state can have more hydrogen bonds than they have in the solid state. Water molecules are further apart in the solid state than they are in the liquid state. These two statements appear to contradict the concept that molecules are more tightly held together in the solid state than in the liquid state. Explain how and why the statements can be correct and not contradict the concepts about solids and liquids.

4. How can a temporary dipole induce a dipole in another molecule?

5. Describe the two factors that affect the formation and strength of London (dispersion) forces.

6. Why do intermolecular interactions have such a significant influence on the structure of matter?

7. Explain the relationship between the strength of intermolecular forces and the physical state of a substance.

8. Prepare a multimedia presentation about intermolecular forces and their influence on the physical state of a compound. (ICT)

Relating Structures and Properties

Section Outcomes

In this section, you will:

- **relate** structures of ionic lattices to their properties
- **describe** how an understanding of electronegativity contributes to the knowledge of melting and boiling points
- **perform** an investigation to illustrate the properties of ionic compounds
- **analyze** experimental data to determine the properties of ionic compounds
- **relate** melting and boiling points, and enthalpies of fusion and vaporization to predicted intermolecular bonding
- **analyze** data for trends and patterns on the melting and boiling points of a related series of molecular substances

Key Terms

melting point
enthalpy of fusion
boiling point
enthalpy of vaporization

Figure 2.31 Try to relate the structures of the book and the box to the property of weight of the objects as it appears to be from the photograph.

What's wrong with this picture? The structures of the book and the box do not have the properties that you would expect. The book should not be as heavy as the student's posture indicates. The box should be too heavy to carry as easily as the student appears to be carrying it. We often find humour in conflicting images, such as in Figure 2.31. In chemistry, however, when the properties of a compound do not seem to fit its structure, chemists are perplexed. When chemists encounter these situations, they keep searching. In this section, you will discover how chemists propose theories to explain the properties and correlate these properties with the structures of many substances.

States of Matter

The state of a pure substance—solid, liquid, or gas—depends on the strengths of the attractive forces between its particles. These forces can be ionic bonds, covalent bonds, or intermolecular or intramolecular forces. In solids, the attractive forces are relatively strong. The kinetic energy of the particles is not great enough to break the bonds between particles. Each particle remains bonded to its adjacent particles. As energy, in the form of heat, is added to the solid substance, the average kinetic energy of the particles increases. Because the temperature of a substance is directly related to the average kinetic energy of the particles, the temperature of the substance also increases. Eventually, the particles reach a temperature at which the kinetic energy of the particles is great enough to break the bonds with neighbouring particles. Individual particles break away from their adjacent particles but immediately form new bonds with other particles. The substance is now in its liquid form. Particles can move past each other, but they are constantly breaking and forming bonds. You could say that particles in a liquid are constantly "changing partners." In liquid water, for example, the hydrogen bonds between water molecules last about 10^{-11} s. In other words, water molecules "change partners" 10^{11} times every second. The temperature at which the change from a solid to a liquid occurs is the **melting point** of the substance. Because pressure affects the melting points of substances, the values listed in tables in chemistry books or on the Internet are melting points for standard atmospheric pressure.

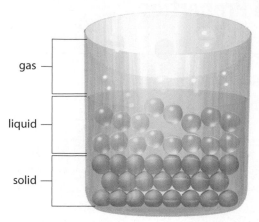

gas

liquid

solid

Figure 2.32 The particles in a solid are rigidly attached to one another and often form a symmetrical structure. In a liquid, the particles are always in contact but can readily move past one another. In a gas, the only contacts between particles are collisions. The particles remain free of one another after the collisions.

Table 2.3 Melting Points and Boiling Points of Some Common Compounds at Standard Atmospheric Pressure.

If you continue to add energy to the substance after it starts to melt, the temperature remains the same until the entire sample has melted. All the incoming energy is used to break the intermolecular bonds. You might recall from previous science courses that the amount of energy required to melt one mole of a substance is called the **enthalpy of fusion** of the substance.

If you continue to add energy to the liquid form of a substance, the average kinetic energy of the particles increases until the particles have enough energy to break all the intermolecular bonds and remain free of any other particles. The substance becomes a gas. In a gas, particles have enough kinetic energy to break any bonds that might form. The temperature at which a liquid becomes a gas is the **boiling point** of the substance. Once again, boiling points are affected by pressure. The boiling points recorded in tables in chemistry books or on the Internet are those for conditions of standard atmospheric pressure.

If you continue to add energy to a substance after it starts to boil, the temperature will remain at the boiling point until the entire sample is transformed into a gas. The amount of energy required for one mole of a liquid to become a gas is called the **enthalpy of vaporization** of the substance. Figure 2.32 illustrates the particles in solid, liquid, and gaseous forms.

Melting Points and Boiling Points

The previous discussion explained how bonds are related to the melting and boiling points of compounds. When metals melt or boil, metallic bonds must be broken. When ionic compounds melt or boil, ionic bonds must be broken. When molecular compounds melt or boil, intermolecular forces must be overcome. The covalent bonds between atoms within a molecule need *not* break when the compound melts or boils. With these concepts in mind, examine the melting and boiling points of a few examples of metals and of ionic and molecular compounds in Table 2.3. Look for trends among the different classes of compounds and within the three classes of compounds.

Metals			Ionic compounds			Molecular substances		
Substance	Melting point (°C)	Boiling point (°C)	Substance	Melting point (°C)	Boiling point (°C)	Substance	Melting point (°C)	Boiling point (°C)
$Li(s)$	+181	+1342	$CsBr(s)$	+636	+1300	$H_2(g)$	−259	−253
$Sn(s)$	+232	+2602	$NaI(s)$	+661	+1304	$Cl_2(g)$	−102	−34
$Al(s)$	+660	+2519	$MgCl_2(s)$	+714	+1412	$H_2O(\ell)$	0	+100
$Ag(s)$	+962	+2162	$NaCl(s)$	+801	+1465	$C_6H_6O(\ell)$	+41	+182
$Cu(s)$	+1085	+2562	$MgO(s)$	+2825	+3600	$C_6H_{12}O_6(s)$	+146	decomposes

From the table you can infer that the melting points of metals and ionic compounds are in about the same order of magnitude, while those of molecular compounds are much lower. The pattern is the same for boiling points. Metallic bonds and ionic bonds are formed by the electrostatic attractive forces between whole positive and negative charges, which create very strong bonds. As well, electrostatic forces are relatively long range forces. Although the strength of the force decreases significantly with distance, each ion attracts oppositely charged ions throughout the crystal or metal. The intermolecular forces between molecules are a result of the interaction between dipoles (either permanent or induced) and are much weaker than ionic or metallic bonds. Much less energy is required to break the weaker bonds between molecules in molecular compounds. Therefore, you would expect them to have lower melting and boiling points, which they do.

Focus, now, on the differences in the melting and boiling points of the ionic compounds. Notice that the MgO(s) has much higher melting and boiling points than the other ionic compounds. The magnesium and oxide ions are slightly smaller than most of the other ions and are, therefore, closer together than the ions in the other compounds, creating a somewhat larger attractive force. The major reason for the higher melting and boiling points is, however, the magnitude of the charges. Magnesium (2+) and oxide (2−) ions have larger charges than most of the other ions in the list, which have a charge of 1+ or 1−. Larger charges create a much stronger attractive force between the ions.

Intermolecular bonds are much weaker than ionic or metallic bonds. Nevertheless, there is a significant amount of variability of melting and boiling points among molecular compounds. Consider several different categories of compounds in relation to the types of intermolecular forces that predominate. First, examine the elements that are found as diatomic molecules in nature. Table 2.4 lists their melting and boiling points, as well as the number of electrons in the molecules.

Table 2.4 Melting Points and Boiling Points of Diatomic Molecules

Substance	Number of electrons in a molecule	Melting point (°C)	Boiling point (°C)
$H_2(g)$	2	−259	−253
$N_2(g)$	14	−210	−196
$O_2(g)$	16	−219	−183
$F_2(g)$	18	−220	−188
$Cl_2(g)$	34	−102	−34
$Br_2(\ell)$	70	−7.2	+59
$I_2(s)$	106	+114	+184

All the molecules in the table are symmetrical, non-polar molecules. Therefore, the only interactions between the molecules in these pure substances are London (dispersion) forces. Because the shapes of these molecules are basically the same, the only factor that should affect the attraction between molecules is the number of electrons. Notice that those substances that have nearly the same number of electrons have very similar melting and boiling points. For the remainder of the substances, both the melting and the boiling points increase as the number of electrons in the molecules increases. The observed data are in agreement with the results predicted by the theory.

Water, $H_2O(\ell)$, has only 10 electrons, yet its melting and boiling points are much higher than those molecules in Table 2.4, which have a similar number of electrons. Compare the data shown in Table 2.5 for water; hydrogen sulfide, $H_2S(g)$; and carbon dioxide, $CO_2(g)$—two other substances that have molecules with three atoms and a similar number of electrons.

Table 2.5 Melting and Boiling Points of Molecules of Similar Size

Substance	Number of electrons in a molecule	Melting point (°C)	Boiling point (°C)
$H_2O(\ell)$	10	0	+100
$H_2S(g)$	18	−86	−60
$CO_2(g)$	22	(sublimation point) −78 °C	

For the compounds in Table 2.5, the number of electrons per molecule is similar and therefore cannot account for the extreme differences in melting and boiling points. As well, all the bonds in all three compounds are polar bonds. Notice that under atmospheric pressure, carbon dioxide neither melts nor boils—it sublimes or goes directly to a gas from a solid.

That carbon dioxide becomes a gas at a temperature below the boiling points for the other two compounds indicates that it has the weakest intermolecular interactions. You have probably already recalled that the shape of carbon dioxide is linear and that of water is bent. The polarities of the $C \!=\! O$ bonds in carbon dioxide cancel each other and make carbon dioxide a non-polar molecule. Because water is bent, it is polar. Even more importantly, the hydrogen atoms are attracted to the very electronegative oxygen, and thus hydrogen bonds form between water molecules. As a result, much more energy is required to convert ice into liquid water and to convert liquid water into a gas than to convert solid carbon dioxide into a gas. Therefore, the melting and boiling points of water are much higher than the sublimation point for carbon dioxide.

Hydrogen sulfide has a bent shape very much like that of water, yet its melting and boiling points are much lower than those of water. To determine why water and hydrogen sulfide have such different physical properties, study the graph in Figure 2.33. The graph compares the boiling points of many binary hydrides—compounds that consist of hydrogen and one other element.

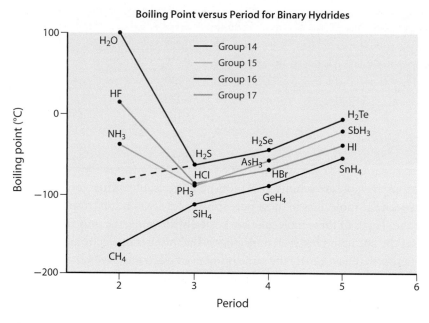

Figure 2.33 The dotted line shows what the boiling point of water would probably be if it had no hydrogen bonds.

Start analyzing the graph by focussing on the hydrides of Group 14 elements (black line). All the molecules are symmetrical because they have four hydrogen atoms bonded to one central atom. Therefore, these hydrides are non-polar. As the period increases, the molecules become larger and have more electrons. The boiling points increase as the number of electrons increase, just as you would expect for London (dispersion) forces.

Inspection of the lines on the graph reveals that for the other three groups (15, 16, and 17) the trend is the same as it is for Group 14, with the exception of the Period 2 elements. In each case, the boiling point of the hydride of the Period 2 elements is much higher than it would be if only London (dispersion) forces were acting. These Period 2 elements (N, F, and O) are all small and have high electronegativity values, making the bonds highly polar. As well, all three hydrides have one or more lone pairs. These conditions are the basis for strong hydrogen bonds. In summary, London (dispersion) forces are largely responsible for the intermolecular interactions of all the hydrides other than water, hydrogen fluoride, and ammonia.

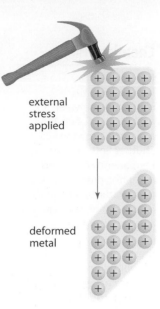

Figure 2.34 Metals deform easily because the layers of positive metal ions can slide over one another while surrounded by delocalized free electrons (yellow cloud).

8. In general, why do molecular substances typically have lower melting points and boiling points than ionic substances?

9. For non-polar molecular substances, predict how the melting point would change as the molar mass of the molecule increases.

10. Would substances stabilized by London (dispersion) forces have higher or lower melting points than substances stabilized by hydrogen bonds? Explain your reasoning.

Mechanical Properties of Solids

You probably learned in previous science courses that metals are malleable and ductile, while ionic compounds are hard and brittle. You now have the knowledge to explain these properties on the basis of bonding. Examine Figure 2.34 and envision the free-electron model of bonding in metals. You can explain the malleability of metals by observing that metallic bonds are non-directional. The somewhat fixed array of positive ions is surrounded by a "sea" of electrons; thus, an attractive force acts in all directions. When stress is applied to a metal, one layer of positive ions can slide over another layer. The layers move without breaking the array because the delocalized, freely moving valence electrons continue to exert a uniform attraction on the positive ions. Metals deform but do not shatter. This property allows many metals, such as copper, to be hammered into ornate designs, as shown in Figure 2.35.

You can explain the hard, brittle properties of ionic compounds by visualizing the ionic bonding in crystals. When stressed or hammered, like charges can become aligned, as shown in Figure 2.36. The crystals break along the line at which the like charges are repelling each other.

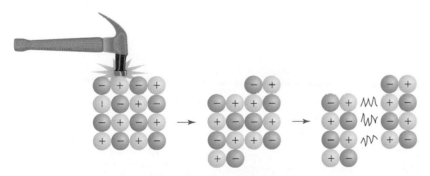

Figure 2.36 Ionic crystals often break along very smooth planes. This diagram shows that, when stressed and when the ions move very slightly, opposite charges can become aligned and the repulsion forces are all on the same plane.

Figure 2.35 Many artists choose to work with copper because of its malleability and attractive colour.

Non-polar, molecular compounds can be solid at standard temperature, but they are often very soft. Paraffin, $C_{25}H_{52}(s)$, is an excellent example. Such solids are stabilized by only London (dispersion) forces arising from induced dipoles. These attractive forces form randomly and are continually changing position and direction. Consequently, the molecules can easily slide along beside each other, making the substance soft and easily broken.

Conductivity

Electric current is the directional flow of charged particles—usually electrons or ions. If charged particles are free to move independently, they can be caused to do so by placing them between oppositely charged electrodes (see Figure 2.37). The electrodes are charged by a battery or other power source. You can determine whether an electric current is flowing if you place a meter or light bulb in the circuit. The conductivity meters that you use in the laboratory contain these components.

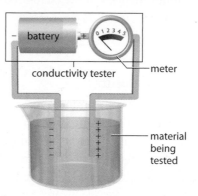

Figure 2.37 If charges are free to move independently of one another, an electric current will flow and register on the meter.

Metals are good conductors of electric current because the valence electrons are delocalized and are, therefore, free to move through the metal. They are, in a sense, moving from one positive ion to the next. They reach the positive electrode and then pass through the circuit in the conductivity tester and register a current.

If you place the electrodes of a conductivity tester on a piece of a solid ionic compound, no current will flow. The attractive forces between the oppositely charged ions are so much stronger than electrical forces exerted by the electrodes in the conductivity tester that the ions in the compound will not move. A solid ionic compound does not conduct electric current.

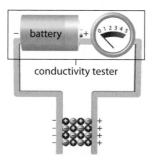

Figure 2.38 Both the positively and negatively charged ions remain in place in a solid ionic compound. No current will register on a conductivity testing meter.

If an ionic compound is dissolved in water, the ions are free to move past one another and will migrate to the electrodes as shown in Figure 2.39. As well, if an ionic compound is heated to the melting point and the ions have enough kinetic energy to break the ionic bonds, the ions will be able to move past one another and move toward the oppositely charged electrode. In either case, a current would register on the meter in the circuit.

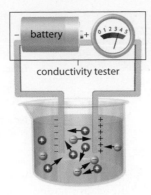

Figure 2.39 An aqueous solution of an ionic compound will conduct electric current because the ions are free to move toward the oppositely charged electrode.

When you read about network solids, you learned that diamond and graphite both consist of pure carbon, but diamond is hard and graphite is slippery. The differences in their properties do not end there. Diamond cannot conduct electric current but graphite can. In diamonds, valence electrons are all involved in very strong covalent bonds. In graphite, three of the four valence electrons of each carbon atom are involved in forming covalent bonds to other carbon atoms. The other electrons of each carbon atom are delocalized throughout the layers of carbon atoms, in a way similar to the delocalization of electrons in metals. These electrons can move parallel to the layers of carbon atoms, and, therefore, graphite can conduct electric current.

In neutrally charged molecular compounds, all the valence electrons are involved in covalent bonds. Even if the molecule is polar, the charges cannot leave the molecule. Molecular compounds cannot conduct electric current either in pure form or in solution in water (see Figure 2.40).

Figure 2.40 If a molecular compound is polar, the positive end will orient toward the negative electrode and the negative end will orient toward the positive electrode. A non-polar molecule will be induced to form a dipole. However, the charges cannot leave the molecule and thus no electric current will flow.

The properties that account for electrical conductivity also account for thermal conductivity. If valence electrons are free to conduct electric current, they are free to move. Then they can easily collide with particles of adjacent hot objects and receive kinetic energy in the collisions. The electrons then move freely throughout the substance and pass their kinetic energy on to other particles by colliding with them. This is the general process by which thermal energy is conducted.

Properties of Substances

In this investigation, you will study the properties of five different types of solids: non-polar covalent, polar covalent, ionic, network, and metallic. You will be asked to identify each substance as one of the five types of solids on the basis of its properties. In some cases, you will make inferences by drawing on past knowledge and experience. In other cases, you will use the process of elimination. The emphasis is on the skills and understanding you use to make your decisions.

Question

Is the unknown solid non-polar covalent, polar covalent, ionic, network, or metallic?

Safety Precautions

- Tie back long hair and any loose clothing.
- Before you light the candle or Bunsen burner, check that there are no flammable solvents anywhere in the laboratory.
- Avoid touching hot surfaces.
- Wash your hands when you have completed the investigation.

Materials

- distilled water
- 5 unknown solids
- 100 mL beaker and stirring rod
- conductivity tester
- metal plate (iron or aluminium)
- ring stand with clamp
- candle
- Bunsen burner
- match, lighter, or striker
- timer
- tongs

Procedure

1. Read the entire Procedure before you begin. Design a table in which to record your observations.

2. Rub each solid on the surface of each other solid. Rank the relative hardness of each solid on a scale of 1 to 5. A solid that receives no scratch marks when rubbed by the other four is the hardest (5). A solid that can be scratched by the other four is the softest (1).

3. One at a time, place a small quantity (about 0.05 g) of each solid into 25 mL of distilled water in a 100 mL beaker. Observe and record the solubility of each solid.

4. For any solids that dissolved in Procedure step 3, test the solution for electrical conductivity.

5. Test each of the five large solid samples for electrical conductivity.

6. One at a time, put a small sample of each solid on the metal plate using tongs. Place the plate on the ring stand and heat it with the burning candle. The plate should be just above the flame. Observe how soon the solid melts.

7. Repeat Procedure step 6 with the flame of a Bunsen burner. Observe how soon the solid melts.

Analysis

1. Based on what you know about bonding, classify each solid as non-polar covalent, polar covalent, ionic, network, or metallic. Give reasons to support your decision.

Conclusion

2. Based on the properties you observed, write a working definition of each type of solid based on its properties.

Mane Products

Walk down the hair-care aisle at any store and you will be confronted with an incredible selection of products that seems to grow daily. You can volumize, strengthen, colour, hydrate, or repair your hair; tame, release, or unfrizz your curls. How do these products supposedly turn your hair, each shaft of which is composed of dead proteins, into a lively, thick, shiny, manageable mane? It is all about scientific research and technology.

Background

Different brands of shampoo include similar ingredients whose properties have been well established based upon knowledge of their chemistry. Scientists and product testers have developed this knowledge over many years. A typical shampoo may contain:

- water, a solvent that keeps the other ingredients in solution
- ammonium lauryl sulfate and ammonium laureth sulfate, anionic surfactants that remove oils and grease from hair
- olealkonium chloride, a cationic surfactant that conditions hair and boosts the viscosity of the shampoo
- dimethicone, a conditioning ingredient that softens dry hair
- ammonium xylenesulfonate and sodium chloride thickeners
- cocamide mea, a lather builder
- hydrolyzed collagen and tricetylmonium chloride, conditioners that control static and reduce tangles
- octyl salicylate, a sunscreen
- cetyl alcohol and stearyl alcohol, thickeners and fatty acids
- disodium EDTA (ethylenediaminetetraacetic acid), a preservative
- methylchloroisothiazolinone, a fragrance
- citric acid, which makes shampoos acidic

Testing an Idea

Surfactants are foaming and cleansing components that clean your hair by stripping sebum, a natural oily coating that collects dirt and styling products. Surfactant molecules have a dual nature. One end is non-polar (uncharged) and thus soluble in dirt and grease, which are also non-polar. The other end is polar (charged) and therefore soluble in water. When you wash your hair, the surfactant acts as a bridge between the sebum and the water. Sebum is attracted to the surfactant and washed away when you rinse. Foaming occurs when surfactant molecules gather around air instead of oil, causing millions of tiny bubbles. So do not judge a shampoo's effectiveness by the amount of lather it produces. These bubbles are using the surfactants that should be removing dirt and oil.

The correct pH (about 5) is important because hair cuticles exposed after the sebum is stripped away are covered with overlapping scales that become smooth in a properly acidic environment. Rough scales do not overlap well and can snag on neighbouring shafts, causing tangles. Conditioners also reduce tangles by leaving a waxy coating on the hair to smooth rough edges.

Ongoing research provides new information that leads to the development of better hair-care products; hence you see new products advertised all the time. Increased scientific knowledge spurs technology. As scientific knowledge increases, technology advances.

• • •

1. Find out how chemistry can allow you to have green or pink hair for a day or several months.

2. Formulators can tweak hair-care products in many ways, but it is the marketing experts who create an image for their products that make consumers choose them. Compare the packaging of several brands of a specific hair product. What criteria will you use?

3. Is your shampoo toxic? Research two or more of the main ingredients in the shampoo you use. Is there any evidence that they could be harmful?

www.albertachemistry.ca
WWW

Section 2.3 Summary

- Stronger bonds require more energy for them to be broken.
- Because the kinetic energy of molecules is directly related to temperature, the temperatures of the melting and boiling points of substances are measures of the strength of their bonds.
- Ionic and metallic bonds are stronger than van der Waal's forces, causing melting and boiling points of ionic and metallic compounds to be higher than those of molecular compounds.
- Trends in melting and boiling points can be explained by the type of bonding in substances.
- Metallic bonds cause metals to be malleable, and ionic bonds cause ionic compounds to be brittle.
- Electrons or ions must be free to move past one another for a substance to conduct electric current or thermal energy.

SECTION 2.3 REVIEW

1. What determines the state of a pure substance at room temperature?

2. The following compounds are all solids at room temperature. Describe the nature of the attractive forces that are broken when the following compounds are melted:
   a) sodium chloride, $NaCl(s)$ c) paraffin wax, $C_{20}H_{42}(s)$
   b) silver, $Ag(s)$ d) graphite

3. Chloroform, $CHCl_3(\ell)$, has the same shape as methane, $CH_4(g)$. The boiling point of methane is $-182\ °C$, and the boiling point of chloroform is $61\ °C$. Explain this difference.

4. Describe the relationship between the strength of the attractive forces between the particle's and compound's melting and boiling points.

5. A solid substance is found to be soluble in water and has a melting point of $140\ °C$. To classify this solid as ionic, covalent, metallic, or network, what additional test(s) should be carried out?

6. Ionic compounds are extremely hard. They hold their shape extremely well. Explain these properties. Give two reasons why it is not practical to make tools out of ionic compounds.

7. A chemist analyzes a white, solid compound and finds that it does not dissolve in water. When the compound is melted, it does not conduct electricity.
   a) What would you predict about the melting point of this compound?
   b) Are the atoms that make up this compound joined by covalent or ionic bonds?

8. Imagine that you have been given a solid piece of graphite. You test the electrical conductivity of the sample in one orientation and find that it does conduct a current. When you test it in a direction that is perpendicular to the first, the conductivity is much lower. Provide a possible explanation for these observations.

9. Design an experiment that would allow you to determine the melting points of solid, crystalline substances. If necessary, look for information on the Internet or in advanced textbooks. Draw and describe the apparatus that you would use and write a detailed procedure. ICT

10. The alkanes are a family of compounds containing carbon and hydrogen. The data table below contains the melting points and boiling points for alkanes having one to 10 carbon atoms. Using an appropriate software package, draw a graph representing this data. Explain the trends in melting and boiling points based on the structures of the compounds.

Name	Formula	Melting point °C	Boiling point °C
Methane	CH_4	−182	−161
Ethane	C_2H_6	−183	−89
Propane	C_3H_8	−188	−42
Butane	C_4H_{10}	−138	−1
Pentane	C_5H_{12}	−130	+36
Hexane	C_6H_{14}	−95	+69
Heptane	C_7H_{16}	−91	+98
Octane	C_8H_{18}	−57	+126
Nonane	C_9H_{20}	−53	+151
Decane	$C_{10}H_{22}$	−30	+174

ICT

In crystals of ionic compounds, the ions align in the way that will bring the oppositely charged ions as close together as possible. The size and charge of the ions determine what the pattern, or crystal lattice, will be. Each positively charged ion is attracted to all negatively charged ions in close proximity to it and vice-versa. Consequently, you cannot identify any specific combinations of positively and negatively charged ions that could be considered to be a molecule. A formula unit is simply the lowest whole number ratio of ions in the crystal. Because each ion is strongly attracted to many others, the crystals are very stable.

In molecular compounds, each atom is covalently bonded to specific atoms forming a unit which is correctly described as a molecule. Molecular compounds have a wide variety of chemical and physical properties. Valence-shell electron-pair repulsion (VSEPR) theory allows you to predict the three dimensional structure around a central atom of a molecule based on its Lewis structure. When you know the three dimensional shape of a molecule, you can analyze the bonds for polarity and then determine whether the molecule is polar.

The forces that act between molecules are called intermolecular forces. These forces vary greatly in strength, giving molecular compounds their wide variety of properties. Most of the forces between molecules in a pure substance can be classed as dipole-dipole attractions or London (dispersion) forces. Both classes of forces vary in strength depending of the properties of the molecules in the substance but dipole-dipole forces are the stronger. In every case, however, the basis of the force is the attraction between positive and negative charges. In dipole-dipole forces, each molecule has a slightly negatively charged end and a slightly positively charged end. These molecules have permanent dipoles. The negative end of one molecule is attracted to the positive end of other molecules. One type of dipole-dipole attraction is so strong that it is given a special category— hydrogen bonds. When a hydrogen atom is bonded to a very electronegative atom such as oxygen, nitrogen, or flourine, the bond is very polar. The positively charged hydrogen atom is strongly attracted to the electronegative atom of another molecule. Hydrogen bonding is critical to the structure and function of many biological molecules.

London (dispersion) forces are the result of temporary charges that form and disappear in molecules. In non-polar molecules, London forces are the only intermolecular forces. They are very weak forces but when molecules can form a large number of these attractions, the resulting combination of forces is very significant.

The physical state of a pure substance is indicative of the strength of the bonds among the atoms. High melting and boiling points indicate that the bonds are very strong. The strength of ionic bonds results in the solid state of ionic compounds at normal temperatures. The fact that the intermolecular forces of molecular compounds vary in strength explains why different compounds can be solid, liquid, and gaseous at normal temperatures.

The alignment of charges in ionic crystals explains why they are brittle. Metals are malleable because their negative charges are randomly distributed and not aligned. The freedom of movement of electrons or ions explains why substances do or do not conduct electrical current or heat.

Concept Organizer   Structure Determines Properties

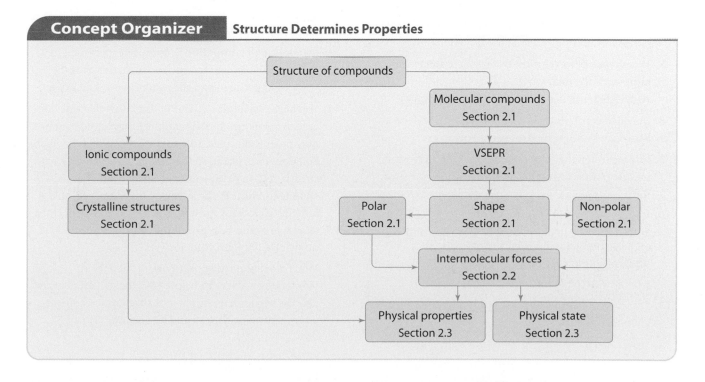

Understanding Concepts

1. Use a diagram to illustrate each of the following ideas:
   a) Water is a molecule that has a bent shape.
   b) Each oxygen atom in carbon dioxide has two bonding pairs and two non-bonding pairs.
   c) Silicon tetrachloride is a tetrahedral molecule.
   d) Carbon tetrachloride contains polar bonds but is a non-polar molecule.
   e) The water molecules in a glass of water are held together by both intermolecular and intramolecular forces.
   f) Hydrogen fluoride, HF(g) is held together by hydrogen bonds but CH_3F(g) is not.

2. Diatomic molecules, such as oxygen, nitrogen and chlorine, tend to exist as gases at room temperature. Explain this observation using bonding theory.

3. A solid molecular compound has both intermolecular and intramolecular forces. Do solid ionic compounds contain these types of forces? Explain your answer.

4. Describe the intermolecular forces between the molecules of hydrogen halides, HF(g), HCl(g), HBr(g), and HI(g), and explain the differences in their boiling points.

5. What is the difference between a permanent molecular dipole in a polar molecule and an induced dipole in a non-polar molecule? What is the effect on melting and boiling points?

6. Cooking oil, a non-polar liquid, has a boiling point in excess of 200 °C. Water boils at 100 °C. How can you explain these facts, given the strength of the hydrogen bonds in water?

7. What types of forces must be broken to melt solid samples of the following?
   a) NH_3(g) b) NaI(s) c) Fe(s) d) CH_4(g)

8. In which compound, H_2O(ℓ) or NH_3(g), will the hydrogen bonding be stronger? Explain.

9. List the following substances in order of increasing boiling points: C_2H_5OH, SiO_2, C_3H_8. Give a reason for your answer.

10. What are the molecular shapes of CCl_4(ℓ), CH_3Cl(ℓ), and $CHCl_3$(ℓ)? In liquid samples of these compounds, which would have dipole-dipole forces between molecules? Predict the order of their boiling points from lowest to highest.

11. Distinguish between dipole-dipole attractive forces and an ionic bond.

12. How can a molecule with polar covalent bonds be non-polar?

13. Predict which substances will have the higher boiling point in each of the following pairs. Explain your predictions using bonding theories.
   a) NH_2Cl or PH_2F
   b) SiC or AsH_3
   c) Ne or Xe
   d) CH_3OH or $C_2H_5NH_2$
   e) CH_3F or F_2
   f) $AlCl_3$ or $AsCl_2$
   g) C_4H_{10} or Cl_2
   h) NH_3 or PH_3
   i) C_5H_{12} or C_4H_9F
   j) ammonia or methane
   k) carbon dioxide or silicon dioxide
   l) neon or krypton
   m) KCl or ICl
   n) KBr or ClBr
   o) NH_2Cl or PH_2Cl
   p) NH_4Cl or CH_3Br
   q) C_2H_5F or CH_3Cl
   r) CH_3F or CH_3Br
   s) C_2F_2 or C_2HCl
   t) H_2O or H_2S
   u) SO_2 or SiO_2
   v) NaCl or CO

14. Explain each of the following using bonding theory:
   a) Glycerol, $C_3H_5(OH)_3$(ℓ), flows much more slowly than water, H_2O(ℓ).
   b) $CaCl_2$(s) has no molecules.
   c) Silver can be bent to make chains.
   d) Graphite is used as a high-temperature lubricant.
   e) Diamonds are not conductors of electric current.
   f) Molten potassium chloride conducts electric current.
   g) Metals have variable melting points.
   h) Diamond is the hardest known substance.
   i) A water-filled bottle breaks when placed in a freezer overnight.
   j) Graphite is used to make tennis rackets and golf clubs.

Applying Concepts

15. Suppose that you have two colourless compounds. You know that one is an ionic compound and the other is a molecular compound. Design an experiment to determine which compound is which. Describe the tests you would perform and the results you would expect.

16. You have two liquids, A and B. You know that one liquid contains polar molecules and the other liquid contains non-polar molecules. You do not know which is which. You pour each liquid so that it falls in a steady, narrow stream. As you pour, you hold a negatively charged ebonite rod near to the stream. The stream of liquid A is deflected toward the rod. The rod does not affect the stream of liquid B. Which liquid is polar and which is non-polar? Explain your answers.

17. Water reaches its maximum density at 4 °C. At temperatures above or below 4 °C, the density is lower. Discuss what is happening to the water molecules at temperatures just above and just below 4 °C.

18. In which liquid, $HF(\ell)$ or $H_2O(\ell)$, would the hydrogen bonding be stronger? Based on your prediction, which of these two liquids would have a higher boiling point? Refer to a reference book or the Internet to find the boiling points of these two liquids to check your predictions. Account for any difference between your predictions and the actual boiling points. **ICT**

19. Compare the molecules $SF_4(g)$ and $SiF_4(g)$ with respect to molecular shape and molecular polarity.

20. Compare the forces of attraction that must be overcome to melt samples of NaI and HI.

21. a) Contrast the physical properties of diamond, a covalent network solid, with the molecular compound 2,2-dimethylpropane shown here.

$$
\begin{array}{c}
\text{H} \\
| \\
\text{H---C---H} \\
\\
\text{H} \qquad\qquad \text{H} \\
| \qquad\qquad\quad | \\
\text{H---C--------C--------C---H} \\
| \qquad\qquad\quad | \\
\text{H} \qquad\qquad \text{H} \\
\\
\text{H---C---H} \\
| \\
\text{H}
\end{array}
$$

  b) In diamond, covalent bonds join the carbon atoms. In a molecule of 2,2-dimethylpropane, a central carbon atom is covalently bonded to four $-CH_3$ groups. How are the covalent bonds in the two substances different from each other?

  c) How are the physical properties of these two substances related to their covalent bonds?

22. At room temperature, carbon dioxide, $CO_2(g)$, is a gas, while silica, $SiO_2(s)$, is a hard solid. Compare the bonding in these two compounds to account for this difference in physical states.

23. Compare the bonding and molecular polarity of $SeO_3(s)$ and $SeO_2(s)$.

24. Discuss the intermolecular and intramolecular forces in $N_2H_4(g)$ and $C_2H_4(g)$. Based on the bonding between molecules, which of these two compounds would have a lower boiling point?

25. The boiling points of argon gas ($-186\,^\circ$C) and fluorine gas ($-188\,^\circ$C) are quite similar. Explain the similarity using bonding theory.

26. Krypton gas, $Kr(g)$ (boiling point $-152\,^\circ$C), hydrogen bromide, $HBr(g)$ (boiling point $-66\,^\circ$C), and hydrogen selenide, $H_2Se(g)$ (boiling point $-41\,^\circ$C), all contain the same number of electrons. Explain the differences in boiling point using bonding theory.

27. Using the following data, classify substances A to D as molecular, ionic, metallic, or network.

Substance	Melting point (°C)	Boiling point (°C)	Conductivity of solid	Conductivity of liquid	Solubility of water
A	+1420	+2435	Poor	Poor	Insoluble
B	+700	+1525	Poor	Good	Soluble
C	−225	−55	Poor	Poor	Low solubility
D	−40	+355	Good	Good	Insoluble

Making Connections

28. Many advanced materials, such as KEVLAR®, have applications that depend on their chemical inertness. What hazards do such materials pose for the environment? In your opinion, do the benefits of using these materials outweigh the long-term risks associated with their use? Give reasons to justify your answer.

29. Considering the changes of states that occur with water in the environment, suggest how intermolecular forces of attraction influence the weather.

30. One form of lightning is cloud-to-ground lightning. This form of lightning originates from clouds that have become negatively charged. The actual mechanism of this charge formation continues to be a hotly debated topic among scientists. The lightning strike consists of two parts, a step leader, which is a path of charged air that leads down from the cloud, and a positive streamer that leads up from the ground. When the leader and streamer meet, a lightning strike (electron flow) will occur. In the picture below, a lightning strike has formed between the cloud and the tree. Many additional step leaders can be seen in the sky.

  a) Compare and contrast the principle behind London (dispersion) forces and the formation of a lightning strike.

  b) People in the immediate area of a lightning strike have reported that their hair stands on end. Explain this phenomenon using your knowledge of London (dispersion) forces.

Career Focus: Ask a Nanotechnologist

Courtesy of the National
Research Council – NINT

The right chemistry makes for effective teamwork and exciting results, say world-renown nanotechnologists Drs. Jillian Buriak and Hicham Fenniri. The pair collaborates on a personal front as parents and increasingly at work as scientists. Each heads a research lab at the National Research Council's National Institute of Nanotechnology (NINT) at the University of Alberta, Edmonton. They are also university chemistry professors. Their work making structures just over a billionth of a metre in size is being used in everything from devising new medical treatments to shrinking the size of computers.

Q What is nanotechnology?

A (JB) Nanotechnology is the application of nanoscience, and nanoscience is the science of materials that are between 1 and 100 nm in size. This is the same size as the basic machinery of cells.

Q What makes nanoscale materials so special?

A (JB) The properties of nanoscale materials can be very different from that of the bulk parent material. For example, if you take a chunk of silicate (a compound of silicon and oxygen), it is a terrible light emitter. But if you take silicate and cram it into a silicate nanoparticle, it is a great light emitter. What is so neat about nanoscale science is that you can get old materials to do wacky things!

Q How do you make materials on the nanoscale?

A (JB) Chemists have been doing nano before nano was nano. Most of the molecules that people make are a nanometre in diameter or less, like Aspirin™, or huge polymers, like plastic bags. But how do you make the intermediate size? Well, that requires new approaches. We admire the way nature builds things, through self-assembly, and processes that occur spontaneously. We try and use the natural ability of things to aggregate …. It is like getting two people to fall in love: You don't put them in a room with fluorescent lights. You choose the reaction conditions—you provide a candlelight dinner.

Q How can you find out which reaction conditions to use?

A (JB) We fool around with thermodynamics and kinetics, and get rates of reaction that will lead to certain sizes. **(HF)** The advantage of nanoscale assembly is that if the pieces don't fit, they don't assemble. Under specific thermodynamic conditions, however, they will result in a well-defined structure. It's an adaptive system, so it can react to its environment. If the environment (temperature, ionic strength, and pH) is not right, it's not going to form … but once you learn the design rules, the sky is the limit.

Q What can you do once you know the design rules?

A (HF) Suppose you have one molecule that has the ability to capture an electron. If you have many of those molecules lined up in a self-organized fashion, that electron can hop from one molecule to another. So, if you design a molecule that can self-assemble in one dimension, you have essentially created a molecular wire.

Q What problem can molecular wires solve?

A (JB) Computers are getting smaller, so that means feature sizes on chips are getting very small. Well, there's an end of the road. The silicon transistors are going to get so small that they just won't function like silicon anymore. And so, can we make computers any faster after that point? What's the new technology going to be? Molecular electronics could be it.

Q How can nanoscale materials be used in medicine?

A (HF) We are very interested in bone-related diseases and bone implants. If you look at the structure of bone, you will see that it is made up of nanoscale materials, such as collagen fibres. The cells that are responsible for building bone recognize these nanostructures. Our group has shown that bone-forming cells stick to nanostructure materials that we have made in the lab. Therefore, we can make nanostructure materials to mimic the nanostructure of bone, so implants won't be rejected. Furthermore, we can choose the components so that the material displays a number of functionalities: It can stimulate growth and repair, inhibit bone resorption, and have an antibiotic function.

We are also making self-assembled nanotubes. In principle, we should be able to load them with specific medications, which they can deliver to specific areas of the body.

Q Does it help to discuss ideas with each other or collaborate on research?

A (HF) Absolutely. You cannot do science alone. (JB) If I see something that he's doing that I think would be useful for me, or vice versa, then we start to talk. But it has to be spontaneous—like any collaboration, it has to be appealing to both.

Materials Chemist Scientists or chemical engineers who design useful organic, inorganic, or organo-metallic materials and study their properties are known as materials chemists. Inorganic chemists study, design, and synthesize molecules and materials containing metals, sometimes called hard materials. Organic chemists similarly work with carbon-based materials, which are known as soft materials. Program leaders in materials chemistry research have post-doctoral training and extensive supervisory experience.

Theoretical Chemist Scientists with postgraduate education in theoretical chemistry predict how a given structure or material could function. Theoretical chemists develop computational methods to predict and explain the behaviour and properties of molecules and materials. Mechanical engineers and materials chemists draw on the work of theoretical chemists.

Biomedical Engineer Specialized engineers design and test materials for use in living systems. In addition to designing and manufacturing devices, such as implants, pacemakers, contact lenses, and diagnostic tools, biomedical engineers create software for analyzing medical data. Biomedical engineers with undergraduate degrees in engineering work in research and industrial laboratories. Post-doctoral training is needed to lead a research team or become a university professor.

Instrumentation Technician Inquiry into nanoscience requires the use of specialized imaging equipment, such as scanning tunnelling microscopes and nanomaterial fabrication equipment. Skilled technicians with college diplomas are needed to operate and maintain laboratory equipment.

Clinician Trained medical doctors with an interest in nanomedicine work with patients in research hospitals to try different treatments, including medications. Some clinicians perform bone implant surgeries and monitor how body tissues respond to the implants. Clinical research is needed to find out if nanotechnology-based treatments are working.

Molecular and Cellular Biologist Biologists who study cells and subcellular structures investigate how nanomaterials interact with cells. Molecular and cellular biologists examine how cells react to various triggers, including which genes are turned "on" or "off," which cellular reactions occur, and how cells interact. Scientists in this field have undergraduate and graduate degrees and work at research institutions, such as universities.

Go Further...

1. How does nanoscience challenge basic knowledge about the structure of matter?

2. Using silicon (Si) as an example, describe how the properties of individual atoms or molecules differ from the properties of nanomaterials. How do silicon nanomaterials differ from larger amounts of silicon?

3. Research the use of light-emitting nanomaterials in sensors. How do these materials react to specific particles in the air?

4. Research the use of nanotechnology in drug delivery. List three medications that are delivered throughout the body by nanotechnology. What is the advantage of using nanomaterials to deliver the medication in each case?

www.albertachemistry.ca
WWW

Understanding Concepts

1. The electrons in the outer energy level of an atom are called:

a) unpaired electrons

b) valence electrons

c) bonding pairs

d) lone pairs

e) octet of electrons

2. Ionic compounds consist of:

a) metal atoms only

b) non-metals only

c) Group 1 and Group 17 only

d) Group 18 atoms

e) metal and non-metal atoms

3. Which of the following are examples of a network solid?

a) sodium, $Na(s)$

b) sucrose, $C_{12}H_{22}O_{12}(s)$

c) graphite, $C(s)$

d) silica, $SiO_2(s)$

e) magnesium fluoride, $MgF_2(s)$

4. The two electrons in a bonding pair are:

a) always donated by the same atom

b) never donated by the same atom

c) usually donated by two different atoms

d) always donated by two different atoms

e) never donated by two different atoms

5. A crystalline substance that is hard, unmalleable, has a high melting point, and is a non-conductor of electric current could be

(I) an ionic crystal   (III) a metal

(II) a polar covalent solid   (IV) a network solid

a) I or II   c) III or IV   e) I or IV

b) I or III   d) II or IV

6. Which of the following molecules does not have a linear shape?

a) $Cl_2O(g)$ c) $SeF_2(g)$ e) $BeF_2(g)$

b) $CO_2(g)$ d) $OCS(g)$

7. The bond between atoms having an electronegativity difference of 1.0 is

a) ionic   d) slightly polar covalent

b) mostly ionic   e) non-polar covalent

c) polar covalent

8. A substance that has a melting point of 1850 °C and a boiling point of 2700 °C is insoluble in water and is a good insulator of electricity. The substance is most likely

a) an ionic solid   d) a network solid

b) a polar covalent solid   e) a molecular solid

c) a metal

9. The main factor that leads to the formation of a chemical bond between atoms is

a) the lowering of the melting point and boiling point when a compound is formed

b) the formation of the most stable electron configuration

c) the formation of a shape consistent with VSEPR theory

d) a tendency to make the attractions equal to the repulsions between atoms

e) a tendency to reduce the number of existing particles

10. Which of the following is *not* a physical property?

a) melting point   d) malleability

b) electronegativity   e) boiling point

c) electrical conductivity

11. List three properties of ionic and molecular compounds that differ. Explain how they differ.

12. Indicate whether the following statement is correct. If not, restate the sentences correctly. "In electron dot diagrams of atoms, some electrons are drawn in pairs when the atom has more than four valence electrons. This method of drawing electrons is simply for convenience. There is no significance in drawing two electrons as a pair."

13. Explain how you can predict whether a compound is ionic or molecular by looking at a formula, such as MgO or NH_3.

14. Use electron dot diagrams to illustrate the formation of potassium iodide and magnesium bromide.

15. Name the following compounds.

a) $NiO(s)$ d) $P_4O_{10}(s)$ g) $N_2O(g)$

b) $SO_3(g)$ e) $Na_3PO_5(s)$ h) $NaOCl(s)$

c) $K_2CO_3(s)$ f) $NH_4HCO_3(s)$ i) $SiO_2(s)$

16. Write the chemical formula for the following compounds.

a) mercury (II) sulfide   f) ammonium nitrite

b) calcium hydroxide   g) manganese (IV) oxide

c) chlorine gas   h) lead (IV) sulfite

d) sulfur dioxide   i) calcium oxalate

e) cesium fluoride

17. What is the difference between *intra*molecular forces and *inter*molecular forces? Provide an example to illustrate the difference.

18. VSEPR theory predicts that the bond angles in a trigonal planar molecule should be 120° and in a bent molecule, 109.5°. Explain how these predicted bond angles are determined and why the actual bond angles may differ significantly from the predicted values.

19. Alkanes are compounds that contain carbon and hydrogen atoms bonded with only single bonds.

a) Describe the shape of an alkane.

b) Describe the nature of the intramolecular forces found in alkanes.

c) Are alkanes polar or non-polar? Explain.

d) Alkanes with 25 carbons have the chemical formula, $C_{25}H_{52}(s)$. Although they all have the same chemical formula, they may have different structural formulas. Molecules with the same chemical formula but a different structural formula are called isomers. Some isomers will be long, straight chains while others will be highly branched. How do you expect the boiling point of these isomers to differ, if at all? Explain your answer.

20. Use bonding theory to explain the following.

a) Radon gas, $Rn(g)$ boils at a much higher temperature than krypton gas, $Kr(g)$.

b) Hydrogen chloride, $HCl(g)$ boils at a much higher temperature than argon gas, $Ar(g)$, even though they have the same number of electrons.

c) Methane, $CH_4(g)$, is non-polar even though its bonds are polar.

d) Hydrogen sulfide, $H_2S(g)$, is a bent molecule.

21. How do the lone pairs of electrons around the central atom of a molecule affect the bond angle between two bonding pairs of electrons?

22. Why is it incorrect to refer to an ionic compound as a "molecule"?

23. Chlorine molecules, $Cl_2(g)$, have a single covalent bond; oxygen molecules, $O_2(g)$, have one double bond; and nitrogen molecules, $N_2(g)$, have a triple bond. Explain why these diatomic molecules have different types of bonds.

24. In both metallic and molecular compounds, atoms share electrons. Why was it necessary to develop a different model of bonding for the two types of substances?

25. Explain why, at room temperature, $CO_2(g)$ is a gas but $CS_2(\ell)$ is a liquid.

26. Explain the process by which you can use the electronegativities of two atoms to determine the type of bond that will form between the atoms.

27. If you read that a certain bond had 70 percent ionic character, what properties would you expect that bond to have? What would be the percent covalent character of the bond?

28. Distinguish between dipole-dipole attractive forces and London (dispersion) forces. Is one type of intermolecular force stronger than the other? Explain.

29. Describe the properties of graphite that make it a good lubricant at high temperatures.

30. Describe some similarities and differences between metallic compounds and network solids.

31. Explain why the boiling point of a pure molecular compound is an indication of the strength of the intermolecular attractive forces among the molecules.

32. State the two factors that affect the strength of London (dispersion) forces between two molecules.

33. What are the requirements for the formation of hydrogen bonds?

34. Under what conditions are hydrogen bonds at their maximum strength?

35. Why does water have a much higher melting point and boiling point than other molecules of similar size, such as ammonia and methane?

36. Explain the reason for the malleability of metals.

37. Ionic crystals can be grown over a long period of time by evaporating the solvent from a concentrated solution of an ionic compound.

a) Describe the two factors that determine the shape of an ionic crystal.

b) List three physical properties associated with ionic crystals.

c) Provide an explanation for these physical properties.

38. How does melting point relate to the strength of intermolecular forces? Explain.

39. Why do ionic compounds conduct electric current when in a molten state but not when they are solid?

Applying Concepts

40. Describe the nature of the chemical bonds in the following compounds. (Identify them as mostly ionic, polar-covalent, non-polar covalent, or metallic.)

a) calcium chloride, $CaCl_2(s)$

b) carbon dioxide, $CO_2(g)$

c) brass (an alloy of copper and zinc)

d) ozone, $O_3(g)$

41. Draw Lewis structures to illustrate how the element bromine can take part in an ionic bond, a non-polar covalent bond, and a polar covalent bond.

42. Which ionic solid would you expect to have a higher melting point, $LiI(s)$ or $LiBr(s)$? Justify your answer.

43. Use electronegativity values to indicate the polarity of the bond in each of the cases below.

a) N—H **b)** F—N **c)** I—Cl

44. Arrange the following sets of bond pairs in order of increasing polarity. Identify the direction of the bond polarity for the bond pairs in each set.
  a) Cl—F, Br—Cl, Cl—Cl
  b) Si—Cl, P—Cl, Si—Si

45. Formulate a procedure by which you could experimentally determine whether an unknown solid was a network solid, an ionic solid, a molecular solid, or a metallic solid.

46. The ammonium ion, $NH_4^+(aq)$, is charged but not polar. The ammonia molecule, $NH_3(g)$, is polar but not charged. Explain why these statements are correct.

47. "If there were no intermolecular forces, all molecular compounds would be gases." Do you agree or disagree with this statement? Explain why.

48. The group trend for boiling point is the same as the trend for atomic radius. For the compounds formed between hydrogen and the first three elements of Group 16, $H_2S(g)$ has a lower boiling point than both $H_2O(\ell)$ and $H_2Se(g)$. Explain this diversion from the trend you would expect.

49. Define the following terms and identify the term that best describes the H—O bond in a water molecule:
  a) non-polar covalent
  b) polar covalent
  c) ionic

50. Define the following terms and identify the term that best describes the forces **between** the water molecules in a glass of water:
  a) London (dispersion) forces
  b) dipole-dipole forces
  c) hydrogen bonds
  d) intermolecular forces
  e) intramolecular forces

51. For a solid metallic element:
  a) identify four physical properties
  b) predict and explain how any of the properties listed would vary from left to right across a period in the periodic table
  c) identify two chemical properties that would enable you to classify the element as metallic

52. You can use electronegativity differences to determine percent ionic or percent covalent character of chemical bonds. The following graph plots percent ionic character versus ΔEN for a number of gaseous binary molecules. Use this graph to answer the following questions:

Ionic Character versus Electronegativity Difference

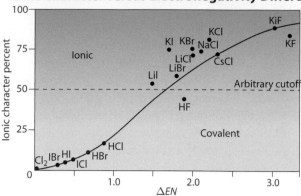

a) Describe the ionic character of the molecules as a function of ΔEN.
b) Which molecule has a zero percent ionic character? What can you infer about the interactions among the electrons of this molecule?
c) Do any molecules have 100 percent ionic character? What can you infer about the interactions among the electrons of ionic compounds?
d) Chemists often assign the value of 50 percent as an arbitrary cutoff for separating ionic compounds from covalent compounds. Based on your answer in part (c), what does the use of the term "arbitrary" suggest about the nature of chemical bonds?

53. In 1906, the Nobel Prize in Chemistry was awarded to a French chemist, Henri Moissan, for isolating fluorine in its pure elemental form. Why would this achievement be deserving of such a prestigious honour? Use your understanding of atomic properties, as well as of chemical bonds, to explain your answer.

54. Use a graphic organizer, such as a Venn diagram, to compare ionic bonding with metallic bonding.

Solving Problems

55. Draw electron dot diagrams for the following neutral atoms:
  a) neon   **c)** phosphorous   **e)** lithium
  b) beryllium **d)** iodine

56. Draw an electron diagram for the ionic compound $Ba(OH)_2(s)$.

57. Draw Lewis structures for each of the following molecules:
  a) $OCF_2(g)$ **c)** $HCN(g)$ **e)** $HCO_3^-(aq)$
  b) $H_2NNH_2(\ell)$ **d)** $Cl_2O(g)$ **f)** $NO_3^-(aq)$

58. Use VSEPR theory to predict the shape of the following molecules. Draw a three-dimensional diagram for each of the molecules to support your decision.
  a) $NO_2(g)$ **b)** $COCl_2(\ell)$

c) NOF(g) **d)** $CH_3Cl(g)$

e) $PCl_3(g)$ **f)** $H_2S_2(g)$

g) $N_2H_2(g)$

59. Which of the molecules above are polar?

60. Use VSEPR theory to predict the molecular shape of $CH_2Cl_2(\ell)$. Make a sketch to indicate the polarity of the bonds around the central atom to verify that this is a polar molecule.

61. Use VSEPR theory to predict the molecular shape and draw a three-dimensional diagram for the compound $C_2H_2Cl_2(\ell)$. Indicate the polarity of the bonds on your diagram. According to your sketch, is this molecule polar or non-polar? In reality, this molecule may be either polar or non-polar. Draw an additional diagram to illustrate this point.

Making Connections

62. Some of the most important discoveries of the last few decades have been made serendipitously, that is, by chance. For example, when the principle behind lasers was discovered and lasers were first built, no one had any idea that there would be a use for them. Such is the case with fullerenes or "buckyballs." Do research in print or on the Internet to find out how fullerenes were discovered. Then find out about the types of ongoing research to learn about possible applications for fullerenes. Focus on a type of research and application that you would enjoy pursuing. What application would you like to develop? What reason do you have to believe that this application is feasible? **ICT**

63. a) Iron is a hard substance, but it is easily corroded by oxygen in the presence of water to form iron(III) oxide, commonly called rust. Rust is chemically very stable but structurally, it has very little strength. Explain this difference in properties of iron and iron(III) oxide, in terms of what you have learned about the type of bonding in these substances.

  b) Aluminium reacts with oxygen in air to form aluminium oxide. Unlike rust, which flakes off, aluminium oxide remains bound to the aluminium metal, forming a protective layer that prevents further reaction. What might account for the difference in behaviour between iron(III) oxide and aluminium oxide and their respective metals?

64. Sour gas is a mixture of methane gas, hydrogen sulfide gas, and water vapour. The methane is a desirable fuel source, but first the dangerous gas, hydrogen sulfide, must be removed. The process of refining sour gas is called "sweetening."

Sour Gas Plant in Alberta

  a) Illustrate the Lewis structure diagrams for methane gas, hydrogen sulfide gas, and water.

  b) Analyze the data below and explain why there is a difference in the boiling points for the three gases.

Chemical name	Boiling point °C
Methane	−161
Hydrogen sulfide	−59.55
Water	+100.0

  c) Predict the boiling points of hydrogen selenide gas ($H_2Se(g)$) and hydrogen telluride gas ($H_2Te(g)$). Justify your answer.

  d) Explain why hydrogen sulfide is considered to be miscible (soluble) in water.

  e) Design a procedure that could be used to "sweeten" sour gas.

65. As you learned in Section 2.2, the double helix of DNA, deoxyribonucleic acid, is held together by both covalent bonds and hydrogen bonds, as shown here.

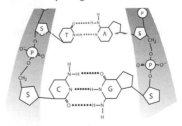

The nucleotides, composed of a phosphate ion, deoxyribose sugar molecule, and one of the four bases are all held together with covalent bonds. The two DNA strands are connected by hydrogen bonds between the bases.

  a) When DNA is analyzed, the strands must first be separated. Using bonding theory, describe a procedure that could be used to separate the strands of DNA without breaking the bonds within the strands themselves.

  b) Proteins are also held together with hydrogen bonds. Using bonding theory, explain why proteins only function properly within a narrow temperature range.

UNIT
2

Forms of Matter: Gases

General Outcome

- explain molecular behaviour using models of the gaseous state of matter

Unit Contents

Focussing Questions

1. How do familiar observations of gases relate to specific scientific models describing the behaviour of gases?

2. What is the relationship between the pressure, temperature, volume, and amount of a gas?

3. How is the behaviour of gases used in various technologies?

Unit PreQuiz ?

www.albertachemistry.ca

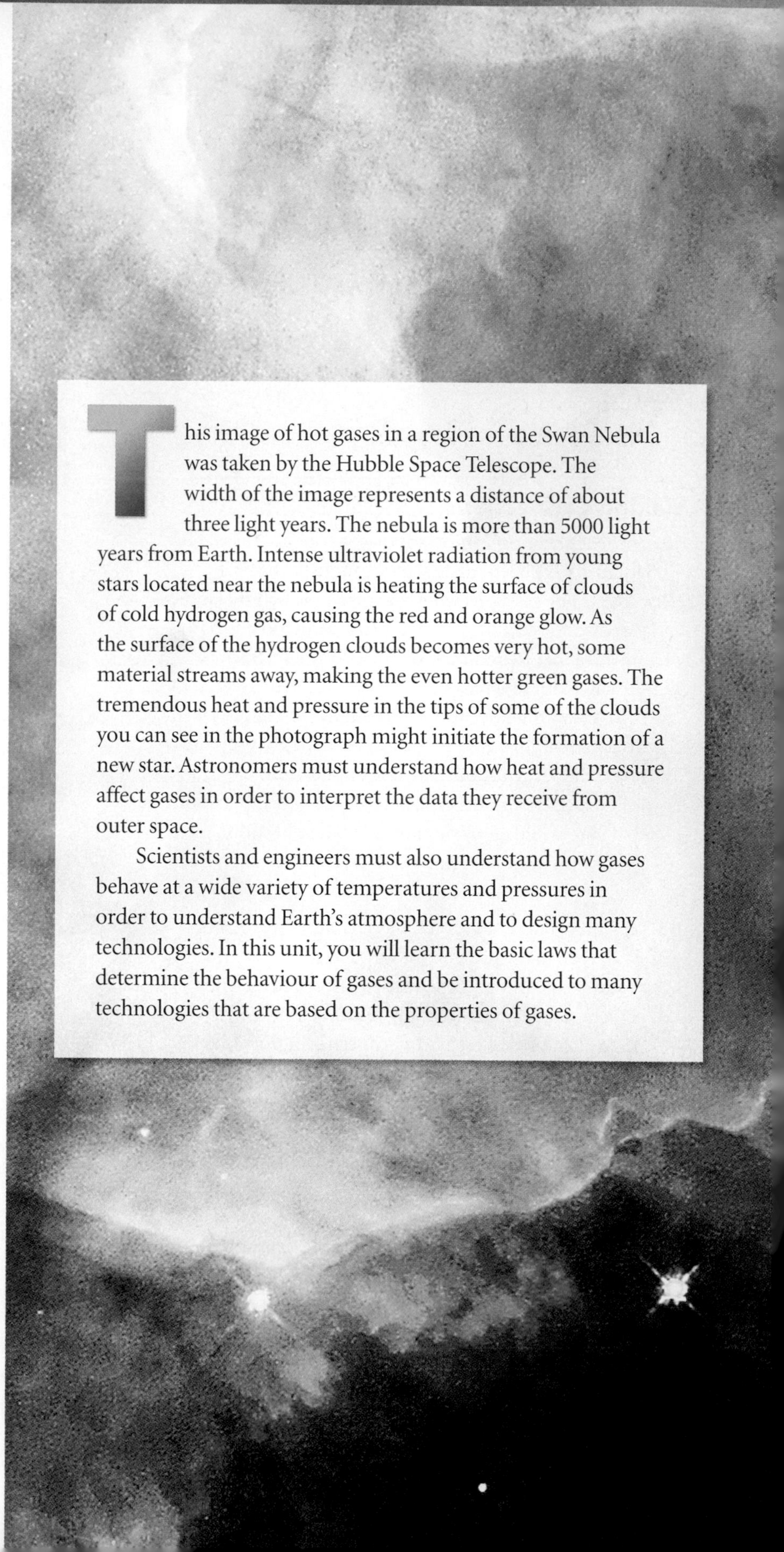

This image of hot gases in a region of the Swan Nebula was taken by the Hubble Space Telescope. The width of the image represents a distance of about three light years. The nebula is more than 5000 light years from Earth. Intense ultraviolet radiation from young stars located near the nebula is heating the surface of clouds of cold hydrogen gas, causing the red and orange glow. As the surface of the hydrogen clouds becomes very hot, some material streams away, making the even hotter green gases. The tremendous heat and pressure in the tips of some of the clouds you can see in the photograph might initiate the formation of a new star. Astronomers must understand how heat and pressure affect gases in order to interpret the data they receive from outer space.

Scientists and engineers must also understand how gases behave at a wide variety of temperatures and pressures in order to understand Earth's atmosphere and to design many technologies. In this unit, you will learn the basic laws that determine the behaviour of gases and be introduced to many technologies that are based on the properties of gases.

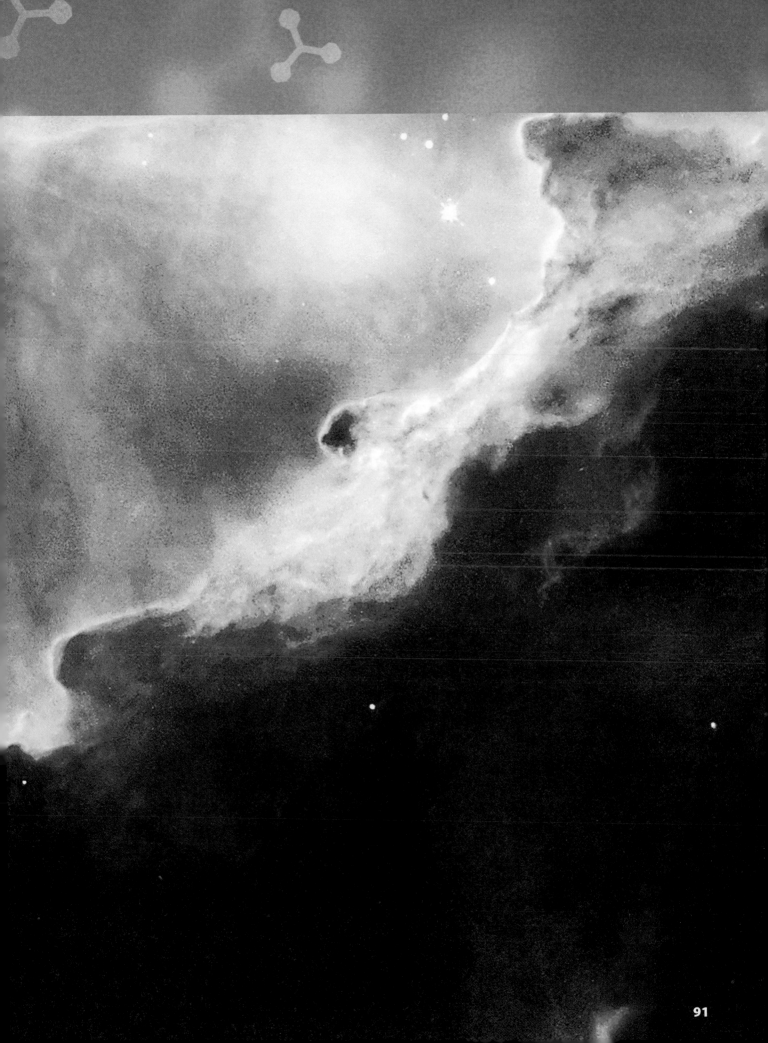

Figure P2.1 How can this heavy, iron bolt float on the surface of liquid mercury? (Note: Because mercury vapours are toxic, the surface is coated with a layer of oil.)

Perhaps you have seen the photograph in Figure P2.1. What physical principles does this photograph illustrate?

The iron bolt floating on liquid mercury illustrates the concept that any object, no matter how heavy, will float on a liquid that has a higher density than that of the object itself.

Density

In previous science courses, you learned about density and buoyancy. You will need to use the concept of density in your study of gases. Density is defined as the mass of an object divided by its volume. You can express the definition of density in mathematical form as shown below.

$$\text{density} = \frac{\text{mass}}{\text{volume}} \qquad D = \frac{m}{V}$$

In SI units, mass is expressed in kilograms (kg) and volume in cubic metres (m^3), so density has units of kilogram per cubic metre (kg/m^3). However, in chemistry, the density of a solid is often expressed in units of gram per cubic centimetre (g/cm^3). Fluid density is often reported in units of gram per litre (g/L) or gram per millilitre (g/mL). You might recall that the density of water (at 15 °C) is 1.0 g/mL.

Pressure

The concept of pressure that you have also studied previously is a concept that you will encounter in this unit. You might recall that pressure is defined as the force exerted on a surface divided by the area of the surface over which the force acts. You can express the definition of pressure mathematically as shown below.

$$\text{pressure} = \frac{\text{force}}{\text{area}} \qquad P = \frac{F}{A}$$

The derived SI unit of force is the newton (N) and the SI unit of area is the square metre (m^2). Therefore, the unit of pressure is newton per square metre (N/m^2). This unit is also called a pascal (Pa), in honour of Blaise Pascal (1623–1662) for his work on concepts involving pressure. When studying gases, you will frequently use the unit kilopascal (kPa), which is equal to 1000 Pa.

Figure P2.2 Syrup is much more viscous than milk.

Viscosity

Do you recall the technical term that describes the resistance to flow of different fluids such as the syrup and the milk shown in Figure P2.2? The viscosity of a fluid is a measure of its resistance to flow. You will not need to use a formula for viscosity in this course. Just remember the definition of viscosity when you are learning about the properties of gases.

Moles and Mass

What property is the same for each of the compounds in the photograph in Figure P2.3? The colours of the substances are not the same. The states and the masses are not the same. However, the numbers of particles—molecules or formula units—in each item are the same. There is one mole of particles of each substance in the containers.

When carrying out reactions between atoms or molecules, they combine in specific ratios. For example, when hydrogen and oxygen combine to form water, two atoms of hydrogen combine with one atom of oxygen. However, one oxygen atom has a much larger mass than do two atoms of hydrogen. Chemists have long realized that it is much more convenient to combine the correct ratios of atoms or molecules in a reaction than to combine the same mass. Because molecules are very small, chemists work with extremely large numbers of molecules. For convenience, the concept of a mole was developed. You probably recall from *Science 10* that a mole is defined as the number of atoms in 12 g of the isotope carbon-12 (^{12}C). The resulting number of atoms, to three significant digits, is 6.02×10^{23} atoms. You can apply the concept of a mole to any particle or object such as atoms, molecules, ions, or even pennies. As you continue your study of chemistry, you will frequently need to convert amount of a substances in moles to its mass and also to convert from mass to moles. The mathematical relationship between mass and moles is shown below.

Figure P2.3 Each container holds 6.02×10^{23} molecules or formula units.

$$\text{number of moles} = \frac{\text{mass}}{\text{molar mass}} \qquad n = \frac{m}{M}$$

The number of moles is called the amount of a substance and the symbol is n. Mass, m, is expressed in units of grams and the molar mass, M, is expressed in grams per mole.

$$n = \frac{m}{M}$$

After you have reviewed some mathematical techniques, you will practice using the formulas.

Solving Equations

When solving problems in chemistry, you are often asked to solve for a variable in an equation that is not isolated. For example, you might need to find the value of force in the pressure equation ($P = \frac{F}{A}$). You might want to solve for mass in the equation for number of moles ($n = \frac{m}{M}$). You will need to rearrange the formula so that the variable that you need to find is isolated. The rules of algebra apply to equations with only variables in exactly the same way that they apply to equations with numerical values. If you remember two rules, you will be able to solve any problem you need to solve in this chemistry course.

Rule 1: *You can perform the same operation on both sides of an equation without changing the equality.*

Rule 2: *To isolate a desired variable, perform the opposite operation to that acting on the variable on both sides of the equation.*

The following steps show you how the rules can be applied. Suppose you are told that the pressure inside a tank of pressurized gas is 525 kPa and the area of the base of the tank was 0.55 m². You are asked to find the force on the bottom of the tank due to the pressure of the gas.

- Write the formula for pressure, P:

$$P = \frac{F}{A}$$

- To isolate the force, F, perform the opposite operation. Because F is divided by A, multiply both sides of the equation by A. Cancel variables were possible:

$$PA = \left(\frac{F}{\cancel{A}}\right)\cancel{A}$$

- F is isolated, so now you can substitute in the numerical values and solve the equation:

$$F = PA$$
$$F = (525 \text{ kPa})(0.55 \text{ m}^2)$$

- To simplify the units, use the definition of the kPa. Recall that Pa = N/m² so kPa = kN/m²:

$$F = 288.75 \frac{\text{kN}}{\cancel{\text{m}^2}} \cancel{\text{m}^2}$$

- Now determine the number of significant digits and round the answer:

$$F = 288.75 \text{ kN}$$

- The area of the tank, 0.55 m², has the fewest significant digits so the answer must have two significant digits:

- There are three places to the left of the decimal point, but you must write the answer with two significant digits. Therefore, write it in scientific notation. Remember that a kN is 1000 N. You can also express the answer in units of N:

$$F = 2.9 \times 10^2 \text{ kN} \text{ or}$$
$$F = 2.9 \times 10^5 \text{ N}$$

Sample Problem

Problem
Find the mass of 12.6 mol of magnesium chloride.

What Is Required?
You need to find the mass, m, of 12.6 mol of magnesium chloride.

What Is Given?
You know the number of moles and the identity of the substance.
$n = 12.6$ mol
The substance is $MgCl_2$.

Plan Your Strategy
First find the molar masses of Mg and Cl from a periodic table. Then determine the molar mass of one formula unit of $MgCl_2$.

Write the formula for the relationship among mass, molar mass, and amount of mass in moles.

Solve the equation for mass, m, using the rules of algebra.

Substitute the known values into the equation and solve it.

Act on Your Strategy
Molar masses:

Mg: $M_{Mg} = 24.31$ g/mol

Cl: $M_{Cl} = 35.45$ g/mol

$$M_{MgCl_2} = 24.31 \frac{\text{g}}{\text{mol}} + 2\left(35.45 \frac{\text{g}}{\text{mol}}\right) = 95.21 \frac{\text{g}}{\text{mol}}$$

$$n = \frac{m}{M}$$

$$nM = \left(\frac{m}{\cancel{M}}\right)\cancel{M}$$

$$nM = m$$

$$m = (12.6 \cancel{\text{mol}})\left(95.21 \frac{\text{g}}{\cancel{\text{mol}}}\right)$$

$$m = 1199.646 \text{ g}$$

$$m \approx 1200 \text{ g or } 1.20 \times 10^3 \text{ g}$$

Solution
The mass of 12.6 mol of magnesium chloride is 1200 g.

Check Your Solution
The unit of mass is grams and the units cancelled correctly to give grams. The value 12.6 is slightly higher than 12 and 95.21 is slightly less than 100. Thus you would expect the value to be near 12 × 100 or 1200, which it is.

Sample Problem

Problem

What is the volume of an object with a mass of 765 g and a density of 2.27 g/cm³?

What Is Required?

You need to find the volume of an object given its mass and density.

What Is Given?

You know the mass and density of an object.

$m = 765$ g

$D = 2.27$ g/cm³

Plan Your Strategy

Use the formula for density and isolate the variable V.

Because V is in the denominator, multiply both sides of the equation by V to place it in the numerator.

Now, to isolate V, you need to perform the opposite operation.

Because V is now multiplied by D, you must divide both sides of the equation by D.

Now you can substitute the numerical values into the equation and solve it

Determine the number of significant digits and round the final answer.

Act on Your Strategy

$$D = \frac{m}{V}$$

$$DV = \left(\frac{m}{\cancel{V}}\right)\cancel{V}$$

$$DV = m$$

$$\frac{\cancel{D}V}{\cancel{D}} = \frac{m}{D}$$

$$V = \frac{m}{D}$$

$$V = \frac{765\,\cancel{g}}{2.27\,\dfrac{\cancel{g}}{cm^3}}$$

$$V = 337.004 \text{ cm}^3$$

$$V \approx 337 \text{ cm}^3$$

Solution

The volume of the object is 337 cm³.

Check Your Solution

The units cancel to give cubic centimetres, which is correct. You could get a very rough estimate of the numerical value by dividing 700 by 2, which is 350. This is close to 337 so a value of 337 cm³ is logical.

Practice Problems

1. Lead has a density of 11.34 g/cm³. What would be the mass of a block of lead if its volume was 225 cm³?

2. The gas pressure in a nitrogen tank is 4.1×10^4 kPa. The nitrogen exerts a force of 4.5×10^7 N on the base of the tank. What is the area of the base of the tank?

3. Water is poured into a graduated cylinder. If the pressure on the bottom of the cylinder, due to the water, is 3.25 kPa and the cross-sectional area of the bottom of the cylinder is 53 cm², what is the weight of the water? (Remember that weight is the force that an object exerts on the surface that is supporting it.)

4. Helium has a density of 0.179 g/L under conditions of standard temperature and pressure. What would be the volume of a sample of helium that has the same mass as the block of lead in question 1? (**Hint:** 1 L = 1000 cm³)

5. You need to add 2.75 moles of potassium carbonate (K_2CO_3) to a quantity of water to make a solution. What will be the mass of the potassium carbonate?

6. You have determined that there are 4.005×10^{-4} mol of a gas in a container. The mass of the gas is 16.00 mg. What is the molar mass of the gas?

In Unit 2, you will build on the concepts you have reviewed here. You will also use some skills that are listed here.

Prerequsite Skills

- **Draw** and **interpret** straight line graphs
- **Collect** and **interpret** numerical data in the laboratory

CHAPTER 3

Properties of Gases

Chapter Concepts

3.1 Gases and Kinetic Molecular Theory

- The properties of gases are important for many common technologies.

- Gases have unique characteristics that distinguish them from liquids and solids.

- The kinetic molecular theory, based on the concept of an ideal gas, can explain the properties of gases.

3.2 Gases and Pressure

- Atmospheric gases exert pressure on Earth's surface.

- The units for pressure reflect the historical development of knowledge about the gases.

- The pressure on a gas is inversely proportional to its volume as described by Boyle's law, as long as the temperature and amount of gas remain constant.

3.3 Gases and Temperature

- As described by Charles's law, when the pressure and amount of gas remain constant the temperature of a gas is directly proportional to its volume.

- The concept of absolute zero determines an absolute temperature scale.

A constant supply of oxygen is a matter of life or death. Nevertheless, people usually take the oxygen in air for granted. In some circumstances, however, it is a challenge to obtain enough oxygen to sustain life. For example, medical emergencies, chronic illnesses, and adventurous pursuits, such as mountain climbing and underwater diving (shown in the photograph) can create situations in which artificial devices are needed to provide the life-sustaining oxygen. Designing and building technologies to manage these situations depends on understanding the properties of gases and predicting the way they will behave. In this chapter, you will learn how to predict and explain changes in the volume, temperature, pressure, and amount of gases. You will learn how these properties can be explained by the kinetic molecular theory. Use the following Launch Lab to reacquaint yourself with some properties of gases.

Balloon in a Bottle

You cannot see most gases, but you can learn about some of the properties of gases by creating situations that you can visualize. In this activity, you will deduce some properties of gases by blowing up a balloon inside a bottle.

Safety Precautions

- Be careful with the sharp point of the scissors when piercing the plastic bottle.
- Your teacher might choose to do this activity as a demonstration.

Materials

- 1 L or 2 L clean plastic soft drink or juice bottle
- round balloon
- pointed scissors, or sharp object

Procedure

1. Hold the open end of the balloon while you insert the closed end into the bottle. Stretch the open end of the balloon over the lip of the bottle as shown in the photograph.

2. Predict how much you will be able to inflate the balloon inside the bottle, relative to the size of the bottle. Record your prediction in your notebook.

3. Inflate the balloon inside the bottle. How large did it get? Record your observations in your notebook. Allow the balloon to deflate.

4. Using the sharp end of a pair of scissors, puncture a hole in the middle of the bottom of the plastic bottle. Inflate the balloon again. Record your observations in your notebook.

Analysis

1. Was your prediction in step 2 verified when you carried out step 3? What happened to the air already inside the bottle when you blew up the balloon inside the bottle?

2. Was there a difference in the volume to which you were able to inflate the balloon before and after you punctured a hole in the bottle? If there was a difference, explain why.

3. Describe one property of gases that was demonstrated in this activity.

Gases and Kinetic Molecular Theory

Blowing up a balloon inside a bottle is just one activity that allows you to observe the behaviour of gases. Adding more gas to a flexible balloon makes the volume of the balloon expand. However, a rigid container, such as an automobile tire, does not expand when more air is added to it. What happens to a tire when you add more air but the volume doesn't change? Why is this result useful? Consider some more technologies that require a knowledge of the properties of gases to see why such knowledge is important.

Using the Properties of Gases—Gas Technologies

Have you ever watched a hot-air balloon drifting through the sky? Sometimes, you can hear the sound of the propane heaters that the pilot uses to warm the air in the balloon. Perhaps you have heard the phrase "hot air rises," but do you know why it does? How does warming the air lift the balloon and its attached basket high above the ground? What properties of gases must you understand in order to design a hot-air balloon?

Look at the picture of a scuba diver on page 96. The diver is carrying a tank of compressed air on her back. You can see bubbles of exhaled air escaping into the water. What properties of gases are critical to know in order to design a "self-contained underwater breathing apparatus" (SCUBA)? How long can a person survive underwater with scuba gear? How is breathing air from the tanks affected by the pressure deep under water, which can be many times greater than its pressure at the surface? What design features must the scuba system incorporate to ensure the diver can breathe the air from the tank without damage to the lungs?

Figure 3.1 The balloon operator controls the temperature of the air in the balloon by regulating the propane burner.

As shown in Figure 3.2, the force of compressed air drives the chuck of the jackhammer, giving it enough force to break through rock and concrete. How much does air have to be compressed to drive the jackhammer? Air compressors also drive dental drills and nail guns. How have these technologies improved your life?

The preceding examples show you how important it is to know about the properties of gases in order to understand how a large number of commonly used products and processes work. In this section, you will read about properties of gases, and then you will be able to answer many of the questions asked above. You will also have the skills to research the answers to many more questions. In addition, you will learn about the scientific theory used to explain these properties.

Properties of Gases

When you read about solids, liquids, and gases, you probably have a clear picture in your mind of the differences among these three states of matter. However, could you give a detailed description of these differences? What are the characteristic physical properties of gases? How do they differ from solids or liquids? To be categorized a gas, the substance must have the following characteristics.

- *Gases are compressible.* The volume of a gas decreases dramatically when pressure is exerted on the gas. In contrast, the volumes of liquids and solids remain almost constant during pressure changes.

- *Gases expand as the temperature is increased* if the pressure remains constant. The volumes of liquids and solids can expand with temperature but to a much smaller extent than can a gas.

- *Gases have very low resistance to flow*—a property called **viscosity**. The viscosity of water is approximately 55 times greater than the viscosity of air. Air and all other gases flow through pipes more freely than do liquids such as water. The low viscosity of gases also enables them to escape quickly through small openings in their containers.

- *Gases have much lower densities* than solids or liquids. The density of water vapor is approximately 1/1000 the density of liquid water.

- *Gases mix evenly and completely* when put in the same container. Substances that mix completely with each other are said to be **miscible**. All gases are miscible. Some liquids, such as water and alcohol are also miscible. However, water and oil are not.

Gases are sometimes defined as fluids that have no shape or volume of their own, but take on the shape and volume of the container in which they are confined. Gases expand to fill any container.

All of these properties of gases are **macroscopic**. That is, they can be directly observed using your senses or a measuring instrument. Scientific knowledge, however, does more than simply provide an accurate description of natural phenomena. It also provides a theoretical basis that enables you to predict, interpret, and explain these phenomena. To accomplish these goals, you must focus on the behaviour of gases at the level of individual molecules.

Figure 3.2 How can air provide a great enough force on the chuck to cause the chuck to break concrete?

Chemistry File

Web Link
If air at the pressure that is found in a scuba tank went directly into a diver's lungs, the lungs would be damaged. What device in the scuba gear prevents this damage? How does this device operate?

www.albertachemistry.ca
WWW

Kinetic Molecular Theory

The kinetic molecular theory provides a scientific model for explaining the behaviour of gases. To develop the theory, scientists defined a hypothetical substance called an ideal gas. Although no gas is ever ideal, the theory quite accurately describes the behaviour of real gases at ordinary temperatures and pressures. Modifications of the theory can account for some of the non-ideal behaviour of real gases. An **ideal gas** is defined by the following characteristics.

- *The gas molecules are in constant, random motion. The molecules travel in straight lines until they collide with other gas particles or the walls of the container* (see Figure 3.3).

- *The molecules of an ideal gas are "point masses."*

- *The gas molecules interact with one another and with the walls on a container only through elastic collisions. The molecules do not exert attractive or repulsive forces on one another.*

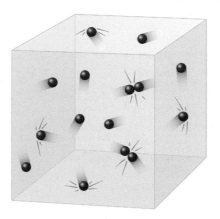
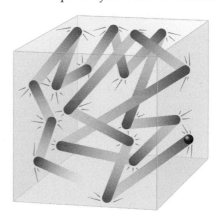

A   **B**

Figure 3.3 (A) Ideal gas molecules move with random motion, colliding with one another and with the walls of the container. **(B)** The path of an individual molecule follows straight lines between collisions. Such a path is sometimes called "random walk."

The definition of a **point mass** is a mass that takes up no space—it has no volume. Of course, such a mass cannot exist, but real gas molecules approximate point masses when the volume of the container is much greater than the volume of the gas molecules themselves.

In an **elastic collision**, kinetic energy is conserved. Molecules can exchange kinetic energy with one another in a collision but the total kinetic energy of all of the molecules remains constant. Contrary to ideal gas molecules, that interact only through elastic collisions, real gas molecules do interact with one another. There is often an attractive force among the molecules. Nevertheless, under standard conditions, the interactions are so small that the gases behave in a nearly ideal manner.

Using the concepts of point masses and elastic collisions, as well as some advanced math, James Clerk Maxwell (1831–1879) developed an equation that showed that the average kinetic energy of the molecules in a gas is directly proportional to temperature of the gas, or $\overline{E}_k \propto T$ (Note: The bar over the E is a symbol for "average.") The relationship was later supported by experimental observations.

Using these basic properties of an ideal gas, you can explain many properties of gases. You can account for the miscibility of gases, for example, by considering the large amount of space available between the molecules of a gas. The molecules of a second gas should readily fit into the spaces between the molecules of the first gas, because both types of gas molecules have negligible volume. Thus, two gases should mix evenly and completely as the molecules continue their constant motion.

ChemistryFile

FYI
Under standard conditions of temperature and pressure, gas molecules have an average speed of around 500 m/s. They can experience more than one billion collisions every second. The average distance between collisions is around 5×10^{-8} m.

ChemistryFile

FYI
The hypothetical point masses of an ideal gas have only translational kinetic energy, or energy of motion in a straight line. Real gas molecules have size and shape and, therefore, vibrate and rotate—motions that also require energy. Real gas molecules have translational, rotational, and vibrational kinetic energy.

Section 3.1 Summary

- Many technologies such as hot-air balloons, SCUBA tanks, dental drills, and nail guns are based on the properties of gases.
- The following properties distinguish gases from other states of matter:
  - compressible
  - expand with an increase in temperature
  - low viscosity
  - low density
  - miscible
- Properties of temperature, pressure, miscibility, viscosity, and density are macroscopic properties.
- Kinetic molecular theory of gases is a model that explains the macroscopic properties of gases based on the behaviour of individual particles (atoms or molecules).
- Molecules of an ideal gas:
  - are in constant motion
  - are point masses
  - collide with walls of a container and each other with elastic collisions
- A point mass is an ideal particle that has no volume.
- In elastic collisions, kinetic energy is conserved.
- The average kinetic energy of gas molecules is proportional to the temperature of the gas.

SECTION 3.1 REVIEW

1. Review the characteristics of gases and select one that you think is important in hot-air ballooning. Explain why you think it is important.

2. Review the characteristics of gases and select one that you think is important in SCUBA diving. Explain why you think it is important.

3. The macroscopic properties of gases are listed on page 99. Identify which macroscopic property explains the following real-life situations.
   a) A full propane tank can provide enough fuel for an entire season of barbecues.
   b) The label on a can of hairspray contains the warning, "caution may explode when heated."
   c) A carbon monoxide leak in the basement spreads quickly throughout the house.
   d) Forced air heating is often a better choice for home heating than hot water.
   e) A bicycle tire develops a small hole and very rapidly becomes flat. Use one of the characteristics of gases to explain why the tire deflates so quickly.

4. Explain the meaning of point mass.

5. How does an elastic collision differ from an inelastic collision? To picture an inelastic collision, imagine throwing a ball of putty against a wall.

6. Why is it important that the molecules of an ideal gas have only elastic collisions?

7. Under what conditions might real gases not behave like ideal gases?

8. The following figure shows three possible paths for a gas molecule moving inside a filled volleyball. Which of these diagrams represents the most likely path of the gas molecule? Justify your choice in terms of the kinetic molecular model of gases.

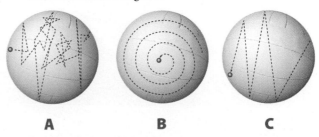

A   B   C

9. Use kinetic molecular theory to explain:
   a) the miscibility of gases
   b) why gases expand to fill the size of their container
   c) why gases can be easily compressed

Gases and Pressure

Section Outcomes

In this section, you will:
- **perform** investigations to determine the quantitative relationships between pressure and the volume of an ideal gas
- **express** atmospheric pressure using mmHg, atm, bar, and kPa
- **use** a broad range of tools and techniques to gather and record data

Key Terms

standard atmospheric pressure
Boyle's law

Have you ever been driving in the mountains and felt your ears pop? If you have flown in an airplane, you have probably felt discomfort in your ears when taking off or landing. As soon as your ears popped, they probably felt better. The reason that your ears become blocked and then pop is due to changes in the atmospheric pressure. Although you are usually unaware that the atmosphere is having any effect on you, it is always exerting a large amount of pressure on you from all directions. Developing an understanding of atmospheric pressure gave scientists the concepts and tools necessary to develop a more general model that applies to all forms of gas pressure.

Atmospheric Pressure

An atmosphere is an envelope of gases extending outward from the surface of any planet. Although gas molecules have very little mass, the gravitational attraction of the planet nevertheless acts on the molecules, keeping them near the surface of the planet. For example, most of the mass of Earth's atmosphere lies within 11 km of its surface.

In general, pressure is defined as force per unit area $\left(P = \frac{F}{A}\right)$. Thus, atmospheric pressure can be described as the force that a column of air exerts on Earth's surface divided by the area of Earth's surface at the base of the column (see Figure 3.4). That column of air extends from Earth's surface to the top of the atmosphere. The force that the air exerts is its weight. An important difference between pressure and force is that pressure is exerted in all directions to the same extent. For example, when you blow up a balloon, the pressure on all parts of the inside of the balloon is the same.

As you read in Section 3.1, gases are highly compressible. Each layer of atmosphere exerts a force on the layer below. Because the weight of the entire column exerts a force on the bottom layer, it is compressed the most. As the altitude increases, the amount of air above that level becomes smaller and, therefore, exerts a smaller force on the air just below it. The higher layers of air are compressed less than are the lower layers. Figure 3.5 illustrates the decrease in the density of the air as the altitude increases.

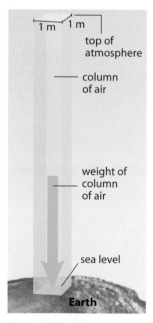

Figure 3.4 A column of air extending from sea level to the top of the atmosphere and having a cross-sectional area of 1 m² weighs 101 325 N. Its mass is 10 329 kg. A column of water with the same mass would be more than 10 m high.

Figure 3.5 The dots represent air molecules. You can see how rapidly the air thins as the altitude increases. For example, the atmospheric pressure on top of Mt. Everest is only about 30% as large as the atmospheric pressure at sea level.

Discovery of Atmospheric Pressure

Early scientists and natural philosophers, such as Aristotle, were unaware of the existence of atmospheric pressure. In fact, a technology that made use of atmospheric pressure—the water pump—was developed before anyone could understand why it worked. By the late sixteenth century, suction water pumps had been invented to pump water out of mine shafts. Galileo suggested that the pump created a vacuum that "pulled" the water up. However, no one, including Galileo, could understand why these pumps could not "lift" water more than 10 m.

In 1634, Evangelista Torricelli (1608–1647), a pupil of Galileo's, worked out the solution. He used mercury, however, because it is 13.6 times more dense than water. Therefore, he could use a column of mercury that is 13.6 times shorter than the 10 m column of water to test his hypothesis. Torricelli designed an experiment that would use an apparatus like the one shown in Figure 3.6. He proposed that if a long tube was filled with mercury and inverted into a dish of mercury, the mercury in the tube would drop down and leave a vacuum in the closed end of the tube. Toricelli hypothesized that the pressure that the column of mercury would exert on the mercury in the dish would be equal to the pressure that the atmosphere was exerting on the surface of the mercury in the dish outside of the tube. These experiments were carried out by a colleague of Torricelli's and his hypothesis was verified.

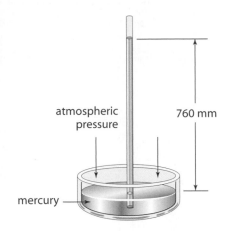

Figure 3.6 Toricelli's apparatus provided the basic design for mercury barometers. As well, this apparatus was the first to produce and sustain a vacuum. The sealed end of the inverted tube contains a vacuum.

Measuring Gas Pressure

In 1648, Blaise Pascal (1623–1662) planned an experiment to test the hypothesis that atmospheric pressure decreased with altitude. His plan was to compare the length of the mercury column at the base of the mountain and on the top of a mountain, about 500 m higher than the base. Because he was ill, Pascal did not perform the experiment himself but convinced his brother-in-law, Florin Perier, to do so. In September 1648, Perier took the apparatus designed by Pascal up to the top of Puy de Dome (see Figure 3.7), the highest mountain in the region. He measured the length of the column of mercury at the base of the mountain, at three locations while climbing the mountain, and at three locations on the top of the mountain. Perier verified that as he ascended the mountain, the column of mercury became shorter. At the top of the mountain, in all three locations, the column of mercury was 76 mm shorter than it was at the base of the mountain.

In addition to demonstrating the existence of atmospheric pressure and its changes with altitude, the combined efforts of Torricelli, Pascal, and Perier resulted in an instrument for measuring atmospheric pressure—the mercury barometer. Barometers designed on the principle developed by Torricelli have been in use ever since Torricelli designed the first one. Although barometers based on newer technologies have been developed, many mercury barometers are still used. Your physician might have measured your blood pressure with a mercury barometer. Many classrooms and laboratories have mercury barometers hanging on a wall.

Chemistry File

FYI
A unit of pressure that is occasionally used is the torr. The torr is equal to the pressure created by 1 mmHg. It was established in honour of Evangelista Torricelli.

Figure 3.7 Puy de Dome is a volcanic crater. Today, modern communication equipment is located on the top of the mountain. This mountain made history when Perier, brother-in-law to Pascal, demonstrated the change in atmospheric pressure with altitude.

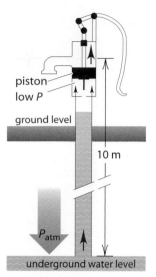

Figure 3.8 The pump creates a partial vacuum in the pipe and the atmospheric pressure pushes water up the pipe. Atmospheric pressure can push water up to a height of only 10 m.

Because mercury barometers, were used extensively, the common unit of pressure became the millimetre of mercury, or mmHg. **Standard atmospheric pressure**, defined as the atmospheric pressure in dry air at 0 °C at sea level, is 760 mmHg. Because standard atmospheric pressure is a common reference point, the unit "atmospheres" (atm) is also commonly used. When SI units became the accepted units, the combined SI unit of pressure (force per unit area) was the Newton per square metre (N/m^2). This derived unit was given the name pascal (Pa) in honour of Blaise Pascal. Standard atmospheric pressure is 101 325 Pa. Because this is such a large number, the unit kPa (kilopascal or 1000 Pa) is commonly used. Standard atmospheric pressure is, thus, 101.325 kPa. As you can see, standard atmospheric pressure, or 1 atm, is close to 100 kPa. Another unit that is often used in chemistry, as well as in meteorology, is called the "bar" and is equal to 100 kPa. Thus, 1 atm is equal to 1.01325 bar. The following expression summarizes the various units of pressure.

$$1 \text{ atm} = 760 \text{ mmHg} = 101\ 325 \text{ Pa} = 101.325 \text{ kPa} = 1.01325 \text{ bar}$$

The development of the barometer and the knowledge that it is atmospheric pressure that pushes the mercury up into the tube answered another question that had been puzzling scientists and inventors of that time. Scientists could now explain why the crude suction pumps that were used to pump water out of mines could not lift water more than 10 m. Now it is clear that the pump is not pulling the water up but atmospheric pressure is pushing it up (see Figure 3.8). Standard atmospheric pressure is about the same as the pressure exerted by 10 m of water.

• • •

1. Explain why the density of the atmosphere decreases as the altitude increases.

2. How does a barometer determine atmospheric pressure?

3. Explain how you would convert pressure given in units of mmHg to units of kPa.

4. How is standard atmospheric pressure defined?

• • •

Chemistry File

FYI
In 1998, a weather balloon carrying instruments to measure the ozone layer drifted off course. It veered into transatlantic air routes, where it posed a serious danger. By this time, the balloon had expanded in size to a volume larger than Commonwealth Stadium. Two Canadian Air Force CF-18 jets directed more than 1000 rounds of cannon fire at the balloon, but could not bring it down. The balloon finally landed on an island off the coast of Finland. University of Toronto physicists have developed a new mechanism to prevent such an event from recurring.

Gas Pressure and Volume

Meteorologists use weather balloons, such as the one shown in Figure 3.9, to carry instrument packages called radiosondes to high altitudes. When the weather balloon reaches the upper atmosphere, the instruments collect data about temperature, pressure, humidity, and wind speeds and send the data back to ground stations. The balloon is partially inflated with helium or hydrogen gas because, at the same pressure, hydrogen and helium are less dense than air. Therefore, the weather balloon rises to very high altitudes. As the altitude increases, the atmospheric pressure decreases and the balloon expands. Eventually, the balloon bursts and a parachute opens to bring the radiosonde safely to the ground.

How can meteorologists predict the altitude at which a weather balloon will burst? If the balloon is designed to burst when it reaches three times the volume to which it was inflated before its release, how do they know the altitude at which that will occur? What is the mathematical relationship between the volume of a gas and its pressure when the temperature remains constant?

Figure 3.9 Weather balloons are designed to reach a pre-determined altitude and then burst.

Chemistry File

Web Link
What do weather balloons carry? How many are released every day? How is their movement tracked?

www.albertachemistry.ca
WWW

Irish scientist, Robert Boyle (1627–1691), studied the relationship between pressure and volume of gases at constant temperatures. By making careful measurements of the volume of a trapped gas, he was able to describe what happened when the pressure exerted on the gas was increased. Figure 3.10 shows an apparatus similar to the one Boyle used in his experiment. He measured the length of the column of trapped air and the length of the column of mercury. Because the length of the mercury column is directly related to the pressure it exerts on the trapped air, Boyle was able to deduce the relationship between the pressure on the air and its volume. You will perform an experiment similar to Boyle's in Investigation 3.A. Because mercury vapours are toxic, you will use a different method to examine the relationship between the pressure and the volume of a gas.

Chemistry File

Try This
In your notebook, write a short paragraph, using kinetic molecular theory, to predict what should happen to the volume of a weather balloon when the atmospheric pressure suddenly increases.

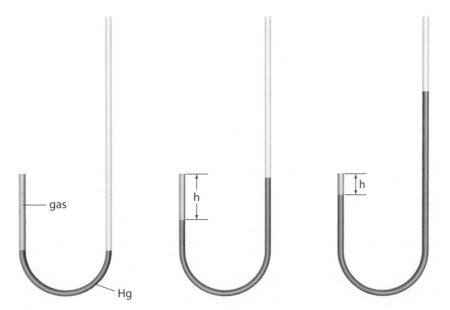

Figure 3.10 The left side of the U-tube is sealed and mercury has trapped air in the tube. As mercury is added to the right side of the tube, the mercury increases the pressure on the trapped gas, causing its volume to decrease.

The Relationship between the Pressure on and the Volume of a Gas

In this investigation, you will observe the change in the volume of air trapped inside a syringe when you apply pressure to the plunger of the syringe and thus to the trapped air. You will use your data to determine the relationship between the pressure on the trapped air and its volume.

Prediction

Predict what will happen to the air inside the syringe when you place weights on top of the plunger of the syringe.

Safety Precautions

- Be careful to centre all of the books that are used for weights so they do not fall.
- Your teacher might choose to do this investigation as a demonstration.

Materials

- 60 mL syringe
- square piece of plexiglass (about 15 to 20 cm on a side)
- glue (strong)
- retort stand
- 3 clamps
- rubber stopper
- scale (with range up to 100 N)
- weights (such as heavy books) totaling a mass of at least 6 kg
- barometer

Procedure

1. Obtain a 60 mL syringe and measure the internal diameter. Calculate the radius. From the radius, calculate the cross-sectional area of the syringe ($A = \pi r^2$).

2. Ensure that the plunger is airtight but slides freely. You might have to lubricate the plunger.

3. Glue the plexiglass platform onto the top of the plunger. Be sure to centre the platform.

4. Determine the weight of the plunger-platform assembly. If the scale or balance that you are using reports mass in kilograms, convert to weight in newtons (N) by multiplying the mass by 9.81.

5. Insert the tip of the syringe into a small hole that has been drilled into a rubber stopper. The hole does NOT penetrate the stopper. It should be just deep enough to fit the tip of the syringe. The fit must be airtight.

6. Assemble the apparatus as shown in the diagram. Notice that the rubber stopper is placed firmly against the base of the retort stand. The clamps on the stopper and the syringe are tight. However, the uppermost clamp is not touching the plunger. It is in place to prevent the plunger from falling, in the event that it should begin to tip over. When you insert the plunger into the syringe, trap as much air as possible.

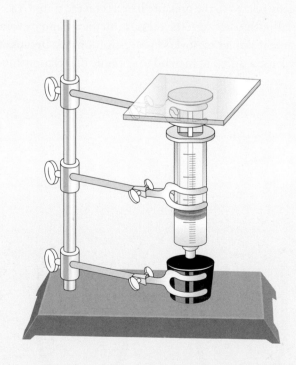

7. Make a data table similar to the one below in which to record your data.

Atmospheric Pressure _____
Cross-sectional area of syringe in m^2 ($A = \pi r^2$) _____

Number of objects	Weight of added object (N)	Total weight on platform (N)	Total pressure (atmospheric pressure plus pressure due to objects) (kPa)	Inverse of pressure, 1/P (1/kPa)	Volume (mL)
1 (plunger and platform)	0 N				
2					
3					
4					

8. Read the current atmospheric pressure from the barometer and record the value. (Atmospheric pressure should not be corrected for altitude.)

9. Record the weight of the plunger-platform apparatus and the volume of air in the syringe with the plunger in place.

10. Calculate the pressure caused by the plunger-platform apparatus by dividing the weight (force) by the cross-sectional area of the plunger.

11. Calculate the total pressure by adding the atmospheric pressure to the pressure due to the weight of the plunger-platform apparatus. Record the total pressure.

12. Determine the weight of the object (such as a heavy book) that you will be adding to the platform. Record the weight in your table. (Recall: weight is mass times 9.81.)

13. Very carefully place the object on the platform. Be sure to centre the object on the platform. Observe and record the volume of the air in the syringe.

14. Calculate the total weight of platform plus object, the pressure due to the weight of the platform plus object (similar to step 10), and total pressure on the air in the syringe (similar to step 11). Record the data.

15. Repeat steps 12 to 14 with more objects until you have added at least 60 N of weight to the platform.

16. Plot a graph of total volume, V (y-axis) versus pressure, P (x-axis).

17. Plot a graph of total volume, V (y-axis) versus the inverse of the pressure, $1/P$ (x-axis).

Analysis

1. What is the manipulated (independent) variable in this investigation? What is the responding (dependent) variable?

2. Which graph, V versus P or V versus $1/P$, appears to give the straightest line?

3. If the plunger and the rubber stopper had not given airtight seals, how would this have affected your data?

4. How might a change in temperature have affected your results?

5. What do you call the variables of amount of air and temperature in this experiment?

Conclusion

6. Describe the type of relationship that exists between volume and pressure.

Technology and the Development of Barometers

Although the mercury barometer invented by Evangelista Torricelli in 1643 is still used today, these instruments tend to be large and easily broken. In 1843, French inventor Lucien Vidie (1805–1866) invented a much smaller and more robust type of barometer that could be used in a greater variety of situations. His aneroid barometer consists of a sealed metal cell with flexible sides. ("Aneroid" means "without fluid.") Air inside the metal cell has been pumped out to create a partial vacuum. As a result, any change in atmospheric pressure outside the cell causes the cell to expand or contract. The aneroid cell in the barometer shown here is shaped like a bellows and moves up and down with changes in air pressure. A pointer connected to the top of the cell moves against a dial or recording device to record the movement.

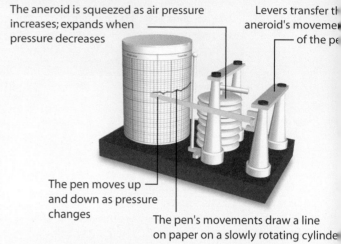

The aneroid is squeezed as air pressure increases; expands when pressure decreases

Levers transfer the aneroid's movement of the pen

The pen moves up and down as pressure changes

The pen's movements draw a line on paper on a slowly rotating cylinder

This aneroid barometer is connected to a recording device called a barograph.

Barometers and Weather Forecasting

Barometers measure air pressure, and changes in air pressure occur when the weather is changing. Atmospheric pressure at Earth's surface increases wherever a mass of air is slowly descending. The increasing pressure warms the air. For this reason, increasing pressure generally means that settled, sunny weather is on the way. In other regions of Earth, air masses will be rising and the atmospheric pressure decreasing. Decreased pressure cools the air and may lead to rain showers or stormy weather.

As you learned on page 103, atmospheric pressure also changes with altitude. No matter what weather systems are doing, air pressure decreases with increasing height above sea level. Why? Because the air pressure at a given point on Earth is produced by the mass of the atmosphere above it. As you climb higher, there is less atmosphere above you.

How to Read a Barometer

Suppose you are taking readings from barometers at different locations to prepare a weather map. To directly compare one barometer reading with another, you must adjust their readings to allow for the effect of elevation on air pressure. To do this, you need to calculate the pressure each barometer would record if it were at sea level directly below the place where it is situated. As a rule of thumb, air pressure decreases by about 1 millibar for each 8 metres of altitude. The official barometric readings broadcast by some radio and television stations and published in newspapers are all corrected to sea level values.

Most household barometers have two pointers. One moves in response to changes in local air pressure. The other can be moved manually by adjusting a screw on the barometer. Once a day, you should set the adjustable pointer to line up with the other pointer. Later in the day, you can see whether the pressure has risen or fallen in comparison with the initial conditions when the hands were together. The direction and amount of change will tell you what weather conditions are on the way.

• • •

1. If you live at 304 metres above sea level, how many millibars should you add to your barometer reading to adjust it to sea level value?

2. Because air masses in the atmosphere are always moving and, thus, air pressure is always changing, a barometer reading in one location can help predict weather for only a few hours ahead. Conduct research to learn how new technologies allow meteorologists to forecast weather changes for up to a week ahead. Make a report describing these techniques and explaining how they work. ICT

Boyle's Law

The relationship that you developed in Investigation 3.A is known as Boyle's law. As you observed, when the pressure on the gas was increased, the volume of the gas decreased. Mathematically, this type of a relationship is called an inverse proportion and can be expressed as $V \propto \frac{1}{P}$ (volume is inversely proportional to pressure). To develop the mathematical equation that is commonly used for calculations involving Boyle's law, use the graphs in Figure 3.11.

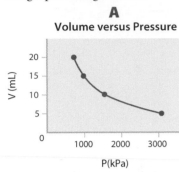

A
Volume versus Pressure

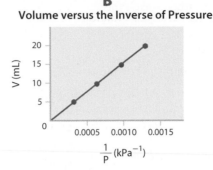

B
Volume versus the Inverse of Pressure

Figure 3.11 The graph for volume versus pressure **(A)** shows an inverse relationship. When you plot volume versus the inverse of pressure **(B)**, you get a straight line.

Graph A in Figure 3.11 represents volume (V) versus pressure (P), and is typical of an inversely proportional relationship. If the relationship is in fact an inverse proportion, you should get a straight line by plotting volume (V) versus the inverse of pressure $\left(\frac{1}{P}\right)$. Graph B, the plot of V versus $\frac{1}{P}$, is linear thus verifying that the data represent an inverse proportion. Now you can use Graph B to write a linear relationship relating volume and pressure. As you learned in math, the general expression for a straight line is $y = mx + b$ in which m is the slope of the line and b is the y-intercept. This information allows you to develop Boyle's law in mathematical form by taking the following steps.

- The general expression for a straight line is: $\qquad y = mx + b$

- In Graph B (Figure 3.11) the y-axis represents volume, $\qquad V = m\frac{1}{P} + b$
  V, and the x-axis represents $\frac{1}{P}$, so you could write:

- The symbol, m, represents the slope of the line and b is the $\qquad V = m\frac{1}{P}$
  y-intercept. From the graph, you can see that the line passes through the origin, thus b is zero leaving:

- Multiplying both sides of the equation by P shows you that the $\qquad PV = \cancel{P}m\frac{1}{\cancel{P}}$
  product PV is equal to a constant, which is the slope of the line: $\qquad PV = m$

- In a given experiment, you always use the same sample of $\qquad P_1V_1 = m$ or $P_2V_2 = m$
  gas and ensure that the temperature is constant. Under these conditions, the slope and, therefore, the constant in the equation is always the same. You could state this mathematically as:

  Where 1 and 2 represent the pressure and volume at any two data points on the graph.

- Because the product of PV for any point on the graph or $\qquad P_1V_1 = P_2V_2$
  any measurement in the experiment is equal to the same constant, m, the products are equal to each other. Thus:

Boyle's Law

The volume of a fixed amount of gas at a constant temperature is inversely proportional to the applied (external) pressure on the gas.

$$P_1V_1 = P_2V_2$$

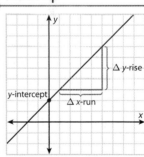

Math Tip

Any straight line can be described by the equation $y = mx + b$. The slope, m, is defined as the rise over the run or $m = \frac{\Delta y}{\Delta x}$. The symbol b represents the y-intercept, which is the value of y at the point at which the line crosses the y-axis. The coordinates of any point on the line satisfy the equation.

Using Boyle's Law to Calculate Volume

Problem

A weather balloon with a volume of 2000 L at a pressure of 96.3 kPa rises to a height of 1000 m, where the atmospheric pressure is measured to be 60.8 kPa. Assuming there is no change in temperature, what is the final volume of the weather balloon?

What Is Required?

You need to find the volume, V_2, after the pressure on the balloon has decreased.

What Is Given?

You know the pressure and volume for the first set of conditions and the pressure for the second set of conditions.

$P_1 = 96.3$ kPa

$V_1 = 2000$ L

$P_2 = 60.8$ kPa

Temperature does not change.

Plan Your Strategy

Because temperature is constant and pressure and volume have been given, you will need to use Boyle's law.

You can solve for the unknown, V_2, and then substitute numbers and units for the variables in the formula.

Act on Your Strategy

$$P_1 V_1 = P_2 V_2$$

$$\frac{P_1 V_1}{P_2} = \frac{\cancel{P_2} V_2}{\cancel{P_2}}$$

$$V_2 = \frac{P_1 V_1}{P_2}$$

$$V_2 = \frac{(96.3 \ \cancel{kPa})(2000 \ L)}{60.8 \ \cancel{kPa}}$$

$$V_2 = 3167.76 \ L$$

$$V_2 \approx 3.17 \times 10^3 \ L$$

Solution

Because the smallest number of significant digits in the question is three, the answer is 3.17×10^3 L or 3.17 kL.

Check Your Solution

The unit for volume is litres. When the other units cancel out, L remains. You would expect the volume to increase when the pressures decrease. As expected, the volume increased.

1. A sample of gas in a flexible container has a volume of 6.9 L after its pressure has been increased from 1.0 atm to 3.5 atm. What was the initial volume of the gas?

2. A flexible container holding 3.50 L of hydrogen gas at standard atmospheric pressure has to be compressed into a volume of 1.75 L. If there is no change in temperature, what pressure is required?

3. A sample of neon gas at room temperature is collected in a 2.50 L balloon at standard atmospheric pressure. The balloon is then submerged into a tub of water, also at room temperature, so that the external pressure is increased to 112.5 kPa. What will be the final volume of the balloon?

4. A flexible container holds 4.0 L of air at 22 °C. If the temperature of the air remains constant, what will be the volume of the air if the pressure doubles?

5. The volume of gas in a 25 mL syringe attached to a pressure gauge is 2.50 mL when the pressure gauge reads 8.26 bar. If there has been no change in temperature when the plunger of the syringe is pulled back to allow the gas to occupy 20.0 mL, what does the pressure gauge read?

6. A 2.5 L container is filled with helium gas at a pressure of 3.5 atm. If the temperature remains constant, and the volume increases to 9.0 L, what is the final pressure on the gas?

Kinetic Molecular Theory and Boyle's Law

Boyle had carried out his experiments and concluded that the pressure on a gas was inversely proportional to its volume at a constant temperature about 200 years before the kinetic molecular theory was developed. Now you can apply the kinetic molecular theory to understand why pressure and volume are related by an inverse proportion.

Pressure on the walls of a container of gas is caused by the collisions of gas molecules with the walls of the container. Every time a molecule collides with the wall, it exerts a force on the wall. Thus, the average force exerted by all the gas molecules divided by the surface area of the container is equivalent to the pressure on the walls of the container. Examine Figure 3.12 to see what happens when you change the external pressure on the gas.

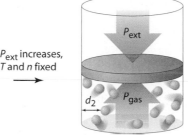

$P_{gas} = P_{ext}$ · P_{ext} increases, T and n fixed · Higher P_{ext} causes lower V, which causes more collisions, increasing the pressure until $P_{gas} = P_{ext}$

Figure 3.12 If the temperature and the amount of gas remain constant and the external pressure increases, it will be larger than the internal pressure. Therefore, the piston will begin to go down and decrease the volume. As the volume decreases, the molecules will be closer together and will collide with the walls of the container and with one another more frequently. (The arrows labelled d_1 and d_2 represent the average distance between the molecules and the walls of the container.)

The containers in Figure 3.12 have movable pistons. The pistons will move until the external pressure and the internal pressure are equal. If you increase the external pressure, the piston will move down thus reducing the volume available to the gas molecules. The gas molecules are now closer together and they collide with one another and the walls of the container more frequently. As the number of collisions over a given time increases, the average force exerted by all the molecules increases while the volume of the container decreases. Consequently, the gas pressure increases. If the temperature remains constant and no gas escapes or enters the container, the decrease in the volume of the container will be inversely proportional to the increase in the pressure.

Section 3.2 Summary

- Atmospheric pressure decreases as altitude increases.
- Torricelli proposed the existence of atmospheric pressure and designed the mercury barometer to demonstrate it.
- Pascal designed the experiment that demonstrated the decrease in atmospheric pressure with altitude.
- A vacuum created in a straw or a pump does not "pull" the water up. Atmospheric pressure pushes the water up where a partial vacuum was created.
- 1 atm = 760 mmHg = 101 325 Pa = 101.325 kPa = 1.01325 bar
- Boyle demonstrated that the volume of a given amount of gas is inversely proportional to the external pressure exerted on it when the temperature is constant.
- $P_1V_1 = P_2V_2$ (Boyle's law)
- As the external pressure on a gas increases, the volume begins to decrease. As the volume decreases, the molecules become closer together causing the collision frequency to increase, increasing the pressure.

1. Discuss the principle on which a mercury barometer is based. How does this principle explain why a water pump cannot bring water up to a height greater than 10 m.

2. Convert the following into kPa:
   a) 3.58 bar   b) 850 mmHg   c) 1.75 atm

3. For each of the following parts, determine which of the following indicates the larger pressure?
   a) 1.25 atm or 101.325 kPa   b) 1.5 bar or 740 mmHg
   c) 1 bar or 105 kPa   d) 800 mmHg or 1.25 atm

4. What characteristics of a graph indicate that there is an inverse relationship between the two variables represented on the graph? How would you determine whether the relationship was, in fact, an inverse relationship?

5. Predict what would happen to a 2.5 L helium balloon if it were taken by a scuba diver 20 m underwater (where pressure is 3 atm). Using kinetic molecular theory, explain this change in volume, assuming there is no temperature change.

6. Raja embeds the end of a 20 mL plastic syringe into a rubber stopper when the plunger is at the 10 mL line. Raja finds that he has to exert considerable force to push the plunger to the 5 mL line, but he can move the plunger easily to the 15 mL line. Using kinetic molecular theory, explain Raja's findings.

5 mL   15 mL

7. A gas with an original volume of 5.0 L at a pressure of 95.8 kPa is allowed to expand until the pressure drops to 20.6 kPa. What is the volume of the expanded gas if the temperature of the gas is unchanged?

8. A 2.5 L sample of gas is trapped at 100 kPa in a cylinder with a moveable piston. If the pressure rises to 3.35 atm while the temperature is kept constant, what is the volume of the sample in the cylinder?

9. State Boyle's law. Describe at least one "everyday observation" that could be explained by Boyle's law.

10. A scuba tank with a volume of 10 L holds air at a pressure of 1.75×10^4 kPa. What volume of air at standard atmospheric pressure was compressed into the tank if the temperature of the air in the tank is the same as the air temperature before it was compressed?

11. Use kinetic molecular theory to explain what would happen if you reduced the external pressure on a sample of a gas.

12. Sara sets up an experiment to collect hydrogen gas in a balloon that is fitted over the top of an Erlenmeyer flask. She records the atmospheric pressure reading from the barometer as 98.5 kPa. When she returns after lunch, she notices that the volume of the balloon has noticeably decreased. Initially, Sara thinks that some of the hydrogen gas has escaped from the balloon. Sara decides she needs to collect more data to account for her observations. What data would you advise Sara to collect? Why?

before lunch   after lunch

13. A student collects oxygen gas in an inverted cylinder and records the volume at 27.9 mL. Later that evening he realizes that he has forgotten to record the pressure. He returns the next day and finds the pressure is 102.1 kPa, but the volume of the gas in the cylinder is now 27.3 mL. How can the student find the pressure of the gas on the previous day? What assumptions must the student make?

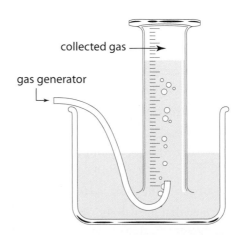

collected gas

gas generator

14. Would it be easier to drink water from a straw at the top of Mt. Everest or at sea level? Explain why.

Gases and Temperature

Section Outcomes

In this section, you will:

- **perform** an investigation to determine the effects of temperature changes on the volume of gases
- **convert** between Celsius and Kelvin temperature scales
- **draw** and **interpret** graphs of experimental data relating temperature to volume

Key Terms

absolute zero
Charles's law

In Section 3.2, you learned about the relationship between pressure and the volume of a gas when the temperature is constant. Most observations and laboratory experiments, however, are carried out at or near standard atmospheric pressure. At constant pressure, what is the relationship between the volume of a sample of gas and its temperature?

Observing the Relationship between Volume and Temperature

Two prominent French scientists Jacques Charles (1746–1823) and Joseph Louis Gay-Lussac (1778–1850) were intrigued by an event in which heated air played a critical role. While Frenchman and papermaker Joseph Montgolfier was watching the fire in his fireplace, he began to wonder what "force" was causing the smoke and flecks of ashes to rise. He and his brother, Etienne, made a small silk bag and placed it above a fire with the open end pointing downward. The bag began to rise. Soon, they made larger bags and studied the way they rose through the air. In 1783, the brothers made a bag 11.5 m in diameter out of linen and lined it with paper. In a public demonstration (see Figure 3.13), they positioned it above a fire and the "balloon" rose to nearly 2000 m and travelled nearly 2 km before it landed. By November of the same year, they had built a balloon that carried a basket with two men riding in it. Their demonstrations created so much interest that they hired Jacques Charles to continue researching balloons. No one yet understood why the heated air caused the balloon to rise.

Chemistry File

FYI
Canada's Aboriginal peoples had long known that heated air would rise and carry smoke and ashes with it. They used this phenomenon to communicate over long distances with smoke signals.

Figure 3.13 There was a flurry of excitement throughout France when the Montgolfier brothers first demonstrated "hot-air ballooning."

When Charles began to study "lighter than air" flight, he knew that an object filled with a gas that was less dense than air would rise in the air. Because carrying straw or wood for fuel in a balloon was difficult and dangerous, Charles began to study balloons filled with hydrogen gas. Although hot-air balloons rapidly disappeared and were replaced with hydrogen-filled balloons, Charles was still interested in the scientific basis for hot-air balloons. He began to study the relationship between temperature and the volume of gases. Gay-Lussac was also a balloon enthusiast and travelled to an altitude of approximately 7000 m in a hydrogen-filled balloon to study gases in the upper atmosphere. Independently, Gay-Lussac also studied the relationship between temperature and the volume of gases. Before examining the results of Charles and Gay-Lussac, complete the following investigation to learn about the relationship.

INVESTIGATION 3.B

Target Skills

Predicting the relationship between the temperature and volume of gas

Performing an experiment to illustrate Charles's law

Drawing and **interpreting** graphs that relate temperature and the volume of a gas

The Relationship between Temperature and Volume of a Gas

In this investigation, you will examine the effect of temperature changes on the volume of a gas.

Question

What is the relationship between the temperature and volume of a fixed amount of gas at a constant external pressure?

Prediction

Predict what the relationship between temperature and volume of a gas will be.

Safety Precautions

• Use care when sealing the end of the plastic pipette with a flame. The plastic will melt and may begin to burn. Hot, molten plastic can burn your skin.

• Do not inhale any of the fumes from the plastic. If a hood is available, seal the pipette in the hood.

• Before lighting the Bunsen burner, check that there are no flammable solvents in the laboratory.

Materials

• coloured water
• tap water
• ice
• thin stem plastic pipette
• match or Bunsen burner
• scissors
• metric ruler (with mm calibrations)
• 2 rubber bands
• 400 mL beaker
• Celsius thermometer
• hot plate
• marker

Procedure

1. Squeeze the pipette bulb and draw enough water into the pipette to form a small plug that is about 0.5 cm long. The rest of the pipette should contain air.

2. Using the flame from a match or a small flame from a Bunsen burner, carefully seal the open end of the pipette completely. Allow the pipette to cool for at least three minutes before proceeding to step 3.

3. Using scissors, cut off a small section at the top of the bulb of the pipette.

4. Carefully attach the pipette to the ruler, using a rubber band, so that the bottom of the tube is even with the 10 mm (1.0 cm) mark of the ruler, as shown in the diagram.

5. Fill a 400 mL beaker about two thirds full of tap water and add three or four ice cubes. Place the thermometer in the water. Then put the ruler with the attached pipette into the water. Allow the ruler and pipette to sit for five minutes.

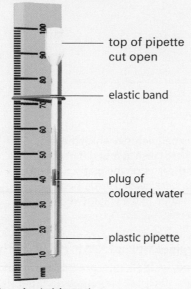

- top of pipette cut open
- elastic band
- plug of coloured water
- plastic pipette

metric ruler (with mm)

6. In your notebook, prepare a data table with the headings shown below. Your table should have at least eight rows for data.

T (°C)	V (mm × A)

7. After five minutes, measure the length of the trapped gas in mm. (Recall that the length is proportional to the volume because $V = A \times L$.) Remember that the bottom of the pipette stem is set at the 10 mm (1.0 cm) mark. Record these values in your data table.

8. Place the beaker on the hot plate and **slowly** heat the water in the beaker. Measure the length and temperature of the trapped gas at every 10 °C to 15 °C. Measure the length of the column of air and temperature of the water to a maximum of 60 °C.

9. Clean the apparatus and dispose of the pipette as directed.

Analysis

1. What is the manipulated (independent) variable? What is the responding (dependent) variable? Name two controlled variables.

2. Plot a graph of volume, in terms of length times constant area (mm × A) versus temperature. Choose your scale so that the temperature axis extends from −300 °C to 100 °C.

3. Draw a line of best fit through your data points. Extrapolate this line until it reaches the x-axis (temperature axis).

4. What relationship between volume and temperature does your graph suggest?

5. Speculate about the significance of the temperature where the line reaches the temperature axis (which indicates zero volume).

6. Put your data table and graph into your notebook because you will use the data again in another investigation.

Conclusion

7. Write a paragraph summarizing your conclusions about the relationship between temperature and the volume of a gas.

Interpreting the Volume versus Temperature Relationship

In Investigation 3.B, did you find that the volume of the gas increased when the temperature was increased? Was this relationship linear? If these were your findings, your results agree with those of Charles and Gay-Lussac. The two scientists performed many experiments with a variety of gases and at several pressures. As long as the amount of gas and the pressure on the gas were constant for a specific experiment, they obtained a linear plot. In addition, they observed another a very consistent result in all their experiments. When they extrapolated their linear plots down to zero volume, all the lines converged at one value of temperature (see Figure 3.14). Using more recent technology to collect data than was available to Charles and Gay-Lussac, the temperature for zero volume of a gas was found to be −273.15 °C. Of course, no real gas could have a volume of zero, but you can show that the volume approaches zero as the temperature approaches −273.15 °C.

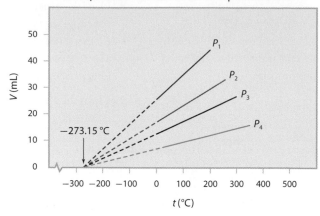

Extrapolated Volume versus Temperature Data

Figure 3.14 This graph represents four experiments in which the same amount of gas was used and data were taken at four different pressures (P_1 to P_4). The solid portions of the lines on the graph represent the temperatures at which data were taken. The dashed portions of the line represent the extrapolation of the volume versus temperature plots. All of the plots intersect at −273.15 °C.

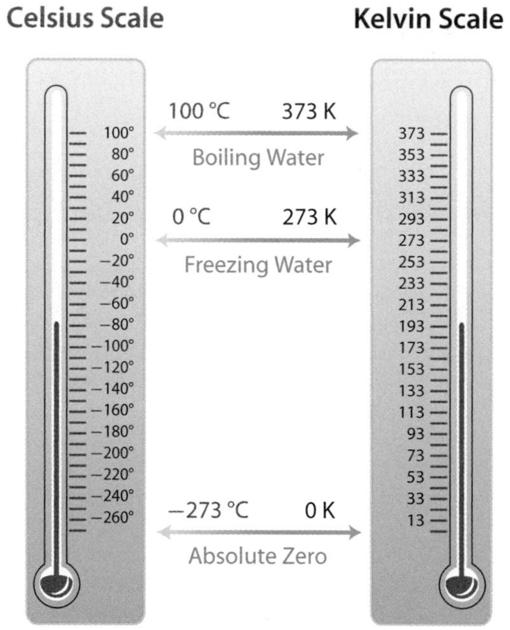

Celsius Scale

Kelvin Scale

100 °C   373 K
Boiling Water

0 °C   273 K
Freezing Water

−273 °C   0 K
Absolute Zero

Figure 3.15 There are 273 temperature units (kelvins or degrees) between absolute zero and the freezing temperature of water on the Celsius and Kelvin scales. There are also 100 temperature units between the freezing and boiling temperatures of water on both scales. (The number 273.15 is often rounded to 273 to simplify the comparison of the temperature scales.)

It was Scottish physicist Lord Kelvin (William Thomson, 1824–1907) who, in 1848, interpreted the significance of this temperature of −273.15 °C. He suggested that this was the theoretically lowest possible temperature or **absolute zero**. He then established a new temperature scale based on absolute zero as the starting point on the scale. The temperature scale was named the Kelvin scale in his honour. The size of a unit of temperature on the Kelvin scale is the same as the size of a degree on the Celsius scale. Only the starting points for the temperature scales are different. The name of a unit in the Kelvin scale is the kelvin (K). Notice that the term "degree" does not appear in the name of the unit for the Kelvin scale. The Celsius and Kelvin scales are compared in Figure 3.15.

For ease of distinguishing between the temperature scales, the symbol t is sometimes used to represent temperature on the Celsius scale and T on the Kelvin scale. (These symbols will be used in the remainder of this textbook.) You can convert between the two scales by using the following formulas.

Celsius to Kelvin	Kelvin to Celsius
$T = t + 273.15$	$t = T - 273.15$

• • •

5. What is the basis of the Kelvin temperature scale?

6. Explain why it is theoretically possible for an ideal gas to have a volume of zero but it is not possible for a real gas to have a volume of zero.

7. How is it possible to determine the value of absolute zero when you cannot measure it?

8. Covert 37 °C to the Kelvin temperature.

9. Convert 77 K to the Celsius temperature.

• • •

Charles's Law

Before completing the development of Charles's law, perform the following Thought Lab. The results of this lab will give you some very important information about the importance of using the Kelvin temperature scale for the development of Charles's law.

Thought Lab **3.1**   **The Importance of the Kelvin Temperature Scale**

In Investigation 3.B, you collected volume and temperature data and completed the first stage of the analysis of the data. In this Thought Lab, you will use the data table and the graph you constructed to complete your interpretation of the relationship between the temperature of a sample of a gas and its volume.

Target Skills

Drawing a graph to relate temperature to the volume of a gas

Determining absolute zero

Identifying the importance of an absolute temperature scale

Procedure

1. Make a new data table like the one below. Include at least eight rows for data.

t (°C)	T (K)	V (mm × A)	$\dfrac{V\,(mm \times A)}{t(°C)}$	$\dfrac{V\,(mm \times A)}{T(K)}$

2. To fill in columns 1 and 3, copy the data from your data table for Investigation 3.B.

3. Convert the Celsius temperature data in column 1 to the Kelvin temperature and record the result in column 2.

4. Calculate the quotient of $\frac{V}{t}$ and record it in column 4.

5. Calculate the quotient of $\frac{V}{T}$ and record it in column 5.

Analysis

1. Describe any trend that you see in the data for quotients of volume to Celsius temperature.

2. Describe any trend that you see in the data for quotients of volume to Kelvin temperature.

3. Propose an explanation for the differences in your descriptions of the trends.

4. What is the x-intercept of the extrapolation on the volume versus temperature graph that you completed in Investigation 3.B?

5. What is the significance of your answer to question 4?

6. In your own words, state the relationship between the volume and temperature of a gas.

To analyze your results from the Thought Lab in more detail, examine the graphs in Figure 3.16. The graphs represent the results of measuring the volume of a gas at various temperatures. The measurements were made for three different values of pressure. In Graph A, the values for volume were plotted against the Celsius temperature, and in Graph B, they were plotted against the Kelvin temperature.

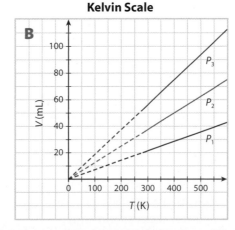

Figure 3.16 In graph **(A)**, using the Celsius temperature scale, each straight line has a different y-intercept, b_1, b_2, and b_3. In graph **(B)**, using the Kelvin temperature scale, the y-intercept of all the straight lines is zero.

Chemistry File

Try This
Obtain a hard-boiled egg and remove the shell. Be sure that the egg is cool and moist. Find a bottle or an Erlenmeyer flask with an opening that is just a little too small for the egg to go through. Set up two containers that will hold the bottle. Place hot tap water in one container and cold tap water in the other. Place the egg, tapered end down, in the opening of the bottle. Place the bottle in the hot water for several minutes. Move the bottle to the cold water. Watch and wait. Explain why the result that you observed occurs.

Chemistry File

FYI

Although it is theoretically impossible to reach absolute zero, some chemists have achieved temperatures of within a small fraction of a kelvin from absolute zero using very advanced equipment.

In both graphs, the relationship between the volume of a gas and its temperature is linear and can thus be described by an equation in the form of $y = mx + b$. However, one temperature scale will generate a much simpler overall equation than the other, as you can see by analyzing the steps below.

	Kelvin scale	Celsius scale
• Write the general equation for a straight line:	$y = mx + b$	$y = mx + b$
• Because V is on the y-axis, substitute V for y. Because temperature is on the x-axis, substitute T (Kelvin) or t (Celsius) for x:	$V = mT + b$	$V = mt + b$
• Examine the graphs and identify the y-intercept for both graphs. On the graph for the Celsius scale, the y-intercept is different for each different pressure. Use b_n to represent the intercepts. On the graph for the Kelvin scale, the y-intercept is zero for all pressures. Eliminate b from the equation:	$V = mT$	$V = mt + b_n$

When temperature is expressed in degrees Celsius, it is not possible to write a general equation that represents all possible conditions of the controlled variables of pressure and amount of gas. Therefore, develop any further equations from the equation that uses the Kelvin scale as follows.

- Divide both sides of the equation by T: $\frac{V}{T} = \frac{m\cancel{T}}{\cancel{T}}$

  $\frac{V}{T} = m$

- For any value of pressure and amount of gas, as long as they are held constant for the experiment, the slope of the line, m, will be the same. Thus, the quotient of $\frac{V}{T}$ for any point on the graph, will be the same constant: $\frac{V_1}{T_1} = m$ or $\frac{V_2}{T_2} = m$
- Because both quotients are equal to the same constant, they are equal to each other: $\frac{V_1}{T_1} = \frac{V_2}{T_2}$

Charles's Law

The volume of a fixed amount of gas at a constant pressure is directly proportional to the absolute (Kelvin) temperature of the gas. Mathematically, Charles's law can be expressed as:

$$\frac{V_1}{T_1} = \frac{V_2}{T_2}$$

where T is the Kelvin temperature.

10 Solve the equation, $V = mt + b_n$, for the Celsius temperature, t.

11 Use your answer to question 10 to explain why the Kelvin temperature scale must always be used when using Charles's law. (Remember, b_n has a different value for each experiment.)

Using Charles's Law to Calculate Volume

Problem

A balloon inflated with air in a room at 22.00 °C has a volume of 650 mL. The balloon is put into a freezer of a refrigerator at 0.00 °C and left long enough for the air in the balloon to reach the same temperature. Predict the volume of the balloon at the end of the two hours, assuming that air pressure in the room and the freezer are the same.

What Is Required?

You need to find the volume, V_2, of the balloon after it has been cooled to 0.00 °C.

What Is Given?

You know the volume and temperature of the air sample for the first set of conditions.

$V_1 = 650$ mL
$t_1 = 22.00$ °C

You know the temperature for the second set of conditions.

$t_2 = 0.00$ °C

You know that the pressure and the amount of gas do not change.

Plan Your Strategy

Use Charles's law to calculate the second volume, V_2. Because temperature is given in degrees Celsius, you will have to convert it to kelvins.

Act on Your Strategy

$T_1 = 22.00$ °C $+ 273.15$ $T_2 = 0.00$ °C $+ 273.15$
$T_1 = 295.15$ K $T_2 = 273.15$ K

$$\frac{V_1}{T_1} = \frac{V_2}{T_2}$$

To isolate V_2, you need to multiply both sides of the equation by T_2.

$$\left(\frac{V_1}{T_1}\right)T_2 = \left(\frac{V_2}{T_2}\right)T_2$$

$$V_2 = \frac{V_1 T_2}{T_1}$$

$$V_2 = \frac{(650 \text{ mL})(273.15 \text{ K})}{295.15 \text{ K}}$$

$$V_2 = 601.55 \text{ mL}$$

$$V_2 \approx 602 \text{ mL}$$

Since the smallest number of significant digits in the question is three, the final volume will be reported to three significant digits.

Solution

The final volume of the gas in the balloon is 602 mL or 0.602 L.

Check Your Solution

The units for the answer are volume units. Volume units remain when the other units cancel out. Because the temperature decreases, you would expect the volume to decrease, which it does.

7. Convert the following Celsius temperatures to the Kelvin scale:

a) 27.3 °C   **d)** −25 °C

b) 37.8 °C   **e)** −40 °C

c) 122.4 °C

8. Convert the following Kelvin temperatures to the Celsius scale:

a) 373.2 K   **d)** 23.5 K

b) 275 K   **e)** 873 K

c) 173 K

9. A 75 mL balloon immersed in liquid nitrogen at −196 °C is lifted out and left in a room at 22.3 °C. What is the final volume of the balloon?

10. A child's balloon is filled to a volume of 3.0 L with room temperature air (22 °C). The balloon will burst if it reaches a volume of 3.5 L. The child takes the balloon with her in the car when she goes shopping with her mother on a hot day. They leave the balloon in the closed car while they are shopping. The temperature of the air in the car reached a temperature of 38 °C. Did the balloon burst? Support your answer with calculations.

Using Charles's Law to Calculate Temperature

Problem
A birthday balloon is filled to a volume of 1.50 L of helium gas in an air-conditioned room at 21.00 °C. The balloon is then taken outdoors on a warm sunny day. The volume of the balloon expands to 1.55 L. Assuming the pressure remains constant, what is the Celsius temperature outdoors?

What Is Required?
You need to find the outdoor temperature, t_2, in degrees Celsius.

What Is Given?
You know the volume and temperature of the air sample for the first set of conditions.
$V_1 = 1.50$ L
$t_1 = 21.00$ °C

You know the volume for the second set of conditions.
$V_2 = 1.55$ L

You know the pressure and the amount of gas do not change.

Plan Your Strategy
Use Charles's law to calculate the second temperature, T_2. Because the temperature is given in degrees Celsius, you will have to convert it to kelvins. After calculating T_2, convert it back to Celsius.

Act on Your Strategy
$T_1 = 21.00$ °C $+ 273.15$
$T_1 = 294.15$ K

$$\frac{V_1}{T_1} = \frac{V_2}{T_2}$$

Solve the equation for T_2 and substitute the values into the rearranged equation.

$$\left(\frac{V_1}{T_1}\right)T_2 = \left(\frac{V_2}{T_2}\right)T_2$$

$$\left(\frac{V_1}{T_1}\right)T_2 = V_2$$

$$\left(\frac{V_1}{T_1}\right)\left(\frac{T_1}{V_1}\right)T_2 = V_2\left(\frac{T_1}{V_1}\right)$$

$$T_2 = V_2\left(\frac{T_1}{V_1}\right)$$

$$T_2 = \frac{(1.55\,L)(294.15\text{ K})}{(1.50\,L)}$$

$$T_2 = 303.955\text{ K}$$
$$T_2 \approx 304\text{ K}$$

Convert the Kelvin temperature to the Celsius temperature.
$t_2 = 303.96$ K $- 273.15$
$t_2 = 30.81$ °C

Because the smallest number of significant digits in the question is three, the outside temperature will be reported to three significant digits.

$t_2 \approx 30.8$ °C

Check Your Solution
The unit for the answer is kelvins. When the other units cancel out, kelvins remain. Because the volume of the balloon increased, you would expect that the temperature had increased. The results showed an increase in temperature as expected.

11. A sealed syringe contains 37.0 mL of trapped air. The temperature of the air in the syringe is the same as room temperature or 295 K. The Sun shines on the syringe causing the temperature of the air inside to increase. If the volume increases to 38.6 mL, what is the new temperature of the air in the syringe?

12. The volume of a 1.5 L balloon at room temperature increases by 25 percent of its volume when it is placed in a hot-water bath. How does the temperature of the water bath compare with room temperature?

13. A birthday balloon is filled with 1.80 L of helium gas at 20.0 °C. The balloon expands to a volume of 5.40 L. If the pressure remains constant, what is the final Celsius temperature of the gas in the balloon?

14. Compressed gases can be condensed when they are cooled. A 500 mL sample of carbon dioxide gas at room temperature (assume 25.0 °C) is compressed by a factor of four, and then is cooled so that its volume is reduced to 25.0 mL. What must the final temperature be (in °C)? (**Hint:** Try using both Boyle's law and Charles's law to answer the question.)

12 Explain the difference between an inverse proportion and a direct proportion. Use the volume versus pressure and volume versus temperature relationships as examples.

13 Imagine that a classmate was sick when you studied Charles's law in class. Explain to this classmate why it is necessary to use Kelvin temperatures and not Celsius temperatures when doing problems involving Charles's law.

Kinetic Molecular Theory and Charles's Law

Charles and Gay-Lussac developed their laws based on observations. They had no theory to explain their result. Now, however, you can use kinetic molecular theory to explain why a gas expands when it is heated.

As you read earlier, the Kelvin temperature of a gas is directly proportional to the average kinetic energy of the gas molecules. You probably recall that an object's kinetic energy is related to its speed $(E_k = \frac{1}{2}mv^2)$. As the temperature of a gas increases, the molecules move at higher speeds. As a result, they collide with the walls of the container and one another more frequently and with a greater force. Therefore, they exert a greater pressure on the walls of the container. If, however, the external pressure on the gas remains the same, the gas pressure causes the container to increase in size. As the volume becomes larger, the gas molecules must travel farther to collide with the walls of the container and one another. As the collisions become less frequent, the pressure drops. The process continues until the pressure inside the container is, once more, equal to the external pressure. The process is depicted in Figure 3.17.

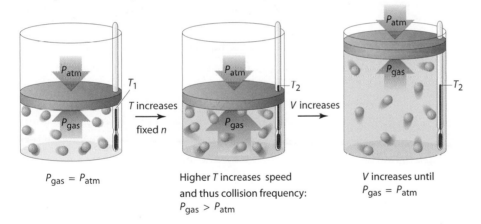

$P_{gas} = P_{atm}$

Higher T increases speed and thus collision frequency:
$P_{gas} > P_{atm}$

V increases until
$P_{gas} = P_{atm}$

Figure 3.17 When a gas is heated, the speed of the molecules increases. They collide with the walls of the container more frequently, increasing the pressure. If the external pressure remains the same, the gas pushes the piston up and increases the volume of the container.

Section 3.3 Summary

• The introduction of hot-air balloons piqued the interest of Charles and Gay-Lussac, and they began to study the relationship between the temperature and volume of gases.

• Plots of volume versus temperature of a gas always extrapolate to a temperature of −273.15 °C.

• Lord Kelvin proposed that −273.15 °C was the theoretically lowest possible temperature. It is now called absolute zero.

• The Kelvin temperature scale is based on a temperature of 0 K at absolute zero. The size of a kelvin (unit of temperature) is the same as the size of a degree on the Celsius scale.

• The symbol T is used for temperature on the Kelvin scale, and t is used for temperature on the Celsius scale.

- $T = t + 273.15$ and $t = T - 273.15$
- Charles's law states that the volume of a fixed amount of gas at a constant pressure is directly proportional to the absolute (Kelvin) temperature of the gas.
- $\dfrac{V_1}{T_1} = \dfrac{V_2}{T_2}$ (Charles's law)
- The average kinetic energy of the molecules in a gas, and thus the speed of the molecules, is directly related to the temperature of the gas. When the temperature increases, the speed of the molecules increases and, therefore, the collision frequency increases.

SECTION 3.3 REVIEW

1. Your friend's little brother is showing you his helium balloon when he looses control of the string. The balloon floats over the barbecue, where some food is cooking. The balloon bursts. Using kinetic molecular theory, explain what has happened.

2. Your driver's licence test is scheduled for the middle of July. Just when you thought things were going fine with your driving lessons, you notice that the car tires seem to be thumping along the road. Your dad tells you to check the air pressure of the tires. Why might he think that there could be something wrong with the air pressure in the tires?

3. The propellant in an aerosol can is pressurized gas. Once the pressure of the gas in the can drops to atmospheric pressure, it can no longer be used to deliver the product. Explain why it is dangerous to throw a used aerosol can into a fire where the temperature might exceed 500 °C.

4. The table below shows the volume of a sample of air at different temperatures and a constant pressure. Use these data to determine the relationship of the volume of a gas and its Celsius temperature. (**Hint:** What is the proportionality constant? What is the y-intercept?)

Temperature (°C)	Volume (mL)
100	126
75	119
50	109
25	102
0	92

5. The temperature of a 6.0 L sample of gas increases from 200 K to 450 K. If the atmospheric pressure is constant, what is the final volume of the gas?

6. When soldering circuits boards, technicians often work with the board in a nitrogen atmosphere. A sample of nitrogen gas occupies a volume of 300 mL at 17 °C. When the technician is welding, the temperature of the nitrogen increases to 100.0 °C. If the pressure remains constant, what volume will the nitrogen occupy?

7. One cool morning, when the temperature is 6.00 °C, a balloon containing 2.00 kL of helium is used as a promotional attraction over a shopping centre. The noon day sun heats the gas. If the atmospheric pressure remains constant and the final volume of the balloon is 2.14 kL, what is the Celsius temperature?

8. A 2.5 L balloon is completely filled with helium indoors at a temperature of 24.2 °C. The balloon is taken out on a cold winter day. If the final volume of the balloon is 2.0 L, what is the Celsius temperature outdoors?

9. Methane gas can be condensed by cooling and increasing the pressure. A 600 L sample of methane gas at 25 °C and 100 kPa is cooled to −20 °C at a constant pressure. In a second step, the gas is compressed until the pressure is quadrupled. What will the final volume be?

10. When you increase the temperature of the air inside a hot-air balloon, the volume of the balloon does not increase. Assuming the pressure remains constant, explain what does happen and why that causes the balloon to rise in the air. If you are not sure of the answer, research hot-air balloons.

11. Which of the following questions is invalid? Explain why. Answer the valid question.
    a) If the Kelvin temperature of a quantity of gas is doubled and the volume is held constant, by what factor is the pressure changed?
    b) If the Celsius temperature of a quantity of gas is tripled and the volume is held constant, by what factor is the pressure changed?

Chapter 3

SUMMARY

The properties of gases are critical to life itself. As well, understanding the properties of gases allows people to develop technologies that can enhance the quality of life and even save lives. Gases are compressible, have a low density, are all miscible, and expand as their temperature is increased. These properties were observed before a model explaining them was developed. The kinetic molecular theory of gases is a model that helps to explain the observable properties.

Curiosity about phenomena that are affected by atmospheric pressure led to detailed studies on the relationship between the volume of a fixed amount of gas and the external pressure applied to it. Galileo did not understand atmospheric pressure and wondered why a vacuum pump could not "pull" water up more than 10 m. Torricelli worked out the answer. He proposed that water was not "pulled" up by a vacuum pump but, instead, was pushed up by the atmosphere. He developed the first mercury barometer to measure atmospheric pressure.

Pascal used a mercury barometer to design an experiment that measured the change of atmospheric pressure with altitude. Boyle did detailed measurements to determine the exact relationship between volume and pressure, which resulted in Boyle's law, $P_1V_1 = P_2V_2$.

Hot-air ballooning motivated Charles and Gay-Lussac to study the relationship between the volume of a fixed amount of gas and its temperature. They both determined that the relationship was linear. When their straight lines are extrapolated to zero volume, the lines all met at one temperature. Kelvin proposed that this value of temperature was absolute zero, meaning that the temperature could not go below that value. When using absolute temperature (the Kelvin scale), the mathematical relationship between the volume of a fixed amount of gas and its temperature at constant pressure can be written as $\dfrac{V_1}{T_1} = \dfrac{V_2}{T_2}$. The Celsius temperature scale cannot be used in calculations involving Charles's law.

Concept Organizer   Developing the Gas Laws

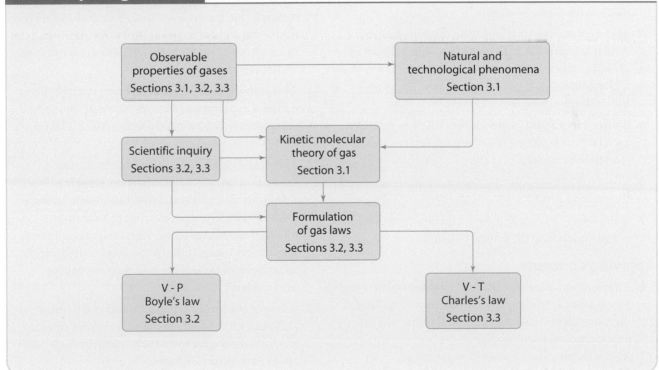

Understanding Concepts

Use the following descriptions of daily events to answer questions 1 to 3.

   a. A scuba diver takes about 2250 L of air underwater in a cylinder that he carries on his back.

   b. When cool dry air over a mountain range reaches the plains, it drops rapidly in altitude. As a result of the altitude change, the air mass is compressed and the temperature increases.

   c. The pilot of a hot-air balloon opens the vent to allow hot air to escape in order to make the balloon "sink" down to the ground.

   d. A weather balloon bursts when it reaches a sufficiently high altitude.

   e. A helium balloon shrinks when it is taken outdoors in winter temperatures below $-10\ °C$.

1. Which of the examples illustrates Boyle's law?

2. Which of the examples illustrates Charles's law?

3. Which of the examples can be explained by a combination of the two laws?

4. a) Describe what will happen to a column of mercury that is inverted and the end is placed in a container of mercury if it is taken from sea level to the top of a mountain.
   b) Explain the changes in this mercury column.

5. Is there a maximum length for which a straw would be useful for drawing up a refreshing drink? How could you test your answer?

6. How does kinetic molecular theory explain a temperature of 0 K?

7. Using kinetic molecular theory, explain why the density of gases is less than the density of liquids.

Applying Concepts

8. A student was asked to write four statements describing temperature conditions of common situations. Read the four statements and identify any statements that provide accurate temperatures.
   a) The freezing temperature of water is 0 K.
   b) On a very hot day, the ambient temperature can reach 313 K.
   c) On a very cold day the ambient temperature can reach 0 K.
   d) A life-threatening fever is at a temperature of 310 K.

9. Correct any of the statements that you found inaccurate in question 8.

10. Which of the following curves accurately communicate the relationship between pressure and the volume of a fixed amount of gas at constant temperature? Explain.

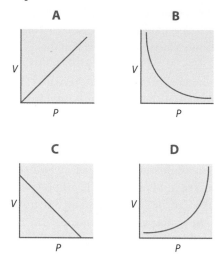

Solving Problems

11. A bicycle pump contains 50 mL of air at 101 kPa. When the handle of the pump is pushed down, the volume of the air drops to 14 mL. If the temperature is constant, what is the pressure of the air?

12. A 50 g sample of dry air in a large advertising balloon occupies a volume of 27 L at 20 °C. What volume will the balloon occupy when the temperature is increased to 35 °C at constant pressure?

13. A student receives a 1.5 L helium balloon at a graduation party in Banff, where the elevation is 1386 m above sea level and the atmospheric pressure is only 89 kPa. The student drives to Vancouver, which is located at sea level, where the atmospheric pressure is 1.0 atm. If the temperature is the same in both places, what is the volume of the balloon in Vancouver?

14. A sample of nitrogen gas has a volume of 10 mL at 22 °C and 1.013 bar. If the pressure remains constant, to what temperature must the nitrogen be cooled to achieve a volume of 2.8 mL?

15. Ammonia gas is used in the production of fertilizers. A sample of ammonia gas is found to be under a pressure of 7.5 atm. What pressure must be exerted on the ammonia gas to reduce its volume to one fifth of its original volume, assuming that the temperature remains constant?

16. A 5.0 L tank of helium, which is at a pressure of 15.0 MPa in a room kept at a temperature of 22.0 °C, is used to blow up balloons. The balloons, when inflated, have a volume of 2.0 L and maintain a pressure of 115 kPa without breaking. If the temperature remains constant, what is the maximum number of balloons that can be filled with this tank?

17. A 19.0 L fire extinguisher contains 18.0 L of water and 1.0 L of compressed air. To operate efficiently, the extinguisher must expel the last drop of water at a pressure of 110 kPa. Assuming that the temperature remains constant, what must be the original pressure of the compressed air in the fire extinguisher?

18. Halogen lamp bulbs are made of quartz to withstand high pressure that develops when they heat up to temperatures of 1000 °C during operation. If a halogen bulb is filled with iodine vapor at 5.00 atm of pressure, what will be the pressure in the bulb when it reaches its operating temperature, if room temperature is 22 °C?

19. The Environment Canada web site listed today's barometric pressure as 102.3 kPa. What would the pressure be in atmospheres? How tall a column of mercury could be supported by the air pressure today?

20. A teacher is doing a demonstration of Boyle's experiments using U-tubes similar to those shown in the diagram. In part A, the mercury levels on the two sides of the U-tube are the same. The column of gas is 15 mm high. To what height must mercury be added to the right side of the U-tube to reduce the height of the gas to 5 mm (as in part B)? (**Hint:** When the gas is compressed, it is replaced with mercury. Also, the right side of the tube is open to the atmosphere. Assume that the conditions of the experiment are at standard atmospheric pressure.)

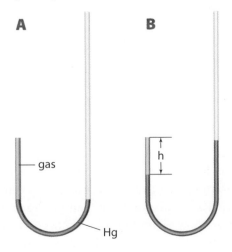

Making Connections

21. a) Vapour pressure can be defined as the degree to which liquid molecules are escaping into the gaseous phase. Water boils when the vapour pressure of the water exceeds the atmospheric pressure of the surrounding air. Using this background information, explain the following data.

Boiling Point of Water at Various Locations in Alberta

Location	Altitude (m)	Boiling point (°C)
Calgary, AB	1048	96.3
Edmonton, AB	668	97.5
Fort McMurray, AB	369	98.5

b) Use your calculator or a spreadsheet package to predict the boiling point of water at the top of Sunshine, a popular ski destination in Banff, Alberta (altitude 2729 m). **ICT**

22. Red blood cells carry the oxygen necessary to sustain life. Using your knowledge of atmospheric pressure, speculate as to the reason that people living in Calgary (altitude 1048 m) have significantly more red blood cells than people living at sea level (altitude 0 m).

23. The transportation of gases poses some unique challenges given their macroscopic properties. In addition to these properties, which are common to all gases, many gases are asphyxiates (cause unconsciousness or death due to lack of oxygen), toxic, and flammable. Design a container that would be suitable to transport these hazardous substances economically, efficiently, and safely. Describe how your container features address these concerns.

24. Design an experiment to generate results that will produce the relationship between the pressure and volume of a fixed mass of gas at a constant temperature given the following list of equipment.
- 25 mL syringe
- plastic tubing
- gas pressure gauge

Sketch a graph of the relationship you expect to see.

25. Using the following equipment, build a thermometer and describe how it would be calibrated.
- Erlenmeyer flask
- water
- one-holed rubber stopper
- glass tubing

Would you expect your thermometer to be accurate on a day-to-day basis? Explain why or why not.

CHAPTER 4

Exploring Gas Laws

Chapter Concepts

4.1 The Combined Gas Law

- Gases can undergo changes in temperature, pressure, and volume simultaneously.

- The combined gas law summarizes a combination of temperature, pressure, and volume changes.

- The volumes of gaseous reactants and products in chemical reactions are in whole-number ratios.

- Equal volumes of gases at the same temperature and pressure contain equal number of molecules (or moles of molecules).

- Molar volume is the same for all gases under the same conditions.

4.2 The Ideal Gas Law

- The ideal gas law includes Avogadro's law and the combined gas law.

- The universal gas constant allows you to use the ideal gas law for any set of conditions.

- You can find the molar mass of a substance by applying the ideal gas law.

- Real gases approximate ideality at standard temperature and pressure.

- The behaviour of real gases diverges from ideality at high pressures and at low temperatures.

Imagine what it would be like to drive the car shown in the photograph. The ride is smooth and luxurious. The engine runs so quietly that it almost purrs. Can you feel the wind in your hair as you glide along an open highway, effortlessly whisked to your destination? It is easy to forget that it takes repeated, violent explosions inside the engine to power this smooth, quiet ride.

Look more closely at the inset image. The pistons in the cylinders of the engine compress an air-and-gasoline mixture that explodes when ignited, producing a large amount of gas at a high temperature that forces the piston back down. The explosive production of gases also contributes to safety. An air bag is folded into the steering wheel, dashboard, or door. When its sensor detects a collision, a chemical reaction produces nitrogen gas that almost instantaneously inflates the air bag.

Gases play a major role in the technologies used to make automobiles functional and safe. Developing these technologies requires a combination of knowledge about the properties of gases and the chemical reactions of gases. In this chapter, you will learn how to combine the gas laws to solve many problems. First, complete the Launch Lab to see how a combination of gas behaviours produces a dramatic result.

Changing Gas Temperature, Pressure, and Volume at the Same Time

An empty container, open to the atmosphere, is bombarded by air molecules from the outside and the inside simultaneously. The volume of the container remains constant because the pressure on the two sides is the same. As you carry out this activity, recall what you have learned about the relationships among changes in gas volume, pressure, and temperature.

Safety Precautions

- Use safety goggles while performing this activity.
- Always use beaker tongs to handle the heated can.
- Your teacher may choose to carry out this investigation as a demonstration.

Materials

- 5 mL of water
- large beaker of ice water
- empty, clean soft drink can
- hot plate
- beaker tongs
- 10 mL graduated cylinder

Procedure

1. Measure 5 mL of water with the graduated cylinder and pour it into the soft drink can.

2. Heat the can on the hot plate until steam begins rising from the opening of the can.

3. Using the beaker tongs, quickly invert the can into the large beaker of ice water so that the opening of the can is just under the surface of the water. Carefully observe the effects on the can.

Analysis

1. What happened to the water molecules inside the can when the water was heated?

2. What happened to the air that was initially present inside the can when it was heated?

3. How can you account for the changes in the can when it was placed in the ice water?

The combustion of fuel in the cylinder causes a sudden change in the temperature and pressure of the gases. The pressure pushes the piston down, resulting in a change in the volume of the gases.

The Combined Gas Law

Section Outcomes

In this section, you will:
- **solve** problems using the combined gas law
- **use** appropriate SI notation, fundamental and derived units, and significant digits when performing gas law calculations
- **predict** volumes of gaseous reactants and products in chemical reactions
- **calculate** the molar volume of gases at STP and SATP

Key Terms

law of combining volumes
Avogadro's law
molar volume
standard temperature and pressure (STP)
standard ambient temperature and pressure (SATP)

In an automobile accident, air pressure inside an air bag can save your life. How can enough pressure be generated quickly enough to prevent your body from hitting the windshield? An automobile air bag must be fully pressurized within milliseconds after the collision. A sensor turns on an electrical circuit that triggers a chemical reaction. The chemical reaction produces about 65 L of nitrogen gas that nearly instantaneously inflates the air bag to cushion your body. As soon as your body hits the air bag, it starts to deflate, cushioning the impact. The nitrogen gas escapes through vents in the bag, and within two seconds, the pressure inside the bag returns to atmospheric pressure.

If you analyze the process that occurs in an air bag, you will discover that the temperature, pressure, and volume of the system changed considerably, all at the same time. In Chapter 3, you considered only two variables at a time. In the laboratory, chemists must make a great effort to control one of the variables to test the other two. In typical situations such as air bags and other industrial processes, all three variables change together. To explain and make predictions about natural and industrial processes involving gases, you need a method for working with changes in temperature, pressure, and volume at the same time.

Combining the Gas Laws

In Chapter 3, you learned about Boyle's law and Charles's law. You can combine these laws as shown below.

- Boyle's law: $V \propto \dfrac{1}{P}$

- Charles's law: $V \propto T$

- If V is proportional to $\dfrac{1}{P}$ and also proportional to T, then V must be proportional to $\dfrac{T}{P}$: $V \propto \dfrac{T}{P}$

- Convert this proportionality to an equality by inserting a proportionality constant: $V = k\dfrac{T}{P}$

- Rearrange the expression algebraically to bring all of the variables to the left side of the equation: $\dfrac{PV}{T} = k$

- For a specific amount of a gas, the pressure on the gas times its volume divided by its temperature is always equal to the same constant: $\dfrac{P_1V_1}{T_1} = k \quad \dfrac{P_2V_2}{T_2} = k$

- Therefore, you can write the combined gas law as shown: $\dfrac{P_1V_1}{T_1} = \dfrac{P_2V_2}{T_2}$

Figure 4.1 Air bags inflate in about one twentieth of a second, significantly reducing injury to the driver or passenger of a car in a collision. People protected by air bags have walked away from collisions that, without the air bag, would have caused severe injury or even death.

The last equation is the most useful statement of the combined gas law.

Combined Gas Law

$$\dfrac{P_1V_1}{T_1} = \dfrac{P_2V_2}{T_2}$$

Note: T is the Kelvin temperature.

Combined Gas Law Calculations

The combined gas law is a valuable tool when you want to make predictions about any of the three variables—temperature, pressure, or volume. For example, imagine that you prepared a gas in a laboratory apparatus that has a specific volume and has sensors to detect the temperature and pressure. You might want to know what the volume would be at typical room temperature and atmospheric pressure. You would mathematically rearrange the equation in the box on page 128 to solve for V_2 by following the steps shown here.

- Write the combined gas law equation:

$$\frac{P_1V_1}{T_1} = \frac{P_2V_2}{T_2}$$

- To eliminate T_2 from the right side of the equation, multiply both sides of the equation by T_2:

$$\frac{P_1V_1}{T_1}(T_2) = \frac{P_2V_2}{\cancel{T_2}}(\cancel{T_2})$$

$$\frac{P_1V_1T_2}{T_1} = P_2V_2$$

- To eliminate P_2 from the right side of the equation, divide both sides by P_2:

$$\frac{P_1V_1T_2}{T_1}\left(\frac{1}{P_2}\right) = \cancel{P_2}V_2\left(\frac{1}{\cancel{P_2}}\right)$$

- Rewrite the equation with the sides exchanged:

$$V_2 = \frac{P_1V_1T_2}{T_1P_2}$$

Sample Problem

Finding Volume Using the Combined Gas Law

Problem
A small balloon contains 275 mL of helium gas at a temperature of 25.0 °C and a pressure of 350.0 kPa. What volume would this gas occupy at 10.0 °C and 101 kPa?

What Is Required?
You need to find the volume, V_2, of the balloon under the new conditions of temperature and pressure.

What Is Given?
You know initial pressure, volume, and temperature:
$P_1 = 350.0$ kPa
$V_1 = 275$ mL
$t_1 = 25.0$ °C

You know the final temperature and pressure.
$t_2 = 10.0$ °C
$P_2 = 101$ kPa

Plan Your Strategy
Use the combined gas law relationship. Remember to convert from the Celsius scale (t) to the Kelvin scale (T) for temperature. Rearrange the equation to solve for V_2 and substitute the known values for variables in the equation.

Act on Your Strategy
$T_1 = 25.0$ °C $+ 273.15$
$T_1 = 298.15$ K
$T_2 = 10.0$ °C $+ 273.15$
$T_2 = 283.15$ K

$$\frac{P_1V_1}{T_1} = \frac{P_2V_2}{T_2}$$

$$\frac{P_1V_1}{T_1}\left(\frac{T_2}{P_2}\right) = \frac{\cancel{P_2}V_2}{\cancel{T_2}}\left(\frac{\cancel{T_2}}{\cancel{P_2}}\right)$$

$$\frac{P_1V_1T_2}{T_1P_2} = V_2$$

$$\frac{(350\text{ kPa})(275\text{ mL})(283.15\text{ K})}{(298.15\text{ K})(101\text{ kPa})} = V_2$$

$V_2 = 905.026$ mL
$V_2 \approx 905$ mL

Solution
The final volume of the gas is 905 mL.

Check Your Solution
The volume unit should be mL and it is. The fewest significant digits in the measured values is three, so when you multiply or divide values, the answer should have only three significant digits. The large decrease in pressure would cause a large increase in volume. The temperature decreased slightly, which should cause a slight decrease in volume. Thus, you would expect an overall increase in volume, which is in agreement with the answer.

Finding Temperature Using the Combined Gas Law

Problem

A volume of 25 mL of gas is produced in a laboratory experiment at a temperature of $-15\,°C$ and a pressure of 700 mmHg. Predict the Celsius temperature of the gas when its volume is reduced to 20 mL and the pressure is increased to 820 mmHg.

What Is Required?

You need to find the final Celsius temperature, t_2, of the gas after the volume and the pressure were changed.

What Is Given?

You know the initial volume, temperature, and pressure of the gas:

$V_1 = 25$ mL

$t_1 = -15\,°C$

$P_1 = 700$ mmHg

You know the final volume and pressure of the gas:

$V_2 = 20$ mL

$P_2 = 820$ mmHg

Plan Your Strategy

Remember to convert initial temperature from Celsius to Kelvin. Rearrange the equation to solve for T_2 and substitute known values into the combined gas law. Convert the temperature back to Celsius. (**Note:** it is not necessary to convert the pressure to kPa or the volume to L because the units will give the correct ratios and will cancel.)

Act on Your Strategy

$T_1 = -15.0\,°C + 273.15$

$T_1 = 258.15$ K

$$\frac{P_1V_1}{T_1} = \frac{P_2V_2}{T_2}$$

$$\frac{\cancel{P_1}\cancel{V_1}}{\cancel{T_1}}\left(\frac{\cancel{T_1}}{\cancel{P_1}\cancel{V_1}}\right)(T_2) = \left(\frac{T_1}{P_1V_1}\right)\frac{P_2V_2}{\cancel{T_2}}(\cancel{T_2})$$

$$T_2 = \frac{T_1 P_2 V_2}{P_1 V_1}$$

$$T_2 = \frac{(258.15\text{ K})(820\text{ }\cancel{mmHg})(20\text{ }\cancel{mL})}{(700\text{ }\cancel{mmHg})(25\text{ }\cancel{mL})}$$

$T_2 = 241.92$ K

$t_2 = 241.92$ K $- 273.15$

$t_2 = -31.23°C$

$t_2 \approx -31°C$

Solution

The final temperature of the gas is $-31\,°C$.

Check Your Solution

The answer is in units of degrees Celsius, as it should be. The volume decreased, which should indicate a temperature decrease. The pressure increased, which should cause the temperature to increase. Because the percentage decrease in volume was greater than the percentage increase in pressure, the overall change in the temperature should be a decrease, which it is.

1. A sample of gas has a volume of 525 mL at 300 K and 746 mmHg. Find the volume if the temperature increases to 350 K and the pressure increases to 780 mmHg.

2. A sample of gas has a volume of 75 mL at $19.0\,°C$ and 120 kPa. Predict what its volume would be at $25\,°C$ and 100 kPa.

3. A chemical researcher produces 15 mL of a new gaseous substance in a laboratory at a temperature of $25\,°C$ and pressure of 100 kPa. Predict the volume of this gas if the temperature was changed to $0.0\,°C$ and the final pressure was 101.325 kPa.

4. A 2.7 L sample of nitrogen gas is collected at a temperature of $45.0\,°C$ and a pressure of 0.92 atm. What pressure would have to be applied to the gas to reduce its volume to 2.0 L at a temperature of $25.0\,°C$?

5. A sample of argon gas occupies a volume of 2.0 L at $-35\,°C$ and standard atmospheric pressure. What would its Celsius temperature be at 2.0 atm if its volume decreased to 1.5 L?

6. A 500 mL sample of oxygen is kept at 950 mmHg and $21.5\,°C$. The oxygen is expanded to a volume of 700 mL and the temperature is adjusted until the pressure is 101.325 kPa. Predict the final temperature of the oxygen gas.

Technology and the Process of Discovery

Science seeks to help us understand why nature and technology act the way that they do. The discoveries of the gas laws are highlights of science in action. Knowledge about gases, however, is closely tied to the instruments that exist to measure their properties. The more you can know about gases, the more questions you can answer, and, in turn, the more new questions you can ask. As more and more sensitive tools are developed, scientists use them to probe deeper into the secrets of our natural world.

Measurements and the Gas Laws

The gas laws are the end result of countless experiments over hundreds of years. Accurate measurements of pressure, volume, and temperature allowed scientists to make precise predictions about the behaviour of gases.

The Bourdon tube pressure gauge, shown below, was patented in 1849 by a French watchmaker. This gauge allows scientists and engineers to make extremely accurate measurements of pressure. The gauge itself consists of a C-shaped tube that is closed at one end and attached to the source of the gas at the other. The tube uncurls slightly when the pressure inside it increases. This movement is reflected on a circular scale by changes in a pointer. The Bourdon gauge is still one of the most widely used instruments for measuring the pressure of liquids and gases of all kinds, up to pressures of 7×10^5 kPa.

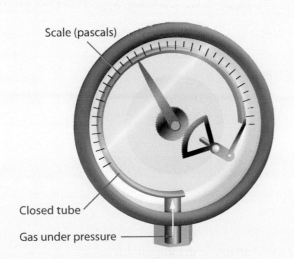

Scale (pascals)

Closed tube

Gas under pressure

Measuring Temperature

The thermocouple and the thermistor are two of the most common temperature measuring devices used in science and engineering today. The detector for the thermistor is a metal conductor that changes resistance with temperature. The thermistor measures the resistance of the conductor and converts the reading to temperature.

The thermocouple is based on the effect of a temperature difference between junctions at the ends of two different types of metal conductors. The ends of the two different wires are welded together, creating two junctions. When the temperatures of the junctions are different, a voltage is generated between the junctions. To measure a temperature, one junction is held at a known temperature and the other is placed in the sample to be measured. A voltmeter in the device measures the voltage between the junctions, as shown in the diagram.

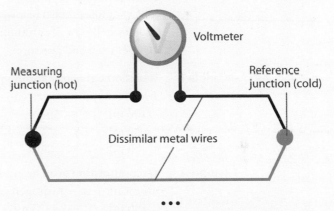

Voltmeter

Measuring junction (hot)

Reference junction (cold)

Dissimilar metal wires

• • •

1. There are many different types of thermocouples and thermistors. Choose one thermocouple or one thermistor and describe in detail how it works. Use diagrams or models to support your work.

2. The currently accepted value of Avogadro's number is $6.022\,136\,7 \times 10^{23}$. Avogadro proposed his hypothesis, now known as Avogadro's law, in 1811. At that time, there were no data on the number of particles in a mole. Conduct research to find at least two steps between the introduction of Avogadro's law and finding a value for Avogadro's number. What technologies were used? ICT

3. Think of a natural phenomenon that interests you and describe a tool that has not been invented yet that would give you scientific insight into that phenomenon. For example, if waves in the air could signal an avalanche, describe a tool that could measure those airwaves.

Combining Volumes of Gases

Early work on gases focussed on the relationships among the volume, temperature, and pressure for a particular mass of a sample of a gas. By the early 1800s, scientists were also exploring chemical reactions between gases. Gay-Lussac and a colleague, Alexander von Humboldt (1769–1859), performed experiments to determine the number of volumes of hydrogen and of oxygen that would react to form a volume of water. They needed this information to determine the percentage of oxygen in the atmosphere. Gay-Lussac made precise measurements and determined that two volumes of hydrogen combine with one volume of oxygen to form two volumes of water. (For example, 2 L of hydrogen combine with 1 L of oxygen to form 2 L of water vapour.) Gay-Lussac continued to study the volumes of other gases in other chemical reactions.

By 1808, Gay-Lussac had completed the analyses of the volumes of gaseous reactants and products that he and other scientists had collected. He concluded that, in chemical reactions, gases combine in very simple proportions. Gay-Lussac's conclusion is known as the **law of combining volumes** and is stated in the following box.

> **Law of Combining Volumes**
>
> When gases react, the volumes of the gaseous reactants and products, measured at the same conditions of temperature and pressure, are always in whole-number ratios.

The examples shown below, of gas volumes before and after chemical reactions, illustrate the law of combining volumes. All gas volumes are measured at the same temperature and pressure.

Example 1   **Reaction between hydrogen gas and oxygen gas**

hydrogen gas + oxygen gas → water vapour

100 mL	+	50 mL	→	100 mL
50 mL	+	25 mL	→	50 mL
8 mL	+	4 mL	→	8 mL

Example 2   **Decomposition of ammonia gas**

ammonia gas → hydrogen gas + nitrogen gas

150 mL	→	225 mL	+	75 mL
50 mL	→	75 mL	+	25 mL
8 mL	→	12 mL	+	4 mL

Sample Problem

Applying the Law of Combining Volumes

Problem

What volume of nitrogen forms when 300 mL of ammonia decomposes?

Solution

Determine the whole-number ratio of ammonia gas to hydrogen gas to nitrogen gas in the reaction by using one of the trials in Example 2, above. You can accomplish this by dividing each of the volumes by the smallest volume.

For example:

$$\frac{8 \text{ mL(ammonia)}}{4 \text{ mL}} : \frac{12 \text{ mL(hydrogen)}}{4 \text{ mL}} : \frac{4 \text{ mL(nitrogen)}}{4 \text{ mL}}$$

2(ammonia) : 3(hydrogen) : 1(nitrogen)

The ratio of ammonia to nitrogen is 2:1.

Therefore: $\dfrac{\text{ammonia}}{\text{nitrogen}} = \dfrac{2}{1}$

$$\frac{300 \text{ mL (ammonia)}}{x \text{ mL (nitrogen)}} = \frac{2}{1}$$

$$x \text{ mL (nitrogen)} = \frac{300 \text{ mL (ammonia)}}{2}$$

$x = 150$ mL (nitrogen)

When 300 mL of ammonia decomposes, 150 mL of nitrogen is formed.

7. What volume of water vapour forms when 250 mL of hydrogen combines with 125 mL of oxygen?

8. What volume of nitrogen gas forms when the decomposition of ammonia gas produces 15 mL of hydrogen gas?

9. A chemist performed an experiment and determined that 125 mL of nitrogen gas reacted with 250 mL of oxygen gas and formed 250 mL of nitrogen dioxide gas. What volume of nitrogen dioxide gas is formed when 350 mL of nitrogen gas reacts with an excess of oxygen gas?

Q1 What laws does the combined gas law combine?

Q2 To what type of process does the law of combining volumes apply?

Avogadro's Law

A few years before Gay-Lussac's description of combining gas volumes, John Dalton (1766–1844) had published his ideas about atoms combining in simple, whole-number ratios to form compounds. Empirical work on the percentage mass of elements present in compounds suggested that chemical combinations of the same elements always occurred in fixed mass ratios or multiples of those ratios. For example, water is always 88.89 percent oxygen and 11.11 percent hydrogen by mass. Thus, the mass of oxygen in a water molecule is eight times the mass of hydrogen in a water molecule. Chemists, at that time, could not understand how one volume of oxygen could combine with two volumes of hydrogen to form two volumes of water vapour if the mass of oxygen in water was eight times the mass of hydrogen in water.

Amadeo Avogadro (1776–1856) resolved the apparent conflict between Gay-Lussac's and Dalton's observations. In 1811, he proposed that the law of combining volumes could be explained if equal volumes of gases contained the same number of particles, regardless of their mass. Avagadro's proposal was ignored for 50 years because scientists could not accept such a revolutionary idea. Eventually, scientists realized the significance of the proposal and more observations supported it. Avogadro's proposal is now known as **Avogadro's law**.

> **Avogadro's Law**
>
> Equal volumes of all ideal gases at the same temperature and pressure contain the same number of molecules.

Chemists now understand that the volume of a gas is directly proportional to the number of molecules of the gas, when the pressure and temperature are constant.

As you might have learned from your earlier studies in chemistry, a large number of molecules are required to measure volumes and masses of substances in the laboratory. Recall that Avogadro's number, 6.02×10^{23} particles, represents one mole of those particles. When the particles are molecules, the ratio of the amounts of the substances present, expressed in moles, is equal to the ratio of the coefficients of the substances in a balanced chemical equation. Examine Figures 4.2 and 4.3 on the next page to see how amounts of gases, expressed in moles, are related to volumes of gases in chemical reactions.

Chemistry File

Try This
- Calculate the mass of oxygen and hydrogen present in 100 g of water.
- Find the ratio of the mass of oxygen for 1 g of hydrogen in water.
- Calculate the mass of oxygen and hydrogen present in 100 g of hydrogen peroxide, $H_2O_2(\ell)$.
- Find the ratio of the mass of oxygen for 1 g of hydrogen in hydrogen peroxide.
- Compare the two ratios.

Chemistry File

FYI
How big is a mole? If pennies were made of pure copper, about 25 pennies would contain one mole of copper. How big is a mole of pennies? If you spread a mole of pennies over the entire land area of Canada, the blanket of pennies would be about 2.5 km deep.

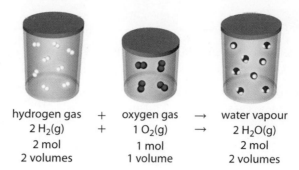

$$\begin{array}{ccccc}
\text{hydrogen gas} & + & \text{oxygen gas} & \rightarrow & \text{water vapour} \\
2\,H_2(g) & + & 1\,O_2(g) & \rightarrow & 2\,H_2O(g) \\
2\,\text{mol} & & 1\,\text{mol} & & 2\,\text{mol} \\
2\,\text{volumes} & & 1\,\text{volume} & & 2\,\text{volumes}
\end{array}$$

Figure 4.2 When hydrogen and oxygen gases combine to form water vapour, the ratios of their volumes is the same as the ratios of the coefficients of the balanced equation.

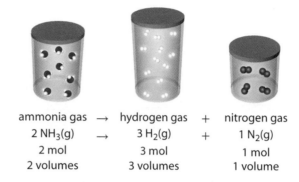

$$\begin{array}{ccccc}
\text{ammonia gas} & \rightarrow & \text{hydrogen gas} & + & \text{nitrogen gas} \\
2\,NH_3(g) & \rightarrow & 3\,H_2(g) & + & 1\,N_2(g) \\
2\,\text{mol} & & 3\,\text{mol} & & 1\,\text{mol} \\
2\,\text{volumes} & & 3\,\text{volumes} & & 1\,\text{volume}
\end{array}$$

Figure 4.3 When ammonia gas decomposes into hydrogen gas and nitrogen gas, the ratios of their volumes is the same as the ratios of the coefficients of the balanced equation.

Because the number of moles of a substance consists of a specific number of molecules, Avogadro's law can be expressed in moles.

The following steps develop a useful form of Avogadro's law in which n represents the number of moles of the gas.

- When the temperature and pressure of a gas are constant, you can express Avogadro's law mathematically as a proportionality.

  $$n \propto V$$

- By using a proportionality constant, you can express it as an equality.

  $$n = kV$$

- As long as the temperature and pressure remain constant, any combination of number of moles divided by the volume of the gas is equal to the same constant.

  $$\frac{n_1}{V_1} = k \quad \text{and} \quad \frac{n_2}{V_2} = k$$

- Therefore,

  $$\frac{n_1}{V_1} = \frac{n_2}{V_2}$$

The last equation is the mathematical statement of Avogadro's law.

Avogadro's Law

$$\frac{n_1}{V_1} = \frac{n_2}{V_2}$$

Figure 4.4 illustrates Avogadro's law by showing that an increase in the amount (n moles) of a gas at a constant temperature and pressure causes the volume to increase.

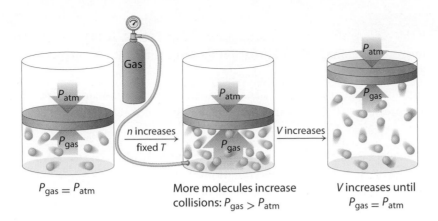

Figure 4.4 When more gas enters a container, the increase in the number of molecules causes the pressure to increase. Because the pressure inside the vessel is greater than the external pressure while the temperature remains constant, the volume will increase. The volume will continue to increase until the internal pressure caused by the gas becomes equal to the external pressure.

Molar Volume of Gases

According to Avogadro's law, the volume of one mole of any gas should be the same as the volume of one mole of any other gas, if the conditions of temperature and pressure are the same. Thus, it is possible to calculate the **molar volume** (*v*) of a gas, or the volume of one mole of a gas. Molar volume is expressed in units of L/mol, which can be found by simply dividing the volume, *V*, by the number of moles, *n*, present.

$$v = \frac{V}{n}$$

The molar volume will, of course, vary with different temperatures and pressures. On Earth, atmospheric pressure and ambient temperature (temperature of the surrounding air) can vary from one place to another and can also vary in one location at different times. Therefore, chemists have agreed on specific sets of conditions under which to report gas volumes. **Standard temperature and pressure (STP)** conditions are defined as 0 °C (273.15 K) and a pressure of 1 atm (101.325 kPa). These values, approximating the freezing temperature of water and atmospheric pressure at sea level, are not the most comfortable conditions in which to make measurements. **Standard ambient temperature and pressure (SATP)** conditions, 25 °C (298.15 K) and 100 kPa, provide a second standard that more closely parallels the conditions in a normal environment, especially in a laboratory. The following table summarizes these conditions.

Table 4.1 Standard Conditions of Temperature and Pressure

Conditions	Pressure	Celsius temperature	Kelvin temperature	Molar volume
STP	101.325 kPa	0 °C	273.15 K	22.4 L/mol
SATP	100.0 kPa	25 °C	298.15 K	24.8 L/mol

Math Tip

- Remember to carry all digits in your calculations through to the final step before rounding to the correct number of significant digits. In your calculations, record at least two more significant digits than you will keep after rounding the answer.
- When using the answer from one step of a calculation in the next step, use the unrounded value.

Analyze the Sample Problem on the next page to see how these tips are applied.

• • •

3 When a formation reaction occurs, such as the reaction of nitrogen and oxygen to form nitrogen dioxide ($NO_2(g)$), the masses of the compounds always react in the same ratios. For example, the ratio of masses of nitrogen, oxygen, and nitrogen dioxide are always approximately 7:16:23. In a chemical reaction between gaseous reactants, the ratio of the volumes is also always the same. However, the ratio of the volumes differs from the ratio of the masses for the same reaction. In the reaction above, the ratio of the volumes of nitrogen, oxygen, and nitrogen dioxide would be 1:2:2 Explain how Avogadro resolved this apparent contradiction.

The following example uses the combined gas law and Avogadro's law to calculate the molar volume of a gas at the standard conditions that are shown in Table 4.1. After studying the example, practise this type of calculation by completing the Thought Lab.

Sample Problem

Calculating the Molar Volume of Nitrogen

Problem

An empty, sealed vacuum container with a volume of 0.652 L is found to have a mass of 2.50 g. When filled with nitrogen gas, the container has a mass of 3.23 g. The pressure of the nitrogen in the container is measured and found to be 97.5 kPa when the temperature is 21.0 °C. Calculate the molar volume of nitrogen gas at STP.

What Is Required?

You need to find the volume, v, of one mole of nitrogen gas at STP.

What Is Given?

You know the temperature, pressure, and the volume of the gas, as well as the mass of the container when empty and when filled with nitrogen:

$T_1 = 21.0\,°C + 273.15\,K$
$T_1 = 294.15\,K$
$P_1 = 97.5\,kPa$
$V_1 = 0.652\,L$
Molar mass of $N_2(g) = 28.02\,\dfrac{g}{mol}$
Mass of container under a vacuum $= 2.50\,g$
Mass of container + nitrogen $= 3.23\,g$
You know the conditions of STP:
$T_2 = 0\,°C + 273.15$
$T_2 = 273.15\,K$
$P_2 = 101.325\,kPa$

Plan Your Strategy

Find the mass of the nitrogen in the container by subtracting the mass of the empty container from the mass of the container filled with nitrogen gas.

Calculate the number of moles of nitrogen by using the formula $n = \frac{m}{M}$, where $m =$ mass of nitrogen and $M =$ molar mass of nitrogen.

Find the volume that the nitrogen would occupy at STP by using the combined gas law.

Find the volume of one mole of nitrogen at STP by dividing the moles of nitrogen by the volume of nitrogen at STP.

Act on Your Strategy

$m_{nitrogen} = m_{container} - m_{vacuum}$
$m_{nitrogen} = 3.23\,g - 2.50\,g$
$m_{nitrogen} = 0.73\,g$

$n = \dfrac{m}{M}$

$n = \dfrac{0.73\,\cancel{g}}{28.02\,\dfrac{\cancel{g}}{mol}}$

$n = 0.026053\,mol$
$n \approx 0.0261\,mol$

$$\dfrac{P_1 V_1}{T_1} = \dfrac{P_2 V_2}{T_2}$$

$$\left(\dfrac{P_1 V_1}{T_1}\right)\left(\dfrac{T_2}{P_2}\right) = \left(\dfrac{\cancel{P_2} V_2}{\cancel{T_2}}\right)\left(\dfrac{\cancel{T_2}}{\cancel{P_2}}\right)$$

$$V_2 = \dfrac{P_1 V_1 T_2}{T_1 P_2}$$

$$V_2 = \dfrac{(97.5\,\cancel{kPa})(0.652\,L)(273.15\,\cancel{K})}{(294.15\,\cancel{K})(101.325\,\cancel{kPa})}$$

$V_2 = 0.582597\,L$
$V_2 \approx 0.583\,L$

$v = \dfrac{V}{n}$

$v = \dfrac{0.582597\,L}{0.026053\,mol}$

$v = 22.36199\,\dfrac{L}{mol}$

$v \approx 22.4\,\dfrac{L}{mol}$

Solution

The molar volume of nitrogen at STP is 22.4 L/mol.

Thought Lab 4.1 Molar Volumes of Gases

Target Skills

Solving problems involving molar volumes of gases

Comparing experimental values with theoretical values of molar volumes

Determining experimental error

Two students decided to calculate the molar volume of carbon dioxide, oxygen, and methane gases. First, they measured the mass of an empty 150 mL syringe with the entire syringe and balance in a vacuum. This ensured that the syringe did not contain any air. Next, they filled the syringe with 150 mL of carbon dioxide gas. They measured and recorded the mass of the syringe plus the gas. The students repeated their procedure for oxygen gas and for methane gas.

Finally, a thermometer in the room registered 23.0 °C (296.15 K) and the barometer registered 98.7 kPa. The students used these values for the temperature and pressure of the three gases. The students' results are given in the table below.

Three Gases at 296 K and 98.7 kPa

Gas	Carbon dioxide	Oxygen	Methane
Volume of gas (V)	150 mL	150 mL	150 mL
Mass of empty syringe	25.08 g	25.08 g	25.08 g
Mass of gas + syringe	25.34 g	25.27 g	25.18 g
Mass of gas (m)			
Molar mass of the gas (M)			
Number of moles of gas $\left(n = \frac{m}{M}\right)$			
Calculations for STP			
Volume of gas STP (273.15 K and 101.325 kPa)			
at STP $v = \frac{V}{n}$			
Calculations for SATP			
Volume of gas at SATP (298.15 K and 100 kPa)			
at SATP $v = \frac{V}{n}$			

Procedure

1. Copy the table into your notebook.
2. Calculate the molar volume for the carbon dioxide, oxygen, and methane gases at STP. Write your calculations and answers in rows indicated.
3. Calculate the molar volume for the carbon dioxide, oxygen, and methane gases at SATP. Write your calculations and answers in the rows indicated.

Analysis

1. Compare the three molar volumes at STP. What do you observe?
2. The accepted molar volume of a gas at STP is 22.4 L/mol. Use this value to calculate the percentage experimental error in the data for each gas.
3. Compare the three molar volumes at SATP. What do you observe?
4. The accepted molar volume of a gas at SATP is 24.8 L/mol. Use this value to calculate the percentage experimental error in the data for each gas.

Section 4.1 Summary

- $\frac{P_1V_1}{T_1} = \frac{P_2V_2}{T_2}$ (combined gas law)
- The law of combining volumes states that, when gases react, the volumes of the gaseous reactants and products, measured at the same conditions of temperature and pressure, are always in whole-number ratios.
- Avogadro proposed that at the same conditions of temperature and pressure, equal volumes of all ideal gases contain the same number of molecules, even though they do not have the same mass.
- $\frac{n_1}{V_1} = \frac{n_2}{V_2}$ (Avogadro's law)
- STP is defined as 0 °C and 1 atm of pressure. SATP is defined as 25 °C and 100 kPa of pressure.
- The molar volume of an ideal gas at STP is 22.4 L/mol.
- The molar volume of an ideal gas at SATP is 24.8 L/mol.

1. Using kinetic molecular theory, explain what happens when air is heated in an open container at atmospheric pressure.

2. Jane was trying to show her friend the collapsing soft drink can that she saw in the Launch Lab. She heated the can for 10 minutes on the stove burner and filled a large bowl with cold water and ice. She lifted the heated can from the stove with tongs and inverted it into the cold water and nothing happened. Explain, using kinetic molecular theory, why the can did not collapse.

3. Hydrogen gas ($H_2(g)$) combines with chlorine gas ($Cl_2(g)$) to form hydrogen chloride gas ($HCl(g)$). What is the whole-number ratio between the volumes of the reactants and the products?

4. How many moles are present in 6.98 g of chlorine gas? What volume would 6.98 g of chlorine occupy at STP?

5. Find the volume of water vapour produced when 2.3 L of hydrogen gas and an excess of oxygen gas (plenty of oxygen to burn all the hydrogen) are put into a reaction vessel and ignited. Perform your calculations, assuming that the gases have all returned to the original temperature and atmospheric pressure.

6. Find the pressure of a 60 mL sample of gas that is heated from SATP conditions to 55.0 °C if its volume expands to 120 mL.

7. A 6.98 g sample of chlorine gas has a volume of 2.27 L at 0 °C and 1.0 atm. Find the molar volume of the chlorine gas at 25 °C and 100 kPa.

8. A large syringe was filled with 48 mL of ammonia gas at STP. If the gas was compressed to 24 mL with a pressure of 110 kPa, what was the final temperature?

9. Jessica's friends are having a birthday party for her on a mild winter day. When the temperature was −2 °C and the atmospheric pressure was 100.8 kPa, her friends tied 4.2 L balloons in front of the house. A sudden cold front passed through and the atmospheric pressure increased to 103.0 kPa and the temperature dropped to −25 °C. What was the final volume of the balloons?

10. Use the data provided in the table below to develop a presentation for your class in which you show the relationships between temperature and the molar volume of an ideal gas, and between pressure and the molar volume of an ideal gas. **ICT**

Molar Volume of an Ideal Gas (L/mol) at Various Temperatures and Pressures

| Temperature | Pressure | | | |
	740 mmHg	750 mmHg	760 mmHg	770 mmHg
30 °C	25.5	25.2	24.9	24.6
25 °C	25.1	24.8	24.5	24.1
20 °C	24.7	24.4	24.1	23.7
15 °C	24.3	24.0	23.6	23.3

a) Identify the manipulated, responding, and controlled variables for each relationship you are planning to develop.

b) Use your calculator, computer, or any other technology available to you to develop graphs and mathematical relationships between the variables.

c) Your presentation should include visual components of the presentation, with the explanation of the relationships in terms of kinetic molecular theory from the text.

11. In Chapter 3, you considered the expansion of weather balloons as they rise to higher altitudes and the atmospheric pressure decreases. You assumed that the temperature remained constant. However, the temperature actually decreases. Discuss, based on the combined gas laws, all of the changes that you predict will take place as a weather balloon is rising upward through the atomsphere.

SECTION 4.2 Ideal Gas Law

Section Outcomes

In this section, you will:

- **explain** Boyle's law, Charles's law, and combined gas law and their relationship to the ideal gas law
- **solve** problems based on the ideal gas law
- **determine** molar mass from gaseous volume
- **describe** and **compare** the behaviour of ideal gases and real gases
- **design** an investigation to determine the universal gas constant, *R*

Key Terms

ideal gas law
partial pressure
Dalton's law of partial pressures

Math Tip

When using the combined gas law, it was not necessary to convert pressure to kPa or volume to L because the variables were used in ratios and the units always cancelled. When using the ideal gas law, however, you must express temperature in K, pressure in kPa, and the volume in L because the value of the universal gas constant was determined using these units. If you always use units in your calculations and ensure that they all cancel properly, you will avoid many common errors.

In the previous section, you related the combined gas law to Avogadro's volume–mole gas relationship using two sets of conditions. This method enabled you to make calculations about pressure, temperature, volume, and amount variables by alternately holding two of the variables constant while manipulating the third and calculating the fourth. The ideal gas law combines all four variables into a single relationship.

Developing the Ideal Gas Law

Recall that Boyle's and Charles's laws were used to produce the combined gas law. The same method can be used to summarize mathematically the relationships expressed by the combined gas law and Avogadro's law for ideal gases, as shown below.

- Avogadro's law — $V \propto n$ — T and P are constant
- The combined gas law — $V \propto \frac{T}{P}$ — n is constant
- Combine the two proportionalities. — $V \propto \frac{nT}{P}$
- Convert the proportionality to an equation by applying a proportionality constant, *R*. — $V = R\frac{nT}{P}$
- Multiply both sides by *P*. — $PV = \frac{RnT}{\cancel{P}}\cancel{P}$

$$PV = nRT$$

The relationship is known as the **ideal gas law**.

Ideal Gas Law

$$PV = nRT$$

The proportionality constant, *R*, is called the **universal gas constant** and applies to all gases.

In the previous section, you learned that the molar volume of any gas at STP is 22.4 L/mol. Thus, when $n = 1.00$ mol, $T = 273.15$ K, $P = 101.325$ kPa, and $V = 22.4$ L, you can calculate *R* as follows.

- Write the ideal gas law. — $PV = nRT$
- Solve for *R* in the ideal gas law. — $\frac{PV}{nT} = \frac{\cancel{nT}R}{\cancel{nT}}$, $R = \frac{PV}{nT}$
- Substitute in numerical values. — $R = \frac{(101.325 \text{ kPa})(22.4 \text{ L})}{(1.00 \text{ mol})(273.15 \text{ K})}$
- Calculate. — $R = 8.31 \frac{\text{kPa} \cdot \text{L}}{\text{mol} \cdot \text{K}}$

When the universal gas constant is calculated using four significant digits, the value is $8.314 \frac{\text{kPa} \cdot \text{L}}{\text{mol} \cdot \text{K}}$. This is the value you will use when making calculations with the ideal gas law.

Ideal Gas Law Calculations

The ideal gas law equation enables you to calculate any one of the four gas variables simply and quickly if you have information about the other three. Importantly, the equation does not require that you compare two sets of conditions for the same gas sample. Examine the following set of Sample Problems to find out how to apply the ideal gas law to a variety of problems.

Sample Problem

Finding Volume Using the Ideal Gas Law

Problem

Find the volume of 100.0 g of oxygen gas at SATP.

Solution

Known variables:

$T = 298.15$ K

$P = 100.0$ kPa

$M_{O_2} = 32.00 \dfrac{\text{g}}{\text{mol}}$

Find the number of moles in 100.0 g of oxygen.

$n_{O_2} = \dfrac{m}{M}$

$n_{O_2} = \dfrac{100.0\,\cancel{\text{g}}}{32.00\,\dfrac{\cancel{\text{g}}}{\text{mol}}}$

$n_{O_2} = 3.125$ mol

Find the volume of 3.125 moles of oxygen using the ideal gas law.

$PV = nRT$

$\cancel{P}V\left(\dfrac{1}{\cancel{P}}\right) = nRT\left(\dfrac{1}{P}\right)$

$V = \dfrac{nRT}{P}$

$V = \dfrac{(3.125\,\cancel{\text{mol}})\left(8.314\,\dfrac{\text{kPa}\cdot\text{L}}{\cancel{\text{mol}}\cdot\cancel{\text{K}}}\right)(298.15\,\cancel{\text{K}})}{(100.0\,\cancel{\text{kPa}})}$

$V = 77.463$ L

$V \approx 77.46$ L

Sample Problem

Finding Temperature Using the Ideal Gas Law

Problem

Find the Celsius temperature when 2.50 moles of a gas occupies a volume of 56.5 L under a pressure of 1.20 atm.

Solution

Known variables:

$n = 2.50$ moles

$V = 56.5$ L

$P = (1.20\,\cancel{\text{atm}})\left(\dfrac{101.325\ \text{kPa}}{\cancel{\text{atm}}}\right) = 121.59$ kPa

Solve for T in the ideal gas law.

$PV = nRT$

$\dfrac{PV}{nR} = \dfrac{\cancel{n}\cancel{R}T}{\cancel{n}\cancel{R}}$

$T = \dfrac{PV}{nR}$

$T = \dfrac{(121.59\,\cancel{\text{kPa}})(56.5\,\cancel{\text{L}})}{(2.50\,\cancel{\text{mol}})\left(\dfrac{8.314\,\cancel{\text{kPa}}\cdot\cancel{\text{L}}}{\cancel{\text{mol}}\cdot\text{K}}\right)}$

$T = 330.5189$ K

Convert to the Celsius scale.

$t = 330.519$ K $- 273.15$

$t = 57.369\ ^\circ$C

$t \approx 57.4\ ^\circ$C

Finding Molar Mass Using the Ideal Gas Law

Problem
Find the molar mass of a gas if a 1.58 g sample occupies a volume of 500.0 mL at STP.

Solution
Known variables:

$V = 0.5000$ L

$T = 273.15$ K

$P = 101.325$ kPa

$m = 1.58$ g

Use the ideal gas law to determine the number of moles in the sample.

$$PV = nRT$$

$$PV\left(\frac{1}{RT}\right) = n\cancel{R}\cancel{T}\left(\frac{1}{\cancel{R}\cancel{T}}\right)$$

$$\frac{PV}{RT} = n$$

$$n = \frac{(101.325\text{ kPa})(0.500\text{ L})}{\left(8.314\frac{\text{kPa}\cdot\text{L}}{\text{mol}\cdot\text{K}}\right)(273.15\text{ K})}$$

$$n = 0.02231 \text{ mol}$$

Rearrange the following formula to determine the molar mass.

$$n = \frac{m}{M}$$

$$\cancel{n}\left(\frac{M}{\cancel{n}}\right) = \frac{m}{M}\left(\frac{M}{n}\right)$$

$$M = \frac{m}{n}$$

$$M = \frac{1.58\text{ g}}{0.02231\text{ mol}}$$

$$M = 70.820\frac{\text{g}}{\text{mol}}$$

$$M \approx 70.8\frac{\text{g}}{\text{mol}}$$

The molar mass is 70.8 g/mol.

Alternate Method
You can build an equation that will enable you to solve the problem in one step. Substitute the value for $\left(n = \frac{m}{M}\right)$ directly into the equation for the ideal gas law ($PV = nRT$). Then solve for M.

$$PV = \frac{m}{M} RT$$

$$PV\left(\frac{M}{PV}\right) = \frac{m}{M} RT\left(\frac{M}{PV}\right)$$

$$M = \frac{mRT}{PV}$$

$$M = \frac{(1.58\text{ g})\left(8.314\frac{\text{kPa}\cdot\text{L}}{\text{mol}\cdot\text{K}}\right)(273.15\text{ K})}{(101.325\text{ kPa})(0.500\text{ L})}$$

$$M = 70.82\frac{\text{g}}{\text{mol}}$$

$$M \approx 70.8\frac{\text{g}}{\text{mol}}$$

Practice Problems

10. What is the pressure on the gas when 3.25 mol of hydrogen gas occupies a volume of 67.5 L at a temperature of 295 K?

11. What is the volume of 5.65 mol of helium gas at 98 kPa of pressure and a temperature of 18.0 °C?

12. How many moles of ammonia are present in a 250 mL container at 25.0 °C and 0.100 bar?

13. Find the volume of 1.87 g of methane gas (CH_4) at 20.0 °C and 780 mmHg.

14. What is the Kelvin temperature of 0.063 mg of argon gas at 1.25 atm of pressure if its volume is 31.5 mL?

15. Find the Celsius temperature of nitrogen gas if a 5.60 g sample occupies 2400 mL at 3.00 atm of pressure.

16. A sample of a gas with a mass of 0.571 g has a volume of 375 mL at 99.0 kPa and 23.8 °C. Find the molar mass of the gas.

• • •

6. Start with the ideal gas law equation and solve for the universal gas constant, R. Show all steps.

7. Explain why you must use units of kPa, K, L, and mol when using the ideal gas law for calculations.

• • •

Sample Problem

Finding the Density of a Gas Using the Ideal Gas Law

Problem
Find the density of helium gas at SATP, accurate to two significant digits.

Solution
Recall the formula for density, $D = \frac{m}{V}$.
For gases, the density is usually given in grams per litre.

Known variables:

$V = 1.0\ \text{L}$
$T = 298.15\ \text{K}$
$P = 100\ \text{kPa}$
$M_{He} = 4.00\ \dfrac{\text{g}}{\text{mol}}$

Start with the ideal gas law.
$PV = nRT$

Because $n = \frac{m}{M}$, you can substitute $\frac{m}{M}$ for n in the ideal gas law.

$$PV = \frac{mRT}{M}$$

Since density is defined as the mass per unit volume, solve for $\frac{m}{V}$ in the above equation.

$$P\cancel{V}\left(\frac{M}{RT}\right)\left(\frac{1}{\cancel{V}}\right) = \frac{m\cancel{RT}}{\cancel{M}}\left(\frac{\cancel{M}}{\cancel{RT}}\right)\left(\frac{1}{V}\right)$$

$$\frac{PM}{RT} = \frac{m}{V}$$

$$\frac{(100\ \cancel{\text{kPa}})\left(4.00\ \dfrac{\text{g}}{\cancel{\text{mol}}}\right)}{\left(8.314\ \dfrac{\cancel{\text{kPa}}\cdot \text{L}}{\cancel{\text{mol}}\cdot \cancel{K}}\right)(298.15\ \cancel{K})} = \frac{m}{V} = D$$

$$D = 0.16137\ \frac{\text{g}}{\text{L}}$$

$$D \approx 0.16\ \frac{\text{g}}{\text{L}}$$

The density of helium gas at SATP is 0.16 g/L.

Practice Problems

17. Calculate the density of carbon dioxide gas at SATP to three significant digits.

18. What is the density of helium gas at −25 °C and a pressure of 90 kPa?

19. What is the pressure on water vapour at 150 °C if its density is 0.500 g/L?

Dalton's Law of Partial Pressures

You have completed calculations that show that a mole of any gas will have the same volume when the temperature and pressure are constant. Because gases are completely miscible and, ideally, do not interact with each other, any mixture of gases will have the same molar volume as any pure gas. Is there a way to describe the pressure of one particular type of gas when it is mixed with other gases? This property is very important when you are analyzing the atmosphere and the availability of oxygen for breathing.

This question was answered by John Dalton when he was studying humidity in the atmosphere. He observed that after he added water vapour to dry gases, the total gas pressure was equal to the sum of the pressure of the dry gas and the pressure of the water vapour on the walls of the container.

$$P_{total} = P_{dry\ air} + P_{added\ water\ vapour}$$

Further studies showed that this phenomenon is true for the addition of any type of gases. When gases are mixed, the pressure that any one gas exerts on the walls of the container is called the **partial pressure** of that gas. Thus, **Dalton's law of partial pressures** can be expressed as shown in the box.

Dalton's Law of Partial Pressures

In a mixture of gases that do not react chemically, the total pressure is the sum of the partial pressures of each individual gas.

Dalton's law is illustrated in Figure 4.5. When two separate gases, originally at the same temperature, are mixed, the temperature—and thus the average speeds of the gases—does not change. Only the total number of molecules in the container increases. As a result, there are more collisions with the walls of the container and thus the pressure is higher.

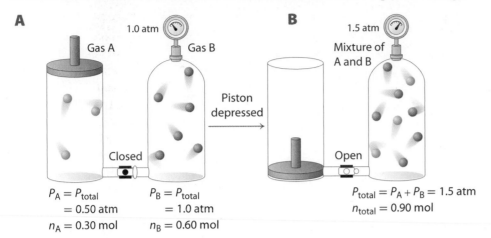

A

1.0 atm

Gas A   Gas B

Piston depressed

Closed

$P_A = P_{total}$ $P_B = P_{total}$
$\quad = 0.50$ atm $\quad = 1.0$ atm
$n_A = 0.30$ mol $n_B = 0.60$ mol

B

1.5 atm

Mixture of A and B

Open

$P_{total} = P_A + P_B = 1.5$ atm
$n_{total} = 0.90$ mol

Figure 4.5 Molecules of each gas collide with the walls of the container as many times and with the same force when both gases are in the same container as the molecules did when the gases were in separate containers. Therefore, the pressure on the wall of the container of mixed gases is the sum of the pressures of the gases in separate containers.

● **End extension material**

Collecting a Gas in the Laboratory

One of the safest ways to collect a gas in the laboratory is by downward displacement of water. A container, such as a graduated cylinder, is filled with water and inverted into a beaker of water. Care must be taken to avoid letting any air into the inverted cylinder. Tubing from the source of the gas is placed in the water and directed up into the inverted cylinder, as shown in Figure 4.6. The gas bubbles up through the water and collects in the closed end of the cylinder. The gas pushes the water down and out of the cylinder, leaving the gas sample trapped above the water. You can adjust the cylinder so that the water level inside the cylinder is even with the water level in the beaker. You can read the volume of the gas from the scale on the cylinder, and you know that the pressure on the gas is the same as the atmospheric pressure. Many laboratories have a barometer in the room from which you can read the barometric (atmospheric) pressure. You can measure the ambient temperature with a laboratory thermometer.

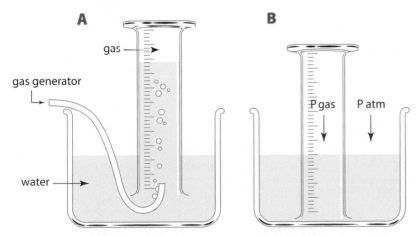

A

gas

gas generator

water

B

P gas   P atm

Figure 4.6 (A) Because gases have much lower density than water does, they float to the top of the container. **(B)** When you adjust the position of the container so that the water levels inside and outside the container are the same, the gas pressures above the water are the same inside and outside the container.

When you collect gases over water, there is one factor that you must take into account even if the gas is not soluble in water. The molecules of water vapour mix with the molecules of the gas. The total pressure of the gas trapped in the cylinder, according to Dalton's law of partial pressures, is then equal to the sum of the pressure exerted by each component of the gas mixture. Mathematically, $P_{total} = P_{gas} + P_{water vapour}$. To determine the pressure of the gas, the pressure of the water vapour must be subtracted from the total pressure. Table 4.2 provides the partial pressures of water vapour at different temperatures.

Table 4.2 Partial Pressure of Water Vapour

Temperature (°C)	Pressure (kPa)	Temperature (°C)	Pressure (kPa)
15	1.71	23	2.81
16	1.81	24	2.99
17	1.93	25	3.17
18	2.07	26	3.36
19	2.20	27	3.56
20	2.33	28	3.77
21	2.49	29	4.00

• • •

8 Ignore all gases in the atmosphere except nitrogen and oxygen. On a day when the percent humidity is zero and the atmospheric pressure is 102 kPa, what is the partial pressure of oxygen if the partial pressure of nitrogen is 71 kPa? How would the partial pressures of nitrogen and oxygen change if the humidity became high?

9 Name two factors that you must take into account when you are collecting a gas by water displacement.

10 A sample of hydrogen gas is collected by water displacement at 20 °C when the atmospheric pressure, as measured by a barometer, is 99.8 kPa. What is the pressure of the "dry" hydrogen?

• • •

INVESTIGATION 4.A

Target Skills

Performing an experiment to find the molar mass of a gas

Controlling variables in an experiment

Using thermometers, balances, and other devices to collect data on a gas

Finding the Molar Mass of a Gas

In this investigation, you will measure the volume and mass of a sample of a hydrocarbon gas to find its molar mass. The purpose of the investigation is to test water displacement as a method for collecting gas in the laboratory. Butane, $C_4H_{10}(g)$, is suggested for use because it is readily available in butane lighters. However, any other available hydrocarbon gas may be substituted.

Question

How can you find the molar volume of a gas when it is combined with water vapour?

Prediction

Predict the molar mass of butane.

Safety Precautions

• Remember that hydrocarbon gases are flammable. Before beginning, check to ensure that there are no open flames in the laboratory.

• If water is spilled on the floor, wipe it up immediately so that no one steps in it.

• Release all the collected gas into an operating fume hood after the investigation. Return the lighter to your classroom teacher after you have completed collecting all your data.

Materials

- tap water
- 4 L beaker or plastic pail
- balance
- disposable butane lighter
- 500 mL graduated cylinder
- blow dryer
- parafilm or plastic wrap
- thermometer (alcohol or digital)
- barometer
- small beaker
- masking tape
- paper towel

Procedure

1. Fill the 4 L beaker (or pail) about two thirds full of tap water at a comfortable temperature to keep your hand immersed in for about 15 minutes.

2. Determine the initial mass of the lighter and record it in your notebook.

3. Fill the graduated cylinder with water. Cover the cylinder tightly with a piece of parafilm. With your hand over the parafilm, place the cylinder upside down into the beaker. Make sure that no air bubbles are trapped in the cylinder. Slide the parafilm away from the mouth of the cylinder.

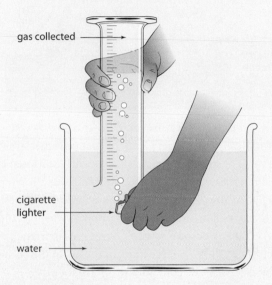

gas collected

cigarette lighter

water

4. As shown in the diagram, hold the lighter underwater, below the graduated cylinder in the beaker. Depress the button on the lighter to release gas into the cylinder. Collect approximately 400 mL of the gas. You may find that you need to have one of the other group members take a turn after you have collected about 150 mL of gas.

5. Adjust the position of the cylinder so that the water inside is at exactly the same level as the water in the beaker. You might have to add water to the beaker. In your notebook, record the volume of the gas that you collected when the water levels are equal inside and outside the cylinder.

6. In your notebook, record the temperature of the water and the atmospheric pressure in the room.

7. Place the lighter on a paper towel and dry it with the blow dryer.

8. Take your cylinder to the fume hood to release all the gas you have collected.

9. After the lighter has dried, determine the final mass and record it in your notebook. Place the lighter in the small beaker and allow it to continue drying overnight. Determine the mass of the lighter the next day. If the mass indicates that no water has evaporated overnight, return the lighter to your teacher. If the lighter has lost mass, indicating that more water has evaporated, correct your calculations.

Analysis

1. Find the mass of the gas in your sample by subtracting the mass of the lighter from its initial mass.

2. Find the partial pressure of the water vapour from Table 4.1. Use that value to find the partial pressure of the butane gas.

3. Use the ideal gas equation to calculate the molar mass of the gas.

4. Determine the theoretical molar mass of butane. (You might have done this calculation for your prediction.)

Conclusion

5. Calculate your percentage experimental error.

6. Discuss the accuracy of your determination of the molar mass of butane. List all possible sources of experimental error.

7. Critically assess the experimental design. What changes in the design do you think might enable you to obtain more accurate results?

Chinook Winds and the Gas Laws

The residents of Pincher Creek, Alberta were in for a surprise on January 6, 1966. At 7:00 A.M., the temperature was a cold −24.4 °C and at 8:00 A.M., the temperature was +0.6 °C. During a four minute period within that hour, the temperature rose by 21 °C. By 9:00 A.M., the temperature had dropped back down to −21.7 °C and remained there until 3:00 P.M. when it once again rose to +2.2 °C and remained there for the rest of the day. What weather condition could possibly cause such dramatic changes in temperature in such a short period of time?

Creating a Chinook Wind

You have probably already guessed that the dramatic temperature variations in Pincher Creek were caused by a chinook wind. Do you know, however, what produces a chinook? An understanding of the gas laws allows you to understand several important factors involved in the production of a chinook wind.

First, a moist air mass forms over the Pacific Ocean and west coast of Canada as shown in the figure. A strong westerly wind pushes the air up the Costal Mountains. Because the atmospheric pressure decreases as the altitude increases, the air mass expands as it rises. As the pressure decreases and the air expands, the temperature drops. As you know, cool air cannot contain as much water vapour as can warm air, so the water condenses and falls as rain or snow. When water vapour condenses, it releases heat (heat, or enthalpy, of vaporization). The decrease in the temperature of the air is much less than it would be for dry air. The large air mass continues moving rapidly eastward until it reaches the Rocky Mountains. It rises higher and the atmospheric pressure drops again, cooling the air more and releasing more condensed water in the form of snow. By the time the air mass reaches the eastern slopes of the Rocky Mountains, it is very dry. It drops down to the Prairies and, as it drops, the atmospheric pressure increases significantly. The pressure rapidly compresses the air mass and with the increase in pressure and thus decrease in volume, the temperature rises. The air is dry now so there is no moisture to evaporate back into vapour. Thus, the heat that was added to the air when the water vapour condensed cannot be removed by evaporation. The temperature rises much more than it decreased when the air was moving up the western slopes of the Costal Mountains. The temperature increases by about 9.8 °C for every 1000 metres that it drops. The compressed warm air tends to expand and the combination of expansion and the prevailing winds cause the air to shoot out across the Prairies as a warm, dry wind—the chinook wind. The wind speed can be as great as 65 to 95 km/h with gusts of more than 160 km/h. The warm winds can remove as much as 30 cm of snow in one hour.

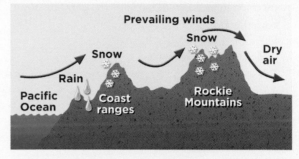

Why did the temperature oscillate up and down so dramatically in Pincher Creek of that unforgettable day? If a mass of cold night air is sitting nearly stationary over the Prairies at the base of the mountains, as shown in the figure, the warm, dry chinook wind can hit the cold air like a rock hitting water in a pond. A wave of cold air moves away but soon returns, like waves in a small pond. Eventually, the very large mass of the air in the chinook wind can push the cold air away permanently.

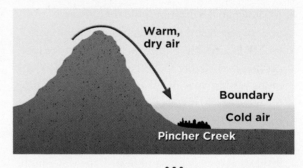

1. How frequent are chinook winds in southern Alberta? How long do they normally last?

2. Some people are physically affected by chinook winds. What physical ailments do these people experience?

3. Chinook winds are usually welcomed in the winter in Southern Alberta. However, at other times and in other parts of the world, such winds can be devastating. What can occur as a result of a chinook-type wind?

www.albertachemistry.ca
WWW

Ideal Gases and Real Gases

The gas laws that you have been using pertain to ideal gases. How well does the behaviour of real gases follow the laws for ideal gases? Examine the data in the first two columns of Table 4.3, the molar volumes of some common gases at STP. A quick calculation shows that maximum deviation of the molar volume of these gases is less than two percent from the ideal molar volume.

Table 4.3 Molar Volume of Some Common Gases at STP (0°C and 1 atm)

Gas	Molar volume (L/mol)	Condensation point (°C)
He	22.435	−268.9
H_2	22.432	−252.8
Ne	22.422	−246.1
Ideal gas	**22.414**	——
Ar	22.397	−185.9
N_2	22.396	−195.8
O_2	22.390	−183.0
CO	22.388	−191.5
Cl_2	22.184	−34.0
NH_3	22.079	−33.4

Under standard conditions of temperature and pressure, most gases behave like ideal gases. To begin to analyze the reasons for deviation from ideal behaviour when the temperature and pressure of molecules of real gas diverge from standard conditions, review the characteristics of ideal gases:

• move in straight lines at speeds determined by their temperature

• collide with one another in elastic collisions

• are point masses that have no volume

• exert no attractive or repulsive forces on other molecules in the container

At STP, gas molecules are moving rapidly and are very far apart, which makes their interactions and volumes insignificant. Real gases begin to deviate from ideal behaviour when they begin to move slowly and are relatively close together. Low temperatures and high pressures create these conditions.

Low-Temperature Effects

In Chapter 2, you learned that intermolecular forces of attraction exist among real molecules of the same kind. These attractive forces do not significantly affect the behaviour of real gas molecules at standard temperatures because the molecules are moving at high speeds. When a molecule has a high speed, it has a large amount of kinetic energy. When molecules collide, their kinetic energy allows them to easily break the attractive interactions. When the temperature, and thus the speed and kinetic energy of the molecules decreases, they cannot readily break the attractive interactions and eventually these attractive interactions cause the gas to condense into a liquid. The third column in Table 4.3 shows the temperatures at which the gases condense. Molecules with stronger attractive forces condense at higher temperatures. If you analyze the sequence of the gases in the table, you can see some trends. Consider, first, the noble gases; helium, neon, and argon. As the molar mass increases, the temperature at which they condense rises. As you learned in Unit 1, as the number of electrons increases (related to molar mass) the strength of the London (dispersion) forces increases.

Chemistry File

FYI

In Chapter 2, you read that Johannes van der Waals studied the interactions between molecules. In conjunction with those studies, van der Waals modified the ideal gas law to account for these intermolecular attractive forces. His equation can be written as $P + \frac{n^2a}{V^2}(V - nb) = nRT$. The "a" and "b" are constants that must be experimentally determined for each individual type of gas. In a very general way, you could say that "a" corrects for the interactions between molecules and "b" corrects for the volume of real gas molecules.

Figure 4.7 Under high pressures, gas molecules are close enough together to interact with one another when they are about to collide with the wall of the container. These interactions reduce the force of the collisions with the wall.

Those gas molecules that exert stronger forces on one another condense at higher temperatures. That is, more energy is needed to separate the molecules and keep them in a gaseous state. The same trend can be seen with the diatomic gases, hydrogen, nitrogen, and oxygen. Finally, consider ammonia. It has a smaller molar mass than all of the other gases except helium and hydrogen, yet it has the highest condensation temperature. Recall that ammonia is a polar molecule. Dipole-dipole interactions are much stronger than London (dispersion) forces and thus these forces require more energy (higher temperature) to separate the molecules from one another and prevent condensation.

High-Pressure Effects

Under standard atmospheric pressure, gas molecules are so far apart that interactions are not frequent enough to cause gases to behave non-ideally. When the external pressure on gases increases to many atmospheres of pressure, the molecules are pushed closer together and they interact frequently. As shown in Figure 4.7, when a molecule is about to collide with the walls of the container, the molecules nearby are exerting attractive forces on the molecule, pulling it away from the wall. Therefore, the force of the collision with the wall is reduced. The gas thus exerts less pressure on the walls of the container than it would in the absence of the intermolecular forces. If you measured the pressure with a pressure gauge, the pressure would be lower than it would for an ideal gas under the same conditions. Thus, calculations of other variables would be incorrect.

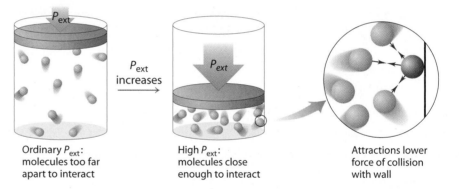

Ordinary P_{ext}: molecules too far apart to interact

High P_{ext}: molecules close enough to interact

Attractions lower force of collision with wall

Another effect of high pressure is a change in the apparent volume of the gases. Under standard atmospheric pressure, the percentage of the total volume of a container that is taken up by gas molecules is insignificant. Gas molecules are behaving as though they were point masses. When high pressures reduce the volume of the container, the percentage of the total volume taken up by the gas molecules becomes significant. According to kinetic molecular theory, the V in the ideal gas law must represent the empty space between molecules. When you measure the volume of a gas at high pressures, the measured value of V is larger than the actual volume of empty space as shown in Figure 4.8. If you use this measured volume to calculate other variables, the calculations of other variables will be incorrect.

Figure 4.8 At standard atmospheric pressure, gas molecules are so far apart that they take up a very small percentage of the volume of the container. At high pressures, the actual volume of the gas molecules is a significant percentage of the volume of the container.

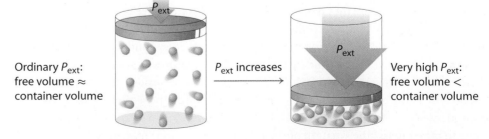

Ordinary P_{ext}: free volume ≈ container volume

P_{ext} increases

Very high P_{ext}: free volume < container volume

In the following investigation, you will draw from the knowledge about gases that you have gained in this unit and design an experiment to determine the universal gas constant.

Finding the Value of the Universal Gas Constant, R

In this investigation, you will work in groups to design an investigation to determine the value of the universal gas constant. After you have agreed on a procedure and safety precautions, you will present your plans to your teacher. When your procedure is approved, you will carry out your experiment.

Question

How can you obtain experimental data that will enable you to calculate the value of the universal gas constant, R?

Safety Precautions

After you have designed your experiment, list the safety precautions you must take while you are performing the experiment.

Materials

With your group, discuss possible types of gas that you could use. Remember that you must have a method for determining the amount of gas in your sample. In Investigation 4.A, you used a butane lighter as a source of gas. What other sources of specific gases are available. Be sure that the gas you choose to use has the properties that are necessary for the experiment. For example, carbon dioxide gas is relatively soluble in water. If you used water displacement to determine a volume, you would not be able to get an accurate value for carbon dioxide.

Make a list of materials that you will need for your experiment. Consider all the materials and equipment that you have used in other investigations. If you believe that you need types of equipment that you have not yet used, find out whether this equipment will be available.

Experimental Plan

With your group, discuss the type of data that you will need to calculate R. Discuss the methods you will need to use to obtain the data. When you have agreed on the general methods, write a detailed, step-by-step procedure for obtaining the data. Remember to include safety precautions. Present your plan to your teacher. When you have received permission, carry out your procedure.

Data and Observations

Based on your experimental plan, decide exactly what type of data you will need to obtain in your investigation. Determine the type of calculations that you will need to make and ensure that you will have the necessary data to make these calculations. Before you start your investigation, prepare data tables in which you will record your data.

Each member of the group should record all the data in his or her notebook. Be sure to include units in your recorded data. Convert your data to kPa for pressure, K for temperature, and L for volume. Use your data to calculate the universal gas constant, R. Record your calculations in your notebook.

Analysis

1. Compare your results with the accepted value of R, $8.314 \frac{kPa \cdot L}{mol \cdot K}$.

2. Calculate your percentage error.

3. Now that you have carried out your procedure, discuss with your group some possible ways to improve the methods of data collection.

Conclusions

4. Evaluate your experimental method. List the parts of your procedure that worked well and those that need improvement.

5. What do you think were the most significant sources of error in your procedure? How could they be avoided or modified?

6. Were your safety precautions adequate?

7. Write a summary of the overall experience of designing and carrying out this investigation.

Section 4.2 Summary

- The ideal gas law equation is $PV = nRT$.
- R is the universal gas constant with a value of $8.314 \dfrac{\text{kPa} \cdot \text{L}}{\text{mol} \cdot \text{K}}$.
- Dalton's law states that in a mixture of gases that do not react chemically, the total pressure is the sum of the partial pressures of each individual gas.
- When you collect a gas by displacement of water, you must account for the partial pressure of the water vapour when determining the pressure of the collected gas by using the equation $P_{\text{total}} = P_{\text{gas}} + P_{\text{water vapour}}$.
- Real gases behave ideally under standard conditions of temperature and pressure.
- Real gases behave non-ideally at low temperatures because the molecules do not have enough kinetic energy to overcome the attractive forces among the molecules.
- Real gases behave non-ideally under pressures of many atmospheres because the molecules are so close together that collisions occur frequently and cause the measured pressure to be low.
- Real gases behave non-ideally under pressures of many atmospheres because the molecules are so close together that they take up a significant percentage of the volume, making volume measurements too large.

SECTION 4.2 REVIEW

1. Explain how the ideal gas law can apply to any gas, even when the masses of individual molecules of various gases are different.

2. Describe two different ways for finding the molar mass of a gas.

3. Would water displacement be a good laboratory procedure for collecting hydrogen sulfide gas? Explain your answer.

4. At STP, a container holds 14.01 g of nitrogen gas, 16.00 g of oxygen gas, 66 g of carbon dioxide gas, and 17.04 g of ammonia gas. What is the volume of the container?

5. A volume of 240 mL of oxygen gas is collected by water displacement when the atmospheric pressure is measured at 100.2 kPa and the temperature is 20.5 °C. What is the pressure exerted by the oxygen gas?

6. Use the following diagram to explain Dalton's law of partial pressures.

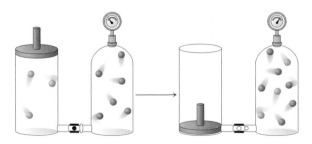

7. What volume would 5.25×10^{27} molecules of xenon gas occupy at SATP?

8. Find the molar mass of an unknown gas if a 0.87 g sample of the gas occupies a volume of 325 mL at 21.5 °C and 102 kPa.

9. Hanna and Josh collected a gas by displacement of water. They measured the volume of gas in the inverted cylinder, as shown in the diagram. Explain what measurements would be inaccurate and why. If they attempted to calculate the number of moles in the gas using $n = \frac{PV}{RT}$, would their calculated value be larger or smaller than the correct value. Explain why.

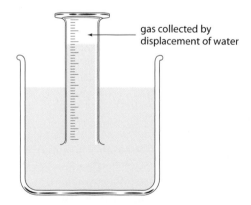

gas collected by displacement of water

10. Calculate the volume of 1.2 mol of carbon dioxide gas at SATP.

Rarely do processes take place with one of the three variables of pressure, volume, or temperature being constant. To solve many real problems, you can combine Boyle's law and Charles's law to produce the combined gas law, $\frac{P_1V_1}{T_1} = \frac{P_2V_2}{T_2}$.

When gases react chemically, the volumes of the gases, measured at the same temperature and pressure, are always whole-number ratios. When compounds react chemically, the masses also combine with the constant ratios. However, the ratio of masses that combine in a chemical reaction differ from the ratio of the volumes of gases for the same reaction. Avogadro resolved this apparent contradiction by proposing that equal volumes of gases under the same conditions of temperature and pressure contain the same number of molecules, regardless of the masses of the molecules. Consequently, the quotient of the number of moles of a gas and the volume of the gas, at constant temperature and pressure, is always the same. You can write his conclusion mathematically as $\frac{n_1}{V_1} = \frac{n_2}{V_2}$.

For consistency, chemists have agreed to report molar volumes of gases under standardized temperature and pressure, either at STP or SATP.

By combining Avogadro's law and the combined gas law, you can produce the ideal gas law, $PV = nRT$, where R is the universal gas constant. This law can be applied to a single gas or mixtures of gases because when gases are mixed each contributes its own pressure to the total pressure. When doing experiments with gases, it is convenient to collect gases by displacement of water. When you do so, you must account for the water vapour that has mixed with the gas on which you are experimenting.

Real gases behave ideally at standard conditions of temperature and pressure but behave non-ideally at low temperatures and very high pressures. At low temperatures, the attractive interactions among gas molecules become significant. At high pressures, the volume of the gas molecules becomes significant. At high pressures, the interactions among the gas molecules prevent molecules from hitting the walls of the container with the expected force.

Concept Organizer — The Ideal Gas Law and Related Concepts

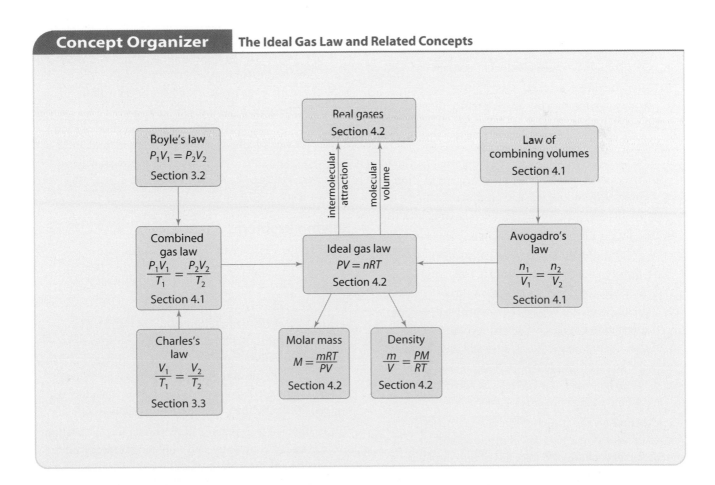

Understanding Concepts

1. Explain why the combined gas law is more practical than either Boyle's law or Charles's law.

2. In your own words, explain the meaning of the law of combining volumes.

3. Use the following diagram to explain the meaning of Avogadro's law.

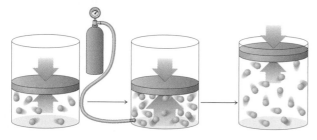

4. Explain why different gases can have different densities if they all have the same molar volume at the same temperature and pressure.

5. Explain why the law of combining volumes stipulates that the gases have to be measured at the same temperature and pressure.

6. How does kinetic molecular theory describe an ideal gas?

7. Using kinetic molecular theory, explain why the pressure of a mixture of gases is equal to the sum of the pressures each gas would exert if it were placed in the container alone.

8. Which aspect of the ideal gas equation enables you to make calculations about the pressure, temperature, volume, and amount of a gas without collecting data about the same mass of gas at different temperatures?

9. Usually, the universal gas constant is given in units of $\frac{kPa \cdot L}{mol \cdot K}$. Convert the universal gas constant to units of

a) $\dfrac{atm \cdot L}{mol \cdot K}$ **b)** $\dfrac{mmHg \cdot L}{mol \cdot K}$

10. Which two attributes of real gases primarily account for the deviation of real gases from ideal behaviour?

11. Draw a diagram to show what happens to the "free space" in a gas sample at high pressure. How does the diagram help explain why real gases do not exhibit ideal behaviour at high pressure?

12. Explain how the attractive interactions between molecules of real gases affect the force with which molecules collide with the walls of the container. How does this non-ideal interaction affect the value of n when substituting observed values into the equation, $n = \frac{PV}{RT}$, to calculate n?

13. A weather balloon is released into the atmosphere. At higher altitudes both the temperature and the pressure of the atmosphere decrease. Predict what will happen to the volume of the balloon. Use the combined gas law to provide reasons for your answer.

Applying Concepts

14. Explosions often occur when liquids or solids are changed into gases. What conditions of volume, temperature, and pressure might generate an explosion? In gasoline engines, a spark plug ignites the fuel. What ignites the fuel in a diesel engine?

15. Explain the apparent contradiction between Gay-Lussac's law of combining volumes and Dalton's ideas about the atoms combining in whole-number ratios to form compounds. How did Avogadro resolve this apparent contradiction?

16. A glass container is filled with helium gas in a chemistry laboratory. An identical container, which is sitting on the counter, is filled with oxygen gas.
  a) Compare the volumes of the helium gas sample with the oxygen gas sample.
  b) Compare the amount (moles) of the helium gas with the amount of oxygen gas.
  c) Compare the mass of the helium gas with the mass of the oxygen gas.
  d) The container with the oxygen gas is placed in an ice-water bath. Compare the amount, pressure, and average kinetic energy of the helium gas with the oxygen gas.

17. What does the expression "dry gas" mean?

Solving Problems

18. A sample of gas has a volume of 532 mL at a temperature of 31.4 °C and a pressure of 1.33 atm. Predict the volume the sample would occupy at STP.

19. A scuba diver is swimming 30.0 m below the ocean surface where the water pressure is 4.0 atm and the temperature is 8.0 °C. A bubble of air with a volume of 5.0 mL is emitted from the breathing apparatus. What will the volume of the air bubble be when it is just below the surface of the water, where the pressure is 101.3 kPa and the water temperature is 24.0 °C?

20. Helium gas is stored in a steel cylinder having a volume of 100 L. The pressure gauge on the cylinder indicates a pressure of 25 atm at 20 °C. The cylinder is used to blow up a weather balloon at 25 °C. If the final pressure in the cylinder and the balloon is 1.05 atm, how large will the balloon be?

21. A mass of 265 g of oxygen is stored in a 10.0 L container at 20.0 atm. What is the temperature of the gas in the container?

22. Find the mass of oxygen in a 5.4 L cylinder of compressed oxygen gas filled at 200 bar and stored at 22.0 °C.

23. Dentists sometimes use laughing gas (dinitrogen monoxide, $N_2O(g)$) as a mild anaesthetic to keep patients relaxed during dental procedures. A cylinder of laughing gas has a diameter of 23.0 cm and a height of 140 cm. The pressure is 108 kPa and the temperature is 294 K. How many grams of laughing gas are in the cylinder?

24. What is the volume occupied by a 1.25 g sample of helium at STP? What is the density of helium at STP?

25. The atmosphere of the imaginary planet, Xylo, consists entirely of chlorine gas ($Cl_2(g)$). The atmospheric pressure of this inhospitable planet is 155.0 kPa and the temperature is 89 °C. What is the density of the atmosphere?

26. Propane gas, $C_3H_8(g)$, in a barbecue undergoes combustion with oxygen gas to produce carbon dioxide gas and water vapour.
    a) Write the balanced equation for this reaction.
    b) What is the total volume of gaseous products, measured at ambient conditions, that are produced when 1 L of propane burns?

27. An 800 mL container holds 2.366 g of an unknown gas at 78.0 °C and a pressure of 103 kPa. Is the gas most likely to be bromine, krypton, neon, or fluorine?

28. A 4.00 L container holds 2.17 mol of ammonia gas at a pressure of 206 kPa. What is the temperature of the ammonia inside the container?

29. A sample of gas is collected at 25 °C. If the Celsius temperature of the gas is tripled and the pressure on the gas is doubled, the final volume of the gas will be what fraction of the original volume?

30. In a study of oxygen uptake by muscle at high altitude, a physiologist prepares an atmosphere consisting of 79% $N_2(g)$, 17% $^{16}O_2(g)$, and 4.0% $^{18}O_2(g)$ by volume. The $^{18}O_2(g)$ is used by the body exactly the same as $^{16}O_2(g)$ but it can be detected and measured by special instruments. The pressure of the mixture will be maintained at 0.75 atm to simulate high altitude. Calculate the partial pressure of $^{18}O_2(g)$ in the mixture.

31. A student collects 55.0 mL of hydrogen gas over water at 23.0 °C and 750 mmHg. What volume will the dry hydrogen occupy at 40.0 °C and 775 mmHg?

32. When 0.600 L of argon at 1.20 atm and 277 °C is mixed with 0.200 L of oxygen at 501 mmHg and 127 °C in a 400 mL flask at 27 °C, what is the pressure in the flask?

33. Acetylene ($C_2H_2(g)$) is an important fuel in welding. A sample is produced in a laboratory by a reaction between calcium carbide and water. The gas was collected by displacement of water. The volume collected was 523 mL when the total gas pressure was 738 mmHg and the temperature was 23 °C. How many grams of acetylene were collected?

34. A sample of an unknown gas was collected in an empty syringe of known mass under vacuum conditions. The gas was then transferred to a flexible container and heated so that the volume of the gas could be measured more accurately.
    a) Use the following data to find the molar mass and the molar volume of the unknown gas:
       Mass of syringe and gas = 2.201 g
       Mass of empty syringe = 2.150 g
       Pressure = 102 kPa
       Temperature = 50.0 °C
       Volume of heated gas = 664 mL
    b) What is the likely identity of the gas?

Making Connections

35. Nearly 2000 people and thousands of livestock in Cameroon, Africa, were killed by carbon dioxide asphyxiation on August 21, 1986. They lived in a valley near Lake Nyos, which is fed by volcanic springs of carbonated ground water. Suddenly, possibly due to a small earthquake, the lower-level carbonated water rose through the fresher surface water, resulting in a decrease in pressure and an increase in temperature of the dissolved carbon dioxide. Calculate the density of carbon dioxide gas at SATP. Can you explain how the carbon dioxide caused so many deaths?

36. Compressed gases, such as oxygen, are used throughout a hospital. What safety precautions should be employed to ensure these gases are used safely? Explain using the macroscopic properties of gases.

37. Although atmospheric pressure is related to altitude, it varies considerably day to day and is associated with certain weather conditions. Low pressure is often associated with stormy weather. As you learned in previous science courses, precipitation is a result of moist air rising and cooling, which leads to the condensation of the water vapour in the air. Provide an explanation to connect these facts.

Career Focus: Ask an Exercise Physiologist

Dr. John Kolb

Jon Kolb took up mountaineering when he was in high school, and he later became a world-class gymnast. Today, as a professor of kinesiology with the Human Performance Lab at the University of Calgary, Dr. Kolb combines his love of sport with his love of science. As a kinesiologist, he studies human movement. He uses his expertise, combined with his knowledge of gas pressure chemistry, to help athletes keep moving. He has designed and tested a hyperbaric tent to help prevent acute mountain sickness in climbers. He and his colleagues are also helping elite athletes to safely and legally improve their endurance through high-altitude training.

Q Why are you interested in the effects of altitude on athletes?

A Being an athlete myself, I was interested in how we could use altitude to help athletes, especially in endurance sports, to maximize their performance legally.

Q Is there less oxygen in the atmosphere at higher altitudes?

A The percentage of oxygen in the atmosphere remains the same. It's 21 percent in Vancouver and 21 percent on the top of Mount Everest. As you go up in altitude, however, the total atmospheric pressure drops and, therefore, so does the partial pressure of every gas in the atmosphere.

Q Why would athletes have difficulty performing at higher altitudes than they are used to?

A The pressure of gas is the driving force behind the diffusion gradient for oxygen to get through the alveoli (of the lungs) and into the bloodstream. Here in Calgary, we're at a substantial altitude—a little more than 1000 m. In Vancouver, at sea level, where the air pressure is 760 mmHg (1.0 atm), the partial pressure of oxygen in the blood is 100 mmHg (0.13 atm). By the time you get to Calgary, the partial pressure of oxygen in the blood has dropped to 88 mmHg (0.12 atm). This small change in the partial pressure of gases is enough to reduce the concentration of oxygen that is in the blood.

Q What is the basis for high-altitude training?

A If you're training at an altitude of, say 2000 m, that creates what's called a hypoxic environment. In other words, there is a reduced partial pressure of oxygen.

If you're hypoxic, the kidneys sense that and release a hormone called erythropoietin (EPO), which travels to the bone marrow and stimulates red blood cell production. After a three- or four-week period—that's the time it takes for red blood cells to mature—there is an increase in the red blood cell count. This allows the same volume of blood to pick up more oxygen.

Q Why not train at even higher altitudes (3000 m or more)?

A If you go too high, there's a really significant increase in red blood cell mass. You can get such an increase that the blood becomes quite viscous, which is difficult on the heart and it doesn't pump that well. This is linked to a higher incidence of stroke and cardiovascular disease.

Q How do you carry out high-altitude training?

A One of the problems that occurs when you go to a high altitude is that you can't train as hard because you're becoming hypoxic. There is less oxygen available to the working muscles.

One of the things our lab has been doing is using the concept of living or sleeping "high" and then training "low." Experiments I've done show that a person can get the same EPO response from living at a high altitude for 12 hours a day for five days as he or she would living there 24 hours a day for five days. So we have athletes "sleep" or "live high" in hypobaric tents overnight, and then train at their normal altitude, which is quite a bit lower, enabling them to train hard. They get the benefits of a natural increase in red blood cell mass, but they don't have to train slower.

Q **You have also designed and used a hyperbaric tent. What is this for?**

A I've done some work on preventing acute mountain sickness by changing the partial pressure of gas. Acute mountain sickness occurs when an individual goes too high too quickly, and the body is not able to adapt to the low oxygen content. This generally happens to mountaineers at altitudes above 3000 m, when there's a rapid ascent. This can have deleterious effects on the brain and lungs. And that's very dangerous.

On a climb in the Andes, I had this big bubble tent. I was getting climbers to go inside it periodically as we went up ... it's like getting into a giant beach ball, and you pump it up with air and the pressure increases. What that does is simulate a very significant decrease in altitude, of about 1000 m. The result is an increase in the oxygen content of the blood, relatively speaking. This kind of tent has been used to treat people with acute mountain sickness, but in my study I used it to prevent individuals from getting sick.

Aerospace Engineer Engineers with various specialties design, create, and test airplanes, spacecraft, and missiles. Wing design, propulsion systems, and cabin air pressure in airplanes and space shuttles are some design features involving gas pressures that aerospace engineers must consider. Junior aerospace engineers have undergraduate degrees in engineering, while senior engineers generally have graduate degrees.

Design Engineer Mechanical engineers specializing in product design make use of the properties of gases to create various tools: nail and staple guns, dental drills, refrigerators, tire-pressure gauges, air bags for cars, camping stoves, and medical devices are some of many examples. Design engineers require undergraduate degrees in mechanical engineering.

Respiratory Physiologist Physiologists study how organisms function, especially with regard to the human body, and may combine research with clinical work in hospitals. Respiratory physiologists specialize in how people breathe, including breathing difficulties and how to overcome them. An undergraduate or graduate degree is usually needed to work in a research laboratory. A research team leader may be a PhD, MD (Doctor of Medicine), or both.

Respiratory Therapist Therapists trained to help people with breathing problems work with patients in hospitals and health clinics or at patients' homes. Respiratory therapists perform breathing tests and educate patients about their respiratory health. They also help patients use breathing-assistance equipment, such as oxygen tanks and mechanical ventilation masks. Therapists must have a three-year college diploma in respiratory therapy, including a work-practicum.

Go Further...

1. Why might someone who has taken a plane from Vancouver to Calgary suffer similar but milder symptoms of acute mountain sickness?

2. Would you expect the partial pressure of oxygen to be higher or lower on hot days (32 °C or more) compared with cold days (−20 °C or below)?

3. A danger that deep-sea divers face when ascending from a dive is called "the bends" or decompression sickness. Research why decompression sickness is a problem and what can be done to prevent it.

Understanding Concepts

1. Two samples of gas have identical volumes. The temperature of Gas A is 22.8 °C. The temperature of Gas B is 296 K. In which sample do the molecules have the larger average kinetic energy?

2. How does kinetic molecular theory explain the pressure exerted by a gas trapped inside a container?

3. Use kinetic molecular theory to explain why the pressure of a gas increases when the temperature is increased and its volume is kept constant.

4. Sometimes molecules of gases are referred to as "point masses." Which characteristic of an ideal gas does this term describe?

5. What variables must be controlled in an investigation to determine the relationship between the pressure and temperature of a gas?

6. What must you know to be able to determine the volume of a gaseous product from the volume of a gaseous reactant, at the same temperature and pressure?

7. What effect does temperature have on the vapour pressure of water?

Applying Concepts

8. What would happen to the volume of a dry gas in a closed container if the pressure exerted on it is tripled at constant temperature?

9. Will the volume of a gas triple when its Celsius temperature is tripled at a constant pressure? Explain your answer.

10. A volume of Gas A of 25 mL combines with 100 mL of Gas B and produces 50 mL of Gas C. All gas volumes are measured at the same temperature and pressure. Write the balanced equation for this chemical reaction.

11. Will the volume of one mole of any gas always be 22.4 L? Give reasons for your answer.

12. Use the data provided in the table below to develop a presentation for your class to show the relationships between the temperature and the density of an ideal gas, and between the pressure and the density of an ideal gas. **ICT**

Density of an Ideal Gas (g/L) at Various Temperatures and Pressures

Temperature (°C) \ Pressure	98.7 kPa	100.0 kPa	101.3 kPa	102.7 kPa
30	1.13	1.14	1.16	1.18
25	1.15	1.16	1.18	1.20
20	1.17	1.18	1.20	1.22
15	1.19	1.20	1.22	1.24

Use your calculator and any other technology available to you to develop graphs and mathematical relationships between the variables. Your presentation should integrate these visual components with text that explains the relationships in terms of kinetic molecular theory.

Solving Problems

13. Convert the following temperatures as indicated:
a) −100 °C to K   **b)** 0 K to °C   **c)** 243.5 °C to K

14. Convert the following pressure values as indicated:
a) 250 kPa to bar   **c)** 978 mmHg to atm
b) 3.69 atm to kPa

15. Predict the volume of a gas, according to Boyle's law, if the value of "k" is 42.5 kPa · L and the pressure is 99.5 kPa.

16. A bottle of carbonated water has a volume of 330 mL and pressure of 105 kPa at 22.0 °C. Find the pressure inside the bottle if it sits in the sun for 2 hours and the temperature of the water rises to 50.0 °C. (Neglect any solubility of CO_2 in water at the high temperature.)

17. A gas balloon containing 4.35 L of helium is cooled to 0 °C from its initial temperature of 19.5 °C at constant pressure. What is the volume of the balloon at the lower temperature?

18. A sample of dry ice, $CO_2(s)$, with a mass of 11.0 g is sealed into a plastic bag and allowed to sublime to $CO_2(g)$. What volume would the plastic bag occupy at SATP?

19. What volume of ammonia gas can be produced by reacting 2.8 L of hydrogen gas with an excess nitrogen gas, when all gas volumes are measured at the same temperature and pressure?

20. Predict the volume of a 2.5 L gas sample when the pressure is tripled and the Kelvin temperature is quadrupled.

21. A sample of hydrogen sulfide gas has a mass of 8.52 g. The volume of the sample is measured to be 5.13 L at −10.0 °C and 97.2 kPa. Calculate the experimental value for the molar volume of hydrogen sulfide at SATP.

22. The volume of your lungs is about 5.0 L. How many moles of air do you hold in your lungs at 100 kPa and a normal body temperature of 37.0 °C?

23. Lindsay fills a birthday balloon with 0.80 g of helium gas to a volume of 5.0 L. Before she can tie the mouth of the balloon, 0.20 g of the helium escapes. What is the new volume of the balloon?

24. Two samples of gases are connected to each other as shown in the figure below. When the clamp is removed and the gases are allowed to mix, what is the pressure in container A? What is the pressure in container B?

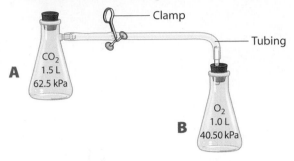

25. An unknown colourless gas is brought to an analytical laboratory for identification. The chemist performs several tests on different samples of the gas and records the following data:

Mass of sample 1	2.51 g
Volume of sample 1	3.06 L
Pressure of sample 1	100.5 kPa
Temperature of sample 1	22.4 °C
Solubility of sample 2	highly soluble in water and ethanol
Litmus test of water solution	litmus paper turns red

a) Find the molar mass of the gas.
b) What is the identity of the gas?
c) Find the density of the gas at SATP.

26. What is the density of argon gas at STP?

27. In a lab group working on designing an investigation to determine the universal gas constant, David suggests collecting the gas in a plastic bag and finding the mass of the plastic bag before and after by using an electronic balance. Anecia wants to find the mass of a pressurized container before and after collecting the gas. With your lab partner, evaluate these two methods. Which method would you choose? Justify your choice.

Making Connections

28. As you have learned in this unit, hydrogen, helium, and hot air may all be used to lift a balloon. Which of these gases would you choose to lift a balloon that would carry you and your family across Alberta? Use your knowledge of gas properties to support you answer.

29. Most cars use an internal combustion engine that involves a four-stroke combustion cycle, as shown in the figure below. In the intake stroke, the piston moves down the combustion cylinder and takes in a mixture of gasoline and air. The piston then moves back up the cylinder, compressing the fuel mixture, which is ignited by a spark from the sparkplug at the top of the stroke. The chemical reaction that occurs is represented by the following unbalanced equation:

$$C_8H_{18}(g) + O_2(g) \rightarrow CO_2(g) + H_2O(g) + energy$$

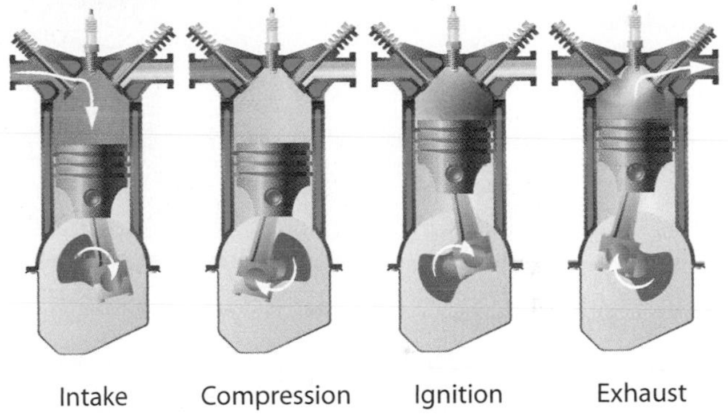

Intake   Compression   Ignition   Exhaust

a) Balance the equation for the combustion of octane, $C_8H_{18}(\ell)$ (a major component in gasoline).
b) Use the law of combining volumes and the combined gas law to explain the explosion that results at the top of the compression stroke.
c) As the piston is driven down to the bottom of the cylinder during the ignition stroke, shown in the figure, the exhaust valve opens and the gaseous products are exhausted out of the tail pipe. One of the exhaust gases is nitrogen monoxide, $NO(g)$, which is partly responsible for the production of smog and acid rain in the atmosphere. Write the balanced chemical equation for the production of nitrogen monoxide. How does the nitrogen become available in the cylinder?

30. As you learned in the Chapter 4 opener, air bags are inflated with nitrogen gas produced by a chemical reaction when a sensor detects a collision. Which gas law(s) is/are important in the operation of an airbag? Do you expect air bags to operate equally well during a winter day in Fort McMurray as during a summer day in Lethbridge? Why or why not?

Matter as Solutions, Acids, and Bases

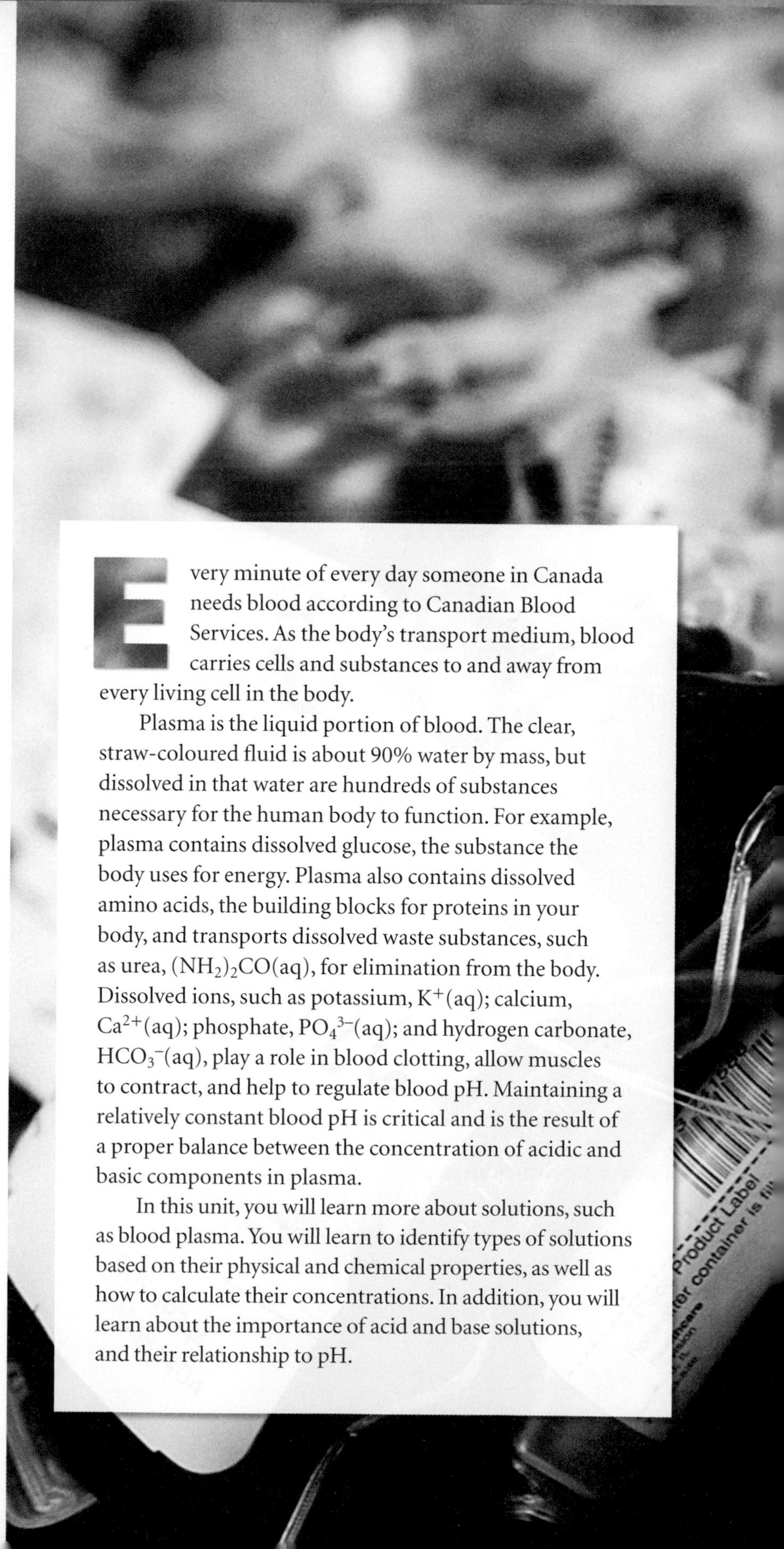

General Outcomes

- investigate solutions and describe their physical and chemical properties

- describe acid and base solutions qualitatively and quantitatively

Unit Contents

Focussing Questions

1. How can you differentiate among matter as solutions, acids, and bases, using theories, properties, and scientific evidence?

2. Why is it important to understand the roles of solutions, acids, and bases in your daily life and in the environment?

Unit PreQuiz ⦿

www.albertachemistry.ca

Every minute of every day someone in Canada needs blood according to Canadian Blood Services. As the body's transport medium, blood carries cells and substances to and away from every living cell in the body.

Plasma is the liquid portion of blood. The clear, straw-coloured fluid is about 90% water by mass, but dissolved in that water are hundreds of substances necessary for the human body to function. For example, plasma contains dissolved glucose, the substance the body uses for energy. Plasma also contains dissolved amino acids, the building blocks for proteins in your body, and transports dissolved waste substances, such as urea, $(NH_2)_2CO(aq)$, for elimination from the body. Dissolved ions, such as potassium, $K^+(aq)$; calcium, $Ca^{2+}(aq)$; phosphate, $PO_4^{3-}(aq)$; and hydrogen carbonate, $HCO_3^-(aq)$, play a role in blood clotting, allow muscles to contract, and help to regulate blood pH. Maintaining a relatively constant blood pH is critical and is the result of a proper balance between the concentration of acidic and basic components in plasma.

In this unit, you will learn more about solutions, such as blood plasma. You will learn to identify types of solutions based on their physical and chemical properties, as well as how to calculate their concentrations. In addition, you will learn about the importance of acid and base solutions, and their relationship to pH.

Figure P3.1 What can you determine about a clear liquid by looking at it? Can you tell if it is acidic, basic, or neutral? Could you tell if it contains a dissolved ionic or molecular compound? What tests would you have to do to identify the liquid?

What is in the flask? Is it a mixture or a pure substance? If it is a mixture, is it a heterogeneous or homogeneous mixture? If it is a pure substance, is it an element or a compound? How could you determine the identity of the liquid?

First, you have to know the definitions of all of the terms. Then you would have to know some of the properties of pure substances and of mixtures, as well as the properties of heterogeneous and homogeneous mixtures. Once you know the properties, you would perform tests to determine the nature of the liquid. How well do you remember all of the definitions and properties that you learned about in previous science courses? Figure P3.2 should help you remember.

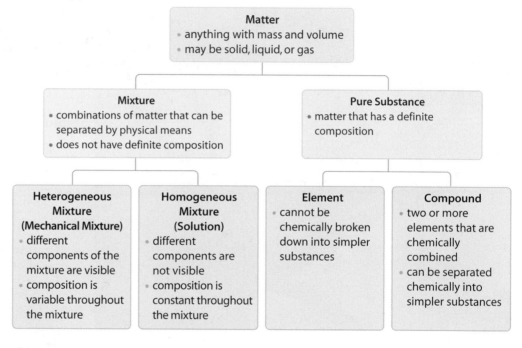

Figure P3.2 Classifying forms of matter makes it easier to understand and remember the definitions and properties of many chemicals.

In this unit, you will be focussing on solutions. Recalling the properties of compounds likely to be dissolved in a solvent will provide a solid foundation on which to study solutions.

Properties of Ionic and Molecular Compounds

Ionic compounds consist of positively and negatively charged ions combined in a ratio that results in a net zero charge. Many ionic compounds are soluble in water. You can predict whether a specific ionic compound is soluble by consulting Table P3.1. The following steps guide you through the use of the table. Then, the example using the table will refresh your memory.

• Find the negatively charged ion in the top row of the table. (Notice that there are three positively charged ions in the second column. All compounds containing these ions are soluble.)

• Look for the positively charged ion in the column below the negatively charged ion.

• If the positively charged ion is in the second row, the compound is soluble in water.

• If the positively charged ion is in the third row, the compound is insoluble in water.

Table P3.1 Solubility of Some Common Ionic Compounds in Water at 298.15 K

Ion / Solubility	$H^+, Na^+, NH_4^+, NO_3^-, ClO_3^-, ClO_4^-, CH_3COO^-$	F^-	Cl^-, I^-, Br^-	SO_4^{2-}	$CO_3^{2-}, PO_4^{3-}, SO_3^{2-}$	$IO_3^-, C_2O_4^{2-}$	S^{2-}	OH^-
Solubility greater than or equal to 0.1 mol/L (**very soluble**)	most Except: $RbClO_4$, $CsClO_4$, $AgCH_3COO$, $Hg_2(CH_3COO)_2$	most	most	most	H^+, Na^+, K^+, NH_4^+ Except: Li_2CO_3	$H^+, Na^+,$ $K^+, NH_4^+,$ $Li^+, Ni^{2+},$ Zn^{2+} Except: $Co(IO_3)_2,$ $Fe_2(C_2O_4)_3,$	H^+, Na^+, K^+, NH_4^+, Li^+, Mg^{2+}, Ca^{2+}	H^+, Na^+, K^+, NH_4^+, Li^+, Sr^{2+}, Ca^{2+}, Ba^{2+}
Solubility less than 0.1 mol/L (**slightly soluble**)	none	Li^+, Mg^{2+}, Ca^{2+}, Sr^{2+}, Ba^{2+}, Fe^{2+}, Hg_2^{2+}, Pb^{2+}	Cu^+, Ag^+, Hg_2^{2+}, Hg^{2+}, PbI_2	Ca^{2+}, Sr^{2+}, Ba^{2+}, Hg_2^{2+}, Pb^{2+}, Ag^+	most	most	most	most

Example

Determine the solubility of silver chloride.

Solution

The chloride ion is in the fourth column. Silver is in the third row below chloride and lists ions that, when combined with chloride (or bromide or iodide), are not soluble. Therefore, silver chloride is insoluble in water and you could write it as $AgCl(s)$.

• • •

The physical state of a compound, at standard temperature, is an indication of the strength of the forces that attract the particles to one another. If the attractive forces are very large, the compounds will be solid. Ionic compounds have very high melting points and are, therefore, solid at room temperature. This property demonstrates that the bonds between ions are very strong.

If a compound can conduct electric current, the positive and negative charges must be able to move freely past one another. Such compounds are called electrolytes. In the solid state, ionic compounds do not conduct electric current. However, in the molten state or dissolved in water, ionic compounds conduct electric current and are thus electrolytes. You have probably tested the conductivity of a solution with an instrument similar to the one in Figure P3.3.

Molecular compounds consist mainly of non-metal elements covalently bonded to one another. Molecular compounds vary in their physical state at standard temperatures. Some are solid, some are liquids, and some are gases. In general, the bonds between molecules are not as strong as those between the ions of ionic compounds. In addition, molecular compounds are non-electrolytes. They do not ionize and, when dissolved in water, they do not conduct electric current. In fact, many molecular compounds are not soluble in water.

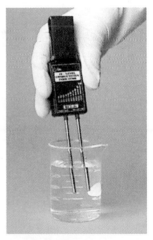

Figure P3.3 By placing the electrodes of this conductivity meter in a solution, you can determine whether the solution contains a dissolved electrolyte. If the meter registers any electric current, an electrolyte is present in the solution.

Acids and Bases

The first chemist to propose a theory of acids and bases was Svanté Arrhenius

ChemistryFile

Web Link

Blood and the solutions in and around living cells must be near pH 7. Have you wondered how the lining of the stomach can tolerate a pH of 3? Find out.

www.albertachemistry.ca
WWW

(1859–1927). An Arrhenius acid is a compound that, when dissolved in water, produces a hydrogen ion, $H^+(aq)$. For example, when hydrogen chloride gas dissolves in water, it ionizes into a hydrogen ion and a chloride ion.

$$HCl(g) \rightarrow H^+(aq) + Cl^-(aq)$$

As the concentration of hydrogen ions in water increases, the solution becomes more acidic.

An Arrhenius base is a compound which, when dissolved in water, dissociates to release a hydroxide ion, $OH^-(aq)$. For example, when potassium hydroxide is dissolved in water it dissociates into a potassium ion and a hydroxide ion.

$$KOH(s) \rightarrow K^+(aq) + OH^-(aq)$$

As the concentration of hydroxide ions in water increases, the solution becomes more basic.

pH Scale

The degree of acidity or basicity of a solution is reported in units of pH. The pH scale goes from 0 to 14, where 7 is neutral—neither acidic nor basic. If a solution has a pH below 7, it is acidic. If the pH is above 7, the solution is basic. You probably recall having seen diagrams that give examples of the acidity of some common solutions.

Measuring pH

Many compounds, when dissolved in water, will be one colour (or colourless) over one range of pH and a different colour over another range of pH values. For example, phenolphthalein is colourless from pH 0 to pH 8.2. At pH 8.2, it begins to turn pink. As the pH increases to 10, the pink colour becomes stronger. Between pH 10 and 14, the colour remains a strong pink. Compounds such as phenolphthalein, which change colour over a specific range of pH values, are called indicators. The colours and pH ranges for several other indicators are shown in Appendix G.

One of the most common indicators is litmus paper. An absorbent paper is soaked in a solution of litmus and dried. When you place a drop of the solution that you are testing onto the pH paper, it changes colour. When an acidic solution touches blue litmus paper, it turns red. When a basic solution touches red litmus paper, it turns blue.

Modern pH paper is made by soaking the paper in several different indicators and drying it. The paper can then assume many different colours, depending on the pH of the solution that moistens the paper. Figure P3.4 shows you how to use pH paper.

As you can see in Figure P3.4, using pH paper gives you a rough estimate of the pH of a solution. If you want a precise measurement, you need to use a pH meter such as the one in Figure P3.5. A good pH meter can give you an accurate reading to two decimal places.

ChemistryFile

FYI

Litmus is a compound extracted from lichens, the organisms that you sometimes see attached to rocks. Lichens are actually two organisms, fungus and algae, living in a symbiotic relationship. Lichens can survive in environments where few other forms of life can exist. The litmus used for pH paper is almost exclusively prepared in the Netherlands from a few specific species of lichens that grow there.

Figure P3.4 Place the pH paper on a surface such as glass. With a dropper, drop a small amount of the unknown solution of the paper. Compare the colour of the moist litmus paper with a scale such as the one in this photograph.

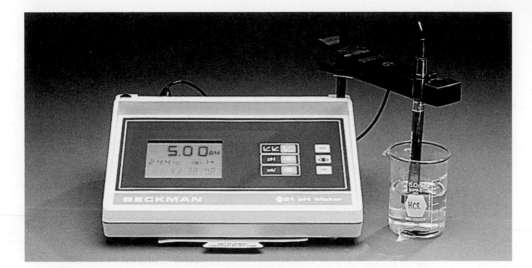

Figure P3.5 Place the probe of the pH meter in the unknown solution. The tip of the probe is sensitive to the hydrogen ion concentration and will allow a current to flow that is proportional to the pH of the solution. The current is registered on a meter or digitally as in this example. pH meters must be calibrated by placing the probe in solutions of known pH and adjusting the meter reading to that pH.

Naming Acids and Bases

You have probably heard the names hydrochloric acid and sulfuric acid, two of the most common acids. Do you remember what the names mean and how to assign names to acids when you are given a formula? The following is a review of the naming of acids and bases that you learned about in previous science courses.

The acids with which you will be working will always consist of a hydrogen ion and a negatively charged ion. The negatively charged ion might consist of a single non-metal atom or a polyatomic ion that contains a non-metal and one or more oxygen atoms. The root of the name of the acid is derived from the non-metal element that is involved. For example, the root name for acids containing chlorine atoms is "chlor." First, you identify the name of the negatively charged ion and then follow the rules in the following table.

Table P3.2 Naming Acids

Name of pure substance (not dissolved in water)	Name of acid	Example
hydrogen (root) ide	hydro (root) ic acid	hydrogen bromide hydrobromic acid
hydrogen hypo (root) ite	hypo (root) ous acid	hydrogen hypochlorite hypochlorous acid
hydrogen (root) ite	(root) ous acid	hydrogen nitrite nitrous acid
hydrogen (root) ate	(root) ic acid	hydrogen sulfate sulfuric acid
hydrogen per (root) ate	per (root) ic acid	hydrogen periodate periodic acid

The names of bases are uncomplicated. Arrhenius bases contain a metal ion or a positively charged polyatomic ion and a hydroxide ion. (The only positively charged polyatomic ion on your list is ammonium.) The name of the base starts with the name of the metal (or ammonium) and ends with "hydroxide." For example, the name of $KOH(aq)$ is potassium hydroxide. The name of $NH_4OH(aq)$ is ammonium hydroxide.

Prerequsite Skills

- **calculate** mass using the molar mass of compounds

- **demonstrate** procedures for safe handling, storing, and disposal of acids and bases

- **demonstrate** proper use of a pH meter and laboratory glassware

Solutions

Chapter Concepts

5.1 Classifying Solutions

- The process of dissolving can be endothermic or exothermic.

- Substances can be classified as electrolytes or non-electrolytes.

- Dissolving substances in water is often a prerequisite for chemical change.

5.2 Solubility

- Solubility is affected by several factors and can be determined in a laboratory.

- Solutions can be described as saturated, unsaturated, or supersaturated.

5.3 The Concentration of Solutions

- The concentration of a solution and its ions can be calculated and expressed in a variety of ways.

- Concentration values can be used in a risk–benefit analysis to determine safe limits of particular solutes.

5.4 Preparing and Diluting Solutions

- Solutions can be prepared and diluted to specific concentrations with proper procedures and calculations.

The environment is made up of many important solutions. The air you breathe is a solution. Many of the metallic objects you use every day are also solutions. Even the water you drink is a solution. As water runs through soil and rocks, it dissolves substances, such as iron, calcium, and magnesium. The Bow River, shown in the photograph, contains many such dissolved substances. The water from your tap contains small amounts of dissolved substances. Some, such as iron or calcium, occur naturally. Some, such as chlorine, have been added to keep the water free of bacteria and safe to drink.

The difference between drinkable water and undrinkable water depends on what is dissolved in it and how much. For example, most tap water contains small amounts of fluoride, which has been added to help keep your teeth healthy. Water with a high amount of fluoride, however, is poisonous. In the Launch Lab on the facing page, you will observe how the identity and amount of substances in a solution affect that solution.

Launch Lab

Sink or Float?

A can of cola and a can of diet cola share many physical properties. What properties do they share? What properties make them different from each other? What do you think will happen when a can of cola is placed in water? What will happen to a can of diet cola placed in water?

Materials

- 1 can of diet cola (unopened)
- 1 can of regular cola (unopened)
- aquarium or large tub
- water
- 500 mL beakers (2)

Procedure

1. Observe the two cans of cola. Compare their physical properties. Read Procedure step 2 and predict what will happen.

2. Watch as your teacher places the two unopened cans in the tub of water.

3. Record your observations.

4. Wait as your teacher opens each can and pours the contents into a separate beaker.

   a) Compare the empty cans. Are there any differences in material or construction? Record your observations.

   b) Compare the volume of the contents of the cans. Record your observations.

Analysis

1. What difference in physical properties could explain your observations when the two cans were placed in water?

2. Account for the difference. How does this difference relate to what you already know about solutions?

Classifying Solutions

In previous courses, you have learned that matter can be classified as either mixtures or **pure substances** and that there can be either heterogeneous or homogeneous mixtures. While the composition of **heterogeneous mixtures** varies throughout the mixture, **homogeneous mixtures** have a uniform composition and are more often called **solutions**. The simplest solutions contain two substances. Most common solutions are mixtures of many substances. Examples of solutions include tap water, filtered air, and iced tea.

A solution consists of a solute or solutes dissolved in a solvent:

- The substance that is present in the largest quantity (whether by volume, mass, or amount) is usually called the **solvent**.
- The substances that are dissolved in the solvent are called the **solutes**.

Both solutions and pure substances can be described as "uniform throughout," so what is the difference between them? Pure substances, such as distilled water, $H_2O(\ell)$, have fixed composition. Distilled water is always 11% hydrogen and 89% oxygen by mass. Solutions, on the other hand, have variable composition. You can mix solutes and solvents together in varying ratios. For example, Figure 5.1 shows two different solutions of iced-tea mix and water. The ratio of solute to solvent in solution A is higher than the ratio of solute to solvent in solution B. Each solution, however, is uniform throughout.

Figure 5.1 How can solutions have variable compositions yet be uniform throughout?

Types of Solutions

When you think of a solution, you probably think of a solid such as salt or sugar dissolved in water. Certainly, water solutions are by far the most common. However, solutions encompass a much wider variety of substances. You might recall from Unit 2 that one property of gases is miscibility. All gases are miscible, meaning that they will all mix together. A homogeneous mixture of gases is a form of a solution. For example, air is a gaseous solution. The main components of dry air (having no water vapour) are nitrogen (78%), oxygen (21%), argon (0.9%), and carbon dioxide (0.03%). Various other gases account for the remaining 0.07% of air. Gases can also dissolve in liquids.

The fizzing that you hear when you open a bottle of a soft drink is the carbon dioxide that is coming out of solution in the water. Solids can also dissolve in solids. Many of the metals that you use every day, such as knives, forks, and spoons, are uniform mixtures of two or more metals. Steel is a solid solution of carbon and other trace elements in iron. Any combination of gases, liquids, and solids can form solutions in each other. Each of the images in Figure 5.2 represents a solution. Examine each image and try to determine what substance is dissolved in what other substance.

Figure 5.2 Try to identify the solutes and solvents in these solutions.

Table 5.1 Types of Solutions

Original state of solute	State of solvent	Examples
gas	gas	air (oxygen, argon, carbon dioxide, and water vapour in nitrogen) natural gas (ethane, propane, and other trace gases in methane) thermogenic gas also contains butane, n-butane, pentane, and n-pentane welding fuel (oxygen and acetylene)
gas	liquid	carbonated drinks (carbon dioxide dissolved in water) river water (contains dissolved oxygen)
gas	solid	hydrogen in platinum
liquid	liquid	alcohol in water ethylene glycol (antifreeze) in water
liquid	solid	amalgams, such as mercury in silver
solid	liquid	sugar in water brine (table salt in water)
solid	solid	copper-nickel alloy (used in coins) brass (a copper-zinc alloy)

Chemistry File

FYI
Colloids, such as milk and fog, may appear to be solutions. They are actually heterogeneous mixtures, but the clumps of particles are so small that you cannot see them without a magnifying glass or a microscope. Other examples of colloids include mayonnaise, marshmallows, shaving foam, and smoke.

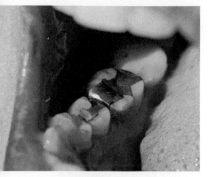

Figure 5.3 Fillings were once made of amalgam. Today, due in part to toxicity studies performed at the University of Calgary in the late 1980s, most dental fillings are made of other substances like ceramics.

Solid Solutions: Alloys and Amalgams

An alloy is a solid solution of metals. Adding even a small quantity of another element to a metal changes the properties of the metal. For example, bronze is an alloy of copper and tin. Although bronze is only about 10 percent tin, it is much stronger than copper and more resistant to corrosion.

Amalgams are solutions of mercury in other metals. Since mercury is a liquid at room temperature, an alloy of mercury in other metals is classified as a liquid in a solid solution. Amalgams were once commonly used for dental fillings, as shown in Figure 5.3, and to coat the back of mirrors.

• • •

1 Classify the following as elements, compounds, heterogeneous mixtures, or solutions (homogeneous mixtures):

a) air	**c)** distilled water	**e)** oxygen
b) brass	**d)** tap water	**f)** concrete

2 Identify the solvent and at least one solute in the following solutions:

a) soda water	**c)** steel
b) antifreeze	**d)** natural gas

• • •

Solutions in Water

You are probably most familiar with liquid solutions, especially aqueous solutions. An **aqueous solution** is any solution for which water is the solvent. Water is sometimes called the universal solvent because so many substances, both ionic and molecular, dissolve in water. Most solutions used in chemistry and most liquid solutions you encounter every day are aqueous solutions. Because aqueous solutions are so important, you will focus on them for the remainder of this chapter and the next.

Dissolving and Forces of Attraction

When a solid solute dissolves in a liquid solvent, such as water, it may appear as if a chemical reaction occurs, since the solute seems to "disappear." In reality, the change is physical, not chemical. The molecules or ions of a solid solute are held together by bonds. When a solute dissolves, these bonds are broken and the molecules or ions of solute become attracted to the particles of the solvent (water molecules, for example). The solute particles then disperse throughout the solution, and your unaided eye cannot distinguish solute from solvent. For example, when you add iced-tea mix to water and stir, at first you can see the granules of sugar and flavouring swirling in the water. Soon, however, they dissolve, and all you can see is a uniform, translucent, tea-coloured solution.

Consider the three processes that occur when a solid solute dissolves in a solvent. Figure 5.4 models the processes for an ionic solid and for a molecular solid dissolving in water. Notice that the positively charged ends of the water molecules are attracted to the negatively charged chloride ions. Eventually, water molecules surround the chloride ions and draw them away from the salt crystal. Likewise, the negatively charged ends of the water molecules are attracted to the positively charged sodium ions and surround them. Through this process, ionic substances are dissolved in water. The sugar in Figure 5.4 B is not ionic but sugar molecules have many polar groups within each molecule. The partially charged ends of the water molecules are attracted to the oppositely charged poles of the sugar molecules. Water molecules form hydrogen bonds with polar atoms on the sugar molecules and eventually surround the molecules, drawing them into solution. The dissolving process is summarized in the following box.

ChemistryFile

Web Link
Using mercury for dental fillings may seem strange, since mercury is such a toxic compound, and the safety of using amalgam was a hotly debated topic in the 1990s. What argument did each side make? What role did researchers at the University of Calgary play?

www.albertachemistry.ca
WWW

Processes of Dissolving

- The forces between the particles in the solute break. This process always *requires* energy. Recall that processes that require energy are called **endothermic** processes. In an ionic solid, the forces holding the ions together are ionic bonds. In a molecular solid, the forces holding the molecules together are intermolecular forces, such as dipole-dipole bonds.

- Some of the intermolecular forces between the particles in the solvent (often water) also break. For example, in an aqueous solution, some of the hydrogen bonds between water molecules break. Because this process involves breaking bonds, it is an *endothermic* process.

- The attraction between the particles of solute and the particles of solvent result in the formation of new chemical bonds. This step always *releases* energy. Processes that release energy are called **exothermic** processes. For aqueous solutions, the dissolving process is sometimes called *solvation* or *hydration*.

A Sodium chloride, NaCl(s) dissolves   **B** Sucrose, $C_{12}H_{22}O_{11}$(s), dissolves

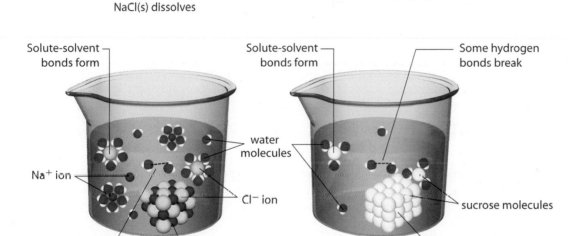

Figure 5.4 When a solute dissolves in a solvent, such as water, the three processes of dissolving occur simultaneously.

Energy and Dissolving for Aqueous Solutions

Using the law of conservation of energy, the overall energy change observed when a solute dissolves in a solvent is the sum of the energy changes in the three steps.

	Energy required to break bonds in solute
+	Energy required to break bonds in solvent
−	Energy released from bonds forming between solute and solvent
=	Total energy change

If the amount of energy required to break the bonds is more than the amount of energy released when the bonds are formed, the overall change is endothermic. If, however, more energy is released when solvent–solute bonds form than is required to break solute–solute and solvent–solvent bonds, the overall change is exothermic. For most substances that dissolve in water, the process is exothermic. For example, when sodium hydroxide, NaOH(s), dissolves in water, the solution becomes quite hot because the process is releasing energy to the solution. For other substances that dissolve in water, however, the process is endothermic. When ammonium nitrate, $NH_4NO_3(s)$, dissolves in water, the solution cools down significantly because the dissolving process is absorbing energy from the solution.

A homemade or commercially available cold pack can be made using ammonium nitrate, $NH_4NO_3(s)$, in a plastic sack within a bag of water. Examine Figure 5.5. When the ammonium nitrate sack is broken and the solute dissolves in the water, the energy required for this endothermic physical change is absorbed from the solution, cooling it down.

Figure 5.5 A typical cold pack contains two chambers. One chamber contains a salt, such as ammonium nitrate, that dissolves endothermically. The other chamber contains water. Crushing the pack allows the salt to dissolve in the water, cooling it down.

• • •

3 List the three processes that occur when a solute is dissolving. Classify each process as endothermic or exothermic.

4 Why are some overall dissolving processes exothermic while others are endothermic?

5 What is one possible reason why one substance dissolves in water and another substance does not? Use the term "forces of attraction" in your answer.

• • •

Properties of Aqueous Solutions

It is often difficult to discriminate visually between aqueous ionic solutions and aqueous molecular solutions. For example, an aqueous solution of sugar—sucrose, $C_{12}H_{22}O_{11}(aq)$—or rubbing alcohol—propan-1-ol, $C_3H_7OH(aq)$—looks the same as a solution of table salt—sodium chloride, NaCl(aq)—or lye—sodium hydroxide, NaOH(aq). They are all clear and colourless. Although these solutions look alike, they have very different properties. In Investigation 5.A on page 174, you will design a procedure to classify solutions as ionic or molecular based on physical properties.

Electrolytes and Non-electrolytes

As you read in the preparation for this unit, ionic compounds conduct electric current while molecular ones do not. When these substances are dissolved in water to make aqueous solutions, the solution has the same conductivity property as the solute. Thus, ionic compounds, when dissolved, form solutions that conduct electric current. Most molecular compounds, when dissolved, form solutions that do not conduct electric current. As you know, substances that dissolve in water to form solutions that conduct electric current are called **electrolytes**. Substances that do not conduct electric current when they dissolve in water are called **non-electrolytes**.

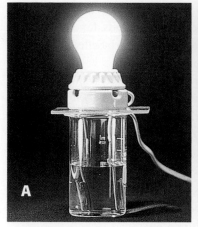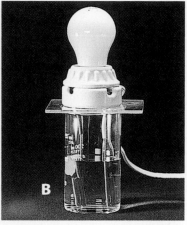

Figure 5.6 A solution of table salt in water conducts electric current **(A)**. Table salt is an electrolyte. A solution of sugar and water does not conduct electric current **(B)**. Sugar is a non-electrolyte.

Ionic Compounds in Solution

Look back at Figure 5.4. When an ionic compound dissolves in water, ionic bonds break, freeing ions. Because these free ions are charged and can move around in solution, they are able to conduct electric current. Chemical equations called dissociation equations represent this process. A **dissociation equation** shows what happens when compounds dissolve. For example, when an ionic compound—such as sodium chloride, $NaCl(s)$, or barium hydroxide, $Ba(OH)_2(s)$—dissolves in water, both positive and negative ions are freed, as shown in the dissociation equations below:

$$NaCl(s) \xrightarrow{\text{in water}} Na^+(aq) + Cl^-(aq)$$

$$Ba(OH)_2(s) \xrightarrow{\text{in water}} Ba^{2+}(aq) + 2OH^-(aq)$$

Ionic compounds separate into charged particles when they dissolve. Ionic compounds that dissolve in water, therefore, are electrolytes.

$Ba(OH)_2(s)$ is comprised of one Ba^{2+} ion for every two OH^- ions. When dissolved, the ions will be produced in this ratio. Notice that the charges on each side of the equations balance.

Soluble and Insoluble Ionic Compounds

All ionic compounds that dissolve in water conduct electric current. Not all ionic compounds dissolve appreciably, however. Compounds that dissolve in a given solvent are **soluble**. Compounds that do not dissolve to any great extent in a given solvent are **insoluble** (or **slightly soluble**). Soluble ionic compounds, such as sodium chloride, $NaCl(s)$, and barium hydroxide, $Ba(OH)_2(s)$, dissociate in water and form electrolytes, as described above. Insoluble ionic compounds, such as silver chloride, $AgCl(s)$, and magnesium hydroxide, $Mg(OH)_2(s)$, do not dissolve appreciably in water. They do not dissociate and, therefore, do not form ions in solution.

$$AgCl(s) \xrightarrow{\text{in water}} AgCl(s)$$

$$Mg(OH)_2(s) \xrightarrow{\text{in water}} Mg(OH)_2(s)$$

When insoluble ionic compounds are mixed with water, they do not dissolve. The solid state symbol, (s), does not change, and the resulting heterogeneous mixture does not conduct electric current.

Solubility Guidelines

When you write a dissociation equation or any equation involving ionic compounds, you need to know whether a compound is soluble in order to include the proper state—solid (s) or aqueous (aq)—in the equation. As you know, when some ionic compounds are mixed together, one of the products might form a precipitate because it is not soluble in water. For example, in the following double-replacement reaction, would either of the products form a precipitate? If so, it should be labelled "s" for solid.

$$AgNO_3(aq) + NaCl(aq) \longrightarrow NaNO_3(?) + AgCl(?)$$

The chart on page 161, that you reviewed in the Unit 3 Preparation, provides you with guidelines that allow you to predict the state of many ionic compounds. If you consult Table P3.1, you will find that sodium nitrate ($NaNO_3(aq)$) should be labelled aqueous and that silver chloride ($AgCl(s)$) should be labelled solid.

If you look closely, you will notice that the label on the lower left says "slightly soluble." To be precise, all ionic compounds are soluble to some extent. However, for many compounds, the amount of the substance that will go into solution is so small that it can often be ignored and the compound can be considered insoluble. There is no clear distinction between "very soluble," "slightly soluble," and "insoluble." The data in Table P3.1 must be considered guidelines and not definitive. In fact, if you refer to other textbooks or sources in the Internet, you will find some differences in the charts. Chemists have not reach a firm agreement about the cutoff limit for calling a compound "soluble." Table 5.2 lists some examples of data from three sources.

Table 5.2 Arbitrary Cutoff Limits for Solubility from Three Sources of Data

| Source | Degree of solubility | | |
Solubility	Table P3.1	Source A	Source B
Soluble	at least 0.1 mol/L	at least 0.1 mol/L	> 1g/100 mL
Slightly soluble	less than 0.1 mol/L	0.1 mol/L to 0.01 mol/L	0.01 g/100 mL to 1 g/100 mL
Insoluble	—	less than 0.01 mol/L	< 0.01 g/100 mL

Inspection of Table P3.1 reveals some trends that simplify making predictions. Nearly all sources of solubility data agree with the following general statements.

- All compounds containing hydrogen, sodium, potassium, or ammonium ions are soluble.

- All compounds containing nitrates are soluble.

These generalizations should help you determine the solubilities of many compounds.

• • •

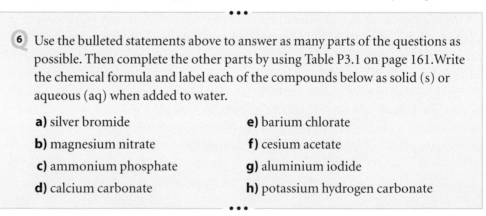

6 Use the bulleted statements above to answer as many parts of the questions as possible. Then complete the other parts by using Table P3.1 on page 161. Write the chemical formula and label each of the compounds below as solid (s) or aqueous (aq) when added to water.

a) silver bromide   **e)** barium chlorate

b) magnesium nitrate   **f)** cesium acetate

c) ammonium phosphate   **g)** aluminium iodide

d) calcium carbonate   **h)** potassium hydrogen carbonate

• • •

Molecular Compounds in Solution

Most molecular compounds are non-electrolytes. Molecular compounds consist of molecules held together by intermolecular forces, such as London (dispersion) forces or dipole-dipole forces. When a molecular compound dissolves, these bonds break and the solute molecules disperse throughout the solvent. For example, when molecular compounds, such as propan-1-ol, $C_3H_7OH(\ell)$, or sucrose, $C_{12}H_{22}O_{11}(s)$, dissolve in water, the solute molecules form intermolecular bonds with water molecules and disperse throughout the solution.

Figure 5.7 Waxed paper is coated with paraffin wax. Waxed paper does not fall apart when it gets damp the way that unprotected paper does, because paraffin wax is insoluble in water.

When molecular compounds dissolve, molecules separate in water. This is indicated by the notation (aq). Only the state changes.

Insoluble molecular compounds, such as paraffin wax, $C_{25}H_{52}(s)$, do not dissolve in water. The molecules do not separate, and the state does not change.

$$C_3H_7OH(\ell) \xrightarrow{\text{in water}} C_3H_7OH(aq)$$

$$C_{12}H_{22}O_{11}(s) \xrightarrow{\text{in water}} C_{12}H_{22}O_{11}(aq)$$

Molecules are not charged and, therefore, do not conduct electric current. As a result, most molecular compounds are non-electrolytes when dissolved in water. Molecular elements, such as $H_2(g)$ and $O_2(g)$, are also non-electrolytes.

As with ionic compounds, not all molecular compounds are soluble in water. In fact, many are not. Paraffin wax, $C_{25}H_{52}(s)$, is one example (see Figure 5.7).

Acids: Molecular Electrolytes

Not all molecular compounds are non-electrolytes, however. For example, hydrogen chloride, $HCl(g)$, is a molecular compound, yet an aqueous solution of hydrogen chloride conducts electric current. That is because hydrogen chloride ionizes (forms ions) in water:

$$HCl(g) \xrightarrow{\text{in water}} H^+(aq) + Cl^-(aq)$$

You probably recognize $HCl(aq)$ as hydrochloric acid. Acids are electrolytes because they ionize (form ions) in water.

Electrolytes in Sports Drinks

You may have heard the term *electrolytes* used in advertising for sports drinks (see Figure 5.8). Your body requires electrolytes, such as sodium, potassium, and calcium ions, for muscle contraction, nerve conduction, waste removal, blood pressure maintenance, and many other functions. Your body also requires the most important non-electrolyte, glucose, as an energy source. If a person loses a significant amount of fluid in a short time through sweating, for example, during intense exercise, the balance of electrolytes and glucose is upset. Sports drinks are designed to provide both the solutes (electrolytes and glucose) and solvent (water) required by the body. Bleeding, vomiting, or diarrhea can upset the electrolyte and non-electrolyte balance of the body, too. For this reason, sports drinks can help a person who has the flu or other illness to replenish fluids, glucose, and electrolytes.

Solutions and Chemical Reactions

Water, the solvent in our body, not only acts as a transport medium for electrolytes and glucose but is also often required for chemical reactions to occur. Solute particles separate, disperse, and collide with particles of other solutes when they dissolve. Therefore, reactions that would occur only very slowly with the undissolved solutes can and do happen. Both double-replacement and single-replacement reactions take place in aqueous solutions (see Figure 5.9). In Chapter 7, you will learn more about reactions in solution.

Figure 5.8 Sports drinks not only provide the water to replace the fluids lost during sweating, but many also contain a variety of vitamins, minerals, and sugars.

Chemistry File

Web Link
Use the Internet to investigate the ingredients in sports drinks and their use in the body. Present your findings as a consumer report. What ingredients do sports drink share? How are the drinks different from one another? Is it possible or advisable to mix homemade sports drink? What are the pros and cons of drinking sports drinks compared with drinking water?

www.albertachemistry.ca
WWW

Figure 5.9 In the solid state, silver nitrate and sodium chromate do not react when mixed. When dissolved, however, both form solutions that when mixed together react to produce a solid precipitate, silver chromate.

Classifying Solutions

How can you classify a solution by studying its physical properties? Ionic and molecular compounds have distinct properties, as you learned in Chapters 1 and 2. What about their properties in solution? In this investigation, you will examine several unknown solutions to classify them as either ionic or molecular based on their conductivity properties.

Question

How do the properties of aqueous solutions of ionic and molecular compounds differ?

Prediction

Before carrying out the procedure, classify the solutes of the solutions below as ionic or molecular. From your knowledge of the forces of attraction in ionic and molecular compounds, predict the difference in conductivity properties of the solutions they form.

Safety Precautions

• Since the solutions are unlabelled, treat them all as though they were toxic and corrosive. Wash any spills on skin or clothing with plenty of cool water and inform your teacher immediately.

• If you are using a conductivity tester with two separate electrodes, be extremely careful to keep the electrodes well separated while you perform your tests.

• When you have completed this investigation wash your hands.

Materials

• distilled water
• unidentified solutions, labelled one to five, of:
  – 1-propanol, $C_3H_7OH(aq)$ Ⓐ Ⓣ
  – hydrochloric acid, $HCl(aq)$ Ⓣ ⊡
  – sodium chloride, $NaCl(aq)$
  – sodium hydroxide, $NaOH(aq)$ Ⓣ ⊡
  – sucrose, $C_{12}H_{22}O_{11}(aq)$
• conductivity tester
• other apparatus and materials as required
• 50 mL beakers (6)

Procedure

1. Construct a table, similar to the one below, to record your results.

Solution	Conductivity
Distilled water	
1	
2	
3	

2. Label the beakers one to five. Use one beaker for distilled water, which you will use as a control.

3. Add 25 mL of unknown solution into the appropriately numbered beaker.

4. Test the distilled water control with the conductivity tester and record your results.

5. Test each of the unknown solutions with the conductivity tester and record your results.

Analysis

1. Classify the five solutions based on their conductivity properties.

2. Explain your observations using your knowledge from Chapters 1 and 2 of ionic and molecular compounds and the properties of solutions they form.

3. Why do you think you needed to use distilled water as a control?

Conclusion

4. **a)** Did your results match your prediction? Suggest explanations for unexpected results.

   b) Using your results, describe the properties of solutions of molecular compounds and ionic compounds.

Extensions

5. Perform Internet research to find two uses for each solution in this investigation. ICT

6. In this investigation, you classified a solution as ionic or molecular. What test or tests would you need to perform to identify the solutions of sodium chloride, $NaCl(aq)$; lye, $NaOH(aq)$; 1-propanol, $C_3H_7OH(aq)$; hydrochloric acid, $HCl(aq)$; and sucrose, $C_{12}H_{22}O_{11}(aq)$?

7 Are the following substances electrolytes or non-electrolytes? Write equations to illustrate your answers.

a) KCl(s) in water

b) AgCH$_3$COO(s) in water

c) C$_2$H$_5$OH(ℓ) in water

d) Na$_2$SO$_4$(s) in water

e) Na$_2$S(s) in water

f) O$_2$(g) in water

g) Mg(OH)$_2$(s) in water

h) HCl(g) in water

8 Explain why a reaction between silver nitrate, and sodium chromate is very slow when the reactants are in the solid state, but almost instantaneous when they are dissolved in water.

Section 5.1 Summary

- Solutions are homogenous mixtures composed of solutes and solvents, which have a uniform composition throughout.

- There are various types of solutions that can exist as or be composed of solids, liquids, or gases.

- Dissolving involves three processes: (1) the breaking of bonds in the solute; (2) the breaking of bonds in the solvent; and (3) the formation of bonds between the solute and solvent. The overall process of dissolving can be endothermic (if breaking bonds requires more energy than making bonds releases) or exothermic (if making bonds releases more energy than breaking bonds requires).

- Soluble ionic compounds dissolve in water to produce solutions that conduct electric current. These solutes are called electrolytes.

- Most molecular compounds that dissolve in water do not produce solutions that conduct electric current and are called non-electrolytes. Acids are an exception to this because they ionize in solution and, therefore, can act as electrolytes.

SECTION 5.1 REVIEW

1. "Natural gas" is actually a solution of many different gases. The table below shows ranges of composition for a typical sample of natural gas.

  a) What compound is the solvent in natural gas?

  b) Name three solutes in natural gas.

Typical Composition of Natural Gas

Component	Formula	Percentage by mass
methane	CH$_4$(g)	70%–90%
ethane	C$_2$H$_6$(g)	0%–20%
propane	C$_3$H$_8$(g)	0%–20%
butane	C$_4$H$_{10}$(g)	0%–20%
carbon dioxide	CO$_2$(g)	0%–8%
oxygen	O$_2$(g)	0%–0.2%
nitrogen	N$_2$(g)	0%–5%
hydrogen sulfide	H$_2$S(g)	0%–5%
other gases	Ar, He, Ne, Xe	less than 1%

2. Explain why the moth pin in Figure 5.2 can be considered to be an example of a solution.

3. Describe the processes that take place when a teaspoon of sugar dissolves in a cup of coffee.

4. When crystalline urea is dissolved in water, the solution becomes very cold. Explain what happens when urea dissolves in water that causes it to become cold.

5. Explain why dissolving substances in water is often a prerequisite for chemical change.

6. Acids such as HCl(g) are molecular compounds yet when dissolved in water they can conduct electric current. How are they different from the majority of molecular compounds?

7. You are given two jars that both contain a white, crystalline solid. You are told that one of them is table sugar (sucrose) and one is table salt (sodium chloride). Without tasting them, what test could you perform to determine which to put in the sugar bowl and which to put in the salt shaker?

Solubility

Section Outcomes

In this section, you will
- **define** solubility and the factors that affect it
- **explain** a saturated solution in terms of equilibrium (equal rates of dissolving and crystallization)
- **perform** an investigation to determine the solubility of a solute in a saturated solution

Key Terms

solubility
saturated
unsaturated
equilibrium

In Section 5.1, you learned that some substances, such as sodium chloride, NaCl(s), are soluble in water, while others, such as paraffin, $C_{25}H_{52}$(s), are insoluble. The **solubility** of a solute is the amount of a solute that dissolves in a given quantity of solvent, at a given temperature. For example, the solubility of sodium chloride, NaCl(s), in water at 20 °C is 36 g/100 mL. In other words, 36 g of sodium chloride can dissolve in 100 mL of water at 20 °C.

A **saturated** solution is a solution that contains the maximum amount of dissolved solute at a given temperature in the presence of undissolved solute. For example, 100 mL of a saturated solution of sodium chloride at 20 °C contains 36 g of dissolved sodium chloride. If more sodium chloride is added to the solution, it will not dissolve. The solution may still be able to dissolve other solutes, however. Notice that for a solution to be called saturated, some undissolved solute must be present. You will find out why later in this section.

An **unsaturated** solution is a solution that is not saturated and, therefore, can dissolve more solute at that particular temperature. For example, a solution that contains 20 g of sodium chloride dissolved in 100 mL of water at 20 °C is unsaturated. This solution has the potential to dissolve another 16 g of sodium chloride, as shown in Figure 5.10.

A *supersaturated solution* is a solution that contains more dissolved solute than its solubility at a given temperature.

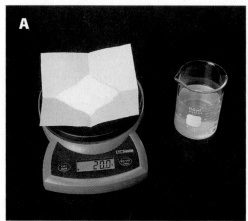

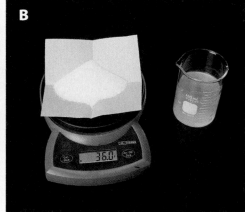

Figure 5.10 At 20 °C, the solubility of NaCl(s) in water is 36 g/100 mL.

A 20 g of NaCl(s) dissolve in 100 mL water to form an unsaturated solution.

B 36 g of NaCl(s) dissolve in 100 mL water to form a saturated solution. (In fact, a tiny amount of excess solute must also be present in the beaker.)

C 40 g of NaCl(s) are added; 36 g dissolve in 100 mL water to form a saturated solution, and 4 g of undissolved solute remain.

9 a) If the solubility of potassium carbonate, $K_2CO_3(s)$, is 94 g/100 mL of water at 20 °C, how much of this solute is dissolved in 200 mL of a saturated solution at 20 °C?

b) How would you describe 500 mL of a solution that contained 94 g of potassium carbonate at 20 °C? (Is it saturated or unsaturated?)

c) A solution of potassium carbonate has a volume of 100 mL and contains 50.0 g of solute at 20 °C. How much more potassium carbonate can dissolve?

d) How much potassium carbonate, in total, is in a beaker containing 100 mL of saturated solution and 10 g of undissolved solid at 20 °C?

Range of Solubility

When you look at the photograph in Figure 5.11, and see the oil from a spill floating on top of the water, you might think of the statement, "Oil and water do not mix." You would conclude that oil is insoluble in water. Is it? In Section 5.1, you read that all ionic compounds dissolve, at least to some extent, in water. Examine Table 5.3 below and decide if the same statement can be said of molecular compounds.

Table 5.3 Solubility of Some Selected Molecular Compounds

Compound	Solubility (g/100 mL) at 25 °C
sucrose ($C_{12}H_{22}O_{11}(s)$)	67.1
glucose ($C_6H_{12}O_6(s)$)	45.0
creatine ($C_4H_9N_3O_2(s)$)	1.6
codeine ($C_{18}H_{21}NO_3(s)$)	0.79
quinine ($C_{20}H_{24}N_2O_2(s)$)	0.057
hexane ($C_6H_{14}(\ell)$)	0.0011
captan ($C_9H_8Cl_3NO_2S(s)$)	0.00005
docosane ($C_{22}H_{46}(s)$)	0.0000006

Figure 5.11 If oil were soluble in water, then oil spills would not form a "slick" on the surface of the ocean. Do you think this would make an oil spill more or less damaging to the environment?

As you can see in Table 5.3, there is a tremendous range of solubilities for molecular substances, even for those of similar molar masses. Nevertheless, all compounds, even the oil in the oil "slick" on the water, dissolve to some extent.

How can you account for the range of solubilities of the compounds listed in Table 5.3? For example, sucrose is nearly 100 times as soluble as codeine (a narcotic) but similar in size. Sucrose has eight −OH groups that can form hydrogen bonds with water molecules. Codeine has one −OH group, but much of the molecule is non-polar. The energy required to break the hydrogen bonds between water molecules cannot be recovered by forming hydrogen bonds with codeine. Similar reasoning explains the extremes in solubility of other compounds listed in the table.

Gases and Solubility

Table 5.4 lists the solubility of several common gases in water. At a glance, you can see that, under nearly any criteria for ionic compounds, gases would have to be considered insoluble. However, you know that aquatic animals survive on the oxygen that is dissolved in lakes, rivers, and oceans. Anything that would reduce the solubility of oxygen in water would be a threat to aquatic animals.

Gas	Solubility (g/100 mL) at 0 °C
He	0.00017
Ne	0.0013
N_2	0.0029
CO	0.0044
O_2	0.0070
NO	0.0098

Table 5.4 Solubility of Some Common Gases in Water at 101.325 kPa

A Closer Look at a Saturated Solution

As you learned in Section 5.1, when a solute dissolves in a solvent, bonds between the solute particles (and some solvent particles) break and bonds between solute and solvent particles form. It may seem reasonable to assume that once the solute–solvent bonds form, they do not break again. Evidence supports another theory, however. Examine Figure 5.12 below, which shows a stoppered flask containing a saturated solution of copper(II) sulfate pentahydrate, $CuSO_4 \cdot 5H_2O(s)$, and undissolved solute.

Figure 5.12 The solubility of copper(II) sulfate pentahydrate, $CuSO_4 \cdot 5H_2O(s)$, is 31.6 g/100 mL of water at room temperature. The solution in photograph **A** was made by adding 40 g of solute to 100 mL of water. Photograph **B** shows the same flask, undisturbed, two weeks later. What differences do you observe?

By examining the saturated solution that has been allowed to sit undisturbed for several weeks, you can see that the undissolved solute at the bottom of the flask has changed in appearance. The powder that was initially present is now in a crystalline form characteristic of that solute.

If you were to separate the undissolved solute from the solution, you would find that the mass remains constant at 8.4 g. Although the same *amount* of solute remains undissolved, the crystals that are present on the bottom of the flask have actually grown from the dissolved solute. Over time, powdered solute dissolves while dissolved solute crystallizes.

Saturated Solutions and Equilibrium

As you learned in the previous section, a dissociation equation can be written for the dissolving of an ionic compound, such as copper(II) sulfate, $CuSO_4(s)$:

$$CuSO_4(s) \xrightarrow{\text{in water}} Cu^{2+}(aq) + SO_4{}^{2-}(aq)$$

Crystallization is the reverse of dissolving and can be written as an equation:

$$Cu^{2+}(aq) + SO_4{}^{2-}(aq) \rightarrow CuSO_4(s)$$

In a saturated solution, which contains both dissolved and undissolved solute, both of these physical changes are taking place at the same time. Figure 5.13 on the next page represents dissolving and crystallization in a saturated solution.

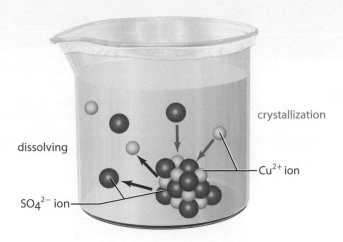

dissolving

crystallization

Cu^{2+} ion

SO_4^{2-} ion

Figure 5.13 Both dissolving and crystallization occur continuously in a saturated solution.

The solubility of copper(II) sulfate in water is constant at a specific temperature, so in a saturated solution the amount of undissolved solute will remain constant. But because both dissolving and crystallization are continually taking place, the rate at which the solute is dissolving must be equal to the rate at which the solute is crystallizing. (Otherwise, the amount of undissolved solute would change.)

A saturated solution is considered to be in a state of equilibrium. **Equilibrium** occurs when a process and the reverse of the process take place at the same rate in a closed system. In this case, dissolving is the process and crystallization is the reverse of the process.

Representing Equilibrium for a Saturated Solution

An equilibrium reaction can be represented by combining both the forward and the reverse processes into one equation with a double-headed arrow, as shown below:

$$CuSO_4(s) \underset{}{\overset{\text{in water}}{\rightleftharpoons}} Cu^{2+}(aq) + SO_4^{2-}(aq)$$

You will learn more about equilibrium in Chapter 6 and in Unit 8.

The double-headed arrow signifies that both the forward and the reverse reactions are taking place at the same rate and at the same time.

• • •

10 **a)** State the conditions necessary for a system to be considered at equilibrium.

b) Explain why a saturated solution (which, by definition, is in the presence of undissolved solute) is a system at equilibrium.

11 Write chemical equations representing the following processes:

a) Sodium chloride, $NaCl(s)$, dissolves to form an unsaturated solution.

b) Calcium chloride, $CaCl_2(s)$, dissolves to form an unsaturated solution.

c) Ammonium carbonate, $(NH_4)_2CO_3(s)$, dissolves to form a saturated solution.

d) Magnesium phosphate, $Mg_3(PO_4)_2(s)$, dissolves to form a saturated solution.

• • •

You may have noticed that solubility data always include temperature, which implies that temperature affects solubility. What do you think the effect is? Find out in the next investigation.

Plotting Solubility Curves

In this investigation, you will determine the temperature at which a certain mass of potassium nitrate is soluble in water. By combining your data with other students' data, you will be able to plot a solubility curve that shows how temperature affects solubility.

Question

How does temperature affect the solubility of potassium nitrate, $KNO_3(aq)$?

Hypothesis

How do you think temperature will affect the solubility of potassium nitrate?

- Write a hypothesis that predicts how temperature will affect the solubility of $KNO_3(s)$ and proposes an explanation.

Prediction

- Draw a sketch to show the shape of the curve you expect for the solubility of $KNO_3(s)$ in water at temperatures ranging from about 20 °C to 80 °C. Plot solubility on the vertical axis and temperature on the horizontal axis.

Safety Precautions

- Potassium nitrate is toxic and reactive. Wash your hands thoroughly after this investigation and inform your teacher of any spills.

Materials

- distilled water
- potassium nitrate, $KNO_3(s)$ 🔥 Ⓣ
- large test tube
- 300 mL tap water
- 400 mL beaker
- hot plate
- retort stand
- test tube clamp
- thermometer clamp
- thermometer
- balance
- scoopula
- stirring rod
- graduated cylinder
- graduated pipette
- pipette bulb

Procedure

1. Read the steps in this Procedure. Prepare a data table to record the mass of the solute, the initial volume of water, and the temperatures at which the solutions begin to crystallize.

2. Your teacher will assign you to one of the following volumes: 10.0 mL, 12.0 mL, 14.0 mL, 16.0 mL, 18.0 mL, 20.0 mL, 22.0 mL, 24.0 mL, 26.0 mL, 28.0 mL, or 30.0 mL. Use the graduated pipette to measure and add the assigned volume of distilled water into the test tube.

3. Pour about 300 mL of tap water into the beaker. Set up a hot water bath using a hot plate, retort stand, and clamp.

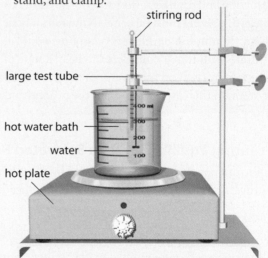

4. Measure 14.0 g of potassium nitrate. Add the potassium nitrate to the test tube.

5. Place the test tube in the beaker of tap water. Secure the test tube and thermometer to the retort stand using clamps. Begin heating the water bath gently.

6. Using the stirring rod, stir the mixture until the solute completely dissolves. Turn the heat source off, and allow the solution to cool.

7. Continue stirring as the solution cools. Record the temperature at which crystals begin to appear in solution.

8. Dispose of the aqueous solutions of potassium nitrate in the labelled waste container provided by your teacher.

Analysis

1. Use the volume of water assigned by your teacher to calculate how much solute dissolved in 100 mL of water. Use the following equation to help you:

$$\text{solubility} = \frac{x\,\text{g}}{100\ \text{mL}} = \frac{14.0\,\text{g}}{\text{your volume}}$$

This equation represents the solubility of $KNO_3(s)$ at the temperature at which you recorded the first appearance of crystals. Your teacher will collect and display the class data for this investigation.

2. Compare and discuss class data and decide whether any solubility data should be discarded.

3. Determine the average temperatures at which crystal formation occurs for solutions that contain the same volume of water. Plot these data on graph paper. Plot solubility on the vertical axis. (The units are g/100 mL.) Plot temperature on the horizontal axis.

4. Draw the best-fit curve through the points. (Do not simply join the points.) Label each axis. Give the graph a suitable title.

Conclusion

5. a) Compare the shape of the sketch you made at the start of the investigation to the graph you created using data from the experiment.

b) According to your results, how does temperature affect the solubility of potassium nitrate?

c) Do your results in this investigation support your hypothesis? Explain your answer.

Applications

6. Use your graph to interpolate the solubility of potassium nitrate at:

a) 60 °C

b) 40 °C

7. Use your graph to extrapolate the solubility of potassium nitrate at:

a) 80 °C

b) 20 °C

Temperature and Solubility

The solubility of a solute in water is usually given as the mass of solute that dissolves in 100 mL of water at a specific temperature, often 20 °C. Specifying temperature with solubility data is essential, since the solubility of a substance is very different at different temperatures. When a solid dissolves in a liquid, energy is needed to break the bonds holding the solid together. At higher temperatures, the particles of the solute and solvent have more energy, on average. Therefore, *the solubility of most solids increases with temperature.*

Because temperature usually increases solubility, boiling leaves, such as mint leaves or tea leaves, to make tea is an effective way to extract tasty or useful compounds from them. For example, some Cree peoples make a drink made from the boiled leaves of the Labrador tea plant (see Figure 5.14). This drink is a traditional remedy for various ailments, including chest pains and colds. Cree Elders possess traditional knowledge of how much Labrador tea to use to make the drink and how often it is safe to drink it. Like most prescription medicines and over-the-counter pills, the correct amount of Labrador tea is medicinal, but too much can be harmful.

How does temperature affect other types of solutions? The bonds between particles in a liquid are not as strong as the bonds between particles in a solid. When a liquid dissolves in a liquid, additional energy is not needed. Therefore, *the solubility of most liquids is not greatly affected by temperature.*

Gas particles move quickly and have a great deal of kinetic energy. When a gas dissolves in a liquid, the gas particles lose some of their energy. At higher temperatures, the dissolved gas gains energy again. As a result, the gas comes out of solution and is less soluble. Therefore, *the solubility of gases decreases at higher temperatures.*

Figure 5.14 Cree peoples use the leaves from this Labrador tea plant to make a medicinal tea.

Thermal Pollution: A Solubility Problem

Figure 5.15 Wabamun Lake, 70 km west of Edmonton, is home to a power plant that released warm water into the lake. The plant is due to be shut down completely by 2010.

Figure 5.16 What happens to the carbon dioxide gas in a soft drink bottle when the pressure is released? The solubility of the gas in the soft drink solution decreases.

Gases are less soluble at higher temperatures. This is why a refrigerated soft drink tastes fizzier than the same drink at room temperature. The warmer drink contains less dissolved carbon dioxide than does the cooler drink.

This property of gases makes heat pollution a serious problem. Many industries and power plants use water to cool overheated machinery. The resulting hot water is then returned to local rivers or lakes. Why is adding warm water into a river or lake a form of pollution? The heat from the hot water increases the temperature of the body of water. As the temperature increases, the dissolved oxygen in the water decreases. Fish and other aquatic wildlife may not have enough oxygen to breathe.

A second problem arising from heat pollution is that harmful solutes, such as mercury compounds and pesticides, may become more soluble. Figure 5.15 shows an Alberta lake affected by heat pollution.

Pressure and Solubility

Another factor that affects solubility is pressure. Changes in pressure have very little effect on the solubility of solids and liquids. Pressure changes do affect the solubility of a gas in a liquid solvent, however. The solubility of the gas is directly proportional to the partial pressure of that gas above the liquid. For example, the solubility of oxygen in lake water depends on the partial pressure of the oxygen in the air above the lake.

When you open a soft drink bottle, you can observe the effect of pressure on solubility. Figure 5.16 shows this effect. Inside a soft drink bottle, the pressure is very high. When you open the bottle, you hear the sound of escaping gas as the pressure decreases, becoming equal to the air pressure outside the bottle. The solubility of the carbon dioxide in the liquid soft drink decreases due to the decrease in pressure. Bubbles begin to rise in the liquid as gas comes out of solution and escapes. It takes time for all of the gas to leave the solution, so you have time to enjoy the taste of the soft drink before it goes flat.

Figure 5.17 illustrates another example of dissolved gases and pressure. The deeper a scuba diver swims underwater, the more the water pressure increases. The solubility of nitrogen gas, which is present in air and is inhaled into the lungs, increases as well. Nitrogen gas dissolves to a greater extent in the diver's blood. As the diver returns to the surface, the pressure acting on the diver decreases. The solubility of the nitrogen gas in the blood decreases, and the gas comes out of the solution. If the diver surfaces too quickly, the effect is similar to opening a soft drink bottle. Bubbles of nitrogen gas form in the blood, which obstructs the blood flow to muscles and the brain. This leads to a painful and sometimes fatal condition known as "the bends."

Section 5.2 Summary

- Solubility represents the quantity of a solute that can be dissolved in a given quantity of solvent. It can be affected by both temperature and pressure.

- An unsaturated solution can dissolve more solute at a particular temperature. A saturated solution cannot dissolve more solute at a given temperature. A supersaturated solution is one which contains more dissolved solute than its solubility would indicate. They can be made by preparing a solution at a high temperature and allowing it to cool undisturbed. As it cools, it will reach a temperature at which it will contain more dissolved solute that it should at that temperature.

- A saturated solution is in equilibrium. The rate at which solid dissolves is equal to the rate at which dissolved solute crystallizes.

- The solubility of most solids in liquids increases with temperature, because the bonds in the solute and solvent are strong and require more energy to break them. The solubility of liquids in another liquid is not greatly affected by temperature.

- At higher temperatures, gases increase in energy and can escape from the liquid in which they are dissolved. Therefore, the solubility of gases decreases at higher temperatures.

- Although changes in pressure have little effect on the solubility of a solid in liquid, it does affect the solubility of a gas in liquid. The solubility of the gas will decrease as the pressure decreases, causing the gas to 'bubble' out of the solution.

Figure 5.17 Scuba divers must heed the effects of decreasing water pressure on dissolved nitrogen gas in their blood. They must surface slowly to avoid "the bends."

SECTION 5.2 REVIEW

1. Define the following terms:
   a) solubility **b)** unsaturated **c)** saturated

2. The solubility of sodium nitrate, $NaNO_3(s)$, is 87.6 g/100 mL in water at 20 °C. Using the terms *unsaturated* and *saturated*, how would you describe the following systems at 20 °C?
   a) 20.0 g of $NaNO_3(s)$ in 100 mL of water
   b) 100 g of $NaNO_3(s)$ in 100 mL of water

3. Write equations that represent saturated solutions of the following compounds:
   a) $LiCl(s)$ **b)** $NH_4F(s)$ **c)** $Na_2S(s)$ **d)** $(NH_4)_2SO_4(s)$

4. The solubility of potassium chloride, $KCl(s)$, is 34.7 g/100 mL of water at 20 °C. Suppose 40.0 g of solute was added to 100 mL of water at 20 °C.
   a) Describe the solution as either saturated or unsaturated.
   b) How much (if any) solute would remain undissolved?
   c) Write a chemical equation to represent the solution.

5. Would you expect to find more mineral deposits near a thermal spring or near a cool mountain spring? Explain.

6. A saturated solution is a solution that contains the maximum amount of dissolved solute at a given temperature *in the presence of undissolved solute*. Why is the presence of undissolved solute necessary for the solution to be considered saturated?

7. The graph below shows the solubility of various substances plotted against the temperature of the solution.
   a) Which substance decreases in solubility as the temperature increases?
   b) Which substance is least soluble at room temperature? Which substance is most soluble at room temperature?
   c) The solubility of which substance is least affected by a change in temperature?
   d) At what temperature is the solubility of potassium chlorate equal to 40 g/100 mL of water?
   e) 20 mL of a saturated solution of potassium nitrate at 50 °C is cooled to 20 °C. Approximately what mass of solid comes out of from the solution?

The Effect of Temperature on Solubility

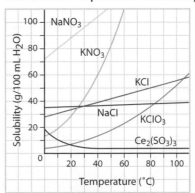

Compare the solubility of these various ionic compounds.

The Concentration of Solutions

Pure phenol is a volatile chemical, which means it evaporates easily at room temperatures and pressures. This volatility is dangerous because inhaling phenol harms the central nervous system and can lead to a coma. Inhalation is not the only way that phenol can affect you. People have died or become comatose within 10 minutes of phenol contacting their skin. As little as 1 g of phenol can be fatal if swallowed.

Are you surprised that such a hazardous chemical is found in over-the-counter medications, such as the product shown in Figure 5.18? If you check your medicine cabinet at home, you may find phenol listed as an ingredient in throat sprays and in lotions to relieve itching. Phenol is also contained in antiseptic or disinfectant products. Its anaesthetic properties soothe a sore throat while combating the harmful bacteria and fungi that cause the infection.

Material Safety Data Sheet

Component Name	CAS Number
PHENOL, 100% Pure	108952

SECTION III: Hazards Identification

- Very hazardous in case of ingestion, inhalation, skin contact, or eye contact.
- Product is corrosive to internal membranes when ingested.
- Inhalation of vapours may damage central nervous system. Symptoms: nausea, headache, dizziness.
- Skin contact may cause itching and blistering.
- Eye contact may lead to corneal damage or blindness.
- Severe over-exposure may lead to lung-damage, choking, or coma.

Figure 5.18 Should medication containing phenol be banned from drugstores?

Is phenol a hazard or a beneficial medicine? The answer depends on concentration. **Concentration** is the quantity of solute per quantity of solution. At high concentrations, phenol can kill. At low concentrations, it is a safe component of many medicines.

Expressing Concentration

The concentration of solutions can be expressed in several ways, both qualitative and quantitative. Qualitatively, solutions can be described as **concentrated** or **dilute**. A concentrated solution contains a large ratio of solute to solvent, whereas a dilute solution contains a small ratio of solute to solvent. The drawback to describing solutions qualitatively is that the terms are relative and do not describe how much more concentrated one solution is than another. It is far more useful to have a quantitative description of solutions.

In the next Thought Lab, you will investigate the variety of ways in which concentrations are stated for various purposes, such as labels for household products or data in environmental reports.

Knowing the concentration of fluoride in two different toothpastes can help you make a buying decision or help you decide whether you should take a fluoride supplement. Measuring and setting safe limits for the concentration in water or air of substances, such as pesticides and heavy metals, is essential for monitoring and maintaining a healthy environment. How are concentrations stated on the labels of household products? How are concentrations stated for environmental data?

Procedure

Part 1: Household Solutions and Medications

1. At home, go through your cupboards to find five labels that include concentration information. Record the information and bring it to class. (Remember that concentration is the quantity of solute per quantity of solution.)

2. Share your results with your classmates.

   a) Which way of expressing concentration is the most common?

   b) What do you think is the meaning of each different way of expressing concentration?

Part 2: Environmental Studies

1. Choose one of the following environmental issues to research:
   • heavy metals
   • pesticides
   • air pollutants

2. Perform Internet research to find a compound that fits into one of the above categories and investigate it in detail. Check with your teacher before going on. **ICT**

3. Use Internet and print resources to find the following information about your compound:
   • original use
   • how its concentration in water or air is measured
   • safe limits for concentration
   • current concentration in the environment or in medication
   • hazards if concentration rises beyond safe values
   • legislation and other controls to ensure concentrations remain within safe limits

4. Share your findings as a presentation or as a poster.

Analysis

1. Discuss your findings as a class.

   a) What are the most common ways of expressing concentration?

   b) What are some advantages and disadvantages of having so many ways to express concentration?

2. Much of what we use at home ends up in a landfill or down the drain. Given what you have learned during this Thought Lab, do you think disposal methods should change? Explain your answer.

Concentration as Percent by Mass

The concentration of a solution can be described as a mass of solute dissolved in a mass of solution. This is usually expressed as a percentage. **Percent by mass** gives the mass of a solute divided by the mass of solution, expressed as a percentage. Percent by mass is often referred to as percent m/m, or mass percent.

$$\text{Percent by mass} = \frac{\text{Mass of solute (g)}}{\text{Mass of solution (g)}} \times 100\%$$

For example, 100 g of seawater contains 0.129 g of magnesium ions (along with many other dissolved substances). The concentration of $Mg^{2+}(aq)$ in seawater is 0.129% m/m. Notice that the number of grams of solute per 100 g of solution is numerically equal to the percent by mass.

The concentration of a solid solution, such as an alloy, is usually expressed as percent by mass. Often, the concentration of a particular alloy varies. Figure 5.19 on the following page shows two objects made from brass that have distinctly different colours. The difference in colours reflects the varying concentrations of copper and zinc that make up the objects. Table 5.5 gives typical percent by mass compositions of some common alloys.

Figure 5.19 Brass contains different ratios of copper and zinc. As a result, two brass objects may look very different.

Table 5.5 The Composition of Some Common Alloys

Alloy	Uses	Typical percent m/m composition
brass	ornaments, musical instruments	Cu (85%); Zn (15%)
bronze	statues, castings	Cu (80%); Zn (10%); Sn (10%)
stainless steel	cutlery, knives	Fe (78%); Cr (15%); Ni (7%)
sterling silver	jewellery	Ag (92.5%); Cu (7.5%)

Sample Problem

Solving for Percent by Mass

Problem
Calcium chloride, $CaCl_2(s)$, can be used instead of road salt to melt the ice on roads during the winter. To determine how much calcium chloride had been used on a nearby road, a student took a sample of slush to analyze. The sample had a mass of 23.47 g. When the solvent was evaporated, the residue had a mass of 4.58 g. (Assume that no other solutes were present.) What was the percent by mass of calcium chloride in the slush?

What Is Required?
You need to calculate the percent by mass of calcium chloride in the slush (that is, how much calcium chloride, in grams, is in a 100 g sample of slush).

What Is Given?
The mass of the solution is 23.47 g. The mass of calcium chloride that was dissolved in the solution (the slush) is 4.58 g.

Plan Your Strategy
Percent by mass is just the quantity of solute per solution expressed as a percent. Use the following equation:

$$\text{Percent by mass} = \frac{\text{Mass of solute (g)}}{\text{Mass of solution (g)}} \times 100\%$$

Act on Your Strategy

$$\text{Percent by mass} = \frac{\text{Mass of solute (g)}}{\text{Mass of solution (g)}} \times 100\%$$

$$= \frac{4.58\ \cancel{g}}{23.47\ \cancel{g}} \times 100\%$$

$$= 19.5\%$$

Solution
The slush was 19.5% calcium chloride by mass.

Check Your Solution
The answer has the correct number of significant digits (three) and the correct units for concentration.

Practice Problems

1. Calculate the percent by mass for each of the following solutes in solution:
  a) An aqueous solution with a mass of 82.0 g contains 17.0 g of sulfuric acid, $H_2SO_4(aq)$.
  b) An aqueous solution with a mass of 110.6 g contains 18.37 g of sodium chloride, $NaCl(aq)$.
  c) A benzene solution with a mass of 85.4 g contains 12.9 g of carbon tetrachloride, $CCl_4(\ell)$.

2. If 55.0 g of potassium hydroxide, $KOH(s)$, is dissolved in 100.0 g of water, what is the concentration of the solution expressed as a percent by mass? (**Hint:** Remember to use the mass of the *solution*, not the mass of the *solvent* in your calculation.)

3. Steel contains about 98.3% iron and about 1.7% carbon. It also contains very small amounts of other materials, such as manganese and phosphorus. What mass of carbon, in grams, is needed to make a 5.0 kg sample of steel?

4. Most cutlery is made of stainless steel, which is a variety of steel that resists corrosion. Stainless steel contains at least 10.5% chromium. What is the minimum mass of chromium needed to make a stainless steel fork with a mass of 60.5 g?

5. Eighteen-carat white gold is an alloy. It contains 75% gold, 12.5% silver, and 12.5% copper. A piece of jewellery made of 18-carat white gold has a mass of 20 g. What mass of pure gold (in grams) does it contain?

Concentration in Parts per Million

Carbon monoxide, CO(g), is an odourless gas that, if absorbed into the human bloodstream, impedes the delivery of oxygen to body tissues. Exposure to carbon monoxide can cause illness or even death. Fortunately, effective carbon monoxide detectors are available, as shown in Figure 5.20. Most carbon monoxide detectors are calibrated to sound an alarm when the concentration of the deadly gas reaches 30 ppm. Most people get a headache when the concentration of carbon monoxide gas in the air reaches 400 ppm. What does ppm mean?

Figure 5.20 A carbon monoxide detector

Very small concentrations of a substance in the human body, or in the environment, are often best expressed in **parts per million (ppm)** or even **parts per billion (ppb)**. Both parts per million and parts per billion are usually mass/mass relationships. They describe the mass of a solute that is present in a certain solution. For example, air that contains 30 ppm carbon monoxide contains 30 g of carbon monoxide per 1 000 000 g (1 million grams) of air. In other words, each gram of air contains 0.000 03 g of carbon monoxide. Notice that parts per million and parts per billion do not refer to number of particles but to the mass of the solute compared with the mass of the solution.

$$\text{Concentration in ppm} = \frac{\text{Mass of solute (g)}}{\text{Mass of solution (g)}} \times 10^6 \text{ ppm}$$

Parts per million can be expressed in a variety of equivalent units.
$$1 \text{ ppm} = 1 \text{ g}/1 \times 10^6 \text{ g}$$
$$= 1 \text{ mg/kg}$$

For aqueous solutions, you can express parts per million using volume units because 1 mL of water has a mass of approximatley 1 g.
$$1 \text{ ppm} = 1 \text{ g/kL}$$
$$= 1 \text{ mg/L}$$

Similarly, for parts per billion:

$$\text{Concentration in ppb} = \frac{\text{Mass of solute (g)}}{\text{Mass of solution (g)}} \times 10^9 \text{ ppb}$$

Figure 5.21 After hydrogen sulfide is extracted from sour gas, it is processed to obtain the useful element, sulfur.

Hydrogen Sulfide: A Hazard at Low Concentrations

The unit ppm is often used to express the concentration of environmental pollutants that are dangerous even in extremely small quantities. Workers in the oil and gas industry are well aware of one such pollutant, hydrogen sulfide, H_2S(g). Hydrogen sulfide is a component of sour gas and can be processed into sulfur, the yellow solid shown in Figure 5.21.

Your nose can detect the characteristic rotten-egg smell of hydrogen sulfide at a concentration as low as 1 ppm in air. At levels of 200 ppm and above, however, the gas causes a loss of sense of smell, so those exposed may not realize they are still being exposed. If the concentration of hydrogen sulfide reaches 700 ppm, those exposed will immediately lose consciousness. At worksites where there is a danger of H_2S(g) exposure, the air is continuously monitored and alarms sound if the concentration reaches a dangerous level. Breathing apparatus stations are located throughout the worksite so all employees will be able to quickly put on the lifesaving equipment.

Calculating Parts per Million

Problem

"Hard" water contains a relatively large quantity of dissolved minerals, one of which is calcium carbonate. Hard water in Alberta may contain up to 300 ppm calcium carbonate. Older models of toilets use up to 20 L per flush. How much calcium carbonate could be crystallized from the water in one flush of an older model of toilet?

What Is Required?

You need to find the mass of calcium carbonate in 20 L of water.

What Is Given?

You know that the concentration of calcium carbonate in the water is 300 ppm and you have 20 L of solution.

Plan Your Strategy

Convert the volume of the solution into mass in grams. You can assume that 1 g of solution has a volume of 1 mL because it is a very dilute aqueous solution. Then rearrange the ppm formula to solve for the mass of the solute.

Act on Your Strategy

$$\text{Mass of solution (g)} = 20\,\cancel{L} \times \frac{1000\,\cancel{mL}}{1.0\,\cancel{L}} \times \frac{1\,g}{1\,\cancel{mL}}$$

$$= 20\,000\,g$$

$$\text{Conc. ppm} = \frac{\text{Mass of solute (g)}}{\text{Mass of solution (g)}} \times 10^6\,\text{ppm}$$

$$\text{Mass of solute (g)} = \frac{\text{Conc. in ppm} \times \text{Mass of solution (g)}}{10^6\,\text{ppm}}$$

$$= 300\,\cancel{ppm} \times \frac{20\,000\,g}{10^6\,\cancel{ppm}}$$

$$= 6.0\,g$$

Solution

From a 20 L flush, 6.0 g of calcium carbonate is present and thus could be crystallized.

Check Your Solution

The answer has the correct number of significant digits (two) and the correct units for mass.

6. Symptoms of mercury poisoning become apparent after a person has accumulated 20 mg of mercury in his or her body.
   a) Express this concentration as parts per million for a 60 kg person.
   b) Express this concentration as parts per billion for the same person.
   c) Express this amount as a percent by mass for the same person.

7. A concentration of 700 ppm hydrogen sulfide in air will cause a person to lose consciousness. Express this concentration as percent by mass.

8. The use of the pesticide DDT has been banned in Canada since 1969 because of its damaging effect on wildlife. In 1967, the concentration of DDT in an average lake trout, taken from Lake Simcoe in Ontario, was 16 ppm. Today it is less than 1 ppm. What mass of DDT would have been present in a 2.5 kg trout with a DDT concentration of 16 ppm?

9. The concentration of chlorine in a swimming pool is generally kept between 1.4 mg/L and 4.0 mg/L. What is the concentration range in ppm? (Assume 1.0 L of water has a mass of 1.0 kg.)
   Hint:
$$\text{ppm} = \frac{1\,g}{1 \times 10^6\,g} = \frac{1000\,\text{mg}}{1000\,\text{kg}} = \frac{1\,\text{mg}}{1\,\text{kg}} = \frac{1\,\text{mg}}{L}$$

10. The drinking water in Stratford, Ontario has a fluoride concentration of 1.6 ppm. A long-term study showed that Stratford residents experienced a much lower incidence of tooth decay than did their neighbours in Woodstock, whose tap water was non-fluoridated. If the average adult consumes 2.0 L of water daily, what mass of fluoride is consumed? (Assume 1.0 mL of tap water has a mass of 1.0 g.)

Biomagnification and Bioaccumulation

Human activities, such as mining and burning of fossil fuels, release significant quantities of toxic metals such as mercury, $Hg(\ell)$, into ecosystems. In living organisms, these elements can build up to unsafe levels that can cause birth defects, brain or kidney damage, and cancer. What processes allow toxic chemicals to build up in living organisms?

Bioaccumulation of Mercury

Industrial activities release various forms of mercury, including elemental mercury, $Hg(\ell)$, into the environment. Elemental mercury is not very soluble in water. However, elemental mercury vaporizes easily. In sunlight, some elemental mercury in the atmosphere forms the mercury(II) ion, $Hg^{2+}(aq)$, which is soluble in water. Rain and snow deposit ionic and some elemental mercury onto land, lakes, and oceans.

Some aquatic bacteria can convert inorganic mercury (Hg or Hg^{2+}) into forms that easily build up in animal tissues. In aquatic sediments, bacteria add a methyl group ($-CH_3$) to mercury ions, creating methylmercury, CH_3Hg^+. Methylmercury forms strong bonds with protein molecules. Therefore, it builds up in protein-rich body tissues when organisms ingest contaminated food or water. For example, you would absorb about 95% of the methylmercury in the fish that you eat.

Bioaccumulation refers to the build-up of substances to higher concentrations in the tissues of living organisms *compared to the non-living environment* (water or soil). Bioaccumulation can occur at the start of a food chain, in plants and algae.

Biomagnification of Mercury

Biomagnification occurs when animals accumulate higher concentrations of substances *compared with their food*. Predatory fish, for example, contain about ten times the concentration of methylmercury found in their prey. Since methylmercury becomes more concentrated with each step in the food chain, even low concentrations of methylmercury in water can build up to dangerous levels in top predators. As a result, some fish contain so much methylmercury that they are unsafe to eat.

Mercury can travel long distances in the atmosphere, and under certain environmental conditions, fall on regions far from the source. Mercury contamination is therefore a major problem in the Arctic, even though few human activities in the region add mercury to the environment.

In order to study the extent of mercury biomagnification in Arctic ecosystems, scientists working in the Northwest Territories collected data on the concentration of mercury in several Arctic species, shown below.

Total Mercury in the Body Tissues of Selected Arctic Animals from the Northwest Territories

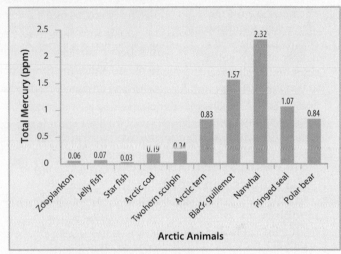

Source: *Can. J. Fish. Aquat. Sci.* Vol. 55, 1998

• • •

1. How much mercury (Hg) would there be in:

   a) a 180 g Arctic cod fish?

   b) a 625 kg polar bear?

2. Alberta uses about 26 million tonnes of coal annually to generate electricity. Assuming the coal contains a minimum of 70 ng/g of mercury, how much mercury would Alberta generate from coal-powered electrical energy every year?

3. Hydrogen sulfide will cause mercury(II) to precipitate out of solution as mercury(II) sulfide:

   $$Hg^{2+}(aq) + H_2S(aq) \rightleftharpoons HgS(s) + 2H^+(aq)$$

   How could hydrogen sulfide be used to help clean up mercury-contaminated environments?

Chemistry File

FYI

You may see the symbol "M" used to represent "molarity" in older textbooks and on the Internet. This symbol is no longer considered acceptable by IUPAC.

Molar Concentration

The most commonly used unit of concentration in chemistry is molar concentration. **Molar concentration** is the amount (number of moles) of solute dissolved in 1 L of solution. Notice that the volume of the *solution* in *litres* is used. Molar concentration is also known as *amount concentration* and *molarity*.

$$\text{Molar concentration (mol/L)} = \frac{\text{Amount of solute (mol)}}{\text{Volume of solution (L)}}$$

You can express this relationship as a shorthand formula, where:

c = concentration (in mol/L)
n = amount of solute (in mol)
V = volume of solution (in L)

$$c = \frac{n}{V}$$

Math Tip

For multistep problems such as those that follow, remember: keep all the digits in your calculator until you have completed the final step of the problem. Only round for significant digits at the final step.

Chemists use a special shorthand for molar concentration because it is used so often. When you see square brackets around a chemical formula, it means "concentration in mol/L." For example: $[NaCl(aq)] = 0.10$ mol/L means "the molar concentration of NaCl(aq) is 0.10 mol/L."

Molar concentration is particularly useful to chemists because it is related to the number of particles in a solution. (The other units you have learned so far are related to mass.) Knowing the number of particles in solution allows you to solve problems involving chemical reactions in which substances react in fixed mole ratios.

Sample Problem

Calculating the Concentration of a Solution in mol/L

Problem

A 100 mL sample of saline solution contains 0.90 g of sodium chloride, NaCl(aq). What is the molar concentration of NaCl(aq)?

What Is Required?

You need to find the molar concentration of the solution.

What Is Given?

You know that 0.90 g of sodium chloride is dissolved in 100 mL of solution.

Plan Your Strategy

To find the amount of sodium chloride (in mol), first find the molar mass of NaCl. Using the molar mass and the volume of the solution, determine the concentration of the solution in mol/L by using the formula $c = \frac{n}{V}$.

Act on Your Strategy

$$\begin{aligned}\text{Molar mass of NaCl} &= 22.99 \text{ g/mol} + 35.45 \text{ g/mol} \\ &= 58.44 \text{ g/mol}\end{aligned}$$

$$\begin{aligned} n &= \frac{m}{M} \\ &= \frac{0.90 \text{ g NaCl}}{58.44 \frac{\text{g NaCl}}{\text{mol NaCl}}} = 0.0154 \text{ mol NaCl}\end{aligned}$$

$$\begin{aligned} c &= \frac{n}{V} \\ &= \frac{0.0154 \text{ mol NaCl}}{100 \text{ mL}\left(\frac{1\text{L}}{1000 \text{ mL}}\right)} = 0.15 \text{ mol/L NaCl}\end{aligned}$$

Solution

The concentration of sodium chloride in the saline solution is 0.15 mol/L.

Check Your Solution

The answer has the correct number of significant digits (two) and the correct units for concentration.

Using Dimensional Analysis

$$\begin{aligned}[NaCl(aq)] &= \frac{0.90 \text{ g NaCl}}{1} \times \frac{1 \text{ mol NaCl}}{58.44 \text{ g NaCl}} \times \frac{1}{100 \text{ mL}} \times \frac{1000 \text{ mL}}{1\text{L}} \\ &= 0.15 \text{ mol/L NaCl}\end{aligned}$$

11. Calculate the molar concentration of the following solutions:
   a) 0.50 mol of NaCl(s) dissolved in 0.30 L of solution
   b) 0.289 mol of iron(III) chloride, $FeCl_3$(s), dissolved in 120 mL of solution

12. Calculate the molar concentration of the following solutions:
   a) 4.63 g of sucrose, $C_{12}H_{22}O_{11}$(s), dissolved in 16.8 mL of solution
   b) 1.2 g of sodium nitrate, $NaNO_3$(s), dissolved in 80 mL of solution

13. If 1.37×10^{-2} g of ammonium phosphate is dissolved in enough water to make 125 mL of solution, what is the molar concentration of the solution?

14. What is the molar concentration of a solution in which 2.1 mg of copper(II) sulfate pentahydrate is dissolved to make 1500 mL of solution?

15. What is the molar concentration of a solution with a concentration of 200 ppm fluoride ions? Remember that 1 mL of water has a mass of 1 g.

16. What is the molar concentration of a solution that is 10% m/m methanol, CH_3OH(aq)? Assume 1 g of CH_3OH solution has a volume of 1 ml.

Hint: use $n = \dfrac{m}{M}$

Molar Concentration of Ions in Solution

In solution, ionic compounds separate by dissociation into individual ions. You may have to calculate the concentration of the individual *ions* in solution, not just the solute. Writing a dissociation equation will help you relate the concentration of ions to the concentration of solute. For example, suppose the concentration of a solution of potassium sulfate, K_2SO_4(aq), is 0.125 mol/L. To determine the concentration of potassium and sulfate ions, first write a dissociation equation.

$$K_2SO_4(s) \xrightarrow{\text{in water}} 2\,K^+(aq) + SO_4{}^{2-}(aq)$$

There are 2 mol K^+(aq) ions produced for every 1 mol K_2SO_4(s), as represented in Figure 5.22. Therefore, you would calculate the concentration of K^+(aq) ions as follows:

$$[K^+(aq)] = \frac{0.125\ \cancel{\text{mol } K_2SO_4(s)}}{L} \times \frac{2\ \text{mol } K^+(aq)}{1\ \cancel{\text{mol } K_2SO_4(s)}}$$
$$= 0.250\ \text{mol/L } K^+(aq)$$

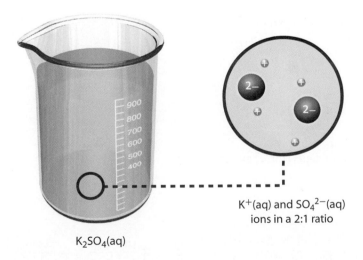

K^+(aq) and $SO_4{}^{2-}$(aq) ions in a 2:1 ratio

K_2SO_4(aq)

Figure 5.22 The concentration of K^+(aq) ions is double the concentration of K_2SO_4(s) and $SO_4{}^{2-}$(aq).

Calculating Ion Concentration in mol/L

Problem

Ammonium phosphate is most commonly used in the treatment of industrial effluent before the effluent is released into lakes or rivers (see figure below). The compound is produced from mined phosphate ore that is reacted with water and sulfuric acid.

Ammonium phosphate is one of many products made by Alberta industries and shipped throughout North America.

A solution is made by dissolving 1.35 mg of ammonium phosphate in enough water to make 250 mL of solution. Calculate the concentration of ammonium ions in mol/L.

What Is Required?

First, you need to calculate the molar concentration of ammonium phosphate solution. Then, you need to use a dissociation equation to determine a mole ratio that will allow you to calculate the concentration of ammonium ions.

What Is Given?

The mass of ammonium phosphate, $(NH_4)_3PO_4(s)$, is 1.35 mg. The volume of the solution is 250 mL.

Plan Your Strategy

Write a dissociation equation and then find the concentration of ammonium phosphate (in mol/L) to solve for the concentration of ammonium ions. Use conversion factors to convert mg into g and mL into L.

• • •

Act on Your Strategy

$$(NH_4)_3PO_4(s) \xrightarrow{\text{in water}} 3NH_4^+(aq) + PO_4^{3-}(aq)$$

$$n = \frac{m}{M} = \frac{1.35 \, \cancel{mg} \, (NH_4)_3PO_4(s) \left(\dfrac{1 \, \cancel{g}}{1000 \, \cancel{mg}}\right)}{149.12 \, \cancel{g}/mol}$$

$$= 9.053 \times 10^{-6} \, mol \, (NH_4)_3PO_4(s)$$

$$c = \frac{n}{V} = \frac{9.053 \times 10^{-6} \, mol \, (NH_4)_3PO_4(s)}{250 \, \cancel{mL}\left(\dfrac{1 \, L}{1000 \, \cancel{mL}}\right)}$$

$$= 3.621 \times 10^{-5} \, mol/L \, (NH_4)_3PO_4(aq)$$

$$[NH_4^+(aq)] = \frac{3.621 \times 10^{-5} \, \cancel{mol \, (NH_4)_3PO_4(aq)}}{L} \times \frac{3 \, mol \, NH_4^+(aq)}{1 \, \cancel{mol \, (NH_4)_3PO_4(aq)}}$$

$$= 1.09 \times 10^{-4} \, mol/L \, NH_4^+(aq)$$

Solution

The concentration of ammonium ions is 1.09×10^{-4} mol/L.

Check Your Solution

The answer has the correct number of significant digits (three) and the correct units for concentration.

Using Dimensional Analysis

$$[(NH_4)_3PO_4(s)]$$

$$= \frac{1.35 \, \cancel{mg \, (NH_4)_3PO_4(s)}}{1} \times \frac{1 \, \cancel{g}}{1000 \, \cancel{mg}} \times \frac{1 \, \cancel{mol \, (NH_4)_3PO_4(s)}}{149.12 \, \cancel{g \, (NH_4)_3PO_4(s)}} \times \frac{1}{250 \, \cancel{mL}} \times \frac{1000 \, \cancel{mL}}{1 \, L} \times \frac{3 \, mol \, NH_4 \, (aq)}{1 \, \cancel{mol \, (NH_4)_3PO_4(aq)}}$$

$$= 1.09 \times 10^{-4} \, mol/L \, NH_4^+(aq)$$

17. Calculate the molar concentration of sodium ions if 2.7 g of sodium carbonate, $Na_2CO_3(s)$, is dissolved in 175 mL of water.

18. What is the molar concentration of calcium ions if 1.3×10^{-4} g of calcium phosphate, $Ca_3(PO_4)_2(s)$, is dissolved to make 3.95 L of solution?

19. Determine the molar concentration of the cation (positive ion) if 0.000 453 g of strontium nitrate is dissolved to make 1 L of solution.

20. Calculate the molar concentration of fluoride ions if 1.45 µg of sodium fluoride, NaF(s), is contained in 100 mL of aqueous solution (µg means 10^{-6} g).

21. What is the molar concentration of nitrate ions if 9.0 g of aluminium nitrate is dissolved to make 5.3 L of solution?

22. Determine the molar concentration of the anion (negative ion) if 4.34 kg of lithium phosphate is dissolved in 3.75 kL of water.

Sample Problem

Calculating Mass from Concentration in mol/L

Problem

At 20 °C, a saturated solution of calcium sulfate, $CaSO_4(aq)$, has a concentration of 0.0153 mol/L. A student takes 65 mL of this solution and evaporates the solvent. What mass (in g) is left in the evaporating dish?

What Is Required?

You need to find the mass of $CaSO_4(aq)$ that remains after evaporation.

What Is Given?

The concentration of the solution is 0.0153 mol/L. The volume of the solution is 65 mL.

Plan Your Strategy

Convert the volume from mL to L and rearrange the formula $c = \frac{n}{V}$ to find the amount of solute in the solution (in moles). Rearrange the formula $n = \frac{m}{M}$ to determine the mass of solute.

• • •

Act on Your Strategy

$$c = \frac{n}{V}$$

$$n = cV = \left(0.0153\, \frac{\text{mol } CaSO_4(aq)}{\cancel{L}}\right)(65\,\cancel{mL})\left(\frac{1\,\cancel{L}}{1000\,\cancel{mL}}\right)$$

$$= 0.000\ 995\ \text{mol } CaSO_4(aq)$$

$$n = \frac{m}{M}$$

$$m = n \times M = (0.000\ 995\ \cancel{\text{mol } CaSO_4(s)})\left(\frac{136.15\ \text{g } CaSO_4(s)}{1\,\cancel{\text{mol } CaSO_4(s)}}\right)$$

$$= 0.14\ \text{g } CaSO_4(s)$$

Check Your Solution

The answer has the correct units (g) and the correct number of significant digits (two).

Using Dimensional Analysis

$$m\ CaSO_4(s) = \frac{65\,\cancel{mL}}{1} \times \frac{1\,\cancel{L}}{1000\,\cancel{mL}} \times 0.0153\,\frac{\cancel{\text{mol } CaSO_4(s)}}{1\,\cancel{L}} \times \frac{136.15\ \text{g } CaSO_4(s)}{1\,\cancel{\text{mol } CaSO_4(s)}}$$

$$= 0.14\ \text{g } CaSO_4(s)$$

Therefore, 0.14 g of calcium sulfate remains in the evaporating dish.

23. Calculate the mass of solute in the following solutions:

 a) 125 mL of 0.200 mol/L NaCl(aq)

 b) 1.35 L of 1.25 mol/L NH_4NO_3(aq)

24. What mass of solute could be obtained by evaporation from these solutions?

 a) 0.38 L of 4.25×10^{-3} mol/L ammonium carbonate

 b) 2.0 L of 1.25 mol/L magnesium hydroxide

25. A 100 mL bottle of skin lotion contains a number of solutes. One of these solutes is zinc oxide, ZnO(s). The concentration of zinc oxide in the skin lotion is 0.915 mol/L. What mass of zinc oxide is present in the bottle?

26. Formalin is an aqueous solution of formaldehyde, HCHO(aq), used to preserve biological specimens. What mass of formaldehyde is needed to prepare 1.5 L of formalin with a concentration of 10 mol/L?

Calculating Volume from Concentration in mol/L

Problem

What volume of a 0.25 mol/L solution of sodium sulfate could be made from 3.5 g of solute?

What Is Required?

You need to find the volume of Na_2SO_4(aq) solution that can be prepared.

What Is Given?

The concentration of the solution to be made is 0.25 mol/L. The mass of $NaSO_4$(s) is 3.5 g.

Plan Your Strategy

Use the formula $n = \dfrac{m}{M}$ to find the equivalent molar amount of 3.5 g of $NaSO_4$(s).

Rearrange the formula $c = \dfrac{n}{V}$ to determine what volume of solution can be made from the solute.

• • •

Solution

$$n = \frac{m}{M} = \frac{3.5 \text{ g } \cancel{Na_2SO_4(s)}}{142.05 \dfrac{\cancel{\text{g } Na_2SO_4(s)}}{\text{mol } Na_2SO_4(s)}}$$

$$= 0.0246 \text{ mol } Na_2SO_4(s)$$

$$c = \frac{n}{V}$$

$$V = \frac{n}{c} = \frac{0.0246 \text{ } \cancel{\text{mol } Na_2SO_4(s)}}{0.25 \dfrac{\cancel{\text{mol } Na_2SO_4(aq)}}{L}}$$

$$= 0.099 \text{ L}$$

Check Your Solution

The answer has the correct units (L) and the correct number of significant digits (two).

Using dimensional analysis

$$x\text{L} = \frac{3.5 \text{ g } \cancel{Na_2SO_4(s)}}{1} \times \frac{1 \text{ } \cancel{\text{mol } Na_2SO_4(s)}}{142.05 \text{ g } \cancel{Na_2SO_4(s)}} \times \frac{1 \text{ L}}{0.25 \text{ } \cancel{\text{mol } Na_2SO_4(s)}}$$

$$= 0.099 \text{ L}$$

Therefore, 0.099 L of a 0.25 mol/L solution of sodium sulfate could be made from 3.5 g of solute.

27. Calculate the volume of solution that could be prepared from the following masses of solutes and concentrations:

    a) 1.65 mol/L solution beginning with 3.3 g of NaCl(s)

    b) 0.225 mol/L solution beginning with 2.0 g of $AgNO_3(s)$

28. What volume of solution could be prepared from the following masses?

    a) 0.398 mol/L solution beginning with 10.0 g of potassium dichromate, $K_2Cr_2O_7(s)$

    b) 4.25 mmol/L solution beginning with 4.5 g of sucrose, $C_{12}H_{22}O_{11}(s)$

29. Intravenous solutions are commonly 0.28 mol/L glucose. What volume of a standard intravenous solution, measured in litres, could be made from 2.5 kg of glucose, $C_6H_{12}O_6(s)$?

30. Household vinegar is a 0.016 mol/L solution of acetic acid, $CH_3COOH(aq)$. How much acetic acid solution with a concentration of 0.016 mol/L could be made from 1.0 g of pure (glacial) acetic acid? (The IUPAC name for acetic acid is ethanoic acid.)

Section 5.3 Summary

- There are a number of ways that the concentration of a solution can be expressed quantitatively. These include as percent by mass, in parts per million or parts per billion (ppm or ppb), and as a molar concentration.

- Percent by mass is the mass of solute divided by the mass of a solution, expressed as a percentage. Because 1 mL of water has a mass of 1 g, mass percent of a water solution is sometimes reported in units of g/100 mL.

$$\text{Percent by mass (m/m)} = \frac{\text{Mass of solute (g)}}{\text{Mass of solution (g)}} \times 100\%$$

- Parts per million is the mass in grams of a solute in one million grams of a solution.

$$\text{Concentration in ppm} = \frac{\text{Mass of solute (g)}}{\text{Mass of solution (g)}} \times 10^6 \text{ ppm}$$

- Parts per billion is the mass in grams of a solute in one billion grams of a solution.

$$\text{Concentration in ppb} = \frac{\text{Mass of solute (g)}}{\text{Mass of solution (g)}} \times 10^9 \text{ ppb}$$

- Molar concentration is the amount in moles of solute dissolved in 1 L of solution. This is also known as molarity and has the units mol/L

$$\text{Molar concentration (mol/L)} = \frac{\text{Amount of solute (mol)}}{\text{Volume of solution (L)}} \quad \text{or} \quad c = \frac{n}{V}$$

- The concentration of individual ions formed when an ionic compound dissolves in solution can be calculated. To perform the calculation, you need to write a dissociation equation for the compound and then use the coefficient of each ion to determine how many moles of that ion will be released when one mole of the compound dissolves in water.

1. Ammonium chloride, $NH_4Cl(s)$, is a very soluble salt. If 300 g of ammonium chloride are dissolved in 600 mL of water, what is the concentration of the solution in percent by mass?

2. A 275 g sample of brass was found to contain 75 g of zinc.
   a) What is the concentration of zinc in percent by mass?
   b) What is the other element found in brass?
   c) What is the concentration of the other element in brass in percent by mass?

3. To prevent bacterial and algal growth in a swimming pool, chlorine concentration should be maintained between 1.5 ppm and 3.0 ppm. If your pool has a volume of 7.00×10^3 L, what is the minimum mass of chlorine that should be in your pool?

4. A contaminant is present in a lake at a concentration of 0.036 ppm. Express the concentration in ppb.

5. The maximum allowable level of selenium in drinking water is 50 ppb. A chemist found 35 µg of selenium in 500 mL of lake water. Would you drink that lake water? Why or why not?

6. Calculate the concentration (in mol/L) of a solution made by dissolving 37.3 g of magnesium nitrate in enough water to make 2.50 L of solution.

7. Given the data in the table below, determine the concentration of the solution (in mol/L).

   Data Obtained in the Crystallization of a Solution of Copper(II) Nitrate

	Results
Initial volume of solution (mL)	100
Mass of solute remaining (g)	0.45

8. A solution is made by dissolving 2.5 g of potassium iodide in enough water to make 375 mL of solution. What is the concentration of the solution (in mol/L)?

9. If 3.75 µg of magnesium phosphate are dissolved in 1.25 L of water, determine the concentration of each of the ions present in mol/L.

10. What mass of potassium chloride, $KCl(s)$, is used to make 25.0 mL of a solution with a concentration of 2.00 mol/L?

11. What volume of a 0.100 mol/L solution of silver nitrate, $AgNO_3(aq)$, could be prepared if you have 2.25 g of solute?

12. Which method of reporting concentration would you expect to be used in the following situations? If you are not sure, make a guess and then explain what type of information would help you to decide.
   a) mercury contamination in a lake, such as the one shown below

   b) a standard solution of potassium permanganate in a chemistry stock room
   c) radon gas in a cellar
   d) carbon in steel
   e) copper in bronze
   f) ethanoic acid in vinegar

13. Your teacher dissolves several lumps of a white compound in some water. She gives you a sample of the water and asks you to determine the percent by mass of the solid dissolved in the solution.
   a) Write a detailed experimental method that you could use to find the answer. Explain what measurements you would take and calculations that you would perform.
   b) The teacher then asks you to determine the molar concentration of the compound in the solution. First, you may ask one question and she will give you the correct answer. What question would you ask? Explain how you would use the answer and the percent by mass that you calculated in part (a) to find the molar concentration.

14. You are told that the concentration of an ionic compound in a solution is 0.275 mol/L. There is no other compound in the solution. You are also told that the concentration of potassium in the same solution is 0.825 mol/L. Explain how both concentrations could be correct. What might the ionic compound be?

15. A technician is asked to prepare a stock solution of 1.5 mol/L sodium sulfate for a class experiment. There will be five groups of students performing the experiment and each group will need to use 25 mL of the stock solution. Describe, in detail, how the technician should prepare the solution.

Preparing and Diluting Solutions

Section Outcomes

In this section, you will:

- **use** a balance and volumetric glassware to prepare solutions of specified concentration
- **describe** the procedures and calculations required for preparing and diluting solutions
- **calculate** concentrations and/or volumes of diluted solutions, and calculate the quantities of a diluted solution and water to use when diluting
- **assess** the impact of industrial effluent on water quality

Key Terms

standard solution

What do the effectiveness of a medicine, the safety of a chemical reaction, the cost of an industrial process, and the taste of a soft drink have in common? They all depend on solutions that are made carefully with known concentrations. A solution with a known concentration is called a **standard solution**. There are two ways to prepare an aqueous solution with known concentration. You can make a solution by dissolving a measured mass of pure solute in a certain volume of solution. Alternatively, you can dilute another standard solution by adding a known volume of additional solvent.

Dilution Calculations

Suppose you want to dilute a standard solution to make a new solution of known concentration. How do you know how much standard solution to use? The amount of solute you draw into the pipette is the same as the amount of solute in the final diluted solution; only the amount of solvent changes. This allows you to set up an equation to perform dilution calculations:

Let c = concentration
$\quad V$ = volume

moles of solute (before dilution) = moles of solute (after dilution)

$$c\left(\frac{\text{mol}}{\text{L}}\right) \times V(\text{L}) \text{ (before dilution)} = c\left(\frac{\text{mol}}{\text{L}}\right) \times V(\text{L}) \text{ (after dilution)}$$

Let c_1 = concentration before dilution
V_1 = volume before dilution
c_2 — concentration after dilution
V_2 = volume after dilution

The formula can then be shortened to give: $c_1 V_1 = c_2 V_2$

First, calculate the amount (number of moles) of one solution and use it to determine the quantity in question. Read the Sample Problem and then try the Practice Problems that follow.

Sample Problem

Diluting a Standard Solution

Problem

For a class experiment, your teacher must make 2.0 L of 0.10 mol/L sulfuric acid, H_2SO_4(aq). This acid is usually sold as an 18 mol/L concentrated solution. How much of the concentrated solution should be used to make a new solution with the correct concentration?

What Is Required?

You need to find the volume of concentrated solution to be diluted.

What Is Given?

Initial concentration (c_1) = 18.0 mol/L
Concentration of diluted solution (c_2) = 0.10 mol/L
Volume of diluted solution (V_2) = 2.0 L

Plan Your Strategy

Once you have assigned your variables, rearrange the equation to solve for the unknown, V_1.

Act on Your Strategy

$$c_1 V_1 = c_2 V_2$$

$$V_1 = \frac{c_2 V_2}{c_1} = \left(\frac{\left(0.10 \, \frac{\text{mol } H_2SO_4(aq)}{\cancel{L}} \right)(2.0L)}{18.0 \, \frac{\text{mol } H_2SO_4(aq)}{\cancel{L}}} \right)$$

$$= 0.011 \text{ L}$$

Solution

Therefore, 0.011 L of the concentrated 18.0 mol/L solution should be diluted to 2.0 L to make 0.10 mol/L $H_2SO_4(aq)$. Always add acid to water when diluting acid, not the other way around!

Check Your Solution

The units are correct. The final solution (0.10 mol/L) is far less concentrated than the original solution (18 mol/L). It is, therefore, reasonable that only a small volume of concentrated solution is needed.

Using Dimensional Analysis

Use the two values that will allow you to solve for moles, then use the other value to solve for the volume of concentrated solution.

$$x\text{L (concentrated)} = \frac{0.10 \, \text{mol } H_2SO_4(aq)}{\cancel{L} \, \text{(dilute)}} \times \frac{2.0 \, \cancel{L} \, \text{(dilute)}}{1} \times \frac{1 \text{ L (concentrated)}}{18.0 \, \text{mol } H_2SO_4(aq)}$$

$$= 0.011 \text{ L}$$

Practice Problems

31. Suppose that you are given a 1.25 mol/L standard aqueous solution of sodium chloride, NaCl(aq). What volume of standard solution must you use to prepare the following solutions?
   a) 50 mL of 1.00 mol/L NaCl(aq)
   b) 200 mL of 0.800 mol/L NaCl(aq)

32. What concentration of solution is obtained by diluting 50.0 mL of 0.720 mol/L aqueous sodium nitrate, $NaNO_3(aq)$, to each of the following volumes?
   a) 120 mL
   b) 400 mL
   c) 5.00 L

33. A solution is prepared by *adding* 600 mL of distilled water to 100 mL of 0.15 mol/L ammonium nitrate. Calculate the molar concentration of the solution. Assume that the volume quantities can be added together.

Now that you understand how to perform calculations for standard solutions and dilutions, it is time for you to try them yourself. In Investigation 5.D on page 200, you will prepare and dilute standard solutions. You will then use these solutions to estimate the concentration of another solution.

The Responsible Care® Program

Although a significant effort is being made globally to minimize the impact of chemical use, chemicals are regularly released into the environment. The Responsible Care® Program, a global initiative of the International Council of Chemical Associations, represents companies that produce, transport, and use chemicals. The goal of this program is to improve the health, safety, and environmental performance of these organizations.

In the next investigation, you will research and analyze the effects of releasing chemicals from an industry into waterways.

INVESTIGATION 5.C

Target Skills

Researching collectively the risk–benefit issue of pollution of waterways by the release of effluents

Proposing a plan for reducing the impact on the environment

Pollution of Waterways: A Risk–Benefit Analysis

In this investigation, you and your group will research the effect of releasing chemicals (called effluent) into waterways and propose a plan to reduce the impact on the surrounding ecosystems.

Pulp and paper mills are a source of effluent. What measures do pulp and paper mills take to reduce their impact on the environment?

Issue

What are the risks and benefits of an effluent-producing industry in Alberta? Create a plan to reduce the impact of the effluent on the environment.

Gathering Data and Information

1. As a class, brainstorm industries in Alberta that release effluent into waterways.

2. Divide into groups of three or four students. In this small group, research one of the situations uncovered in your brainstorming session. Your research should include: **ICT**

- Background on the origin of the effluent: who produces it, why is it produced, what benefits are involved?

- Background on the negative effects of the effluent: what potential harm could the effluent cause to the ecosystem and human health?

- Technology, including both processes and equipment, that has been developed to reduce the risk and how it reduces the risk.

Organizing Findings

3. Perform a risk–benefit analysis on the production of the effluent.

4. In your group, propose a plan that would reduce the impact of this effluent on the ecosystem.

5. Share your findings in a presentation or as a poster.

Opinions and Recommendations

6. Discuss the groups' findings as a class.

a) Are the safety, environmental, and health concerns of chemical use in Alberta being addressed adequately?

b) In general, do the benefits of producing and using chemicals in Alberta and their subsequent appearance in our waterways outweigh the risks?

c) Do you think your proposal to reduce the environmental impact of the effluent you researched is reasonable? Explain your answer in detail.

Preparing and Diluting a Standard Solution

Copper(II) sulfate, $CuSO_4(s)$, is a soluble ionic compound. It is sometimes added to decorative ponds to control the growth of fungi. Solutions of copper(II) sulfate are blue. The intensity of the colour increases with increased concentration. In this investigation, you will prepare copper(II) sulfate solutions with known concentrations. Then you will dilute the solution to a specific concentration and verify your accuracy by comparing its colour intensity with the colour intensities of known solutions. You teacher will provide you with step-wise instructions for preparing a diluted solution of known concentration from a standard solution.

Copper(II) sulfate pentahydrate, $CuSO_4 \cdot 5H_2O(s)$, is a *hydrated salt*. Hydrated salts are ionic compounds that have a specific number of water molecules associated with each formula unit.

Question

How can you use colour to compare the concentration of solutions?

Safety Precautions

- Copper(II) sulfate is poisonous. Wash your hands at the end of this investigation.
- If you spill any solution on your skin, wash it off immediately with copious amounts of cool water.

Materials

- copper(II) sulfate pentahydrate, $CuSO_4 \cdot 5H_2O(s)$ ☠ Ⓣ
- standard solutions of $CuSO_4 \cdot 5H_2O(s)$ in large test tubes labelled 1–6 ☠ Ⓣ
- 100 mL volumetric flask
- 100 mL beaker
- 250 mL beakers (2)
- light box, overhead projector, or spectrophotometer
- distilled water
- medicine dropper
- paper towel
- electronic balance
- scoopula
- pipette bulb
- large test tube
- grease pencil or marker
- pipette (or graduated cylinder)
- stirring rod
- funnel

Procedure

Part 1: Preparing a Standard Solution

1. With your partner, calculate the mass of copper(II) sulfate pentahydrate required to make 100 mL of a 0.500 mol/L aqueous solution. *Include the water molecules that are hydrated to the crystals, as given in the chemical formula, in your calculation of molar mass.* Show all of your calculations.

2. Using an electronic balance, measure the mass of solute you calculated in step 1 into a 100 mL beaker.

3. Add approximately 50 mL of distilled water to the $CuSO_4 \cdot 5 H_2O(s)$. Stir with the stirring rod until the solute is completely dissolved.

4. Transfer your solution to a 100 mL volumetric flask using a funnel.

5. Rinse the beaker and stirring rod into the funnel with distilled water. Rinse the funnel into the volumetric flask.

6. Add distilled water until the bottom of the meniscus reaches the etched line on the volumetric flask.

7. Stopper the flask, keep your thumb on the stopper, and invert several times to mix the solution.

Part 2: Diluting the Standard Solution

In this part of the investigation, you will be diluting your solution to obtain 100 mL of a 0.0500 mol/L solution of copper (II) sulfate pentahydrate.

1. With your partner, calculate the volume of the standard solution that must be diluted to obtain 100 mL of a 0.0500 mol/L solution.

2. Transfer your standard solution to a clean 250 mL beaker.

3. Rinse a pipette or graduated cylinder with a small amount of your standard solution. Discard the rinse solution to the waste container.

4. Transfer the volume of the standard solution calculated in step 1 using either the pipette or graduated cylinder to a clean volumetric flask. (See Appendix for instructions on the proper use of a pipette.)

5. Fill the volumetric flask to the 100 mL line with distilled water.

6. Invert the flask several times to mix.

7. Label a large test tube "standard solution" using a grease pencil or marker. Transfer some of your diluted solution to the large test tube to compare with the standard solutions. The height of the solution in all of the test tubes should be the same. Use a medicine dropper to add or take away solution from your labelled test tube, as needed.

8. The best way to compare colour intensity is by looking down through the test tube. Wrap each test tube with a small piece of paper towel to stop light from entering the side.

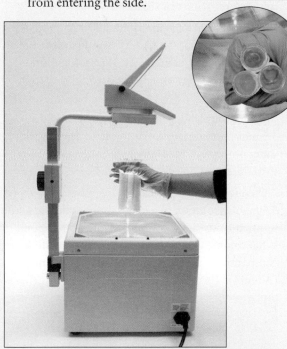

9. Place the solutions over a diffuse light source, such as a light box or overhead projector, as shown in the photograph. Compare the intensity of the colour of your solution with the standards. Identify which standard most closely matches your solution.

10. Dispose of the solutions as directed by your teacher.

Analysis

1. Describe any possible sources of error for this investigation.

Conclusion

2. Obtain the concentrations of the labelled standards from your teacher. Using the concentration of the solution in the test tube that most closely matched yours and the theoretical concentration of your diluted solution (0.0500 mol/L), calculate the percentage error in this experiment.

Extensions

3. This experiment may also be performed with coloured sodium chloride (sodium chloride that has been treated with food colouring). Explain why this choice is a responsible choice and still achieves the objectives of the experiment.

4. Suppose you have 250.0 mL of standard solution of 0.500 mol/L $CuSO_4 \cdot 5H_2O(s)$. You are given a solution of $CuSO_4 \cdot 5H_2O(s)$ of unknown concentration. (Its colour is less intense than the colour of the standard.) Write a procedure that uses the technique of dilution to estimate the concentration of the unknown solution.

Intensity of colour can be used to compare the concentration of solutions. The more intense the colour, the more concentrated the solution.

Section 5.4 Summary

• There are numerous applications that require the use of solutions of known concentration. These can be prepared by either dissolving a pure solid of known mass in a known volume of solution or diluting a more concentrated standard solution.

• Dilution calculations allow you to determine how much of a standard solution you should use to prepare a new diluted solution. The equation $c_1V_1 = c_2V_2$ can be used to calculate the amount of standard required for a diluted solution of known concentration and volume.

1. What is a standard solution? Explain how a standard solution is used in a laboratory.

2. List the steps in preparing a standard solution.

3. List the steps in performing a dilution.

4. A student made the following errors while preparing solutions. Describe how the errors will affect the resulting solution. Will it be more concentrated or more dilute than expected? Explain what the student should have done in each case.

   a) The student dissolves the solute in a beaker and transfers the solution to a volumetric flask but forgets to rinse the beaker and add the rinse to the volumetric flask.

   b) The student fills the volumetric flask so that the *top* of the meniscus is at the fill line.

   c) Before using a pipette to draw up the standard solution, the student rinses the pipette with distilled water but not with standard solution.

   d) The student uses the pipette bulb to force out the final drop of solution from the pipette.

5. The figure below shows three pieces of glassware.

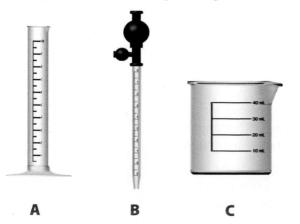

A   **B**   **C**

a) Give the name for each piece of glassware.

b) Which piece of glassware measures 10 mL of solution most precisely? Explain your answer.

c) Which piece of glassware measures 10 mL of solution least precisely? Explain your answer.

6. In Investigation 5.D, you made serial dilutions of a standard solution. Serial dilutions are made by diluting a standard solution then using a sample of the diluted solution to make another dilution. Why might it be better to make serial dilutions, than one complete dilution of a solution, when you want to make 100 mL of a 0.001 mol/L solution from a standard solution that had a concentration of 5.0 mol/L?

7. What is the final concentration of a solution if 15.0 mL of a 3.75 mol/L solution of potassium chloride, KCl(aq), were diluted to 4.50 L?

8. How much of a 1.3 mol/L solution could be made from 350 mL of a 5.7 mol/L standard solution? Describe the steps in preparing this solution.

9. What volume of a 5.0 mol/L standard solution of calcium nitrate, $Ca(NO_3)_2(aq)$, would have to be diluted to make 150 mL of a 0.175 mol/L solution? Approximately how much water would have to be added?

10. You and your lab partners are given 25 mL of a 1.5 mol/L solution of sodium sulfate. How much water must you add in order to dilute the solution to a concentration of 0.050 mol/L?

11. A chemical company accidentally released 475 L of a 5.50 mol/L solution of a toxic compound. The solution went into a stream that emptied into a small pond. What volume would the pond have to be to dilute the compound to a safe concentration of 0.35 μmol/L? (A μmol is 10^{-6} mol.)

Many chemical reactions occur in solution. While there are various types of solutions, all have a uniform composition throughout (homogeneous mixture). Each solution is a particular combination of solute dissolved in solvent, and can be in the form of solid, liquid, or gas.

Aqueous solutions are the most common in chemistry. Typically, these involve a solid solute dissolving in water, which occurs according to basic physical principles. Dissolving involves the breaking of bonds in the solute, the breaking of bonds in the solvent, and the formation of bonds between the solute and solvent. The overall process of dissolving can be endothermic or exothermic.

Soluble ionic compounds dissolve in water to produce solutions that conduct electric current and are referred to as electrolytes. A dissociation equation indicates what ions are in solution when an ionic compound is dissolved in water. Most molecular compounds that dissolve in water do not produce solutions that conduct electric current and are called non-electrolytes. Acids are an exception to this because they ionize in solution and, therefore, can act as electrolytes.

Solubility represents the quantity of a solute that can be dissolved in a given quantity of solvent at a specific temperature. It can be affected by both temperature and pressure. An unsaturated solution can dissolve more solute at a particular temperature. A saturated solution cannot dissolve more solute at a given temperature. A saturated solution is in equilibrium, because the rate at which the solute dissolves is equal to the rate at which the solute crystallizes. A supersaturated solution is one which contains more dissolved solute than its solubility would indicate. They can be made by preparing an solution at high temperature and allowing it to cool undisturbed.

As the solution cools, it will reach a temperature at which the solution contains more dissolved solute that it should at that temperature.

The solubility of most solids in liquids increases with temperature, because the bonds in the solute and solvent have more energy. The solubility of a liquid in another liquid is not greatly affected by temperature. At higher temperatures, gases increase in energy and can come out of the liquid in which they are dissolved. Therefore, the solubility of gases decreases at higher temperatures. Although changes in pressure have little effect on the solubility of solids in liquid, it does affect the solubility of a gas in liquid solvent. The solubility of the gas will decrease as the pressure decreases, causing the gas to 'bubble' out of the solution.

There are a number of ways that the concentration of a solution can be expressed quantitatively. The formulas are summarized on page 195. The concentration of individual ions formed when an ionic compound dissolves in solution can be calculated. This calculation requires knowledge of the concentration of the ionic substance dissolved and writing a dissociation equation for the reaction to know the moles of ions produced for every mole of ionic compound dissolved.

Solutions of known concentration can be prepared by either dissolving a pure solid of known mass in a known volume of solution or diluting a more concentrated standard solution. Dilution calculations allow you to determine how much of a standard solution you should use to prepare a new diluted solution. The equation $c_1V_1 = c_2V_2$ can be used to calculate the amount of stock required for a diluted solution of known concentration and volume.

Concept Organizer — Properties of Solutions

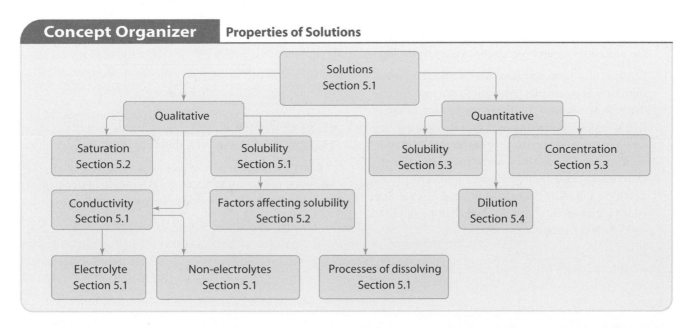

Chapter 5

Understanding Concepts

1. Identify at least two solutions in your home that are
   a) beverages
   b) found in the bathroom or medicine cabinet
   c) solids

2. How is a solution different from a pure substance? A mechanical mixture? Give specific examples.

3. Mixing 2 mL of linseed oil and 4 mL of turpentine makes a binder for oil paint. Which liquid is the solvent? Why?

4. Why does an aqueous solution of an electrolyte conduct electric current but an aqueous solution of a non-electrolytes does not?

5. Explain why, when some substances such as sodium hydroxide dissolve in water, energy is released (the reaction is exothermic), whereas when other substances such as ammonium nitrate dissolve in water, energy is consumed (the reaction is endothermic).

6. Why is it correct to say that nothing is really insoluble?

7. Explain the difference(s) among an unsaturated solution, a saturated solution, and a supersaturated solution.

8. Describe the concept of equilibrium using a saturated solution.

9. What two factors affect solubility? How do they affect solids, liquids, and gases in solution differently?

Applying Concepts

10. Suppose that your teacher gives you three test tubes. Each test tube contains a clear, colourless liquid. One liquid is an aqueous solution of an ionic compound. Another liquid is an aqueous solution of a molecular compound. The third liquid is distilled water. Outline the procedure for an experiment to identify the liquids.

11. Fertilizers for home gardens may be sold as aqueous solutions. Suppose that you want to start a company that sells an aqueous solution of potassium nitrate, $KNO_3(aq)$, fertilizer. You need a solubility curve (a graph of solubility versus temperature) to help you decide what concentration to use for your solution. Describe an experiment that you might perform to develop a solubility curve for potassium nitrate. State which variables are controlled, which is manipulated, and which is responding.

12. Potassium alum, $KAl(SO_4)_2 \cdot 12H_2O(aq)$, is used to stop bleeding from small cuts. The solubility of potassium alum, at various temperatures, is given below.

Solubility of Potassium Alum

Temperature (°C)	Solubility (g/100 mL water)
0	4
10	10
20	15
30	23
40	31
50	49
60	67
70	101
80	135

a) Plot a graph of solubility versus temperature.

b) From your graph, interpolate the solubility of potassium alum at 67 °C.

c) By extrapolation, estimate the solubility of potassium alum at 82 °C.

d) Look at your graph. At what temperature will 120 g of potassium alum form a saturated solution in 100 mL of water?

Solubility of Four Salts at Various Temperatures

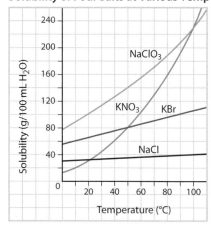

Use the graph above to answer questions 13 and 14.

13. At 80 °C, what mass of sodium chloride dissolves in 1.0 L of water?

14. What minimum temperature is required to dissolve 24 g of potassium nitrate in 40 g of water?

204 MHR · Unit 3 Matter as Solutions, Acids, and Bases

Solving Problems

15. Boric acid solution, $H_3BO_3(aq)$, is used as eyewash. What mass of boric acid is present in 250 g of a solution that is 2.25% m/m acid in water?

16. Sodium hydroxide solution, $NaOH(aq)$, with a concentration of 10% m/m is used to break down wood fibre to make paper.
   a) What mass of solute is needed to make 250 mL of 10% m/m solution?
   b) What mass of solvent is needed?
   c) What is the molar concentration of the solution?

17. Some municipalities add sodium fluoride, $NaF(s)$, to drinking water to help protect children's teeth. If a municipality wants to maintain the concentration of sodium fluoride at 2.9×10^{-5} mol/L, what mass (in mg) of sodium fluoride is dissolved in 1 L of solution? Express this concentration as ppm.

18. A gardening "recipe" for fertilizer suggests adding 15 g of Epsom salts, magnesium sulfate heptahydrate, $MgSO_4 \cdot 7H_2O(s)$, to enough water to make 4.0 L of solution. What will be the concentration of magnesium ions (in mol/L)?

19. Calculate the concentration (in mol/L) of each of the following aqueous solutions:
   a) 7.37 g of table sugar, $C_{12}H_{22}O_{11}(s)$, dissolved in 125 mL of solution
   b) 15.5 g of ammonium phosphate, $(NH_4)_3PO_4(s)$, dissolved in 180 mL of solution
   c) 76.7 g of glycerol, $C_3H_8O_3(\ell)$, dissolved in 275 mL of solution

20. Calculate the concentration (in mol/L) of each of the ions in the following solutions:
   a) 2.45 mg of magnesium hydroxide, $Mg(OH)_2(s)$, in 275 mL of solution
   b) 1.9 g of aluminium nitrate, $Al(NO_3)_3(s)$, in 1.2 L of solution
   c) 0.225 g of sodium phosphate, $Na_3PO_4(s)$, in 750 mL of solution

21. A saturated solution of sodium acetate, $NaCH_3COO(aq)$, can be prepared by dissolving 4.65 g in enough water to make 10.0 mL of solution at 20 °C. What is the molar concentration of the solution?

22. Calculate the mass of solute required to prepare each solution below:
   a) 250 mL of 0.250 mol/L calcium acetate, $Ca(CH_3COO)_2(aq)$

   b) 1.8 L of 0.35 mol/L ammonium sulfate, $(NH_4)_2SO_4(aq)$

23. Calculate the volume of solution that could be prepared of the following:
   a) 0.205 mol/L sodium fluoride, $NaF(aq)$, from 2.0 g of solute
   b) 1.3 mmol/L ammonium bromide, $NH_4Br(aq)$, from 1.3 g of solute

24. Calculate the molar concentration of each solution after dilution:
   a) 20 mL of 6.0 mol/L hydrochloric acid, $HCl(aq)$, diluted to 70 mL
   b) 300 mL of 12.0 mol/L ammonia, $NH_3(aq)$, diluted to 2.50 L

25. Calculate the volume of the standard solution required to prepare the following solutions:
   a) 500 mL of 2.50 mol/L sulfuric acid, $H_2SO_4(aq)$, from a 6.0 mol/L standard
   b) 1.50 L of 0.00350 mol/L copper (II) sulfate, $CuSO_4(aq)$, from a 1.25 mol/L standard

Making Connections

26. Suppose that you made a pot of hot tea. Later, you put a glass of the clear tea in the refrigerator to save it for a cool drink. When you take it out of the refrigerator hours later, you notice that it is cloudy. How could you explain this to a younger sibling?

27. "Everyone knows that oil doesn't dissolve in water." Is this statement true or false? Explain.

28. A sour gas plant is being proposed only a few kilometres from your school. What concerns would you have? Write a short speech that you could give to city council.

29. List ten activities you performed yesterday that used water. Which activity required the most water? Estimate the volume of water you used. Would it make sense to have two supplies of water to your home, one for drinking and a second of lower purity for other activities that use water? Why or why not? Why do you think our water system is not set up in this way?

30. A concentrated standard solution of ammonium sulfate sat open and undisturbed in the stock room of a chemistry lab for several days. When the teacher noticed it and picked it up to put it away, crystals suddenly started to form. Provide a logical explanation for the crystal formation.

Acids and Bases

Chapter Concepts

6.1 Theories of Acids and Bases

- Acids and bases have been used throughout history to solve technological problems.

- Acids produce $H_3O^+(aq)$ in aqueous solutions. Bases produce $OH^-(aq)$ in aqueous solutions.

6.2 Strong and Weak Acids and Bases

- Acid or base properties are determined by both the concentration and the identity of the acid or base.

- The relative strength of acids and bases can be explained by the degree of ionization or dissociation.

- Some acid molecules have more than one hydrogen atom that can ionize and react with water.

6.3 Acids, Bases, and pH

- Indicators enable you to estimate the hydronium ion concentration in a solution; pH meters enable you to precisely measure the hydronium ion concentration in a solution.

- $H_3O^+(aq)$ and $OH^-(aq)$ concentrations, and pH and pOH of acid and base solutions can be calculated based on logarithmic expressions.

For many people, the word "acid" evokes the image of a dangerous, fuming liquid. Many acids are indeed corrosive and reactive, and they can be hazardous to work with. Hydrofluoric acid, for example, although it is a weak acid, can cause deep, slow-healing tissue burns if handled carelessly. Its corrosiveness can also be useful, however. Artisans use this acid to etch delicate patterns on glass. In this chapter, you will discover why the very properties that make acids and bases dangerous also make them extremely useful for a variety of applications. In the Launch Lab on the facing page, observe an acid–base reaction.

The Colour of Your Breath

Bromothymol blue is an acid–base indicator, meaning it turns different colours in solutions of different pH. It is used in products to test the pH of swimming pools. As you carry out this activity, recall what you have learned about acids, bases, pH, and indicators.

Safety Precautions

- Bromothymol blue is harmful if swallowed. It may cause irritation to your skin, eyes, and respiratory tract.
- Sodium hydroxide is corrosive. If you spill any on your skin, immediately rinse with plenty of cold water.

Materials

- tap water
- bromothymol blue indicator
- 0.10 mol/L NaOH(aq) in a dropper bottle
- 100 mL graduated cylinder
- 250 mL Erlenmeyer flask
- straws
- stopwatch or clock with second hand

Procedure

1. Using a graduated cylinder, measure 100 mL of tap water and pour it into the Erlenmeyer flask.

2. Add five drops of bromothymol blue indicator. Record your observations.

3. Add drops of NaOH(aq), swirling the solution after each addition, until the solution turns blue.

4. Start the stopwatch. Using the straw, blow into the solution until the solution changes colour. **Caution** Do not draw the solution into the straw.

5. Observe how long it takes for a colour change to occur.

6. Repeat steps 3 to 5, but hold your breath for 30 s before step 4.

7. Observe how long it takes for a colour change to occur.

8. Dispose of the waste as directed by your teacher.

Analysis

1. Based on your prior knowledge, what type of substance was produced when you blew into the solution? Explain your answer.

2. How did holding your breath for 30 s affect your observations, if at all? Explain your observations.

3. As a class, compare and discuss your results. Did your classmates see similar results? Discuss why or why not.

Hydrofluoric acid, HF(aq), is especially dangerous because the fluoride ion, F⁻(aq), which is tiny relative to other ions, easily penetrates skin, damaging tissue and bone.

Theories of Acids and Bases

Section Outcomes

In this section, you will:
- **recall** nomenclature of acids and bases
- **design** a procedure to determine the properties of acids and bases
- **describe** acids and bases empirically, using indicators, pH, and conductivity
- **define** Arrhenius (modified) acids as substances that produce $H_3O^+(aq)$ in aqueous solution
- **define** Arrhenius (modified) bases as substances that produce $OH^-(aq)$ in aqueous solution

Key Terms

neutral solution
Arrhenius theory of acids and bases
Arrhenius acid
Arrhenius base
ionize
dissociate
hydronium ion, $H_3O^+(aq)$
modified Arrhenius theory of acids and bases
modified Arrhenius acid
modified Arrhenius base

Figure 6.1 Saskatoon berries, native to Alberta, can be dried and used to make pemmican.

At room temperature, most food spoils quickly, which is a problem if you need to carry food with you on a long journey or if you want to store it for a long time. We can slow the spoilage by storing food in a refrigerator or freezer. However, these are relatively recent technologies. Throughout human history, people have developed many ways to halt or delay food spoilage. Ancient technologies still used today include drying, smoking, or pickling food in vinegar.

For example, pemmican, a dried meat product, is a stable trail food that North American Aboriginal peoples have made and used for thousands of years. The word "pemmican" is derived from the word *pemikan* (pay-me-kan) from the Abnaki language and the word *pimikan* (pe-me-kan) from the Cree language. The original word describes the process of bone marrow grease preparation but later became the name of the final product.

Consisting of pounded dried meat, bone marrow grease, and locally available berries, pemmican remains unspoiled and edible for a long time—up to 30 years. The citric acid and ascorbic acid in the berries (Figure 6.1) add a pleasantly tangy flavour and, more importantly, help to preserve the meat.

Acids and Bases: Problem Solvers

The longevity of pemmican is an example of a food-preservation technology that depends on the properties of acids. There are countless other technologies that depend on the properties of acids and bases, some of which you probably encounter frequently. For example, does your dinner of halibut taste too fishy? Squeeze some acid on it. Is your drain clogged? Use some base to unclog it. Is there a wart on your hand? Apply some acid to remove it. Do you have indigestion? Consume a tablet containing base to stop the burn. Table 6.1 shows some products and the acids or bases they contain.

Table 6.1 Common Products Containing Acids or Bases

Acids	
Product	**Acid(s) contained in the product**
vinegar	acetic (ethanoic) acid, $CH_3COOH(aq)$
wart remover	salicylic acid, $C_6H_4(OH)CO_2H(aq)$
soft drinks	carbonic acid, $H_2CO_3(aq)$ phosphoric acid, $H_3PO_4(aq)$
rust remover	glycolic acid, $CH_2(OH)COOH(aq)$ phosphoric acid, $H_3PO_4(aq)$
lemon juice	citric acid, $C_6H_8O_7(aq)$ ascorbic acid, $C_6H_8O_6(aq)$
Bases	
Product	**Base(s) contained in the product**
oven cleaner	aqueous sodium hydroxide, $NaOH(aq)$
drain cleaner	aqueous sodium hydroxide, $NaOH(aq)$
antacids (some brands)	magnesium hydroxide, $Mg(OH)_2(s)$
glass cleaner (some brands)	aqueous ammonia, $NH_3(aq)$

Reviewing Naming Rules for Acids and Bases

As you can see in Table 6.1 on the previous page, bases are not named according to any special rules. Sodium hydroxide, $NaOH(s)$, when dissolved in water, is called aqueous sodium hydroxide, $NaOH(aq)$. Acids, however, can be named according to two different systems—the classical system and the International Union of Pure and Applied Chemistry (IUPAC) guidelines. Using the IUPAC system, you name acids as you would any solution in water. Add "aqueous" to the beginning of the name of the compound in solution. The classical system names acids differently, according to whether they contain oxygen. The classical system is still widely used, so it is important to know both systems. Use Table 6.2 to help you recall how to name acids.

Table 6.2 Naming Acids

Acids that do not contain any oxygen			
Pure substance (ends in -ide)	**IUPAC name for acid**	**Classical name for acid**	**Formula for acid**
hydrogen chloride	aqueous hydrogen chloride	hydrochloric acid	$HCl(aq)$
hydrogen cyanide	aqueous hydrogen cyanide	hydrocyanic acid	$HCN(aq)$
hydrogen sulfide	aqueous hydrogen sulfide	hydrosulfuric acid	$H_2S(aq)$
Acids that contain oxygen atoms			
Pure substance (ends in -ate)	**IUPAC name for acid**	**Classical name for acid**	**Formula for acid**
hydrogen chlorate	aqueous hydrogen chlorate	chloric acid	$HClO_3(aq)$
hydrogen perchlorate	aqueous hydrogen perchlorate	perchloric acid	$HClO_4(aq)$
Pure substance (ends in -ite)	**IUPAC name for acid**	**Classical name for acid**	**Formula for acid**
hydrogen chlorite	aqueous hydrogen chlorite	chlorous acid	$HClO_2(aq)$
hydrogen hypocholorite	aqueous hydrogen hypochlorite	hypochlorous acid	$HClO(aq)$

Naming acids containing sulfur or phosphorus is analogous to naming acids containing chlorine. Add the ending -ic or -ous to sulfur- and phosphor-. For example, $H_2SO_2(aq)$ (from hydrogen hyposulfite) is hyposulfurous acid, and $H_3PO_5(aq)$ (from hydrogen perphosphate) is perphosphoric acid.

1 Name the following acids using classical and IUPAC systems:

a) $HF(aq)$
b) $HSCN(aq)$
c) $H_2CO_3(aq)$
d) $HNO_2(aq)$
e) $H_2SO_4(aq)$
f) $H_2SO_5(aq)$

2 Provide formulas for the following acids:

a) hydroiodic acid
b) aqueous hydrogen bromide
c) dichromic acid
d) aqueous hydrogen permanganate
e) hypocarbonous acid
f) aqueous hydrogen nitrate

ChemistryFile

Web Link

There is now considerable concern over the use of many commercially available cleaning products. Most are highly toxic and harmful to the environment. However, there are numerous non-toxic acids and bases that can be used instead, which are common to most households. Research the effects that commercial cleaning products have on the environment and what some of the non-toxic acid and base alternatives are. Try comparing how well a natural product works versus a commercial one that is sold for the same use. You can perform this in groups and present your findings to the class as a poster.

www.albertachemistry.ca
WWW

ChemistryFile

FYI

When naming acids, it is helpful to name the pure substance first. For example, to name HSCN(aq), first name HSCN(g). You will need to find the name of the SCN^- ion: use the list of polyatomic ions in Appendix I to help you.

Figure 6.2 Sulfuric acid, $H_2SO_4(aq)$, conducts electric current very well; that's one reason why it is used in car batteries.

Empirical Definitions of Acids and Bases

Many solutions of acids and bases are clear and colourless. However, **neutral solutions**, which are neither acidic nor basic, can also be clear and colourless. Therefore, determining whether a solution is acidic, basic, or neutral can be done using experimental tests that are based on their characteristic properties.

Empirical definitions of acids and bases, which are derived from their properties, have been developed over many centuries of experience and experimentation. The observable properties of acids and bases, most of which you have already studied, are listed in Table 6.3 below.

Table 6.3 Observable Properties of Acids and Bases

Test	Acids	Bases
litmus paper	turn blue litmus red	turn red litmus blue
pH paper	pH less than 7	pH greater than 7
electrical conductivity in solution	conduct electric current (electrolyte)	conduct electric current (electrolyte)
reaction with active metals, such as Mg(s) and Zn(s)	react to produce $H_2(g)$	do not react with active metals to produce $H_2(g)$
taste **Caution:** Never taste anything in the lab.	taste sour	taste bitter
feel **Caution:** Never deliberately touch chemicals. Many acids and bases can burn your skin.	have no characteristic feel	feel slippery

INVESTIGATION | 6.A

Target Skills

Designing a procedure to determine the properties of acids and bases

Conducting investigations into relationships and **using** a broad range of tools and techniques

An Empirical Definition for Acids and Bases

Many useful household substances are acids or bases. In this investigation, you will use known acidic and basic substances to determine the properties of acids and bases and to develop an empirical definition for each.

Question

What are the properties of acids and bases?

Safety Precautions

- Household solutions containing acids and bases are often toxic or corrosive. Wash any spills on skin or clothing with plenty of cool water. Inform your teacher immediately.

- If you are using a conductivity tester with two separate electrodes, keep the electrodes well separated while you perform your tests.

- Do not taste any of the substances you are working with.

- When you have completed the investigation, wash your hands.

Materials

- solutions of:
  - hand soap
  - laundry detergent
  - glass cleaner
  - antacid
  - milk
  - sour milk
  - carbonated water
  - soda pop (carbonated)
  - soda pop (flat)
  - apple juice
  - vinegar
  - baking soda
  - water and table salt solution
- red and blue litmus paper
- pH paper and/or pH meter
- short strips of magnesium ribbon
- spot plate
- dropper
- conductivity tester
- beakers

Procedure

1. Read the procedure steps below and construct an appropriate table to record your results.

2. Perform a litmus test on the solutions by placing a small amount on red and blue litmus paper. Record your observations.

3. Determine the pH of the solutions using either pH paper or a pH meter. Record your observations.

4. Determine whether each solution conducts electric current by placing 25 mL of a solution into a 50 mL beaker and using the conductivity tester. Record your observations.

5. Determine the reactivity of each solution with Mg(s) metal by placing 5 mL of the solution into a test tube and adding a small piece of the Mg(s). Record your observations.

6. Dispose of all materials as directed by your teacher.

Analysis

1. Use your observations to design a chart or concept map to compare the properties of acids and bases.

2. Categorize your solutions by their use. What is one common use for bases? What is one common use for acids?

3. What happens to the pH of milk as it sours?

Conclusion

4. As a class, write an empirical definition for acids and an empirical definition for bases.

Extension

5. **a)** How did the pH of the carbonated soda pop compare with the pH of the flat soda pop?

   b) (Perform this activity at home.) How does carbonated water taste compared with regular water? How does carbonated soda pop taste compared with flat soda pop? What might account for the difference in taste?

6. What are the names of the acids and bases in each of the products you tested? You will find some of the answers in Table 6.1. For other answers, examine the labels of the products, and use the Internet to conduct research. ICT

Explaining the Properties of Acids and Bases

To solve many everyday problems, it is often enough to know about the properties of acids and bases. For example, people were making and using soap and pemmican long before any theories of acids and bases had been developed. You read about pemmican on page 208. For hundreds, possibly even thousands, of years, people have been making lye soap. They poured water slowly through wood ashes and filtered the solution through straw. The resultant liquid, called lye, was a strongly basic solution. They then boiled the lye with animal fat that they had saved. The boiling liquid eventually became a thick, foamy soap they called lye soap.

So why develop theories to explain the properties of acids and bases if you already know how to make use of these properties? The goal of technology (solving problems) and the goal of science (explaining phenomena) are related. Understanding the science behind acids and bases can help people not only design new technologies but also improve existing technologies. In addition, an understanding of the chemistry of acids and bases allows chemists to develop the best methods for cleaning spills and preventing accidents with acids and bases.

Sometimes, the development of a new technology allows scientists to gather evidence to explain phenomena. For example, the discovery of electric current at the beginning of the nineteenth century and the technologies developed to measure it allowed scientists to study solutions, and therefore acids and bases, in terms of their conductivity.

Acids, Bases, and Conductivity

Svante August Arrhenius (1859–1927), shown in Figure 6.3, theorized about why some solutions conduct electric current while others do not. In 1887, Arrhenius proposed that certain substances break apart in water to form ions, producing a solution that can then conduct electric current.

Arrhenius's work on the conductivity of solutions led him to think about the properties of acids and bases. He knew that solutions of both acids and bases conduct electric current, as shown in Figure 6.4.

Arrhenius proposed that acids contained hydrogen atoms that formed hydrogen ions ($H^+(aq)$) in solution. He also proposed that bases contained hydroxide ions ($OH^-(aq)$) that could dissociate in solution.

Figure 6.3 Arrhenius's ideas about what happens to solutes in water were considered so radical that he was reluctantly awarded the lowest possible pass on his PhD thesis—a fourth-class degree. Ironically, he later won the 1913 Nobel Prize for Chemistry for the same work.

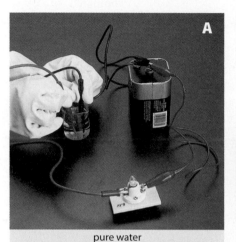

pure water

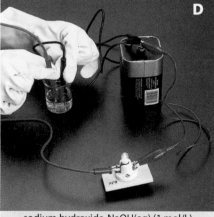

hydrochloric acid, HCl(aq) (1 mol/L)

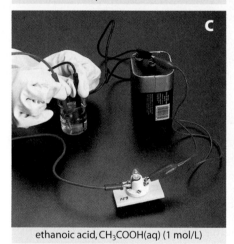

ethanoic acid, CH_3COOH(aq) (1 mol/L)

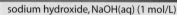

sodium hydroxide, NaOH(aq) (1 mol/L)

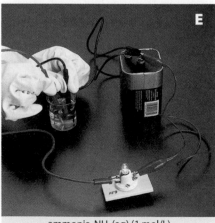

ammonia, NH_3(aq) (1 mol/L)

Figure 6.4 A conductivity tester enables you to decide whether a solution conducts electric current and, therefore, whether it contains ions. Notice that the bulb glows more brightly in some of the photographs. How might the brightness of the bulb relate to the concentration of ions in solution?

Acids and Bases in Solution

Hydrogen chloride, $HCl(g)$, is a molecular substance. According to Arrhenius, this molecular substance, when dissolved in water, forms $H^+(aq)$ and $Cl^-(aq)$ ions:

$$HCl(g) \xrightarrow{\text{in water}} H^+(aq) + Cl^-(aq)$$

Sodium hydroxide is an ionic substance. When dissolved in water, the ions break apart to form $Na^+(aq)$ and $OH^-(aq)$ ions:

$$NaOH(s) \xrightarrow{\text{in water}} Na^+(aq) + OH^-(aq)$$

Arrhenius's explanation of the properties of acids and bases is the **Arrhenius theory of acids and bases**.

The Arrhenius Theory of Acids and Bases

- An **Arrhenius acid** is a substance that ionizes to form hydrogen ions, $H^+(aq)$, in aqueous solution.

- An **Arrhenius base** is a substance that dissociates to produce hydroxide ions, $OH^-(aq)$, in aqueous solutions.

When acidic molecular substances dissolve in water, they **ionize** or form ions. When ionic substances break apart into separate ions, they **dissociate**. They do not ionize, because they are already made of ions. (This distinction was not yet made when Arrhenius developed his theory, and he used the term "dissociate" for all substances that formed ions in solution.)

According to the Arrhenius theory, acids increase the concentration of $H^+(aq)$ in aqueous solutions. Therefore, it can be inferred that an Arrhenius acid must contain hydrogen as the source of $H^+(aq)$, as illustrated in the ionization of $HCl(g)$ above. Bases, conversely, increase the concentration of $OH^-(aq)$ in aqueous solutions. An Arrhenius base must similarly contain the hydroxide ion, $OH^-(aq)$, as the source of $OH^-(aq)$, as shown above in the dissociation of $NaOH(aq)$.

Using the Arrhenius theory, you could infer that acids must contain a source of $H^+(aq)$ and bases must contain a source of $OH^-(aq)$. Does the Arrhenius theory successfully predict acidic or basic properties in all cases? In Investigation 6.B on the following page, you will test the Arrhenius theory of acids and bases.

• • •

3 Classify the following solutions as acidic, basic, or neutral using the Arrhenius theory of acids and bases. Provide a name for each solution:

a) $HNO_3(aq)$ d) $LiOH(aq)$

b) $Ca(OH)_2(aq)$ e) $CaCl_2(aq)$

c) $NaNO_3(aq)$ f) $CH_3COOH(aq)$
                       (**Hint:** Name the pure substance first.)

4 List three properties for each of acidic, basic, and neutral solutions.

• • •

Testing the Arrhenius Theory of Acids and Bases

Using the Arrhenius theory of acids and bases, predict whether the following solutions are acidic, basic, or neutral. Then measure pH and verify your predictions. Recall from your previous science courses that acids have a pH less than 7, and bases have a pH greater than 7. Neutral solutions have a pH of approximately 7.

Question

Does the Arrhenius theory of acids and bases successfully predict and explain the properties of acids and bases?

Prediction

Predict whether each solution will be acidic, basic, or neutral. Explain your predictions.

Safety Precautions

- Hydrochloric acid, sodium hydroxide, and ammonia are toxic and corrosive. Wash any spills on skin or clothing with plenty of cool water. Inform your teacher immediately.
- When you have completed this investigation, wash your hands.

Materials

- distilled water
- solutions of:
  - 0.10 mol/L NaOH(aq)
  - 0.10 mol/L NaCl(aq)
  - 0.10 mol/L $C_{12}H_{22}O_{11}$(aq)
  - 0.10 mol/L HCl(aq)
  - 0.10 mol/L NH_3(aq)
  - 0.10 mol/L CH_3COOH(aq)
  - 0.10 mol/L $NaHCO_3$(aq)
- pH paper/pH meter

Procedure

1. Construct a table to record your data. Include a column for your predictions.

2. Measure the pH of the distilled water, which will act as a control.

3. Measure the pH of each solution using a pH meter or pH paper.

4. Record your results.

5. Dispose of your materials as directed by your teacher.

Analysis

1. Provide the IUPAC and classical name for each of the substances in solution.

2. All the solutions were of equal concentration. Did all the acids have the same pH? What about all the bases? Explain your answers.

Conclusions

3. Re-examine your predictions.

   a) Did any results surprise you? Explain your answer.

   b) Explain your observations using the Arrhenius theory of acids and bases. Did any of your observations appear to contradict the Arrhenius theory of acids and bases? If so, specify.

Extension

4. For each of the solutions that produced unexpected results, describe two tests that you could perform to verify your observations. With your teacher's permission, perform these tests. Did your additional results support the pH you measured? What do these results suggest?

A Limitation of the Arrhenius Theory of Acids and Bases

If you carried out Investigation 6.B, you were probably able to successfully predict and explain the acid or base properties of most of the solutions by using the Arrhenius theory of acids and bases.

What happened when you tested aqueous ammonia, $NH_3(aq)$? A 0.10 mol/L solution of ammonia conducts electric current (see Figure 6.4), turns red litmus blue, and has a pH greater than 7. These results would probably lead you to conclude that aqueous ammonia is a base. Similarly, aqueous sodium hydrogen carbonate, $NaHCO_3(aq)$, also exhibits the properties of a base. How do you explain these results using the Arrhenius theory of acids and bases? You cannot. By definition, an Arrhenius base must contain a hydroxide ion that can be released during dissociation. Ammonia and sodium hydrogen carbonate contain no such ion.

What happens, however, if you modify the theory slightly? A modified theory recognizes that substances like ammonia *react* with water to produce an $OH^-(aq)$ ion:

$$NH_3(g) + H_2O(\ell) \rightleftharpoons NH_4^+(aq) + OH^-(aq)$$

Thus, although ammonia does not contain an $OH^-(aq)$ ion, it can react with water and produce an $OH^-(aq)$ ion. (Notice the equilibrium arrow, $\rightleftharpoons$, which you first saw in Chapter 5. You will learn more about equilibrium and acids and bases on page 219.)

The Hydronium Ion

A more recent discovery necessitates another modification to the Arrhenius theory of acids and bases. This discovery involves the ion that is responsible for acidity, $H^+(aq)$. Look again at the equation for the ionization of hydrochloric acid:

$$HCl(g) \xrightarrow{\text{in water}} H^+(aq) + Cl^-(aq)$$

This ionization occurs in aqueous solution, but chemists often leave out water as a component of the reaction. They simply assume it is there. What happens if you put water into the equation as a reactant? According to the Arrhenius theory, this is what happens:

$$HCl(aq) + H_2O(\ell) \rightarrow H^+(aq) + Cl^-(aq) + H_2O(\ell)$$

Notice that the water is unchanged when the reaction is represented in this way. However, since water is a polar molecule, the oxygen atom has a partial negative charge, and the hydrogen atoms have partial positive charges. Thus, water likely interacts in some way with hydrogen chloride as it ionizes.

When analytical technology was developed early in the twentieth century, scientists were able to gather evidence that a hydrogen ion does not exist in isolation in aqueous solution. Instead, it exists as the **hydronium ion, $H_3O^+(aq)$**. Hydronium ions and water molecules in solution are attracted by hydrogen bonds, as shown in Figure 6.5.

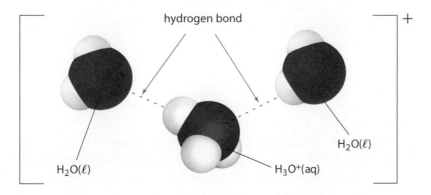

Figure 6.5 Although $H_3O^+(aq)$ is currently accepted as the form in which $H^+(aq)$ exists in solution, scientific debate continues about whether this is the best representation.

To account for the hydronium ion, you can modify the Arrhenius theory slightly to recognize that when acids ionize, the hydrogen ion exists in solution as the hydronium ion. In other words, in solution, acids react with water to form hydronium ions. For hydrochloric acid, the formation of the hydronium ion can be illustrated as shown in Figure 6.6 below.

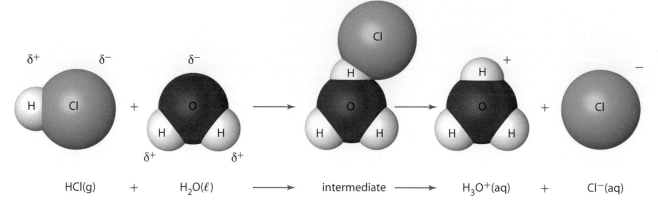

A Modified Theory

Figure 6.6 The polar ends of the water molecule attract the oppositely charged polar ends of the hydrogen chloride molecule. The molecular bond between the hydrogen and chloride ions is broken, producing hydronium, $H_3O^+(aq)$, and chloride, $Cl^-(aq)$, ions.

The Arrhenius theory of acids and bases becomes even more useful to explain the properties of acids and bases if you make a few modifications, shown below as the **modified Arrhenius theory of acids and bases**.

The Modified Arrhenius Theory of Acids and Bases

- An **Arrhenius acid** reacts with water to produce $H_3O^+(aq)$ in aqueous solution.
- An **Arrhenius base** dissociates or reacts with water to produce $OH^-(aq)$ in aqueous solution.

According to the modified theory, when neutral substances dissolve in water they produce neither $H_3O^+(aq)$ ions nor $OH^-(aq)$ ions.

Look back at the original Arrhenius definitions for acids and bases and compare them with the ones presented in the modified theory.

The reaction equations below represent the reaction of an Arrhenius acid with water to produce $H_3O^+(aq)$.

$$HI(aq) + H_2O(\ell) \rightarrow H_3O^+(aq) + I^-(aq)$$
$$HNO_3(aq) + H_2O(\ell) \rightarrow H_3O^+(aq) + NO_3^-(aq)$$
$$HClO_4(aq) + H_2O(\ell) \rightarrow H_3O^+(aq) + ClO_4^-(aq)$$

The reaction equations

$$LiOH(aq) \rightarrow Li^+(aq) + OH^-(aq)$$
$$Ba(OH)_2(aq) \rightarrow Ba^{2+}(aq) + 2OH^-(aq)$$

represent the dissociation of an Arrhenius base to produce $OH^-(aq)$. These reactions are consistent with the non-modified theory you saw earlier.

For the reaction

$$NH_3(aq) + H_2O(\ell) \rightleftharpoons NH_4^+(aq) + OH^-(aq)$$

ammonia, $NH_3(aq)$, is considered an Arrhenius base according to the modified theory because it *reacts* with water to produce $OH^-(aq)$.

Non-metallic oxides (substances made from a non-metal and oxygen), such as carbon dioxide, $CO_2(g)$, and metallic oxides (substances made from a metal and oxygen), such as barium oxide, $BaO(s)$, will undergo a two-step reaction to form acids and bases, as shown on the next page:

Step 1 $CO_2(g) + H_2O(\ell) \rightleftharpoons H_2CO_3(aq)$
Step 2 $H_2CO_3(aq) + H_2O(\ell) \rightleftharpoons H_3O^+(aq) + HCO_3^-(aq)$ } Non-metallic oxides form acids in water

Step 1 $SO_2(g) + H_2O(\ell) \rightleftharpoons H_2SO_3(aq)$
Step 2 $H_2SO_3(aq) + H_2O(\ell) \rightleftharpoons H_3O^+(aq) + HSO_3^-(aq)$

Step 1 $Na_2O(s) + H_2O(\ell) \rightarrow 2NaOH(aq)$

Step 2 $2NaOH(aq) \xrightarrow{\text{in water}} 2Na^+(aq) + 2OH^-(aq)$ } Metallic oxides form bases in water

Step 1 $BaO(s) + H_2O(\ell) \rightarrow Ba(OH)_2(aq)$

Step 2 $Ba(OH)_2(aq) \xrightarrow{\text{in water}} Ba^{2+}(aq) + 2OH^-(aq)$

Note how step 1 is a reaction with water, while step 2 is a dissociation in water to produce the $OH^-(aq)$.

Section 6.1 Summary

- Indicators, pH, conductivity, reactivity with metals, taste, and feel are all properties that can define an acid, base, or neutral solution.

- Acids and bases are used for a variety of things that include food preservatives, batteries, cleaning products, and the treatment of accidental spills.

- The original Arrhenius theory of acids and bases defined an acid as a substance that ionizes to form hydrogen ions, $H^+(aq)$ and a base as a substance that dissociates to produce hydroxide ions, $OH^-(aq)$.

- The original theory that Arrhenius proposed did not allow for the recognition of substances, such as ammonia, which are bases that do not dissociate to produce the hydroxide ion. In addition, it was discovered that the hydrogen ion does not exist in isolation but instead as the hydronium ion, $H_3O^+(aq)$.

- The modified Arrhenius theory defines an acid as a substance that reacts with water to produce $H_3O^+(aq)$ in aqueous solution and a base as a substance that dissociates or reacts with water to produce $OH^-(aq)$ in aqueous solution.

- Non-metallic and metallic oxides can undergo two-step reactions to form acids and bases, respectively.

SECTION 6.1   REVIEW

1. a) List three characteristic properties of an acid.
  b) List three characteristic properties of a base.

2. Acids and bases must be treated with care and caution in the laboratory.
  a) Explain why acids and bases can be dangerous.
  b) List the safety precautions you should take when working with acids and bases.

3. a) Define an acid and a base according to Arrhenius.
  b) List two observations that can be explained using the Arrhenius theory.
  c) Provide one example of an observation that *cannot* be explained using the Arrhenius theory.
  d) How does the modified Arrhenius theory help explain the observation you described in part (c)?

4. Write a chemical equation to show how the following acids would react in water, according to the modified Arrhenius theory of acids and bases. Name each acid:
  a) $HCl(aq)$ **d)** $HNO_3(aq)$
  b) $HBr(aq)$ **e)** $CO_2(g)$ (a non-metallic oxide)
  c) $HClO_4(aq)$

5. Write a chemical equation to show how the following bases would dissociate or react in water, according to the modified Arrhenius theory of acids and bases. Name each base:
  a) $NH_3(aq)$ **d)** $KOH(aq)$
  b) $NaOH(aq)$ **e)** $SrO(s)$ (a metallic oxide)
  c) $Ca(OH)_2(aq)$

Strong and Weak Acids and Bases

Section Outcomes

In this section, you will:
- **distinguish** between strong acids and bases and weak acids and bases based on ionization
- **design** an experiment to differentiate between weak and strong acids and between weak and strong bases
- **compare** the ionization of monoprotic acids with the ionization of polyprotic acids
- **compare** monoprotic and polyprotic bases
- **define** neutralization as a reaction that occurs between hydronium and hydroxide ions
- **assess** the risks and benefits of producing, using, and transporting acidic and basic substances

Key Terms

strong acid
weak acid
strong base
weak base
monoprotic acids
polyprotic acids
monoprotic base
polyprotic base
salt
neutralization reaction

Would you be more concerned if you spilled 0.10 mol/L hydrochloric acid or 0.10 mol/L ethanoic acid on your new pair of jeans? Do you think 1.0 mol/L hydrochloric acid would cause more damage than 0.10 mol/L hydrochloric acid? Examine the four test tubes in Figure 6.7. Each test tube contains an acid and a small piece of magnesium. Which two factors appear to affect the vigour of the reaction?

1.0 mol/L HCl(aq)   1.0 mol/L CH₃COOH(aq)   0.10 mol/L HCl(aq)   0.10 mol/L CH₃COOH(aq)

Figure 6.7 Reactions of different solutions with magnesium.

From Figure 6.7, you can infer that changing the concentration of an acid changes the acidic properties of a solution, because the reaction is more vigorous in the solution of 1.0 mol/L HCl(aq) than in the solution of 0.10 mol/L HCl(aq). The difference that concentration makes is even more noticeable when you compare the reaction in the solution of 1.0 mol/L CH₃COOH(aq) to the reaction in the solution of 0.10 mol/L CH₃COOH(aq).

It appears, however, that different acids with the same concentration can also have different acidities (or acidic properties)—the reaction is more vigorous in the solution of 1.0 mol/L HCl(aq) than it is in the solution of 1.0 mol/L CH₃COOH(aq). Similarly, the reaction is more vigorous in the solution of 0.10 mol/L HCl(aq) than it is in the solution of 0.10 mol/L CH₃COOH(aq).

You can see further evidence that different acids with the same concentration have different acidic properties if you refer back to Figure 6.4 on page 212. Table 6.4 below summarizes the observations and inferences you could make by testing the conductivity of acidic and basic solutions.

Table 6.4 Conductivity of Acids and Bases

Solution	Observation	Inference
Acidic solutions		
1.0 mol/L HCl(aq)	bulb glows brightly	high concentration of ions; more acidic
1.0 mol/L CH₃COOH(aq)	bulb glows weakly	low concentration of ions; less acidic
Basic solutions		
1.0 mol/L NaOH(aq)	bulb glows brightly	high concentration of ions; more basic
1.0 mol/L NH₃(aq)	bulb glows weakly	low concentration of ions; less basic

Figure 6.4 and Figure 6.7 provide evidence that *the acidic or basic properties of a solution are affected by both the concentration of the solution and the identity of the acid or base.* Why is a solution of 1.0 mol/L HCl(aq) a better conductor of electricity than a solution of 1.0 mol/L CH₃COOH(aq)?

Strong Acids and Weak Acids

According to the modified Arrhenius theory of acids and bases, all acids produce hydronium ions in water and all bases produce hydroxide ions in water. Not all acids, however, ionize to the same degree. Acids and bases are further defined as strong or weak. An acid that ionizes nearly 100% in water is known as a **strong acid**.

For example, hydrochloric acid is a strong acid. Nearly 100% of the molecules of hydrochloric acid in an aqueous solution react with water to form H_3O^+(aq) and Cl^-(aq) ions (see Figure 6.8).

$$HCl(g) + H_2O(\ell) \rightarrow H_3O^+(aq) + Cl^-(aq) \qquad (\sim100\% \text{ reaction})$$

Thus, a 1.0 mol/L solution of hydrochloric acid contains 1.0 mol/L of hydronium ions.

There are only six strong acids, all of which are listed in Table 6.5.

Table 6.5 The Strong Acids

Perchloric acid (aqueous hydrogen perchlorate), $HClO_4$(aq)
Hydroiodic acid (aqueous hydrogen iodide), HI(aq)
Hydrobromic acid (aqueous hydrogen bromide), HBr(aq)
Hydrochloric acid (aqueous hydrogen chloride), HCl(aq)
Sulfuric acid (aqueous hydrogen sulfate), H_2SO_4(aq)
Nitric acid (aqueous hydrogen nitrate), HNO_3(aq)

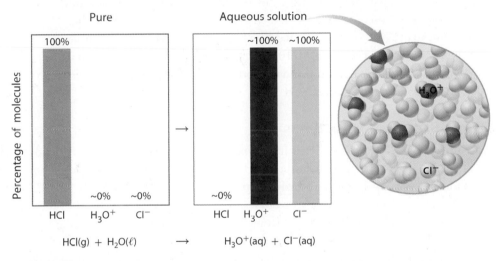

Figure 6.8 In an aqueous solution, close to 100% of hydrogen chloride molecules ionize to form H_3O^+(aq) and Cl^-(aq).

A **weak acid** reacts very little in water. In other words, as shown in Figure 6.9 on the following page, only a small percentage of the acid molecules form ions in water. Most of the acid molecules remain intact. For example, ethanoic acid, CH₃COOH(aq), is a weak acid. Only about 1% of the ethanoic acid molecules react with water in a 0.10 mol/L solution of ethanoic acid at 25 °C.

$$CH_3COOH(aq) + H_2O(\ell) \rightleftharpoons CH_3COO^-(aq) + H_3O^+(aq) \qquad (\sim1\% \text{ reaction})$$

Notice the double-headed equilibrium arrow ⇌ in the equation above. As you learned in Chapter 5, this arrow indicates that the solution is in equilibrium. In other words, in solution, ethanoic acid molecules are constantly reacting with water to form CH₃COO⁻(aq) and H_3O^+(aq), and those ions are constantly reacting to form ethanoic acid molecules and water. (Contrast this with the ionization of a strong acid in water.) *On average, however, at any one time, only about 1% of ethanoic acid molecules are ionized.*

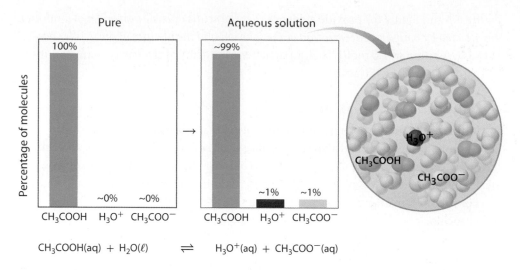

$$CH_3COOH(aq) + H_2O(\ell) \rightleftharpoons H_3O^+(aq) + CH_3COO^-(aq)$$

Figure 6.9 In a 0.10 mol/L solution of $CH_3COOH(aq)$, the concentration of $H_3O^+(aq)$ is only about 0.0010 mol/L, and the concentration of $CH_3COO^-(aq)$ is also only about 0.0010 mol/L.

So how can you explain why the identity of an acid affects its acidic properties? A 0.10 mol/L solution of a strong acid, such as hydrochloric acid, has a higher concentration of $H_3O^+(aq)$ ions than does a 0.10 mol/L solution of a weak acid, such as ethanoic acid. Therefore, a 0.10 mol/L solution of hydrochloric acid reacts more vigorously with magnesium and is a better conductor of electric current than a 0.10 mol/L solution of ethanoic acid.

Strong Bases and Weak Bases

A **strong base** dissociates completely into ions in water. All oxides and hydroxides of the alkali metals—Group 1—are strong bases. The oxides and hydroxides of the alkaline earth metals—Group 2—below beryllium are also strong bases.

Table 6.6 lists some common strong bases. Barium hydroxide, $Ba(OH)_2(s)$, and strontium hydroxide, $Sr(OH)_2(s)$, are strong bases that are soluble in water. Many strong bases, such as magnesium hydroxide, $Mg(OH)_2(s)$, have very low solubility in water. However, the small amount of these compounds that does dissolve dissociates completely. Therefore, these compounds are considered to be strong bases.

A very low percentage of molecules of a **weak base** react in water to produce $OH^-(aq)$. One common weak base is aqueous ammonia, $NH_3(aq)$. Only about 1 percent of the ammonia molecules react in water to form hydroxide ions at 25 °C. This reaction is represented in Figure 6.10. Note again the equilibrium arrow.

Table 6.6 Common Strong Bases

Aqueous sodium hydroxide, NaOH(aq)
Aqueous potassium hydroxide, KOH(aq)
Aqueous calcium hydroxide, $Ca(OH)_2(aq)$
Aqueous strontium hydroxide, $Sr(OH)_2(aq)$
Aqueous barium hydroxide, $Ba(OH)_2(aq)$

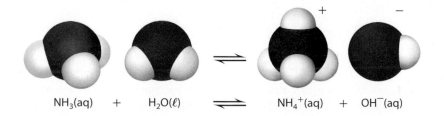

$$NH_3(aq) + H_2O(\ell) \rightleftharpoons NH_4^+(aq) + OH^-(aq)$$

Figure 6.10 About 1% of ammonia molecules react in water to produce $OH^-(aq)$. In a 0.10 mol/L solution of ammonia, the concentration of $OH^-(aq)$ is only about 0.0010 mol/L.

Differentiating between Weak and Strong Acids and Bases

Given six unknown solutions of equal concentration, you will carry out a procedure that will identify the solutions as being a strong acid, strong base, weak acid, weak base, neutral ionic solution, or molecular solution.

Question

What laboratory tests will allow you to identify weak and strong acids and basis?

Safety Precautions

- Acids and bases are often both toxic and corrosive. Wash any spills on skin or clothing with plenty of cool water and inform your teacher immediately.

- When you have completed this investigation, wash your hands.

Materials

- unknown solutions of:
  - 0.10 mol/L molecular solution
  - 0.10 mol/L neutral ionic solution
  - 0.10 mol/L strong base
  - 0.10 mol/L strong acid
  - 0.10 mol/L weak base
  - 0.10 mol/L weak acid
- conductivity testers
- pH paper or pH meter
- short strips of magnesium ribbon

Procedure

1. Read the procedure steps below and construct an appropriate table to record your results.

2. Determine the pH of the solutions, using either pH paper or a pH meter. Record your observations.

3. Determine the conductivity of each solution by placing 25 mL of a solution into a 50 mL beaker and using the conductivity tester. Record your observations.

4. Determine the reactivity of each solution with Mg(s) metal by placing 5 mL of the solution into a test tube and adding a small piece of Mg(s).

   Record your observations, making note of whether some solutions reacted more vigorously with the metal than others.

5. Dispose of all materials as directed by your teacher.

Analysis

1. Using your data, classify each of the six unknown solutions. Explain your classifications.

Conclusion

2. Provide empirical definitions for each of strong acid, weak acid, strong base, weak base, neutral ionic solution, and molecular solution.

Weak Acids and Weak Bases in Living Organisms

When you think of acids and bases, you probably think of serious burns and other dangers. However, weak acids and weak bases perform many important functions in your body and other living things. Many weak acids are intermediates in the breakdown of glucose for energy. For example, citric acid, the weak acid that is responsible for the flavour of citrus fruits, is an intermediate in cellular respiration. Your body also forms ammonia, a weak base, in the process of breaking down proteins in your food. Ammonia is toxic so the body converts it into other, weaker bases in the process of making the urea that is eliminated from your body. Many of the amino acids that make up the proteins in your body are weak acids or weak bases. Their chemical nature is critical to the proper functioning of all proteins. Your body is unable to produce one very important weak acid that it critical to the proper functioning of your body. You must obtain this weak acid in your diet. Its name is ascorbic acid but you know it as vitamin C. These are just a few examples of the weak acids and weak bases that keep your body functioning correctly.

● ● ●

5 All the following solutions are acidic. Using the modified Arrhenius theory of acids, classify the following as strong or weak acids:

a) $HNO_3(aq)$

c) $HI(aq)$

b) $C_2H_5COOH(aq)$

d) $HF(aq)$

6 All the following solutions are basic. Using the modified Arrhenius theory of bases, classify the following as strong or weak bases:

a) $NH_3(aq)$

c) $Na_2CO_3(aq)$

b) $KOH(aq)$

d) $Mg(OH)_2(aq)$

7 Consider 0.10 mol/L $HCl(aq)$ and 0.10 mol/L $HF(aq)$. Which would:

a) have a higher concentration of hydronium ions?

b) have a lower conductivity?

● ● ●

Monoprotic and Polyprotic Acids

The prefix *mono-* means "one." The prefix *di-* means "two." The prefix *tri-* means "three." The prefix *poly-* means "many." The root *-protic* refers to "proton," meaning $H^+(aq)$.

Some acids react in a 1:1 mole ratio with water. These acids, called **monoprotic acids**, have only one hydrogen atom per molecule that ionizes in water. Hydrochloric acid, $HCl(aq)$, hydrobromic acid, $HBr(aq)$, and hydroiodic acid, $HI(aq)$, are all strong monoprotic acids. Hydrofluoric acid, $HF(aq)$, and ethanoic acid, $CH_3COOH(aq)$, are both weak monoprotic acids.

Many acids contain two or more hydrogen atoms that can ionize. These acids are called **polyprotic acids.** For example, sulfuric acid, $H_2SO_4(aq)$, has two hydrogen atoms that can ionize. As you know from Table 6.5, sulfuric acid is a strong acid. This is true only for its first ionization, however.

$$H_2SO_4(aq) + H_2O(\ell) \rightarrow H_3O^+(aq) + HSO_4^-(aq) \qquad (\sim100\% \text{ reaction})$$

The resulting aqueous hydrogen sulfate ion, $HSO_4^-(aq)$, is also an acid. A low percentage of $HSO_4^-(aq)$ reacts with water, however, and it is therefore a weak acid.

$$HSO_4^-(aq) + H_2O(\ell) \rightleftharpoons H_3O^+(aq) + SO_4^{2-}(aq) \qquad (\text{minimal reaction})$$

Since sulfuric acid contains two hydrogen atoms that can ionize and react with water, it is also known as a *diprotic acid*.

Acids that have three hydrogen atoms that can ionize are called *triprotic acids*. Phosphoric acid, $H_3PO_4(aq)$, is a weak triprotic acid. It gives rise to three anions as follows:

$$H_3PO_4(aq) + H_2O(\ell) \rightleftharpoons H_3O^+(aq) + H_2PO_4^-(aq) \quad (\text{weak acid})$$

$$H_2PO_4^-(aq) + H_2O(\ell) \rightleftharpoons H_3O^+(aq) + HPO_4^{2-}(aq) \quad (\text{weaker acid})$$

$$HPO_4^{2-}(aq) + H_2O(\ell) \rightleftharpoons H_3O^+(aq) + PO_4^{3-}(aq) \quad (\text{weakest acid})$$

Chemistry File

FYI
Sulfurous acid, $H_2SO_3(aq)$, is a weak diprotic acid produced when $SO_2(g)$ reacts with $H_2O(\ell)$ in the atmosphere. Sulfurous acid is one of the components of acid rain.

Here again, the acid in the first ionization, $H_3PO_4(aq)$, reacts more completely with water than does the second acid, $H_2PO_4^-(aq)$. The second acid, in turn, reacts more completely with water than the acid in the third ionization, $HPO_4^{2-}(aq)$. With each hydrogen ionized, the strength of a polyprotic acid decreases. Keep in mind, however, that all three of these acids are weak because only a very small proportion of each one ionizes.

Monoprotic and Polyprotic Bases

Just as acids can be classified as monoprotic or polyprotic, so can bases. The term **monoprotic base** is used to describe a base that dissociates or reacts with water in one step, forming hydroxide ions. A **polyprotic base** is a base that reacts with water in two or more steps, forming hydroxide ions. One example of a polyprotic base is hydrazine, $N_2H_4(\ell)$, which is most often used in rocket fuel. Hydrazine gives rise to two ions as follows:

$$N_2H_4(\ell) + H_2O(\ell) \rightleftharpoons N_2H_5^+(aq) + OH^-(aq)$$
$$N_2H_5^+(aq) + H_2O(\ell) \rightleftharpoons N_2H_6^{2+}(aq) + OH^-(aq)$$

As with polyprotic acids, the base that ionizes in the first step, $N_2H_4(\ell)$, ionizes more completely than the base in the second ionization step, $N_2H_5^+(aq)$.

Another polyprotic base is sodium carbonate, $Na_2CO_3(aq)$. In water, sodium carbonate dissociates, forming sodium ions and carbonate ions:

$$Na_2CO_3(s) \xrightarrow{\text{in water}} 2Na^+(aq) + CO_3^{2-}(aq)$$

The carbonate ion acts as a base, reacting with water in two steps as follows:

$$CO_3^{2-}(aq) + H_2O(\ell) \rightleftharpoons HCO_3^-(aq) + OH^-(aq)$$
$$HCO_3^-(aq) + H_2O(\ell) \rightleftharpoons H_2CO_3(aq) + OH^-(aq)$$

Acids and Bases in Industry

Sulfuric acid, a strong diprotic acid, is a very important industrial acid. It is so important that the mass of sulfuric acid used by a nation is a measure of the strength of a country's economy. Figure 6.11 shows some uses of sulfuric acid.

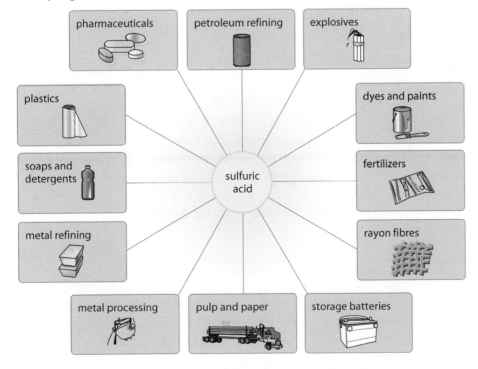

Figure 6.11 Sulfuric acid is used in all of these products and processes.

Chemistry File

Try This
Today, sulfuric acid is produced by the contact process, which was first used commercially in 1900. Find out more about the history of sulfuric acid production and its link to the Industrial Revolution. With a partner or group, create a multimedia presentation to share your findings. ICT

Bases are also important industrial chemicals. For example, one of the most important manufactured bases in Canada is ammonia, which is used to produce fertilizers, explosives, refrigerants, and textiles. Canada is the fifth-largest producer of ammonia in the world.

Clearly, there are many benefits to manufacturing and using acids and bases. However, due to their reactivity and corrosiveness, there are also risks involved in producing and transporting them, especially in large quantities.

Preventing and Dealing with Acid and Base Spills

The problems of safely making, transporting, and using acids or bases require technological solutions. Many such technologies focus on developing processes, materials, and vehicles that minimize the chances of a spill. For example, rail cars are specially designed with rubber linings, acid-resistant coatings, and special fittings to make loading and unloading acids safer. Spills do occur, however, and knowledge of acid and base chemistry is very helpful in minimizing their impact.

The two key ways to deal with an acid or base spill are as follows.

1. **Dilute it:** Diluting an acid (or base) decreases the concentration of $H_3O^+(aq)$ or $OH^-(aq)$ ions, minimizing the harmful properties of the solution. For example, eyewash stations minimize any damage to eyes caused by an acid or base by immediately diluting the solution with copious amounts of cool water.

2. **Neutralize it:** Neutralizing an acid spill involves adding a weak base to reduce the acidity of the solution. Similarly, neutralizing a base spill involves adding a weak acid. This technique can be used on a very small or a very large scale. For example, the train wreck in Figure 6.12A caused the spillage of acid, which was neutralized by the addition of enormous quantities of a solid mixture that included a weak base. Figure 6.12B shows a student neutralizing an acid spill using a spill mix containing baking soda. You must not use a strong base to neutralize an acid spill—this would result in another hazardous spill, because you are likely to add excess base, leaving you with a hazardous basic mixture.

Figure 6.12 Neutralization reactions can help clean up acid spills on both a large scale (**A**) and a small scale (**B**).

Neutralization

The modified Arrhenius theory of acids and bases helps to explain neutralization. The reaction between an acid and a base produces an ionic compound (a salt) and water:

$$acid + base \rightarrow a\ salt + water$$

A **salt** is an ionic compound comprising an anion from an acid and a cation from a base. For example, sodium nitrate, a salt, can be added to processed meat to preserve the colour and to slow the rate of spoiling by inhibiting bacterial growth. Sodium nitrate can be prepared in a laboratory by reacting nitric acid with sodium hydroxide, as shown below.

Why does adding acid to base diminish or remove altogether the harmful properties of both substances?

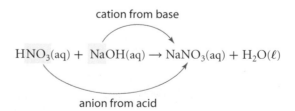

cation from base

$$HNO_3(aq) + NaOH(aq) \rightarrow NaNO_3(aq) + H_2O(\ell)$$

anion from acid

Notice that the products of the reaction are water and sodium nitrate, which is a neutral salt. In other words, a strong acid and a strong base, both of which are corrosive and harmful, form a neutral solution of sodium nitrate in water.

A similar reaction occurs between all kinds of strong acids and bases when they are mixed. You can think of neutralization in a general way if you remember that acids produce hydronium ions in solution and bases produce hydroxide ions. When solutions containing hydronium ions and hydroxide ions are mixed, the ions react to form water, as shown:

$$H_3O^+(aq) + OH^-(aq) \rightarrow 2H_2O(\ell)$$

In other words, a **neutralization reaction**, also called an acid–base reaction, involves hydronium and hydroxide ions reacting to form water. As neutralization reactions remove both hydronium and hydroxide ions from solution, producing only water and a salt, the substances produced during neutralization do not have the acidic or basic properties of the original acid and base.

Applications of Neutralization Reactions

Neutralization reactions have many applications. For example, neutralization reactions can be used to determine the concentration of an acid or base in pharmaceuticals, cosmetics, or swimming pools (see Figure 6.13). Pool technicians use neutralization reactions to keep [$H_3O^+(aq)$] high enough to maintain bacteria-killing reactions but low enough so that the acidity does not irritate the skin of swimmers.

As you have learned, neutralization also plays an important role in dealing with some of the hazards of acids and bases. In Thought Lab 6.1, you and your group will take a closer look at technologies developed to deal with these hazards.

Figure 6.13 The world's largest indoor wave pool is at the West Edmonton Mall.

Thought Lab | 6.1 | Risks and Benefits of Transporting Acids and Bases

The acids and bases produced and used in Alberta pose numerous challenges for the organizations responsible for their safe handling and transport, including Alberta Environment, Alberta Occupational Health and Safety, and the Canadian Association of Petroleum Producers. These groups ensure that safety standards are met and develop new technologies to decrease the likelihood of spills and workplace accidents.

Target Skills

Assessing, qualitatively, the risks and benefits of producing, using, and transporting acidic and basic substances

Procedure

1. Divide into groups of three or four students. In this small group, choose an acid or base produced or used in Alberta. Some examples are listed below. You may also perform an Internet search to find other examples.
   - hydrochloric acid
   - phosphoric acid
   - nitric acid
   - ammonia
   - sodium hydroxide (caustic soda)

2. Record your ideas about the use and safety issues with the production or use of this acid or base.

3. Research issues surrounding the handling or transportation of your chosen acid or base. Ensure that your group uses a variety of sources to find data. Be aware of any bias in your sources. Your research should include:
   - background on how the acid or base is produced and used
   - background on safety issues, based on WHMIS guidelines and Transportation of Dangerous Goods Regulations (Transport Canada)

   - descriptions of technology (processes and equipment) that has been developed to reduce the risk and explanations of how it reduces the risk

4. Present your findings to the class using a computer slide show presentation, and summarize your findings in a report. **ICT**

Analysis

1. As a class, discuss the findings of each group.
   a) Are the safety, environmental, and health concerns of acid–base production and transportation in Alberta being addressed adequately?
   b) Do the benefits of producing and using acids and bases in Alberta outweigh the risks of their use?
   c) Look back at your original ideas about the use and risk of acids and bases. Which ideas were supported by your research? Which findings surprised you?
   d) Do you think the media adequately reports the risks involved in using acids and bases?

Section 6.2 Summary

- Strong acids and bases ionize or dissociate completely in water, whereas weak acids and bases only partially ionize in water. As a result, characteristic properties, such as pH, conductivity, and reactivity with metals, are altered for weak acids and bases.

- Monoprotic acids react in a 1:1 molar ratio with water because they only have one hydrogen per molecule that ionizes in water. Polyprotic acids have more than one hydrogen that can ionize in water. Diprotic and triprotic acids are polyprotic acids that have two or three hydrogens, respectively, that ionize in water.

- Monoprotic bases are those that dissociate or react with water in one step to form hydroxide ions. Polyprotic bases are those that react with water in more than one step to form hydroxide ions.

- Acids and bases have numerous industrial uses. Some of the most highly used are sulfuric acid and ammonia. An important aspect to industrial uses of acids and bases is developing safe methods for use and transport. When accidents do occur, methods for neutralization have been developed to minimize the detrimental effects. These neutralization reactions are simply reactions of an acid with an appropriate amount of base to produce water and a salt.

SECTION 6.2   REVIEW

1. Use the modified Arrhenius theory of acids and bases to explain the following observations:
   a) A conductivity tester shows that a 1.0 mol/L solution of hydrochloric acid is a better conductor of electric current than is a 0.1 mol/L solution of hydrochloric acid.
   b) A conductivity tester shows that a 1.0 mol/L solution of sodium hydroxide is a better conductor of electric current than is a 1.0 mol/L solution of ammonia.

2. Give one example of each of the following:
   a) a monoprotic acid
   b) a polyprotic acid
   c) a monoprotic base
   d) a polyprotic base

3. Give one example of each of the following:
   a) a weak acid
   b) a strong acid
   c) a strong base
   d) a weak base
   e) a metallic oxide that forms a base in water
   f) a non-metallic oxide that forms an acid in water

4. Formic acid, $HCOOH(aq)$, is responsible for the painful bites of fire ants. Is formic acid a strong acid or a weak acid?

5. Your friend states that a 0.10 mol/L solution of sulfuric acid, $H_2SO_4(aq)$, must have twice the concentration of hydronium ions compared to a 0.10 mol/L solution of hydrochloric acid, $HCl(aq)$. Do you agree? Why or why not?

6. Read the article below and answer the questions that follow.

Neutralization on a Large Scale

In October 2002, more than 150 L of hydrochloric acid was spilled as it was being unloaded in Red Deer's industrial park. Eight minutes after the spill occurred, the Red Deer response team arrived as the two employees were neutralizing the acid with a base. The Red Deer team added foam (to decrease the vaporization of the acid) and soda ash to further neutralize the acid. The residue was later sucked up by a vacuum truck. Although 15 people came in contact with the spill, none required medical treatment.

   a) What technique was used to minimize the danger from the spilled acid? Why does this method work?
   b) According to the article, is soda ash an acid or a base? Explain your answer.
   c) Soda ash is the common name for sodium carbonate. Write the formula for solid sodium carbonate.
   d) Write chemical equations to show what happens when sodium carbonate is added to water.

Acids, Bases, and pH

Section Outcomes

In this section, you will:

- **calculate** $H_3O^+(aq)$ and $OH^-(aq)$ concentrations, pH, and pOH based on logarithmic expressions
- **compare** magnitude changes in pH and pOH with changes in concentration for acids and bases
- **explain** how indicators, pH meters, or pH paper can be used to measure $H_3O^+(aq)$ concentration
- **relate** pH and hydronium ion concentration
- **use** indicators, pH paper, and a pH meter to determine acidity and alkalinity of a solution
- **use** appropriate SI units to express pH and concentration to the correct number of significant digits

Key Terms

pH
acid–base indicator
universal indicator
pOH

You are probably familiar with the term "pH" from a variety of sources. Advertisers talk about the "pH balance" of products such as soaps, shampoos, and skin creams (Figure 6.14). Gardeners and farmers monitor the pH of the soil, as most plants and food crops grow best within a narrow pH range. Your own health depends on your body maintaining its fluids at certain pH levels. For example, the blood, saliva, and spinal fluid of a healthy person have a pH of about 7.4. Meanwhile, stomach fluid has a pH ranging from 1.0–3.0.

You probably know that a solution with a pH of 1.0 is "more acidic" than a solution with a pH of 7.4. But what exactly is pH? How is it measured? To answer these questions, consider a familiar substance—water.

Figure 6.14 Skin has a pH between 4.5 and 6.0. The slight acidity helps to kill bacteria that might cause infection. Products called toners neutralize the basic residue left behind by soaps and restore the pH of the skin.

Ion Concentration in Water

Water is a molecular substance. In previous science classes, you may have learned and observed that water does not conduct electric current. Actually, water *does* conduct electric current, though *very* weakly. From this observation, scientists infer that pure water self-ionizes, meaning that two molecules of water can react to form ions as shown below.

$$H_2O(\ell) + H_2O(\ell) \rightleftharpoons H_3O^+(aq) + OH^-(aq)$$

On average, at 25 °C, only about two water molecules in every billion are ionized at any given moment. Since so very few ions are produced, very little electric current is conducted. This is why pure water is such a poor conductor.

Chemists have determined that the concentration of the hydronium ions in pure water at 25 °C is only 1.0×10^{-7} mol/L. The ionization of water produces the same low concentration of hydroxide ions. Thus, the concentration of hydronium ions and hydroxide ions in pure water, which is a neutral substance, can be written as follows:

$$[H_3O^+(aq)] = [OH^-(aq)] = 1.0 \times 10^{-7} \text{ mol/L}$$

Acids produce hydronium ions in water, therefore acidic solutions contain a higher concentration of hydronium ions than hydroxide ions. Bases, conversely, produce hydroxide ions in water. Therefore, basic solutions contain a higher concentration of hydroxide ions than hydronium ions, as shown in Figure 6.15.

acidic solution	neutral solution	basic solution
$[H_3O^+] > [OH^-]$	$[H_3O^+] = [OH^-]$	$[H_3O^+] < [OH^-]$

Figure 6.15 The relationship between the concentrations of hydronium ions and hydroxide ions in a solution determines whether the solution is acidic, basic, or neutral.

ChemistryFile

FYI

Sørenson was studying enzymes involved in beer-making when he proposed the method of pH. He wrote pH as P$_H$, and called his way of reporting pH a *wasserstoffionexponent*, meaning "hydrogen ion exponent" in German.

The pH Scale: Measuring by Powers of Ten

In the late nineteenth and early twentieth centuries, scientists gathered evidence that the concentration of hydrogen (hydronium) ions ranged from about 10 mol/L for a concentrated strong acid to about 1.0×10^{-15} mol/L for a concentrated strong base. This huge range of concentrations and the negative powers of 10 are not very convenient to work with. In 1909, a Danish biochemist, Søren Sørenson (1868–1939), suggested a method for reporting hydrogen ion concentrations using more convenient numbers. His method used the numerical system of base 10 logarithms.

The base 10 logarithm of a number is the power to which you must raise 10 to equal that number. For example, the logarithm of 10 is 1, because $10^1 = 10$. The logarithm of 100 is 2, because $10^2 = 100$. (See Appendix E for more information about exponents and logarithms.)

Sørenson defined pH as $-\log[H^+(aq)]$. Since Sørenson did not know about hydronium ions, his definition of pH is based on the Arrhenius theory. As you know, scientists have since gathered evidence that $H^+(aq)$ exists in an aqueous solution as $H_3O^+(aq)$, and the modified Arrhenius theory takes this evidence into account. Therefore, according to the modified Arrhenius theory, **pH** is the negative base-ten logarithm of the molar concentration of hydronium ions in a solution.

$$pH = -\log[H_3O^+(aq)]$$

Figure 6.16 shows a way to remember the meaning of pH.

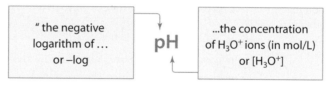

Figure 6.16 The concept of pH makes working with very small values, such as 1.0×10^{-7} mol/L (0.000 000 10 mol/L), much easier.

The concept of pH allows hydronium ion concentration to be expressed as positive numbers, usually between 0 and 14, rather than very small numbers. For example, recall that $[H_3O^+(aq)]$ of pure (neutral) water at 25 °C is 1.0×10^{-7} mol/L.

$$
\begin{aligned}
pH &= -\log[H_3O^+(aq)] \\
&= -\log(1.0 \times 10^{-7}) \\
&= -(-7.00) \\
&= 7.00
\end{aligned}
$$

MathTip

To find the log of a number on newer models of calculators, use the log button [LOG]. Press [LOG], enter the value for concentration, and then press [=]. For more information on these calculations, see Appendix D.

You will probably agree that the number 7.00 is easier to work with than 1.0×10^{-7}.

All acidic solutions have a pH less than 7 and all basic solutions have a pH greater than 7. You can calculate the pH of any solution by finding the negative base-ten logarithm of the concentration of hydronium ions in the solution.

Significant Digits and pH

In the calculation above, notice that the value for concentration has two significant digits (1.0×10^{-7}). The calculated pH, however, is reported with three digits (7.00). This is not an error. *When reporting pH values, only the digits to the right of the decimal point are considered significant.* For example, a pH of 8.29 has two significant digits. A pH of 3 has zero significant digits.

Use the Sample Problems and Practice Problems on the following pages to practise calculating pH.

MathTip

The digits to the left of the decimal point in a pH value refer to the exponent portion of the number in scientific notation. For example, the $[H_3O^+(aq)]$ of 1.0×10^{-7} mol/L (0.000 000 10 mol/L), a number with two significant digits, corresponds to a pH of 7.00. The 7 tells you where to place the decimal point, and is therefore not considered a significant digit.

Sample Problem

Calculating the pH of a Solution from $[H_3O^+(aq)]$

Problem

Calculate the pH of a solution with $[H_3O^+](aq) = 3.8 \times 10^{-3}$ mol/L. Is the solution acidic or basic?

Solution

Use the equation $pH = -\log [H_3O^+(aq)]$ to solve for the unknown.

$$\begin{aligned} pH &= -\log(3.8 \times 10^{-3}) \\ &= -(-2.42) \\ &= 2.42 \end{aligned}$$

The pH of the solution is 2.42.

Check Your Solution

$[H_3O^+(aq)]$ is greater than 1.0×10^{-7} mol/L. Therefore, the pH should be less than 7.00. The solution is acidic, as you would expect. The answer has the correct number of significant digits (two).

Sample Problem

Calculating the pH of a Solution

Problem

Calculate the pH of a solution made by dissolving 3.75 g of HCl(g) in enough water to make 250 mL of solution. Is the solution acidic or basic?

What Is Required?

You need to determine the pH of the solution and if the solution is acidic or basic.

You need to calculate [HCl(aq)] from the mass of HCl(g) and the volume of solution.

What Is Given?

Mass HCl(g)	$m = 3.75$ g
Volume of solution	$V = 250$ mL

Plan Your Strategy

Calculate the molar mass of HCl(g), calculate [HCl(aq)] in mol/L using the given mass of HCl(g), the molar mass of HCl(g), and the volume of the solution. Since HCl(aq) is a strong acid, write an ionization equation and use a mole ratio to determine $[H_3O^+(aq)]$. Finally, use the equation $pH = -\log [H_3O^+(aq)]$ to solve for pH.

Act on Your Strategy

$$\begin{aligned} \text{Molar mass of HCl(g)} &= 1.01 \text{ g/mol} + 35.45 \text{ g/mol} \\ &= 36.46 \text{ g/mol} \end{aligned}$$

$$n = \frac{m}{M} = \frac{3.75 \text{ g HCl(g)}}{36.46 \frac{\text{g HCl(g)}}{\text{mol HCl(g)}}}$$

$$= 0.10285 \text{ mol HCl(g)}$$

$$c = \frac{n}{V} = \frac{0.10285 \text{ mol HCl(g)}}{250 \text{ mL}\left(\frac{1 \text{L}}{1000 \text{ mL}}\right)}$$

$$= 0.4114 \text{ mol/L HCl(g)}$$

The ionization equation is:

$$HCl(aq) + H_2O(\ell) \rightarrow H_3O^+(aq) + Cl^-(aq)$$

Since HCl(aq) is a strong acid and ionizes ~100%

$$[H_3O^+(aq)] = [HCl(aq)]$$

$$[H_3O^+(aq)] = \frac{0.4114 \text{ mol HCl(aq)}}{\text{L}} \times \frac{1 \text{ mol } H_3O^+(aq)}{1 \text{ mol HCl(aq)}}$$

$$= 0.4114 \text{ mol/L } H_3O^+(aq)$$

$$\begin{aligned} pH &= -\log[H_3O^+(aq)] \\ &= -\log(0.4114) \\ &= 0.386 \end{aligned}$$

The pH of the solution is 0.386.

Check Your Solution

HCl(g) is an acid, so it makes sense that the pH is less than 7. The answer has the correct number of significant digits (three).

Using Dimensional Analysis

$$[HCl(aq)] = \frac{3.75 \text{g HCl(g)}}{1} \times \frac{1 \text{ mol HCl(g)}}{36.46 \text{ g HCl(g)}} \times \frac{1}{250 \text{ mL}} \times \frac{1000 \text{ mL}}{1.000 \text{ L}}$$

$$= 0.4114 \text{ mol/L HCl(aq)}$$

1. Calculate the pH of each solution, given the following information:
   a) $[H_3O^+(aq)] = 0.0027$ mol/L
   b) $[H_3O^+(aq)] = 5.20$ mol/L
   c) $[H_3O^+(aq)] = 8.27 \times 10^{-12}$ mol/L
   d) $[HI(aq)] = 9.7 \times 10^{-5}$ mol/L
   e) 1.25×10^{-2} g $HClO_4(g)$ in 770 mL of solution

2. $[H_3O^+(aq)]$ in a cola drink is about 5.0×10^{-3} mol/L. Calculate the approximate pH of the drink. Is the drink acidic or basic?

3. A glass of orange juice has $[H_3O^+(aq)]$ of 2.9×10^{-4} mol/L. Calculate the pH of the drink. Is the drink acidic or basic?

4. $[H_3O^+(aq)]$ of a solution of sodium hydroxide is 6.59×10^{-10} mol/L. Calculate the pH of the solution.

5. The concentration of a dilute solution of nitric acid, $HNO_3(aq)$, is 6.30×10^{-3} mol/L. Calculate the pH of the solution.

6. A hydrobromic acid solution was made by dissolving 1.36 g of the gas in enough water to make 3.50 L of solution. Determine the pH.

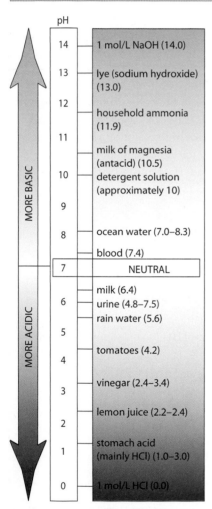

Figure 6.17 The pH scale is a logarithmic scale. The Richter scale and the decibel scale are two other examples of logarithmic scales. You can find out more about logarithmic scales in Appendix D and online.

www.albertachemistry.ca
WWW

The pH Scale

Notice that the pH for question 1(b) has a negative value. How is this possible if the pH scale ranges from 0 to 14? In fact, the pH scale was defined so that most acidic and basic solutions fall within this range at 25 °C. It is possible, however, for concentrated, strong bases to have a pH value greater than 14 and for concentrated, strong acids to have a negative pH value. Still, most familiar solutions fall between pH 0.0 and pH 14.0, as shown in Figure 6.17.

Measuring pH

For centuries, people have known that certain vividly coloured substances change colour depending on whether they are in acidic or basic conditions. Dyers and painters used this property to their advantage, adding acid or base to a dye or pigment to achieve a different colour.

Scientists used and continue to use these substances to determine whether a substance is acidic or basic. An **acid–base indicator** is any chemical substance that changes colour in an acidic or basic solution. Litmus is a famous example.

Once chemists learned how to measure the exact concentration of hydronium ions in solution, they could link an indicator's colour change with a specific pH range. Some common acid–base indicators and their colour changes are shown in Table 6.7.

Table 6.7 Some Common Acid–Base Indicators

Indicator	pH range	Colour change as pH increases
methyl violet	0.0–1.6	yellow to blue
orange IV	1.4–2.8	red to yellow
methyl orange	3.2–4.4	red to yellow
bromocresol green	3.8–5.4	yellow to blue
methyl red	4.8–6.0	red to yellow
bromothymol blue	6.0–7.6	yellow to green to blue
phenol red	6.6–8.0	yellow to red
phenolphthalein	8.2–10.0	colourless to pink
indigo carmine	11.4–13.0	blue to yellow

Indicators and Indicator Paper

Figure 6.18 shows several of the indicators listed in Table 6.7 in solutions of different pH.

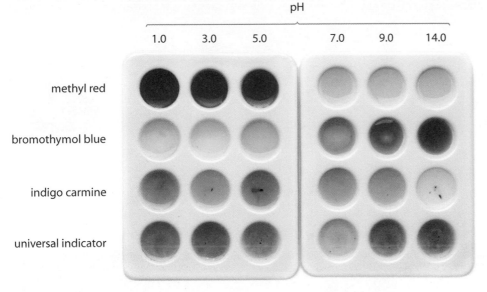

Figure 6.18 How is the universal indicator solution different from the other indicators shown here?

Acid–base indicators can be used as solutions that are added to the test solution, or they can be soaked onto paper and then dried. (Litmus paper is a familiar example.) Acid–base indicators can be used one at a time, as in Investigation 6.D on the following page, or they can be combined to create a **universal indicator**, which contains a number of indicator substances. The indicator has different characteristic colours over the range of pH values and can thus provide a more accurate pH estimate than one indicator alone can. Figure 6.19 shows an example of universal indicator in use. Universal indicator is available as a liquid or as universal indicator paper, also called pH paper.

Figure 6.19 Compare the colour of the indicator paper with the indicator colour key. What is the approximate pH of each of these three solutions?

There are also many naturally occurring indicators. For example, tea, grape juice, the juice of red cabbage (Figure 6.20 on the following page), and blueberries all change colour in the presence of acids and bases, as do many other plant extracts.

Chemistry File

FYI
Robert Boyle (1627–1691), famous for his work on gas laws, is also credited with the discovery of the acid–base indicator litmus, which he isolated from lichen.

Chemistry File

TryThis
If you put strong-smelling onion or vanilla extract into a basic solution like baking soda, they lose their characteristic smell. Onion and vanilla are known as *olfactory indicators* because you can detect whether a solution is basic by using your olfactory sense—your sense of smell. These indicators change in odour because the molecules, which you can smell, ionize in basic solutions to produce an unscented ion.

Figure 6.20 Red cabbage is a particularly useful natural indicator as it changes colour with relatively small changes in pH.

INVESTIGATION | 6.D

Determining the pH of an Unknown Solution with Indicators

In this investigation, you will determine the pH of four unknown solutions by using four acid–base indicators.

Question

How can you use indicators to determine the pH of a solution?

Safety Precautions

- Solutions of acids and bases can be toxic and corrosive. Wash any spills on skin or clothing with plenty of cool water. Inform your teacher immediately.
- Dispose of all materials as directed by your teacher

Materials

- 4 solutions of unknown pH
- 4 indicator solutions
  - methyl orange
  - methyl red
  - bromothymol blue
  - phenolphthalein
- spot plates or small test tubes
- medicine droppers

Procedure

1. Place two drops of methyl orange indicator into four wells on the spot plate (or in four small test tubes).

2. Add five drops of each of the unknown solutions to the indicator.

3. Record the colours.

4. Repeat with the other three indicators.

Analysis

1. Using Table 6.7 on page 230 , estimate the pH of each of the four solutions.

Conclusion

2. Your teacher will use a pH meter to determine the pH of the solutions. How do your results, obtained using indicators, compare with the results obtained using a pH meter?

Extension

3. With the permission of your teacher, repeat the investigation using cabbage juice indicator. Use the Internet to find a procedure for preparing an indicator by using cabbage juice. Which method, a combination of indicators, the use of a universal indicator (such as cabbage juice), or a pH meter, appears to be the most practical? **ICT**

8 A solution of unknown pH is yellow in orange IV and red in methyl red. Approximate its pH.

9 Two drops of phenol red are added to a solution with a pH of 5.6. What colour will be observed?

10 What colour is indigo carmine in a solution with a pH of 12.0?

pH Meters

A pH meter is the most precise way to measure the pH of a solution. It uses an ion-specific electrode that compares the concentration of $H_3O^+(aq)$ in the solution with a reference electrode. In many modern pH meters, the electrodes are combined in a single probe. Before use, a pH meter should be calibrated (adjusted for accuracy) using solutions of known pH.

The first commercial pH meter, shown in Figure 6.21A, was developed in response to the technological problem of monitoring the quality of citrus fruit. A citrus grower's company was working on a fast, simple way to test the pH of citrus fruit, and Professor Arnold Beckman developed an "acidimeter" in response. Soon, chemists were using the pH meters too, and they became a staple of analytical laboratories. Today, you can attach a pH probe directly to a computer, a specialized readout unit, or even a calculator for data collection and analysis, as shown in Figure 6.21B.

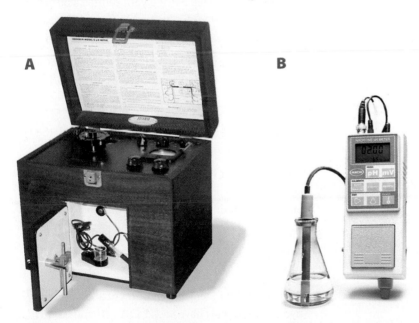

A

B

Figure 6.21 The Beckman Model G pH meter **(A)** had a mass of nearly 7 kg. The earliest meters had a design glitch, in that the pH reading changed with the depth of immersion of the electrodes. Beckman fixed this problem by sealing the glass bulb of the electrode. Modern pH meters include both electrodes inside a single, sealed probe, connected to the meter readout or a computer **(B)**.

You learned earlier that the concentration of an acid affects its pH. In the next investigation, you will practise using pH paper or a pH meter. You will predict how diluting an acid will change its pH, then test your prediction.

The Effect of Dilution on the [H₃O⁺(aq)] and pH of an Acid

In this investigation, you will predict the pH of a strong acid as it is diluted by a series of ten-fold dilutions.

Question

How does diluting a strong acid affect its pH?

Prediction

Calculate $[H_3O^+(aq)]$ and predict the pH for:

a) the original 0.10 mol/L hydrochloric acid

b) the hydrochloric acid after one ten-fold dilution

c) the hydrochloric acid after each of six more ten-fold dilutions

Safety Precautions

- Hydrochloric acid is corrosive. Wash any spills on skin or clothing with plenty of cool water. Inform your teacher immediately.

- Dispose of all materials as directed by your teacher.

Materials

- 0.10 mol/L hydrochloric acid
- distilled water
- 10 mL graduated cylinders or graduated pipettes (2)
- 100 mL beaker
- small beaker or large test tube
- universal indicator paper (pH paper)
- pH meter (optional)
- stirring rod

Procedure

1. Create a table with the following headings: [HCl(aq)], $[H_3O^+(aq)]$, predicted pH, pH measured with a universal indicator, and pH measured with pH meter (optional). Record your predictions in the table.

2. Pour about 10 mL of 0.10 mol/L hydrochloric acid into a clean, dry 100 mL beaker. Dip a piece of universal indicator paper into the acid. Compare the colour against the colour chart to determine the pH and record it in your table.

3. If possible, measure and record the pH of the acid using a pH meter. Rinse the electrode with distilled water after making the measurement.

4. Measure 1.0 mL of the acid from Procedure step 2 and dilute to 10.0 mL with water. The resulting 10 mL of solution is one-tenth as concentrated as the acid from Procedure step 2. Mix thoroughly.

5. Use universal indicator paper and a pH meter (if available) to measure the pH. Record your results.

6. Dispose of the more concentrated acid solution as directed by your teacher and rinse and dry the beaker or test tube so you can use it for the next dilution.

7. Repeat Procedure step 3 using your diluted acid instead of the 0.10 mol/L hydrochloric acid. Repeat steps 4 through 6.

8. Make further dilutions and pH measurements until the hydrochloric acid solution is 1.0×10^{-9} mol/L

Analysis

1. Which do you think gave the more accurate pH: the universal indicator paper or the pH meter? Explain.

2. Compare the pH values you predicted with the measurements you made. If there were differences for the first few dilutions, how can you explain them?

3. What was the pH of the solution that had a concentration of 1.0×10^{-8} mol/L and 1.0×10^{-9} mol/L? Explain the pH you obtained.

4. What effect does a ten-fold dilution of a strong acid (hydrochloric acid) have on $[H_3O^+(aq)]$ and the pH of the acid?

5. Use a spreadsheet program to construct a graph (use an *xy* scatter plot) to compare pH and hydronium ion concentration and describe the relationship. **ICT**

Conclusions

6. Write a statement or statements to explain how changes in the concentration of $H_3O^+(aq)$ relate to changes in the pH of the solution.

Diluting an Acid

You may have noticed in Investigation 6.E that diluting an acid will increase its pH (since $[H_3O^+(aq)]$ is decreased), but only until a pH of 7.0 is reached. In calculating the pH of the acid during dilution, you considered only $[H_3O^+(aq)]$ from the hydrochloric acid. As long as the concentration of hydrochloric acid was high enough to increase the concentration of hydronium ion beyond the concentration of $H_3O^+(aq)$ found in pure water, the pH remained below 7.0. Once the HCl(aq) was diluted to a concentration below 1.0×10^{-7} mol/L, the $H_3O^+(aq)$ concentration in water became significant, and the pH of the solution remained at 7.0. Similarly, the pH of a base can be decreased only to a pH of 7.0 through dilution—it will not fall below 7.0.

Chemistry File

FYI
Although it is perfectly acceptable to use pOH to describe the basicity or acidity of a solution, people more often use pH. That is, we usually think of acids having a low pH and bases having a high pH rather than bases having a low pOH and acids having a high pOH.

Calculating pOH

So far, you have explored how measuring the hydronium ion concentration of a solution allows you to determine the basicity or acidity—the pH—of a solution. What about the hydroxide ion concentration, however? Just as pH refers to the exponential power of the hydronium ion concentration in a solution, **pOH** refers to the power of hydroxide ion concentration. You can calculate the pOH of a solution from the $[OH^-(aq)]$, as shown below:

$$pOH = -\log[OH^-(aq)]$$

For example, recall that $[OH^-(aq)]$ of pure (neutral) water at 25 °C is 1.0×10^{-7} mol/L.

$$\begin{aligned}
pOH &= -\log[OH^-(aq)] \\
&= -\log(1.0 \times 10^{-7}) \\
&= -(-7.00) \\
&= 7.00
\end{aligned}$$

If $[OH^-(aq)]$ is greater than 1.0×10^{-7} mol/L, the solution will be basic. If $[OH^-(aq)]$ is less than 1.0×10^{-7} mol/L, the solution will be acidic. Therefore, if pOH is less than 7.00, the solution will be basic. If pOH is greater than 7.00, the solution will be acidic. In this sense, pOH is the opposite of pH.

Every magnitude change of 1.0 in pOH represents a ten-fold change in concentration of hydroxide ions.

Sample Problem

Calculating the pOH of a Solution from $[OH^-(aq)]$

Problem
Calculate the pOH of a solution with $[OH^-] = 6.8 \times 10^{-5}$ mol/L. Is the solution acidic or basic?

Solution
Use the equation $pOH = -\log[OH^-(aq)]$ to solve for the unknown.

$$\begin{aligned}
pOH &= -\log(6.8 \times 10^{-5}) \\
&= -(-4.17) \\
&= 4.17
\end{aligned}$$

The pOH of the solution is 4.17. The solution is basic.

Check Your Solution
$[OH^-(aq)]$ is greater than 1.0×10^{-7} mol/L. Therefore, the pOH should be less than 7.00. The solution is basic, as you would expect. The answer has the correct number of significant digits (two).

Calculating the pOH of a Solution from Mass and Volume

Problem
Calculate the pOH of a solution of barium hydroxide, $Ba(OH)_2(aq)$, if 0.5 g of solid are dissolved in enough water to make 375 mL of solution.

What Is Required?
You need to calculate pOH using $pOH = -\log[OH^-(aq)]$.

You need to calculate $[Ba(OH)_2(aq)]$ from the mass of $Ba(OH)_2(s)$ and the volume of solution.

You need to write a dissociation equation to determine $[OH^-(aq)]$.

What Is Given?
You have the following data:
Mass $Ba(OH)_2(aq) = 0.5$ g
Volume of solution = 375 mL

Plan Your Strategy
Calculate the concentration of $Ba(OH)_2(s)$ using the given mass of $Ba(OH)_2(s)$, molar mass of $Ba(OH)_2(s)$, and the volume of the solution (converted to L using the conversion factor 1.000 L = 1000 mL).

Write a dissociation equation to determine the relationship between $[Ba(OH)_2(s)]$ and $[OH^-(aq)]$.

Solve for $[OH^-(aq)]$.

Use the equation $pOH = -\log[OH^-(aq)]$ to solve for pOH.

• • •

Act on Your Strategy
Molar mass of $Ba(OH)_2(s) = 137.33$ g/mol $+ 2(16.00$ g$)$/mol $+ 2(1.01$ g$)$/mol
$$= 171.35 \text{ g/mol}$$

$$n = \frac{m}{M} = \frac{0.5 \text{ g } Ba(OH)_2(s)}{171.35 \dfrac{\text{g } Ba(OH)_2(s)}{\text{mol } Ba(OH)_2(s)}}$$

$$= 0.002\,918 \text{ mol } Ba(OH)_2(s)$$

$$c = \frac{n}{V} = \frac{0.002\,918 \text{ mol } Ba(OH)_2(s)}{375 \text{ mL} \left(\dfrac{1 \text{ L}}{1000 \text{ mL}}\right)}$$

$$= 7.78 \times 10^{-3} \text{ mol/L } Ba(OH)_2(aq)$$

$Ba(OH)_2(aq)$ is a strong base that produces 2 mol $OH^-(aq)$ for every 1 mol $Ba(OH)_2(s)$, as shown in the following dissociation equation:

$$Ba(OH)_2(aq) \rightarrow Ba^{2+}(aq) + 2OH^-(aq)$$

So, calculating $[OH^-(aq)]$

$$[OH^-(aq)] = \frac{7.78 \times 10^{-3} \text{ mol } Ba(OH)_2(aq)}{L} \times \frac{2 \text{ mol } OH^-(aq)}{1 \text{ mol } Ba(OH)_2(aq)}$$

$$= 1.56 \times 10^{-2} \text{ mol/L } OH^-(aq)$$

$$pOH = -\log[OH^-(aq)]$$
$$= -\log(1.56 \times 10^{-2})$$
$$= 1.8$$

The pOH of the solution is 1.8 (one significant digit).

Check Your Solution
$Ba(OH)_2(aq)$ is a base, so the pOH should be less than 7.
The answer has the correct number of significant digits (one).

Using Dimensional Analysis

$$[Ba(OH)_2(aq)] = \frac{0.5 \text{ g } Ba(OH)_2(s)}{1} \times \frac{1 \text{ mol } Ba(OH)_2(s)}{171.35 \text{ g } Ba(OH)_2(s)} \times \frac{1}{375 \text{ mL}} \times \frac{1000 \text{ mL}}{1.000 \text{ L}}$$

$$= 7.78 \times 10^{-3} \text{ mol/L } Ba(OH)_2(aq)$$

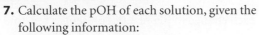

7. Calculate the pOH of each solution, given the following information:
   a) $[OH^-(aq)] = 0.0062$ mol/L
   b) $[OH^-(aq)] = 3.95$ mol/L
   c) $[OH^-(aq)] = 5 \times 10^{-11}$ mol/L
   d) $[NaOH\ (aq)] = 2.4 \times 10^{-3}$ mol/L
   e) 2.95 g of KOH(s) dissolved to make 100 mL of solution
   f) 0.42 g of $Sr(OH)_2$(s) dissolved to make 600 mL of solution

8. The $[OH^-(aq)]$ in seawater is about 2.0×10^{-6} mol/L. Calculate the pOH of seawater. Is the seawater acidic or basic?

9. A harsh liquid soap has $[OH^-(aq)]$ of 2.5×10^{-4} mol/L. Calculate the pOH of the soap. Is the soap acidic or basic?

10. $[NaOH\ (aq)]$ in a dilute solution is 3.47×10^{-3} mol/L. Calculate the pOH of the solution.

11. $[OH^-(aq)]$ of a solution of hydrochloric acid, HCl(aq), is 9.6×10^{-11} mol/L. Calculate the pOH of the solution.

12. If 9.20 mg of cesium hydroxide is dissolved in 225 mL of solution, determine its pOH.

The Relationship Between pH and pOH

Notice that for water at 25 °C, pH + pOH = 7.00 + 7.00 = 14.00. This relationship holds for all dilute aqueous solutions at 25 °C, so you can use the following formula:

$$pH + pOH = 14.00$$

In Investigation 6.E, you learned about the relationship between pH and $H_3O^+(aq)$. This relationship is summarized below in Table 6.8, with the addition of corresponding pOH values for comparison.

Table 6.8 Understanding pH and pOH

Range of acidity and basicity	$[H_3O^+(aq)]$ (mol/L)	Exponential notation (mol/L)	pH ($-\log$ $[H_3O^+(aq)]$)	pOH (14 − pH)
very acidic	1	1×10^0	0.0	14.0
	0.1	1×10^{-1}	1.0	13.0
	0.01	1×10^{-2}	2.0	12.0
	0.001	1×10^{-3}	3.0	11.0
	0.0001	1×10^{-4}	4.0	10.0
	0.000 01	1×10^{-5}	5.0	9.0
	0.000 001	1×10^{-6}	6.0	8.0
neutral $[H_3O^+(aq)]$ = $[OH^-(aq)]$ = 1.0×10^{-7} mol/L	0.000 000 1	1×10^{-7}	7.0	7.0
	0.000 000 01	1×10^{-8}	8.0	6.0
	0.000 000 001	1×10^{-9}	9.0	5.0
	0.000 000 000 1	1×10^{-10}	10.0	4.0
	0.000 000 000 01	1×10^{-11}	11.0	3.0
	0.000 000 000 001	1×10^{-12}	12.0	2.0
	0.000 000 000 000 1	1×10^{-13}	13.0	1.0
very basic	0.000 000 000 000 01	1×10^{-14}	14.0	0.0

You can make two observations by examining Table 6.8:

1. As $[H_3O^+(aq)]$ increases, pH decreases (an inverse relationship) and pOH increases.

2. Because pH is a logarithmic scale, each change in pH of 1.0 represents a ten-fold change in $H_3O^+(aq)$ concentration. In other words, a solution with pH 8.00 is ten times more concentrated than a solution with pH 9.00. The $[H_3O^+(aq)]$ in a solution with pH 3.00 is 100 times greater than it is in a solution with pH 5.00.

pH and Acid Deposition

Normal rainwater has a pH of approximately 5.6. Due to acid deposition caused by industrial pollutants, however, the pH of rain in Ontario has an average pH of just more than 4.0. Although the difference in pH seems small (just 1.4), the acidity of Ontario's rain is actually $10^{1.4}$ times (25 times) as acidic as normal rain.

Acid deposition, a problem for all Canadians, is a result of human activities, such as burning fossil fuels and smelting metal. The problem was first recognized in the late nineteenth century in England, but it was not considered critical until the 1970s. Since then, people have developed technologies to reduce both the quantity and the effects of acid deposition.

Oxides of sulfur, produced by burning fossil fuels that contain sulfur, are one of the major causes of acid rain. These oxides react with water in the air to form acids. For example, sulfur dioxide reacts with water in the air to form sulfurous acid:

$$SO_2(aq) + H_2O(\ell) \longrightarrow H_2SO_3(aq)$$

Four ways to prevent this reaction from occurring are as follows:

- Removing sulfur from the fuel before it is burned
- Using alternative forms of energy
- Reducing consumption of energy
- Adding scrubbers to smokestacks (the scrubbers remove oxides that lead to acid formation)

Figure 6.22 below shows how scrubbers remove sulfur from fuels such as coal. The key to "scrubbing" exhaust gases is the addition of calcium oxide, $CaO(s)$. Calcium oxide reacts with sulfur dioxide gas, $SO_2(g)$, to form calcium sulfite, $CaSO_3(s)$. The calcium sulfite can then be washed away by water.

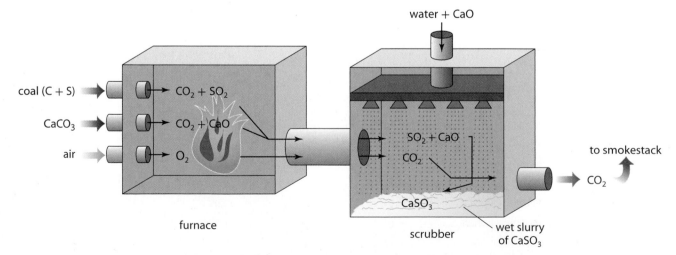

Figure 6.22 Nearly all sources of coal contain sulfur as a contaminant. A scrubber removes most of the sulfur via a reaction of sulfur dioxide with calcium oxide.

Drain Cleaners

Have you ever seen the black "goop" that comes out of a clogged sink drain? This is biofilm, a combination of bacteria, food particles, soap, cosmetics, toothpaste, hair, and body oils. Biofilm is sticky on the surface and hardens as layers build. When drains get clogged, most people think powerful chemicals are required and turn to commercial drain cleaners, some of the most dangerous chemicals found in the home. These products can harm our health, our plumbing pipes, and the wastewater stream.

Chemical Cleaners

Chemical drain cleaners can be acids (e.g., sulfuric acid, $H_2SO_4(aq)$, hydrochloric acid, $HCl(aq)$) or bases (e.g., sodium hydroxide, $NaOH(aq)$, also known as lye). Both strong acids and strong bases are reactive.

The hydronium ions, $H_3O^+(aq)$, in acidic drain cleaners react exothermically with the blockage and water in the pipes. The heat generated is usually sufficient to liquefy the blockage. Effective acidic drain cleaners have high concentrations of hydronium ion and, therefore, a very low pH.

Basic drain cleaners work in a similar way, except it is their hydroxide ions, $OH^-(aq)$, that react exothermically with water and the organic materials in the blockage. A higher pH means a higher concentration of hydroxide ions and a more effective reaction. Lye drain cleaners, for example, have a pH of 12 to 14.

Hazards of Chemical Cleaners

What are some dangers of using commercial drain cleaners? These products rely on heat-generating reactions that liberate potentially toxic gases. The addition of water to a concentrated base or acid can also result in the solution boiling and spattering. This solution will attack human tissue in a similar manner as it does the organic materials in drain blockages. Chemical drain cleaners can corrode metal plumbing. Plastic (PVC) pipes will not corrode, but the heat liberated by chemical drain cleaners may soften them. Chemical drain cleaners can also corrode stainless steel or aluminium sinks and fixtures, and the heat produced may crack porcelain. Empty containers and unused products must be taken to a hazardous waste disposal centre.

Drain-Cleaning Alternatives

Enzymatic drain cleaners work similarly to the enzymes that digest waste in your body. Because they do not produce significant thermal effects, enzymatic drain cleaners reduce the potential for the hazards associated with chemical cleaners. They may not be as effective at dissolving the blockage, however. Enzymatic reactions are much slower than acid or alkaline chemical reactions are and work best on organic materials. Although they are non-corrosive, most labels warn of harm from swallowing, skin contact, and inhalation.

A still gentler approach is to use simple household ingredients and tools. If the drain is sluggish but not completely clogged, turn on the hot water tap and let the hot water run for five to ten minutes. Then try an old-fashioned plunger. If the drain is still clogged, get your parent or guardian's permission to try this. Pour $\frac{1}{4}$ cup of baking soda and $\frac{1}{2}$ cup of vinegar into the drain. The resulting reactions are shown below.

$$NaHCO_3(s) + HC_2H_3O_2(aq) \rightarrow NaC_2H_3O_2(aq) + H_2CO_3(aq)$$

$$H_2CO_3(aq) \rightarrow H_2O(\ell) + CO_2(g)$$

The $CO_2(g)$ is released quickly. Its bubbles and fizzing loosen clogs in the drain.

Wait at least ten minutes and then flush with boiling water. The boiling water helps dissolve soap and grease. Do not try this if you have recently used any chemical drain cleaner.

• • •

1. Survey the drain-cleaning products in your home or school. Ask the people who buy these products how they decide what product to buy.

2. Design a controlled experiment to test the effectiveness of several drain cleaners. List the safety precautions necessary. Obtain permission from your teacher before performing such an experiment.

Math Tip

To find the antilog of a number, use the antilog button on your calculator [10^x]. If your calculator does not have a [10^x], you can also use the [y^x]. Enter 10, then press the [y^x], enter the negative value for pH (or pOH), and then press [=]. For more information on these calculations, see Appendix D.

Calculating $[H_3O^+(aq)]$ and $[OH^-(aq)]$ from pH and pOH

A pH meter will give you a precise and fairly accurate reading for the pH of a substance. Suppose you used a pH meter to determine the pH of a solution of HCl(aq). How would you convert pH back to $[H_3O^+(aq)]$? You can calculate $[H_3O^+(aq)]$ or $[OH^-(aq)]$ by finding the *antilog* of the pH or pOH, using the following relationships:

$$[H_3O^+(aq)] = 10^{-pH}$$

$$[OH^-(aq)] = 10^{-pOH}$$

Work through the next Sample Problem to see how to find the antilog of pH and determine $[H_3O^+(aq)]$. The following Sample Problem will show you how to use pH to find mass.

Sample Problem

Calculating $[H_3O^+(aq)]$ from pH

Problem

Calculate the $[H_3O^+(aq)]$ of a solution at 25°C with pH = 2.47.

Solution

$$\begin{aligned}[H_3O^+(aq)] &= 10^{-pH} \\ &= 10^{-2.47} \\ &= 0.0034 \text{ mol/L} \\ &= 3.4 \times 10^{-3} \text{ mol/L}\end{aligned}$$

Check Your Solution

- Since pH = 2.47, which is less than 7.00, the solution should have a hydronium ion concentration that is greater than 1.0×10^{-7} mol/L, and it does.
- Since you know pH 2.00 has $[H_3O^+(aq)] = 1.0 \times 10^{-2}$ and pH 3.00 has $[H_3O^+(aq)] = 1.0 \times 10^{-3}$, you would expect $[H_3O^+(aq)]$ for pH 2.47 to lie somewhere in between, and it does.
- The answer has the correct number of significant digits (two).

Sample Problem

Using pH to Find Mass

Problem

Calculate the mass of sodium hydroxide required to make 200 mL of solution with a pH of 11.50 at 25°C.

What Is Required?

You need to find the mass (in grams) of the solute, sodium hydroxide, NaOH(s), required to make the solution.

What Is Given?

pH = 11.50

V = 200 mL

Plan Your Strategy

Step 1 Calculate pOH from the given pH using pH + pOH = 14.00.
Rearrange to get pOH = 14.00 – pH.

Step 2 Find $[OH^-(aq)]$ using the formula $[OH^-(aq)] = 10^{-pOH}$

Step 3 Determine [NaOH (aq)].
NaOH(aq) dissociates in water:

$NaOH(s) \rightarrow Na^+(aq) + OH^-(aq)$ (a 1:1 molar ratio)

$$[NaOH(aq)] = [OH^-(aq)] \times \frac{1 \text{ mol NaOH(aq)}}{1 \text{ mol OH}^-\text{(aq)}}$$

Step 4 Calculate the molar mass of NaOH.

Step 5 Calculate the mass of NaOH(s) required using the concentration of the solution, the volume (in L using the conversion factor 1000 mL = 1.000 L), and the molar mass of NaOH(s).

Math Tip

Remember that only the digits following the decimal point in pH (or pOH) are significant. Since 2.47 has only two digits after the decimal point, the final answer in mol/L should have only two significant digits.

Act on Your Strategy

Step 1 Convert pH into pOH:

$$\begin{aligned} pOH &= 14.00 - pH \\ &= 14.00 - 11.50 \\ &= 2.50 \end{aligned}$$

Step 2 Solve for $[OH^-(aq)]$:

$$\begin{aligned} [OH^-(aq)] &= 10^{-pOH} \\ &= 10^{-2.50} \\ &= 0.00316 \text{ mol/L} \end{aligned}$$

Step 3 Determine $[NaOH(aq)]$:

$NaOH(s) \rightarrow Na^+(aq) + OH^-(aq)$ (a 1:1 molar ratio)

$$\begin{aligned} [NaOH(aq)] &= \frac{0.00316 \cancel{\text{mol } OH^-(aq)}}{L} \times \frac{1 \text{ mol NaOH(aq)}}{1 \cancel{\text{mol } OH^-(aq)}} \\ &= 0.00316 \text{ mol/L NaOH(aq)} \end{aligned}$$

Step 4 Calculate the molar mass of $NaOH(s)$:

$$\begin{aligned} \text{Molar mass of NaOH(s)} &= 22.99 \text{ g/mol} + 16.00 \text{ g/mol} + 1.01 \text{ g/mol} \\ &= 40.00 \text{ g/mol} \end{aligned}$$

Step 5 Determine the mass of $NaOH(s)$ needed.

$$c = \frac{n}{V}$$

$$n = c \times V = \frac{0.00316 \text{ mol NaOH(aq)}}{\cancel{L}} \times \frac{200 \cancel{\text{mL}}}{\left(\frac{1000 \cancel{\text{mL}}}{\cancel{L}}\right)} = 6.325 \times 10^{-4} \text{ mol NaOH(s)}$$

$$n = \frac{m}{M}$$

$$m = n \times M = (6.325 \times 10^{-4} \cancel{\text{mol NaOH(s)}})\left(\frac{40.00 \text{ g NaOH(s)}}{\cancel{\text{mol NaOH(s)}}}\right) = 0.025 \text{ g NaOH(s)}$$

Therefore, 2.5×10^{-2} g (or 25 mg) of $NaOH(s)$ are required to make 200 mL of solution with a pH of 11.50.

Check Your Solution

The answer has the correct units and the correct number of significant digits (two). Remember that all the digits must be kept in the calculator throughout the problem to obtain the correct answer.

Practice Problems

13. Calculate the concentration of hydronium ion or hydroxide ion. Is the solution acidic or basic?

a) $[H_3O^+(aq)]$ if pH = 3.9

b) $[H_3O^+(aq)]$ if pOH = 5.22

c) $[OH^-(aq)]$ if pOH = 5.422

d) $[OH^-(aq)]$ if pH = 2.65

14. The lowest recorded pH of rain falling in Fort McMurray, Alberta was measured at 4.80. Calculate $[H_3O^+(aq)]$.

15. The pOH of spaghetti sauce is 9.79. Determine $[H_3O^+(aq)]$.

16. One brand of glass cleaner has a pH of 11.18. Determine the concentration of hydroxide ions.

17. The pOH of oven cleaner is often as low as 0.50. What mass of sodium hydroxide would have to be dissolved to make 100 mL of a solution with the same pOH?

18. If lemon juice has a pOH of 11.80, determine the mass of $HCl(g)$ that would have to be dissolved in 50 mL of solution to make a solution of the same pOH.

Section 6.3 Summary

- The pH of a solution is an indication of the concentration of hydronium ions, $H_3O^+(aq)$. The pH can be calculated using the equation $pH = -\log[H_3O^+(aq)]$. The $[H_3O^+(aq)]$ can be calculated from pH using $[H_3O^+(aq)] = 10^{-pH}$.

- The pOH of a solution is an indication of the concentration of hydroxide ions, $OH^-(aq)$. The pOH of a solution can be calculated using the equation $pOH = -\log[OH^-(aq)]$. The $[OH^-(aq)]$ can be calculated from pOH using the equation $[OH^-(aq)] = 10^{-pOH}$.

- For every change of 1.0 in pH or pOH, there is a 10-fold change in $[H_3O^+(aq)]$ or $[OH^-(aq)]$, respectively.

- For all dilute aqueous solutions at 25 °C, $pH + pOH = 14.00$.

- The pH scale was defined so that most acidic and basic solutions fall within the range of 0–14 at 25 °C. All acidic solutions have a pH that is less than 7 and all basic solutions have a pH that is greater than 7.

- The pH of a solution can be measured using an indicator. Indicators change colour in acid and basic solutions, and have now been standardized so that a particular colour of an indicator is associated with a particular pH. There are also universal indicators that will be different colours over a particular pH range. An example of this is pH paper.

- pH meters can also be used to determine the pH of a solution. A pH meter uses an ion-specific electrode to measure the $[H_3O^+(aq)]$ of the solution.

- Acid deposition from human activities such as burning fossil fuels and smelting has resulted in a decrease in the pH of rainwater in Canada. This is particularly troublesome in areas such as Ontario, where rain has an average pH of as low as just over 4.0. Methods are now being developed to help reduce the effects of acid deposition.

SECTION 6.3 REVIEW

1. Explain why pH was adopted as a way to express $[H_3O^+]$.

2. Arrange the following foods in order of increasing acidity: beets, pH = 5.0; camembert cheese, pH = 7.4; egg white, pH = 8.0; sauerkraut, pH = 3.5; yogurt, pH = 4.5

3. Using Table 6.7 on page 230, determine an approximate pH (a range) given the following indicator colours:
   a) methyl orange is yellow and methyl red is red
   b) bromothymol blue is blue and phenol red is orange
   c) methyl violet is green

4. Calculate the pH of these two body fluids, given the concentration of hydronium ions. Classify the solutions as acids or bases.
   a) tears: $[H_3O^+(aq)] = 4.0 \times 10^{-8}$ mol/L
   b) stomach acid: $[H_3O^+(aq)] = 4.0 \times 10^{-2}$ mol/L

5. Calculate the pOH of the following drinks, given the concentration of hydroxide ions. Classify the solutions as acids or bases.
   a) table wine: $[OH^-(aq)] = 3.1 \times 10^{-11}$ mol/L
   b) milk: $[OH^-(aq)] = 3.89 \times 10^{-7}$ mol/L

6. Calculate the pH of a solution made by dissolving 4.22 g of HI(g) in enough water to make 175 mL of solution.

7. Calculate the pH of a solution made by dissolving 1.25 kg of NaOH(s) in enough water to make 1000 L of solution.

8. Calculate $[H_3O^+(aq)]$ given the following:
   a) pH = 2.77   b) pOH = 10.432

9. Calculate $[OH^-(aq)]$ given the following:
   a) pOH = 3.96   b) pH = 9.9

10. What is the concentration of hydrogen nitrate, $HNO_3(aq)$, in an aqueous solution with a pH of 2.27?

11. What mass of potassium hydroxide, KOH(aq), would have to be dissolved in enough water to make 500 mL of solution with a pH of 10.20?

12. Make a chart to compare $[OH^-(aq)]$, $[H_3O^+(aq)]$, pOH, and pH (use Table 6.7 as a guide). If you have access to a spreadsheet program, also construct a graph to compare these four values. Include proper headings and labels for your chart and graph. (ICT)

13. Construct a set of formulas in a spreadsheet or write a program for a graphing calculator that will calculate pH, pOH, $[H_3O^+(aq)]$, and $[OH^-(aq)]$ when one value is entered. Submit your program or spreadsheet to your teacher. (ICT)

Acid and base solutions have a variety of essential applications that range from food preservation to industrial processes. Their properties, such as reactivity to metal, conductivity, pH, taste, and feel, have been used to develop empirical definitions for acidic, basic, and neutral solutions.

In addition, theories have been developed to help define acids and bases. The original Arrhenius theory of acids and bases defined an acid as a substance that ionizes to form hydrogen ions, $H^+(aq)$, and a base as a substance that dissociates to produce hydroxide ions, $OH^-(aq)$. However, the original theory that Arrhenius proposed did not allow for the recognition of substances, such as ammonia, which are bases that do not dissociate to produce hydroxide ions. In addition, it was discovered that the hydrogen ion does not exist in isolation but instead as the hydronium ion, $H_3O^+(aq)$. The modified Arrhenius theory defines an acid as a substance that reacts with water to produce $H_3O^+(aq)$ in aqueous solution, and a base as a substance that dissociates or reacts with water to produce $OH^-(aq)$ in aqueous solution.

The properties of an acid or base are determined by both its concentration and identity. Strong acids and bases ionize or dissociate completely in water, whereas weak acids and bases only partially ionize in water. As a result, characteristic properties, such as pH, conductivity, and reactivity with metals, are altered for weak acids and bases. Monoprotic acids are those that react in a 1:1 molar ratio with water because they only have one hydrogen per molecule that ionizes in water. Polyprotic acids have more than one hydrogen that can ionize in water. Monoprotic bases are those that dissociate or react with water in one step to form hydroxide ions. Polyprotic bases are those that react with water in more than one step to form hydroxide ions.

Acids and bases have numerous industrial uses. Some of the most highly used are sulfuric acid and ammonia. An important aspect to the industrial use of acids and bases is ensuring their safe production, transport, and use. When accidents do occur, methods for neutralization have been developed to minimize the detrimental effects. These neutralization reactions are simply reactions of an acid with an appropriate amount of base to produce water and a salt.

The pH of a solution is an indication of the concentration of hydronium ions, $H_3O^+(aq)$, which can be calculated using the equation $pH = -\log[H_3O^+(aq)]$. The pOH of a solution is an indication of the concentration of hydroxide ions, $OH^-(aq)$. The pOH of a solution can be calculated using the equation $pOH = -\log[OH^-(aq)]$. For every change of 1.0 in pH or pOH, there is a ten-fold change in $[H_3O^+(aq)]$ or $[OH^-(aq)]$, respectively. For all dilute aqueous solutions at 25 °C, $pH + pOH = 14.00$

The pH scale was defined so that most acidic and basic solutions fall within the range of 0–14 at 25 °C. All acidic solutions have a pH that is less than 7 and all basic solutions have a pH that is greater than 7. The pH of a solution can be measured using an indicator. Indicators change colour in acid and basic solutions, and have now been standardized so that a particular colour of an indicator is associated with a particular pH range. There are also universal indicators that will be different colours over a particular pH range. An example of this is pH paper. pH meters can also be used to determine the pH of a solution. In this case, it uses an ion-specific electrode to measure the $[H_3O^+(aq)]$ of the solution.

Acid deposition from human activities such as burning fossil fuels and smelting has resulted in a decrease in the pH of rainwater in Canada, particularly in Ontario. Considerable effort is now being made to reduce the detrimental effects of acid deposition in Canada.

Concept Organizer | Acids and Bases

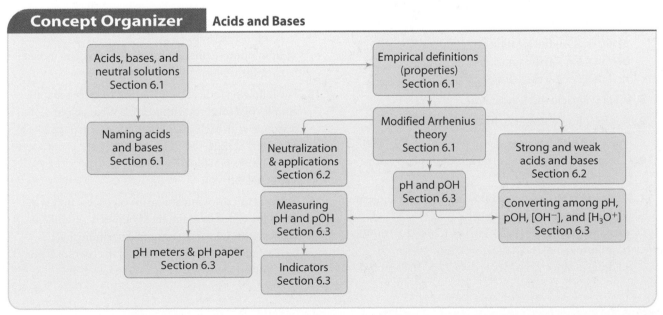

Understanding Concepts

1. List three properties of acids and three properties of bases.

2. Use the modified Arrhenius theory to describe the following acid–base concepts:
   a) composition of an acid and a base
   b) conductivity of an acidic or basic solution
   c) interaction between an acid and water
   d) interaction between a base and water
   e) strong and weak acids and bases
   f) the pH of a solution
   g) the pOH of a solution

3. Write an equation representing the neutralization of aqueous sodium hydroxide with aqueous nitric acid. What are the products of neutralization reactions?

4. Classify each compound as a strong acid, strong base, weak acid, or weak base.
   a) phosphoric acid, $H_3PO_4(aq)$, (used in cola beverages and rust-proofing products)
   b) chromic acid, $H_2CrO_4(aq)$, (used in the production of wood preservatives)
   c) barium hydroxide, $Ba(OH)_2(aq)$, a white, toxic base (can be used to de-acidify paper)
   d) $CH_3NH_2(aq)$, commonly called methylamine (is responsible for the characteristic smell of fish that are no longer fresh)

5. Carbonic acid is a diprotic acid. Explain the term "diprotic."

6. Compare the degree of ionization for $H_2SO_4(aq)$ and $HSO_4^-(aq)$. Which is a stronger acid?

7. Complete the following statements using the words "acidic," "basic," or "neutral": (do not write in this textbook)
   a) if pH > 7.0 the solution is _____
   b) if pOH = 7.0 the solution is _____
   c) if pOH > 7.0 the solution is _____

8. What is an acid–base indicator?

9. Describe how a universal indicator differs from bromothymol blue.

10. Compare the use of universal indicator paper with a pH meter.

11. Determine the pH of the following solutions and state whether they are acidic or basic:
    a) cranapple juice: $[H_3O^+(aq)] = 1.26 \times 10^{-3}$ mol/L
    b) mouthwash: $[H_3O^+(aq)] = 3.55 \times 10^{-6}$ mol/L
    c) drain cleaner: $[H_3O^+(aq)] = 9.77 \times 10^{-13}$ mol/L

12. Determine the pOH of the following solutions and state whether they are acidic or basic:
    a) green tea: $[OH^-(aq)] = 7.94 \times 10^{-7}$ mol/L
    b) root beer: $[OH^-(aq)] = 2.34 \times 10^{-7}$ mol/L
    c) orange punch: $[OH^-(aq)] = 5.13 \times 10^{-10}$ mol/L

13. From the following pH and pOH, determine $[H_3O^+(aq)]$ or $[OH^-(aq)]$:
    a) $[H_3O^+(aq)]$ if pH = 4.97
    b) $[H_3O^+(aq)]$ if pH = 10.5
    c) $[H_3O^+(aq)]$ if pOH = 3.22
    d) $[H_3O^+(aq)]$ if pOH = 2.1
    e) $[OH^-(aq)]$ if pOH = 12.56
    f) $[OH^-(aq)]$ if pOH = 2.2
    g) $[OH^-(aq)]$ if pH = 7.3
    h) $[OH^-(aq)]$ if pH = 11.3

Applying Concepts

14. In the laboratory, you have samples of three different solutions of equal concentration: a 1.0 mol/L solution of acetic acid, a 1.0 mol/L solution of hydrochloric acid, and a 1.0 mol/L solution of ammonia. How would the pH of each solution compare? Explain.

15. Codeine, a compound extracted from opium, is used for pain relief. The pOH of a solution of codeine is 3.74. Is codeine an acid or a base? Explain.

16. How does diluting an acidic or basic solution ten-fold affect the pH of a solution? What if you diluted it one hundred–fold? What happens to the pH as you continue to dilute the acid?

17. Determine a possible range of pH values for solutions A, B, and C using Table 6.7 on page 230.
    a) In solution A, bromocresol green is blue and methyl red is orange.
    b) In solution B, phenol red is red and phenolphthalein is colourless.
    c) In solution C, phenolphthalein is pink and indigo carmine is blue.

18. The following indicators are added to three separate samples of a solution with pH 6.2. What colour would you observe in each case? Use Table 6.7 on page 230.
    a) methyl orange
    b) bromothymol blue
    c) phenol red

19. Harmful micro-organisms grow on cheese unless the pH of the cheese is maintained between 5.5 and 5.9. You test a sample of cheese and determine that $[H_3O^+(aq)]$ is 1.0×10^{-7} mol/L. Is the cheese safe to eat? Why or why not?

20. You have six unknown solutions, all of equal concentration—a strong base, a strong acid, a weak base, a weak acid, a neutral ionic solution, and a neutral molecular solution. Design a procedure to identify each solution. Summarize the expected results in a data table.

21. Explain the rules for significant digits to use when working with pH values.

22. "A strong acid is more dangerous than a weak acid." Is this statement true or false? Explain your answer using the terms "concentration" and "strength."

23. How is a 1.0 mol/L solution of hydrochloric acid different from a 1.0 mol/L solution of acetic acid? If you added a strip of magnesium metal to each acid, would you observe any differences in the reactions? Explain so that Grade 9 students could understand your answer.

Solving Problems

24. Calculate the pH of a solution made by dissolving 0.001 23 g of HCl(g) in enough water to make 375 mL of solution.

25. Calculate the pH of a solution made by dissolving 9.99 mg of strontium hydroxide in enough water to make 500 mL of solution.

26. How much hydrogen iodide (in grams) would have to be dissolved in enough water to make 2.0 L of solution with a pH of 3.45?

27. How much nitric acid could be isolated from 600 mL of a solution with a pH of 4.0?

28. Calculate the mass of potassium hydroxide that would have to be dissolved to make 195 mL of solution with a pOH of 0.07.

Making Connections

29. Alberta has a large amount of limestone in geological formations. Limestone consists of calcium carbonate. As a result of the presence of the limestone, bodies of water such as lakes and ponds are less likely to become acidic due to acid rain. Use the information that you learned about acids and bases to explain why limestone might protect lakes from acidity caused by acid rain.

30. "Normal" rainwater is slightly acidic because it contains dissolved carbon dioxide, CO_2(g), which occurs naturally in the air. Carbon dioxide reacts with water to form aqueous hydrogen carbonate, a weak diprotic acid.

  a) Write an equation to show how carbon dioxide reacts with water to form aqueous hydrogen carbonate.

  b) If normal rainwater is slightly acidic, why is acid precipitation such a concern? Explain your answer.

  c) Why is acid precipitation currently more of a concern in Eastern Canada than in Western Canada? Explain your answer.

  d) Explain why removing sulfur from fuels helps reduce acid deposition.

  e) Explain in detail how a scrubber removes sulfur from coal as it is burned.

  f) List three ways in which you could reduce your use of fossil fuels. How does reducing your use of fossil fuels help prevent acid deposition?

How does the choice to ride a bike instead of drive affect the environment?

31. Look back to the Launch Lab at the begining of this chapter. If you carried out this activity, you observed that a solution containing sodium hydroxide, NaOH(aq), and bromothymol blue is coloured blue. If you blow into the solution, it turns green, then yellow. Based on your answer to part (a) of question 30 and your knowledge of acids and bases, provide a detailed analysis of these observations.

Career Focus: Ask a Marine Conservationist

Dr. Farideh Jalilehvand
(Courtesy of the National Research Council – NINT)

For 333 years, the Swedish warship *Vasa*, who sunk on her maiden voyage in 1628, lay in Stockholm's harbour. Pollution made the water around the ship almost oxygen free (anoxic), preventing organisms that consume wood from decomposing the ship. As a result, the ship was very well preserved. In 1961 the ship was raised to dry land, refurbished, and placed on display in a museum. But by 2000, acidic reactions within the hull (the ship's body) threatened to disintegrate the *Vasa*. The decomposition had to be halted. Dr. Farideh Jalilehvand, an assistant chemistry professor at the University of Calgary, is one of the minds behind saving the *Vasa*.

Q How did you get involved in the *Vasa* case?

A I lived in Stockholm, Sweden, while working on my second PhD at the Royal Institute of Technology. One of my professors there got me involved in the effort to save the ship.

Q What caused the damage to the ship?

A As the *Vasa* lay underwater, increasing pollution from the growing city gradually reduced the oxygen content of the water. In order to gain oxygen, bacteria feeding on the ship would convert sulfate (SO_4^{2-}(aq)) ions, which are present in seawater, to oxygen. As a byproduct of this reaction, toxic hydrogen sulfide (H_2S(aq)) was created. The dissolved hydrogen sulfide reacted with the *Vasa's* wood compounds. The reactions gradually converted the wood to sulfides, elemental sulfur and organic sulfur compounds, which accumulated in the wood. When the ship was salvaged and brought to dry land, the sulfur compounds began to react with oxygen in the air, forming sulfuric acid together with sulfate salts. The resulting acids ate away at the ship's wood. The iron ions from the ship's rusting iron bolts and the humid conditions within the *Vasa* Museum speeded the disintegration.

Q What steps do scientists and museum curators take to reduce the damage from the acid?

A To neutralize the acid, we treat the wood with cloths (called *poultices*) soaked in a solution of sodium bicarbonate, $NaHCO_3$(aq), and sodium carbonate decahydrate, $Na_2CO_3 \cdot 10H_2O$(aq). This treatment produces a solution that reduces the acidity of the wood. This helps prevent the wood's cellulose from breaking down. The harmless products of carbon dioxide, CO_2(g), and sodium sulfate, Na_2SO_4(aq), are formed in the reactions. But, this treatment is only a "band-aid" solution. The mass of acid neutralised in this way is only about 50 kg, and it is estimated that about 2 t of sulfuric acid built up in the wood. A few weeks after treatment with poultices, treated areas become acidic again.

A more efficient way of neutralizing the acid would be to treat the wood with ammonia gas, NH_3(g). That's a less likely solution, since it would require constructing a gas-tight tent around the whole ship! Scientists are researching more lasting treatments to reduce the rate of acid-producing reactions, such as spraying treatments to wash away acid, and methods to remove the rust particles that speed up harmful reactions in the wood.

Q What is the state of the *Vasa* today?

A Since her salvage, the *Vasa's* massive oak timbers only have decayed by a few centimetres. This is not cause for alarm yet, and it will take a long time before the whole ship is softened by disintegration. We have also discovered that the wood is also unevenly acidic, with low pH (less than pH = 3) in several hundred areas.

Q What lessons can we draw from the *Vasa*?

A Many shipwrecks are still buried in the seabed under anoxic conditions. We should resist salvaging them until a solution is found for the acidity problem caused by sulfur accumulation.

Q **You've achieved an extensive education, earning two PhD degrees in chemistry. What interests you about chemistry?**

A Chemicals and chemical reactions are everywhere. To start it was the colours! In high school, I was fascinated to see how the two colourless solutions reacted to form a blood red solution of a complex ion $[Fe(SCN) \cdot 5(H_2O)]^{2+}(aq)$. I wanted to understand how reactions work.

Art conservationist Art conservationists work to preserve cultural artifacts such as musical instruments, textiles, paintings, archival materials, such as letters and books, and museum displays. They test and analyze the best conditions for storage, display, and conservation. They also devise treatments for cleaning and preserving to prevent further damage, such as corrosion from acid damage to art and artifact. Art conservationists need a strong background in organic and inorganic chemistry, as well as art history.

Architectural conservationist Cleaning and restoring building exteriors is a key part of keeping them functional and looking as they were intended to look. Architectural conservationists often employ chemical cleaning with bases and acids to refurbish buildings made of marble, limestone, dolomite, and travertine. Reversing damage from acid precipitation is a common goal of architectural conservationists, many of whom work on buildings in cities with acid rain problems.

Archaeological conservationist Work for an archaeological conservationist often begins on-site at an archaeological dig, where they consult with excavators on how best to remove artifacts from the ground or water in which the artifacts have been preserved. When an object moves to a museum or university lab, conservators work to clean and chemically stabilize objects, in order to keep the object from deterioration. Undergraduate courses in archeology and chemistry are good preparation for graduate study in the field, which is usually required for entrance into the profession.

Marine archaeological conservationist Scientists and explorers continue to locate preserved shipwrecks and marine artifacts. Marine archaeological conservationists preserve organic and inorganic marine artifacts alike—from iron, copper, brass, wood, and leather, to whole ships. They also work to restore artifacts for display and storage. Marine archaeological conservations usually study underwater archaeology after completing an education in marine biology, chemistry, archaeology, history, or art history.

Go Further...

1. The *Mary Rose* was a warship raised after hundreds of years spent underwater. What are the similarities between the preservation of the *Mary Rose* and the *Vasa*?

2. Acid rain is a common problem in North American cities. Why does this phenomenon occur? Which materials are most resistant to acid rain damage and why?

3. Research three preservation techniques used to halt acid damage in marine artifacts.

www.albertachemistry.ca
WWW

Understanding Concepts

1. How can a homogeneous mixture be distinguished from a heterogeneous mixture? Give one example of each.

2. **a)** List the steps in the dissolving process.

   b) Classify each step as either endothermic or exothermic.

   c) What determines whether the dissolving process is endothermic or exothermic overall?

3. List three different ways in which the concentration of a solution could be described. Provide an example of each and where it could be used.

4. Is a saturated solution always a concentrated solution? Give an example to illustrate your answer.

5. Define the term "solubility." What factors affect the solubility of solids, liquids, and gases in other liquids?

6. Iron concentration of 0.2 to 0.3 parts per million in water can cause fabric staining when washing clothes. A typical wash uses 12.0 L of water. What is the maximum mass of iron that can be present without staining the clothes?

7. High levels of phosphorus are not toxic but can cause digestive problems. The allowable drinking water concentration is 0.05 ppm. What is the maximum mass of phosphorus that could be present in a 250 mL glass of tap water? What is the concentration in mol/L?

8. Is a 1% m/m solution of table salt, $NaCl(aq)$, more concentrated, less concentrated, or at the same concentration as a 1% solution of table sugar, $C_{12}H_{22}O_{11}(aq)$? Explain.

9. What are the concentrations of each anion and cation in the solutions below?

   a) 4.12 mol/L $NH_4Cl(aq)$

   b) 0.275 mol/L $Ba(OH)_2(aq)$

   c) 0.543 mol/L $(NH_4)_3PO_4(aq)$

10. Calculate the new concentration when the following dilutions are performed:

    a) 25.0 mL of 1.50 mol/L $HCl(aq)$ diluted to 150 mL of solution

    b) 1.3 L of 1.2 mol/L $NaNO_3(aq)$ diluted to 7.3 L of solution

    c) 500 mL of 3.75 mol/L $(NH_4)_3PO_4(aq)$, with 2.75 L of water added

11. Vinegar, which contains acetic acid, $CH_3COOH(aq)$, is added to a kettle with a build-up of scale (a base). What type of reaction would you expect to observe? What other changes could be noted? Explain.

12. A sample of lemon juice has a pH of 2. A sample of water has a pH of 5. How much more concentrated is the hydronium ion in the lemon juice than in the water?

13. If you did not know that acids and bases are electrolytes, using the Arrhenius theory, could you infer this fact? Why or why not?

14. The pH of pure water is 7.00. Explain why.

15. **a)** Acids and bases can be described as strong or weak. What is the difference? Give an example of a strong base, a strong acid, a weak base, and a weak acid.

    b) Acids and bases can be described as concentrated or dilute. Explain how these terms are different from those used in part (a).

16. Which of the solutions in each pair has a higher pH? Explain why.

    a) 0.10 mol/L solution of a weak acid;
    0.010 mol/L solution of the same acid

    b) 0.10 mol/L solution of an acid;
    0.10 mol/L solution of a base

    c) a solution with a pOH of 7.0;
    a solution with a pOH of 8.0

    d) 0.10 mol/L solution of a strong acid;
    0.10 mol/L solution of a weak acid

    e) 0.10 mol/L solution of a strong base;
    0.10 mol/L solution of a weak base

    f) 0.10 mol/L solution of $H_2SO_4(aq)$;
    0.10 mol/L solution of $HSO_4^-(aq)$

17. Calculate the pH of the following solutions:

    a) $[H_3O^+(aq)] = 2.3$ mol/L

    b) $[H_3O^+(aq)] = 1.43 \times 10^{-4}$ mol/L

    c) pOH = 12.3

Applying Concepts

18. Design a procedure to distinguish among the following eight solutions. All common laboratory apparatus is available:

    - 0.10 mol/L $NaOH(aq)$
    - 0.010 mol/L $NaOH(aq)$
    - 0.10 mol/L $CH_3COOH(aq)$
    - 0.10 mol/L $HCl(aq)$
    - 0.10 mol/L $NaCl(aq)$
    - 0.10 mol/L $Na_3PO_4(aq)$
    - 0.10 mol/L $C_{12}H_{22}O_{11}(aq)$
    - distilled water

19. Use Table 6.7 on page 230 to answer the following questions:
  a) If a vinegar solution has a pH of 5, what colour would you expect the following indicators to show if placed into separate samples of the vinegar?
   • bromocresol green
   • orange IV
   • phenolphthalein
  b) An aqueous solution of sodium acetate, $NaCH_3COO(aq)$, used in photographic development, makes phenol red indicator red and phenolphthalein pink. What is the pH of this solution?

20. Design an experiment to collect data on the pH of a stream over one year. Why might the pH vary at different times of the year?

21. Design a procedure to provide evidence to support the following claim: "Hydrobromic acid is a strong acid."

22. "A saturated solution is in equilibrium." Explain what this statement means. Draw a diagram to illustrate this equilibrium that could be used to explain saturation to a junior high school student.

23. Preparing and diluting solutions requires using precision glassware and following a series of specific steps. Create a poster that could be displayed in your classroom that outlines the processes to be followed in the preparation of a standard solution and dilution to a specific concentration. Include descriptions and labelled pictures of the glassware and equipment involved.

24. The concentration of solutions can be expressed in a variety of ways. Describe each of the ways in which concentration can be expressed and include an example of an appropriate use of that measure. Which method of expressing concentration made sense to you at the beginning of this unit? Has your opinion changed over the course of this unit?

25. You have a sample of acetic acid, $CH_3COOH(aq)$, with a concentration of 0.10 mol/L and a sample of hydrochloric acid, $HCl(aq)$, with a concentration of 0.00010 mol/L. Both samples have nearly the same pH.
  a) Explain how this is the case.
  b) Are these two particular samples equally "strong"? Justify your answer.

26. Sketch a line graph of $[H_3O^+(aq)]$ versus pH. Your graph should cover the range from 1.0 mol/L to 1.0×10^{-14} mol/L. Explain the shape of your curve. Repeat the process with $[OH^-(aq)]$ and pOH.

27. Science and technology have both intended and unintended consequences for humans and the environment. Using specific examples, address both positive and negative consequences that the use and production of acids and bases has had through time.

28. Using a spreadsheet package, create an appropriate line graph representing the following solubility data. From your graph, answer the questions below. **ICT**

Solubility Data at Various Temperatures

Temperature (°C)	Ammonium chloride (g/100 mL)	Potassium chloride (g/100 mL)	Sodium chloride (g/100 mL)
0	29.4	28.1	35.7
10	33.3	31.2	35.8
20	37.2	34.2	36.0
40	45.8	40.0	36.6
60	55.2	45.8	37.3
80	65.6	51.3	38.4
100	77.3	56.3	39.8

  a) What is the solubility of potassium chloride at 35 °C?
  b) What is the solubility of ammonium chloride at 90 °C?
  c) Which salt has the highest solubility overall?
  d) Which of the three ionic compounds' solubility is most affected by an increase in temperature?

29. Using the data in the table above, describe the procedure you would use to make 200 mL of a saturated solution of ammonium chloride at 40 °C. Describe the appearance of this solution.

Solving Problems

30. Calculate the molar concentration of a solution containing 3.75 g of ammonium phosphate, $(NH_4)_3PO_4(aq)$, dissolved in enough water to make 500 mL of solution. Determine the concentration of each ion in solution.

31. What volume of 5.00×10^{-2} mol/L $Ca(NO_3)_2(aq)$ solution will contain 2.50×10^{-2} mol of nitrate ions?

32. What volume of a 0.100 mol/L solution of potassium carbonate, $K_2CO_3(aq)$, could be made from 25 mg of solute. Describe the procedure you would follow to make this solution.

33. How much (in grams) solute is required to make 1.0 L of 0.500 mol/L copper(II) sulfate pentahydrate, $CuSO_4 \cdot 5\ H_2O(aq)$.

34. How much of a 1.75 mol/L solution would be required to make 750 mL of a 0.250 mol/L solution? How much water should be added?

35. Calculate the pH of the following solutions:
a) 2.24 g of hydrogen chloride, $HCl(g)$, dissolved to make 150 mL of solution
b) 175 mg of hydrogen perchlorate, $HClO_4(aq)$, dissolved to make 25 mL of solution
c) $[OH^-(aq)] = 2.47 \times 10^{-9}$ mol/L

36. A 250 mL sample of a 0.25 mol/L solution of hydrochloric acid, $HCl(aq)$, is diluted to 500 mL. What is the effect on pH? What is the effect on the sample's acid properties?

37. Sort the following in order of increasing pH:
- 0.10 mol/L $NaOH(aq)$
- 0.10 mol/L $Ba(OH)_2(aq)$
- 0.10 mol/L $HCl(aq)$
- 0.010 mol/L $HF(aq)$
- 0.10 mol/L $HF(aq)$
- 0.10 mol/L $NaCl(aq)$
- 0.10 mol/L $NH_3(aq)$

38. Calculate the pOH of the following solutions:
a) 5.6×10^{-2} g of sodium hydroxide, $NaOH(s)$, dissolved to make 750 mL of solution
b) 2.35 g of barium hydroxide, $Ba(OH)_2(s)$, dissolved to make 450 mL of solution
c) $[H_3O^+(aq)] = 2.00 \times 10^{-2}$ mol/L

39. What mass of hydrogen bromide, $HBr(g)$, is required to produce 250.0 mL of a solution with a pH of 4.5?

40. What mass of strontium hydroxide, $Sr(OH)_2(s)$, would be required to produce 375 mL of a solution with a pH of 9.45?

Making Connections

41. Vitamin A is a compound that is soluble in fats but not in water. It is found in certain foods, including yellow fruit and green vegetables. In parts of Central Africa, children frequently show signs of vitamin A deficiency, although their diet contains a good supply of necessary fruits and vegetables. Why? What implication does this have on how we usually take vitamin supplements?

42. Lead is highly toxic when absorbed into the body, especially for young children. A level of 10 μg of lead per decilitre of blood is cause for concern. Convert 10 μg per decilitre to concentration in mol/L and to ppm by mass.

43. The following data were assembled by Parks Canada to compare three hot springs in Alberta. No two hot springs are the same. Differences in temperature and the variety of mineral content vary due to factors such as the type of rock in the area.

Hot Spring Mineral Content (in ppm) and Other Data

	Banff Upper Hot Springs	Radium Hot Springs	Miette Hot Springs
sulfate ion, $SO_4^{2-}(aq)$	572.0	302.0	1130.0
aqueous hydrogencarbonate ion, $HCO_3^-(aq)$, (base)	134.0	100.8	124.3
aqueous hydrogen sulfide, $H_2S(aq)$ (acid)	2.0	n/a	6.0
total dissolved solids	1677	706	1798
pH	7.15	7.05	6.8
temperature (°C)	47.3	45.5	53.9

a) Account for the differences in pH among the three hot springs.
b) Account for the difference in total dissolved solids measured at the three hot springs. What fluctuations would you expect to see during the summer months?
c) Given the relatively low pH at Miette Hot Springs, should swimmers be concerned? Justify your answer.

44. Think about a drive you take frequently along a major highway. Likely, hazardous materials, including acids and bases along with other toxic substances, are transported along the route. What concerns would you have if you lived along a hazardous-goods route? How could your city or town reduce the risk and your level of concern?

45. Are sports drinks worth the expense? What do they contain, and what is the importance of each component? Could you make your own sports drinks? If so, what and how much of each component should be included? Why? Write a letter to the editor of a health magazine that answers these questions.

46. Acids and bases are electrolytes and their strengths in both "roles" are related. Explain whether the electrical conductivity of 0.10 mol/L $HCl(aq)$ is higher, lower, or the same as that of 0.10 mol/L $CH_3COOH(aq)$. Is the conductivity of 1.0×10^{-7} mol/L $HCl(aq)$ higher, lower, or the same as that of 1.0×10^{-7} mol/L $CH_3COOH(aq)$? Defend your answer.

47. Hydrangea is a shrub that boasts showy flowers that change colour depending on soil pH. Using the Internet for background research, put together a pamphlet for aspiring gardeners on how they can turn their garden into a showpiece by changing the colour of their hydrangea. Your report should include the chemical species responsible for the colour changes, the pH at which these colour changes occur, pictures of the hydrangea flowers in both acidic and basic conditions, and ways that gardeners can manipulate their soil pH to achieve the desired results. **ICT**

48. In this unit you learned how acid rain forms from oxides of non-metals, such as carbon, sulfur, and nitrogen. These oxides are a part of specific biogeochemical cycles: the carbon cycle, the sulfur cycle, and the nitrogen cycle. Write chemical equations to show how these oxides form acids.

49. The damage that strong acids and bases can have on the environment when improperly handled or disposed of is well known and understood.

  a) What strong acids and bases do you and your family use on a regular basis?

  b) Could any of them be easily substituted with less harmful products?

  c) Why do you and others resist this change? Justify your position.

50. In the past, scurvy was a disease that killed many sailors. James Lind (1716–1794) discovered that eating citrus fruits prevents scurvy. What do citrus fruits have in common? How might they help prevent scurvy?

James Lind, a naval physician, used controlled scientific studies to determine that eating citrus fruits, such as lemons and limes, prevented scurvy.

51. The people in the helicopter are spraying ground limestone onto the lake. Limestone is mostly calcium carbonate. For what reason might they be "liming" the lake? What do you think "liming" will accomplish? Explain the chemistry of "liming."

Quantitative Relationships in Chemical Changes

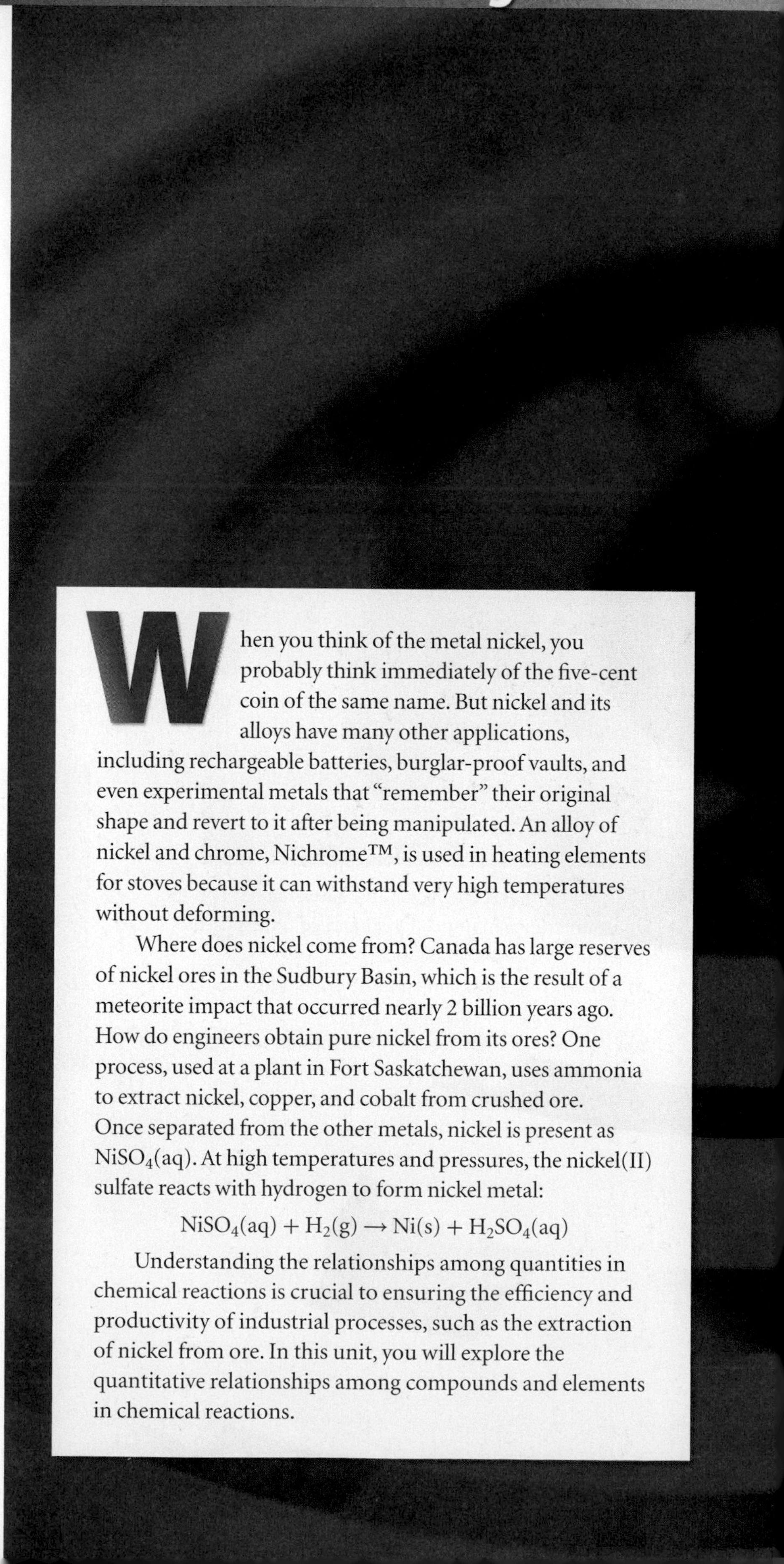

General Outcomes

- explain how balanced chemical equations show the quantitative relationships among the reactants and products involved in chemical changes

- use stoichiometry in quantitative analysis

Unit Contents

Focussing Questions

1. How do scientists and engineers use mathematics to analyze chemical change?

2. How are balanced chemical equations used to predict yields in chemical reactions?

Unit PreQuiz ⑦
www.albertachemistry.ca

W hen you think of the metal nickel, you probably think immediately of the five-cent coin of the same name. But nickel and its alloys have many other applications, including rechargeable batteries, burglar-proof vaults, and even experimental metals that "remember" their original shape and revert to it after being manipulated. An alloy of nickel and chrome, Nichrome™, is used in heating elements for stoves because it can withstand very high temperatures without deforming.

Where does nickel come from? Canada has large reserves of nickel ores in the Sudbury Basin, which is the result of a meteorite impact that occurred nearly 2 billion years ago. How do engineers obtain pure nickel from its ores? One process, used at a plant in Fort Saskatchewan, uses ammonia to extract nickel, copper, and cobalt from crushed ore. Once separated from the other metals, nickel is present as $NiSO_4(aq)$. At high temperatures and pressures, the nickel(II) sulfate reacts with hydrogen to form nickel metal:

$$NiSO_4(aq) + H_2(g) \rightarrow Ni(s) + H_2SO_4(aq)$$

Understanding the relationships among quantities in chemical reactions is crucial to ensuring the efficiency and productivity of industrial processes, such as the extraction of nickel from ore. In this unit, you will explore the quantitative relationships among compounds and elements in chemical reactions.

Each of the images in Figure P4.1 shows a chemical reaction in progress. How are these reactions similar? How do they differ? What is the mass of each reactant and each product? What amount of each reactant is involved? What is the difference between mass and amount when describing a chemical reaction? These questions all involve the topics you have studied in previous science courses that you will need to understand as you study Unit 4.

Figure P4.1 **A** Magnesium is burning in air. **B** Sodium azide is decomposing into gases that are rapidly expanding the air bag. **C** Silver nitrate is being added to sodium chloride.

Conservation of Mass

Figure P4.2 The mass of the products is the same as the mass of the reactants. As well, the number of sodium atoms and the number of chlorine atoms is the same before and after a chemical reaction. Is the number of moles of product the same as the number of moles of reactant?

Law of Conservation of Mass

After a chemical reaction, the total mass of the products is always equal to the total mass of the reactants.

One of the most fundamental laws, not only of chemistry, but of all nature, is the law of conservation of mass. Through chemical reactions, matter can be transformed from one compound to another and can take on many different properties. However, the total mass of the products is the same as the mass of the reactants. In fact, the total number of atoms of each type of element is the same after the reaction as it was before the reaction. Figure P4.2 shows how the law of conservation of mass helps you balance chemical equations.

$Na(s) + Cl_2(g) \longrightarrow NaCl(s)$

1 The equation is unbalanced. The mass of the reactants is greater than the mass of the products.

$Na(s) + Cl_2(g) \longrightarrow 2NaCl(s)$

2 Add a coefficient of 2 in front of NaCl. The equation is still unbalanced. The mass of the products is greater than the mass of the reactants.

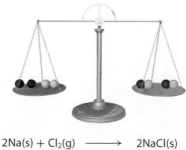

$2Na(s) + Cl_2(g) \longrightarrow 2NaCl(s)$

3 Add a coefficient of 2 in front of Na. The equation is now balanced.

Classifying Chemical Reactions

Learning to classify chemical reactions helps you understand the equations and make predictions about the products. As you might expect, there are several ways to classify chemical reactions but those used in this textbook cover most of the types of reactions that you will ever encounter. The names of the classes of reactions provide clues about the nature of the reactions. A general equation and one or more examples will be given for each class.

Formation Reactions

In a formation reaction, two or more reactants combine to produce one new product.

$$X + Y \longrightarrow XY \qquad\qquad N_2(g) + 3H_2(g) \longrightarrow 2NH_3(g)$$
$$2H_2(g) + O_2(g) \longrightarrow 2H_2O(g)$$

Decomposition Reactions

In a decomposition reaction, one compound breaks down into two or more simpler compounds or elements.

$$XY \longrightarrow X + Y \qquad\qquad 2NaN_3(s) \longrightarrow 3N_2(g) + 2Na(s)$$
$$NH_4NO_3(s) \longrightarrow N_2O(g) + 2H_2O(g)$$

Single-Replacement Reactions

In a single-replacement reaction, an atom of one element takes the place of an atom of another element. These reactions typically involve ionic compounds. Since either the positively charged ion or the negatively charged ion can be replaced, two general equations are often used to represent single-replacement reactions. In the general equations below, A and B represent the positively charged ions (or atom that can become positively charged) and X and Y represent the negatively charged ions (or atoms that can become negatively charged).

$A + BX \longrightarrow AX + B \qquad Cu(s) + 2AgNO_3(aq) \longrightarrow Cu(NO_3)_2(aq) + 2Ag(s)$

$AX + Y \longrightarrow AY + X \qquad 2NaBr(aq) + Cl_2(g) \longrightarrow 2NaCl(aq) + Br_2(aq)$

Double-Replacement Reactions

In a double-replacement reaction, the positively charged ions of two different ionic compounds exchange places, forming two new compounds. Neutralization reactions represent a special case of double-replacement reactions in which one of the compounds is an acid and the other is a base. The products are a salt and water.

$WX + YZ \longrightarrow WZ + YX \qquad AgNO_3(aq) + NaCl(aq) \longrightarrow NaNO_3(aq) + AgCl(s)$

$HA + BOH \longrightarrow AB + H_2O \qquad H_2SO_4(aq) + 2KOH(aq) \longrightarrow K_2SO_4(aq) + 2H_2O(\ell)$

Complete Combustion Reactions

A complete combustion reaction occurs when a hydrocarbon burns in a plentiful supply of oxygen and the only products are carbon dioxide and water. Many compounds can burn in the presence of oxygen, such as the magnesium in Figure P4.1. However, the term *complete combustion* is often reserved for the burning of hydrocarbons.

hydrocarbon + oxygen $\longrightarrow$ carbon dioxide + water $\quad CH_4(g) + 2O_2(g) \longrightarrow CO_2(g) + 2H_2O(g)$

Predicting the Products of a Reaction

When you understand the classes of reactions, you can usually predict the products of a reaction when given the reactants. The following steps provide a guide for predicting the products of a reaction.

1. First, identify what class of reaction best fits the reactants. For example:

a) $Na_2S(aq) + FeSO_4(aq) \rightarrow$
The reactants are two soluble ionic compounds, each with cations and anions that could exchange. The reaction is likely a double-replacement reaction.

b) $O_2(g) + CH_4(g) \rightarrow$
The reactants are oxygen and a hydrocarbon. The reaction is likely complete hydrocarbon combustion.

c) $MgO(s) \rightarrow$
There is only one reactant. The reaction is likely decomposition.

2. Next, predict the products. Remember to include states.

a) In a double-replacement reaction, the ions exchange places. The products are Na_2SO_4 and FeS. For a reaction to occur, one of the products must be an insoluble solid precipitate, molecular, or decompose to form a gas. According to the solubility guidelines, sodium sulfate is soluble and iron(II) sulfide is insoluble. Therefore, the unbalanced equation is:
$Na_2S(aq) + FeSO_4(aq) \rightarrow Na_2SO_4(aq) + FeS(s)$

b) The products of complete hydrocarbon combustion are water and carbon dioxide. Therefore, the unbalanced equation is:
$O_2(g) + CH_4(g) \rightarrow H_2O(g) + CO_2(g)$

c) Magnesium oxide decomposes to form magnesium metal, a solid, and oxygen, a gas. The unbalanced equation is:
$MgO(s) \rightarrow Mg(s) + O_2(g)$

Balanced Chemical Equations

Balanced chemical equations convey a lot of information. Examine Figure P4.3 to review the information that you can derive from some typical chemical equations. Notice that the coefficients of the each chemical formula represent the relative number of atoms, formula units, molecules, or ions involved in a reaction. The subscripts of the individual atoms (or polyatomic ions) tell you the number of atoms of that element that are in one molecule or formula unit.

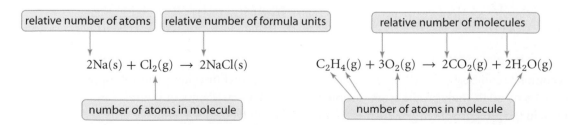

Figure P4.3 When no number is present as a coefficient or a subscript, it is assumed to be one.

The following guidelines will help you recall the methods that you learned for balancing chemical equations.

Guidelines for Balancing Chemical Equations

The process of balancing an equation by inspection involves trial and error. You may find helpful the systematic approach outlined below.

1. Compare the number of reactant atoms and polyatomeic ions on both sides of the unbalanced equation.

$$Cu(NO_3)_2(aq) + KOH(aq) \longrightarrow Cu(OH)_2(s) + KNO_3(aq)$$

Find the number of Cu^{2+}, NO_3^-, K^+, and OH^- ions on each side of the equation.

2. Identify the atoms or polyatomic ions that are present in different numbers on the two sides of the equation.

There is one hydroxide ion on the left and two on the right. There are two nitrate ions on the left and one on the right.

3. Choose one of the ions mentioned in step 2 and write an integer coefficient in front of the compound that will balance the number of ions on the two sides of the equation.

For example, choose the hydroxide ion. Place a 2 in front of the KOH term.
$$Cu(NO_3)_2(aq) + 2KOH(aq) \longrightarrow Cu(OH)_2(s) + KNO_3(aq)$$

4. Identify another atom or polyatomic ion that is different in number on the two sides of the equation.

There are still two nitrate ions on the left and one on the right side of the equation.

5. Determine the integer coefficient for the compound containing that atom or polyatomic ion that will balance it on the two sides of the equation.

Place a 2 in front of the KNO_3 compound on the right side of the equation.
$$Cu(NO_3)_2(aq) + 2KOH(aq) \longrightarrow Cu(OH)_2(s) + 2KNO_3(aq)$$

6. Once more, compare the number of each type of atom or polyatomic ion on each side of the equation. If the addition of coefficients has cause another atom or polyatomic ion to become unbalanced, repeat steps 4 and 5. If the numbers of each of the atoms and polyatomic ions is now the same on both sides of the equation, the equation is balanced.

The equation, $Cu(NO_3)_2(aq) + 2KOH(aq) \longrightarrow Cu(OH)_2(s) + 2KNO_3(aq)$, was balanced in two steps. (For some equations, more steps will be required.)

Balancing Chemical Equations

Problem

Octane, $C_8H_{18}(\ell)$, burns in an oxygen atmosphere to produce carbon dioxide and water vapour. This complete hydrocarbon combustion reaction takes place when automobile gasoline burns in an engine. Balance the chemical equation:

$$\_\_ C_8H_{18}(\ell) + \_\_ O_2(g) \rightarrow \_\_ CO_2(g) + \_\_ H_2O(g)$$

Solution

The unbalanced chemical equation is a hydrocarbon combustion reaction:

$$\_\_ C_8H_{18}(\ell) + \_\_ O_2(g) \rightarrow \_\_ CO_2(g) + \_\_ H_2O(g)$$

Since there are 18 hydrogen atoms on the reactant side and 2 hydrogen atoms on the product side, the hydrogen atoms are the most different in number (18 in $C_8H_{18}(\ell)$ and 2 in $H_2O(g)$).

To balance the hydrogen atoms, one $C_8H_{18}(\ell)$ and nine $H_2O(g)$ are needed:

$$C_8H_{18}(\ell) + \_\_ O_2(g) \rightarrow \_\_ CO_2(g) + \mathbf{9H_2O(g)}$$

(The 1 is not written in the equation.)

Now there are eight carbon atoms on the reactant side and one carbon atom on the product side. The carbon atoms are the most different in number in this partially balanced equation. To balance the carbon atoms, $8CO_2(g)$ are needed to balance the eight carbons in $C_8H_{18}(\ell)$:

$$C_8H_{18}(\ell) + \_\_ O_2(g) \rightarrow \mathbf{8}CO_2(g) + \mathbf{9}H_2O(g)$$

Now there remain 2 oxygen atoms on the reactant side and 25 oxygen atoms on the product side (16 in the $8CO_2(g)$ and 9 in the $9H_2O(g)$). There is no integer that you can put in front of $O_2(g)$ that will result in an odd number of oxygen atoms and you need 9 oxygen atoms. Therefore, double all of the coefficients in the equation but leave the coefficient of oxygen blank.

$$2C_8H_{18}(\ell) + \_\_ O_2(g) \rightarrow 16CO_2(g) + 18H_2O(g)$$

Once again, determine the number of oxygen atoms on each side of the equation. There are still two oxygen atoms on the reactant side and now there are 50 oxygen atoms on the products side. ($16 \times 2 + 18 = 50$) You can now balance the number of oxygen atoms in the equation by placing a coefficient of 25 in front of the $O_2(g)$ term.

$$2C_8H_{18}(\ell) + 25O_2(g) \rightarrow 16CO_2(g) + 18H_2O(g)$$

The equation is now balanced.

Check Your Solution

Tally the atoms of each type on each side of the chemical equation.

$$2C_8H_{18}(\ell) + 25O_2(g) \rightarrow 16CO_2(g) + 18H_2O(g)$$

Reactant side		Product side	
Atom	Number	Atom	Number
C	16	C	16
H	36	H	36
O	50	O	50

1. Balance the following equations and classify each reaction:
   a) $CH_4(g) + O_2(g) \rightarrow CO_2(g) + H_2O(g)$
   b) $P_4(s) + I_2(s) \rightarrow P_2I_4(s)$
   c) $Cl_2(g) + CsBr(aq) \rightarrow Br_2(aq) + CsCl(aq)$
   d) $Ba(ClO_3)_2(aq) + Na_3PO_4(aq) \rightarrow$
   $\qquad\qquad\qquad Ba_3(PO_4)_2(s) + NaClO_3(aq)$
   e) $Li_3N(s) \rightarrow Li(s) + N_2(g)$
   f) $C_6H_{12}O_6(aq) \rightarrow C_2H_5OH(aq) + CO_2(g)$

2. Predict the products for the following incomplete chemical equations. Then balance the equations:
   a) $C_4H_{10}(g) + O_2(g) \rightarrow$
   b) $Zn(s) + Pb(NO_3)_2(aq) \rightarrow$
   c) $Mg(s) + S_8(s) \rightarrow$
   d) $Sr(OH)_2(aq) + H_2SO_4(aq) \rightarrow$
   e) $LiN_3(s) \rightarrow$

Calculating Amounts of Substances

As you read above, the coefficients of the chemical formulas in a chemical equation represent the relative amounts of the compounds that react with each other. You probably recall that chemists find it very convenient to use the concept of moles instead of trying to describe the number of molecules or formula units taking part in a reaction. You have encountered several different formulas that include n, the amount of a substance in units of moles. Before learning about the quantitative calculations you will be performing in conjunction with chemical equations, it will be helpful to summarize the different methods for determining the amount of a substance in moles.

In previous science courses, you learned about the definition of the mole. In the Unit 2 Preparation, you reviewed that information and performed calculations using the formula $n = \frac{m}{M}$, where n is the number of moles, m is the mass, and M is the molar mass of a substance.

In Unit 2, you learned about the ideal gas law and you developed the formula, $PV = nRT$. You can rearrange this formula to solve for n, which will allow you to calculate the number of moles of a gas when you are given data for the pressure, temperature, and volume of the gas.

- Write the ideal gas law equation.

$$PV = nRT$$

- Since the variable that you want to isolate is multiplied by RT, perform the opposite operation and divide both sides by RT. Cancel like symbols.

$$\frac{PV}{RT} = \frac{n\cancel{RT}}{\cancel{RT}}$$

- Simplify.

$$n = \frac{PV}{RT}$$

In Unit 3, you learned to express the concentration of a solution in terms of moles per litre. You can write this definition in mathematical terms as $c = \frac{n}{V}$. You can rearrange this equation to solve for the amount of the substance in moles.

- Write the definition of concentration.

$$c = \frac{n}{V}$$

- Since the variable that you want to isolate is divided by V, perform the opposite operation and multiply both sides of the equation by V. Cancel like symbols.

$$cV = \frac{n}{\cancel{V}}\cancel{V}$$

- Simplify.

$$n = cV$$

The following table summarizes the different methods for calculating the amount of a substance in moles.

Condition of substance	Formula
When working with solid substances and the mass is known, use:	$n = \frac{m}{M}$
When working with gases and the pressure, temperature, and volume data are known, use:	$n = \frac{PV}{RT}$
When working with dissolved substances and the concentration and volume data are known, use:	$n = cV$

Prerequisite Skills

- **determine** molar mass of a compound
- **convert** mass to amount and amount to mass
- **recall** lab safety procedures

CHAPTER 7

Stoichiometry

Chapter Concepts

7.1 Reactions in Aqueous Solution

- Complete balanced, ionic, and net ionic equations are all ways to represent reactions in solution.

- Qualitative tests, such as flame tests and precipitation reactions, test for the presence of dissolved ions.

7.2 Stoichiometry and Quantitative Analysis

- The coefficients of a balanced chemical reaction can be used to predict the masses of products and reactants, given the mass of another product or reactant.

- Stoichiometric techniques can be used to determine the concentration and volume of aqueous solutions.

- Precipitation reactions are useful for determining the concentration of an unknown solution.

- Stoichiometric techniques can be used to determine the volume, temperature, mass, or pressure of gaseous reactants or products.

- Industrial processes use stoichiometric principles to maximize yield and minimize loss.

Plants, such as the wheat plants shown in the photograph, require nitrogen to grow. Nitrogen is the most abundant gas in our atmosphere (79%) but the supply of *usable nitrogen* is limited, because molecular nitrogen is unreactive. To enrich soil for commercial crops, industrially produced nitrogen compounds are often added to the soil as fertilizer.

Liquid ammonia, $NH_3(\ell)$, is a commonly used nitrogen fertilizer. Ammonia is also a starting material for the manufacture of other nitrogen-containing fertilizers.

For example, the fertilizer ammonium sulfate, $(NH_4)_2SO_4(s)$, is made at a chemical plant in Fort Saskatchewan, Alberta, according to the following chemical equation:

$$2NH_3(g) + H_2SO_4(aq) \rightarrow (NH_4)_2SO_4(s)$$

How do chemical engineers know what quantities of ammonia and sulfuric acid to use to make the fertilizer? In this chapter, you will learn how to use a balanced chemical equation to calculate the amounts of reactants needed or products expected in any chemical reaction.

Launch Lab

The Thermal Decomposition of Baking Soda

Baking soda and sodium bicarbonate are common names for sodium hydrogen carbonate ($NaHCO_3$). When baking soda is heated, such as in baking, it decomposes (thermal decomposition) to produce sodium carbonate, $Na_2CO_3(s)$, carbon dioxide, $CO_2(g)$, and water, $H_2O(g)$.

Observe the thermal decomposition of $NaHCO_3$ and compare the mole relationship between reactant and products in this reaction.

Safety Precautions

Make sure to allow the crucible to cool after the reaction before you touch it.

Materials

- baking soda
- crucible
- iron ring
- clay triangle
- Bunsen burner or alcohol burner
- retort stand
- electronic balance

Procedure

1. Read the entire procedure and construct a data table to record your observations.

2. Carefully weigh a dry empty crucible. Place 2 to 3 g of baking soda in the crucible and reweigh it. Record the weight of the empty crucible and the crucible containing the baking soda.

3. Set the crucible on a clay triangle supported by an iron ring attached to a retort stand. Heat gently with a Bunsen burner (or alcohol burner) for 5 or 6 minutes, then increase the heat for an additional 3 or 4 minutes. Be sure that the crucible is in the flame of the burner. Record your observations of what occurs during the reaction.

4. Allow the crucible and contents to cool to room temperature. Reweigh the crucible containing the product, Na_2CO_3, and record your data.

Analysis

1. Write a balanced chemical equation for the thermal decomposition of sodium hydrogen carbonate.

2. Calculate the amount (in moles) of sodium hydrogen carbonate used and the amount of sodium carbonate produced.

3. Determine the ratio of $\dfrac{\text{amount of sodium hydrogen carbonate (mole)}}{\text{amount of sodium carbonate (mole)}}$

4. What do you think is the significance of the ratio that you calculated in question 3?

Liquid ammonia, shown here, is used as a fertilizer. Ammonia is also an important raw material for the manufacture of other fertilizers, such as ammonium sulfate.

Reactions in Aqueous Solution

Key Terms

spectator ions
complete balanced equation
ionic equation
net ionic equation
qualitative analysis
quantitative analysis
flame test
precipitation reaction

Figure 7.1 Most of the time, spectators at a sports event have no net impact on the result of the game. Similarly, spectator ions are "passive onlookers" in a chemical reaction.

Reactions in aqueous solution are relevant to numerous chemical, biological, and industrial processes. For example, many single-replacement and double-replacement reactions, which you reviewed in the Unit Preparation, occur with water as the solvent. For these reactions, many have dissolved ions that are present in *both the reactants and the products*. In other words, although these ions are important components of the reactants, they remain unchanged when the reaction is over. For example, the following equation represents the reaction of silver nitrate and sodium chloride:

$$AgNO_3(aq) + NaCl(aq) \rightarrow NaNO_3(aq) + AgCl(s)$$

The silver ions and chloride ions react to form a solid. Sodium ions, $Na^+(aq)$, and nitrate ions, $NO_3^-(aq)$, remain dissolved in both the reactants and the products. Dissolved ions that are in the reactants and in the products once a reaction is complete are called **spectator ions** (see Figure 7.1).

Chemists use the concept of spectator ions to write specialized equations for reactions occuring in aqueous solutions.

Representing Aqueous Reactions with Net Ionic Equations

Chemists use three types of equations to represent chemical reactions in aqueous solution.

- The **complete balanced equation**: This equation shows all the reactants and products as if they were intact compounds that had not dissociated in solution. This is the type of equation that you are accustomed to writing. For example, the complete balanced equation for the reaction of aqueous solutions of silver nitrate and sodium chromate is written as follows:

$$2AgNO_3(aq) + Na_2CrO_4(aq) \rightarrow Ag_2CrO_4(s) + 2NaNO_3(aq)$$

- The **ionic equation**: This equation is a more accurate representation of a reaction in solution, because it shows all the high-solubility ionic compounds dissociated into ions. For the reaction of aqueous silver nitrate and sodium chromate, the ionic equation is written as follows:

$$2Ag^+(aq) + \mathbf{2NO_3^-(aq)} + \mathbf{2Na^+(aq)} + CrO_4^{2-}(aq) \rightarrow Ag_2CrO_4(s) + \mathbf{2Na^+(aq)} + \mathbf{2NO_3^-(aq)}$$

Notice that the coefficients and formulas of the sodium ions, $Na^+(aq)$, and the nitrate ions, $NO_3^-(aq)$, are identical on both sides of the equation. The sodium and nitrate ions are the *spectator ions*. Notice, too, that the charges on both sides of the equation are balanced. There are equal numbers of positive and negative charges on both sides of the equation resulting in a total charge of zero on each side.

- The **net ionic equation**: This equation shows only the actual chemical change that occurs. To write a net ionic equation, you cancel the spectator ions from each side of the equation, as shown below:

$$2Ag^+(aq) + \cancel{2NO_3^-(aq)} + \cancel{2Na^+(aq)} + CrO_4^{2-}(aq) \rightarrow Ag_2CrO_4(s) + \cancel{2Na^+(aq)} + \cancel{2NO_3^-(aq)}$$

For the reaction of aqueous solutions of silver nitrate and sodium chromate, the net ionic equation is as follows:

$$2Ag^{2+}(aq) + CrO_4^{2-}(aq) \rightarrow Ag_2CrO_4(s)$$

Figure 7.2 shows what the dramatic double-replacement reaction between silver nitrate and sodium chromate looks like in the laboratory. The pale-coloured ions represent the spectator ions, which, though present in the reactants, are not involved in the reaction.

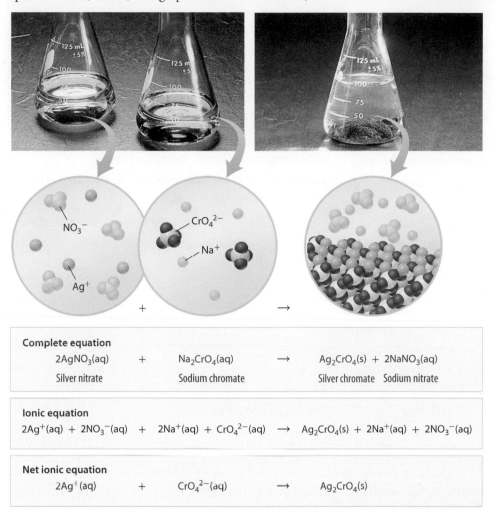

Figure 7.2 When you mix a clear solution of silver nitrate with a yellow solution of sodium chromate, you get a clear solution of sodium nitrate and a red precipitate of silver chromate.

Complete equation

$$2AgNO_3(aq) \quad + \quad Na_2CrO_4(aq) \quad \longrightarrow \quad Ag_2CrO_4(s) + 2NaNO_3(aq)$$

Silver nitrate   Sodium chromate   Silver chromate   Sodium nitrate

Ionic equation

$$2Ag^+(aq) + 2NO_3^-(aq) \quad + \quad 2Na^+(aq) + CrO_4^{2-}(aq) \quad \longrightarrow \quad Ag_2CrO_4(s) + 2Na^+(aq) + 2NO_3^-(aq)$$

Net ionic equation

$$2Ag^+(aq) \quad + \quad CrO_4^{2-}(aq) \quad \longrightarrow \quad Ag_2CrO_4(s)$$

Stoichiometry and Industry

The Connections feature on page 320 describes how natural gas is prepared for distribution. The harmful by-product gas, hydrogen sulfide ($H_2S(g)$), is removed and converted to the useful substance, sulfur. Using by-products to produce useful substances (instead of disposing of them) is one way that companies minimize waste and maximize efficiency and profit (Figure 7.3).

All companies in the chemical industry need to obtain the highest possible quantity of product from their starting materials. Stoichiometric analysis (gravimetric, solution, and gas) can help them to do this. Successful companies manufacture substances as efficiently as possible, with the highest yield possible and with the highest purity possible. In Investigation 7.C, you will choose an Alberta-related industrial process to research in-depth.

Figure 7.3 A responsible chemical company will minimize waste. Not only does this improve efficiency, but it also minimizes the effect of the company's activities on the environment.

Writing a Net Ionic Equation

Problem

When aqueous sodium iodide and aqueous lead(II) nitrate are mixed, a vivid yellow precipitate of lead(II) iodide is produced. Write a net ionic equation to represent this reaction.

What Is Required?

You need to write a net ionic equation for the reaction between aqueous lead(II) nitrate and aqueous sodium iodide to form solid lead(II) iodide.

What Is Given?

You know that the reactants are aqueous lead(II) nitrate and sodium(II) iodide.

You know that one of the products is solid lead(II) iodide.

Plan Your Strategy

Write formulas for all compounds involved.

Write the complete balanced equation.

Rewrite the equation and dissociate all the high-solubility ionic compounds. This is the ionic equation.

Look for ions that are unchanged in the reactants and the products—the spectator ions.

Cancel the spectator ions and write the net ionic equation.

• • •

Act on Your Strategy

The reactants in the problem are $Pb(NO_3)_2(aq)$ and $NaI(aq)$.

The product in the problem is $PbI_2(s)$. Another product must be formed of sodium and nitrate ions: $NaNO_3(aq)$. The *complete balanced equation*:

$$Pb(NO_3)_2(aq) + 2NaI(aq) \rightarrow PbI_2(s) + 2NaNO_3(aq)$$

According to the solubility table, $Pb(NO_3)_2(aq)$, $NaI(aq)$, and $NaNO_3(aq)$ are all high-solubility ionic compounds. Show how they dissociate into their constituent ions to give a *ionic equation*:

$$Pb^{2+}(aq) + 2NO_3^-(aq) + 2Na^+(aq) + 2I^-(aq) \rightarrow PbI_2(s) + 2Na^+(aq) + 2NO_3^-(aq)$$

The $NO_3^-(aq)$ and $Na^+(aq)$ ions are the same on both sides of the equation. Therefore, they are the *spectator ions*.

Eliminate the spectator ions from both sides of the equation:

$$Pb^{2+}(aq) + \cancel{2NO_3^-(aq)} + \cancel{2Na^+(aq)} + 2I^-(aq) \rightarrow PbI_2(s) + \cancel{2Na^+(aq)} + \cancel{2NO_3^-(aq)}$$

The *net ionic equation* for the reaction of $Pb(NO_3)_2(aq)$ and $NaI(aq)$ is

$$Pb^{2+}(aq) + 2I^-(aq) \rightarrow PbI_2(s)$$

Check Your Solution

The atoms and the charges in the final equation are balanced. The product is an insoluble salt. The spectator ions that were cancelled form a soluble compound.

Practice Problems

1. Write net ionic equations for the following *balanced* single-replacement reactions:
   a) $Cl_2(g) + 2RbBr(aq) \rightarrow Br_2(\ell) + 2RbCl(aq)$
   b) $Cu(s) + 2AgNO_3(aq) \rightarrow Cu(NO_3)_2(aq) + 2Ag(s)$
   c) $2Al(s) + 3CuCl_2(aq) \rightarrow 3Cu(s) + 2AlCl_3(aq)$
   d) $Zn(s) + Pb(NO_3)_2(aq) \rightarrow Zn(NO_3)_2(aq) + Pb(s)$
   e) $H_2(g) + Na_2SO_4(aq) \rightarrow 2Na(s) + H_2SO_4(aq)$

2. Write balanced net ionic equations for the following *unbalanced* double-replacement reactions:
   a) $Ba(ClO_3)_2(aq) + Na_3PO_4(aq) \rightarrow$
      $\quad\quad Ba_3(PO_4)_2(s) + NaClO_3(aq)$
   b) $Na_2SO_4(aq) + Sr(OH)_2(aq) \rightarrow$
      $\quad\quad SrSO_4(s) + NaOH(aq)$
   c) $Al_2(SO_4)_3(aq) + (NH_4)_2Cr_2O_7(aq) \rightarrow$
      $\quad\quad Al_2(Cr_2O_7)_3(s) + (NH_4)_2SO_4(aq)$
   d) $NaOH(aq) + HCl(aq) \rightarrow NaCl(aq) + H_2O(\ell)$
   e) $MgCl_2(aq) + NaOH(aq) \rightarrow$
      $\quad\quad Mg(OH)_2(s) + NaCl(aq)$

Qualitative versus Quantitative Analysis

The identity and concentration of metal ions in water can be determined by qualitative and quantitative analyses, respectively. One important application of this is the testing of drinking water. Health Canada specifies the allowable limits for ions of the metal elements cadmium, chromium, copper, lead, and zinc in drinking water. (See Table 7.1.)

Table 7.1 Limits for Metal Ions in Drinking Water

Metal	Allowable limits for ions
cadmium	5 ppb
chromium	0.05 ppm
copper	1 ppm
lead	0.01 ppm
zinc	5.0 ppm

Source: Summary of the Guidelines for Canadian Drinking Water Quality (04/04) Safe Environments Programme

Qualitative analysis involves determining by experiment whether a certain substance is present in a sample. By contrast, **quantitative analysis** allows you to determine *how much* of a certain substance is present in a sample.

Techniques of Qualitative Analysis

Advanced techniques, such as atomic absorption spectroscopy (AAS), can be used to carry out both qualitative and quantitative analysis. AAS uses the absorption of light to measure the concentration of metal atoms in a hot flame. However, other more conventional experimental techniques for qualitative analysis are still used today, which include observing solution colour, flame tests, and precipitation reactions.

Ion Colour in Solution

Aqueous solutions of ionic compounds of certain cations and anions have distinct solution colours. For example, most solutions that contain aqueous copper(II) ions are blue. Therefore, the colour of a solution provides evidence for the presence of certain ions. The colours in solution of some common ions are provided in Table 7.2.

Table 7.2 The Colour of Some Common Ions in Aqueous Solution

	Ions	Symbol	Colour
Cations	chromium(II) copper(II)	$Cr^{2+}(aq)$ $Cu^{2+}(aq)$	blue
	chromium(III) copper(I) iron(II) nickel(II)	$Cr^{3+}(aq)$ $Cu^+(aq)$ $Fe^{2+}(aq)$ $Ni^{2+}(aq)$	green
	iron(III)	$Fe^{3+}(aq)$	pale yellow
	cobalt(II) manganese(II)	$Co^{2+}(aq)$ $Mn^{2+}(aq)$	pink
Anions	chromate	$CrO_4^{2-}(aq)$	yellow
	dichromate	$Cr_2O_7^{2-}(aq)$	orange
	permaganate	$MnO_4^-(aq)$	purple

Chemistry File

WebLink
You have probably seen examples of qualitative analysis while watching forensics shows on television. The investigators spray a liquid they call Luminol on a floor then shine ultraviolet light on it in a darkened room. If hemoglobin is present, it will fluoresce (glow). Also, an investigator might have swabbed a spot that is suspected to contain human blood. They drop a liquid on the swab and if it turns blue, they have evidence of human blood. Research these qualitative analysis techniques on the Internet or in print resources to find out how they work.

www.albertachemistry.ca
WWW

Figure 7.4 The bright colours of fireworks result from the explosive ignition of metals with distinctive flame colours, such as strontium and copper.

Figure 7.5 Heated strontium **(A)** produces a red flame, while heated copper **(B)** produces a bluish-green flame.

Flame Tests

Many metal ions produce a distinct colour of flame when they are heated. Therefore, one way to test for the presence of metal ions in solution is to heat a drop of the solution in a hot flame and observe the colour. This qualitative procedure is called a **flame test**.

Why do different metal ions produce different colours when they are heated? Electrons within heated ions (or atoms) absorb thermal energy. This energy allows them to occupy higher electron energy levels within the atom or ion. These electrons are said to be *thermally excited*. The characteristic colours appear when the thermally excited electrons emit energy as visible light as they move back down to lower energy levels within the ion.

Fireworks, shown in Figure 7.4, are a familiar example of the different flame colours produced by heated metal ions. Strontium and copper ions are two examples, shown separately in Figure 7.5. The flame colours of some common ions are provided in Table 7.3.

Table 7.3 The Flame Colour of Selected Metallic Ions

Ion	Symbol	Colour
lithium	$Li^+(aq)$	red
sodium	$Na^+(aq)$	yellow
potassium	$K^+(aq)$	violet
cesium	$Cs^+(aq)$	violet
calcium	$Ca^{2+}(aq)$	yellowish-red
strontium	$Sr^{2+}(aq)$	red
barium	$Ba^{2+}(aq)$	yellowish-green
copper	$Cu^{2+}(aq)$	bluish-green
boron	$B^{2+}(aq)$	yellowish-green
lead	$Pb^{2+}(aq)$	bluish-white

Precipitation Reactions

A **precipitation reaction** is another term for a double-replacement reaction in solution that produces a solid product. The solid product is called a **precipitate**. Predicting the outcome of precipitation reactions can help in qualitative analysis. Chemists add dissolved substances to unknown solutions and observe whether a precipitate forms. Their observations help them infer which ions must have been present in the unknown solution. Figure 7.6 shows a general scheme for identifying ions in this way, and it illustrates how this is often done as a series of reactions to identitfy the different ions that are present. At each stage, the chemist may perform flame tests on the precipitate and observe the colour of the solution.

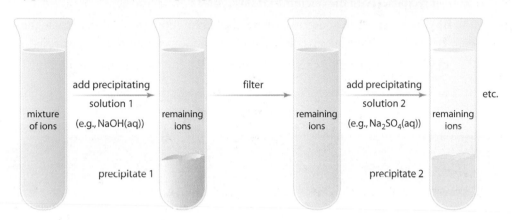

Figure 7.6 This illustration shows the idea behind using precipitation reactions for identifying ions in an aqueous solution. At each stage, the resulting precipitate is removed.

In the Thought Lab and questions that follow, interpret observations of flame tests, solution colour, and precipitation reactions to determine the identity of dissolved metal ions.

Thought Lab 7.1 | Identifying Unknown Aqueous Solutions

Target Skills

Predicting the identity of an ion based on qualitative tests.

A student performed a series of precipitation reactions, flame tests, and observations of solution colour tests on a solution known to contain two metal ions. The student does not know the identities of the ions. Use the data collected by the student and the information provided in Table 7.2, and Table 7.3 to identify the two metal ions present in the aqueous solution.

Procedure

1. Examine the observations in the Table of Evidence and then answer Analysis questions 1–4.

Table of Evidence: Testing a Solution of Unknown Metal Ions

Test	Observation
1 solution colour	solution is colourless
2 effect of adding NaOH(aq) to solution	white precipitate is produced. Mixture is filtered and remaining solution (filtrate) is colourless
3 flame test on precipitate from Test 2	flame colour is red
4 effect of adding Na₂SO₄(aq) to filtrate from Test 2	a second white precipitate is produced. Mixture is filtered and remaining solution is colourless
5 flame test on precipitate from Test 4	flame colour is red but different from the colour in Test 3

Analysis

1. List all the possible ions that give a red flame test and a precipitate in the presence of hydroxide ions.

2. List all the possible cations that give a red flame test and precipitate in the presence of sulfate ions.

3. If all traces of the two metal cations are removed from the solution at Test 4, what might the flame colour be when a sample of the solution is tested? Explain your prediction.

4. Write predictions for what you would observe if you used solution colour and precipitation reaction tests to identify the metal ion in the following unknown solutions:

   Solution 1: contains Na⁺(aq) only
   Solution 2: contains Cu²⁺(aq) only
   Solution 3: contains Na⁺(aq) and Ag⁺(aq)
   Solution 4: contains Cu²⁺(aq) and Ag⁺(aq)

1 Which ion or ions may be present, according to the following observations?

a) A solution is orange.

b) A blue solution gives a bluish-green flame.

c) A colourless solution gives a bright yellow flame.

2 You are given two unlabelled colourless solutions: $AgNO_3(aq)$ and $Ca(NO_3)_2(aq)$.

a) Which ion, when added to each unlabelled solution, would produce a precipitate with the calcium and silver ions?

b) Which ion, when added to each unlabelled solution, would differentiate the solution containing silver ions from the solution containing calcium ions?

c) What test, other than checking for precipitates, might identify the solutions?

INVESTIGATION | 7.A

Qualitative Analysis

In this investigation, apply your knowledge of chemical reactions and solubility to identify ions in solution.

Question

What ions are in the unknown solutions A, B, C, and D?

Predictions

Read the entire Procedure. Can you predict the results of any steps? Write your predictions in your notebook. Justify each prediction.

Safety Precautions

- Be careful not to contaminate the dropper bottles. The tip of a dropper should not make contact with either the plate or another solution. Put the cap back on the bottle immediately after use.

- Hydrochloric acid and sulfuric acid are corrosive. Wash any spills on your skin with plenty of cool water. Inform your teacher immediately.

- Part 2 of this investigation requires an open flame. Tie back long hair, and confine any loose clothing.

- When you have completed the investigation, wash your hands.

Materials

Part 1

- unknowns: four dropper bottles (labelled A, B, C, and D) of solutions, each containing one of the following ions: $Na^+(aq)$, $Ag^+(aq)$, $Ca^{2+}(aq)$, and $Cu^+(aq)$

- reactants: two labelled dropper bottles containing dilute $HCl(aq)$ and dilute $H_2SO_4(aq)$

- 12-well or 24-well plate, or spot plate

Part 2

- unknowns: solutions A, B, C, and D from Part 1

- reactants: four labelled dropper bottles, each containing one of the following cations: $Na^+(aq)$, $Ag^+(aq)$, $Ca^{2+}(aq)$, and $Cu^{2+}(aq)$

- laboratory burner

- heat-resistant pad

- cotton swabs

Procedure

Part 1 Using Acids to Identify Cations

1. Read Procedure steps 2 and 3 below. Design a suitable table for recording your observations.

2. Place one or two drops of each unknown solution into its own well or spot. Add one or two drops of hydrochloric acid to each unknown. Record your observations.

3. Repeat Procedure step 2, but this time, test each unknown solution with one or two drops of sulfuric acid. Record your observations.

4. Answer Analysis questions 1 to 5.

Part 2 Using Flame Tests to Identify Cations
Note: Your teacher may demonstrate this part.

5. Read Procedure steps 6 to 10. Design tables to record your observations.

6. Observe the appearance of each known solution. Record your observations. Repeat for each unknown solution. If you think that you can identify one of the unknowns, record your identification (refer to Table 7.4).

7. Flame tests can identify some cations. Set up the laboratory burner and heat-resistant pad. Light the burner. Adjust the air supply to produce a hot flame with a blue cone.

8. Place a few drops of solution containing $Na^+(aq)$ on one end of a cotton swab. Carefully hold the swab so that the saturated end is just in the laboratory burner flame, near the blue cone. You may need to hold it in this position for as long as 30 s to allow the solution to vaporize and mix with the flame. Record the colour of the flame.

9. Sodium is often present in solution as a contaminant. For a control, repeat Procedure step 8 with water and record your observations. You can use the other end of the swab for a second test. Dispose of used swabs in the container your teacher provides.

10. Repeat the flame test for each of the other known solutions. Then test each of the unknown solutions.

11. Answer Analysis question 6.

Analysis

1. a) Which of the cations you tested should form a precipitate with hydrochloric acid? Write the net ionic equation. (See Table P3.1 on page 161.)

   b) Did your results support your predictions? Explain.

2. a) Which cation(s) should form a precipitate when tested with sulfuric acid? Write the net ionic equation.

   b) Did your results support your predications? Explain.

3. Which cation(s) should form a soluble chloride and a soluble sulfate?

4. Which cation(s) has a solution that is not colourless?

5. Based on your analysis so far, tentatively identify each unknown solution.

6. Use your observations of the flame tests to confirm or refute the identifications you made in Analysis question 5. If you are not sure, check your observations and analysis with other students' results. If necessary, repeat some of your tests.

Conclusion

7. Identify the unknown cations in this investigation. Explain why you do, or do not, have confidence in your decisions. What could you do to be more confident?

Section 7.1 Summary

- Reactions in aqueous solution often contain spectator ions, which are ions present in both the reactants and products that do not undergo change in the reaction. Aqueous reactions can be depicted using complete balanced, ionic, and net ionic equations.

- The complete balanced equation shows all reactants and products as intact compounds, and not as if they had dissociated in solution. This is the type of equation with which you are already familiar.

- The ionic equation is a more accurate representation of a reaction in aqueous solution. It shows all of the ions that are formed in solution, including the spectator ions. Both the number of atoms and the charges must be balanced on both sides of the equation.

- The net ionic equation is the same as the ionic equation except that it lacks the spectator ions. Therefore, it only contains the ions that undergo chemical change in the reaction.

- Qualitative and quantitative analyses can be used to determine the identity and concentration, respectively, of ions in solution. Common experimental techniques for qualitative analysis include observation of solution colour, flame tests, and precipitation reactions.

- The colour of an aqueous solution may indicate the presence of a particular ion. Some dissolved ions cause a distinct colouring of the solution. Table 7.2 in this section provides a summary of ions and their associated colours when dissolved in solution.
- Flame tests involve heating a sample in a flame and observing the resultant colour of the flame. When metal ions are heated in this manner they produce distinctly coloured flames. Table 7.3 in this section provides a summary of flame colours that are associated with certain metal ions.
- Precipitation reactions take advantage of the solubility characteristics of substances. Solutions that are known to cause the precipitation of certain ions are added to a sample and the formation of a solid precipitate helps infer the presence of those ions.
- The qualitative tests discussed in this section are often used in combination to provide the strongest supporting evidence for the identity of an ion.

SECTION 7.1   REVIEW

1. Explain what is meant by spectator ions. What characteristics make an ion a spectator ion?
2. Write a net ionic equation for each double-displacement reaction in aqueous solution.
   a) tin(II) chloride with potassium phosphate
   b) nickel(II) chloride with sodium carbonate
   c) chromium(III) sulfate with ammonium sulfide
3. For each reaction in question 2, identify the spectator ions.
4. State the name and formula of the precipitate that forms when aqueous solutions of copper(II) sulfate and sodium carbonate are mixed. Write the net ionic equation for the reaction. Identify the spectator ions.
5. Which two soluble salts, when mixed together, result in each of the following net ionic equations? (There is more than one solution to each.)
   a) $3Ba^{2+}(aq) + 2PO_4^{3-}(aq) \rightarrow Ba_3(PO_4)_2(s)$
   b) $Mg^{2+}(aq) + 2OH^-(aq) \rightarrow Mg(OH)_2(s)$
   c) $2 Al^{3+}(aq) + 3Cr_2O_7^{2-}(aq) \rightarrow Al_2(Cr_2O_7)_3(s)$
6. Why might a chemist need to carry out qualitative analysis on a solution? Give an example.
7. Would you expect a qualitative analysis of a solution to give you information about the amount of each ion present? Explain your answer.
8. A solution of limewater, $Ca(OH)_2(aq)$, is basic. It is used to test for the presence of carbon dioxide, $CO_2(g)$. When carbon dioxide is bubbled through an aqueous solution, it reacts with the water to form carbonic acid, $H_2CO_3(aq)$. Carbon dioxide bubbled through limewater results in a milky-white precipitate.
   a) Write two balanced chemical equations to show what happens when carbon dioxide is bubbled through limewater.
   b) Classify the reactions.
   c) Is this an example of a qualitative test or a quantitative test? Explain your answer.

9. An ion in solution forms a precipitate when NaI(aq) is added to it. The precipitate produces a blue-white colour when heated in a flame.
   a) Suggest the formula of the ion and its low-solubility salt.
   b) Write a net ionic equation to represent the reaction.
10. To answer the following questions, refer to the solubility table (Appendix G or Unit 3 Preparation on page 161).
   a) What aqueous solution will precipitate $Pb^{2+}(aq)$ but not $Eu^{3+}(aq)$ or $Cu^+(aq)$ ions?
   b) What aqueous solution will precipitate $Cu^+(aq)$ ions but not $Eu^{3+}(aq)$ ions?
   c) Using your answers to (a) and (b), suggest a procedure that would allow you to first precipitate the $Pb^{2+}(aq)$ ions, followed by the $Cu^+(aq)$ ions, followed by the $Eu^{3+}(aq)$ ions.
   d) How would you test a solution for the presence of $Pb^{2+}(aq)$, $Cu^+(aq)$, and $Eu^{3+}(aq)$ ions?
11. All the solutions in the figure below have the same concentration: 0.1 mol/L. Use Table 7.2 to infer which ion causes the colour in each solution. How much confidence do you have in your inferences? What could you do to increase your confidence?

What do the colours tell you about the solutes of these solutions?

Stoichiometry and Quantitative Analysis

Section Outcomes

In this section, you will:

- **analyze** data and apply mathematical and conceptual models to develop and assess possible solutions to problems by using stoichiometric ratios from chemical equations
- **write** balanced ionic and net ionic equations, including identification of spectator ions for reactions taking place in aqueous solution
- **calculate** the quantities of reactants and products involved in chemical reactions using gravimetric, solution, and gas stoichiometry
- **use** appropriate SI notation, fundamental and derived units, and significant digit rules when performing stoichiometry calculations
- **explain** that the focus of technology is on the development of answers to problems, involving devices and systems that meet a given need within the constraints of a problem

Key Terms

mole ratio
stoichiometry
gravimetric stoichiometry
solution stoichiometry
gas stoichiometry

In section 7.1, you learned about qualitative analysis: determining by experiment whether a certain substance is present in a sample. By contrast, quantitative analysis allows you to determine *how much* of a certain substance is present in a sample. Quantitative analysis is used for many crucial applications, from determining the amount of pollutants in drinking water to ensuring that pharmaceutical products contain the correct amount of medicine.

Balanced chemical equations are essential for doing calculations and making predictions related to quantities in chemical reactions. To understand why, consider the following analogy. Suppose you work in a deli and your manager insists you prepare clubhouse sandwiches using three slices of toast, two slices of turkey, and four strips of bacon. Figure 7.7 shows how you can express this sandwich recipe as an equation.

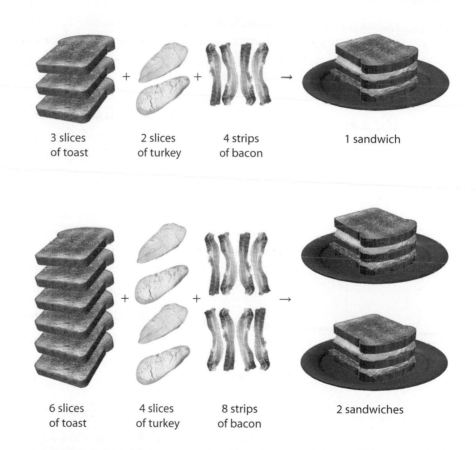

| 3 slices of toast | 2 slices of turkey | 4 strips of bacon | 1 sandwich |

| 6 slices of toast | 4 slices of turkey | 8 strips of bacon | 2 sandwiches |

Figure 7.7 Chemical equations, like sandwich recipes, can be multiplied. How much of each ingredient would you need to make three sandwiches?

If you are making two sandwiches, you need twice the amount of ingredients that you used to make the first sandwich. If you are making three sandwiches, you would need three times the amount of each ingredient.

You can get the same kind of information from a balanced chemical equation.

WebLink

Ammonia is produced using the Haber process, which uses the reaction shown on the right. A fertilizer plant in Redwater, Alberta, produces 960 000 Mg of ammonia every year via the Haber process. What are some of the key industrial uses of ammonia? Find out more about ammonia and the Haber process and present your findings as a brochure.

www.albertachemistry.ca
WWW

Mole Relationships in the Balanced Chemical Equation

The coefficients in front of the formulas for any chemical equation specify the relative number of particles of each chemical involved. For example, when ammonia is produced, we know that for every two molecules of $NH_3(g)$ formed, one molecule of $N_2(g)$ and three molecules of $H_2(g)$ have been consumed, as shown in the following equation:

$$N_2(g) + 3H_2(g) \rightarrow 2NH_3(g)$$

At the molecular scale, therefore, the ratio of the reaction components is

1 molecule of $N_2(g)$:3 molecules of $H_2(g)$:2 molecules of $NH_3(g)$

This knowledge may not seem useful, because molecules are too small to count. Recall, however, that chemists use the mole to compare the number of molecules with amounts that you can see and measure. At the molar scale (1 mole = 6.022×10^{23} molecules, atoms, or ions) the ratio of the reaction components is

1 mole of $N_2(g)$:3 moles of $H_2(g)$:2 moles of $NH_3(g)$

Mole Ratios

The ratios between the molar amounts of any *two* elements, ions, and compounds in a chemical equation are called **mole ratios**. For example, the mole ratio

1 mol $N_2(g)$: 2 mol $NH_3(g)$

can be used to predict how many moles of $N_2(g)$ would be produced by the decomposition of a certain amount of $NH_3(g)$. It can also be used to determine how many moles of $NH_3(g)$ would be produced by the consumption of a certain amount of $N_2(g)$.

The number of moles of each reactant or product in a mole ratio is identical to the coefficient in the balanced chemical equation you are considering. For example, the ratio of the components of the decomposition of potassium carbonate,

$$K_2CO_3(s) \rightarrow K_2O(s) + CO_2(g)$$

is 1 mol of $K_2CO_3(s)$: 1 mol of $K_2O(s)$: 1 mol of $CO_2(g)$.

The ratio of the components of the complete combustion of octane

$$2C_8H_{18}(\ell) + 25O_2(g) \rightarrow 16CO_2(g) + 18H_2O(g)$$

is 2 mol of $C_8H_{18}(\ell)$:25 mol of $O_2(g)$:16 mol of $CO_2(g)$:18 mol of $H_2O(g)$.

• • •

3 Consider the following chemical equation:
$$2H_2(g) + O_2(g) \rightarrow 2H_2O(\ell)$$
a) Write the ratio for all components of the reaction.
b) Write the mole ratio for $O_2(g)$ and $H_2O(\ell)$.
c) Write the mole ratio for $H_2(g)$ and $O_2(g)$.

4 Consider the following chemical equation:
$$2C_2H_6(g) + 7O_2(g) \rightarrow 4CO_2(g) + 6H_2O(g)$$
a) Write the ratio for all components of the reaction.
b) Write the mole ratio for ethane, $C_2H_6(g)$, and carbon dioxide, $CO_2(g)$.
c) Write the mole ratio for ethane and oxygen.

• • •

Using Mole Ratios

The mass of a reactant available for most reactions in the chemistry laboratory rarely corresponds to the precise number of moles specified by the balanced chemical equation. For one thing, it is impractical and often expensive to investigate the reactions of large quantities of chemicals.

For example, suppose a student carries out an investigation to observe the thermal decomposition of malachite, $Cu_2(CO_3)(OH)_2(s)$, shown in the following chemical equation:

$$Cu_2(CO_3)(OH)_2(s) \rightarrow CO_2(g) + H_2O(g) + 2CuO(s)$$

The balanced chemical equation states that 1 mol of malachite decomposes to form 1 mol of carbon dioxide, 1 mol of water, and 2 mol of copper(II) oxide. The student is unlikely to use 1 mol of malachite, however—this corresponds to 221.13 g of malachite, quite a large mass. Instead, the student might decompose about 1 g (0.0045 mol) of malachite, a less expensive quantity that is also easier to handle.

How would the student predict the amount of copper(II) oxide that is likely to be produced from 0.0045 mol of malachite? Examine the following Sample Problem to find out.

Sample Problem

Mole Ratios

Problem
What amount (in mol) of copper(II) oxide, $CuO(s)$, could be produced if 0.0045 mol of malachite, $Cu_2(CO_3)(OH)_2(s)$, decomposes completely according to the following equation?

$$Cu_2(CO_3)(OH)_2(s) \rightarrow CO_2(g) + H_2O(g) + 2CuO(s)$$

What Is Required?
You need to determine the amount (in mol) of copper(II) oxide that would be produced by the decomposition of 0.0045 mol of malachite.

What Is Given?
You know the balanced chemical equation for the reaction.

You know you have 0.0045 mol of malachite.

Plan Your Strategy
Use the balanced chemical equation to write the mole ratio.

Convert moles of malachite into moles of copper oxide by multiplying the moles of malachite by the mole ratio. Keep track of your units.

• • •

Act on Your Strategy
The ratio of malachite to copper(II) oxide is 2 mol $CuO(s)$:1 mol $Cu_2(CO_3)(OH)_2$

$$\frac{n(CuO)}{n(Cu_2(CO_3)(OH)_2)} = \frac{2 \text{ mol CuO}}{1 \text{ mol } Cu_2(CO_3)(OH)_2}$$

So

$$n(CuO) = n(Cu_2(CO_3)(OH)_2) \times \frac{2 \text{ mol CuO}}{1 \text{ mol } Cu_2(CO_3)(OH)_2}$$

$$= 0.0045 \text{ mol } Cu_2(CO_3)(OH)_2 \times \frac{2 \text{ mol CuO}}{1 \text{ mol } Cu_2(CO_3)(OH)_2}$$

$$= 0.0090 \text{ mol CuO}$$

Therefore, 0.0090 mol of copper(II) oxide is produced from the complete decomposition of 0.0045 mol of malachite.

Check Your Solution
Since the ratio of copper(II) oxide to malachite is two to one, the answer is as expected: there is twice the amount of copper(II) oxide as of malachite.

3. For the following reaction:
$$2AgNO_3(aq) + Na_2CrO_4(aq) \rightarrow$$
$$Ag_2CrO_4(s) + 2NaNO_3(aq)$$
  a) Write the ratio of all the components.
  b) Write the mole ratio for silver nitrate, $AgNO_3(aq)$, and silver chromate, $Ag_2CrO_4(s)$.
  c) What amount (in mol) of $Ag_2CrO_4(s)$ would be produced from 0.5 mol of $AgNO_3(aq)$?

4. For the following reaction:
$$2NH_3(g) + CO_2(g) \rightarrow NH_2CONH_2(s) + H_2O(g)$$
  a) Write the ratio of all the components of the reaction.
  b) What amount (in mol) of ammonia, $NH_3(g)$, is required to prepare 1.30 mol of urea, $NH_2CONH_2(s)$?
  c) What amount (in mol) of water is formed when 6.00 mol of carbon dioxide is consumed in the reaction?

5. The fertilizer ammonium sulfate, $(NH_4)_2SO_4(s)$, is made at Sherrit International Corporation's plant in Fort Saskatchewan. The following chemical equation shows the reaction:
$$2NH_3(g) + H_2SO_4(aq) \rightarrow (NH_4)_2SO_4(s)$$
  a) Write the ratio of all the components of the reaction.
  b) What amount of ammonia is necessary to prepare 20 000 mol of ammonium sulfate?

  c) What amount of ammonium sulfate fertilizer is formed when 3.28 mol of ammonia is consumed in the reaction?

6. At 400 °C, xenon and fluorine react to produce colourless crystals of xenon tetrafluoride, $XeF_4(s)$, as shown below:
$$Xe(g) + 2F_2(g) \rightarrow XeF_4(s)$$
  a) Write the ratio of all the components of the reaction.
  b) What amount of $F_2(g)$ is necessary to prepare 2.35 mol of $XeF_4(s)$?
  c) What amount of $F_2(g)$ is required to react with 12.2 mmol of xenon?
  (**Note:** 1 mol = 1000 mmol.)

7. A wide variety of products are possible from the reaction of nitrogen with oxygen. The equations below show two possible reactions:
$$2N_2(g) + O_2(g) \rightarrow 2N_2O(g)$$
$$N_2(g) + 2O_2(g) \rightarrow 2NO_2(g)$$
  a) For each equation, write the ratio of all the components of the reaction.
  b) What amount of $O_2(g)$ reacts with 0.0935 mol of nitrogen to form $N_2O(g)$?
  c) What amount of $O_2(g)$ reacts with 93.5 mmol of nitrogen to form $NO_2(g)$?
  (**Note:** 1 mol = 1000 mmol.)

Mass Relationships in Chemical Equations

Although the *mole* ratios of any two substances in any balanced reaction equation are simple whole number ratios, the *mass* ratios of any two chemicals in a balanced chemical reaction equation are rarely, if ever, so neat. For example, in the synthesis of ammonia,

$$N_2(g) + 3H_2(g) \rightleftharpoons 2NH_3(g)$$

the mole ratio of nitrogen to ammonia is:

$$1 \text{ mol } N_2(g):2 \text{ mol } NH_3(g)$$

Recall that you can use molar mass (M) and the equation $m = n \times M$ to find the corresponding mass of each compound as follows:

$$m = 1 \cancel{\text{ mol }} N_2(g) \times 28.02 \text{ g/}\cancel{\text{mol }} N_2(g) = 28.02 \text{ g } N_2(g)$$
$$m = 2 \cancel{\text{ mol }} NH_3(g) \times 17.04 \text{ g/}\cancel{\text{mol }} NH_3(g) = 34.08 \text{ g } NH_3(g)$$

Therefore, the mass ratio of nitrogen to ammonia is:

$$28.01 \text{ g } N_2(g):34.08 \text{ g } NH_3(g)$$

Although in theory you could use a "mass ratio" to determine the masses of other chemicals in a reaction, it is more convenient to use mole ratios for the conversion because you can obtain them directly from the balanced chemical equation.

Table 7.4 shows how amount and mass are related for the synthesis of ammonia. Notice that the mass of the reactants equals the mass of the products, as you would predict based on the law of conservation of mass.

Table 7.4 What a Balanced Chemical Equation Tells You

Balanced equation	$N_2(g) + 3H_2(g) \rightleftharpoons 2NH_3(g)$
Number of particles (molecules)	1 molecule $N_2(g)$ + 3 molecules $H_2(g)$ $\rightleftharpoons$ 2 molecules $NH_3(g)$

Amount (mol)	1 mol $N_2(g)$ + 3 mol $H_2(g)$ $\rightleftharpoons$ 2 mol $NH_3(g)$
Mass (g)	28.02 g $N_2(g)$ + 6.06 g $H_2(g)$ $\rightleftharpoons$ 34.08 g $NH_3(g)$
Total mass (g)	34.08 g reactants $\rightleftharpoons$ 34.08 products

• • •

5 What equation is used to calculate the number of moles of a reactant or product based on its mass? Define each term.

6 The previous Sample Problem demonstrated how 0.0090 mol of copper(II) oxide, CuO(s), is produced from the complete decomposition of 0.0045 mol of malachite. Calculate the mass of CuO(s) produced.

• • •

Chemistry File

FYI
The word *stoichiometry* is derived from two Greek words: *stoicheion* ("element") and *metron* ("measure"). Stoichiometry involves performing calculations using the masses (or the volumes) of reactants and products involved in a chemical reaction.

Stoichiometric Calculations

Stoichiometry is a method of predicting the quantity of a reactant or product in a chemical reaction based on the quantity of another reactant or product in the same reaction. In general, stoichiometry is the study of the relative quantities of reactants and products in chemical reactions.

Stoichiometric calculations are useful for many applications. For example, chemical engineers at NASA use stoichiometric calculations to determine the masses of the liquid hydrogen, liquid oxygen, and solid fuel needed for launching a craft, such as the space shuttle. Stoichiometric calculations can also help keep the air in the space shuttle safe for breathing (see Figure 7.8). Astronauts produce carbon dioxide as they breathe. To maintain a low level of carbon dioxide, air in the cabin is passed through canisters of solid lithium hydroxide. The lithium hydroxide reacts with carbon dioxide according to the following equation:

$$CO_2(g) + 2LiOH(s) \rightarrow Li_2CO_3(s) + H_2O(g)$$

Engineers use stoichiometric calculations to determine how much lithium hydroxide is needed to maintain a safe level of carbon dioxide, based on how much carbon dioxide an average astronaut exhales in a day.

Figure 7.8 A spacecraft is a closed system. If there was not a mechanism in place to remove carbon dioxide, the air would soon become unbreathable.

Chemistry File

FYI

According to IUPAC definitions, the term *stoichiometry* refers to the relationship between the amounts of substances that react together in a particular chemical reaction and the amounts of products that are formed. The general stoichiometric equation, $aA + bB + ... \rightarrow ...yY + zZ$, provides the information that a moles of A reacts with b moles of B to produce y moles of Y and z moles of Z.

Gravimetric Stoichiometry

Stoichiometric analysis involving mass is called **gravimetric stoichiometry**. To convert the mass of one component (reactant or product) of a reaction into the mass of another component (reactant or product) we must know the molar mass of each substance and the mole ratio of the two components in the reaction under investigation. Guidelines for solving these types of problems are provided below. The following Sample Problems show examples of gravimetric stoichiometry calculations.

Process for Solving Gravimetric Stoichiometry Problems

1. Write the complete balanced equation for the reaction and calculate the molar masses for the appropriate components (M_{given} and $M_{required}$).

2. Convert the mass of a *given substance* into a molar amount, n_{given}, using its molar mass (M_{given}) and the equation $n = \dfrac{m}{M}$.

3. Determine the molar amount of the *required substance*, $n_{required}$, using the mole ratio $\dfrac{n_{required}}{n_{given}}$ from the balanced equation, and the n_{given} you calculated in 2. The equation for this calculation is $n_{required} = n_{given} \times$ mole ratio.

4. Convert the molar amount of the *required substance*, $n_{required}$, into a mass, m, using its molar mass, $M_{required}$, and the equation $m = n \times M$.

Sample Problem

Gravimetric Stoichiometry: Reactant to Reactant

Problem

The exhaled carbon dioxide, $CO_2(g)$, produced by astronauts can be removed from the air in the spacecraft using lithium hydroxide, $LiOH(s)$. The reaction produces lithium carbonate, $Li_2CO_3(s)$, and water, $H_2O(g)$. An astronaut exhales an average of 1.00×10^3 g of carbon dioxide each day. What mass of lithium hydroxide should engineers place on board a spacecraft, per astronaut, per day?

What Is Required?

You need to determine the mass of $LiOH(s)$ required to react completely with 1.00×10^3 g $CO_2(g)$.

What Is Given?

You know the reaction between carbon dioxide and lithium hydroxide produces lithium carbonate and water. You know the mass of carbon dioxide is 1.00×10^3 g.

Plan Your Strategy

Step 1 Write the balanced chemical equation and determine M for each substance.

Step 2 Convert the mass of $CO_2(g)$ into a molar amount using $n = \dfrac{m}{M}$.

Step 3 Determine the molar amount of $LiOH(s)$ based on the mole ratio $\dfrac{n_{required}}{n_{given}}$.

Step 4 Determine the mass of $LiOH(s)$ using $m = n \times M$.

Act On Your Strategy

Step 1 Write the balanced chemical equation and determine M for each substance.

$$CO_2(g) + 2LiOH(s) \rightarrow Li_2CO_3(s) + H_2O(g)$$

$$m_{CO_2} = 1.00 \times 10^3 \text{ g} \qquad m_{LiOH} = ?$$

$$M_{CO_2} = 44.01 \text{ g/mol} \qquad M_{LiOH} = 23.95 \text{ g/mol}$$

Step 2 Convert the mass of $CO_2(g)$ into moles of $CO_2(g)$.

$$n_{CO_2} = \frac{m_{CO_2}}{M_{CO_2}}$$

$$= \frac{1.00 \times 10^3 \text{ g } CO_2(g)}{44.01 \text{ g/mol}}$$

$$= 22.72 \text{ mol } CO_2(g)$$

Step 3 Convert the amount of $CO_2(g)$ in moles into the amount of $LiOH(s)$ in moles.

$$n_{LiOH} = n_{CO_2} \times \frac{2 \text{ mol } LiOH(s)}{1 \text{ mol } CO_2(g)}$$

$$= 22.72 \text{ mol } CO_2(g) \times \frac{2 \text{ mol } LiOH(s)}{1 \text{ mol } CO_2(g)}$$

$$= 45.44 \text{ mol } LiOH(s)$$

Step 4 Use $m = n \times M$ to convert the moles of LiOH(s) into the mass of LiOH(s).

$$m_{LiOH} = n_{LiOH} \times M_{LiOH}$$

$$= 45.4 \text{ mol LiOH(s)} \times 23.95 \text{ g/mol}$$

$$= 1.09 \times 10^3 \text{ g LiOH(s)}$$

To react completely with 1.00×10^3 g CO_2(g), 1.09×10^3 g LiOH(s) is required.

Check Your Solution

Lithium hydroxide needs to be present in twice the amount (in mol) of carbon dioxide (the molar ratio is 2:1). The molar mass of lithium hydroxide, however, is about half the molar mass of carbon dioxide. Therefore, it makes sense that the mass of carbon dioxide (1000 g) would be approximately equal to the mass of the reactant (1090 g). The number of significant digits is correct (three).

• • •

Using Dimensional Analysis: A one-step way to check the answer to this problem is to link these three conversion steps into a single calculation. Set up the equation so that all terms cancel except the required term (in this case, g LiOH(s)). Consider trying this method once you feel comfortable doing the calculations in steps.

$$m_{LiOH} = 1.00 \times 10^3 \text{ g } CO_2\text{(g)} \times \frac{\text{mol } CO_2\text{(g)}}{44.01 \text{ g } CO_2\text{(g)}} \times \frac{2 \text{ mol LiOH(s)}}{1 \text{ mol } CO_2\text{(aq)}} \times \frac{23.95 \text{ g LiOH(s)}}{\text{mol LiOH(s)}}$$

$$= 1.09 \times 10^3 \text{ g LiOH(s)}$$

Sample Problem

Gravimetric Stoichiometry: Reactant to Product

Problem

The intense yellow colour of chrome yellow is the main pigment in the paint used to create the yellow lines on Canadian roads and highways. Chrome yellow—lead(II) chromate, $PbCrO_4$(s)—is the precipitate produced when aqueous solutions of lead(II) nitrate, $Pb(NO_3)_2$(aq), and potassium chromate, K_2CrO_4(aq), are mixed. What mass of chrome yellow is produced from the reaction of 5.10g of lead(II) nitrate with sufficient potassium chromate in aqueous solution?

What Is Required?

You need to determine the mass of $PbCrO_4$(s) produced from 5.10 g of $Pb(NO_3)_2$(aq) and sufficient K_2CrO_4(aq).

What Is Given?

You know that lead(II) nitrate and potassium chromate react to form a precipitate lead(II) chromate (and another compound) because the reaction is a double displacement reaction. You know the mass of $Pb(NO_3)_2$(aq) is 5.10 g.

Plan Your Strategy?

Step 1 Write the balanced chemical equation and determine M for each substance.

Step 2 Convert the mass of $Pb(NO_3)_2$(aq) into a moles amount using $n = \frac{m}{M}$.

Step 3 Determine the molar amount of $PbCrO_4$(aq) based on the mole ratio $\frac{n_{required}}{n_{given}}$.

Step 4 Determine the mass of $PbCrO_4$(aq) using $m = n \times M$.

Act on Your Strategy

Step 1 Write the balanced chemical equation and determine M for each substance.

$$Pb(NO_3)_2\text{(aq)} + K_2CrO_4\text{(aq)} \rightarrow PbCrO_4\text{(s)} + 2KNO_3\text{(aq)}$$

$$m_{Pb(NO_3)_2} = 5.10 \text{ g} \qquad m_{PbCrO_4} = ?$$

$$M_{Pb(NO_3)_2} = 331.23 \text{ g/mol} \quad M_{PbCrO_4} = 323.21 \text{ g/mol}$$

Step 2 Convert the mass of $Pb(NO_3)_2$(aq) into moles of $Pb(NO_3)_2$(aq).

$$n_{Pb(NO_3)_2} = \frac{m_{Pb(NO_3)_2}}{M_{Pb(NO_3)_2}}$$

$$= \frac{5.10 \text{ g } Pb(NO_3)_2\text{(aq)}}{331.23 \text{ g/mol}}$$

$$= 0.0154 \text{ mol } Pb(NO_3)_2\text{(aq)}$$

Step 3 Convert the amount of $Pb(NO_3)_2$(aq) in moles into the amount of $PbCrO_4$(s) in moles.

$$n_{PbCrO_4} = n_{Pb(NO_3)_2} \times \frac{1 \text{ mol } PbCrO_4\text{(s)}}{1 \text{ mol } Pb(NO_3)_2\text{(aq)}}$$

$$= 0.0154 \text{ mol } Pb(NO_3)_2\text{(aq)} \times \frac{1 \text{ mol } PbCrO_4\text{(s)}}{1 \text{ mol } Pb(NO_3)_2\text{(aq)}}$$

$$= 0.0154 \text{ mol } PbCrO_4\text{(s)}$$

Step 4 Use $m = n \times M$ to convert the moles of $PbCrO_4(s)$ into the mass of $PbCrO_4(s)$.

$m_{PbCrO_4} = n_{PbCrO_4} \times M_{PbCrO_4}$

$= 0.0154 \text{ mol } PbCrO_4(s) \times 323.21 \text{ g/mol}$

$= 4.98 \text{ g of } PbCrO_4(s)$

From 5.10 g of $Pb(NO_3)_2(aq)$, 4.98 g of $PbCrO_4(s)$ will be produced.

Check Your Solution

The product (lead(II) nitrate) and reactant (lead(II) chromate) exist in a 1:1 molar ratio. The molar masses of both substances are very similar. Therefore, it makes sense that the mass of the product (4.98 g) would be approximately equal to the mass of the reactant (5.10 g). The number of significant digits is correct (three).

• • •

Using Dimensional Analysis: A one-step way to check the answer to this problem is to link these three conversion steps into a single calculation. Set up the equation so that all terms cancel except the required term (in this case, g $PbCrO_4(s)$). Consider trying this method once you feel comfortable doing the calculations in steps.

$m_{PbCrO_4} = 5.10 \text{ g } Pb(NO_3)_2(aq) \times \dfrac{\text{mol } Pb(NO_3)_2(aq)}{33.123 \text{ g } Pb(NO_3)_2(aq)} \times \dfrac{1 \text{ mol } PbCrO_4(s)}{1 \text{ mol } Pb(NO_3)_2(aq)} \times \dfrac{323.21 \text{ g } PbCrO_4(s)}{\text{mol } PbCrO_4(s)}$

$= 4.98 \text{ g } PbCrO_4(s)$

Practice Problems

8. Ethanoic acid, $CH_3COOH(\ell)$, is produced according to the following chemical equation:

$CH_3OH(\ell) + CO(g) \rightarrow CH_3COOH(\ell)$

Calculate the mass of ethanoic acid that would be produced by the reaction of 6.0×10^4 g of $CO(g)$ with sufficient $CH_3OH(\ell)$.

9. For the following chemical reaction:

$2NaCl(aq) + 2H_2O(\ell) \rightarrow$
$\qquad\qquad 2NaOH(aq) + Cl_2(g) + H_2(g)$

Calculate the mass of sodium chloride that must react with water to provide 400 kg of sodium hydroxide.

10. As an experiment, a student performed a single-replacement reaction by dipping a strip of copper metal, $Cu(s)$, into an aqueous solution of silver nitrate, $AgNO_3(aq)$, to produce silver, $Ag(s)$.

   a) Write the balanced equation for this reaction.

   b) Calculate the mass of $Ag(s)$ that would be produced if the copper strip had a mass of 1.00 g and was completely consumed in the reaction.

11. At a cement plant in Edmonton, limestone, $CaCO_3(s)$, decomposes into calcium oxide, $CaO(s)$ (lime), and carbon dioxide, $CO_2(g)$, when it is heated to about 900 °C. Calculate the mass of lime that would be produced by heating 200 g of limestone to 900 °C.

12. What mass of magnesium oxide, $MgO(s)$, would be produced by the reaction of 4.86 g of magnesium metal, $Mg(s)$, in a copious supply of oxygen gas, $O_2(g)$?

13. Calculate the mass of liquid metal element produced by the decomposition of a 23.3 g mass of mercury(II) sulfide.

14. The compound cisplatin, $Pt(NH_3)_2Cl_2(s)$, is commonly administered in combination with other chemotherapy drugs to treat cancers of the reproductive tracts, head, neck, bladder, esophagus, and lung. Cisplatin is prepared from potassium tetrachloroplatinate, $K_2PtCl_4(aq)$, by reaction with ammonia, $NH_3(aq)$, according to the following reaction:

$K_2PtCl_4(aq) + 2NH_3(aq) \rightarrow 2KCl(aq) + Pt(NH_3)_2Cl_2(s)$

What mass of cisplatin would result from the reaction of 55.8 g of $K_2PtCl_4(s)$ in aqueous solution?

15. The molecular compound phosphorus trichloride, $PCl_3(\ell)$, is a commercially important compound used in the manufacture of pesticides. It is prepared by the direct combination of phosphorus, $P_4(s)$, and chlorine, $Cl_2(g)$, according to the following *unbalanced* reaction equation:

$P_4(s) + Cl_2(g) \rightarrow PCl_3(\ell)$

What mass of $PCl_3(\ell)$ forms when 323 g of $Cl_2(g)$ reacts completely with $P_4(s)$?

Waste Water Treatment

Inadequately treated waste water can cause major environmental problems. Primary and secondary sewage treatment removes solids from the water. However, though the water may look clean at this stage, it still contains large amounts of dissolved phosphate, $PO_4^{3-}(aq)$, and nitrogen-containing compounds. Excess phosphate and nitrogen in rivers, lakes, and oceans causes algae to grow out of control, killing off other aquatic life. To prevent this problem, many cities use a third treatment stage to remove phosphate and nitrogen from waste water.

Chemical Treatment of Sewage Waste Water

Adding certain ionic compounds can remove some of the phosphate and nitrogen-containing compounds in waste water. For example, to remove phosphate, water treatment operators can add alum, a hydrate of aluminium sulfate ($Al_2(SO_4)_3 \cdot 14H_2O$). The reaction of alum with dissolved phosphate produces a precipitate, which settles out of the water:

$$Al_2(SO_4)_3 \cdot 14H_2O(aq) + 2PO_4^{3-}(aq) \rightarrow$$
$$2AlPO_4(s) + 3SO_4^{2-}(aq) + 14H_2O$$

Another way to precipitate the phosphate is by adding calcium chloride ($CaCl_2$):

$$3CaCl_2(aq) + 2PO_4^{3-}(aq) \rightarrow Ca_3(PO_4)_2(s) + 6Cl^-(aq)$$

Waste water contains various forms of nitrogen. When biological material breaks down, it releases ammonia (NH_3), which converts to ammonium ($NH_4^+(aq)$) in solution. Ammonium can be removed from water by adding the base sodium hydroxide (NaOH):

$$NaOH(aq) + NH_4^+(aq) \rightarrow Na^+(aq) + NH_3(g) + H_2O(\ell)$$

As the equation shows, ammonium gets converted to ammonia, which comes out of the water as a gas. The gas is collected and used to make fertilizers.

This treatment centre cleans water for the city of Calgary.

Biological Treatment Methods

An alternative to chemically treating waste water to remove phosphate and nitrogen is to get micro-organisms to do the work. After secondary treatment, waste water can be fed into large containers called biodigestors. Inside the biodigestors, bacteria take up the phosphate and nitrogen. The bacteria convert some of the nitrogen-containing compounds into nitrogen gas, $N_2(g)$, which is released into the atmosphere.

The phosphate, however, becomes part of the bacteria. The resulting phosphate-rich sludge is removed from the water and used to fertilize farmers' fields.

• • •

1. Another chemical method to remove phosphate from waste water is to add iron(III) chloride ($FeCl_3$). Provide a balanced equation to show how adding iron(III) chloride could be used to precipitate phosphate from waste water.

2. How many grams of alum would you need to add to 1 L of waste water to reduce the concentration of phosphate from 3 mg/L to 1 mg/L?

3. The city of Calgary treats enough water to fill 128 Olympic-sized swimming pools daily. This amounts to a lot of phosphate. The city's waste-water system includes biodigestors to remove excess phosphate from waste water. The resulting sludge is used to fertilize certain grain crops. However, the sludge is not used to grow vegetables. Why not?

4. Suggest one benefit and one drawback to removing phosphate or nitrogen from waste water using

   a) chemical treatment methods

   b) biological treatment methods

Chemistry File

Web Link

Police officers who suspect a person of drunk driving will ask them to take a Breathalyser™ test by blowing into a small hand-held device. This device estimates a driver's blood alcohol concentration (BAC), which is based on a colour change that occurs when alcohol in the driver's breath reacts with an acidic aqueous solution of potassium dichromate. What is the balanced equation for this reaction and why are stoichiometry calculations important? Research the answers to these questions and develop a diagram illustrating how the Breathalyser™ device works. Present this information as a poster. You may also want to include some statistics such as what the BAC legal limit is for driving in Alberta, and how many people are killed or injured due to drunk driving every year.

www.albertachemistry.ca
WWW

Solution Stoichiometry

Solution stoichiometry, like gravimetric stoichiometry, allows you to determine the quantity of one reactant or product when you know the quantity of another reactant or product. In solution stoichiometry, however, the volumes and concentrations of reactants or products are used to determine the concentrations, volumes, or masses of other reactants or products.

A common problem in solution stoichiometry is calculating how much of a substance must be added to precipitate a particular ion from solution. For example, suppose you wanted to remove a known quantity of silver ions, $Ag^+(aq)$, from an aqueous solution. To do this, you could add some aqueous magnesium chloride, $MgCl_2(aq)$. The $Ag^+(aq)$ and the $Cl^-(aq)$ ions would form a precipitate of $AgCl(s)$ according to the following net ionic equation:

$$Ag^+(aq) + Cl^-(aq) \rightarrow AgCl(s)$$

What volume of magnesium chloride solution would you need to add to precipitate all the silver ions? If the concentration and volume of the $Ag^+(aq)$ solution are known, a stoichiometric calculation can be used to determine the volume of the magnesium chloride solution that is required. Guidelines for solving these types of problems are provided below. Examine the Sample Problem that follows to see an example of this calculation.

Process for Solving Solution Stoichiometry Problems

1. Write the complete balanced equation for the reaction.

2. Determine the molar amount of a *given substance*, n_{given}, using the equation $n_{given} = V_{given} \times c_{given}$. This equation is simply a rearrangement of $c = \frac{n}{V}$, where c is the concentration (mol/L) and V is the volume (L).

3. Determine the molar amount of the *required substance*, $n_{required}$, using the mole ratio $n_{required}/n_{given}$ based on the balanced equation, and the n_{given} you calculated in 2. The equation for this calculation is $n_{required} = n_{given} \times$ mole ratio.

4. Convert the molar amount of the *required substance*, $n_{required}$, into a volume using the equation $V = \frac{n}{c}$.

• • •

7 How many moles of NaCl are in a 25 mL sample of a 1 mol/L NaCl(aq) solution?

• • •

Sample Problem

Solution Stoichiometry

Problem

Aqueous solutions of silver ions are usually treated with solutions containing chloride ions so that the silver can be recovered as solid silver chloride, $AgCl(s)$. This reaction, shown in the figure on the next page, can be represented by the following balanced net ionic equation:

$$Ag^+(aq) + Cl^-(aq) \rightarrow AgCl(s)$$

What is the minimum volume of 0.250 mol/L $MgCl_2(aq)$ needed to precipitate all the silver ions in 60 mL of 0.30 mol/L $AgNO_3(aq)$? (Assume that the silver chloride is completely insoluble in water.)

What Is Required?

You need to calculate the *volume* of 0.250 mol/L $MgCl_2(aq)$ that will completely react with all the silver ions in the 0.30 mol/L $AgNO_3(aq)$ solution.

What Is Given?

The volume of $AgNO_3(aq)$ is 60 mL.

$c_{AgNO_3} = 0.30$ mol/L

$c_{MgCl_2} = 0.250$ mol/L

Adding chloride ions to a solution is a common way to test for the presence of silver. If silver is present, a white precipitate of silver chloride, AgCl(s), forms.

Plan Your Strategy

Step 1 Write a complete balanced equation for this reaction.

Step 2 Determine the amount of $AgNO_3(aq)$ in moles using $n = c \times V$.

Step 3 Determine the amount of $MgCl_2(aq)$ in moles based on the mole ratio.

Step 4 Determine the volume of $MgCl_2(aq)$ using $V = \frac{n}{c}$.

• • •

Act on Your Strategy

Step 1 Write the complete balanced equation is as follows:

$$2AgNO_3(aq) + MgCl_2(aq) \rightarrow Mg(NO_3)_2(aq) + 2AgCl(s)$$

$c_{AgNO_3} = 0.30 \text{ mol/L}$ $\quad c_{MgCl_2} = 0.250 \text{ mol/L}$

$V_{AgNO_3} = 60 \text{ mL}$ $\quad V_{MgCl_2} = ?$

Step 2 Determine the moles of $AgNO_3(aq)$

$$n_{AgNO_3} = c_{AgNO_3} \times V_{AgNO_3}$$

$$= 0.30 \frac{\text{mol}}{\cancel{L}} \times 0.060 \cancel{L} \, AgNO_3(aq)$$

$$= 0.018 \text{ mol } AgNO_3(aq)$$

Step 3 Determine the moles of $MgCl_2(aq)$

$$n_{MgCl_2} = 0.018 \cancel{\text{mol } AgNO_3(aq)} \times \frac{1 \text{ mol } MgCl_2(aq)}{2 \cancel{\text{mol } AgNO_3(aq)}}$$

$$= 0.0090 \text{ mol } MgCl_2(aq)$$

Step 4 Determine the volume of $MgCl_2(aq)$

$$V_{MgCl_2} = \frac{n_{MgCl_2}}{c_{MgCl_2}}$$

$$= \frac{0.0090 \cancel{\text{mol}} \, MgCl_2(aq)}{0.250 \cancel{\text{mol}}/L}$$

$$= 0.036 \text{ L } MgCl_2(aq)$$

The minimum volume of $MgCl_2(aq)$ solution required is 0.036 L (or 36 mL).

Check Your Solution

The answer has the correct number of significant digits (two). The volume of $MgCl_2(aq)$ required is reasonable. The two solutions are of comparable concentration and 1 mol of magnesium chloride is required for every 2 mol of silver nitrate. Therefore, the volume of magnesium chloride solution required should be close to half the volume of silver nitrate solution.

Using Dimensional Analysis

$$V_{MgCl_2} = 0.060 \cancel{\text{L } AgNO_3(aq)} \times \frac{0.30 \cancel{\text{mol } AgNO_3(aq)}}{1 \cancel{\text{L } AgNO_3(aq)}} \times \frac{1 \cancel{\text{mol } MgCl_2(aq)}}{2 \cancel{\text{mol } AgNO_3(aq)}} \times \frac{1 \text{ L } MgCl_2(aq)}{0.250 \cancel{\text{mol } MgCl_2(aq)}}$$

$$= 0.036 \text{ L } MgCl_2(aq)$$

16. What minimum volume of 0.50 mol/L aqueous magnesium chloride do you need to add to 60 mL of 0.30 mol/L aqueous silver nitrate in order to remove all the chloride ions?

17. Sulfuric acid, $H_2SO_4(aq)$, can be neutralized by reacting it with aqueous barium hydroxide, $Ba(OH)_2(aq)$. The reaction is:
$H_2SO_4(aq) + Ba(OH)_2(aq) \rightarrow BaSO_4(s) + 2H_2O(\ell)$
What volume of 0.676 mol/L $H_2SO_4(aq)$ can be neutralized by 22.7 mL of 0.385 mol/L $Ba(OH)_2(aq)$?

18. When solutions of lead(II) nitrate and sodium iodide are mixed, a bright yellow precipitate appears.
   a) Write the complete balanced equation for this double-replacement reaction.
   b) What volume of 0.125 mol/L $NaI(aq)$ is necessary to precipitate all the aqueous lead(II) ions in 25.0 mL of 0.100 mol/L $Pb(NO_3)_2(aq)$?
   c) What mass of precipitate is formed in this reaction?

19. The cells lining your stomach secrete hydrochloric acid, with a typical concentration of 0.030 mol/L $HCl(aq)$. Antacid tablets are used to relieve the pain of heartburn, caused by excess stomach acid irritating the lining of the esophagus just above your stomach. One Brand X antacid tablet contains 500 mg of $CaCO_3(s)$ and 110 mg of $Mg(OH)_2(s)$.
   a) Calculate the volume of stomach acid neutralized by the $CaCO_3(s)$ in one Brand X antacid tablet according to the following equation:
$2HCl(aq) + CaCO_3(s) \rightarrow CaCl_2(aq) + CO_2(g) + H_2O(\ell)$
   b) Calculate the volume of stomach acid neutralized by the $Mg(OH)_2(s)$ in one Brand X tablet according to the following equation:
$2HCl(aq) + Mg(OH)_2(s) \rightarrow MgCl_2(aq) + 2H_2O(\ell)$
   c) What total volume of stomach acid is neutralized by one Brand X antacid tablet?

INVESTIGATION 7.B

Target Skills

Evaluating an experiment based on a precipitation reaction to determine the concentration of a solution

Determining the Concentration of a Solution

You will now use your knowledge of how to analyze stoichiometric data of precipitation reactions to determine the concentration of a solution. Your teacher will give you a sample of a solution, $Mg(NO_3)_2(aq)$. You will need to express the concentration of your solution as

   a) a mass of $Mg(NO_3)_2(aq)$ dissolved in 100 mL of solution
   b) a molar concentration

Question

What is the molar concentration of a $Mg(NO_3)_2(aq)$ solution?

Safety Precautions

If you spill any $Mg(NO_3)_2(aq)$ or $Na_3PO_4(aq)$ on your skin, flush with plenty of cool water and inform your teacher immediately. Once you have completed this investigation, wash your hands.

Materials

- 50 mL of a $Mg(NO_3)_2(aq)$ solution of unknown concentration
- 50 mL of 0.200 mol/L $Na_3PO_4(aq)$
- deionized water
- 150 mL beaker
- 50 mL volumetric pipette
- 250 mL Erlenmeyer flask
- funnel
- retort stand
- wash bottle
- drying oven (if available)
- stirring rod
- ring clamp or funnel rack
- filter paper
- large watch glass
- electronic balance

Procedure

1. Copy a table similar to the one below in your notebook.

Mass of filter paper	
Mass of filter paper and precipitate	
Mass of precipitate	

Observations

1. Using a volumetric pipette, measure 50.0 mL of the $Mg(NO_3)_2(aq)$ solution into a 100 mL beaker.

2. Slowly add the $Na_3PO_4(aq)$ to the solution in small amounts while swirling the beaker. Continue until no more precipitate is formed upon adding small amounts of $Na_3PO_4(aq)$.

3. Set up your filtration apparatus, as shown below. Record the mass of your filter paper. Be sure to record the mass of the filter paper before folding and wetting it.

4. Filter the mixture through the filter paper and record the color of the precipitate. Wash the precipitate with small amounts of deionized water.

5. Remove the filter paper with the precipitate and place it on a watch glass.

6. Leave the sample to dry (one hour in an oven at 70 °C or overnight on the counter).

7. When the sample is dry, determine the mass of the filter paper with precipitate. Record the mass.

Analysis

1. Use the data you collected to calculate the following values. Show all your calculations:

  a) the mass of dissolved $Mg(NO_3)_2$ in the solution you were given

  b) the concentration of the solution in mass of solute per 100 mL of solution

  c) the molar concentration of the solution (mol/L)

2. Identify at least two significant sources of error in your measurements.

3. Describe at least two improvements that you could make to your procedure.

4. Did your observations suggest that any of the $Mg(NO_3)_2$ decomposed, reacted incompletely, or became otherwise lost as a result of the design of your investigation method? If so, how do you think this affected the results of your experiment? Explain your answer.

Conclusions

5. State the molar concentration of your solution.

---

1 Fold a piece of fluted filter paper.

  a) Fold the filter paper in half.

  b) Make creases in the half to divide it into eight sections of equal size.

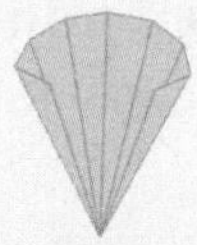

  c) Flip the piece over. Make a fan shape by folding each section in the direction opposite to the previous direction.

  d) Open up the two halves. You have now "fluted" your filter paper.

2 Place your fluted filter paper in the plastic funnel. Use your wash bottle to add a little distilled water to the centre of the filter paper so that it will stay in place.

3 Set up the filtration apparatus as shown. The diagram also shows how to pour the liquid down a stirring rod to ensure no product is lost.

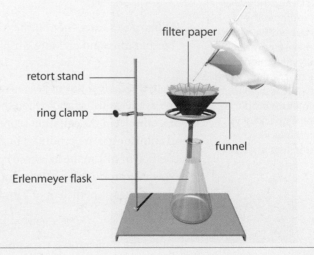

filter paper

retort stand

ring clamp

funnel

Erlenmeyer flask

Gas Stoichiometry

Gas stoichiometry, like gravimetric stoichiometry and solution stoichiometry, allows you to determine the quantity of one reactant or product if you know the quantity of another reactant or product. In gas stoichiometry, the volumes, temperatures, and pressures of gaseous reactants or products are involved in determining the quantities of other reactants or products.

For example, a chemical plant in Fort Saskatchewan, Alberta, uses the chlor-alkali process, shown in Figure 7.9. This process produces chlorine gas, $Cl_2(g)$, hydrogen gas, $H_2(g)$, and sodium hydroxide, $NaOH(aq)$, according to the following chemical equation:

$$2NaCl(aq) + H_2O(\ell) \xrightarrow{\text{electrical energy}} Cl_2(g) + 2NaOH(aq) + H_2(g)$$

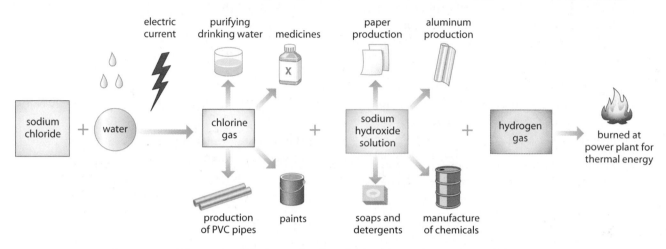

Figure 7.9 Chlorine and sodium hydroxide are the products of the chlor-alkali process. Hydrogen is a by-product. The hydrogen is not discarded, however, but is used for energy or to make other products.

Chlorine is a toxic gas. Hydrogen is an explosive gas. Therefore, for safety as well as for efficiency, engineers must be able to predict the volumes of chlorine and hydrogen gas that will be produced by a given quantity of brine.

To solve gas stoichiometry problems, you will use the concepts and skills you learned in Chapters 3 and 4, including the law of combining volumes and the ideal gas equation.

Stoichiometry and the Law of Combining Volumes

Recall from Chapter 3 that the law of combining volumes can be stated as follows:

> When gases react, the volumes of reactant and product gases, at the same temperature and pressure, are always in whole number ratios.

In fact, the mole ratios of gaseous species in a balanced chemical equation are the same as the volume ratios when the gases have the same temperature and pressure. For example, in the equation representing the chlor-alkali process, hydrogen gas and chlorine gas are produced in a 1:1 volume ratio. This is because they both have a coefficient of 1 in the balanced equation. In the electrolysis of water, hydrogen and oxygen are produced in a 2:1 volume ratio. As you can see in the balanced equation, hydrogen has a coefficient of 2 and oxygen has an implied coefficient of 1.

$$2H_2O(\ell) \xrightarrow{\text{electrical energy}} 2H_2(g) + O_2(g)$$

Because the mole ratio is the same as the volume ratio for gaseous reactions at constant temperatures and pressures, you can use volume ratios to solve stoichiometric problems.

Chemistry File

FYI
The chloro-alkali process results in the production of chemicals that have many commercial and industrial uses. For example, the chlorine gas produced is used to make a polymer called polyvinyl chloride, or PVC. Due to its strength and durability, PVC is used to make numerous products such as pipes, packaging, and even things like whitewater rafts.

8 For the reaction $N_2(g) + 2O_2(g) \longrightarrow N_2O_4(g)$:

   a) What is the mole ratio of $O_2(g)$ and $N_2O_4(g)$?

   b) What is the volume ratio of $O_2(g)$ and $N_2O_4(g)$?

   c) If 1 mol of $N_2O_4(g)$ is produced, how many moles of $O_2(g)$ must be consumed?

   d) If 1L of $N_2O_4(g)$ is produced, what volume of $O_2(g)$ must be consumed?

Sample Problem

Gas Stoichiometry Using the Law of Combining Volumes

Problem
Propane gas reacts with oxygen via complete combustion to produce carbon dioxide gas and water vapour. Suppose 1.50 L of propane gas, $C_3H_8(g)$, is consumed through complete combustion.

   a) What volume of oxygen gas is consumed? (The propane and oxygen are at the same temperature and pressure.)

   b) What volume of carbon dioxide gas is produced? (The propane and carbon dioxide are at the same temperature and pressure.)

What Is Required?
   a) You need to calculate the volume of oxygen gas required for complete combustion of 1.50 L of propane.

   b) You need to calculate the volume of carbon dioxide gas produced by the complete combustion of 1.50 L of propane.

What Is Given?
All the gases mentioned are at the same temperature and pressure, and 1.50 L of propane is consumed.

Plan Your Strategy
According to the law of combining volumes, the mole ratios of oxygen to propane and of carbon dioxide to propane from the balanced chemical equation will be the same as the volume ratios *when the gases have the same temperature and pressure.*

Write a balanced chemical equation.

   a) Use a mole ratio to convert the volume of propane into a volume of oxygen.

   b) Use a mole ratio to convert the volume of propane into a volume of carbon dioxide.

Act on Your Strategy
The balanced chemical equation for this reaction and the given data are as follows:

$$C_3H_8(g) + 5O_2(g) \rightarrow 3CO_2(g) + 4H_2O(g)$$
$$V_{C_3H_8} = 1.50 \text{ L}$$

   a) The mole ratio, and therefore the volume ratio, of $O_2(g)$ to $C_3H_8(g)$ is 5:1.

$$V_{O_2} = 1.50 \text{ L } C_3H_8(g) \times \frac{5 \text{ L } O_2(g)}{1 \text{ L } C_3H_8(g)}$$
$$= 7.50 \text{ L } O_2(g)$$

When 1.50 L of propane is consumed by complete combustion, 7.50 L of oxygen gas is consumed (at the same temperature and pressure).

   b) The mole ratio, and thus the volume ratio, of $CO_2(g)$ to $C_3H_8(g)$ is 3:1.

$$V_{CO_2} = 1.50 \text{ L } C_3H_8(g) \times \frac{3 \text{ L } CO_2(g)}{1 \text{ L } C_3H_8(g)}$$
$$= 4.50 \text{ L } CO_2(g)$$

When 1.50 L of propane is consumed by complete combustion, 4.50 L of carbon dioxide gas is produced (at the same temperature and pressure).

Check Your Solution
The solutions have the correct number of significant digits (three). The answers are reasonable.

20. For the reaction $2H_2(g) + O_2(g) \rightarrow 2H_2O(g)$
   a) What is the mole ratio of oxygen gas to water vapour?
   b) What is the volume ratio of hydrogen gas to water vapour?
   c) Calculate the volume of $H_2O(g)$ produced if 20 L of hydrogen gas reacts with the appropriate amount of oxygen.

21. Suppose that 12.0 L of nitrogen gas reacts with excess hydrogen gas to make ammonia gas, all at the same temperature and pressure. What volume of ammonia is expected from this reaction?

22. If 15 L of methane gas burns in a hot-water heater, what volume of oxygen gas at the same temperature and pressure is required for the methane to undergo complete combustion?

23. In the gas phase, 2.0 L of element A reacts with 1.0 L of element B to make 1.0 L of compound C. All the gasses are at the same temperature and pressure.
   a) Write a balanced chemical equation for this reaction.
   b) Each molecule of element A is actually made of two A atoms—it is really $A_2(g)$. Each molecule of element B is actually made of two B atoms—it is really $B_2(g)$. What is the formula of compound C in terms of A and B atoms?

Using the Ideal Gas Law in Stoichiometry Calculations

When the volumes of gaseous reactants or products are not under the same conditions of temperature and pressure, or when the reaction occurs under non-standard conditions, you will need to use the ideal gas law to determine the amounts and volumes of gases in a chemical reaction. Recall from Chapter 4 that the ideal gas law is represented by the following equation:

$$PV = nRT$$

Recall from Chapter 4 that: $n =$ amount of gas (mol)

$$P = \text{pressure of gas (kPa)}$$
$$V = \text{volume of gas (L)}$$
$$T = \text{temperature of gas (K)}$$
$$R = 8.314 \ \frac{\text{kPa} \cdot \text{L}}{\text{mol} \cdot \text{K}}$$

• • •

9 What is the volume of 3.0 mol of hydrogen gas at 100 kPa and 25 °C?

• • •

The following Sample Problem shows how to use a balanced chemical equation, a stoichiometric calculation, the mole ratio, and the ideal gas law to determine the volume of gas produced in a reaction.

Sample Problem

Gas Stoichiometry Using the Ideal Gas Law

Problem
In a laboratory investigation, what volume of hydrogen gas is likely to be produced, at 93.0 kPa and 23 °C from the reaction of 33 mg of magnesium metal with sufficient hydrochloric acid?

What Is Required?
You must calculate the volume of hydrogen gas produced at 93.0 kPa and 23 °C from the reaction of hydrochloric acid with magnesium.

What Is Given?
The mass of magnesium that reacted is 33 mg.
The temperature is 23 °C.
The pressure is 93.0 kPa.

Plan Your Strategy

Step 1 Write the complete balanced chemical equation.

Step 2 Use $n = \dfrac{m}{M}$ to calculate the moles of magnesium, n_{Mg}.

Step 3 Determine the moles of hydrogen, n_{H_2}, using the mole ratio from the balanced equation and the n_{Mg} you calculated in 2. The equation for this calculation is $n_{H_2} = n_{Mg} \times$ mole ratio.

Step 4 Use the ideal gas law to determine the volume of hydrogen produced.

Act on Your Strategy

Step 1 The balanced chemical equation for this reaction and the given data are as follows:

$$Mg(s) + 2HCl(aq) \rightarrow MgCl_2(aq) + H_2(g)$$

$m_{Mg} = 33$ mg $\qquad V_{H_2} = ?$

$M_{Mg} = 24.31$ g/mol

Also,

$P = 93.0$ kPa

$T = 23\,°C + 273.15 = 296.15$ K

Step 2 Determine the moles of Mg(s).

$$n_{Mg} = \frac{m_{Mg}}{M_{Mg}}$$

$$= \frac{0.033\ \cancel{g}\ Mg}{24.31\ \cancel{g}/mol}$$

$$= 0.001375\ \text{mol Mg}$$

Step 3 Determine the moles of $H_2(g)$.

$$n_{H_2} = 0.001375\ \cancel{\text{mol Mg}} \times \frac{1\ \text{mol}\ H_2}{1\ \cancel{\text{mol Mg}}}$$

$$= 0.001375\ \text{mol}\ H_2$$

Step 4 Use n_{H_2} to determine the volume of $H_2(g)$ at the given temperature and pressure.

$$PV = nRT$$

Therefore,

$$V = \frac{nRT}{P}$$

$$= \frac{0.001375\ \cancel{\text{mol}} \times 8.314\ \dfrac{L \cdot \cancel{kPa}}{\cancel{K} \cdot \cancel{mol}} \times 296.15\ \cancel{K}}{93.0\ \cancel{kPa}}$$

$$= 0.036\ L$$

The volume of hydrogen gas expected from this reaction is 36 mL.

Check Your Solution

The units of the answer are reasonable. The answer has the correct number of significant digits (two).

Note: If you were reporting the number of moles of magnesium in Step 2 or number of moles of hydrogen in Step 3, you would round it to 0.0014 mol. However, if you used 0.0014 mol in Step 4, your final answer would be incorrect. Always use the unrounded number for further calculations.

Remember: Report values in the correct number of significant digits but perform calculations with unrounded numbers.

Practice Problems

24. a) What is the volume of 2.0 mol of chlorine gas at 75 kPa and 27 °C?

   b) What amount, in moles, of an unknown gas occupies 3.2 L at 16.6 kPa and 127 °C?

   c) What amount, in moles, of $C_2F_6(g)$ occupies 12.5 mL at 600 mmHg pressure and −23 °C? (Note that 760 mmHg pressure is equivalent to 1 atm, or 101.325 kPa.)

25. Oxygen gas and magnesium react to form 2.43 g of magnesium oxide. What volume of oxygen gas at 94.9 kPa and 25.0 °C would be consumed to produce this mass of MgO(s)?

26. A 3070 kg load of coal is 90% carbon by mass. The coal burns to produce carbon dioxide. What volume of carbon dioxide gas is produced at 100 kPa and 25 °C, once it has cooled? (**Hint:** Use the percentage given to determine how much carbon udergoes combustion.)

27. In the semiconductor industry, hexafluoroethane, $C_2F_6(g)$, is used to remove silicon dioxide, $SiO_2(s)$, from apparatus as silicon tetrafluoride, $SiF_4(g)$, according to the following chemical equation:

$$2SiO_2(s) + 2C_2F_6(g) + O_2(g) \rightarrow$$
$$2SiF_4(g) + 2COF_2(g) + 2CO_2(g)$$

What mass of $SiO_2(s)$ reacts with 1.270 L of $C_2F_6(g)$ at a pressure of 0.200 kPa and a temperature of 400 °C?

Computer chips are made of silicon dioxide. A very clean room is necessary for building delicate computer components.

Analyzing Industrial Processes

Alberta's chemical and petrochemical industry is the second-largest manufacturing industry in the province, currently producing more than $6 billion worth of products annually. About half these products are exported to the American and Asian markets. Some of Alberta's major players in the chemical industry include:

Agrium
Albchem
Borden
BP Amoco
Canadian Fertilizers
Celanese Canada
Degussa-Huls Canada Inc.
Dow Chemical Canada Inc.

Laporte Performance
  Chemicals
Nexen Chemicals
Nova Chemicals
OXY Vinyl
Shell Canada Chemical
  Company
Sterling Chemical

Issue

Working in groups, choose an Alberta-related industrial process to research in depth. How does the process maximize yield, minimize waste, and minimize environmental impact? Could the company that uses the process improve on any of these areas?

Gathering Data and Information

1. Choose a chemical company and find out what chemical processes it uses. Choose one process.

2. Use the Internet and print resources to research the process. **ICT**

Organizing Findings

3. Use a word-processing program to prepare a report. Your report must

   - provide the details of your process from the raw materials to a finished product

   - include a list of by-products produced and how they are used (are they sold? discarded? burned at the plant for energy?)

   - include a schematic diagram, flowchart, or 3-D model to illustrate your chosen process

   - provide the details of the balanced chemical equations for all the reactions in your chosen process, including the details of the conditions used

   - list the ways that the overall yield of the process is maximized

   - list the ways that wastes are minimized

   - specify the ways that environmental impacts are minimized

   - include a properly cited list of references consulted

Your final report must incorporate the appropriate diagrams and charts into the body of the report.

Opinions and Recommendations

1. Using facts from your research, state an opinion on how well the process you researched maximizes yield, minimizes waste, and minimizes environmental impact.

2. Do you think the process should be improved in any of these areas? For example, could a by-product that is currently discarded as waste be used to create something useful? Could waste be reduced further? Explain your recommendations. Why do you think the company has not improved their performance in these areas?

Section 7.2 Summary

- Quantitative analysis of a reaction involves determining how much of a substance is present or expected in a reaction. Predicting the quantities of reactants or products in chemical reactions is done using stoichiometric calculations.

- Stoichiometric calculations involve the use of balanced chemical equations and the coefficients of the reactants and products, which indicate the relative amount of reactant(s) required or product(s) expected. This can also be viewed as the relative number of moles of required or expected components and can be represented using the mole ratio. The reaction $A + 2B \longrightarrow C$ can be thought of as 1 mole of A required to react with 2 moles of B, and the mole ratio of A to B is 1:2.

- The three types of stoichiometric calculations discussed in this section are gravimetric (using mass of reactants and products), solution (using concentration and volume of reactants and products), and gas (using volume, temperature, and pressure of reactants and products).

- The process for solving gravimetric stoichiometry problems is:
  1. Write the complete balanced equation for the reaction and calculate the appropriate molar masses (M_{given} and $M_{required}$).
  2. Determine n_{given} using M_{given} and the equation $n = \frac{m}{M}$.
  3. Determine $n_{required}$ using the mole ratio from the balanced equation, n_{given} from step 2, and $n_{required} = n_{given} \times$ mole ratio.
  4. Convert $n_{required}$ into a mass, m, using $M_{required}$ and $m = n \times M$.

- The process for solving solution stoichiometry problems is:
  1. Write the complete balanced equation for the reaction.
  2. Determine n_{given} using the equation $n_{given} = V_{given} \times c_{given}$.
  3. Determine $n_{required}$ using the mole ratio from the balanced equation, n_{given} from step 2, and $n_{required} = n_{given} \times$ mole ratio.
  4. Convert $n_{required}$ into a volume using the equation $V = n/c$.

- Gas stoichiometric calculations take advantage of the fact that, for gaseous reactions at constant temperature and pressure, the mole ratio is the same as the volume ratio. Therefore, given the volume of one component, the volume of another can be calculated. In addition, when the reactants and products are not under the same conditions or the reaction occurs under non-standard conditions, the ideal gas law can be used to calculate the amounts of gases produced. The ideal gas law is represented as $PV = nRT$.

- Stoichiometry has many important industrial applications that include removal of sulfur from natural gas. Stoichiometric calculations can help chemical industries to obtain the highest possible yield from starting materials, which is essential for their prosperity.

SECTION 7.2 REVIEW

1. Explain the difference between qualitative analysis and quantitative analysis.

2. What mass of precipitate will form when 3.66 g of potassium hydroxide in aqueous solution reacts with sufficient aqueous copper(II) nitrate solution?

3. What mass of each product would be formed by the complete combustion of 1.00 kg of methane, $CH_4(g)$, to give carbon dioxide and water vapour?

4. The Thermite® reaction, shown in the figure on the right, produces molten iron from the reaction of iron(III) oxide with excess aluminium powder. Predict the mass of molten iron that would be produced by the reaction of 28 kg of iron(III) oxide with excess aluminium.

The reaction of iron(III) oxide and aluminium generates enough thermal energy to melt the iron produced.

5. What mass of solid product is expected when 16.8 g of sodium hydrogen carbonate is heated until it fully decomposes to sodium carbonate, carbon dioxide, and water vapour?

6. A 13.2 g bar of zinc reacted with a silver nitrate solution. When the reaction was complete, 11.6 g of zinc *remained*. What mass of silver nitrate was in aqueous solution?

7. Suppose you have an aqueous solution of lead(II) nitrate. You do not know how much lead(II) nitrate is in your solution and want to find out. Given that you also have some sodium iodide solution, design an experiment that would allow you to determine how much lead(II) nitrate was in the solution.

8. Use a spreadsheet program to create a spreadsheet that carries out gravimetric stoichiometry calculations automatically when you fill in the variables. Test your spreadsheet by using it to solve problems you have already solved using a pen, paper, and a calculator. ICT

9. You intend to remove all of the barium ions in 60.0 mL of a 0.100 mol/L aqueous barium nitrate solution by precipitation followed by filtration. What is the minimum mass of sodium carbonate that you should add to the solution to accomplish this goal?

10. Hydrochloric acid reacts with pure magnesium carbonate, $MgCO_3(s)$, according to the following balanced equation:

$$MgCO_3(s) + 2HCl(aq) \rightarrow MgCl_2(aq) + H_2O(\ell) + CO_2(g)$$

What volume of 0.383 mol/L hydrochloric acid is needed to react with 1.62 g of pure magnesium carbonate?

11. Dissolved chlorine, $Cl_2(aq)$, is toxic to fish. At fairly low concentrations, it stresses fish by damaging their gills. A solution of sodium thiosulfate, $Na_2S_2O_3(aq)$, is used to remove chlorine from 54 L of tap water needed to fill an aquarium. The reaction of sodium thiosulfate and chlorine is as follows:

$$2Na_2S_2O_3(aq) + Cl_2(aq) \rightarrow 2NaCl(aq) + Na_2S_4O_6(aq)$$

What is the minimum volume of 0.170 mol/L $Na_2S_2O_3(aq)$ necessary to remove all the $Cl_2(aq)$, concentration 2.8×10^{-5} mol/L, from the tap water? (This is about 2.0 ppm, the residual chlorine concentration in municipal tap water.)

12. Calculate the volume of 0.150 mol/L sodium sulfate solution, $Na_2SO_4(aq)$, necessary to precipitate all the lead(II) ions in 100 mL of a 0.100 mol/L lead(II) nitrate solution, $Pb(NO_3)_2(aq)$.

13. What volume of 0.100 mol/L aqueous strontium hydroxide will completely neutralize 5.00 mL of household vinegar (essentially 0.833 mol/L ethanoic acid in water)?

14. The last step in manufacturing nitric acid, $HNO_3(aq)$, involves this reaction :

$$3NO_2(g) + H_2O(\ell) \rightarrow 2HNO_3(aq) + NO(g)$$

What volume of 15.4 mol/L $HNO_3(aq)$ is prepared from the reaction of 2.31×10^4 mol of $NO_2(g)$?

15. Wine becomes sour and "vinegary" when the ethanol it contains, $C_2H_5OH(aq)$, reacts with oxygen and is converted to ethanoic acid, $CH_3COOH(aq)$, as follows:

$$C_2H_5OH(aq) + O_2(g) \rightarrow CH_3COOH(aq) + H_2O(\ell)$$

Determine the concentration of ethanoic acid, $CH_3COOH(aq)$, that would be produced if 6.00 g of $C_2H_5OH(aq)$ in 1.00 L of wine were converted into $CH_3COOH(aq)$.

16. If all the reactants and products are at the same temperature and pressure, what volume of hydrogen chloride gas is produced from the reaction of 2.00 L of hydrogen gas with sufficient chlorine gas?

17. When it is disturbed by the shock wave of a detonator, nitroglycerin, $C_3H_5(NO_3)_3(\ell)$, decomposes into several gaseous products, as follows:

$$4C_3H_5(NO_3)_3(\ell) \rightarrow 6N_2(g) + O_2(g) + 12CO_2(g) + 10H_2O(g)$$

For every 4 mol of nitroglycerin, 29 mol of gas are produced. Under SATP conditions what total volume of gas is expected from the decomposition of 4.00 mol of nitroglycerin?

18. Nitrogen triodide, $NI_3(s)$, is a highly unstable compound. When crystals of this compound are touched very lightly, they decompose explosively to produce nitrogen, $N_2(g)$, and iodine, $I_2(g)$. Calculate total volume of each gas produced at STP when 395 mg of $NI_3(s)$ decomposes.

19. Chlorine gas can be made in the laboratory from the reaction of hydrochloric acid and manganese(IV) oxide. The balanced reaction equation is

$$MnO_2(s) + 4HCl(aq) \rightarrow MnCl_2(aq) + 2H_2O(\ell) + Cl_2(g)$$

Calculate the volume of chlorine gas produced at SATP from the reaction of 100 mL of 0.104 mol/L HCl(aq) and 0.250 g of $MnO_2(s)$.

20. A 2.00 L sample of ethane, $C_2H_6(g)$, at 1.00 atm and 25.0 °C, was burned in 2.00 L of oxygen at the same temperature and pressure to form carbon dioxide gas and liquid water. What is the volume of the carbon dioxide gas produced at the same temperature and pressure?

Qualitative analysis determines whether particular ions are present in a solution. Three common experimental techniques are observation of solution colour, flame tests, and precipitation reactions. Because some dissolved ions produce a distinct colour solution, the solution colour may indicate the presence of a particular ion. Flame tests involve heating a sample in a flame and observing the resultant flame colour. Finally, precipitation reactions rely on ion solubility when a particular precipitating solution is added. The formation of a precipitate supports the presence of certain ions in the solution being tested. Information from multiple qualitative tests provides strong supporting evidence for the identity of an ion.

Quantitative analysis determines how much of a substance is present or expected using stoichiometric calculations. These calculations use balanced chemical equations and coefficients of the reactants and products, which indicate the relative amount of reactant(s) required or product(s) expected. This can also be viewed as the relative number of moles of required or expected components and can be represented using the mole ratio. Three common types of stoichiometric analyses are gravimetric, solution, and gas stoichiometry.

Gravimetric stoichiometric calculations determine the mass of one component based on the mass of another. This requires the determination of the molar mass of each component and their mole ratio based on the balanced chemical equation. By converting the known mass to moles and using the mole ratio to determine the expected moles of the second component, the expected mass of that component can be calculated.

Solution stoichiometry calculations determine the concentration, volume, or mass of one component based on the volume or concentration of another in solution. For these calculations, the number of moles of a known reactant are determined based on its concentration and volume used in the reaction. Using this value and the mole ratio of the two reactants, the moles of precipitate formed can be calculated and, therefore, its mass also determined.

Gas stoichiometry calculations use the volumes, temperatures, and pressures of gaseous reactants and products to determine the amount of other reactants or products. These calculations use the fact that, for gaseous reactions at constant temperature and pressure, the mole ratio is the same as the volume ratio. Therefore, given the volume of one component, the volume of another can be calculated. When the reactants and products are not under the same conditions or the reaction occurs under non-standard conditions, the ideal gas law can be used to calculate the amounts of gases produced.

There are numerous important commercial, health, and industrial applications of stoichiometry, which include the development of the Breathalyser™ test, the production of pharmaceuticals, and the removal of sulfur from natural gas.

Concept Organizer Qualitative and Quantitative Analysis

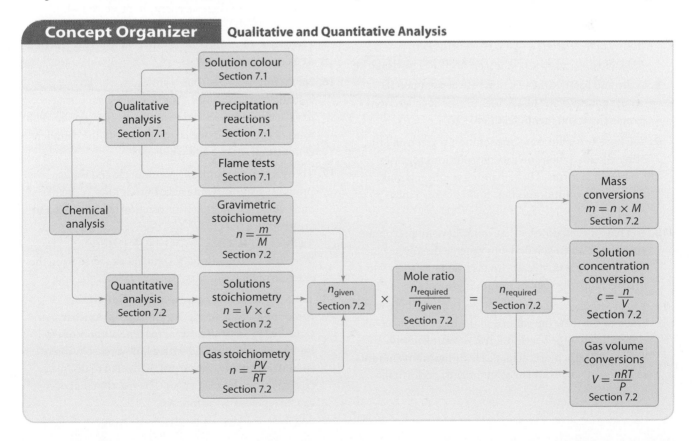

Understanding Concepts

1. Explain the differences among complete balanced equations, ionic equations, and net ionic equations.

2. What is happening in a flame when a metal ion emits a specific colour?

3. How do you determine the mole ratio of two compounds, either reactants or products, in a reaction?

4. Give a practical example of an instance in which a stoichiometric calculation would be important.

5. Distinguish among gravimetric, solution, and gas stoichiometry.

Applying Concepts

6. Identify the ion in each solution based on the evidence.
   a) The solution is purple.
   b) The solution is green. It reacts with $NaCl(aq)$ to produce a precipitate.
   c) The solution is colourless. It gives a bright yellow colour to a flame.
   d) The solution is blue. It gives a bluish-green colour to flame.

7. First balance and then write net ionic equations for the following reactions:
   a) $Ba(ClO_4)_2(aq) + Na_3BO_3(aq) \rightarrow$
   $$Ba_3(BO_3)_2(s) + NaClO_4(aq)$$
   b) $Cr(s) + AgNO_3(aq) \rightarrow Cr(NO_3)_2(aq) + Ag(s)$
   c) $Al(s) + ZnCl_2(aq) \rightarrow Zn(s) + AlCl_3(aq)$
   d) $Br_2(\ell) + NaI(aq) \rightarrow I_2(s) + NaBr(aq)$
   e) $H_2SO_4(aq) + Sr(OH)_2(aq) \rightarrow SrSO_4(s) + 2H_2O(\ell)$

8. Write and balance the reaction between copper(II) nitrate and sodium sulfate. Identify the spectator ions. Explain how you identified them.

9. You have two solutions. One solution is green and the other is pink. When you mix the solutions together, a precipitate forms. The remaining solution is pink. Suggest two compounds that could give you these results.

10. You have a solid powder that you believe might contain strontium. When you add it to water, it dissolves. Provide a series of at least two tests that will allow you to determine whether strontium is present.

11. You are told what the mass of a piece of beryllium is and asked to calculate the volume of the hydrogen that will be produced if the beryllium reacts complete with an excess of hydrochloric acid. List the minimum amount of data that you would need in order to perform the calculation.

Solving Problems

12. Methanol is made from synthesis gas, which is a mixture of carbon monoxide and hydrogen:
   $$2H_2(g) + CO(g) \rightarrow CH_3OH(g)$$
   a) Write the ratio of all the components of the reaction.
   b) What amount of hydrogen is required to prepare 130 kmol of methanol?
   c) What amount of hydrogen is required for every 6.00 mol of carbon monoxide consumed in the reaction?

13. The 2-methylpropane, $C_4H_{10}(g)$, that fuels a portable camping stove burns in oxygen to produce carbon dioxide, water vapour, and, of course, heat:
   $$2C_4H_{10}(g) + 13O_2(g) \rightarrow 8CO_2(g) + 10H_2O(g)$$
   a) Write the ratio of all the components of the reaction.
   b) What amount of oxygen is required to consume all 8.0 mol of 2-methylpropane in a small canister of the stove fuel?
   c) What amount of water vapour is produced when 3.50 mol of 2-methylpropane is consumed in the reaction?

14. What mass of metal oxide is formed when 50 g of magnesium in a road flare reacts with excess oxygen?

15. If a sample of sodium is lowered into a beaker filled with excess chlorine, it "burns" brightly with the release of a great amount of thermal energy to produce a white solid:
   $$Cl_2(g) + 2Na(s) \rightarrow 2NaCl(s)$$
   What mass of white solid is expected from the reaction of 4.6 g of sodium metal?

16. MSR camping stoves burn white gas, which is essentially octane, $C_8H_{18}(\ell)$. The white gas burns in oxygen to form carbon dioxide and water vapour. What mass of carbon dioxide is expected from the complete combustion of 72.1 g of white gas?

17. The reaction of titanium(IV) chloride and water produces a smoky white haze of titanium(IV) oxide and hydrogen chloride (this reaction has been used to create smoke screens):
   $$TiCl_4(\ell) + 2H_2O(\ell) \rightarrow TiO_2(s) + 4HCl(g)$$
   What mass of titanium(IV) oxide is expected from the reaction of 85.6 g of titanium(IV) chloride?

18. The tarnish on silverware is almost entirely the compound silver sulfide, $Ag_2S(s)$. If you touch a tarnished piece of silverware to some aluminium foil lining a plastic tub filled with a hot water solution of sodium carbonate, the tarnish will seem to disappear (see figure on the next page). The reaction that occurs can be represented as follows:
   $$3Ag_2S(s) + 2Al(s) \rightarrow 6Ag(s) + Al_2S_3(aq)$$

The tarnish on this silver spoon can be removed using a reaction with aluminium.

What mass of tarnish is removed if 6.4 g of silver metal is produced using this method?

Making Connections

19. You have been given the task of removing mercury(II) ions present in the waste water of an industrial facility—mostly as mercury(II) nitrate. You decide to use an aqueous solution of sodium sulfide to remove the ions. Craft a short essay that addresses the following points and is accompanied by well-organized stoichiometric calculation(s).

   a) Include a complete balanced and a net ionic equation for the reaction you propose to use to remove the mercury(II) nitrate.

   b) Explain why the reactant you chose is suitable for removing the mercury(II) ions from the waste water. What laboratory techniques and tests must be used to ensure that this reaction is as effective as possible for removing the mercury ions?

   c) What property of mercury(II) sulfide makes it less of an environmental concern than is mercury(II) nitrate?

   d) What assumptions are you making with regard to the toxicities of sodium sulfide or sodium nitrate relative to either mercury(II) nitrate or mercury(II) sulfide?

   e) If every litre of the waste water contains 0.03 g of mercury(II) nitrate, what mass of sodium sulfide will be required to remove all the soluble mercury compound from 10 000 L of waste water?

20. The labels have fallen off five bottles containing aqueous solutions. The labels indicate that the bottles contain $CaCl_2(aq)$, $NiSO_4(aq)$, $AgNO_3(aq)$, $LiOH(aq)$, and $NH_4Cl(aq)$. Write a procedure that would enable you to collect the data necessary to identify each solution. Prepare an evidence table including predicted evidence and diagnostic tests, and the identification of the solutions.

21. An automobile air bag system relies on the decomposition of sodium azide. The system is actually more complex than a simple decomposition, and contains a mixture of sodium azide, $NaN_3(s)$; potassium nitrate, $KNO_3(s)$; and silicon dioxide, $SiO_2(s)$. The potassium nitrate and silicon dioxide are present so that sodium metal will not be a product of the reaction. This mixture reacts rapidly when electrically heated, to provide a large volume of nitrogen gas at an appropriate temperature and pressure in a time of 0.03 s! The air bag reaction is

$$10NaN_3(s) + 2KNO_3(s) + 6SiO_2(s) \rightarrow$$
$$K_2SiO_3(s) + 5Na_2SiO_3(s) + 16N_2(g)$$

   a) What mass of $NaN_3(s)$ reacts with an excess of $KNO_3(s)$ and $SiO_2(s)$ to produce 60.0 L of $N_2(g)$ at 115 kPa and 30.0 °C?

   b) What would happen if there was not an excess amount of potassium nitrate and silicon dioxide?

   c) How important is it to design a test to analyse the safety of a reaction like this one?

22. The compound potassium superoxide, $KO_2(s)$, has been investigated as an air purifier for scrubbing carbon dioxide from the air in submarines. $KO_2(s)$ was used for this very purpose in the *Salyut-4* space station some time ago. The reaction of potassium superoxide with carbon dioxide is represented by the following unbalanced chemical equation:

$$\_\_KO_2(s) + \_\_CO_2(g) \rightarrow \_\_K_2CO_3(s) + \_\_O_2(g)$$

   a) Calculate the mass of potassium superoxide necessary to react with 50.0 L of carbon dioxide gas at 25.0 °C and 1.00 atm.

   b) Lithium hydroxide canisters have long been used to scrub carbon dioxide from the atmospheres of American space vehicles. Lithium hydroxide reacts according to the following balanced chemical equation:

$$2LiOH(s) + CO_2(g) \rightarrow Li_2CO_3(s) + H_2O(g)$$

   The water produced by the reaction can be recycled. Suggest one reason why potassium superoxide may be a better air purifier than is lithium hydroxide.

Applications of Stoichiometry

Chapter Concepts

8.1 Limiting and Excess Reactants

- Reactants are not usually present in the exact amounts required by the reaction.

- One reactant limits the reaction, while the other reactants are left over after the reaction is complete.

8.2 Predicted and Experimental Yield

- Chemical reactions rarely produce the predicted amount of product.

- Percentage yield compares the experimental yield with the predicted yield.

8.3 Acid–Base Titration

- You can determine the concentration of a strong acid by adding a strong base or the concentration of strong base by adding a strong acid.

- Indicators and pH meters allow you to determine when neutralization is complete.

In January 2004, two weeks apart, the twin exploration rover robots, *Spirit* and *Opportunity*, landed on Mars. Their successful landings were due in large part to a sophisticated air bag landing system. As a lander enters the atmosphere, the air bags inflate to surround it. The inflated air bags cushion the lander as it hits the planet's surface at 100 km/h. Nestled in the air bag cushion, the lander bounces off the rocky Mars terrain several times before coming safely to a halt. The air bags deflate and the robot within rolls out to begin exploring.

An artist's rendering of this air bag system, first developed in 1996 for use in the *Mars Pathfinder* mission, is shown on the right. The system uses technology that is similar to automobile air bag technology.

In air bags for both automobiles and rovers, the key reaction that inflates the bag is the fast decomposition of sodium azide, $NaN_3(s)$, to produce nitrogen gas:

$$2NaN_3(s) \rightarrow 2Na(s) + 3N_2(g)$$

The reaction is triggered by an impact or a rapid decrease in speed. In Chapter 7 you learned how stoichiometric calculations could be used to determine the required mass of sodium azide.

In this chapter, we will further explore stoichiometry by looking at factors that can influence the extent of a reaction and how it can be quantitated or monitored.

The Model Air Bag

Household vinegar and baking soda react to produce carbon dioxide gas:

$$CH_3COOH(aq) + NaHCO_3(s) \rightarrow NaCH_3COO(aq) + CO_2(g) + H_2O(\ell)$$

You can use this reaction to create a model for an air bag.

Safety Precautions

-  If the bag bursts, the solution mixture could spatter.

Materials

- household white vinegar (dilute ethanoic acid, $CH_3COOH(aq)$)
- baking soda (sodium hydrogencarbonate, $NaHCO_3(s)$)
- resealable sandwich bag
- barometer or the latest air pressure (uncorrected)
- thermometer (alcohol or digital)
- electronic balance
- 50 mL graduated cylinder
- 50 mL beaker

Procedure

1. Working in groups, use what you know about gas volume at a given pressure and temperature to calculate the amount, in moles, of $CO_2(g)$ needed to fill the sandwich bag you are using. Your teacher will provide the volume of the sandwich bag. Use this result to estimate how much baking soda and vinegar you will need to just fill the sandwich bag with gas. Note that household vineger in typically 5% w/v ethanoic acid in water.

2. Measure and record the air pressure with a barometer.

3. Use the thermometer to measure and record the temperature of the air.

4. Use the electronic balance to measure the mass of baking soda you predicted you would need. Put the baking soda in the sandwich bag.

5. Use the graduated cylinder to obtain the volume of household vinegar you predicted you would need.

6. While you pinch off the corner of the bag containing the $NaHCO_3(s)$, add the vinegar to the other corner and, with the aid of another student, seal the sandwich bag so that it contains as little extra air as possible.

7. Place the sealed bag in the sink and allow the reactants to mix. When the reaction stops, record your observations. Is the bag partly full, full, or did it burst?

8. Dispose of the reaction mixtures down the sink with lots of water.

Analysis

1. Share your results with the class. Which group or groups succeeded in filling their air bag and why? What errors were made by other groups?

2. Suggest two reasons why the reaction between vinegar and baking soda is not used to fill air bags for automobiles or Mars landers.

The rover robot shown in the large picture carries delicate scientific instruments to investigate the geology and chemistry of the surface of Mars. Air bag technology allowed this robot to make a soft landing and carry out its mission of scientific discovery.

SECTION 8.1

Limiting and Excess Reactants

Section Outcomes

In this section, you will:
- **calculate** the quantities of reactants and products involved in chemical reactions using gravimetric and solution stoichiometry
- **interpret** stoichiometric ratios from chemical reaction equations
- **perform** stoichiometry calculations to determine predicted yields
- **identify** limiting and excess reactants in chemical reactions

Key Terms

stoichiometric coefficients
stoichiometric amounts
limiting reactant
excess reactant

The coefficients of a balanced chemical reaction equation are often called the **stoichiometric coefficients** because they are used in stoichiometric calculations. As you know, stoichiometric coefficients are used to determine the mole ratios for the reactants and products of a reaction. If the reactants are present in the amounts that correspond exactly to the mole ratios, they are said to be present in **stoichiometric amounts**. When reactants are present in stoichiometric amounts, then absolutely no trace of any the reactants will be left when the chemical reaction has ceased. In practice, it is very rare that reactants are present in stoichiometric amounts. One or more of the reactants are likely to be left unreacted when the reaction is over.

Recall the clubhouse sandwich "equation" from section 7.2:

$$3 \text{ slices of toast} + 2 \text{ slices of turkey} + 4 \text{ strips of bacon} \rightarrow 1 \text{ sandwich}$$

What is the "limiting ingredient" if you have 6 slices of toast, 12 slices of turkey, and 20 strips of bacon? Examine Figure 8.1 to find out.

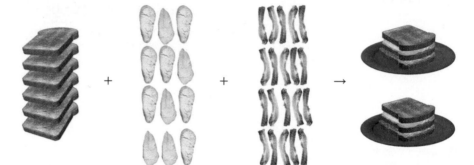

Figure 8.1 Once you make two sandwiches you have run out of toast, but you still have turkey and bacon left over. Toast is the "limiting ingredient."

Try the following Thought Lab to practise identifying the limiting item.

Target Skills

Identify limiting and excess "reactants"

Suppose that you are in charge of a company that is manufacturing widgets. A widget is a contraption that, in this case, consists of two nuts, a washer, and a bolt.

The word equation for making one widget is

$$\text{one bolt} + \text{one washer} + \text{two nuts} \rightarrow \text{one widget}$$

- a bolt has the symbol Bt
- a washer has the symbol Wa
- a nut has the symbol Nu

Using the above symbols, we can write a "chemical equation" for the formation of one widget:

$$Bt + Wa + 2Nu \rightarrow BtWaNu_2$$

$BtWaNu_2$ is the "formula" for a widget.

1 bolt + 1 washer + 2 nuts → 1 widget

Analysis

1. If you are given 75 bolts (75 Bt), 100 washers (100 Wa), and 100 nuts (100 Nu), how many widgets can you make if they must all have the formula $BtWaNu_2$?

   a) Which "reactant" limits the number of widgets you can make?

   b) Which "reactants" were present in excess?

   c) How much of each excess reactant remains?

   d) What is the "mole ratio" of Nu to Bt?

   e) What is the "mole ratio" of Nu to Wa?

2. How does the amount of a reactant that is present in excess affect the quantity of product that is obtained?

3. In this case, there were fewer Bt than Nu or Wa. Explain why the bolts are not the limiting reactant, despite being present in the least number of the three reactants.

Limiting and Excess Reactants in Chemical Reactions

In the Thought Lab, you saw that one component ("reactant") can limit the number of widgets that can be constructed. In the same way, the quantity of product that results from a chemical reaction is limited by the reactant that is used up or completely consumed first. It is routine laboratory practice to use excess amounts of the least expensive or most readily available reactant or reactants to ensure that the more expensive or scarcer reactant is consumed.

The **limiting reactant**, also called the limiting reagent, of a chemical reaction is the reactant that is completely consumed in the reaction. The reactant (or reactants) that remains after a reaction is over is called the **excess reactant**. The limiting reactant determines the maximum quantity of product or products that can arise from a chemical reaction. The limiting reactant does not need to be the reactant present in fewer moles. Rather, it is the reactant that will form fewer moles of product(s). Consider, for example, the reaction of 1.5 mol of hydrogen with 1.0 mol of oxygen to produce water, according to the following equation:

$$2H_2(g) + O_2(g) \rightarrow 2H_2O(\ell)$$

Given an excess of oxygen, 1.5 mol of hydrogen would produce 1.5 mol of water. Given an excess of hydrogen, however, 1.0 mol of oxygen would produce 2.0 mol of water. Therefore, hydrogen limits the amount of water produced and is the limiting reactant. Likewise, the limiting reactant is not necessarily the lowest mass reactant but it is the reactant that gives the lowest mass of product(s).

Excess Oxygen for Motor Home Heaters

Not only is the maximum quantity of product determined by the limiting reactant but in some significant cases, the composition of the product can also be affected. Ensuring proper ventilation in a motor home (see Figure 8.2) is essential for the proper functioning of a propane heater. Good ventilation ensures that an excess supply of oxygen is available for complete combustion of the propane as follows:

$$C_3H_8(g) + 5O_2(g) \rightarrow 3CO_2(g) + 4H_2O(g)$$

If the supply of oxygen dips below limiting quantities for the complete combustion of propane, then dangerous quantities of $CO(g)$ may be produced. One possible reaction for incomplete combustion of the propane is shown by the following equation:

$$2C_3H_8(g) + 7O_2(g) \rightarrow 6CO(g) + 8H_2O(g)$$

In fact, reactions that produce both carbon dioxide and carbon monoxide are more likely; however, any amount of carbon monoxide is dangerous.

Figure 8.2 Without proper ventilation, oxygen becomes the limiting reactant for the combustion of propane, a reaction used to heat motor homes. Incomplete combustion results, producing poisonous carbon monoxide.

Q 1 a) Define the term limiting reactant.

b) Is the limiting reactant always the chemical present in the fewest number of moles? Explain your answer.

Identifying the Limiting Reactant

Once the limiting reactant in a chemical reaction is consumed, no more product can be made. Therefore, to determine the limiting reactant, find out which reactant would give the smallest amount of product. In general, limiting reactant problems are easy to identify because you are given enough information to determine the amounts of two reactants, only one of which is the limiting reactant. Examine the following Sample Problem as an example of how this is done for gravimetric stoichiometry calculations.

Sample Problem

Gravimetric Stoichiometry with a Limiting Reactant

Problem
Acetylene, $C_2H_2(g)$, also called ethyne, is used in welding torches. Acetylene reacts explosively with itself and is therefore hazardous to transport. In most welding shops, acetylene is produced from the reaction of water and calcium carbide, $CaC_2(s)$, as needed, according to the following equation:

$$CaC_2(s) + 2H_2O(g) \rightarrow C_2H_2(g) + Ca(OH)_2(s)$$

If 1.00×10^2 g of calcium carbide is mixed with 1.00×10^2 g of water, what mass of acetylene is produced?

What Is Required?
You need to determine which of $CaC_2(s)$ or $H_2O(g)$ is the limiting reactant. You need to use the limiting reactant to calculate the mass of $C_2H_2(g)$ produced.

What Is Given?
You know the chemical reaction.
You know the mass of calcium carbide: 100 g
You know the mass of water: 100 g

Plan Your Strategy
Complete a gravimetric stoichiometry calculation for each reactant. Whichever reactant produces the lowest mass of $C_2H_2(g)$ is the limiting reactant.

• • •

Act on Your Strategy

Step 1 Write the balanced chemical equation and determine M for each substance:

$$CaC_2(s) \quad + \quad 2H_2O(\ell) \quad \rightarrow \quad Ca(OH)_2(s) \quad + \quad C_2H_2(g)$$

$m_{CaC_2} = 1.00 \times 10^2$ g $\qquad m_{H_2O} = 1.00 \times 10^2$ g $\qquad m_{C_2H_2} = ?$

$M_{CaC_2} = 64.10$ g/mol $\qquad M_{H_2O} = 18.02$ g/mol $\qquad M_{C_2H_2} = 26.04$ g/mol

Step 2 Calculate the mass of $C_2H_2(g)$ that could be produced from 100 g of $CaC_2(s)$.
Note: the calculations are shown all in one step. You may choose to carry out the calculations in multiple steps, as shown in section 7.2.

$$m_{C_2H_2} = 100 \text{ g } CaC_2(s) \times \frac{\text{mol } CaC_2(s)}{64.10 \text{ g } CaC_2(s)} \times \frac{1 \text{ mol } C_2H_2(g)}{1 \text{ mol } CaC_2(s)} \times \frac{26.04 \text{ g } C_2H_2(g)}{\text{mol } C_2H_2(g)}$$

$$= 40.6 \text{ g } C_2H_2(g)$$

Step 3 Calculate the mass of $C_2H_2(g)$ that could be produced from 100 g of $H_2O(\ell)$.

$$m_{C_2H_2} = 100 \text{ g } H_2O(\ell) \times \frac{\text{mol } H_2O(\ell)}{18.02 \text{ g } H_2O(\ell)} \times \frac{1 \text{ mol } C_2H_2(g)}{2 \text{ mol } H_2O(s)} \times \frac{26.04 \text{ g } C_2H_2(g)}{\text{mol } C_2H_2(g)}$$

$$= 72.3 \text{ g } C_2H_2(g)$$

Since the mass of product from 100 g of $CaC_2(s)$ is less than the mass of product from 100 g $H_2O(\ell)$, $CaC_2(s)$ is the limiting reactant.

The maximum mass of $C_2H_2(g)$ expected from this reaction is 40.6 g.

Check Your Solution
The number of significant digits is correct (three). From the balanced equation roughly 64 g of $CaC_2(s)$ (1 mol) is required to produce about 26 g of $C_2H_2(g)$ (1 mol). So, 40.6 g of $C_2H_2(g)$ is a suitable amount of product from the reaction of 100 g of $CaC_2(s)$.

1. What would be the limiting reactant if 4.0 g of Cr(s) and 6.0 g of $Fe(NO_3)_2(aq)$ were used in the reaction below?

   $Cr(s) + Fe(NO_3)_2(aq) \rightarrow Fe(s) + Cr(NO_3)_2(aq)$

2. What would be the expected mass of Fe(s) formed in the reaction described in question 1.

3. Predict the mass of zinc sulfide produced from the reaction of 15 g of zinc with 15 g of yellow sulfur ($S_8(s)$).

4. Hydrogen fluoride, HF(g), is produced when concentrated sulfuric acid, $H_2SO_4(\ell)$, reacts with the mineral fluorspar, $CaF_2(s)$:

   $CaF_2(s) + H_2SO_4(\ell) \rightarrow 2HF(g) + CaSO_4(s)$

   Calculate the mass of HF(g) produced when 15 g of calcium fluoride is mixed with 20 g of $H_2SO_4(\ell)$.

5. If 15 g of zinc powder is stirred into an aqueous solution of copper(II) nitrate that contains 50 g of solute, does the zinc react completely?

6. Acrylonitrile, $C_3H_3N(g)$, is the starting material for acrylic fibres. It is produced using the gas phase reaction of ammonia, $NH_3(g)$; oxygen from the air, $O_2(g)$; and propene, $C_3H_6(g)$, according the following chemical equation:

   $C_3H_6(g) + 2NH_3(g) + 2O_2(g) \rightarrow$
   $\qquad\qquad C_3H_3N(g) + HCN(g) + 4H_2O(g)$

   What is the limiting reactant if 1.0 kg of $C_3H_6(g)$ reacts with 600 g of $NH_3(g)$?

The Limiting Reactant: Testing Your Prediction

In the laboratory, you can carry out a reaction and observe which reactant has been completely consumed and which reactant or reactants are left over. In Investigation 8.A, the colour of one of the reactants will allow you to easily observe whether or not it is still present after the reaction is complete.

Limiting Reactant Problems in Aqueous Solution

Limiting reactant stoichiometric calculations for solution stoichiometry problems are essentially the same as they are for gravimetric stoichiometric situations. You will often be asked to determine the maximum mass of precipitate that can be formed when two reactant solutions combine. Guidelines for solving these types of problems are outlined below. Note, however, that the calculations and equations are all familiar because you used them in Chapter 7.

Process for Solving Limited Reactant Solution Stoichiometry Problems

1. Write the complete balanced equation for the reaction.

2. Determine the molar amount of each reactant, $n_{reactant}$, using the equation $n = V \times c$, where c is concentration (mol/L) and V is volume (L).

3. Determine the molar amount of the product formed using the mole ratio $\dfrac{n_{product}}{n_{reactant}}$ based on the balanced equation and $n_{reactant}$ for EACH reactant. The equation for this calculation is $n_{product} = n_{reactant} \times$ mole ratio.

4. Determination of the masses of reactants and product can also be calculated using the equation $m = n \times M$, with the appropriate n and M values for each reactant and product.

5. The reactant that results in the production of the LEAST amount of product is the limiting reactant. It will limit the amount of product that is formed.

INVESTIGATION 8.A

Target Skills

Identifying limiting and excess reactants in chemical reactions

The Limiting Reactant

In this investigation, you will predict the limiting reactant in a chemical reaction and test your prediction. You will use the single-replacement reaction of aluminium with aqueous copper(II) chloride:

$$2Al(s) + 3CuCl_2(aq) \rightarrow 3Cu(s) + 2AlCl_3(aq)$$

Question

How can observations tell you which is the limiting reactant in the reaction of aluminium with aqueous copper(II) chloride?

Prediction

Your teacher will give you a flask that contains a 0.25 g piece of aluminium foil and 0.51 g of copper(II) chloride. Use stoichiometric calculations to predict which one of these reactants will be the limiting reactant.

Safety Precautions

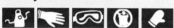

The reaction mixture may get hot. Do not hold the flask as the reaction proceeds.

Copper(II) chloride is toxic. Inform your teacher immediately of any spills.

Materials

- 0.51 g $CuCl_2(s)$
- 0.25 g aluminium foil
- water
- 125 mL Erlenmeyer flask
- stirring rod

Procedure

1. To begin the reaction, add about 50 mL of water to the flask that contains the aluminium foil and copper(II) chloride.
2. Record the colour of the solution and any metal that is present at the beginning of the reaction.
3. Record any colour changes as the reaction proceeds. Stir occasionally with the stirring rod.
4. When the reaction is complete, return the flask, with its contents, to your teacher for proper disposal. Do not pour anything down the drain.

Analysis

1. According to your observations, which reactant was the limiting reactant? Explain your answer.
2. According to your observations, which reactant was the excess reactant? Explain your answer.
3. Did your predictions match your observations? Explain your answer.

Conclusion

4. Summarize your findings. Are stoichiometric calculations an effective way to predict limiting and excess reactants?

Extension

5. Magnesium, Mg(s), and hydrochloric acid, HCl(aq), react according to the following equation:

$$Mg(s) + 2HCl(aq) \rightarrow MgCl_2(aq) + H_2(g)$$

a) Examine the equation carefully. What would you expect to observe if the magnesium were the limiting reactant? What would you expect to observe if the hydrochloric acid were the limiting reactant?

b) Suppose you have a piece of magnesium and a beaker containing some hydrochloric acid of unknown concentration. Design an experiment to determine which reactant is the limiting reactant. If your teacher approves, carry out your procedure.

Precipitating Mercury: A Limiting Reactant Problem

Problem

A 0.221 mol/L aqueous solution of mercury(II) chloride, $HgCl_2(aq)$, is used to preserve and harden tissue samples for bone marrow and lymph node biopsies. This solution is called B5. The mercury(II) ions must be removed from leftover B5 solution for safe disposal as solid, toxic waste. What mass of mercury(II) sulfide, $HgS(s)$, precipitates from 200.0 mL of B5 solution when it is treated with 100.0 mL of 0.50 mol/L aqueous sodium sulfide, $Na_2S(aq)$? (Assume that the mercury(II) sulfide is completely insoluble in water.)

What Is Required?

You need to determine the mass of $HgS(s)$ that precipitates, assuming that it is completely insoluble in water. You must first determine the limiting reactant.

What Is Given?

$V_{HgCl_2} = 200.0$ mL $\qquad$ $V_{Na_2S} = 100$ mL

$c_{HgCl_2} = 0.221$ mol/L $\qquad$ $c_{Na_2S} = 0.50$ mol/L

Plan Your Strategy

Step 1 Write the balanced chemical equation.

Step 2 Perform two stoichiometric calculations to identify the limiting reactant:

1. Convert the amount of mercury(II) chloride into an amount of mercury(II) sulfide and ultimately into a mass of that precipitate.

2. Convert the amount of sodium sulfide into an amount of mercury(II) sulfide and ultimately into another mass of precipitate.

The calculation that gives the lower mass of precipitate identifies the limiting reactant and represents the correct answer.

• • •

Act on Your Strategy

Step 1 Write the balanced chemical equation for this precipitation reaction and M for $HgS(s)$:

$$HgCl_2(aq) \quad + \quad Na_2S(aq) \quad \rightarrow \quad HgS(s) \quad + \quad 2NaCl(aq)$$

$V_{HgCl_2} = 200.0$ mL $\qquad$ $V_{Na_2S} = 100$ mL $\qquad$ $m_{HgS} = ?$

$c_{HgCl_2} = 0.221$ mol/L $\qquad$ $c_{Na_2S} = 0.50$ mol/L $\qquad$ $M_{HgS} = 232.66$ g/mol

Step 2 Determine the limiting reactant and corresponding mass of $HgS(s)$.

Assuming the $HgCl_2(aq)$ is the limiting reactant, the mass of $HgS(s)$ is calculated as follows:

$$n_{HgS} = c_{HgCl_2} \times V_{HgCl_2} \times \frac{1 \text{ mol HgS(s)}}{1 \text{ mol HgCl}_2\text{(aq)}} = \frac{0.221 \text{ mol HgCl}_2\text{(aq)}}{1 \text{ L}} \times 0.200 \text{ L} \times \frac{1 \text{ mol HgS(s)}}{1 \text{ mol HgCl}_2\text{(aq)}}$$

$$= 0.0442 \text{ mol HgS(s)}$$

$$m_{HgS} = n_{HgS} \times M_{HgS} = 0.0442 \text{ mol HgS(s)} \times 232.66 \text{ g/mol}$$
$$= 10 \text{ g HgS(s)}$$

The mass of $HgS(s)$ predicted to precipitate if $HgCl_2(aq)$ is the limiting reactant is 10.3 g.

Assuming the $Na_2S(aq)$ is the limiting reactant, the mass of $HgS(s)$ is calculated as follows:

$$n_{HgS} = c_{Na_2S} \times V_{Na_2S} \times \frac{1 \text{ mol HgS(s)}}{1 \text{ mol Na}_2\text{S(aq)}} = \frac{0.50 \text{ mol Na}_2\text{S(aq)}}{1 \text{ L}} \times 0.100 \text{ L} \times \frac{1 \text{ mol HgS(s)}}{1 \text{ mol Na}_2\text{S(aq)}}$$

$$= 0.050 \text{ mol HgS}$$

$$m_{HgS} = n_{HgS} \times M_{HgS} = 0.050 \text{ mol HgS} \times 232.66 \text{ g/mol}$$
$$= 12 \text{ g HgS}$$

The mass of $HgS(s)$ predicted to precipitate if $Na_2S(aq)$ is the limiting reactant is 12 g.

Mercury(II) chloride is the limiting reactant. The mass of mercury(II) sulfide that will precipitate in this procedure is 10 g.

Check Your Solution

The answer has the correct number of significant digits compared with the data that was used to calculate it (two).

The next Sample Problem shows an alternative method for solving limiting reactant problems. In this method, you calculate the amount (in mol) of product that would be produced by each reactant. The smaller result indicates the limiting reactant. Use this amount to calculate the mass of product. You can choose to use either method to solve the Practice Problems that follow.

Sample Problem

Precipitating Silver Chromate

Problem

Silver chromate, $Ag_2CrO_4(s)$, is an insoluble, brick-red compound. Calculate the mass of silver chromate that forms when 50.0 mL of 0.100 mol/L silver nitrate, $AgNO_3(aq)$, reacts with 25.0 mL of 0.150 mol/L sodium chromate, $Na_2CrO_4(aq)$.

What Is Required?

You need to determine the mass of silver chromate that precipitates. Because two reactants are provided, you must first determine the limiting reactant.

What Is Given?

$V_{AgNO_3} = 50.0$ mL $V_{Na_2CrO_4} = 25.0$ mL

$c_{AgNO_3} = 0.100$ mol/L $c_{Na_2CrO_4} = 0.150$ mol/L

Plan Your Strategy

Step 1 Write a balanced equation for the reaction.

Step 2 Find the amount (in mol) of silver nitrate that would be precipitated by each reactant. Whichever produces the least amount of product is the limiting reactant.

Step 3 Calculate the mass of silver chromate that precipitates based on the limiting reactant.

• • •

Act on Your Strategy

Step 1 Write the balanced chemical reaction equation for this precipitation reaction and calculate the M for $Ag_2CrO_4(s)$.

$$2AgNO_3(aq) \quad + \quad Na_2CrO_4(aq) \quad \longrightarrow \quad Ag_2CrO_4(s) + 2NaNO_3(aq)$$

$V_{AgNO_3} = 50.0$ mL $V_{Na_2CrO_4} = 25.0$ mL $m_{Ag_2CrO_4} = ?$

$c_{AgNO_3} = 0.100$ mol/L $c_{Na_2CrO_4} = 0.150$ mol/L $M_{Ag_2CrO_4} = 331.74$ g/mol

Step 2 Determine the limiting reactant.

For $AgNO_3(aq)$:

$n_{AgNO_3} = 0.0500$ L $AgNO_3(aq) \times 0.100$ mol/L

$\qquad = 5.00 \times 10^{-3}$ mol $AgNO_3(aq)$

$n_{Ag_2CrO_4} = 5.00 \times 10^{-3}$ ~~mol $AgNO_3(aq)$~~ $\times \dfrac{1 \text{ mol } Ag_2CrO_4(s)}{2 \text{ mol } AgNO_3(aq)}$

$\qquad = 2.50 \times 10^{-3}$ mol $Ag_2CrO_4(s)$

For $Na_2CrO_4(aq)$, the calculation is shown in one step (you can do the calculation either way):

$n_{Ag_2CrO_4} = 0.0250$ ~~L $Na_2CrO_4(aq)$~~ $\times 0.150$ ~~mol/L~~ $\times \dfrac{1 \text{ mol } Ag_2CrO_4(s)}{1 \text{ mol } Na_2CrO_4(aq)}$

$\qquad = 3.75 \times 10^{-3}$ mol $Ag_2CrO_4(s)$

Step 3 Calculate the mass of $Ag_2CrO_4(s)$ based on the limiting reactant.

Since the given quantity of $AgNO_3(aq)$ would give a smaller amount of $Ag_2CrO_4(s)$, $AgNO_3(aq)$ is the limiting reactant. Use $n_{Ag_2CrO_4}$, which you calculated using silver nitrate, to determine $m_{Ag_2CrO_4}$.

$m_{Ag_2CrO_4} = n_{Ag_2CrO_4} \times M_{Ag_2CrO_4}$

$\qquad = 2.50 \times 10^{-3}$ mol $Ag_2CrO_4(s) \times 331.74$ g/mol

$\qquad = 0.829$ g $Ag_2CrO_4(s)$

The mass of silver chromate that precipitates is 0.829 g.

Check Your Solution

The answer has the correct number of significant digits (three). The answer has the correct units (grams).

7. What would be the limiting reactant if 100.0 mL of a 0.5 mol/L solution of $Mg(NO_3)_2(aq)$ and 125.0 mL of a 1.2 mol/L solution of $Na_3PO_4(aq)$ were used in the reaction below?

$$3Mg(NO_3)_2(aq) + 2Na_3PO_4(aq) \rightarrow$$
$$Mg_3(PO_4)_2(s) + 6Na(NO_3)(aq)$$

8. What would be the expected mass of $Mg_3(PO_4)_2(s)$ formed in the reaction described in question 7.

9. Would 600 mL of 0.085 mol/L $Na_2S(s)$ be sufficient to remove all the mercury(II) ions as $HgS(s)$ from 200 ml of B5 solution (0.221 mol/L $HgCl_2(aq)$)? Explain your answer.

10. If 250 mL of 0.400 mol/L aqueous lead(II) nitrate, $Pb(NO_3)_2(aq)$, is mixed with 300 mL of 0.22 mol/L aqueous potassium iodide, $KI(aq)$, what is the maximum mass of precipitate that would be formed in the resulting reaction?

11. Priti mixed 25.0 mL of 0.320 mol/L aqueous copper(II) sulfate, $CuSO_4(aq)$, with 29.7 mL of 0.270 mol/L aqueous strontium nitrate, $Sr(NO_3)_2(aq)$.

   a) Write the balanced reaction equation for this precipitation reaction.

b) The precipitate that Priti collected by filtration through paper and dried in an oven was a white, powdery solid. What colour was the filtrate after the reaction mixture was filtered? Explain your answer.

c) Calculate the mass of precipitate that Priti can expect from this reaction.

d) Priti collected 1.432 g of precipitate. Calculate the percentage yield of precipitate.

12. Silicate salts of most transition metal ions, many of which are too toxic for disposal down the drain, have very low aqueous solubilities. It is common laboratory practice to add $Na_2SiO_3(aq)$ to remove these ions for disposal as solid waste. Suppose 150 mL of 0.250 mol/L aqueous sodium silicate, $Na_2SiO_3(aq)$, are mixed with 950 mL of a 0.035 mol/L silver nitrate solution, $AgNO_3(aq)$.

   a) Write the balanced chemical equation for this precipitation reaction.

b) Which reactant is the limiting reactant? Explain your answer.

c) Calculate the predicted yield of precipitate.

d) Are all the silver ions removed from the solution? Explain briefly.

Section 8.1 Summary

- When reactants are present in amounts that correspond to the mole ratio from the balanced chemical equation, they are considered to be present in stoichiometric amounts and are completely consumed in a reaction. However, that is rarely the case. Usually, there is one reactant that is completely consumed while some of the others remain.

- The limiting reactant in a chemical reaction is the reactant that is completely consumed in a reaction. It is given this name because once it is consumed the reaction no longer continues. Therefore, the limiting reactant dictates the amount of product that will be formed. It is not necessarily the component that is present in the least number of moles or mass. It is the reactant that will produce the least amount of product, based on stoichiometric calculations.

- Excess reactants are not completely consumed and are still present, to some extent, once the reaction is complete. For reactions such as the combustion of propane in heaters, the presence of the excess reactant oxygen is important in making sure that the correct products (carbon dioxide and water) are formed as much as possible and production of a toxic side-product (carbon monoxide) is limited.

- Gravimetric, solution, and gas stoichiometric calculations can be done to determine the limiting reactant in chemical reactions by identifying the one that produces the least amount of product.

1. Why do you not need to consider reactants that are present in excess amounts when performing a stoichiometric calculation? Use an everyday analogy to explain the idea of excess quantity of reactant.

2. **a)** When a small quantity of phosphorus, $P_4(s)$, reacts with oxygen, $O_2(g)$, in open air, which reactant do you think will be in excess? Explain your answer.

   b) Gold is quite an unreactive metal, but it does react with a mixture of hydrochloric and nitric acids called *aqua regia*. The reaction is shown below:
   $$Au(s) + 3NO_3^-(aq) + 4Cl^-(aq) + 6H_3O^+(aq) \rightarrow$$
   $$AuCl_4^-(aq) + 3NO_2(g) + 9H_2O(\ell)$$
   This reaction is always performed with the *aqua regia* in excess. Why would a chemist not have the gold as the excess reactant?

   c) In general terms, what are the characteristics or properties of a reactant that would make it suitable as an excess reactant?

3. Copper is more reactive than gold. Copper reacts with nitric acid to give a blue solution:
   $$Cu(s) + 4HNO_3(aq) \rightarrow$$
   $$Cu(NO_3)_2(aq) + 2NO_2(g) + 2H_2O(\ell)$$

   a) When 57.4 g of $Cu(s)$ react with 140 g of $HNO_3(aq)$, which reactant is in excess?

   b) What mass of $Cu(NO_3)_2(aq)$ could be obtained from the product solution?

4. Coal contains sulfur. This sulfur is present either as metal sulfides or as organic compounds. The sulfur dioxide, $SO_2(g)$, produced when coal burns in coal-fired power plants is the major source of polluting gases that cause acid rain. One way of scrubbing sulfur dioxide from emissions of coal-fired plants is to pass the gases through a mixture of calcium carbonate and water so that the following reaction occurs:
   $$2CaCO_3(s) + 2SO_2(g) + O_2(g) \rightarrow 2CaSO_4(s) + 2CO_2(g)$$
   The resulting $CaSO_4(s)$, also called gypsum, can then be used to make wallboard. If 1000 kg of $SO_2(g)$ passes through a slurry of 2000 kg of $CaCO_3(s)$, will all the $SO_2(g)$ be removed?

5. A waste laboratory solution contained 7.16 g of lead(II) nitrate, $Pb(NO_3)_2(aq)$. This solution was mixed with an aqueous solution containing 20 g of sodium silicate, $Na_2SiO_3(aq)$, to precipitate the lead(II) ions as lead(II) silicate, $PbSiO_3(s)$. Estimate the mass of precipitate likely to be collected.

6. A student adds some zinc powder to a beaker containing an aqueous solution of copper(II) chloride, which has a deep blue colour. The reaction mixture is stirred using a magnetic stirrer. After several hours, the student notes that the solution colour has faded but has not completely disappeared.

   a) Write a balanced chemical reaction equation to describe the reaction.

   b) List the measurements and observations that the student should record to fully analyze this reaction.

   c) According to the initial and final solution colours, which reactant was likely in excess?

   d) The beaker contained 3.12 g of copper(II) chloride. What does this tell you quantitatively about the mass of zinc that was added?

7. A container of dry sodium hydrogen carbonate is always on a certain shelf in the laboratory in case there are any acid spills. One day, your lab partner knocked over a graduated cylinder containing 35.0 mL of 0.05 mol/L sulfuric acid. You reached for the sodium hydrogen carbonate and discovered there was not much there so you sprinkled all of it on the sulfuric acid spill. Assume that there had been 0.640 g of sodium hydrogen carbonate in the container. The unbalanced equation for the reaction is:
   $$H_2SO_4(aq) + NaHCO_3(s) \rightarrow$$
   $$Na_2SO_4(aq) + CO_2(g) + H_2O(\ell)$$

   a) Was the sulfuric acid completely neutralized?

   b) How much of the excess reactant remained?

   c) What volume of carbon dioxide gas was produced if the conditions in the room were 100.0 kPa and 25 °C (SATP)?

   d) Why is it better to use sodium hydrogen carbonate than sodium hydroxide to neutralize an acid?

8. Write a procedure for recovering all of the silver from 355 mL of a 0.75 mol/L silver nitrate solution. Include the volume and concentration of all other chemicals that you would use. Include a list of all apparatus that you would need. Include safety precautions. How much silver would you expect to recover?

Predicted and Experimental Yield

Key Terms

predicted yield
experimental yield
mechanical losses
percentage yield

When you write a test, the highest grade that you can earn is usually 100%. Most people, however, do not regularly earn a grade of 100%. Similarly, in baseball, a batter does not get a hit with every swing. A batter's success rate is expressed as a decimal fraction (see Figure 8.3). In this section, you will learn how chemists predict and express the "success" of chemical reactions.

Figure 8.3 A baseball player's batting average is calculated as hits/attempts. For example, a player with 6 hits for 21 times at bat has a batting average of 6/21 = .286. This represents a success rate of 28.6%.

Predicted Yield and Experimental Yield

Chemists use stoichiometric calculations to determine the maximum quantity of product expected from a reaction. The expected quantity of product is called the **predicted yield**, also known as the *theoretical yield*. The predicted yield is calculated on the assumption that all the limiting reactant reacts in the proportions described by the balanced chemical equation. Predicted yield is analogous to the maximum number of marks you can get on a test or the number of times at bat for a baseball player.

The quantity of product actually obtained from a reaction is called the **experimental yield**, sometimes called the *actual yield*. For example, complete decomposition of 7.955 g copper(II) oxide, CuO(s), should produce 6.355 g of copper metal, Cu(s), if all the CuO(s) reacts as expected. In reality, when you perform this experiment, you will probably obtain less than this mass of copper. Experimental yield is analogous to the marks you actually get on a test or the number of successful hits for a baseball player.

Predicted yield and experimental yield are usually different. Just as you rarely score 100% on a test and baseball players never have 1.000 batting averages (at least, not for long), predicted yield and experimental yield are almost never the same. In practice, several factors usually contribute to the difference.

Competing Reactions

In certain circumstances, the same two chemicals can react to give different products. For example, when carbon burns in a plentiful supply of oxygen, it reacts to produce carbon dioxide, $CO_2(g)$, as expected in the presence of a plentiful supply of oxygen, according to the following chemical equation:

$$C(s) + O_2(g) \rightarrow CO_2(g)$$

In fact, carbon monoxide, $CO(g)$, is also produced to a small extent even when carbon burns with an excess of oxygen available:

$$2C(s) + O_2(g) \rightarrow 2CO(g)$$

This secondary reaction is an example of a *competing reaction*. Since some of the carbon reacts to form carbon monoxide in the competing reaction, the experimental yield of carbon dioxide is always less than predicted.

Other Factors That Limit Experimental Yield

For most reactions, lower experimental yields are found for a variety of reasons other than competing reactions.

Slow Reaction
- If a reaction is slow, or if insufficient time has been allowed for the complete conversion of the reactants into products, the mass of product is often less than that predicted from the stoichiometric calculation.

Collection and Transfer Methods
- The method used to collect and transfer the product often results in some small losses. For example, if a low-solubility product is isolated by filtration, some of it will remain dissolved in the filtrate.

- When a low-solubility product is collected in a filter paper, it is common practice to rinse it with solvent. This treatment is used to remove traces of the reactants or the other products from the precipitate. Some of the precipitate will, however, also dissolve in the solvent used for rinsing.

- Handling solid products when transferring them from a filter paper to a sample bottle always results in **mechanical losses**. Mechanical losses are the small amounts of product that are lost when they remain stuck to glassware or filter paper as they are transferred in the lab. It is to be expected that some of your product will adhere to even the cleanest of spatulas. When you pour finely powdered solids from, or into, plastic containers, or if you use plastic utensils, it is not uncommon for static to cause some solids to adhere to the lips of containers.

Reactant Purity
- Another common source of reduced yields is reactant purity. Look at the label on any reactant-grade chemical and you will see that, although they are close to 100% pure, they do contain contaminants (see Figure 8.4).

Reactions That Do Not Proceed to Completion
- Many reactions reach a point at which the reaction appears to stop, although less than 100% of the reactants have been converted into products. These reactions have reached equilibrium and the products are reacting to form the reactants at the same rate as the reactants are reacting to form products. For example, under most conditions, only a small percentage of hydrogen and iodine molecules have reacted to form hydrogen iodide at any one time:

$$H_2(g) + I_2(g) \rightleftharpoons 2HI(g)$$

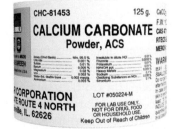

Figure 8.4 Most laboratory reactants contain small amounts contaminants—and not all of them are inert!

2 **a)** Briefly describe 5 factors that can limit or reduce the yield of a chemical reaction.

b) Is the predicted yield higher or lower than the experimental yield for a reaction? Explain why.

Calculating the Percentage Yield of a Reaction

The **percentage yield** of a product from a reaction is the quantity of product actually obtained, the *experimental yield*, expressed as a percentage of the *predicted yield*:

$$\text{Percentage yield} = \frac{\text{Experimental yield}}{\text{Predicted yield}} \times 100\%$$

For example, when magnesium metal, $Mg(s)$, is heated vigorously in air, it reacts with oxygen to form magnesium oxide, $MgO(s)$. If 2.43 g of $Mg(s)$, is heated in air, we would expect 4.03 g of $MgO(s)$ to be produced. If the mass of product is measured and found to be 3.95 g, the percentage yield is calculated as follows:

$$\text{Percentage yield of } MgO(s) = \frac{3.95\ \text{g}}{4.03\ \text{g}} \times 100\% = 98.0\% \text{ of } MgO(s)$$

3 A student performs a chemical reaction that produces silver chromate, $Ag_2CrO_4(s)$. Using stoichiometric calculations, she expected to obtain a mass of 1.40 g of $Ag_2CrO_4(s)$. When she weighed her product it had a mass of 0.90 g. What is the percentage yield of this student's chemical reaction?

To calculate the percentage yield of a particular product when the predicted mass of product is not provided, we must first calculate the predicted yield using a gravimetric stoichiometry calculation. The Sample Problem on the next page illustrates how this is accomplished.

Percentage Yield and Industry

The percentage yield of chemical reactions is extremely important in industrial chemistry. For example, the synthesis of certain drugs involves many sequential chemical reactions. The product of one reaction is the reactant of the next reaction. To determine the overall percentage yield, you must multiply the yields of each step in their decimal form. For example, imagine that a chemist synthesizes a drug in five steps. The percentage yield of the first step is 86%, of the second step is 91%, of the third step is 77%, of the fourth step is 82%, and of the last step is 74%. Each yield is reasonably good. Now calculate the overall yield:

$$0.86 \times 0.91 \times 0.77 \times 0.82 \times 0.74 = 0.366$$

The overall yield is only about 37%. As you can see, even when yields look reasonable, when many steps are necessary the yield can be very low.

Research chemists, who generally work with small quantities, may not need to worry about small yields. Chemical engineers, conversely, work with very large quantities. A difference of 1% in the yield of a reaction can translate into thousands of dollars.

In Investigation 8.B on page 309, you can determine the percentage yield of a reaction on a smaller scale.

Calculating Percentage Yield

Problem

When 1.84 g of strontium nitrate is mixed with excess copper(II) sulfate in aqueous solution, a precipitate of strontium sulfate is produced:

$$Sr(NO_3)_2(aq) + CuSO_4(aq) \rightarrow SrSO_4(s) + Cu(NO_3)_2(aq)$$

Calculate the percentage yield of this investigation if 1.50 g of strontium sulfate is obtained by filtration.

What Is Required?

You need to determine the predicted yield of strontium sulfate and use that mass, along with the experimental yield, to calculate a percentage yield.

What Is Given?

You know the mass of strontium nitrate used: 1.84 g.
You know the mass of strontium sulfate obtained: 1.50 g.

Plan Your Strategy

Step 1 Write a balanced equation for the reaction, and calculate M for the product and reactant.

Step 2 Complete a gravimetric stoichiometry calculation to get the predicted yield/mass of strontium sulfate from the reaction of 1.84 g of strontium nitrate.

Step 3 Determine percentage yield according to the equation:

$$\text{Percentage yield} = \frac{\text{Experimental yield}}{\text{Predicted yield}} \times 100\%$$

• • •

Act on Your Strategy

Step 1 Write a balanced chemical equation and calculate M for product and reactant as follows:

$$Sr(NO_3)_2(aq) \quad + \quad CuSO_4(aq) \rightarrow SrSO_4(s) \quad + \quad Cu(NO_3)_2(aq)$$

$m_{Sr(NO_3)_2} = 1.84 \text{ g}$ predicted $m_{SrSO_4} = ?$

$M_{Sr(NO_3)_2} = 211.64 \text{ g/mol}$ $M_{SrSO_4} = 183.69 \text{ g/mol}$

Step 2 Calculate the predicted yield of $SrSO_4(s)$.

$$m_{SrO_4} = 1.84 \text{ g } Sr(NO_3)_2(aq) \times \frac{\text{mol } Sr(NO_3)_2(aq)}{211.64 \text{ g } Sr(NO_3)_2(aq)} \times \frac{1 \text{ mol } SrSO_4(s)}{1 \text{ mol } Sr(NO_3)_2(aq)} \times \frac{183.69 \text{ g } SrSO_4(s)}{\text{mol } SrSO_4(s)}$$

$$= 1.597 \text{ g } SrSO_4(s)$$

The predicted yield of precipitate, $SrSO_4(s)$, is 1.60 g.

Step 3 Calculate percentage yield.

$$\text{Percentage yield} = \frac{\text{Experimental yield}}{\text{Predicted yield}} \times 100\%$$

$$= \frac{1.50 \text{ g}}{1.597 \text{ g}} \times 100\%$$

$$= 93.9\%$$

The percentage yield for this experiment is 93.9%.

Note: You would always report the predicted mass of strontium sulfate in the correct number of significant digits, or 1.60 g. In this example, the unrounded number is used to make it easier to follow the calculations in Step 3.

Check Your Solution

The experimental yield is less than the predicted yield, which makes sense. From the balanced equation, roughly 212 g of $Sr(NO_3)_2(s)$ (one mole) is required to produce about 184 g of $SrSO_4(s)$ (one mole). Therefore, a 1.60 g predicted yield of $SrSO_4(s)$ from 1.84 g of $Sr(NO_3)_2(s)$ is reasonable. The significant digits are correct (three).

Practice Problems

13. For the following reaction:

$$Fe(s) + CuCl_2(aq) \rightarrow FeCl_2(aq) + Cu(s)$$

  a) What is the predicted yield if 10.0 g of $Fe(s)$ is reacted with an excess amount of $CuCl_2(aq)$?

  b) What would be the percentage yield if 9.0 g of $Cu(s)$ is actually obtained?

14. Calculate the percentage yield for the following reaction if 60 g of $SO_2(s)$ is produced using 50 g of $S_8(s)$.

$$S_8(s) + 8O_2(g) \rightarrow 8SO_2(s)$$

15. A chemical engineering student performs the following reaction:

$$Fe_2O_3(s) + 3CO(g) \rightarrow 2Fe(s) + 3CO_2(g)$$

The student carries out three trials with the same mass of $Fe_2O_3(s)$ and obtains the following masses of $Fe(s)$.

Trial 1: 23.5 g   Trial 2: 23.2 g   Trial 3: 23.9 g

The predicted yield of $Fe(s)$ is 24.6 g. Calculate the percentage yield for each trial and the average percentage yield for this series of trials.

16. Prairie Chemical in northeast Edmonton manufactures household and industrial bleach from liquid chlorine using the following process:

$$Cl_2(\ell) + 2NaOH(aq) \rightarrow$$
$$NaOCl(aq) + NaCl(aq) + H_2O(\ell)$$

For every 0.150 kg of liquid chlorine consumed, 0.150 kg of $NaOCl(aq)$ is produced. Calculate the percentage yield of $NaOCl(aq)$.

17. The fermentation enzymes of baker's yeast convert a solution of glucose, $C_6H_{12}O_6(aq)$, to ethanol, $C_2H_5OH(aq)$, and carbon dioxide, $CO_2(g)$:

$$C_6H_{12}O_6(aq) \xrightarrow{\text{baker's yeast}} 2C_2H_5OH(aq) + 2CO_2(g)$$

If 223 g of ethanol is obtained from the fermentation of 1.63 kg of glucose, what is the percentage yield of the reaction?

18. When a 1 g to 5 g sample of the pale-green mineral malachite, $Cu_2(CO_3)(OH)_2(s)$, is heated vigorously over a Bunsen burner flame for about 20 minutes, it is transformed into black copper(II) oxide:

$$Cu_2(CO_3)(OH)_2(s) \rightarrow 2CuO(s) + CO_2(g) + H_2O(g)$$

a) Calculate the predicted yield of $CuO(s)$ from vigorous heating of 4.00 g of malachite.

b) If 2.80 g of $CuO(s)$ remains after the $CO_2(g)$ and $H_2O(g)$ have been burned off, what is the percentage yield of the solid product?

19. The reaction of toluene, $C_7H_8(\ell)$, with potassium permanganate, $KMnO_4(aq)$, proceeds with significantly less than 100% yield under most conditions:

$$C_7H_8(\ell) + 2KMnO_4(aq) \rightarrow$$
$$KC_7H_5O_2(aq) + 2MnO_2(s) + KOH(aq) + H_2O(\ell)$$

a) If 8.60 g of toluene reacts with excess potassium permanganate, what is the predicted yield, in grams, of potassium benzoate, $KC_7H_5O_2(aq)$?

b) If the percentage yield is 70.0%, what mass of potassium benzoate would you expect to be produced?

c) What mass of toluene is needed to produce 13.4 g of potassium benzoate if the percentage yield is 60.0%?

INVESTIGATION 8.B

Determining the Percentage Yield of a Chemical Reaction

The percentage yield of a reaction is determined by numerous factors: the nature of the reaction itself, the conditions under which the reaction was carried out, the purity of the reactants, and the skill of the experimenter.

In this investigation, you will determine the percentage yield of the following chemical reaction:

$$Fe(s) + CuCl_2(aq) \rightarrow FeCl_2(aq) + Cu(s)$$

You will use steel wool as your source of iron.

Question

What is the percentage yield of the reaction of iron and copper(II) chloride when steel wool and copper(II) chloride dihydrate are used as reactants?

Predictions

While waiting for the filtration in Procedure step 7, calculate the predicted yield. Assume the steel wool is 100% iron. Assume the iron reacts completely with a solution containing excess $CuCl_2(aq)$. Then predict the percentage yield and actual yield, giving reasons for your predictions.

Safety Precautions

Hydrochloric acid is corrosive. If either $CuCl_2(aq)$ or $HCl(aq)$ spill on your skin, wash immediately with plenty of cold water and inform your teacher. When you have completed the investigation, wash you hands.

continued ...

Materials

- distilled water
- 5.00 g copper chloride dihydrate, $CuCl_2 \cdot 2H_2O(s)$ ⊗ Ⓣ 🔥
- 20 mL 1 mol/L hydrochloric acid, HCl(aq) ⊗ 🔥
- 250 mL beaker
- stirring rod
- electronic balance
- 1.00 g–1.20 g rust-free, degreased steel wool
- Erlenmeyer flask
- plastic funnel
- retort stand
- wash bottle
- ring clamp
- filter paper
- watch glass
- drying oven (if available)

Procedure

1. Copy the table below into your notebook.

Observations

Mass of filter paper	
Mass of steel wool	
Mass of filter paper and solid product	

2. Place about 50 mL of distilled water in a 250 mL beaker. Add 5.00 g of $CuCl_2 \cdot 2H_2O(s)$ to the water. Stir to dissolve.

3. Determine the mass of your sample of steel wool. Record the mass in your table.

4. Add the steel wool to the solution in the beaker. Allow the mixture to sit until the reaction is complete. The reaction could take up to 20 minutes.

5. While the reaction is proceeding, set up your filtration apparatus as you did for Investigation 7.B. Be sure to determine the mass of your piece of filter paper before folding and wetting it. Record the mass of the filter paper in your table.

6. When you believe that the reaction is complete, carefully decant most of the liquid in the beaker through the filter paper. Pouring the liquid down a stirring rod helps avoid losing any solid.

7. Pour the remaining liquid and solid through the filter paper. While you are waiting for the filtration to be completed, calculate the predicted yield (see Predictions).

8. Rinse the beaker and stirring rod several times with small quantities of water. Pour the rinse water through the filter paper. Ensure there is no solid product remaining in the beaker or on the stirring rod.

9. Rinse the filter paper and solid with about 10 mL of 1 mol/L HCl(aq). Then rinse the solid with water to remove the acid.

10. After all the liquid has drained from the funnel, carefully remove your paper and place it on a labelled watch glass. Be careful not to lose any solid product.

11. Place the watch glass in a drying oven overnight. If no drying oven is available, allow the solid to dry in a safe place for several days. Dispose of the material in the flask as your teacher directs.

12. Determine the mass of the dried filter paper and product. Dispose of the filter paper and product as your teacher directs.

Analysis

1. Based on the amount of iron that you used, prove that the 5.00 g of $CuCl_2 \cdot 2H_2O(s)$ was the excess reactant.

2. Suggest sources of error for this reaction.

3. How might you attain an improved percentage yield if you performed this reaction again? Consider materials and technique.

Conclusion

4. Calculate the percentage yield for this reaction. How does the percentage yield compare with the percentage yield you predicted?

Extension

5. Consider the precision and accuracy of your results.

   a) How precise was your determination of the maximum percentage yield of the reaction of iron with copper(II) chloride? Explain your answer.

   b) Suggest how you could have improved the precision of your determination.

   c) How accurate was your determination of the maximum percentage yield of the reaction of iron with copper(II) chloride? Explain your answer and list factors that affected the accuracy of your determination.

   d) Suggest how you could have improved the accuracy of your determination.

Section 8.2 Summary

- Optimizing the efficiency of a chemical reaction is very important, particularly in industrial chemistry. Stoichiometric calculations can be used to determine the maximum amount of product that is expected from a reaction based on the amounts of reactants present. This value is called the predicted yield. The amount of product that is actually obtained from a reaction is referred to as the experimental yield. Most often, the experimental yield is lower than the predicted.

- If a competing reaction occurs, some of the reactant(s) may be used up in the formation of that product and no longer be available for the primary product formation. In addition, if the reactant is not very pure then a considerable proportion of the amount added will be the impurity, which will not contribute to forming the desired product.

- If a reaction is very slow or the reaction has not been allowed enough time to react, the amount of product will not be optimized. There are also some reactions that never undergo complete conversion of reactants because they have reached an equilibrium, in which the products react to form the reactants at the same rate as the reactants react to form the product.

- Techniques for isolating and purifying the product can result in reducing its experimental yield. For example, filtration can result in losses due to some solubility in solvents used and losses during the handling and transfer of any solids.

- The efficiency of chemical reactions is expressed as the percentage yield.

- The percentage yield can be calculated by the equation $\dfrac{\text{experimental yield}}{\text{predicted yield}} \times 100\%$

SECTION 8.2 REVIEW

1. Would the units you use make any difference when calculating a percentage yield of a product? Explain your answer.

2. If the limiting reactant in a certain chemical reaction is impure, what might happen to the percentage yield of the products?

3. Methyl salicylate, $C_8H_8O_3(\ell)$, one of the aromatic ingredients used in products for soothing sore muscles, is made by gently boiling salicylic acid, $C_7H_6O_3(s)$, in excess methanol at 70 °C for 30 minutes:

   $$C_7H_6O_3(s) + CH_3OH(\ell) \rightarrow C_8H_8O_3(\ell) + H_2O(\ell)$$

   a) Predict the yield of $C_8H_8O_3(\ell)$ from the reaction of 3.50 g of $C_7H_6O_3(s)$.

   b) If 2.84 g of $C_7H_6O_3(s)$ are recovered unreacted after 30 minutes, what is the maximum possible yield of $C_8H_8O_3(\ell)$?

   c) Calculate the percentage yield of the reaction.

4. A quantity of sodium chloride, 0.58 g, dissolved in water, reacts with 3.50 g of lead(II) nitrate in aqueous solution.

   a) What mass of precipitate is expected?

   b) If 1.22 g of precipitate are obtained from the reaction, what is the percentage yield?

5. Molten chromium, $Cr(\ell)$, can be obtained from the reaction of chromium(III) oxide with powdered aluminium.

   a) What mass of molten chromium do you predict will be produced when 100 g of chromium(III) oxide are reacted with 54 g of aluminium powder?

   b) A 60 g chunk of chromium was obtained from the reaction. Calculate the percentage yield.

6. You want to determine the percentage yield of the reaction of copper with silver nitrate. Design an experiment to find the percentage yield.

7. When product collected on filter paper is rinsed with solvent, a small amount of the product is lost as it dissolves and is washed away by the solvent. One way to reduce the amount of product lost in this way is to cool the solvent used for rinsing in an ice bath. Explain why using cold solvent reduces the amount of product lost during rinsing.

8. Explain why it is important for chemists to know the purity of the reactants that they are using.

Acid–Base Titration

When solving stoichiometric problems, you have thus far been provided with the information you needed to solve the problem. What if you did not have some of this information? You would need to carry out a quantitative analysis. One example of a quantitative analysis is performing a precipitation reaction to determine the concentration of an aqueous solution, as you saw in section 7.2. Another technique for obtaining quantitative results is titration.

Titration

Chemists often need to know the concentration of a solution. To acquire this information, they may use an experimental procedure called a titration. In a **titration**, the concentration of one solution is determined by quantitatively observing its reaction with a solution of known concentration. Figure 8.5 shows a typical titration apparatus. As you learned in Chapter 6, a solution of precisely known concentration is called a *standard solution*. In a titration, the solution that is being added and its volume measured is called the **titrant**. The titrant solution is the standard solution, which is added to the unknown solution. Using a standard solution to determine the concentration of another solution by titration is often called standardization. You will learn more about standardization in the following pages.

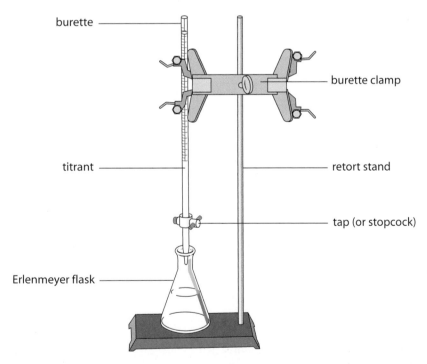

burette

burette clamp

titrant

retort stand

tap (or stopcock)

Erlenmeyer flask

Figure 8.5 A typical titration set-up is shown here. The titrant may be an acid or a base.

The stage of the titration at which the reaction is complete is called the **equivalence point**. The equivalence point occurs when stoichiometrically equivalent amounts of reactant have been consumed. Often a third chemical is present to provide visual evidence of the end of a reaction by means of a dramatic colour change. These chemicals are called indicators about which you learned in Chapter 6. The point at which the indicator changes colour is called the **endpoint**.

Standardizing Acids

Why titrate? The concentrations of strong monoprotic acids, such as HCl(aq), need to be standardized periodically because, over time, the acid concentrations change. **Standardizing** a solution means establishing its concentration by analyzing its reaction with a substance of known purity—a standard. For example, over time HCl(g) escapes as a gas from solution. The result is that the pH of the acid increases. A popular primary standard for standardizing acids is solid sodium carbonate, $Na_2CO_3(s)$, shown in Figure 8.6. The equation representing the reaction when aqueous sodium carbonate is used to standardize a nitric acid solution is as follows:

$$2HNO_3(aq) + Na_2CO_3(aq) \rightarrow 2NaNO_3(s) + H_2O(\ell) + CO_2(g)$$

Standardizing Bases

Strong bases also need to be standardized frequently. The concentrations of strong bases change over time because they tend to absorb carbon dioxide from the air. In solution, the carbon dioxide reacts to form insoluble carbonates, which often cause glass lids on bottles of this strong base to stick. The reactions also change the pH of the base. A common standard for standardizing strong bases, such as sodium hydroxide, NaOH(aq), is potassium hydrogen phthalate, $KO_2CC_6H_4CO_2H(s)$, in aqueous solution.

The Equivalence Point

In an acid–base titration, an acid titrant is added to a base sample, or vice versa. For monoprotic acids and bases, the point at which equal moles of reactant acid and base combine is called the equivalence point of the titration. An example of this type of neutralization reaction is as follows:

$$HCl(aq) + NaOH(aq) \rightarrow H_2O(\ell) + NaCl(aq)$$

The mole ratio of HCl(aq):NaOH(aq) is 1:1. Therefore, the equivalence point occurs when an equal amount (in mol) of HCl(aq) has been added to NaOH(aq).

When a strong monoprotic acid (e.g., $HClO_4(aq)$, HCl(aq), HI(aq), HBr(aq), $HNO_3(aq)$) is neutralized by a strong base (any Group I metal hydroxide, $Ba(OH)_2(aq)$, or $Sr(OH)_2(aq)$), the equivalence point has a pH of 7 at 25 °C. This is because all hydronium ions from the acid have been neutralized by an equal amount of hydroxide ions and no amount of other significant acid or base substance remains in solution.

Using Indicators to Observe the Equivalence Point

For most neutralization reactions, there are no visible signs that a reaction is occurring. How do you know when a neutralization reaction is complete? One way is to use an acid–base indicator. You learned about indicators in Chapter 6. Recall that an indicator is a substance that changes colour over a given pH range. Usually indicators are weak monoprotic acids. They do not ionize much in acidic solutions, but they ionize a great deal in basic solutions. The molecular and ionized forms of the indicator have different colours, and different indicators change colour at different pH levels.

$$\underset{\text{colour 2}}{\overset{\text{colour 1}}{HIn(aq)}} \rightleftharpoons H^+ + In^-(aq)$$

For example, phenolphthalein is a commonly used indicator for acid–base reactions. It is colourless between pH 0 and pH 8. It turns pink between pH 8 and pH 10. Figure 8.7 shows phenolphthalein and its colour change.

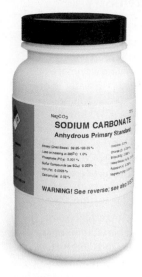

Figure 8.6 Very pure sodium carbonate samples are relatively inexpensive, are stable, are highly soluble in water, and have a long shelf life.

Figure 8.7 A good indicator changes colour dramatically just after the neutralization reaction is complete.

FYI

A standard solution is usually prepared by dissolving a primary standard in water to give an aqueous solution of precisely known concentration. A *primary standard* is a stable, high-purity solid compound with a known, preferably high, molar mass. (When the molar mass of the primary standard is high, errors in mass measurement are minimized.) Alternatively, another solution may be standardized by titrating it against a known volume of primary standard solution. This type of solution is called a *secondary standard*.

4 A student performs a titration in which hydrochloric acid, HCl(aq), of known concentration is added from a burette to a sample of sodium hydroxide, NaOH(aq), of unknown concentration. Which solution is the titrant?

5 **a)** What is standardization?

   b) Why do strong acids need to be standardized?

   c) Why do strong bases need to be standardized?

6 What is the purpose of an indicator in a titration?

7 **a)** How many moles of HCl(aq) are required to neutralize 1.5 moles of KOH(aq)? Explain.

   b) If you had 1 mL of a 1.5 mol/L solution of HCl(aq) how many mL of a 2.0 mol/L solution of KOH(aq) would be required to neutralize the acid solution?

Titrations can be a very valuable method to determine the unknown concentration of acid or base in a sample. The following Sample Problem demonstrates how the unknown concentration of an acid in a solution can be determined based on the known amount of base with which it is titrated.

Sample Problem

Determining Acid Concentration by Titration

Problem

A student titrates 20.00 mL of a solution of hydrochloric acid, HCl(aq), with 0.1000 mol/L NaOH(aq). The initial reading of the burette is 0.20 mL. At the endpoint, the burette reads 36.30 mL. What is the concentration of the hydrochloric acid?

What Is Required?

You need to find the concentration of the hydrochloric acid.

What Is Given?

Volume of NaOH = $V_{final} - V_{initial}$ = 36.30 mL − 0.20 mL
$$= 36.10 \text{ mL}$$

Volume of HCl(aq) = 20.00 mL
Concentration of NaOH(aq) = 0.1000 mol/L

Plan Your Strategy

Step 1 Write the balanced chemical equation for the reaction.

Step 2 Calculate the amount (in mol) of sodium hydroxide present.

Step 3 Use the mole ratio from the balanced chemical equation to determine the amount (in mol) of hydrochloric acid needed to neutralize the sodium hydroxide.

Step 4 Use $c = n/V$ to determine the concentration of the hydrochloric acid.

Act on Your Strategy

Step 1 Write the balanced chemical equation.

$$\text{NaOH(aq)} + \text{HCl(aq)} \rightarrow \text{NaCl(aq)} + \text{H}_2\text{O}(\ell)$$

$V_{NaOH} = 36.10$ mL $V_{HCl} = 20.00$ mL

$c_{NaOH} = 0.1000$ mol/L $c_{HCl} = ?$

Step 2 $n_{NaOH} = c_{NaOH} \times V_{NaOH}$

$$= 0.1000 \text{ mol/}\cancel{L} \times 0.03610 \cancel{L} \text{ NaOH(aq)}$$

$$= 0.003610 \text{ mol NaOH(aq)}$$

Step 3 $n_{HCl} = 0.003610 \cancel{\text{mol NaOH(aq)}} \times \dfrac{1 \text{ mol HCl(aq)}}{1 \cancel{\text{mol NaOH(aq)}}}$

$$= 0.003610 \text{ mol HCl(aq)}$$

Step 4 $c_{HCl} = \dfrac{n_{HCl}}{V_{HCl}}$

$$= \dfrac{0.003610 \text{ mol HCl(aq)}}{0.02000 \text{ L}}$$

$$= 0.1805 \text{ mol/L HCl(aq)}$$

The concentration of the hydrochloric acid is 0.1805 mol/L.

Check Your Solution

The concentration of HCl(aq) is nearly twice the concentration of NaOH(aq). This makes sense because the volume of HCl(aq) was about half the volume of NaOH(aq) used, and the mole ratio is 1:1. The number of significant digits is correct (four).

Using Dimensional Analysis

$$c_{HCl} = 0.03610 \text{ L NaOH(aq)} \times \dfrac{0.1000 \text{ mol NaOH(aq)}}{1 \text{ L NaOH(aq)}} \times \dfrac{1 \text{ mol HCl(aq)}}{1 \text{ mol NaOH(aq)}} \times \dfrac{1}{0.02000 \text{ L HCl}}$$

$$= 0.1805 \text{ mol/L HCl(aq)}$$

Practice Problems

20. If the pH of a 50 ml solution of NaOH(aq) reaches 7 after you have added 36.2 mL of a 1.5 mol/L solution of acetic acid, CH_3COOH(aq), what is the concentration of the NaOH(aq) solution?

21. A student titrates 20.00 mL of HCl(aq) with 0.150 mol/L NaOH(aq). The initial reading on the burette is 1.50 mL. The final reading on the burette is 29.51 mL. What is the concentration of the HCl(aq)?

22. A student uses 0.2030 mol/L NaOH(aq) to titrate 40.00 mL of HNO_3(aq). The initial reading on the burette is 0.050 mL. The final reading on the burette at endpoint is 38.00 mL. What is the concentration of the HNO_3(aq)?

23. A student performs a series of titrations in order to determine the concentration of a KOH(aq) solution. The titrant used is a 0.50 mol/L solution of HNO_3(aq). The following is a table summarizing the student's experimental data for the 3 trials that were performed:

Trial #	1	2	3
Final burette volume	5.35	5.42	6.58
Initial burette volume	0.15	0.20	1.43
Volume of KOH(aq)	10.0 mL	10.0 mL	10.0 mL

Calculate the concentration of the KOH(aq) solution for each trial. What is the average value for the concentration of KOH(aq)?

24. A student is about to titrate 25.00 mL of HCl(aq) using 0.1500 mol/L of NaOH(aq). The teacher tells the class that the concentration of the acid is approximately 0.2 mol/L.

  a) Using this information, estimate what volume of sodium hydroxide will be required to titrate the hydrochloric acid. (**Hint:** Use $V = n/c$.)

  b) Why might it be useful to perform this type of calculation before a titration?

25. Geologists carry small bottles of hydrochloric acid in the field to identify carbonate rocks and minerals. When a few drops of the acid are applied to a carbonate containing rock or mineral, the geologist observes fizzing. The test geologists perform can be recreated on a larger scale in the chemistry laboratory. It takes 125.0 mL of hydrochloric acid to react with 3.28 g of calcium carbonate according to the following chemical equation:

$$CaCO_3(s) + 2HCl(aq) \rightarrow CO_2(g) + H_2O(\ell) + CaCl_2(aq)$$

What is the molar concentration of the hydrochloric acid?

Standardizing a Hydrochloric Acid Solution

Hydrochloric acid, HCl(aq), has the common name of muriatic acid. It is used for cleaning concrete and removing stains on driveways. In industry, hydrochloric acid is often used to neutralize basic waste waters. Metal electroplating facilities use it to clean rust from steel. In this investigation, you will determine the concentration of a sample of HCl(aq) by titrating it with an aqueous sodium carbonate standard solution in the presence of a few drops of the indicator, phenolphthalein.

Question

What is the molar concentration of the HCl(aq) samples that are to be standardized?

Prediction

Your teacher will tell you the approximate molar concentration of the acid. Once you have completed Procedure step 1, calculate the approximate volume of base that you expect to add to reach the endpoint.

Safety Precautions

Hydrochloric acid and sodium carbonate solutions are corrosive. Wash any spills on your skin or clothing with plenty of cool water, and inform your teacher immediately.

Materials

- 100 mL of HCl(aq), concentration unknown 😵 ☣
- samples of Na₂CO₃(s), no more than 2 g needed 😵 ☣
- dropper bottle of phenolphthalein
- wash bottle of deionized or distilled water
- 50 mL burette
- 10 mL volumetric pipette
- 50 mL beaker
- 250 mL beaker
- 125 mL Erlenmeyer flasks (2)
- 100 mL volumetric flask
- scoopula or spatula
- glass stirring rod
- funnel
- meniscus reader
- pipette bulb
- retort stand
- burette clamp
- electronic balance

Procedure

1. Use the electronic balance to obtain 0.90 g–1.20 g of $Na_2CO_3(s)$ in a clean, dry 50 mL beaker. Record the mass of the sodium carbonate.

2. With the aid of the stirring rod, dissolve the $Na_2CO_3(s)$ in a minimum of distilled water.

3. Transfer the solution from Procedure step 2 to the 100 mL volumetric flask and prepare a standard 100.00 mL solution. (Be sure to use a funnel and add all the washing of the glassware you have used to the volumetric flask.) Calculate the molar concentration of $Na_2CO_3(aq)$.

4. Copy the table below into your notebook to record your observations.

Volumes of Standard $Na_2CO_3(aq)$ Solution Added to 10.00 mL Samples of HCl(aq)

Trial #	1	2	3	4
Final burette volume (± 0.1 mL)				
Initial burette volume (± 0.1 mL)				
Indicator colour at endpoint				
Mass of $Na_2CO_3(s)$ standard =	Total volume of $Na_2CO_3(aq)$ standard =			

5. Clean and rinse the burette with the $Na_2CO_3(aq)$, and then fill it with fresh $Na_2CO_3(aq)$, making sure there is no air bubble in the burette tip. Record the initial volume of titrant in the burette to the nearest 0.05 mL. (It need not be 0.00 mL!)

6. With a clean, dry 10.00 mL volumetric pipette, obtain a 10.00 mL sample of the HCl(aq).

7. Deliver the 10.00 mL HCl(aq) sample to an Erlenmeyer flask and add two drops of indicator.

8. Position the flask under the burette as shown in Figure 8.5 on page 312. With a sheet of white paper under the flask, slowly titrate the $Na_2CO_3(aq)$ standard into the HCl(aq) sample with gentle swirling. As the endpoint is approached, you will see a slight colour change that will disappear with swirling. Now add the $Na_2CO_3(aq)$ drop by drop. You have reached the endpoint when the indicator permanently changes colour. Record the final volume of titrant in the burette and the indicator colour.

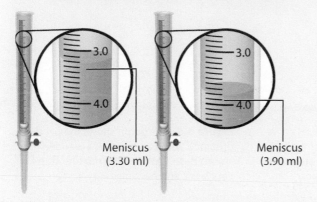

Meniscus
(3.30 ml)

Meniscus
(3.90 ml)

9. Repeat Procedure steps 5–8 three more times or until you have several results in which the volumes of titrant added are the same to within ± 0.2 mL. (The volume of titrant added is the difference between the initial and final burette volumes.)

10. When you have completed at least four titrations, dispose of all the excess chemicals as directed by your teacher. Clean and rinse all the glassware thoroughly and leave the burette, tap open, and the pipette upside down in the burette clamp to dry.

Analysis

1. a) Using the trials that agree to within ± 0.2 mL, calculate the average volume of $Na_2CO_3(aq)$ added.

  b) Why is it advisable to carry out multiple trials?

2. Write the balanced chemical equation for the reaction of $Na_2CO_3(aq)$ and HCl(aq).

3. Perform a stoichiometric calculation to determine the concentration of HCl(aq).

4. Post your results. As a class, determine the average [HCl(aq)]. (Discard any results that seem unreasonable.)

Conclusion

5. Report the average [HCl(aq)].

6. Compare [HCl(aq)] to the value provided by your teacher.

7. List several likely sources of error in this investigation.

Extension

8. In addition to being useful as a primary standard, sodium carbonate is also the principal ingredient of most dishwasher detergents. Use the Internet to find the common name of sodium carbonate and suggest why it is used as a dishwashing detergent. **ICT**

Acid–Base Titration Curves

As you have learned, you can determine the equivalence point in an acid–base titration by using an indicator that changes colour near the equivalence point. You can also determine when this happens by using a pH meter to measure pH as the reaction progresses.

Strong monoprotic acids have pH values below 7 (at 25 °C). Strong monoprotic bases have pH values above 7 (at 25 °C). For a titration in which you are using a pH meter to determine the equivalence point, the pH of the reaction mixture is monitored as an acid (or base) titrant is added to a base (or acid) sample. As you add the solution, you record the pH versus the volume of titrant added.

A plot of the pH of the reaction mixture versus the volume of titrant added is called a **pH titration curve**. If a strong base titrant is added to a strong acid sample, the pH titration curve will start at the low pH of the acid sample and rise as more titrant is added. Eventually, if you continue to add titrant past the equivalence point, the solution reaches a pH close to that of the base titrant (see Figure 8.8). Conversely, when a strong acid titrant is added to a strong base sample, the pH titration curve starts at the high pH of the base sample and decreases as more titrant is added, to a pH close to that of the acid titrant (see Figure 8.9).

As you can see from Figures 8.8 and 8.9, pH titration curves are not linear but are in fact S-shaped. There is a steep rise accompanying the complete neutralization of the strong acid sample and a steep drop accompanying the complete neutralization of the strong base sample.

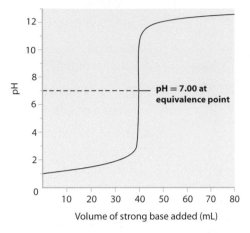

Figure 8.8 This graph shows the pH curve obtained when a strong base is added to a strong acid. The reaction mixture starts off very acidic and becomes very basic.

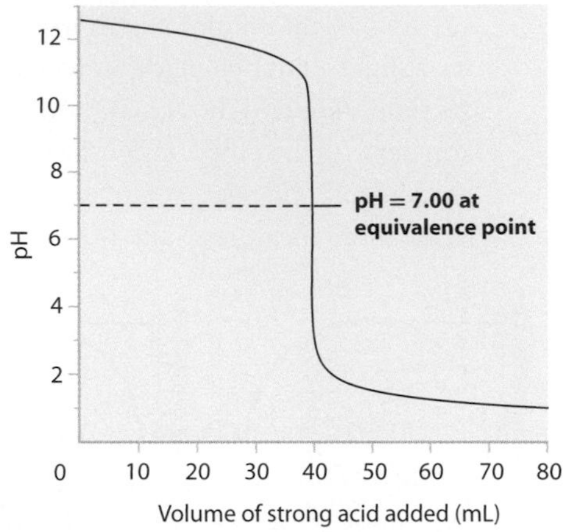

Figure 8.9 This graph shows the pH curve obtained when a strong acid is added to a strong base. The reaction mixture starts off very basic and becomes very acidic.

The middle of the steep rise, or steep drop, in a pH titration curve corresponds to the pH of the reaction mixture at the equivalence point of the reaction. The equivalence point, when a monoprotic strong acid is neutralized by a monoprotic strong base, is at a pH of 7. You will notice that in the vicinity of the equivalence point, the reaction mixture pH changes very rapidly. Therefore, as the equivalence point approaches, a very small amount of added titrant will result in a large change in pH.

Choosing an Indicator

It is often more convenient to use an indicator than to use a pH meter to carry out a titration. Indicators are inexpensive, and multiple titration trials are easy to perform and improve the accuracy of results. In contrast, pH probes and meters are often expensive. Also, pH titration analyses are time consuming unless a data logging device is available.

How do you know what indicator to use for a given titration? When you are choosing an indicator for a titration, the endpoint pH for the indicator must be within the steep rise or drop in the titration curve. Ideally, the endpoint of the indicator would occur right at the equivalence point of the reaction. Indicators that are commonly used in the labratory include phenolphthalein and bromothymol blue. Table 8.1 lists a variety of indicators that are suitable for strong acid–strong base titrations.

Table 8.1 Indicators Suitable for Strong Acid–Strong Base Titrations

Indicator	pH range	Colour change as pH increases
bromocresol green	3.8–5.4	yellow to blue
methyl red	4.8–6.0	red to yellow
chlorophenol red	5.2–6.8	yellow to red
bromothymol blue	6.0–7.6	yellow to blue
phenol red	6.6–8.0	yellow to red
phenolphthalein	8.2–10.0	colourless to pink

8. Explain how you would tell the difference between an acid–base titration curve in which acid is the titrant and an acid–base titration curve in which base is the titrant. Sketch an example of each titration curve.

9. How do you identify the equivalence point of the reaction on an acid–base titration curve for a strong acid and a strong base?

10. What does the steep slope within a titration curve tell you about the volume of titrant that is required to change the pH near the equivalence point of the reaction?

In the following Thought Lab, you will analyze pH data from an acid–base titration to determine the equivalence point, suggest a suitable indicator, and determine the concentration of an acid solution. In the investigation that follows, you will write and carry out a procedure to determine the concentration of a sample of sodium hydroxide, NaOH(aq).

Thought Lab 8.2 Plotting a Titration Curve

Target Skills

Creating and interpreting a titration curve graph for an acid–base experiment

A student carried out the titration of 50.00 mL of aqueous potassium hydroxide, KOH(aq), with 0.100 mol/L nitric acid, HNO_3(aq). The student used a pH meter to measure the changing pH of the reaction mixture. The data are in the table below.

Titration pH Data

Volume of 0.100 mol/L HNO_3(aq) added (mL)	Reaction mixture pH
0.00	13.00
10.00	12.82
20.00	12.63
30.00	12.40
40.00	12.05
45.00	11.72
49.00	10.00
50.00	7.00
51.00	4.00
55.00	2.32
60.00	2.04
70.00	1.78
80.00	1.64
90.00	1.54
100.00	1.48

Procedure

1. Use a spreadsheet program to plot these data as an x-y scatter graph. Include all appropriate labels on the axes and a title for the graph. Select the graph display mode that does not include a line through the data points. (Alternatively, create your graph on graph paper.) **ICT**

Analysis

1. Print your graph and draw a smooth line of best fit through the data points.

2. Locate and then label the equivalence point pH on the graph.

3. Suggest three suitable indicators for this titration and list the colour change you are likely to see for each. Explain why the indicators you selected are appropriate.

4. How does the shape of this pH titration curve compare with the shape of the curve in Figure 8.8? How does it compare with the shape of the curve in Figure 8.9?

5. Use your graph to determine the concentration of the KOH(aq) sample.

Extension

6. The indicator methyl violet changes colour from yellow to blue over a pH range of 0.0–1.6. Would methyl violet be a suitable indicator to use for this titration? Explain.

Sulfur from Sour Gas

Natural gas is one of the most widely used fuels in North America. This clean-burning fuel consists mostly of methane, $CH_4(g)$, a colourless, odourless gas. Many households and industries in Canada use natural gas for heat and power, and 80% of the gas used in Canada comes from Alberta.

When it is pumped from the well, natural gas is not ready to use. In its crude (unprocessed) form it contains water, oil, and other gases such as hydrogen sulfide, $H_2S(g)$. Hydrogen sulfide smells bad, is corrosive, and is highly toxic. Natural gas that contains large amounts of hydrogen sulfide is called sour gas. Before it is used for heating, natural gas must be processed to remove the hydrogen sulfide. This process is called "sweetening" the gas. The natural gas companies can process the hydrogen sulfide by-product to obtain sulfur.

Stacked sulfur at a sour gas processing facility

Chemical Sweetening of Sour Gas

Sweetening natural gas takes place at several plants in Alberta. The most common way of removing sulfur from sour gas uses amine solutions, such as monoethanolamine, $(CH_2)_2OHNH_2(\ell)$. Amines are a class of compounds that contain nitrogen. In their liquid form, they absorb $H_2S(g)$ from sour gas.

The $H_2S(g)$ is then fed to a Claus processing unit, where it is converted to elemental sulfur in two steps:

1. In the *thermal step*, the H_2S is combined with oxygen in a furnace at high temperatures. This oxidizes the H_2S, removing some of the sulfur and using some of it to create sulfur dioxide (SO_2):

$$6H_2S(g) + 3O_2(g) \rightarrow 4H_2S(g) + 2SO_2(g) + 2H_2O(g)$$

2. The remaining gas goes through a *catalytic* stage, in which chemicals called catalysts speed up the reaction process. The $H_2S(g)$ reacts with the $SO_2(g)$ that formed in the thermal stage:

$$2H_2S(g) + SO_2(g) \xrightarrow{\text{catalyst}} 3S(\ell) + 2H_2O(g)$$

Engineers repeat this process using different catalysts so that they can extract as much H_2S as possible. Eventually, they capture approximately 97% of the sulfur from the sour gas. When burned, $H_2S(g)$ creates $SO_2(g)$ in the atmosphere, which leads to acid rain. To minimize this effect, the gas goes through more processing and incineration to extract another 1% of H_2S from the gas and to make the gas that consumers burn as pure as possible. The solid sulfur that is produced from this process can be used to produce sulfuric acid.

• • •

1. **a)** How many grams of sulfur would you obtain from 100 L of $H_2S(g)$ at SATP extracted from a batch of sour gas? Assume 100% extraction. (**Hint:** There is 1 mol of liquid sulfur, $S(\ell)$, for every 1 mol of $H_2S(g)$.)

   b) Sulfur is more widely used as sulfuric acid, $H_2SO_4(aq)$, than in its elemental form. According to the following equation, what volume of 6.0 mol/L $H_2SO_4(aq)$ could you obtain from 1.00 kg of $S(\ell)$?

   $$2S(\ell) + 2H_2O(\ell) + 3O_2(g) \rightarrow 2H_2SO_4(aq)$$

2. After it goes through the Claus treatment, natural gas is about 97% free of hydrogen sulfide. Why must it go through further processing?

Claus extraction

furnace   catalytic section

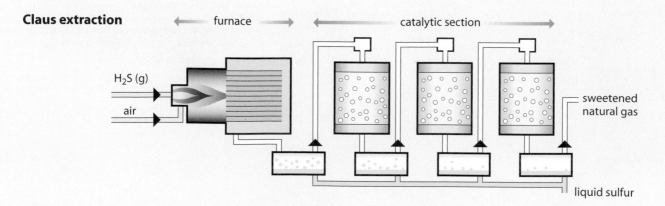

H_2S (g)

air

sweetened
natural gas

liquid sulfur

Titrating a Strong Base with a Strong Acid

Sodium hydroxide, NaOH(s), is a raw material for the pulp and paper industry. It is used to neutralize acidic waste water and to manufacture soap from animal fats and vegetable oils, and it turns up in your local supermarket as the principal ingredient of basic drain cleaners. In this investigation, you will design a procedure to determine the concentration of an aqueous sodium hydroxide solution by titrating it with standardized hydrochloric acid and using an appropriate indicator.

Question

What is the molar concentration of a sample of NaOH(aq)?

Prediction

Your teacher will provide you with the approximate concentration of the base. Using this concentration, predict how much HCl(aq) solution you will need to reach the equivalence point.

Safety Precautions

Include all appropriate safety precautions in your experimental design.

Materials

You may specify any appropriate, readily available apparatus and reactants.

Experimental Plan

1. Write a logically sequenced, easily understood, experimental procedure that any chemistry student could use to record good-quality data with which to calculate the concentration of the NaOH(aq). Your procedure must be accompanied by an example of the kind of table necessary to record the data from the titration. Include a materials list and safety precautions. Use Investigation 8.C as a guideline for how to design this experiment. (This experiment is basically the reverse of what was done in Investigation 8.C.)

2. Submit your procedure to your teacher for approval.

Data and Observations

3. Carry out your procedure and record all observations.

4. Clean up and dispose of all materials as directed by your teacher.

Analysis

1. For trials that agree to within ± 0.2 mL, calculate the average volume of titrant used in your experiment.

2. Write the balanced chemical equation for the reaction of NaOH(aq) and HCl(aq).

3. Perform a stoichiometric calculation to determine [NaOH(aq)].

Conclusion

4. Compare your base concentration with the values obtained by your fellow student experimenters.

5. List several likely sources of error in this investigation.

6. If you were to repeat this experiment, would you modify your procedure in any way? What would you change? Explain your answer.

Extension

7. If you wanted to analyze the NaOH(aq) content of a concentrated, basic drain cleaner, how would you prepare the sample? Explain your answer. (Do not carry out this experiment without the permission of your teacher.)

8. Suggest two household bases and two household acids that you could analyze by titration. (Do not carry out this analysis without the permission of your teacher.)

Section 8.3 Summary

- Acid–base titrations can be used to determine the concentration of a strong acid or strong base in aqueous solution. The titrant, a solution of known concentration, is added to a sample of unknown concentration until the equivalence point is reached.

- The equivalence point of an acid–base titration occurs when stoichiometrically equal amounts of acid and base have reacted. For strong monoprotic acids and bases, it is the point at which equal moles of each have reacted and results in neutralization (pH 7 at 25 °C).

- Acid–base titrations are useful for standardizing acids and bases, which tend to change in concentration over time.

- Indicators can be used to monitor acid–base titrations. An indicator will undergo a distinct colour change within a certain pH range. For titrations of strong monoprotic acids and bases, the indicator of choice should be one that changes colour at a pH near the equivalence point.

- A pH meter can also be used to monitor an acid–base titration. The data collected is plotted as pH of the reaction mixture versus the volume of titrant added and is referred to as a pH titration curve. These curves are S-shaped, with a steep slope near the equivalence point.

SECTION 8.3   REVIEW

1. Write a chemical equation for each of the following neutralization reactions:
   a) $KOH(aq)$ with $HClO_3(aq)$
   b) $HI(aq)$ with $NaOH(aq)$
   c) $HBr(aq)$ with $RbOH(aq)$
   d) $NaOH(aq)$ with $HNO_3(aq)$
   e) $HCl(aq)$ with $Na_2CO_3(aq)$

2. a) Explain the difference between the endpoint of an acid–base titration and the equivalence point of an acid–base reaction.
   b) When an indicator is chosen for a strong acid–strong base titration, do the pH of the equivalence point and the endpoint pH range of the indicator need to coincide exactly? Explain your answer.

3. A student used 0.118 mol/L H_2SO_4 to titrate 10.00 mL of $NaOH(aq)$. The initial volume reading on the burette was 1.05 mL. The final volume reading on the burette was 23.00 mL. What is the concentration of the $NaOH(aq)$, according to these data?

4. The set of data in the table below was collected during the titration analysis of an unknown hydrochloric acid sample with a sodium hydroxide titrant. Use the data to calculate the concentration of the acid. (**Hint:** Would you discard any data?)

5. It requires 18.2 mL of 0.100 mol/L $KOH(aq)$ to neutralize all the citric acid, $H_3C_6H_5O_7(aq)$, in a 10.00 mL sample. The balanced chemical equation for this reaction is as follows:
$3KOH(aq) + H_3C_6H_5O_7(aq) \rightarrow K_3C_6H_5O_7(aq) + 3H_2O(g)$
What is the concentration of the citric acid?

6. While performing the first titration of a set of four, a student does an excellent job of locating the endpoint very close to the equivalence point. It takes only one drop of $NaOH(aq)$ titrant to cause the phenolphthalein indicator to change colour in the hydrochloric acid sample. Subsequent analysis of this trial gives an acid concentration that is 10% higher than the correct value. The student suspects that a mistake has occurred during this titration. Discuss the following possible mistakes. In each case state whether the mistake would give an artificially high or a low concentration for the acid. (One "mistake" would not affect the result at all.)
   a) The pipette used to collect the acid sample was wet and this diluted the acid sample.
   b) The Erlenmeyer flask to which the measured acid sample was added, though clean, was wet.
   c) The tip of the burette was filled with air and not titrant when the first trial was performed.
   d) A large drop of titrant was on the very tip of the burette when the first trial was performed.

Volumes of 0.06649 mol/L NaOH(aq) Added to 10.00 mL Samples of HCl(aq)

Trial #	1	2	3	4
Final burette volume ($\pm$ 0.1 mL)	24.06	46.06	23.08	45.12
Initial burette volume ($\pm$ 0.1 mL)	0.10	24.06	1.06	23.08
Indicator colour at endpoint (methyl red)	deep red	pale red	pale red	pale red

An important component to stoichiometric analyses of reactions is considering how the relative amounts of components can influence the extent of a reaction. When reactants are present in amounts that correspond to the mole ratio from the balanced chemical equation, they are considered to be present in stoichiometric amounts and are completely consumed in a reaction. However, that is rarely the case. Usually, there is one reactant that is completely consumed before the others. This reactant is referred to as the limiting reactant. It dictates the amount of product formed since, once it is consumed, the reaction cannot continue. Stoichiometric calculations are used to determine the limiting reactant in a reaction by identifying the one that would produce the least amount of product. It is important to remember that it is not necessarily the component that is present in the least amount. Reactants that remain upon completion of a reaction are referred to as the excess reactants. The safe combustion of propane in heaters relies on the presence of excess oxygen in order for the production of a toxic side-product, carbon monoxide, to be minimized.

Optimizing the efficiency of a chemical reaction is very important, particularly in industrial chemistry. Stoichiometric calculations can be used to determine the maximum amount of product that is expected from a reaction based on the amounts of reactants present. This value is called the predicted yield. The amount of product that is actually obtained from a reaction is referred to as the experimental yield. Most often, the experimental yield is lower than the predicted. The efficiency of a chemical reaction is expressed as the percentage yield. The percentage yield can be calculated by the equation

$$\frac{\text{experimental yield}}{\text{predicted yield}} \times 100\%$$

There are several factors that can contribute to limiting the experimental yield of a product. Competing reactions can cause the consumption of some of the reactants and, therefore, a decrease in the amount available for the formation of the primary product. The presence of considerable amounts of impurities in reactants can also reduce the amount of product expected since a considerable proportion of the amount of reactant added will not contribute to the reaction. In addition, if a reaction is very slow or the reaction has not been allowed enough time to react, the amount of product formed will not be optimized. There are also some reactions that never undergo complete conversion of reactants because they have reached an equilibrium, whereby the products react to form the reactants at the same rate as the reactants reacting to form the product. Finally, some techniques used for isolating and purifying products can result in reduction of the experimental yield. For example, if isolating a solid product by filtration is required there can be losses due to solubility in the solvents used and improper handling and transfer of the solid.

Acid–base titrations are a form of quantitative analysis that can be used to determine the concentration of a strong acid or strong base in aqueous solution. A practical example of this is the standardization of stock acid and base solutions, which are known to change in concentration over time. In these titrations, the titrant with known concentration is added to a sample of unknown concentration until the equivalence point is reached. This equivalence point occurs when stoichiometrically equal amounts of acid and base have reacted. For strong monoprotic acids and bases, it is the point at which equal moles of each have reacted and results in neutralization (pH 7 at 25 °C).

Acid–base titrations can be monitored using indicators, which undergo a distinct colour change within a certain pH range and signal that the equivalence point has been reached. A pH meter can also be used to monitor acid–base titrations. The data collected is plotted as pH of the reaction mixture versus the volume of titrant added and is referred to as a pH titration curve. These curves are S-shaped with a steep slope near the equivalence point.

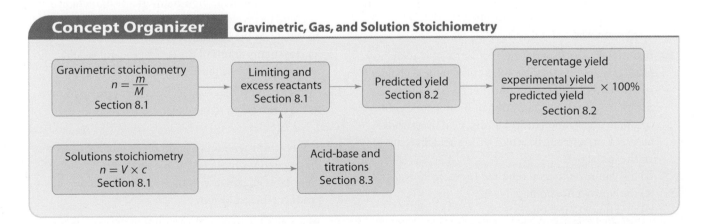

Concept Organizer — **Gravimetric, Gas, and Solution Stoichiometry**

Gravimetric stoichiometry
$$n = \frac{m}{M}$$
Section 8.1

→ Limiting and excess reactants
Section 8.1

→ Predicted yield
Section 8.2

→ Percentage yield
$$\frac{\text{experimental yield}}{\text{predicted yield}} \times 100\%$$
Section 8.2

Solutions stoichiometry
$$n = V \times c$$
Section 8.1

→ Acid-base and titrations
Section 8.3

Understanding Concepts

1. Explain the meaning of the term, "limiting reactant."

2. What does it mean when it is said that the reactants are present in stoichiometric amounts?

3. Explain why, in a chemical reaction in which a company is preparing an industrial chemical, they might choose to use a large excess of one reactant instead of using stoichiometric amounts of all reactants.

4. List and briefly explain three factors that cause the yield of a chemical reaction to be lower than the predicted amount.

5. In industry, several steps are often required to produce a chemical for market. How does the number of steps affect the percentage yield of the final product? Explain.

6. During an acid-base titration experiment, which is reached first, the end point or the equivalence point? Explain why.

7. What is the term that describes the solution that is placed in the burette during an acid base titration?

Applying Concepts

8. A student used the endpoint from an indicator that changes colour between pH 10.1 and pH 12.0, to measure the volume of a standardized NaOH(aq) titrant necessary to neutralize a sample of HCl(aq). Use the pH curve in Figure 8.8 to decide if this student's calculated acid concentration is likely to be higher, lower, or the same as the acid's true concentration.

9. Some of the glassware used in a titration analysis can be wet (with water) and have no effect on the results of the analysis. Other glassware if wet, will lead to unreliable results.
   a) Which of the burette, the pipette, or the Erlenmeyer flask will spoil the data obtained if they are wet? Explain the effect(s) on the results.
   b) Which of the burette, the pipette, or the Erlenmeyer flask can be wet without affecting the data? Explain why.

10. In an acid–base titration, an acid (or base) is added to a base (or acid) and the pH of the initial sample and the reaction mixture is monitored after each addition of titrant. For equal concentration of monoprotic entities, compare the titration of (i) a strong acid by a strong base and (ii) a strong base by a strong acid, in terms of
    a) the initial pH
    b) the volume of titrant required to reach equivalence
    c) the equivalence point pH
    d) the choice of indicator
    e) the final pH

Solving Problems

11. A chemist adds some zinc shavings to a beaker containing a blue solution of copper(II) chloride. The contents of the beaker are stirred. After several hours, the chemist observes that the blue colour has almost, but not completely, disappeared.
    a) Write a balanced chemical equation to describe this reaction.
    b) What other observations would you expect the chemist to make?
    c) According to the chemist's observations, which reactant was the limiting reactant?
    d) The beaker contained 3.12 g of copper(II) chloride dissolved in water. What does this tell you, quantitatively, about the amount of zinc that was added?

12. 20.8 g of calcium phosphate, 13.3 g of silicon dioxide, and 3.9 g of carbon react according to the following equation:
$$2Ca_3(PO_4)_2(s) + 6SiO_2(s) + 5C(s) \rightarrow P_4(s) + 6CaSiO_3(s) + 5CO_2(g)$$
    a) What is the limiting reactant?
    b) Determine the mass of calcium silicate that is produced.

13. 2.85×10^2 g of pentane, $C_5H_{12}(\ell)$ reacts with 3.00 g of oxygen gas according to the following equation:
$$C_5H_{12}(\ell) + 8O_2(g) \rightarrow 5CO_2(g) + 6H_2O(\ell)$$
    a) What is the limiting reactant?
    b) What mass of carbon dioxide is produced?
    c) What mass of the excess reactant remains after the reaction is complete?

14. Zinc reacts with carbon monoxide according to the following equation:
$$ZnO(s) + CO(g) \rightarrow Zn(s) + CO_2(g)$$
    a) If you combined 40.7 g of zinc oxide and 28.0 g of carbon monoxide, how many grams of metallic zinc could be produced in the reaction?
    b) What was the limiting reactant?
    c) How many grams of the excess reactant would remain after the reaction was complete?

15. Nitroglycerin ($C_3H_5N_3O_9(\ell)$) is a powerful explosive. Its decomposition may be represented by
$$4C_3H_5N_3O_9(\ell) \rightarrow 6N_2(g) + 12CO_2(g) + 10H_2O(g) + O_2(g)$$
    This reaction generates a large amount of heat and many gaseous products. It is the sudden formation of these gases, together with their rapid expansion, that produces the explosion.
    a) What is the maximum amount of oxygen, in grams, that can be obtained from the decomposition of

2.00×10^2 g of nitroglycerin?

b) Calculate the percentage yield in this reaction if the amount of oxygen generated is found to be 6.55 g.

c) Based on your answer to part b), what mass of nitrogen was formed in the reaction?

16. The complete combustion of octane, the major component of gasoline for automobiles, can be described by the equation below.

$$2C_8H_{18}(\ell) + 25O_2(g) \rightarrow 16CO_2(g) + 18H_2O(g)$$

If automobile engines are not properly tuned and the air filter replaced, there is often not sufficient oxygen for complete combustion. Under these conditions, a competing reaction occurs along with the complete combustion reaction. The equation for this reaction is shown below. This reaction is the source of carbon monoxide in automobile exhaust.

$$2C_8H_{18}(\ell) + 17O_2(g) \rightarrow 16CO(g) + 18H_2O(g)$$

a) What is the maximum amount of carbon dioxide that can be produced by the complete combustion of 750 g octane?

b) If 350 g of carbon monoxide are produced by the burning of 750 g of octane, how many grams of carbon dioxide would you predict were generated?

17. Ammonium nitrate is used as a fertilizer. It is synthesized from ammonia and nitric acid.

a) Write a balanced equation for the reaction between ammonia and nitric acid to produce ammonium nitrate.

b) What is the predicted yield of ammonium nitrate if 17 t of ammonia is consumed in the reaction?

c) If 63 t of fertilizer is obtained from the reaction that consumed 17 t of ammonia, what is the percentage yield?

18. In a titration experiment, 50.0 mL of 0.010 mol/L hydrochloric acid was required to neutralize 25.0 mL of an aqueous solution of sodium hydroxide. What was the molar concentration of the sodium hydroxide solution?

19. You measured 25.0 mL of a nitric acid solution and poured it into an Erlenmeyer flask and added a drop of phenolphthalein solution. You then filled a burette with 0.115 mol/L sodium hydroxide solution. Before beginning your titration, you observed that reading on the burette was 1.7 mL. As you titrated the nitric acid, you began to see a pink colour when the burette read 19.4 mL. However, when you swirled the flask, the pink colour disappeared. When the burette reading was 20.0 mL, the pink colour remained after mixing. What was the concentration of the nitric acid solution?

20. If it takes 18.0 mL of 0.500 mol/L aqueous sodium hydroxide to neutralize a 15.0 mL sample of sulfuric acid, what is the acid's concentration?

21. Commercial processors of potatoes remove the skins by briefly soaking the potatoes in a solution of caustic soda (sodium hydroxide). The potatoes are soaked in the solution for a short time at 60 °C–70 °C, after which the peel is easily removed with a spray of fresh water. As a technician at a large food-processing facility, you are responsible for analyzing a batch of NaOH(aq) solution. You titrate 25.00 mL samples of the base and find that, on average, it requires 30.21 mL of 1.986 mol/L HCl(aq) to reach a satisfactory endpoint.

a) Calculate the molar concentration of the sodium hydroxide.

b) If the sodium hydroxide concentration drops below 10 g/100 mL, it cannot be used for the skin-removing process. Calculate the concentration of the solution in grams of NaOH per 100 mL and determine whether the solution can be re-used.

c) What safety gear is likely worn by workers at the potato-processing plant? Explain your answer.

22. The arsenic in a 1.22 g sample of pesticide was converted by a chemical reaction to aqueous arsenate ion, $AsO_4^{3-}(aq)$. This aqueous arsenate solution was titrated with an aqueous solution of silver ions, $Ag^+(aq)$, to give a precipitate of silver arsenate.

a) Write the net ionic equation for the formation of the precipitate from arsenate and silver ions.

b) It took 25.0 mL of a 0.102 mol/L $Ag^+(aq)$ titrant to precipitate all the arsenate ions in the solution. What mass of arsenate ion was in the solution?

c) What is the percentage by mass of arsenic in the pesticide?

23. A 2.50 mL sample of pickling vinegar (a solution of ethanoic acid, $CH_3COOH(aq)$, in water) requires 34.9 mL of 0.0960 mol/L sodium hydroxide to reach the equivalence point when titrated. How many grams of ethanoic acid are in 2.00 L of this pickling vinegar?

Making Connections

24. Phosphate ions promote algal growth in rivers and lakes. Phosphates enter rivers and lakes in run-off from heavily fertilized fields or in untreated waste water that contains phosphate detergents.

a) How could phosphate-contaminated water be treated to remove the phosphate ions? Describe what steps would be involved.

b) In addition to treating the water, what are some other options for reducing the concentration of phosphates in rivers and lakes?

Career Focus: Ask a Pharmaceutical Chemist

Dr. Lee Wilson

Dr. Lee Wilson made history in 1998 when he became the first Métis student to achieve a PhD in chemistry from the University of Saskatchewan, where he is now a researcher and assistant professor in the Department of Chemistry. In his research, he studies the uses for container molecules: hollow structures (or "hosts") that can hold other molecules (or "guests"). These container molecules can be developed to improve the delivery of drugs to the body's systems. Dr. Wilson uses stoichiometry in his research to predict the yield of complex reactions between different molecules.

Cyclodextrin, shown here, is an example of a "container molecule."

medicine molecule ("guest") + container molecule ("host") ⇌ host-guest complex

A container molecule holds a medicine molecule, forming a host–guest complex.

Q Can you tell us a bit about your background? How did you first become interested in chemistry?

A My interest in chemistry began in high school. I had a chemistry teacher who told wonderfully inspiring stories and got me interested in the subject. I chose chemistry as a major area of study in university because I thought it would provide more job options once I graduated. During my undergraduate degree at the University of Winnipeg, I was fortunate to do a major research project in my fourth year that inspired me to carry on to graduate school. I had a good thesis supervisor, and he was very energetic and enthusiastic: Professor Alaa Abd-El-Aziz. I would have to say that the influential people who mentored and supervised me have given me strength and inspiration to continue in my profession.

Q Why is the pharmaceutical industry interested in your research?

A At first, my research was aimed at understanding container molecules that carry medicines. These are called transporter devices. My PhD research unravelled some of the complexities in what was once considered a simple topic. Currently, my group is designing transporter devices that can be used in new pills for diabetes, cancer, AIDS, and heart disease. Using these devices makes the dosages more efficient. Since many drugs are toxic at high concentrations, this will ultimately reduce dangerous side effects.

Q What are some other ways your research might be used?

A Other applications of this research include new low-energy methods of separating molecules in chemical mixtures, ways to clean up contaminated drinking water, designs for new sensors, and ways to store gases for fuel. Container molecules can also provide ways to produce proteins more effectively. Insulin, used for the treatment of diabetes, is an example of an important protein that biotechnology helps produce on a large scale.

Q How do you develop container molecules?

A We use anything that helps us work with chemicals at the molecular level. For example, we use different types of modern equipment to study the structure of molecules and how they interact with each other under a variety of conditions. We use computers to process our data and to predict what will occur during our experiments.

Q How do you use stoichiometry in your work?

A Stoichiometry is ever-present in my work. Nature is very complex and adaptive. We often create experimental conditions in which the stoichiometry of a drug and transporter device is not known. Quite often, this will cause interactions to occur that we couldn't have predicted.

For example, consider the interaction between a drug (D) and a carrier molecule (C). In a simple case, the drug and carrier may bind together in a 1:1 stoichiometry, as follows:

$$D(aq) + C(aq) \rightleftharpoons D{-}C(aq)$$

Depending on what the carrier is and how concentrated D and C are, the equations may balance differently, resulting in different molar ratios:

$$2D(aq) + C(aq) \rightleftharpoons D_2{-}C\,(aq)$$
$$D(aq) + 2C(aq) \rightleftharpoons D{-}C_2\,(aq)$$

The latter two examples are simplistic but illustrate that drug delivery systems can follow complex stoichiometry pathways. Improving our understanding of stoichiometry has contributed immensely to solving important problems in science.

Q What advice would you give to young people who are considering a career in chemistry?

A Choose a career path that you enjoy and that you are passionate about! Work hard toward your goal. Study and discipline will provide the push to allow you to overcome obstacles that you encounter; the love and passion of the subject will provide the pull to overcome the more difficult obstacles.

Other Chemistry Careers Related to Health

Pharmacologist Pharmacologists study how chemical substances, like drugs, affect the body. They test new drugs to make sure that they are effective and safe to use, secure approval for new types of medicine, and carry out safety tests to discover potentially dangerous side effects. They often work in pharmaceutical companies, research firms and institutes, universities, and hospitals. The basic requirement for a career as a pharmacologist is a four-year Bachelor of Science (BSc) degree in pharmacology or a related biomedical science. Because pharmacology involves the use of principles from many different fields, there is more than one education route to becoming a pharmacologist.

Pharmacist Pharmacists take prescriptions for medicine from physicians and dispense the medication to patients. Often, pharmacists own or manage the pharmacy in which they work. Historically, they were required to complete an undergraduate degree in pharmaceutical chemistry, but today, the basic requirement for a career as a pharmacist is an undergraduate degree in pharmacy from a recognized university. Pharmacists play key roles in health care beyond dispensing medicine. They give advice on common ailments, like colds and allergies, and optimal use of medications, such as whether you should take a pill before or after eating. They review possible conflicts among different types of medications; for example, if you take warfarin, which prevents blood clots, you shouldn't take Aspirin™. They refer patients to other health professionals and monitor patients' general health.

Medicinal Chemist Medicinal chemists (or pharmaceutical chemists) apply chemical research techniques to study existing natural and synthetic drugs and to design and develop new drugs suitable for therapeutic use. This interdisciplinary field combines organic chemistry, biochemistry, computational chemistry, pharmacology, molecular biology, statistics, and physical chemistry. Medicinal chemists work in academic institutions, pharmaceutical companies, and government institutes, both inside and outside laboratories. Chemists with BSc degrees typically work as research technicians, while research careers are open to PhD chemists. Pharmaceutical companies usually look for people who have research experience, advanced degrees (particularly in organic chemistry), and at least two years of post-doctoral laboratory experience.

Go Further...

1. Why is it important for drugs to be "targeted" so that they reach different systems in the human body or are released into the bloodstream at specific times?

2. Describe how container molecules are used in one of the other applications suggested by Dr. Wilson, such as water purification or energy processing.

www.albertachemistry.ca
WWW

Understanding Concepts

1. Explain what the coeffiecients of a balanced chemical equation mean in terms of
   a) atoms, molecules, and ions
   b) molar amounts

2. Explain how a balanced chemical equation refects the law of conservation of mass.

3. Classify each of the following unbalanced equations. Then balance the equations.
   a) $H_2(g) + CuO(s) \rightarrow Cu(s) + H_2O(g)$
   b) $Ag(s) + S_8(s) \rightarrow Ag_2S(s)$
   c) $C_4H_8(g) + O_2(g) \rightarrow CO_2(g) + H_2O(g)$
   d) $MgO(s) \rightarrow Mg(s) + O_2(g)$
   e) $Al_2(SO_4)_3(aq) + K_2CrO_4(aq) \rightarrow$
   $K_2SO_4(aq) + Al_2(CrO_4)_3(s)$

4. Balance and then classify each of the reactions represented by the following chemical equations:
   a) $CH_3OH(\ell) + O_2(g) \rightarrow CO_2(g) + H_2O(g)$
   b) $P_4(s) + O_2(g) \rightarrow P_4O_{10}(s)$
   c) $CaCO_3(s) \rightarrow CaO(s) + CO_2(g)$
   d) $Sr(NO_3)_2(aq) + K_3PO_4(aq) \rightarrow$
   $Sr_3(PO_4)_2(s) + KNO_3(aq)$
   e) $Li_3N (s) \rightarrow Li(s) + N_2(g)$
   f) $Ca(s) + H_2O(\ell) \rightarrow Ca(OH)_2(s) + H_2(g)$

5. Use the solubility chart to decide which of the following salts are soluble or insoluble in aqueous solution:
   a) $Eu(CH_3COO)_3(s)$ **d)** $TlBr(s)$
   b) $KSCN(s)$ **e)** $PbSO_4(s)$
   c) $(NH_4)_2CrO_4(s)$

6. Predict the products of and balance the following equations:
   a) $C_2H_2(g) + O_2(g) \rightarrow$
   b) $Cd(s) + Pb(NO_3)_2(aq) \rightarrow$
   c) $Al(s) + O_2(g) \rightarrow$
   d) $Sr(OH)_2(aq) + H_2SO_3(aq) \rightarrow$
   e) $HgS(s) \rightarrow$ two products (a silvery liquid and yellow crystals)

7. Write a dissociation equation to represent what happens when aluminium sulfate dissolves in water.

8. Consider the reaction of lead(II) nitrate with sodium bromide.
   a) Write a complete balanced equation for this reaction.
   b) Write an ionic equation for this reaction.
   c) Identify the spectator ions.
   d) Write a net ionic equation for this reaction.

9. Write a net ionic equation to represent what happens when silver nitrate, $AgNO_3(aq)$, and ammonium chromate, $(NH_4)_2CrO_4(aq)$ are mixed.

10. Equal molar amounts of hydrochloric acid and barium hydroxide are mixed according to the following unbalanced chemical equation:
    $$\_\_ HCl(aq) + \_\_ Ba(OH)_2(aq) \rightarrow$$
    $$\_\_ BaCl_2(aq) + \_\_ H_2O(\ell)$$
    Identify the limiting and excess reactant for this reaction. (If the reactants are present in stoichiometric amounts, state that there is no limiting or excess reactant.)

11. **a)** List the laboratory apparatus you would need to carry out an acid–base titration.
    b) Draw a simple labelled diagram to show how you would set up the apparatus.

12. What is the difference between the equivalence point of a reaction and the endpoint of an indicator? How are they related?

13. Explain briefly how to clean a burette before using it.

14. Explain the difference between an ionic equation and a net ionic equation.

15. Provide a balanced chemical equation with reactants of your choice to demonstrate the neutralization of a strong acid by a strong base.

Applying Concepts

16. When 100 molecules of hydrogen gas and 200 molecules of chlorine gas are mixed, certain amounts of the elements react completely. How many molecules of hydrogen chloride gas are produced? Which gas is present in excess? How many molecules of that gas are in excess?

17. A high solubility sulfate salt produces a red colour when heated in a Bunsen burner flame. Suggest the identity of the cation of this salt and provide its chemical formula.

18. What aqueous solutions, added in sequence, would allow you to precipitate the ions $Ag^+(aq)$, $Cu^+(aq)$, and $Sr^{2+}(aq)$ from a contaminated aqueous solution one at a time for disposal as solid waste?

19. Outline an experimental design to test the effectiveness of four common antacid medications using criteria of your own choosing. The concentration of hydrochloric acid in the stomach is 0.030 mol/L, on average.

20. The presence of copper(II) ions in aqueous solution can be tested by adding an aqueous solution of sodium sulfide. If a black precipitate forms, the test is positive for the presence of copper(II) ions. What precipitate is formed? Write the net ionic equation for a positive test.

21. Use labelled sketches to compare the titration curves for
  a) a sample of a strong monoprotic acid titrated with a strong monoprotic base
  b) a sample of a strong monoprotic base titrated with a strong monoprotic acid

22. Design an experiment to measure the concentration of strong base in a liquid drain "unclogging" solution. Be sure to incorporate all the necessary steps to make the data you collect as reliable as possible. Include safety precautions. Do not carry out this investigation without the permission of your teacher.

Solving Problems

23. Phosphoric acid is produced from the reaction of "phosphate rock," $Ca_3(PO_4)_2(s)$, and sulfuric acid to produce a slurry of phosphoric acid and gypsum.

$$3H_2SO_4(aq) + Ca_3(PO_4)_2(s) + 6H_2O(\ell) \rightarrow$$
$$2H_3PO_4(aq) + 3CaSO_4 \cdot 2H_2O(s).$$

The slurry is pumped over filters that separate the acid from the gypsum. What mass of gypsum, $3CaSO_4 \cdot 2H_2O(s)$, is produced for every 1.00 kg of phosphate rock consumed?

24. At the Dow chemical plant in Fort Saskatchewan, ethane, $C_2H_6(g)$, is decomposed into ethene, $C_2H_4(g)$, and hydrogen gas at high temperature. The Ethane Cracker at Dow is the world's second largest, producing 1.1 Tg (teragram, $1\,Tg = 10^{12}$ g) of ethene per year. What mass of ethane is required to produce 1.1 Tg of ethene?

25. When a strip of lead metal is dipped in an aqueous copper(II) nitrate solution, a single-replacement reaction occurs. Predict the mass of the element produced if a 1.0 g strip of lead metal is completely consumed in an aqueous copper(II) nitrate solution.

26. A simple way to remove lime scale (mostly calcium carbonate) from the inside of a kettle is to rinse the kettle with household vinegar. The chemical equation for the net reaction that occurs is

$$CaCO_3(s) + 2CH_3COOH(aq) \rightarrow$$
$$Ca(CH_3COO)_2(aq) + CO_2(g) + H_2O(\ell)$$

Will the 12.5 g of ethanoic acid in 250 mL of household vinegar be sufficient to remove 20 g of $CaCO_3(s)$ from your kettle?

27. When aqueous solutions of lead(II) acetate and ammonium chloride are mixed, a white precipitate forms.
  a) Write the balanced chemical equation for this reaction.
  b) If 21.0 mL of 0.125 mol/L ammonium chloride reacts with excess lead(II) acetate, what mass of precipitate forms?

28. What amount of potassium chloride forms from the reaction of 2.68 L of 2.11 mol/L hydrochloric acid with 3.17 L of 2.29 mol/L potassium hydroxide?

29. Calculate the concentration of a sulfuric acid solution if a 25.0 mL sample of it requires 38.93 mL of 4.50 mmol/L potassium hydroxide for complete neutralization.

30. When 50.0 mL of 0.200 mol/L calcium nitrate are mixed with 200 mL of 0.180mol/L sodium sulfate, what is the molar concentration of sulfate ions remaining in the final solution?

31. When 25.0 mL of aqueous silver nitrate solution, $AgNO_3(aq)$, react with excess potassium chloride solution, $KCl(aq)$, a precipitate with a mass of 0.842 g is produced. Calculate the molar concentration of the original $AgNO_3(aq)$ solution.

32. In a sulfur-burning sulfuric acid plant, the net reaction for the production of the acid is

$$S_8(s) + 12O_2(g) + 8H_2O(\ell) \rightarrow 8H_2SO_4(\ell)$$

What mass of sulfuric acid is produced for every 1.00 kg of sulfur consumed?

33. The unbalanced chemical equation below describes what happens when a match is struck against a rough surface, producing a flame.

$$P_4S_3(s) + O_2(g) \rightarrow P_4O_{10}(g) + SO_2(g)$$

  a) Balance this chemical equation.
  b) If 5.3 L of oxygen gas at STP were consumed, what volume of sulfur dioxide at STP would be produced?

c) What mass of $P_4S_3(s)$ would be consumed in the same reaction described in part (b)?

34. The production of ammonia is described by the following chemical equation:

$$3CH_4(g) + 6H_2O(\ell) + 4N_2(g) \rightarrow 8NH_3(g) + 3CO_2(g)$$

a) For every 1000 kg of methane that reacts, what volume of gaseous ammonia is produced once it has been cooled to 27 °C and a pressure of 100 kPa?

b) Ammonia is stored in its liquid form at −33 °C. What mass of liquid ammonia could be produced for every 1000 kg of methane that reacts?

c) If the density of liquid ammonia is 681.91 kg/m³, at −33 °C, what volume of liquid ammonia is produced for every 1000 kg of methane that reacts?

35. Aspirin™ tablets contain the analgesic (pain-killing medication) acetylsalicylic acid or ASA ($CH_3CO_2C_6H_4CO_2H(s)$). Aspirin™ tablets can be analyzed by titration with sodium hydroxide according to the following balanced chemical equation:

$$CH_3CO_2C_6H_4CO_2H(aq) + NaOH(aq) \rightarrow$$
$$NaCH_3CO_2C_6H_4CO_2(aq) + H_2O(\ell)$$

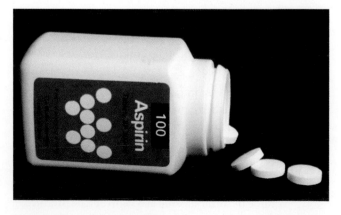

a) What mass of ASA is present in one analgesic tablet if it takes 13.4 mL of standardized 0.132 mol/L sodium hydroxide to reach an endpoint indicated by phenolphthalein?

b) With few exceptions, regular-strength analgesic tablets contain 325 mg of ASA. What is the percentage yield of ASA in your analysis relative to this mass?

c) Suggest a connection between the discrepancy in your analysis and the fact that many bottles of ASA-containing analgesic tablets have a large cotton ball in the bottle.

36. The concentration of $Ca^{2+}(aq)$ in blood plasma is 2.5 mmol/L. Calcium ions control a wide range of cellular functions, most significantly heart muscle contraction. Sources of fluoride ions, such as hydrofluoric acid, HF(aq), are especially toxic because they can precipitate calcium ions throughout the body. Deaths have occurred from concentrated HF(aq) burns to as little as 2.5% of a person's body surface area. What quantity of 1.5 mol/L commercial hydrofluoric acid could combine with all the calcium ions in 2.8 L of blood plasma?

37. The disagreeable odour of skunks is caused by a thiol. Thiols are sulfur-containing organic molecules that are structurally similar to alcohols. The odour of a butanethiol, $C_4H_9SH(\ell)$, can be deodorized with household bleach, NaOCl(aq), as follows:

$$2C_4H_9SH(\ell) + NaOCl(aq) \rightarrow$$
$$(C_4H_9)_2S_2(\ell) + NaCl(aq) + H_2O(\ell)$$

What mass of butanethiol can be deodorized with 5.00 mL of 0.0985 mol/L NaOCl(aq)?

38. One option for removing a rust stain from a piece of clothing is to soak the stain in a weak oxalic acid solution, HOOCCOOH(aq). The reaction that occurs is

$$Fe_2O_3(s) + 6HOOCCOOH(aq) \rightarrow$$
$$2Fe(OOCCOO)_3(aq) + 3H_2O(\ell) + 6H^+(aq)$$

Will 500 mL of 0.11 mol/L HOOCCOOH(aq) remove a stain caused by 1.0 g of red rust, $Fe_2O_3(s)$?

39. A 10.0 g sample of a commercial washing powder contains ammonium sulfate. When the sample of washing powder is treated with excess sodium hydroxide, ammonia gas is generated as follows:

$$2NaOH\ (aq) + (NH_4)_2SO_4(aq) \rightarrow$$
$$Na_2SO_4(aq) + 2NH_3(g) + 2H_2O(\ell)$$

The ammonia that forms in this reaction is reacted with 50.0 mL of 0.250 mol/L of sulfuric acid—an excess. The balanced equation for this second reaction is

$$2NH_3(g) + H_2SO_4(aq) \rightarrow (NH_4)_2SO_4(aq)$$

The excess sulfuric acid that remains requires 27.9 mL of 0.230 mol/L sodium hydroxide to reach a phenolphthalein endpoint. Use the information to determine what percentage of the 10.0 g of washing powder is actually $(NH_4)_2SO_4(aq)$. Assume the other ingredients are unreactive in this series of reactions.

40. Silica (also called silicon dioxide), along with other silicates, makes up approximately 95% of Earth's crust—the outermost layer of rocks and soil. Silicon dioxide is also used to manufacture transistors. Silicon reacts with hydrofluoric acid (HF) to produce silicon tetrafluoride and water vapour.

$$SiO_2(s) + 4HF(aq) \rightarrow SiF_4(g) + 2H_2O(g)$$

a) 12.2 g of SiO_2 is reacted with a small excess of HF. What is the theoretical yield, in grams, of H_2O?

b) If the actual yield of water is 2.50 g, what is the percentage yield of the reaction?

c) Assuming the yield obtained in part (b), what mass of SiF_4 is formed?

41. In the reaction of 4.36 g of sodium sulfate, $Na_2SO_4(aq)$, with excess barium chloride, $BaCl_2(s)$, 2.62 g barium sulfate, $BaSO_4(s)$, is filtered and dried.

$$BaCl_2(s) + Na_2SO_4(aq) \rightarrow BaSO_4(s) + 2NaCl(aq)$$

Determine the percentage yield of the barium sulfate?

42. Benzene, $C_6H_6(\ell)$, reacts with bromine, $Br_2(\ell)$, to form bromobenzene, $C_6H_5Br(\ell)$.

$$C_6H_6(\ell) + Br_2(\ell) \rightarrow C_6H_5Br(\ell) + HBr(g)$$

a) What is the maximum amount of C_6H_5Br that can be formed from the reaction of 7.50 g of C_6H_6 with excess Br_2?

b) A competing reaction is the formation of dibromobenzene, $C_6H_4Br_2$.
$$C_6H_6(\ell) + 2Br_2(\ell) \rightarrow C_6H_4Br_2(\ell) + 2HBr(g)$$
If 1.25 g of $C_6H_4Br_2$ was formed by the competing reaction, how much C_6H_6 was not converted to C_6H_5Br?

c) Based on your answer to part (b), what was the actual yield of C_6H_5Br? Assume that all the C_6H_5Br that formed was collected.

d) Calculate the percentage yield of C_6H_5Br.

Making Connections

43. Air bags can save lives, but they can also cause injuries. Make a list of the pros and cons of using air bags in automobiles. Do you think the benefits of air bags outweigh the potential harm? Explain your answer.

44. By-products are substances that are formed in industrial processes that are not the intended product of the process. Your friend thinks that for maximum efficiency, all by-products should just be disposed of. Do you agree or disagree? Provide examples to support your answer.

45. Scurvy is a disease that killed many early settlers during long Canadian winters. This disease is caused by a deficiency of vitamin C (ascorbic acid, $H_2C_6H_6O_6$ (s)). Today, scurvy is much less common due to the yearlong availability of citrus fruits and vitamin supplements. One good source of vitamin C is fresh orange juice. Design an experiment that would allow you to calculate the mass of vitamin C in a 250 mL glass of orange juice.
- List the manipulated, responding, and controlled variables.
- Construct a data table to record all required data.
- Draw a diagram of the necessary equipment.
- Provide a sample calculation.

UNIT 5
Thermochemical Changes

General Outcomes

- determine and interpret energy changes in chemical reactions

- explain and communicate energy changes in chemical reactions

Unit Contents

Focussing Questions

1. How does our society use the energy of chemical changes?

2. What are the impacts of energy use on society and the environment?

3. How do chemists determine how much energy will be produced or absorbed for a given chemical reaction?

Unit PreQuiz ?

www.albertachemistry.ca

For generations, dog teams like this one were essential to the Inuit way of life in the Canadian North. Capable of pulling heavy loads for long distances, sled dogs were vital for transportation and hunting over snow and ice. Although such dogs as these can survive for a long time without food, like all animals, they get their energy from their food. The chemical reactions related to digestion break down protein and fat molecules from meat, releasing the energy that keeps the dogs warm and able to run in the bitter cold. Energy not used immediately is stored as fat, to be converted back to energy by chemical reactions when needed.

Today, the snowmobile has largely replaced sled dogs as a way for people in Northern Canada to travel and to hunt. Like sled dogs, snowmobiles require energy, but snowmobiles use gasoline, not food, as the source of energy-releasing chemical reactions. Snowmobiles and other vehicles that run on fossil fuels have improved life in some ways for the Inuit and other people around the world by offering fast, comfortable transportation. That improvement, however, has come at a cost: the burning of fossil fuels contributes to environmental problems, such as acid precipitation and the greenhouse effect.

How do chemical reactions release energy? What are the similarities and differences between the breakdown of food in the body and the combustion of gasoline in an engine? What are the issues related to the use of fossil fuel? In this unit, you will find out about the science and technology of the energy changes caused by chemical reactions and how they affect society.

There are many ways to describe the contents and the processes occurring in test tubes, flasks, and beakers such as the ones in Figure P5.1. A review of some of the methods you learned in previous science courses will help you to understand the new concepts that you will learn in this unit.

Describing Chemical Reactions

Possibly the first thing you think of when you read "chemical reaction" is a balanced equation such as those in Figure P5.2. A balanced equation gives you many details about chemical reactions. Some of these details are listed here.

- The formulas of each reactant and product tell you exactly how many atoms of each element comprise one molecule or formula unit of the compound.

- The symbol in brackets tells you the state of the compound when it is undergoing the reaction.

- The coefficients of each reactant and product tell you the mole ratios of the compounds in the reaction. For example, in the second equation you know that one mole of ethene gas ($C_2H_4(g)$) reacts with three moles of oxygen gas to produce two moles of carbon dioxide gas and two moles of water vapour.

Figure P5.1 When you add chemicals to a water solution, chemical reactions might occur. In some cases, the solution becomes warmer or cooler. How do you analyze and describe the processes occurring in the beakers?

Figure P5.2 You will find a lot of information about a chemical reaction in a balanced equation.

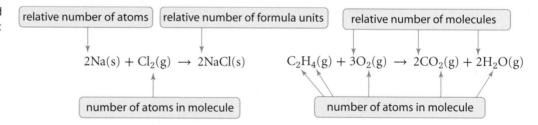

Another way to describe a chemical reaction involves energy. As you know, when some reactions occur in a test tube, the test tube feels warmer than it did before you added the reactants. The reaction generated heat thus raising the temperature of the solution. Such a reaction is called *exothermic*. Conversely, some reactions make the solution cooler by absorbing heat. These reactions are called *endothermic*.

Thermodynamics

The mention of thermal energy leads to the topic of thermodynamics, the movement of thermal energy or heat. You have probably heard and read the terms "system" and "surroundings" in relation to the study of thermodynamics. To describe the thermodynamics of a solution in a beaker or flask, you first need to define exactly what "system" you are studying. For a process occurring in a solution, the solution itself is usually defined as the system. Everything else in the universe including the container is considered the surroundings. To properly analyze a system, you need to classify it among the following three categories. These categories are illustrated in Figure P5.3.

- *Open system:* Both matter and energy are allowed to enter or leave the system.

- *Closed system:* Energy can enter or leave the system but matter cannot.

- *Isolated system:* Neither matter nor energy is allowed to enter or leave the system.

open system   closed system

isolated system

Figure P5.3 An open pot of potatoes boiling on the stove represents an open system, because heat is entering the pot and water vapour along with heat are leaving the system. A pressure cooker prevents any matter from escaping but heat is entering, so the pressure cooker represents a closed system. If the pot is placed inside a perfect insulator, neither heat nor water can enter or leave the system, making it an isolated system.

Many chemists and physicists will say that no system is truly isolated except the universe itself. However, a well-insulated system can often be considered to be isolated because the amount of heat entering or leaving the system is so small that it would not affect any analysis or calculations involving the process occurring in the system.

The laws of thermodynamics govern the processes involving systems. The first law of thermodynamics states that:

The total energy of the universe is constant.

You will often hear the law stated, "Energy cannot be created or destroyed but can be transformed from one form to another or transferred from one object to another." This law tells you that if heat leaves your system, the same amount of heat must enter the surroundings. Since you can measure the heat that enters the surroundings, you can determine exactly how much heat left your system.

How can you be sure that the heat generated in the system will actually leave the system? Why doesn't it just stay in the system? The first law of thermodynamics does not say that thermal energy must leave a system, even if it is a closed system that allows heat to leave. The movement of heat is dictated by the second law of thermodynamics which states:

In the absence of energy input, a system becomes more disordered.

How does disorder relate to the movement of thermal energy or heat? Imagine a system that consists of two objects in thermal contact (objects can exchange energy) but at different temperatures. Recall that temperature is a measure of the average kinetic energy of the particles (atoms or molecules) of a system. This system can be considered orderly because particles having a higher kinetic energy are on one side of the system and particles having a lower kinetic energy are on the other side. Imagine putting red marbles and white marbles in a box and shaking the box. When you open the box would you expect it to be orderly, that is, would you expect the red marbles to be on one side and the white ones on the other side? Common sense says you would not. The second law of thermodynamics says you could not expect that. Therefore, when two objects are in thermal contact and are at different temperatures, the thermal energy must be transferred from the system at a higher temperature to the one at a lower temperature so that the system becomes disorderly. In this case, disorderly means that particles having different kinetic energies are completely mixed.

In relation to a system that generated thermal energy, if it is a closed system, the second law of thermodynamics says that the thermal energy must leave the system until the system and its surroundings are at the same temperature.

Measuring Thermal Energy Changes

You cannot measure thermal energy directly, but you can measure temperature. As you know, temperature is a measure of the average kinetic energy of the particles of a substance. A change in temperature indicates that energy has entered or left a substance. However, the amount of energy required to change the temperature of different substances by the same amount varies with the substance. For example, 4.19 J of energy are required to change the temperature of one gram of water by one degree Celsius while only 0.897 J of energy are required to change the temperature of one gram of aluminium by one degree Celsius. Figure P5.4 shows you another example.

You probably recall that the amount of energy necessary to increase one gram of a substance by one degree Celsius is called the *specific heat capacity* of a substance. The specific heat capacities of some common substances are listed in Table P5.1. Notice, in the table, that the state of the substance affects its specific heat capacity. For example, the specific heat capacity of liquid ammonia is 4.70 J/g • °C whereas for gaseous ammonia it is 2.06 J/g • °C.

Figure P5.4 This wooden bench is comfortably warm to sit on in the warm Sun. However, an iron bench would be uncomfortably hot. The specific heat capacity of wood is significantly larger than that of iron.

Table P5.1 Specific Heat Capacities of Some Common Substances

Substance	Specific heat capacity (J/g • °C at SATP)
Elements	
aluminium	0.897
carbon (graphite)	0.709
copper	0.385
gold	0.129
hydrogen	14.304
iron	0.449
Compounds	
ammonia (liquid)	4.70
ammonia (gas)	2.06
ethanol	2.44
water (solid)	2.00
water (liquid)	4.19
water (gas)	2.02
Other materials	
air	1.01
concrete	0.88
glass	0.84
granite	0.79
wood	1.26

In addition to the specific heat capacity and the temperature change of a substance, you also need to know the mass of the substance before you can calculate the amount of heat entering or leaving a system. Knowing these three values, you can calculate the amount of heat absorbed or released by a substance by using the formula shown in Figure P5.5.

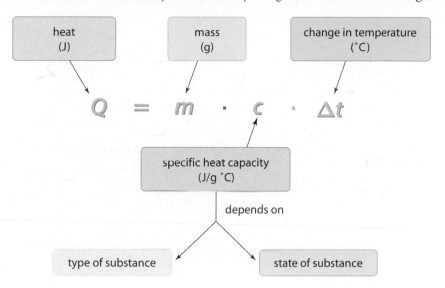

Figure P5.5 This formula allows you to calculate the amount of heat absorbed or released by a substance.

The temperature change, Δt, in the formula is always defined as the final temperature minus the initial temperature.

$$\Delta t = t_f - t_i$$

If the temperature change is negative, the temperature decreased. As you can see, a negative Δt will make Q negative, indicating that heat left the system. Therefore, the equation not only tells you how much heat was exchanged between a system and its surroundings, it also tells you the direction of the heat transfer. Renew your skills for using this formula by completing the following Practice Problems.

Prerequisite Skills

- **constructing** data tables

- **drawing** and **interpreting** graphs

- **using** appropriate tools and techniques to gather experimental data

Practice Problems

1. 100 g of ethanol at 25 °C is heated until it reaches 50 °C. How much thermal energy does the ethanol gain?

2. Beaker A contains 50 g of liquid at room temperature. The beaker is heated until the liquid increases in temperature by 10 °C. Beaker B contains 100 g of the same liquid at room temperature. The beaker is also heated until the liquid increases in temperature by 10 °C. In which beaker does the liquid absorb the most heat? Explain.

3. How much heat is released when the temperature of 789 g of liquid ammonia decreases from 82.7 °C to 25.0 °C?

4. A solid substance has a mass of 250.00 g. It is cooled by 25.00 °C and loses 4.937 kJ of heat. What is the specific heat capacity? Identify the substance using the values in Table P5.1.

Energy and Chemical Reactions

Chapter Concepts

9.1 Enthalpy and Thermochemical Equations

- Energy stored in the chemical bonds of hydrocarbons originated from the Sun.

- Enthalpy change (ΔH) refers to the potential energy change of a system during a process.

- Enthalpy changes can be communicated in chemical equations using ΔH notation and potential energy diagrams.

- Chemical reactions can be classified as endothermic or exothermic.

9.2 The Technology of Energy Measurement

- Calorimetery is used to determine enthalpy changes in chemical reactions.

- A calorimeter works on the principle that the heat lost (or gained) by the system is equal to the heat gained (or lost) by the surroundings.

Every May, runners meet at the Banff Recreational Centre for a 136.2 km relay race from Banff to the Eau Clair Market in Calgary. To fuel their bodies to run long distances, and to help their bodies recover afterward, runners eat before, during, and after the race. Whether competing in a race or just walking to school, your body has a lot in common with a coal-burning power plant or the engine in your car. It needs fuel to keep going.

Hydrocarbon combustion, the chemical reaction that takes place in a car's engine, is in some ways quite similar to cellular respiration, the chemical reaction that takes place in our cells. Both processes "burn" fuel (either gasoline or food) in the presence of oxygen by breaking and forming chemical bonds, and both processes release energy. In this chapter, you will learn more about the energy involved in chemical reactions and how to measure it. In the Launch Lab that follows, you will create a hot pack or cold pack, which could come in handy when the race is over!

Launch Lab

Hot Packs and Cold Packs

A cold pack helps reduce swelling in sore muscles after a long-distance race. Commercially available hot and cold packs are useful for first-aid treatment when ice and hot water are not readily available. You can make simple hot or cold packs using only a soluble salt (ionic compound), water, and a suitable container. In this Launch Lab, you and your group will recall what you know about endothermic and exothermic processes to design a hot or cold pack using only these materials.

Safety Precautions

- Some salts are toxic and corrosive. If you spill any on your skin, immediately rinse with plenty of water.
- Large amounts of heat may be generated if large quantities of salts are used.
- Dispose of your solutions as instructed by your teacher.
- Your teacher may choose to carry out this investigation as a demonstration.

Materials

- a variety of soluble ionic compounds
- water
- common household containers and resealable bags
- balance

Procedure

1. With your group, come up with an initial design for your hot or cold pack that will allow you to scientifically test the salts for their effectiveness at absorbing or releasing heat.

2. Construct a data table to record appropriate data.

3. Choosing one of the salts, test your design.

4. Refine your design if necessary.

5. Using your refined design, test one of the salts for suitability as a hot or cold pack.

6. Repeat Procedure step 5 for the other available salts.

7. Dispose of the waste as directed by your teacher.

Analysis

1. Based on your data, which salt would make the best hot pack? Why? Which would make the best cold pack? Why? Use the terms endothermic or exothermic in your answers.

2. Are there any safety concerns you should consider? Explain.

3. If you were to prepare these hot or cold packs commercially, what design changes would you like to make? Explain.

Edible gels are transportable and easily digestible forms of concentrated energy—perfect for runners who need to refuel but don't want to slow down.

Enthalpy and Thermochemical Equations

All chemical reactions are accompanied by changes in energy. These energy changes are crucial to life on Earth. For example, through photosynthesis green plants and blue-green algae absorb energy from sunlight and store it in the chemical bonds of the carbohydrates they produce. Animals that eat the plants use the stored energy through the process of cellular respiration. This energy generates heat to help regulate body temperature and fuels growth and movement. After plants and animals die, their remains—over millions of years—are eventually transformed into the hydrocarbons that make up the fossil fuels that provide energy to power today's society. See Figure 9.1 below.

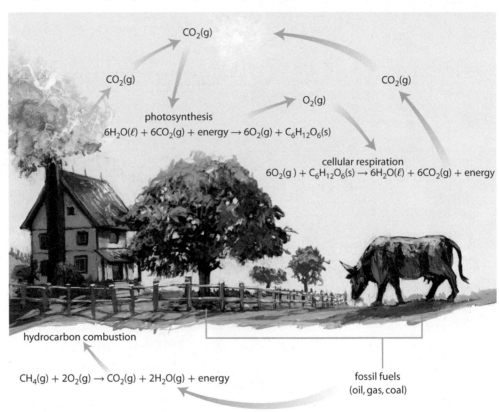

$CO_2(g)$

$CO_2(g)$

$CO_2(g)$

$O_2(g)$

photosynthesis
$6H_2O(\ell) + 6CO_2(g) + energy \rightarrow 6O_2(g) + C_6H_{12}O_6(s)$

cellular respiration
$6O_2(g) + C_6H_{12}O_6(s) \rightarrow 6H_2O(\ell) + 6CO_2(g) + energy$

hydrocarbon combustion

$CH_4(g) + 2O_2(g) \rightarrow CO_2(g) + 2H_2O(g) + energy$

fossil fuels
(oil, gas, coal)

Figure 9.1 Sunlight is the original source of energy for the formation of fossil fuels.

Studying Energy Changes

The study of energy transfer and energy changes is known as **thermodynamics**. Chemists are interested in a branch of thermodynamics called **thermochemistry**—the study of the energy changes involved in physical and chemical changes. The discussion of energy and its interconversions involves several terms and definitions that you will learn while studying the next few pages. Then you will examine the energy changes that accompany chemical reactions.

The law of conservation of energy states that the total energy of the universe is constant. In other words, energy can be neither destroyed nor created. This idea is expressed by the following equation where $E_{universe}$ is the total energy of the universe.

$$\Delta E_{universe} = 0$$

Energy can, however, be transformed from one form to another or transferred from one substance to another. Chemists study chemical and physical changes; thus thermochemists study energy transfers that accompany chemical and physical processes. To analyze such a process, the chemist must carefully describe the process involved in the study. For example, chemical reactions often occur in solutions in laboratory glassware. Chemists would call the solution the **system** and everything else in the entire universe would be the **surroundings**. In a practical sense, the surroundings may be considered as anything that might exchange energy with the system such as the flask and the surrounding air, as illustrated in Figure 9.2. Now you can apply the conservation of energy to a system. First, you can express the definition of the system and surroundings as:

$$\text{Universe} = \text{System} + \text{Surroundings}$$

The law of conservation of energy says that:

$$\Delta E_{\text{universe}} = 0$$

You know that the energy of the system can change because heat can leave the system—the solution—and enter the surroundings—the flask and air. Given that the system and the surroundings comprise the universe, you can now state:

$$\Delta E_{\text{universe}} = \Delta E_{\text{system}} + \Delta E_{\text{surroundings}} = 0$$

or

$$\Delta E_{\text{system}} = -\Delta E_{\text{surroundings}}$$

The relationship in the box is known as the **first law of thermodynamics**. It tells you that any energy that enters the system must come from the surroundings or any energy that leaves the system must enter the surroundings. Therefore, if a chemist can measure the energy that enters the surroundings, the chemist knows that the same amount of energy was lost by the system being studied. Thermochemists use this concept to great advantage.

Figure 9.2 The solution in the flask is the system. The flask, the laboratory, and the student are the surroundings.

Types of Energy

You may recall from earlier science courses that energy can be classified into two fundamental types:

- **kinetic energy (E_k):** the energy of motion (in chemistry, this usually means the energy of the motion of particles or thermal energy)

- **potential energy (E_p):** energy that is stored (in chemistry, potential energy is usually energy stored in chemical bonds)

In chemistry, energy changes involved in chemical and physical processes fit into one or both of these two energy categories. The SI-derived unit for both kinetic and potential energy is the joule (symbol J).

Heat Transfer and Temperature Change

The absolute, or Kelvin, temperature, T, is a measure of the average kinetic energy of the particles that make up a substance or system. A change in temperature signals a change in kinetic energy. The temperature variable used is the change in temperature, ΔT ("delta tee"). The size of a temperature unit in the Kelvin and Celsius systems is the same so the *change* in the temperature is the same in both systems, or $\Delta T = \Delta t$. Because you often take measurements and report temperature in Celsius, you will usually use Δt when working with a temperature change.

A positive value for Δt indicates an increase in temperature. A negative value for Δt indicates a decrease in temperature.

Chemistry File

FYI
The joule is named for James Prescott Joule (1818–1889). He studied heat and its relationship to work, which led to the development of the law of conservation of energy. The joule is derived from other SI units: one joule is equal to $1\,\dfrac{\text{kg} \cdot \text{m}^2}{\text{s}^2}$. 1000 J, or 1 kJ, is enough energy to melt about 3 g of ice.

Transfer of Kinetic Energy

Heat, Q, refers to the transfer of thermal (kinetic) energy between objects with different temperatures. Heat, therefore, has the same units as energy—joules (J) or kilojoules (kJ).

According to the particle model of matter, matter is made up of particles that are constantly in motion. When a substance absorbs thermal energy as heat, the average speed of the particles in the object increases. Therefore, the temperature of the substance increases. Figure 9.3 models what happens when hot chocolate is heated in a pot on a stove.

Figure 9.3 The length of each arrow represents the speed of the particle. As the hot chocolate absorbs heat, the average speed of the particles increases.

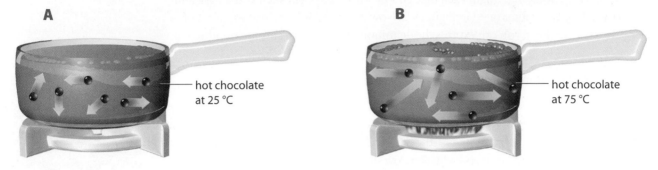

A

hot chocolate at 25 °C

B

hot chocolate at 75 °C

When substances with different temperatures come into contact, kinetic energy is transferred as heat from the particles of the warmer substance to the particles of the cooler substance.

For example, if you hold a mug of hot chocolate, it feels warm because the particles in the hot chocolate are transferring thermal energy to the mug, and the mug is transferring thermal energy to your hand. (See Figure 9.4.)

Figure 9.4 As the average kinetic energy of the particles in the mug and your hand increases, the average kinetic energy of the hot chocolate decreases until they reach the same temperature.

• • •

1. Why is it a bad idea to leave the front door open on a hot day when the air conditioning is on in your home?

2. Why do you feel cold after you touch an object that has a lower temperature than your skin, such as a concrete wall on a winter day?

• • •

Energy Changes in Chemical Reactions

Thermal energy is released or absorbed by chemical reactions when chemical bonds are broken and new ones are formed. Chemical bonds are sources of stored potential energy. Recall that *breaking a bond is a process that requires energy, while creating a bond is a process that releases energy*. Figure 9.5 represents these ideas. Bond breakage and formation is shown here in two steps for clarity. In a real chemical reaction, however, bond breakage and formation occur simultaneously.

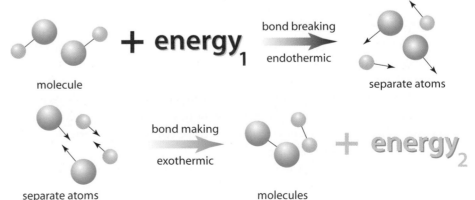

Figure 9.5 This illustration shows bonds being broken and formed during a chemical reaction.

Recall that when a reaction results in a net absorption of energy, it is called an **endothermic reaction**. When a reaction results in a net release of energy, it is called an **exothermic reaction**. In Figure 9.5, if "energy$_1$" is greater than "energy$_2$", the reaction will be endothermic. If "energy$_2$" is greater than "energy$_1$," the reaction will be exothermic. Figure 9.6 illustrates the concepts.

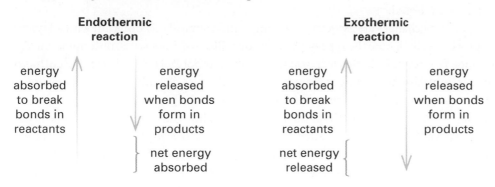

Endothermic reaction

energy absorbed to break bonds in reactants

energy released when bonds form in products

net energy absorbed

Exothermic reaction

energy absorbed to break bonds in reactants

energy released when bonds form in products

net energy released

Figure 9.6 The difference between the amount of energy needed to break bonds and the energy released when bonds form determines whether a reaction is endothermic or exothermic.

Consider for example, the reaction that takes place when nitrogen reacts with oxygen:

$$N_2(g) + O_2(g) \rightarrow 2NO(g)$$

In this reaction, one mole of nitrogen–nitrogen triple bonds and one mole of oxygen–oxygen double bonds are broken. Two moles of nitrogen–oxygen bonds are formed. Because more energy is required to break one nitrogen–nitrogen bond and one oxygen–oxygen bond than is released when two nitrogen–oxygen bonds are formed, the overall process absorbs energy. This is an endothermic reaction.

Energy and Enthalpy

The total internal energy of a system is the sum of all of the forms of potential energy and kinetic energy of the atoms and molecules of the system. It is not possible to determine the total internal energy of a system. In fact, it is often very difficult to determine the change in the internal energy of a system during a process. Chemists have defined another property of systems that is related to energy but for which it is easier to measure changes. That property is **enthalpy**, *H*. As is the case with energy, it is not possible to determine the absolute enthalpy of a system. However, when a process takes place under conditions of constant pressure, the **enthalpy change**, ΔH, is equal to the heat, *Q*, absorbed or released by a system. The enthalpy change of a chemical reaction is directly related to the change in the potential energy of the components of the system—the reactants and products.

Enthalpy change is a measure of *relative* enthalpy. This is like saying that the distance between your home and your school is 2 km, instead of stating their positions in terms of their latitude, longitude, and elevation. You usually talk about the relative positions of two locations, not their absolute positions. Enthalpy changes may be expressed as either an enthalpy of reaction (in kJ) or as a molar enthalpy (in kJ/mol). **Molar enthalpy** is the change in enthalpy when one mole of the substance undergoes a process. Both enthalpy and molar enthalpy changes are represented by the symbol ΔH. Therefore, when you see or use the symbol ΔH, you need to pay close attention to the context. The units "kJ" always indicate an enthalpy change and the units "kJ/mol" always indicate a molar enthalpy change. In addition, when discussing the enthalpy change of a reaction, you might read the phrase "the enthalpy change of the reaction *as written* is −802.5 kJ." (The negative value of the enthalpy change means that thermal energy is released.) Then you will see an equation such as:

$$CH_4(g) + 2O_2(g) \rightarrow CO_2(g) + 2H_2O(g)$$

Chemistry File

FYI
The biggest explosion created by humans before the development of the atomic bomb was caused by exothermic reactions that occurred on December 6, 1917, in Halifax, Nova Scotia. The reactions occurred in the harbour when two ships collided. One ship was carrying 300 tons of picric acid (2,4,6-trinitrophenol), 200 tons of TNT (trinitrotoluene), 10 tons of gun cotton (nitrocellulose), and 35 tons of benzol. The blast killed more than 1900 people. Its effects were later studied by J. Robert Oppenheimer, one of the scientists involved in developing the atomic bombs dropped on Hiroshima and Nagasaki.

The phrase, "as written," means that the coefficients in the written equation represent the number of moles of each substance that participate in the reaction which releases 802.5 kJ of energy. The negative value of the enthalpy change means that thermal energy is released. That is to say, when one mole of methane ($CH_4(g)$) reacts with two moles of oxygen and produces one mole of carbon dioxide and two moles of water, 802.5 kJ of energy are released.

Enthalpy changes are associated not only with chemical reactions but with all types of processes. To denote the type of process, you will sometimes see a subscript in the ΔH symbol. For example, in this text, you will see the symbol $\Delta_r H$ to represent the **enthalpy of reaction**. Also, as you might suspect, enthalpy changes depend on the temperature and pressure under which a process occurs. As long as the pressure does not change during a chemical reaction, the enthalpy change will be equal in magnitude to the heat exchanged between the system and the surroundings. However, the enthalpy change will be different at different pressures. Therefore, chemists often talk about the **standard enthalpy of reaction, $\Delta_r H°$**—the enthalpy change of a chemical reaction that occurs at SATP (25 °C and 100 kPa). The superscript "°" indicates SATP conditions.

Enthalpy Changes of Exothermic Reactions

In Alberta, people commonly burn natural gas to heat their homes or as fuel for gas stoves. Natural gas is mostly methane, $CH_4(g)$. Methane is also the main gas found in "firedamp," the dangerous mixture of gases released by coal mining (see Figure 9.7). As you read above, when one mole of methane burns in an excess of oxygen, carbon dioxide gas and water vapour are produced along with 802.5 kJ of energy. The combustion of methane is an example of hydrocarbon combustion. How can you represent the enthalpy change for exothermic reactions, such as hydrocarbon combustion reactions?

Figure 9.7 In the early days of coal mining, canaries were used to detect methane. If the canary died, the miners would know the methane was at a dangerous level, and they would leave. Today, methanometers, which continually record and display methane levels, help keep miners safe.

The simplest way to represent an exothermic reaction is to use a **thermochemical equation**—a balanced equation that includes the change in enthalpy for the reaction it represents. The thermochemical equation for the complete combustion of one mole of methane is shown below:

$$CH_4(g) + 2O_2(g) \rightarrow CO_2(g) + 2H_2O(g) + 802.5 \text{ kJ}$$

Notice that the enthalpy term is included with the products because thermal energy is released in the reaction. You can also indicate the enthalpy of reaction using ΔH **notation,** in which the enthalpy term is written as a separate expression beside the chemical equation. For exothermic reactions, ΔH is always negative:

$$CH_4(g) + 2O_2(g) \rightarrow CO_2(g) + 2H_2O(g) \quad \Delta_r H = -802.5 \text{ kJ}$$

A final way to represent the enthalpy of reaction is to use a simple **potential energy diagram**. A potential energy diagram is a graphical representation of the change in enthalpy during a chemical or physical change. Examine Figure 9.8 to see how this is done.

Enthalpy Change for Combustion of Methane

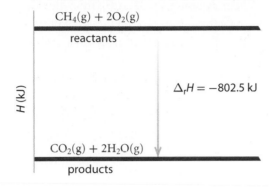

Figure 9.8 In an exothermic reaction, the enthalpy of the system decreases as energy is released to the surroundings.

Enthalpy Changes of Endothermic Reactions

Photosynthesis is the endothermic process by which green plants absorb energy from the Sun and produce glucose and oxygen. For each mole of glucose produced, 2802.5 kJ of energy from the Sun are absorbed. As with exothermic reactions, there are three different ways to represent the enthalpy change of an endothermic reaction.

You can include the enthalpy of reaction as an enthalpy term in the chemical equation. Because energy is absorbed in an endothermic reaction, the enthalpy term is included on the reactant side of the equation:

$$6H_2O(\ell) + 6CO_2(g) + 2802.5 \text{ kJ} \rightarrow 6O_2(g) + C_6H_{12}O_6(s)$$

You can also indicate the enthalpy of reaction as a separate expression beside the chemical reaction. For endothermic reactions, the enthalpy of reaction is always positive:

$$6H_2O(\ell) + 6CO_2(g) \rightarrow 6O_2(g) + C_6H_{12}O_6(s) \qquad \Delta_r H = 2802.5 \text{ kJ}$$

Finally, you can use a diagram to show the enthalpy of reaction. Figure 9.9 shows how the process of photosynthesis can be represented graphically. (The photosynthesis equation represents a long series of reactions rather than a single reaction.)

Enthalpy Change For Photosynthesis

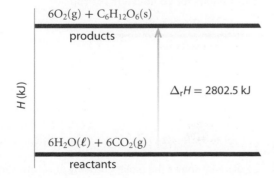

Figure 9.9 In an endothermic reaction, the enthalpy of the system increases as energy is absorbed from the surroundings.

Figure 9.10 Rock climbers use magnesium carbonate, "chalk," to absorb sweat and prevent their hands from becoming slippery. Magnesium carbonate is used in a similar way by gymnasts and weightlifters.

Figure 9.11 A gas barbecue is an open system because the gaseous products are allowed to escape, and thermal energy is transferred to the surrounding air. The reaction occurs at constant pressure.

• • •

3 When methane is formed in the chemical reaction shown below at standard conditions, 74.6 kJ of energy is produced:

$$C(s) + 2H_2(g) \rightarrow CH_4(g)$$

Express the energy change:
  a) as part of a thermochemical equation
  b) using ΔH notation
  c) as an enthalpy, or potential energy, diagram

4 The endothermic decomposition of solid magnesium carbonate (see Figure 9.10) produces solid magnesium oxide and carbon dioxide gas. For each mole of magnesium carbonate that decomposes, 117.3 kJ of thermal energy is absorbed. Write a balanced chemical equation and express the enthalpy change as a term in the equation:
  a) as a reactant   **b)** as a separate expression
  c) as an enthalpy, or potential energy, diagram

• • •

Molar Enthalpy of Combustion

The molar enthalpy of combustion, $\Delta_c H$, is the enthalpy change associated with the complete combustion of *one mole* of a given substance. The change in enthalpy is measured for the products and reactants as they would be found under the combustion conditions. There are two sets of conditions in which combustion occurs. The most common set of conditions are described as an **open system**, in which the products of combustion (thermal energy and gases) are allowed to escape, as in a car engine or a barbecue, shown in Figure 9.11. Combustion can also take place in an **isolated system**, such as a *bomb calorimeter*, in which the products of combustion (both substances and thermal energy) are contained. The enthalpy change associated with combustion in an open system and a closed system differs slightly. In an open system, the process occurs at constant pressure. Therefore, the enthalpy change is equal to the heat exchanged between the system and the surroundings. In a closed system, the pressure can change and the heat cannot be used to determine the enthalpy change. For example, in the combustion of propane, $C_3H_8(g)$, a common barbecue fuel, $\Delta_c H = -2043.9$ kJ/mol in an open system. You can represent the molar enthalpy of combustion using a thermochemical equation, using ΔH notation, or using an enthalpy diagram (Figure 9.12):

$$C_3H_8(g) + 5O_2(g) \rightarrow 3CO_2(g) + 4H_2O(g) \qquad \Delta_c H = -2043.9 \text{ kJ/mol } C_3H_8(g)$$

Molar Enthalpy Change for Combustion of Propane

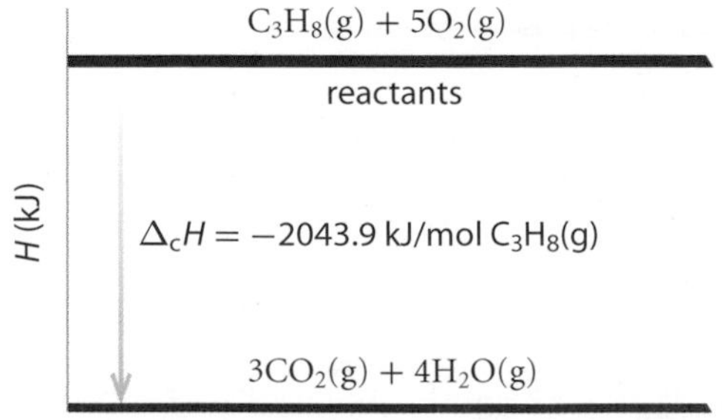

$C_3H_8(g) + 5O_2(g)$

reactants

H (kJ)

$\Delta_c H = -2043.9$ kJ/mol $C_3H_8(g)$

$3CO_2(g) + 4H_2O(g)$

Figure 9.12 This enthalpy diagram shows the molar enthalpy of combustion of propane in an open system.

Cellular respiration, which is another important series of reactions that are sometimes classified as combustion reactions, take place in the cells of all living things. The chemical equation for cellular respiration is shown below:

$$C_6H_{12}O_6(s) + 6O_2(g) \rightarrow 6CO_2(g) + 6\,H_2O(\ell) \qquad \Delta_cH = -2802.5 \text{ kJ/mol glucose}$$

Although this is a combustion reaction as written, the water that is produced is in liquid form because the reactions are taking place at body temperature.

Table 9.1 lists selected molar enthalpies of combustion in an open system for the first eight straight-chain alkanes. Alkanes are compounds that contain hydrogen and carbon and in which all carbon–carbon bonds are single bonds.

Table 9.1 Molar Enthalpies of Combustion for Alkanes in an Open System*

Alkane name	Formula	Δ_cH (kJ/mol)
methane	$CH_4(g)$	−802.5
ethane	$C_2H_6(g)$	−1428.4
propane	$C_3H_8(g)$	−2043.9
butane	$C_4H_{10}(g)$	−2657.3
pentane	$C_5H_{12}(\ell)$	−3244.8
hexane	$C_6H_{14}(\ell)$	−3854.9
heptane	$C_7H_{16}(\ell)$	−4464.7
octane	$C_8H_{18}(\ell)$	−5074.1

*products are $H_2O(g)$ and $CO_2(g)$

• • •

5. Write a balanced thermochemical equation using ΔH notation to represent the molar enthalpy of combustion in an open system of each of the following alkanes (see Table 9.1). Recall that water is produced as a vapour and that all reactions must be balanced to show the complete combustion of one mole of fuel.

   a) pentane   **c)** hexane

   b) butane   **d)** octane

6. Draw a potential energy diagram to represent the molar enthalpy of combustion of heptane in an open system.

• • •

Reactant Amounts and Enthalpy of Reaction

You know intuitively that a roaring bonfire releases more thermal energy than does a burning matchstick. The reaction is the same, but the quantities involved are different. The enthalpy change associated with a reaction depends on the amounts of the reactants involved. For example, the thermochemical equation for photosynthesis indicates that 2802.5 kJ of energy is absorbed when one mole, or 180.18 g, of glucose is formed. The formation of two moles of glucose requires twice as much energy, or 5605.0 kJ:

$$6H_2O(\ell) + 6CO_2(g) + 2802.5 \text{ kJ} \rightarrow 6O_2(g) + C_6H_{12}O_6(s)$$
$$12H_2O(\ell) + 12CO_2(g) + 5605.0 \text{ kJ} \rightarrow 12O_2(g) + 2C_6H_{12}O_6(s)$$

Enthalpy of reaction is *directly proportional* to the amounts of the substances that react. That is, if the amounts of the reactants double, the enthalpy change also doubles. In other words, when you multiply the stoichiometric coefficients of a thermochemical equation by any factor, you must multiply the enthalpy term by the same factor.

Chemistry File

Web Link
Although biological processes, such as cellular respiration and photosynthesis, are written as single reactions, they are, in fact, composed of numerous steps. Find out more about the multiple steps and chemical reactions they involve. Why is it important that these reactions occur over multiple steps? What would happen if cellular respiration occurred all in one step?

www.albertachemistry.ca
www

Calculating Enthalpy Changes

If you know the thermochemical equation for a reaction, you can multiply the enthalpy term and the appropriate stoichiometric factor to predict the enthalpy change associated with the reaction for any amounts of products or reactants.

The enthalpy change for a reaction is equal to the number of moles of a specified reactant or product times the molar enthalpy change for the specified reactant or product.

$$\Delta H = n\Delta_r H$$

Similarly, if you know the enthalpy change of a reaction, you can use it to determine the amounts of the reactants or products involved as shown below.

Sample Problem

Predicting an Enthalpy Change

Problem
a) What is the enthalpy change when 50.00 g of methane undergoes complete combustion in an open system, according to the equation below?

$$CH_4(g) + 2O_2(g) \rightarrow CO_2(g) + 2H_2O(g)$$

b) What is the enthalpy of the reaction per mole of O_2, $CO_2(g)$, and $H_2O(g)$?

What Is Required?
You need to determine the enthalpy change, ΔH, for the combustion reaction that consumes 50.00 g of methane.

What Is Given?
Mass of methane $(m) = 50.00$ g

$$CH_4(g) + 2O_2(g) \rightarrow CO_2(g) + 2H_2O(g)$$

$\Delta_c H = -802.5$ kJ/mol $CH_4(g)$ (Table 9.1)

Plan Your Strategy
a) Determine the amount of methane using its mass and its molar mass. Multiply the molar enthalpy of combustion by the amount of methane to determine the enthalpy change associated with the combustion of 50.00 g of methane:

$$\Delta H = n\Delta_c H$$

b) Use the mole ratio of $O_2(g)$, $CO_2(g)$, and $H_2O(g)$ to $CH_4(g)$ to find the enthalpy of reaction per mole of the three compounds.

Act on Your Strategy
a) $n = \dfrac{m}{M}$

$$= \frac{50.00 \text{ g } CH_4(g)}{16.05 \text{ g/mol}}$$

$$= 3.115 \text{ mol } CH_4(g)$$

$$\Delta H = n\Delta_c H$$

$$= (3.115 \text{ mol } CH_4(g))\left(\frac{-802.5 \text{ kJ}}{1 \text{ mol } CH_4(g)}\right)$$

$$= -2.500 \times 10^3 \text{ kJ}$$

When 50.00 g of methane undergoes complete combustion, the enthalpy change is -2.500×10^3 kJ. If you assume the enthalpy change is released as thermal energy, you would predict that -2.500×10^3 kJ of heat would be produced.

b) $\Delta H(\text{per mol } O_2) = \left(\dfrac{\text{mol } CH_4}{\text{mol } O_2}\right)\left(\dfrac{\Delta H}{\text{mol } CH_4}\right)$

$$\Delta H(\text{per mol } O_2) = \left(\frac{1 \text{ mol } CH_4}{2 \text{ mol } O_2}\right)\left(\frac{-802.5 \text{ kJ}}{\text{mol } CH_4}\right)$$

$$\Delta H(\text{per mol } O_2) = -401.2 \frac{\text{kJ}}{\text{mol } O_2}$$

$$\Delta H(\text{per mol } CO_2) = \left(\frac{\text{mol } CH_4}{\text{mol } CO_2}\right)\left(\frac{\Delta H}{\text{mol } CH_4}\right)$$

$$\Delta H(\text{per mol } CO_2) = \left(\frac{1 \text{ mol } CH_4}{1 \text{ mol } CO_2}\right)\left(\frac{-802.5 \text{ kJ}}{\text{mol } CH_4}\right)$$

$$\Delta H(\text{per mol } CO_2) = -802.5 \frac{\text{kJ}}{\text{mol } CO_2}$$

$$\Delta H(\text{per mol } H_2O) = \left(\frac{\text{mol } CH_4}{\text{mol } H_2O}\right)\left(\frac{\Delta H}{\text{mol } CH_4}\right)$$

$$\Delta H(\text{per mol } H_2O) = \left(\frac{1 \text{ mol } CH_4}{2 \text{ mol } H_2O}\right)\left(\frac{-802.5 \text{ kJ}}{\text{mol } CH_4}\right)$$

$$\Delta H(\text{per mol } H_2O) = -401.2 \frac{\text{kJ}}{\text{mol } H_2O}$$

Check Your Solution
The units are correct. The sign of the enthalpy change is negative, which makes sense because a hydrocarbon combustion reaction is always exothermic. The enthalpy change for the reaction is about three times as great as the molar enthalpy of combustion, which makes sense because about 3 mol of methane was burned.

Using Dimensional Analysis

$$\Delta H = 50.00 \text{ g } CH_4(g) \times \frac{1 \text{ mol } CH_4(g)}{16.05 \text{ g } CH_4(g)} \times \frac{-802.5 \text{ kJ}}{1 \text{ mol } CH_4(g)}$$

$$= -2.500 \times 10^3 \text{ kJ}$$

Using Enthalpy Data to Determine the Mass of Products

Problem

When methane is burned, along with the heat produced, oxygen is consumed. Determine what mass of oxygen is consumed if the change in enthalpy is −250 kJ, given the following thermochemical equation:

$$CH_4(g) + 2O_2(g) \rightarrow CO_2(g) + 2H_2O(g)$$
$$\Delta_c H = -802.5 \text{ kJ/mol } CH_4(g)$$

What Is Required?

You need to determine the mass of oxygen that is used when the enthalpy change for the reaction is −250 kJ.

What Is Given?

You know the chemical equation and the enthalpy change for 1 mol of $CH_4(g)$ and 2 mol of $O_2(g)$:

$$CH_4(g) + 2O_2(g) \rightarrow CO_2(g) + 2H_2O(g)$$
$$\Delta_c H = -802.5 \text{ kJ/mol } CH_4(g)$$

You also know the enthalpy change for the specific reaction:

$$\Delta H = -250 \text{ kJ}$$

Plan Your Strategy

You need to determine the number of moles of oxygen consumed. You can use the formula $\Delta H = n\Delta_c H$.
Note that:

$$\Delta H_c = \frac{-802.5 \text{ kJ}}{1 \text{ mol } CH_4(g)} = \frac{-802.5 \text{ kJ}}{2 \text{ mol } O_2(g)}$$

Then you need to find the mass by using molar mass.

$$n = \frac{m}{M}$$

Act on Your Strategy

$$\Delta H = n\Delta_c H$$

$$n = \frac{\Delta H}{\Delta_c H} = \frac{-250 \text{ kJ}}{\left(\dfrac{-802.5 \text{ kJ}}{2 \text{ mol } O_2(g)}\right)}$$
$$= 0.623 \text{ mol } O_2(g)$$

$$n = \frac{m}{M}$$

$$m = n \times M$$
$$= (0.623 \text{ mol } O_2(g))(32.00 \frac{g}{mol})$$
$$= 19.9 \text{ g } O_2(g)$$

Therefore, 19.9 g of oxygen gas are consumed when the enthalpy change is −250 kJ.

Check Your Solution

The units are correct. The mass of oxygen produced is less than the mass of 2 mol $O_2(g)$, which is 64 g. This answer makes sense because the enthalpy change of the reaction was less than $\Delta_c H$.

Using Dimensional Analysis

$$m_{O_2(g)} = \frac{-250 \text{ kJ}}{1} \times \frac{2 \text{ mol } O_2(g)}{-802.5 \text{ kJ}} \times \frac{32.00 \text{ g } O_2(g)}{1 \text{ mol } O_2(g)}$$
$$= 19.9 \text{ g } O_2(g)$$

1. Hydrogen gas and oxygen gas react to form 0.534 g of *gaseous* water. What is the enthalpy change?
$$\Delta_r H = -241.8 \text{ kJ/mol}$$

2. Pentane reacts with an excess of oxygen to produce carbon dioxide and water. What is the enthalpy change of the reaction per mole of each of the following?
   a) oxygen
   b) carbon dioxide
   c) water

3. Determine the thermal energy released by the combustion of each of the following samples of hydrocarbons in an open system:
   a) 56.78 g pentane, $C_5H_{12}(\ell)$
   b) 1.36 kg octane, $C_8H_{18}(\ell)$
   c) 2.344×10^4 g heptane, $C_7H_{16}(\ell)$

4. What is the enthalpy change for the combustion of a 1.00 g sample of methane, $CH_4(g)$, in an open system?

5. The molar enthalpy of combustion of methanol, $CH_3OH(\ell)$, is −725.9 kJ. What mass of methanol must be burned to generate 2.34×10^4 kJ of thermal energy?

6. The enthalpy change for the following reaction, as written is −906 kJ.
$$4NH_3(g) + 5O_2(g) \rightarrow 4NO(g) + 6H_2O(g)$$
What is the molar enthalpy change of the reaction for each of the following?
   a) ammonia
   b) oxygen
   c) nitrogen monoxide
   d) water

Section 9.1 Summary

- The law of conservation of energy states that there is no change in the total energy of the universe.
- A result of the law of conservation of energy, also called the first law of thermodynamics, is that when thermal energy leaves one system it must enter another.
- The two types of energy are kinetic energy (the energy of motion) and potential energy (stored energy).
- Temperature is a measure of the average kinetic energy of the particles of a system.
- For a chemical reaction to occur, some bonds must break and others must form.
- Energy is required to break bonds. Energy is released when bonds are formed.
- Chemical reactions are either exothermic (release energy) or endothermic (require energy).
- Enthalpy is a property of systems. When a process takes place under conditions of constant pressure, the change in the enthalpy of a system is equal to the amount of heat gained or lost by the system.
- The enthalpy change of an exothermic reaction is negative and the enthalpy change of an endothermic reaction is positive.
- The enthalpy change of a reaction as written ($\Delta_r H$) is the enthalpy change for the number of moles of each reactant and product as determined by the coefficient of the term representing the reactant or product.
- The enthalpy of combustion ($\Delta_c H$) is the enthalpy change that occurs when one mole of the compound reacts completely with oxygen.

SECTION 9.1 REVIEW

1. In your own words, state the first law of thermodynamics. Then express the first law of thermodynamics as an equation.

2. If a chemical reaction is endothermic, which process involves a larger enthalpy change: breaking the reactant bonds or forming the product bonds? Explain your answer.

3. Rewrite each of the following thermochemical equations using ΔH notation. Use the most specific form(s) of ΔH. Remember to pay attention to the sign of ΔH:
   a) $Ag(s) + \frac{1}{2} Cl_2(g) \rightarrow AgCl(s) + 127.0 \text{ kJ}$
      (at 25 °C and 100 kPa)
   b) $C_2H_4(g) + 3O_2(g) \rightarrow 2CO_2(g) + 2H_2O(g) + 1322.9 \text{ kJ}$
   c) $NaCl(s) \rightarrow Na^+(aq) + Cl^-(aq) + 44.2 \text{ kJ}$

4. Acetylene, $C_2H_2(g)$, undergoes complete combustion in oxygen. Carbon dioxide and water in its liquid form are produced. The molar enthalpy of the complete combustion of acetylene is $-1.25 \times 10^3 \text{ kJ/mol}$.
   a) Write a thermochemical equation for this reaction.
   b) Draw a potential energy diagram to represent the thermochemical reaction.
   c) What is the change in enthalpy during the complete combustion of 2.17 g of acetylene?

5. In your own words, explain why exothermic reactions have $\Delta H < 0$.

6. The standard molar enthalpy of formation of sucrose, $C_{12}H_{22}O_{11}(s)$, is -2226.1 kJ/mol. Write a thermochemical equation representing this reaction
   a) with the enthalpy term as a product
   b) using ΔH notation

7. Hydrogen sulfide gas, $H_2S(g)$, has a distinct, powerful smell of rotten eggs. The gas undergoes a combustion reaction with oxygen to produce gaseous sulfur dioxide and gaseous water. The molar enthalpy of combustion of hydrogen sulfide gas is $-519 \text{ kJ/mol } H_2S$.
   a) Write a balanced thermochemical equation to represent the combustion of hydrogen sulfide gas.
   b) What is the change in enthalpy when 15.0 g of hydrogen sulfide gas undergoes combustion?
   c) A sample of hydrogen sulfide gas undergoes combustion, and the enthalpy change is -47.2 kJ. What volume of sulfur dioxide at STP is produced by the reaction?

The Technology of Energy Measurement

Section Outcomes

In this section, you will:
- **determine** enthalpy changes using calorimetry data
- **design** a method to compare molar enthalpies when burning two or more fuels
- **compare** energy changes associated with a variety of chemical reactions through the analysis of data and energy diagrams
- **use** appropriate SI units to express enthalpy changes and molar enthalpies

Key Terms

calorimeter
calorimetry
simple calorimeter
thermal equilibrium
bomb calorimeter
heat capacity, C

Why is it important to know how to determine the energy changes associated with chemical and physical changes? Engineers need to know how much energy is released from a fuel when they design an engine to use that fuel. Firefighters need to know how much thermal energy can be released by the combustion of different materials so they can choose the best way to fight a fire. Dietitians need to know how much energy is released by the digestion of certain foods so that they can plan balanced diets that provide the right amount of energy for healthy living (see Figure 9.13).

Figure 9.13 Which food releases more energy when your body digests it, a hamburger or an apple? How do you know? How do food scientists quantify the energy in foods?

How do you determine the thermal energy absorbed or released by chemical and physical processes? In this section, you will learn some ways to determine experimentally the enthalpy changes of various processes based on a system of heat exchange.

Calorimetry

A **calorimeter** is a device that is used to measure the heat released or absorbed by a physical or chemical process taking place within it. **Calorimetry** is the use of a calorimeter to study the energy changes associated with physical and chemical processes. Ideally, a calorimeter is an isolated system, which does not allow products to escape or thermal energy to be transferred to the air or other materials surrounding it.

Figures 9.14 and 9.15 model how a calorimeter works.

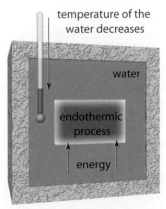
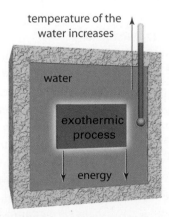

Figure 9.14 An endothermic process reduces the temperature of the water in the calorimeter.

Figure 9.15 An exothermic process increases the temperature of water in the calorimeter.

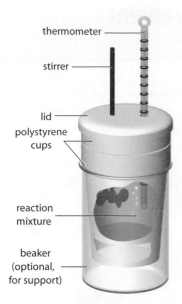

thermometer

stirrer

lid
polystyrene
cups

reaction
mixture

beaker
(optional,
for support)

Figure 9.16 A simple calorimeter, made from nested polystyrene cups, is most often used in high school laboratories. Although it is "low tech," this type of calorimeter can yield very accurate results if used carefully. It is helpful to place the calorimeter in a beaker to reduce the chance that it will tip over.

Figure 9.17 A hot water bottle warms a cold bed because the thermal energy is transferred from the hot water to its surroundings—the sheets, the air trapped under the sheets, and the person trying to stay warm.

Using a Simple Calorimeter

Water, a thermometer (alcohol or digital), nested polystyrene cups, and a plastic lid are the basic components of a **simple calorimeter**, shown in Figure 9.16. In a simple calorimeter, a known mass of water is inside two nested polystyrene cups covered with a lid. The polystyrene cups are excellent insulators and prevent thermal energy from being transferred to the air outside the calorimeter. The nested cups also keep gaseous and liquid products and reactants inside. However, gases can escape because the lid has holes for the thermometer and stirrer. Because the calorimeter allows gases to escape, this calorimeter measures the thermal energy changes of processes at a constant pressure.

Steps for Using a Simple Calorimeter

1. Measure the initial temperature of the water in the calorimeter. (If you are mixing two solutions in the calorimeter, measure the initial temperature of the solutions.)

2. Add the reactants to the calorimeter. The water surrounds, and is in direct contact with, the chemical reaction (or physical change) that releases or absorbs thermal energy.

3. Allow the reaction to proceed. Stir the solution to ensure an even temperature throughout. The system is at a constant pressure because it is open to the air. Record the changing temperature of the water as the reaction proceeds.

4. For an exothermic reaction, the maximum temperature is used to calculate the thermal energy released. For an endothermic reaction, the minimum temperature is used to calculate the thermal energy absorbed.

A simple calorimeter can measure thermal energy changes during chemical changes such as neutralization.

Why Calorimetry Works

Calorimetry is based on two fundamental laws—the law of conservation of energy and the second law of thermodynamics. Recall from Section 9.1 that the law of conservation of energy states that the energy of the universe (system plus surroundings) is constant. This law is also known as the first law of thermodynamics.

The second law of thermodynamics can be expressed in many different statements but they are all based on the same principle. As it relates to thermodynamics and calorimetry, the most useful statement of the second law is: *Thermal energy is spontaneously transferred from an object at a higher temperature to an object at a lower temperature until the two objects reach the same temperature.* This law applies to processes as familiar as a pot sitting on the burner of a stove and warming a cold bed with a hot water bottle as shown in Figure 9.17.

Energy cannot be created or destroyed, and heat flows from a hot object to a cold one until they reach the same temperature. It follows, then, that as energy is released (or absorbed) by a chemical reaction in a calorimeter, it will result in a change in the temperature of the calorimeter surrounding it. Therefore, the enthalpy change of a reaction can be calculated based on the thermal energy transfer in the calorimeter.

When using a simple calorimeter, you are measuring the temperature change of the water in which a process such as a chemical reaction is occurring. To calculate the amount of heat transferred from the chemical reaction to the water in which the reaction is occurring, you must use the specific heat capacity of the water. Therefore, the solution must be dilute so that the presence of the reactants and products do not significantly change the specific heat capacity of the water.

When using a simple calorimeter, you must make the following assumptions:
- The system is isolated. (Any thermal energy that is exchanged with the surroundings outside the calorimeter is small enough to be ignored.)
- The thermal energy that is exchanged with the calorimeter polystyrene cups, thermometer, lid, and stirring rod is small enough to be ignored.
- If something dissolves or reacts with the water in the calorimeter, the resulting solution retains the properties of water. (For example, density and specific heat capacity remain the same.)
- The process takes place under constant pressure.

Once you make those assumptions, the following equation applies:

thermal energy released by system = −thermal energy absorbed by surroundings

heat lost (or gained) by the system = −heat gained (or lost) by the surroundings

The system is the chemical reaction (or physical change) that you are studying. The surroundings consist of the water or solution in the calorimeter. When a chemical reaction releases or absorbs thermal energy in a calorimeter, the change in temperature is measured by a thermometer in the water. If you know the mass of the water and its specific heat capacity, you can calculate the change in thermal energy caused by the process using the equation $Q = mc\Delta t$, which you reviewed in the Unit 5 Preparation.

Using a Calorimeter to Determine the Enthalpy Change of Reaction

A simple calorimeter is well suited to determine the enthalpy changes of reactions in dilute aqueous solutions. The water in the calorimeter absorbs (or provides) the energy that is released (or absorbed) by a chemical reaction. When carrying out an experiment in a dilute solution, *the solution itself* absorbs or releases the energy. You can calculate the quantity of thermal energy that is absorbed or released by the solution using $Q = mc\Delta t$. The mass, m, is the mass of the water or solution. If the solution is dilute, you can assume it has the same specific heat capacity as water.

$Q = m_w c_w \Delta t_w$ is used to determine the quantity of thermal energy absorbed or released by the water or solution in the calorimeter. To find the quantity of thermal energy released or absorbed by the system, just change the sign of the calculated value for Q.

You can use the data collected from a calorimetry experiment to determine the enthalpy of reaction, $\Delta_r H$, for a reaction that takes place in the calorimeter. In a simple calorimeter, the process takes place at constant pressure thus the heat, Q, exchanged by the system and the calorimeter is equal to the enthalpy change of the system, ΔH. To find the enthalpy change of the reaction as written, solve for $\Delta_r H$ in the equation:

$$\Delta H = n\Delta_r H$$

$$\Delta_r H = \frac{\Delta H}{n}$$

Recall that n is the number of moles of the specified reactant or product.

The following Sample Problem shows how calorimetry can be used to determine the enthalpy change of a chemical reaction in solution.

Notice that all the materials in the calorimeter in the following Sample Problem have the same final temperature. A system is said to be at **thermal equilibrium** when all its components have the same temperature.

Determining the Enthalpy Change of a Reaction

Problem

Aqueous copper(II) sulfate, $CuSO_4(aq)$, reacts with sodium hydroxide, $NaOH(aq)$, in a double-replacement reaction. A precipitate of copper(II) hydroxide, $Cu(OH)_2(s)$, and aqueous sodium sulfate, $Na_2SO_4(aq)$, is produced:

$$CuSO_4(aq) + 2NaOH(aq) \rightarrow Cu(OH)_2(s) + 2Na_2SO_4(aq)$$

50.00 mL of 0.300 mol/L $CuSO_4(aq)$ is mixed with 50.00 mL containing an excess of $NaOH(aq)$. The initial temperature of both solutions is 21.40 °C. After mixing the solutions in a simple calorimeter, the highest temperature reached is 24.60 °C. Determine the enthalpy change for the reaction as written and write a thermochemical equation.

What Is Required?

You need to calculate $\Delta_r H$ for the reaction as written.

What Is Given?

Volume of $CuSO_4(aq) = 50.00$ mL
Concentration of $CuSO_4(aq) = 0.300$ mol/L
Volume of $NaOH(aq) = 50.00$ mL
Concentration of $NaOH(aq)$ excess
Initial temperature, $t_i = 21.40$ °C
Final temperature, $t_f = 24.60$ °C
Specific heat capacity of $H_2O(\ell) = 4.19 \dfrac{J}{g\,°C}$ (see Table P5.1)

Plan Your Strategy

The total volume of the solution is the volume of the water in the calorimeter which is 50.00 mL plus 50.00 mL. Assuming the density is 1.00 g/mL, you can determine the mass of water. Determine the amount of $CuSO_4(aq)$.

Use the formula $Q = m_{solution}\, c_{solution}\, \Delta t_{solution}$ to calculate the change in thermal energy of the surroundings (water). Change the sign of Q to find the change in thermal energy of the system (the reaction). This value is equivalent to ΔH of the system. Use the formula $\Delta H = n\Delta_r H$ and solve for $\Delta_r H$. Use the number of moles of $CuSO_4(aq)$.

Act on Your Strategy

Find the total volume of the solution:

$V_{total} = 50.00$ mL $+ 50.00$ mL $= 100.0$ mL

Assume the solutions have the same density as water:

$$m_{total} = 100.0 \text{ mL} \times 1.00 \frac{g}{mL} = 100.0 \text{ g}$$

The mass of the reaction mixture is 100.0 g.

Determine the amount of copper sulfate present:

$$n_{CuSO_4(aq)} = c \times V$$
$$= 0.300 \frac{mol}{L} \times 50.00 \text{ mL} \times \frac{L}{1000 \text{ mL}}$$
$$= 0.0150 \text{ mol}$$

Determine the change in thermal energy of the water:

$$Q = m_{solution} c_{solution} \Delta t_{solution}$$
$$= (100.0 \text{ g})\,(4.19 \frac{J}{g\,°C})\,(24.60\,°C - 21.40\,°C)$$
$$= 1.341 \times 10^3 \text{ J}$$
$$= 1.341 \text{ kJ}$$

The change in the thermal energy of the surroundings is 1.341 kJ. Therefore, the change in thermal energy of the system is -1.341 kJ.

This is the same value as ΔH of the system. Thus $\Delta H_{system} = -1.341$ kJ

Determine the enthalpy change for the reaction as written.

$$\Delta H = n\Delta_r H$$

$$\Delta_r H = \frac{\Delta H}{n}$$
$$= \frac{-1.341 \text{ kJ}}{0.0150 \text{ mol}}$$
$$= -89.4 \text{ kJ}$$

$CuSO_4(aq) + 2NaOH(aq) \rightarrow$
$$Cu(OH)_2(s) + 2Na_2SO_4(aq) + 89.4 \text{ kJ}$$

Since the coefficient of $CuSO_4(aq)$ is one, $\Delta_r H$ represents the enthalpy change when one mole of copper(II) sulfate reacts with an excess of sodium hydroxide.

Check Your Solution

The solution has the correct number of significant digits. The units are correct. You know that the reaction was exothermic, because the temperature of the solution increased. The calculated value of ΔH is negative, which is correct for an exothermic reaction.

Using Dimensional Analysis

You can determine the enthalpy change in one step as follows:

$$\Delta_r H = -\frac{100.0 \text{ g}}{1} \times \frac{4.19 \text{ J}}{g\cdot°C} \times \frac{(24.6\,°C - 21.4\,°C)}{1} \times \frac{1}{0.300 \text{ mol/L}} \times \frac{1000 \text{ mL/L}}{50.00 \text{ mL}}$$
$$= -8.94 \times 10^4 \text{ J/mol } CuSO_4(aq)$$
$$= -89.4 \text{ kJ/mol } CuSO_4(aq)$$

7. A chemist wants to determine the enthalpy of neutralization for the following reaction:

$$HCl(aq) + NaOH(aq) \rightarrow NaCl(aq) + H_2O(\ell)$$

The chemist uses a simple calorimeter to neutralize completely 61.1 mL of 0.543 mol/L HCl(aq) with 42.6 mL of sufficiently concentrated NaOH(aq). The initial temperature of both solutions is 17.80 °C. After neutralization, the highest recorded temperature is 21.60 °C. Calculate the enthalpy change of neutralization in kJ/mol of HCl(aq). Assume that the density of both solutions is 1.00 g/mL.

8. A chemist wants to determine empirically the enthalpy change for the following reaction as written:

$$Mg(s) + 2HCl(aq) \rightarrow MgCl_2(aq) + H_2(g)$$

The chemist uses a simple calorimeter to react 0.50 g of magnesium ribbon with 100 mL of 1.00 mol/L HCl(aq). The initial temperature of the HCl(aq) is 20.40 °C. After the reaction, the highest recorded temperature is 40.70 °C.

   a) Calculate the enthalpy change, in kJ/mol of Mg(s), for the reaction as written.

   b) State all assumptions that you made in order to determine the enthalpy change.

9. Sodium reacts violently to form sodium hydroxide when placed in water, as shown in the equation below:

$$2Na(s) + 2H_2O(\ell) \rightarrow 2NaOH(aq) + H_2(g)$$

Determine an experimental value for the molar enthalpy of reaction for sodium given the following data:

mass of sodium, Na(s): 0.37 g
mass of water in calorimeter: 175 g
initial temperature of water: 19.30 °C
final temperature of mixture: 25.70 °C

10. Nitric acid is neutralized with potassium hydroxide in the following reaction:

$$HNO_3(aq) + KOH(aq) \rightarrow KNO_3(aq) + H_2O(\ell)$$
$$\Delta_r H = -53.4 \text{ kJ/mol } HNO_3$$

55.0 mL of 1.30 mol/L solutions of both reactants, at 21.40 °C, are mixed in a calorimeter. What is the final temperature of the mixture? Assume that the density of both solutions is 1.00 g/mL. Also assume that the specific heat capacity of both solutions is the same as the specific heat capacity of water. No heat is lost to the calorimeter itself.

11. A student uses a simple calorimeter to determine the enthalpy change of reaction for hydrobromic acid and aqueous potassium hydroxide. The student mixes 100.0 mL of 0.50 mol/L HBr(aq) at 21.00 °C with 100.0 mL of 0.50 mol/L KOH(aq), also at 21.00 °C. The highest temperature that is reached is 24.40 °C. Write a thermochemical equation for the reaction.

12. In a simple calorimeter, 150 mL of 1.000 mol/L NaOH(aq) is mixed with 150 mL of 1.000 mol/L HCl(aq). If both solutions were initially at 25.00 °C and after mixing the temperature rose to 30.00 °C, what is the enthalpy change of reaction as written?

In the following investigation, you will construct a simple calorimeter and use it to determine the enthalpy of a neutralization reaction.

• • •

7 Define calorimetry.

8 State the first law of thermodynamics and explain how it forms a basis for calorimetry.

9 Explain the relationship between the second law of thermodynamics and the theory of calorimetry.

10 What conditions allow you to determine the enthalpy change of a reaction by measuring the change of the temperature of the water in a calorimeter during a chemical reaction?

• • •

Determining the Enthalpy of a Neutralization Reaction

The neutralization of hydrochloric acid with sodium hydroxide solution is represented by the following equation:

$$HCl(aq) + NaOH(aq) \rightarrow NaCl(aq) + H_2O(\ell)$$

Using a simple calorimeter, determine the enthalpy change for this reaction.

Question

What is the enthalpy of neutralization for a hydrochloric acid and sodium hydroxide solution?

Prediction

Will the neutralization reaction be endothermic or exothermic? Record your prediction, and give reasons.

Safety Precautions

- If you get any hydrochloric acid or sodium hydroxide solution on your skin, flush your skin with plenty of cold water.

Materials

- 1.00 mol/L HCl(aq)
- 1.00 mol/L NaOH(aq)
- polystyrene cups that are the same size (2)
- plastic lid for cup
- 100 mL graduated cylinder
- 400 mL beaker
- thermometer (alcohol or digital)
- stirring rod

Procedure

1. Your teacher will allow the hydrochloric acid and sodium hydroxide solution to come to room temperature overnight.

2. Read the rest of this Procedure carefully before you continue. Set up a graph to record your temperature observations.

3. Build a simple calorimeter, using Figure 9.16 as a guide. You will need to make two holes in the lid—one for the thermometer and one for the stirring rod. The holes should be as small as possible to minimize thermal energy exchange with the surroundings.

4. Rinse the graduated cylinder with a small quantity of 1.00 mol/L NaOH(aq). Use the cylinder to add 50.0 mL of 1.00 mol/L NaOH(aq) to the calorimeter. Record the initial temperature of the NaOH(aq). This will also represent the initial temperature of the HCl(aq).

5. Rinse the graduated cylinder with tap water. Then rinse it with a small quantity of 1.00 mol/L HCl(aq). Quickly and carefully, add 50.0 mL of 1.00 mol/L HCl(aq) to the NaOH(aq) in the calorimeter.

6. Cover the calorimeter. Record the temperature every 30 s, stirring gently and continuously.

7. When the temperature levels off, record the final temperature, t_f.

8. If time permits, repeat Procedure steps 4 to 7.

Analysis

1. Determine the quantity of thermal energy absorbed by the solution in the calorimeter.

2. Determine the quantity of thermal energy released by the reaction.

3. Determine the amount (in mol) of HCl(aq) and NaOH(aq) that were involved in the reaction.

4. Use your knowledge of solutions to explain what happens during a neutralization reaction. Use equations in your answer. Was thermal energy absorbed or released during the neutralization reaction? Explain your answer.

Conclusion

5. Use your results to determine the enthalpy change of the neutralization reaction in kJ/mol of NaOH(aq). Write the thermochemical equation for the neutralization reaction.

6. When an acid touches your skin, why must you flush the area with plenty of water rather than neutralizing the acid with a base?

7. Suppose that you had added solid sodium hydroxide pellets to hydrochloric acid, instead of adding hydrochloric acid to sodium hydroxide solution:

   a) Would you have obtained a different enthalpy change? Explain your answer.

b) If so, would the enthalpy change have been higher or lower?

c) How can you test your answer? Design an investigation, and carry it out with the permission of your teacher.

d) What change do you need to make to the thermochemical equation if you perform the investigation using solid sodium hydroxide?

Determining Enthalpy of Combustion

In the next investigation, you will determine the molar enthalpy of combustion of paraffin wax, $C_{25}H_{52}(s)$, more commonly known as candle wax. To determine the enthalpy of combustion, a flame calorimeter is commonly used in a school setting (see Figure 9.18). Unlike simple calorimeters, flame calorimeters are fire-resistant. They do, however, absorb a significant amount of energy; therefore, the heat capacity of the calorimeter must be included in energy calculations. The fuel is lit under the small can, heating both the can and the water inside.

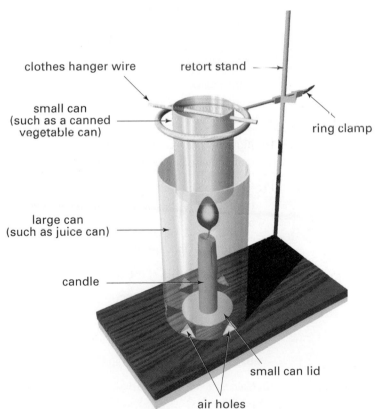

clothes hanger wire

retort stand

small can (such as a canned vegetable can)

ring clamp

large can (such as juice can)

candle

small can lid

air holes

Figure 9.18 A flame calorimeter is commonly made from a small can suspended on a ring stand by a stirring rod or wire. A large can with air holes surrounds the fuel source to direct the heat upward to minimize the loss of thermal energy to the surroundings.

In the second part of the investigation, you will design a procedure to compare the molar enthalpy of combustion when burning two or more fuels. Knowing how much energy can be harnessed from an energy source not only helps engineers choose a fuel for the engines they design and build but also helps us decide what to bring on a long hike. Our bodies, through cellular respiration, "burn" the foods we eat to generate the energy we need to live.

Molar Enthalpy of Combustion

You have probably gazed into a candle flame, a roaring fire, or your gas barbecue without thinking about chemistry! Now, however, you will use the combustion of candle wax to gain insight into the measurement of heat exchanges. With this experience, you will then design a procedure to compare the molar enthalpies of paraffin and two other fuels. You will also evaluate the design of this investigation and make suggestions for improvements.

Question

1. What is the molar enthalpy of combustion of paraffin?

2. How can you compare the molar enthalpies of combustion for several different fuels?

Safety Precautions

- Tie back long hair and secure any loose clothing. Before you light any fuel source, check that there are no flammable solvents nearby.

Part 1: The Molar Enthalpy of Combustion of Paraffin

Materials

- candle
- water
- matches
- balance
- calorimeter
- apparatus: includes retort stand, large can with air holes, small can, ring clamp, wire or stirring rod
- thermometer (alcohol or digital)
- stirring rod
- can lid or cardboard base

Procedure

1. Light the candle to melt some wax. Drip the wax onto the base and attach the candle to it. Blow out the candle.

2. Set up the apparatus as shown in Figure 9.18, but do not include the large can yet. Adjust the ring stand so that the small can is about 5 cm above the wick of the candle. When the candle is lit, the tip of the flame should just touch the bottom of the small can.

3. Measure the mass of the candle and the base.

4. Measure the mass of the small can. Measure the mass of the hanger.

5. Place the candle inside the large can on the retort stand.

6. Fill the small can about two thirds full of cold water (10 °C to 15 °C). You will measure the mass of the water later.

7. Stir the water in the can. Measure the temperature of the water.

8. Light the candle. Quickly place the small can in position over the candle. **Caution** Be careful of the open flame.

9. Continue stirring. Monitor the temperature of the water until it has reached 10 °C to 15 °C above room temperature.

10. Blow out the candle. Continue to stir. Monitor the temperature until you observe no further change.

11. Record the highest temperature reached. Examine the bottom of the container and record your observations.

12. Measure the mass of the small can and the water.

13. Measure the mass of the candle, base, and any drops of candle wax.

Analysis

1. **a)** Calculate the mass of the water.

   b) Calculate the mass of candle wax that was burned.

2. Calculate the molar enthalpy of combustion of paraffin wax.
   - Assume that the candle wax is pure paraffin wax.
   - Include the mass of the small can in your calculations as part of the calorimeter. Ask your teacher for the composition of the can.

heat lost by system = −heat gained by surroundings

= −(heat gained by water

+ heat gained by can)

Conclusion

3. **a)** List some possible sources of error that may have affected the results you obtained.

   b) Evaluate the design and the procedure of this investigation. Think about the assumptions that are made in calorimetry calculations. Consider the apparatus, the combustion, and anything else you can think of. Make suggestions for possible improvements.

4. What if soot (unburned carbon) accumulated on the bottom of the small can? Would this produce a greater or lower heat value than you expected? Explain.

Part 2: Comparing the Molar Enthalpies of Combustion for Three Fuels

Materials

- water
- two fuels other than paraffin
- matches
- balance
- calorimeter apparatus:
  includes retort stand, large can with air holes, small can, ring clamp, wire or stirring rod
- thermometer (alcohol or digital)
- stirring rod
- can lid or cardboard base

Experimental Plan

1. List the variables for your investigation: manipulated, responding, and controlled.

2. Write a procedure that you will carry out for each fuel.

3. Design a data table to record your data.

4. Think about the calculations that you will have to perform to calculate molar enthalpy of combustion. Are you collecting enough data?

5. You have to be able to compare the three fuels (the two other fuels plus the paraffin) fairly. Are you controlling all the variables that should be controlled?

Data and Observations

6. Perform your investigation. Record all data.

Analysis

1. Calculate the molar enthalpy of combustion for the other two fuels.

Conclusion

2. **a)** Write a balanced thermochemical equation for the complete combustion of paraffin wax and of the other two fuels you studied.

   b) Draw an enthalpy diagram for each fuel.

3. Compare the three fuels. Your answer should include:
   - a statement about the relative molar enthalpies of combustion
   - the environmental impact of each fuel
   - the appropriate uses of each fuel and why they are appropriate

   Use an Internet search to obtain information to satisfy the last two bulleted points. ICT

Extension

4. Repeat this experiment using food instead of fuel. What aspects other than enthalpy of combustion are important in evaluating the appropriateness of food that may be taken on a survival hike? Explain.

Bomb Calorimetry

In Investigation 9.B, you constructed a calorimeter to determine the molar enthalpy of combustion of paraffin wax and two other fuels. Your calorimeter was more practical to use with combustion reactions than was a simple calorimeter, but some thermal energy was transferred to the air and to the large metal can. To measure precisely and accurately the enthalpy changes of combustion reactions, chemists use a calorimeter called a **bomb calorimeter**, shown in Figures 9.19 and 9.20. A bomb calorimeter measures enthalpy changes during combustion reactions at a constant volume.

stirrer

ignition terminals

thermometer

water

insulated container

steel chamber holding oxygen and the sample

Figure 9.19 A bomb calorimeter is more sophisticated than a simple calorimeter or a flame calorimeter.

Figure 9.20 A chemist prepares a sample for testing in a bomb calorimeter.

The bomb calorimeter works on the same general principle as the simple calorimeter. The reaction, however, takes place inside an inner metal chamber, called a *bomb*. This bomb contains pure oxygen. The reactants are ignited using an electric coil. A known quantity of water surrounds the bomb and absorbs the energy released by the reaction.

A bomb calorimeter has many more parts than a polystyrene calorimeter has. All these parts can absorb or release small quantities of energy. Therefore, you cannot assume that the heat lost to these parts is small enough to be negligible. To obtain precise heat measurements, you must know or find out the **heat capacity, C**, of the entire bomb calorimeter. The heat capacity of a substance, measured in J/°C or kJ/°C, is the quantity of energy required to raise its temperature by one degree Celsius. The heat capacity of a bomb calorimeter takes into account the heat that *all* parts of the calorimeter can lose or gain (see Figure 9.21).

Figure 9.21 The heat capacity of a bomb calorimeter incorporates the heat capacity of all its components.

heat capacity of calorimeter

$C_{total} = C_{water} + C_{thermometer} + C_{stirrer} + C_{container}$

A bomb calorimeter is calibrated for a constant mass of water. Since the mass of the other parts remains constant, there is no need for mass units in the heat capacity value. The manufacturer usually includes the heat capacity value in the instructions for the calorimeter.

Another aspect of a bomb calorimeter is that it is a closed system under pressure. In a bomb calorimeter, processes occur at constant volume but not constant pressure. Since the pressure can change significantly, the amount of heat transferred from the system to the calorimeter is not the same as the change in enthalpy of the system. Chemists must make corrections in their calculations to account for the constant volume conditions.

Open system:

$C_3H_8(g) + 5O_2(g) \rightarrow 3CO_2(g) + 4H_2O(g)$ $\Delta_cH = -2043.9$ kJ/mol C_3H_8

Bomb calorimeter:

$C_3H_8(g) + 5O_2(g) \rightarrow 3CO_2(g) + 4H_2O(\ell)$ $\Delta_cH = -2219.9$ kJ/mol C_3H_8

Thermal energy calculations must be done differently when the heat capacity of a calorimeter walls is included. The Sample Problem that follows the special feature illustrates how to use the heat capacity of a calorimeter in your calculations.

Chemistry File

FYI

In the Sample Problem, you will notice that the enthalpy of combustion of peanut butter is expressed in kJ/g, not kJ/mol. Molar enthalpy of combustion is used for pure substances. Peanut butter, however, like most food, is not a pure substance. Another appropriate way to express the enthalpy of combustion for food is in kJ/serving.

Energy for Living: How Food Fuels You

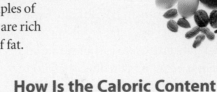

Like any machine that does work, our bodies need fuel. Carbohydrates, proteins, and fats are classes of food compounds that become oxidized in the body to provide energy through cellular respiration. Although many foods contain a mix of all three nutrients, one type usually dominates. Such foods as bread, rice, and potatoes are examples of carbohydrate-rich foods. Fish, beans, poultry, and red meat are rich in protein. Oil, butter, and nuts contain a high proportion of fat.

Calories in Food and Cellular Respiration

Cellular respiration is similar to the release of energy from burning a piece of wood, but the "burning" process in respiration occurs more gradually and through multiple steps. Human cells obtain most of their energy from carbohydrates. Before your body uses the energy in carbohydrates, the digestive system breaks them down into glucose ($C_6H_{12}O_6$). The glucose then undergoes cellular respiration, which consists of many steps. The overall reaction of cellular respiration is shown below:

$$C_6H_{12}O_6(aq) + 6O_2(g) \rightarrow 6CO_2(g) + 6H_2O(g)$$

glucose   oxygen   carbon   water
                        dioxide

The complete reaction of 1 mol of glucose in the body releases about 2870 kJ of energy.

The energy released by cellular respiration is the same as would be released by burning the food in a calorimeter, assuming that the sample is completely digested.

How do we quantify the energy we get from food? You are probably used to seeing the unit "Calorie" on food labels. One Calorie is the quantity of thermal energy required to raise the temperature of 1 kg of water by one degree Celsius:

$$1 \text{ Calorie} = 4.18 \text{ kJ}$$

(The unit Calorie (with a capital C) comes from the historical unit calorie (with a lowercase c), which is equal to 4.18 J. Therefore, 1 Calorie = 1 kilocalorie = 1000 calories.)

The body uses fats, proteins, and carbohydrates for energy. In general, fats have nine Calories per gram, while proteins and carbohydrates have four Calories per gram. Caloric requirements vary with the individual; active people require more energy than inactive people do. Excess food energy is stored as glycogen or, if glycogen stores are full, as fat.

How Is the Caloric Content of Food Determined?

There are three ways to estimate the caloric content of food products:

1. The 4-9-4 rule is most often used for nutritional labelling purposes. First, protein, fat, moisture, and ash/mineral content are determined by extracting them using various methods. Then the carbohydrate (CHO) content is calculated by subtracting their mass from the total mass. The final step involves another calculation:

   Calories/serving = (mass protein/serving × 4 Cal/g)
   + (mass fat/serving × 9 Cal/g)
   + (mass CHO/serving × 4 Cal/g)

   Thus, the label on an energy bar that contains 10 g of protein, 9 g of fat, and 20 g of carbohydrate would list 201 Calories.

2. Food scientists also use the 4-9-4 rule, but they deduct Calories contributed by insoluble fibre. The insoluble fibre portion of the CHO is subtracted from the total CHO difference value before multiplying by 4 Cal/g.

3. Using bomb calorimetry data and information about how much of a food your body digests, the Atwater System tabulates data for particular foods or ingredients. For example, to estimate the Calories obtained from 1 gram of potato, bomb calorimetry experiments determine the heat of combustion of potato protein. This number, 3.75 Cal/g, is multiplied by 74%, which represents the percentage of potato protein that your body digests, giving a value of 2.78 Cal/g. The process is repeated for potato fat and potato CHO, and then all three values are added together and divided by three to estimate the total calories per gram of potato consumed.

....*continued*

Nutrition Facts
Valeur nutritive
Per 3 Crackers (21 g)
Pour 3 craquelins (21 g)

Amount Teneur	% Daily Value % valeur quotidienne
Calories / Calories 90	
Fat / Lipides 2.0 g	3 %
Saturated / saturés 0.5 g + Trans / trans 0 g	3 %
Cholesterol / Cholestérol 0 mg	0 %
Sodium / Sodium 95 mg	4 %
Carbohydrate / Glucides 16 g	5 %
Fibre / Fibres 1 g	4 %
Sugars / Sucres 0 g	
Protein / Protéines 2 g	
Vitamin A / Vitamine A	0 %
Vitamin C / Vitamine C	0 %
Calcium / Calcium	0 %
Iron / Fer	6 %

New Challenges in Determining Caloric Content

Accurately determining a food product's Caloric content is becoming increasingly difficult with the trend toward reduced-fat and non-fat products. When food additives, such as fibre, gums, bulking agents, and such specially engineered ingredients as salatrim and olestra, are used, they affect Calorie estimation. For example, laboratory experiments show that polydextrose, a carbohydrate and a popular bulking agent used in many reduced-fat foods, contributes only 1 Cal/g. Similarly, salatrim, though a real fat, contributes only 5 Cal/g as opposed to the usual 9 Cal/g because only 55% of it is useable by the body. The continuing challenge is to develop fast and accurate tests that allow scientists to determine how much energy you get from the food you eat.

• • •

1. Using an online Calorie chart or food labels, compare the caloric values of your five favourite foods. For mixed dishes, estimate the amounts of the major ingredients. **ICT**

2. Choose one food with a food label that includes Calorie, fat, protein, and carbohydrate information. Estimate the Calorie content using the 4-9-4 rule and compare your estimate with the listed Calorie content.

3. What percentage of the food you eat should be carbohydrates? proteins? fats? Compare these recommended percentages with your diet for three days.

4. Many mammals rely on fat that is stored in their bodies. By surviving on fat reserves that are stored during the autumn, bears can hibernate throughout the winter without eating. Why is excess energy in the body stored primarily as fat, not as protein or carbohydrates?

Grizzly bears need to eat constantly during the summer in order to survive a winter of hibernation.

To learn more about how food fuels you, go to the McGraw-Hill Ryerson web site.

www.albertachemistry.ca
WWW

Sample Problem

Calculating Thermal Energy in a Bomb Calorimeter

Problem

A laboratory was contracted to test the energy content of Brand P peanut butter. A technician placed a 16.0 g sample of peanut butter in the steel bomb of a bomb calorimeter, along with sufficient oxygen to burn the sample completely. She ignited the mixture and took temperature measurements. The heat capacity of the calorimeter was calibrated at 8.28 kJ/°C. During the experiment, the temperature increased by 50.5 °C.

a) What was the amount of thermal energy released by the sample of peanut butter?

b) What is the enthalpy of combustion of the peanut butter per gram of sample?

What Is Required?

a) You need to calculate the thermal energy lost by the peanut butter.

b) You need to calculate the enthalpy change per gram of peanut butter.

What Is Given?

You know the mass of the peanut butter, the heat capacity of the calorimeter, and the change in the temperature of the system:

Mass of peanut butter, $m = 16.0$ g
Heat capacity of calorimeter, $C = 8.28$ kJ/ °C
Change in temperature, $\Delta t = 50.5$ °C

Plan Your Strategy

a) The heat capacity of the calorimeter takes into account the specific heat capacities and masses of all the parts of the calorimeter. Calculate the amount of heat absorbed by the calorimeter, Q, using the equation

$$Q = C\Delta t$$

Note: C is the heat capacity of the calorimeter in J/°C or kJ/°C. It replaces the m and c in other calculations involving specific heat capacity.

First calculate the thermal energy gained by the calorimeter. When the peanut butter burns, the thermal energy released by the peanut butter sample equals the thermal energy absorbed by the calorimeter.

b) To find the heat of combustion per gram, divide the thermal energy by the mass of the sample.

Act on Your Strategy

a) $Q = C\Delta t$

$$= (8.28\frac{kJ}{°C})(50.5\,°C)$$
$$= 418.14\ kJ$$

The calorimeter gained 418 kJ of thermal energy. Therefore, the sample of peanut butter released 418 kJ of thermal energy. For the combustion of peanut butter, $Q = -418$ kJ.

b) Heat of combustion per gram $= Q/m$

$$= \frac{-418\ kJ}{16.0\ g}$$
$$= -26.1\ kJ/g$$

The heat of combustion per gram of peanut butter is -26.2 kJ/g.

Check Your Solution

Thermal energy was released by the peanut butter, so it makes sense that the thermal energy value is negative.

$$\frac{Q}{m} = \frac{\left(-8.28\frac{kJ}{°C}\right)(50.0\,°C)}{16\ g}$$
$$= -26.1\frac{kJ}{g}$$

Practice Problems

13. Predict the final temperature of a 500 g iron ring that is initially at 25.0 °C and is heated by combusting 1.95 g of ethanol, $\Delta_c H = -1234.8$ kJ/mol $C_2H_5OH(\ell)$ in an open system.

14. Calculate the molar enthalpy of combustion of octane if 0.53 g of the fuel increased the temperature of a coffee can calorimeter (13 g of aluminium and 250 mL of water) by 17.2 °C. Remember to include the heat gained by not only the water but also by the aluminium can.

15. How much propane (in grams) would have to be burned in an open system to raise the temperature of 300 mL of water from 20.00 °C to its boiling point? (The molar enthalpy of combustion of propane may be found in Table 9.1 on page 347).

16. A lab technician places a 5.00 g food sample into a bomb calorimeter that is calibrated at 9.23 kJ/°C. The initial temperature of the calorimeter system is 21.0 °C. After burning the food, the final temperature of the system is 32.0 °C. What is the enthalpy of combustion of the food in kJ/g?

17. Determine the enthalpy of combustion of an unknown fuel if a 2.75 g sample increased the temperature of 500 mL of hot chocolate ($c = 3.75$ J/g • °C) in a 150 g glass mug ($c = 0.84$ J/g • °C) from 10.00 °C to 45.00 °C. Express the value for enthalpy of combustion in appropriate units.

When you begin your study of chemistry, you usually think only about the chemical compounds that comprise the reactants and products of a chemical reaction. You often think of the study of chemistry as the study of matter and changes that occur in matter. Concepts involving energy seem to belong to the study of physics. In this chapter, you have begun to see the union between chemistry and physics. The precise measurement of the energy changes that occur in chemical reactions have provided chemists with a wealth of information about chemical reactions. In the next chapter, you will discover more ways to analyze the role of energy in chemical changes and how this information can be applied to some very practical situations.

Section 9.2 Summary

- A calorimeter is an instrument that measures the amount of heat that enters or leaves a system during a process such as a chemical reaction.
- Calorimetry is based on the first law of thermodynamics that states that the change in the energy of the universe during any process is zero. Therefore, any thermal energy that leaves a system must enter its surroundings.
- When using a simple calorimeter, the process takes place at constant pressure. Therefore, the amount of heat that is exchanged between the calorimeter and the system is equal to the change in the enthalpy of the system.
- If the process that takes place in the calorimeter is a chemical reaction, you can use the data from the calorimetry experiment to calculate the enthalpy of the reaction.
- In a bomb calorimeter, the process takes place in a sealed container so the process occurs at constant volume and not constant pressure. The heat that is exchanged between the calorimeter and the system is *not* equal to the enthalpy change of the system. The enthalpy change must be corrected for a changing pressure.

SECTION 9.2 REVIEW

1. List two characteristics of a calorimeter that are necessary for successful determination of thermal energy change.

2. A calorimeter is calibrated at 7.61 kJ/°C. When a sample of coal is burned in the calorimeter, the temperature increases by 5.23 °C. What is the enthalpy change of the reaction?

3. A reaction in a calorimeter causes 150 g of water to decrease in temperature by 5.0 °C. What is the thermal energy change of the water?

4. What properties of polystyrene make it a suitable material for a constant-pressure calorimeter?

5. Suppose you use concentrated reactant solutions in an experiment with a simple calorimeter. Should you make the same assumptions that you did when you used dilute solutions? Explain.

6. Concentrated sulfuric acid can be diluted by adding it to water. The "dilution" is actually an extremely exothermic reaction of sulfuric acid with water. In this question, you will design an experiment to measure the enthalpy change (in kJ/mol) for the dilution of concentrated sulfuric acid. Assume that you have access to any equipment in your school's chemistry laboratory. Do not carry out this experiment.
   a) State the equipment and chemicals that you need.
   b) Write a step-by-step procedure.
   c) Set up an appropriate data table.
   d) State any information that you need.
   e) State any simplifying assumptions that you will make.

7. A chemist mixes 100.0 mL of 0.050 mol/L aqueous potassium hydroxide with 100.00 mL of 0.050 mol/L nitric acid in a simple calorimeter. The temperature of the reactants is 21.01 °C. The temperature of the products is 21.34 °C.
   a) Determine the molar enthalpy of neutralization of KOH(aq) with HNO_3(aq).
   b) Write a thermochemical equation and draw an enthalpy diagram for the reaction.
   c) If you performed this investigation, how would you change the procedure? Explain your answer.

8. Scientists can study the thermal energy that is produced by human metabolism reactions using a calorimeter for humans. Based on what you know about calorimetry, how would you design a calorimeter for humans? What variables would you control and study in an investigation using your calorimeter? Write a brief proposal outlining the design of your human calorimeter and the experimental approach you would take.

Which person do you think would release more thermal energy in one hour?

Chapter 9

SUMMARY

Nearly all of the energy for every process on Earth ultimately comes from the Sun. Energy exists in the form of kinetic energy, the energy of motion, or potential energy, which is stored energy. In chemistry, kinetic energy is usually the energy of motion of atoms and molecules, which is called thermal energy. Potential energy is the energy stored in chemical bonds. Energy is required to break bonds and energy is released when bonds are formed. If the energy required to break bonds in reactants is greater than the energy released to form the bonds in the products of a chemical reaction, the reaction is endothermic. If the energy that is released when bonds form in the products is greater than the energy consumed to break the bonds in the reactants, the reaction is exothermic.

Chemists often use a property of systems called enthalpy to analyze chemical reactions. It is not possible to determine the total enthalpy of a system. Only an enthalpy change can be measured. Enthalpy is related to energy and thus the same unit, the joule, is used to report an enthalpy change. When a process such as a chemical reaction takes place at constant pressure, the enthalpy change of the process is equal to the amount of heat exchanged between the system and its surroundings. The enthalpy change is negative for an exothermic reaction and positive for an endothermic reaction.

Calorimetry is a technique by which the amount of heat exchanged between a system and its surroundings can be measured. You can study enthalpy changes for systems by using calorimetry. The processes that are carried out in simple calorimeters take place at constant pressure, thus the measured amount of heat exchanged between the system and the calorimeter is equal to the change in the enthalpy of the system. In a bomb calorimeter, processes take place in a constant volume and the pressure can change. To relate the heat exchanged between the system and the calorimeter to the change in enthalpy of the system, chemists must correct for the change in pressure.

Concept Organizer

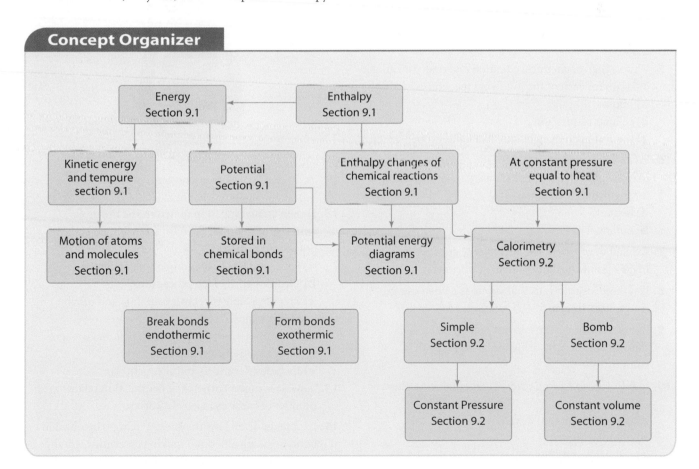

Understanding Concepts

1. State the first law of thermodynamics and express the law as an equation.

2. Define and distinguish between the following terms. Use examples in your answer:
   a) heat capacity, C, and specific heat capacity, c
   b) enthalpy change and molar enthalpy change
   c) formation reactions and combustion reactions

3. Define and distinguish between kinetic energy and potential energy, and give one example of each of the following types of changes:
   a) a process that involves changes in kinetic energy only
   b) a process that involves changes in potential energy only
   c) a process that involves changes in both potential energy and kinetic energy

4. An equal amount of heat is absorbed by a 25 g sample of aluminium ($c = 0.897$ J/g • °C) and a 25 g sample of nickel ($c = 0.444$ J/g • °C). Which metal will show a greater increase in temperature? Explain your answer.

5. Write a thermochemical equation and draw an enthalpy diagram for each of the following reactions:
   a) photosynthesis
   b) cellular respiration
   c) the combustion of butane in a lighter (open system)

6. Write a balanced equation for the combustion of the following substances. Include a separate expression for the enthalpy changes.
   a) heptane
   b) butane

7. Explain why two nested polystyrene cups, with a lid, make a good constant-pressure calorimeter.

8. List three assumptions that you make when using a simple calorimeter.

9. Explain why juice cans do not make a very good calorimeter for measuring molar enthalpy of combustion.

10. Identify the following reactions as either endothermic or exothermic:
    a) the combustion of hexane
    b) photosynthesis
    c) cellular respiration

Applying Concepts

11. Explain in detail why water is more suitable to use in a simple calorimeter than oil.

12. Design an investigation to determine the molar enthalpy of combustion of ethanol, $C_2H_5OH(\ell)$, using a wick-type burner, similar to that in a kerosene lamp, shown below.
    a) Draw and label a diagram of the apparatus.
    b) Write a step-by-step procedure.
    c) Prepare a table to record your data and other observations.
    d) State any assumptions that you will make when carrying out the calculations.
    e) Include relevant chemical equations.

A kerosene lamp

13. Design an investigation to determine the molar enthalpy of neutralization for sulfuric acid and potassium hydroxide.
    a) Draw and label a diagram of the apparatus.
    b) Write a step-by-step procedure.
    c) Prepare a table to record your data and other observations.
    d) State any assumptions that you will make when carrying out the calculations.
    e) Include relevant chemical equations.

14. How does molar enthalpy of combustion relate to the number of carbons in a hydrocarbon?

15. Explain in detail how the energy changes involved in bond breaking and bond making determine whether a chemical reaction is endothermic or exothermic. Use a diagram to illustrate your explanation.

Solving Problems

16. To make four cups of tea, 1.00 kg of water is heated from 22.0 °C to 99.0 °C. How much thermal energy is absorbed by the water?

17. A 10.0 g sample of pure acetic acid, $CH_3COOH(\ell)$, is completely burned in oxygen, and the enthalpy change is -144.77 kJ. What is the molar enthalpy of combustion of acetic acid?

18. The molar enthalpy of combustion of methanol, $CH_3OH(\ell)$, in an open system is -727 kJ/mol.
  a) Write a balanced thermochemical equation to show the complete combustion of methanol to form water and carbon dioxide.
  b) 44.3 g of methanol undergoes complete combustion to form 56.2 L of carbon dioxide at STP. What is the enthalpy change?

19. Calcium oxide, $CaO(s)$, reacts with carbon as shown.
$$CaO(s) + 3C(s) + 462.3 \text{ kJ} \rightarrow CaC_2(s) + CO(g)$$
  a) If the enthalpy change for a reaction is $+246.7$ kJ, what mass of calcium carbide is produced?
  b) What is the enthalpy change if 46.7 g of graphite reacts with excess calcium oxide?
  c) How much energy is required to consume 1.38×10^5 g of calcium oxide?

20. Determine an experimental value for the molar enthalpy of combustion of paraffin wax given the following experimental data:
mass of candle and base before burning = 17.36 g
mass of candle and base after burning = 17.01 g
mass of aluminium can = 35.03 g
mass of water = 125.03 g
initial temperature of water = 14.3 °C
final temperature of water = 35.7 °C
specific heat capacity of aluminium = 0.897 J/g • °C
Perform an Internet search to determine an accepted value for the molar enthalpy of combustion for paraffin wax. Calculate a percentage difference. Is your value lower or higher than the accepted value? Explain. **ICT**

21. How many grams of pentane must be burned to raise the temperature of 500 g of water by 13.5 °C? (See Table 9.1 on page 347 for the molar enthalpy of combustion of pentane.)

22. A chemist wants to calibrate a new bomb calorimeter. She completely burns 0.930 g of carbon in the calorimeter. The temperature of the calorimeter changes from 25.00 °C to 28.15 °C. If the molar enthalpy of combustion of carbon is -394 kJ/mol, what is the heat capacity of the new calorimeter? What experimental evidence shows that this reaction is exothermic?

Making Connections

23. When walking briskly, you use about 20 kJ of energy per minute. Eating one serving of a whole wheat cereal (37.5 g) provides about 527.18 kJ of energy.
  a) How long could you walk after eating one serving of cereal?
  b) What mass of cereal would provide enough energy for you to walk for 4.0 h?
  c) An average apple provides 283 kJ of energy. How many apples would provide enough energy for you to walk for 4.0 h?

24. When one mole of gaseous water forms from the reaction of $H_2(g)$ and $O_2(g)$, the change in enthalpy is -241.8 kJ. In other words, when hydrogen burns in oxygen or air, it produces a great deal of energy relative to its mass. Since the nineteenth century, scientists have been researching the potential of hydrogen as a fuel. One way in which the energy of the combustion of hydrogen has been successfully harnessed is as rocket fuel for spacecraft shown below.
  a) Write a thermochemical equation for the combustion of hydrogen.
  b) Determine the volume of hydrogen gas at STP that would have to be burned completely to result in an enthalpy change of -9.0×10^5 kJ—about the same amount produced by a tank of gas in a small car.
  c) Suggest three reasons why hydrogen gas is a desirable rocket fuel.
  d) What do you think might be some challenges that engineers have had to overcome in order to make hydrogen a workable rocket fuel for spacecraft.

A mixture of liquid hydrogen and liquid oxygen is used as fuel for the space shuttle.

CHAPTER
10

Theories of Energy and Chemical Changes

Chapter Concepts

10.1 Hess's Law

- Hess's law can be used to determine the net enthalpy change for a chemical reaction from a series of reactions.

- Enthalpy change (ΔH) can be predicted from standard enthalpies of formation.

- Calorimetry can be used along with Hess's law to determine enthalpy changes.

10.2 Energy and Efficiency

- The efficiency of a thermal energy conversion system is given by:

  $$\text{Efficiency} = \frac{\text{Energy output}}{\text{Energy input}} \times 100\%$$

- Hess's law and calorimetry data can be combined to determine the efficiency of a thermal energy conversion device.

- Consumer products differ in efficiency, and consumers can choose more energy-efficient products.

10.3 Fuelling Society

- Canadians rely on the potential energy stored in fossil fuels for everyday life.

- Although fossil fuels have many benefits, using them comes at a cost to the environment.

- Fuels can be evaluated based on their efficiency, heat content, and harmful emissions.

The hibernating little brown bats above stay huddled together as they sleep through the long winter months. The thermal energy released by their body processes helps keep the other bats in the colony warm. What do these bats have in common with the chemical plant on the right? Both have found a way to use thermal energy that would otherwise be wasted—released into the surroundings.

The thermal energy of steam is a by-product of the processes that burn fossil fuel to produce electrical energy in power plants. In traditional power plants, the thermal energy is considered a waste product and is released into the atmosphere. In cogeneration plants, however, the thermal energy is captured and used. For example, the Joffre cogeneration facility, pictured on the right, produces electricity by burning natural gas. The thermal energy produced by the plant is not wasted; it is used to heat a nearby chemical plant and increases the overall efficiency of burning the natural gas. Using the thermal energy that would otherwise be released into the surroundings means that the output of useful energy is increased for the same energy input.

In this chapter, you will learn how to calculate the theoretical amount of energy that is absorbed or released by chemical reactions by using known thermochemical equations. You will also examine the efficiency of traditional energy sources, and you will explore the effects of Canadians' dependence on fossil fuels. In the Launch Lab activity on the next page, you will begin your investigation into efficiency.

Bake a Cake

You may have heard the term *efficiency* used when talking about a car, an air conditioner, a factory, or even a person. What does efficiency mean to you? What does it mean when a process or appliance is "energy efficient?" In this Launch Lab, you will take a closer look at energy efficiency by constructing a working oven from household materials and making improvements to increase its efficiency.

Safety Precautions

- Hot and cold materials often look the same. Do not touch the oven with your bare hands.
- Your teacher may choose to carry out this investigation as a demonstration.

Materials

- light bulbs and receptacle
- retort stand
- cake mix
- disposable metal tin pans
- graduated cylinder
- lamp or flashlight
- available household materials

Procedure

1. With your group, determine how you will assess the efficiency of an oven designed to bake a very small cake by using the energy provided by a light bulb.

2. Design a procedure that will allow you to assess efficiency.

3. Construct a data table to record the appropriate data.

4. Build your oven and test the design.

5. Refine your design if necessary.

6. Using your refined design, test again.

7. Dispose of any waste as directed by your teacher.

Analysis

1. List the criteria your group used to determine efficiency.

2. As a class, discuss your criteria and results, and share your designs. What did all the designs have in common? What was unique?

3. In your opinion, is using a light bulb an efficient way to bake a cake? Why or why not?

The Joffre cogeneration facility, located near Red Deer, Alberta, is the largest cogeneration power plant in Canada.

SECTION
10.1

Hess's Law

Section Outcomes

In this section, you will:
- **explain** and use Hess's law to calculate energy changes for a net reaction from a series of reactions
- **predict** the enthalpy change for chemical reactions by using standard heats of formation
- **compare** molar enthalpy data from calorimetry experiments with theoretical values obtained by using Hess's law

Key Terms

Hess's law
thermal stability
Standard molar enthalpy of formation, $\Delta_f H°$

Germain Henri Hess (1802–1850), a Swiss-born Russian chemist, stated his "law of constant heat summation" in 1840. Hess's law, as it is now more commonly known, enabled other scientists to develop theories of chemical thermodynamics.

In Section 9.2, you learned how to use a simple calorimeter to determine the thermal energy released or absorbed during a chemical reaction. Simple calorimeters are generally used only for reactions and changes involving dilute aqueous solutions. There are many non-aqueous chemical reactions, however. Furthermore, there are reactions that occur too slowly for the calorimetric method to be practical. There are also many reactions that release so much energy they are not safe to perform using a simple calorimeter. Imagine trying to determine the enthalpy of reaction for the detonation of nitroglycerin, an unstable and powerfully explosive compound.

Understanding Hess's Law

Chemists can determine the enthalpy change of any reaction by using an important law, known as Hess's law (see Figure 10.1). Which is stated in the box below.

Hess's Law

The enthalpy change of a physical or chemical process depends only on the initial and final states. The enthalpy change of the overall process is the sum of the enthalpy changes of its individual steps.

You can develop an understanding of Hess's law by analyzing the following example. Carbon dioxide can be formed from carbon and oxygen by one step or by two steps as shown below.

Two steps	**One step**
$2C(s) + O_2(g) \rightarrow 2CO(g)$	$C(s) + O_2(g) \rightarrow CO_2(g)$
$2CO(g) + O_2(g) \rightarrow 2CO_2(g)$	

According to Hess's law, the sum of the enthalpy changes for the set of two reactions on the left should be the same as the enthalpy change for the single reaction on the right. Before attempting to apply Hess's law, you must consider two factors. Notice that if you determined the enthalpy changes for the reactions on the left as written, you would obtain the enthalpy change for two moles of carbon dioxide. The enthalpy change for the reaction on the right, as written, would represent the formation of one mole of carbon dioxide. To compare the reactions, you need to compare the same number of moles of product. Therefore, you would divide the coefficients of the two reactions on the left by two.

$$C(s) + \frac{1}{2}O_2(g) \rightarrow CO(g)$$
$$CO(g) + \frac{1}{2}O_2(g) \rightarrow CO_2(g)$$

The next factor to consider is the conditions under which the reactions occur. As you know, the enthalpy change of a reaction is influenced by the temperature and pressure of the system. To compare enthalpy changes of chemical reactions, they must be carried out under the same conditions. Chemists have agreed to use the standard enthalpy change, $\Delta H°$ that implies conditions at SATP. For chemical reactions taking place in solution, the standard state includes a concentration of exactly 1 mol/L. Now you can compare the two pathways by which you could form carbon dioxide from carbon and oxygen.

	Two steps			**One step**

$$C(s) + \frac{1}{2}O_2(g) \rightarrow CO(g) \quad \Delta H° = -110.5 \text{ kJ} \qquad C(s) + O_2(g) \rightarrow CO_2(g) \quad \Delta H° = -393.5 \text{ kJ}$$

$$CO(g) + \frac{1}{2}O_2(g) \rightarrow CO_2(g) \quad \underline{\Delta H° = -283.0 \text{ kJ}}$$

$$\Delta H° = -393.5 \text{ kJ}$$

The standard enthalpy change for the two pathways is the same. You can also represent this comparison on a potential energy diagram as shown in Figure 10.2.

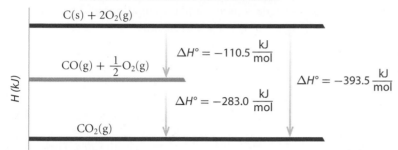

Oxidation of Carbon to Carbon Dioxide

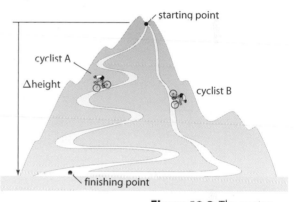

Figure 10.2 Carbon dioxide can be formed by the reaction of carbon with oxygen via two different pathways. The enthalpy change of the overall reaction is the same regardless of the method or pathways that is used.

Two Pathways, One Change

One way to think about Hess's law is to compare the energy changes that occur in a chemical reaction with the changes in the potential energy of a cyclist on hilly terrain. This comparison is shown in Figure 10.3. Hess's law is valid because enthalpy change is determined only by the initial and final conditions of the system. It is not dependent on the path taken by the system. Hess's law allows you to calculate the enthalpy change of a chemical reaction without collecting experimental data by using calorimetry.

Combining Chemical Equations Algebraically

You can use Hess's law to find the enthalpy change of any reaction as long as you know the enthalpy changes for any set of reactions that add to the reaction of interest. For example, the first two steps in the industrial synthesis of sulfuric acid (see Figure 10.4) are shown here.

(1) $S_8(s) + 8O_2(g) \rightarrow 8SO_2(g)$ **(2)** $2SO_2(g) + O_2(g) \rightarrow 2SO_3(g)$

Figure 10.3 The routes that cyclists take to get from the starting point to the finishing point have no effect on the net change in the cyclists' gravitational potential energy.

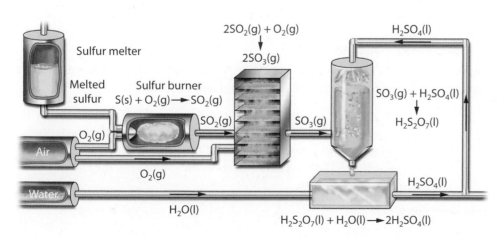

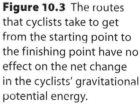

Figure 10.4 The industrial production of sulfuric acid shown here is called the contact process. To speed the reaction of the second step, the oxidation of sulfur dioxide, the gases come in contact with a catalyst illustrated by the box with many layers of the catalyst.

Suppose you would like to find the enthalpy change for the formation of sulfur trioxide from sulfur and oxygen. The overall equation is:

$$S_8(s) + 12O_2(g) \rightarrow 8SO_3(g)$$

You need to examine all of the equations and decide whether you need to divide the coefficients by a constant. Since you want to find the enthalpy change of one mole of sulfur trioxide, you need to divide the overall equation by eight. You must also divide equation (2) by two. Since equation (2) will now have one mole of sulfur dioxide on the left, you need to divide equation (1) by eight so it will produce only one mole of sulfur dioxide. Write equations (1) and (2) with their arrows aligned and put their known standard enthalpy changes beside the equations and add the equations and the standard enthalpy changes.

$$\frac{1}{8} S_8(s) + O_2(g) \rightarrow SO_2(g) \qquad \Delta H° = -296.8 \text{ kJ}$$

$$SO_2(g) + \frac{1}{2}O_2(g) \rightarrow SO_3(g) \qquad \Delta H° = -99.2 \text{ kJ}$$

$$\frac{1}{8} S_8(s) + O_2(g) + SO_2(g) + \frac{1}{2}O_2(g) \rightarrow SO_2(g) + SO_3(g) \qquad \Delta H° = -296.8 \text{ kJ} -99.2 \text{ kJ}$$

Examine the final equation to find any compounds occurring on both sides. These compounds will cancel each other. Also look for any compounds that occur more than once on the same side and add them together.

$$\frac{1}{8} S_8(s) + \left(1 + \frac{1}{2}\right) O_2(g) + \cancel{SO_2(g)} \rightarrow \cancel{SO_2(g)} + SO_3(g) \qquad \Delta H° = -296.8 \text{ kJ} -99.2 \text{ kJ}$$

$$\frac{1}{8}S_8(s) + \frac{3}{2}O_2(g) \rightarrow SO_3(g) \qquad \Delta H° = -396.0 \text{ kJ}$$

The potential energy diagram representing the process of synthesizing sulfur trioxide is shown in Figure 10.5.

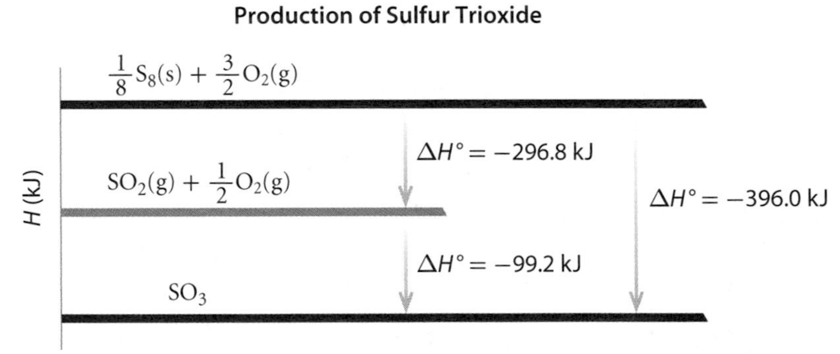

Production of Sulfur Trioxide

Figure 10.5 This potential energy diagram shows the first two steps in the industrial production of sulfuric acid.

$$\frac{1}{8} S_8(s) + \frac{3}{2} O_2(g)$$

H (kJ)

$$SO_2(g) + \frac{1}{2} O_2(g)$$

$$\Delta H° = -296.8 \text{ kJ}$$

$$\Delta H° = -396.0 \text{ kJ}$$

$$\Delta H° = -99.2 \text{ kJ}$$

SO_3

Manipulating Equations

You will often need to manipulate the equations before adding them, as was done above. There are two key ways in which you can manipulate an equation:

1. *Reverse an equation* so that the products become reactants and the reactants become products. When you reverse an equation, you need to change the sign of ΔH (multiply by -1). When a reaction is written in reverse, the enthalpy change has the opposite sign.

2. *Multiply each coefficient* in an equation by the same integer or fraction. When you multiply an equation, you need to multiply ΔH by the same number. As you learned in Chapter 9, enthalpy change is directly related to the amount of the substances involved in a reaction. If two candles are burned instead of one, then twice as much heat and light will be produced.

Examine the following Sample Problems to see how to manipulate equations so that they add to the overall equation. Try the Practice Problems that follow to practice finding the enthalpy change by adding equations.

Sample Problem

Using Hess's Law to Determine Enthalpy Change

Problem
One of the methods that the steel industry uses to obtain metallic iron is to react iron(III) oxide, $Fe_2O_3(s)$, with carbon monoxide, $CO(g)$, as shown in the balanced equation below:

$$Fe_2O_3(s) + 3CO(g) \rightarrow 3CO_2(g) + 2Fe(s)$$

Determine the enthalpy change of this reaction, given the following equations and their enthalpy changes.

(1) $CO(g) + \frac{1}{2}O_2(g) \rightarrow CO_2(g)$ $\Delta H° = -283.0 \text{ kJ}$

(2) $2Fe(s) + \frac{3}{2}O_2(g) \rightarrow Fe_2O_3(s)$ $\Delta H° = -824.2 \text{ kJ}$

What Is Required?
You need to manipulate equations (1) and (2), along with their enthalpy changes, so they add up to the overall equation:

$$Fe_2O_3(s) + 3CO(g) \rightarrow 3CO_2(g) + 2Fe(s)$$

What Is Given?
You know the chemical equations for reactions (1) and (2), and you know their corresponding enthalpy changes.

Plan Your Strategy
Step 1 Examine equations (1) and (2) to see how they compare with the overall equation. Decide how you need to manipulate equations (1) and (2) so that they add to the overall equation. (Reverse the equation, multiply the equation by a coefficient, do both, or do neither.) Remember to adjust $\Delta H°$ accordingly for each equation.

Step 2 Write the manipulated equations so that their equation arrows line up. Add the reactants and products on each side, and cancel substances that appear on both sides in equal amounts.

Step 3 Ensure that you have obtained the overall equation. Add $\Delta H°$ for the combined equation.

Act on Your Strategy
Step 1
- Equation (1) has $CO(g)$ as a reactant and $CO_2(g)$ as a product, as does the overall equation. The stoichiometric coefficients do not match the coefficients of the overall equation, however. To achieve the same coefficients, you must multiply equation (1) by 3.
- Equation (2) has the required stoichiometric coefficients, but $Fe(s)$ and $Fe_2O_3(s)$ are on the wrong sides of the equation. You need to reverse equation (2) and, therefore, change the sign of the $\Delta H°$.
- **Note:** Since oxygen gas, $O_2(g)$, is present in both equations that you are manipulating, do not use $O_2(g)$ to decide on how to manipulate the equations. Always start with a chemical species that is present in only one of the equations and is also present in the overall equation.

Step 2

3 × (1) $3CO(g) + \frac{3}{2}O_2(g) \rightarrow 3CO_2(g)$ $\Delta H° = 3(-283.0 \text{ kJ})$

−1 × (2) $Fe_2O_3(s) \rightarrow 2Fe(s) + \frac{3}{2}O_2(g)$ $\Delta H° = -1(-824.2 \text{ kJ})$

Step 3

$$Fe_2O_3(s) + 3CO(g) \rightarrow 3CO_2(g) + 2Fe(s)$$

$$\Delta H° = 3(-283.0 \text{ kJ}) + (-1(-824.2 \text{ kJ}))$$
$$= -24.8 \text{ kJ}$$

The manipulated equations add to the overall equation. Therefore, the sum of the manipulated enthalpy changes is the enthalpy change of the overall equation:

$$Fe_2O_3(s) + 3CO(g) \rightarrow 3CO_2(g) + 2Fe(s) \quad \Delta H° = -24.8 \text{ kJ}$$

Check Your Solution
The equations added correctly to the overall equation. Check to ensure that you adjusted $\Delta H°$ accordingly for each equation. Because you added the $\Delta H°$ values, the final answer will be as precise as the least precise number used in the calculation. The final answer has one digit after the decimal point, which is correct.

Using Hess's Law to Determine Enthalpy Change for Formation Reactions

Problem

From the following information, calculate the enthalpy of formation of solid phosphorus pentachloride as shown in the equation below:

$$\frac{1}{4}P_4(s) + \frac{5}{2}Cl_2(g) \rightarrow PCl_5(s) \qquad \Delta H° = ?$$

(1) $P_4(s) + 6Cl_2(g) \rightarrow 4PCl_3(\ell) \qquad \Delta H° = -1272 \text{ kJ}$

(2) $PCl_3(\ell) + Cl_2(g) \rightarrow PCl_5(s) \qquad \Delta H° = -125 \text{ kJ}$

What Is Required?

You need to manipulate and add equations (1) and (2) along with their enthalpy changes to obtain the overall equation.

What Is Given?

You know the overall equation and the thermochemical equations for reactions (1) and (2).

Plan Your Strategy

The overall equation contains fractions. To simplify your manipulations of equations (1) and (2), you may find it easier to first find the enthalpy change for the overall equation with whole-number coefficients and then divide to obtain the true overall equation at the end of the problem.

First, balance the overall equation with whole numbers by multiplying the coefficients by 4:

$$P_4(s) + 10Cl_2(g) \rightarrow 4PCl_5(s) \qquad \Delta H° = ?$$

Second, manipulate equations (1) and (2).

- Equation (1) can be used as written to obtain $P_4(s)$ as a reactant.
- Equation (2) has $PCl_5(s)$ on the correct side of the equation but not in the correct stoichiometric quantities. Multiply equation (2) by 4.

Act on Your Strategy

(1) $\qquad P_4(s) + 6Cl_2(g) \rightarrow 4\cancel{PCl_3(\ell)} \quad \Delta H° = -1272 \text{ kJ}$

(2) × 4 $\quad 4\cancel{PCl_3(\ell)} + 4Cl_2(g) \rightarrow 4PCl_5(s) \quad \Delta H° = 4(-125 \text{ kJ})$

$$P_4(s) + 10Cl_2(g) \rightarrow 4PCl_5(s)$$
$$\Delta H° = -1272 \text{ kJ} + 4(-125 \text{ kJ})$$
$$= -1772 \text{ kJ}$$

The desired overall equation is for the formation of only 1 mol of $PCl_5(s)$. Therefore, divide the equation and corresponding $\Delta H°$ by 4, thus giving the following solution.

$$\frac{1}{4}P_4(s) + \frac{5}{2}Cl_2(g) \rightarrow PCl_5(s) \quad \Delta H° = \frac{-1772 \text{ kJ}}{4}$$
$$= -443 \text{ kJ}$$

Therefore, the enthalpy of formation of phosphorus pentachloride is $\Delta H° = -443$ kJ/mol.

Check Your Solution

The equations and the $\Delta H°$ have been adjusted and are added correctly to the overall equation. Substances not appearing in the overall equation have been cancelled. The final answer has the same precision as the $\Delta H°$ used in the calculations (no digits after the decimal point).

---

1. Ethene, $C_2H_4(g)$, reacts with water to form ethanol, $C_2H_5OH(\ell)$, as shown below:

$$C_2H_4(g) + H_2O(\ell) \rightarrow C_2H_5OH(\ell)$$

Determine the enthalpy change of this reaction, given the following thermochemical equations.

(1) $C_2H_5OH(\ell) + 3O_2(g) \rightarrow 3H_2O(\ell) + 2CO_2(g)$
$$\Delta_c H° = -1366.8 \text{ kJ/mol}$$

(2) $\quad C_2H_4(g) + 3O_2(g) \rightarrow 2H_2O(\ell) + 2CO_2(g)$
$$\Delta_c H° = -1411.0 \text{ kJ/mol}$$

2. A typical automobile uses a lead-acid battery. During discharge, the following chemical reaction takes place:

$$Pb(s) + PbO_2(s) + 2H_2SO_4(\ell) \rightarrow$$
$$2PbSO_4(aq) + 2H_2O(\ell)$$

Determine the enthalpy change of this reaction, given the following equations:

(1) $Pb(s) + PbO_2(s) + 2SO_3(g) \rightarrow 2PbSO_4(aq)$
$$\Delta H° = -775 \text{ kJ}$$

(2) $SO_3(g) + H_2O(\ell) \rightarrow H_2SO_4(\ell) \qquad \Delta H° = -133 \text{ kJ}$

3. Mixing household cleansers can result in the production of hydrogen chloride gas, $HCl(g)$. Not only is this gas toxic and corrosive, but it also reacts with oxygen to form poisonous chlorine gas according to the following equation:

$$4HCl(g) + O_2(g) \rightarrow 2Cl_2(g) + 2H_2O(g)$$

Determine the enthalpy change of this reaction, given the following equations:

(1) $H_2(g) + Cl_2(g) \rightarrow 2HCl(g) \ \Delta H° = -184.6 \text{ kJ}$

(2) $H_2(g) + \frac{1}{2}O_2(g) \rightarrow H_2O(\ell) \ \Delta H° = -285.8 \text{ kJ/mol}$

(3) $H_2O(\ell) \rightarrow H_2O(g) \ \Delta_{vap} H° = +40.7 \text{ kJ/mol}$

4. Calculate the enthalpy change of the following reaction between nitrogen gas and oxygen gas, given thermochemical equations (1), (2), and (3):

$$2N_2(g) + 5O_2(g) \rightarrow 2N_2O_5(g)$$

(1) $2H_2(g) + O_2(g) \rightarrow 2H_2O(\ell)$ $\Delta H^\circ = -572$ kJ

(2) $N_2O_5(g) + H_2O(\ell) \rightarrow 2HNO_3(\ell)$ $\Delta H^\circ = -77$ kJ

(3) $\frac{1}{2}N_2(g) + \frac{3}{2}O_2(g) + \frac{1}{2}H_2(g) \rightarrow HNO_3(\ell)$
$$\Delta H^\circ = -174 \text{ kJ/mol}$$

5. Ethene, $C_2H_4(g)$, is used in the manufacture of many polymers, including polyethylene terephthalate (PET), which is used to make pop bottles. Determine the molar enthalpy of formation for ethene, as shown by $2C(s) + 2H_2(g) \rightarrow C_2H_4(g)$, given the following equations:

(1) $C(s) + O_2(g) \rightarrow CO_2(g)$ $\Delta H^\circ = -393.5$ kJ

(2) $H_2(g) + \frac{1}{2}O_2(g) \rightarrow H_2O(\ell)$ $\Delta H^\circ = -285.8$ kJ

(3) $C_2H_4(g) + 3O_2(g) \rightarrow 2CO_2(g) + 2H_2O(\ell)$
$$\Delta H^\circ = -1411.2 \text{ kJ}$$

6. From the following equations, determine the molar enthalpy of formation for $HNO_2(aq)$, as shown below in the overall equation:

$$\frac{1}{2}H_2(g) + \frac{1}{2}N_2(g) + O_2(g) \rightarrow HNO_2(aq)$$

(1) $NH_4NO_2(aq) \rightarrow N_2(g) + 2H_2O(\ell)$
$$\Delta H^\circ = -320.1 \text{ kJ}$$

(2) $NH_3(aq) + HNO_2(aq) \rightarrow NH_4NO_2(aq)$
$$\Delta H^\circ = -37.7 \text{ kJ}$$

(3) $2NH_3(aq) \rightarrow N_2(g) + 3H_2(g)$ $\Delta H^\circ = +169.9$ kJ

(4) $H_2(g) + \frac{1}{2}O_2(g) \rightarrow H_2O(\ell)$ $\Delta H^\circ = -285.8$ kJ

Hess's Law and Calorimetry

Sometimes, it is not practical to use a simple calorimeter to find the enthalpy change of a reaction. You can, however, use the calorimeter to find the enthalpy changes of other reactions. You can combine the equations of these reactions to arrive at the equation of the reaction in which you are interested. In the following investigation, you will apply Hess's law to determine the enthalpy change of a reaction.

INVESTIGATION | 10.A

Target Skills

Performing calorimetry experiments to determine the molar enthalpy change of chemical reactions

Using thermometers or temperature probes appropriately when measuring temperature changes

Plotting enthalpy diagrams indicating energy changes

Hess's Law and the Enthalpy of Combustion of Magnesium

Magnesium ribbon burns in air in a highly exothermic combustion reaction. (See equation (1).) A very bright flame accompanies the production of magnesium oxide, as shown in the photograph on the right. It is impractical and dangerous to use a simple calorimeter to determine the enthalpy change for this reaction:

(1) $Mg(s) + \frac{1}{2}O_2(g) \rightarrow MgO(s)$

Instead, you will determine the enthalpy changes for two other reactions (equations (2) and (3) below). You will use these enthalpy changes, along with the known enthalpy change for another reaction (equation (4) below), to determine the enthalpy change for the combustion of magnesium.

(2) $MgO(s) + 2HCl(aq) \rightarrow MgCl_2(aq) + H_2O(\ell)$

(3) $Mg(s) + 2HCl(aq) \rightarrow MgCl_2(aq) + H_2(g)$

(4) $H_2(g) + \frac{1}{2}O_2(g) \rightarrow H_2O(\ell) + 285.8$ kJ

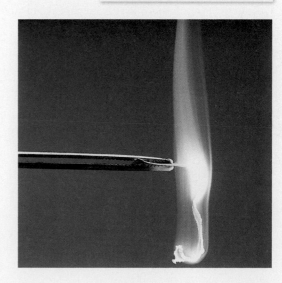

....*continued*

Notice that equations (2) and (3) occur in aqueous solution. You can use a simple calorimeter to determine the enthalpy changes for these reactions. Equation (4) represents the formation of water directly from its elements in their standard state.

Question

What is the molar enthalpy of formation of magnesium oxide?

Predictions

- Predict the molar enthalpy of formation of magnesium oxide.
- Predict whether reactions (2) and (3) will be exothermic or endothermic.

Safety Precautions

- Hydrochloric acid is corrosive. Use care when handling it.
- Be careful not to inhale the magnesium oxide powder.

Materials

- 1.00 mol/L HCl(aq) ☠ ☣
- MgO(s) powder
- Mg ribbon (or Mg turnings) 🔥 ⓣ
- simple calorimeter
- 100 mL graduated cylinder
- scoopula
- electronic balance
- thermometer (alcohol or digital)
- sandpaper or emery paper

Procedure

Part 1: Determining ΔH of Reaction (2)

1. Read the Procedure for Part 1. Prepare a data table to record mass and temperature data.

2. Set up the simple calorimeter (refer to Investigation 9.B on page 358). Using a graduated cylinder, add 100 mL of 1.00 mol/L HCl(aq) to the calorimeter. **Caution** Hydrochloric acid can burn your skin.

3. Record the initial temperature, t_i, of the HCl(aq), to the nearest tenth of a degree.

4. Find the mass of no more than 0.80 g of MgO(s) powder. Record the exact mass.

5. Add the MgO(s) powder to the calorimeter containing the HCl(aq). Swirl the solution gently, recording the highest temperature, t_f, reached.

6. Dispose of the reaction solution as directed by your teacher.

Part 2: Determining ΔH of Reaction (3)

1. Read the Procedure for Part 2. Prepare a data table to record mass and temperature data.

2. Using a graduated cylinder, add 100 mL of 1.00 mol/L HCl(aq) to the calorimeter.

3. Record the initial temperature, t_i, of the HCl(aq) to the nearest tenth of a degree.

4. If you are using magnesium ribbon (as opposed to turnings), sand the ribbon. Accurately determine the mass of no more than 0.50 g of magnesium. Record the exact mass.

5. Add the Mg(s) to the calorimeter containing the HCl(aq). Swirl the solution gently, recording the highest temperature, t_f, reached.

6. Dispose of the solution as directed by your teacher.

Analysis

1. Determine the enthalpy change of reactions (2) and (3). List any assumptions you make and explain why they are valid assumptions.

2. Write thermochemical equations for reactions (2) and (3) using ΔH notation. Ensure the signs you use are correct.

3. **a)** Algebraically combine equations (2), (3), and (4), and their corresponding ΔH values, to arrive at equation (1) and the molar enthalpy of combustion of magnesium.

   b) Draw a potential energy diagram to represent the combining of equations (2), (3), and (4) to obtain equation (1) and the molar enthalpy of combustion of magnesium.

4. **a)** Compare your result with the accepted value of $\Delta_c H$ for magnesium. Calculate your percent error.

   b) Suggest some sources of error in the procedure. In what ways could you improve the procedure?

Conclusion

5. Explain how you used Hess's law to determine ΔH for the combustion of magnesium. State the result you obtained for the thermochemical equation that corresponds to chemical equation (1).

6. Design an investigation to test Hess's law by using the following equations.

(1) $NaOH(s) \rightarrow NaOH(aq)$
(2) $NaOH(s) + HCl(aq) \rightarrow NaCl(aq) + H_2O(\ell)$
(3) $NaOH(aq) + HCl(aq) \rightarrow NaCl(aq) + H_2O(\ell)$

Assume you have a simple calorimeter, $NaOH(s)$, 1.00 mol/L $HCl(aq)$, 1.00 mol/L $NaOH(aq)$, and standard laboratory equipment. Write a step-by-step procedure for the investigation. Then outline a plan for analyzing your data. Be sure to include appropriate safety precautions. If time permits, obtain your teacher's approval and carry out the investigation.

Standard Molar Enthalpies of Formation

You have learned how to add equations for reactions with known standard enthalpy changes to find the enthalpy change of another reaction. As you could imagine, chemists have measured and calculated the standard enthalpy changes for thousands of chemical reactions. If you could find the right set of equations, you could calculate the standard enthalpy change for nearly any reaction. How can you find these known enthalpy changes? Chemists have collected and organized data for one specific type of reaction that allows you to generate equations for almost any reaction that you could imagine. The type of reaction is a specific class of formation reactions in which the compound must be formed from its elements and not from any other compound.

The enthalpy change of formation for a compound at SATP is called the standard molar enthalpy of formation. The **standard molar enthalpy of formation, $\Delta_f H°$,** is the change in enthalpy when one mole of a compound is formed directly from its elements in their most stable state at SATP (25 °C and 100 kPa). Because elements in their most stable state have been selected as a reference, the standard enthalpy change of formation of the elements has been arbitrarily set at zero.

Some molar enthalpies of formation are listed in Table 10.1. Additional molar enthalpies of formation are found in Appendix G and also in your Chemistry Data Booklet.

Table 10.1 Selected Standard Molar Enthalpies of Formation

Compound	$\Delta_f H°$(kJ/mol)	Formation Equation
$CO(g)$	−110.5	$C(s) + \frac{1}{2}O_2(g) \rightarrow CO(g)$
$CO_2(g)$	−393.5	$C(s) + O_2(g) \rightarrow CO_2(g)$
$CH_4(g)$	−74.6	$C(s) + 2H_2(g) \rightarrow CH_4(g)$
$CH_3OH(\ell)$	−239.2	$C(s) + 2H_2(g) + \frac{1}{2}O_2(g) \rightarrow CH_3OH(\ell)$
$C_2H_5OH(\ell)$	−277.6	$2C(s) + 3H_2(g) + \frac{1}{2}O_2(g) \rightarrow C_2H_5OH(\ell)$
$C_6H_6(\ell)$	+49.1	$6C(s) + 3H_2(g) \rightarrow C_6H_6(\ell)$
$C_6H_{12}O_6(s)$	−1273.3	$6C(s) + 6H_2(g) + 3O_2(g) \rightarrow C_6H_{12}O_6(s)$
$H_2O(\ell)$	−285.8	$H_2(g) + \frac{1}{2}O_2(g) \rightarrow H_2O(\ell)$
$H_2O(g)$	−241.8	$H_2(g) + \frac{1}{2}O_2(g) \rightarrow H_2O(g)$
$CaCl_2(s)$	−795.4	$Ca(s) + Cl_2(g) \rightarrow CaCl_2(s)$
$CaCO_3(s)$	−1207.6	$Ca(s) + C(s) + \frac{3}{2}O_2(g) \rightarrow CaCO_3(s)$
$NaCl(s)$	−411.2	$Na(s) + \frac{1}{2}Cl_2(g) \rightarrow NaCl(s)$
$HCl(g)$	−92.3	$\frac{1}{2}H_2(g) + \frac{1}{2}Cl_2(g) \rightarrow HCl(g)$
$HCl(aq)$	−167.5	$\frac{1}{2}H_2(g) + \frac{1}{2}Cl_2(g) \rightarrow HCl(aq)$

You will notice that many of the formation equations include fractions. It is often necessary to use fractions in formation equations, because by definition they show the formation of exactly one mole of the product compound. For example, the following equation shows the formation of calcium oxide under standard conditions:

$$Ca(s) + \frac{1}{2}O_2(g) \rightarrow CaO(s) \qquad \Delta_f H^\circ = -634.9 \text{ kJ/mol CaO(s)}$$

To produce 1 mol of CaO(s), you need 1 mol of Ca(s) and 0.5 mol of O_2(g). If you were to eliminate the fractions by multiplying the coefficients by two, the equation would show the formation of 2 mol of CaO(s), not 1 mol.

• • •

1. Using the Table of Standard Enthalpies of Formation found in Appendix G or your Chemistry Data Booklet, write a formation equation for the following substances. Include energy as a term in this equation:

   a) hydrogen iodide, HI(g)   c) barium oxide, BaO(s)

   b) iron(III) oxide, Fe_2O_3(s)   d) hydrogen perchlorate, $HClO_4(\ell)$

2. Identify each reaction in question Q1 as either endothermic or exothermic.

3. Draw a potential energy diagram to represent the standard molar enthalpy of formation of sodium chloride.

• • •

Some elements exist in more than one form under standard conditions. For example, two forms of carbon are graphite and diamond, as shown in Figure 10.6. Graphite is the standard state of carbon. Therefore, the standard molar enthalpy of formation of graphite carbon is 0 kJ/mol. The standard molar enthalpy of formation of diamond is +1.9 kJ/mol.

Figure 10.6 Two forms of carbon under standard conditions are graphite and diamond. Carbon can, however, have only one standard state. Carbon's standard state is graphite.

$$C_{(graphite)} \longrightarrow C_{(diamond)} \quad \Delta_f H^\circ = + 1.9 \text{ kJ/mol}$$

The following equation and accompanying enthalpy diagram (Figure 10.7) show the formation of liquid water from its elements under standard conditions.

$$H_2(g) + \frac{1}{2}O_2(g) \rightarrow H_2O(\ell) \qquad \Delta_f H^\circ = -285.8 \text{ kJ/mol}$$

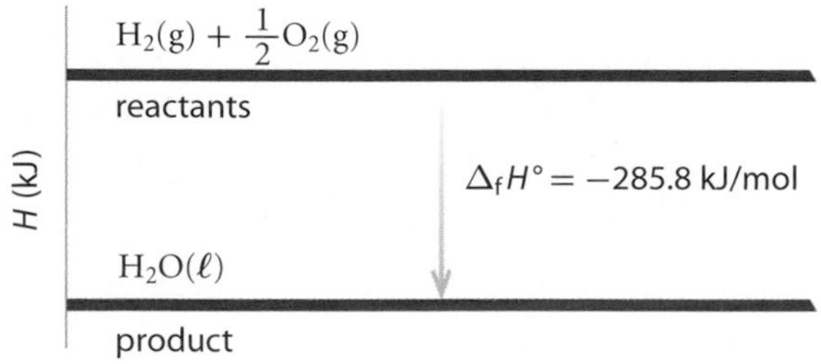

Figure 10.7 As one mole of water is formed from its elements in standard state, 285.8 kJ of energy is released.

$H_2(g) + \frac{1}{2}O_2(g)$

reactants

H (kJ)

$\Delta_f H^\circ = -285.8$ kJ/mol

$H_2O(\ell)$

product

Formation Reactions and Thermal Stability

Formation reactions can also provide information about how stable a substance is. The ability of a substance to resist decomposition when heated is known as its **thermal stability**. The reverse of a formation reaction is a decomposition reaction. The reverse of the enthalpy change of a formation reaction, therefore, is the enthalpy change of a decomposition reaction. The greater the enthalpy change of a decomposition reaction, the more energy input is required to decompose a substance into its elements. In other words, the greater the enthalpy change of decomposition reaction, the greater its thermal stability.

For example, compare the enthalpies of decomposition of calcium carbonate, $CaCO_3(s)$, and methane, $CH_4(g)$. (Use the enthalpies of formation from Table 10.1 and multiply by −1.)

- Methane's enthalpy of decomposition is +74.6 kJ/mol.

- Calcium carbonate's enthalpy of decomposition is +1207.6 kJ/mol.

Calcium carbonate requires a far greater energy input to decompose into its elements; therefore, calcium carbonate is considered more thermally stable.

Try the following problems to practice writing and working with formation reactions.

• • •

4. Write a thermochemical equation for the formation of each substance. Be sure to include the physical state of all the elements and compounds in the equation. You can find the standard enthalpy of formation of each substance in Appendix G.

   a) $CH_4(g)$ **c)** $NaCl(s)$ **e)** $CaCO_3(s)$

   b) $C_6H_6(\ell)$ **d)** $MgO(s)$

5. Arrange the substances in question Q4 in order from most to least thermally stable.

6. Liquid sulfuric acid has a very large negative standard enthalpy of formation (−814.0 kJ/mol). Write an equation to show the formation of liquid sulfuric acid. The standard state of sulfur is orthorhombic sulfur, $S_8(s)$.

7. Write a thermochemical equation for the formation of glucose, $C_6H_{12}O_6(s)$. Compare this with a thermochemical equation for photosynthesis. The enthalpy change of reaction for photosynthesis is +2802.5 kJ/mol of glucose.

• • •

Using Enthalpies of Formation and Hess's Law

A variation of Hess's law allows you to calculate the enthalpy change of a chemical reaction by adding the enthalpies of formation of the products and subtracting the total enthalpies of formation of the reactants. The following equation can be used to determine the enthalpy change of a chemical reaction:

$$\Delta_r H^\circ = \Sigma(n\Delta_f H^\circ \text{ products}) - \Sigma(n\Delta_f H^\circ \text{ reactants})$$

where n represents the stoichiometric coefficient for each substance and Σ means "the sum of."

As usual, you need to begin with a balanced chemical equation. Consider, for example, the complete combustion of methane, $CH_4(g)$, in an open system:

$$CH_4(g) + 2O_2(g) \rightarrow CO_2(g) + 2H_2O(g)$$

Using the equation for the enthalpy change and the standard molar enthalpies of formation in Appendix G, you can calculate the enthalpy change for this reaction.

$$\Delta_r H° = \Sigma(n\Delta_f H° \text{ products}) - \Sigma(n\Delta_f H° \text{ reactants})$$
$$= [(1 \text{ mol})(\Delta_f H° \text{ CO}_2(g)) + (2 \text{ mol})(\Delta_f H° \text{ H}_2O(g))] - [(1 \text{ mol})(\Delta_f H° \text{ CH}_4(g)) + (2 \text{ mol})(\Delta_f H° \text{ O}_2(g))]$$
$$= [(1 \text{ mol})(-393.5 \text{ kJ/mol}) + (2 \text{ mol})(-241.8 \text{ kJ/mol})] - [(1 \text{ mol})(-74.6 \text{ kJ/mol}) + (2 \text{ mol})(0 \text{ kJ/mol})]$$
$$= -877.1 \text{ kJ} - (-74.6 \text{ kJ})$$
$$= -802.5 \text{ kJ}$$

Note: Oxygen gas, $O_2(g)$, at SATP is an element in its most stable state. Therefore, its standard enthalpy of formation is zero.

Therefore,

$$CH_4(g) + 2O_2(g) \rightarrow CO_2(g) + 2H_2O(g) \quad \Delta H° = -802.5 \text{ kJ}$$

Relating Enthalpies of Formation and Hess's Law

Why does this method of adding and subtracting enthalpies of formation work, and how does this method relate to Hess's law? Consider the equations for the formation of each compound that is involved in the reaction of methane with oxygen:

(1) $H_2(g) + \frac{1}{2}O_2(g) \rightarrow H_2O(g)$ $\qquad\qquad \Delta_f H° = -241.8 \text{ kJ}$

(2) $C(s) + O_2(g) \rightarrow CO_2(g)$ $\qquad\qquad \Delta_f H° = -393.5 \text{ kJ}$

(3) $C(s) + 2H_2(g) \rightarrow CH_4(g)$ $\qquad\qquad \Delta_f H° = -74.6 \text{ kJ}$

There is no equation for the formation of oxygen because oxygen is an element in its most stable state.

By adding the formation equations and their enthalpy changes, you can obtain the overall equation. Notice that you need to reverse equation (3) and multiply equation (1) by 2.

2 × (1) $2H_2(g) + O_2(g) \rightarrow 2H_2O(g)$ $\qquad\qquad \Delta H° = 2(-241.8 \text{ kJ})$

(2) $C(s) + O_2(g) \rightarrow CO_2(g)$ $\qquad\qquad \Delta H° = -393.5 \text{ kJ}$

−1 × (3) $CH_4(g) \rightarrow C(s) + 2H_2(g)$ $\qquad\qquad \Delta H° = -1(-74.6 \text{ kJ})$

$$CH_4(g) + 2O_2(g) \rightarrow CO_2(g) + 2H_2O(g) \quad \Delta H° = -802.5 \text{ kJ}$$

This value of $\Delta H°$ is the same as the value you obtained by adding and subtracting enthalpies of formation. Therefore, using enthalpies of formation to determine the enthalpy of a reaction is consistent with Hess's law. In fact, using the formula in the box above makes it unnecessary to look for equations to add and carry out all of the extra steps. Figure 10.8 shows the general process for determining the enthalpy of reaction from enthalpies of formation.

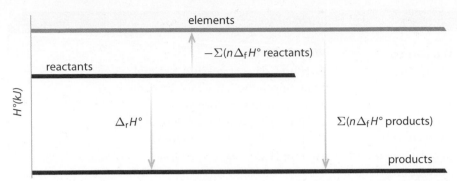

Figure 10.8 To visualize the overall reaction, imagine the reactants decomposing entirely into their elements and then the elements forming the products.

When using enthalpies of formation, remember that for most reactions the *reactants do not actually break down into their elements and then react to form products*. Since there are extensive data about enthalpies of formation, however, it is useful to calculate the overall enthalpy change this way. Remember that, according to Hess's law, the enthalpy change is the same, regardless of the pathway. The first Sample Problem that follows shows how to use enthalpies of formation to determine the enthalpy change of a reaction. Examine the second Sample Problem to learn how to determine enthalpy of formation using a known enthalpy of reaction. Then try the Practice Problems that follow.

Sample Problem

Using Enthalpies of Formation

Problem
Iron(III) oxide reacts with carbon monoxide to produce elemental iron and carbon dioxide. Determine the enthalpy change of this reaction by using known enthalpies of formation:

$$Fe_2O_3(s) + 3CO(g) \rightarrow 3CO_2(g) + 2Fe(s)$$

What Is Required?
You need to find $\Delta_r H°$ for the given chemical equation by using $\Delta_f H°$ data.

What Is Given?
From Appendix G, you can find the molar enthalpies of formation:

$\Delta_f H°$ of $Fe_2O_3(s) = -824.2$ kJ/mol
$\Delta_f H°$ of $CO(g) = -110.5$ kJ/mol
$\Delta_f H°$ of $CO_2(g) = -393.5$ kJ/mol
$\Delta_f H°$ of $Fe(s) = 0$ kJ/mol (by definition, because $Fe(s)$ is elemental iron in its most stable state)

Plan Your Strategy
Multiply the enthalpies of formation by their stoichiometric coefficients. Subtract the total enthalpy of formation for the reactants from the total enthalpy of formation for the products:

$$\Delta_r H° = \Sigma(n\Delta_f H° \text{ products}) - \Sigma(n\Delta_f H° \text{ reactants})$$

$\bullet\bullet\bullet$

Act on Your Strategy
$\Delta_r H° = \Sigma(n\Delta_f H° \text{ products}) - \Sigma(\Delta_f H° \text{ reactants})$

$\quad = [(3 \text{ mol})(\Delta_f H° \, CO_2(g)) + (2 \text{ mol})(\Delta_f H° \, Fe(s))] - [(1 \text{ mol}))(\Delta_f H° \, Fe_2O_3(s)) + (3 \text{ mol})(\Delta_f H° \, CO(g))]$

$\quad = [(3 \text{ mol})(-393.5 \, \dfrac{kJ}{mol}) + (2 \text{ mol})(0 \, \dfrac{kJ}{mol})] - [(1 \text{ mol}))(-824.2 \, \dfrac{kJ}{mol}) + (3 \text{ mol})(-110.5 \, \dfrac{kJ}{mol})]$

$\quad = [-1180.5 \text{ kJ} + 0 \text{ kJ}] - [-824.2 \text{ kJ} - 331.5 \text{ kJ}]$

$\quad = -1180.5 \text{ kJ} + 1155.7 \text{ kJ}$

$\quad = -24.8 \text{ kJ}$

Therefore,
$Fe_2O_3(s) + 3CO(g) \rightarrow 3CO_2(g) + 2Fe(s) \quad \Delta H° = -24.8 \text{ kJ}$

Check Your Solution
A balanced chemical equation was used in the calculation. The answer is correctly expressed to the same precision as $\Delta_f H°$, with one digit after the decimal point. The units are also correct.

Using an Enthalpy of Combustion to Determine an Enthalpy of Formation

Problem

Octane, $C_8H_{18}(\ell)$, one of the major components in gasoline, burns completely in an open system with oxygen, producing carbon dioxide and water vapour. The molar enthalpy of combustion of octane in these conditions is −5074.1 kJ/mol. Determine the molar enthalpy of formation of octane.

What Is Required?

You need to write a balanced chemical equation for the reaction. You need to find $\Delta_fH°$ of octane by using the given molar enthalpy of combustion for octane and $\Delta_fH°$ data.

What Is Given?

$\Delta_cH°$ of $C_8H_{18}(\ell) = -5074.1$ kJ/mol
From Appendix G, you can obtain the molar enthalpies of formation.
$\Delta_fH°$ of $O_2(g) = 0$ kJ/mol (by definition)
$\Delta_fH°$ of $CO_2(g) = -393.5$ kJ/mol
$\Delta_fH°$ of $H_2O(g) = -241.8$ kJ/mol

Plan Your Strategy

Write a chemical equation representing the combustion of octane in an open system. Use the available $\Delta_fH°$ data and the $\Delta_cH°$ (heat of combustion of octane) to substitute into the equation below, multiplying by the molar coefficients. Solve for the unknown $\Delta_fH°$ for octane:

$\Delta_rH° = \Sigma(n\Delta_fH° \text{ products}) - \Sigma(n\Delta_fH° \text{ reactants})$

• • •

Act on Your Strategy

Write and balance the chemical equation for the combustion of 1 mol of octane:

$C_8H_{18}(\ell) + \dfrac{25}{2}O_2(g) \rightarrow 8CO_2(g) + 9H_2O(g)$

$\Delta_rH° = \Sigma(n\Delta_fH° \text{ products}) - \Sigma(n\Delta_fH° \text{ reactants})$

$\Delta_cH° \, C_8H_{18}(\ell) = [(8 \text{ mol})(\Delta_fH° \, CO_2(g)) + (9 \text{ mol})(\Delta_fH° \, H_2O(g))] - [(1 \text{ mol})(\Delta_fH°C_8H_{18}(\ell)) + (\dfrac{25}{2} \text{ mol})(\Delta_fH° \, O_2(g))]$

$-5074.1 \text{ kJ} = [(8 \,\cancel{\text{mol}})(-393.5 \, \dfrac{\text{kJ}}{\cancel{\text{mol}}}) + (9 \,\cancel{\text{mol}})(-241.8 \, \dfrac{\text{kJ}}{\cancel{\text{mol}}})] - [(1 \text{ mol})(\Delta_fH° \, C_8H_{18}(\ell)) + (\dfrac{25}{2} \,\cancel{\text{mol}})(0 \, \dfrac{\text{kJ}}{\cancel{\text{mol}}})]$

$-5074.1 \text{ kJ} = -3148.0 \text{ kJ} - 2176.2 \text{ kJ} - (1 \text{ mol})(\Delta_fH° \, C_8H_{18}(\ell))$

$(1 \text{ mol})(\Delta_cH° \, C_8H_{18}(\ell) = +5074.1 \text{ kJ} - 3148.0 \text{ kJ} - 2176.2 \text{ kJ}$

$\Delta_fH° \, C_8H_{18}(\ell) = \dfrac{-250.1 \text{ kJ}}{1 \text{ mol}}$

$= -250.1 \, \dfrac{\text{kJ}}{\text{mol}}$

The enthalpy of formation of octane, $C_8H_{18}(\ell)$, is −250.1 kJ/mol.

Check Your Solution

A balanced chemical equation was used in the calculation with appropriate molar coefficients. The answer is and should be expressed to the same precision as $\Delta_fH°$, with one digit after the decimal point. The units are also correct.

Practice Problems

Refer to Appendix G or your Chemistry Data Booklet for data on molar enthalpy of formation.

7. Hydrogen can be added to ethene, $C_2H_4(g)$, to obtain ethane, $C_2H_6(g)$:

$C_2H_4(g) + H_2(g) \rightarrow C_2H_6(g)$

Write thermochemical equations for the formation of both ethene, $C_2H_4(g)$, and ethane, $C_2H_6(g)$. Show that these equations can be algebraically combined to obtain the equation for the addition of hydrogen to ethene. Determine the enthalpy change of reaction.

8. Zinc sulfide reacts with oxygen gas to produce zinc oxide and sulfur dioxide:

$2ZnS(s) + 3O_2(g) \rightarrow 2ZnO(s) + 2SO_2(g)$

Calculate the enthalpy change of this reaction by using enthalpies of formation from your Chemistry Data Booklet.

9. Small amounts of oxygen gas can be produced in a laboratory by heating potassium chlorate, $KClO_3(s)$:

$2KClO_3(s) \rightarrow 2KCl(s) + 3O_2(g)$

Calculate the enthalpy change of this reaction by using enthalpies of formation.

10. Use the following equation to answer the questions below:

$$CH_3OH(\ell) + \frac{3}{2}O_2(g) \rightarrow CO_2(g) + 2H_2O(g)$$

a) Calculate the molar enthalpy of combustion of methanol by using enthalpies of formation.

b) Use your answer from (a) to determine how much energy is released when 125 g of methanol undergoes complete combustion.

11. Consider the following equation representing the reaction of methane and chlorine to form chloroform, $CHCl_3(g)$:

$$CH_4(g) + 3Cl_2(g) \rightarrow CHCl_3(g) + 3HCl(g)$$
$$\Delta H° = -305.0 \text{ kJ}$$

Use standard molar enthalpies of formation to determine the molar enthalpy of formation for chloroform, $CHCl_3(g)$.

12. The molar enthalpy of combustion of heptane, $C_7H_{16}(\ell)$, in a bomb calorimeter is -4816.7 kJ/mol of heptane. Using this and $\Delta_fH°$ data, determine the molar enthalpy of formation of heptane.

Section 10.1 Summary

- Hess's Law states that the enthalpy change of a physical of chemical process depends only on the initial and final states. The enthalpy change of the overall process is the sum of the enthalpy changes of its individual steps.

- You can add any number of chemical equations to obtain an equation that you need, and the enthalpy change of the overall reaction is the sum of the enthalpy changes of the individual reactions.

- You can manipulate chemical equations to make them fit into another set of reactions by multiplying the equation by a constant or by reversing the equation. If you multiply by a constant, you must multiply the enthalpy change by that same constant. If you reverse an equation, you must change the sign of the enthalpy change.

- The standard enthalpy of formation of a compound ($\Delta_fH°$) is the enthalpy change for synthesizing the compound from its elements in their most stable state under standard conditions.

- The enthalpy of formation of an element in its most stable state under standard conditions is arbitrarily set at zero.

- You can build any chemical equation by summing the enthalpies of formation for each reactant and product.

- You can calculate the enthalpy change for any reaction by applying the formula:

$$\Delta_rH° = \Sigma(n\Delta_fH° \text{ products}) - \Sigma(n\Delta_fH° \text{ reactants})$$

SECTION 10.1 REVIEW

1. Explain why you need to change the sign of $\Delta H°$ when you reverse an equation. Use an example in your answer.

2. Rank the following compounds from most to least thermally stable by using data on enthalpy of formation found in Appendix G or your Chemistry Data Booklet: $C_2H_4(g)$, $C_2H_2(g)$, $H_2S(g)$, $NaCl(s)$, and $O_2(g)$.

3. Benzene, $C_6H_6(\ell)$, is naturally produced in the intense heat of volcanic eruptions and forest fires.
   a) Write a thermochemical equation for the formation of benzene.
   b) 32.6 g of benzene is formed from elements in their standard state. What is the enthalpy change?

4. Calculate the enthalpy change of the following reaction, given equations (1), (2) and (3).

$$2H_3BO_3(aq) \rightarrow B_2O_3(s) + 3H_2O(\ell)$$

(1) $H_3BO_3(aq) \rightarrow HBO_2(aq) + H_2O(\ell)$ $\Delta H° = -0.02$ kJ
(2) $H_2B_4O_7(s) + H_2O(\ell) \rightarrow 4HBO_2(aq)$ $\Delta H° = -11.3$ kJ
(3) $H_2B_4O_7(s) \rightarrow 2B_2O_3(s) + H_2O(\ell)$ $\Delta H° = 17.5$ kJ

5. "Using enthalpies of formation is like a shortcut for manipulating and adding equations to obtain $\Delta H°$." Do you agree with this statement? Explain your answer.

Energy and Efficiency

Section Outcomes

In this section, you will:
- **determine** the efficiency of thermal energy conversion systems
- **design**, **build**, and **evaluate** a heating device
- **explain** the discrepancy between the theoretical and the actual efficiency of energy conversion systems
- **identify** ways to use energy more efficiently
- **use** appropriate SI units to express enthalpy changes and molar enthalpies

Key Terms

efficiency
energy output
energy input

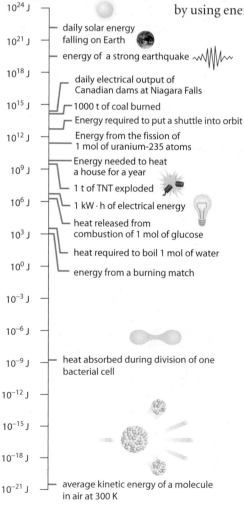

Compared with many other countries, Canada has huge energy requirements per capita. This energy demand is due in part to Canada's vast size compared with its population and the energy required for transportation of goods and people as a result. Also, Canada's northern climate means that, for many months of the year, Canadians rely on natural gas, oil, or electricity to heat their homes.

Where does the energy come from? Most of Canada's energy comes from chemical processes, such as the combustion of petroleum, coal, or natural gas. A significant portion of the energy is derived from nuclear processes. Most power plants, whether nuclear or chemical, use the heat generated by exothermic processes to convert water into steam. The steam moves a turbine, which generates electrical energy.

Energy from Physical, Chemical, and Nuclear Processes

How does the energy from physical, chemical, and nuclear processes compare, and how do the differences affect their use as energy sources?

- Changes of state involve the breaking and forming of intermolecular forces. Physical changes, such as changes of state, usually involve 10^0 to 10^2 (1 to 100) kJ/mol. When water vapour condenses, for example, 40.8 kJ/mol of heat is released. The hydrologic (water) cycle, which involves the cycling of water through its liquid and gaseous phases by using energy input from the sun, is the basis for hydroelectricity.

- Chemical changes involve the breaking and forming of chemical bonds, which are stronger than intermolecular forces. Chemical changes usually involve 10^2 to 10^5 kJ/mol, as you learned in the previous chapter and section. For example, the molar enthalpy of combustion of coal, which is used to produce about half of Alberta's electricity, is about 3900 kJ/mol.

- Nuclear changes involve changes within the nuclei of atoms. These types of changes involve enormous changes in energy. Nuclear changes involve millions and billions of kilojoules per mole. For example, nuclear power plants derive their energy from the fission (splitting) of uranium-235. The fission of 1 mol of uranium-235 releases about 2.1×10^{10} kJ, or 21 billion kilojoules. In Canada, nuclear power is harnessed in plants located in Ontario, Quebec, and New Brunswick.

Figure 10.9 shows the energy changes associated with various processes.

Figure 10.9 Chemical reactions involve energy changes that tend to be greater than those associated with physical changes, but less than those associated with nuclear changes.

Energy and Efficiency

When you think about energy efficiency, what comes to mind? You may think about taking the stairs instead of the elevator, choosing to drive a small car instead of a sport utility vehicle, or turning off the lights when you are not using them. What, however, does efficiency really mean? How did you define "efficiency" in the Launch Lab at the beginning of this chapter? How do you quantify efficiency?

There are several ways to define efficiency. One general definition says that energy efficiency is the ability to produce a desired effect with minimum energy expenditure. For example, suppose that you want to cook a potato. You can use a microwave oven or a conventional oven. Both options achieve the same effect (cooking the potato), but the first option uses less energy. According to the general definition above, using the microwave oven is more energy efficient than using the conventional oven. The general definition is useful, but it is not quantitative.

The following definition quantitatively compares input and output of energy. **Efficiency** is *the ratio of useful energy produced (**energy output**) to energy used in its production (**energy input**), expressed as a percentage*. When you use this definition, however, you need to be clear about what you mean by "energy used." Figure 10.10 shows factors to consider when calculating efficiency or analyzing efficiency data.

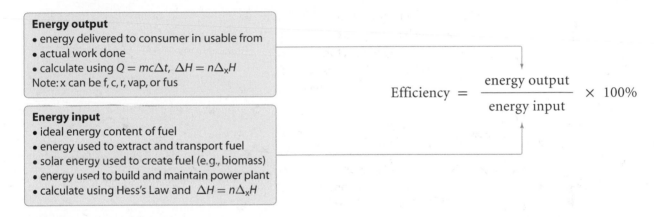

Energy output
- energy delivered to consumer in usable from
- actual work done
- calculate using $Q = mc\Delta t$, $\Delta H = n\Delta_xH$

Note: x can be f, c, r, vap, or fus

Energy input
- ideal energy content of fuel
- energy used to extract and transport fuel
- solar energy used to create fuel (e.g., biomass)
- energy used to build and maintain power plant
- calculate using Hess's Law and $\Delta H = n\Delta_xH$

$$\text{Efficiency} = \frac{\text{energy output}}{\text{energy input}} \times 100\%$$

Figure 10.10 Efficiency is expressed as a percentage. Always specify what is included in the "energy input" part of the ratio.

It is often difficult to determine how much energy is used and exactly how much useful energy is produced. Often, an efficiency percentage only takes into account the ideal energy output of a system, based on the theoretical energy content of the fuel.

The efficiency of a thermal energy conversion (heating) system is often calculated by using calorimetry data (energy output) and a calculation for energy input, which is a theoretical quantity (usually determined by Hess's law). The energy output takes into account only the useful energy for a particular process, which, in the above example, would be in cooking the potato. The energy input includes that energy plus all of the "wasted" energy, which is used to heat the oven, the air in the oven, and your kitchen. Examine the following Sample Problem to learn how to determine energy efficiency. Try the problems that follow to practice making calculations by using the following equation:

$$\text{Efficiency} = \frac{\text{Energy output}}{\text{Energy input}} \times 100\%$$

Sample Problem

The Efficiency of a Propane Barbecue

Problem

Propane, $C_3H_8(g)$, is a commonly used barbecue fuel. Determine the efficiency of the barbecue as a heating device if 5.10 g of propane are required to change the temperature of 250 g of water contained in a 500 g stainless steel pot ($c = 0.503$ J/g • °C) from 25.0 °C to 75.0 °C.

What Is Required?

You need to find how much energy was released by the burning propane. You need to find how much energy was absorbed by the water and pot. You then need to use those values to determine efficiency.

What Is Given?

Energy input

Mass of propane, $C_3H_8(g)$, used = 5.10 g

Energy output

$m_{water} = 250$ g

$m_{steel\ pot} = 500$ g

$c_{water} = 4.19$ J/g • °C

$c_{stainless\ steel} = 0.503$ J/g • °C

Initial temperature, $t_i = 25.0$ °C

Final temperature, $t_f = 75.0$ °C

Plan Your Strategy

Energy input

Write a balanced equation for the complete combustion of propane, and calculate $\Delta_cH°$ by using enthalpies of formation.

Calculate the number of moles of propane burned from the mass and use the equation $\Delta H = n\Delta_cH°$ to determine the theoretical energy content of the fuel.

Energy output

Use the equation $Q = mc\Delta t$ to determine how much energy was absorbed by the pot and water.

Efficiency

Finally, calculate efficiency by using the following equation:

$$\text{Efficiency} = \frac{\text{Energy output}}{\text{Energy input}} \times 100\%$$

Act on Your Strategy

Energy input

The equation for the complete combustion of propane is shown below. Recall that hydrocarbon combustion in an open system, such as a barbecue, produces gaseous water:

$$C_3H_8(g) + 5O_2(g) \rightarrow 3CO_2(g) + 4H_2O(g)$$

$$\begin{aligned}\Delta_cH° &= \Sigma(n\Delta_fH \text{ products}) - \Sigma(n\Delta_fH \text{ reactants})\\ &= [(3\text{ mol})(\Delta_fH°\,CO_2(g)) + (4\text{ mol})(\Delta_fH°\,H_2O(g))] - [(1\text{ mol})(\Delta_fH°\,C_3H_8(g)) + (5\text{ mol})(\Delta_fH°\,O_2(g))]\\ &= [(3\,\cancel{\text{mol}})(-393.5\,\tfrac{\text{kJ}}{\cancel{\text{mol}}}) + (4\,\cancel{\text{mol}})(-241.8\,\tfrac{\text{kJ}}{\cancel{\text{mol}}})] - [(1\,\cancel{\text{mol}})(-103.8\,\tfrac{\text{kJ}}{\cancel{\text{mol}}}) + (5\,\cancel{\text{mol}})(0\,\tfrac{\text{kJ}}{\cancel{\text{mol}}})]\\ &= -2043.9\text{ kJ}\end{aligned}$$

Therefore, the molar enthalpy of combustion of propane is $\Delta_cH° = -2043.9$ kJ/mol.

$$\begin{aligned}n = \frac{m}{M} &= \frac{5.10\,\cancel{\text{g}}\,C_3H_8(g)}{44.11\,\frac{\cancel{\text{g}}}{\text{mol}}}\\ &= 0.116\text{ mol }C_3H_8(g)\end{aligned}$$

$$\begin{aligned}\Delta H = n\Delta_cH° &= 0.116\,\cancel{\text{mol}}\,C_3H_8(g)(-2043.9\,\tfrac{\text{kJ}}{\cancel{\text{mol}}})\\ &= -236\text{ kJ}\end{aligned}$$

Energy output

$$\begin{aligned}Q_{total} &= Q_{water} + Q_{steel}\\ &= mc\Delta t_{water} + mc\Delta t_{steel}\\ &= (250\,\cancel{\text{g}})\left(\frac{4.19\text{ J}}{\cancel{\text{g}}\cdot\cancel{°\text{C}}}\right)(75.0\,\cancel{°\text{C}} - 25.0\,\cancel{°\text{C}}) + (500\,\cancel{\text{g}})\left(\frac{0.503\text{ J}}{\cancel{\text{g}}\cdot\cancel{°\text{C}}}\right)(75.0\,\cancel{°\text{C}} - 25.0\,\cancel{°\text{C}})\\ &= 6.50 \times 10^4\text{ J}\\ &= 65.0\text{ kJ}\end{aligned}$$

Efficiency

$$\text{Efficiency} = \frac{\text{Energy output}}{\text{Energy input}} \times 100\%$$

$$= \frac{65.0 \ \cancel{kJ}}{236 \ \cancel{kJ}} \times 100\%$$

$$= 27.5\%$$

As a heating device, the barbecue was 27.5% efficient.

Check Your Solution

The units are correct. The efficiency is significantly less than 100%, which makes sense. Much of the thermal energy is transferred to the surrounding air and the barbecue itself, rather than to the food being heated.

Practice Problems

13. Using the data for the molar enthalpy of combustion of butane from Table 9.1 (page 347), determine the efficiency of a lighter as a heating device if 0.70 g of butane is required to raise the temperature of a 250 g stainless steel spoon ($c = 0.503$ J/g·°C) by 45 °C.

14. A camping fuel has a posted energy content of 50 kJ/g. Determine its efficiency if a 2.50 g piece was required to raise the temperature of 500 g of soup ($c = 3.77$ J/g·°C) in a 50 g aluminium pot by 45 °C.

15. What mass of pentane, $C_5H_{12}(g)$, would have to be burned in an open system to heat 250 g of hot chocolate ($c = 3.59$ J/g·°C) from 20.0 °C to 39.8 °C if the energy conversion is 45% efficient?

16. Determine the efficiency of a heating device that burns methanol, $CH_3OH(\ell)$, given the following information:

Data for Determining the Efficiency of a Methanol-Burning Heater

Quantity being measured	Data
Initial mass of burner	38.37 g
Final mass of burner	36.92 g
Mass of aluminium can	257.36 g
Mass of aluminium can and water	437.26 g
Initial temperature of water	10.45 °C
Final temperature of water	23.36 °C
$\Delta_c H(CH_3OH)$	$726.1 \frac{kJ}{mol}$

INVESTIGATION 10.B

Target Skills

Designing and **building** a heating device

Determining the efficiency of thermal energy conversion systems

Evaluating a personally designed and constructed heating device

Build a Heating Device

In this investigation, you and the members of your group will design, build, and calculate the efficiency of a heating device. You will have the opportunity to test your design and make modifications to it in order to improve its efficiency.

Problem

How can you build an efficient heating device?

Safety Precautions

• Tie back long hair and secure any loose clothing. Before you light any fuel source, check that there are no flammable solvents nearby. 🔥

Materials

• balance
• thermometer (alcohol or digital)
• stirring rod
• water
• other available materials as dictated by your plan

Design Specifications

1. As a class, develop a list of specifications that each group's heater must meet.

Plan and Construct

2. In your group, put together an initial design for your water heater that meets the class specifications.

....continued

3. Design a data table to record the data necessary to determine efficiency.

4. Think about the calculations that you will have to perform to calculate the efficiency of your heater. Will you be collecting enough data?

5. Check your initial design with your teacher.

6. Build your heater. Carry out your efficiency investigation and record all observations and data.

7. Determine the efficiency of your water heater.

   a) Calculate the energy output (the thermal energy absorbed by the water).

   b) Calculate the theoretical energy input from your energy source.

   c) Calculate efficiency by using this formula:

   $$\text{Efficiency} = \frac{\text{Energy output}}{\text{Energy input}} \times 100\%$$

8. With the members of your group, list sources of energy loss from your system.

9. Revisit your design and make improvements that will decrease energy loss.

10. Repeat the investigation with your improved design.

Evaluate and Communicate

1. Evaluate your original and improved designs. What other design modifications would you like to make? Explain your answer.

2. Perform an Internet search to investigate ways in which the efficiency of water heaters can be improved. How would you apply this knowledge to your classroom water heater? **ICT**

3. Compare your design with the designs of other groups. Write a one-page advertisement for your water heater that highlights its selling points.

ChemistryFile

WebLink

In 1997, representatives from Canada and more than 160 other countries met in Kyoto, Japan, to draft the Kyoto Protocol. They were attempting to address the problem of global warming caused by greenhouse gas emissions. Check the Internet for the latest information on Canada's progress in meeting its Kyoto goals.

www.albertachemistry.ca
WWW

Using Energy Efficiently

What happens to the "wasted" energy from inefficient appliances? As you learned in Chapter 9, according to the law of conservation of energy, energy is not destroyed, only transformed. A water heater that is not 100% efficient does not use all of the input energy for heating the water. Some of the input energy likely heats up the container holding the water as well as the air surrounding it. The challenge in developing more efficient technology is to find ways to convert more of the input energy into a useful form.

New appliances and automobiles sold in Canada all bear an EnerGuide label that provides potential buyers with the item's energy consumption information (see Figure 10.11). EnerGuide labels on appliances state the approximate energy consumption per year and also rate the appliance against others in a given size and type category. An indicator arrow on the left side of the EnerGuide scale means the appliance is more energy efficient and has lower operating costs compared with other, similar appliances. Vehicle EnerGuide labels show city and highway fuel consumption ratings and an estimated annual fuel cost for that particular vehicle.

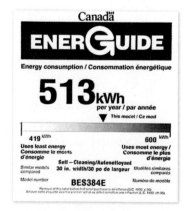

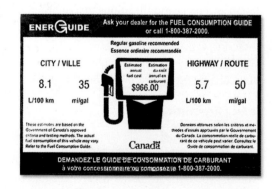

Figure 10.11 EnerGuide labels contain energy consumption information for appliances, heating and cooling systems, cars, and even homes.

You may have also heard about ENERGY STAR®. The internationally used ENERGY STAR® mark (see Figure 10.12) indicates that the product is one of the most energy efficient in its category.

Why is it important to purchase energy-efficient appliances and automobiles? Not only does saving energy save you money, but it also helps to protect the environment. Although the energy that Canadians use come from a variety of sources, the significant majority involves the combustion of fossil fuels. Burning fossil fuels releases greenhouse gases and gases that contribute to acid precipitation. The Canadian government has committed to reducing greenhouse gas emissions by 6% below the 1990 levels between 2008 and 2012, as agreed to in the Kyoto Protocol. One way to achieve this goal is to improve energy efficiency. In the next Thought Lab, you will look at how to improve your personal energy efficiency.

Figure 10.12 When you see the ENERGY STAR® mark, you know that the product is one of the most energy efficient available.

Thought Lab 10.1 Improving Energy Efficiency at Home

In Canada, approximately 80% of the energy used in our homes is used to heat water and the air. The next largest category of energy consumers is appliances, which includes washing machines, dishwashers, and televisions. (The graph below summarizes energy use in homes by category.) Canadians also use a great deal of energy getting from place to place. All forms of transportation account for 25% of all energy use. Automobiles account for 39% of that 25%.

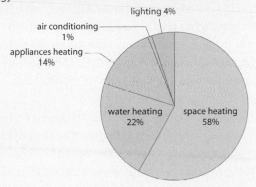

Most of the energy used in our homes is for heating.

The Office of Energy Efficiency (OEE) was established in 1998 as part of Natural Resources Canada. The OEE helps Canadians improve energy efficiency through a variety of programs aimed at both individuals and businesses. These programs include educational programs, such as EnerGuide and ENERGY STAR®, and significant rebates and cash incentives for individuals and businesses that switch to more energy-efficient options. In this Thought Lab, you will take a look at ways to improve how efficiently you use energy.

Procedure

1. Divide into groups of three or four students. Each group will concentrate on one category of residential energy use. Your teacher will assign the category.

2. In your group, brainstorm to obtain a list of all the ways in which you use energy in the assigned category during your day.

3. Perform an Internet search to obtain the following background information:
   - the amount of energy consumed by that activity
   - ideas to help decrease the quantity of energy used in that activity. These ideas should be grouped into two categories:
     - changes in behaviour that would decrease the total amount of energy required
     - technology that can be used to improve energy efficiency while not limiting the activity in any way
   - possible energy and monetary savings: include any special offers and rebates that are currently available

4. Organize your findings and display them on a web page, in a pamphlet, or on a poster that links to or connects with those from other groups and can be used to market this information to the rest of the school.

Analysis

1. Using the results from all the groups, make a list of energy-saving tips that you are willing to implement. Perform a cost–benefit analysis for each tip.

2. What energy-saving tips could be implemented at school?

3. Which energy-saving strategies would you not consider implementing? Support your answer.

Extension

4. This Thought Lab focussed on energy use in the home. Providing our homes with water for bathing, watering our gardens, and cooking also uses energy. List ways in which water, and therefore energy, can be conserved.

Efficient Home Heating

According to the pie chart in Thought Lab 10.1, about 60 percent of the total energy consumed in a typical Canadian home is used for space heating. Another 20 percent is used to heat water. Most Canadian homes are heated by furnaces or boilers that burn fossil fuels. Is it possible to reduce this fossil fuel consumption? By improving the technology that converts fuels to heat, Canadians can not only reduce fuel consumption, but also save money on heating bills and lower the amount of air pollutants created by burning oil and natural gas.

Measuring Energy Efficiency

When a fuel such as oil or natural gas is burned in a home furnace, some of the heat produced is wasted. The extent of this waste determines the efficiency of the furnace, measured as a percentage called the Annual Fuel Utilization Efficiency (AFUE). An AFUE of 80 percent means that 80 percent of the energy in the fuel becomes heat for the home while the other 20 percent escapes up the chimney and elsewhere in the system. By determining their furnace's AFUE, homeowners can tell how much they could reduce their annual heating costs by replacing an older furnace with a higher efficiency furnace. The minimum acceptable AFUE rating for a new fossil-fuelled, warm-air furnace in Canada is 78 percent.

Different Technologies, Different Efficiencies

Table 10.2 on page 391 shows that the most energy-efficient furnaces on the market today are condensing furnaces fuelled by natural gas. They can achieve an amazing efficiency of up to 97 percent by using not only heat from the combustion of natural gas but also heat from the condensation of water vapour produced in the combustion process. When natural gas is burned, it produces a mixture of hot gases and water vapour. The water vapour contains a significant amount of latent heat—about 11 percent of the total energy in the fuel. This heat is extracted from the water vapour by a stainless steel condensing heat exchanger (see furnace diagram). The heat exchanger condenses the water vapour in the combustion gases and releases its latent heat into the house air circulating through the furnace. After this process, the waste gases have cooled to between 40 °C and 50 °C. The cooled gases are vented to the side wall of the house, and the liquid condensate runs to a drain outlet.

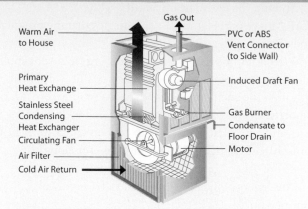

A high-efficiency condensing gas furnace. A furnace's AFUE rating is shown on a label. The higher the AFUE, the greater the efficiency.

The condensing process is efficient at extracting heat from waste vapour, but this technology is not generally used in furnaces that burn oil. Why not? First, the burning of oil produces only half as much water vapour as the burning of gas. In other words, oil has much less of its energy tied up in the form of latent heat. Second, the dew point (temperature at which condensation occurs) is lower for the combustion gases from burning oil, so the furnace must use more energy to cool the gases sufficiently. Third, the combustion gases from burning oil contain higher levels of sulfur and soot, which slowly corrode the heat exchange surfaces and reduce the efficiency of the system. For these reasons, a condenser oil furnace is not significantly more efficient than a well-designed, mid-efficiency oil furnace.

Paying attention to the AFUE rating on a furnace is only one strategy for efficient home heating. AFUE refers only to the efficiency of the furnace—it does not tell homeowners about heat losses that occur in the duct system that carries the warmed air from the furnace throughout the house. Heat inside the house can also be lost through the windows, doors, walls, and roof.

• • •

1. In many older homes, the original furnace had a capacity to produce more than twice the amount of heat needed for the size of the home. Why do you think that was? What improvements could the owner of an older house make to allow the house to have a much smaller furnace while still maintaining a comfortable temperature in winter?

2. Research other technologies and designs that are used to increase the efficiency of home heating, and write a brief report on one of them to explain how it works. (For example, consider programmable thermostats, heat pumps, integrating space and water heating systems, sealed combustion, or variable-speed fan motors.)

Efficiency and Energy Conversions

When assessing the efficiency of devices, such as microwave ovens and light bulbs, manufacturers focus on a single energy conversion—the input of electrical energy versus the output of useful energy.

When looking at the bigger picture of energy efficiency, you also need to think about the source of the electricity. Consider, for example, natural gas. In Alberta, natural gas is used both to heat our homes and to generate electricity in power plants. Natural gas is primarily methane. Therefore, you can estimate an ideal value for energy input by using the enthalpy of combustion of methane, $CH_4(g)$:

$$CH_4(g) + 2O_2(g) \rightarrow CO_2(g) + 2H_2O(g) \qquad \Delta_c H° = -802.5 \text{ kJ/mol}$$

When natural gas is used directly to provide heat, its efficiency can be as high as 90% (the efficiency of a high-efficiency furnace). Thus, for every mole of natural gas burned, you get about 720 kJ of energy (0.90×802.5 kJ). If natural gas is used to produce electric energy in a power plant, however, the efficiency is much lower—about 37%. Why? The heat from burning the natural gas is used to boil water. The kinetic energy of the resulting steam is transformed to mechanical energy for turning a turbine. The turbine generates the electrical energy. Each step has an associated efficiency that is less than 100%. Thus, at each step, the overall efficiency of the fuel decreases. Figure 10.13 shows the difference between the home furnace and the power plant. Table 10.2 lists comparisons among furnace technologies.

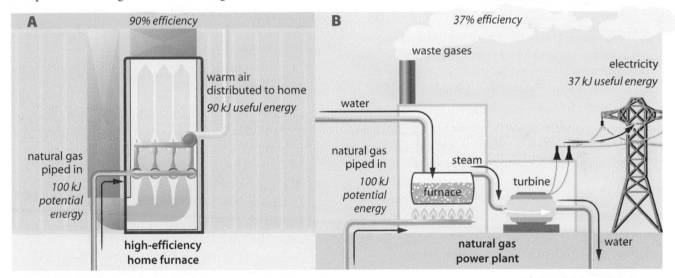

Table 10.2 Comparison of Typical Heating System Efficiencies

Fuel	Furnace technology	AFUE (% efficiency)	Energy savings (% of base)
Oil	Old cast-iron	60	base
	New standard	78–86	23–30
	Mid-efficiency	83–89	28–33
Natural Gas	Conventional	60	base
	Mid-efficiency	78–84	23–28
	High-efficiency condensing	89–97	33–38

Figure 10.13 Natural gas is a far more efficient fuel when used to heat a home furnace than when it is used to generate electricity. There is only one energy conversion step in a high-efficiency furnace. Natural gas is burned, and the energy released heats air, which is distributed throughout the home **(A)**. In a power plant, there are several energy conversions, and "waste" energy is released to the surroundings at each step **(B)**.

Section 10.2 Summary

- The efficiency of a chemical, physical, or nuclear process can be expressed as:

$$\text{Efficiency} = \frac{\text{energy output}}{\text{energy input}} \times 100\%$$

- An EnergyGuide label is found on all new appliances and automobiles sold in Canada. The label provides information about the efficiency of the appliance or automobile.

- An ENERGY STAR® mark on a product means that it is one of the most energy-efficient products in its category.

- Every energy-conversion step in a process reduces its efficiency.

- Energy-conversion processes that involve boiling water into steam and using the steam to turn a turbine have low efficiencies.

SECTION 10.2 REVIEW

1. Compare the energy changes associated with physical, chemical, and nuclear processes. Give one example of each process.

2. Your friend tells you about an energy source that is supposed to be 46% efficient. What questions do you need to ask your friend in order to verify this claim?

3. Design an experiment to determine the efficiency of a laboratory burner. You will first need to decide how to define the efficiency, and you will also need to find out what fuel your burner uses. Include a complete procedure and safety precautions.

4. Some high-efficiency gas furnaces can heat with up to 97% efficiency. These gas furnaces work by allowing the water vapour produced during combustion to condense. Condensation is an exothermic process that releases further energy for heating. Explain why allowing the water to condense increases the efficiency of the furnace.

5. The label on a kettle claims that the kettle is 95% efficient.
   a) What definition of efficiency is the manufacturer using?
   b) Write an expression that shows how the manufacturer might have arrived at an efficiency of 95% for the kettle.
   c) Design a detailed experiment to test the manufacturer's claim. Include safety precautions.

6. The cost of a propane barbecue is 90% of the cost of a natural gas barbecue. A hibachi, which burns charcoal, costs only 10% the cost of the gas barbecue. From an environmental and economic perspective, which type of barbecue would you choose? Justify your answer.

7. A student uses a methane burner to heat a beaker of water. The student's data is below.

Data for Determining the Efficiency of a Methane Burner

Quantity measured	Data
mass of methane consumed	0.37 g
mass of glass beaker	275.38 g
mass of glass beaker and water	427.96 g
initial temperature of water	25.3 °C
final temperature of water	37.9 °C
specific heat capacity of glass	0.84 J/g · °C

   a) Calculate the efficiency of the natural gas (methane) burner in heating the beaker of water.
   b) Account for the wasted energy.
   c) Suggest improvements to increase efficiency and explain them.

8. The car you have been driving has a posted fuel consumption of 10.0 L/100 km (city) and 7.1 L/100 km (highway) according to the EnerGuide information. Overall, your actual fuel consumption has averaged 10.5 L/100 km.
   a) Is this better or worse than the posted fuel consumption?
   b) Provide three reasons why your actual fuel consumption may be different from the posted consumption.
   c) List three ways in which you could improve your fuel consumption (other than buying a new car!).

9. You have probably noticed that a lot of heat is generated by the compressor of your refrigerator. Design a system that could use this and other sources of "waste heat" in your home for some useful purpose.

Fuelling Society

Key Terms

greenhouse gas
global warming
non-renewable energy
   source
renewable
heat content

Figure 10.14 Alberta harnesses more wind power than any other province in Canada. These turbines are part of a wind farm located in Pincher Creek, Alberta.

In the previous section, you learned about conserving energy through maximizing efficiency. You can minimize the energy you use by making smarter choices, such as by using compact fluorescent bulbs instead of incandescent bulbs, or by driving a smaller, more efficient car. Energy can also be saved on a large scale at the source. For example, when generating electricity, hydroelectric power (falling or moving water turns the turbines directly) is more efficient than burning coal or natural gas (the heat from the burning fuel is first used to boil water, then the kinetic energy in the steam is used to turn the turbines). It's because there are fewer energy conversion steps. Efficiency is not the only criterion to use when selecting an energy source, however.

Thinking About the Environment

Since the 1970s, people have become increasingly conscious of the impact of energy technologies on the environment.

Suppose that you want to analyze the environmental impact of an energy source. You can ask the following questions:

- *Are any waste products or by-products of the energy production process harmful to the environment*? For example, any process in which a hydrocarbon is burned produces carbon dioxide, a **greenhouse gas**. Greenhouse gases trap heat in Earth's atmosphere and prevent the heat from escaping into outer space. Scientists theorize that a build-up of greenhouse gases in the atmosphere is leading to an increase in global temperatures, known as **global warming**. Any combustion process provides the heat required to form oxides of nitrogen from nitrogen gas. Nitrogen oxides contribute to acid precipitation. Nuclear fission does not produce greenhouse gases, but it does leave behind radioactive waste that remains a danger for thousands of years.

- *Is obtaining or harnessing the fuel harmful to the environment*? For example, oil wells and strip coal mines destroy habitat. The large wind turbines found in Southern Alberta have been described as noisy and can be a danger to birds (see Figure 10.14). Although hydroelectric power plants produce very little in the way of emissions in the production of power, ecosystems both upstream and downstream of the plant are affected by their construction and presence (see Figure 10.15).

Figure 10.15 The Brazeau Dam, located near Drayton Valley on the North Saskatchewan River, is Alberta's largest hydroelectric power-generating plant.

• *Will using the energy source permanently remove the fuel from the environment?* A **non-renewable energy source** (such as coal, oil, or natural gas) is effectively gone once we have used it up. Non-renewable energy sources take millions of years to form. We use them up at a much faster rate than they can be replenished. Energy sources that are clearly **renewable** include solar and wind energy. The sun will continue to radiate energy toward Earth and produce wind over its lifetime—for many millions of years more. A somewhat renewable energy source is wood. Trees can be grown to replace those cut down. It takes trees a long time to grow, however, and habitat is often destroyed or permanently altered in the meantime by activities involved in logging.

Fossil Fuels in Alberta

Canadians depend on fossil fuels for approximately 80% of their energy requirements. In Alberta, essentially all our energy needs are met by fossil fuels, as 97% of our electricity is produced from either coal or natural gas (see Figure 10.16). We use natural gas for the vast majority of our heating. Not only do Albertans depend on fossil fuels for energy, but the fuels are also a major contributor to Alberta's economy. In 2003, more than 300 000 people were employed directly or indirectly in the energy sector, which amounts to more than 18% of the Alberta workforce. In other words, one in about every six working Albertans is employed by companies related to fossil fuels. Alberta's 2003 oil and mineral exports were valued at almost $40 billion and accounted for 70% of all of Alberta's international exports that year.

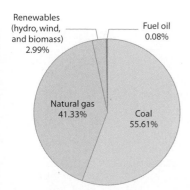

Figure 10.16 Ninety-seven percent of Alberta's electricity requirements are met by burning fossil fuels.

Renewables (hydro, wind, and biomass) 2.99%
Fuel oil 0.08%
Natural gas 41.33%
Coal 55.61%

Chemistry File

FYI

A major source of oil pollution comes from the everyday use of oil by ordinary people. A high school student did a home experiment to discover how much oil remains in "empty" motor oil containers that are thrown out. Calculations showed that nearly five million litres of oil are dumped into landfill sites every year, just in "empty" oil containers!

Fossil Fuels and the Environment

Fossil fuels have changed the way we live in many positive ways. We use them to light and heat our homes, to cook and to keep our food fresh, to supply power for our work and pastimes, and to transport ourselves and the products we make across the country and around the globe. Nearly everything we do that involves technology also involves fossil fuels at some stage. Our dependence on them, however, has affected the world around us. The greenhouse effect, global warming, acid rain, air pollution, and the environmental damage resulting from oil spills, such as occurred in the train derailment at Wabamun Lake in the summer of 2005 (see Figure 10.17), are familiar topics on the news today. Our use, or misuse, of these fuels is linked directly to these problems. Figure 10.18 shows how these issues are related to fossil fuel use.

Figure 10.17 On August 3, 2005, 43 rail cars carrying petroleum products derailed near Wabamun Lake just west of Edmonton. The oil that spilled had an immediate and devastating effect on waterfowl and marine life. The long-term effects are still largely unknown, and scientists will continue to monitor the area.

Figure 10.18 This concept organizer shows the interrelated issues that arise from the use of fossil fuels.

Concept Organizer   Fossil Fuels and the Environment

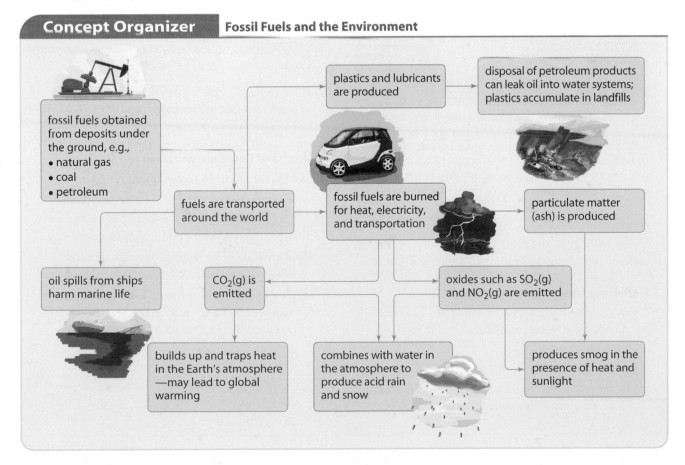

In the next investigation, you will research and analyze the risks and benefits of relying on fossil fuels as an energy source.

INVESTIGATION | 10.C

Target Skills

Assessing qualitatively the risks and benefits of relying on fossil fuels as energy sources

Conducting investigations into relationships and **using** a broad range of tools and techniques to gather and record data and information

Working as a member of a team in addressing problems and applying the skills and conventions of science

Fossil Fuels as Energy Sources: A Risk–Benefit Analysis

In this investigation, you and your group will research the use and production of fossil fuels in Alberta and propose a plan to reduce the impact on the environment.

Would you like to have a coal-burning power plant near your home? Some people might be upset by this idea because they believe that there maybe a health risk caused by pollution from the plant. Other people might think that a coal-burning power plant poses very little threat, and the benefits are worth the risk. Who is right? How do you decide?

Issue

What are the risks and benefits of continuing to depend on fossil fuels as our primary energy source in Alberta?

Gathering Data and Information

1. As a class, produce a list of fossil fuels that are produced and used in Alberta, and the ways in which they are used.

2. Divide into groups of three or four students. In your small group, research one of the fossil fuels and its uses. Your research should include

- benefits: where is it recovered, what is it used for, how does it benefit our economy, how large are the reserves (how long will we be able to use it)?

- risks: what potential harm could recovering and using this fossil fuel have on the environment and human health? What will happen when the reserves are depleted?

- technology, including both processes and equipment, that has been developed to reduce the risk. How does it reduce the risk? **ICT**

Organizing Findings

3. Perform a risk–benefit analysis on our dependency on fossil fuels.

4. In your group, propose a plan that would reduce the impact of our dependency on fossil fuels.

5. Share your proposal and findings in a presentation or as a poster that incorporates data, graphics, and text in a persuasive manner.

Opinions and Recommendations

1. Discuss the groups' findings as a class.

   a) Are the safety, environmental, and health concerns arising from fossil fuel production and use in Alberta being addressed adequately?

   b) In general, do the benefits of producing and using fossil fuels in Alberta outweigh the risks?

   c) Do you think your proposal to reduce the negative impact of fossil fuel production and use is reasonable? Explain your answer in detail.

What Is a "Clean" Fuel?

You may have discovered through your research in the previous investigation that not all fossil fuels have the same impact on the environment, either during their recovery or in their use. Some fossil fuels are considered cleaner than others, meaning that their emissions are lower than others' for an equivalent amount of energy produced. Table 10.3 summarizes the carbon dioxide emissions for a variety of fuels in kilograms of CO_2(g) per kilojoule of energy produced.

Table 10.3 Fuel Source and Carbon Dioxide Emissions

Fuel	Carbon dioxide emissions (kg CO_2(g)/kJ of energy produced)
anthracite coal	108.83
lignite coal	103.08
subbituminous coal	101.79
bituminous coal	98.25
municipal solid waste (landfill gas)	95.64
wood and wood waste (biomass)	93.32
tires/tire-derived fuel	90.71
oil	78.48
kerosene	76.35
gasoline	74.86
jet fuel	74.78
propane	66.61
flare gas	57.77
natural gas (pipeline)	56.03
methane	55.16
nuclear	0.00
renewables (solar thermal and photovoltaic, geothermal, wind, hydroelectric)	0.00

As you can see from Table 10.3, renewable energy sources, such as wind power, produce no carbon dioxide in their operation. However, coal and natural gas, which provide 97% of Alberta's electricity, produce a significant amount of carbon dioxide. Other emissions, such as particulates (ash particles) and oxides that combine with water to produce acid rain and snow, should also be factored in when evaluating a fuel source. From Table 10.4, it is clear that the levels of other pollutants produced in the combustion of three commonly used fossil fuels—coal, oil, and natural gas—varies widely.

Table 10.4 Common Fossil Fuels and Harmful Emission Levels

Pollutant	Emission levels (kg/MJ of energy produced)		
	natural gas	oil	coal
carbon monoxide	19.14	15.79	99.54
nitrogen oxides	44.03	214.40	218.70
sulfur dioxide	0.48	536.95	1239.96
particulates	3.35	40.20	1313.18

The **heat content** of a fuel is the amount of energy released per kilogram of a fuel. To fairly assess a fuel, heat content, along with environmental and economic factors, should be considered. In the next investigation, you will collect additional data and evaluate two fossil fuels.

Chemistry File

FYI
The SI unit for energy is the joule (J). You may have come across other units for energy in your research. One of the most commonly used units is the British thermal unit, Btu. One British thermal unit equals 1.06 kJ and is defined as the quantity of energy required to raise the temperature of one pound of water by one degree Fahrenheit. "Horsepower," which is a unit of power (like the watt), is equal to 2540 Btu/h.

Chemistry File

WebLink
What makes some fuels "cleaner" than others? Find out more about the differences among types of fossil fuels and the way in which the location they are recovered affects emission levels.

www.albertachemistry.ca
WWW

Fuelling Thermal Power Plants

In this investigation, you and your group will compile data on the heat content of fuels used in Alberta's thermal power plants and combine that information with environmental data to assess the fuels.

Target Skills

Evaluating the economic and environmental impact of different fuels by relating carbon dioxide emissions and the heat content of fuels

Using library and electronic research tools to compile information on the energy content of fuels used in Alberta power plants

Assessing whether coal or natural gas should be used to fuel thermal power plants in Alberta

Issue

Is coal or natural gas a wiser choice from thermochemical, environmental, and economic perspectives for Alberta?

Gathering Data and Information

1. Using the Internet or other resources, research the two fuels used in Alberta's thermal power plants. **ICT** Your research should include

   • the heat content of the fuels
   • environmental and economic issues surrounding recovery and use—include quantitative and qualitative data

Organizing Findings

2. Using a spreadsheet package, organize your findings for both fuels. **ICT**

3. Graph numerical data to illustrate quantitative differences between the two fuels.

4. With your group, analyze your data and propose a plan for thermal power plants.

5. Display your findings in a presentation or on a poster.

Opinions and Recommendations

1. Discuss the groups' findings as a class.

   a) Did all groups come to the same conclusions? Why or why not?

   b) Do you think your proposal is reasonable? Why or why not?

   c) What other factors should have been considered that were not? Explain.

Section 10.3 Summary

• Energy efficiency is important because the by-products of current energy conversion processes are harmful to the environment.

• Ninety-seven percent of the resources that are used to produce electric energy in Alberta are fossil fuels. Fossil fuels are non-renewable resources.

• The burning of fossil fuels contributes to global warming, acid rain, and pollution of the environment.

• The heat content of a fuel is the amount of energy produced per unit mass (g or kg) of the fuel.

SECTION 10.3 REVIEW

1. List five activities in your daily life in which fossil fuels are used. Identify the fuel used.

2. What is meant by a "clean" fuel? Are there really any clean fuels? Justify your answer.

3. Identify the products of combustion associated with the following environmental issues:
   a) global warming
   b) acid rain or snow
   c) smog

4. Differentiate between the heat content and heat of combustion of a fuel. Which term is more useful in evaluating fuels? Why?

5. List three benefits and three risks that Albertans face by relying so heavily on fossil fuels.

6. Which, fuel, coal or natural gas, should Albertans use in their thermal power plants? Explain your decision.

Hess's law allows you to calculate enthalpy changes for reactions that cannot be performed in the laboratory. You can do this be summing any combination of reactions that add to the reaction you want and for which the enthalpy changes are known. Chemists have accumulated a database that allows you to use Hess's law to calculate the enthalpy change for nearly any chemical reaction. That database consists of the standard enthalpy changes of formation of a very large number of compounds.

It is important to be able to determine the enthalpy change for many reactions so you can determine the efficiency of industrial processes that use that reaction. Developing energy efficient procedures is critical for the economic use of resources, the reduction in environmental pollution, and preserving non-renewable energy resources. Meantime, scientists and engineers are studying ways to economically use renewable resources. The general public can contribute to the efficient use of resources by choosing to purchase automobiles and appliances that are energy efficient as shown on the EnerGuide labels.

Although Alberta produces more electrical energy from wind than another province, 97% of Alberta's electricity is generated by the burning of fossil fuels. Burning fossil fuels contributes to global warming, acid rain, and environmental pollution as well as using up non-renewable resources.

Concept Organizer   Hess's Law, Efficiency, and the Environment

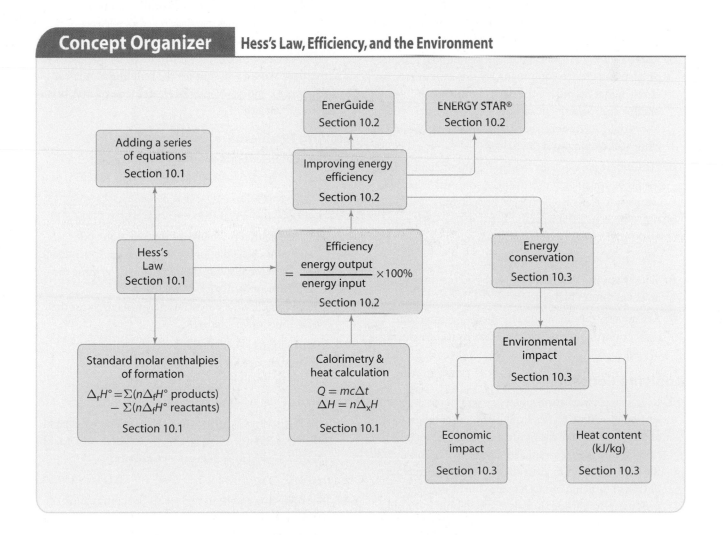

Understanding Concepts

1. A given chemical equation representing a chemical reaction is tripled and then reversed. What effect, if any, will there be on the enthalpy change of the reaction? On the molar enthalpy change of the reaction?

2. **a)** Write a balanced thermochemical equation for the formation of each of the following substances:
   i) $AgCl(s)$
   ii) $C_2H_5OH(\ell)$
   iii) $NH_4NO_3(s)$
   b) Rank the substances from least to most stable.

3. If the enthalpy of formation of an element in its standard state is equal to zero, explain why the enthalpy of formation of iodine gas, $I_2(g)$, is $+21$ kJ/mol.

4. Define *efficiency*. Which is more efficient: driving a car to school or taking the bus? Explain using your definition.

5. Approximately 5% of Canada's air pollution is produced by lawn and garden equipment, such as gas mowers. An older-model gas mower produces as much VOC (volatile organic compounds) in an hour as driving a car 500 km! How could you reduce your impact on the environment while still keeping your lawn tidy?

6. Hair dryers are typically less than 50% efficient. What would be considered useful energy output? What happens to the wasted energy? Be specific.

7. If you were shopping for a new appliance, what two labels should you look for when assessing energy efficiency? What do they mean?

8. What steps can you take to ensure your car is operating at its maximum fuel efficiency?

9. Explain why producing electric energy by using wind power or hydro is more efficient than by using natural gas or coal.

Applying Concepts

10. In an oxygen-rich atmosphere, carbon burns to produce carbon dioxide, $CO_2(g)$. Both carbon monoxide, $CO(g)$, and carbon dioxide are produced when carbon is burned in an oxygen-deficient atmosphere. This makes the direct measurement of the enthalpy of formation of $CO(g)$ difficult, but $CO(g)$ also burns in oxygen to produce pure carbon dioxide. Explain how you would experimentally determine the enthalpy of formation of carbon monoxide.

11. Your friend says, "I want to help conserve energy and decrease my greenhouse gas emissions, but I don't want to carpool or take the bus." What energy-conservation suggestions you would make to this person?

12. Acetylene (ethyne), $C_2H_2(g)$, and ethylene (ethene), $C_2H_4(g)$, are both used as fuels. Acetylene also reacts with hydrogen to produce ethylene, as shown:
   $$C_2H_2(g) + H_2(g) \rightarrow C_2H_4(g) \quad \Delta H° = -175.1 \text{ kJ}$$
   a) Without doing any calculations, explain how you know that $C_2H_2(g)$ has a less negative enthalpy of formation than does $C_2H_4(g)$.
   b) Which do you think produces more energy per mole when combusted: acetylene or ethylene? Explain.

13. "Hydroelectric power produces no emissions and does not harm any ecosystems during power generation." Do you agree or disagree with this statement? Explain.

14. Canada produces 28% of the world's supply of uranium from mines. Nuclear reactors release no greenhouse gases. Why, then, do you think that none of Alberta's electric energy is produced by nuclear reactors? Would you support the building of nuclear reactors for the generation of electric energy in Alberta? Why or why not?

Solving Problems

15. The complete combustion of 1.00 mol of sucrose, $C_{12}H_{22}O_{11}(s)$, releases 5641 kJ of energy (at 25 °C and 100 kPa):
   $$C_{12}H_{22}O_{11}(s) + 12O_2(g) \rightarrow 12CO_2(g) + 11H_2O(\ell)$$
   a) Use the enthalpy change of this reaction and standard enthalpies of formation from Appendix G or your Chemistry Data Booklet to determine the enthalpy of formation of sucrose.
   b) Draw and label a potential energy diagram for the combustion of sucrose.

16. Use equations (1), (2), and (3) below to find the enthalpy change for the production of methane, $CH_4(g)$ from chloroform, $CHCl_3(\ell)$:
   $$CHCl_3(\ell) + 3H_2(g) \rightarrow CH_4(g) + 3HCl(g) \quad \Delta H° = ?$$
   (1) $\frac{1}{2}H_2(g) + \frac{1}{2}Cl_2(g) \rightarrow HCl(g) \qquad \Delta H° = -92.3 \text{ kJ}$
   (2) $C(s) + 2H_2(g) \rightarrow CH_4(g) \qquad \Delta H° = -74.6 \text{ kJ}$
   (3) $C(s) + \frac{1}{2}H_2(g) + \frac{3}{2}Cl_2(g) \rightarrow CHCl_3(\ell)$
   $$\Delta H° = -134.5 \text{ kJ}$$

17. The following equation represents the combustion of ethylene glycol, $(CH_2OH)_2(\ell)$:
   $$(CH_2OH)_2(\ell) + \frac{5}{2}O_2(g) \rightarrow 2CO_2(g) + 3H_2O(\ell)$$
   $$\Delta H° = -1178 \text{ kJ/mol}$$
   Use known enthalpies of formation and the given enthalpy change to determine the molar enthalpy of formation of ethylene glycol.

18. Hydrogen peroxide, $H_2O_2(\ell)$, is a strong oxidizing agent.

   a) Write a balanced chemical equation for the formation of $H_2O_2(\ell)$.

   b) Using the following equations, determine the enthalpy of formation of $H_2O_2(\ell)$.

   (1) $2H_2O_2(\ell) \rightarrow 2H_2O(\ell) + O_2(g)$ $\Delta H^\circ = -196$ kJ
   (2) $H_2(g) + \frac{1}{2}O_2(g) \rightarrow H_2O(\ell)$ $\Delta H^\circ = -286$ kJ

19. Hydrogen cyanide is a highly poisonous gas. It is produced from methane, $CH_4(g)$, and ammonia, $NH_3(g)$:

   $CH_4(g) + NH_3(g) \rightarrow HCN(g) + 3H_2(g)$

   Find the enthalpy change of this reaction by using the following thermochemical equations.

   (1) $H_2(g) + 2C(s) + N_2(g) \rightarrow 2HCN(g)$ $\Delta H^\circ = +270$ kJ
   (2) $N_2(g) + 3H_2(g) \rightarrow 2NH_3(g)$ $\Delta H^\circ = -92$ kJ
   (3) $C(s) + 2H_2(g) \rightarrow CH_4(g)$ $\Delta H^\circ = -75$ kJ

20. The following equation represents the complete combustion of butan-1-ol, $C_4H_9OH(\ell)$:

   $C_4H_9OH(\ell) + 6O_2(g) \rightarrow 4CO_2(g) + 5H_2O(g)$

   a) Using known standard enthalpies of formation, calculate the molar enthalpy of combustion for butan-1-ol. (The standard enthalpy of formation for butan-1-ol is −244.4 kJ/mol.)

   b) Draw a potential energy diagram for this reaction.

   c) If 10.0 g of butan-1-ol is combusted in an alcohol burner, determine the quantity of energy released.

21. Determine the enthalpy of combustion in kilojoules per gram for an alcohol burner containing 30% butan-1-ol and 70% ethanol by mass.

22. How much energy is produced or absorbed when 7.9 g of methane, $CH_4(g)$, is produced in the following reaction?

   $3H_2(g) + CO(g) \rightarrow CH_4(g) + H_2O(g)$

23. White gas is approximately 60% hexane and 40% heptane by mass. Design an investigation to determine the enthalpy of combustion of white gas.

   a) Draw and label a diagram of the apparatus.

   b) Write a step-by-step procedure.

   c) Prepare a table to record your data and other observations.

   d) The accepted value for the molar enthalpy of combustion of camping fuel in an open system is 44.6 kJ/g. Do you expect your experimental value to be higher or lower than this value? Explain.

24. Calculate the efficiency of a camp stove if 3.25 g of fuel used to heat a 150 g aluminium pot with enough water to make one cup (250 mL) of hot chocolate. The water

was initially at 2.5 °C and was heated to 85.0 °C. Use the enthalpy of combustion value from question 23.

25. How much water could you heat from 10 °C to its boiling point by using 4.75 g of methane in a natural gas–burning barbecue if the barbecue is 47% efficient?

26. How much propane would have to be combusted to preheat a 5.0 kg cast iron barbecue to a temperature of 175 °C (from 25 °C) if the barbecue is 50% efficient and cast iron has a specific heat capacity of 0.46 J/g • °C?

Making Connections

27. Which of your pastimes involve energy consumption? You can calculate your energy consumption for any activity by using the information on the appliance and two mathematical formulas. The power (in watts) needed to run the appliance is usually on an attached plate. With that number, and how many hours you use that appliance each month, you can determine your consumption:

   Consumption in kilowatt hours $= \dfrac{\text{watts} \times \text{hours used}}{1000}$

   To calculate the monthly cost, multiply the consumption in kilowatt hours (kW • h) by the cost per kilowatt hour, which is on your energy bill. **ICT**

   a) List your energy-consuming pastimes.

   b) Calculate the monthly energy cost for these activities.

   c) Use Excel to construct a spreadsheet for your data and draw a graph to illustrate your findings.

   d) Which activity is the most energy consuming? The least? Do the results surprise you?

   e) Create a spreadsheet that will allow you to quickly calculate energy consumption or activity expenditure.

28. Compare methane, $CH_4(g)$, and hydrogen, $H_2(g)$.

   a) Write the chemical equation for the complete combustion of each fuel. Then, find the enthalpy of combustion, $\Delta_c H^\circ$, of each fuel. Express your answers in kilojoules per mole and in kilojoules per gram. Assume that water vapour, rather than liquid water, is formed in both reactions.

   b) Which fuel has the higher heat content (in kilojoules per gram)?

   c) Consider a fixed mass of each fuel. Which fuel would allow you to drive a greater distance? Explain briefly.

   d) Describe how methane and hydrogen could be obtained. Which of these methods is less expensive? Explain.

   e) Which fuel do you think is more environmentally friendly? Explain.

Activation Energy and Catalysts

Chapter Concepts

11.1 Reaction Pathways

- Activation energy is the energy barrier that must be overcome for reactions to occur.

- You can use energy diagrams that include both enthalpy change and activation energy data for a chemical reaction.

11.2 Catalysts and Reaction Rates

- Catalysts increase reaction rates by providing alternative pathways for change.

- Technologies using catalysts are used to reduce air pollution caused by the burning of hydrocarbons.

- Catalysts in the human body are called enzymes and are essential for life-sustaining processes.

In Alberta, more than 40% of forest fires are caused by people. The remaining 60% are caused mostly by lightning strikes. To reduce the risk of wildfires, Alberta Forest Protection recommends that campers keep their campfires small and soak them thoroughly with water to extinguish them. Fire bans are issued during dry spells, restricting the types of fires and locations of fires people are allowed to have. Despite these precautions, an average of more than 1000 forest fires occurred each year between 1995 and 2005. These fires consumed more than 120 000 ha of land each year.

Cellulose, the main component of wood in trees, is a highly flammable material, and the combustion of cellulose is a highly exothermic reaction. Without being ignited by a carelessly tossed cigarette, a bolt of lightning, or a spark from a campfire, however, a tree will grow peacefully for many, many years—never catching fire.

If the combustion of wood is so exothermic, why does it take a spark to start the reaction? Why does the reaction of cellulose with oxygen, once started, proceed so quickly, while the reaction of iron with oxygen (rusting) proceeds very slowly? How can some substances increase the speed of a reaction without affecting the products or reactants?

In this chapter, you will find out why some reactions, such as campfires or forest fires, need an initial input of energy to begin. You will discover one reason why some reactions are fast while others are slow. You will also explore an important class of substances that change the speed of reactions and find out how they work. You will begin to learn about this class of substances in the Launch Lab on the next page.

Does It Gel?

Gelatin is a protein obtained as a by-product from the meat and leather industry. Gelatin has a number of food and non-food uses, such as gelatin desserts, chewy candies, and marshmallows. It is also used in margarine and ice cream. In this activity, you will explore how gelatin dessert can be quickly broken down by substances in some familiar foods.

Safety Precautions

- Your teacher may do this lab as a demonstration.

Materials

- gelatin (cut into 3 cm × 3 cm squares)
- variety of fresh fruits (cut into 1 cm squares): apple, banana, kiwi, orange, pineapple, strawberry
- balance
- a beaker for each fruit and one for just gelatin

Procedure

1. Copy the data table below. In the first column, list all the fruits you will be testing.

The Effect of Fruit on Gelatin

Fruit	Mass of gelatin before (g)	Mass of gelatin after (g)	Other observations
apple			
banana			

2. Determine the initial mass of each square of gelatin and place each square in its own beaker.

3. Place three squares of one type of fruit on the gelatin in all but one beaker. Each beaker should have a different fruit. The beaker that contains a square of gelatin only is your control.

4. Leave your set-up on the lab bench overnight.

5. Determine the final mass of each square of solid gelatin. Observe and record any other changes in the gelatin or fruit that you observed.

6. Dispose of the waste as directed by your teacher.

Analysis

1. Which fruit or fruits had an effect on the gelatin?

2. Without the addition of the fruit, do you think the gelatin would deteriorate? Explain.

3. Compare the deterioration of gelatin with the combustion of cellulose described in the chapter opener.

Rumours are sometimes said to spread "like wildfire." Like a forest fire, however, a rumour does not start by itself—someone has to be the "spark" that starts the rumour.

Reaction Pathways

Figure 11.1 Gasoline combustion occurs very quickly in the engines of these race cars.

The combustion of gasoline in the race cars shown in Figure 11.1 is a very fast and highly exothermic reaction. Conversely, the rusting of the steel body of the automobile proceeds quite slowly. The speed at which a reaction occurs, or the **reaction rate**, is the change in the amount of reactants consumed or products generated over time. Why do some reactions proceed slowly while others seem to take place instantaneously? Why do some reactions, such as the combustion of gasoline, require an initial input of energy while others do not? In this section, you will explore these ideas.

Collision Theory and Beyond

What causes a reaction to occur? One obvious answer is that a reaction occurs when two reactant particles collide with each other. Chemists theorize that for a reaction to occur, reacting particles (atoms, molecules, or ions) must collide with one another. This theory is known as **collision theory**.

Does every collision result in a reaction? Consider a 1 mL sample of gas at room temperature and atmospheric pressure. In the sample, about 10^{28} collisions take place among gas molecules every second. If each collision resulted in a reaction, *all* gas phase reactions would be complete in about a nanosecond (10^{-9} s)! In reality, gas phase reactions can occur quite slowly, which suggests that *not every collision between reactants results in a reaction*. For a collision between reactants to result in a reaction, the collision must be *effective*. An effective collision—one that results in the formation of products—must satisfy the following two criteria.

For a collision to be effective, it must satisfy *both* of the following criteria:
- the correct orientation of reactants (collision geometry)
- sufficient collision energy (activation energy, E_a)

Orientation of Reactants

Reacting particles must collide with the proper orientation relative to one another. This is also known as having the correct *collision geometry*. The importance of proper collision geometry can be illustrated by the following reaction:

$$NO(g) + NO_3(g) \rightarrow NO_2(g) + NO_2(g)$$

Figure 11.2 shows five of the many possible ways in which $NO(g)$ and $NO_3(g)$ can collide. Only *one* of the five possibilities has the correct collision geometry for a reaction to occur. As shown in the figure, only a specific orientation of reactants before collision leads to the formation of two molecules of nitrogen dioxide.

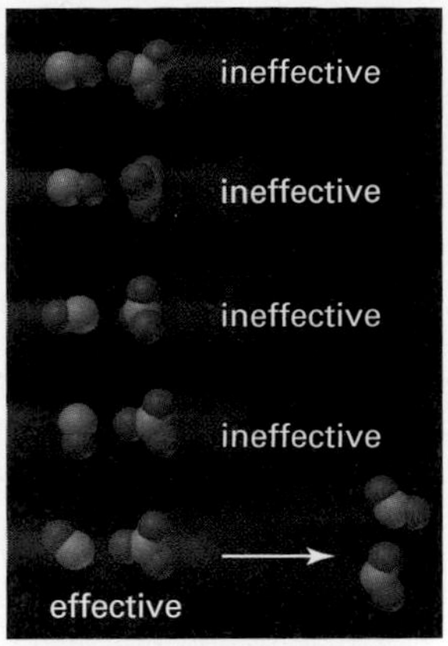

Activation Energy

In addition to collision geometry, the energy of the collision also determines whether a reaction can take place. The reactant particles must collide with sufficient energy to begin to break the bonds in the reactants and begin to form the bonds in the products. In most reactions, only a small fraction of collisions have sufficient energy for a reaction to occur. The **activation energy, E_a,** of a reaction is the minimum collision energy required for a successful reaction.

The collision energy depends on the kinetic energy of the colliding particles. As you know, temperature is a measure of the average kinetic energy of the particles in a substance. If you plot the number of collisions in a substance at a given temperature against the kinetic energy of each collision, you get a curve like the one in Figure 11.3. The type of distribution shown by this curve is known as a *Maxwell-Boltzmann distribution*. The dotted line indicates the activation energy. The shaded part of the graph indicates the collisions with energy equal to or greater than the activation energy.

Temperature and Activation Energy

How does the distribution of kinetic energy change as the temperature of a substance increases? Figure 11.4 shows the distribution of kinetic energy in a sample of reacting gases at two different temperatures, T_1 and T_2, where $T_2 > T_1$. The activation energy is indicated by the dashed vertical line. Two observations are apparent from the graph:

- At both temperatures, a relatively small fraction of collisions have enough energy for a collision to result in a reaction. ("Enough energy" means greater than or equal to activation energy.)
- As the temperature of a sample increases, the fraction of collisions with an energy equal to or greater than the activation energy increases significantly.

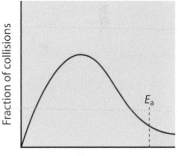

Figure 11.2 Only one of these five possible orientations of $NO(g)$ and $NO_3(g)$ will lead to the formation of a product.

End extension material

Effective Collisions

Figure 11.3 The area under a Maxwell-Boltzmann distribution graph represents the distribution of the kinetic energy of collisions at a constant temperature. At any given temperature, only a certain fraction of the molecules in a sample have enough kinetic energy to react.

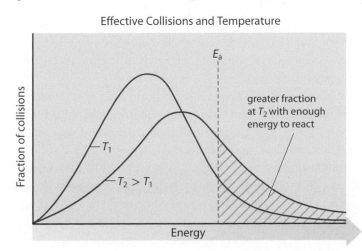

Figure 11.4 At increased temperatures, more particles collide with enough energy to react.

These observations explain why, for most reactions, the reaction rate will increase at higher temperatures. At higher temperatures, more collisions are successful.

A Closer Look at a Molecular Collision

You can represent the changes in potential energy during a chemical reaction using a potential energy diagram, as shown in Figures 11.5 and 11.6. These potential energy diagrams are similar to the ones you examined and drew in Chapters 9 and 10, with one difference. In addition to the enthalpy change of the reaction, these diagrams also show the activation energy of the reaction. The "hill" in each diagram illustrates the energy barrier for the reaction. Reactions that have a large activation energy will occur more slowly than will reactions that have a smaller activation energy, as relatively few reactants will have sufficient kinetic energy for a successful reaction. Reactions with low activation-energy barriers will be fast and in some cases even explosive.

Exothermic Reaction

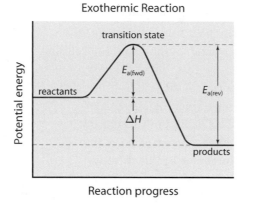

Endothermic Reaction

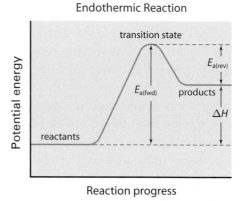

Figure 11.5 A potential energy diagram for an exothermic reaction

Figure 11.6 A potential energy diagram for an endothermic reaction

Activation Energy and Enthalpy

It is tempting to think that the activation energy for a reaction can be determined using enthalpy change, ΔH. This is not the case, however. *There is no way to predict the activation energy of a reaction from its enthalpy change.* A highly exothermic reaction can have a very high activation energy and, therefore, be very slow at room temperature. Conversely, a reaction may release very little heat, or even be endothermic, and still take place quite rapidly at room temperature. As you learned in Chapter 10, the enthalpy change of a reaction is determined from the difference in the potential energy of the reactants and products and is independent of the pathway.

The activation energy of a reaction is determined by analyzing the reaction rate at various temperatures. In general, reactions with low activation energies tend to proceed quickly at room temperature, regardless of whether they are endothermic or exothermic. Exothermic or endothermic reactions with high activation energies tend to proceed slowly at room temperature.

As you know, gasoline is highly flammable and burns in an exothermic combustion reaction. Why, then, does gasoline not burst spontaneously into flames? You know it does not, because it is safe to store gasoline in the tanks of cars and at gas stations. The answer is that gasoline, which is mostly octane, $C_8H_{18}(\ell)$, requires a spark—a small energy input—for the combustion reaction to begin. The spark provides a few octane and oxygen molecules with enough energy to overcome the energy barrier—the activation energy. Once ignited, the gasoline will continue to burn because the energy released during combustion provides the activation energy for other molecules of octane and oxygen. Figure 11.7 illustrates the difference in potential energy diagrams representing these two reactions.

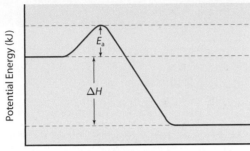

E_p Diagram for Combustion of Gasoline

Figure 11.7 Gasoline requires a spark to begin the combustion process. The spark allows nearby molecules to reach or exceed activation energy.

Activation Energy for Reactions in Both Directions

You already know that many reactions can proceed in two directions. For example, hydrogen and oxygen gases react to form water. Water, however, can also undergo electrolysis, forming hydrogen and oxygen gases. Electrolysis is the reverse of the first reaction.

$$H_2(g) + \frac{1}{2}O_2(g) \rightarrow H_2O(\ell) \qquad \Delta_r H = -285.8 \text{ kJ}$$
$$H_2O(\ell) \rightarrow H_2(g) + \frac{1}{2}O_2(g) \qquad \Delta_r H = 285.8 \text{ kJ}$$

The enthalpy change of the first reaction is the same in magnitude as the enthalpy change of the second reaction, with the opposite sign. (You can show this using Hess's law.) How are the activation energies of the forward and the reverse reactions related? For an exothermic reaction, as shown in Figure 11.5, the activation energy of the reverse reaction, $E_{a(rev)}$, is equal to the activation energy of the forward reaction plus the magnitude of the enthalpy change. For an endothermic reaction, as shown in Figure 11.6, the activation energy of the reverse reaction is equal to the activation energy of the forward reaction minus the magnitude of the enthalpy change. In other words,

- For an exothermic reaction: $E_{a(rev)} = E_{a(fwd)} + \Delta_r H$
- For an endothermic reaction: $E_{a(rev)} = E_{a(fwd)} - \Delta_r H$

The top of the activation energy barrier on a potential energy diagram represents the changeover point of the reaction. The chemical species that exists at the top of the activation energy barrier is referred to as the **activated complex**. The activated complex is a transitional species that is neither product nor reactant. It has partial bonds and is highly unstable.

Tracing a Reaction with a Potential Energy Diagram

Consider the substitution reaction between a hydroxide ion, $OH^-(aq)$, and methyl bromide:

$$BrCH_3(aq) + OH^-(aq) \rightarrow CH_3OH(aq) + Br^-(aq)$$

Figure 11.8 is a potential energy diagram for this reaction. It includes several "snapshots" of the reactants, activation complex, and products as the reaction proceeds.

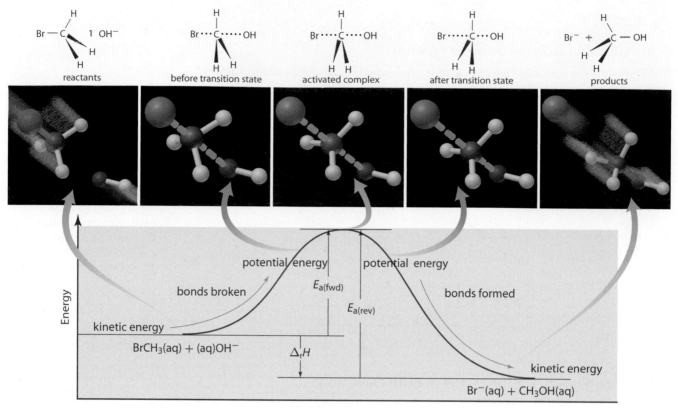

reactants   before transition state   activated complex   after transition state   products

potential energy   potential energy

$E_{a(fwd)}$

bonds broken   bonds formed

$E_{a(rev)}$

kinetic energy

Energy

$BrCH_3(aq) + (aq)OH^-$ Δ_rH kinetic energy

$Br^-(aq) + CH_3OH(aq)$

Reaction Progress

Figure 11.8 As the reactants collide, chemical bonds break and form.

For a successful reaction to take place, $BrCH_3(aq)$ and $OH^-(aq)$ must collide. If the collision occurs in a favourable orientation with sufficient kinetic energy, the kinetic energy of the colliding molecules is converted to potential energy, which is stored in the partial bonds of the activated complex. The activated complex is said to be in its **transition state**.

Because the activated complex contains partial bonds, it is highly unstable. It can either break down to form products, or it can decompose to re-form reactants. The activated complex is like a rock teetering on top of a mountain. It could fall either way.

When the partial bonds of the activated complex re-form as chemical bonds, the potential energy that was stored is converted back into kinetic energy as the molecules again separate. This conversion results in a decrease in potential energy. Use the Sample Problem and Practice Problems that follow to practise labelling and sketching potential energy diagrams.

Sample Problem

Drawing a Potential Energy Diagram

Problem

Carbon monoxide, $CO(g)$, reacts with nitrogen dioxide, $NO_2(g)$. Carbon dioxide, $CO_2(g)$, and nitrogen monoxide, $NO(g)$, are formed. Draw a potential energy diagram to illustrate the progress of the reaction. (You do not need to draw your diagram to scale). Label the axes, the transition state, and the activated complex. Indicate the activation energy of the forward reaction, $E_{a(fwd)} = 134$ kJ, as well as $\Delta_rH = -226$ kJ. Calculate the activation energy of the reverse reaction, $E_a(rev)$, and show it on the graph. Because the reaction is going in the reverse direction, Δ_rH would be positive.

Solution

The activation energy of the reverse reaction may be determined using the formula described on page 407, or it can be determined from the potential energy diagram. The reaction is exothermic therefore the reverse reaction is endothermic.

$$E_{a(rev)} = E_{a(fwd)} + \Delta_rH$$
$$= 134 \text{ kJ} + 226 \text{ kJ}$$
$$= 360 \text{ kJ}$$

The activation energy of the reverse reaction is 360 kJ.

Potential Energy Diagram for CO + NO$_2$(g)

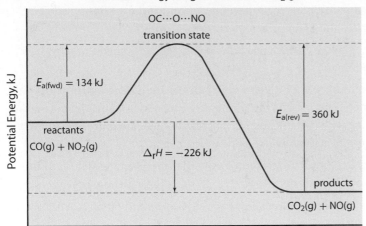

Using the potential energy diagram, you can confirm that the activation energy of the reverse reaction is 360 kJ.

Check Your Solution
Look carefully at the potential energy diagram. Check that you have labelled it completely. Since the forward reaction is exothermic, the reactants should be shown at a higher energy level than the products, and they are. The value of $E_{a(rev)}$ is reasonable.

Practice Problems

1. Complete the potential energy diagram by adding the following labels. Is the reaction endothermic or exothermic?

$E_{a(fwd)}$ $E_{a(rev)}$ $\Delta_r H$

appropriate x-axis label   appropriate y-axis label

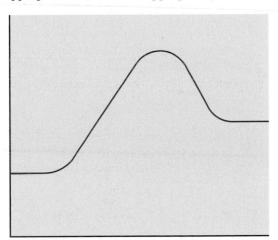

2. Consider the following reaction:
AB + C → AC + B $\Delta_r H = 65$ kJ, $E_{a(rev)} = 34$ kJ
Draw and label a potential energy diagram for this reaction. Calculate and label $E_{a(fwd)}$.

3. Consider the reaction below:
C + D → CD $E_{a(fwd)} = 61$ kJ, $E_{a(rev)} = 150$ kJ
Draw and label a potential energy diagram for this reaction. Calculate and label ΔH.

4. Using the potential energy diagram below, estimate the values at $E_{a(fwd)}$, $E_{a(rev)}$ and $\Delta_r H$. Is the reaction endothermic or exothermic?

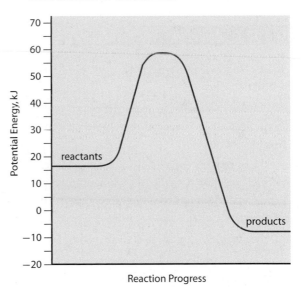

5. In the upper atmosphere, oxygen exists in forms other than O$_2$(g). For example, it exists as ozone, O$_3$(g), and as individual oxygen atoms, O(g). Ozone and atomic oxygen react to form two molecules of oxygen gas. For this reaction, the enthalpy change is −392 kJ and the activation energy is 19 kJ. Draw and label a potential energy diagram. Include a value for $E_{a(rev)}$.

Section 11.1 Summary

- According to collision theory, a reaction will occur only if the reactants collide with the correct orientation and sufficient energy.
- The energy equal to or above which reactants must collide is called the activation energy and symbolized, E_a.
- At any given temperature, the atoms and molecules of a substance are moving with a large range of kinetic energies. As the temperature increases, there are more atoms or molecules moving with higher energies. Therefore, there are more atoms or molecules with energies at or above the activation energy.
- You can illustrate the reaction progress on a potential energy curve. The amount of energy above the potential energy of the reactants is the activation energy.
- Reactions having a large activation energy proceed more slowly than reactions with a smaller activation energy.
- The potential energy curve for the reverse of a reaction is the mirror image of the curve for the forward reaction.
- For an exothermic reaction, $E_{a(rev)} = E_{a(fwd)} + \Delta_r H$
- For an endothermic reaction, $E_{a(fwd)} = E_{a(rev)} - \Delta_r H$
- When reactants reach the top of the potential energy curve they are said to be in a transition state, which is an unstable compound that is between the reactants and produces.

1. In your own words, describe the criteria necessary for a reaction to occur.

2. How does an increase in temperature affect the rate of a chemical reaction? Explain your answer in terms of activation energy.

3. Describe the energy conversions that take place during a chemical reaction. Use the terms kinetic energy, potential energy, bonds breaking, and bonds forming in your answer.

4. Describe what is meant by an activated complex.

5. An activated complex has been compared with a boulder sitting on the peak of a mountain. Explain why this is an appropriate comparison.

6. Consider the following reaction:
   $A_2(g) + B_2(g) \rightarrow 2AB(g)$ $E_{a(fwd)} = 143$ kJ; $E_{a(rev)} = 75$ kJ
   a) Is the reaction endothermic or exothermic in the forward direction?
   b) Draw and label a potential energy diagram. Include a value for ΔH.

7. Consider two exothermic reactions. Reaction (1) has a much smaller activation energy than Reaction (2).
   a) Sketch a potential energy diagram for each reaction, showing how the difference in activation energy affects the shape of the graph.

   b) How do you think the rates of reactions (1) and (2) compare? Explain your answer.

8. Answer the following questions by giving the number(s) of the reaction represented by the graph labelled with the number.

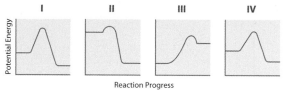

Which reaction has:
   a) largest $E_{a(fwd)}$
   b) smallest ΔH
   c) endothermic
   d) fastest reaction

9. Your friend is confused about the difference between the enthalpy change and the activation energy of a chemical reaction. Write a few paragraphs in which you define each term and distinguish between them. Use potential energy diagrams to illustrate your answer.

Catalysts and Reaction Rates

Section Outcomes

In this section, you will:
- **explain** that catalysts increase reaction rates by providing alternative pathways
- **design** and carry out an investigation to illustrate the effect of a catalyst on a chemical reaction
- **explain** how enzymes function as biological catalysts
- **explain** how catalysts reduce air pollution caused by the burning of hydrocarbons

Key Terms

catalyst
enzyme
active site
substrate

As you learned in the previous section, the rate of a chemical reaction is affected by the activation energy, or energy barrier, for that reaction. For some chemical reactions, that energy barrier is so high that the reaction will either not occur at all or will occur very slowly, over days or even years. Speeding up reactions to make them practical and profitable for industrial applications is important, and it is achieved using catalysts.

Catalysts

A **catalyst** is a substance that increases the rate of a chemical reaction without being consumed by the reaction. Catalysts are enormously important in industry and have a wide variety of applications (see Figure 11.9). Well over three million tonnes of catalysts are produced annually in North America. Newly discovered catalysts are patented and closely guarded because they increase productivity and give an organization a competitive edge.

Figure 11.9 Catalysts have many different industrial applications—from the production of sulfuric acid to the making of margarine, and from the refining of petroleum to the cleaning of car exhaust.

How a Catalyst Works

A catalyst works by providing an alternative pathway for a reaction to occur. The alternative pathway has a smaller activation energy than the uncatalyzed reaction; therefore, a larger fraction of the reactants have sufficient energy to react. The potential energy diagram in Figure 11.10 shows the activation energy for an uncatalyzed reaction and the activation energy for the same reaction with the addition of a catalyst.

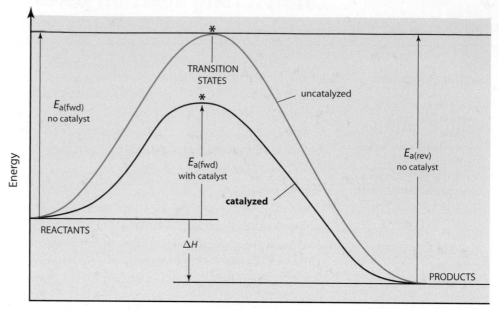

Figure 11.10 A catalyst provides an alternative pathway for a chemical reaction. The alternative pathway has a lower activation energy. A catalyst also increases the rate of the reverse reaction. What effect does a catalyst have on ΔH?

ChemistryFile

Web Link
You probably know that compounds called chlorofluorocarbons (CFCs) are responsible for depleting the ozone layer in Earth's stratosphere. Did you know, however, that CFCs do their destructive work by acting as catalysts? How do CFCs catalyze the decomposition of ozone in the stratosphere? Use the Internet to find out and communicate your findings as a two-page press release.

www.albertachemistry.ca
WWW

Although the catalyzed reaction has the same reactants, products, and enthalpy change (ΔH) as the uncatalyzed reaction, the catalyzed reaction pathway has a lower activation energy and is therefore faster. The catalyst takes part in the reaction and is changed during the reaction, but it is regenerated unchanged at the end of the overall reaction.

Figure 11.11 shows the reaction of sodium potassium tartrate, $NaKC_4O_6H_4(aq)$, with hydrogen peroxide, $H_2O_2(aq)$, catalyzed by cobalt(II) chloride, $CoCl_2(aq)$. The $CoCl_2(aq)$ is pink in solution, while the reactants are colourless. Notice that the contents of the beaker briefly change colour to dark green.

There are several possible products of the reaction—including oxygen, carbon dioxide, and oxalic acid, $HOOCCOOH(aq)$—and all the possible products are colourless. This observation means that the green colour is due to the interaction of the catalyst with the reactants during the reaction. The catalyst, in other words, changes during the reaction. The pink colour of the solution after the reaction is over, however, provides evidence that the catalyst is regenerated afterward.

Figure 11.11 The reaction of Rochelle's salt, sodium potassium tartrate, with hydrogen peroxide is catalyzed with $CoCl_2(aq)$ at 70 °C. **(A)** shows the reactants before mixing. **(B)** was taken immediately after $CoCl_2(s)$ was added. Notice the pink colour. **(C)** was taken after about 20 s. **(D)** was taken after about 2 min, when the reaction was complete.

• • •

4 How does the addition of a catalyst affect the following?

a) $E_{a(\text{fwd})}$ **d)** the speed of the forward reaction

b) $E_{a(\text{rev})}$ **e)** the speed of the reverse reaction

c) ΔH

5 What happens to a catalyst when the reaction is complete?

• • •

Catalysts in Industry

Many reactions that produce useful compounds proceed too slowly to be used in industries. Some reactions need to be carried out at high temperatures or pressures to proceed quickly without a catalyst. These conditions, however, are expensive to maintain. Therefore, when it is possible, chemists and engineers use catalysts to speed up the reactions and obtain products at a reasonable rate and under reasonable conditions.

Sulfuric Acid

For example, sulfur extracted from sour gas or oil sands can be converted into sulfuric acid, an enormously important industrial acid. This process, called the contact process, uses vanadium(V) pentoxide, $V_2O_5(s)$, as a catalyst to speed up one step in the process. The chemical reactions involved in the contact process, beginning with elemental sulfur, are shown below:

(1) $S_8(s) + 8O_2(g) \rightarrow 8SO_2(g)$

(2) $2SO_2(g) + O_2(g) \xrightarrow{V_2O_5(s)} 2SO_3(g)$

(3) $SO_3(g) + H_2SO_4(aq) \rightarrow H_2S_2O_7(aq)$

(4) $H_2S_2O_7(aq) + H_2O(\ell) \rightarrow 2H_2SO_4(aq)$

As you can see, step (2) involves the use of the catalyst. This step would otherwise be very slow, reducing the overall rate of the production of sulfuric acid. Notice that the catalyst is written above the reaction arrow and not as a reactant because the catalyst is not consumed during the reaction.

Nitric Acid

Another important industrial acid, nitric acid, $HNO_3(aq)$, is produced using the Ostwald process. In Alberta, one major use of nitric acid is as a starting material for the production of ammonium nitrate, $NH_4NO_3(s)$, a fertilizer.

The Ostwald process is shown below. The first step is catalyzed using a platinum-rhodium catalyst (Pt-Rh) at 800 °C. A photograph of the catalyst is shown in Figure 11.12. Notice that both nitric acid and nitrous acid are formed in step (3). With heating, the nitrous acid is converted to nitric acid:

(1) $4NH_3(g) + 5O_2(g) \xrightarrow{\text{Pt-Rh, 800° C}} 4NO(g) + 6H_2O(g)$

(2) $2NO(g) + O_2(g) \rightarrow 2NO_2(g)$

(3) $2NO_2(g) + H_2O(\ell) \rightarrow HNO_2(aq) + HNO_3(aq)$

(4) $3HNO_2(aq) \rightarrow HNO_3(aq) + H_2O(\ell) + 2NO(g)$

The catalyst industry, although consistently strong, has seen incredible growth because of environmental regulations surrounding acceptable levels of air pollution. Read the Connections feature on the next page to learn more about how catalysts are used in automobiles to reduce air pollution caused by the burning of hydrocarbons.

Figure 11.12 The platinum-rhodium catalyst, shown here, is in the form of gauze or mesh to maximize its surface area. Why do you think this is important?

Chemistry File

FYI

The same technology that was used on the space shuttle for plant-growth experiments is now used to extend the life of fruits, vegetables, and cut flowers. Ethene, a plant hormone responsible for the natural ripening of fruits and vegetables, also leads to their spoilage. A catalyst, titanium dioxide, $TiO_2(s)$, can be used to treat the air in spaces where fresh fruits, vegetables, or flowers are stored. Titanium dioxide speeds up the breakdown of ethene into carbon dioxide and water vapour, lengthening the shelf life of the produce.

Car Pollution Solution? Inside a Catalytic Converter

Since their introduction in 1975, catalytic converters have become mandatory for all new vehicles to reduce the level of harmful emissions in vehicle exhaust. When car engines burn fuel, they produce exhaust, made up primarily of nitrogen gas, $N_2(g)$, carbon dioxide, $CO_2(g)$, and water vapour, $H_2O(g)$. Although carbon dioxide is a greenhouse gas that contributes to global warming, it is not the most harmful emission produced by a car engine. Because the combustion process in an engine is not perfect, car engines also produce smaller amounts of the poisonous gas carbon monoxide, $CO(g)$, as well as hydrocarbons and nitrogen oxides, $NO(g)$ and $NO_2(g)$, which together are called $NO_x(g)$. These substances, together, contribute to the formation of smog and acid rain.

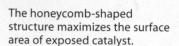

The honeycomb-shaped structure maximizes the surface area of exposed catalyst.

Three Stages of Conversion

Most modern vehicles are equipped with a three-way catalytic converter (TWC) that is located in the exhaust system and exposed to the exhaust stream.

Most TWCs consist of a honeycomb-shaped structure coated with the metal catalysts platinum, $Pt(s)$, palladium, $Pd(s)$, and rhodium, $Rh(s)$. As engine exhaust gases flow through the exhaust pipe into the honeycomb passageways of the catalytic converter, they come into contact with the metal catalysts.

TWCs reduce $CO(g)$, hydrocarbon, and $NO_x(g)$ emissions in three stages:

Stage 1: The $Pt(s)$ and $Rh(s)$ catalyze reactions that convert $NO_x(g)$ and $CO(g)$ into $N_2(g)$ and $CO_2(g)$.

Stage 2: The $Pt(s)$ and $Pd(s)$ catalyze the complete combustion of the hydrocarbons and $CO(g)$, producing $H_2O(g)$ and $CO_2(g)$.

Stage 3: For the complete combustion in stage 2 to take place, there must be the correct concentration of oxygen in the exhaust. During the third stage, an oxygen sensor mounted between the car's engine and the TWC determines the amount of oxygen in the exhaust and relays this information to the engine computer. The engine computer can control the amount of oxygen in the exhaust by adjusting the air-to-fuel ratio.

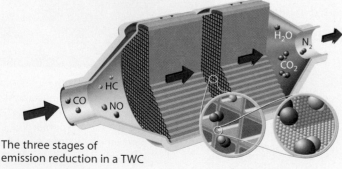

The three stages of emission reduction in a TWC

A Perfect Solution?

The catalytic converter can eliminate up to 95% of hydrocarbons, $CO(g)$, and $NO_x(g)$, but it is not perfect. The exhaust still contains $CO_2(g)$, which contributes to global warming. Also, the TWC only begins to work at a relatively high temperature. It begins to operate at around 288 °C, and efficient conversion starts at 399 °C. So when you start your car on a cold day, harmful gases escape with the exhaust until the catalytic converter heats up. Using a block heater in vehicles during the colder months of the year helps combat this problem.

• • •

1. Write a chemical equation to show the complete combustion of carbon monoxide.

2. Researchers have recently discovered that catalytic converters can produce dinitrogen monoxide, $N_2O(g)$. Commonly known as "laughing gas," $N_2O(g)$ makes up about 7.2% of greenhouse gases. Although the auto industry proposes redesigning the catalytic converter, environmentalists argue that this is just another reason to move away from gasoline-powered cars to electric or hybrid cars. Which do you think is the better solution to the problem? Explain your answer.

3. Fuel emission levels are connected to fuel consumption. The more fuel you use, the more emissions you produce.
   a) What are some ways to reduce the fuel used by an existing vehicle?
   b) How can you minimize your fuel consumption when choosing a new vehicle to buy? Search the Internet to provide examples of cars of varying fuel efficiency. Go to the web site below to get started.

www.albertachemistry.ca
WWW

Enzymes: Biological Catalysts

Your body depends on reactions that are catalyzed by amazingly efficient and specific biological catalysts. Biological catalysts are enormous protein molecules called **enzymes**. The molar masses of enzymes range from 15 000 to 1 000 000 g/mol. In industrial applications, chemists will often use an increase in temperature to increase the rate of a chemical reaction, with or without a catalyst. In the body, however, all chemical reactions must take place at body temperature, 37 °C. Without enzymes, chemical reactions in the body would occur far too slowly to maintain life.

Only a small portion of the enzyme, called the **active site**, is actually involved in the catalysis reaction. Compared with the enzyme's overall shape, the active site is like a nook or a fold in the surface. The reactant molecule, called the **substrate** in an enzyme reaction, binds to the active site. The enzyme works by stabilizing the reactant molecule(s) during the reaction. Figure 11.13 models the binding of a substrate to an enzyme.

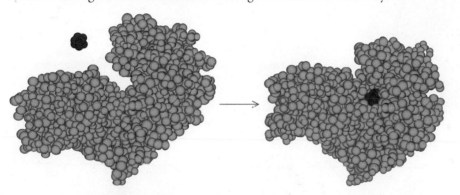

Figure 11.13 A glucose molecule binds to the enzyme hexokinase. Notice the enormous size of the enzyme relative to the substrate.

An early model of the interaction between an enzyme and its substrate was called the lock and key model. This model suggested that an enzyme is like a lock, and its substrate is like a key. The shape of the active site on the enzyme exactly fits the shape of the substrate. The current model is called the induced fit model or the hand-in-glove model. This model suggests that the active site of an enzyme changes its shape to fit its substrate. Figure 11.14 shows both models.

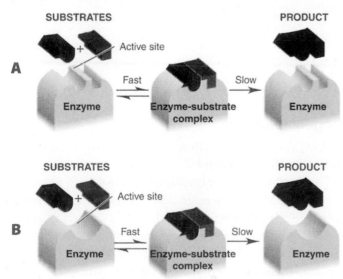

Figure 11.14 Diagram A shows the lock and key model of enzyme function. Diagram B shows the induced fit model of enzyme function.

In both models, the fit between the substrate and the active site is critical to the functioning of the enzyme. If the enzyme's shape changes in any way, the enzyme will cease to function. The enzyme's overall shape is maintained by hydrogen bonds, dipole-dipole interactions, and London (dispersion) forces. These attractive forces or bonds can break when exposed to high temperatures or extremes of pH.

If you carried out the Launch Lab at the beginning of this chapter, you discovered that pineapple effectively broke down, or digested, the previously firm gelatin. Pineapple contains bromelain, a collection of enzymes that break down protein. Bromelain, which is concentrated in the pineapple's stem, has a number of medicinal uses, including the treatment of athletic injuries (as an anti-inflammatory), digestive problems, and blood clots, among many others. How do you think cooking the pineapple would affect the action of bromelain? Design an experiment to test your prediction and carry it out at home.

For example, the enzyme pepsin, produced by cells in the stomach, catalyzes the breakdown of protein molecules into smaller units in the acidic environment of the stomach. At a higher, more neutral pH, pepsin's shape is changed, rendering the enzyme inactive. The cells of the stomach lining itself are protected from the acid by a mucous layer.

Digestion involves many other enzymes in addition to pepsin. Most of the digestive enzymes are active in the small intestine. For example, the enzyme lactase is responsible for catalyzing the breakdown of lactose, a sugar found in milk. People who are lactose intolerant usually do not produce lactase, or they have insufficient amounts. If you are lactose intolerant, you can take commercially produced supplements that contain lactase (Figure 11.15).

Figure 11.15 People who are lactose intolerant can still enjoy a glass of milk or cheese on a pizza by taking a supplement containing lactase.

Catalase and Hydrogen Peroxide

Hydrogen peroxide, $H_2O_2(aq)$, is a by-product in several reactions in living cells, especially liver and kidney cells. Hydrogen peroxide is also toxic to cells. Catalase, an enzyme found in living cells, is responsible for breaking down hydrogen peroxide and, therefore, protecting the cells. Catalase is also responsible for the fizzing you see when you use a dilute hydrogen peroxide solution to disinfect a cut, as shown in Figure 11.16. Catalase speeds up the decomposition of hydrogen peroxide into water and oxygen. Hydrogen peroxide works as an antiseptic because many types of bacteria lack the enzyme catalase and succumb to the toxicity of the hydrogen peroxide.

Figure 11.16 When you put hydrogen peroxide on a cut, it kills bacteria. Then the catalase in your tissues breaks down the peroxide and you see bubbles of oxygen.

Have you ever opened an old bottle of hydrogen peroxide and noticed that it had lost its potency? Over time, hydrogen peroxide decomposes, leaving you with a bottle of water. At room temperature, in the absence of a catalyst, the decomposition of $H_2O_2(aq)$ occurs very slowly, over months or years. In the presence of catalase, this reaction occurs in seconds. In Investigation 11.A, you will examine the effect of an inorganic catalyst, $I^-(aq)$ ion, on the rate of decomposition of hydrogen peroxide.

The Effect of a Catalyst on the Decomposition of Hydrogen Peroxide, $H_2O_2(aq)$

Hydrogen peroxide, $H_2O_2(aq)$, can be purchased as a dilute solution in a pharmacy or supermarket. It is used as a topical antiseptic for minor cuts, among other things. In this investigation, you will investigate the effect of the $I^-(aq)$ ion as a catalyst on the decomposition of $H_2O_2(aq)$. Hydrogen peroxide decomposes as shown in the equation below:

$$2H_2O_2(aq) \xrightarrow{I^-(aq)} 2H_2O(\ell) + O_2(g)$$

The gas produced can be measured as a decrease in mass of the system (caused by the release of oxygen gas) or by the downward displacement of water, as shown.

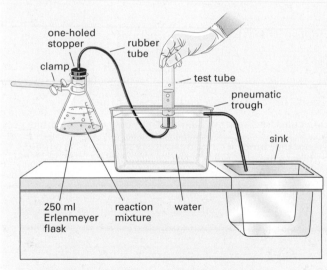

The downward displacement of water: To invert a test tube filled with water, place a piece of paper over the mouth of the filled test tube before inverting it.

Prediction

Predict how the addition of a catalyst will affect the rate of decomposition of hydrogen peroxide.

Safety Precautions

- The reaction mixture will get hot. Handle all glassware with care.
- 6% hydrogen peroxide, $H_2O_2(aq)$, is an irritant. Wear safety glasses, gloves, and a laboratory apron.

Materials

- 60 mL 6% (m/v) $H_2O_2(aq)$ 🔥
- 60 mL 1.0 mol/L NaI(aq) Ⓣ
- masking tape or grease pencil
- 100 mL beakers
- 10 mL graduated cylinders
- 250 mL Erlenmeyer flask
- clock with a second hand or stopwatch
- water
- one-holed stopper, fitted with a piece of glass tubing (must be airtight)
- rubber tubing to fit glass tubing (must be airtight)
- large test tube
- pneumatic trough or large beaker
- electronic balance accurate to 0.001 g

Experimental Plan

1. From your prediction, provide a list of variables (manipulated, responding, and controlled).

2. Using these variables, write a detailed procedure to test your prediction. Your procedure should
   - control all necessary variables
   - provide sufficient data for analysis
   - use no more than the quantities of materials provided

3. Design a data table to record your observations.

4. Check your plan to ensure that all safety requirements are addressed.

5. Ensure that the manipulated variable is the only one that changes during the experiment.

6. Ensure that the responding variable is measurable.

7. Check to see that the variables that should be controlled are kept constant throughout the experiment.

8. If you need a control experiment, ensure that it is included in the plan.

9. Be sure to collect sufficient data (perform enough trials).

10. Be sure that you have sufficient materials to complete the trials.

Data and Observations

11. Perform the experiment and record your data.

Analysis

1. How did the addition of a catalyst affect the volume or mass of gas produced (as measured by a decrease in the mass of the system)? How does this relate to the rate of reaction?

2. Construct a graph illustrating the effect of a catalyst on the rate of reaction.

3. What changes or additions would you make to your procedure if you were to repeat this experiment? Why?

Conclusion

4. Describe the effect of iodide on the rate of the decomposition of hydrogen peroxide. Be as specific as possible.

Application

5. The rate of decomposition of $H_2O_2(aq)$ can be increased using other catalysts, such as $Mn^{2+}(aq)$ and $Fe^{2+}(aq)$. How do you expect ΔH for the decomposition reaction to change if different catalysts are used?

Extension

6. Perform an Internet search to verify your findings.
ICT

Section 11.2 Summary

- A catalyst increases the rate of a reaction but is unchanged at the end of the reaction.
- Catalysts reduce the activation energy of the reaction thus creating a condition in which a larger percentage of reactants have kinetic energy at or above the activation energy.
- Catalysts are widely used in industry to make the rate of reactions economical.
- Biological catalysts are called enzymes.
- Enzymes are very specialized to catalyze only one or a few specific reactions.
- In an enzyme-catalyzed reaction, the reactants are called the substrate.
- Enzymes are large protein molecules that are much larger than the substrate molecules.
- The part of the enzyme molecule that binds to the substrate is called the active site.

SECTION 11.2 REVIEW

1. Explain why a catalyst has no effect on the change in total energy from reactants to products. Illustrate your answer with a potential energy diagram.

2. Why is only a small amount of catalyst necessary to speed up a chemical reaction?

3. List two similarities and two differences between industrial catalysts and enzymes.

4. **a)** What catalysts are present in a catalytic converter?
   b) What components of air pollution are removed by a catalytic converter? What are the eventual products?

5. You have read that the chlorine atoms released from CFCs in the upper atmosphere act as catalysts in the breakdown of ozone. Explain why the damage to the ozone layer is much greater than it would be if the chlorine atoms were reactants.

6. Use the following diagram to estimate the answers to the questions that follow.

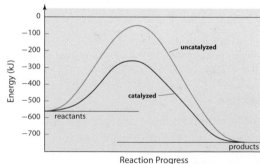

Reaction Progress

a) What is the $E_{a(fwd)}$ of the uncatalyzed reaction?
b) What is the $E_{a(fwd)}$ of the catalyzed reaction?
c) Which reaction is faster? Why?
d) What is the $E_{a(rev)}$ of the uncatalyzed reaction?
e) What is the $E_{a(rev)}$ of the catalyzed reaction?
f) What is the ΔH of the forward reaction?
g) What is the ΔH of the reverse reaction?
h) Explain your answers to (f) and (g) using Hess's law.

Chemical reactions occur only when the reactant atoms or molecules collide with the correct orientation and with sufficient energy to overcome the energy barrier. As you know, temperature is a measure of the average kinetic energy of atoms and molecules. At any given temperature, the kinetic energy of individual atoms or molecules varies over a wide range as described by a Maxwell-Boltzmann distribution. The fraction of reactant atoms or molecules that have sufficient energy, that is, equal to or greater than the activation energy, is determined by the temperature of the system.

A potential energy graph describes the progress of a reaction. When reactant atoms or molecules collide in the correct orientation and with sufficient kinetic energy, they convert to an intermediate form called the transition state. Transition state compounds are unstable and can go either forward to form products or backward to the reactants.

Catalysts are substances that increase the rate of a reaction without being permanently changed by the reaction. They are often changed temporarily but return to their original form by the end of a reaction. Catalysts reduce the activation energy but have no effect on the enthalpy change of the reaction. Catalysts are important to industry because in their absence most reactions would occur so slowly that they would not be economically feasible.

The chemical reactions that occur in living systems must occur at body temperatures. In humans, this is 37 °C. Since the reactions that occur in living systems would be too slow without a catalyst, all chemical reactions in living systems must be catalyzed. Biological catalysts are called enzymes. Reactants that enzymes act on are called substrates. Enzymes are very large protein molecules that are specialized to catalyze one or a few specific reactions. The region of the enzyme that binds the substrate and catalyzes the reaction is called the active site.

Concept Organizer   Activation Energy and Catalysts

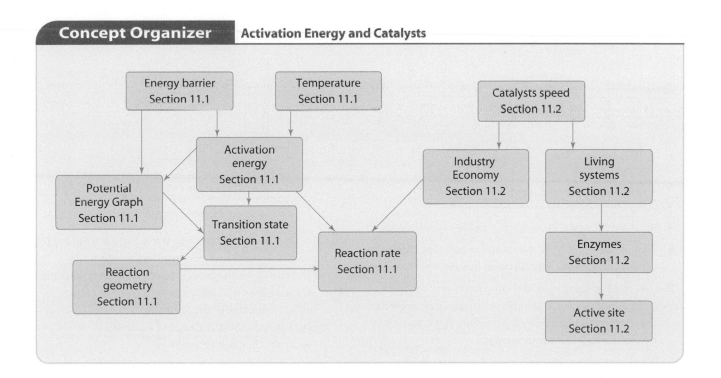

Understanding Concepts

1. For a reaction to take place, what three things must happen?

2. Use the lighting of a match as an example to illustrate the term *activation energy*.

3. What is the activated complex? Why is it so unstable? Use the term *energy* in your answer.

4. Why does increasing the temperature often increase the rate of a chemical reaction? Use a diagram in your answer.

5. Where does the energy required to reach activation energy come from? Explain your answer.

6. Complete the potential energy diagram below with the labels provided. Is the reaction endothermic or exothermic?
   a) $E_{a(fwd)}$
   b) $E_{a(rev)}$
   c) $\Delta_r H$
   d) reactants
   e) products
   f) activated complex
   g) reaction progress
   h) potential energy (kJ/mol)

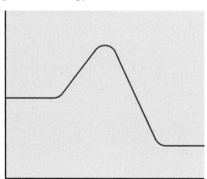

7. What are catalysts? How do they work?

8. What is the effect of a catalyst on the enthalpy change of a chemical reaction? Explain your answer.

9. Describe the energy conversions that take place during a chemical reaction as bonds are broken and reformed. Use the terms *kinetic energy*, *potential energy*, and activated complex in your answer.

10. Why are catalysts usually shown above the reaction arrow in a chemical reaction, instead of being included with the products or reactants?

11. a) Give one example of an industrial catalyst and explain its function.
   b) Give one example of an enzyme and explain its function.

c) While pepsin has a pH optimum around pH 3, the enzyme, trypsin, has a pH optimum closer to pH 7. Why do you think the pH optima for the two enzymes are so different?

12. Most reactions, including reactions catalyzed by enzymes, proceed faster at higher temperatures. Above a certain temperature, however, the rate for a given enzyme-catalyzed reaction will decrease dramatically. Explain why.

Applying Concepts

13. Consider the following reaction:
   $$A_2(g) + B_2(g) \rightarrow 2AB(g)$$
   $$E_{a(fwd)} = 143 \text{ kJ}, E_{a(rev)} = 75 \text{ kJ}$$
   a) Is the reaction endothermic or exothermic in the forward direction? How do you know?
   b) Draw and label a potential energy diagram. Include a value for $\Delta_r H$.

14. Use the potential energy diagram below to answer the following questions:
   a) Estimate the values for $\Delta_r H$ for the forward reaction, $E_{a(fwd)}$, and $E_{a(rev)}$ for the uncatalyzed reaction.
   b) What is the $\Delta_r H$, $E_{a(fwd)}$, and $E_{a(rev)}$ for the catalyzed reaction?

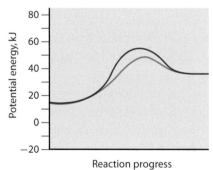

15. Consider two exothermic reactions. Reaction (1) has a much smaller activation energy than reaction (2).
   a) Sketch a potential energy diagram for each reaction, showing how the difference in activation energy affects the shape of the graph.
   b) How do you think the rates of reactions (1) and (2) compare? Explain your answer.

16. Why can't the activated complex be isolated during a chemical reaction?

17. The burning of gasoline in air is very exothermic and is used to provide the energy for many of our methods of transportation, yet a mixture of gasoline and air can be kept safely. Explain.

18. Consider the following general endothermic reaction:

$$A(s) + BC(aq) + energy \rightarrow B(s) + AC(aq)$$

   a) Explain why a catalyst has no effect on the change in total energy from reactants to products.

   b) Illustrate your answer with a potential energy diagram.

19. For the following reaction, for which $E_{a(fwd)} = 15$ kJ:

$$A(s) + B(g) \rightarrow AB(g) \quad \Delta H = -25 \text{ kJ}$$

   The addition of a catalyst decreases the $E_{a(rev)}$ to 32 kJ.

   a) What is the value for $E_{a(fwd)}$ of the catalyzed reaction?

   b) Illustrate your answer with a potential energy diagram.

20. For the following chemical reactions, suggest ways by which you could measure a change in rate of reaction when a catalyst is added. (**Hint:** See Investigation 11.A on page 417).

   a) $Zn(s) + 2 HCl(aq) \rightarrow ZnCl_2(aq) + H_2(g)$

   b) $H_2(g) + I_2(g) \rightarrow 2HI(g)$ (iodine gas is violet in colour)

   c) $2NOBr(g) \rightarrow 2NO(g) + Br_2(g)$ (both $NOBr(g)$ and $Br_2(g)$ are brown)

   d) $4Fe(s) + 3O_2(g) \rightarrow 2Fe_2O_3(s)$

21. Inhibitors slow down the rate of chemical reactions. Some inhibitors will actually prevent any reaction from occurring at all. How do you think inhibitors affect the E_a of a chemical reaction?

22. Hydrogen gas and oxygen gas are mixed in a balloon. No reaction occurs until the mixture is sparked or a small amount of powdered metal is added, at which point, the mixture explodes.

   a) Is the spark acting as a catalyst? Explain your answer.

   b) Is the metal acting as a catalyst? Explain your answer.

Making Connections

23. For years, tetraethyl lead was added to gasoline to prevent "knocking." Lead has since been removed from gasoline. Why? Is this lead additive still used in other parts of the world? Why or why not? ICT

24. Why are catalytic converters not very effective immediately after starting a vehicle in the winter? Suggest a way to correct this problem.

25. Aspirin™, or acetylsalicylic acid, inhibits pain by affecting an enzyme in the body. One of the classes of compounds in the body that is responsible for inflammation and pain sensations is the class called prostaglandins. Propose a mechanism by which aspirin might reduce pain.

26. Enzymes are sometimes used in industrial applications. You have probably read or heard in advertisements that some products contains enzymes. Think of one of those products and suggest a reason that enzymes might be useful in that product.

27. The toxic substance ethanol affects the nervous system, which can impair heart function and breathing. In humans, ethanol is broken down by an enzyme called alcohol dehydrogenase to form acetaldehyde. This substance causes headaches and other symptoms of a hangover.

   a) Why do you think consuming too much alcohol too fast (binge drinking) can be fatal?

   b) Methanol, which is even more toxic than ethanol, is processed by the same enzyme. The product is formaldehyde, which can lead to blindness and even death. An antidote to methanol poisoning is ethanol. Why do you think this works?

28. Historians suspect that many ancient Romans suffered from lead poisoning. The ancient Romans used lead for their plumbing system (hence the Latin name for the element) and also used lead acetate to sweeten wine. Lead is a poison because it can form strong bonds with proteins, including enzymes. Lead and other heavy metals react with enzymes by binding to functional groups on their active sites. When lead binds to a protein, such as an enzyme, the protein often precipitates out of solution.

   a) Based on your understanding of enzymes and the information above, explain why lead's ability to form strong bonds with the active sites of enzymes makes lead toxic to humans.

   b) Egg whites and milk are used as antidotes for heavy metal poisoning. The victim must ingest the egg white or milk soon after the lead has been ingested. Then the victim's stomach must be pumped. Explain why this antidote works. Why must the stomach be pumped?

29. Birds and mammals regulate their body temperature so the enzymes in their bodies are always functioning at the same temperature. Other families of animals do not generate enough heat in their bodies to maintain a constant body temperature. Such animals are very slow and sluggish when their body temperatures become very low. Provide a possible explanation for this sluggishness.

Career Focus: Ask a Supramolecular Chemist

In high school, Felaniaina Rakotondradany loved her chemistry labs because they were the only classes where she could use the experiments to create something useful and concrete. She pursued her interest in university, where she completed an undergraduate degree in chemical engineering, followed by a PhD in chemistry. Today, as a research manager in the Department of Chemical and Materials Engineering at the University of Alberta, she works to understand complex hydrocarbons like bitumen and heavy oils.

Felaniaina Rakotondradany

Q What kind of research are you doing now?

A The type of research I do relates to the fundamental properties of bitumen—organic liquids such as asphalt and tar that are very viscous; and heavy oils— petroleum deposits that are too thick to be extracted by conventional oil pumping methods. We know bitumen contains many different molecules, but we don't have a formula or know what many of those molecules are. This makes it harder to know how bitumen behaves. In order to get a better understanding of the compound, I synthesize and study hydrocarbons that we think are similar to those in bitumen.

Q What led you to this particular field?

A My PhD thesis is related to supramolecular chemistry, which is a relatively new field. Supramolecular chemists use weak interactions between molecules (i.e., monomers) to design new, larger supermolecules (i.e., aggregates). To better control the size and shape of these supermolecules, a supramolecular chemist needs to understand the nature of the monomers and the weak interactions that act as a glue between the monomers. Bitumen and heavy oil are extremely complex mixtures. We know some of the components of bitumen and heavy oil also form aggregates. I decided to make use of my experience with supramolecular structures to understand these natural mixtures.

Q What are some practical applications of your work on hydrocarbons?

A My work has implications for the petroleum industry. Understanding the structure of bitumen and heavy oils will help solve problems like fouling—the build-up of highly viscous materials in places such as chemical reactors or pipes. When a build-up occurs, it can plug pipes and reactor components, increasing pressure and causing explosions.

Q What is bitumen and heavy oil upgrading? Why is this process necessary?

A Bitumen and heavy oils are crude oils composed of hydrocarbons varying in size and nature. The goal of bitumen upgrading is to reduce the viscosity of bitumen, making it easier to handle. Upgrading transforms this tar-like mixture into useful materials, by breaking up the large compounds (cracking) and removing sulphur, nitrogen, and other non-hydrocarbon compounds.

Q What kinds of products can be made from hydrocarbons?

A Bitumen upgrading leads to "synthetic crude oil," which can subsequently be refined by distillation to produce kerosene, gas, oil, diesel, and fuel gas. During upgrading, small molecules like ammonia and ethylene are created. Ethylene is a widely used raw material for polymers and plastics. Ammonia is used in the pharmaceutical industry to make drugs.

Related careers

Petroleum Engineer Petroleum engineers analyze data about oil and gas wells. They predict the immediate and long-term production capabilities of wells to decide the quantities of hydrocarbons that should be extracted. Petroleum engineers also help design health, safety, and environmental controls on oil and gas extraction operations. A bachelor's degree in petroleum engineering or chemical engineering is good education for this type of work.

Petroleum Engineer Technologist Petroleum engineer technologists work with the technical aspects of oil and gas extraction. They supervise well equipment servicing, calculate hydrocarbon reserves to evaluate a well's economic potential, and construct maps based on geological data. They may also be involved in monitoring occupational health, safety, and environmental programs. They can enter the workforce after two years of post-secondary training in petroleum engineering technology.

Hydrocarbon Data Analysts Hydrocarbon mud loggers, also called hydrocarbon data analysts, test mud from wells to see whether or not oil or gas is present. Their work helps determine the rate at which a well should be drilled, and the tools a drilling team should use to extract oil from a well. You can be trained on the job for this kind of work.

Petroleum Exploration Geophysicist Geophysics is the study of the earth's composition and structure through physical methods, such as measuring seismic reflection, gravity, electromagnetic, radioactive, and magnetic activity. Petroleum exploration geophysicists use their knowledge about how to study the earth to figure out where oil and gas deposits are most likely to be found. These geophysicists work in teams with petroleum engineers, technicians, and other oil and gas professionals, and usually have a bachelor's degree in geophysics.

Go Further...

1. Bitumen is often used in roofing and highway construction. What properties of bitumen make it suitable for these uses?

2. Like fouling, corrosion of pipelines can cause dangerous problems, including spills and explosions. What is corrosion? How might understanding the behaviour and properties of bitumen help prevent corrosion?

3. Waste plastic can be mixed with bitumen and laid down to make roads. Research the environmental benefits of such a mixture.

www.albertachemistry.ca
WWW

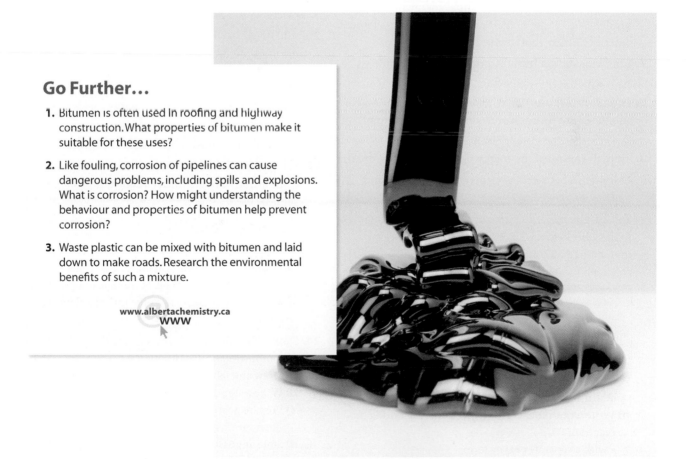

Understanding Concepts

1. Describe an endothermic reaction in terms of the energy required to break bonds compared with the energy released when bonds are made.

2. If beaker A contains 500 mL of a liquid and beaker B contains 1000 mL of the same liquid, what happens when 200 kJ of energy is absorbed by the liquid in each beaker?

3. Which of the following processes are exothermic?
a) boiling water
b) forming water
c) photosynthesis
d) combustion of methane
e) cellular respiration

4. What is the standard molar enthalpy of formation for all elements in their most stable state under standard conditions? Why?

5. Illustrate the difference between the terms *system* and *surroundings*.

6. "The reactants have more potential energy than the products do." What kind of reaction does this statement describe? Justify your answer.

7. Compare and contrast the process of photosynthesis with that of respiration.

8. How does hydrocarbon combustion in an open system compare with hydrocarbon combustion in a closed system, such as a bomb calorimeter?

9. Trace the origin of the energy stored in chemical bonds.

10. Distinguish between the terms *enthalpy change* and *molar enthalpy change*.

11. You are given a 70 g sample of each of the metals shown in the table below, all at 25 °C. You heat each metal under identical conditions. Which metal will be first to reach 30 °C? Which will be last? Explain your reasoning.

Metal	Specific heat capacity (J/g • °C)
platinum	0.133
titanium	0.523
zinc	0.388

12. When ammonia is formed from its elements in standard state, 45.9 kJ of energy is released.
a) Write a thermochemical equation representing this reaction:
• with a separate energy term
• with the energy as a product
b) Draw a potential energy diagram for this reaction.

13. Define *calorimetry*. Describe two commonly used calorimeters.

14. When taking calorimetric measurements, why do you need to know the heat capacity of the calorimeter?

15. Compare the following terms: *specific heat capacity* and *heat capacity*.
a) What are their units?
b) In which situations would they be used?

16. Define *activation energy*.

17. For a reaction to occur when molecules collide, what two criteria must be satisfied?

18. How does a catalyst help achieve the requirements in question 17?

19. Draw and label a potential energy diagram for the following exothermic reaction:

$$A(s) + B(g) \rightarrow AB(s) + 397 \text{ kJ} \qquad E_{a(fwd)} = 27 \text{ kJ}$$

Include activation energy (forward and reverse), the activated complex, and the enthalpy change for the reaction.

20. Add the following labels to the potential energy diagram you drew for question 19:
• bonds being broken
• bonds being formed
• kinetic energy → potential energy
• potential energy → kinetic energy

21. Compare and contrast catalysts and enzymes.

22. What is the difference between heat content and heat of combustion? Under which circumstances should each value be used?

Applying Concepts

23. Some solid sodium hydroxide, $NaOH(s)$, is added to some dilute hydrochloric acid, $HCl(aq)$, in a simple calorimeter at room temperature. The temperature of the solution increases significantly. Explain this observation.

24. Design an experiment to determine the molar enthalpy of combustion of two different fuels.
a) Draw and label a diagram of the apparatus.
b) Write a step-by-step procedure.
c) Prepare a table to record your data and other observations.
d) State any assumptions that you will make when carrying out the calculations.

25. Write balanced thermochemical equations for the formation of following substances at standard conditions:
   a) $CH_3COOH(\ell)$
   b) $BaSO_4(s)$
   c) $C_2H_2(g)$
   d) $NO_2(g)$

26. Rank the substances in question 25 from least thermally stable to most thermally stable. Justify your answer.

27. In a short paragraph, explain why it is more efficient to produce electricity from flowing water than from burning a fossil fuel.

28. Use the EnerGuide label below to answer the following questions:

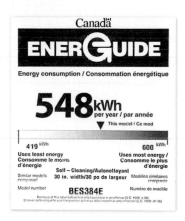

   a) What is the expected yearly consumption of energy for this model of dishwasher?
   b) How does this model compare with others in its class?
   c) What mark signifies that an appliance is among the most energy efficient in its class?

29. The decomposition of aqueous hydrogen peroxide, $H_2O_2(aq)$, into water, $H_2O(\ell)$, and oxygen gas, $O_2(g)$, can be catalyzed by different substances. One such substance is aqueous sodium iodide, $NaI(aq)$; another is aqueous iron(II) nitrate, $Fe(NO_3)_2(aq)$.
   a) The enthalpy change of this reaction is the same, regardless of the catalyst. Explain why, with the help of a potential energy diagram.
   b) Design an investigation to verify your explanation in part (a). Do not attempt to carry out the investigation without the supervision of your teacher.
   c) What enthalpy change would you predict for the decomposition of 1 mol of hydrogen peroxide? Show your work.

Solving Problems

30. a) Use the following data to determine an experimental value for the molar enthalpy of combustion for ethanol, $C_2H_5OH(\ell)$:

   initial mass of alcohol burner = 47.39 g

   final mass of alcohol burner = 47.19 g

   mass of aluminium can = 84.37 g

   mass of aluminium can and water = 239.36 g

   initial temperature of water = 15.7 °C

   final temperature of water = 18.3 °C

   b) The accepted value for the molar enthalpy of combustion is −507 kJ/mol. Calculate the percentage error.
   c) What happened to the "lost" energy?

31. Consider the following chemical equations and their enthalpy changes:

$$CH_4(g) + 2O_2(g) \rightarrow CO_2(g) + 2H_2O(g) \quad \Delta_rH = -8.0 \times 10^2 \text{ kJ}$$
$$CaO(s) + H_2O(\ell) \rightarrow Ca(OH)_2(s) \quad \Delta_rH = -65 \text{ kJ}$$

   What volume of methane at STP would have to be burned to release the same amount of energy as the reaction of 1.0×10^2 g of $CaO(s)$ with sufficient water?

32. The complete combustion of 1.00 mol of sucrose (table sugar), $C_{12}H_{22}O_{11}(s)$, yields 5.65×10^3 kJ of energy.
   a) Write a balanced thermochemical equation for the combustion of sucrose in an open system.
   b) Calculate the amount of energy that is released when 5.00 g of sucrose (about one teaspoon) is burned.
   c) How much water (in grams) could be warmed by 5.0 °C when 5.00 g of sucrose is completely burned?

33. When 1 mol of potassium bromide, $KBr(s)$, decomposes to its elements, 394 kJ of heat is absorbed.
   a) Write a thermochemical equation for this reaction.
   b) Draw and label a potential energy diagram.
   c) How much heat is released when 1.37 g of $KBr(s)$ forms from its elements in standard state?

34. Rust, $Fe_2O_3(s)$, forms when the iron in steel is exposed to water. A simplified reaction (without water) is shown below:

$$4Fe(s) + 3O_2(g) \rightarrow 2Fe_2O_3(s) \quad \Delta_rH = -1.65 \times 10^3 \text{ kJ}$$

   a) How much energy is released when 1.37 kg of iron rusts?
   b) How much rust forms along with the release of 3.75 MJ of energy?

35. Four fuels are listed in the table below. Equal masses of these fuels are burned completely.

Fuel	Enthalpy of combustion at SATP (kJ/mol)
octane, $C_8H_{18}(\ell)$	$-5\,513$
methane, $CH_4(g)$	-890
ethanol, $C_2H_5OH(\ell)$	$-1\,367$
hydrogen, $H_2(g)$	-285

a) Put the fuels in order, from greatest to least amount of energy provided per mol.

b) Write a balanced chemical equation for the complete combustion of each fuel.

c) Determine the heat content of each fuel (in kJ/g). Rank the fuels again from greatest to least amount of energy provided per gram.

d) Which method of ranking fuels do you think is most appropriate? Why?

36. A group of students tested two white, crystalline solids, A and B, to determine their enthalpies of solution. The students dissolved 10.00 g of each solid in 100.0 mL of water in a polystyrene calorimeter and collected the following temperature data:

Time (min)	Temperature of A (°C)	Temperature of B (°C)
0.0	15.0	25.0
0.5	20.1	18.8
1.0	25.0	16.7
1.5	29.8	15.8
2.0	31.9	15.2
2.5	32.8	15.0
3.0	33.0	15.0
3.5	33.0	15.2
4.0	32.8	15.5
4.5	32.5	15.8
5.0	32.2	16.1
5.5	31.9	16.4

a) Using a spreadsheet package, graph the data in the table above, placing time on the x-axis and temperature on the y-axis. Use your graph to answer parts (b) to (d). **ICT**

b) Classify the enthalpy of solution of each solid as exothermic or endothermic.

c) From the data given, calculate the enthalpy of solution for each solid. Your answer should be in kJ/g.

d) Is there any evidence from the data that the students' calorimeter could have been more efficient? Explain your answer.

37. Carbon monoxide reacts with hydrogen gas to produce a mixture of methane, carbon dioxide, and water. (This mixture is known as substitute natural gas.) This reaction is summarized below:

$$4CO(g) + 8H_2(g) \rightarrow 3CH_4(g) + CO_2(g) + 2H_2O(\ell)$$

Use the following thermochemical equations to determine the enthalpy change of the reaction.

$$C(s) + 2H_2(g) \rightarrow CH_4(g) + 74.8 \text{ kJ}$$
$$CO(g) + \tfrac{1}{2}O_2(g) \rightarrow CO_2(g) \quad \Delta H° = -283.1 \text{ kJ}$$
$$H_2(g) + \tfrac{1}{2}O_2(g) \rightarrow H_2O(g) + 241.8 \text{ kJ}$$
$$C(s) + \tfrac{1}{2}O_2(g) \rightarrow CO(g) \quad \Delta H° = -110.5 \text{ kJ}$$
$$H_2O(\ell) + 44.0 \text{ kJ} \rightarrow H_2O(g)$$

38. Sulfur hexafluoride, $SF_6(g)$, one of the densest known gases, is used to insulate electrical equipment, including high-voltage wires. Determine the enthalpy change of reaction (1) below using reactions (2), (3), and (4).

(1) $H_2S(g) + 4F_2(g) \rightarrow 2HF(g) + SF_6(g) \quad \Delta_r H° = ?$

(2) $\tfrac{1}{2}H_2(g) + \tfrac{1}{2}F_2(g) \rightarrow HF(g) \quad \Delta_r H° = -273 \text{ kJ}$

(3) $\tfrac{1}{8}S_8(s) + 3F_2(g) \rightarrow SF_6(g) \quad \Delta_r H° = -1220 \text{ kJ}$

(4) $H_2(g) + \tfrac{1}{8}S_8(s) \rightarrow H_2S(g) \quad \Delta_r H° = -21 \text{ kJ}$

39. Using standard enthalpies of formation, determine the enthalpy of reaction for the following:

a) $CaO(s) + CO_2(g) \rightarrow CaCO_3(s)$

b) $C_2H_2(g) + 2H_2(g) \rightarrow C_2H_6(g)$

c) $4NH_3(g) + 5O_2(g) \rightarrow 4NO(g) + 6H_2O(g)$

40. Given the following thermochemical equation, determine the molar enthalpy of formation for silicon tetrafluoride, $SiF_4(g)$:

$$SiO_2(s) + 4HF(g) \rightarrow SiF_4(g) + 2H_2O(\ell)$$
$$\Delta H° = -182.6 \text{ kJ}$$

41. Nitroglycerine, $C_3H_5(NO_3)_3(\ell)$, an explosive, detonates to produce nitrogen, water vapour, carbon dioxide, and oxygen gases, releasing 1.15×10^4 kJ/mol of nitroglycerine.

a) Write a balanced chemical equation for the reaction.

b) Using standard molar enthalpies of formation, determine the molar enthalpy of formation for nitroglycerine.

42. Use the figure below to answer the following questions:
   a) Determine the enthalpy change when 3.7 g of ethanol, $C_2H_5OH(\ell)$, is formed from its elements in its standard state.
   b) How much oxygen is consumed if the enthalpy change is -100 kJ?
   c) Write a thermochemical equation for the combustion of ethanol in an open system ($\Delta_cH = -1234.8$ kJ/mol).
   d) Determine the mass of ethanol that would have to be combusted to result in an enthalpy change of -550 kJ.

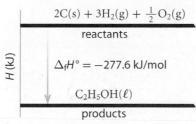

$$2C(s) + 3H_2(g) + \tfrac{1}{2}O_2(g)$$

reactants

H (kJ)

$\Delta_fH° = -277.6$ kJ/mol

$C_2H_5OH(\ell)$

products

43. The standard molar enthalpy of formation of poisonous hydrogen cyanide gas, $HCN(g)$, is 135 kJ/mol.
   a) Write a balanced thermochemical equation to show the formation of 1 mol of hydrogen cyanide from its elements.
   b) Draw an enthalpy diagram for this reaction.
   c) What would the change in enthalpy be to form 50.0 L of $HCN(g)$ at STP from its elements?

44. Determine the efficiency of a camp stove given the following data:

mass of propane burned = 4.48 g
mass of steel pot = 575.00 g
mass of soup = 275.37 g
initial temperature of soup = 4.5 °C
final temperature of soup = 75.4 °C
specific heat capacity of steel = 0.503 J/g • °C
specific heat capacity of soup = 3.57 J/g • °C

What improvements could be made to a camp stove to improve its efficiency?

45. For the reaction $A(g) + B(g) \rightarrow AB(g)$, $E_{a(fwd)} = 25$ kJ and $\Delta_rH = +13$ kJ,
   a) Sketch a potential energy diagram for this reaction.
   b) Determine the activation energy for the reverse reaction, and add it to the diagram.
   c) Sketch the effect of a catalyst on the energy pathway.

Making Connections

46. How efficient is your dream car?
   a) How do you think that your car compares with others in its class?
   b) What concessions could you make to choose a vehicle that is more fuel efficient?
   c) What driving and maintenance habits could you adopt to ensure you achieve the maximum fuel efficiency for that vehicle?

47. Suppose that you are having a new home built in a rural area, where natural gas is not available. You have two choices for fuelling your furnace:
   • propane, $C_3H_8(g)$, delivered as a liquid under pressure and stored in a tank
   • home heating oil, delivered as a liquid (not under pressure) and stored in a tank.
   What factors do you need to consider to choose the best fuel? What assumptions do you need to make? Are there any other alternatives that you would like to consider? Explain your response.

48. Suppose that you read the following statement in a magazine: "0.95 thousand cubic feet of natural gas is equal to a gigajoule, GJ, of energy." Being a media-literate student, you are sceptical of this claim and want to verify it. The following assumptions and information might be useful:
   • Natural gas is pure methane.
   • Methane undergoes complete combustion.
   • $H_2O(\ell)$ is the product of combustion, not $H_2O(g)$.
   • 1.00 mol of any gas occupies 24 L at 20 °C and 100 kPa.
   • 1 foot = 12 inches; 1 inch = 2.54 cm; 1 L = 1 dm^3

49. Pepsin, chymotrypsin, and lipase are digestive enzymes that require very different conditions in order to be effective. Pepsin and chymotrypsin hydrolyze proteins while lipase breaks down lipids. Lipids are biological molecules that are not soluble in water. Fats are a type of lipid. Chymotrypsin and lipase are effective at neutral pH while pepsin is active at a low pH. Provide a logical explanation for the properties of these enzymes.

UNIT 6

Electrochemical Changes

General Outcomes

- explain the nature of oxidation-reduction reactions

- apply the principles of oxidation-reduction to electrochemical cells

Unit Contents

Focussing Questions

1. What is an electrochemical change?

2. How have scientific knowledge and technological innovation been integrated in the field of electrochemistry?

Unit PreQuiz ?

www.albertachemistry.ca

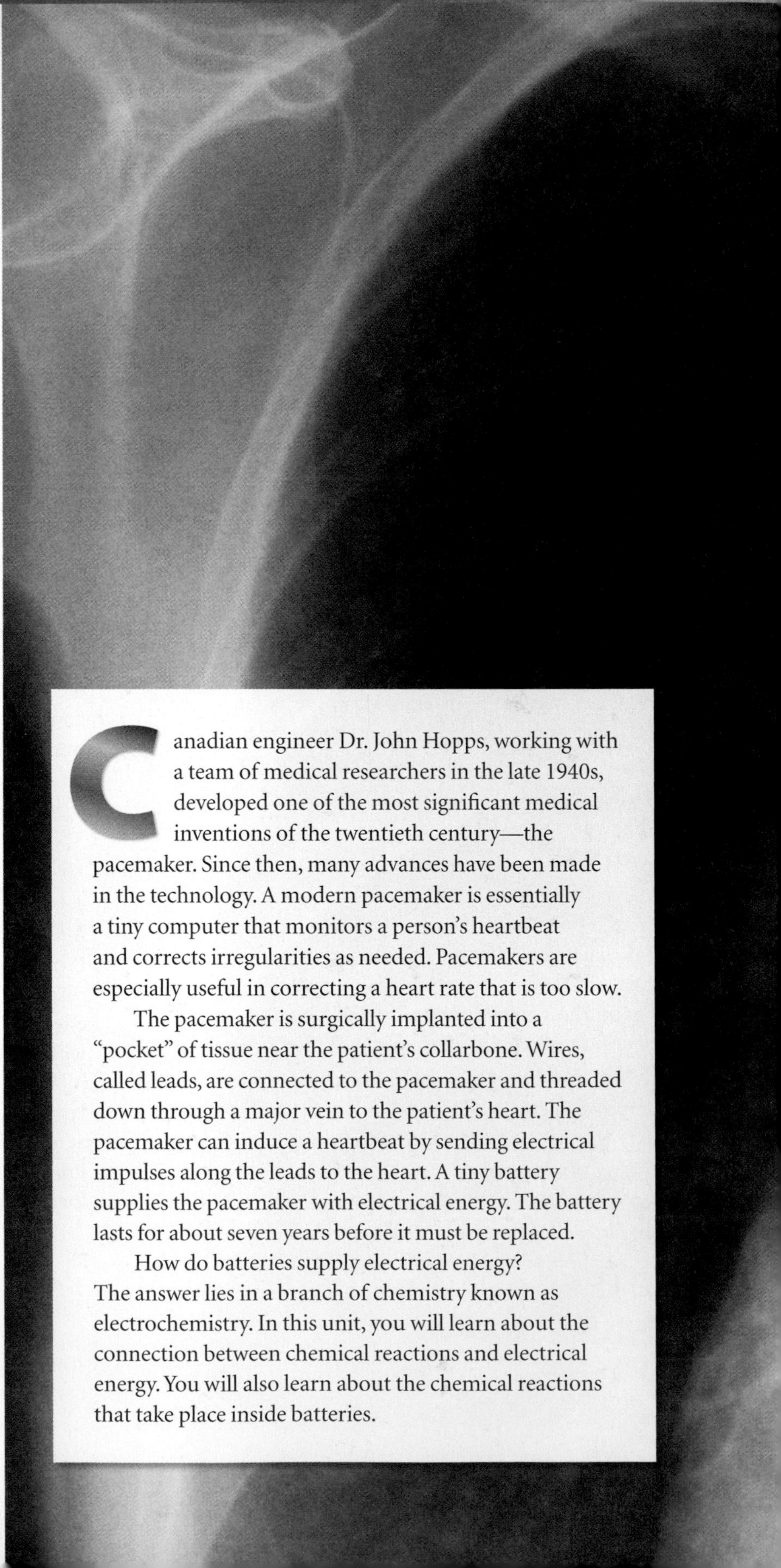

Canadian engineer Dr. John Hopps, working with a team of medical researchers in the late 1940s, developed one of the most significant medical inventions of the twentieth century—the pacemaker. Since then, many advances have been made in the technology. A modern pacemaker is essentially a tiny computer that monitors a person's heartbeat and corrects irregularities as needed. Pacemakers are especially useful in correcting a heart rate that is too slow.

The pacemaker is surgically implanted into a "pocket" of tissue near the patient's collarbone. Wires, called leads, are connected to the pacemaker and threaded down through a major vein to the patient's heart. The pacemaker can induce a heartbeat by sending electrical impulses along the leads to the heart. A tiny battery supplies the pacemaker with electrical energy. The battery lasts for about seven years before it must be replaced.

How do batteries supply electrical energy? The answer lies in a branch of chemistry known as electrochemistry. In this unit, you will learn about the connection between chemical reactions and electrical energy. You will also learn about the chemical reactions that take place inside batteries.

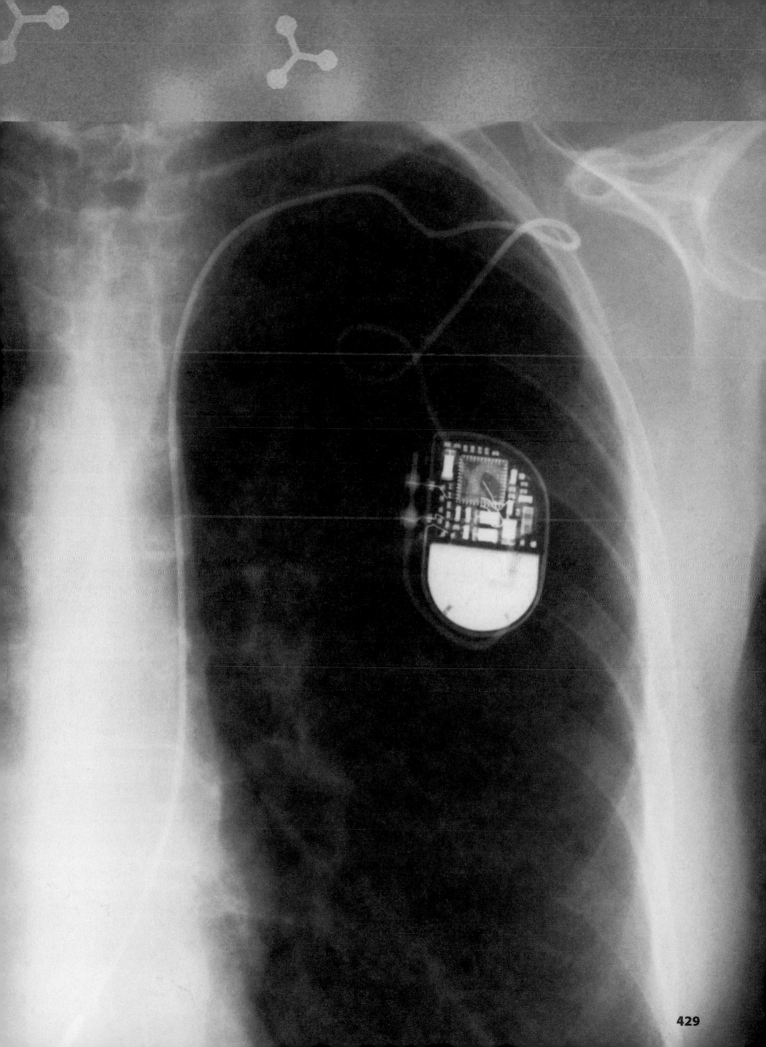

In chemistry, concepts build on one another. When you begin to learn about a new concept such as electrochemistry, you need to reach back into your memory and pull out concepts that you learned in previous science courses. The following topics are some of the concepts that you have "stored" in your memory and will be using in this unit.

Types of Chemical Reactions

Although there are millions of possible chemical reactions, many of them can be classified within five different types of reactions. These five types of reactions are:

Formation reactions: Reactions in which two reactants combine to form one product.

Decomposition reactions: Reactions in which one reactant breaks down into more than one product.

Single-replacement reactions: Reactions in which an atom of one element takes the place of an atom of another element in a compound.

Double-replacement reactions: Reactions in which two ionic compounds "change partners."

Complete combustion reactions: Reactions in which a hydrocarbon reacts with oxygen to produce carbon dioxide and water.

In this unit, you will be working predominantly with single-replacement reactions and double-replacement reactions. You will also consider complete combustion reactions. Analyze the equation below.

$$2Al(s) + 3NiSO_4(aq) \longrightarrow Al_2(SO_4)_3(aq) + 3Ni(s)$$

Notice that the aluminium atoms replaced the nickel atoms in the nickel sulfate. Since atoms of one element replaced atoms of another element and no other changes occurred, this is a single-replacement reaction. In Unit 6, you will learn how to predict whether a reaction such as this will proceed spontaneously.

Consider, now, the following equation.

$$Ba(NO_3)_2(aq) + Na_2SO_4(aq) \longrightarrow BaSO_4(s) + 2NaNO_3(aq)$$

In this reaction, the barium ion replaced the sodium ion in the sodium sulfate but the sodium ion also replaced the barium ion in the barium nitrate. Since there were two "replacements," this is a double-replacement reaction. When working with a reaction such as this, always remember to consult your solubility table to find out whether a compound is soluble or insoluble. You need this information to determine whether to label the compound aqueous (aq) for soluble compounds or solid (s) for insoluble compounds.

Another concept that you might need to review is how to determine the subscripts for ionic compounds. You know that an ionic compound must have an equal number of positive and negative charges so that it will have a net zero charge. For individual atoms, consult the periodic table and look for the "most stable ion charges" for the atom. For polyatomic ions, you will find a table of common polyatomic ions with their charges in Appendix G and also in your data booklet. For example, if you look for barium in your periodic table, you will find that it ionizes to Ba^{2+}. The nitrate polyatomic ion (NO_3^-) has a charge of 1− and, therefore, you need two nitrate ions to balance the charge of one barium ion.

Types of Equations

The two equations that are used as examples are called complete balanced equations because everything that is involved in the original compounds is present. However, in some situations, the actual changes that are taking place are easier to visualize by using ionic or net ionic equations. In ionic equations, you write each ion and its charge individually. The ionic forms of the two equations are written as follows.

$$2Al(s) + 3Ni^{2+}(aq) + 3SO_4^{2-}(aq) \longrightarrow 2Al^{3+}(aq) + 3SO_4^{2-}(aq) + 3Ni(s)$$

$$Ba^{2+}(aq) + 2NO_3^{-}(aq) + 2Na^{+}(aq) + SO_4^{2-}(aq) \longrightarrow BaSO_4(s) + 2Na^{+}(aq) + 2NO_3^{-}(aq)$$

If you analyze the first equation, you will see that three sulfate ions are present on both sides of the equation. Since no change occurred in the sulfate ions in this reaction, they are called spectator ions. In net ionic equations, the spectator ions are omitted. The net ionic form of the first equation would thus be:

$$2Al(s) + 3Ni^{2+}(aq) \longrightarrow 2Al^{3+}(aq) + 3Ni(s)$$

This net ionic form makes it clear that the aluminium atom is losing three electrons and is becoming ionized. The nickel ion is gaining two electrons and is becoming a nickel atom.

In the second ionic equation, you can see that both sodium and nitrate are spectator ions. The net ionic equation would thus be:

$$Ba^{2+}(aq) + SO_4^{2-}(aq) \longrightarrow BaSO_4(s)$$

This form of the equation makes it very clear that the barium and sulfate ions are forming a solid precipitate.

Electronegativity

All of the compounds in the example equations above are ionic compounds. Do you recall how to determine whether a compound is held together by ionic bonds or covalent bonds? You probably recall that ionic compounds consist of a metal and a non-metal ion and that molecular compounds consist of only non-metal ions. This is correct but the reasoning does not stop there. The type of bond is determined by the electronegativity of the elements that combine to form a compound. If the difference in the electronegativities of two elements is large (over 1.7) the bonds are "mostly ionic." If the electronegativity difference between the elements in a compound is low (less than 1.7) but not zero, the bonds are polar covalent. The electrons in a bond are not shared equally by the two atoms. One nucleus usually exerts a greater attractive force on the electrons. The electronegativity of an element is a measure of the strength of the attractive force that atoms of that element exert on the electrons in the bond. In general, the electronegativities of the elements tends to increase as you go from left to right and as you go up the periodic table. For example, the element in Period 2 and Group 17 is fluorine with an electronegativity of 4.0. The element in Period 7, Group 1 is francium with an electronegativity of 0.7. Figure P6.1 is one square from a periodic table and shows you that the electronegativity for the element is listed in the centre of the left side of the square. Understanding electronegativity will help you when you are balancing equations involving electrochemical reactions.

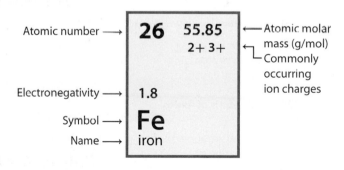

Figure P6.1 Each element on the periodic table except the noble gases (which do not normally form bonds at all) and the very rare elements, have the value of their electronegativity in the centre of the left side of the square. As shown here, the electronegativity of iron is 1.8.

Oxidation-Reduction Reactions

Chapter Concepts

12.1 Characterizing Oxidation and Reduction

- Oxidation is a loss of electrons. Reduction is a gain of electrons.

- When one atom is oxidized, another must be reduced.

- You can experimentally determine the relative strength of atoms or ions as reducing or oxidizing agents.

12.2 Redox Reactions Involving Ionic Compounds

- An equation for a half-reaction includes only those compounds that are reduced or only those that are oxidized.

- A set of rules guides you through the process of balancing half-reactions and total reactions.

12.3 Redox Reactions Involving Molecular Compounds

- Oxidation numbers describe the oxidation state of atoms in both ionic and molecular compounds.

- A set of rules allows you to assign oxidation numbers to all atoms in a compound and balance equations.

12.4 Quantitative Analysis of Redox Reactions

- Because some elements change colour with a change in oxidation state, you can titrate redox reactions.

- Quantities measured in redox titrations allow you to perform quantitative calculations on redox reactions.

The chemical reactions that take place in burning logs, contracting muscles, and rusting metals are very similar. In each case, a compound reacts with oxygen and the reaction releases heat. In some reactions, large amounts of heat are released rapidly. In the reactions that occur in muscles, the overall reaction with oxygen takes place in many small steps; therefore, heat is released evenly over time. When metals rust, the reactions occur so slowly that it is difficult to detect the heat that is released. Because oxygen is a reactant in all these reactions, chemists originally called them oxidation reactions.

In this chapter, you will begin your study of oxidation and reduction reactions. You will learn the theoretical definitions of oxidation and reduction. You will also learn about one way of predicting whether these reactions will be spontaneous.

Penny Chemistry

Oxidation and reduction reactions are responsible for many familiar occurrences, such as the tarnishing of coins and other objects. In this activity, you will use some common items to clean tarnished pennies.

Safety Precautions

- If you spill any vinegar on your skin, wash it with water.
- Wash your hands when you have completed the lab.

Materials

- white vinegar
- table salt
- measuring cup
- paper towel
- marker
- stirring rod
- measuring spoons
- 1 clean steel screw or nail
- 1 small, clear glass or plastic bowl
- 6 pennies

Procedure

1. Measure 1 tsp of salt and $\frac{1}{4}$ cup of vinegar and pour them into the bowl. Stir until the salt dissolves.

2. Hold a penny so that half of the penny is in the solution. After about 15 s, remove the penny. Record the appearance of the penny.

3. Put all the pennies in the solution and leave them for about 5 min.

4. Remove the pennies but leave the solution in the bowl for use in step 7.

5. Label one piece of paper towel "not rinsed" and lay half the pennies on it to dry.

6. Label a second piece of paper towel "rinsed." Rinse the remaining pennies thoroughly under running water then lay them on the paper towel to dry.

7. Position the screw (or nail) so that about half of it is in the solution and the other half is in air. Secure it in this position.

8. Observe the screw for a few minutes. Look for any evidence of a chemical reaction occurring. Record your observations but leave the screw in the solution.

9. At the end of the class period (about an hour) examine the screw. Observe and record the appearance of the screw.

Analysis

1. What do you think causes pennies to become dull?

2. Speculate about the chemical reaction that occurred when you placed the pennies in the vinegar-and-salt solution.

3. What chemical reaction might have caused the change in the pennies that dried on the paper towel without being rinsed?

4. Provide a possible explanation for the change in the appearance of the screw that was placed into the solution after the pennies had been removed.

5. After you have completed your study of this chapter, review your answers to these Launch Lab questions. If any of your answers need to be corrected, make those corrections.

Characterizing Oxidation Reduction

In chemistry, the terms oxidation and reduction are often used together to describe reactions. What is the origin of the term *reduction* and what is its meaning as it relates to chemical reactions?

Figure 12.1 Today, the process of reducing ore is called smelting.

When metal is extracted from raw ore, as shown in Figure 12.1, the mass of the metal is much smaller than the mass of the mined ore. Thus, the process of extracting the metal was historically called reduction (a reduction in the amount of mass). The chemical reactions that remove the metals from the compounds were, therefore, called reduction reactions. How are these reduction reactions related to oxidation reactions? To answer that question and to see how chemists now define the terms oxidation and reduction, look more closely at some examples of oxidation and reduction reactions.

Oxidation

As you read in the introduction to this chapter, chemists originally defined oxidation as any chemical reaction in which an atom or a compound reacted with molecular oxygen. You would call this the operational definition of oxidation. As chemists made more observations and analyzed more data, they began to see similarities between reactions of atoms and compounds with oxygen and reactions of the same atoms or compounds with elements other than oxygen.

Consider the reaction between magnesium and oxygen. You might have observed this reaction in a laboratory activity. Magnesium burns very rapidly and emits a very bright light, as shown in Figure 12.2. The product of the reaction is magnesium oxide. The chemical equation is shown below:

$$2Mg(s) + O_2(g) \rightarrow 2MgO(s)$$

Figure 12.2 Magnesium burns in air with a bright white light and forms magnesium oxide.

Magnesium oxide, as you probably recall, is an ionic compound containing magnesium ions (Mg^{2+}) and oxide ions (O^{2-}). Magnesium atoms lose two electrons and become positively charged, while oxygen atoms gain two electrons and become negatively charged. Even in solid form, the elements exist as separate magnesium ions and oxide ions. You could write the ionic equation as shown here:

$$2Mg(s) + O_2(g) \rightarrow 2Mg^{2+}(s) + 2O^{2-}(s)$$

Now compare the reaction between magnesium and oxygen with the reaction between magnesium and chlorine. The total ionic equation is shown here:

$$Mg(s) + Cl_2(g) \rightarrow Mg^{2+}(s) + 2Cl^-(s)$$

The product, magnesium chloride, is also an ionic compound. In this reaction, magnesium atoms, $Mg(s)$, lose two electrons and become ionized (Mg^{2+}). As you can see, when solid magnesium reacts with either oxygen or chlorine, it loses two electrons and becomes a magnesium ion. This example illustrates the modern, or theoretical, definition of **oxidation**, which is stated in the box below.

> **Oxidation**
>
> Oxidation is the loss of electrons.

Reduction

If one atom, ion, or molecule in a reaction loses electrons, then another atom, ion, or molecule must gain electrons, because electrons cannot exist free in a solution. If oxidation is the loss of electrons, what is the gain of electrons called?

To answer this question, consider the reactions that occur when ores are reduced. Iron ore, for example, usually contains magnetite, $Fe_3O_4(s)$, or hematite, $Fe_2O_3(s)$, both of which are iron oxides. The process of converting the iron ions (Fe^{3+} or Fe^{2+}) in the oxides to metallic iron involves the addition of electrons to the ions. Since this process was historically called reduction, chemists now apply the term **reduction** to all cases in which an atoms or ions gain electrons in a reaction.

> **Reduction**
>
> Reduction is the gain of electrons.

Redox Reactions

If one atom or ion is oxidized in a chemical reaction, another must be reduced. Therefore, reactions in which electrons are gained by one atom or ion and lost by another are called **oxidation-reduction reactions** or, more briefly, **redox reactions**. Redox reactions do not comprise a separate set of reactions. On the contrary, redox reactions fit into several basic categories of reactions. The first reaction that you considered above, the oxidation of magnesium, is a formation reaction. Decomposition reactions are basically the reverse of formation reactions; thus, they are usually redox reactions. Since combustion reactions are defined as reactions in which a compound reacts with oxygen, they are also redox reactions. Single-replacement reactions, which you will observe in Investigation 12.A, are always redox reactions. The reactants and products in single-replacement reactions are often visibly different from each other, as shown in Figure 12.3. The first photograph shows a zinc strip that has just been placed in a copper(II) sulfate solution. The second photograph shows the same zinc strip and copper(II) sulfate solution several hours later.

Figure 12.3 A solid zinc strip reacts with a solution that contains blue copper(II) ions. The black material is Cu(s). When copper atoms deposit on the zinc strip, they do not form a smooth, shiny surface.

Chemistry File

FYI
When writing equations for reactions that take place in an aqueous solution, refer to your solubility table in Appendix G to determine whether the salts are soluble. If they are soluble, the state will be aqueous (aq). If they are insoluble, the state will be solid (s).

The single-replacement reaction that is taking place between zinc and copper(II) sulfate is represented by the following complete balanced equation:

$$Zn(s) + CuSO_4(aq) \longrightarrow Cu(s) + ZnSO_4(aq)$$

You can write this equation as an **ionic equation**:

$$Zn(s) + Cu^{2+}(aq) + SO_4^{2-}(aq) \longrightarrow Cu(s) + Zn^{2+}(aq) + SO_4^{2-}(aq)$$

The sulfate ions are **spectator ions**, meaning they are ions that are not involved in the chemical reaction. By omitting the spectator ions (SO_4^{2-}), you obtain the following **net ionic equation**:

$$Zn(s) + Cu^{2+}(aq) \longrightarrow Cu(s) + Zn^{2+}(aq)$$

The following equation shows you how to track the electrons. In the reaction of zinc atoms with copper(II) ions, the zinc atoms *lose* electrons and undergo oxidation—the zinc atoms are *oxidized*. The copper(II) ions *gain* electrons and undergo reduction—the copper(II) ions are *reduced*.

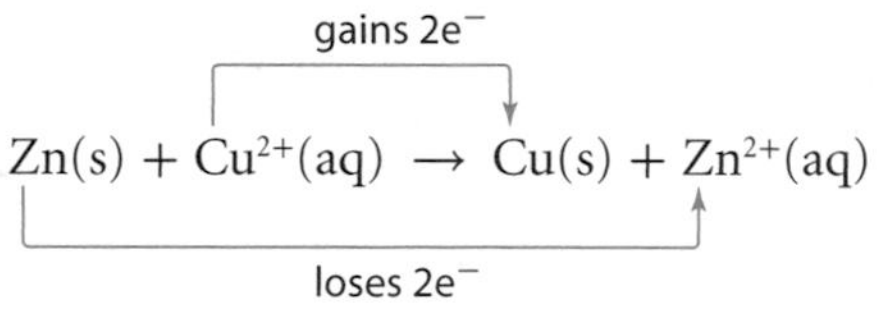

$$\overbrace{Zn(s) + Cu^{2+}(aq)}^{\text{gains } 2e^-} \longrightarrow Cu(s) + Zn^{2+}(aq)$$
$$\underbrace{}_{\text{loses } 2e^-}$$

Since electrons are transferred from zinc atoms to copper(II) ions, the copper(II) ions are responsible for the oxidation of the zinc atoms. A reactant that oxidizes another reactant is called an **oxidizing agent**. The oxidizing agent receives electrons in a redox reaction and is therefore reduced. In this reaction, copper(II) is the oxidizing agent. The zinc atoms are responsible for the reduction of the copper(II) ions. A reactant that reduces another reactant is called a **reducing agent**. The reducing agent donates or loses electrons in a redox reaction and therefore is oxidized. In this reaction, zinc is the reducing agent.

A redox reaction can also be defined as a reaction between an oxidizing agent and a reducing agent, as illustrated in Figure 12.4.

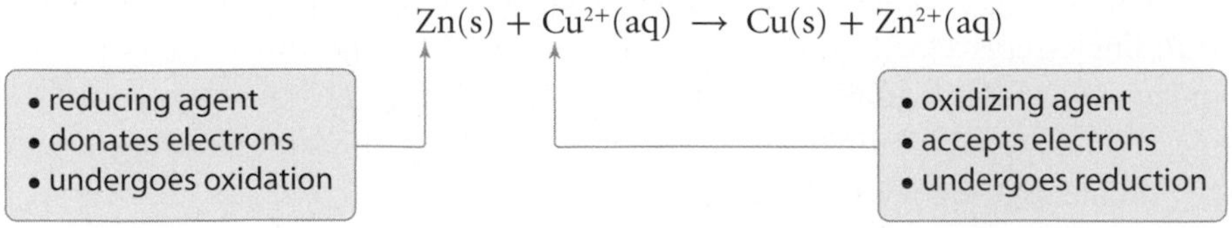

$$Zn(s) + Cu^{2+}(aq) \longrightarrow Cu(s) + Zn^{2+}(aq)$$

Figure 12.4 In a redox reaction, the reducing agent is oxidized and the oxidizing agent is reduced.

- reducing agent
- donates electrons
- undergoes oxidation

- oxidizing agent
- accepts electrons
- undergoes reduction

1. Explain how a compound or atom can become oxidized in a reaction that does not involve oxygen.

2. What is the original (operational) definition for reduction? What is the theoretical definition of reduction? How do the operational and theoretical definitions of reduction differ? How are they the same?

3. Analyze the following single-replacement reaction and identify the oxidizing agent and the reducing agent:

$$2Al(s) + 3Fe(SO_4)(aq) \rightarrow Al_2(SO_4)_3(aq) + 3Fe(s)$$

4. Analyze the following double-replacement reaction equation. Using this reaction as a model, do you think that double-replacement reactions are typically redox reactions or not? Explain why.

$$BaCl_2(aq) + K_2CO_3(aq) \rightarrow BaCO_3(s) + 2KCl(aq)$$

ChemistryFile

FYI

Try using a mnemonic to remember the definitions for oxidation and reduction. For example, in "LEO the lion says GER," LEO stands for "**l**oss of **e**lectrons is **o**xidation," and GER stands for "**g**ain of **e**lectrons is **r**eduction." More appropriate for Alberta, the mnemonic "OIL RIG" stands for "**o**xidation **i**s **l**oss; **r**eduction **i**s **g**ain."

Spontaneity of Redox Reactions

The reaction of zinc with copper sulfate was spontaneous—it proceeded with no addition of energy or any other stimulus. When zinc replaced copper in the copper sulfate compound, zinc acted as a reducing agent and reduced the copper(II) ion, while the zinc itself became oxidized. When metallic copper formed, it did not replace the zinc ions that had formed. It appears that copper atoms cannot reduce zinc ions. Zinc atoms are a stronger reducing agent than are copper atoms, and copper(II) ions are a stronger oxidizing agent than are zinc ions. These concepts are expressed in table form in Table 12.1.

Table 12.1 Strength of Oxidizing and Reducing Agents

Oxidizing agent	Reduction → ← Oxidation	Reducing agent
Stronger oxidizing agent		
Cu^{2+}		Cu
Zn^{2+}		Zn
		Stronger reducing agent

In 12.A Investigation, you will test several more metals and their soluble salts to determine the relative ability of each to oxidize or reduce the others. You will use the data to extend Table 12.1. To decide where the additional metals and their ions fit in the table, remember that any ion in the left column is a stronger oxidizing agent than any ion below it. Any atom in the right column is a stronger reducing agent than any atom above it.

To prepare for the Investigation, be sure that you understand the definitions for *oxidizing agent, reducing agent, ionic equation,* and *net ionic equation*. Because you will be writing ionic and net ionic equations, be sure that you can do so and also identify the atoms and ions that gain or lose electrons. Be able to relate the gain or loss of electrons to oxidation and reduction. If you are not certain about all of these concepts, review page 436.

Testing Relative Oxidizing and Reducing Strengths of Metal Atoms and Ions

By observing whether reactions occur between solid metals and metal ions in solution, you can determine the order of oxidizing and reducing agents according to strength.

Question

How can the presence or absence of a reaction provide information about the relative strength of oxidizing and reducing agents?

Safety Precautions

- Wear goggles, gloves, and an apron for all parts of this investigation.

- If you spill any solution on your skin, wash it off with large amounts of water.

- Wash your hands when you have completed the investigation.

Materials

- 4 small pieces of each of the following metals:
  - aluminium foil
  - thin copper wire or tiny copper beads
  - iron filings
  - magnesium
  - zinc

- dropper bottles containing dilute solutions of
  - aluminium sulfate
  - copper(II) sulfate
  - iron(II) sulfate
  - magnesium sulfate
  - zinc nitrate
- well plate
- white paper

Procedure

1. Place the well plate on a piece of white paper. Label the paper to match the data table below.

Metal \ Compound	$Al_2(SO_4)_3$(aq)	$CuSO_4$(aq)	$FeSO_4$(aq)	$MgSO_4$(aq)	$Zn(NO_3)_2$(aq)
Al(s)					
Cu(s)					
Fe(s)					
Mg(s)					
Zn(s)					

2. Place the four small pieces of each metal, about the size of a grain of rice, into the well plate. Use the data table as a guide for placement of metals. Cover each piece of metal with a few drops of the appropriate solution. Wait 3–5 min to observe whether a reaction occurs.

3. Look for evidence of a chemical reaction in each mixture. Record the results by using a "y" for a reaction and an "n" for no reaction. If you are unsure, repeat the process on a larger scale in a small test tube.

4. Discard the mixtures into the waste beaker supplied by your teacher. Do not pour anything down the drain.

Analysis

1. For each single-replacement reaction that proceeded spontaneously, write
   a) a complete balanced equation
   b) an ionic equation
   c) a net ionic equation

2. Identify the oxidizing agent and the reducing agent in each of the reactions that proceeded spontaneously.

3. Make a simple redox table similar to Table 12.1 that contains all the metal atoms and metal ions that you analyzed in this investigation. Note that the ion that was able to oxidize all other metal atoms is placed at the top of the left column. In the next row, place the ion that oxidized all but the first metal atom. Complete the table.

Conclusion

4. Based on your observations, do you think that there are any properties of metal atoms or ions that would allow a chemist to predict whether one would be a better oxidizing or reducing than another? Explain your reasoning.

Predicting the Spontaneity of Redox Reactions

An analysis of the redox table that you created in Investigation 12.A shows you that metal ions can act as oxidizing agents because they can remove electrons from certain metal atoms. Likewise, metal atoms act as reducing agents because they can donate electrons to certain ions. By performing many experiments similar to those in Investigation 12.A, chemists have determined the relative strengths of many ions as oxidizing agents and atoms as reducing agents. Table 12.2 lists the results of some of those experiments. Notice that the strength of the ions as oxidizing agents increases as you go up the list on the left. The strength of metal atoms as reducing agents increases as you go down the list on the right.

Table 12.2 Oxidation-Reduction Table

Strongest Oxidizing Agent	Weakest Reducing Agent
$Au^+(aq)$	$Au(s)$
$Pt^{2+}(aq)$	$Pt(s)$
$Ag^{2+}(aq)$	$Ag(s)$
$Hg^{2+}(aq)$	$Hg(s)$
$Cu^{2+}(aq)$	$Cu(s)$
$Sn^{2+}(aq)$	$Sn(s)$
$Ni^{2+}(aq)$	$Ni(s)$
$Co^{2+}(aq)$	$Co(s)$
$Tl^+(aq)$	$Tl(s)$
$Cd^{2+}(aq)$	$Cd(s)$
$Fe^{2+}(aq)$	$Fe(s)$
$Cr^{3+}(aq)$	$Cr(s)$
$Zn^{2+}(aq)$	$Zn(s)$
$Al^{3+}(aq)$	$Al(s)$
$Mg^{2+}(aq)$	$Mg(s)$
$Ca^{2+}(aq)$	$Ca(s)$
$Ba^{2+}(aq)$	$Ba(s)$
Weakest Oxidizing Agent	Strongest Reducing Agent

You can use Table 12.2 to predict whether a reaction between atoms of one element and ions of another element will spontaneously undergo a redox reaction. Remember that a stronger reducing agent loses electrons more readily than does a weaker reducing agent. A stronger oxidizing agent gains electrons more readily than does a weaker oxidizing agent.

The following steps and examples in Figure 12.5 will help you predict the spontaneity of redox reactions.

- Write the net ionic equation.
- Add arrows to indicate the gain and loss of electrons.
- Identify the stronger reducing and oxidizing agents.
- If the stronger reducing agent is losing electrons and stronger oxidizing agent is gaining electrons, the reaction will proceed spontaneously.
- If the stronger reducing agent has gained electrons and stronger oxidizing agent has lost electrons, the reaction will NOT proceed spontaneously.

Example 1

Does the reaction between metallic tin and platinum nitrate proceed spontaneously?

$$\text{loses } 2e^-$$
$$Sn(s) + Pt^{2+}(aq) \longrightarrow Sn^{2+}(aq) + Pt(s)$$
$$\text{gains } 2e^-$$

stronger reducing agent   stronger oxidizing agent

The stronger reducing agent is losing electrons and the stronger oxidizing agent is gaining electrons. Therefore, the reaction will proceed spontaneously.

Example 2

Does the reaction between metallic iron(II) and magnesium nitrate proceed spontaneously?

$$\text{loses } 2e^-$$
$$Fe(s) + Mg^{2+}(aq) \longrightarrow Fe^{2+}(aq) + Mg(s)$$
$$\text{gains } 2e^-$$

stronger oxidizing agent   stronger reducing agent

The stronger reducing agent has gained electrons and the stronger oxidizing agent has lost electrons. Therefore, the reaction will NOT proceed spontaneously.

Figure 12.5 These examples show you how to apply the steps listed above.

Notice that, when the stronger reducing and oxidizing agents are on the left of the equation, the reaction proceeds spontaneously. When the stronger reducing and oxidizing agents are on the right of the equation, the reaction does NOT proceed spontaneously. Use this information to help you answer the following questions.

Chemistry File

FYI

Be cautious when using oxidation-reduction tables or activity series in other chemistry texts or tables books. Tables are not always written in the same direction as those in this textbook. Be sure to identify the direction of higher and lower activities or of stronger versus weaker oxidizing or reducing agents. Use the concepts of activity and of oxidation and reduction instead of memorizing "up" or "down."

Use Table 12.2 to answer the following questions.

5. Which of the following reactions will proceed spontaneously?
   a) solid aluminium and aqueous copper sulfate
   b) aqueous calcium nitrate and solid nickel
   c) solid chromium and aqueous silver nitrate
   d) aqueous barium sulfate and solid tin

6. If you wanted to demonstrate the appearance of a reaction between a solid and an ionic solution and the solid you wanted to use was cobalt, which of the following aqueous solutions would you use? Explain why.
   a) silver nitrate
   b) zinc sulfate

7. List three solid metals that, when placed in an aqueous solution of cadmium chloride, will react spontaneously. Write net ionic equations for the reactions.

8. List three soluble salts that would react spontaneously with solid tin. Write net ionic equations for the reactions.

Section 12.1 Summary

- The operational definition of oxidation is any reaction involving molecular oxygen. The theoretical definition of oxidation is the loss of electrons.
- The operational definition of reduction is the extracting of metals from metal ore. The theoretical definition of reduction is the gain of electrons.
- If one atom is oxidized in a reaction, another atom must be reduced.
- An oxidizing agent accepts electrons from another compound and becomes reduced.
- A reducing agent donates electrons to another compound and becomes oxidized.
- You can experimentally determine whether one element is a stronger reducing agent or a stronger oxidizing agent than the other. From accumulated data, you can predict whether a given reaction will proceed spontaneously.

SECTION 12.1 REVIEW

1. Explain why oxidation and reduction reactions always occur together. Use the theoretical definitions of oxidation and reduction in your answer.
2. Explain why, in a redox reaction, the oxidizing agent undergoes reduction.
3. Write equations that represent three different spontaneous reactions in which calcium is oxidized.
4. When metallic lithium is a reactant in a formation reaction, does it act as an oxidizing agent or a reducing agent? Explain.
5. Predict whether each of the following single-replacement reactions will proceed spontaneously. For those that will, write a complete balanced equation, an ionic equation, and a net ionic equation. Use the redox table that you generated in Investigation 12.A.
   a) aqueous silver nitrate and metallic cadmium
   b) solid gold and aqueous copper(II) sulfate
   c) solid aluminium and aqueous mercury(II) chloride
6. Write a net ionic equation for a spontaneous reaction in which
   a) $Fe^{2+}(aq)$ acts as an oxidizing agent
   b) $Al(s)$ acts as a reducing agent
   c) $Au^{3+}(aq)$ acts as an oxidizing agent
   d) $Cu(s)$ acts as a reducing agent
7. Potassium is made industrially through the single-displacement reaction of molten sodium with molten potassium chloride.
   a) Write a net ionic equation for the reaction, assuming that all reactants and products are in the liquid state.
   b) Identify the oxidizing agent and the reducing agent in the reaction.

Section Outcomes

In this section, you will:
- **define** half-reaction and disproportionation
- **write** and balance equations for redox reactions using the half-reaction method
- **develop** equations for half-reactions from information about redox changes
- **perform** investigations on redox reactions

Key Terms

half-reactions
disproportionation reaction
smelting
refining

Redox reactions are an important part of compact disc (CD) manufacturing. A master disc is made of glass coated with silver, as shown in Figure 12.6. Silver is deposited onto the glass disc by the reduction of silver ions with methanal, $HCHO(\ell)$ (commonly known as formaldehyde). In the reaction, methanal is oxidized to methanoic acid, $HCOOH(\ell)$ (commonly known as formic acid). The redox reaction occurs under acidic conditions. Text or music is first recorded on the master disc. Then copies are made on discs made of Lexan, the same plastic used to make riot shields and bulletproof windows. The Lexan is coated with a thin film of aluminium.

Figure 12.6 The production of CDs depends on a redox reaction used to coat the master disc with silver.

Most of the balanced chemical equations that you have seen represent redox reactions that can be balanced by inspection. Some reactions, however, require specific techniques for balancing the equations, particularly those reactions that take place under acidic or basic conditions, such as the acidic conditions used in coating a glass CD with silver. A common technique for balancing redox reactions uses **half-reactions**, which are equations that describe the changes in only the compound that is oxidized or only the compound that is reduced.

Writing Balanced Half-Reactions

Consider, again, the reaction of metallic zinc with aqueous copper(II) sulfate. Analyze the net ionic equation below:

$$Zn(s) + Cu^{2+}(aq) \rightarrow Cu(s) + Zn^{2+}(aq)$$

Try This
You can write separate oxidation and reduction half-reactions to represent a redox reaction, but a half-reaction cannot occur on its own. Explain why this statement is true.

You can see that each neutral Zn atom is oxidized to form a Zn^{2+} ion. Thus, each Zn atom must lose two electrons. You can write an oxidation half-reaction to show this change:

$$\textit{Oxidation half-reaction}: Zn(s) \rightarrow Zn^{2+}(aq) + 2e^-$$

Each Cu^{2+} ion is reduced to form a neutral Cu atom. Thus, each ion must gain two electrons. You can write a reduction half-reaction to show this change:

$$\textit{Reduction half-reaction}: Cu^{2+}(aq) + 2e^- \rightarrow Cu(s)$$

Look again at each half-reaction above. Notice that the atoms and the charges are balanced. Like other types of balanced equations, half-reactions are balanced using the smallest possible whole-number coefficients.

• • •

9 Write balanced half-reactions from the net ionic equation for the reaction between solid aluminium and aqueous iron(III) sulfate. The sulfate ions are spectator ions and are not included:

$$Al(s) + Fe^{3+}(aq) \rightarrow Al^{3+}(aq) + Fe(s)$$

10 Write balanced half-reactions from the following net ionic equations:

a) $Fe(s) + Cu^{2+}(aq) \rightarrow Fe^{2+}(aq) + Cu(s)$

b) $Cd(s) + 2Ag^+(aq) \rightarrow Cd^{2+}(aq) + 2Ag(s)$ (**Hint:** Remember to use the smallest possible whole-number coefficients in the reduction half-reaction.)

11 Write balanced half-reactions for each of the following equations:

a) $Sn(s) + PbCl_2(aq) \rightarrow SnCl_2(aq) + Pb(s)$

b) $Au(NO_3)_3(aq) + 3Ag(s) \rightarrow 3AgNO_3(aq) + Au(s)$

c) $3Zn(s) + Fe_2(SO_4)_3(aq) \rightarrow 3ZnSO_4(aq) + 2Fe(s)$

• • •

Balancing Equations Using Half-Reactions

For simple equations, such as the one for the reaction between zinc and copper(II) sulfate, you would simply add the half-reactions together. Since the number of electrons on each side of the equation are the same, they can be cancelled. Finally, if you choose to do so, you can add back the spectator ions (SO_4^{2-}) as shown.

$$Zn(s) \rightarrow Zn^{2+}(aq) + 2e^-$$
$$Cu^{2+}(aq) + 2e^- \rightarrow Cu(s)$$
$$\overline{Zn(s) + Cu^{2+}(aq) + 2e^- \rightarrow Cu(s) + Zn^{2+}(aq) + 2e^-}$$
$$Zn(s) + Cu^{2+}(aq) \rightarrow Cu(s) + Zn(aq)$$
$$Zn(s) + Cu^{2+}(aq) + SO_4^{2-}(aq) \rightarrow Cu(s) + Zn^{2+}(aq) + SO_4^{2-}(aq)$$

One additional step is needed when the numbers of electrons in the two half-reactions are not the same. For example, consider the synthesis of potassium chloride from its elements. Metallic potassium is oxidized to form potassium ions, and gaseous chlorine is reduced to form chloride ions (see Figure 12.7). Each half-reaction can be balanced by writing the correct formulas for the reactant and product, balancing the numbers of atoms, and then adding the correct number of electrons to balance the charges. The oxidation half-reaction is as follows:

$$\textit{Oxidation half-reaction}: K(s) \rightarrow K^+(s) + e^-$$

The atoms are balanced. The net charge on each side is zero.

The reduction half-reaction is

Reduction half-reaction: $Cl_2(g) + 2e^- \rightarrow 2Cl^-(s)$

The atoms are balanced and the net charge on each side is -2. There is one electron in the oxidation half-reaction and two electrons in the reduction half-reaction. Before you can add the equations, you must multiply the entire oxidation half-reaction by two and then proceed as before. There are no spectator ions to add to this equation.

$$K(s) \rightarrow K^+(s) + e^-$$
$$2[K(s)] \rightarrow 2[K^+(s) + e^-]$$
$$2K(s) \rightarrow 2K^+(s) + 2e^-$$
$$Cl_2(g) + 2e^- \rightarrow 2Cl^-(s)$$
$$\overline{2K(s) + Cl_2(g) + 2e^- \rightarrow 2K^+(s) + 2e^- + 2Cl^-(s)}$$
$$\text{or}$$
$$2K(s) + Cl_2(g) \rightarrow 2KCl(s)$$

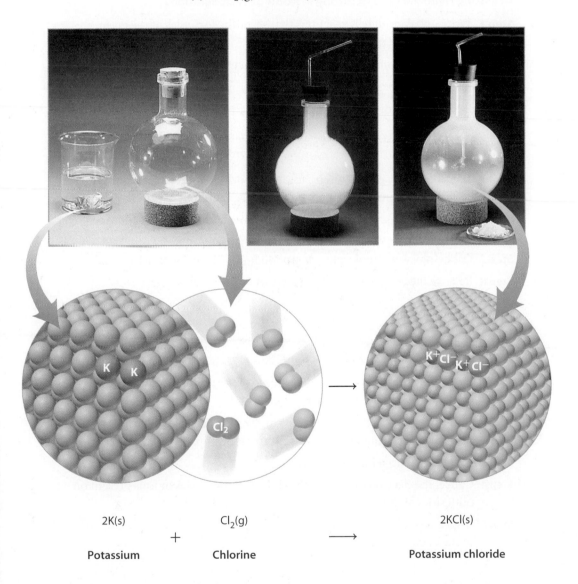

2K(s)   Cl₂(g)   2KCl(s)

        +   ⟶

Potassium   Chlorine   Potassium chloride

Figure 12.7 Grey potassium metal, which is stored under oil, reacts very vigorously with greenish-yellow chlorine gas to form white potassium chloride.

Balancing Equations for Reactions That Occur in Acidic or Basic Solutions

You could have balanced the two equations discussed above by inspection. However, when balancing equations for reactions that take place under acidic or basic conditions, you must account for the hydrogen, $H^+(aq)$, or hydroxide, $OH^-(aq)$, ions present in the solutions. The overall approach is similar, but a few additional steps are necessary. The first box below lists the steps to follow when balancing the half-reactions. The second box lists the steps to follow when combining the half-reactions to produce a balanced equation. Read through the steps and then study the Sample Problems that follow to learn how to apply the rules. Note that these steps allow you to balance equations but do not represent the actual mechanisms that occur in the solutions. Your final result does, however, represent the products of the reaction.

Balancing Half-Reactions Occurring in Acidic or Basic Solutions

Step 1 Write unbalanced half-reactions that show the formulas of the given reactant(s) and product(s).

Step 2 Balance any atoms other than oxygen and hydrogen first.

Step 3 Balance any oxygen atoms by adding water molecules.

Step 4 Balance any hydrogen atoms by adding hydrogen ions.

If your reaction is taking place in an acidic solution, skip to step 8.
If your reaction is taking place in a basic solution, proceed to step 5.

Step 5 Adjust for basic conditions by adding to both sides the same number of hydroxide ions as the number of hydrogen ions already present.

Step 6 Simplify the equation by combining hydrogen ions and hydroxide ions that appear on the same side of the equation into water molecules.

Step 7 Cancel any water molecules present on both sides of the equation.

Step 8 Balance the charges by adding electrons.

Balancing Equations Using Half-Reactions

Step 1 Determine the lowest common multiple of the numbers of electrons in the oxidation and reduction half-reactions.

Step 2 Multiply one or both half-reactions by the number that will bring the number of electrons to the lowest common multiple.

Step 3 Add the balanced half-reactions.

Step 4 Cancel the electrons and any other identical molecules or ions present on both sides of the equation.

Step 5 If spectator ions were removed when forming half-reactions, add them back to the equation.

While this many steps may seen intimidating, you will find that you can combine some of these as you become more comfortable writing balanced equations.

Balancing an Equation for a Reaction That Occurs in an Acidic Solution

Problem

Sulfur is oxidized by nitric acid in an aqueous solution, producing sulfur dioxide, nitrogen monoxide, and water, as shown by the unbalanced equation. Use the half-reaction method to balance the equation:

$$S(s) + HNO_3(aq) \rightarrow SO_2(g) + NO(g) + H_2O(\ell)$$

(**Note**: To reduce the complexity, the subscript, 8, has been omitted from the sulfur. After the final step, you could add the subscript to the sulfur and multiply all other coefficients by 8.)

Solution

First, write the total ionic equation:

$$S(s) + H^+(aq) + NO_3{}^-(aq) \rightarrow SO_2(g) + NO(g) + H_2O(\ell)$$

Step 1 Write the unbalanced half-reactions. To write the unbalanced half-reactions, write only those compounds that contain the atom that is oxidized or the atom that is reduced. For this step, ignore the fact that one side of the equation might have oxygen atoms and the other side has none.

Oxidation half-reaction: $S(s) \rightarrow SO_2(g)$

Reduction half-reaction: $NO_3{}^-(aq) \rightarrow NO(g)$

Step 2 Balance atoms other than oxygen and hydrogen.

The sulfur and nitrogen atoms are already balanced.

Step 3 Balance oxygen atoms by adding water molecules.

Oxidation half-reaction: There are two oxygen atoms on the right side of the equation, so you must add two water molecules to the left side:

$$S(s) + 2H_2O(\ell) \rightarrow SO_2(g)$$

Reduction half-reaction: There are three oxygen atoms on the left side of the equation and one on the right. Add two water molecules to the right side.

$$NO_3{}^-(aq) \rightarrow NO(g) + 2H_2O(\ell)$$

Step 4 Balance hydrogen atoms by adding hydrogen ions.

Oxidation half-reaction: There are four hydrogen atoms in the water on the left side of the equation, so add four hydrogen ions to the right side:

$$S(s) + 2H_2O(\ell) \rightarrow SO_2(g) + 4H^+(aq)$$

Reduction half-reaction: There are four hydrogen atoms in the water molecules on the right side of the equation, so add four hydrogen ions to the left side:

$$NO_3{}^-(aq) + 4H^+(aq) \rightarrow NO(g) + 2H_2O(\ell)$$

The nitric acid makes this an acid solution, so skip to step 8 of the steps for balancing half-reactions.

Step 8 Balance the charges by adding electrons.

Oxidation half-reaction: There are zero net charges on the left side of the equation and four positive charges on the right. Therefore, give the right side a net charge of zero by adding four electrons to the right side:

$$S(s) + 2H_2O(\ell) \rightarrow SO_2(g) + 4H^+(aq) + 4e^-$$

Reduction half-reaction: There are four positive charges and one negative charge on the left side of the equation, giving it a net positive charge of three. The net charge on the right side of the equation is zero. Add three electrons to the left side to give it a net zero charge:

$$NO_3{}^-(aq) + 4H^+(aq) + 3e^- \rightarrow NO(g) + 2H_2O(\ell)$$

The half-reactions are balanced. Now balance the entire equation by following the steps in the shaded box.

Step 1 Determine the lowest common multiple of the numbers of electrons in the oxidation and reduction half-reactions.

There are four electrons in the oxidation half-reaction and three electrons in the reduction half-reaction. The lowest common multiple of 4 and 3 is 12.

Step 2 Multiply one or both half-reactions by the number that will bring the number of electrons to the lowest common multiple.

Oxidation half-reaction:

$3[S(s) + 2H_2O(\ell) \rightarrow SO_2(g) + 4H^+(aq) + 4e^-]$
$3S(s) + 6H_2O(\ell) \rightarrow 3SO_2(g) + 12H^+(aq) + 12e^-$

Reduction half-reaction:

$4[NO_3{}^-(aq) + 4H^+(aq) + 3e^- \rightarrow NO(g) + 2H_2O(\ell)]$
$4NO_3{}^-(aq) + 16H^+(aq) + 12e^- \rightarrow 4NO(g) + 8H_2O(\ell)$

Step 3 Add the balanced half-reactions:

$3S(s) + 6H_2O(\ell) + 4NO_3{}^-(aq) + 16H^+(aq) + 12e^- \rightarrow$
$\qquad 3SO_2(g) + 12H^+(aq) + 12e^- + 4NO(g) + 8H_2O(\ell)$

Step 4 Cancel the electrons and any other identical molecules or ions present on both sides of the equation:

$3S(s) + 6H_2O(\ell) + 4NO_3{}^-(aq) + 16H^+(aq) + 12e^- \rightarrow$
$\qquad 3SO_2(g) + 12H^+(aq) + 12e^- + 4NO(g) + 8H_2O(\ell)$

Subtract $6H_2O(\ell)$ and 12 $H^+(aq)$ from both sides of the equation, as well as the 12 electrons:

$3S(s) + 4NO_3{}^-(aq) + 4H^+(aq) \rightarrow$
$\qquad\qquad 3SO_2(g) + 4NO(g) + 2H_2O(\ell)$
$3S_8(s) + 32NO_3{}^-(aq) + 32H^+(aq) \rightarrow$
$\qquad\qquad 24SO_2(g) + 32NO(g) + 16H_2O(\ell)$

In the previous Sample Problem, you were told that nitric acid was the oxidizing agent. In some problems, you might not know which compound is the oxidizing agent and which is the reducing agent. The next Sample Problem will demonstrate that you do not need to know. If you follow the steps, the identity of the agents will be revealed. Examine the following Sample Problem to see how to balance equations in which some information appears to be missing.

Sample Problem

Balancing an Equation for a Reaction That Occurs in an Acidic Solution

Problem

The unbalanced equation for the reaction between permanganate ions, $MnO_4^-(aq)$, and oxalate ions, $C_2O_4^{2-}(aq)$, in an acidic solution is shown below. Use the half-reaction method to balance the equation:

$$MnO_4^-(aq) + C_2O_4^{2-}(aq) \rightarrow Mn^{2+}(aq) + CO_2(g)$$

Solution

Step 1 Write the unbalanced half-reactions.

Carbon half-reaction: $C_2O_4^{2-}(aq) \rightarrow CO_2(g)$

Manganese half-reaction: $MnO_4^-(aq) \rightarrow Mn^{2+}(aq)$

Step 2 Balance any atoms other than oxygen and hydrogen.

Carbon half-reaction: There are two carbon atoms on the left side of the equation and one carbon on the right. Multiply the right side by two to make the number of carbon atoms the same:

$$C_2O_4^{2-}(aq) \rightarrow 2\,CO_2(g)$$

Manganese half-reaction: Already balanced for Mn.

Step 3 Balance oxygen atoms by adding water molecules.

Carbon half-reaction: Already balanced for oxygen.

Manganese half-reaction: There are four oxygen atoms on the left side of the equation and zero on the right. Add four water molecules to the right side:

$$MnO_4^-(aq) \rightarrow Mn^{2+}(aq) + 4H_2O(\ell)$$

Step 4 Balance hydrogen atoms by adding hydrogen ions.

Carbon half-reaction: There are no hydrogen atoms.

Manganese half-reaction: There are eight hydrogen atoms in the water molecules on the right side of the equation so add eight hydrogen ions to the left side:

$$MnO_4^-(aq) + 8H^+(aq) \rightarrow Mn^{2+}(aq) + 4H_2O(\ell)$$

The reaction takes place in an acidic solution, so skip to step 8.

Step 8 Balance the charges by adding electrons.

Carbon half-reaction: There are two negative charges on the left and zero on the right, so add two electrons to the right side. Since the oxalate is losing electrons, it is oxidized, making this the oxidation half-reaction:

$$C_2O_4^{2-}(aq) \rightarrow 2CO_2(g) + 2e^-$$

Manganese half-reaction: There are seven positive charges on the left and two positive charges on the right. Reduce the charge on the left side to two positive charges by adding five electrons to the left side. Since the permanganate is gaining electrons, it is being reduced, making this the reduction half-reaction:

$$MnO_4^-(aq) + 8H^+(aq) + 5e^- \rightarrow Mn^{2+}(aq) + 4H_2O(\ell)$$

The half-reactions are balanced. Now balance the entire equation.

Step 1 Determine the lowest common multiple of the numbers of electrons in the two half-reactions.

There are two electrons in the oxidation half-reaction and five electrons in the reduction half-reaction. The lowest common multiple of 2 and 5 is 10.

Step 2 Multiply one or both half-reactions by the number that will bring the number of electrons to the lowest common multiple:

Oxidation half-reaction:
$$5\,[C_2O_4^{2-}(aq) \rightarrow 2CO_2(g) + 2e^-]$$
$$5C_2O_4^{2-}(aq) \rightarrow 10CO_2(g) + 10e^-$$

Reduction half-reaction:
$$2\,[MnO_4^-(aq) + 8H^+(aq) + 5e^- \rightarrow$$
$$Mn^{2+}(aq) + 4H_2O(\ell)]$$
$$2MnO_4^-(aq) + 16H^+(aq) + 10e^- \rightarrow$$
$$2Mn^{2+}(aq) + 8H_2O(\ell)$$

Step 3 Add the balanced half-reactions.
$$2MnO_4^-(aq) + 16H^+(aq) + 10e^- + 5C_2O_4^{2-}(aq) \rightarrow$$
$$2Mn^{2+}(aq) + 8H_2O(\ell) + 10CO_2(g) + 10e^-$$

Step 4 Remove the electrons and any other identical molecules or ions present on both sides of the equation:
$$2MnO_4^-(aq) + 16H^+(aq) + 10e^- + 5C_2O_4^{2-}(aq) \rightarrow$$
$$2Mn^{2+}(aq) + 8H_2O(\ell) + 10CO_2(g) + 10e^-$$
$$2MnO_4^-(aq) + 16H^+(aq) + 5C_2O_4^{2-}(aq) \rightarrow$$
$$2Mn^{2+}(aq) + 8H_2O(\ell) + 10CO_2(g)$$

The atoms of each element are balanced and there is a net charge of plus four on each side of the equation. The equation is balanced.

Balancing an Equation for a Reaction That Occurs in a Basic Solution

Problem

Cyanide, $CN^-(aq)$, is oxidized by permanganate, $MnO_4^-(aq)$, in a basic solution, as shown in the following unbalanced equation. Use the half-reaction method to balance the equation:

$$CN^-(aq) + MnO_4^-(aq) \rightarrow CNO^-(aq) + MnO_2(s)$$

Solution

Step 1 Write the unbalanced half-reactions:

Oxidation half-reaction: $CN^-(aq) \rightarrow CNO^-(aq)$

Reduction half-reaction: $MnO_4^-(aq) \rightarrow MnO_2(s)$

Step 2 Balance any atoms other than oxygen and hydrogen.

Oxidation half-reaction: The carbon and nitrogen atoms are already balanced.

Reduction half-reaction: The manganese atoms are already balanced.

Step 3 Balance oxygen atoms by adding water molecules.

Oxidation half-reaction: There is one oxygen atom in the cyanate ion on the right side of the equation. Add one water molecule to the left side to balance the number of oxygen atoms:

$$CN^-(aq) + H_2O(\ell) \rightarrow CNO^-(aq)$$

Reduction half-reaction: There are four oxygen atoms in the permanganate ion on the left side and two oxygen atoms in the manganese oxide on the right. Therefore, add two water molecules to the right side:

$$MnO_4^-(aq) \rightarrow MnO_2(s) + 2H_2O(\ell)$$

Step 4 Balance hydrogen atoms by adding hydrogen ions.

Oxidation half-reaction: There are two hydrogen atoms in the water molecule on the left side of the equation and zero on the right. Therefore, add two hydrogen ions to the right side:

$$CN^-(aq) + H_2O(\ell) \rightarrow CNO^-(aq) + 2H^+(aq)$$

Reduction half-reaction: There are four hydrogen atoms in the two water molecules on the right side of the equation and zero on the left. Therefore, add four hydrogen ions to the left side:

$$MnO_4^-(aq) + 4H^+(aq) \rightarrow MnO_2(s) + 2H_2O(\ell)$$

Step 5 Adjust for basic conditions by adding to both sides the same number of hydroxide ions as the number of hydrogen ions already present.

Oxidation half-reaction: There are two hydrogen ions on the right side of the equation, so add two hydroxide ions to both sides:

$$CN^-(aq) + H_2O(\ell) + 2OH^-(aq) \rightarrow$$
$$CNO^-(aq) + 2H^+(aq) + 2OH^-(aq)$$

Reduction half-reaction: There are four hydrogen ions on the left side of the equation so add four hydroxide ions to both sides:

$$MnO_4^-(aq) + 4H^+(aq) + 4OH^-(aq) \rightarrow$$
$$MnO_2(s) + 2H_2O(\ell) + 4OH^-(aq)$$

Step 6 Simplify the equation by combining the hydrogen ions and hydroxide ions that appear on the same side of the equation into water molecules:

Oxidation half-reaction:

$$CN^-(aq) + H_2O(\ell) + 2OH^-(aq) \rightarrow$$
$$CNO^-(aq) + 2H_2O(\ell)$$

Reduction half-reaction:

$$MnO_4^-(aq) + 4H_2O(\ell) \rightarrow$$
$$MnO_2(s) + 2H_2O(\ell) + 4OH^-(aq)$$

Step 7 Remove any water molecules present on both sides of the equation.

Oxidation half-reaction:

$$CN^-(aq) + H_2O(\ell) + 2OH^-(aq) \rightarrow$$
$$CNO^-(aq) + 2H_2O(\ell)$$
$$CN^-(aq) + 2OH^-(aq) \rightarrow CNO^-(aq) + H_2O(\ell)$$

Reduction half-reaction:

$$MnO_4^-(aq) + 4H_2O(\ell) \rightarrow$$
$$MnO_2(s) + 2H_2O(\ell) + 4OH^-(aq)$$
$$MnO_4^-(aq) + 2H_2O(\ell) \rightarrow MnO_2(s) + 4OH^-(aq)$$

Step 8 Balance the charges by adding electrons.

Oxidation half-reaction: There are three negative charges on the left side of the equation and one on the right. Balance the charges by adding two electrons to the right side:

$$CN^-(aq) + 2OH^-(aq) \rightarrow CNO^-(aq) + H_2O(\ell) + 2e^-$$

Reduction half-reaction: There is one negative charge on the left side of the equation and four on the right. Balance the charges by adding three electrons to the left side:

$$MnO_4^-(aq) + 2H_2O(\ell) + 3e^- \rightarrow MnO_2(s) + 4OH^-(aq)$$

The half-reactions are balanced, so proceed to balancing the entire equation.

Step 1 Determine the lowest common multiple of the numbers of electrons in the oxidation and reduction half-reactions.

There are two electrons in the oxidation half-reaction and three electrons in the reduction half-reaction. The lowest common multiple of 2 and 3 is 6.

Step 2 Multiply one or both half-reactions by the number that will bring the number of electrons to the lowest common multiple.

Oxidation half-reaction:

$$3[CN^-(aq) + 2OH^-(aq) \rightarrow CNO^-(aq) + H_2O(\ell) + 2e^-]$$
$$3CN^-(aq) + 6OH^-(aq) \rightarrow 3CNO^-(aq) + 3H_2O(\ell) + 6e^-$$

Reduction half-reaction:

$$2[MnO_4^-(aq) + 2H_2O(\ell) + 3e^- \rightarrow$$
$$MnO_2(s) + 4OH^-(aq)]$$
$$2MnO_4^-(aq) + 4H_2O(\ell) + 6e^- \rightarrow 2MnO_2(s) + 8OH^-(aq)$$

Step 3 Add the balanced half-reactions:

$$2MnO_4^-(aq) + 4H_2O(\ell) + 6e^- + 3CN^-(aq) + 6OH^-(aq)$$
$$\rightarrow 2MnO_2(s) + 8OH^-(aq) + 3CNO^-(aq) + 3H_2O(\ell) + 6e^-$$

Step 4 Remove the electrons and any other identical molecules or ions present on both sides of the equation:

$$2MnO_4^-(aq) + 4H_2O(\ell) + 6e^- + 3CN^-(aq) + 6OH^-(aq)$$
$$\rightarrow 2MnO^2(s) + 8OH^-(aq) + 3CNO^-(aq) +$$
$$3H_2O(\ell) + 6e^-$$

$$2MnO_4^-(aq) + H_2O(\ell) + 3CN^-(aq) \rightarrow$$
$$2MnO_2(s) + 2OH^-(aq) + 3CNO^-(aq)$$

The atoms of each element are balanced, and there is a net charge of minus five on each side of the equation. The equation is balanced.

Practice Problems

1. Balance each of the following ionic equations for acidic conditions. Identify the oxidizing agent and the reducing agent in each case:
   a) $MnO_4^-(aq) + Ag(s) \rightarrow Mn^{2+}(aq) + Ag^+(aq)$
   b) $Hg(\ell) + NO_3^-(aq) + Cl^-(aq) \rightarrow$
   $$HgCl_4^{2-}(s) + NO_2(g)$$
   c) $AsH_3(s) + Zn^{2+}(aq) \rightarrow H_3AsO_4(aq) + Zn(s)$
   d) $I_2(s) + OCl^-(aq) \rightarrow IO_3^-(aq) + Cl^-(aq)$

2. Balance each of the following ionic equations for basic conditions. Identify the oxidizing agent and the reducing agent in each case:
   a) $MnO_4^-(aq) + I^-(aq) \rightarrow MnO_4^{2-}(aq) + IO_3^-(aq)$
   b) $H_2O_2(aq) + ClO_2(aq) \rightarrow ClO^-(aq) + O_2(g)$
   c) $ClO^-(aq) + CrO_2^-(aq) \rightarrow CrO_4^{2-}(aq) + Cl_2(g)$
   d) $Al(s) + NO^-(aq) \rightarrow NH_3(g) + AlO_2^-(aq)$

In the next investigation, you will carry out several redox reactions, including reactions of acids with metals and the combustion of hydrocarbons.

INVESTIGATION 12.B

Target Skills

Describing procedures for the safe handling and disposal of acids and products of hydrocarbon combustion

Analyzing data from an experiment on redox reactions

Using appropriate numeric and symbolic representations to communicate equations for redox reactions

Redox Reactions and Balanced Equations

You have practised balancing equations for redox reactions, but can you predict the products of a redox reaction? Can you determine whether a reaction has occurred and, if so, whether it was a redox reaction? In this investigation, you will develop these skills.

Question

How can you tell whether a redox reaction occurs when reactants are mixed? Can you observe the transfer of electrons in the mixture?

Predictions

- Predict which metals of magnesium, zinc, copper, and aluminium can be oxidized by aqueous hydrogen ions (acidic solution). Explain your reasoning.
- Predict whether metals that cannot be oxidized by hydrogen ions can be dissolved in acids. Explain your reasoning.

- Predict whether the combustion of a hydrocarbon is a redox reaction. What assumptions have you made about the products?

Safety Precautions

- The acid solutions are corrosive. Handle them with care.
- If you accidentally spill a solution on your skin, wash the area immediately with copious amounts of cool water. If you get any acid in your eyes, use the eye wash station immediately. Inform your teacher.

- Before lighting a Bunsen burner or candle, make sure that there are no flammable liquids in the laboratory. Tie back long hair, and confine any loose clothing.
- Wash your hands when you have completed the investigation.

Materials

- small pieces of each of the following metals:
  - magnesium
  - zinc
  - copper
  - aluminium
- hydrochloric acid (1 mol/L)
- sulfuric acid (1 mol/L)
- well plate
- 4 small test tubes
- test-tube rack
- Bunsen burner
- candle

Procedure

Part 1 Reactions of Acids with Metals

1. Place one small piece of each metal into a well on the well plate. Add a few drops of hydrochloric acid to each metal. Record your observations. If you are unsure of your observations, repeat the procedure on a larger scale in a small test tube.

2. Place another small piece of each metal into clean sections of the well plate. Add a few drops of sulfuric acid to each metal. Record your observations. If you are unsure of your observations, repeat the procedure on a larger scale in a small test tube.

3. Dispose of the mixtures into the beaker supplied by your teacher.

Part 2 Combustion of Hydrocarbons

4. Light a Bunsen burner and observe the combustion of natural gas. Adjust the colour of the flame by varying the quantity of oxygen admitted to the burner. Describe how the colour depends on the quantity of oxygen.

5. Light a candle and observe its combustion. Compare the colour of the flame with the colour of the Bunsen burner's flame. Which adjustment of the burner makes the colours of the two flames most similar?

Analysis

Part 1 Reactions of Acids with Metals

1. Write a complete balanced equation for each of the reactions of an acid with a metal.

2. Write each equation from question 1 in net ionic form.

3. Determine which of the reactions from question 1 are redox reactions.

4. Write each redox reaction from question 3 as two half-reactions.

5. Describe and explain any trends that you see in your answers to question 4.

6. In the reactions you observed, are the hydrogen ions acting as an oxidizing agent, a reducing agent, or neither?

7. In the neutralization reaction of hydrochloric acid and sodium hydroxide, do the hydrogen ions behave in the same way as you described in question 6? Explain.

8. Your teacher might demonstrate the reaction of copper with concentrated nitric acid to produce copper(II) ions and brown, toxic nitrogen dioxide gas. Write a balanced net ionic equation for this reaction. Do the hydrogen ions behave in the same way as you described in question 7? Identify the oxidizing agent and the reducing agent in this reaction.

9. From your observations of the reaction of copper with hydrochloric acid in your investigation and the reaction of copper with nitric acid in the demonstration described above, can you tell whether hydrogen ions or nitrate ions are the better oxidizing agent? Explain.

Part 2 Combustion of Hydrocarbons

10. The main component of natural gas is methane, $CH_4(g)$. The products of the combustion of this gas in a Bunsen burner depend on how the burner is adjusted. A blue flame indicates complete combustion. What are the products of complete combustion? Write a balanced chemical equation for this reaction.

11. A yellow or orange flame from a Bunsen burner indicates incomplete combustion and the presence of carbon in the flame. Write a balanced chemical equation for this reaction.

12. Name another possible carbon-containing product from the incomplete combustion of methane. Write a balanced chemical equation for this reaction.

13. Assume that the fuel in the burning candle is paraffin. Although paraffin is a mixture of hydrocarbons, you can represent it by the formula $C_{25}H_{52}(s)$. Write a balanced chemical equation for the complete combustion of paraffin.

14. Write two different balanced equations that both represent a form of incomplete combustion of paraffin.

15. How do you know that at least one of the incomplete combustion reactions is taking place when a candle burns?

16. Are combustion reactions also redox reactions? Does your answer depend on whether the combustion is complete or incomplete? Explain.

Conclusion

17. How could you tell whether a redox reaction occurred when the reactants were mixed? Could you observe the transfer of electrons in the mixture?

Applications

18. Gold is very unreactive and does not dissolve in most acids. However, it does dissolve in *aqua regia* (Latin for "royal water"), which is a mixture of concentrated hydrochloric acid and nitric acid. The unbalanced ionic equation for the reaction is as follows:

$$Au(s) + NO_3^-(aq) + Cl^-(aq) \rightarrow AuCl_4^-(s) + NO_2(g)$$

Balance the equation, and identify the oxidizing agent and reducing agent.

19. Natural gas is burned in gas furnaces. Give at least three reasons why this combustion reaction should be as complete as possible. How would you try to ensure complete combustion?

Disproportionation Reactions

In most redox reactions, atoms of one element are oxidized and atoms of a different element are reduced. It is possible, however, for some atoms of an element to undergo oxidation and other atoms of the same element to undergo reduction in a single reaction called a **disproportionation reaction**. Examine the copper atoms and ions in the following equations.

complete balanced equation $Cu_2O(aq) + H_2SO_4(aq) \rightarrow Cu(s) + CuSO_4(aq) + H_2O(\ell)$
total ionic equation $2Cu^+(aq) + O^{2-}(aq) + 2H^+(aq) + SO_4^{2-}(aq) \rightarrow$
$$Cu(s) + Cu^{2+}(aq) + SO_4^{2-}(aq) + H_2O(\ell)$$
net ionic equation $2Cu^+(aq) + O^{2-}(aq) + 2H^+(aq) \rightarrow Cu(s) + Cu^{2+}(aq) + H_2O(\ell)$
(**Note:** Copper(II) oxide (Cu_2O) is nearly insoluble in water but it will react with sulfuric acid.)

Of the two copper(I) ions in the reactant, one is reduced to elemental copper, $Cu(s)$, by gaining an electron and the other is oxidized to copper(II), $Cu^{2+}(aq)$, by losing an electron. You can track the electrons as shown below.

The two half-reactions are as follows.

Oxidation half-reaction: $Cu^+(aq) \rightarrow Cu^{2+}(aq) + e^-$
Reduction half-reaction: $Cu^+(aq) + e^- \rightarrow Cu(s)$

Disproportionation Reaction

In a disproportionation reaction, some atoms of an element are oxidized and other atoms of the same element are reduced.

Disproportionation reactions are balanced in the same way as any other redox reaction is. Study the Sample Problem and then complete the Practice Problems to extend your problem-solving skills.

Balancing a Disproportionation Reaction

Problem

Balance the following unbalanced equation for the disproportionation, in an acidic solution, of nitrous acid, $HNO_2(aq)$, forming nitric acid, $HNO_3(aq)$; nitrogen monoxide, $NO(g)$; and water:

$$HNO_2(aq) \rightarrow HNO_3(aq) + NO(g) + H_2O(\ell)$$

Solution

First, write the ionic and then the net ionic equations.

$$H^+(aq) + NO_2^-(aq) \rightarrow H^+(aq) + NO_3^-(aq) + NO(g) + H_2O(\ell)$$

$$NO_2^-(aq) \rightarrow NO_3^-(aq) + NO(g) + H_2O(\ell)$$

Step 1 Write the unbalanced half-reactions that show the formulas of the given reactant(s) and product(s).

Oxidation half-reaction: $NO_2^-(aq) \rightarrow NO_3^-(aq)$

Reduction half-reaction: $NO_2^-(aq) \rightarrow NO(g)$

Step 2 Balance any atoms other than oxygen and hydrogen first.

Oxidation half-reaction: The nitrogen atoms are already balanced.

Reduction half-reaction: The nitrogen atoms are already balanced.

Step 3 Balance any oxygen atoms by adding water molecules.

Oxidation half-reaction: There are two oxygen atoms in the nitrite ion on the left side of the equation and three oxygen atoms in the nitrate ion on the right side. Therefore, add one water molecule to the left side of the equation:

$$NO_2^-(aq) + H_2O(\ell) \rightarrow NO_3^-(aq)$$

Reduction half-reaction: There are two oxygen atoms in the nitrite ion on the left side of the equation and one oxygen atom in the nitrogen monoxide on the right side. Therefore, add one water molecule to the right side:

$$NO_2^-(aq) \rightarrow NO(g) + H_2O(\ell)$$

Step 4 Balance any hydrogen atoms by adding hydrogen ions.

Oxidation half-reaction: There are two hydrogen atoms in the water molecule on the left side of the equation and zero on the right. Therefore, add two hydrogen ions to the right side:

$$NO_2^-(aq) + H_2O(\ell) \rightarrow NO_3^-(aq) + 2H^+(aq)$$

Reduction half-reaction: There are two hydrogen atoms in the water molecule on the right side of the equation and zero on the left. Therefore, add two hydrogen ions to the left side:

$$NO_2^-(aq) + 2H^+(aq) \rightarrow NO(g) + H_2O(\ell)$$

The solution is acidic, so skip to step 8.

Step 8 Balance the charges by adding electrons.

Oxidation half-reaction: There is one negative charge on the left side of the equation and a net charge of plus one on the right. To balance the charges, add two electrons to the right side:

$$NO_2^-(aq) + H_2O(\ell) \rightarrow NO_3^-(aq) + 2H^+(aq) + 2e^-$$

Reduction half-reaction: There is a net charge of plus one on the left side of the equation and zero on the right. To balance the charges, add one electron to the left side:

$$NO_2^-(aq) + 2H^+(aq) + e^- \rightarrow NO(g) + H_2O(\ell)$$

The half-reactions are balanced. Now balance the entire equation.

Step 1 Determine the lowest common multiple of the numbers of electrons in the oxidation and reduction half-reactions. There were two electrons lost in the oxidation half-reaction and one electron was gained in the reduction half-reaction. The lowest common multiple of 1 and 2 is 2.

Step 2 Multiply one or both half-reactions by the number that will bring the number of electrons to the lowest common multiple.

Oxidation half-reaction:
$$NO_2^-(aq) + H_2O(\ell) \rightarrow NO_3^-(aq) + 2H^+(aq) + 2e^-$$

Reduction half-reaction:
$$2[NO_2^-(aq) + 2H^+(aq) + e^- \rightarrow NO(g) + H_2O(\ell)]$$
$$2NO_2^-(aq) + 4H^+(aq) + 2e^- \rightarrow 2NO(g) + 2H_2O(\ell)$$

Step 3 Add the balanced half-reactions:

$$NO_2^-(aq) + H_2O(\ell) + 2NO_2^-(aq) + 4H^+(aq) + 2e^- \rightarrow$$
$$NO_3^-(aq) + 2H^+(aq) + 2e^- + 2NO(g) + 2H_2O(\ell)$$

Step 4 Remove the electrons and any other identical molecules or ions that are present on both sides of the equation:

$$NO_2^-(aq) + H_2O(\ell) + 2NO_2^-(aq) + 4H^+(aq) + 2e^- \rightarrow$$
$$NO_3^-(aq) + 2H^+(aq) + 2e^- + 2NO(g) + 2H_2O(\ell)$$

$$NO_2^-(aq) + 2NO_2^-(aq) + 2H^+(aq) \rightarrow$$
$$NO_3^-(aq) + 2NO(g) + H_2O(\ell)$$

$$3NO_2^-(aq) + 2H^+(aq) \rightarrow NO_3^-(aq) + 2NO(g) + H_2O(\ell)$$

Step 5 If spectator ions were removed when forming half-reactions, add them back to the equation. One hydrogen ion was removed when forming half-reactions:

$$3NO_2^-(aq) + 3H^+(aq) \rightarrow$$
$$H^+(aq) + NO_3^-(aq) + 2NO(g) + H_2O(\ell)$$

There are three nitrogen atoms, three hydrogen atoms, and six oxygen atoms on each side of the equation. The net charge on each side is zero. The equation is balanced.

Balance the following disproportionation reactions:

3. $PbSO_4(aq) \rightarrow Pb(s) + PbO_2(aq) + SO_4^{2-}(aq)$
(acidic solution; $PbSO_4(aq)$ and $PbO_2(aq)$ are
soluble in an acidic solution)

4. $NO_2(g) + H_2O(\ell) \rightarrow HNO_3(aq) + NO(g)$ (acidic
solution)

5. $Cl_2(g) \rightarrow ClO^-(aq) + Cl^-(aq)$ (basic conditions)

6. $I_3^-(aq) \rightarrow I^-(aq) + IO_3^-(aq)$ (acidic conditions)

Reducing Iron Ore

The term *reduction* originally came from the treatment of ore to produce metal. Today, however, the terms *smelting* and *refining* are often applied to processing and purifying ore. How are these terms related to the original term *reduction* and to the modern meaning of chemical reduction? A brief glimpse into the historical development of metal implements will tie these concepts together.

Copper artefacts, similar to those in Figure 12.8, that date back to about 3600 B.C.E. show that crude copper was being extracted from copper ore, probably malachite, $CuCO_3 \cdot Cu(OH)_2$, and used to make utensils. To extract the copper from the ore, the ore had to be exposed to very high temperatures, higher than a camp fire could provide. Since the only way to generate such high temperatures at that time was by using the ovens in which clay pots were fired, archaeologists speculate that potters might have discovered the process of preparing copper. If ore had been placed in high-temperature ovens, the copper would have melted and thus separated from the other substances in the ore. The copper product produced during that era was quite crude but could, nevertheless, be shaped into very effective tools and other utensils.

For about 1000 years after people had learned how to prepare copper, the only source of iron for tool making was meteorites. Archaeologists have evidence that ancient peoples began to process iron ore about 2500 B.C.E. However, it was not a common practice until about 1200 B.C.E. By then, people had discovered that heating iron ore, probably hematite, Fe_2O_3, to very high temperatures and in contact with charcoal produced crude but useful iron. The fact that the mass of iron obtained from ore was much smaller than the mass of the original ore was the basis of the term *reduction*.

Today, we are still heating iron ore with a form of charcoal to extract metallic iron. The process is called **smelting** and it involves the chemical reduction of the iron ions to iron atoms with carbon as the reducing agent. The process takes place in several steps inside a large *blast furnace*, as shown in Figure 12.9.

The chemistry of the reactions is somewhat complex, but it can be summarized by the equations in Figure 12.9. A mixture, called the *charge*, containing pulverized iron ore—usually hematite, $Fe_2O_3(s)$—limestone, $CaCO_3(s)$, and coke, $C(s)$, is poured into the top of the blast furnace. Blasts of hot air travel up through the particles as they fall.

At the lowest level of the furnace, where oxygen from injected air is available, carbon in the form of coke burns, heating the air to about 2000 °C. By controlling the amount of air and coke, the burning of the coke can be made incomplete, producing carbon monoxide according to the reaction below:

$$2C(s) + O_2(g) \rightarrow 2CO(g) + heat$$

Carbon monoxide is the compound that acts as the reducing agent for the ore. As the carbon monoxide travels up the furnace, it reacts with the compounds from the iron ore that are falling. A variety of reduction reactions occur throughout the furnace. The important reactions are shown in Figure 12.9, along with the temperature range in which they occur. As you can see, complete reduction of the hematite occurs in three steps. Finally, the metallic iron melts and pools in the bottom of the furnace.

Figure 12.8 These tools were made from raw copper. Knives and utensils much older than these have also been discovered.

ChemistryFile

FYI
When ancient peoples were just learning how to prepare iron, it was five times as expensive as gold. Like gold, iron was used in jewellery and ornaments.

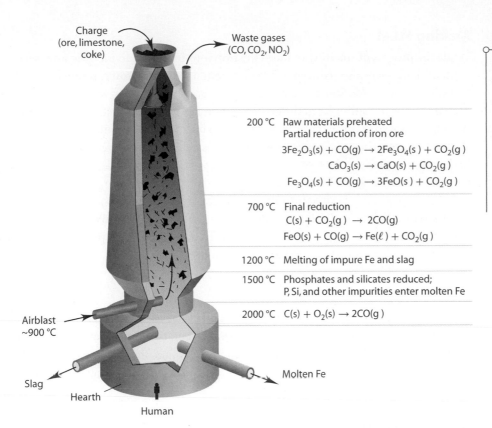

Charge (ore, limestone, coke)

Waste gases (CO, CO_2, NO_2)

200 °C Raw materials preheated
Partial reduction of iron ore

$$3Fe_2O_3(s) + CO(g) \rightarrow 2Fe_3O_4(s) + CO_2(g)$$
$$CaO_3(s) \rightarrow CaO(s) + CO_2(g)$$
$$Fe_3O_4(s) + CO(g) \rightarrow 3FeO(s) + CO_2(g)$$

700 °C Final reduction
$$C(s) + CO_2(g) \rightarrow 2CO(g)$$
$$FeO(s) + CO(g) \rightarrow Fe(\ell) + CO_2(g)$$

1200 °C Melting of impure Fe and slag

1500 °C Phosphates and silicates reduced;
P, Si, and other impurities enter molten Fe

2000 °C $C(s) + O_2(s) \rightarrow 2CO(g)$

Airblast ~900 °C

Slag

Hearth

Human

Molten Fe

Chemistry File

FYI
Coke is prepared by heating coal in the absence of oxygen, which drives off most of the impurities. When an inexpensive process to prepare coke from coal was developed, it made iron smelting much less expensive and more efficient.

Figure 12.9 Notice the size of the person relative to the blast furnace.

The purpose of the lime, $CaO(s)$, which is produced by heating the limestone, $CaCO_3(s)$, in the charge, is to react with impurities. Significant amounts of silicon dioxide and aluminium(III) oxide are found in iron ore. The lime reacts with these compounds, according to the equations below, to produce a liquid called *slag*:

$$CaO(s) + SiO_2(s) \rightarrow CaSiO_3(\ell)$$
$$CaO(s) + Al_2O_3(s) \rightarrow Ca(AlO_2)_2(\ell)$$

The slag is less dense than the molten iron and thus floats on top. As shown in Figure 12.9, the slag and molten iron are drawn off at different levels.

The iron still contains contaminants—mainly carbon (about 5%), silicon, phosphorus, manganese, and sulfur. This crude molten iron is poured into long narrow moulds to form bars. In this form, the metal is called *pig iron* (see Figure 12.10). The pig iron is then transported to other locations, where most of it is converted into steel. Some, however, is used for cast iron and wrought iron.

Figure 12.10 Pig iron was given its name because the bars cooling and solidifying in their moulds made the workers think of piglets nursing.

Making Steel

While the process of smelting reduced the iron ions in the ore to metallic iron, several contaminants were also reduced to their elemental form. Pig iron contains about 5% carbon from the coke. Some of the silicon and phosphorus, along with small amounts of manganese and sulfur from the ore, remain in the pig iron. Most of these impurities must be removed because they cause the iron to be brittle and granular. By the process of purifying or **refining**, pig iron is converted into steel. Many of the contaminants remain in the pig iron because they were reduced along with the iron. Thus, the reasonable way to remove them is by oxidation.

Molten pig iron is poured into an upright vessel. Oxygen gas is pumped into the vessel over the iron. Lime, CaO(s), called *flux*, is poured into the mixture. The impurities oxidize much more readily than does the iron. These oxidized compounds then react with the lime, according to the reactions shown here:

$$SiO_2(s) + CaO(s) \rightarrow CaSiO_3(\ell)$$
$$P_4O_{10}(\ell) + 6CaO(s) \rightarrow 2Ca_3(PO_4)_2(\ell)$$

The reactions produce a liquid called slag that is similar to the slag in the smelting process. The slag is less dense than the molten iron and therefore floats on it. When the reactions are complete, the vessel is tilted on its hinges, the slag is decanted (poured off the top), and the refined steel is recovered. Steel still contains some carbon (0.03% to 1.4%) but much less than does pig iron.

Section 12.2 Summary

- Oxidation half-reactions include only those atoms that become oxidized in a redox reaction.

- Reduction half-reactions include only those atoms that become reduced in a redox reaction.

- You can use a set of rules to balance oxidation and reduction half-reactions. You can then add the half-reactions to obtain a balanced redox reaction.

- In a disproportionation reaction, one or more atoms of an element are oxidized and one or more atoms of the same element are reduced in the same reaction.

- Iron ore is reduced in a series or reactions involving carbon monoxide to form pig iron.

- Pig iron is refined by oxidizing the impurities and reacting them with calcium oxide to form a liquid that floats on the molten iron.

SECTION 12.2 REVIEW

1. Balance each half-reaction. Identify each reaction as an oxidation or reduction half-reaction:
   a) $Cr_2O_7^{2-}(aq) \rightarrow Cr^{3+}(aq)$ (acidic conditions)
   b) $S_2O_3^{2-}(aq) \rightarrow S_4O_6^{2-}(aq)$
   c) $AsO_4^{3-}(aq) \rightarrow As_4O_6(aq)$ (acidic conditions)
   d) $Br_2(g) \rightarrow BrO_3^-(aq)$ (basic conditions)

2. Balance each equation:
   a) $Au^{3+}(aq) + Co(s) \rightarrow Au(s) + Co^{3+}(aq)$
   b) $Cu(s) + NO_3^-(aq) \rightarrow Cu^{2+}(aq) + NO(g)$ (acidic conditions)
   c) $NO_3^-(aq) + Al(s) \rightarrow NH_3(g) + AlO_2^-(aq)$ (basic conditions)

3. Write the net ionic equation and the half-reactions for the disproportionation of mercury(I) ions in aqueous solution to produce liquid mercury and aqueous mercury(II) ions. Assume that mercury(I) ions exist in solution as $2Hg^+(aq)$.

4. Balance the following equations that describe the processes occurring in an iron-ore smelting furnace:
   a) $Fe_2O_3(s) + CO(g) \rightarrow Fe_3O_4(s) + CO_2(g)$
   b) $CaCO_3(s) \rightarrow CaO(s) + CO_2(g)$
   c) $Fe_3O_4(s) + CO(g) \rightarrow FeO(s) + CO_2(g)$
   d) $FeO(s) + CO(g) \rightarrow Fe(\ell) + CO_2(g)$

5. Explain how and why both reduction and oxidation processes are used in the smelting and refining of steel.

Redox Reactions Involving Molecular Compounds

Section Outcomes

In this section, you will:
- **differentiate** between redox reactions and other reactions using oxidation numbers
- **write** and **balance** redox reactions by assigning oxidation numbers

Key Term

oxidation numbers

The light emitted by these fireflies (see Figure 12.11) is produced by a unique redox reaction. The chemical reaction that produces light in fireflies is shown in the inset of Figure 12.11. Notice that there are no metal ions or atoms in the reaction. How would you write half-reactions for a reaction like this one?

Figure 12.11 When organisms, such as these fireflies, use a chemical reaction to produce light, the process is called *bioluminescence*. These flashes of light, produced by the redox reactions shown in the inset, attract a mate. The reactant in the chemical equation is called luciferin and the enzyme that catalyzes the reaction is luciferase.

In addition to the few redox reactions that emit light, there many more redox reactions for which it would be difficult to write half-reactions. For example, many of the reactions that occur in living organisms involve molecular compounds and no ions or metal atoms. You have probably seen the summary reactions for cellular respiration. The unbalanced reaction is shown here:

$$C_6H_{12}O_6(aq) + O_2(g) \rightarrow CO_2(g) + H_2O(\ell)$$

In the cell, the reaction takes place in many separate steps, some of which are redox reactions. How could you determine whether such a reaction as this, that involves only molecular compounds, is a redox reaction? How would you write half-reactions for the cellular respiration summary equation? A method that makes balancing redox reactions that involve only molecular compounds much simpler is called the *oxidation number method*. To use this method, you need to assign an oxidation number to each atom in the equation.

Assigning Oxidation Numbers

Oxidation numbers are, in general, the same as the charge that an atom in a compound would have if no electrons were shared but if, instead, the electrons were completely held by the atom with the greatest electronegativity (see the periodic table). For example, you learned that, in a water molecule, oxygen is more electronegative than hydrogen, and thus the oxygen atoms attract the shared electrons more strongly than do the hydrogen atoms.

Therefore, to determine oxidation numbers, you would assign all the valence electrons to oxygen and none to hydrogen. Since the oxygen atom would then possess one extra electron from each of the two hydrogen atoms in addition to its own valence electrons, it would have an oxidation number of -2. The hydrogen atoms would each have an oxidation number of $+1$ because their valence electrons are considered to belong to the oxygen atom. It is important to remember that oxidation numbers do *not* represent actual charges on atoms. Oxidation numbers are a way to describe some properties of atoms in a compound.

When electrons are equally shared by two identical atoms, such as chlorine atoms in a chlorine molecule, $Cl_2(g)$, half of the shared electrons are considered to be held by each of atoms. In the chlorine molecule, $Cl_2(g)$, neither atom "gained" or "lost" an electron, so each has an oxidation number of 0. To assign all oxidation numbers by using such a reasoning process would be very time consuming. You can assign oxidation numbers by following the rules listed in Table 12.3 below.

Table 12.3 Oxidation Number Rules

Rules	Examples
1. A pure element has an oxidation number of 0.	Na in Na(s), Br in $Br_2(\ell)$, and P in $P_4(s)$ all have an oxidation number of 0.
2. The oxidation number of an element in a monatomic ion equals the charge on the ion.	The oxidation number of aluminium in $Al^{3+}(aq)$ is $+3$. The oxidation number of selenium in $Se^{2-}(aq)$ is -2.
3. The oxidation number of hydrogen in compounds is $+1$, except in metal hydrides, where the oxidation number of hydrogen is -1.	The oxidation number of hydrogen in $H_2S(g)$ or $CH_4(g)$ is $+1$. The oxidation number of hydrogen in NaH(s) or in $CaH_2(s)$ is -1.
4. The oxidation number of oxygen in compounds is usually -2, but there are exceptions. These include peroxides, such as H_2O_2, superoxides, and the compound OF_2.	The oxidation number of O in Li_2O or KNO_3 is -2.
5. In molecular compounds that do not contain hydrogen or oxygen, the more electronegative element is assigned an oxidation number equal to the negative charge it usually has when it is in ionic compounds.	The oxidation number of Cl in PCl_3 is -1. The oxidation number of S in CS_2 is -2.
6. The sum of the oxidation numbers of all the atoms in a neutral compound is 0.	In CF_4, the oxidation number of F is -1 and the oxidation number of C is $+4$. The oxidation numbers of $$1\text{ C atom} + 4\text{ F atoms} = 0$$ $$(+4) + 4(-1) = 0$$ $$+4 - 4 = 0$$
7. The sum of the oxidation numbers of all the atoms in a polyatomic ion equals the charge on the ion.	In NO_2^-, the oxidation number of O is -2 and the oxidation number of N is $+3$. The oxidation numbers of $$1\text{ N atom} + 2\text{ O atoms} = -1$$ $$(+3) + 2(-2) = -1$$ $$+3 - 4 = -1$$

Problem Tip

When finding the oxidation numbers of elements in ionic compounds, you can work with the ions separately. For example, $Na_2Cr_2O_7(aq)$ contains two $Na^+(aq)$ ions and sodium has an oxidation number of $+1$. The oxidation numbers of chromium and oxygen in the dichromate ion, $Cr_2O_7^{2-}$, can then be calculated as shown in part (c) of the Sample Problem.

In the following Sample Problem, you will find out how to apply these rules to covalent molecules and polyatomic ions.

Assigning Oxidation Numbers

Problem

Assign an oxidation number to each atom in the following compounds:

a) $SiBr_4(\ell)$ **b)** $HClO_4(aq)$ **c)** $Cr_2O_7^{2-}(aq)$

Solution

a) • Because the compound $SiBr_4(\ell)$ does not contain hydrogen or oxygen, rule 5 from Table 12.2 applies.
   • Because $SiBr_4(\ell)$ is a neutral compound, rule 6 also applies.
   • Silicon has an electronegativity of 1.9. Bromine has an electronegativity of 3.0. From rule 5, therefore, you can assign bromine an oxidation number of −1.
   • The oxidation number of silicon is unknown, so call it x. You know from rule 6 that the sum of the oxidation numbers is 0. Therefore, the oxidation numbers of

   1 Si atom + 4 Br atoms = 0
   $$x + 4(-1) = 0$$
   $$x - 4 = 0$$
   $$x = +4$$

   • The oxidation number of silicon is +4. The oxidation number of bromine is −1.

b) • Because the compound $HClO_4(aq)$ contains hydrogen and oxygen, rules 3 and 4 apply.
   • Because $HClO_4(aq)$ is a neutral compound, rule 6 also applies.
   • Hydrogen has its usual oxidation number of +1.

   • Oxygen has its usual oxidation number of −2.
   • The oxidation number of chlorine is unknown, so call it x.
   • You know from rule 6 that the sum of the oxidation numbers is 0. Therefore, the oxidations numbers of 1 H atom + 1 Cl atom + 4 O atoms = 0
   $$(+1) + x + 4(-2) = 0$$
   $$1 - 8 + x = 0$$
   $$x - 7 = 0$$
   $$x = +7$$

   • The oxidation number of hydrogen is +1.
   • The oxidation number of chlorine is +7.
   • The oxidation number of oxygen is −2.

c) • Because the dichromate, $Cr_2O_7^{2-}(aq)$, contains oxygen, rule 4 applies.
   • Because it is a polyatomic ion, rule 7 also applies.
   • Oxygen has its usual oxidation number of −2.
   • The oxidation number of chromium is unknown, so call it x.
   • You know from rule 7 that the sum of the oxidation numbers is −2. Therefore, the oxidation numbers of

   2 Cr atoms + 7 O atoms = −2
   $$2x + 7(-2) = -2$$
   $$2x - 14 = -2$$
   $$2x = 14 - 2$$
   $$2x = 12$$
   $$x = +6$$

   • The oxidation number of chromium is +6. The oxidation number of oxygen is −2.

7. Determine the oxidation number of the atoms of the specified element in each of the following:
   a) N in $NF_3(g)$ **d)** P in $P_2O_5(s)$
   b) S in $S_8(s)$ **e)** C in $C_{12}H_{22}O_{11}(s)$
   c) Cr in $CrO_4^{2-}(aq)$ **f)** C in $CHCl_3(g)$

8. Determine the oxidation number of each of the atoms in each of the following compounds:
   a) $H_2SO_3(aq)$
   b) $OH^-(aq)$
   c) $HPO_4^{2-}(aq)$

9. As stated in rule 4, oxygen does not always have its usual oxidation number of −2. Determine the oxidation number of oxygen in each of the following:
   a) the compound oxygen difluoride, $OF_2(g)$
   b) the peroxide ion, $O_2^{2-}(aq)$

10. Determine the oxidation number of each element in each of the following ionic compounds by considering the ions separately. (**Hint:** One formula unit of the compound in part (c) contains two identical monatomic ions and one polyatomic ion.)
   a) $Al(HCO_3)_3(s)$
   b) $(NH_4)_3PO_4(aq)$
   c) $K_2H_3IO_6(aq)$

Begin extension material

When you apply these rules, you will discover that some oxidation numbers do not appear to be integers. For example, an important iron ore called magnetite has the formula $Fe_3O_4(s)$. Using the oxidation number rules, you can assign oxygen an oxidation number of -2 and calculate an oxidation number of $(+\frac{8}{3})$ for iron. However, oxidation numbers must be integers. When an oxidation number appears to be a fraction, it must be an average of the oxidation numbers for the atoms in the compound. Magnetite actually contains iron(III) ions and iron(II) ions in a 2:1 ratio. The formula of magnetite is sometimes written as $Fe_2O_3 \cdot FeO(s)$ to indicate that there are iron atoms with two different oxidation numbers present in the compound. The value of $(+\frac{8}{3})$ for the oxidation number of iron is an average value as shown.

$$\frac{2(\text{iron ions with oxidation number } +3) + 1(\text{iron ion with oxidation number } +2)}{3 \text{ iron ions}}$$

$$\frac{2(+3) + (+2)}{3} = +\frac{8}{3}$$

Now consider acetone, a compound often found in nail polish remover. The formula is $H_6C_3O(\ell)$. The oxidation number for each of the six hydrogen atoms is $+1$, and the oxidation number of the oxygen atom is -2. Since acetone is a neutral molecule, the oxidation numbers must add to zero. Let x be the average oxidation number for each of the three carbon atoms:

$$6(+1) + 3x + (-2) = 0$$
$$6 + 3x - 2 = 0$$
$$3x = -4$$
$$x = -\frac{4}{3}$$

Thought Lab | 12.1 | Oxidation Numbers and Lewis Structures

Target Skills

Analyze Lewis structures and use electronegativities to determine oxidation numbers for atoms in compounds

You read that, for some compounds, the rules for assigning oxidation numbers give you only an average value for the oxidation numbers of atoms of a particular element. In some cases, you might want to know the oxidation number of each individual atom instead of an average value. Lewis structures can sometimes help you determine individual oxidation numbers. The two examples below show you the basic method for using Lewis structures to assign oxidation numbers.

	Chlorine molecule	Cyanide ion
• Draw the Lewis structure.	:Cl:Cl:	[:C:::N:]⁻
• Find electronegativities for each atom in your periodic table.	:Cl:Cl: 3.2 3.2	[:C:::N:]⁻ 2.6 3.0
• Circle the electrons that "belong" to each atom. Remember that the electrons in a bond belong to the atom with the largest electronegativity. For bonds between atoms of the same element, the electrons are shared equally between the atoms.	3.2 3.2	 2.6 3.0
• Count the valence electrons inside of each circle.	chlorine atoms: 7 electrons each	carbon atom: 2 electrons nitrogen atom: 8 electrons
• Subtract the number of electrons in the circles from the number of valance electrons that the neutral atom would have. The answer is the oxidation number of the atom.	chlorine: $7 - 7 = 0$ The oxidation number for each chlorine atom is 0.	carbon: $+4 - 2 = +2$ The oxidation number for the carbon atom is $+2$. nitrogen: $+5 - 8 = -3$ The oxidation number for the nitrogen atom is -3.

Procedure

1. Study the following example for using Lewis structures to find the oxidation numbers for each atom in acetone.

- Draw the Lewis structure $H_6C_3O(\ell)$.

$$H \quad \overset{..}{\underset{..}{O}} \quad H$$
$$H : \overset{H}{\underset{H}{C}} : \overset{..}{\underset{..}{C}} : \overset{H}{\underset{H}{C}} : H$$

- Find and record the electronegativities for each atom.

- Circle the electrons that "belong" to each atom.

- Count the electrons that are inside each circle.

- Subtract the number of valance electrons for each neutral atom. These results are the oxidation numbers of the atoms.

all H atoms: $+1 - 0 = +1$

O atom: $+6 - 8 = -2$

central C atom: $+4 - 2 = +2$

two outer C atoms: $+4 - 7 = -3$

Notice that the oxidation numbers of the carbon atoms are $+2$, -3, and -3.

The average is $\dfrac{+2 - 3 - 3}{3} = \dfrac{-4}{3}$, which agrees with the average value calculated above.

2. Use Lewis structures and the rules for assigning oxidation numbers to determine the oxidation numbers of all atoms in the compounds below:

   a) methane, $CH_4(g)$

   b) carbon dioxide, $CO_2(g)$

   c) ammonium ion, $NH_4^+(g)$

   d) $CHCl_3(\ell)$

3. The following is the Lewis structure for glucose. Assign oxidation numbers to each atom. Use the rules for assigning oxidation numbers to glucose using the formula $C_6H_{12}O_6(s)$.

Analysis

1. Describe what you can learn by using Lewis structures that you cannot determine by using the rules for assigning oxidation numbers.

— End extension material

Applying Oxidation Numbers to Redox Reactions

You can use oxidation numbers to identify redox reactions and to identify the oxidizing and reducing agents. First, apply oxidation numbers to familiar equations to see how they can be used in more challenging equations.

The oxidation number for balancing equations can be applied to reactions involving ionic compounds as well as molecular compounds. You have seen that the single-displacement reaction of zinc with copper(II) sulfate is a redox reaction, represented by the following complete balanced equation and net ionic equation:

$$Zn(s) + CuSO_4(aq) \rightarrow Cu(s) + ZnSO_4(aq)$$
$$Zn(s) + Cu^{2+}(aq) \rightarrow Cu(s) + Zn^{2+}(aq)$$

Chapter 12 Oxidation-Reduction Reactions · **MHR** **459**

You can assign an oxidation number to each atom or ion in the net ionic equation, as follows:

- Solid Zn has an oxidation number of 0.
- Cu^{2+} has an oxidation number of +2.
- Solid Cu has an oxidation number of 0.
- Zn^{2+} has an oxidation number of +2.

Thus, there are changes in oxidation numbers in this reaction. The oxidation number of zinc increases, while the oxidation number of copper decreases:

oxidation number increases
(loss of electrons)

$$Zn + Cu^{2+} \longrightarrow Zn^{2+} + Cu$$
$$0 \quad\quad +2 \quad\quad\quad +2 \quad\quad\quad 0$$

oxidation number decreases
(gain of electrons)

In the oxidation half-reaction, zinc undergoes an increase in its oxidation number from 0 to +2:

$$Zn \rightarrow Zn^{2+} + 2e^-$$
$$0 \quad\quad +2$$

In the reduction half-reaction, copper undergoes a decrease in its oxidation number from +2 to 0:

$$Cu^{2+} + 2e^- \rightarrow Cu$$
$$+2 \quad\quad\quad\quad 0$$

Therefore, you can describe redox reactions, oxidation, and reduction as follows. (See also Figure 12.12.)

- A *redox reaction* is a reaction in which the oxidation numbers of at least two atoms change.
- *Oxidation* is an increase in oxidation number.
- *Reduction* is a decrease in oxidation number.

You can also monitor changes in oxidation numbers in reactions that involve covalent molecules. For example, oxidation number changes occur in the reaction of hydrogen and oxygen to form water:

$$2H_2(g) + O_2(g) \rightarrow 2H_2O(\ell)$$
$$0 \quad\quad\quad\quad 0 \quad\quad\quad\quad +1 \,\, -2$$

Because hydrogen combines with oxygen in this reaction, hydrogen undergoes oxidation, according to the operational definition. Hydrogen also undergoes oxidation according to the modern definition, because the oxidation number of hydrogen increases from 0 to +1. Hydrogen is the reducing agent in this reaction. The oxygen undergoes reduction, because its oxidation number decreases from 0 to −2. Oxygen is the oxidizing agent in this reaction. The reaction is a redox reaction, because the oxidation numbers of three atoms—two hydrogen atoms and one oxygen atom—change.

The following Sample Problem illustrates how to use oxidation numbers to identify redox reactions, oxidizing agents, and reducing agents.

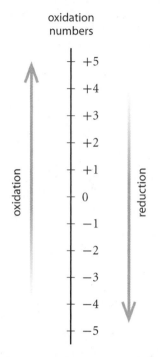

oxidation numbers

+5
+4
+3
+2
+1
0
−1
−2
−3
−4
−5

oxidation

reduction

Figure 12.12 Oxidation and reduction are directly related to changes in oxidation numbers.

Problem Tip

- Use the rule that the sum of the oxidation numbers in a molecule is zero to check the assignment of the oxidation numbers.
- Make sure that a reaction does not include only a reduction or only an oxidation. Oxidation and reduction must occur together in a redox reaction.

Identifying Redox Reactions

Problem

Determine whether each of the following reactions is a redox reaction. If so, identify the oxidizing agent and the reducing agent:

a) $CH_4(g) + Cl_2(g) \rightarrow CH_3Cl(g) + HCl(g)$

b) $CaCO_3(s) + 2HCl(aq) \rightarrow$
$$CaCl_2(aq) + H_2O(\ell) + CO_2(g)$$

c) $3HNO_2(aq) \rightarrow HNO_3(aq) + 2NO(g) + H_2O(\ell)$

Solution

Find the oxidation number of each atom in the reactants and products. Identify any atoms that undergo an increase or a decrease in oxidation number during the reaction.

a) The oxidation number of each element in the reactants and products is as shown:

$$CH_4(g) + Cl_2(g) \rightarrow CH_3Cl(g) + HCl(g)$$
$${-4\ +1}0{-2\ +1\ -1}{+1\ -1}$$

- The oxidation number of hydrogen is $+1$ on both sides of the equation, so hydrogen is neither oxidized nor reduced.
- Both carbon and chlorine undergo changes in oxidation number, so the reaction is a redox reaction.
- The oxidation number of carbon increases from -4 to -2. The carbon atoms on the reactant side exist in methane molecules, $CH_4(g)$, so methane is oxidized. Therefore, methane is the reducing agent.
- The oxidation number of chlorine decreases from 0 to -1, so elemental chlorine, $Cl_2(g)$, is reduced. Therefore, elemental chlorine is the oxidizing agent.

b) Because this reaction involves ions, write the equation in its ionic form:

$$CaCO_3(s) + 2H^+(aq) + 2Cl^- \rightarrow$$
$$Ca^{2+}(aq) + 2Cl^-(aq) + H_2O(\ell) + CO_2(g)$$

The chloride ions are spectator ions, which do not undergo oxidation or reduction. The net ionic equation is as follows:

$$CaCO_3(s) + 2H^+(aq) \rightarrow Ca^{2+}(aq) + H_2O(\ell) + CO_2(g)$$

For the net ionic equation, the oxidation number of each atom in the reactants and products is as shown:

$$CaCO_3(s) + 2H^+(aq) \rightarrow Ca^{2+}(aq) + H_2O(\ell) + CO_2(g)$$
$${+2\ +4\ -2}{+1}{+2}\phantom{Ca^{2+}(aq) +}{+1\ -2}{+4\ -2}$$

No atoms undergo changes in oxidation numbers, so the reaction is *not* a redox reaction.

c) Because this reaction involves ions, write the equation in its ionic form. The oxidation number of each atom or ion in the reactants and products is as shown:

$$3H^+(aq) + 3NO_2^-(aq) \rightarrow$$
$${+1}{+3\ -2}$$
$$H^+(aq) + NO_3^-(aq) + 2NO(g) + H_2O(\ell)$$
$${+1}{+5\ -2}{+2\ -2}{+1\ -2}$$

- The oxidation number of hydrogen is $+1$ on both sides of the equation, so hydrogen is neither oxidized nor reduced.
- The oxidation number of oxygen is -2 on both sides of the equation, so oxygen is neither oxidized nor reduced.
- The only remaining atoms are nitrogen atoms. Nitrogen atoms are found in two different compounds among the products, and the nitrogen atoms in these compounds have different oxidation numbers. Therefore, this is a disproportionation reaction.
- The oxidation number of one nitrogen atom increases from $+3$ to $+5$; therefore, it is oxidized.
- The oxidation number of two of the nitrogen atoms decreases from $+3$ to $+2$; therefore, they are reduced.
- Nitrous acid, $HNO_2(aq)$, is both the oxidizing agent and the reducing agent.

11. Which of the following are redox reactions? Identify any disproportionation reactions:
a) $H_2O_2(aq) + 2Fe(OH)_2(s) \rightarrow 2Fe(OH)_3(s)$
b) $PCl_3(\ell) + 3H_2O(\ell) \rightarrow H_3PO_3(aq) + 3HCl(aq)$
c) $2C_2H_6(g) + 7O_2(g) \rightarrow 4CO_2(g) + 6H_2O(\ell)$
d) $3NO_2(g) + H_2O(\ell) \rightarrow 2HNO_3(aq) + NO(g)$

12. Identify the oxidizing agent and the reducing agent for the redox reaction(s) in the previous question.

13. For the following balanced net ionic equation, identify the reactant that undergoes oxidation and the reactant that undergoes reduction:

$$Br_2(\ell) + 2ClO_2^-(aq) \rightarrow 2Br^-(aq) + 2ClO_2(aq)$$

14. Nickel and copper ores usually contain the metals as sulfides, such as $NiS(s)$ and $Cu_2S(s)$. Does the extraction of these pure elemental metals from their ores involve redox reactions? Explain your reasoning.

Balancing Equations Using the Oxidation Number Method

Now that you know how to assign oxidation numbers to atoms in compounds and you can identify redox equations, you are ready to learn how to balance redox reactions by using oxidation numbers. The critical step in balancing redox equations is to ensure that the total increase in the oxidation number of the oxidized element or elements equals the total decrease in the oxidation number of the reduced element or elements. This statement is the same as saying that the total number of electrons lost by the reducing agent must be the same as the total number of electrons gained by the oxidizing agent.

For example, the combustion of ammonia in oxygen produces nitrogen dioxide and water:

$$NH_3(g) + O_2(g) \rightarrow NO_2(g) + H_2O(\ell)$$
$$\;\;{\scriptstyle -3\;+1}\qquad{\scriptstyle 0}\qquad{\scriptstyle +4\;-2}\qquad{\scriptstyle +1\;-2}$$

The oxidation number of nitrogen increases from -3 to $+4$, an increase of 7. The oxidation number of oxygen decreases from 0 to -2, a decrease of 2. When you conclude that the oxidation number of nitrogen increased by 7, you know that the nitrogen atom effectively lost 7 electrons. Similarly, when you know that the oxidation number of the oxygen atoms decreased by 2, you know that each oxygen atom in the oxygen molecule effectively gained 2 electrons for a total of 4 electrons. You can then write the equation with lines connecting the oxidized and reduced atoms and write the number of electrons gained or lost above or below the lines as shown:

$$\overset{\displaystyle -7e^-}{NH(g) + O_2(g) \rightarrow NO_2(g) + H_2O(\ell)}$$
$$+4e^- \qquad (2e^-\text{ per O atom})$$

Next, you balance the number of electrons gained and lost by determining the lowest common multiple of 7 and 4, which is 28. Multiply each number of electrons by the number that will bring it to the lowest common multiple, 28, as shown:

$$\overset{\displaystyle 4(-7e^-) = -28e^-}{NH_3(g) + O_2(g) \rightarrow NO_2(g) + H_2O(\ell)}$$
$$7(+4e^-) = +28e$$

Insert the numbers by which you multiplied the electrons as coefficients in the equation.

$$4NH_3(g) + 7O_2(g) \rightarrow NO_2(g) + H_2O(\ell)$$

Finally, finish balancing the equation by inspection.

$$4NH_3(g) + 7O_2(g) \rightarrow 4NO_2(g) + 6H_2O(\ell)$$

A summary of the steps of the oxidation number method is given below. The Sample Problem that follows shows how to apply these steps.

Balancing Equations Using the Oxidation Number Method

Step 1 Write an unbalanced equation, if it is not given.

Step 2 Assign an oxidation number to each atom in the equation to determine whether it is a redox reaction.

Step 3 If the reaction is a redox reaction, identify the atom or atoms that undergo an increase in oxidation number and the atom or atoms that undergo a decrease in oxidation number.

Step 4 Find the numerical values of the increase and the decrease in oxidation numbers.

Step 5 Determine the lowest common multiple of the number of electrons lost by the reducing agent (increase in oxidation number) and the number of electrons gained by the oxidizing agent (decrease in oxidation number).

Step 6 Apply the numbers found in step 5 to balance the number of atoms oxidized and the number of atoms reduced.

Step 7 Balance the other elements by inspection, if possible.

Step 8 For reactions that occur in acidic or basic solutions, include water molecules, hydrogen ions, or hydroxide ions as needed to balance the equation.

Sample Problem

Balancing a Redox Equation in Basic Solution

Problem

Write a balanced net ionic equation to show the formation of iodine by bubbling oxygen gas through a basic solution that contains iodide ions.

Solution

Step 1 Write an unbalanced equation from the given information:
$$O_2(g) + I^-(aq) \rightarrow I_2(s)$$

Step 2 Assign an oxidation number to each atom in the equation to determine whether it is a redox reaction.
$$\underset{0}{O_2}(g) + \underset{-1}{I^-}(aq) \rightarrow \underset{0}{I_2}(s)$$

Because iodide is oxidized to iodine, the reaction is a redox reaction. The product that contains oxygen is unknown at this stage. However, oxygen must be reduced.

Step 3 Because the reaction is a redox reaction, identify the atom or atoms that undergo an increase in oxidation number and the atom or atoms that undergo a decrease in oxidation number.

Iodine is the element that undergoes an increase in oxidation number. Oxygen is the element that undergoes a decrease in oxidation number.

Step 4 Find the numerical values of the increase and the decrease in oxidation numbers.

Iodine undergoes an increase in its oxidation number from -1 to 0, an increase of 1. Assume that the oxidation number of oxygen after reduction is its normal value, which is -2. Thus, oxygen undergoes a decrease in its oxidation number from 0 to -2, a decrease of 2.

Step 5 Determine the lowest common multiple of the increase in oxidation number and the decrease in oxidation number.

Step 6 Apply the numbers found in step 5 to balance the number of atoms oxidized (iodine) and the number of atoms reduced (oxygen). Make sure there are two iodine atoms for every oxygen atom:
$$O_2(g) + 4I^-(aq) \rightarrow 2I_2(s)$$

Step 7 Balance the other elements by inspection, if possible.

No other reactants or products can be balanced by inspection.

Step 8 For reactions that occur in acidic or basic solutions, include water molecules, hydrogen ions, or hydroxide ions as needed to balance the equation.

The reaction occurs in a basic solution. As you learned in section 12.2, for basic conditions, start by assuming that the conditions are acidic. Add water molecules and hydrogen ions as necessary to balance the atoms:
$$O_2(g) + 4I^-(aq) \rightarrow 2I_2(s) + 2H_2O(\ell)$$
$$O_2(g) + 4I^-(aq) + 4H^+(aq) \rightarrow 2I_2(s) + 2H_2O(\ell)$$

Add hydroxide ions to adjust for basic conditions. Simplify the resulting equation:
$$O_2(g) + 4I^-(aq) + 4H^+(aq) + 4OH^-(aq) \rightarrow$$
$$2I_2(s) + 2H_2O(\ell) + 4OH^-(aq)$$
$$O_2(g) + 4I^-(aq) + 4H_2O(\ell) \rightarrow$$
$$2I_2(s) + 2H_2O(\ell) + 4OH^-(aq)$$
$$O_2(g) + 4I^-(aq) + 2H_2O(\ell) \rightarrow 2I_2(s) + 4OH^-(aq)$$

15. Use the oxidation number method to balance the following equation for the combustion of carbon disulfide:

$CS_2(g) + O_2(g) \rightarrow CO_2(g) + SO_2(g)$

16. Use the oxidation number method to balance the following equations:

a) $B_2O_3(aq) + Mg(s) \rightarrow MgO(s) + Mg_3B_2(aq)$

b) $H_2S(g) + H_2O_2(aq) \rightarrow S_8(s) + H_2O(\ell)$

17. Use the oxidation number method to balance each ionic equation in an acidic solution:

a) $Cr_2O_7^{2-}(aq) + Fe^{2+}(aq) \rightarrow Cr^{3+}(aq) + Fe^{3+}(aq)$

b) $I_2(g) + NO_3^-(aq) \rightarrow IO_3^-(aq) + NO_2(g)$

c) $PbSO_4(aq) \rightarrow Pb(s) + PbO_2(aq) + SO_4^{2-}(aq)$

18. Use the oxidation number method to balance each ionic equation in a basic solution:

a) $Cl^-(aq) + CrO_4^{2-}(aq) \rightarrow ClO^-(aq) + CrO_2^-(aq)$

b) $Ni(s) + MnO_4^-(aq) \rightarrow NiO(s) + MnO_2(s)$

c) $I^-(aq) + Ce^{4+}(aq) \rightarrow IO_3^-(aq) + Ce^{3+}(aq)$

Concept Organizer | **The Oxidation Number Method for Balancing Redox Equations**

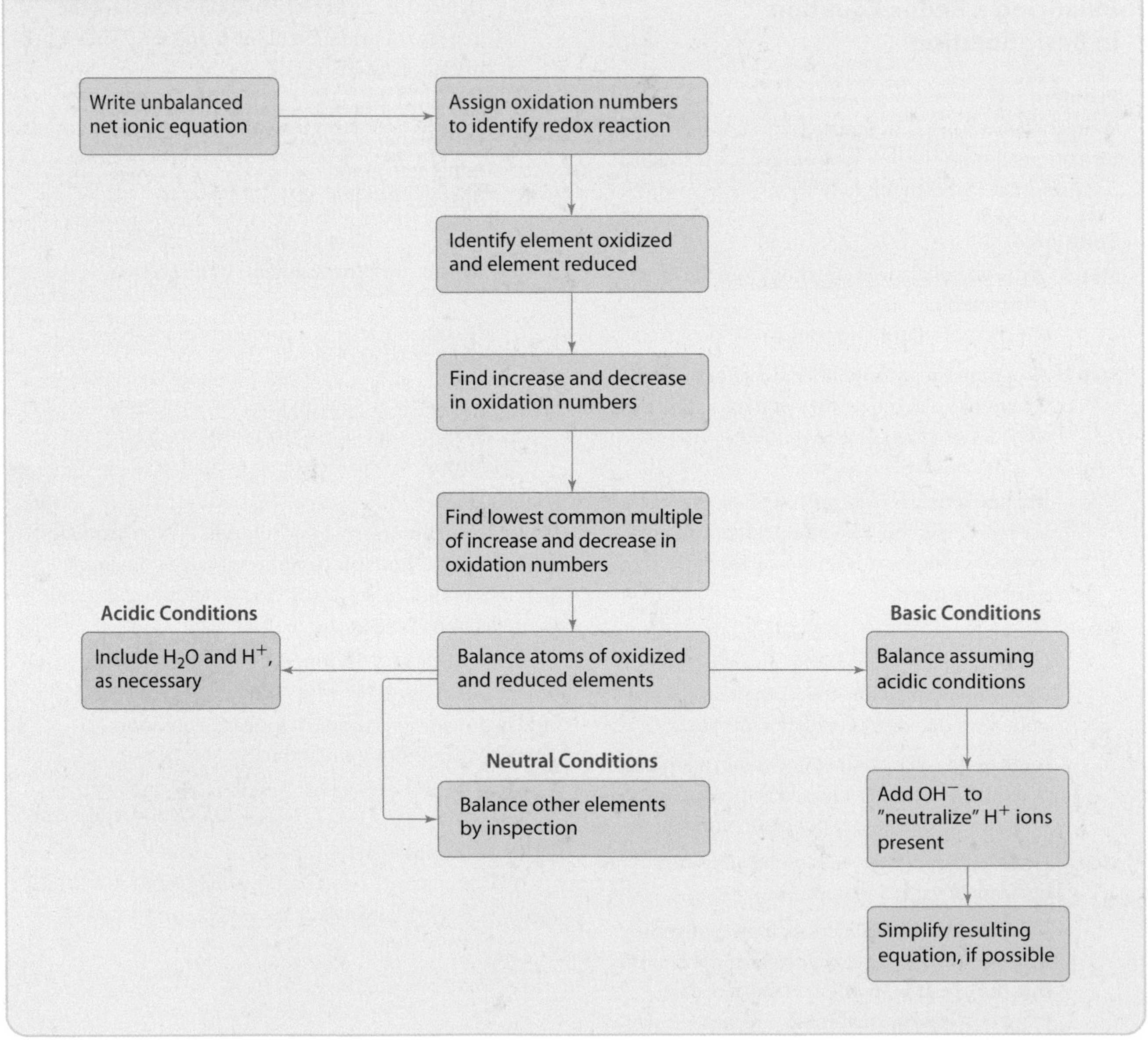

| Write unbalanced net ionic equation | → | Assign oxidation numbers to identify redox reaction |

↓

Identify element oxidized and element reduced

↓

Find increase and decrease in oxidation numbers

↓

Find lowest common multiple of increase and decrease in oxidation numbers

↓

Acidic Conditions

Include H_2O and H^+, as necessary ← Balance atoms of oxidized and reduced elements → **Basic Conditions** Balance assuming acidic conditions

Neutral Conditions

Balance other elements by inspection

Add OH^- to "neutralize" H^+ ions present

↓

Simplify resulting equation, if possible

How Green Is White Paper?

Wood pulp that is to be used for making paper is usually bleached not only to make it white but to make it more durable. The North American Paper Industry is enormous, producing 45–55 million tonnes of paper every year, about half of which is bleached. Canada produces 29% of the world's wood pulp and 49% of its newsprint. Alberta is a major player in the manufacture of paper and paper products because the slow-growing trees in Alberta produce a very high-quality, high-strength fibre.

Wood Pulp Bleaching
TAML® catalysts activate hydrogen peroxide to provide a superfast totally chlorine-free method for bleaching wood pulp with very high selectivity. "Bleaching" is breaking the coloured lignin away from the solid pulp to leave white cellulose.

Wood pulp is a mixture of cellulose and lignin (a complex phenolic polymer)

Dominant Bleaching Technology
"Elemental Cl-free" – ClO_2

Little lignin fragments are destroyed by bacteria in treatment lakes.

ClO_2

Cellulose

Lignin fragments and chlorinated polutants

Big lignin fragments carry "colour" and resist biological degradation.

Bleaching Pulp

Bleaching not only whitens but also improves absorption, softness, flexibility, and resistance to aging. Bleaching involves the removal of a polymer, called lignin, from the pulp. The conventional method of bleaching pulp is called the Kraft process. It involves a series of processes using alkali, acid, hydrogen and sodium peroxide, oxygen, dithionite salts, sodium bisulfite, and a wash water process. This is followed by chlorinating treatments to remove any residual lignin.

Chlorine gas was used to bleach pulp until scientists discovered that toxic by-products called dioxins were formed during the bleaching. Currently, the primary bleaching agent is chlorine dioxide, $ClO_2(g)$, which has completely different properties than chlorine. It can be produced by reducing sodium chlorate in a strong acid with a suitable reducing agent:

$$2ClO_3^-(aq) + 2Cl^-(aq) + 4H^+(aq) \rightarrow$$
$$2ClO_2(g) + Cl_2(g) + 2H_2O(\ell)$$

Chlorine dioxide is an oxidant with a low reduction potential of +0.96 compared to chlorine's reduction potential of +1.36. It is a yellow-to-brown gas at standard temperature and pressure. Because it is a strong oxidant, chlorine dioxide must be generated on site. It is explosive as a gas, but is stable in water when kept in the dark.

The chlorine dioxide removes lignin from the cellulose fibers in wood pulp. Lignin is a polymer that binds the cellulose fibers together in wood. The lignin fragments are placed in artificial oxidation lakes, where they are digested by bacteria before being returned to natural bodies of water. A problem arises because the bacteria cannot easily digest the larger fragments. When these fragments are returned to bodies of water as effluent, they retain their dark color and thus stain natural waterways. The use of chlorine dioxide in place of chlorine lessens, but does not eliminate, the problem. The industry's move to Elemental Chlorine Free (ECF) processes opens up new opportunities for biotechnology and other competing technologies.

Environmentally-Friendly Bleaching

The pulp and paper industry has been examining the use of ozone, a very powerful oxidant, for many years. Ozone can be used in the bleaching process and also applied as a final polishing treatment to the waste effluent. Domtar's mill in Espanola, Ontario, is the first in Canada to produce hardwood pulp using the ozone bleaching process.

Work performed in the laboratories of Terrence Collins at Carnegie Mellon University has produced a series of oxidant activators referred to as "tetraamido-macrocyclic ligand activators" (TAML™). TAML™ catalysts are catalysts that can improve peroxide pulp bleaching.

Enzymatic bleaching methods using oxidative enzymes have recently drawn much attention as being environmentally friendly. Investigations in electrochemical bleaching take this process further by replacing enzymes as oxidants with electrodes. Ultimately this research may lead to novel, cost-effective, bleaching strategies that are environmentally benign.

Making Connections

1. Pulp is converted to paper on a paper machine, for example the Fourdrinier machine. Find out how it works.

2. What are some advantages of using the TAML™/ hydrogen peroxide bleaching system over chlorine systems?

3. Contact an Alberta pulp and paper mill. Ask how they are addressing the use of elemental chlorine free (ECF) processes.

To learn more about making paper, go on-line.

www.albertachemistry.ca
WWW

Section 12.3 Summary

- Some redox equations cannot be written as half-reactions. You can use the method of oxidation numbers to balance these equations.
- An oxidation number of an atom in a compound is the charge that the atom would have if, instead of sharing electrons, the electrons were held by the atom having the greatest electronegativity.
- You can assign oxidation numbers by following a set of rules.
- Sometimes the rules lead to fractional oxidation numbers. In these cases, you can determine the oxidation numbers for individual atoms in a compound by using Lewis structures.
- If the oxidation number of an atom increases in a reaction, it indicates a loss of electrons. If the oxidation number of an atom decreases in a reaction, it indicates a gain of electrons.
- You can balance an equation by finding the coefficients that make the number of electrons lost by one atom of one element equal to the number of electrons gained by atoms of another element. You then balance the rest of the equation by inspection.

SECTION 12.3 REVIEW

1. Determine whether each of the following reactions is a redox reaction:
    a) $H_2(g) + I_2(s) \rightarrow 2HI(aq)$
    b) $2NaHCO_3(aq) \rightarrow Na_2CO_3(aq) + H_2O(\ell) + CO_2(g)$
    c) $2HBr(aq) + Ca(OH)_2(s) \rightarrow CaBr_2(aq) + 2H_2O(\ell)$
    d) $PCl_5(\ell) \rightarrow PCl_3(\ell) + Cl_2(g)$

2. Write three different definitions for a redox reaction.

3. Explain why fluorine has an oxidation number of -1 in all its compounds other than fluorine gas, $F_2(g)$.

4. When atoms of one element combine with atoms of another element, is the reaction a redox reaction? Explain your answer.

5. **a)** Use the oxidation number rules to find the oxidation number of sulfur in a thiosulfate ion, $S_2O_3{}^{2-}(aq)$.
    b) The Lewis structure of a thiosulfate ion is given here. Use the Lewis structure to find the oxidation number of each sulfur atom.

$$\left[\begin{array}{c} :\!\ddot{O}\!: \\ :\!\ddot{O}\!:\!\ddot{S}\!:\!\ddot{S}\!: \\ :\!\ddot{O}\!: \end{array} \right]^{2-}$$

    c) Compare your results from parts (a) and (b) and explain any differences.
    d) What are the advantages and disadvantages of using Lewis structures to assign oxidation numbers?
    e) What are the advantages and disadvantages of using the oxidation number rules to assign oxidation numbers?

6. **a)** The Haber process for the production of ammonia from nitrogen gas and hydrogen gas is a very important industrial process. Write a balanced chemical equation for the reaction. Use oxidation numbers to identify the oxidizing agent and the reducing agent.
    b) When ammonia reacts with nitric acid to make the common fertilizer ammonium nitrate, is the reaction a redox reaction? Explain. (**Hint:** Consider the two polyatomic ions in the product separately.)

7. Balance each equation by the method of your choice. Explain your choice of method in each case:
    a) $CH_3COOH(aq) + O_2(g) \rightarrow CO_2(g) + H_2O(\ell)$
    b) $O_2(g) + H_2SO_4(aq) \rightarrow HSO_4{}^-(aq)$ (acidic conditions)

8. Use the oxidation number method to balance the following equations:
    a) $NH_3(g) + Cl_2(g) \rightarrow NH_4Cl(aq) + N_2(g)$
    b) $Mn_3O_4(aq) + Al(s) \rightarrow Al_2O_3(aq) + Mn(s)$

9. Explain why, in redox reactions, the total increase in the oxidation numbers of the oxidized elements must equal the total decrease in the oxidation numbers of the reduced elements.

10. The combustion of ammonia in oxygen to form nitrogen dioxide and water vapour involves covalent molecules in the gas phase. Use the oxidation number method for balancing the equation.

Quantitative Analysis of Redox Reactions

Section Outcomes

In this section, you will:

- **describe calculations** to determine quantities of substance involved in redox titrations
- **perform** a titration experiment
- **analyze** data from a titration experiment
- **calculate** the concentration of a reducing agent from titration data

Redox titrations are an important application of redox chemistry and stoichiometry. In an acid-base titration, a base is used to find the concentration of an acid, or an acid is used to find the concentration of a base. Similarly, in a redox titration, a known concentration of an oxidizing agent can be used to find the unknown concentration of a reducing agent, or a known concentration of a reducing agent can be used to find the unknown concentration of an oxidizing agent. Redox titrations are used in a wide range of situations, including the measuring of the iron content in drinking water (Figure 12.13) and the vitamin C content in foods or vitamin supplements.

Figure 12.13 The dripping water contained iron, which stained this fountain.

Stoichiometry and Redox Titrations

The permanganate ion, $MnO_4^-(aq)$ and dichromate ion, $Cr_2O_7^{2-}(aq)$, are commonly used as an oxidizing agents in redox titrations. For example, the permanganate ion has a strong purple colour, meaning that no additional indicator is needed to determine the endpoint of the titration. Figure 12.14 shows the titration of a solution of sodium oxalate, $Na_2C_2O_4(aq)$, with a solution of potassium permanganate, $KMnO_4(aq)$.

As long as there are oxalate ions present in solution, they reduce the manganese in the purple permanganate ions (oxidation number +7) to nearly colourless manganese(II) ions, $Mn^{2+}(aq)$ (oxidation number +2). Once all oxalate ions have been oxidized, the next drop of potassium permanganate solution turns the solution light purple. This purple colour signals the endpoint of the titration.

Figure 12.14 The photo on the right shows the endpoint of the titration of a solution containing oxalate ions with a solution containing permanganate ions. In the balanced equation, the ratio of $MnO_4^-(aq)$ to $OOCCOO^{2-}(aq)$ is 2:5. This ratio means that at the endpoint, 2 mol of $MnO_4^-(aq)$ have been added for every 5 mol of $C_2O_4^{2-}(aq)$ that were present initially.

$2MnO_4^-(aq)$		$5OOCCOO^{2-}(aq)$		$16H^+$		$2Mn^{2+}(aq)$		$10CO_2(g)$		$8H_2O(\ell)$
+7 −2	+	+3 −2	+	+1	→	+2	+	+4 −2	+	+1 −2
permanganate		oxalate		hydrogen ion		manganese (II)		carbon dioxide		water

The permanganate ion can also be used to oxidize hydrogen peroxide, $H_2O_2(aq)$. The following equation shows the redox reaction in acidic conditions:

$$5H_2O_2(aq) + 2MnO_4^-(aq) + 6H^+(aq) \rightarrow 5O_2(g) + 2Mn^{2+}(aq) + 8H_2O(\ell)$$

Aqueous solutions of hydrogen peroxide sold in pharmacies are often about 3% $H_2O_2(aq)$ by mass. In solution, however, hydrogen peroxide decomposes steadily to form water and oxygen.

Suppose that you need to use a hydrogen peroxide solution with a concentration of at least 2.5% by mass for a certain experiment. Your 3% $H_2O_2(aq)$ is not fresh, so it might have decomposed significantly. How do you find out whether your $H_2O_2(aq)$ solution is concentrated enough? The following Sample Problem shows how to solve this problem by using data from the titration of the hydrogen peroxide solution with a solution of potassium permanganate.

Sample Problem

Redox Titrations

Problem

You are using a 0.011 43 mol/L $KMnO_4(aq)$ solution to determine the percentage by mass of an aqueous solution of $H_2O_2(aq)$. You know that the peroxide solution is about 3% $H_2O_2(aq)$ by mass.

You prepare the sample by adding 1.423 g of the hydrogen peroxide solution to an Erlenmeyer flask. (Although the hydrogen peroxide is in aqueous solution, you can determine the mass by placing an empty flask on a balance and adding peroxide with a pipette.) You add about 75 mL of water to dilute the solution. You also add some dilute sulfuric acid to acidify the solution.

You reach the light-purple-coloured endpoint of the titration when you have added 40.22 mL of the $KMnO_4(aq)$ solution. What is the percent, by mass, of the peroxide solution?

(Percent by mass is defined as: $\dfrac{\text{mass of peroxide}}{\text{total mass of solution}} \times 100\%$)

What Is Required?

You need to determine the mass of $H_2O_2(aq)$ in the sample. You need to express your result as a mass percent.

What Is Given?

Concentration of $KMnO_4(aq)$ = 0.011 43 mol/L
Volume of $KMnO_4(aq)$ = 40.22 mL
Mass of $H_2O_2(aq)$ solution = 1.423 g

Plan Your Strategy

Step 1 Write the balanced chemical equation for the reaction.

Step 2 Calculate the amount (in mol) of permanganate ion added, based on the volume and concentration of the potassium permanganate solution.

Step 3 Determine the amount (in mol) of hydrogen peroxide needed to reduce the permanganate ions.

Step 4 Determine the mass of hydrogen peroxide, based on the molar mass of hydrogen peroxide. Finally, express your answer as a mass percent of the hydrogen peroxide solution, as the question directs.

Act on Your Strategy

Step 1 The redox equation was provided on the previous page. It is already balanced:

$$5H_2O_2(aq) + 2MnO_4^-(aq) + 6H^+(aq) \rightarrow$$
$$5O_2(g) + 2Mn^{2+}(aq) + 8H_2O(\ell)$$

Step 2 The concentration of $MnO_4^-(aq)$ is the same as the concentration of $KMnO_4(aq)$:

$$n = cV$$
$$n = \left(0.011\,43\,\frac{mol}{L}\right)(0.040\,22\,L)$$
$$n = 4.597 \times 10^{-4}\,mol$$

Step 3 Permanganate ions react with hydrogen peroxide in a 2:5 ratio, as shown by the coefficients in the balanced equation:

Amount (in moles) $H_2O_2(aq)$

$$n = \left(\frac{5\,mol\,H_2O_2(aq)}{2\,mol\,MnO_4^-(aq)}\right)(4.597 \times 10^{-4}\,mol\,MnO_4^-(aq))$$
$$= 1.149 \times 10^{-3}\,mol\,H_2O_2(aq)$$

Step 4

$$M_{H_2O_2(aq)} = 2\left(1.01\,\frac{g}{mol}\right) + 2\left(16\,\frac{g}{mol}\right)$$
$$= 34.02\,\frac{g}{mol}$$

$$m_{H_2O_2(aq)} = nM_{H_2O_2(aq)}$$
$$= (1.149 \times 10^{-3}\,mol)\left(34.02\,\frac{g}{mol}\right)$$
$$= 0.039\,09\,g$$

$$H_2O_2\,percent\,(m/m) = \left(\frac{0.03909\,g}{1.432\,g}\right)100\%$$
$$= 2.730\%$$

Check Your Solution

The units are correct. The value for the mass of pure $H_2O_2(aq)$ that you obtained is less than the mass of the $H_2O_2(aq)$ sample solution, as you would expect. The mass percent you obtained for the solution is close to the expected value. It makes sense that the value is somewhat less than 3%, since $H_2O_2(aq)$ decomposes in solution, forming water and oxygen.

Using Dimensional Analysis:

$$m_{H_2O_2(aq)} = \left(0.011\,43\,\frac{mol}{L}\,MnO_4^-(aq)\right)\left(0.040\,22\,L\right)\frac{(5\,mol\,H_2O_2(aq)}{2\,mol\,MnO_4^-(aq)}\right)\left(34.02\,\frac{g}{mol}\right)$$
$$= 0.039\,09\,g$$

Practice Problems

19. An analyst prepares a $H_2O_2(aq)$ sample solution by placing 1.284 g of $H_2O_2(aq)$ solution in a flask then diluting it with water and adding sulfuric acid to acidify it. The analyst titrates the $H_2O_2(aq)$ sample solution with 0.020 45 mol/L $KMnO_4(aq)$ and determines that 38.95 mL of $KMnO_4(aq)$ is required to reach the endpoint.

  a) What is the mass of pure $H_2O_2(aq)$ that is present in the sample solution?

  b) What is the mass percent of pure $H_2O_2(aq)$ in the sample solution?

20. A forensic chemist wants to determine the level of alcohol in a sample of blood plasma. The chemist titrates the plasma with a solution of potassium dichromate. The balanced equation is

$$16H^+(aq) + 2Cr_2O_7^{2-}(aq) + C_2H_5OH(aq) \rightarrow$$
$$4Cr^{3+}(aq) + 2CO_2(g) + 11H_2O(\ell)$$

If 32.35 mL of 0.050 23 mol/L $Cr_2O_7^{2-}(aq)$ is required to titrate 27.00 g plasma, what is the mass percent of alcohol in the plasma?

21. An analyst titrates an acidified solution containing 0.153 g of purified sodium oxalate, $Na_2C_2O_4(aq)$, with a potassium permanganate solution, $KMnO_4(aq)$. The light purple endpoint is reached when the chemist has added 41.45 mL of potassium permanganate solution. What is the molar concentration of the potassium permanganate solution? The balanced equation is

$$2MnO_4^-(aq) + 5Na_2C_2O_4(aq) + 16H^+(aq) \rightarrow$$
$$10Na^+ + 2Mn^{2+}(aq) + 10CO_2(g) + 8H_2O(\ell)$$

22. 25.00 mL of a solution containing iron(II) ions was titrated with a 0.020 43 mol/L potassium dichromate solution. The endpoint was reached when 35.55 mL of potassium dichromate solution had been added. What was the molar concentration of iron(II) ions in the original, acidic solution? The *unbalanced* equation is

$$Cr_2O_7^{2-}(aq) + Fe^{2+}(aq) \rightarrow Cr^{3+}(aq) + Fe^{3+}(aq)$$

In the following investigation, you will apply the concepts that you have just learned about redox titrations. First, you will standardize an iodine solution. Then you will use the standardized solution to determine the concentration of vitamin C in a commercial orange juice. In a reaction of vitamin C with iodine, the vitamin C becomes oxidized. You will determine the amount of iodine solution required to oxidize all the vitamin C.

INVESTIGATION | 12.C

Measuring the Concentration of Vitamin C in Orange Juice

Vitamin C is vital to the production of collagen, and it also protects the fat-soluble vitamins A and E and fatty acids from oxidation (vitamin C is an antioxidant). Vitamin C is a weak acid (ascorbic acid) and is stable in weak acids. Bases, such as baking soda, destroy vitamin C. It is also easily oxidized in air and sensitive to heat and light. Since it is contained in the watery part of fruits and vegetables, it is easily lost during cooking in water. Loss is minimized when vegetables, such as broccoli or Brussels sprouts, are cooked over water in a double boiler instead of directly in water. Any copper in the water or in the cookware will also diminish the vitamin C content of foods, because copper binds to vitamin C. In this investigation, you will determine the concentration of vitamin C (ascorbic acid) in a sample of orange juice.

Question

What is the molar concentration of vitamin C in a sample of orange juice?

Prediction

Predict the approximate concentration of vitamin C in the orange juice from the information given on the label. Your teacher will help you with the calculation.

Safety Precautions

- The iodine solution can cause stains on clothing and skin. Wash any spills on your skin or clothing with plenty of cool water.
- Dispose of the iodine solution used to rinse the burette and any waste iodine solution into the container provided.
- Sodium thiosulfate should be available to remove any stains that occur.

- Wash your hands thoroughly when you have completed the investigation.

Materials

- iodine solution
- 5.68 mmol/L ascorbic acid standard solution
- dropper bottle containing starch indicator
- orange juice, pulp removed
- deionized water
- burette
- burette clamp
- ring stand
- meniscus reader
- volumetric pipette (10.00 mL)
- suction bulb
- 3 Erlenmeyer flasks (125 mL)
- 3 beakers (250 mL)
- labels
- sheet of white paper
- funnel

Procedure

1. Copy the tables below into your notebook to record your observations.

Volumes of Iodine Solution Required to React with 10.00 mL Samples of a 5.68 mmol/L Standard Solution of Ascorbic Acid

	Trial 1	Trial 2	Trial 3	Trial 4
Final buret reading (mL)				
Initial buret reading (mL)				
Volume added (mL)				
Indicator colour at endpoint (starch)				

Volumes of Iodine Solution Required to React with 10.00 mL Samples of Orange Juice

	Trial 1	Trial 2	Trial 3	Trial 4
Final buret reading (mL)				
Initial buret reading (mL)				
Volume added (mL)				
Indicator colour at endpoint (starch)				

2. Clean and rinse the burette with the iodine solution, and then fill it with fresh iodine solution. Ensure that there is no air bubble in the burette tip. Record the initial volume of titrant to the nearest 0.05 mL.

3. Rinse the 10.00 mL pipette with standard solution and then use it to obtain a 10.00 mL sample of the 5.68 mmol/L aqueous ascorbic acid and deliver it to a 125 mL Erlenmeyer flask. Add 4 drops of starch indicator.

4. Titrate the sample with the iodine solution with a gentle swirling. The endpoint occurs when the indicator changes to blue. Record the final volume of titrant in the burette and the indicator colour.

5. Repeat Procedure steps 3 and 4 until you have three results in which the volumes of titrant added are the same to within ±0.10 mL. (Fill the burette with fresh iodine solution as needed.)

6. Fill the burette with fresh iodine solution and record the initial volume of titrant to the nearest 0.05 mL.

7. Rinse the 10.00 mL pipette with orange juice and then use it to obtain a 10.00 mL sample of the orange juice and deliver it to a 125 mL Erlenmeyer flask. Add 4 drops of starch indicator.

8. Titrate the sample with the iodine solution with a gentle swirling. The endpoint occurs when the indicator changes to blue; however, since the orange juice is orange, the endpoint will appear as a greenish hue in the orange juice. Record the final volume of titrant in the burette and the indicator colour.

9. Repeat Procedure steps 7 and 8 until you have three results in which the volumes of titrant added are the same to within ±0.10 mL. (Fill the burette with fresh iodine solution as needed.)

10. Dispose of all excess chemicals as directed by your teacher. Rinse the pipette and burette with deionized water. Leave the burette tap open.

Analysis

1. Balance the equation for the reaction of ascorbic acid with iodine. Use the method of your choice:

$$C_6H_8O_6(aq) + I_2(aq) \rightarrow C_6H_6O_6(aq) + I^-(aq) \text{ (acidic)}$$

2. Calculate the concentration of iodine in the iodine solution by using the average volume of iodine used in the first series of titrations.

3. Using your calculation from question 2, calculate the concentration of vitamin C in the orange juice. Use the average volume of iodine used in the second series of titrations.

Conclusion

4. Compare your results with the results of other students.

5. Compare your results with the concentration you calculated as a prediction.

6. List several possible sources of error in this investigation.

7. Manufacturers will often add more vitamin C to a product than what is listed on the label. Why might the manufacturer do this?

Extension

8. Vitamin C is an antioxidant commonly found in foods. Use the Internet to find other antioxidants found in foods. Determine how antioxidants work and how they are beneficial to the human body. ICT

Section 12.4 Summary

- You can perform titrations on redox reactions by using compounds that have different colours in their reduced and oxidized forms.

- The permanganate ion (MnO_4^-(aq), oxidized form) is purple and the manganese ion (Mn^{2+}, reduced form) is colourless.

- You can determine molar concentrations or mass percent of a compound using data from redox titrations.

- To apply stoichiometry to redox reactions, you must account for the number of moles of electrons that are exchanged by the reactants while forming products.

SECTION 12.4 REVIEW

1. The permanganate ion is an oxidizing agent. Explain why it is a good choice of oxidizing agent with which to perform a redox titration experiment.

2. Starch binds an iodine-iodide mixture to give a blue-black colour. Why is starch added to a solution of vitamin C when titrating the vitamin C with an iodine solution?

3. If you titrate orange juice that has been exposed to the air for a week, will the vitamin C concentration be different from the vitamin C concentration in fresh juice? If so, will it decrease or increase? Explain your prediction, in terms of redox reactions.

4. A chemist adds a few drops of deep violet-red iodine to a vitamin C tablet. The iodine solution quickly becomes colourless. Then the chemist adds a solution that contains chlorine, Cl_2(aq). The chemist observes that the violet-red colour of the iodine reappears. Explain the chemist's observations, in terms of redox reactions.

5. Oxygen and other oxidizing agents can react harmfully in your body. Vitamin C acts as an antioxidant: Because it is very easily oxidized, it reacts with oxidizing agents, preventing them from reacting with other important molecules in the body. Vitamin C is relatively stable when oxidized, therefore, it does not propagate a harmful series of oxidation reactions. One way to determine the vitamin C content of a sample is to titrate it with an iodine-iodide solution. The diagram below shows the reaction involved:

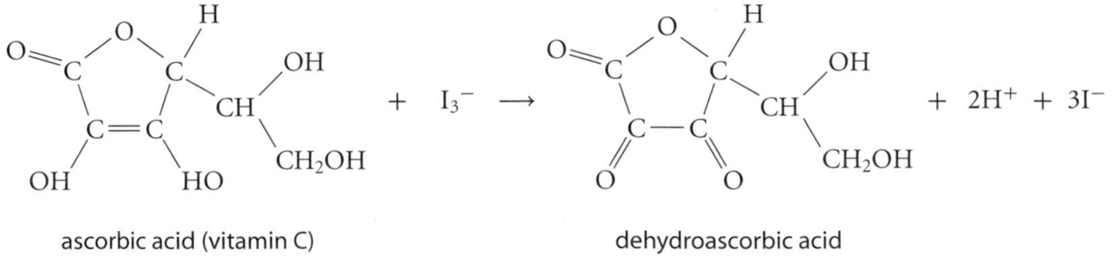

In the iodine solution, iodine, I_2(s), and iodide ions, I^-(aq), are in equilibrium with triiodide, as shown below:

$$I_2(aq) + I^-(aq) \rightleftharpoons I_3^-(aq)$$

Molecular iodine is a deep violet-red colour. Iodide ions are colourless. Thus, when an antioxidant reduces the iodine molecules in a solution, the iodine colour disappears completely.

a) In storage, the concentration of iodine in solution decreases fairly quickly over time. Why do you think this happens?

b) Because the iodine solution's concentration is unstable, it should be standardized frequently. To standardize an iodine solution, use it to titrate a solution that contains a known quantity of vitamin C. Explain how you would standardize a solution of iodine using vitamin C tablets from a pharmacy.

c) To standardize iodine using vitamin C tablets, you should use fresh tablets. Explain why.

ascorbic acid (vitamin C)   dehydroascorbic acid

Combustion reactions occur when atoms or compounds react violently with molecular oxygen. It is said that the compound is oxidized. However, other compounds or atoms can react with oxygen very slowly and generate imperceptible amounts of heat. These are also oxidation reactions. Chemists realized that the effects of oxidation on an atom could be achieved by reactions with elements other than oxygen. The modern definition of oxidation is therefore, the loss of electrons. The term reduction was originally applied to the reduction in the mass when ore was converted into pure metals. To extract a pure metal from ore, electrons must be added to the metal ions to convert them into atoms. Today, reduction of an atom means to add electrons to it.

Because electrons cannot exist free of atoms for any significant period of time, they must be passed from one atom to another. Therefore, if one atom is oxidized (loses electrons) in a chemical reaction, another atom must be reduced (gain electrons). Thus such reactions are called oxidation-reduction reactions or redox reactions.

Some redox reactions such as the reaction between zinc metal and copper(II) sulfate proceed spontaneously. Since the zinc atoms lost electrons they were oxidized by the copper(II) ions. Also, the copper(II) ions gained electrons and were reduced by the zinc atoms. Since the reaction was spontaneous, the zinc must be a stronger reducing agent than copper atoms and copper(II) ions must be a stronger oxidizing agent than zinc ions. By experimentally observing which reactions proceed spontaneously, you can list elements and ions in the order of their strength as reducing agents. This order allows you to predict the results of other reactions.

Many redox equations can be balanced by inspection but others, especially those that take place in acidic or basic conditions, are more difficult to balance by inspection. You can balance these reactions either by the half-reaction method or by the oxidation number method. The half-reaction method involves separating the reactions into oxidizing half-reactions and reduction half reactions, balancing the half-reactions according to a list of rules, and then adding the half-reactions. The oxidation number method involves assigning oxidation numbers to each element in each compound and then determining which atoms increased or decreased in oxidation number during the reactions. You use the increase and decrease in oxidation number to balance the number of electrons that were exchanged by the atoms. The remainder of the equations can often be balanced by inspection.

Iron ore is converted into partially pure pig iron by several reduction reactions. The pig iron is then refined by an oxidation reaction. The contaminants in the pig iron, such as silicon and phosphorous, are oxidized more easily than the iron, making it possible to remove them.

You can often determine the concentration of a compound in a sample of unknown concentration by performing a redox titration. To perform a redox titration, you react the compound of the unknown concentration with another compound that will either oxidize or reduce the unknown. Some compounds have different colours in their oxidized versus reduced forms and these colours provide you with a visible endpoint.

Concept Organizer · Identifying and Balancing Redox Reactions

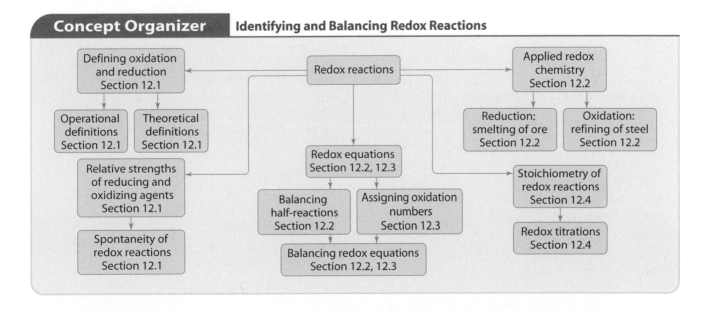

Understanding Concepts

1. For each reaction described below:
   a) write a balanced chemical equation by inspection
   b) write the ionic and net ionic equations
   c) identify the oxidizing agent and the reducing agent
   d) write the two half-reactions
      i) zinc metal with aqueous silver nitrate
      ii) aqueous cobalt(II) bromide with aluminium metal
      iii) metallic cadmium with aqueous tin(II) chloride

2. When a metallic element reacts with a non-metallic element, which reactant is
   a) oxidized?   c) the oxidizing agent?
   b) reduced?   d) the reducing agent?

3. Determine the oxidation number of each element present in the following substances:
   a) $BaCl_2(s)$ f) $S_8(s)$
   b) $Al_4C_3(s)$ g) $AsO_3^{3-}(s)$
   c) $KCN(s)$ h) $VO_2^+(s)$
   d) $LiNO_2(s)$ i) $XeO_3F^-(s)$
   e) $(NH_4)_2C_2O_4(s)$ j) $S_4O_6^{3-}(s)$

4. Explain why the historical use of the word *reduction*, that is, the production of a metal from its ore, is consistent with the modern definitions of reduction.

5. Determine which of the following balanced chemical equations represent redox reactions. For each redox reaction, identify the oxidizing agent and the reducing agent:
   a) $2C_6H_6(\ell) + 15O_2(g) \rightarrow 12CO_2(g) + 6H_2O(\ell)$
   b) $CaO(s) + SO_2(g) \rightarrow CaSO_3(aq)$
   c) $H_2(g) + I_2(s) \rightarrow 2HI(aq)$
   d) $KMnO_4(aq) + 5CuCl(s) + 8HCl(aq) \rightarrow$
      $KCl(aq) + MnCl_2(aq) + 5CuCl_2(s) + 4H_2O(\ell)$
   e) $2Ag^+(aq) + Cu(s) \rightarrow 2Ag(s) + Cu^{2+}(aq)$
   f) $Pb^{2+}(aq) + S^{2-}(aq) \rightarrow PbS(s)$
   g) $2Mn^{2+}(aq) + 5BiO_3^-(aq) + 14H^+(aq) \rightarrow$
      $2MnO_4^-(aq) + 5Bi^{3+}(aq) + 7H_2O(\ell)$

6. a) Examples of molecules and ions composed only of vanadium and oxygen are listed below. In this list, identify molecules and ions in which the oxidation number of vanadium is the same:
   $V_2O_5(s)$ $VO(s)$ $VO_3^-(s)$
   $V_2O_3(s)$ $VO_2^+(s)$ $VO_4^{3-}(s)$
   $VO_2(s)$ $VO^{2+}(s)$ $V_3O_9^{3+}(s)$
   b) Is the following reaction a redox reaction?
   $2NH_4VO_3(aq) \rightarrow V_2O_5(s) + 2NH_3(g) + H_2O(\ell)$

Applying Concepts

7. Identify a polyatomic ion in which chlorine has an oxidation number of +3.

8. If possible, give an example of each of the following:
   a) a formation reaction that is a redox reaction
   b) a formation reaction that is not a redox reaction
   c) a decomposition reaction that is a redox reaction
   d) a decomposition reaction that is not a redox reaction
   e) a double-replacement reaction that is a redox reaction
   f) a double-replacement reaction that is not a redox reaction

9. Phosphorus, $P_4(s)$, reacts with hot water to form phosphine, $PH_3(g)$, and phosphoric acid, $H_3PO_4(aq)$.
   a) Write a balanced chemical equation for this reaction.
   b) Is the phosphorus oxidized or reduced? Explain your answer.

10. The thermite reaction, which is highly exothermic, can be used to weld metals. In the thermite reaction, aluminium reacts with iron(III) oxide to form iron and aluminium oxide. The temperature becomes so high that the iron is formed as a liquid.
    a) Write a balanced chemical equation for the reaction.
    b) Is the reaction a redox reaction? If so, identify the oxidizing agent and the reducing agent.

11. Describe a laboratory investigation that you could perform to decide whether tin or nickel is the better reducing agent. Include in your description all the materials and equipment you would need, and describe the procedure you would follow.

12. The following table shows the average formation, by volume, of the air we inhale and exhale as part of a biochemical process called respiration. (The values are rounded.) How do the data indicate that at least one redox reaction is involved in respiration?

Gas	Inhaled Air (% by volume)	Exhaled Air (% by volume)
oxygen	21	16
carbon dioxide	0.04	4
nitrogen and other gases	79	80

13. Explain why you would not expect sulfide ions to act as an oxidizing agent.

Solving Problems

14. Use the half-reaction method to balance each of the following equations:

a) $MnO_2(s) + Cl^-(aq) \rightarrow Mn^{2+}(aq) + Cl_2(g)$
(acidic conditions)

b) $NO(g) + Sn(s) \rightarrow NH_2OH(aq) + Sn^{2+}(aq)$
(acidic conditions)

c) $Cd^{2+}(aq) + V^{2+}(aq) \rightarrow Cd(s) + VO_3^-(aq)$
(acidic conditions)

d) $Cr(s) \rightarrow Cr(OH)_4^-(aq) + H_2(g)$ (basic conditions)

e) $S_2O_3^{2-}(aq) + NiO_2(s) \rightarrow Ni(OH)_2(s) + SO_3^{2-}(aq)$
(basic conditions)

f) $Sn^{2+}(aq) + O_2(g) \rightarrow Sn^{4+}(aq)$ (basic conditions)

15. Use the oxidation number method to balance each of the following equations:

a) $SiCl_4(aq) + Al(s) \rightarrow Si(s) + AlCl_3(aq)$

b) $PH_3(aq) + O_2(g) \rightarrow P_4H_{10}(s) + H_2O(\ell)$

c) $I_2O_5(s) + CO(g) \rightarrow I_2(s) + CO_2(g)$

d) $SO_3^{2-}(aq) + O_2(g) \rightarrow SO_4^{2-}(aq)$

16. Complete and balance a net ionic equation for each of the following disproportionation reactions:

$NO_2(g) \rightarrow NO_2^-(aq) + NO_3(g)$ (acidic conditions)

$Cl_2(g) \rightarrow ClO^-(aq) + Cl^-(aq)$ (basic conditions)

a) Which is one of the reactions involved in acid rain formation?

b) Which is one of the reactions involved in the bleaching action of chlorine in a basic solution?

17. Balance each of the following net ionic equations. Then include the named spectator ions to write a balanced chemical equation:

a) $Co^{3+}(aq) + Cd(s) \rightarrow Co^{2+}(aq) + Cd^{2+}(aq)$
(spectator ions are $NO_3^-(aq)$)

b) $Ag^+(aq) + SO_2(g) \rightarrow Ag(s) + SO_4^{2-}(aq)$ (acidic conditions; spectator ions are $NO_3^-(aq)$)

c) $Al(s) + CrO_4^{2-}(aq) \rightarrow Al(OH)_3(s) + Cr(OH)_3(s)$
(basic conditions; spectator ions are $Na^+(aq)$)

18. Iodine reacts with concentrated nitric acid to form iodic acid, gaseous nitrogen dioxide, and water.

a) Write the balanced chemical equation.

b) Calculate the mass of iodine needed to produce 28.0 L of nitrogen dioxide at STP.

19. Highly toxic phosphine gas, $PH_3(g)$, is used in industry to produce flame retardants. One way to make phosphine on a large scale is by heating elemental phosphorus with a strong base.

a) Balance the following net ionic equation for the reaction under basic conditions:
$P_4(s) \rightarrow H_2PO_2^-(aq) + PH_3(g)$

b) Show that the reaction in part (a) is a disproportionation reaction.

c) Calculate the mass of phosphine that can theoretically be made from 10.0 kg of phosphorus by this method.

Making Connections

20. The compound $NaAl(OH)_2CO_3(s)$ is a component of some common stomach acid remedies.

a) Determine the oxidation number of each element in the compound.

b) Predict the products of the reaction of the compound with stomach acid (hydrochloric acid), and write a balanced chemical equation for the reaction.

c) Were the oxidation numbers from part (a) useful in part (b)? Explain your answer.

d) What type of reaction is this?

e) Check your medicine cabinet at home for stomach acid remedies. If possible, identify the active ingredient in each remedy.

21. Two of the substances on the head of a safety match are potassium chlorate and sulfur. When the match is struck, the potassium chlorate decomposes to give potassium chloride and oxygen. The sulfur then burns in the oxygen and ignites the wood of the match.

a) Write balanced chemical equations for the decomposition of potassium chlorate and for the burning of sulfur in oxygen.

b) Identify the oxidizing agent and the reducing agent in each reaction in part (a).

c) Does any element in potassium chlorate undergo disproportionation in the reaction? Explain your answer.

d) Research the history of the safety match to determine when it was invented, why it was invented, and what it replaced.

22. Ammonium ions, from fertilizers or animal waste, are oxidized by atmospheric oxygen. The reaction results in the acidification of soil on farms and the pollution of ground water with nitrate ions.

a) Write a balanced net ionic equation for this reaction.

b) Why do farmers use fertilizers? What alternative farming methods have you heard of? Which farming method(s) do you support and why?

CHAPTER
13

Cells and Batteries

Chapter Concepts

13.1 Voltaic Cells
- Voltaic cells transform chemical energy into electrical energy.
- The chemical reaction determines the cell voltage.

13.2 Applications of Voltaic Cells
- The common flashlight battery is a dry cell.
- Fuel cells offer a potentially clean form of energy for the future if they can be produced economically.

13.3 Electrolytic Cells
- Electrolysis is the process in which electrical energy is used to drive non-spontaneous redox reactions.
- You can use standard reduction potentials to calculate the potential difference needed to drive a specific reaction.
- Electrolysis is used to produce many industrially important products such as chlorine, sodium hydroxide, and aluminium.

13.4 Stoichiometry and Faraday's Law
- You can apply stoichiometric calculations to redox reactions after determining the charge on a mole of electrons.
- You can calculate the amount of time that a specific current is needed in an electrochemical cell to deposit a known amount of mass on an electrode.
- Some metals are refined by electrolysis of molten salts or salt solutions.

Batteries come in a wide range of sizes and shapes, from batteries large enough to power these golf carts to the tiny button batteries in watches. Many golf cart batteries generate 6 V, while watch batteries usually generate 1.5 V. Automobile batteries generate 12 V. Flashlight batteries come in many sizes (AAA to D) but they all provide 1.5 V. The small rectangular batteries that are used in some smoke alarms and in some small portable radios are 9 V batteries. There is no relationship between the voltage a battery generates and its size. What is the difference between a large and a small battery that provide the same voltage? What determines the voltage a battery generates? What is the relationship between batteries and the redox reactions you studied in Chapter 12? This chapter will help you find answers to these questions.

What Determines Voltage?

You might have seen or even built a lemon battery similar to the one in the photograph. In this activity, you will test some of the characteristics of lemon batteries. You will then use your observations to make predictions about commercial batteries.

Materials

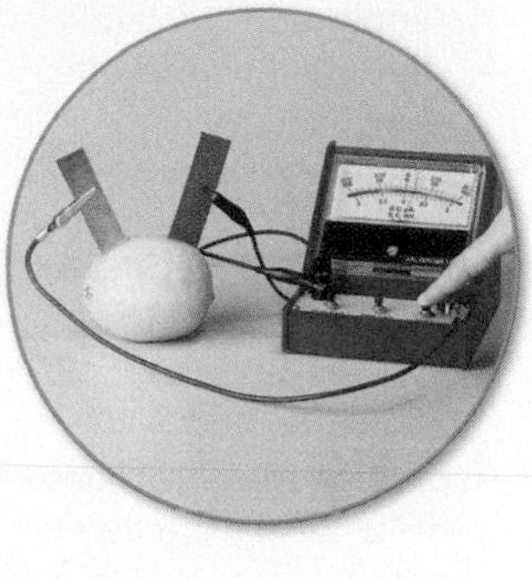

- 4 zinc strips (1 cm × 5 cm)
- 4 copper strips (1 cm × 5 cm)
- 4 lemons
- fine sandpaper
- 8 electrical leads with alligator clips
- voltmeter (high sensitivity)
- small flashlight bulb

Procedure

1. Clean the zinc and copper strips with the sandpaper. Roll each lemon on the table with your hand on top, pressing down on the lemon to break open the pockets of juice inside.

2. Insert one zinc strip and one copper strip into one lemon, as shown in the photograph. Attach one electrical lead to each strip by using the alligator clips, as shown. Connect the other end of each lead to the voltmeter. Read and record the voltage displayed by the voltmeter.

3. Disconnect the leads from voltmeter and connect them to the flashlight bulb. If you see any light, describe its intensity.

4. Insert one zinc strip and one copper strip into each of the lemons. Connect the strips as shown in the diagram labelled "series."

5. Connect the final leads to the voltmeter. Read and record the voltage. Repeat step 3.

6. Disconnect then reconnect the lemons as shown in the diagram labelled "parallel." Connect them to the voltmeter and then the light bulb, as shown. Record the voltage and describe the light intensity.

Analysis

1. Speculate about answers to the following questions.

   a) Which connection—single lemon, series lemons, or parallel lemons—produced the highest voltage?

   b) In which case (if any), did the system cause the flashlight bulb to produce the brightest light?

2. Speculate about the difference between a cell and a battery.

4. Why do you think that some batteries are larger than others?

5. As you study this chapter, look for answers to these questions. Compare your speculations with the answers that you find. Correct your answers.

Series

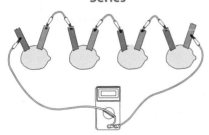

Parallel

Voltaic Cells

You learned in Chapter 12 that a zinc strip reacts with a solution containing copper(II) ions to form zinc ions and metallic copper. The reaction is spontaneous, meaning that occurs by itself, without an ongoing input of energy. In fact, the reaction releases energy in the form of heat—it is exothermic:

$$Zn(s) + Cu^{2+}(aq) \rightarrow Zn^{2+}(aq) + Cu(s) + heat$$

While analyzing your observations in the Launch Lab, you probably realized that similar reactions were taking place in the "solutions" inside the lemons. However, the copper and zinc strips were not in contact with each other inside the lemons. Also, the electrons appeared to be passing through the electrical conductors. You probably also know that a flow of electric charges, such as electrons, is called an **electric current**. These two concepts, redox reactions and electric current, form the basis of the field of **electrochemistry**: the study of the processes involved in converting chemical energy to electrical energy and in converting electrical energy to chemical energy.

The Voltaic Cell

A **voltaic cell** is a device that uses redox reactions, such as the reaction between copper(II) ions and zinc atoms, to transform chemical potential energy into electrical energy. The key to the operation of the voltaic cell is the prevention of direct contact between the reactants in a redox reaction. Instead of allowing the electrons to be transferred directly from zinc to copper, for example, the electrons must flow through an **external circuit**—a circuit outside the reaction vessel. The energy carried by the electric current can then be used for other purposes, such as lighting a light bulb.

The Daniell Cell

One example of a voltaic cell, called the Daniell cell, is shown in Figure 13.1. One half of the cell consists of a piece of zinc placed in a zinc sulfate solution. The other half of the cell consists of a piece of copper placed in a copper(II) sulfate solution. In the photograph, the zinc sulfate solution is in a porous cup that is sitting in the copper(II) sulfate solution. The cup forms a *porous barrier* that prevents the copper(II) ions from coming into direct contact with the zinc electrode. However, ions can gradually diffuse through the porous cup.

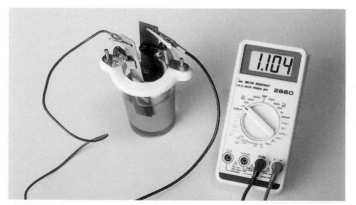

Figure 13.1 The Daniell cell is named after its inventor, the English chemist John Frederic Daniell (1790–1845). In the photograph shown here, the zinc sulfate solution is placed inside a porous cup, which is placed in a larger container of copper sulfate solution. The cup acts as the porous barrier.

In a Daniell cell, the strips of metallic zinc and copper act as electrical conductors. Conductors that carry electrons into and out of a cell are called **electrodes**. The zinc sulfate and copper(II) sulfate solutions act as electrolytes. **Electrolytes** are substances that, when dissolved in water, conduct electric current. The electric charges that flow in an aqueous solution are ions. The terms *electrode* and *electrolyte* were coined by the leading pioneer of electrochemistry, Michael Faraday (1791–1867).

As you know, if the zinc atoms in the zinc electrodes and the copper(II) ions in the copper sulfate solution were in direct contact, zinc atoms would spontaneously transfer electrons to the copper(II) ions. In the Daniell cell, the contact is established through the external circuit and the copper electrode. Zinc atoms still transfer electrons to the copper(II) ions but the electrons must travel through the circuit to do so. The electrons that are travelling through the circuit carry energy that can be transformed into other forms of energy rather than into heat. For example, the energy can be used to light a light bulb. The oxidation half-reactions occur in one half-cell and the reduction half-reactions occur in the other half-cell. The following half-reactions occur in the Daniell cell:

$$\text{oxidation (loss of electrons):} \quad Zn(s) \rightarrow Zn^{2+}(aq) + 2e^-$$
$$\text{reduction (gain of electrons):} \quad Cu^{2+}(aq) + 2e^- \rightarrow Cu(s)$$

In any voltaic cell, the electrode at which oxidation occurs is called the **anode**. In this Daniell cell, zinc atoms undergo oxidation at the zinc electrode; therefore, it is the anode. The electrode at which reduction occurs is called the **cathode**. Since copper(II) ions undergo reduction at the copper electrode in the Daniell cell, it is the cathode.

To visualize the resultant motion of the ions and electrons in a Daniell cell, examine Figure 13.2. This Daniell cell is somewhat different from the one in the photograph in Figure 13.1. Instead of separating the copper(II) sulfate and zinc sulfate solutions with a porous barrier, this cell separates the solutions with a **salt bridge**. The open ends of the salt bridge are plugged with a porous material, such as cotton or glass wool, to stop the electrolyte from flowing out. However, the ions can diffuse out slowly. The electrolyte solution in the salt bridge—in this case, it is KCl(aq)—is selected so that it does not interfere in the reaction. For example, if one of the solutions contained silver nitrate, you would not use KCl(aq) in the salt bridge because, when the chloride ions diffused into the silver nitrate solution, silver chloride would precipitate out of the solution.

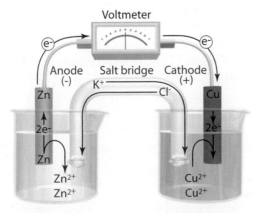

Figure 13.2 By convention, voltaic cells, such as this Daniell cell, are usually drawn with the anode on the left and the cathode on the right.

Chemistry File

Web Link
Voltaic cells are named after the Italian physicist Count Alessandro Giuseppe Antonio Anastasia Volta (1745–1827), who built the first chemical batteries. These cells are also called galvanic cells, after the Italian physician, Luigi Galvani (1737–1798). How did Galvani "discover" electrical cells while dissecting frog legs? How did Galvani's and Volta's interpretation of these observations differ?

www.albertachemistry.ca
WWW

Chemistry File

FYI
You may have noticed that biting on a piece of aluminum foil can be painful if it has come in contact with any metal, such as a filling, in your mouth. By having two dissimilar metals present in the moist, salty environment that exists in your mouth, you have set up a battery and an electric current is generated. This current gets conducted into the tooth's root, which causes the sensation of pain.

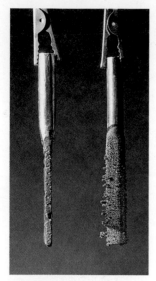

Figure 13.3 The mass of the zinc anode decreases as the zinc is oxidized and the ions go into solution. The mass of the copper cathode increases as the copper(II) ions are reduced and become copper atoms that become part of the electrode.

To track the motion of charges in the cell, start at the zinc electrode on the left in Figure 13.2. When the zinc atoms are oxidized, the electrons remain in the electrode while the zinc ions go into the solution of zinc sulfate. Consequently, the zinc electrode (anode) becomes negatively charged and the zinc sulfate solution near the electrode becomes positively charged. At the same time, at the copper electrode, copper(II) ions in the solution are being reduced by combining with electrons in the copper electrode. The newly formed copper atoms remain as part of the electrode. As a result, the copper electrode (cathode) is positively charged, while the copper(II) sulfate solution near the electrode becomes negatively charged through the loss of the positive ion. Now the electrical conductor—consisting of the zinc electrode, connecting wire conductor, and the copper electrode—is negatively charged on one end and positively charged on the other end. Consequently, electrons flow away from left to right—from the negatively charged anode to the positively charged cathode.

The copper and zinc sulfate solutions are also oppositely charged and are connected by a salt bridge containing an electrolyte. The negative chloride ions are attracted to the positively charged zinc sulfate solution and the positively charged potassium ions are attracted negatively charged copper(II) sulfate solution.

Current continues to flow until the concentrations of the solutions and the changes in the electrodes are so great that the processes can no longer continue. For example, as copper accumulates on the copper electrode, the concentrations of copper(II) ions in the solution decreases. As well, potassium ions are accumulating in the solution, and the zinc electrode is dissolving as zinc ions are formed and go into solution. Figure 13.3 shows electrodes after they have been used for a long period.

• • •

Q 1 How can zinc atoms reduce copper(II) ions when the two metals are not in contact?

Q 2 Explain the function of the porous cup or the salt bridge in a voltaic cell.

Q 3 How do the anode and cathode in a voltaic cell differ?

• • •

Types of Electrodes

Voltaic cells can be made with a wide variety of electrodes other than the copper and zinc of the Daniell cell. For example, Figure 13.4 shows a voltaic cell with chromium and silver electrodes. Notice that the electrolyte in the salt bridge is potassium nitrate rather than potassium chloride. Although the chloride ions would, in general, migrate to the chromium(III) nitrate cell, a small amount could diffuse into the silver nitrate cell and form a precipitate with the silver ions. It is always best to avoid such possibilities.

Figure 13.4 Chromium is the anode and silver is the cathode because chromium is a stronger reducing agent than silver.

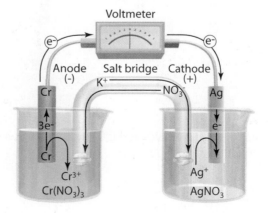

Inert Electrodes

The zinc anode and copper cathode of a Daniell cell and the silver and chromium electrodes in Figure 13.4 are all metals and can act as electrical conductors. However, some redox reactions involve oxidizing and reducing agents that are not solid metals but, instead, are dissolved electrolytes or gases and, therefore, cannot be used as electrodes. To construct a voltaic cell that will use these oxidizing and reducing agents, you have to use inert electrodes. An **inert electrode** is an electrode made from a material that is neither a reactant nor a product of the redox reaction. Instead, the inert electrode can carry a current and provide a surface on which redox reactions can occur. Figure 13.5 shows a cell that contains one example of an inert electrode—a platinum electrode. The complete balanced equation, net ionic equation, and half-reactions for this cell are given below.

complete balanced equation:	$Pb(s) + 2FeCl_3(aq) \rightarrow 2FeCl_2(aq) + PbCl_2(aq)$
net ionic equation:	$Pb(s) + 2Fe^{3+}(aq) \rightarrow 2Fe^{2+}(aq) + Pb^{2+}(aq)$
oxidation half-reaction:	$Pb(s) \rightarrow Pb^{2+}(aq) + 2e^-$
reduction half-reaction:	$Fe^{3+}(aq) + e^- \rightarrow Fe^{2+}(aq)$

The anode is the lead electrode. Lead atoms lose electrons that remain in the electrode while the lead(II) ions dissolve in the solution in the same way that the anode did in previous example. However, the reduction half-reaction involves iron(III) ions that accept an electron from the platinum inert electrode and become iron(II) ions. The platinum atoms in the electrode (cathode) remain unchanged.

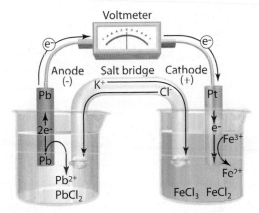

Figure 13.5 This cell uses an inert electrode to conduct electrons. Why do you think that platinum is often chosen as an inert electrode? Another common choice is graphite.

Cell Notation

Sketching entire cells helps you to visualize and learn about the movement of ions and electrons in a voltaic cell. However, it becomes tedious when you are trying to communicate quickly. Chemists have therefore developed a shorthand method for representing voltaic cells called a **cell notation**. The cell notation of a Daniell cell is as follows:

$$Zn(s) \mid Zn^{2+}(aq) \parallel Cu^{2+}(aq) \mid Cu(s)$$

Chemistry File

FYI

The cell notation is correctly called a cell diagram. However, some people confuse the term cell diagram with a drawing of a complete cell. Therefore, this text uses the term cell notation.

In the cell notation, as in the sketch of the entire cell, the anode is usually shown on the left and the cathode on the right. Each single vertical line, |, represents a phase boundary between the electrode and the solution in a half-cell. For example, the first single vertical line shows that the solid zinc and aqueous zinc ions are in different phases or states—solid and aqueous. The double vertical line, ||, represents the porous barrier or salt bridge between the half-cells.

Inert electrodes, such as the platinum electrode in the cell shown in Figure 13.5, do not appear in the chemical equation or in half-reactions. However, they are included in the cell notation. The cell notation for the cell in Figure 13.5 is shown here.

$$Pb(s) \mid Pb^{2+}(aq) \parallel Fe^{3+}(aq), Fe^{2+}(aq) \mid Pt(s)$$

A comma separates the formulas Fe^{3+} and Fe^{2+} ions involved in the reduction half-reaction to indicate that they are in the same phase—the aqueous phase.

4 **a)** If the reaction of zinc with copper(II) ions is carried out in a test tube, what is the oxidizing agent and what is the reducing agent?

b) In a Daniell cell, what is the oxidizing agent and what is the reducing agent? Explain your answer.

5 Sketch voltaic cells for each of the following cell notations. Label all parts of the sketches. Below the sketch, write the oxidation half-reaction, the reduction half-reaction, and the overall cell reaction. Identify the anode and the cathode in each case. Identify any inert electrodes. Explain why there is a line instead of a comma between the symbols for the hydrogen ions and the hydrogen gas in reaction (b).

a) $Sn(s) \mid Sn^{2+}(aq) \parallel Tl^{+}(aq) \mid Tl(s)$

b) $Cd(s) \mid Cd^{2+}(aq) \parallel H^{+}(aq) \mid H_2(g) \mid Pt(s)$

6 A voltaic cell involves the overall reaction of iodide ions with acidified permanganate ions to form manganese(II) ions and iodine. The salt bridge contains potassium nitrate. Both electrodes are inert and are made of graphite.

a) Write the half-reactions and the overall cell reaction.

b) Identify the oxidizing agent and the reducing agent.

c) Solid iodine forms on one of the electrodes. Does it form on the anode or the cathode? Explain.

d) Sketch and label the entire voltaic cell.

7 As you saw earlier, pushing a zinc electrode and a copper electrode into a lemon makes a "lemon cell." In the following representation of the cell, $C_6H_8O_7$ is the formula for citric acid. Explain why the representation does not include a double vertical line:

$$Zn(s) \mid C_6H_8O_7(aq) \mid Cu(s)$$

Cell Potentials

You read that the oxidation reaction at the anode causes the anode to become negatively charged and the reduction reaction that occurs at the cathode causes the cathode to become positively charged. Whenever a separation of charge exists, a force acts on charged particles. Therefore, they have potential energy. The amount of energy conferred on a unit of charge when it moves from one point to another is called the **electrical potential difference** between those two points. In the case of voltaic cells, these two points are the two electrodes. In the sketches of the voltaic cells, you saw a voltmeter. A voltmeter measures the potential difference (also called **voltage**). The magnitude of the electrical potential difference between the electrodes in a voltaic cell is an important property of a cell because it determines how much energy can be conferred on the moving charges. For any given cell, the electrical potential difference between the electrodes is called the **cell potential**.

The cell potential depends on many factors, including the nature of the oxidizing and reducing agents, the concentrations of the salt solutions in the half-cells, the temperature of the solutions, and the atmospheric pressure. Chemists have agreed on a set of standard conditions for reporting cell potentials. The **standard cell potential**, symbolized by $E°_{cell}$, is the potential difference between the electrodes of a voltaic cell when the concentrations of the salt solutions are 1.0 mol/L, the atmospheric pressure is 1.01325×10^5 Pa (1.0 atm), the electrodes are pure metals, the temperature is 25 °C, and there is no current flowing in the cell.

You can measure a standard cell potential for any combination of oxidizing or reducing agents. However, it would be convenient to be able to describe half-cell potentials. Because any type of electrical potential difference always reports a *difference* between two points, you can define a half-cell potential only by choosing an arbitrary reference point. This situation is analogous to defining an altitude. When you note the altitude at which an airplane is flying, you are using sea level as a reference point. You are not referring to the distance that the airplane is above the ground. For example, when you land at the Calgary airport, you are at an altitude of more than 1000 m. The altitude at sea level is *defined* as zero and all other altitudes are measured in relation to sea level. Chemists have thus agreed on a reference half-cell to which all other half-cells are related. This reference point is defined as the half-reaction between hydrogen gas and hydrogen ions, as shown:

$$H_2(g) \rightarrow 2H^+(aq) + 2e^-$$

Since hydrogen is a gas, standard conditions are defined as a platinum electrode immersed in a 1.0 mol/L solution of a monoprotic acid with hydrogen gas being bubbled past the electrode at 1.0 atm pressure (see Figure 13.6). The half-cell potential of the hydrogen half-cell is defined as zero. The half-cell potential of all other cells is defined as the potential difference between the hydrogen half-cell and the chosen cell.

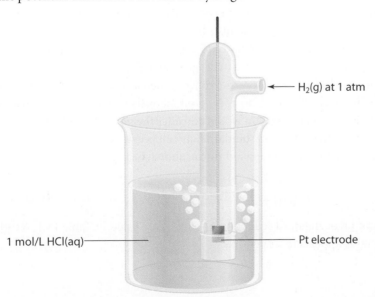

H₂(g) at 1 atm

1 mol/L HCl(aq)

Pt electrode

Figure 13.6 The potential of this standard hydrogen half-cell is arbitrarily assigned to be zero. This half-cell is the reference against which all other half-cell potentials are measured.

The hydrogen half-cell can be either the reducing half-cell or the oxidizing half-cell, depending on whether hydrogen is a stronger or weaker reducing agent than the compound in the opposite half-cell. Figure 13.7 shows that the hydrogen half-cell is the reducing half-cell (cathode) when connected to a zinc half-cell and is the oxidizing half-cell (anode) when connected to the copper half-cell.

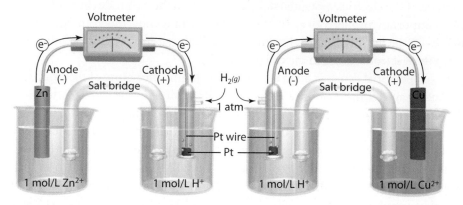

Figure 13.7 Zinc is a stronger reducing agent than hydrogen is, therefore, the hydrogen electrode is the cathode when connected to a zinc electrode. Hydrogen is a stronger reducing agent than copper is, so the hydrogen electrode is the anode when connected to the copper electrode.

If you used an apparatus, such as the one in Figure 13.7, and measured the standard cell potentials of the two cells, you would find that the cell potential for the first cell is 0.76 V. The cell notation for that cell is shown here:

$$Zn(s) \mid Zn^{2+}(aq) \parallel H^+(aq) \mid H_2(g) \mid Pt(s)$$

Since the hydrogen half-cell potential is defined as zero, the zinc half-cell potential is 0.76 V. Similarly, if you measured the cell potential of the second cell, you would find a potential difference of 0.34 V. The cell notation is shown here:

$$H^+(aq) \mid H_2(g) \mid Pt(s) \parallel Cu^{2+}(aq) \mid Cu(s)$$

Notice that, in the first cell, zinc is being oxidized and, in the second, cell copper(II) ions are being reduced. To be consistent and to make calculations more uniform, half-cell potentials should all be reported for either oxidation or reduction.

Standard Reduction Potential

The standard cell potentials for a large number of compounds have been measured and the values listed in tables. Chemists have generally agreed to list the half-reactions as reduction reactions. How, then, would you write a reduction half-cell potential for a compound that is not reduced by a hydrogen half-cell, such as the zinc half-cell?

As you saw in Figure 13.7, zinc is oxidized when a zinc half-cell is connected to a hydrogen half-cell. The reaction goes spontaneously in a direction opposite to the reduction of zinc. Therefore, you would consider the potential difference to be negative. You would write the reduction standard half-cell potential, called the **standard reduction potential**, for copper and zinc as shown below:

$$Zn^{2+}(aq) + 2e^- \rightarrow Zn(s) \qquad E° = -0.76\,V$$
$$Cu^{2+}(aq) + 2e^- \rightarrow Cu(s) \qquad E° = +0.34\,V$$

More examples of standard reduction potentials are listed in Table 13.1. A more complete table appears in Appendix G.

Table 13.1 Standard Reduction Potentials (298 K, 1 atm)

	Half-reaction		$E°$ (V)
	$F_2(g) + 2e^- \rightarrow 2F^-(aq)$		+2.87
	$Br_2(\ell) + 2e^- \rightarrow 2Br^-(aq)$		+1.07
	$Ag^+(aq) + e^- \rightarrow Ag(s)$		+0.80
	$I_2(s) + 2e^- \rightarrow 2I^-(aq)$		+0.54
	$Cu^{2+}(aq) + 2e^- \rightarrow Cu(s)$		+0.34
	$2H^+(aq) + 2e^- \rightarrow H_2(g)$		0.00
	$Fe^{2+}(aq) + 2e^- \rightarrow Fe(s)$		−0.45
	$Cr^{3+}(aq) + 3e^- \rightarrow Cr(s)$		−0.74
	$Zn^{2+}(aq) + 2e^- \rightarrow Zn(s)$		−0.76
	$Al^{3+}(aq) + 3e^- \rightarrow Al(s)$		−1.66
	$Na^+(aq) + e^- \rightarrow Na(s)$		−2.71

Increasing strength as oxidizing agent (left margin, arrow pointing up)

Increasing strength as reducing agent (middle, arrow pointing down)

If you compare the sequence of the metals in Table 13.1 with the table of oxidizing and reducing agents in Table 12.2, you will discover that the sequences are the same. Table 12.2 showed the relative strengths of the ions and atoms as oxidizing or reducing agents. In this table, however, you have numerical values that tell you how much stronger the atoms, ions, or compounds are as oxidizing or reducing agents.

Another important application of these data is in choosing the salts for a salt bridge in a voltaic cell. Re-examine the Daniell cell as it is reproduced in Figure 13.8. Suppose, for example, that you had chosen silver nitrate, $AgNO_3(aq)$, for the salt bridge. Silver, rather than potassium, would diffuse into the copper(II) sulfate solution. Silver ions are stronger oxidizing agents than are copper(II) ions, so the silver ions would migrate to the copper anode and accept the electrons in place of copper(II) ions. As a result, the reading on the voltmeter would be incorrect.

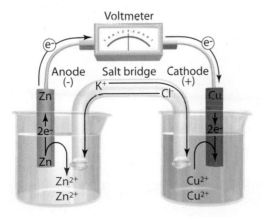

Figure 13.8 When potassium diffuses into the copper(II) sulfate solution, it has no effect on the reduction reaction of copper. However, any ion that has a more positive standard reduction potential than does copper(II) will displace the copper(II) ions at the cathode.

Calculating Standard Cell Potentials

You can use Table 13.1 to calculate the standard cell potential, from the formula below, of any cell for which the standard reduction potentials of the oxidizing and reducing agents are listed on a table.

$$E°_{cell} = E°_{cathode} - E°_{anode}$$

$E°_{cell}$: standard cell potential
$E°_{cathode}$: standard reduction potential for the reduction (cathode) half-reaction
$E°_{anode}$: standard reduction potential for the oxidation (anode) half-reaction

A positive standard cell potential tells you that the reaction will proceed spontaneously, in the direction indicated. A reaction will always proceed spontaneously if the reducing agent on the left is stronger than the reducing agent on the right. These concepts are summarized below, using the reaction between the zinc and copper(II) ion.

$Zn(s)$	+	$Cu^{2+}(aq)$	$\rightarrow$	$Zn^{2+}(aq)$	+	$Cu(s)$
stronger reducing agent than $Cu(s)$		stronger oxidizing agent than $Zn^{2+}(aq)$		weaker oxidizing agent than $Cu^{2+}(aq)$		weaker reducing agent than $Zn(s)$

If your calculated standard cell potential is negative, the reaction will not proceed spontaneously in the direction indicated. In fact, it will go in the reverse direction.

Study the following Sample Problems for calculating standard cell potentials and then complete the Practice Problems that follow.

Sample Problem

Calculating a Standard Cell Potential, Given a Net Ionic Equation

Problem

Calculate the standard cell potential for the voltaic cell in which the following reaction occurs:

$2I^-(aq) + Br_2(\ell) \rightarrow I_2(s) + 2Br^-(aq)$

What Is Required?

You need to find the standard cell potential for the reaction above between iodide ions and liquid bromine.

What Is Given?

You have the balanced net ionic equation and a table of standard reduction potentials.

Plan Your Strategy

- Write the oxidation and reduction half-reactions.

- Locate the relevant reduction potentials in a table of standard reduction potentials.

- Subtract the reduction potentials to find the cell potential by using the formula

$$E^{\circ}{}_{cell} = E^{\circ}{}_{cathode} - E^{\circ}{}_{anode}$$

Act on Your Strategy

Oxidation half-reaction (occurs at the anode):

$2I^-(aq) \rightarrow I_2(s) + 2e^-$

Reduction half-reaction (occurs at the cathode):

$Br_2(\ell) + 2e^- \rightarrow 2Br^-(aq)$

The relevant reduction potentials in the table of standard reduction potentials are as follows:

$I_2(s) + 2e^- \rightarrow 2I^-(aq)$	$E^{\circ}{}_{anode} = +0.54\ V$
$Br_2(\ell) + 2e^- \rightarrow 2Br^-(aq)$	$E^{\circ}{}_{cathode} = +1.07\ V$

Calculate the cell potential.

$E^{\circ}{}_{cell} = E^{\circ}{}_{cathode} - E^{\circ}{}_{anode}$
$E^{\circ}{}_{cell} = 1.07\ V - 0.54\ V$
$E^{\circ}{}_{cell} = +0.53\ V$

Sample Problem

Calculating a Standard Cell Potential, Given a Chemical Reaction

Problem

Calculate the standard cell potential for the voltaic cell in which the following reaction occurs:

$2Na(s) + 2H_2O(\ell) \rightarrow 2NaOH(aq) + H_2(g)$

(**Note:** Sodium reacts violently with water so these reactants could not be used to construct an electrochemical cell. Nevertheless, you can still calculate a "cell" potential for the chemical reaction.)

Solution

Write the equation in ionic form to identify the half-reactions:

$2Na(s) + 2H_2O(\ell) \rightarrow 2Na^+(aq) + 2OH^-(aq) + H_2(g)$

Write the oxidation and reduction half-reactions.

Oxidation half-reaction (occurs at the anode):

$Na(s) \rightarrow Na^+(aq) + e^-$

Reduction half-reaction (occurs at the cathode):

$2H_2O(\ell) + e^- \rightarrow 2OH^-(aq) + H_2(g)$

Locate the relevant reduction potentials in a table of standard reduction potentials.

$Na^+(aq) + e^- \rightarrow Na(s)$ $\quad\quad\quad E^\circ_{anode} = -2.71$ V

$2H_2O(\ell) + 2e^- \rightarrow 2OH^-(aq) + H_2(g)$ $\; E^\circ_{cathode} = -0.83$ V

Subtract the standard reduction potentials to calculate the cell potential.

$E^\circ_{cell} = E^\circ_{cathode} - E^\circ_{anode}$

$E^\circ_{cell} = -0.83$ V $- (-2.71$ V$)$

$E^\circ_{cell} = +1.88$ V

The standard cell potential is $+1.88$ V.

Practice Problems

(**Note:** Obtain the necessary standard reduction potential values from Table 13.1 or the table in Appendix G.) Write the two half-reactions for the following redox reactions. Find the standard cell potentials for voltaic cells in which these reactions occur.

1. $Cl_2(g) + 2Br^-(aq) \rightarrow 2Cl^-(aq) + Br_2(\ell)$
2. $Mg(s) + 2AgNO_3(aq) \rightarrow Mg(NO_3)_2(aq) + 2Ag(s)$
3. $Sn(s) + 2HBr(aq) \rightarrow SnBr_2(aq) + H_2(g)$
4. $Cr(s) + 3AgCl(s) \rightarrow CrCl_3(aq) + 3Ag(s)$

You have learned that the standard hydrogen electrode has an assigned standard reduction potential of exactly 0 V and is the reference for all half-cell standard reduction potentials. What would happen to cell potentials if a different reference were used? You will address this question in the following Thought Lab.

Thought Lab 13.1 Assigning Reference Values

Target Skill

Explaining that values of standard reduction potentials are all relative

Many scales of measurement have zero values that are arbitrary. For example, on Earth, average sea level is often assigned as the zero of altitude. In this Thought Lab, you will investigate what happens to calculated cell potentials when the reference half-cell is changed.

Procedure

1. Copy the following table of reduction potentials into your notebook. Choose the half reaction for Al^{3+} and Al as your reference point and assign a value of 0 V for this half reaction. To make the standard cell potential for the Al^{3+}/Al half-reaction equal to zero, you would have to add 1.66 V to the accepted standard reduction potential. To adjust all the reduction potentials to the new reference, you add 1.66 V to each value.

Reduction half-reaction	Accepted E° (V)	Adjusted E°(V) [+ 1.66 (V)]
$F_2(g) + 2e^- \rightarrow 2F^-(aq)$	+2.87	
$Fe^{3+}(aq) + e^- \rightarrow Fe^{2+}(aq)$	+0.77	
$2H^+(aq) + 2e^- \rightarrow H_2(g)$	0.00	
$Al^{3+}(aq) + 3e^- \rightarrow Al(s)$	−1.66	
$Li^+(aq) + e^- \rightarrow Li(s)$	−3.04	

2. Use the given standard reduction potentials to calculate the standard cell potentials for the following redox reactions:

a) $2Li(s) + 2H^+(aq) \rightarrow 2Li^+(aq) + H_2(g)$

b) $2Al(s) + 3F_2(g) \rightarrow 2Al^{3+}(aq) + 6F^-(aq)$

c) $2FeCl_3(aq) + H_2(g) \rightarrow 2FeCl_2(aq) + 2HCl(aq)$

d) $Al(NO_3)_3(aq) + 3Li(s) \rightarrow 3LiNO_3(aq) + Al(s)$

3. Repeat your calculations using the new, adjusted reduction potentials.

Analysis

1. Compare your calculations from Procedures steps 2 and 3. What effect does changing the zero on the scale of reduction potentials have on

a) reduction potentials?

b) cell potentials?

2. Find the difference between the temperatures at which water boils and freezes on the following scales (assume that a difference is positive, rather than negative):

a) the Celsius temperature scale

b) the Kelvin temperature scale

3. What do your answers to question 2 tell you about these two temperature scales?

4. How are temperature scales and reduction potentials similar?

5. The zero on a scale of masses is not arbitrary. Why not?

Measuring Cell Potentials of Voltaic Cells

In this investigation, you will build some voltaic cells and measure their cell potentials

Question

What factors affect the cell potential of a voltaic cell?

Prediction

Predict the values of the cell potentials for the cells that you build. Give reasons for your predictions. For those measured cell potentials that are not approximately the same as your predictions, do you think the values will be larger or smaller than your predicted values? Explain your reasoning.

Safety Precautions

- Handle the nitric acid solution with care.
- Immediately wash any spills on your skin with copious amounts of water. Inform your teacher.
- Wash your hands when you have completed the investigation.

Materials

- 1 Styrofoam™ or clear plastic egg carton with 12 wells
- 5 mL of 0.1 mol/L solutions of each of the following:
  - $Mg(NO_3)_2(aq)$
  - $Cu(NO_3)_2(aq)$
  - $Al(NO_3)_3(aq)$
  - $Zn(NO_3)_2(aq)$
  - $SnSO_4(aq)$
  - $Fe(NO_3)_3(aq)$
  - $AgNO_3(aq)$
  - $HNO_3(aq)$ 🔥 ☣
- 5 mL of saturated NaCl(aq) solution
- 5 cm strip of Mg ribbon
- 1 cm × 5 cm strips of each of the following: Cu(s), Al(s), Zn(s), Sn(s), Fe(s), and Ag(s)
- 15 mL of 1.0 mol/L KNO_3
- 5 cm of thick graphite pencil lead or a graphite rod
- sandpaper
- 25 cm clear aquarium rubber tubing (Tygon®; internal diameter: 4–6 mm)
- cotton batting
- disposable pipette
- black and red electrical leads with alligator clips
- voltmeter set to a scale of 0 V to 20 V
- paper towel
- marker

Procedure

1. Use tape or a permanent marker to label the outside of nine wells of your egg carton with the nine different half-cells. Each well should correspond to one of the eight different metal|metal ion pairs: $Mg(s)|Mg^{2+}(aq)$, $Cu(s)|Cu^{2+}(aq)$, $Al(s)|Al^{3+}(aq)$, $Zn(s)|Zn^{2+}(aq)$, $Sn(s)|Sn^{2+}(aq)$, $Fe(s)|Fe^{3+}(aq)$, and $Ag(s)|Ag^+(aq)$. Label the ninth well $H^+(aq)|H_2(g)$.

2. Prepare an 8 × 8 grid in your notebook. Label the nine columns to match the nine half-cells. Label the nine rows in the same way. You will use this chart to record the cell potentials you obtain when you connect two half-cells to build a voltaic cell. You will also identify which half-cell contains the anode and which half-cell contains the cathode for each voltaic cell you build. (You may not need to fill out the entire chart.)

3. Sand each of the metals to remove any oxides.

4. Pour 5 mL of each metal salt solution into the appropriate well of the egg carton. Pour 5 mL of the nitric acid into the well labelled $H^+|H_2$.

5. Prepare your salt bridge as follows.

   a) Roll a small piece of cotton batting so that it forms a plug about the size of a grain of rice. Place the plug in one end of your aquarium tubing, but leave a small amount of the cotton hanging out, so you can remove the plug later.

   b) Fill a disposable pipette as full as possible with the 1 mol/L $KNO_3(aq)$ electrolyte solution. Fit the tip of the pipette firmly into the open end of the tubing. Slowly inject the electrolyte solution into the tubing. Fill the tubing completely, so that the cotton on the other end becomes wet.

   c) With the tubing completely full, insert another cotton plug into the other end. There should be no air bubbles. (You will have to repeat this step from the beginning if you have air bubbles.)

6. Insert each metal strip into the corresponding well. Place the graphite rod in the well with the nitric acid. The metal strips and the graphite rod are your electrodes. (**Note:** The graphite electrode is very fragile. Be gentle when using it.)

7. Attach the alligator clip on the red lead to the red probe of the voltmeter. Attach the black lead to the black probe.

8. Choose two wells to test. Insert one end of the salt bridge into the solution in the first well. Insert the other end of the salt bridge into the solution in the second well. Attach a free alligator clip to the electrode in each well. You have built a voltaic cell.

9. If you get a negative reading, switch the alligator clips. Once you obtain a positive value, record it in your chart. The black lead should be attached to the anode (electrons flowing into the voltmeter). Record which metal is acting as the anode and which is acting as the cathode in this voltaic cell.

10. Remove the salt bridge and wipe any excess salt solution off the outside of the tubing. Remove the alligator clips from the electrodes.

11. Repeat procedure steps 8 to 10 for all other combinations of electrodes. Record your results.

12. Re-attach the leads to the silver and magnesium electrodes, and insert your salt bridge back into the two appropriate wells. While observing the reading on the voltmeter, slowly add 5 mL of saturated $NaCl(aq)$ solution to the $Ag(s)|Ag^+(aq)$ well to precipitate $AgCl(s)$. Record any changes in the voltmeter reading. Observe the $Ag(s)|Ag^+(aq)$ well.

13. Rinse off the metals and the graphite rod with water. Dispose of the salt solutions into the heavy metal salts container your teacher has set aside. Rinse out your egg carton. Remove and discard the plugs of the salt bridge, and dispose of the $KNO_3(aq)$ solution in the heavy metal salts container because it is contaminated with heavy metals. Return all materials to their appropriate locations.

Analysis

1. For each cell in which you measured a cell potential, identify
   a) the anode and the cathode
   b) the positive and negative electrodes

2. For each cell in which you measured a cell potential, write a balanced equation for the reduction half-reaction, the oxidation half-reaction, and the overall cell reaction.

3. For any one cell in which you measured a cell potential, describe
   a) the direction in which electrons flow through the external circuit
   b) the movements of ions in the cell

4. Use your observations to decide which of the metals used as electrodes is the most effective reducing agent. Explain your reasoning.

5. List all the reduction half-reactions you wrote in question 2 so that the metallic elements in the half-reactions appear in order of their strength as reducing agents. Put the least effective reducing agent at the top of the list and the most effective reducing agent at the bottom.

6. In which part of your list from question 5 are the metal ions that are the best oxidizing agents? Explain.

7. How well did your observed cell potentials fit your predicted cell potentials?

8. Were your measured values of cell potentials larger or smaller than your predictions? Does this observation fit with your predictions?

9. List some possible factors that might have caused your measured values to differ from your predicted values for cell potentials.

10. a) When saturated sodium chloride solution was added to the silver nitrate solution (in Procedure step 12), what reaction took place? Explain.
    b) Does the concentration of an electrolyte affect the cell potential of a voltaic cell? How do you know?
    c) How do your observations about concentration help you to explain your answers to questions 8 and 9?

Conclusion

11. Identify factors that affect the cell potential of a voltaic cell.

12. Explain the difference between "cell potential" and "standard cell potential."

Application

13. Predict any factors, other than those you were able to observe, that you think might affect the cell potential of a voltaic cell. Describe an investigation you could use to test your prediction.

Section 13.1 Summary

- A voltaic cell is a device that uses the energy from spontaneous redox reactions to generate an electrical potential difference, or voltage.

- A Daniell cell is an example of a voltaic cell that uses zinc and copper electrodes in zinc sulfate and copper sulfate electrolytes. Electrons are spontaneously donated by zinc atoms to copper ions. The electrons are caused to flow through an external circuit.

- In any voltaic cell, the oxidation half-reaction occurs at the anode and the reduction half-reaction occurs at the cathode.

- The anode loses mass and the cathode gains mass while the reactions are proceeding.

- If the oxidizing or reducing agent is not a solid, inert electrodes are used to provide a surface on which the reactions can occur and also carry the electrons to and from the external circuit.

- A cell notation is a shorthand method for describing voltaic cells. Atoms or ions that are in different phases (solid, liquid, gas) are separated by one vertical line. The half cells are separated by a double vertical line. Ions in the same phase are separated by a comma in the cell notation.

- The cell potential is the potential difference between the two electrodes that is generated by the chemical reactions.

- The hydrogen half-cell was selected as the reference to which to measure all other half cells.

- The standard reduction potential of a half-cell is the potential difference between any chosen half-cell and the hydrogen half-cell under conditions of one atmosphere of pressure, 25 °C, 1.0 mol/L concentrations of electrolyte solutions and hydrogen gas is pumped into the half-cell at one atmosphere of pressure.

- The standard cell potential can be calculated by using the formula:

$$E°_{cell} = E°_{cathode} - E°_{anode}$$

SECTION 13.1 REVIEW

1. To construct a voltaic cell, why is it necessary to prevent any direct contact between the reactants in a redox reaction?

2. It is possible to measure and describe a voltage between two electrodes. How, then, is it possible to define half-cell potentials?

3. Determine the standard cell potential for each of the following redox reactions:
   a) $3Mg(s) + 2Al^{3+}(aq) \rightarrow 3Mg^{2+}(aq) + 2Al(s)$
   b) $2K(s) + F_2(g) \rightarrow 2K^+(aq) + 2F^-(aq)$
   c) $Cr_2O_7^{2-}(aq) + 14H^+(aq) + 6Ag(s)$
      $\rightarrow 2Cr^{3+}(aq) + 6Ag^+(aq) + 7H_2O(\ell)$

4. Determine the standard cell potential for each of the following redox reactions:
   a) $CuSO_4(aq) + Ni(s) \rightarrow NiSO_4(aq) + Cu(s)$
   b) $4Au(OH)_3(s) \rightarrow 4Au(s) + 6H_2O(\ell) + 3O_2(g)$
   c) $Fe(s) + 4HNO_3(aq) \rightarrow Fe(NO_3)_3(aq) + NO(g) + 2H_2O(\ell)$

5. Look at the half-cells in the table of standard reduction potentials in Appendix G. Could you use two of the standard half-cells to build a voltaic cell with a standard cell potential of 7 V? Explain your answer.

6. Explain the difference between a standard reduction potential and a cell potential.

7. The cell potential for the following voltaic cell is given:
   $Zn(s) | Zn^{2+}(aq)(1mol/L) || Pd^{2+}(aq)(1mol/L) | Pd(s)$
   $E°_{cell} = 1.750$ V
   Determine the standard reduction potential for the following half-reaction:
   $Pd^{2+}(aq) + 2e^- \rightarrow Pd(s)$

Applications of Voltaic Cells

Section Outcomes

In this section, you will:

- **analyze** the relationship between scientific knowledge and the technological development of batteries and fuel cells
- **investigate** the use of technology to solve problems related to corrosion
- **predict** the future of fuel cells in transportation

Key Terms

dry cell
battery
primary battery
secondary battery
alkaline batteries
button battery
fuel cell
corrosion
galvanizing
sacrificial anode
cathodic protection

Voltaic cells are fairly large and filled with liquid. You could not use a voltaic cell to power a wristwatch, a remote control, or a flashlight. How are the common batteries, that you use daily, related to the voltaic cells about which you have been learning?

Dry Cells

A **dry cell** is a voltaic cell in which the electrolyte has been thickened into a paste. The first dry cell was invented in 1866 by the French chemist Georges Leclanché, who used starch to thicken the electrolyte. The cell was called the Leclanché cell.

Modern dry cells are closely modelled on the Leclanché cell and also contain electrolyte pastes. You have probably used inexpensive dry cells in many kinds of applications, such as a flashlight, a remote control, or a portable radio. The lowest priced 1.5-V batteries (sizes AAA, AA, A, C, and D) are dry cells.

A **battery** is defined as a set of voltaic cells connected in series. In a series connection, the negative electrode of one cell is connected to the positive electrode of the next cell, just as the series lemon battery was connected in the Launch Lab. *The voltage of a set of cells connected in series is the sum of the voltages of the individual cells.* Thus, a 9-V battery contains six 1.5-V dry cells connected in series. Often, the term *battery* is (incorrectly) used to describe a single cell. For example, a 1.5-V dry cell "battery" contains only a single cell. It is not, in fact, a battery.

A dry cell stops producing electrical energy when the reactants are used up. For many years, all batteries were disposable and were discarded after they had run down completely. A disposable battery is known as a **primary battery**. Some newer batteries, known as **secondary batteries**, are rechargeable. You will learn more about these batteries in section 13.3.

A typical dry cell contains a zinc anode and an inert graphite cathode, as shown in Figure 13.9. The electrolyte is a moist paste of manganese(IV) oxide, MnO_2; zinc chloride, $ZnCl_2$; ammonium chloride, NH_4Cl; and "carbon black," $C(s)$.

zinc anode

electrolyte paste

graphite cathode

Figure 13.9 The D-size dry cell battery is shown both whole and cut in half. The anode is the zinc container, located just inside the outer paper, steel, or plastic case. The graphite cathode runs through the centre of the cylinder.

You are already familiar with the oxidation half-reaction at the zinc anode:

$$Zn(s) \rightarrow Zn^{2+}(aq) + 2e^-$$

The reduction half-reaction at the cathode is somewhat more complicated. A simplified equation is shown here:

$$2MnO_2(s) + H_2O(\ell) + 2e^- \rightarrow Mn_2O_3(s) + 2OH^-(aq)$$

Therefore, a simplification of the overall cell reaction is

$$2MnO_2(s) + Zn(s) + H_2O(\ell) \rightarrow Mn_2O_3(s) + Zn^{2+}(aq) + 2OH^-(aq)$$

The more expensive alkaline cell, shown in Figure 13.10, is an improved, longer-lasting version of the dry cell.

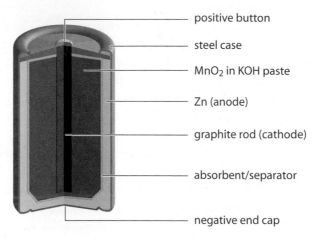

positive button
steel case
MnO_2 in KOH paste
Zn (anode)
graphite rod (cathode)
absorbent/separator
negative end cap

Figure 13.10 The structure of an alkaline cell is similar to the structure of a dry cell. Each type produces a voltage of 1.5 V.

Figure 13.11 Button batteries are small and long lasting.

Billions of **alkaline batteries**, each containing a single alkaline cell, are made every year. The ammonium chloride and zinc chloride used in a dry cell are replaced by strongly alkaline (basic) potassium hydroxide, KOH. The half-reactions and the overall reaction in an alkaline cell are given here.

Oxidation (at the anode): $Zn(s) + 2OH^-(aq) \rightarrow ZnO(s) + H_2O(\ell) + 2e^-$
Reduction (at the cathode): $MnO_2(s) + 2H_2O(\ell) + 2e^- \rightarrow Mn(OH)_2(s) + 2OH^-(aq)$
Overall cell reaction: $Zn(s) + MnO_2(s) + H_2O(\ell) \rightarrow ZnO(s) + Mn(OH)_2(s)$

A **button battery** is much smaller than an alkaline battery. Button batteries are commonly used in watches, as shown in Figure 13.11. Because it is so small, the button battery is also used for hearing aids, pacemakers, and some calculators and cameras. The development of smaller batteries has had an enormous impact on portable devices.

Two common types of button batteries both use a zinc container, which acts as the anode, and an inert stainless steel cathode, as shown in Figure 13.12. In the mercury button battery, the alkaline electrolyte paste contains mercury(II) oxide, HgO. In the silver button battery, the electrolyte paste contains silver oxide, Ag_2O. The batteries have similar voltages of about 1.3 V for the mercury cell and about 1.6 V for the silver cell.

Figure 13.12 A common type of button battery, shown here, contains silver oxide or mercury(II) oxide. Mercury is cheaper than silver, but discarded mercury batteries release toxic mercury metal into the environment.

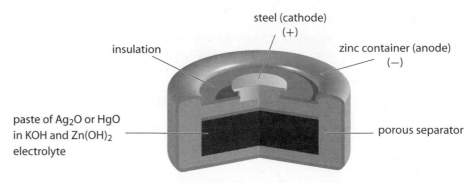

steel (cathode)
(+)
insulation
zinc container (anode)
(−)
paste of Ag_2O or HgO in KOH and $Zn(OH)_2$ electrolyte
porous separator

The reaction products in a mercury button battery are solid zinc oxide and liquid mercury. The two half-reactions and the overall equation are as follows.

Oxidation half-reaction:	$Zn(s) + 2OH^-(aq) \rightarrow ZnO(s) + H_2O(\ell) + 2e^-$
Reduction half-reaction:	$HgO(s) + H_2O(\ell) + 2e^- \rightarrow Hg(\ell) + 2OH^-(aq)$
Overall reaction:	$Zn(s) + HgO(s) \rightarrow ZnO(s) + Hg(\ell)$

• • •

8 Describe the similarities and differences between a Daniell cell and a dry cell.

9 What is the difference between a cell and a battery?

• • •

Fuel Cell History

Fuel cells offer great promise for the future by providing clean electric energy for transportation as well as for industries and homes. A **fuel cell** is, fundamentally, a battery that can be refuelled. Instead of a container with chemical reactants sealed inside, the reactants flow into a fuel cell and the products flow out. Although this sounds similar to an internal combustion engine or a fossil fuel–burning electrical generator, fuel cells convert the energy in the fuel directly into electrical energy, as does a battery, instead of burning the fuel and using the heated gases to drive an engine or generator. Fuel cells are more efficient than are engines or generators that burn fuel and, in addition, they are much cleaner. The major "waste" product is water.

So, if fuel cells are so clean and efficient, why are they not in everyday use? The simple answer is cost. Nevertheless, many government and industry research projects are finding ways to reduce the cost and improve the technology. For example, the bus in Figure 13.14 is part of a testing, demonstration, and development program. More than 20 buses similar to this have been used for public transportation in cities throughout the world.

Considering the stage of the development of fuel cells, you might think that the theory was only recently developed. On the contrary, the first demonstration of a working fuel cell was carried out by Sir William Robert Grove (1811–1896) in 1839. He did not pursue the development of the "gas battery" as he called it. In 1889, Ludwig Mond (1839–1909) and Charles Langer tried to modify Grove's design but were unable to produce a practical device. It was Mond and Langer who coined the term *fuel cell*. In 1932, Dr. Francis Thomas Bacon (1904–1993; direct descendent of Sir Francis Bacon) made modifications to Mond and Langer's fuel cell and created one that was able to power a welding machine. Although a proven concept, the fuel cell had not yet become practical or economical.

Figure 13.13 With small, long-lasting batteries, a pacemaker can now be implanted in a heart patient's chest. Early pacemakers, such as this one developed at the University of Toronto in the 1950s, were so big and heavy that patients had to wheel them around on a cart.

Figure 13.14 This city bus in Vancouver, British Columbia, is part of a fleet of buses in a development program. The buses are powered by fuel cells produced by Ballard Power Systems located in Burnaby, British Columbia. They are fuelled by hydrogen and the only emission is pure water.

In 1960, NASA (National Aeronautics and Space Administration) began to look for a safe, efficient, light-weight source of electrical energy for space flights that had crews. Funded by NASA, several companies carried out intensive research and the first practical—though expensive—fuel cells were developed. Not only could the fuel cells produce electrical energy efficiently but also the "waste" product was pure water that the astronauts could consume (Figure 13.15).

Figure 13.15 This fuel cell provides electrical energy for instruments and equipment on the space shuttle, as well as water for the astronauts to drink.

Fuel Cell Technology

A variety of types of fuel cells are currently being developed but they are all based on the same principle—the principle of the voltaic cell. The main difference is that the chemical reactants are gases that flow past the anode and cathode where the oxidation and reduction half-reactions occur. An electric current then flows through an external circuit. Finally, the gaseous products are eliminated from the cell.

The type of fuel cell that was first used by NASA on space flights for their Gemini program, which is also under development by Ballard Power Systems for use in cars and buses, is called the *proton exchange membrane* (PEM) *fuel cell*. A schematic diagram of a PEM fuel cell is shown in Figure 13.16.

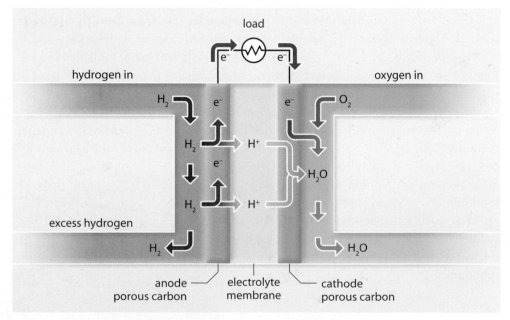

Figure 13.16 Hydrogen gas from tanks and oxygen in air provide a continuous supply of fuel for the fuel cell. The electrolyte is a solid polymer that allows hydrogen ions to pass through.

The anode of the PEM fuel cell is porous carbon coated with platinum to catalyze the oxidation half-reaction. A thin solid polymer acts as the electrolyte, allowing the positively charged hydrogen ions, formed in the oxidation half-reactions, to pass through but blocking the passage of negatively charged electrons. The electrons move from the anode, through an external circuit, to the cathode. The source of oxygen for the cathode is air. As air passes beside the porous carbon cathode, oxygen combines with the electrons returning to the cathode from the external circuit and protons that have passed through the membrane. The oxygen, hydrogen ions, and electrons combine to form water. The reactions that take place at the electrodes are shown below. Notice that the hydrogen ions are in the solid phase because they bind to the solid membrane as they pass through it.

Oxidation half-reaction (anode): $H_2(g) \rightarrow 2H^+(s) + 2e^-$
Reduction half-reaction (cathode): $O_2(g) + 4H^+(s) + 4e^- \rightarrow 2H_2O(\ell)$
Overall reaction: $2H_2(g) + O_2(g) \rightarrow 2H_2O(\ell)$

A complete fuel cell consists of many layers of individual cells that, together, are called a fuel cell stack. PEM fuel cells are about 40% to 50% efficient and operate at temperatures between 60 °C and 100 °C. PEM fuel cells are currently being tested in buses and cars and for electrical energy for homes.

Another type of fuel cell that is being tested for use in public transportation is the *phosphoric acid fuel cell*. The oxidation and reduction half-reactions are the same as those for the PEM fuel cell. However, the electrolyte is phosphoric acid. These fuel cells run at slightly higher temperatures than the PEM fuel cells but can achieve slightly higher efficiencies. Also, they can tolerate fuels that are not as pure as those for the PEM fuel cells.

A third type of fuel cell, called the *alkaline fuel cell*, was developed for NASA for the Apollo program and is still used in space shuttles. In an alkaline fuel cell (see Figure 13.17), the electrolyte is aqueous potassium hydroxide. When hydrogen reaches the anode, it combines with the hydroxide ions from the electrolyte and forms water. The electrons travel along the electrode to the external circuit. At the cathode, oxygen gas combines with the electrons that are returning from the external circuit and with water. The anode, cathode, and overall reactions are shown here.

Oxidation half-reaction (anode): $2H_2(g) + 4OH^-(aq) \rightarrow 4H_2O(\ell) + 4e^-$
Reduction half-reaction (cathode): $O_2(g) + 2H_2O(\ell) + 4e^- \rightarrow 4OH^-(aq)$
Overall reaction: $2H_2(g) + O_2(g) \rightarrow 2H_2O(\ell)$

Figure 13.17 In an alkaline fuel cell, the hydroxide ions that are used in the oxidation half-reaction at the anode are replaced by the reduction half-reaction at the cathode.

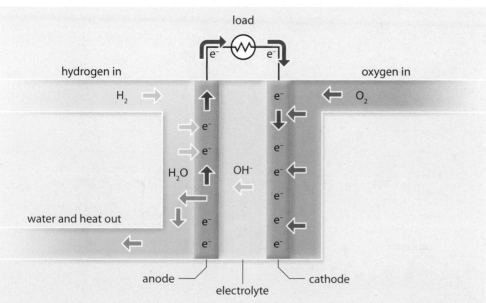

Alkaline fuel cells operate at temperatures around 90 °C to 100 °C and are about 40% to 50% efficient. Their use will probably remain with the space program and the military.

Source of Hydrogen Gas

Are you wondering about the source of the pure hydrogen for these fuel cells? Only trace amounts of hydrogen gas are found in the atmosphere. Hydrogen for use in fuel cells must be extracted from other compounds and any extraction processes requires energy. For example, electrical energy can split water into hydrogen and oxygen. Also, hydrogen can be chemically removed from hydrocarbons by processes called reforming. Once again, all these processes require energy. If the energy used to extract hydrogen is obtained from the burning of fossil fuels, there would be no benefit in using fuel cells. If, however, wind and solar energy are used to produce the hydrogen, the entire process is still environmentally friendly.

Research is underway to develop fuel cells that have internal reformers—systems that remove hydrogen from hydrocarbons. These fuel cells can use a variety of fuels, such as methane or other hydrocarbons. Some carbon dioxide is produced, but for the amount of energy generated, much less carbon dioxide is released than from internal combustion engines or fossil fuel–burning power plants.

Progress is being made in the development of another unique fuel cell—the *direct methanol fuel cell* (DMFC). The reactions that take place at the electrodes are shown below. This fuel cell can use methanol rather than hydrogen (see Figure 13.18).

Oxidation half-reaction (anode): $CH_3OH(\ell) + H_2O(\ell) \rightarrow CO_2(g) + 6H^+ + 6e^-$
Reduction half-reaction (cathode): $O_2(g) + 4H^+ + 4e^- \rightarrow 2H_2O(\ell)$
Overall reaction: $2CH_3OH(\ell) + 3O_2(g) \rightarrow 2CO_2(g) + 4H_2O(\ell)$

Figure 13.18 In addition to this MP3 player, Toshiba has developed a laptop computer that is powered by a direct methanol fuel cell. However, they will not market them unless air travel regulations change. Currently it is against the law to carry methanol on a commercial aircraft.

10. Calculate $E°_{cell}$ for a hydrogen fuel cell.

11. Reactions that occur in fuel cells are sometimes referred to as "flameless combustion reactions." Explain why.

12. What is the source of hydrogen for fuel cells?

Corrosion: Unwanted Voltaic Cells

You have probably seen many examples of rusty objects, such as the one in Figure 13.19. Did you know that rusting costs many billions of dollars per year in prevention, maintenance, and replacement costs? What is the relationship between rust and electrochemical cells? You will now learn what causes rust and how knowledge of electrochemistry can help to prevent it.

Figure 13.19 Rust is a common sight on iron objects.

Rusting is an example of **corrosion**, which is a spontaneous redox reaction of materials with substances in their environment. Figure 13.20 shows an example of the hazards that result from corrosion.

Figure 13.20 On April 28, 1988, this Aloha Airlines aircraft was flying at an altitude of 7300 m when a large part of the upper fuselage ripped off. This accident was caused by undetected corrosion damage. The pilot showed tremendous skill in landing the plane.

Many metals are easily oxidized by a powerful oxidizing agent in the atmosphere—oxygen. Because metals are constantly in contact with oxygen, they are vulnerable to corrosion. In fact, the term *corrosion* is sometimes defined as the oxidation of metals exposed to the environment. In North America, about 20% to 25% of the iron and steel produced is used to replace objects that have been damaged or destroyed by corrosion. However, not all corrosion is harmful. For example, the green layer formed by the corrosion of a copper roof is considered attractive by many people. Once this layer is formed, it protects the copper beneath it from further corrosion.

Rust is a hydrated iron(III) oxide, $Fe_2O_3 \cdot xH_2O$. (Note that the "x" signifies a variable number of water molecules per formula unit.) The surface of a piece of iron behaves as though it consists of many small voltaic cells in which electrochemical reactions form rust, as shown in Figure 13.21. In each small cell, iron acts as the anode. The cathode is inert and may be an impurity that exists in the iron or is deposited onto it. For example, the cathode could be a piece of soot that has been deposited onto the iron surface from the air.

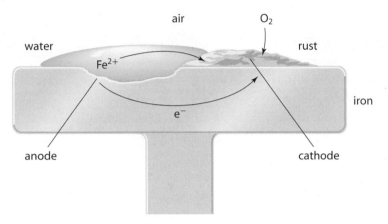

Figure 13.21 The rusting of iron involves the reaction of iron, oxygen, and water in a naturally occurring voltaic cell on the exposed surface of the metal. There are usually many of these small cells on the surface of the same piece of iron.

Water, in the form of rain, is needed for rusting to occur. Carbon dioxide in the air reacts with rainwater to form carbonic acid, $H_2CO_3(aq)$. This weak acid partially dissociates into ions, $H_2CO_3(aq) \rightleftharpoons H^+(aq) + HCO_3^-(aq)$. Thus, the carbonic acid is an electrolyte for the corrosion process. Other electrolytes, such as road salt, may also be involved. The circuit is completed by the iron itself, which conducts electrons from the anode to the cathode.

The rusting process is complex, and the equations can be written in various ways. A simplified description of the half-reactions and the overall cell reaction is given here.

Oxidation half-reaction (occurs at the anode):	$Fe(s) \rightarrow Fe^{2+}(aq) + 2e-$
Reduction half-reaction (occurs at the cathode):	$O_2(g) + 2H_2O(\ell) + 4e^- \rightarrow 4OH^-(aq)$
Overall reaction:	$2Fe(s) + O_2(g) + 2H_2O(\ell) \rightarrow$
	$2Fe^{2+}(aq) + 4OH^-(aq)$

There is no barrier in the cell, so nothing stops the dissolved Fe^{2+} and OH^- ions from mixing. The iron(II) ions produced at the anode and the hydroxide ions produced at the cathode react to form a precipitate of iron(II) hydroxide, $Fe(OH)_2(s)$. Therefore, the overall cell reaction could be written as follows:

$$2Fe(s) + O_2(g) + 2H_2O(\ell) \rightarrow 2Fe(OH)_2(s)$$

The iron(II) hydroxide undergoes further oxidation through reaction with the oxygen in the air to form iron(III) hydroxide:

$$4Fe(OH)_2(s) + O_2(g) + 2H_2O(\ell) \rightarrow 4Fe(OH)_3(s)$$

Iron(III) hydroxide readily breaks down to form hydrated iron(III) oxide, $Fe_2O_3 \cdot xH_2O$, more commonly known as rust:

$$2Fe(OH)_3(s) \rightarrow Fe_2O_3 \cdot 3H_2O(s)$$

Both $Fe(OH)_3$ and $Fe_2O_3 \cdot xH_2O$ are reddish-brown, or "rust-coloured." A rust deposit may contain a mixture of these compounds.

Not all metals corrode to the same extent as iron. In fact, many metals corrode in air to form a surface coating of metal oxide that, in many cases, adheres to the metal surface and forms a protective layer that prevents the metal from further corrosion. For example, aluminium, chromium, and magnesium are readily oxidized in air to form their oxides, Al_2O_3, Cr_2O_3, and MgO. Unless the oxide layer is broken by a cut or a scratch, the layer prevents further corrosion. In contrast, rust easily flakes off from the surface of an iron object and provides little protection against further corrosion.

Corrosion Prevention

Corrosion, and especially the corrosion of iron, can be very destructive. For this reason, a great deal of effort goes into corrosion prevention. The simplest method of preventing corrosion is to paint an iron object. The protective coating of paint prevents air and water from reaching the metal surface. Other effective protective layers include grease, oil, plastic, or a metal that is more resistant to corrosion than is iron. For example, a layer of chromium protects bumpers and metal trim on cars. An enamel coating is often used to protect metal plates, pots, and pans. *Enamel* is a shiny, hard, and very unreactive type of glass that can be melted onto a metal surface. A protective layer is effective as long as it completely covers the iron object. If a hole or scratch breaks the layer, the metal underneath can corrode.

It is also possible to protect iron against corrosion by forming an alloy with a different metal. *Stainless steel* is an alloy of iron that contains at least 10% chromium, by mass, in addition to small quantities of carbon and occasionally metals, such as nickel. Stainless steel is much more resistant to corrosion than is pure iron. Therefore, stainless steel is often used for cutlery, taps, and various other applications where rust resistance is important. However, chromium is much more expensive than iron. As a result, stainless steel is too expensive for use in large-scale applications, such as building bridges.

Galvanizing is a process in which iron is covered with a protective layer of zinc. Galvanized iron is often used to make metal buckets and chain-link fences. Galvanizing protects iron in two ways. First, the zinc acts as a protective layer. If this layer is broken, the iron is exposed to air and water. When this happens, however, the iron is still protected. Zinc is more easily oxidized than is iron. Therefore, zinc, not iron, becomes the anode in the voltaic cell. The zinc metal is oxidized to zinc ions. In this situation, zinc is called a **sacrificial anode**, because it is destroyed (sacrificed) to protect the iron. Iron acts as the cathode when zinc is present. Thus, iron does not become oxidized, until all the zinc has reacted.

Cathodic protection, shown in Figure 13.22, is another method of preventing rusting that is similar to galvanizing, in that a more reactive metal is attached to the iron object. This reactive metal acts as a sacrificial anode, and the iron becomes the cathode of a voltaic cell. Unlike galvanizing, the metal used in cathodic protection does not completely cover the iron. Because the sacrificial anode is slowly destroyed by oxidation, it must be replaced periodically.

Figure 13.22 Magnesium, zinc, or aluminium blocks are attached to ships' hulls, oil and gas pipelines, underground iron pipes, and gasoline storage tanks. These reactive metals provide cathodic protection by acting as a sacrificial anode.

If iron is covered with a protective layer of a metal that is less reactive than iron, there can be unfavourable results. A "tin" can is actually a steel can coated with a thin layer of tin. While the tin layer remains intact, it provides effective protection against rusting. If the tin layer is broken or scratched, however, the iron in the steel corrodes faster in contact with the tin than the iron would on its own. Since tin is less reactive than iron is, tin acts as a cathode in each miniature voltaic cell on the surface of the can. Therefore, the tin provides a large area of available cathodes for the small voltaic cells involved in the rusting process. Iron acts as the anode of each cell, which is its normal role when rusting.

Sometimes, the rusting of iron is promoted accidentally. For example, by connecting an iron pipe to a copper pipe in a plumbing system, an inexperienced plumber could accidentally speed up the corrosion of the iron pipe. Copper is less reactive than iron. Therefore, copper acts as the cathode and iron as the anode in numerous small voltaic cells at the intersection of the two pipes.

Build on your understanding of corrosion by answering the following questions.

• • •

13 **a)** Use the two half-reactions for the rusting process and a table of standard reduction potentials to determine the standard cell potential for this reaction.

b) Do you think that your calculated value is the actual cell potential for each of the small voltaic cells on the surface of a rusting iron object? Explain.

14 Explain why aluminium provides cathodic protection to an iron object.

15 In 2000, Transport Canada reported that thousands of cars sold in the Atlantic provinces between 1989 and 1999 had corroded engine cradle mounts. Failure of these mounts can cause the steering shaft to separate from the car. The manufacturer recalled the cars so that repairs could be made, where necessary. The same cars were sold in other parts of the country but had little corrosion damage. Why do you think that the corrosion problems showed up in the Atlantic provinces?

16 **a)** Use a table of standard reduction potentials to determine whether elemental oxygen, $O_2(g)$, is a better oxidizing agent under acidic conditions or basic conditions.

b) From your answer to part (a), do you think that acid rain promotes or helps prevent the rusting of iron?

• • •

Chemistry File

Web Link
Battery makers have been challenged to make batteries smaller, lighter, longer lasting, and more powerful. In March 2003, Dalhousie University established a Research Chair in Battery and Fuel Cell Materials. The university's new laboratory is one of the few in the world equipped to use a new mode of research developed in 1995, called combinatorial materials synthesis (CMS). CMS rapidly shortens the time needed to test new combinations of materials by uncovering thousands of distinct compositions in a single experiment. Chief researcher Jeffery Dahn is applying CMS methods to improve the safety of lithium-ion batteries and produce cells large enough to power electric vehicles. How does CMS allow for the testing of so many materials at one time?

www.albertachemistry.ca
WWW

Section 13.2 Summary

• Voltaic cells are made more portable and practical by thickening the electrolyte into a paste and sealing the components of the cell. The resulting cell is called a dry cell.

• A battery is a set of cells connected in series. The potential difference, or voltage, between the electrodes of a battery is the sum of the voltages across each dry cell.

• The common dry cell that is often called the "flashlight battery" has a zinc anode and a graphite cathode with manganese oxide in the electrolyte paste.

• An alkaline battery lasts longer than the older batteries and has potassium hydroxide in the electrolyte.

- A fuel cell is based on the concept of a voltaic cell. For most fuel cells, hydrogen is the reducing agent and oxygen is the oxidizing agent. The by-product is pure water. The hydrogen and oxygen flow into the fuel cell and the excess hydrogen and water flow out of the cell.
- Although the only by-product of a hydrogen fuel cell is pure water and seems to be ideal for the environment, there is very little hydrogen available. It must be generated from some source and the process typically uses a significant amount of energy.
- Corrosion is caused by the spontaneous formation of voltaic cells on the surface of metals. A drop of rainwater with some dissolved carbon dioxide or salt along with a spec of contaminant such as a piece of dust can form a voltaic cell in which the iron or other metal is the anode which loses electrons. The iron ion then goes into solution.
- Corrosion can be reduced or eliminated by covering the iron surface with paint or some other material.
- Galvanizing, applying a protective layer of zinc, protects iron from corrosion. Even if the layer of zinc is broken, the zinc protects the iron because the zinc is more easily oxidized than the iron.
- Adding any metal that oxidizes more easily than iron provides protection for the iron. This technique is called cathodic protection.

SECTION 13.2 REVIEW

1. Identify the oxidizing agent and the reducing agent in a dry cell.

2. Explain why the top of a commercial 1.5-V dry cell "battery" is always marked with a plus sign.

3. The reaction products in a silver button battery are solid zinc oxide and solid silver.
   a) Write the two half-reactions and the equation for the overall reaction in the battery.
   b) Name the materials used to make the anode and the cathode.

4. If two 1.5-V D-size batteries, connected in series, power a flashlight, at what voltage is the flashlight operating? Explain.

5. How many dry cells are needed to make a 6-V battery? Explain.

6. Many 9 V batteries are very small while some 1.5 V dry cells are very large. Explain how a small battery can create a large voltage (potential difference) when a large cell produces a small voltage. What is the advantage of a large size for a 1.5 V dry cell?

7. When a dry cell produces an electric current, what happens to the container? Explain.

8. Use the following cell notation to sketch a possible design of the cell. Include as much information as you can. Identify the anode and cathode, and write the half-reactions and the overall cell reaction:

   Fe(s) | Fe^{2+}(aq) || Ag+(aq) | Ag(s)

9. Explain how a fuel cell is similar to a dry cell and how a fuel cell differs from a dry cell.

10. Name three ways in which fuel cells are superior to other possible methods of producing electrical energy for space flights.

11. Make a simple sketch of a PEM fuel cell, label it, and write the reactions that occur at the anode and cathode. Briefly explain how it provides electrical energy.

12. Why does the use of road salt cause cars to rust faster than they otherwise would?

13. Aluminium is a more reactive metal than any of the metals present in steel. However, discarded steel cans disintegrate much more quickly than discarded aluminium cans when both are left open to the environment in the same location. Give an explanation.

14. Explain why zinc acts as a sacrificial anode in contact with iron.

15. a) Identify two metals that do not corrode easily in the presence of oxygen and water. Explain why they do not corrode.
    b) How are these metals useful? How do the uses of these metals depend on their resistance to corrosion?

16. A silver utensil is said to *tarnish* when its surface corrodes to form a brown or black layer of silver sulfide. Research and describe a chemical procedure that can be used to remove this layer. Write balanced half-reactions and a chemical equation for the process. ICT

Electrolytic Cells

In the first section of this chapter, you learned more about redox reactions and how spontaneous redox reactions release energy. In the second section, you learned how voltaic cells harness that energy in the form of electrical energy. Would it be possible to introduce electrical energy from an external source and cause redox reactions to proceed in the opposite direction? If so, what could you accomplish by reversing the direction of redox reactions?

Electrolytic Cells

A cell that uses an external source of electrical energy to drive a non-spontaneous redox reaction is called an **electrolytic cell**. You could describe an electrolytic cell as a cell that converts electrical energy into chemical energy. The process that takes place in an electrolytic cell is called **electrolysis**.

Electrolytic cells can look very much like voltaic cells. In fact, some voltaic cells can be converted into electrolytic cells. Figure 13.23 shows how a Daniell cell can be converted into an electrolytic cell by adding a power source.

In the typical Daniell cell on the left in Figure 13.23, zinc is spontaneously oxidized and electrons flow through the external circuit to the copper electrode, on the surface of which, copper(II) ions are reduced to copper atoms. On the right of the figure, a power source was added to the external circuit, making the cell an electrolytic cell. The power supply "pulls" electrons from the copper electrode, thus oxidizing copper atoms in the electrode to copper(II) ions that go into solution. The copper electrode is, therefore, the anode because oxidation half-reactions occur there. At the same time, the power supply is forcing electrons onto the zinc electrode causing zinc ions in the solution to become reduced to zinc atoms that become part of the zinc electrode. Since zinc ions are becoming reduced at the zinc electrode, that electrode is the cathode. The entire system is operating opposite to the Daniell cell. The processes happening in the voltaic cell and the electrolytic cell are compared in Table 13.2.

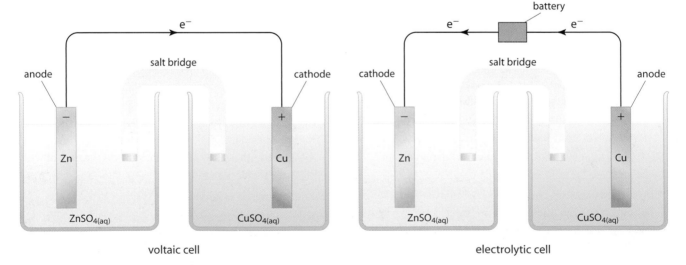

Figure 13.23 Adding an external voltage to reverse the electron flow converts a Daniell cell from a voltaic cell into an electrolytic cell.

Table 13.2 Comparison of Voltaic Cell and Electrolytic Cell

Voltaic cell	Electrolytic cell
spontaneous reaction	non-spontaneous reaction
converts chemical energy to electrical energy	converts electrical energy to chemical energy
anode (negatively charged): zinc electrode	anode (positively charged): copper electrode
cathode (positively charged): copper electrode	cathode (negatively charged): zinc electrode
oxidation (at anode): $Zn(s) \rightarrow Zn^{2+}(aq) + 2e^-$	oxidation (at anode): $Cu(s) \rightarrow Cu^{2+}(aq) + 2e^-$
reduction (at cathode): $Cu^{2+}(aq) + 2e^- \rightarrow Cu(s)$	reduction (at cathode): $Zn^{2+}(aq) + 2e^- \rightarrow Zn(s)$
cell reaction: $Zn(s) + Cu^{2+}(aq) \rightarrow Zn^{2+}(aq) + Cu(s)$	cell reaction: $Cu(s) + Zn^{2+}(aq) \rightarrow Cu^{2+}(aq) + Zn(s)$

Chemistry File

FYI
To remember the sign of the charge on the cathode in the two different types of cells (voltaic cells versus electrolytic cells), remember that electrons always leave the cathode and a reduction half-reaction occurs there. In a voltaic cell, the spontaneous chemical reaction "pulls" the electrons off the cathode leaving it positively charged. In an electrolytic cell, the external source of electrical energy "pushes" electrons onto the cathode making it negatively charged. This negatively charged cathode then "pushes" electrons onto the compound in solution, causing a reduction half-reaction to occur.

Previously, you calculated the standard cell potential for the redox reaction between zinc atoms and copper(II) ions and found it to be 1.10 V.

$$E^\circ_{cell} = E^\circ_{cathode} - E^\circ_{anode}$$
$$E^\circ_{cell} = E^\circ_{copper} - E^\circ_{zinc}$$
$$E^\circ_{cell} = +0.34\,V - (-0.76\,V)$$
$$E^\circ_{cell} = +1.10\,V$$

Similarly, if you calculated the standard cell potential for the reverse reaction, you would obtain $-1.10\,V$.

$$E^\circ_{cell} = E^\circ_{cathode} - E^\circ_{anode}$$
$$E^\circ_{cell} = E^\circ_{zinc} \quad E^\circ_{copper}$$
$$E^\circ_{cell} = -0.76\,V - (+0.34\,V)$$
$$E^\circ_{cell} = -1.10\,V$$

The negative sign means that the reaction is not spontaneous. This negative value represents the minimum potential difference that you would have to apply from the external power source to drive the cell reaction for the electrolytic cell.

Electrolysis in Aqueous Solutions

When you convert some voltaic cells into electrolytic cells, the reactions are effectively the reverse of those in the voltaic cell, as in the example above. However, depending on the nature of the electrolyte that is dissolved in the aqueous solution, the reaction that occurs might not be the reverse of the reaction that would occur in the voltaic cell.

Shortly after Volta invented the voltaic cell (1800), Sir Humphry Davy (1778–1829) built a large voltaic cell and used it to perform electrolysis experiments. He wanted to reduce metal ions in salts to produce the pure metals. At this time, few pure metals had been discovered. Davy first tried to carry out the electrolysis of salts, such as sodium chloride and potassium chloride, in aqueous solutions. The reactions that he hoped would occur when using sodium chloride are shown below in the form in which we would now express them.

Oxidation (anode):	$2Cl^-(aq) \rightarrow Cl_2(g) + 2e^-$
Reduction (cathode):	$Na^+(aq) + e^- \rightarrow Na(s)$

Davy observed that chlorine gas did, in fact, form at the anode. However, hydrogen gas formed at the cathode. The only source of hydrogen in the electrolytic cell was the water in which the sodium chloride was dissolved. Water was being reduced at the cathode rather than sodium ions. Chemists now understand how and why that process occurred. To learn how to predict the products of electrolysis in aqueous solutions, you need to first understand the electrolysis of water itself.

Electrolysis of Water

When water is subjected to electrolysis, some water molecules are oxidized at the anode and other water molecules are reduced at the cathode. The half-reaction equations that occur at each electrode and their standard reduction potentials are shown below. (Because the reaction that occurs at the anode is written as an oxidation half-reaction, a negative sign is needed in front of the standard reduction potential.)

oxidation (anode): $2H_2O(\ell) \rightarrow O_2(g) + 4H^+(aq) + 4e^-$ $-E° = -1.23\,V$
reduction (cathode): $2H_2O(\ell) + 2e^- \rightarrow H_2(g) + 2OH^-(aq)$ $E° = -0.83\,V$

Oxygen gas is generated at the anode and hydrogen gas is generated at the cathode. To prevent the oxygen and hydrogen gases from mixing, the electrodes are usually positioned below or in the closed end of glass tubes, as shown in Figure 13.24.

Figure 13.24 When water is subjected to electrolysis, the volume of the hydrogen gas is twice a great as the volume of the oxygen gas. Can you explain why this is occurs?

Several factors that affect the electrolysis of water must be considered when you are performing an experiment, making predictions about reactions, or carrying out calculations. Pure water is a poor conductor of electric current so the electrolysis of pure water proceeds very slowly. If you use a salt that will conduct current but will not interfere with the reactions, you can increase the rate of the reaction. Sodium sulfate, $Na_2SO_4(aq)$, is often used for this purpose. A second factor that affects the electrolysis of water is the fact that the concentrations of the reactants and products—$H^+(aq)$, $OH^-(aq)$, and $H_2O(\ell)$—are not 1.0 mol/L. The concentrations of hydrogen ions and hydroxide ions are 1.0×10^{-7} mol/L. Water has a concentration of about 55 mol/L. Therefore, standard reduction potentials cannot be used for predictions or calculations. The reduction potentials for the half-reactions of water under the non-standard concentrations are as follows:

oxidation $2H_2O(\ell) \rightarrow O_2(g) + 4H^+(aq) + 4e^-$ $-E = -0.82\,V$
reduction $2H_2O(\ell) + 2e^- \rightarrow H_2(g) + 2OH^-(aq)$ $E = -0.42\,V$

Notice that there are no superscripts on the symbol E for the reduction potentials because the values do not represent standard conditions.

Another important factor becomes obvious when you calculate the cell potential and compare it with the potential that must be applied to cause electrolysis of water. The calculated cell potential is

○ **Begin extension material**

$$E_{cell} = E_{cathode} - E_{anode}$$
$$E_{cell} = -0.42\,V - 0.82\,V$$
$$E_{cell} = -1.24\,V$$

The negative value means that the reaction is not spontaneous. You would probably predict that this value is then the potential that you must apply from an external source to cause the reaction to proceed in the indicated direction. However, if you applied this potential difference to an electrolysis cell, the reaction would not occur. By experimenting, you would discover that you must apply a significantly higher potential difference. This increase in potential difference beyond the calculated value for the cell potential is called the **overpotential**. An overpotential is necessary because, in addition to the energy needed to drive the reactions in the reverse directions, energy is also needed for the gases to form at the electrodes. For hydrogen and oxygen gases to form, the overpotential must be approximately 0.6 V.

Electrolysis of Aqueous Solutions of Sodium Chloride

By understanding the conditions for the electrolysis of water, you can now explain why Davy observed chlorine gas at the anode and hydrogen gas at the cathode of his electrolysis cell. There are two possible half-reactions that could occur at the cathode: the reduction of sodium ions or the reduction of water. The half-reaction that is most likely to occur is the half-reaction that requires the smallest amount of energy—the half-reaction with the less negative reduction potential. The two half-reactions are shown below with their reduction potentials:

$$Na^+(aq) + e^- \rightarrow Na(s) \qquad\qquad E° = -2.71\,V$$
$$2H_2O(\ell) + 2e^- \rightarrow H_2(g) + 2OH^-(aq) \qquad E = -0.42\,V$$

Even if you include the necessary overpotential of 0.6 V for the reduction of water, the reduction potential is about -1.0 V. This value is much less negative than the value of -2.71 V for sodium.

Why was chlorine gas produced at the anode rather than oxygen gas? You can use the same reasoning to answer this. The two possible oxidation half-reactions that could occur at the anode are shown below with their reduction potentials. Because these half-reactions are written in the direction opposite to reduction, you must use the negative of the reduction potentials.

$$2Cl^- \rightarrow 2e^- + Cl_2(g) \qquad\qquad -E° = -1.36\,V$$
$$2H_2O(\ell) \rightarrow O_2(g) + 4H^+(aq) + 4e^- \qquad -E = -0.82\,V$$

Now if you include the 0.6 V overpotential that is needed for oxygen gas to form, the reduction potential becomes $-E = -1.42$ V. The values for the formation of chlorine and for oxygen are similar but the value for chlorine is slightly less negative than for oxygen. Chemists now know that by making the concentration of chloride ion high, the reaction will favour the formation of chlorine gas. The net ionic equation and the complete balanced equation for the reaction that occurs during the electrolysis of an aqueous solution of sodium chloride are shown below:

$$2Cl^-(aq) + 2H_2O(\ell) \rightarrow Cl_2(g) + H_2(g) + 2OH^-(aq)$$
$$2NaCl(aq) + 2H_2O(\ell) \rightarrow Cl_2(g) + H_2(g) + 2NaOH(aq)$$

As you can see in the last equation, common table salt, which is readily available, can be used to produce chlorine gas, hydrogen gas, and sodium hydroxide base—three commercially important products. The electrolysis of sodium chloride is the basis of an important industrial process called the chlor-alkali process.

The Chlor-Alkali Process

Sodium hydroxide and chlorine are two of the most extensively produced commercial chemicals, with billions of kilograms of each produced every year in North America. Chlorine is used to make laundry bleach, to bleach pulp for paper, to make compounds for treating water, to use as a disinfectant, and to make hydrochloric acid. A large portion of the chlorine that is produced every year goes into making polyvinyl chloride (PVC), a type of plastic. Large amounts of sodium hydroxide are used in the pulp and paper industry to break down the lignin in wood that holds fibres together. Sodium hydroxide is also used in making soaps and detergents, in the production of aluminium, and in the manufacturing of many different chemicals.

Most of the chlorine and sodium hydroxide that are used commercially are produced by the **chlor-alkali process**. The redox reactions of the chlor-alkali process are shown on the previous page. In this process, brine (aqueous sodium chloride) is electrolyzed in a cell, like the one shown in Figure 13.25.

End extension material

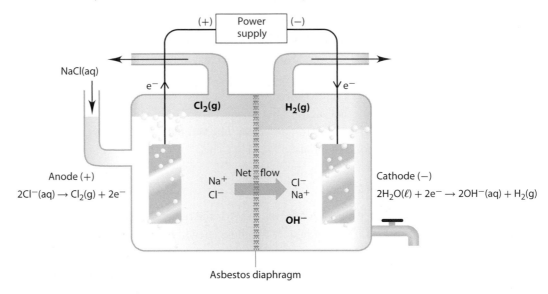

Figure 13.25 The chlor-alkali cell in this diagram electrolyzes an aqueous solution of sodium chloride to produce chlorine gas, hydrogen gas, and aqueous sodium hydroxide.

Chlorine gas, hydrogen gas, and sodium hydroxide are all highly reactive with one another and must be separated as they are being produced. The difference in the liquid level between the compartments keeps a net movement of solution into the cathode compartment in which hydrogen gas is produced. Chlorine gas is produced at the anode, where it is removed and prevented from mixing with the hydroxide ions produced at the cathode. Sodium hydroxide solution is removed from the cell periodically, and fresh brine is added to the cell. The aqueous sodium hydroxide is later dried by evaporation and packaged as a solid.

Hydrogen gas is not as widely used commercially as are chlorine gas and sodium hydroxide. Sometimes hydrogen gas is used as fuel to heat and thus dry the sodium hydroxide. However, if hydrogen fuel cell automobiles and electrical generators become economical, more hydrogen produced in the chlor-alkali industry might be distributed for use in fuel cells.

Predicting the Products of Electrolysis of Aqueous Solutions

Predicting the products of the electrolysis of any aqueous solution is essentially the same as the process used in analyzing the electrolysis of a sodium chloride solution. Write all possible half-reactions that could occur at each of the electrodes. Find the oxidation half-reaction and the reduction half-reaction that require the smallest amount of energy and then put them together. You must always include an overpotential for oxidation and the reduction of water. When you find the amount of energy or potential difference required to drive the half-reaction, you must increase it by the overpotential. Although overpotential will be between 0.4 V and 0.6 V, it is not possible to be precise because the value is affected by the amount of current that runs through the cell. Using the maximum amount of 0.6 V will give you a good approximation. The only way to be certain is to perform the experiment and observe what happens at each electrode. The following Sample Problem and Practice Problems will help you learn how to make your predictions. The investigation that follows the Practice Problems will give you a chance to test one of your predictions.

Sample Problem

Electrolysis of an Aqueous Solution

Problem
Predict the products of the electrolysis of 1.0 mol/L LiBr(aq).

What Is Required?
You need to predict the products of the electrolysis of 1.0 mol/L LiBr(aq).

What Is Given?
This is an aqueous solution. You are given the formula and concentration of the electrolyte. You have a table of standard reduction potentials, and you know the non-standard reduction potentials for water.

Plan Your Strategy
List each possible half-reaction that could occur at each electrode. List the reduction potentials for each half-reaction. Include a 0.6-V overpotential for the half-reactions involving water. Select the oxidation and reduction half-reactions that require the least amount of energy. These are the reactions that have the smallest negative reduction potential. Put the selected half-reactions together.

Act on Your Strategy
The $Li^+(aq)$ and $Br^-(aq)$ concentrations are 1.0 mol/L, so use the standard reduction potentials for the half-reactions that involve these ions. Use the non-standard values for water.

There are two possible oxidation half-reactions at the anode: the oxidation of bromide ion in the electrolyte or the oxidation of water:

$$2Br^-(aq) \rightarrow Br_2(\ell) + 2e^- \qquad -E^\circ_{anode} = -1.07\,V$$
$$2H_2O(\ell) \rightarrow O_2(g) + 4H^+(aq) + 4e^- \quad -E_{anode} = -0.82\,V$$
$$\text{with overpotential} = -1.42\,V$$

There are two possible reduction half-reactions at the cathode: the reduction of lithium ions in the electrolyte or the reduction of water:

$$Li^+(aq) + e^- \rightarrow Li(s) \qquad E^\circ_{cathode} = -3.04\,V$$
$$2H_2O(\ell) + 2e^- \rightarrow H_2(g) + 2OH^-(aq) \quad E_{cathode} = -0.42\,V$$
$$\text{with overpotential} = -1.02\,V$$

The oxidation of the bromide ion has the smallest negative reduction potential of the oxidation half-reactions.

The reduction of water has the smallest negative reduction potential of the reduction half-reactions. Notice that if you ignored the overpotential for the reduction of water, you would have predicted that the oxidation of water would occur instead of the oxidation of the bromide ion.

The combined half-reactions are

$$2Br^-(aq) + 2H_2O(\ell) \rightarrow Br_2(\ell) + H_2(g) + 2OH^-(aq)$$

The cell potential would be
$$E_{cell} = E_{cathode} - E_{anode}$$
$$E_{cell} = -1.02\,V - 1.07\,V$$
$$E_{cell} = -2.09\,V$$

Check Your Solution
By comparing this reaction with the known reaction observed when sodium chloride is electrolyzed, the results are analogous. Lithium has a more negative reduction potential than sodium does and thus is less likely to be reduced. You would expect that water would be reduced instead of the lithium ion. As well, bromide has a smaller reduction potential than does chloride and you know that chloride is oxidized in the electrolysis of aqueous sodium chloride. You would expect that bromide would be oxidized.

5. One half-cell of a voltaic cell has a nickel electrode in a 1 mol/L nickel(II) chloride solution. The other half-cell has a cadmium electrode in a 1 mol/L cadmium chloride solution.
  a) Find the cell potential.
  b) Identify the anode and the cathode.
  c) Write the oxidation half-reaction, the reduction half-reaction, and the overall cell reaction.

6. An external voltage is applied to change the voltaic cell in question 5 into an electrolytic cell. Repeat parts (a) to (c) for the electrolytic cell.

7. Predict the products of the electrolysis of a 1 mol/L solution of cobalt chloride.

8. Predict the products of the electrolysis of a 1 mol/L solution of silver nitrate.

INVESTIGATION 13.B

Target Skills

Identifying the products of an electrolysis cell

Evaluating experimental designs for an electrolytic cell and suggesting alternatives

Using appropriate SI notation and fundamental and derived units to communicate answers to problems related to functioning electrolytic cells

Electrolysis of Aqueous Potassium Iodide

When an aqueous solution is electrolyzed, the electrolyte or water can undergo electrolysis. In this investigation, you will build an electrolytic cell, carry out the electrolysis of an aqueous solution, and identify the products.

Question

What are the products from the electrolysis of a 1 mol/L aqueous solution of potassium iodide? Are the observed products the ones predicted using reduction potentials?

Predictions

Use the relevant standard reduction potentials from the table in Appendix G, and the non-standard reduction potentials you used previously for water, to predict the electrolysis products. Predict which product(s) are formed at the anode and which product(s) are formed at the cathode.

Safety Precautions

Materials

- 10 mL 1 mol/L KI
- 20 gauge platinum wire (2 cm)
- 1 graphite pencil lead, 2 cm long
- 1 drop 1% starch solution
- 1 drop 1% phenolphthalein
- sheet of white paper
- 1 beaker (600 mL or 400 mL)
- 1 elastic band
- 25 cm clear aquarium tubing (Tygon®; internal diameter: 4–6 mm)

- 3 disposable pipettes
- 3 toothpicks
- 2 wire leads (black and red) with alligator clips
- 9-V battery or variable power source set to 9 V

Procedure

1. Fold a sheet of paper lengthwise. Curl the folded paper so that it fits inside the beaker. Invert the beaker on your lab bench.

2. Use the elastic to strap the aquarium tubing to the side of the beaker in a U shape, as shown in the diagram.

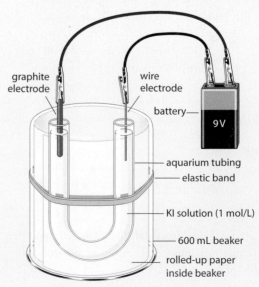

3. Fill a pipette as completely as possible with 1 mol/L KI solution. Insert the tip of the pipette firmly into one end of the aquarium tubing. Slowly inject the solution into the U-tube until the level of the solution is within 1 cm to 2 cm from the top of both ends. If air bubbles are present, try to remove them by poking them with a toothpick. You may need to repeat this step from the beginning to ensure there are no air bubbles.

4. Attach the black lead to the 2 cm piece of wire. Insert the wire into one end of the U-tube. Attach the red electrical lead to the graphite. Insert the graphite into the other end of the U-tube.

5. Attach the leads to the 9-V battery or to a variable power source set to 9 V. Attach the black lead to the negative terminal and the red lead to the positive terminal.

6. Let the reaction proceed for three minutes, while you examine the U-tube. Record your observations. Shut off the power source and remove the electrodes. Determine the product formed around the anode by adding a drop of starch solution to the end of the U-tube that contains the anode. Push the starch solution down with a toothpick if there is an air lock. Determine one of the products around the cathode by adding a drop of phenolphthalein to the appropriate end of the U-tube.

7. Dispose of your reactants and products as instructed by your teacher. Take your apparatus apart, rinse out the tubing, and rinse off the electrodes. Return your equipment to its appropriate location.

Analysis

1. Sketch the cell you made in this investigation. On your sketch, show
   a) the direction of the electron flow in the external circuit
   b) the anode and the cathode
   c) the positive electrode and the negative electrode
   d) the movement of ions in the cell

2. Use your observations to identify the product(s) formed at the anode and the product(s) formed at the cathode.

3. Write a balanced equation for the half-reaction that occurs at the anode.

4. Write a balanced equation for the half-reaction that occurs at the cathode.

5. Write a balanced equation for the overall cell reaction.

6. Calculate the external voltage required to carry out the electrolysis. Why was the external voltage used in the investigation significantly higher than the calculated value?

Conclusion

7. What are the products from the electrolysis of a 1 mol/L aqueous solution of potassium iodide? Are the observed products the same as the products predicted using reduction potentials?

Applications

8. If you repeated the electrolysis using aqueous sodium iodide instead of aqueous potassium iodide, would your observations change? Explain your answer.

9. To make potassium by electrolyzing potassium iodide, would you need to modify the procedure? Explain your answer.

Electrolysis of Molten Salts

Since sodium cannot be purified by electrolysis in an aqueous solution, how is pure sodium obtained? Sir Humphry Davy found the answer. Salts are solid at standard temperatures and will not conduct electric current. So Davy heated the salts until they melted and applied electrolysis to the molten solids. Since there was no water present, it could not interfere with the reactions. Davy was successful in using electrolysis to isolate sodium, potassium, magnesium, calcium, strontium, and barium.

The electrolysis of molten sodium chloride is an important industrial reaction. The half-reactions are shown here.

Reduction half-reaction (cathode): $Na^+(\ell) + e^- \rightarrow Na(\ell)$

Oxidation half-reaction (anode): $2Cl^-(\ell) \rightarrow Cl_2(g) + 2e^-$

Figure 13.26 shows the large electrolytic cell used in the industrial production of sodium and chlorine.

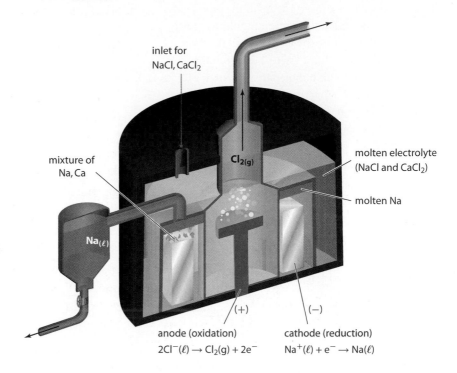

Figure 13.26 The large cell used for the electrolysis of sodium chloride in industry is known as a Downs cell. To decrease heating costs, calcium chloride is added to lower the melting point of sodium chloride from about 800 °C to about 600 °C. The reaction produces sodium and calcium by reduction at the cathode, and chlorine by oxidation at the anode.

Check your understanding of electrolytic cells by completing the following questions.

17 The electrolysis of molten calcium chloride produces calcium and chlorine. Write

   a) the half-reaction that occurs at the anode

   b) the half-reaction that occurs at the cathode

   c) the complete balanced equation for the overall cell reaction

18 For the electrolysis of molten lithium bromide, write

   a) the half-reaction that occurs at the negative electrode

   b) the half-reaction that occurs at the positive electrode

   c) the net ionic equation for the overall cell reaction

19 A voltaic cell produces direct current, which flows in one direction. The main electrical supply at your home is a source of alternating current, which changes direction 120 times a second. Explain why the external electrical supply for an electrolytic cell must be a direct current rather than an alternating current.

20 Suppose a battery is used as the external electrical supply for an electrolytic cell. Explain why the negative terminal of the battery must be connected to the cathode of the cell.

Rechargeable Batteries

In section 13.2, you learned about several primary (disposable) batteries that contain voltaic cells. Now that you have learned about electrolytic cells, you might have figured out how secondary batteries can be recharged. One of the most common secondary (rechargeable) batteries is found in car engines. Most cars contain a lead-acid battery, similar to the one shown in Figure 13.27. When you turn the ignition, a surge of electrical current from the battery starts the engine.

When in use, a lead-acid battery partially discharges. In other words, the cells in the battery operate as voltaic cells and produce electrical energy. The reaction in each cell proceeds spontaneously in one direction. To recharge the battery, an alternator, driven by the car engine, supplies electrical energy to the battery. The external voltage of the alternator reverses the reaction in the cells. The reaction in each cell now proceeds non-spontaneously, and the cells operate as electrolytic cells. All secondary batteries, including the lead-acid battery, operate some of the time as voltaic cells and some of the time as electrolytic cells.

As the name suggests, the materials used in a lead-acid battery include lead and an acid. Figure 13.28 shows that the electrodes in each cell are constructed using lead grids. One electrode consists of powdered lead packed into one grid. The other electrode consists of powdered lead(IV) oxide packed into the other grid. The electrolyte solution is sulfuric acid at a concentration of about 4.5 mol/L.

Figure 13.27 A typical car battery consists of six 2-V cells. The cells are connected in series to give a total potential of 12 V.

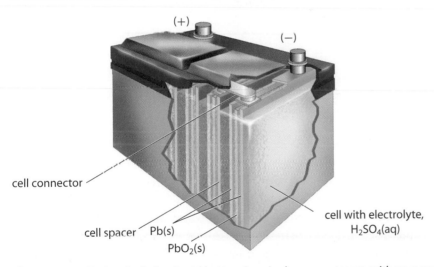

(+)   (−)

cell connector

cell spacer   Pb(s)

PbO₂(s)

cell with electrolyte, $H_2SO_4(aq)$

Figure 13.28 Each cell of a lead-acid battery is a single compartment, with no porous barrier or salt bridge. Fibreglass or wooden sheets are placed between the electrodes to prevent them from touching.

When the battery supplies electrical energy, the half-reactions and overall cell reaction are as follows.

oxidation (at the Pb anode):
$$Pb(s) + SO_4^{2-}(aq) \rightarrow PbSO_4(s) + 2e^-$$
reduction (at the PbO₂ cathode):
$$PbO_2(s) + 4H^+(aq) + SO_4^{2-}(aq) + 2e^- \rightarrow PbSO_4(s) + 2H_2O(\ell)$$
overall cell reaction:
$$Pb(s) + PbO_2(s) + 4H^+(aq) + 2SO_4^{2-}(aq) \rightarrow 2PbSO_4(s) + 2H_2O(\ell)$$

The reaction consumes some of the lead in the anode, some of the lead(IV) oxide in the cathode, and some of the sulfuric acid. A precipitate of lead(II) sulfate forms.

Figure 13.29 Billions of rechargeable nicad batteries are produced every year. They are used in portable devices, such as cordless razors and cordless power tools.

When the battery is recharged, the half-reactions and the overall cell reaction are reversed. In this reverse reaction, lead and lead(IV) oxide are redeposited in their original locations, and sulfuric acid is re-formed.

reduction (at the Pb cathode):
$$PbSO_4(s) + 2e^- \rightarrow Pb(s) + SO_4^{2-}(aq)$$

oxidation (at the PbO_2 anode):
$$PbSO_4(s) + 2H_2O(\ell) \rightarrow PbO_2(s) + 4H^+(aq) + SO_4^{2-}(aq) + 2e^-$$

overall cell reaction:
$$2PbSO_4(s) + 2H_2O(\ell) \rightarrow Pb(s) + PbO_2(s) + 4H^+(aq) + 2SO_4^{2-}(aq)$$

In practice, this reversibility is not perfect. Nevertheless, the battery can go through many charge/discharge cycles before it eventually wears out.

Many types of rechargeable batteries are much more portable than a lead-acid battery. For example, a rechargeable version of the alkaline battery is now available. Another example, shown in Figure 13.29, is the rechargeable nickel-cadmium (nicad) battery. Figure 13.30 shows a nickel-cadmium cell, which has a potential difference of about 1.4 V. A typical nicad battery contains three cells in series to produce a suitable voltage for electronic devices. When the cells in a nicad battery operate as voltaic cells, the half-reactions and the overall cell reaction are as follows.

oxidation (at the Cd anode):
$$Cd(s) + 2OH^-(aq) \rightarrow Cd(OH)_2(s) + 2e^-$$

reduction (at the NiO(OH) cathode):
$$NiO(OH)(s) + H_2O(\ell) + e^- \rightarrow Ni(OH)_2(s) + OH^-(aq)$$

overall cell reaction:
$$Cd(s) + 2NiO(OH)(s) + 2H_2O(\ell) \rightarrow Cd(OH)_2(s) + 2Ni(OH)_2(s)$$

Like many technological innovations, nickel-cadmium batteries carry risks as well as benefits. After being discharged repeatedly, they eventually wear out. In theory, worn-out nicad batteries should be recycled. In practice, however, many end up in garbage dumps. Over time, discarded nicad batteries release toxic cadmium. The toxicity of this substance makes it hazardous to the environment, as cadmium can enter the food chain. Long-term exposure to low levels of cadmium can have serious medical effects on humans, such as high blood pressure and heart disease.

Section 13.3 Summary

- An electrolytic cell is a device that converts electrical energy into chemical potential energy.
- A voltaic cell can be turned into an electrolytic cell by inserting a battery or electrical power source into the external circuit and driving the reactions backwards.

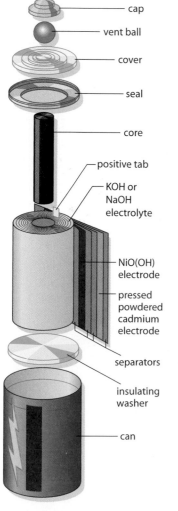

cap
vent ball
cover
seal
core
positive tab
KOH or NaOH electrolyte
NiO(OH) electrode
pressed powdered cadmium electrode
separators
insulating washer
can

Figure 13.30 A nicad cell has a cadmium electrode and another electrode that contains nickel(III) oxyhydroxide, NiO(OH). When the cell is discharging, cadmium is the anode. When the cell is recharging, cadmium is the cathode. The electrolyte is a base—either sodium hydroxide or potassium hydroxide.

- The electrode that would be the anode in a voltaic cell is made to be the cathode in an electrolytic cell because the external source of electrical energy forces the electrons in the direction opposite to which they would go spontaneously. Similarly, the electrode that would be the cathode in the voltaic cell becomes the anode in the electrochemical cell.

- Some reactions in aqueous solutions do not proceed exactly backwards because water can be oxidized or reduced in place of the metal or salt of the voltaic cell.

- Water and the hydrogen and hydroxide ions cannot be present in 1.0 mol/L concentrations so you cannot use the standard reduction potentials for water to calculate the external potential needed to drive a reaction. You must use the reduction potentials for the actual concentrations of water, hydrogen, and hydroxide ions.

- Energy is required to oxidize or reduce water but energy is also needed to convert the hydrogen and oxygen into gases. This extra energy is called the over potential.

- The electrolysis of brine (sodium chloride in water) is used commercially to produce chlorine gas, hydrogen gas, and sodium hydroxide base in a process known as the chlor-alkali process.

- To produce pure metals from ionic compounds, it is often necessary to electrolyze the molten salts in the absence of water.

- Metallic sodium and chlorine gas are produced by electrolysis of molten sodium chloride in a large cell called a Down cell.

- Electrolysis is also used to recharge batteries. By driving the chemical reactions in a battery backwards, the original compounds can be regenerated.

SECTION 13.3 REVIEW

1. Predict the products of the electrolysis of a 1 mol/L aqueous solution of copper(I) bromide.

2. In this section, you learned that an external source of electrical energy can reverse the cell reaction in a Daniell cell so that the products are zinc atoms and copper(II) ions.
   a) What are the predicted products of this electrolysis reaction?
   b) Explain the observed products.

3. Predict whether each of the following reactions is spontaneous or non-spontaneous under standard conditions:
   a) $2FeI_3(aq) \rightarrow 2Fe(s) + 3I_2(s)$
   b) $2Ag^+(aq) + H_2SO_3(aq) + H_2O(\ell)$
   $\rightarrow 2Ag(s) + SO_4^{2-}(aq) + 4H^+(aq)$

4. Write the two half-reactions and the overall cell reaction for the process that occurs when a nicad battery is being recharged.

5. What external voltage is required to recharge a lead-acid car battery?

6. The equation for the overall reaction in an electrolytic cell does not include any electrons. Why is an external source of electrical energy needed for the reaction to proceed?

7. a) Predict whether aluminium will displace hydrogen from water.
   b) Water boiling in an aluminium saucepan does not react with the aluminium. Give possible reasons why.

8. What are the advantages and disadvantages of the lead acid battery used in cars and trucks? Why is it extremely important to use caution when recharging a lead acid car battery?

9.

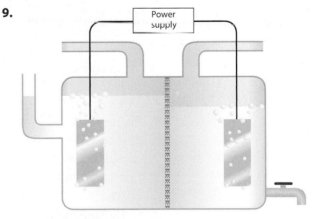

Sketch this chlor-alkali cell in your notebook. Add labels, arrows to indicate flow of matter, and chemical equations to show what is taking place at each electrode. Explain how this chlor-alkali cell works.

Stoichiometry and Faraday's Law

Section Outcomes

In this section, you will:

- **calculate** mass, amounts, current, and time in electrolytic cells by applying Faraday's law and stoichiometry
- **describe** the ways in which scientific knowledge led to the development of electroplating and refining metals from ores

Key Terms

Faraday's law
extraction
refining

You read in section 13.1 that Michael Faraday (1791–1867) was the leading pioneer of electrochemistry. One of Faraday's major contributions was to connect the concepts of stoichiometry and electrochemistry. Using minimal technology, as shown in Figure 13.31, Faraday developed concepts and laws of electricity and magnetism that are still fundamental to these fields today. In this section, you will apply some of the concepts that Faraday developed.

Figure 13.31 Michael Faraday carried out some of the most fundamental experiments in electricity and magnetism is his laboratory, which is depicted here.

Stoichiometry in Electrochemistry

You learned about stoichiometry in Chapter 7, so you know that the coefficients in a balanced chemical equation represent the mole ratios of the reactants and products in the reaction. When you apply stoichiometry to electrochemical equations, you must include an additional factor—the current passing through the external circuit. How can you quantitatively relate current flowing through a circuit to the reactions occurring in the half-cells?

First, you will need to review the definitions and relationships among the electrical quantities. As you know, the flow of electrons through a metal conductor is called current. The current (I) that passes any given point in a conductor is defined as the amount of charge (q) that passes the point in an interval of time (Δt). You can write the relationship mathematically as $I = \frac{q}{\Delta t}$.

The SI unit of electric current is the ampere (A), named after the French physicist André Ampère (1775–1836). The SI unit of electric charge is the coulomb (C), named after another French physicist, Charles Coulomb (1736–1806). The unit of time is the second (s). Therefore, if one ampere of current is flowing in a conductor, one coulomb of charge is passing each point in the conductor every second (A = C/s).

In the half-cells, during a redox reaction, electrons are being donated by a reducing agent and accepted by an oxidizing agent. You can measure the current passing through the external circuit and the time interval during which that current was flowing.

Using these data, you can calculate the total amount of charge, in the form of electrons, that was transferred from the reducing agent to the oxidizing agent. By rearranging the equation for current, you can derive the mathematical equation $q = I\,\Delta t$. When you perform this calculation, the charge will be in units of coulombs. Now you need a way to convert a coulomb of charge into a mole of electrons because you must work in units of moles to use the data in stoichiometric calculations.

You can find the charge on one mole of electrons by using Avogadro's number for the number of particles in a mole. You then use the charge on one electron—a measurement made by Robert Millikan (1868–1953).

Charge on one mole of electrons = charge on one electron × number of electrons in a mole

$$= \left(\frac{1.602 \times 10^{-19}\,\text{C}}{1e^-}\right)\left(\frac{6.022 \times 10^{23}\,e^-}{1\,\text{mol}}\right)$$

$$= 9.647 \times 10^4\,\frac{\text{C}}{\text{mol}}$$

The numerical value of charge on a mole of electrons is called a faraday (1 F), in honour of Michael Faraday, who carried out the original experiments. A rounded value of 9.65×10^4 C/mol is often used in calculations.

The information above provides you with a method to very precisely control electrolysis. For example, suppose you modified a Daniell cell to operate as an electrolytic cell. You want to plate 0.1 mol of zinc atoms onto the zinc electrode. The coefficients in the half-reaction for the reduction represent stoichiometric relationships. Figure 13.32 shows that two moles of electrons are needed for each mole of zinc deposited. Therefore, to deposit 0.1 mol of zinc, you need to use 0.2 mol of electrons:

Zn^{2+}	+	$2e^-$	$\longrightarrow$	Zn
1 ion		2 electrons		1 atom
$1 \times 6.02 \times 10^{23}$ ions		$2 \times 6.02 \times 10^{23}$ electrons		$1 \times 6.02 \times 10^{23}$ atoms
1 mol of ions		2 mol of electrons		1 mol of atoms

Figure 13.32 A balanced half-reaction shows the relationships among the amounts of reactants and products and the amount of electrons transferred.

In the next Sample Problem, you will learn to apply the relationship between the amount of electrons and the amount of an electrolysis product.

Chemistry File

FYI
Considered by many to be the greatest experimental chemist ever, Michael Faraday did not receive any formal education beyond the primary grades. At the age of 14, Faraday worked as an apprentice at a bookbindery in London, where he educated himself by reading many of the books brought there for binding, including the section on electricity in the *Encyclopaedia Britannica*.
   A client of the bookbindery gave Faraday tickets to lectures given by Sir Humphry Davy at the Royal Institution. Faraday eagerly attended the lectures and afterward presented his detailed and precise notes on them to Davy. Impressed by the young Faraday's diligence, Davy hired him as his laboratory assistant in 1813, saying "his disposition is active and cheerful, his manner intelligent." In 1825, Faraday took over from Davy directing the laboratory at the Royal Institution and went on to contribute even more to the study of electricity and its applications than did Davy, himself an eminent figure in the field.

Sample Problem

Calculating the Mass of an Electrolysis Product

Problem
Calculate the mass of aluminium produced by the electrolysis of molten aluminium chloride, if a current of 500 mA passes between the half-cells for 1.50 h.

What Is Required?
You need to calculate the mass of aluminium produced.

What Is Given?
You know the name of the electrolyte, the current, and the time.

electrolyte: $AlCl_3(\ell)$
current: 500 mA
time: 1.50 h
You know the charge on one mole of electrons: 9.65×10^4 C/mol.

Plan Your Strategy
Use the current and the time to find the quantity of electric charge that passed from the anode to the cathode. From the charge, find the amount of electrons that passed through the circuit. Use the stoichiometry of the relevant half-reaction to relate the amount of electrons to the amount of aluminium produced. Use the molar mass of aluminium to convert the amount of aluminium to a mass of aluminium.

Act on Your Strategy

To calculate the quantity of electrical charge in coulombs, convert the data to SI units:

$1000 \text{ mA} = 1 \text{ A}$

$500 \text{ mA} = (500 \text{ mA})\left(\dfrac{1 \text{ A}}{1000 \text{ mA}}\right)$

$= 0.500 \text{ A}$

$1.50 \text{ h} = (1.50 \text{ h})\left(\dfrac{60 \text{ min}}{1 \text{ h}}\right)\left(\dfrac{60 \text{ s}}{1 \text{ min}}\right)$

$= 5.4 \times 10^3 \text{ s}$

$q = I\Delta t$

$q = (0.500 \text{ A})(5400 \text{ s})$

$= 2.700 \times 10^3 \text{ C}$

Find the amount of electrons. One mole of electrons has a charge of 9.65×10^4 C:

Amount of electrons $= 2700 \text{ C} \times \dfrac{1 \text{ mol e}^-}{9.65 \times 10^4 \text{ C}}$

$= 0.0280 \text{ mol e}^-$

The half-reaction for the reduction of aluminium ions to aluminium atoms is

$Al^{3+} + 3e^- \rightarrow Al$

Amount of aluminium formed $= 0.0280 \text{ mol e}^- \times \dfrac{1 \text{ mol Al}}{3 \text{ mol e}^-}$

$= 0.009\,33 \text{ mol Al}$

Convert the amount of aluminium to a mass:

$m_{Al} = nM_{Al}$

$= (0.009\,33 \text{ mol Al})\left(\dfrac{27.0 \text{ g Al}}{1 \text{ mol Al}}\right)$

$= 0.252 \text{ g}$

Check Your Solution

The answer is expressed in units of mass. To check your answer, use estimation. If the current were 1 A, then 1 mol of electrons would pass in 9.65×10^4 s. In this example, the current is less than 1 A, and the time is much less than 9.65×10^4 s. Therefore, much less than 1 mol of electrons would be used, and much less than 1 mol (27 g) of aluminium would be formed.

Using Dimensional Analysis:

$m_{Al} = \left(\dfrac{0.500 \text{ C}}{1 \text{ s}} \times 5400 \text{ s}\right)\left(\dfrac{1 \text{ mol e}^-}{9.65 \times 10^4 \text{ C}}\right)\left(\dfrac{1 \text{ mol Al}}{3 \text{ mol e}^-}\right)\left(\dfrac{27.0 \text{ g}}{1 \text{ mol}}\right)$

$= 0.252 \text{ g}$

Practice Problems

9. Calculate the mass of zinc plated onto the cathode of an electrolytic cell by a current of 750 mA in 3.25 h.

10. How many minutes does it take to plate 0.925 g of silver onto the cathode of an electrolytic cell using a current of 1.55 A?

11. The nickel anode in an electrolytic cell decreases in mass by 1.20 g in 35.5 min. The oxidation half-reaction converts nickel atoms to nickel(II) ions. What is the average current?

12. The following two half-reactions take place in an electrolytic cell with an iron anode and a chromium cathode:
Oxidation: $Fe(s) \rightarrow Fe^{2+}(aq) + 2e^-$
Reduction: $Cr^{3+}(aq) + 3e^- \rightarrow Cr(s)$
During the process, the mass of the iron anode decreases by 1.75 g.
a) Find the change in mass of the chromium cathode.
b) Explain why you do not need to know the electric current or the time to complete part (a).

The preceding Sample Problem gave an example of the mathematical use of **Faraday's law**, which is given in the following box.

> **Faraday's law**
>
> The amount of a substance produced or consumed in an electrolysis reaction is directly proportional to the amount of charge that flows through the circuit.

To illustrate this law, think about changing the amount of charge used in the Sample Problem. Suppose this amount were doubled by using the same current, 500 mA, for twice the time, 3 h. As a result, the amount of electrons passing into the cell would also be doubled.

$$500 \text{ mA} = 0.500 \text{ A}$$
$$3 \text{ h} = 2 \times 1.5 \text{ h}$$
$$= 2 \times 5400 \text{ s}$$
$$q = I\Delta t$$
$$q = (0.500 \text{ A})(2 \times 5400 \text{ s})$$
$$= 2 \times 2700 \text{ C}$$
$$= 5.40 \times 10^3 \text{ C}$$

$$\text{Amount of electrons} = 5.40 \times 10^3 \, \cancel{C} \times \frac{1 \text{ mol e}^-}{9.65 \times 10^4 \, \cancel{C}}$$
$$= 0.0560 \text{ mol e}^-$$

Then, as you can see from the relevant half-reaction, the mass of aluminium produced would be doubled. The mass of aluminium produced is clearly proportional to the amount of charge used.

In Investigation 13.C, you will apply Faraday's law to an electrolytic cell that you construct.

INVESTIGATION | 13.C

Target Skills

Describing procedures for safe handling and disposal of materials used in the laboratory

Identifying limitations of data collected

Comparing predictions with observations of electrochemical cells

Electroplating

You have learned that electroplating is a process in which a metal is deposited, or plated, onto the cathode of an electrolytic cell. In this investigation, you will build an electrolytic cell and electrolyze a copper(II) sulfate solution to plate copper onto the cathode. You will use Faraday's law to relate the mass of metal deposited to the amount of electric charged used.

Question

Does the measured mass of copper plated onto the cathode of an electrolytic cell agree with the mass calculated according to Faraday's law?

Prediction

Predict whether the measured mass of copper plated onto the cathode of an electrolytic cell will be greater than, equal to, or less than the mass calculated using Faraday's law.

Safety Precautions

- Nitric acid is corrosive. Note that the $CuSO_4$ solution contains sulfuric acid and hydrochloric acid. Wash any spills on your skin with plenty of cold water. Inform your teacher immediately.
- Avoid touching the parts of the electrodes that have been washed with nitric acid.

- Make sure your hands and your lab bench are dry before handling any electrical equipment.

Materials

- 3 cm × 12 cm × 1 mm Cu strip
- 150 mL 1.0 mol/L HNO_3 in a 250 mL beaker
- deionized water in a wash bottle
- 50 cm 16-gauge bare solid copper wire
- 120 mL acidified 0.50 mol/L $CuSO_4$ solution (with 5 mL of 6 mol/L H_2SO_4 and 3 mL of 0.1 mol/L HCl added)
- fine sandpaper
- 250 mL beaker
- 2 electrical leads with alligator clips
- adjustable D.C. power supply with ammeter
- drying oven
- electronic balance

Procedure

1. Clean any tarnish off the copper strip by sanding it gently. Working in a fume hood, dip the bottom of the copper strip in the nitric acid for a few seconds, and then rinse the strip carefully with de-ionized water. Avoid touching the section that has been cleaned by the acid.

2. Place the copper strip in the beaker, with the clean part of the strip at the bottom. Bend the top of the strip over the rim of the beaker so that the copper strip is secured in a vertical position, as shown in the diagram. This copper strip will serve as the anode.

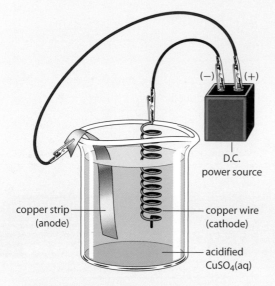

copper strip (anode)
copper wire (cathode)
acidified CuSO₄(aq)
(−) (+)
D.C. power source

3. Wrap the copper wire around a pencil to make a closely spaced coil. Leave 10 cm of the wire unwrapped. Measure and record the mass of the wire. Working in a fume hood, dip the coil in the nitric acid, and rinse the coil with de-ionized water. Use the 10 cm of uncoiled wire to secure the coil on the opposite side of the beaker from the anode, as shown in the diagram. This copper wire will serve as the cathode.

4. Pour 120 mL of the acidified CuSO₄(aq) solution into the beaker. Attach the lead from the negative terminal of the power supply to the cathode. Attach the positive terminal to the anode.

5. Turn on the power supply and set the current to 1.0 A. Maintain this current for 20 min by adjusting the variable current knob as needed.

6. After 20 min, turn off the power. Remove the cathode and rinse it very gently with deionized water. Place the cathode in a drying oven for 20 min.

7. Measure and record the new mass of the cathode.

8. Dispose of all materials as instructed by your teacher.

Analysis

1. Write a balanced equation for the half-reaction that occurs at the cathode.

2. Use the measured current and the time for which the current passed to calculate the amount of charge used.

3. Use your answers to questions 1 and 2 to calculate the mass of copper plated onto the cathode.

4. Compare the calculated mass from question 3 with the measured increase in mass of the cathode. Give possible reasons for any difference between the two values.

Conclusion

5. How did the mass of copper electroplated onto the cathode of the electrolytic cell compare with the mass calculated using Faraday's law? Compare your answer with your prediction from the beginning of this investigation.

Applications

6. Suppose you repeated this investigation using iron electrodes and 0.5 mol/L iron(II) sulfate solution as the electrolyte. If you used the same current for the same time, would you expect the increase in mass of the cathode to be greater than, less than, or equal to the increase in mass that you measured? Explain your answer.

7. Suppose you repeated the investigation with the copper(II) sulfate solution, but you passed the current for only half as long as before. How would the masses of copper plated onto the cathode compare in the two investigations? Explain your answer.

8. Could you build a voltaic cell without changing the electrodes or the electrolyte solution you used in this investigation? Explain your answer.

Industrial Extraction and Refining of Metals

Many metals, and their alloys, are widely used in modern society. The enormous variety of metal objects ranges from large vehicles, such as cars and aircraft, to small items, such as the pop cans shown in Figure 13.33.

Extraction is a process by which a metal is obtained from an ore. Some metals are extracted in electrolytic cells. In Section 13.3, you saw the extraction of sodium from molten sodium chloride in a Downs cell. Other reactive metals, including lithium, beryllium, magnesium, calcium, and radium, are also extracted industrially by the electrolysis of their molten chlorides.

One of the most important electrolytic processes is the extraction of aluminium from an ore called bauxite. This ore is mainly composed of hydrated aluminium oxide, $Al_2O_3 \cdot xH_2O$. In industry, the scale of production of metals is huge. The electrolytic production of aluminium is more than two million tonnes per year in Canada alone. As you know from Faraday's law, the amount of a metal produced by electrolysis is directly proportional to the amount of charge used. Therefore, the industrial extraction of aluminium and other metals by electrolysis requires vast amounts of electrical energy. The availability and cost of electrical energy greatly influence the location of industrial plants.

In industry, the process of purifying a material is known as **refining**. After the extraction stage, some metals are refined in electrolytic cells. For example, copper is about 99% pure after extraction. This copper is pure enough for some uses, such as the manufacture of copper pipes for plumbing. However, the copper is not pure enough for one of its principal uses, electrical wiring. Therefore, some of the impure copper is refined electrolytically, as shown in Figure 13.34. Nickel can be refined electrolytically in a similar way. You refined copper on a small scale in Investigation 13.C.

Figure 13.33 The alloy used to make pop cans contains about 97% aluminium, by mass. The other elements in the alloy are magnesium, manganese, iron, silicon, and copper.

Figure 13.34 This electrolytic cell is used to refine copper. The anode is impure copper, and the cathode is pure copper. During electrolysis, the impure copper anode dissolves, and pure copper is plated onto the cathode. The resulting cathode is 99.99% pure metal. Most impurities that were present in the anode either remain in solution or fall to the bottom of the cell as sludge.

Section 13.4 Summary

- To perform stoichiometric calculations on redox reactions occurring in electrochemical cells, you must account for the electrons that are flowing in the external circuit.

- You can determine the charge per mole of electrons by multiplying the number of coulombs per electron (a quantity measured by Millikan) times the number of electrons in a mole (Avogadro's number). The result, 9.647×10^4 C/mol, is called a faraday (F) after Michael Faraday.

- Faraday's law states that the amount of a substance produced or consumed in an electrolysis reaction is directly proportional to the amount of charge that flows through the circuit.

- You can find the charge that flows through a circuit by multiplying the current in the circuit times the time interval over which the current was flowing.

- Electrolysis is used industrially for electroplating of metals, extracting metals from ore, and for refining metals.

SECTION 13.4 REVIEW

1. In section 12.3, you learned about a redox reaction used in the production of compact discs. In another step of this production process, nickel is electroplated onto the silver-coated master disc. The nickel layer is removed and used to make pressings of the CD onto plastic discs. The plastic pressings are then coated with aluminium to make the finished CDs.

   a) When nickel is plated onto the silver master disc, is the master disc the anode or the cathode of the cell? Explain.

   b) Calculate the amount of charge needed to plate each gram of nickel onto the master disc. Assume that the plating process involves the reduction of nickel(II) ions.

2. Most industrial reactions take place on a much larger scale than reactions in a laboratory or classroom. The voltage used in a Downs cell for the industrial electrolysis of molten sodium chloride is not very high, about 7 V to 8 V. However, the current used is 25 000 A to 40 000 A. Assuming a current of 3.0×10^4 A, determine the mass of sodium and the mass of chlorine made in 24 h in one Downs cell. Express your answers in kilograms.

3. An industrial cell that purifies copper by electrolysis operates at 2.00×10^2 A. Calculate the mass, in tonnes, of pure copper produced if the cell is supplied with raw materials whenever necessary and if it works continuously for a year that is not a leap year.

4. Canada is a major producer of aluminium by the electrolysis of bauxite. However, there are no bauxite mines in Canada, and all the ore must be imported. Explain why aluminium is produced in Canada.

5. Aluminium is the most abundant metal in Earth's crust. However, obtaining pure aluminium from bauxite ore is a very complex process. First aluminium oxide, $Al_2O_3(s)$, is prepared from the bauxite ore by a multistep process involving acid-base reactions and solubility properties. Pure aluminium is prepared from the aluminium oxide by electrolysis. Aluminium is too strong a reducing agent to obtain it by electrolysis from an aqueous solution. However, aluminium oxide has a melting point of 2030 °C making it extremely costly to heat it to its melting point to electrolyze the molten salt. Thus the aluminium oxide is mixed with a compound called cryolite ($Na_3AlF_6(s)$) because the mixture melts at about 1000 °C. Electrolysis of the molten mixture is carried out in a graphite lined furnace. The graphite lining acts as one electrode and graphite rods form the other electrode. The process is usually operated at 4.5 V with a current as high as 2.5×10^5 A. The overall cell reaction is:

$$2Al_2O_3(\ell) + 3C(s) \rightarrow 4Al(\ell) + CO_2(g)$$

Notice that the graphite electrodes also act as one of the reactants and the rods are used up in the process and must periodically be replaced.

In 1998, the world production of aluminium was 22.7 t. For what period of time would one cell have to be operated to produce that amount of aluminium? If electric energy is calculated from $E = qV$, how much energy was used to produce 22.7 t of aluminium?

6. Nickel and copper are both very important to the Canadian economy. Before they can be refined by electrolysis, they must be extracted from their ores. Both metals can be extracted from a sulfide ore, NiS or Cu_2S. The sulfide is roasted to form an oxide, and then the oxide is reduced to the metal. Research the extraction processes for both nickel and copper, and write balanced equations for the redox reactions involved. One product of each extraction process is sulfur dioxide. Research the environmental effects of this compound. Describe any steps taken to decrease these effects. **ICT**

Chapter 13 SUMMARY

In a redox reaction, a reducing agent spontaneously donates electrons to an oxidizing agent. If the oxidizing and reducing agents are separated by an external circuit, the electrons will be forced to travel through the circuit. The electrical energy can then be used for a variety of purposes. Such a system is called a voltaic cell.

A variety of metals can be used as electrodes. When ions or gases act as oxidizing or reducing agents, an inert electrode must be used. The potential difference between the electrodes that has been generated by the chemical reaction is called the cell potential. Chemists have agreed on the standard conditions of a concentration of 1.0 mol/L, an external pressure of $1.013\,25 \times 10^5$ Pa, pure metal electrodes, and the temperature is 25 °C for reporting cell potentials.

Chemists have also agreed to use the hydrogen half-cell as a reference. This consists of hydrogen ions at a concentration of 1.0 mol/L in solution and hydrogen gas at a pressure of $1.013\,25 \times 10^5$ Pa bubbling past a platinum electrode. The standard reduction potential for all other half-cells is the potential difference between the platinum electrode of the hydrogen half-cell and the electrode other half-cell. From known standard reduction potentials, you can calculate the standard cell potential of any cell by using the formula,
$$E°_{cell} = E°_{cathode} - E°_{anode}.$$
When the electrolyte of a voltaic cell is thickened to a paste and the components of the cell are sealed, a dry cell is produced. A battery is a set of cells connected in series. A fuel cell is similar to a "battery" but the oxidizing and reducing agents flow through the cell instead of being sealed in the cell.

Corrosion, or rust, results from the spontaneous formation of many voltaic cells on the surface of metals. Corrosion is costly and thus methods of preventing corrosion are important. A coat of paint, a coat of zinc (galvanizing), or the presence of a sacrificial anode can protect iron. A sacrificial anode is a metal that is oxidized more readily than iron.

An electrolytic cell is an electrochemical cell in which an external source of electrical energy drives non-spontaneous redox reactions. In a solution of a salt in water, you can predict which components of the solution will be oxidized and which will be reduced by comparing their reduction potentials. The reaction that requires the least energy will occur. The principle of electrolysis is used in the chlor-alkali process to produce chlorine gas, hydrogen gas, and sodium hydroxide from brine. To produce metallic sodium, molten sodium chloride must be electrolyzed. Electrolysis is also used to recharge batteries.

You can calculate the extent of a redox reaction that occurs under electrolysis by using the formula, $q = I\,\Delta t$ (the relationship among charge, current, and a time interval) along with the Faraday constant which is the number of coulombs of charge in a mole of electrons. These calculations are important for processes such as electroplating and for the electrolytic extraction of metals from ore and for refining of metals.

Concept Organizer   Electrochemical Cells

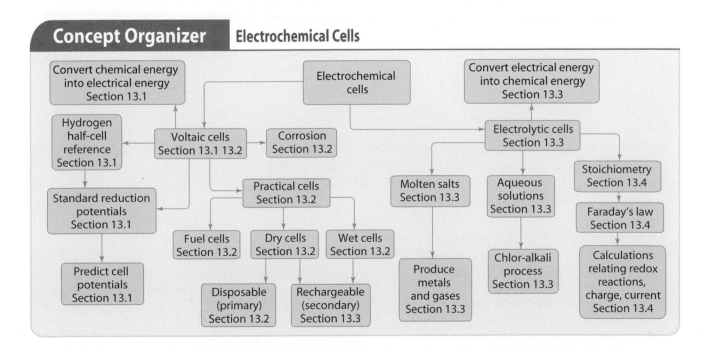

Understanding Concepts

1. Explain the function of the following parts of an electrolytic cell:
   a) electrodes
   b) electrolyte
   c) external voltage

2. In a voltaic cell, one half-cell has a cadmium electrode in a 1 mol/L solution of cadmium nitrate. The other half-cell has a magnesium electrode in a 1 mol/L solution of magnesium nitrate. Write the cell notation for the cell.

3. Write the oxidation half-reaction, the reduction half-reaction, and the overall cell reaction for the following voltaic cell:

   $Pt \mid NO(g) \mid NO_3^-(aq), H^+(aq) \parallel I^-(aq) \mid I_2(s), Pt$

4. What is the importance of the hydrogen electrode?

5. Lithium, sodium, beryllium, magnesium, calcium, and radium are all made industrially by the electrolysis of their molten chlorides. These salts are all soluble in water, but aqueous solutions are not used for the electrolytic process. Explain why.

6. Use the following two half-reactions to write balanced net ionic equations for one spontaneous reaction and one non-spontaneous reaction. State the standard cell potential for each reaction:

   $N_2O(g) + 2H^+(aq) + 2e^- \rightarrow N_2(g) + H_2O(\ell)$
   $$E° = 1.770\ V$$
   $CuI(s) + e^- \rightarrow Cu(s) + I^-(aq) \quad E° = -0.185\ V$

7. Identify the oxidizing agent and the reducing agent in a lead-acid battery that is
   a) discharging
   b) recharging

8. Rank the following in order from most effective to least effective oxidizing agents under standard conditions: $Zn^{2+}(aq), Co^{3+}(aq), Br_2(\ell), H^+(aq)$

9. Rank the following in order from most effective to least effective reducing agents under standard conditions: $H_2(g), Cl^-(aq), Al(s), Ag(s)$

10. The ions $Fe^{2+}(aq), Ag^+(aq)$, and $Cu^{2+}(aq)$ are present in the half-cell that contains the cathode of an electrolytic cell. The concentration of each of these ions is 1 mol/L. If the external voltage is very slowly increased from zero, in what order will the three metals Fe, Ag, and Cu begin to be plated onto the cathode? Explain your answer.

Applying Concepts

11. Write the half-reactions and calculate the standard cell potential for each reaction. Identify each reaction as spontaneous or non-spontaneous:
    a) $Zn(s) + Fe^{2+}(aq) \rightarrow Zn^{2+}(aq) + Fe(s)$
    b) $Cr(s) + AlCl_3(aq) \rightarrow CrCl_3(aq) + Al(s)$
    c) $2AgNO_3(aq) + H_2O_2(aq) \rightarrow$
    $$2Ag(s) + 2HNO_3(aq) + O_2(g)$$

12. a) Describe a method you could use to measure the standard cell potential of the following voltaic cell:

    $Sn(s) \mid Sn^{2+}(aq)(1\ mol/L) \parallel Pb^{2+}(aq)(1\ mol/L) \mid Pb(s)$

    b) Why is this cell unlikely to find many practical uses?

13. The two half-cells in a voltaic cell consist of one iron electrode in a 1 mol/L iron(II) sulfate solution, and a silver electrode in a 1 mol/L silver nitrate solution.
    a) Assume the cell is operating as a voltaic cell. State the cell potential, the oxidation half-reaction, the reduction half-reaction, and the overall cell reaction.
    b) Repeat part (a), but this time, assume that the cell is operating as an electrolytic cell.
    c) For the voltaic cell in part (a), do the mass of the anode, the mass of the cathode, and the total mass of the two electrodes increase, decrease, or stay the same while the cell is operating?
    d) Repeat part (c) for the electrolytic cell in part (b).

14. a) Describe an experiment you could perform to determine the products from the electrolysis of aqueous zinc bromide. How would you identify the electrolysis products?
    b) Zinc and bromine are the observed products from the electrolysis of aqueous zinc bromide solution under standard conditions. They are also the observed products from the electrolysis of molten zinc bromide. Explain why the first observation is surprising.

15. Use the half-cells shown in a table of standard reduction potentials. Could you build a battery with a potential of 8 V? If your answer is yes, give an example.

16. Research the following information. Prepare a short presentation or booklet on the early history of electrochemistry: ICT
    a) the contributions of Galvani and Volta to the development of electrochemistry
    b) how Humphry Davy and Michael Faraday explained the operation of voltaic and electrolytic cells (note: these scientists could not describe them in terms of electron transfers, because the electron was not discovered until 1897)

17. How rapidly do you think iron would corrode on the surface of the moon? Explain your answer.

18. Reactions that are the reverse of each other have standard cell potentials that are equal in size but opposite in sign. Explain why.

19. Use a labelled diagram to represent each of the following:
    a) a voltaic cell in which the hydrogen electrode is the anode
    b) a voltaic cell in which the hydrogen electrode is the cathode

20.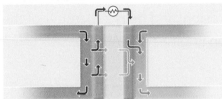

Sketch this PEM fuel cell in your notebook. Add the following labels and explain how it works
anode (porous carbon)
electrolyte membrane
cathode (porous carbon)
hydrogen in
excess hydrogen
oxygen in
load
H_2 (4 times)
e^- (5 times)
H_2O (2 times)
O_2
H^+ (2 times)

Solving Problems

21. Calculate the standard cell potential of a voltaic cell that uses $Ag(s)/Ag^+(aq)$ and $Al(s)/Al^+(aq)$ half-cell reactions. State which half-cell is the oxidation half-cell and which is the reduction half-cell.

22. A voltaic cell using $Mg(s)/Mg^{2+}(aq)$ and $Cu(s)/Cu^{2+}(aq)$ half-cells operates under standard conditions. The cell delivers 0.22 A for 31.6 h. How many grams of $Cu(s)$ are deposited?

23. Calculate the mass of magnesium that can be plated onto the cathode by the electrolysis of molten magnesium chloride by using a current of 3.65 A for 55.0 min.

24. Suppose you produce a kilogram of sodium and a kilogram of aluminium by electrolysis. Compare your electricity costs for these two processes. Assume that electricity is used for electrolysis only and not for heating.

25. What is the hourly production rate of chlorine gas (in kg) from an electrolytic cell using aqueous $NaCl(aq)$ and carrying a current of 1.500×10^3 A?

26. A constant current flows for 3.75 h through an electrolytic cell containing aqueous $AgNO_3(aq)$. During this time, 2.00 g of silver are deposited on the electrode. What is the current flowing through the cell?

Making Connections

27. A D-size dry cell flashlight battery is much bigger than a AAA-size dry cell calculator battery. However, both have cell potentials of 1.5 V. Do they supply the same amount of charge? Explain your answer.

28. a) Would you use aluminium nails to attach an iron gutter to a house? Explain your answer.
    b) Would you use iron nails to attach aluminium siding to a house? Explain your answer

29. Research the aluminium-air battery and the sodium-sulfur battery. Both are rechargeable batteries that have been used to power electric cars. In each case, describe the design of the battery, the half-reactions that occur at the electrodes, and the overall cell reaction. Also, describe the advantages and disadvantages of using the battery as a power source for a car. ICT

30. Explain why the recycling of aluminium is more economically viable than the recycling of many other metals.

31. Suppose you live in a small town with a high rate of unemployment. A company plans to build a smelter there to produce copper and nickel by roasting their sulfide ores and reducing the oxides formed. Would you be in favour of the plant being built or opposed to it? Explain and justify your views.

32. Many metal objects are vulnerable to damage from corrosion. A famous example is the Statue of Liberty. Research the history of the effects of corrosion on the Statue of Liberty. Give a chemical explanation for the processes involved. Describe the steps taken to solve the problem and the chemical reasons for these steps. ICT

33. a) Estimate the number of used batteries you discard in a year. Survey the class to determine an average number. Now estimate the number of used batteries discarded by all the high-school students in your province in a year.
    b) Prepare an action plan suggesting ways of decreasing the number of batteries discarded each year.

Career Focus: Ask a Pipeline Corrosion Expert

Dave Grzyb took a program in Plastics Engineering Technology.

Dave Grzyb is a Technical Specialist with the Alberta Energy and Utilities Board (EUB). The EUB is a provincial, quasi-judicial agency that oversees the discovery, production, and distribution of Alberta's energy resources. Grzyb's team provides advice and recommendations to oil and gas producers and pipeline operators. In Alberta, about 350 000 km of pipelines transport oil and gas from wells, refineries, and plants—that's a lot of pipeline to oversee! Grzyb's specialty is pipeline corrosion, and he is a member of the National Association of Corrosion Engineers (NACE).

Q Why do petroleum pipelines corrode?

A Corrosion has always been an issue with iron and steel products of all types. In Alberta, we have more pipeline carrying raw petroleum and gas than finished product. These unprocessed production streams are corrosive in nature because they often contain water from formation, ionic species, such as chlorides, as well as dissolved gases, such as carbon dioxide and hydrogen sulfide, which make an environment that is very corrosive to steel.

Q What happens when pipeline corrosion is allowed to continue?

A In the event that corrosion is unaddressed, you will eventually end up with a perforated pipe, where at some point it develops a hole and leaks. Occasionally, however—in perhaps 5 percent of cases—you could end up with a pipeline that becomes thinned down to the point that it ruptures and explodes because there is not enough material in the pipe wall to contain the pressure. In this event, not only is there a spill of product, there may be an explosion or fire.

Q Is pipeline corrosion a big problem?

A People just tend to accept corrosion as the way life is— "stuff rusts, what's the big deal?" In fact, it is a big deal. The amount of equipment and material that we lose to corrosion is huge. The cost of corrosion of bridges, railways, automobiles, ships, and pipeline infrastructures is in the billions of dollars. The impact of corrosion on our global economy is vast. Preventing corrosion also has a tremendous social benefit in terms of reducing waste. To make any product requires energy, and if we are losing that product because of corrosion, it is not an efficient use of resources. There is also a risk to the environment and the public if a pipeline ruptures or leaks. The prevention of corrosion and the maintenance of our infrastructure are very important.

Q How can you prevent pipeline corrosion?

A The basic principle of corrosion is that there is an electrochemical cell. One side of the circuit is the steel pipe-wall, the other side of the circuit is the water (electrolyte) in the line, and there is electron flow in the circuit. An iron molecule at surface dissolves, and releases two iron ions into solution, while two released electrons flow through the metal to react with other ionic species at other locations. If you can totally eliminate water you generally eliminate corrosion. So when we look at mitigating pipeline corrosion we usually look at three things: A, the possibility of water removal; B, the implementation of pipeline cleaning, which we call pigging; and C, putting a chemical inhibitor or other barrier on the steel to break the circuit and stop the electrochemical cell.

Q What inhibitors do you use?

A Amines are one of the more common chemicals that are added. They form a thin film on the steel surface inside the pipe, which prevents the water from being in direct physical contact with the steel. These films are quite tenacious and resistant to breaking down. Another method is to use plastic liner and plastic coatings on the inside of pipe or on the inside of vessels and tanks to prevent corrosion. That is how I got started in the corrosion world. My initial specialty was in the field of plastic linings and fibreglass materials for use in corrosion control.

Other Careers Involving Corrosion Management

Auto-Body Technician Technicians repair, replace, and refinish motor vehicle parts, and sometimes make damage appraisals. Auto-body technicians often have to remove rust from vehicle bodies, and prepare surfaces with anti-corrosion coatings. The work is physically demanding. Training is completed through registered apprenticeship programs. High-school students can take apprenticeship programs while completing their high-school diplomas. Businesses usually prefer to hire workers who have completed high school.

Biocorrosion Specialist Pipeline corrosion is often enhanced by the growth of bacteria, which grow in formations called biofilms on the pipe walls. Some corrosion consulting companies hire microbiologists to test samples from pipelines for the presence of corrosion-enhancing bacteria, or provide chemical cleaners called biocides to remove biofilms. This is a very specialized field, although there are many routes of entry. In general, biocorrosion specialists are engineers or environmental microbiologists with graduate degrees.

Home Inspector In addition to inspecting a property's exterior, overall structure, electrical wiring, and insulation, home inspectors visually inspect the plumbing and look for signs of corrosion damage to copper and steel pipes. They prepare unbiased reports about property damage or poorly built structures for people who are buying or selling homes. Most are self-employed and have experience working in construction. They can register with the Canadian Association of Home and Property Inspectors if they have successfully met the association's course and practical requirements.

Pipefitter Also called steamfitter-pipefitters, these workers interpret blueprints to install piping systems in homes or commercial or industrial settings. They install and repair pipelines that carry water, steam, chemicals, or fuel. To become registered to work in Alberta, workers must successfully complete a four-year apprenticeship program, which includes on the job training and some class-work. Steamfitter-pipefitters on construction crews often work overtime to meet construction deadlines.

Restoration Specialist Some restoration experts specialize in restoring metal artifacts (such as musical instruments), antique cars, or metal components on historic buildings. Work on architectural metals often includes gently removing spots of corrosion, identifying its causes, and treating the metal to prevent further corrosion. Other forms of corrosion, such as the blue-green patina of oxidized copper are desirable and therefore preserved. Some restoration specialists are professional architects. Others are professional conservators with undergraduate degrees in arts or sciences.

Go Further...

1. In your notebook, complete the equation for the corrosion of iron (in a steel pipe):

$$Fe^{2+}(aq) + O_2(g) + \underline{\quad} \rightarrow 2Fe_2O_3 \cdot nH_2O + \underline{\quad}$$
$$(\text{rust})$$

2. Write an equation for the corrosion of copper (Cu) used in a decorative roof.

3. Some pipeline corrosion occurs when there is no oxygen available, but hydrogen sulphide ($H_2S(g)$) is present. Suggest how corrosion would occur in this case and name the reducing agent and oxidizing agent in the overall reaction.

4. Based on the interview, how would you define the term, *infrastructure*? Name some examples of what makes up the *infrastructure* of cities and towns.

www.albertachemistry.ca
WWW

Understanding Concepts

1. State the theoretical definition for oxidation reactions. Write ionic equations for the complete balanced equations below. Use the equations to explain why chemists developed the theoretical definition that you stated.

$$4Na(s) + O_2(g) \rightarrow 2Na_2O(s)$$
$$2Na(s) + Cl_2(g) \rightarrow 2NaCl(s)$$

2. Write the following equation as an ionic equation and then as a net ionic equation.

$$Cd(s) + Sn(NO_3)_2(aq) \rightarrow Cd(NO_3)_2(aq) + Sn(s)$$

   a) Use the equation to explain the meaning of "spectator ion."
   b) Identify the reducing agent.
   c) Identify the oxidizing agent.
   d) The atom of which element is oxidized?
   e) The atom of which element is reduced?

3. Which of the following reactions will proceed spontaneously?
   a) aqueous calcium chloride and metallic silver
   b) aqueous nickel sulfate and metallic zinc

4. Describe one step in the production of steel that is a reduction reaction and one step that is an oxidation reaction.

5. Determine the oxidation number of the specified atom in each of the following compounds:
   a) N in N_2O_3
   b) P in $H_4P_2O_7$
   c) Si in $SiF_6{}^{2-}$
   d) each atom in $(NH_4)_2SO_4$

6. Does the fact that you can assign oxidation numbers of +1 to hydrogen and −2 to oxygen in water mean that water is an ionic substance? Explain.

7. Describe a situation in which it is easier to use the oxidation number method for balancing a redox equation than to use the half-reaction method.

8. a) Limewater, $Ca(OH)_2(aq)$, turns cloudy in the presence of carbon dioxide. Write a balanced chemical equation for the reaction. Include the states of the substances.
   b) State whether the reaction is a redox reaction. Explain.

9. Explain why you cannot use the table of standard reduction potentials in Appendix G to calculate the external voltage required to electrolyze molten sodium chloride.

Applying Concepts

10. A student prepared a standard solution by dissolving 0.852 g of potassium permanganate in enough water to make 250.0 mL of solution. The student measured 25.00 mL of a hydrogen peroxide solution of unknown concentration. The student filled a burette with the standard potassium permanganate. The initial reading on the burette was 1.34 mL. When the endpoint was observed, the student determined that the final reading on the burette was 21.54 mL. What was the molar concentration of the hydrogen peroxide solution?

11. Sketch the following voltaic cell in your notebook. Assume that one half-cell consists of a magnesium electrode in 1.0 mol/L magnesium nitrate and the other half-cell consists of a silver electrode in 1.0 mol/L silver nitrate.

   a) Label the electrodes and solutions in the sketch.
   b) Label the anode and the cathode and the direction of the current.
   c) Choose an appropriate solution for the salt bridge and label it.
   d) Write half-reactions for each half-cell.
   e) Write the overall reaction.
   f) Calculate the cell potential.
   g) Write the cell notation for the voltaic cell.

12. What is an "inert electrode" and when must one be used? Name two materials that are commonly used as inert electrodes.

13. Explain the meaning of the standard reduction potentials that are listed in Appendix G. How can a half-cell have a potential when a potential must be measured between two points or two electrodes?

14. What is the difference between a primary battery and a secondary battery?

15. Use the diagram below to explain how iron objects rust. Copy the sketch in your notebook and add labels that show the rusting process.

16. Choose and explain one technique for preventing corrosion.

17. When Sir Humphrey Davy attempted to electrolyze aqueous solutions to reduce the metal ions to their pure metal form, his attempts did not work. Explain why he was unable to obtain the pure metals and what was electrolyzed in his experiments.

18. Explain how electrolysis can be successfully used to refine metals.

19. Why is the density of the electrolyte solution in a lead-acid battery greatest when the battery is fully recharged?

20. Predict the products from the electrolysis of a 1 mol/L solution of hydrochloric acid.

21. Suppose you decide to protect a piece of iron from rusting by covering it with a layer of lead.
 a) Would the iron rust if the lead layer completely covered the iron? Explain.
 b) Would the iron rust if the lead layer partially covered the iron? Explain.

22. Design and describe a procedure you could use to galvanize an iron nail. Explain your choice of materials.

23. Suppose you build four different half-cells under standard conditions in the laboratory.
 a) What is the greatest number of different voltaic cells you could make from the four half-cells?
 b) What is the smallest number of standard cell potentials you would need to measure to rank the half-cells from greatest to least standard reduction potentials? Explain.

24. Describe how you could build a voltaic cell and an electrolytic cell in which the two electrodes are made of lead and silver. Include a list of the materials you would require.

25. It is possible to use the standard reduction potentials for the reduction of hydrogen ions and the reduction of water molecules to show that the dissociation of water molecules into hydrogen ions and hydroxide ions is non-spontaneous under standard conditions. Describe how you would do this. How is this result consistent with the observed concentrations of hydrogen ions and hydroxide ions in pure water?

26. List all the information you can obtain from a balanced half-reaction. Give two examples to illustrate your answer.

27. a) If one metal (metal A) displaces another metal (metal B) in a single-replacement reaction, which metal is the more effective reducing agent? Explain. Give an example to illustrate your answer.

 b) If one non-metal (non-metal A) displaces another non-metal (non-metal B) in a single-replacement reaction, which non-metal is the more effective oxidizing agent? Explain. Give an example to illustrate your answer.

28. a) Sketch a cell that has a standard cell potential of 0 V.
 b) Can the cell be operated as an electrolytic cell? Explain.

29. If a redox reaction cannot create or destroy electrons, how is it possible that the balanced half-reactions for a redox reaction may include different numbers of electrons?

30. a) A friend in your class has difficulty recognizing disproportionation reactions. Write a clear explanation for your friend on how to recognize a disproportionation reaction.
 b) Write a balanced chemical equation that represents a disproportionation reaction.

31. a) An electrolytic cell contains a standard hydrogen electrode as the anode and another standard half-cell. Is the standard reduction potential for the half-reaction that occurs in the second half-cell greater than or less than 0 V? Explain how you know.
 b) Will your answer for part (a) change if the hydrogen electrode is the cathode of the electrolytic cell? Explain.

32. Write a descriptive paragraph to compare the reactions that occur in a Downs cell and a chlor-alkali cell. Describe the similarities and differences.

33. One of your classmates is having trouble understanding some of the main concepts in this unit, listed below. Use your own words to write an explanation for each of the following concepts. Include any diagrams or examples that will help to make the concepts clear to your classmate.
 a) oxidation and reduction
 b) voltaic and electrolytic cells

Solving Problems

34. Balance each of the following half-reactions:
 a) $Hg_2^{2+} \rightarrow Hg$
 b) $TiO_2 \rightarrow Ti^{2+}$ (acidic conditions)
 c) $I_2 \rightarrow H_3IO_6^{3-}$ (basic conditions)

35. Some metals can have different oxidation numbers in different compounds. In the following reactions of iron with concentrated nitric acid, assume that one of the products in each case is gaseous nitrogen monoxide. Include the states of all the reactants and products in the equations.

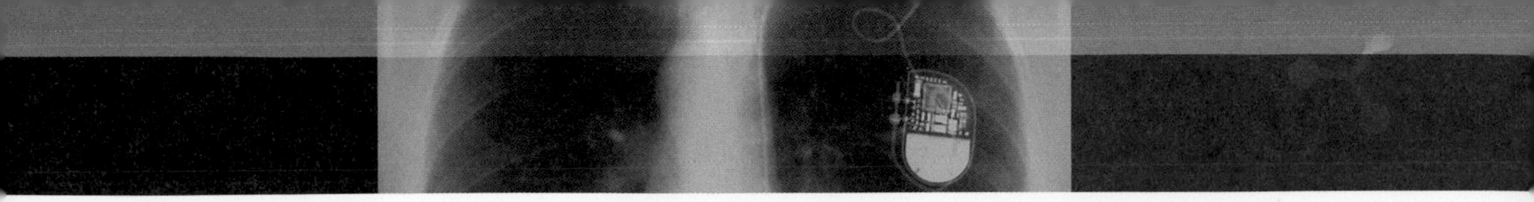

a) Write a balanced chemical equation for the reaction of iron with concentrated nitric acid to form iron(II) nitrate.

b) Write a balanced chemical equation for the reaction of iron(II) nitrate with concentrated nitric acid to form iron(III) nitrate.

c) Write a balanced chemical equation for the reaction of iron with concentrated nitric acid to form iron(III) nitrate.

d) How are the equations in parts (a), (b), and (c) related?

36. For the following voltaic cell,
$$C(s), I_2(s) \mid I^-(aq) \parallel Ag^+(aq) \mid Ag(s)$$
    a) identify the anode, the cathode, the positive electrode, and the negative electrode

    b) write the two half-reactions and the overall cell reaction

    c) identify the oxidizing agent and the reducing agent

    d) determine the standard cell potential

37. State whether the reaction shown in each of the following unbalanced equations is a redox reaction. If so, identify the oxidizing agent and the reducing agent. Balance each equation:

    a) $Cl_2O_7 + H_2O \rightarrow HClO_4$

    b) $I_2 + ClO_3^- \rightarrow IO_3^- + Cl^-$ (acidic conditions)

    c) $S^{2-} + Br_2 \rightarrow SO_4^{2-} + Br^-$ (basic conditions)

    d) $HNO_3 + H_2S \rightarrow NO + S + H_2O$

38. Determine the standard cell potential for each of the following reactions. State whether each reaction is spontaneous or non-spontaneous:

    a) $2Fe^{2+}(aq) + I_2(s) \rightarrow 2Fe^{3+}(aq) + 2I^-(aq)$

    b) $Au(NO_3)_3(aq) + 3Ag(s) \rightarrow 3AgNO_3(aq) + Au(s)$

    c) $H_2O_2(aq) + 2HCl(aq) \rightarrow Cl_2(g) + 2H_2O(\ell)$

39. When aqueous solutions of potassium permanganate, $KMnO_4(aq)$, and sodium oxalate, $Na_2C_2O_4(aq)$, react in acidic solution, the intense purple colour of the permanganate ion fades and is replaced by the very pale pink colour of manganese(II) ions. Gas bubbles are observed as the oxalate ions are converted to carbon dioxide. If the redox reaction is carried out as a titration, with the permanganate being added to the oxalate, the permanganate acts as both reactant and indicator. The persistence of the purple colour in the solution with the addition of one drop of permanganate at the endpoint shows that the reaction is complete.

a) Complete and balance the equation for acidic conditions:
$$MnO_4^-(aq) + C_2O_4^{2-}(aq) \rightarrow Mn^{2+}(aq) + CO_2(g)$$

b) If 14.28 mL of a 0.1575 mol/L potassium permanganate solution reacts completely with 25.00 mL of a sodium oxalate solution, what is the concentration of the sodium oxalate solution?

40. Redox titrations can be used for chemical analysis in industry. For example, the percent by mass of tin in an alloy can be found by dissolving a sample of the alloy in an acid to form aqueous tin(II) ions. Titrating with cerium(IV) ions produces aqueous tin(IV) and cerium(III) ions. A 1.475 g sample of an alloy was dissolved in an acid and reacted completely with 24.38 mL of a 0.2113 mol/L cerium(IV) nitrate solution. Calculate the percent by mass of tin in the alloy.

41. In a voltaic cell, the mass of the magnesium anode decreased by 3.38 g while the cell produced electric current.

    a) Calculate the amount of charge produced by the cell.

    b) If the constant current flowing through the external circuit was 100 mA, for how many hours did the cell produce electric current?

42. Calculate the mass of aluminium plated onto the cathode by a current of 2.92 A that is supplied for 71.0 min to an electrolytic cell containing aqueous aluminium nitrate as the electrolyte.

43. An industrial method for manufacturing fluorine gas is the electrolysis of liquid hydrogen fluoride. If the current supplied to the electrolytic cell is 5000 A, what mass of fluorine, in tonnes, is produced by one cell in one week?

44. A variety of electrolytic cells can be based on the design of the diagram shown here.

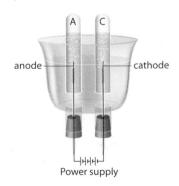

When the electrolyte in another cell is molten sodium chloride and the electrodes are graphite, the cell will produce one gas and another product. Both products will form in collecting tubes, such as those labeled A and C.

a) When the electrolyte is aqueous sodium chloride:

i) identify the gas that is produced in collecting tube A. Justify your answer by writing the half-reaction that occurs on the electrode.

ii) How would you identify the gas that collects in C? Justify your answer by writing the half-reaction that occurs at that electrode.

iii) What other material could replace carbon for the cathode and would not significantly change the operation of the cell?

iv) Determine the minimum voltage required to operate the cell.

v) Compare the contents of collectors A and C after the cell has operated for some time.

vi) Describe the results of a litmus test of the liquids surrounding the anode and cathode, after the cell has operated for some time.

b) When the electrolyte is molten sodium chloride:

i) Explain why gas only collects in A, and not in C. Support your answer with relevant half-reactions.

ii) Determine the minimum voltage required to operate the cell. What assumption must you make?

iii) Contrast the net reaction in the cell containing molten sodium chloride with the net reaction in the cell containing aqueous sodium chloride.

c) When the general cell design is considered:

i) Explain why only one of these designs is able to produce sodium metal.

ii) Identify an oxidizing agent and reducing agent pair that would be likely to produce oxygen and hydrogen gases in A and C.

iii) Describe the direction of the electron flow in all of these cells.

Making Connections

45. Black-and-white photographic film contains silver bromide. When exposed to light, silver bromide decomposes in a redox reaction. Silver ions are reduced to silver metal, and bromide ions are oxidized to bromine atoms. The brighter light that hits the film, the greater is the decomposition. When the film is developed, the parts of the film that contain the most silver metal produce the darkest regions on the negative. In other words, the brightest parts of the photographed scene give the darkest parts of the image on the negative. High-speed black-and-white film uses silver iodide in place of silver bromide. Do you think that silver iodide is more sensitive or less sensitive to light than silver bromide? Explain.

46. The two rocket booster engines used in a space shuttle launch contain a solid mixture of aluminium and ammonium perchlorate. The products of the reaction after ignition are aluminium oxide and ammonium chloride.

a) Write a balanced chemical equation for this reaction, and identify the oxidizing agent and reducing agent.

b) The other four engines used in a space shuttle launch use the redox reaction of liquid hydrogen and liquid oxygen to form water vapour. Write a balanced chemical equation for this reaction.

c) Which of the reactions described in parts (a) and (b) do you think is the more environmentally friendly reaction? Explain.

47. Every year, corrosion is responsible for the failure of many thousands of water mains. These are the pipes that transport water to Canadian homes and businesses. Research and describe the economic, environmental, health, and safety issues associated with rupture and repair of water mains. Describe the methods that are being used to improve the situation. **ICT**

48. Since the corrosion of iron is such an expensive problem, why do you think that iron is still used for so many purposes?

UNIT 7

Chemical Changes of Organic Compounds

General Outcomes

- explore organic compounds as a common form of matter

- describe chemical reactions of organic compounds

Unit Contents

Focussing Questions

1. What are the common organic compounds and what is the system for naming them?

2. How does society use the reactions of organic compounds?

3. How can society ensure that the technical applications of organic chemistry are assessed to ensure future quality of life and a sustainable environment?

Unit PreQuiz ②

www.albertachemistry.ca

At this moment, you are walking, sitting, or standing in an "organic" body. Your skin, hair, muscles, heart, and lungs are all made of *organic compounds* —chemical compounds that are based on the carbon atom. In fact, the only parts of your body that are not mostly organic are your teeth and bones! When you study organic chemistry, you are studying the substances that make up your body and much of the world around you. Medicines, clothing, carpets, curtains, and plastics are all manufactured from organic chemicals. The canola plants in this field as well as the oil that will be extracted from the seeds all consist of organic compounds. The petroleum being pumped from deep beneath the soil is also organic.

Are you having a sandwich for lunch? Bread, butter, meat, and lettuce are made of organic compounds. Will you have dessert? Sugar, flour, vanilla, and chocolate are also organic. What about a drink? Milk and juice are solutions of water in which organic compounds are dissolved.

In this unit, you will study a variety of organic compounds. You will learn how to name them and how to draw their structures. You will also learn how these compounds react, and you will use your knowledge to predict the products of organic reactions. In addition, you will discover the amazing variety of organic compounds in your body and in your life.

Do you recall learning about the types of intermolecular forces that act between molecules? If so, you might remember analyzing the diagrams in Figure P7.1. What type of intermolecular force is acting between the molecules of these two compounds? What other type of intermolecular force can act between molecules? A review of the bonding patterns that determine the shape and polarity of molecules and of intermolecular forces will provide a strong foundation for your study of Unit 7.

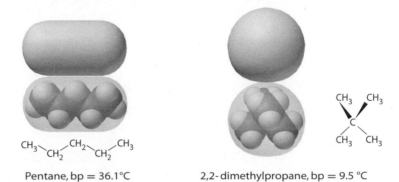

Pentane, bp = 36.1°C   2,2-dimethylpropane, bp = 9.5 °C

Figure P7.1 Both molecules shown here have five carbon atoms and 12 hydrogen atoms. What causes the large difference in their boiling points?

Valence-Shell Electron-Pair Repulsion Theory or VSEPR

The shape of a molecule influences the interactions of that molecule with other molecules of the same kind and of different kinds. VSEPR theory allows you to predict the shape of a molecule around a specific atom. You can apply the theory to any or all of the atoms in a molecule to predict the general shape the molecule. VSEPR is based on the concept that electrons in the valence shell, or outer energy level, repel one another because "like charges repel." Electron groupings will move as far away from each other as possible.

The term "electron grouping" can apply to a single bond, a double bond, a triple bond, or a lone pair. To predict the shape around a single atom, you first identify the number of electron groupings associated with the atom. Then relate that number to one of the three general classes of shapes associated with atoms having eight electrons in their outer shell. Those three shapes are listed in the table below.

Table P7.1 Three General VSEPR Shapes

Number of electron groupings	2	3	4
Shape	linear	trigonal planar	tetrahedral
Diagram			

In all of the diagrams in Table P7.1, the electron groupings are bonding electrons. In trigonal planar shapes, one electron grouping can be a lone pair, and in tetrahedral shapes, one or two electron groupings can be lone pairs. These shapes are shown in Figure P7.2.

bent (trigonal planar)   trigonal pyramidal   bent (tetrahedral)
                          (tetrahedral)

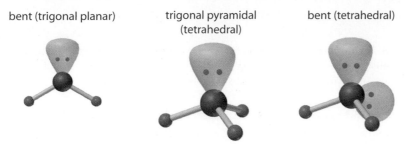

Figure P7.2 Because the lone pairs spread out more than bonding pairs of electrons, the lone pairs exert a greater force on each other and on the bonding pairs than do the bonding pairs. This greater force affects the bond angles.

The simplest way to determine the number of electron groupings around an atom in a molecule is to draw the Lewis structure of the molecule. For example, the Lewis structure of ethyne (commonly called acetylene) is shown in Figure P7.3A. Both carbon atoms have two electron groupings so the shape around each carbon atom must be linear. Because the two carbon atoms share an electron grouping, the entire molecule is linear. The two-dimensional drawing (Figure 7.3B) is accurately showing that the molecule is linear. Often the two-dimensional drawings of molecules do not correctly depict the structure.

A H : C ⫶ C : H   **B** H — C ≡ C — H

Figure P7.3 Ethyne (acetylene) is a completely linear molecule.

A compound called propanone (commonly called acetone) will illustrate both trigonal planar and tetrahedral shapes. Figure P7.4A shows the Lewis structure of propanone.

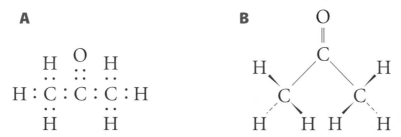

Figure P7.4 **(A)** The Lewis structure of propanone. **(B)** The three carbon atoms and the oxygen atom form a flat plane while the hydrogen atoms extend above and below that plane.

The Lewis structure of propanone shows that the central carbon atom has three electron groupings and no lone pairs. Therefore, the shape around the central carbon atom is trigonal planar. The three carbon atoms and the oxygen atoms lie in a flat plane. The two outer carbon atoms have four electron groupings with no lone pairs. The shape around the two outer carbon atoms is tetrahedral. The hydrogen atoms can lie above or below the plane defined by the three carbon atoms.

You can use your knowledge of polarity of bonds to determine whether a molecule has a polar region. The important polar bonds that you will encounter in Unit 7 are the oxygen–hydrogen bond and the carbon–oxygen bond shown in Figure P7.5.

$$\overset{\longleftarrow\;\;+}{O-H} \qquad\qquad \overset{\overset{+}{\longleftarrow\;\;\longrightarrow}}{C=O}$$

Figure P7.5 The tip of the arrow points to the end of the bond to which the electrons are drawn. Because the electrons spend more time at this end of the bond, it is slightly negative. The tail of the arrow that looks like a plus sign (+) is slightly positive.

When you see these groups in a molecule, you will know that the oxygen atom is slightly negative and the hydrogen or the carbon atom is slightly positive. You will also see many carbon–hydrogen bonds. The polarity of these bonds is negligible in strength so you can consider them non-polar. Given this information, you can see that the ethyne molecule if Figure P7.3 is entirely non-polar and the propanone in Figure P7.4 has a polar region around the C=O bond. If you need a more thorough review of VSEPR, go back to Chapter 2, page 52.

Intermolecular Forces

Intermolecular forces determine some important properties of molecular compounds. If the forces are strong, large amounts of energy will be required to pull the molecules apart. When the intermolecular forces are small, the molecules can be easily separated. This energy is in the form of kinetic energy of the individual molecules. You probably recall that temperature is a measure of the average kinetic energy of the molecules in a substance. Therefore, an increase in the temperature of a substance indicates that the kinetic energy of the molecules is increasing.

Based on the above information, you can infer that the molecules of substances with low boiling points and melting points require less energy to separate the molecules than the molecules of substances with higher melting and boiling points. In summary, if the melting and boiling points of a substance are low, the intermolecular forces are not strong. Higher melting and boiling points indicate that the intermolecular bonds of the substance are strong.

In your previous chemistry course, you learned about dipole-dipole attractions and about London (dispersion) forces. Dipole-dipole attractions, shown in Figure P7.6, are just what the name implies. The positive end of one dipole is attracted to the negative end of another dipole. Many molecules with polar bonds will be able to experience dipole-dipole attractions. (Recall that in some molecules, the polar bonds are oriented in directions that cause the dipoles to cancel each other within the molecule. Carbon dioxide is an example of such a molecule.) Therefore, when you have analyzed the shape and polarity of the molecule of a substance, you can determine whether they will experience dipole-dipole attractions.

Figure P7.6 Many molecules with polar bonds have a partially positive region and a partially negative region. These regions can be called positive and negative "poles," thus the molecules are called dipoles. The dotted lines in the diagram represent electrostatic attractive forces between the positive pole of one molecule and the negative pole of another.

You probably recall the type of dipole-dipole interaction that is so strong that it has a name of its own. Hydrogen bonds are the strongest type of dipole-dipole attraction. Hydrogen bonds can form between molecules that have a hydrogen atom bonded to an electronegative atom such as oxygen or nitrogen. Figure P7.7 shows an example of hydrogen bond forming between an —OH group of one molecule and a C=O group in another molecule.

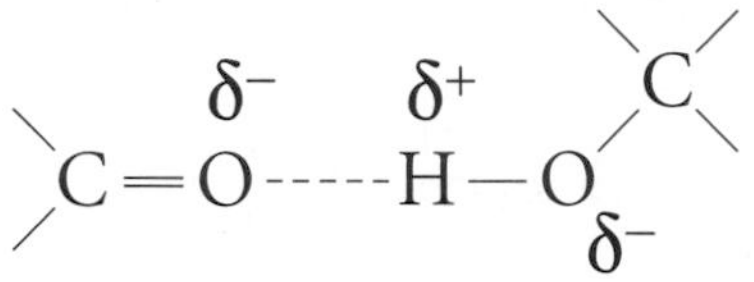

$$\underset{/}{\overset{\backslash}{C}} = \overset{\delta^-}{O} \text{-----} \overset{\delta^+}{H} - \underset{\delta^-}{O} \overset{\backslash}{\underset{/}{C}}$$

Figure P7.7 The carbon-oxygen group on one part of a molecule can form a hydrogen bond (dotted line) with a hydrogen atom bonded to another oxygen atom. The two groups might be on the same type of molecule or on molecules of two different substances.

London (dispersion) forces can act between any two molecules. However, they are so much weaker than dipole-dipole attractions that they are often ignored when dipole-dipole attractions are present. London (dispersion) forces are the only type of intermolecular force that can act between non-polar molecules or non-polar regions of any molecule. Although they are weak, when the contact between molecules is large, so many attractive London (dispersion) forces are acting that they become very significant.

London (dispersion) forces are created by instantaneous formations of temporary dipoles that can form when electrons are momentarily concentrated is one region of a molecule. Each time a temporary dipole forms, the very slightly negative region of the molecule repels the electrons very close to it and attracts positive charges. These forces induce another temporary dipole in the adjacent region. These instantaneous attractions between temporary dipoles are the source of the London (dispersion) forces.

The two factors that affect the strength and number of London (dispersion) forces are the size and shape of the molecules. Larger molecules have more electrons thus increasing the probability that temporary dipoles will form. Molecules that have shapes that allow a lot of surface contact between molecules can experience more London (dispersion) forces. Examine the two molecules in Figure P7.1. These two molecules have exactly the same number of carbon and hydrogen atoms. The more spherical molecules, however, have much less surface area in contact with one another. Therefore, the cylindrical shaped molecules with more area of contact will experience more London (dispersion) forces. This difference in the strength of the London (dispersion) forces is the reason for the difference in the boiling points of the two compounds.

Prerequsite Skills

- **Relating** the valence electron structure of an atom to its ability to form covalent bonds

- **Writing** and balancing chemical equations

Structure and Physical Properties of Organic Compounds

Chapter Concepts

14.1 Introducing Organic Compounds

- The study of traditional sources of medicines has led to important discoveries of organic compounds that can be used to treat illness and disease.

- Organic compounds contain carbon atoms, which bond to one another in chains, rings, and two- and three-dimensional networks to form a variety of structures.

14.2 Hydrocarbons

- The International Union of Pure and Applied Chemistry (IUPAC) system lists rules for naming organic compounds based on structural features.

- Structural models are useful tools to use in predicting the chemical and physical properties of organic compounds.

14.3 Hydrocarbon Derivatives

- Functional groups are responsible for characteristic physical and chemical properties of organic compounds.

- Isomers of compounds contain the same number and type of atoms but are structurally different and have different properties.

14.4 Refining and Using Organic Compounds

- Organic compounds can be separated from natural mixtures, such as petroleum or coal, by physical and chemical processes.

- The enhanced recovery techniques that must be used to extract crude oil from Alberta's bitumen deposits presents technological and environmental challenges.

No one knows when the first person drank sap from the paper birch tree or tea made from its bark and leaves and discovered that these were effective treatments for aches and pains. The paper birch tree was only one source of many such traditional medicines for Canada's Aboriginal peoples. For more than 100 years, chemists have used the valuable knowledge passed on by traditional healers to assist them in chemically isolating and studying many valuable compounds. The chemical isolation of just one organic compound from the paper birch, salicylic acid, forms the basis of a multibillion-dollar global industry.

Salicylic acid, an effective painkiller, tastes horrible and is corrosive to the mouth and stomach. Felix Hoffmann synthesized a much less unpleasant painkiller by adding an acetyl group (CH_3CO—) to form acetylsalicylic acid. In 1893, the Bayer company obtained a patent for acetylsalicylic acid, the first truly synthetic drug. You likely know this drug better by its trade name, Aspirin™, a combination of acetyl and *Spiraea*, a plant genus that was a common source of salicylic acid.

What do chemists need to know in order to modify a natural compound or produce a synthetic compound? What are some of the risks and benefits of producing organic compounds? In this chapter, you will examine the nature of organic compounds and the features that determine their use.

Familiar Organic Compounds

A derivative of salicylic acid called methyl salicylate is easily detected by the characteristic "medicinal" smell it gives to plants that contain it.

In this activity, you will prepare methyl salicylate from salicylic acid and identify the common chemical name for this compound from its characteristic aroma.

Safety Precautions

- Methanol is flammable. Ensure that there are no open flames in the laboratory. Use a hot plate.
- Methanol is toxic. If it gets on your skin, rinse with copious amounts of water and notify your teacher.
- $H_2SO_4(aq)$ is corrosive. Use extreme caution. In the case of accidental spills, rinse skin and clothing with plenty of cold water and notify your teacher.
- The boiling point of methanol is 64.7 °C. When heating, be sure to point the mouth of the test tube away from others. Avoid overheating the methanol.

Materials

- hot plate or electric kettle
- small test tube
- rubber stopper to fit test tube
- test-tube rack
- beaker tongs
- test-tube clamp
- water
- salicylic acid
- methanol ☠
- 3.0 mol/L sulfuric acid, $H_2SO_4(aq)$ ⚠
- 1.0 mol/L sodium carbonate, $Na_2CO_3(aq)$
- 100 mL graduated cylinder
- 250 mL beaker

Procedure

1. Using the graduated cylinder, measure 100 mL of water and pour it into a 250 mL beaker. Place the beaker on a hot plate and bring the water to a gentle boil (or boil water in the kettle).

2. Place one drop of salicylic acid in a test tube. Add three drops of methanol to the test tube, and then add one drop of 3.0 mol/L $H_2SO_4(aq)$. Insert a rubber stopper firmly into the test tube. Place the test tube in the test tube rack.

3. Using tongs, remove the beaker of boiling water from the hot plate. Place the test tube in the hot water and allow it to stand for 8 to 10 minutes. Do not allow water to enter the test tube.

4. Remove test tube from the water bath by using the test tube clamp. Add four drops of 1.0 mol/L $Na_2CO_3(aq)$ to the tube. Swirl gently.

5. Wave the aroma gently toward your nose. Record your observations of the aroma. Dispose of all materials as directed by your teacher.

Analysis

1. Based on the aroma, what commercial products do you think would contain methyl salicylate?

2. What is the common name for methyl salicylate?

More than 70% of the biologically active compounds that chemists have derived from plants have been found by exploring plants, such as the wintergreen plant shown here, that were used in traditional medicines. In some cases, surprises were waiting. The organic compound that gives paper birch its pain-killing properties was isolated more than 100 years ago, but recently chemists have isolated another useful compound from the bark of the paper birch. Betulinic acid shows promise as an effective treatment for both skin cancer and HIV/AIDS.

SECTION 14.1

Introducing Organic Compounds

Section Outcomes

In this section, you will:
- **define** organic compounds and **give examples** of their origin, significance, and applications
- **carry out** an investigation comparing the properties of household organic and inorganic compounds

Key Terms

organic compounds
inorganic compounds
allotrope

Of the many substances that you encounter in your daily life, the vast majority can be classified as organic. The proteins, starches, and sugars you consume with each meal are all organic compounds. Almost all the antibiotics, detergents, solvents, and pesticides you use are also organic compounds. All plastics and fuels, and even the fabrics used to make the clothes you wear, whether they are natural or synthetic, are classified as organic. Considering the remarkable differences among these substances, you might be wondering what, exactly, makes a compound "organic."

Defining Organic Compounds

For early chemists, the answer to that question was simply that organic materials were substances derived from living or once-living organisms. Living plants and animals were thought to possess a "vital force" that gave them the unique ability to produce organic compounds. Inorganic compounds by contrast could be derived from rocks and mineral deposits.

This idea was challenged in 1828 when Friedrich Wöhler (1800–1882) was attempting to produce ammonium cyanate by reacting silver cyanate and ammonium chloride. He was startled to find that he had obtained a white, crystalline substance that showed none of the chemical or physical properties of ammonium cyanate. On testing this mystery substance, Wöhler was surprised to find that it was identical to urea—an organic compound found in the urine of almost all mammals. Clearly, an organic compound could be produced in the absence of a living organism (see Figure 14.1).

Figure 14.1 When Wöhler synthesized urea, he started a major change in the fundamental concepts of organic chemistry.

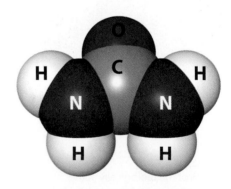

$$AgCNO + NH_4Cl \rightarrow CO(NH_2)_2 + AgCl$$
urea

In 1843, a former student of Wöhler, Adolf Wilhelm Hermann Kolbe (1818–1884), prepared an organic compound from inorganic compounds in a series of reactions, including a substitution reaction. Starting with carbon disulfide, an inorganic compound, Kolbe prepared ethanoic acid, also called acetic acid. Chemists finally had to rethink the definition of organic compounds. Eventually, scientists abandoned the idea of a vital force.

The terms *organic* and *inorganic* are currently used to make a distinction between carbon-based compounds and other compounds. Although some carbon-containing compounds, such as carbonates (CO_3^{2-}), cyanides (CN^-), carbides (C^{2-}), and oxides of carbon (CO_2, CO), are still classed as inorganic, they are exceptions. The modern definition states that **organic compounds** are those in which carbon atoms are nearly always bonded to each other, to hydrogen atoms, and often to atoms of a few specific elements. These other elements are usually oxygen, nitrogen, sulfur, and phosphorus. Some synthetic organic compounds contain halogens. Notice that the exceptions do not contain any carbon–carbon or carbon–hydrogen bonds. These exceptions along with all compounds that contain no carbon atoms are classified as **inorganic compounds.** In the next investigation, you will compare the physical and chemical properties of several common organic and inorganic compounds.

The Special Nature of the Carbon Atom

In Chapter 2, you read that the hardest substance known—diamond—and the slippery substance in pencil lead—graphite—both consist of pure carbon. What is the nature of the carbon atom that allows it to form these widely differing structures and, as well, to form the basis of the thousands of molecules found in living organisms? The chart below outlines three key properties of carbon. Throughout this section and section 14.2, you will explore the consequences of each of these properties.

Properties of Carbon

Carbon has four bonding electrons.
This electron structure enables carbon to form four strong covalent bonds. As a result, carbon may bond to itself and to many different elements (mainly hydrogen, oxygen, and nitrogen but also phosphorus, sulfur, and halogens, such as chlorine).

Carbon can form strong single, double, and triple bonds with itself. This property allows carbon to form long chains of atoms—something that very few other atoms can achieve. In addition, the resulting compounds are fairly stable under standard conditions of temperature and pressure.

Carbon atoms can bond together to form a variety of geometric structures. These structures include straight chains, branched chains, rings, sheets, tubes, and spheres. No other atom can do this.

Thought Lab 14.1 Nanotubes, Buckyballs, and Allotropes

The serendipitous discovery of fullerenes in 1985 revealed the existence of a third form or **allotrope** of pure carbon. Interest in fullerenes has been intense since the discovery of this unusual form of carbon. You are already familiar with the use of graphite in pencils, and you may be aware that, in addition to being used in jewellery, diamonds are used extensively in industrial applications. How much do you know about fullerenes? The physical properties of graphite and diamond are almost the complete opposite of each other. How do the physical properties of fullerenes compare with those of diamond and graphite? Where do nanotubes fit in with these allotropes?

Procedure

Use library and electronic research tools to find answers for the following questions.
1. What is the origin of the name *fullerene* for this third allotrope of carbon?

2. What are the physical properties (electrical, structural, hardness, colour) of graphite and of diamond?

3. What are the electrical properties of fullerenes? How do those properties compare with the electrical properties of diamond and graphite?

4. What are the physical characteristics of a carbon nanotube? What are three technological uses for nanotubes?

Analysis

1. Diamond is one of the hardest known substances, and graphite one of the softest. Yet most of the carbon–carbon bonds in both graphite and diamond are approximately equal in strength. How can diamond be so much harder than graphite?

2. What effect is the physical structure of a nanotube predicted to have on the structural strength of carbon composites?

3. Is it appropriate to describe a nanotube as merely a special case of a fullerene? Explain.

Comparing Organic and Inorganic Compounds

Of the millions and millions of organic compounds that exist, each one is based on a "skeleton" of carbon atoms covalently bonded to one another and to other atoms. Hydrocarbons containing only non-polar covalent bonds between carbon and hydrogen will be somewhat unreactive and insoluble in water. They will tend to interact only with one another. The presence of polar covalent bonds between carbon and oxygen in hydrocarbon derivatives, such as alcohols and carboxylic acids, is expected to produce different characteristics. How will the properties of hydrocarbon derivatives compare with those of hydrocarbons? How will the properties of hydrocarbons and their derivatives compare with those of inorganic compounds? Your teacher will assign either Part I or Parts II and III to each group. You will share data with another group before completing the Analysis section.

Question

Do the observed properties of solubility, conductivity, miscibility, and viscosity of hydrocarbons, hydrocarbon derivatives, and inorganic compounds agree with the expected properties?

Hypothesis

Knowing that water is a substance with polar chemical bonds and using your previous knowledge of chemical bonding, formulate a hypothesis about the solubility of hydrocarbons and of hydrocarbon derivatives in water.

Knowing that viscosity is the resistance of a liquid to a change in its position, formulate a hypothesis about the relative viscosities of liquids (a) in which the molecules have many polar bonds and (b) in which the molecules have very few polar bonds.

Safety Precautions

- Organic solvents are flammable. Extinguish all flames in the laboratory area.
- Propylene glycol is toxic. Do not ingest it. Ⓣ

Materials

- deionized (or distilled) water, H_2O (ℓ)
- ethanoic acid, $C_2H_4O_2$ (ℓ) (acetic acid or vinegar)
- propanol, C_3H_8O (ℓ)
- pentane, C_5H_{12} (ℓ)
- sodium chloride crystals, $NaCl$(s) (table salt)
- sodium hydrogen carbonate, $NaHCO_3$(s) (baking soda)
- calcium carbonate, $CaCO_3$(s)
- salicylic acid crystals, $C_7H_6O_3$(s)
- sucrose crystals, $C_{12}H_{22}O_{11}$(s) (table sugar)
- naphthalene, $C_{10}H_8$(s) (mothball crystals) Ⓣ
- food colouring (red, green, or blue)
- hexane, C_6H_{14} (ℓ) or cyclohexane, C_6H_{12} (ℓ) Ⓣ
- propane-1,2-diol (propylene glycol), $C_3H_8O_2$ (ℓ)
- propane-1,2,3-triol (glycerol), $C_3H_8O_3$ (ℓ)
- 1 mL pipette or medicine dropper
- well plate (12 or 24 wells are preferable)
- conductivity tester
- microspatulas or wood splint
- test tubes
- grease pen or marker
- plastic wrap or Parafilm™
- test tube rack
- four 50 mL burettes
- 2 double burette clamps
- 2-ring stands
- four 250 mL Erlenmeyer flasks
- stopwatch
- 4 funnels

Procedure

Part I: Testing for Solubility and Conductivity

1. Use the pipette or medicine dropper to place 1 mL of each liquid listed below (the solvents) into its own well on your well plate. Label each well:

   A deionized water   **C** propanol
   B ethanoic acid   **D** pentane

2. Measure the conductivity of each solvent using a conductivity tester. Remember to clean the electrical contacts after each test. Record the conductivity of each solvent.

3. Place a small quantity of sodium chloride into each solvent using a microspatula or wood splint. Stir to determine whether the crystals will dissolve. Record your results.

4. Measure the conductivity of each well again using a conductivity tester. Remember to clean the electrical contacts after each test. Record the conductivity of each solution and note any differences from the results in Procedure step 2.

5. Repeat Procedure steps 1–4 using baking soda as the solute.

6. Repeat Procedure steps 1–4 using calcium carbonate as the solute.

7. Repeat Procedure steps 1–4 using salicylic acid as the solute.

8. Repeat Procedure steps 1–4 using table sugar as the solute

9. Repeat Procedure steps 1–4 using naphthalene as the solute.

Part II: Testing for Miscibility

1. Using the pipette or medicine dropper, place 1 mL of deionized water into each of five small test tubes. Label the tubes A to E. Add a drop of food colouring to each tube and swirl gently to mix.

2. Into the test tube labelled with its letter, place 1 mL of the following liquids:

   A deionized water   D pentane
   B ethanoic acid   E hexane or cyclohexane
   C propanol

3. Seal the end of each test tube with plastic wrap or Parafilm™. Place your thumb over the opening and shake the tube several times to mix the contents.

4. Let the test tubes stand undisturbed in the test tube rack for 1 min.

5. Observe each tube for evidence of mixing between the liquids. Use tube A as a standard against which to compare the other tubes.

Part III: Testing for Relative Viscosity

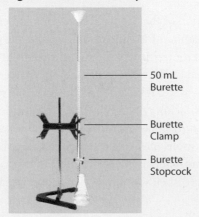

50 mL Burette

Burette Clamp

Burette Stopcock

1. Proceed to the lab station your instructor has set up. There you will find four burettes. Burette 1 contains 25 mL of ethanoic acid, burette 2 contains 25 mL of propanol, burette 3 contains 25 mL of propane-1, 2-diol, and burette 4 contains 25 mL of propane-1,2,3-triol.

2. Position an empty Erlenmeyer flask under each burette. Be ready to start the stopwatch when you open the stopcock. Open the stopcock on burette 1 and start the stopwatch. Stop the watch when 10 mL of liquid has flowed into the Erlenmeyer flask. Record that time.

3. Repeat Procedure step 2 for burette 2, burette 3, and burette 4.

4. Close the stopcock on each burette. Use the funnels to transfer the contents of each Erlenmeyer flask back into their original burettes.

Analysis

1. Using the definition of organic compounds, classify each liquid and solid in this investigation as either organic or inorganic.

2. Use your knowledge of chemical bonding to classify the solids used in this investigation as either ionic or molecular. Repeat for each liquid used.

3. Is it appropriate to conclude that organic solids dissolve in organic solvents and that inorganic solids dissolve in inorganic solvents? Explain.

4. Would it be reasonable to conclude that "like dissolves like" in regard to polar and non-polar substances? Explain.

5. Electrical conductivity is the result of the presence of ions in solution. Another student suggests that only ionic substances can conduct electricity. Do your results support this? Explain.

6. How does the presence of carbon–oxygen polar covalent bonds in an organic liquid affect the viscosity of that liquid? Use your knowledge of hydrogen bonding to explain why.

Conclusion

7. How well do the results you observed in this investigation support your hypotheses?

After inspecting the periodic table, you might expect silicon and carbon atoms to react in much the same way. The silicon–silicon bond is, however, much weaker than the bond between carbon atoms and, therefore, the bond between silicon atoms tends to be unstable. Despite this tendency, scientists have great interest in *silanes*, the silicon analogue to the simplest type of organic carbon compounds. Silanes will bond to organic carbon molecules and bond readily to inorganic substances, such as glass and metals. In medical applications (such as hip replacements), silanes are used to bond organic layers to titanium metal implants to form a "bio-inert" layer that decreases the risk of rejection of that implant by a body's immune system.

This metal femur prosthesis is coated with a silicon-based layer covering the metal.

Classifying Organic Compounds

The ability of carbon atoms to form chains, rings, sheets, tubes, and spheres is an important feature of organic compounds. These structures can consist of hundreds or even thousands of carbon atoms. The addition of each new carbon atom produces a new compound. Is it any wonder that there are so many organic compounds? In Thought Lab 14.1, you learned that the physical properties of carbon-based compounds depend on how those carbon atoms are linked together. Basically, the number of carbon atoms that are linked together, the way in which those atoms are linked, and the other types of atoms that are present determine how a compound behaves chemically. Using structural similarities, chemists organize the various organic compounds into chemical "families" that have the same general chemical reactivity. Knowing the general chemical and physical properties of any chemical family is an important tool for chemists to use in designing a new compound or in determining the hazards associated with any organic compound.

One classification scheme is presented in the concept organizer below. Note how the structure and type of atoms present is the key to classifying compounds into families. Hydrocarbon compounds typically contain only carbon and hydrogen atoms connected by non-polar covalent bonds. The hydrocarbon derivatives will contain atoms, such as oxygen, nitrogen, sulfur, or various halogens, connected to carbon by polar covalent bonds.

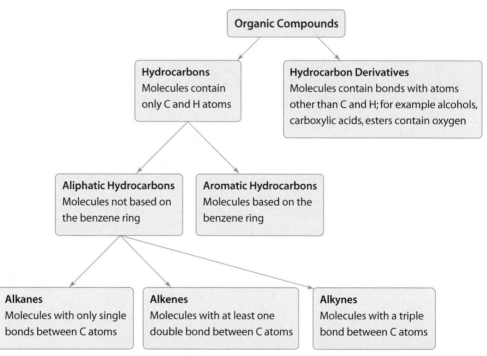

In general, it is the nature and location of the chemical bonds that determine the chemical and physical properties of organic compounds. For example, the presence of polar carbon–oxygen bonds within a molecule will increase the solubility of that molecule in water. Solubility is enhanced by the hydrogen bonding between the polar covalent hydrogen–oxygen bonds of the water molecule and the carbon–oxygen bond of the organic molecule. Hydrocarbons, lacking polar bonds, are insoluble in water. In addition, short-chain molecules containing polar carbon–oxygen bonds can form more hydrogen bonds than can longer-chain molecules. As the length of the carbon chain increases, the number of non-polar bonds increases and the effect of hydrogen bonding becomes less significant, while London (dispersion) forces that act between molecules becomes significant.

Section 14.1 Summary

- The historical definition for "organic compounds" was "substances derived from living or once-living organisms."
- Friedrich Wöhler was the first chemist to synthesize an organic compound. When he synthesized urea, chemists began to rethink their concept of a "vital force," something in living organisms that they believed was responsible for the synthesis of organic compounds.
- The modern definition of organic compounds is a compound in which carbon is nearly always bonded to itself, hydrogen, or a few other elements.
- Any compound that is not organic is called an inorganic compound.
- Carbon atoms can form four strong covalent bonds with other atoms, allowing it to form long chains, rings, sheets, and other structures.
- Organic compounds are classified as either hydrocarbons or hydrocarbon derivatives. Each class is then subdivided into many other categories.

SECTION 14.1 REVIEW

1. Give the names of three organic compounds and three inorganic compounds.

2. The theory of "vitalism" was originally used to separate compounds into organic and inorganic. What was the key idea behind vitalism? What findings led to the downfall of that theory?

3. What are the three properties of carbon that allow it to form such a great variety of compounds?

4. Explain why the physical properties of graphite and diamond are not the same.

5. Silicon lies immediately below carbon on the periodic table in Group 14 and, like carbon, has four valence electrons that may form covalent bonds. Why are there more carbon compounds than silicon compounds?

6. In general, what determines the solubility of any compound?

7. Many sources regarding the nature of organic and inorganic compounds state that organic substances have low solubility in water and don't conduct electricity. What rationale is used to justify this? Based on your experience, is this a true statement? Explain.

SECTION 14.2

Hydrocarbons

Section Outcomes

In this section, you will:

- **name** and **draw** structural, condensed structural, and line diagrams and formulas for saturated and unsaturated aliphatic and aromatic hydrocarbons
- **construct** molecular models to represent the structures of selected organic and inorganic compounds

Key Terms

hydrocarbons
alkanes
saturated hydrocarbons
aliphatic compounds
homologous series
root
prefix
suffix
substituent groups
side groups
alkyl groups
alkenes
unsaturated hydrocarbons
alkynes
cyclic hydrocarbons
aromatic compounds
benzene
phenyl group

Hydrocarbons, compounds that contain only carbon atoms and hydrogen atoms, are part of your everyday life. The gasoline that fuels the school bus and your car is a hydrocarbon. The natural gas that heats your home and school is a hydrocarbon. You might even have a fireplace in your home that burns natural gas (see Figure 14.2). You have also encountered hydrocarbons previously in chemistry. You have balanced equations involving hydrocarbon combustion. You have seen the structure of methane ($CH_4(g)$) in examples of the shapes of molecules. In this section, you will focus on naming and drawing hydrocarbons and learn about their physical properties.

Figure 14.2 Modern gas fireplaces often have artificial logs and burn natural gas.

Alkanes

Alkanes are the simplest hydrocarbons. They contain only *single* covalent bonds. Organic compounds that contain only single bonds, no double or triple bonds, are called **saturated hydrocarbons**. They are saturated because each carbon atom is bonded to as many other atoms as possible. As you saw in the diagram on page 542, alkanes are a subclass of compounds called aliphatic hydrocarbons. **Aliphatic compounds** contain no benzene rings.

Methane is the simplest alkane containing only one carbon atom and four hydrogen atoms. Figure 14.3 shows the structure of methane and the next three members of the alkane family. By examining the four structures, you will see patterns that will be true of all alkanes.

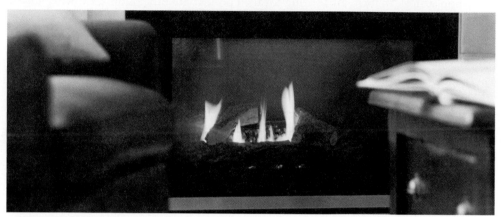

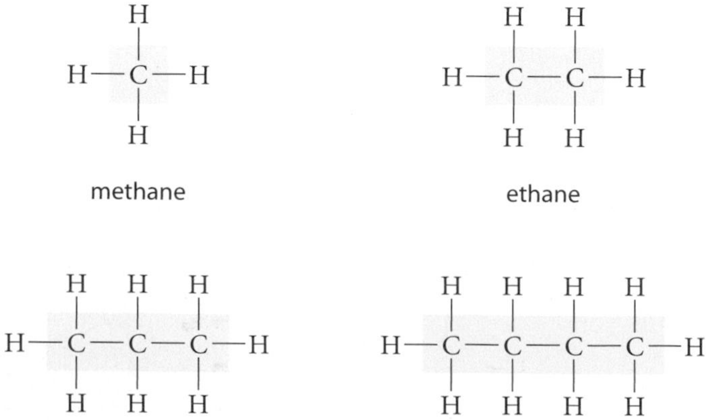

methane ethane propane butane

Figure 14.3 These are the first four alkanes. How are they similar? How do they differ?

As you can see, each molecule differs from the previous one by the addition of the structural unit $-CH_2-$. Each carbon atom in the chain has a minimum of two hydrogen atoms bonded to it. As well, the two carbon atoms at the ends of the chain are bonded to an additional hydrogen atom. You can thus generalize and say that the number of hydrogen atoms is an alkane is two times the number of carbon atoms plus two more. If n represents the number of carbon atoms, then the general formula for all straight or branched chain alkanes is C_nH_{2n+2}. Therefore, if an alkane has five carbon atoms, its formula is $C_5H_{2\times5+2}$ or C_5H_{12}. Any set of molecules that differ by one specific unit such as $-CH_2-$, is called a **homologous series**. Alkanes form a homologous series.

Modelling Alkanes

The formula, C_5H_{12}, tells you only that the molecule has five carbon atoms and twelve hydrogen atoms. It tells you nothing about structure of the molecule. Therefore, chemists have developed several ways to represent organic compounds that give a more complete picture of the molecule. The chart below shows you five different ways to represent the structures of alkanes, using C_5H_{12} as an example. With only slight modifications, these models can be used to represent all other families of hydrocarbons. As you see in the chart, the chain of carbon atoms in alkanes need not be a straight chain. Molecules can have branches in which one carbon atom is bonded to more than two other carbon atoms. Examine the structure to convince yourself that the formula C_nH_{2n+2} applies even when branches are present in the carbon chain.

Empirical molecular formula C_5H_{12}	An empirical formula shows the number and types of atoms present. This formula makes no attempt to specify structure, so it is not very useful for most applications.
Expanded molecular formula $CH_3CH(CH_3)CH_2CH_3$	An expanded molecular formula shows groupings of atoms. Brackets are used to indicate the locations of branched chains. In the formula on the left, a side chain consisting of $-CH_3$ is shown attached to the second carbon from the left end of the molecule. Bonds are assumed to exist between atoms.
Structural formula 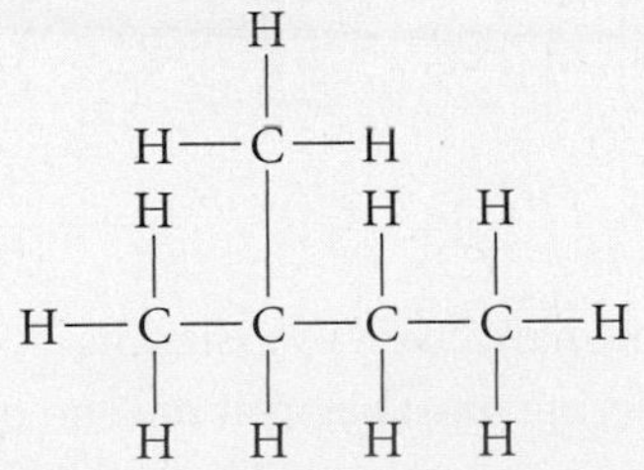 $$\begin{array}{c} H \\ | \\ H-C-H \\ H \quad | \quad H \quad H \\ | \quad | \quad | \quad | \\ H-C-C-C-C-H \\ | \quad | \quad | \quad | \\ H \quad H \quad H \quad H \end{array}$$	The structural formula gives a clear picture of all atoms and the locations of the bonds. Straight lines represent chemical bonds between atoms. Although detailed and accurate, this method requires a lot of space.
Condensed structural formula 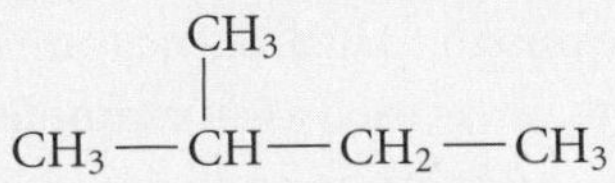 $$\begin{array}{c} CH_3 \\ | \\ CH_3-CH-CH_2-CH_3 \end{array}$$	To save space, the condensed structural formula does not show the carbon–hydrogen bonds (they are assumed to be present). The model does show all other bonds. This method still specifies the location of the side branches but has the advantage of being much cleaner and clearer than a structural formula.
Line structural formula 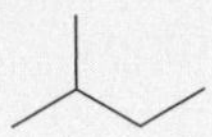 	This model uses lines to represent chemical bonds. Each end of a straight line represents a carbon atom (unless otherwise specified) and each carbon is assumed to have as many hydrogen atoms bonded to it as is necessary to give it four bonds.

Naming Alkanes

You are familiar with the IUPAC system of naming ionic compounds and simple molecular compounds. As you can imagine, there are literally millions of different organic compounds. Nevertheless, the IUPAC system has rules for naming every compound based on its structure. The names are so precise that you can draw any structure from its name. The IUPAC name of any organic compound has three basic parts: a root, a prefix and a suffix.

$$\text{prefix} \quad + \quad \text{root} \quad + \quad \text{suffix}$$

In the box below, each part of the name is explained as it applies to alkanes.

- The **root** denotes the number of carbon atoms in the longest continuous chain of carbon atoms.
- The **prefix** gives the positions and names of any branches from the main chain.
- The **suffix** indicates the series to which the molecule belongs. The suffix for alkanes is "–ane."

There is a root name for every possible chain of carbon atoms. You will only be responsible for learning the root names for chains of up to ten carbon atoms. Those root names are listed in Table 14.1.

Table 14.1 Root Names for First Ten Alkanes

Number of carbon atoms	Root name
1	meth
2	eth
3	prop
4	but
5	pent
6	hex
7	hept
8	oct
9	non
10	dec

ChemistryFile

FYI
The number system for organic compounds is very similar to that for inorganic compounds. Special prefixes, dating back to an earlier system that used descriptive names, describe compounds having one to four carbons. The prefixes used for numbering more than four carbons are based on the Greek numbering system, with the exception of *non* for nine, which is derived from Latin.

The prefix indicates the positions of any branches on the main chain by assigning numbers to the carbon atoms in the main chain. Numbering must begin at the end of the main chain that will give the branches the lowest possible numbers. To name the side groups, think of them as separate groups that have been substituted in the place of a hydrogen atom. They are called **substituent groups** or, more commonly, **side groups**. The side groups on alkanes are basically alkanes that are missing a hydrogen atom. For example, look at the side group on the C_5H_{12} molecule that was used as an example in the chart of formulas in the blue box on page 545. It is a $-CH_3$ group and it looks like methane that is missing one hydrogen atom. These side groups that are based on alkanes are called **alkyl groups**. They are named by using the same root that you would use for a main chain but instead of adding the suffix "–ane" you add "–yl." So the $-CH_3$ group is called a methyl group. Table 14.2 shows you the first four alkyl groups.

Table 14.2 Common Alkyl Groups

methyl	ethyl	propyl	butyl
$-CH_3$	$-CH_2CH_3$	$-CH_2CH_2CH_3$	$-CH_2CH_2CH_2CH_3$

In summary, the prefix consists of the number of the carbon atom on the main chain that is bonded to the side chain, followed by a hyphen, and then the name of the alkyl group. For example, "2-methyl–" is the prefix when a methyl group is on carbon number two. If there is more than one side group, they are named in alphabetically order. If there is more than one of the same side group, use a prefix that indicates the number of that type of alkyl group. For example, if there are two methyl groups, one on carbon atom number 3 and the other on number 5, the prefix would be *3,5-dimethyl–*. Note the comma between the numbers 3 and 5. When you are determining the alphabetical order, ignore the prefixes di–, tri–, and tetra–. For example, *dipropyl–* would come after *ethyl–* because *p* comes after *e* in the alphabet. The steps and examples in Table 14.3 will clarify the process of naming alkanes.

Table 14.3 Steps in Naming Alkanes

- **Identify the root.**

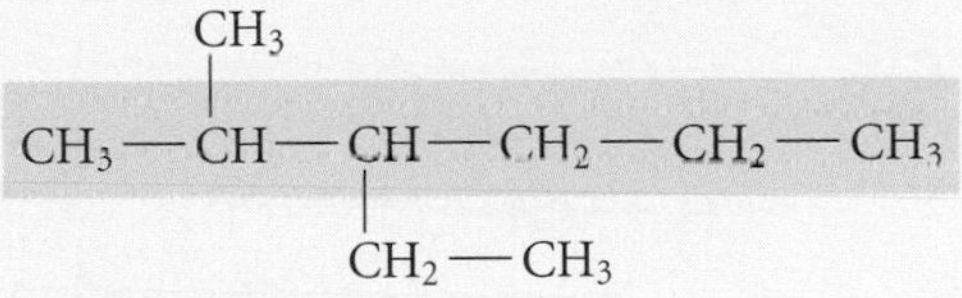

– Find the longest continuous chain.	6 carbon atoms
– Find the root name for the number of carbon atoms in the chain.	Root is hex-.

- **Identify the suffix.**

– For alkanes, the suffix is *–ane*.	Compound is an alkane. hex- plus –ane is hexane

- **Identify the prefix.**

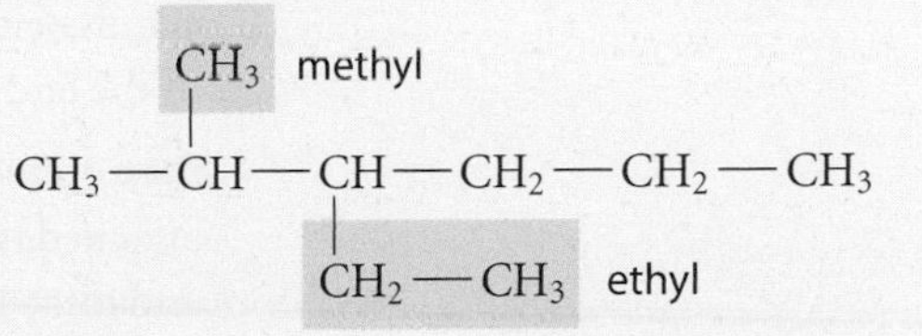

– Find the number of carbon atoms in each side group.	One side group has one carbon so it is a *methyl* group. The other side group has two carbon atoms so it is an *ethyl* group.
– Find the root name for each side group according to the number of carbon atoms. Add *–yl* to the root to name the side group.	
– Determine the alphabetical order of the side groups (if there is more than one).	Alphabetical order is *ethyl methyl*.
– Find the position of each side group.	

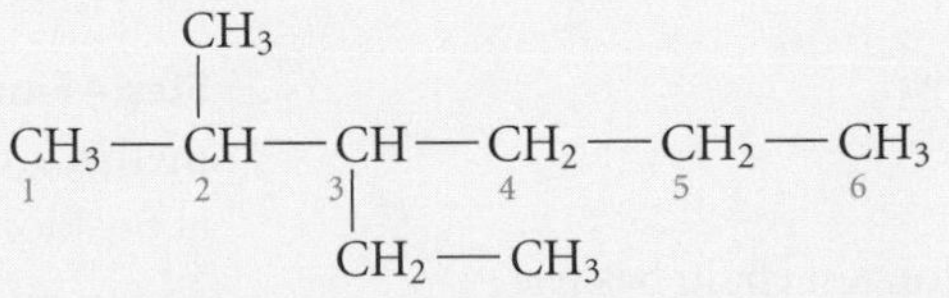

– Precede each side group name with the number of the carbon atom to which it is attached on the main chain.	The ethyl group is on carbon atom 3. The methyl group is on carbon atom 2. The name of the compound is:
– There are hyphens between numbers and words but no hyphen or space between the last prefix and the root.	3-ethyl-2-methylhexane

Figure 14.4 shows one more example that illustrates an alkane that has two of the same type of subgroups. Notice that a comma is placed between two numbers when they appear together. Review the steps for naming alkanes and examine the Sample Problem that follows the figure. Then develop your skills at naming by completing the Practice Problems.

A 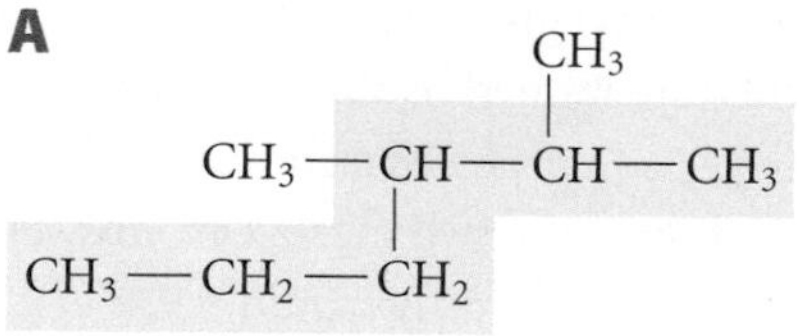 CH_3 CH_3—CH—CH—CH_3 CH_3—CH_2—CH_2	There are six carbons in the longest chain for this molecule, as highlighted. This molecule will be named as a **hex**ane.
B 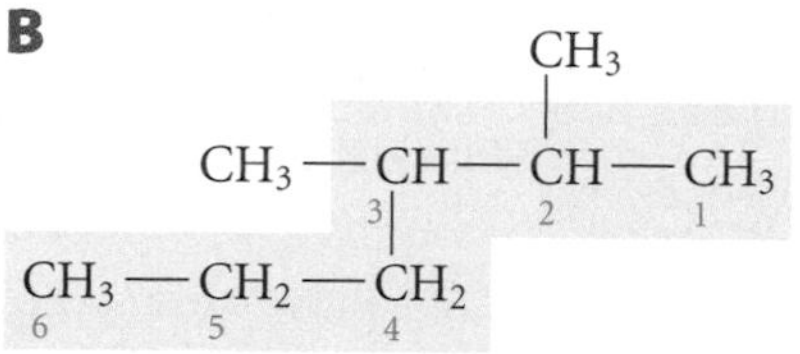 CH_3 CH_3—CH—CH—CH_3 3 2 1 CH_3—CH_2—CH_2 6 5 4	Since there are 2 branches on this molecule, one closest to the right end. Begin numbering the carbons on the main chain from the right end and continue sequentially towards the left.
C 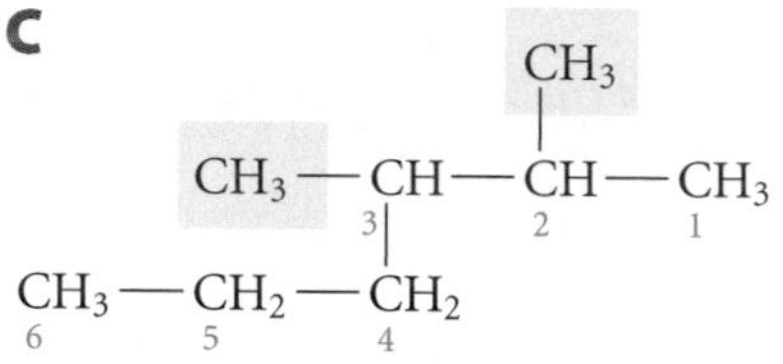 CH_3 CH_3—CH—CH—CH_3 3 2 1 CH_3—CH_2—CH_2 6 5 4	The branches are both single carbon branches. Use the meth– prefix to indicate this. As branches, these are both named as methyl. Since the methyl groups are bonded to carbons 2 and 3, it is necessary to use 2,3-dimethyl to denote this. Remember the di– prefix is necessary to say that there are 2 methyl groups.

Figure 14.4 The name of this molecule will be 2,3-dimethylhexane.

Sample Problem

Naming Alkanes

Problem
Name the following alkanes:

a) 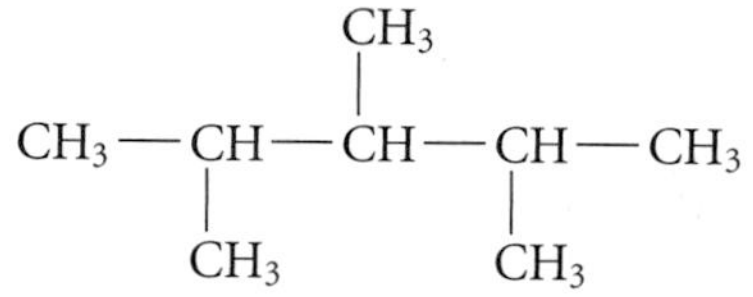
$$CH_3$$
$$CH_3—CH—CH—CH—CH_3$$
$$CH_3 \qquad CH_3$$

b) 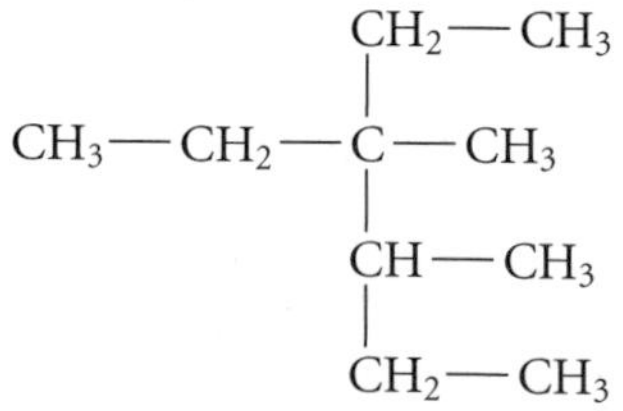
$$CH_2—CH_3$$
$$CH_3—CH_2—C—CH_3$$
$$CH—CH_3$$
$$CH_2—CH_3$$

Solution
a) **Step 1 Find the root:** The longest chain has five carbon atoms, so the root is pent.

Step 2 Find the suffix: Since the bonds are all single bonds, the suffix is –ane.

Step 3 Assign position numbers: You can start numbering at either end because the molecule is symmetrical. The name will be the same regardless of the end at which you start numbering.

Step 4 Find the prefix: Three methyl groups are attached to the main chain at positions two, three, and four. Therefore, the prefix is 2,3,4-trimethyl–.

Step 5: The full name is 2,3,4-trimethylpentane.

b) **Steps 1 and 2 Find the root and suffix:** There are six carbon atoms in the main chain. (Note that the main chain is bent at two places.) The root is hex. The suffix is –ane.

Step 3 Assign position numbers: Number from the left so the branches are at positions three and four.

Step 4 Find the prefix: There is an ethyl group and a methyl group at position 3, and another methyl group at position 4. The prefix is 3-ethyl-3,4-dimethyl–.

Step 5: The full name is 3-ethyl-3,4-dimethylhexane.

Notice that the "di-" in dimethyl is ignored when alphabitizing the prefixes. The "e" in ethyl comes before the "m" in methyl.

1. Provide names for the following molecules:

a) 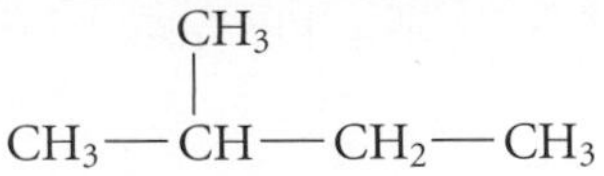
$$CH_3-\underset{\underset{CH_3}{|}}{CH}-CH_2-CH_3$$

b) 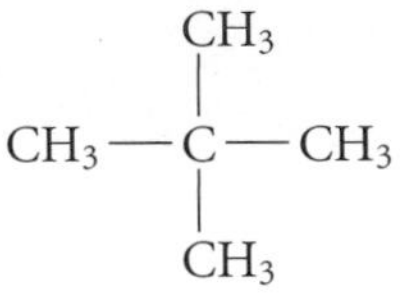
$$CH_3-\underset{\underset{CH_3}{|}}{\overset{\overset{CH_3}{|}}{C}}-CH_3$$

c)
$$CH_3-\underset{\underset{CH_3}{\overset{|}{CH_2}}}{\overset{|}{CH}}-CH_2-\underset{\overset{|}{CH_2}}{CH}-\underset{\overset{|}{CH_3}}{CH}-CH_3$$
with CH_2-CH_3 at top of middle CH.

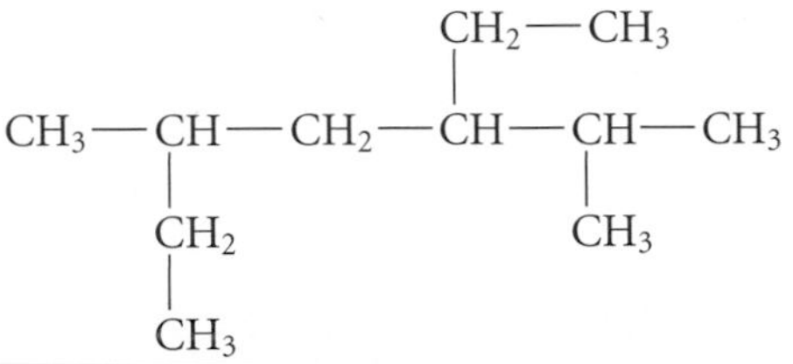

d) 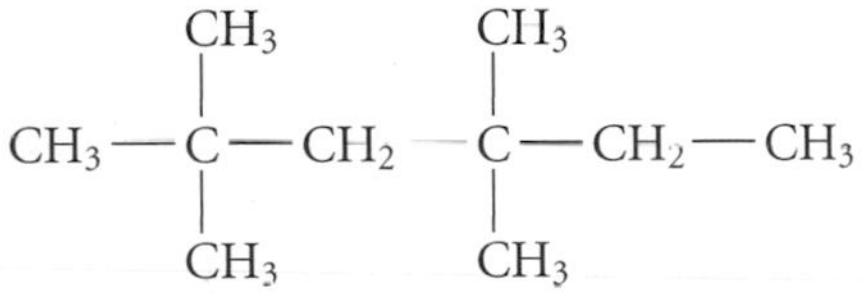
$$CH_3-\underset{\underset{CH_3}{\overset{|}{CH_3}}}{\overset{\overset{CH_3}{|}}{C}}-CH_2-\underset{\underset{CH_3}{\overset{|}{CH_3}}}{\overset{\overset{CH_3}{|}}{C}}-CH_2-CH_3$$

e) 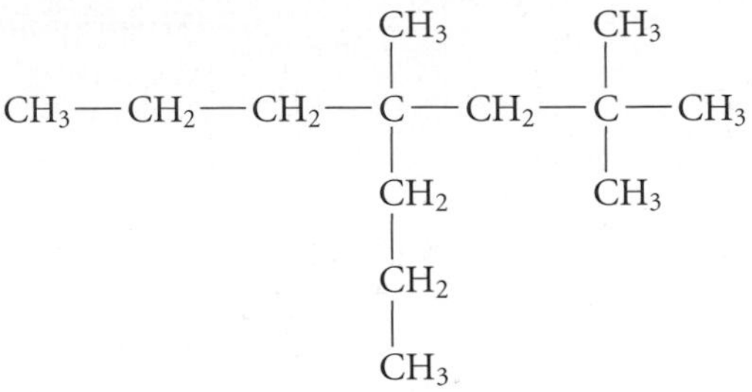
$$CH_3-CH_2-CH_2-\underset{\underset{CH_2}{\overset{|}{CH_2}}}{\overset{\overset{CH_3}{|}}{C}}-CH_2-\underset{\overset{|}{CH_3}}{\overset{\overset{CH_3}{|}}{C}}-CH_3$$
with $CH_2-CH_2-CH_3$ chain descending from fourth carbon.

2. Identify any errors in the name of each of the following hydrocarbons. (**Hint:** Where possible, draw the structures as named. Then examine the structures and rename them.)
   a) 2,2,3-dimethylbutane
   b) 2,4-diethyloctane
   c) 3-methyl-4,5-diethylnonane

3. Name each of the following compounds:

a)

b)

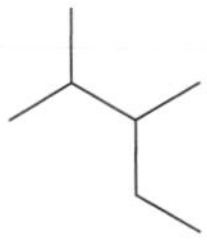

c)

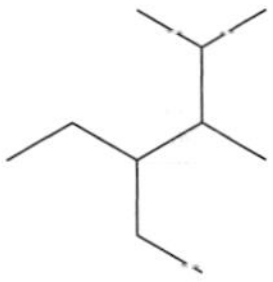

Drawing Alkanes

The IUPAC system also allows you to draw a structural formula of any organic molecule given its name. The logic involved is the same as that used in naming the molecule. For example, when asked to draw the condensed structural formula for 3-ethyl-3-methylpentane, you can do so by following the steps below.

Step 1 Identify the root and the suffix of the name: This step gives you the number of carbons in the main chain and the type of functional group present. For 3-ethyl-3-methylpentane, the root and suffix are –pentane. The root, pent, tells you there are five carbons in the main chain. The suffix, –ane, tells you that the chain has only single carbon–carbon bonds.

Step 2 Draw the main chain first: Draw a straight chain containing five carbon atoms with single bonds between the atoms. Do not include hydrogen atoms yet.

$$C-C-C-C-C$$

Step 3 Choose one end of your carbon chain to be carbon number 1 and number the rest of the chain in sequence. Add the indicated branches to the appropriate carbon: In this case, add a methyl group and an ethyl group both at carbon three. You may place the branches on either side of the main chain.

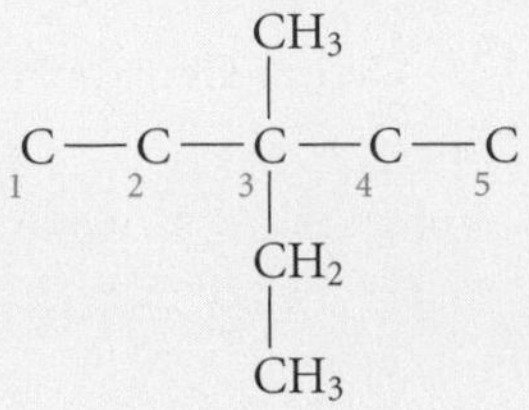

$$\underset{1}{C}-\underset{2}{C}-\underset{3}{\overset{\overset{CH_3}{|}}{\underset{\underset{\underset{CH_3}{|}}{CH_2}}{C}}}-\underset{4}{C}-\underset{5}{C}$$

Step 4 Complete your formula by adding the number of hydrogen atoms beside each carbon that will give each carbon four bonds: Add hydrogen atoms as shown below.

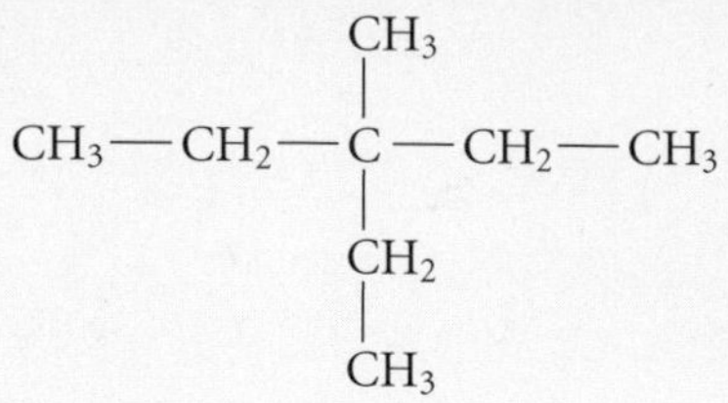

$$CH_3 - CH_2 - \underset{\underset{CH_3}{|}}{\overset{\overset{CH_3}{|}}{C}} - CH_2 - CH_3$$

The following Sample Problem will guide you through another example of drawing a structure from a name. After you familiarize yourself with the process, use the Practice Problems to practise your drawing skills.

Sample Problem

Drawing an Alkane

Problem

Draw a condensed structural formula of 3-ethyl-2-methylheptane.

Solution

Step 1 The root is -hept- indicating that there are seven carbons in the main chain. The suffix is –ane, indicating that this structure is an alkane having only single carbon–carbon bonds.

Step 2 Draw seven carbon atoms in a row, leaving spaces for hydrogen atoms.

Step 3 The number 3 and the -eth- indicate that a side group having two carbon atoms is attached to carbon atom number 3. The number 2 and the meth- indicate that a side group having one carbon atom is attached to carbon atom number 2.

Step 4 Add hydrogen atoms so that each carbon atom will have four bonds.

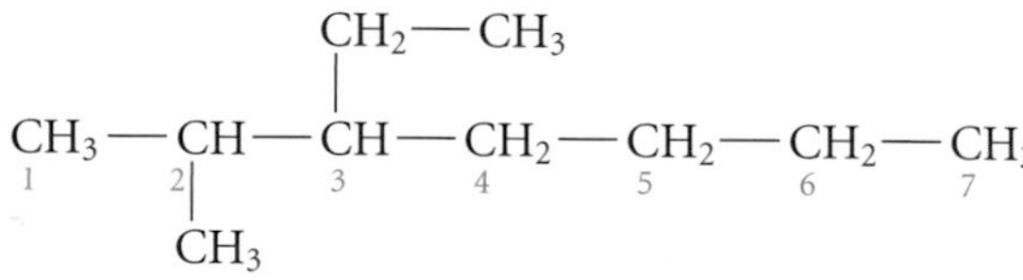

$$CH_3 - \underset{1}{} \underset{2}{\overset{}{CH}} - \underset{3}{\overset{\overset{CH_2 - CH_3}{|}}{CH}} - \underset{4}{CH_2} - \underset{5}{CH_2} - \underset{6}{CH_2} - \underset{7}{CH_3}$$
$$\underset{CH_3}{|}$$

Practice Problems

4. Draw a structural formula for each of the following organic molecules:
   a) propane
   b) 2-methylbutane

5. Draw a condensed structural formula for each of the following:
   a) 2,4,6- trimethyloctane
   b) 4-ethyl-3-methylheptane

6. For each of the molecules listed in question 5, draw an expanded molecular formula.

7. The following names are incorrect. Draw structures that these names describe. Examine your drawing, and rename the hydrocarbon correctly.
   a) 3-propylbutane
   b) 1,3-dimethylhexane
   c) 4-methylpentane

8. Draw a line structural formula for each of the following alkanes:
   a) 3-ethyl-3,4-dimethyloctane
   b) 2,3,4-trimethylhexane
   c) 4-ethyl-3,3-dimethylheptane
   d) 4,6-diethyl-2,5-dimethylnonane

9. Examine the following compounds and their names. Identify any mistakes, and correct the names as necessary.
   a) 4-ethyl-2-methylpentane

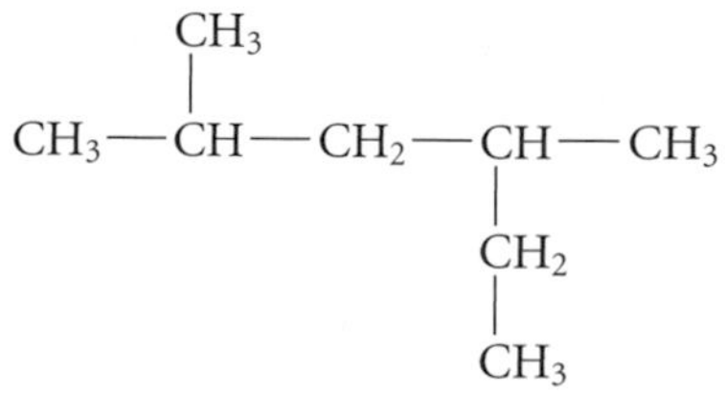

$$CH_3 - CH - CH_2 - \underset{\underset{\underset{CH_3}{|}}{\underset{CH_2}{|}}}{\overset{\overset{CH_3}{|}}{CH}} - CH_3$$

b) 4,5-methylhexane

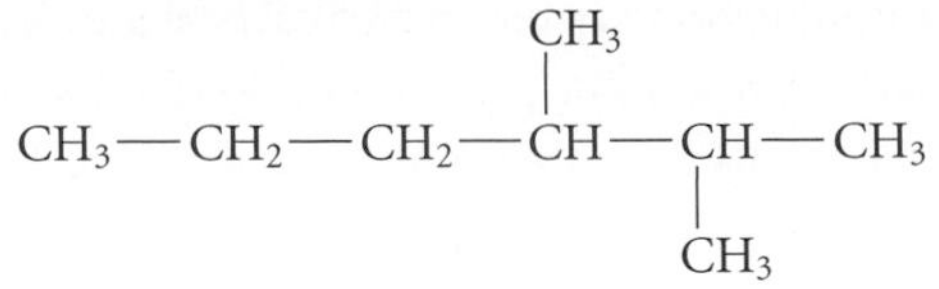

$$CH_3—CH_2—CH_2—CH—CH—CH_3$$

with CH₃ above the fourth carbon (CH) and CH₃ below the fifth carbon (CH)

c) 3-methyl-3-ethylpentane

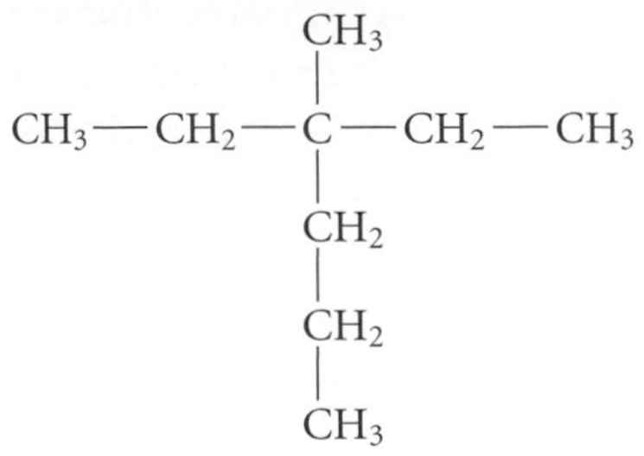

$$CH_3—CH_2—C—CH_2—CH_3$$

with CH₃ above the central carbon, and CH₂, CH₂, CH₃ below it

Physical Properties of Alkanes

You have read, several times, that methane is the major component of natural gas used for home heating. Clearly, it is a gas at standard temperature. As well, you probably know that propane is a commonly used gas for camp stoves, barbeques, and some other types of burners. Thus you know that it is also a gas. You might have used a butane burner in a laboratory experiment. Are all alkanes gases?

All alkanes are non-polar molecules and, therefore, the only intermolecular forces that act among them are London (dispersion) forces. As you know, these are the weakest of the intermolecular forces. Therefore, the boiling points of small alkanes are below 25 °C, making the gases at standard temperature. As the length of the chain increases, however, the sum of the intermolecular forces become stronger and the medium length alkanes are liquids at standard temperature. The very large alkanes are waxy solids. Table 14.4 lists the range of boiling point ranges and some examples of uses for the alkanes. These values are given in ranges because, as you know, the shape of the molecules as well as the size, affects the boiling points. Highly branched-chain molecules would not attract one another as strongly as would straight chain molecules. Being non-polar, alkanes are not soluble in water. However, alkanes are soluble in benzene and other non-polar solvents.

Table 14.4 Sizes and Boiling Points of Alkanes

Size (number of carbon atoms per molecule)	Boiling point range (°C)	Examples of uses
1 to 4	below 30	gases: used for fuels to cook and heat homes
5 to 16	30 to 275	liquids: used for automotive, diesel, and jet engine fuels; also used as raw material for the petrochemical industry
16 to 22	over 250	heavy liquids: used for oil furnaces and lubricating oils; also used as raw materials to break down into smaller molecules
over 18	over 400	semi-solids: used for lubricating greases and paraffin waxes to make candles, waxed paper, and cosmetics
over 26	over 500	solid residues: used for asphalts and tars in the paving and roofing industries

Alkenes

Fresh fruits and vegetables that you find in the supermarket have often been shipped over long distances. If they were fully ripe when they were picked, they would begin to spoil by the time they reached the supermarket. Therefore, fruits and vegetables are usually picked before they ripen. You have probably seen green bananas (see Figure 14.5) in your local store but most fruits and vegetable are ripe when they arrive at the supermarket. Producers refrigerate the fruits and vegetables to slow the ripening during transport. When they reach their destination, ethene, a plant hormone, is pumped into the containers to ripen fruits. Ethene is the simplest alkene.

Figure 14.5 Fruits such as bananas are ripened after transport by exposing them to the plant hormone, ethene. However, they sometimes reach the supermarket slightly green. They will ripen soon.

Alkenes are aliphatic hydrocarbons that have at least one double bond in the carbon chain. Since there must be at least one double bond between two carbon atoms, there are no alkenes consisting of just one carbon atom. The presence of the double bond gives alkenes the ability to bond to more atoms than are present in the molecules. For example, Figure 14.6 shows an alkene reacting with hydrogen and bonding to two more hydrogen atoms than were originally present. The result is the alkane, ethane. Since the carbon atoms in alkenes are not bonded to the maximum number of atoms possible, they are called **unsaturated hydrocarbons**.

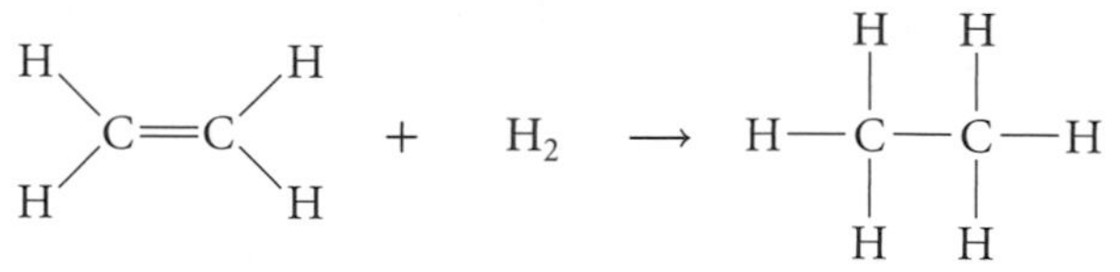

unsaturated
hydrocarbon

saturated
hydrocarbon

Figure 14.6 An unsaturated hydrocarbon bond reacts with a hydrogen molecule and becomes a saturated hydrocarbon.

From Figure 14.6, you can also see that an alkene with one double bond has two fewer hydrogen atoms than the alkane having the same number of carbon atoms. Therefore, the general formula for straight and branched chain alkenes with one double bond is C_nH_{2n}. Check the structures in Figure 14.7 to ensure that they fit the formula. Figure 14.7 shows you the first three alkenes. Examine the figure to find the similarities and differences between alkenes and alkanes.

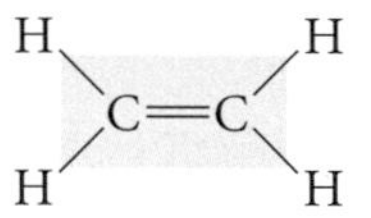

ethene

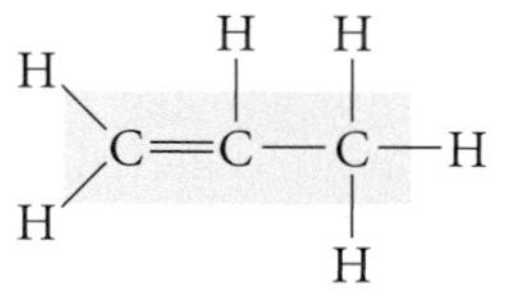

propene

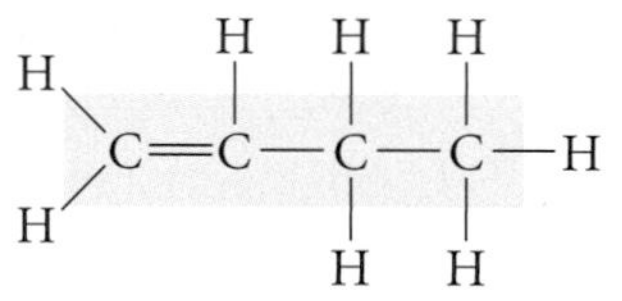

but-1-ene

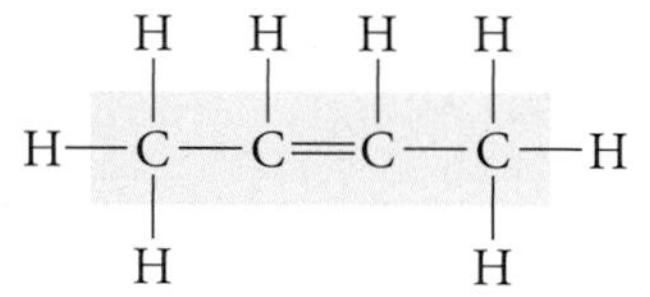

but-2-ene

Figure 14.7 These are the alkenes that consist of two, three, and four carbon atoms. Why are there two different alkenes with four carbon atoms?

As you can see in Figure 14.7, there is only one possible structure for the first two alkenes. If you moved the double bond in the second alkene to the second position and then flipped it over side-to-side, it would not differ from the structure as written. Alkenes with at least four carbon atoms, however, can have more than one structure because the double bond can be in either the first or second position. As more carbon atoms are added, the variety becomes greater.

Modelling Alkenes

You can use the same five types of formulas to model alkenes. The alkene with five carbon atoms, C_5H_{10}, is used as an example in Figure 14.8.

Empirical molecular formula	Expanded molecular formula
C_5H_{10}	$CH_3C(CH_3)CHCH_3$

Structural formula

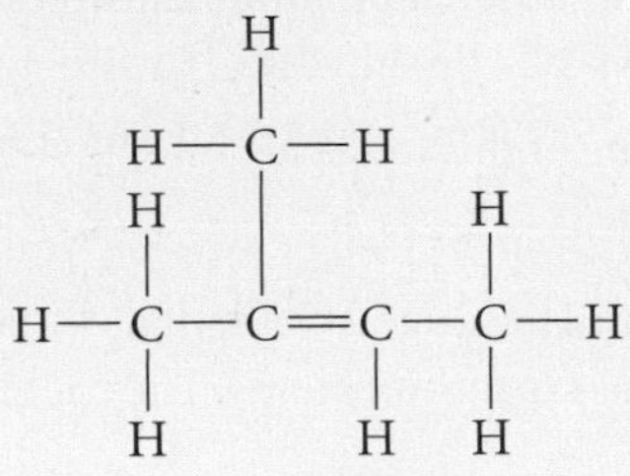

Condensed structural formula

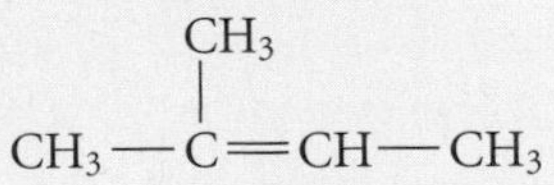

$$CH_3 - \overset{\overset{\displaystyle CH_3}{|}}{C} = CH - CH_3$$

Line structural formula

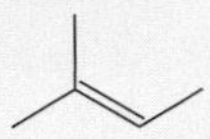

Figure 14.8 This example shows just one of several possible alkenes that have five carbon atoms.

As you can see, you would have to analyze the empirical molecular formula and the expanded molecular formula at length to determine that the molecule is an alkene. It is not possible to determine the position of the double bond from the empirical molecular formula and you would have to analyze the expanded molecular formula to determine where the double bond is located. All of the structural formulas clearly show the double bond but the condensed and line structural formulas are quickest to draw.

Naming Alkenes

The IUPAC name of alkenes has the same three basic parts: a root, a prefix and a suffix.

prefix + root + suffix

Determining the prefix, root, and suffix are very similar to those of alkanes with a few modifications as follows.

- **Identify the root.**
  Find the longest continuous chain *that contains the double bond*. The root name for a given chain of carbon atoms is the same as the root name for alkanes.

- **Identify the suffix.**
  The suffix for all alkenes is *−ene*. However, the suffix of an alkene must also indicate the location of the double bond. The numbering of the main chain must begin at the end of the chain nearest the double bond and continue through the double bond. The position of the double bond is indicated by the number of the carbon atom that *precedes* the double bond. The suffix consists of a hyphen, a number, a hyphen, and *−ene*.

- **Identify the prefix.**
  The rules for naming alkyl groups as side groups on alkenes are the same as they are for alkanes.

Notice that in some molecules the chain containing the double bond is not the longest chain. Nevertheless, the root name must describe the chain that contains the double bond. Also, since there is only one possible position for the double bond in ethene and propene, a number indicating the position is not needed. For longer chains alkenes, the number that describes the position of the double bond is part of the suffix. This rule was recently revised by IUPAC so, in many textbooks, you might see the number in the prefix. Always use the new rule that puts the number in the suffix. For example, a six carbon chain that has a double bond between the second and third carbon atom is named, hex-2-ene. Analyze the following Sample Problem to learn how to apply the rules. Then complete the Practice Problems to develop your skills.

Naming Alkenes

Problem

Name the following alkenes:

a) 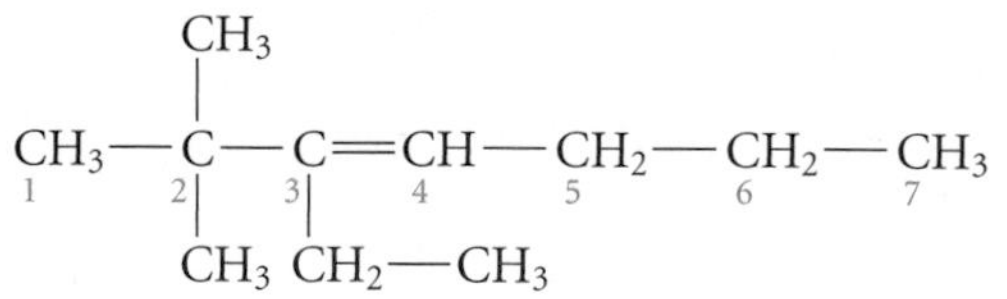
$$CH_3 - \underset{\underset{CH_3}{|}}{\overset{\overset{CH_3}{|}}{\underset{2}{C}}} - \underset{\underset{CH_2-CH_3}{|}}{\overset{}{\underset{3}{C}}} = \underset{4}{CH} - \underset{5}{CH_2} - \underset{6}{CH_2} - \underset{7}{CH_3}$$

b) 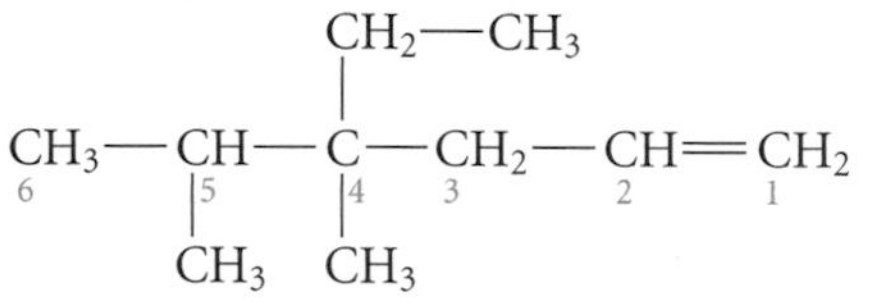
$$CH_3 - \underset{6}{CH} - \underset{\underset{CH_3}{|}}{\overset{\overset{CH_2-CH_3}{|}}{\underset{4}{C}}} - \underset{3}{CH_2} - \underset{2}{CH} = \underset{1}{CH_2}$$
with CH_3 and CH_3 below carbons 5 and 4.

a) Step 1 Find the root: The longest chain that contains the double bond has seven carbon atoms. The root is -hept-.

Step 2 Find the suffix: The molecule has a double bond so the suffix ends with –ene.

Step 3 Assign position numbers: Numbering must begin at the left end because the double bond is closest to that end. The position of the double bond is 3 so the suffix is -3-ene.

Step 4 Find the prefix: Two methyl groups are attached to carbon atom number two. One ethyl group is attached to carbon atom number three. The prefix is 3-ethyl-2,2-dimethyl-.

Step 5 The full name of this molecule is 3-ethyl-2,2-dimethylhept-3-ene.

b) Step 1 Find the root: The longest chain that contains the double bond has six carbon atoms. The root is -hex-.

Step 2 Find the suffix: The molecule has a double bond so the suffix ends with –ene.

Step 3 Assign position numbers: Numbering must begin at the right end of the molecule because it is closest to the double bond. The position of the double bond is 1 so the suffix is -1-ene.

Step 4 Find the prefix: There is one methyl group on carbon atom number 4 and another on carbon atom number 5. There is also an ethyl group on carbon atom number 4. The prefix is 4-ethyl-4,5-dimethyl-.

Step 5 The full name of the compound is 4-ethyl-4,5-dimethylhex-1-ene.

Practice Problems

10. Name each alkene:

a) $CH_3 - CH_2 - CH = CH - CH_2 - CH_3$

b) 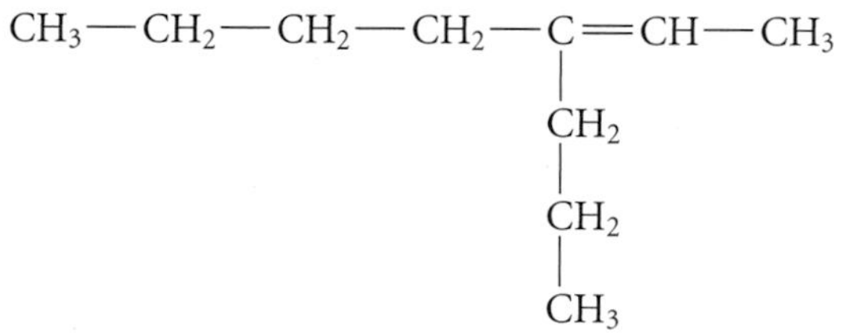 $CH_3 - CH_2 - CH_2 - CH_2 - \underset{\underset{\underset{\underset{\underset{CH_3}{|}}{CH_2}}{|}}{\overset{}{\underset{|}{CH_2}}}}{C} = CH - CH_3$

11. Name this alkene:

$$CH_3 - \underset{\underset{CH_3}{|}}{CH} - \underset{\underset{CH_3}{|}}{CH} - \underset{\underset{CH-CH_2-CH_2-CH_3}{\|}}{C} - CH_2 - CH_3$$

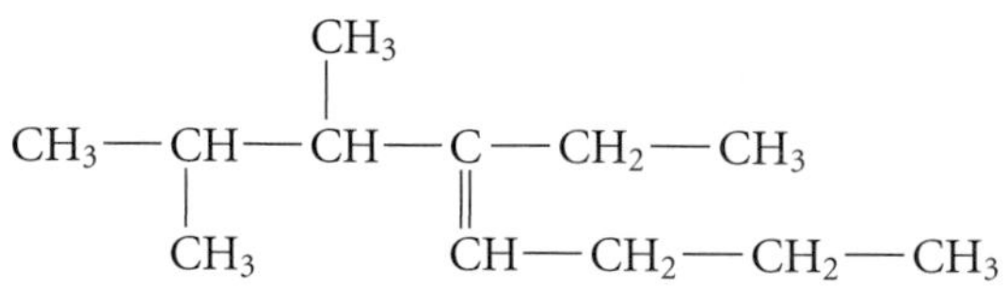

12. Name the following alkene:

$$CH_3 - \underset{\underset{CH_3}{|}}{\overset{\overset{CH_3}{|}}{C}} - CH = CH - CH_3$$

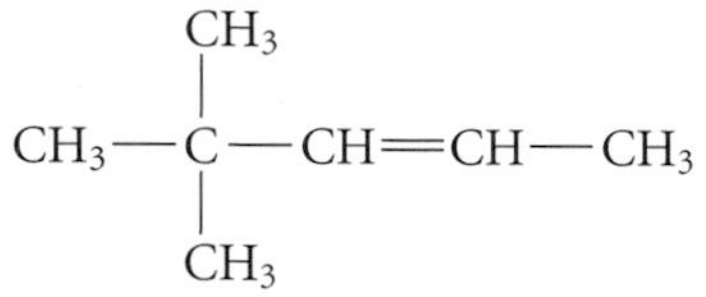

13. A classmate claims that the following alkene is named 3-methyl-4-ethyl-4-propyl-hex-5-ene. Do you agree? If not, give the correct name.

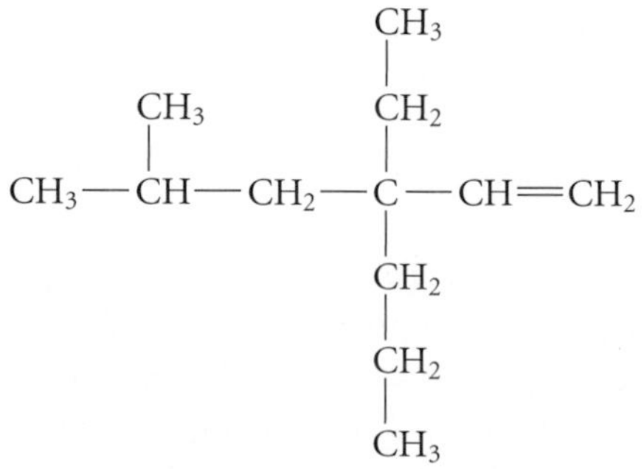

$$CH_3 - \underset{\underset{CH_3}{|}}{CH} - CH_2 - \underset{\underset{\underset{\underset{\underset{CH_3}{|}}{CH_2}}{|}}{\overset{\overset{\overset{\overset{CH_3}{|}}{CH_2}}{|}}{C}}}{C} - CH = CH_2$$

Drawing Alkenes

The method for drawing alkenes is fundamentally the same as that for drawing alkanes. The main difference is in placing the double bond in the correct place. Remember that it comes after the carbon atom with the number stated in the suffix. Also remember that the two carbon atoms that share the double bond have only two other bonds. When you have completed a structure, check to be sure that all carbon atoms have only four bonds.

Sample Problem

Drawing Alkenes

Problem
Draw a structural formula for 2-methylbut-2-ene.

Solution
Step 1 The root is *but-* so there are four carbon atoms in the main chain.

Step 2 The suffix is -2-*ene* so it has a double bond after carbon atom number two.

Step 3 The prefix is 2-methyl- so there is a $-CH_3$ group of the second carbon atom.

Step 4 All carbon atoms must be bonded to enough hydrogen atoms to give them exactly four bonds.

The structural formula of 2-methylbut-2-ene is:

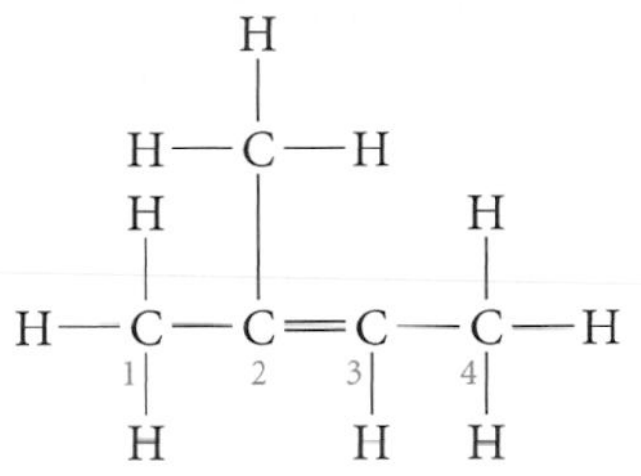

Practice Problems

14. Draw structural formulas for each compound:
- **a)** 3-methylbut-1-ene
- **b)** pent-2-ene
- **c)** 4,4-dimethylpent-1-ene

15. Draw condensed structural formulas for each compound:
- **a)** 5-ethyl-3,4,6-trimethyloct-2-ene
- **b)** 4,5-dimethylhept-2-ene
- **c)** 3-ethyl-4-methylhex-1-ene
- **d)** 2,5,7-trimethyloct-3-ene

Physical Properties of Alkenes

The physical properties of alkenes are very similar to those of alkanes. Alkenes are non-polar and thus do not dissolve in water but do dissolve in non-polar solvents. The first three alkenes—ethene, propene, and butene—are all gases at standard temperature while the intermediate size alkenes are liquids. However, the boiling points of the alkenes are slightly lower than those of the alkanes that have the same number of carbon atoms. For example, the boiling point of ethene is −103.8 °C while that of ethane is −88.6 °C. Likewise, the boiling point of propene is −47.7 °C while that of propane is −42.1 °C. The reason for the slight difference is that the alkenes have fewer atoms and thus fewer electrons. As you reviewed in the Unit Preparation, the number of electrons in a molecule influences the strength of the London (dispersion) forces.

As you know, when the number of carbon atoms in an alkene is four or greater, there is more than one possible position for the double bond. Even the slight differences in the shape of the molecules caused by the location of the double bond make a difference in the boiling point. For example, the boiling point of but-1-ene is −6.3 °C and that of one of the configurations of but-2-ene is +0.88 °C.

Figure 14.9 Acetylene, an alkyne, is commonly used for welding. The IUPAC name for acetylene is ethyne.

Alkynes

One of the most common alkynes, commonly called acetylene, is used by welders as the fuel in their torches, such as the torch shown in Figure 14.9. The tool is often called an oxyacetylene torch because pure oxygen is combined with acetylene. When the mixture of oxygen and acetylene is burning, the temperature can be higher than 3000 °C. The oxyacetylene torch can be used for welding, cutting, or shaping metals.

Alkynes are aliphatic hydrocarbons that have at least one triple bond. Like alkenes, the carbon atoms in alkynes are not bonded to the maximum number of other atoms so they are also unsaturated hydrocarbons. Inspect the examples in Figure 14.10 to determine the general formula for straight or branched-chain alkynes.

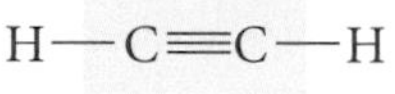

$$H—C\equiv C—H$$

ethyne

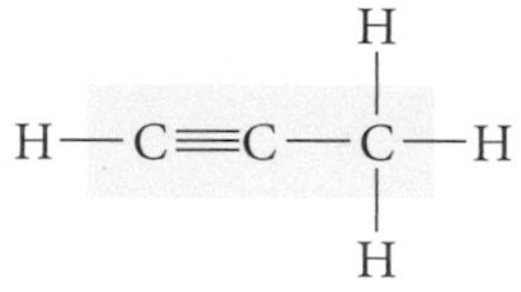

$$H—C\equiv C—\underset{\underset{H}{|}}{\overset{\overset{H}{|}}{C}}—H$$

propyne

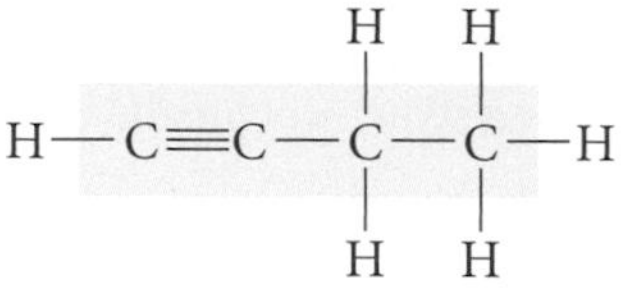

$$H—C\equiv C—\underset{\underset{H}{|}}{\overset{\overset{H}{|}}{C}}—\underset{\underset{H}{|}}{\overset{\overset{H}{|}}{C}}—H$$

but-1-yne

$$H—\underset{\underset{H}{|}}{\overset{\overset{H}{|}}{C}}—C\equiv C—\underset{\underset{H}{|}}{\overset{\overset{H}{|}}{C}}—H$$

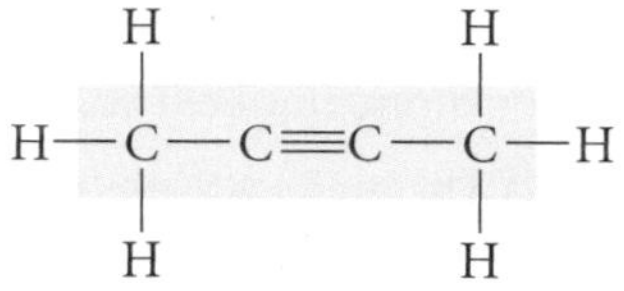

but-2-yne

Figure 14.10 Similar to the alkenes, the alkynes having four or more carbon atoms can have more than one structure because the triple bond can be in either of two different places.

As you can see, for molecules with the same number of carbon atoms, alkynes have two fewer hydrogen atoms than do alkenes. Therefore, the general formula for straight or branched-chain alkynes is C_nH_{2n-2}. For example, the alkyne that has five carbon atoms has a formula of $C_5H_{2\times5-2}$ or C_5H_8.

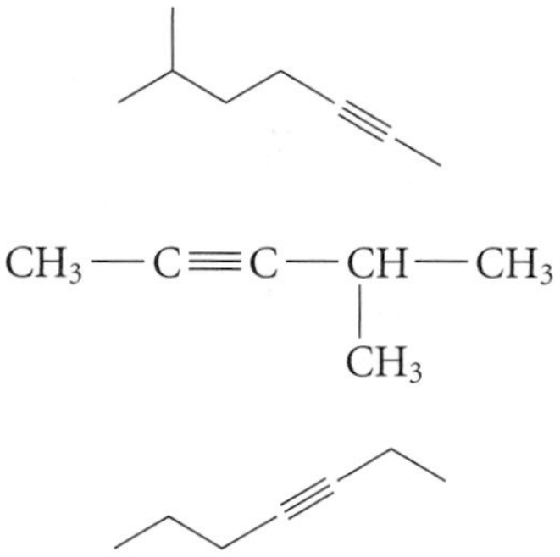

Naming and Drawing Alkynes

The naming of alkynes follow the rules for alkenes with the exception of the suffix which is *–yne* instead of *–ene*. Drawing alkynes is also very similar to drawing alkenes. The only significant difference is that the two carbons on either end of the double bond have only two electron groups and no lone pairs. Recall from the review of VSEPR theory in the Unit Preparation, that central atoms with two electron groupings have a linear structure. Therefore, when drawing alkynes, the lines representing bonds on either side of the triple bond are in line with the triple bond. This linear structure is shown in Figure 14.11. Develop your skills for naming and drawing alkynes in the next set of Practice Problems.

$$CH_3—C\equiv C—\underset{\underset{CH_3}{|}}{CH}—CH_3$$

Figure 14.11 Alkynes are always linear around the triple bond.

Practice Problems

16. Name the following alkynes.

a)
$$CH_3—\underset{\underset{CH_3}{|}}{\overset{\overset{CH_3}{|}}{C}}—C\equiv C—CH_3$$

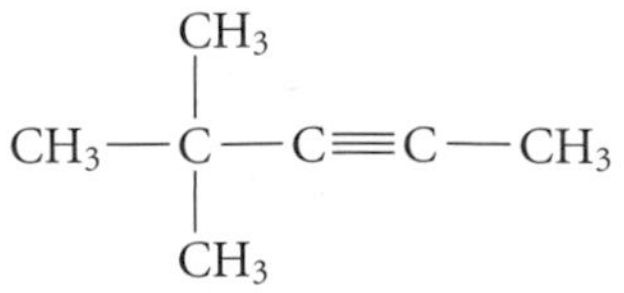

b)
$$CH_3—\underset{\underset{}{|}}{\overset{\overset{CH_3}{|}}{CH}}—CH_2—\underset{\underset{CH_2}{|}}{\overset{\overset{CH_2}{|}}{\underset{\underset{CH_2}{|}}{\underset{\underset{CH_3}{|}}{C}}}}—C\equiv CH$$

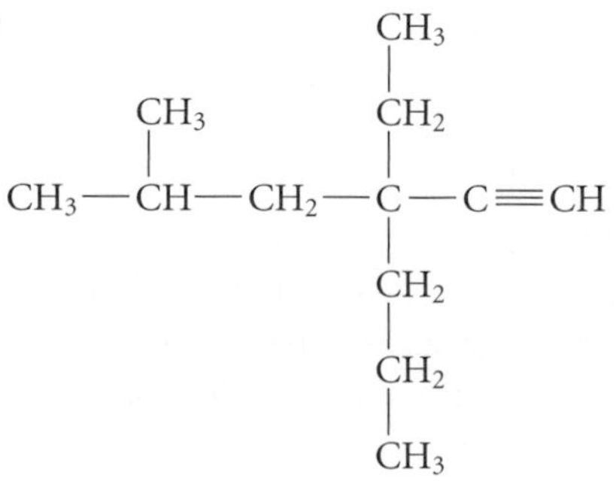

17. Draw a condensed structural formula for each of the following compounds.

a) pent-2-yne

b) 4,5-dimethlyhept-2-yne

c) 3-ethyl-4-methylhex-1-yne

d) 2,5,7-trimethyloct-3-yne

Physical Properties of Alkynes

Like the other aliphatic hydrocarbons, alkynes are non-polar and thus insoluble in water. Also, like the alkanes and alkenes, the first few alkynes exist as gases at standard temperature. An interesting property of the aliphatic hydrocarbons arises when you compare the boiling points of the lower-mass, straight-chain alkanes, alkenes, and alkynes that are listed in Table 14.5. (Note: The multiple bonds are all in position 1.)

Table 14.5 Boiling Points of Small Alkanes, Alkenes, and Alkynes

Alkanes	Boiling point (°C)	Alkenes	Boiling point (°C)	Alkynes	Boiling point (°C)
ethane	−89	ethene	−104	ethyne	−84
propane	−42	propene	−47	propyne	−23
butane	−0.5	butene	−6.3	butyne	8.1
pentane	36	pentene	30	pentyne	39
hexane	69	hexene	63	hexyne	71

As you read previously, the alkenes have lower boiling points that their corresponding alkane. However, alkynes have a higher boiling point than the corresponding alkane. Although the alkynes have fewer electrons than their corresponding alkanes and alkenes, their linear structure and the nature of the triple bond cause them to attract one another more strongly than do the alkenes or alkanes and, thus, it takes more energy to overcome these attractive forces.

Cyclic Hydrocarbons

You have heard and read a lot about cholesterol, the compound that is often a causative factor in heart disease. Cholesterol, shown in Figure 14.12, has four connected hydrocarbon rings. Although an excess of cholesterol can be harmful, it is a necessary part of cell membranes. As well, it is the "raw material" that the body uses to synthesize several different steroid hormones including testosterone and the estrogens, also shown in Figure 14.12. The estrogens consist of three related compounds, 17-β-estradiol, estrone, and estriol. These compounds are members of the class of hydrocarbons called cyclic hydrocarbons.

cholesterol   estrone   testosterone

Figure 14.12 Cyclic hydrocarbons are the foundation of many biologically important molecules. You are probably familiar with one famous (or infamous) class of cyclic hydrocarbons, the steroids, in relation to "doping" scandals in various sports. Steroid hormones help control many functions within cells and are sometimes subject to misuse by bodybuilders and other athletes.

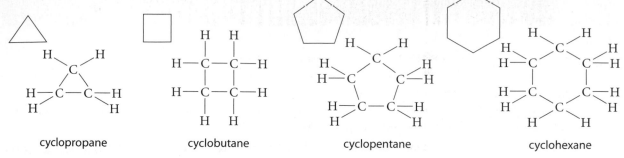

cyclopropane   cyclobutane   cyclopentane   cyclohexane

Figure 14.13 A carbon chain must have at least three carbon atoms to form a ring. Notice that, in the cycloalkanes, there are two hydrogen atoms for each carbon atom. Therefore, the general formula is C_nH_{2n}.

Cyclic hydrocarbons are aliphatic hydrocarbon chains that have no beginning and no end but, instead, form rings. They can be alkanes, alkenes, or alkynes. Cyclic hydrocarbons are non-polar and have physical properties similar to their straight chain counterparts. The first four cyclic alkanes are shown in Figure 14.13 as both expanded structural formulas and line structural formulas.

Naming Cyclic Hydrocarbons

All of the basic rules for naming hydrocarbons apply to cyclic hydrocarbons with a few additional rules. You have probably already noticed that *cyclo* precedes each name. The general rules for naming cyclic aliphatic hydrocarbons are listed below followed by Sample and Practice Problems. Study the Sample Problems to see how the rules are applied. Then develop your skills by completing the Practice Problems.

- **Identify the root**
  Determine the number of carbon atoms in the ring. The root name is the same as the straight chain alkane, alkene, or alkyne, with the same number of carbons atoms preceded by *cyclo*.
- **Identify the suffix**
  Determine whether the molecule has all single bonds, at least one double bond, or at least one triple bond. The suffix is *–ane*, *–ene*, or *–yne*, respectively. There are no numbers to indicate the location of the double or triple bonds because they are always assumed to be between carbon atoms number one and two.
- **Identify the prefix**
  The names for the alkyl prefixes are the same as they are for the straight-chain hydrocarbons. They are given in alphabetical order. However, there are a special set of rules for numbering the carbon atoms in the ring to which side groups are attached.
  - If there are no side groups or only one side group on an alkane, the carbon atoms are not numbered.
  - If the molecule is a cycloalkane and there are two or more side groups, the numbering must result in the lowest possible numbers. The carbon atom to which one of the side groups is attached is carbon atom number one. Numbering goes in the direction to make the numbers of any other side groups as small as possible. You can choose any carbon atom with a side group to start numbering.
  - If the molecule is an alkene or alkyne, the multiple bond takes highest priority. The carbon atom on one side of the multiple bond is carbon atom number one and the one on the other side is number two. If there is a side group, the numbering starts in the position that will make the number of the carbon atom with the side group as small as possible.

Naming Cylic Hydrocarbons

Problem
Name the following cyclic hydrocarbons.

a)

b) CH$_3$

c) CH$_3$

CH$_2$—CH$_3$

d) CH$_3$

Solution

a) Step 1 Find the root: The ring has five carbon atoms so the root is cyclopent–.

Step 2 Find the suffix: There are no multiple bonds so the suffix is –ane.

Step 3 Assign the position numbers: There are no side groups so the carbon atoms are not numbered.

Step 4 Find the prefix: There are no side groups so there is no prefix.

Step 5 The name of the compound is cyclopentane.

b) Step 1 Find the root: The ring has six carbon atoms so the root is cyclohex–.

Step 2 Find the suffix: There are no multiple bonds so the suffix is –ane.

Step 3 Assign the position numbers: There is only one side group so the carbon atoms are not numbered.

Step 4 Find the prefix: The side group has one carbon atom so the prefix is methyl–.

Step 5 The name of the compound is methylcyclohexane.

c) Step 1 Find the root: The ring has six carbon atoms so the root is cyclohex–.

Step 2 Find the suffix: There are no multiple bonds so the suffix is –ane.

Step 3 Assign the position numbers: There are two side groups so the numbering will start at one of the side groups and proceed toward the other. Call the lower right carbon atom number one and number counterclockwise. The side groups will be numbered 1 and 3.

Step 4 Find the prefix: One side group has two carbon atoms so it is an ethyl group. The other side group has one carbon atom so it is a methyl group. Name them in alphabetical order. The prefix is 1-ethyl-3-methyl.

Step 5 The name of the compound is 1-ethyl-3-methylcyclohexane. The numbering is shown here.

CH$_3$

CH$_2$—CH$_3$

d) Step 1 Find the root: The ring has six carbon atoms so the root is cyclohex–.

Step 2 Find the suffix: The ring has one double bond so the suffix is –ene.

Step 3 Assign the position numbers: The carbon atoms beside the double bond must be numbered 1 and 2. The numbering must proceed in the direction of the side group. Therefore, the lower left carbon atom must be number one and the numbering then goes clockwise giving the side group the number 3. Because the numbering of carbon atoms one and two is mandatory, the numbers are not included in the name.

Step 4 Find the prefix: The side group has one carbon atoms so the prefix is methyl.

Step 5 The name of the compound is 3-methylcyclohexene. The numbering is shown here.

CH$_3$

18. Name each of the following cyclic hydrocarbons:

CH$_3$

a)

b) CH$_3$—CH$_2$ CH$_3$

19. Draw a condensed structural formula for each compound:

a) methylcyclopentane

b) 1,2-dimethylcyclobutane

20. Name each of the following cyclic hydrocarbons:

a)

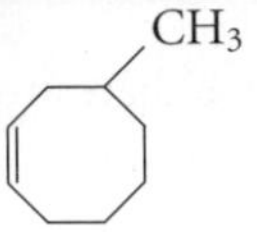

b) CH_3—CH_2—CH_2——CH_3

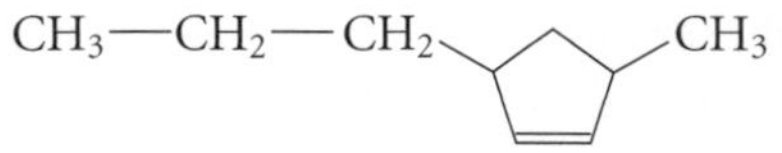

21. Draw a condensed structural formula for each compound:

a) 3-methylcyclopentene

b) 2-ethyl-3-propylcyclobutene

22. Name each of the following cyclic hydrocarbons:

a)

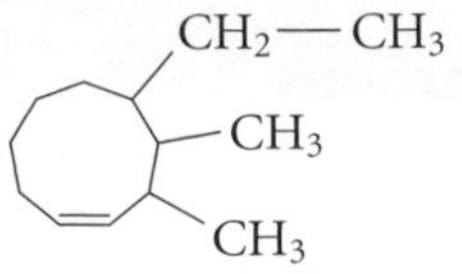

b)

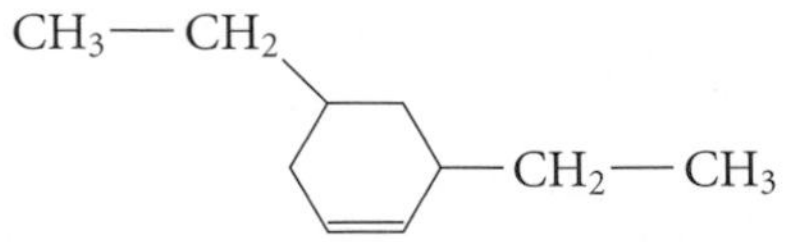

23. Draw a condensed structural formula for each compound:

a) 1,3-diethyl-2-methylcyclopentane

b) 4-butyl-1,3-dimethylcyclohexane

The Aromatic Hydrocarbons

A type of hydrocarbon that looks very much like a typical cyclic hydrocarbon has such unique properties that it forms the basis of an entire class of hydrocarbons—the aromatic hydrocarbons. Originally, the name was applied to naturally occurring compounds that are isolated from plants and characterized by intense aromas. The methylsalicylate that you prepared in the Launch Lab at the beginning of this chapter is one such aromatic compound.

For many years chemists struggled to explain two contradictory properties of aromatic compounds. Aromatic hydrocarbons have a low hydrogen to carbon ratio and unusual stability. A low hydrogen to carbon ratio is usually associated with the presence of multiple bonds. However, multiple bonds tend to make compounds more reactive, not less reactive. How could a low carbon to hydrogen ratio be associated with stability? This question was answered with the discovery that naturally occurring aromatics are all based on the presence of a benzene ring—a six-carbon ring with one hydrogen atom bound to each carbon, thus having the formula C_6H_6. **Aromatic compounds** are those hydrocarbons derived from the **benzene** ring.

Chemists originally drew the benzene ring with alternating single and double bonds as shown in Figure 14.14A. However, experimental observations showed that all six carbon–carbon bonds were identical in length and in other properties. The length of the carbon–carbon bonds is intermediate between a single and a double bond. Chemists then realized that benzene was a resonance hybrid of the two structures shown in Figure 14.14B. You might recall that you first read about resonance hybrids when considering the Lewis structure of sulfur dioxide ($SO_2(g)$). Chemists now realize that the electrons that make up the second bond in the "double bonds" are equally shared by all six carbon atoms. Electrons that behave in this way are called *delocalized electrons*. Benzene is not unique in having delocalized electrons. In any compound that has one single bond between two double bonds, the electrons are delocalized. The double bonds in this combination are called conjugated double bonds. Such structures are very stable structure and thus the electrons are not readily available for chemical reactions. Based on this information about the structure, chemists now prefer drawing the benzene ring with single bonds between the carbon atoms and then a circle within the carbon ring to represent the six delocalized electrons (see Figure 14.14C).

Chemistry File

FYI
Although the empirical formula C_6H_6 for benzene was determined in 1825, it was not until 1865 that German chemist August Kekulé (1829–1896) was able to deduce an appropriate molecular structure. Waking from a dream in which a snake was eating its own tail, Kekulé was prompted to propose that benzene was based on a ring structure of six carbon atoms with alternating single and double bonds.

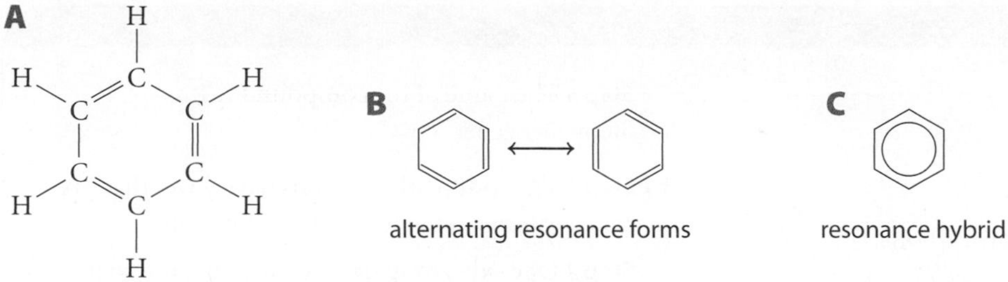

A

B alternating resonance forms

C resonance hybrid

Figure 14.14 Although the double bonds portray the correct number of electrons in the bonds, they do not portray the true chemical and physical properties of benzene.

Naming Aromatic Hydrocarbons

Since benzene forms the basis of all aromatic hydrocarbons, the naming of simple aromatic hydrocarbons uses benzene as the root. The major rules are given below.

Rules

- The root name for the molecule is –benzene.
- The carbons are numbered to locate the presence of more than one side group.
- Numbering starts at the carbon with the highest priority (or most complex) group in the list here.
- Numbering continues in the direction of the nearest group.

Highest priority	–OH
	–NH$_2$
	–F, –Cl, –Br, –I
Lowest priority	alkyl groups with 6 or fewer carbon atoms in alphabetical order

If a benzene ring is attached to a single hydrocarbon chain that is larger than the benzene ring itself (more than six carbon atoms) the benzene ring is considered to be the substituent group. In such cases, the attached benzene ring is called a **phenyl group**.

To draw aromatic hydrocarbons with side groups that have six or fewer carbon atoms in any given chain, simply draw the benzene ring and then add the side groups as indicated. If a benzene ring is attached to a hydrocarbon chain with more than six carbon atoms, name the compound according to the rules for naming aliphatic hydrocarbons and then name the benzene ring as a phenyl group just as you would have named an alkyl group such as "methyl." The Sample and Practice Problems on the next page will help you clarify the naming of aromatic hydrocarbons.

Physical Properties of Aromatic Compounds

Benzene is a liquid at standard temperature. If you compare the boiling points of simple aromatic hydrocarbons, you will see that they are very similar to those of the aliphatic hydrocarbons having the same number of carbon atoms. Also, as the name *aromatic* implies, these compounds often have strong aromas. Examples of aromatic compounds that have aromas with which you might be familiar are shown in Figure 14.15. In the next section, you will learn about some of the side groups that you see in this figure.

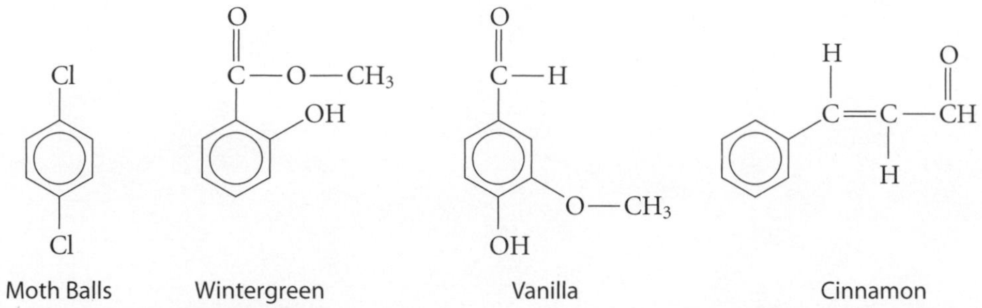

Chemistry File

FYI
Many chemicals were in widespread use in laboratories long before IUPAC naming was instituted. Thus, chemists became familiar with their common chemical names, sometimes called trivial names. Two aromatic compounds that are heavily used in industry are toluene and styrene. In the IUPAC system, their names would be methylbenzene and phenylethene.

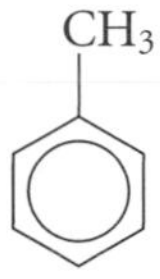

common name: toluene
IUPAC name: methylbenzene

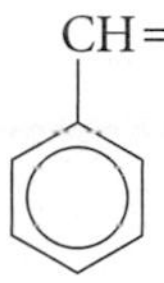

common name: styrene
IUPAC name: phenylethene

Moth Balls   Wintergreen   Vanilla   Cinnamon

Figure 14.15 Substituted benzene rings form the basis of compounds having many familiar odours— some are pleasant, others are not.

Naming and Drawing Aromatic Hydrocarbons

Problem
Name the following aromatic hydrocarbons.

a) CH_2—CH_3

CH_2—CH_3

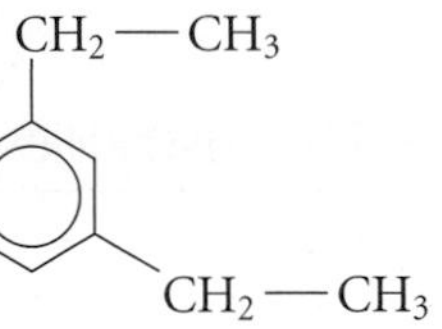

b) CH_3

CH_2—CH_2—CH_3

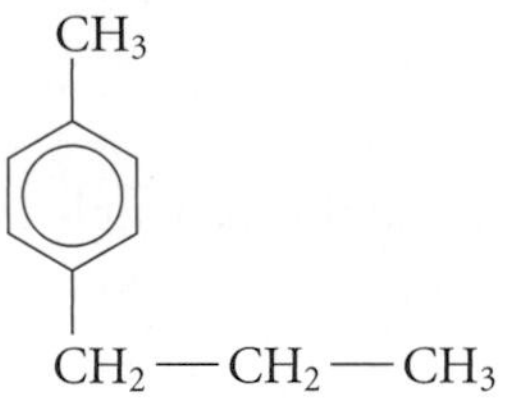

c) CH_3—CH_2—CH_2—CH_2—CH—CH_2—CH_3

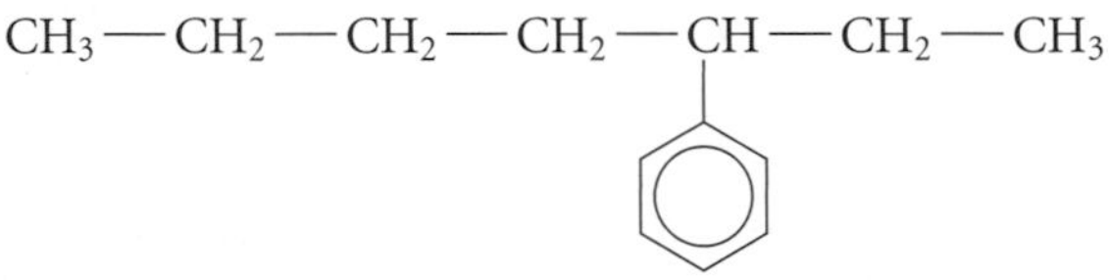

Solution

a) Step 1 The hydrocarbon chains have fewer than six carbon atoms so the root name is benzene.

Step 2 Both of the side groups have two carbon atoms so they are ethyl groups.

Step 3 Either ethyl group can be chosen to be on carbon atom number one. Counting must go in the direction that will give the second group the smaller number.

Step 4 The name of the compound is 1,3-diethylbenzene.

b) Step 1 The hydrocarbon chains have fewer than six carbon atoms so the root name is benzene.

Step 2 One side group has one carbon atom so it is a methyl group. The second side group has three carbon atoms so it is a propyl group.

Step 3 Methyl comes before propyl alphabetically so the carbon atom to which the methyl group is attached is carbon atom number one. The propyl group will be on carbon atom number three regardless of the direction in which you count.

Step 4 The name of the compound is 1-methyl-3-propylbenzene.

c) Step 1 The hydrocarbon chain has seven carbon atoms so the benzene ring is a substituent group. There are no multiple bonds in the seven carbon chain so the root name and suffix becomes heptane.

Step 2 The prefix is phenyl.

Step 3 The phenyl group is nearer to the right end of the chain so it is on carbon atom number three.

Step 4 The name of the compound 3-phenylheptane.

24. Name the following aromatic compound:

CH_3

CH_3 CH_3

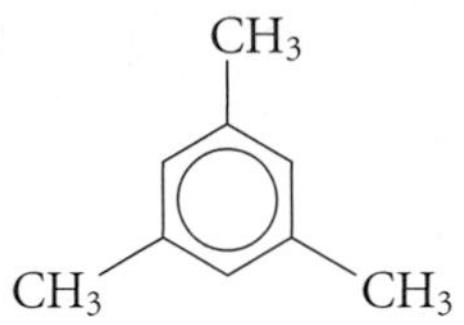

25. Draw a condensed structural formula for each aromatic compound given:

    a) 1-ethyl-3-methylbenzene

    b) 2-ethyl-1,4-dimethylbenzene

26. Draw a structural formula for 2-phenyloctane.

27. Name the following aromatic compound:

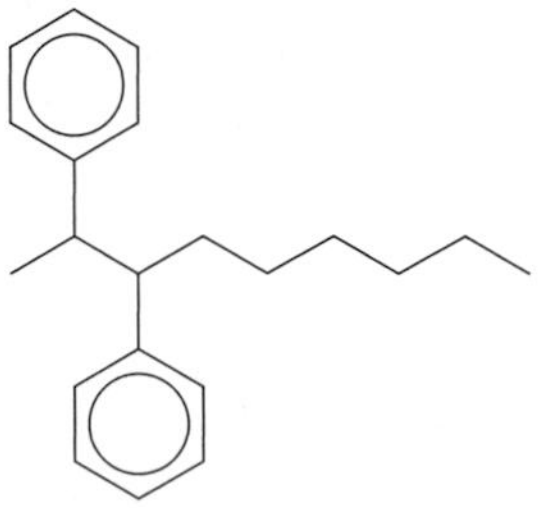

INVESTIGATION 14.B

Target Skills

Building molecular models depicting the structures of selected organic compounds

Analyzing the structure of molecular models

Modelling Organic Compounds

As you probably realize, the two-dimensional structures that you draw do not give you a good image of a molecule. Three-dimensional models help you understand the structure of organic compounds much better than sketches do. Chemists use models to help them predict the various ways in which a molecule might interact with other molecules. In this investigation, you will prepare molecular models to represent several organic compounds. You will also examine the changes to structure that result from the presence of an unsaturated bond. Along the way, you will gain practice in reading and interpreting structural diagrams.

Materials

- paper and pencil
- molecular modelling kit

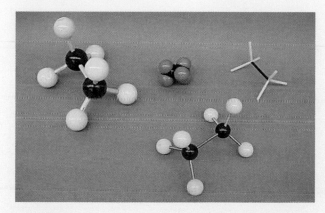

Procedure

1. Try to construct three-dimensional models for each of the indicated series of molecules. As you complete each model, try to rotate each of the bonds in the molecules, and then draw a careful diagram of the structure. Your diagram might be similar to the one shown below.

   a) propane, propene, propyne
   b) methylcyclobutane, 3-methylcyclobutene, 3-methylcyclobutyne
   c) cyclohexane, cyclohexene, benzene
2. Beginning with a three-dimensional model for methane, replace hydrogen atoms with oxygen atoms to form carbon dioxide. Each oxygen atom should have two bonds with the carbon atom. Again, as you complete each model, try to rotate each of the bonds in the molecules and then draw a careful diagram of the structure.
3. Build a model for carbonic acid. (**Hint:** Start with your model of carbon dioxide and add hydrogen atoms to each oxygen, and oxygen to the carbon atom.) Draw a careful diagram of the structure.

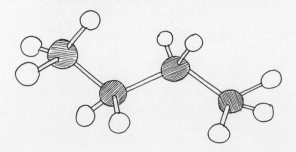

Analysis

1. How did the addition of a multiple bond change the ability of a bond to rotate?
2. How did the addition of a multiple bond change the three-dimensional shape of each series of molecules?
3. Other than the presence of multiple double bonds, how do the structures of cyclohexane and benzene differ?
4. Were any molecules more difficult to construct than the others? Explain.

Conclusion

5. How would you describe to another student the effect on a molecule of an increasing number of multiple bonds?
6. Inspect your drawing of carbonic acid and carbon dioxide. How are these compounds different from the other compounds that you constructed in this exercise?

Section 14.2 Summary

- Hydrocarbons consist of carbon and hydrogen atoms only.
- Aliphatic hydrocarbons contain no benzene rings.
- Aliphatic hydrocarbons include alkanes, alkenes, and alkynes.
- Alkanes have only single bonds and are called saturated compounds.
- Alkenes have at least one double bond and are called unsaturated compounds.
- Alkynes have at least one triple bond and are called unsaturated compounds.
- Cyclic aliphatic hydrocarbons contain at least one chain of carbon atoms that forms a closed ring.
- Aromatic hydrocarbons contain a benzene ring.
- Benzene, C_6H_6, has a six-membered ring structure in which six electrons are shared by all six carbon atoms. These shared electrons are said to be delocalized.
- Names of all hydrocarbons include a root, a prefix, and a suffix. IUPAC rules for naming and drawing hydrocarbons are precise and allow you to construct a name from a structure or a structure from a name.
- Hydrocarbons with four or fewer carbon atoms are gases at standard temperature. Mid-length hydrocarbons are liquid at standard temperature and long-chain hydrocarbons (approximately over 18 carbon atoms) are waxy solids.
- Hydrocarbons are insoluble in water but are soluble in non-polar solvents such as benzene.

SECTION 14.2 REVIEW

1. What is the difference between an aliphatic and an aromatic hydrocarbon?

2. What advantage does a structural formula have over an empirical formula?

3. What advantage does a line structural formula have over the other representations of a molecule?

4. What is the difference between a saturated and an unsaturated hydrocarbon? How does this affect their properties?

5. Name each of the following compounds:

a)

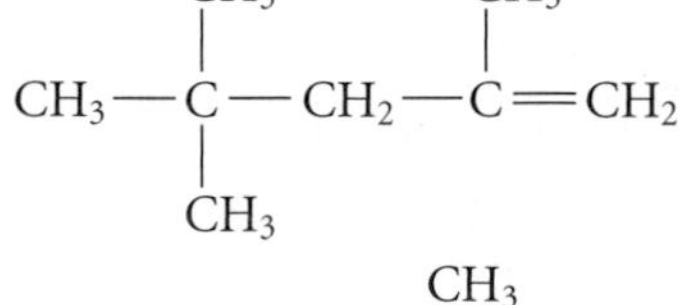

b)

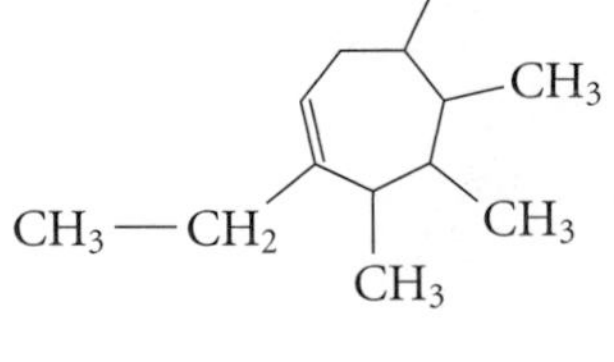

c)

6. Draw a condensed structural formula for each of the following compounds:
   a) 2,4-dimethylhex-3-ene
   b) 5-ethyl-4-propylhept-2-yne
   c) 3,5-diethyl-2,4,7,8-tetramethyl-5-propyldecane
   d) 1,4-dimethylbenzene
   e) 1,3-diethylcyclopentene

7. Draw a line structural formula for each of the following compounds:
   a) pentylcylclohexane
   b) but-2-yne
   c) 3-ethyl-4-methylhexane

8. Identify any mistakes in the name and structure of each compound:
   a) 3-methylbut-2-ene   b) 2-ethyl-4,5-methylhex-1-ene

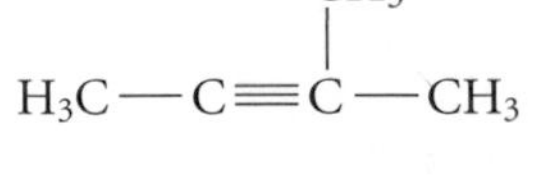

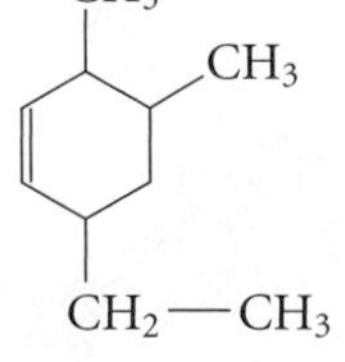

Hydrocarbon Derivatives

Section Outcomes

In this section, you will:

- **identify** types of compounds from the functional groups, given structural formulas
- **define** structural isomerism as compounds having the same empirical formulas but different structural formulas
- **compare** the boiling points and solubility of examples of aliphatics, aromatics, alcohols, and carboxylic acids

Key Terms

functional group
alcohol
parent alkane
alkyl halides
carboxylic acid
carboxyl group
ester
structural isomers

Organic chemists classify chemical compounds according to their functional group. A **functional group** is a special arrangement of atoms that is mainly responsible for the chemical behaviour of the molecule. Functional groups also determine some of the physical properties of the molecule. Table 14.6 lists some of the common chemical families and the characteristic functional group associated with each family. You have already learned how to name and draw alkenes and alkynes as well as learning about their physical properties. Throughout the remainder of this section, you will learn to recognize, name, and draw alcohols, alkyl halides, carboxylic acids, and esters. You will also learn about some of their physical properties.

Table 14.6 Common Chemical Families and Their Functional Groups

Family name	Suffix or prefix	Functional group	Example	Generic formula
alkene	-ene	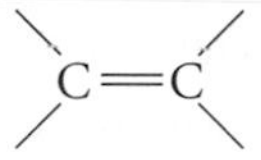 $C=C$	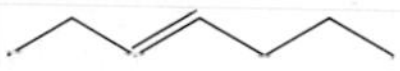 hept-3-ene	
alkyne	-yne	$-C\equiv C-$	 pent-2-yne	
alcohol	–ol	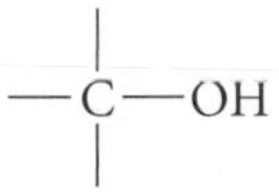 $-C-OH$ (commonly called the hydroxyl group)	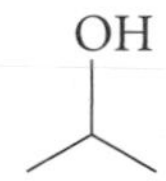 propan-2-ol	$R-OH$
alkyl halide	prefix varies with halogen	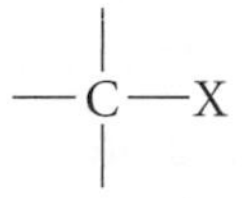 $-C-X$ $x =$ a halogen	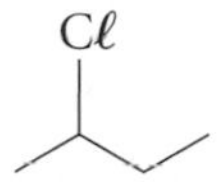 2-chlorobutane	$R-X$
carboxylic acid	-oic acid	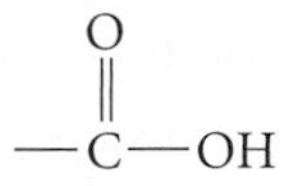 $-C-OH$ (commonly called the carboxyl group)	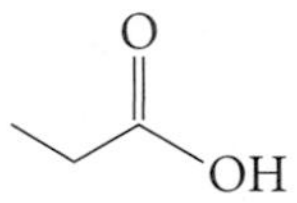 propanoic acid	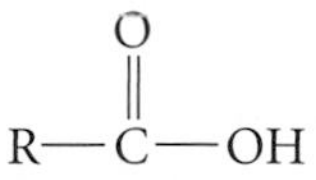 $R-C-OH$
ester	-oate	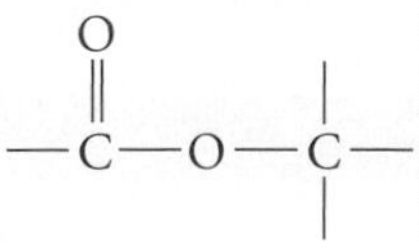 $-C-O-C-$	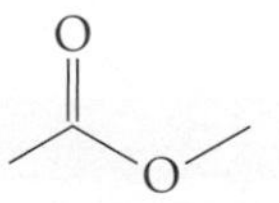 methyl ethanoate	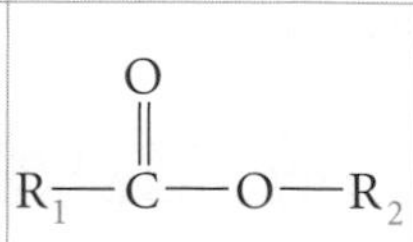 $R_1-C-O-R_2$

Alcohols

An **alcohol** is a hydrocarbon derivative that contains an —OH, or hydroxyl, functional group. Alcohols are components of many commonly used products. The following table summarizes a few common alcohols and some of their uses.

Table 14.7 Common Alcohols and Their Uses

IUPAC name	Common name(s)	Structure	Boiling point	Use(s)
methanol	wood alcohol, methyl alcohol	CH_3-OH	64.6 °C	• solvent in many chemical processes • component of automobile antifreeze • fuel
ethanol	grain alcohol, ethyl alcohol	CH_3-CH_2-OH	78.2 °C	• solvent in many chemical processes • component of alcoholic beverages • antiseptic liquid • additive for fuel
propan-2-ol	rubbing alcohol, isopropyl alcohol, isopropanol	CH_3 $CH-OH$ CH_3	82.4 °C	• antiseptic liquid
ethane-1,2-diol	ethylene glycol	$HO-CH_2-CH_2-OH$	197.6 °C	• main component automobile antifreeze

Naming and Drawing Alcohols

The root of the name of an alcohol is based on the **parent alkane,** which is the alkane having the same basic carbon structure. The following steps will help you name alcohols. Draw alcohols according to the parent alkane and then place the hydroxyl group on the correct carbon atom as determined by the number.

• **Identify the root:**
  Locate the longest chain that includes the hydroxyl group. Name the parent alkane.

• **Identify the suffix:**
  – The suffix always ends with –*ol*. The position of the hydroxyl is indicated by a number in front of the –ol.

  – Numbering of the carbon chain begins with the carbon atom at the end of the chain nearest the hydroxyl group.

  – If there is more than one hydroxyl group, use a prefix (di-, tri-, tetra-) to indicate the number of hydroxyl groups. A number for each hydroxyl is placed at the beginning of the suffix.

  – If the suffix begins with a vowel, drop the –*e* on the end of the parent alkane. If the suffix begins with a consonant, do not drop the –*e*.

• **Identify the prefix:**
  Name and number any alkyl side groups on the main chain.

Chemistry File

FYI
Do not confuse the name of the functional group for alcohols, the *hydroxyl group* (–OH), with the name of the ion containing hydrogen and oxygen, the *hydroxide ion* (OH⁻). The line beside the hydroxyl group represents a covalent bond between the oxygen atom and the remainder of the alcohol molecule. The superscript line on the hydroxide ion indicates that the ion is negatively charged.

Sample Problem

Naming Alcohols

Problem
Name the following alcohols.

a) $CH_3 - CH_2 - CH_2$
                  |
   $CH_3 - CH - CH_2 - CH_2 - OH$

b) OH OH OH
   | | |
   $CH_2 - CH - CH_2$

Solution

a) **Identify the root:** The hydroxyl group is attached to a hydrocarbon chain having six carbon atoms. The root is hexane.

Identify the suffix: The only hydroxyl group is at an end of the chain so start the numbering with the carbon attached to the hydroxyl group. There is only one hydroxyl group so the suffix is –ol. The suffix starts with a vowel so drop the –e at the end of hexane.

Identify the prefix: There is a methyl group on carbon atom number three. The prefix is 3-methyl-.

The full name of the alcohol is: 3-methylhexan-1-ol.

b) Identify the root: The hydroxyl groups are all attached to a carbon chain having three carbon atoms. The root is propane.

Identify the suffix: There are three hydroxyl groups so the suffix is -*triol*. Since the suffix starts with a consonant, do not remove the –*e*. You could start numbering at either end because the molecule is symmetrical. There is a hydroxyl group on each of the three carbon atoms so the suffix must contain -1,2,3-.

Identify the prefix: There are no additional groups so there is no prefix.

The full name of the alcohol is propane-1,2,3-triol.

Practice Problems

28. Name each of the following alcohols:

a) $CH_3—CH_2—CH_2—OH$

b) 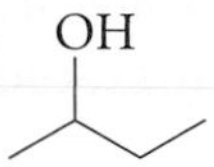

c) 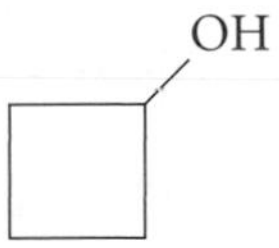

d) 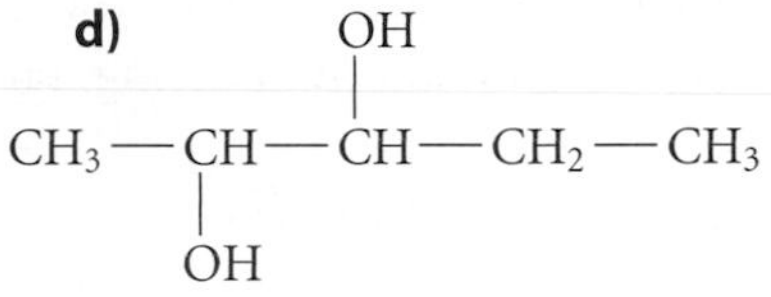

e) 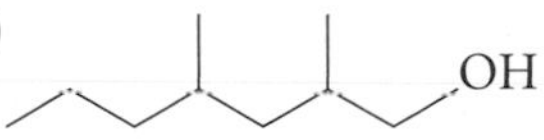

29. Draw each of the following alcohols:

a) methanol
b) propan-2-ol
c) butane-2,2-diol
d) 3-ethyl-4-methyloctan-1-ol
e) 2,4-dimethylcyclopentanol

30. Identify any errors in each name. Give the correct name for the alcohol:

a) heptan-1,3-ol

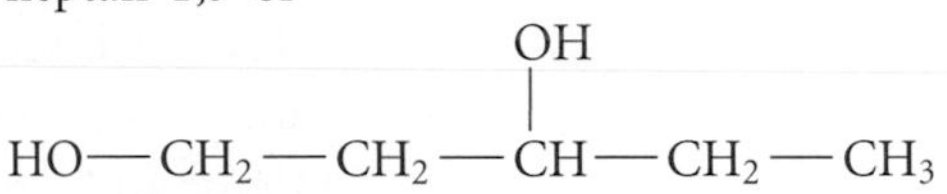

b) 3-ethyl-4-ethyldecan-1-ol

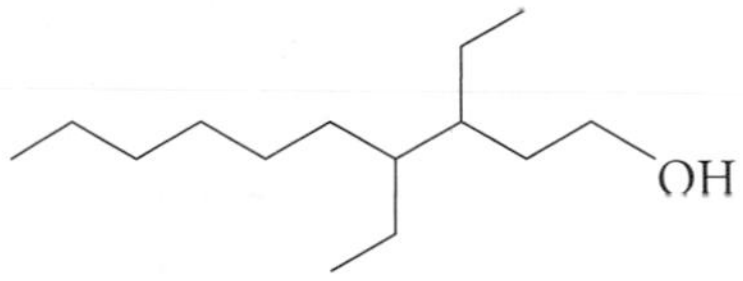

c) 1,2-dimethylbutan-3-ol

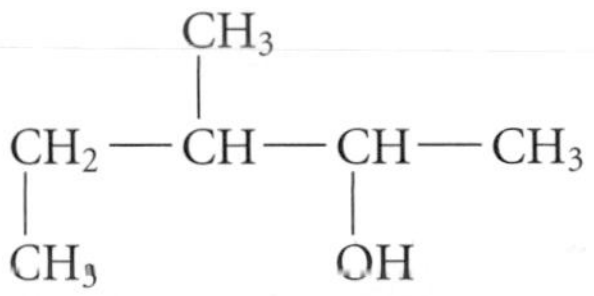

Physical Properties of Alcohols

The hydroxyl group is very polar making the small alcohols polar. Consequently, the smallest alcohols – methanol and ethanol – are miscible with water. As the hydrocarbon chain becomes longer, the non-polar characteristics supersede the polarity of the hydroxyl group and the alcohols become less soluble in water.

The hydroxyl groups also allow the alcohols to hydrogen bond with one another. Therefore, the boiling points of the pure alcohols are much higher than are those of the corresponding alkanes. All straight chain alcohols with fewer than twelve carbon atoms are liquids at standard temperatures.

All alcohols are toxic. Methanol can cause blindness or death. Although ethanol is consumed in large quantities, it, too can cause death if excessive amounts are consumed.

Alkyl Halides

Alkyl halides are hydrocarbons that contain at least one halogen atom. They are not found in living systems but are artificial. Trichloromethane, commonly known as chloroform, was once used as an anaesthetic. It is no longer used as such because it is now considered a possible carcinogen. Another familiar group of alkyl halides are the chlorofluorocarbons (CFCs) that were once used as refrigerants and as propellants in spray cans. In addition to being potent greenhouse gases, CFCs were discovered to be responsible for damage to the ozone layer. The ozone layer in the stratosphere normally absorbs most ultraviolet light before it reaches Earth's surface. The reduction in the ozone layer permits dangerous levels of ultraviolet light from the Sun to strike Earth's surface. Because serious health effects such as skin cancer, as well as crop damage, can result from exposure to ultraviolet light, CFCs are now banned substances. In general, the physical properties of alkyl halides are very similar to the corresponding alcohol.

Naming and Drawing Alkyl Halides

Like alcohols, alkyl halides are named according to the parent alkane. The following steps will help you name and draw alkyl halides. Notice that the suffix is the same as the alkane.

- **Identify the root:**
  Locate the longest chain that includes the halogen atom(s). Name the parent alkane.

- **Identify the prefix:**
  - Number the parent carbon chain starting at the end nearest the halogen atom(s). If the compound is a cycloalkane, numbering starts at the carbon atom bonded to the halogen atom.
  - Name and number any alkyl side groups on the main chain.
  - Insert the number(s) of the carbon atom(s) bonded to the halogen(s).
  - Use the prefix(es) that identify the specific halogen(s) (chloro-, fluoro-, bromo-, iodo-)
  - If there are two or more of the same type of halogen, use a prefix to indicate the number.
  - If there is more than one type of halogen present, write them alphabetically. Any prefixes (di-, tri-) are not considered when alphabetizing the terms.

Sample Problem

Naming Alkyl Halides

Problem

Name the following alkyl halides.

a)

$$CH_3 - CH_2 - \overset{\overset{\displaystyle Br}{|}}{CH} - \overset{\overset{\displaystyle |}{CH}}{\underset{\underset{\displaystyle Cl}{|}}{CH}} - \overset{\overset{\displaystyle Br}{|}}{CH} - CH_3$$

b)

Solution

a) **Identify the root:** The chain has six carbon atoms, therefore, the parent compound is hexane.

Identify the prefix: The carbon atom on the right end is nearest the first halogen so numbering starts at the right end. There is a bromine atom on carbon number two and carbon number four. The prefix will include 2,4-dibromo-. There is a chlorine atom on carbon atom number three so the prefix will include 3-chloro-. Alphabetically, bromo- is before chloro-.

The name of the compound is 2,4-dibromo-3-chlorohexane.

b) **Identify the root:** The parent compound is a cyclohexane.

Identify the prefix: Two bromine atoms are present as well as a methyl group. Numbering begins at a bromine atom and proceeds toward the second bromine and also in a direction that will give the methyl group the lowest possible number. Therefore, numbering begins at the top and proceeds clockwise placing the second bromine atom on carbon atom number three. This places the methyl group on carbon atom number four. The prefix is thus 1,3-dibromo-4-methyl-.

The full name of the compound is 1,3-dibromo-4-methylcyclohexane.

31. Name the following alkyl halides:

a) CH₃ — CH — Br
         |
         CH₃

b)
        Cl   F
        | |
CH₃ — CH — CH — CH — CH₂ — CH₃
              |
              Cl

c)
        CH₃   Br
        | |
CH₃ — CH — CH₂ — C — CH₃
                  |
                  Br

32. Draw condensed structural formulas for the following alkyl halides:

a) bromoethane

b) 2,3,4-triiodo-3-methylheptane

Carboxylic Acids

Do you eat vinegar with your French fries? Maybe you like vinegar and oil dressing (see Figure 14.16) on your salad. If so, you are eating ethanoic acid (acetic acid), a carboxylic acid. In fact, if you eat oranges, you are eating another carboxylic acid called citric acid.

Figure 14.16 Carboxylic acids are found in many foods such as the vinegar shown here. Acids have a sour taste.

A **carboxylic acid** is a compound that contains a **carboxyl group**, or —COOH. Figure 14.17 shows the structural formula for a carboxyl group and two examples of carboxylic acids. Carboxylic acids are weak acids for which the ionization equation can be written as shown below.

$$-COOH(aq) \rightarrow -COO^-(aq) + H^+(aq)$$

            O
            ‖
    — C — OH

carboxyl group

            O
            ‖
    HC — OH

methanoic acid
(common name: formic acid)

            O
            ‖
CH₃ — C — OH

ethanoic acid
(common name: acetic acid)

Figure 14.17 Any compound containing a carboxyl group is a carboxylic acid.

Naming and Drawing Carboxylic Acids

Notice that three bonds on the carbon atom in the carboxyl group are bonded to the other atoms in the group. There is only one position available on the carbon atom to bond to other atoms. Therefore, the carboxyl group must always be on the end of a chain. The carbon atom of the carboxyl group is always given the number one. The following steps will allow you to name and draw simple carboxylic acids.

- **Identify the root:**
  Locate the longest chain that includes the carboxyl group. Name the parent alkane.
- **Identify the suffix:**
  Drop the −*e* at the end of the name of the parent alkane and replace it with −*oic acid*.
- **Identify the prefix:**
  Name and number any alkyl side groups on the main chain. The carbon atom of the carboxyl group is always number one.

Sample Problem

Naming Carboxylic Acids

Problem
Name the following carboxylic acid.

$$CH_3-CH_2-\overset{\overset{\displaystyle CH_3}{|}}{CH}-CH_2-\overset{\overset{\displaystyle O}{\|}}{C}-OH$$

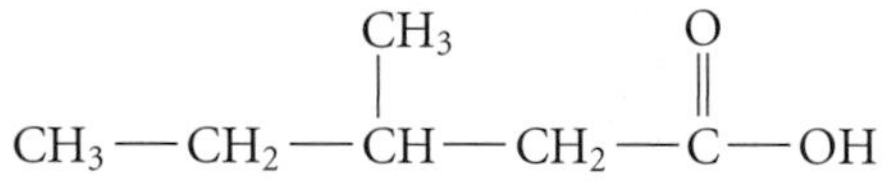

Solution
Identify the root:
The longest chain that includes the carboxyl group has five carbon atoms including the carbon atom in the carboxyl group. The parent alkane is pentane.

Identify the suffix:
Replace the −*e* at the end of pentane with −*oic acid*. The name includes pentanoic acid.

Identify the prefix:
There is a methyl group on carbon atom number three so the prefix is 3-methyl.

The name of the compound is 3-methylpentanoic acid.

Practice Problems

33. Name each of the following carboxylic acids:

a)
$$HO-\overset{\overset{\displaystyle O}{\|}}{C}-CH_2-CH_3$$

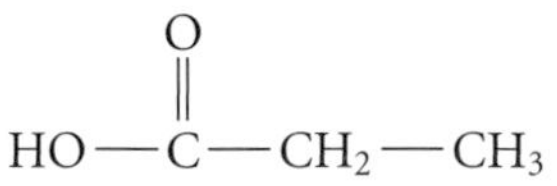

b)
$$CH_3-CH_2-\overset{\overset{\displaystyle CH_3}{|}}{\underset{\underset{\displaystyle CH_3}{|}}{C}}-\overset{\overset{\displaystyle O}{\|}}{C}-OH$$

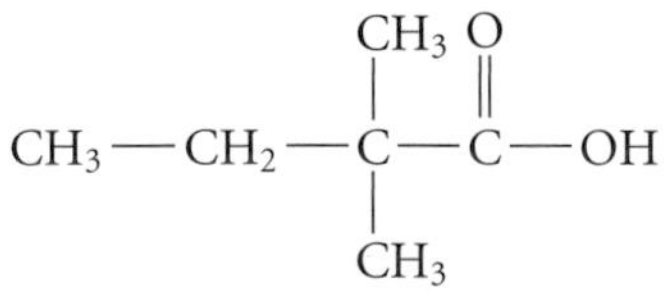

c)

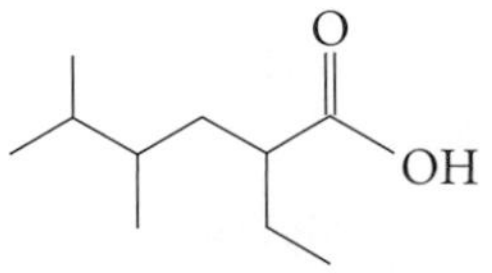

34. Draw a condensed structural formula for each of the following carboxylic acids:
   a) hexanoic acid
   b) 3-propyloctanoic acid
   c) 3,4-diethyl-2,3,5-trimethylheptanoic acid

35. Draw a line structural formula for each compound in question 34.

36. Draw and name two different carboxylic acids that have the molecular formula $C_4H_8O_2$.

Physical Properties of Carboxylic Acids

The presence of both a —C=O group and the —OH group make the carboxyl group very polar, allowing carboxylic acid molecules to form hydrogen bonds with one another. Thus, the boiling points of carboxylic acids are much higher than those of other hydrocarbon and their derivatives with the same number of carbon atoms. For example, examine the boiling points of some four carbon hydrocarbon derivatives listed in Table 14.8.

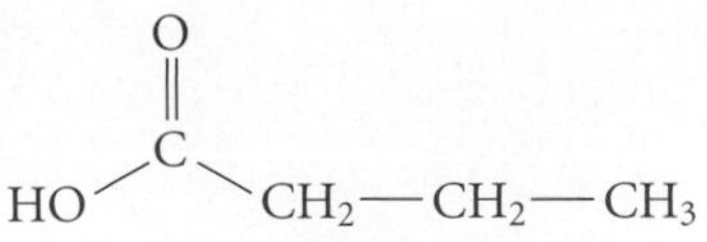

Figure 14.18 Butanoic acid is also known by its more common name, butyric acid. It is found in some animal milk and has been used in disinfectants and pharmaceuticals.

Table 14.8 Boiling Points of Some Four Carbon Compounds

Compound	Boiling point (°C)
butane	−0.5
butan-1-ol	117.2
butanoic acid	165.5

Short-chain carboxylic acids are liquids at standard temperature while those with longer chains are waxy solids.

The polarity of the carboxyl group makes the small carboxylic acids soluble in water. In fact, carboxylic acids with one to four carbon atoms are completely miscible with water. Those with chain lengths of five to nine carbon atoms are less soluble in water, while those with chains longer than ten carbon atoms are insoluble in water.

Carboxylic acids are weak acids and thus turn litmus paper red. As weak acids, they conduct electric current.

Esters

When you smell the aroma of fresh fruit such as the raspberries in Figure 14.19, you probably do not think about what chemical compound might be producing the pleasant fruity odor. In many cases, the fruity odor is caused by a compound called an ester.

An **ester** is a hydrocarbon derivative that contains the following functional group.

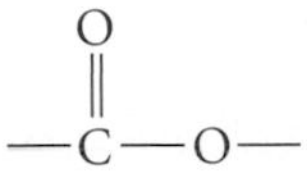

$$-\overset{\displaystyle O}{\overset{\displaystyle \|}{C}}-O-$$

You can write the general formula for an ester as RCOOR′. The symbol, R, represents any hydrocarbon or just a hydrogen atom. The symbol, R′, represents a hydrocarbon that cannot simply be a hydrogen atom. To name an ester, you must think of it as a combination of a carboxylic acid and an alcohol because the names of those compounds are used in the name of the ester. The equation below will illustrate this combination.

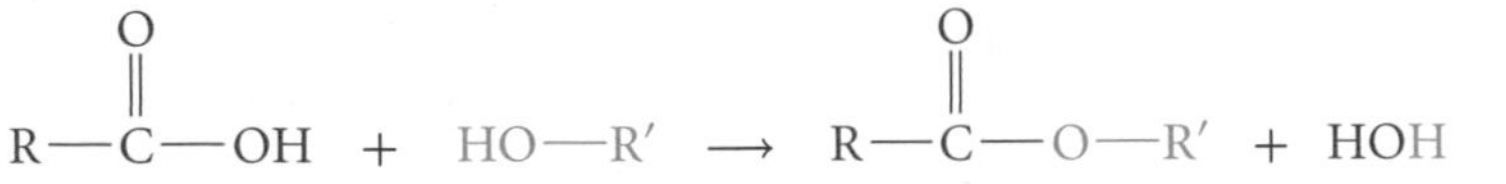

Figure 14.19 The aroma of these raspberries is caused by a hydrocarbon derivative called an ester.

$$\underset{\text{carboxylic acid}}{R-\overset{O}{\overset{\|}{C}}-OH} + \underset{\text{alcohol}}{HO-R'} \longrightarrow \underset{\text{ester}}{R-\overset{O}{\overset{\|}{C}}-O-R'} + \underset{\text{water}}{HOH}$$

Naming and Drawing Esters

Refer to the equation above while you read through the steps for naming an ester. Then examine the Sample Problem and complete the Practice Problems.

- **Identify the root:**

  Identify the part of the ester that contains the C=O group. This is the part of the ester that is red on the previous page and came from the acid. The root of the name of the ester is based on the name of the acid. Determine the name of the parent acid.

- **Identify the suffix:**

  Remove the −*oic acid* from the name of the parent acid and replace it with −*oate*.

- **Identify the prefix:**

  To form the prefix, consider the part of the ester that is associated with the alcohol, the part in blue in the equation on the previous page. Ignore the oxygen atom and use only the alkyl group. Identify the name of the alkyl group. The name of the alkyl group is the prefix for the name of the ester. There is always a space between the name of the alkyl group and the root. For example, the name of the following ester is ethyl butanoate.

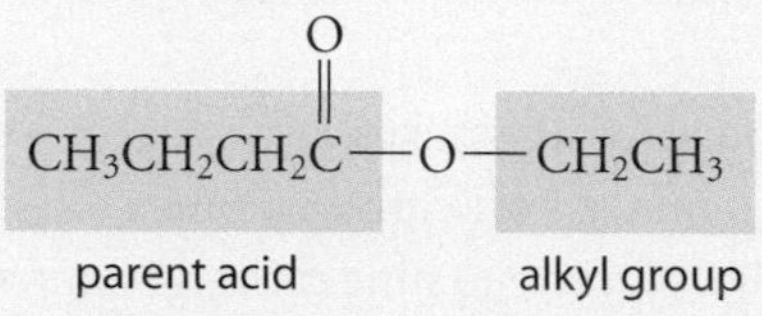

$$CH_3CH_2CH_2C \overset{O}{\overset{\|}{-}} O - CH_2CH_3$$

parent acid   alkyl group

Sample Problem

Naming Esters

Problem

Name the following esters.

a)
$$CH_3CH_2C \overset{O}{\overset{\|}{-}} O - CH_3$$

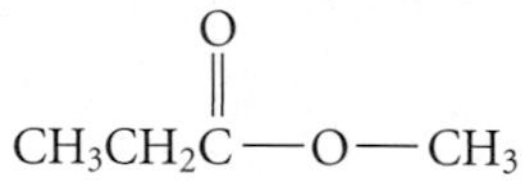

b)
$$H - C \overset{O}{\overset{\|}{-}} O - CH_2CH_2CH_3$$

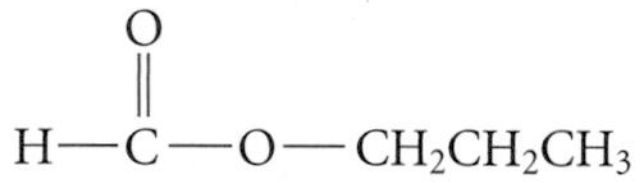

Solution

a) **Identify the root:** The C=O is part of a three carbon group making the parent acid a propanoic acid.

  Identify the suffix: Remove the −*oic acid* from the name of the parent acid and replace it with −*oate*. The root plus the suffix is now propanoate.

Identify the prefix: The part of the ester that is associated with an alcohol has one carbon atom therefore the prefix is methyl.

The full name of the ester is methyl propanoate

b) **Identify the root:** The C=O is bonded to only a hydrogen atom so the parent acid had only one carbon atom. The acid was methanoic acid.

  Identify the suffix: Remove the −*oic acid* from the name of the parent acid and replace it with −*oate*. The root plus suffix is methanoate.

  Identify the prefix: The part of the ester that is associated with an alcohol has three carbon atoms therefore the prefix is propyl.

The full name of the ester is propyl methanoate

Practice Problems

37. Name the following esters:

a)
$$CH_3CH_2 - O \overset{O}{\overset{\|}{-}} CH$$

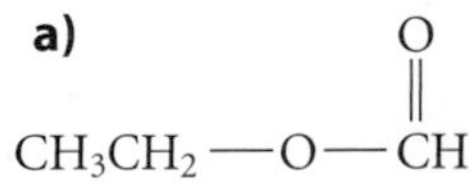

b)
$$CH_3CH_2CH_2C \overset{O}{\overset{\|}{-}} O - CH_3$$

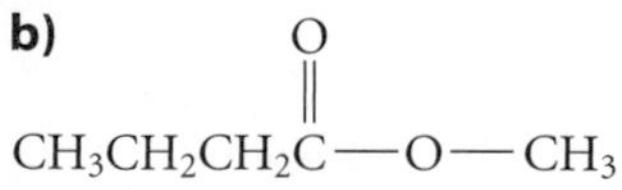

c)
$$CH_3CH_2CH_2CH_2C \overset{O}{\overset{\|}{-}} O - CH_2CH_2CH_2CH_2CH_3$$

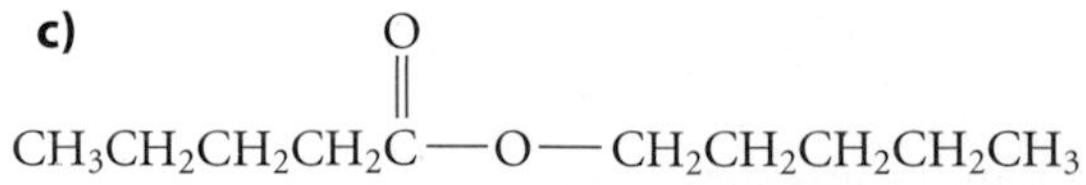

38. Draw the following esters:

a) methyl pentanoate

b) heptyl methanoate

c) butyl ethanoate

d) propyl octanoate

e) ethyl 3,3-dimethylbutanoate

f) ethyl octanoate (found in oranges)

g) methylpropyl methanoate (responsible for the aroma of raspberries)

Physical Properties of Esters

The presence of the $C=O$ group makes esters somewhat polar but without an $-OH$ group, ester molecules cannot form hydrogen bonds with one another. Therefore, the boiling points are lower than the corresponding alcohols and carboxylic acids. The smaller esters are liquids at standard temperature while the longer chain esters waxy solids. Esters with four or fewer carbon atoms are soluble in water while larger esters are insoluble. The most noticeable characteristic of esters is their volatility which allows them to generate aromas.

Structural Isomerism

One observation that puzzled many chemists for years was the fact that compounds having identical empirical formulas could have different properties. For example, they found two different compounds that both have the molecular formula C_3H_8O. One of these compounds has a boiling point of 82.5 °C and the other has a boiling point of 7.4 °C. The first compound must have much stronger intermolecular forces than does the second. This observation would lead you to think that the possibility that one with the higher boiling point should be polar and form hydrogen bonds. Conversely, the lower boiling point compound should experience only London (dispersion) forces. How could this be if the two compounds have the same combination of atoms? The answer is found in the concept of structural isomerism. The molecular formula for the two compounds is identical but the structural formulas are different. Any two compounds that have the same molecular formula but different structural formulas are called **structural isomers**. The structural isomers of the two C_3H_8O compounds are shown in Figure 14.20. One of the structural isomers is an alcohol and the other is an ether. (Ethers are not discussed in this textbook.) The alcohol can form hydrogen bonds but the ether cannot. The alcohol, propan-2-ol, has a boiling point of 82.5 °C while the ether, methoxyethane, commonly called methyl ethyl ether, boils at 7.4 °C.

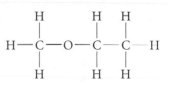

propan-2-ol

methoxyethane

Figure 14.20 These are the two structural isomers of C_3H_8O. There is a third structural isomer. Can you determine what it is?

• • •

1. There are just three ways to connect the five carbon atoms that make up C_5H_{12}. Determine the structure of those three structural isomers and provide the correct name for each.

2. Of two compounds having the molecular formula, $C_5H_{10}O_2$, one has a boiling point that is approximately 85 °C higher than the other. Determine the structures of two different hydrocarbon derivatives that are structural isomers with this molecular formula. State which one would probably have the higher boiling point and explain why.

• • •

Section 14.3 Summary

- Hydrocarbon derivatives have functional groups that confer unique properties on the compounds.
- Alcohols contain a hydroxyl group, $-OH$, that makes the small alcohols polar.
- Small alcohols are liquids at standard temperature while longer-chain alcohols are waxy solids
- Alcohols are used as solvents, antiseptics, and in antifreeze.
- Alkyl halides are artificial hydrocarbon derivatives that contain at least one halogen (fluorine, chlorine, bromine, or iodine,) atom.
- Carboxylic acids contain a carboxyl group, $-COOH$.

- Carboxylic acids have a sour taste and are found in many foods.
- Carboxylic acids form strong hydrogen bonds with one another and thus they have high boiling points relative to other hydrocarbon derivatives with the same number of carbon atoms.
- Esters have the functional group $-COO-$. They are formed by a combination of a carboxylic acid and an alcohol and are named according to the names of the acid and alcohol.
- Esters are responsible for the pleasant aroma of many fruits.
- The names of hydrocarbon derivatives include a root, suffix, and a prefix. IUPAC naming rules are so precise that you can draw the structure given the name or construct the name given the structure.
- Structural isomers are compounds that have the same empirical molecular formula but different structural formulas.
- Structural isomers can have very different physical and chemical properties.

SECTION 14.3 REVIEW

1. What is a functional group?

2. How does the location of the functional group affect the chemical reactivity of a molecule?

3. Define the term isomer.

4. Draw condensed structural formulas and provide names for three structural isomers for $C_4H_8O_2$.

5. What is the most important force when considering the intermolecular interactions between non-polar molecules and between polar molecules?

6. Provide names for each of the following molecules:

a)

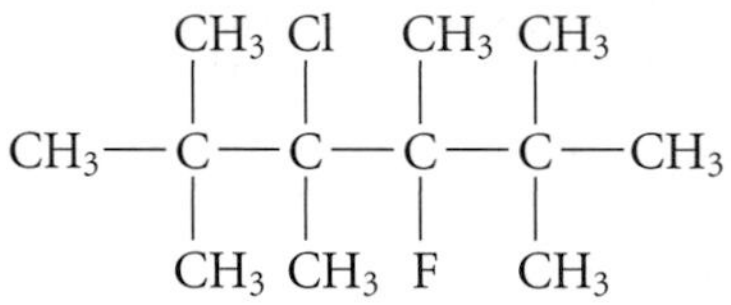

b)

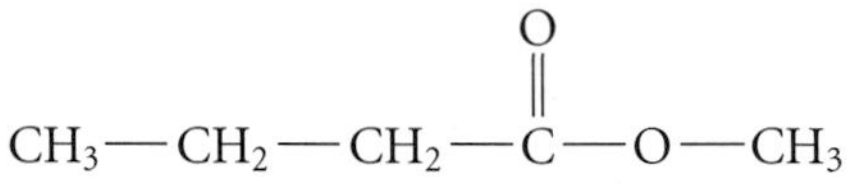

c)

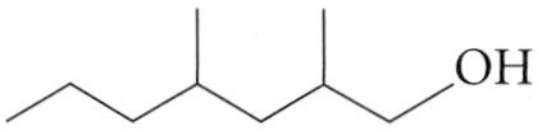

d)

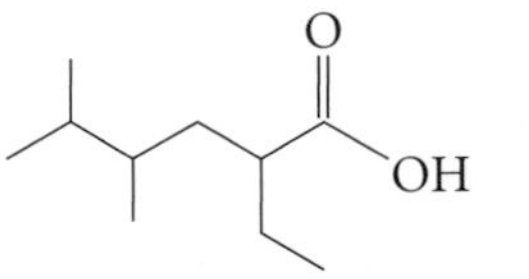

e)

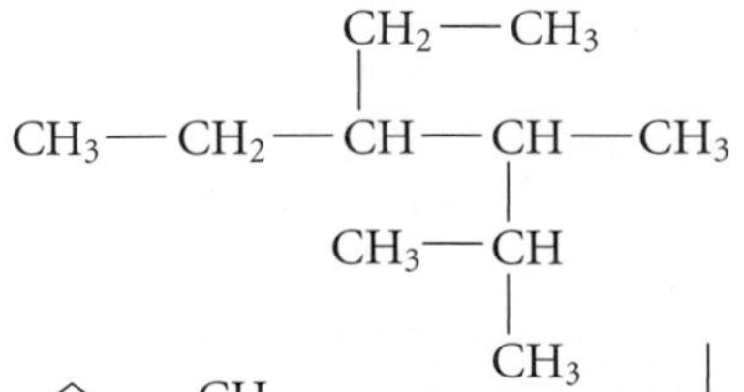

f)

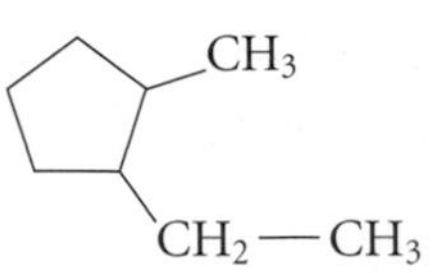

g)

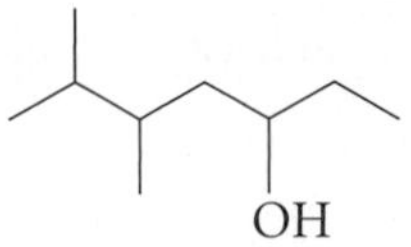

h)

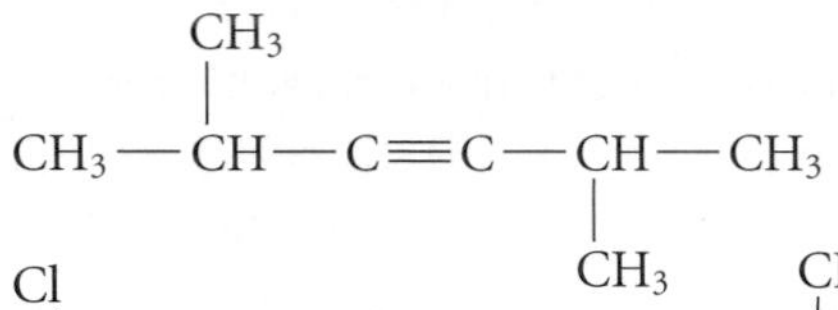

i)

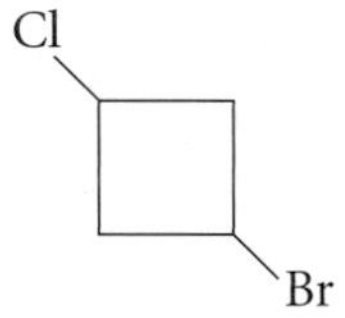

j)

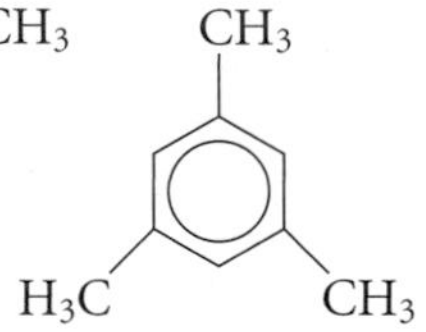

k)

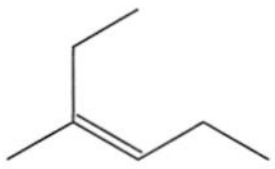

l)

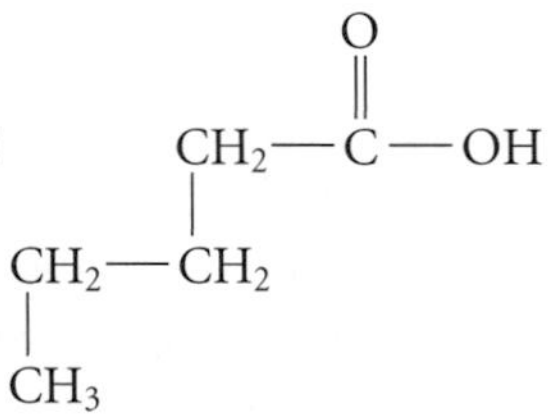

m)

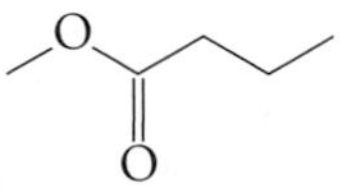

7. Given the following names, provide a condensed structural formula:
   a) butan-2-ol
   b) 1,3-dibromocyclohexane
   c) 3-propyloctanoic acid
   d) butyl ethanoate
   e) but-1-yne
   f) 3-methylhex-2-ene
   g) propanoic acid
   h) heptan-3-ol
   i) propyl octanoate
   j) 4-propylcylcononene
   k) 1,3-dibromo-2-iodopropane

8. Hexane has a boiling point of 68.7 °C. Benzene has a boiling point of 80.1 °C and cyclohexane, 80.7 °C. Since both benzene and cyclohexane have lower molar masses than hexane, they might be expected to also have lower boiling points. Explain why this is not the case.

Refining and Using Organic Compounds

Section Outcomes

In this section, you will:

- **investigate** the physical, chemical, and technological processes used to separate organic compounds from natural mixtures or solutions
- **design** a procedure to separate a mixture of organic compounds by fractional distillation
- **describe** the processes involved in bitumen recovery
- **assess** the risks and benefits and some positive and negative consequences for humans and the environment resulting from the process of bitumen recovery

Key Terms

petrochemicals
fractional distillation
cracking
steam cracking
catalytic cracking
hydrocracking
reforming
alkylation
coke
solvent extraction

You live in a world of synthetics. Most of the items that you use every day were not available before the 1850s. Since then, the number and uses of synthetic organic compounds have been growing exponentially. One major factor in this growth has been access to "chemical feedstocks"—inexpensive raw materials. During the late nineteenth century, these raw materials were derived from coal. Since the early twentieth century, the main source of raw materials has been petroleum. Petroleum is a complex mixture of hydrocarbons—mostly alkanes and alkenes—in the form of natural gas, crude oil, and bitumen. Natural gas and crude oil are often found together in the same underground reservoirs, extracted by using wells drilled into those reservoirs, and then pumped to refineries for processing (see Figure 14.21).

Figure 14.21 Alberta has about 150 operating oil refineries similar to this one.

Alberta has a vast natural wealth of petroleum. Both natural gas and crude oil reserves have been exploited in the south and central regions of the province since the early twentieth century, and they form the mainstay of Alberta's petrochemical industry. In recent years, gigantic bitumen deposits covering more than 140 800 square kilometres in the Athabasca, Peace River, and Cold Lake areas have become important. These gigantic bitumen deposits or "oil sands" were utilized historically by the First Nation Cree people as a source of tar to caulk birch bark vessels and canoes or to burn in smudge pots for mosquito control. The existence of the oil sands was first reported by European explorers in 1719, but early attempts to extract oil from those deposits failed. The development of an efficient process for separating the bitumen from the sand and clay it is mixed with required years of trial and error. In 1967 the first barrel of commercially viable oil was produced. More than 50% of Alberta's total oil production is derived from just one of those deposits.

Most of the petroleum from Alberta's oil, gas, and oil sands is refined and burned as fuel. Approximately 5% is used to synthesize petrochemicals. **Petrochemicals** are the basic hydrocarbon raw materials, such as ethene and propene, that are used to make plastics and other synthetic materials. The petrochemical industry has developed an efficient process for separating the materials; recovering the individual, pure hydrocarbon components; and forming petrochemicals.

Tar Sands and Bitumen

The tar sands, or oil sands, found in Alberta are massive deposits of bitumen that can ultimately yield petroleum products. However, the bitumen must be separated out from the sand and processed into a crude oil product before it can be further refined into a useable product. The deposits in Alberta contain up to 15% bitumen, with the remainder being clay, sand, other minerals, and water.

Extracting Bitumen

The depth of the bitumen deposits determines how bitumen is extracted from the oil sands. If the bitumen deposit is near the surface, it can be recovered by using open pit mining techniques. Giant machines, like the ones shown below, dig up and move the sand to a processing facility where the separation takes place. Once the bitumen has been removed from the sand, the processed sand is returned to the pit and the site is reclaimed.

If the bitumen deposits are found deeper than 75 m below the surface, the bitumen must be mined and processed *in situ*, or at the site of digging. The largest bitumen mining site in Canada is found in Cold Lake, Alberta. Steam is injected into the sands, and this causes the bitumen to rise to the surface. It is turned into a material called condensate by mixing it with substances to keep it liquid, and then it is pumped through pipelines to facilities where it can be processed even further into other petroleum products.

Environmental Costs

There are some critical environmental issues associated with removing the bitumen from the oil sands. The operations emit large amounts of greenhouse gases, up to two-and-a-half times more than traditional oil production, because of the additional technology and machines that are needed to separate the bitumen from the sand. Processing the oil sands also requires water, which is taken from the Athabasca River. Once the bitumen has been removed, the water and remaining mineral material (together called tailings) is diverted into ponds called tailings ponds. These ponds are left alone so the sand can settle to the bottom. Fresh water is added, and scientists believe that after two to three years, aquatic systems will be able to survive in the ponds.

Returning the landscape to its pre-mined natural state is also difficult. Once the sand has been mined and the bitumen removed, the sand is used to fill in the areas that were mined. Topsoil, removed from the original land, is put back on and seeds and vegetation are planted. The process of reclaiming land takes 12 to 15 years, and although the landscape can support wildlife, it will never be the same as before the mining occurred.

1. Create two flowcharts that compare how the bitumen is removed in an open pit mine versus an *in situ* mining operation. Which do you think has less environmental impact? Explain your answer.

2. Oil companies are spending millions of dollars to lessen the environmental impact of their operations. Most of the new environmental work is experimental so there is no way to know whether the impact of oil sand production is really being minimized. Imagine you are a scientist 20 years from now. Design an experiment that would show whether or not environmental damage from past open pit mining is permanent or whether land reclamation was successful.

3. Production in the oil sands is being increased because of a growing global shortage of petroleum. Outline a strategy to reduce your dependence on fossil fuels. Be sure to explain how you would implement this strategy in your home.

A Oil sands mine before reclamation

B Oil sands mine after reclamation

RECLAMATION AREA DO NOT DISTURB

Petroleum, from Latin *petra*, for "rock" and *oleum* for "oil," is the fossilized remains of plankton transformed geologically under extremely high temperatures and pressures into a complex mixture of solid, liquid, and gaseous hydrocarbons. The liquid form is called crude oil and ranges in colour and viscosity. The gaseous portion, primarily methane, is odourless and colourless and is referred to as natural gas. Petroleum naturally seeps to the Earth's surface along fault lines and cracks in rocks from underground reservoirs. It pools in surface deposits as bitumen (more commonly, tar and asphalt). Early cultures used the bitumen seeping out of faults from underground deposits to waterproof objects, to make lubricants, to make adhesives, and to use as fuel for heat and light.

Separating Petroleum Components

The first step in refining petroleum involves the separation of petroleum into its hydrocarbon components by using fractional distillation. **Fractional distillation** involves successive heating, evaporation, cooling, and condensation. This process is energy intensive and can contribute to more than half of a refinery's operating costs. At the refinery, the petroleum is heated to high temperatures inside a large fractionating tower (see Figure 14.22A). Each hydrocarbon component, called *fractions*, has its own range of boiling points and relative densities. The boiling point of the hydrocarbons increases with increasing size of the molecules. When the petroleum is heated to very high temperatures, it vapourizes. As the vapour rises upward in the tower, it gradually cools. As the fractions cool to their own boiling point, they condense and are collected. The process continues with each fraction of the mixture being separated according to differences in boiling point and density.

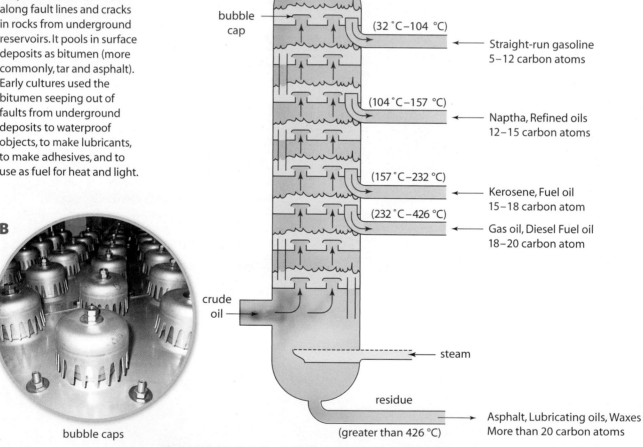

bubble caps

Figure 14.22 Fractional distillation towers can be as high as 60 m.

Perforated plates, often fitted with bubble caps (see Figure 14.22B), are placed at various levels in the tower. As the crude oil vapour rises upward in the tower, it begins to cool. The plates are placed at strategic locations where the temperature is just below the boiling point of a particular fraction. On reaching that plate, the fraction will condense to a liquid. Using a bubble cap to divert rising vapours from below through the liquid on a plate increases the efficiency of capture for each fraction. Pipes channel the liquid away for further treatment. Thus, larger hydrocarbon molecules with higher boiling points condense in the hotter, lower levels of the tower. The smaller, lighter hydrocarbon molecules remain in a gaseous state until reaching the higher, cooler levels.

Cracking and Reforming

Once the fractions are removed from the distillation tower, they may be chemically processed or purified further to make them marketable. There has been a tremendous increase in the demand for a variety of petroleum products since the early twentieth century. The reactive carbon compounds in petrochemicals are in great demand because they can be converted into a wide range of products, from drugs and fertilizers to plastics and synthetic rubber. Crude oil, however, contains relatively small amounts of these compounds, requiring the oil industry to develop techniques to convert less useful crude oil fractions into the needed petrochemicals. These techniques are called cracking and reforming. Figure 14.23 will give you an idea of the technological complexity of the process.

Petroleum Fractionation Products

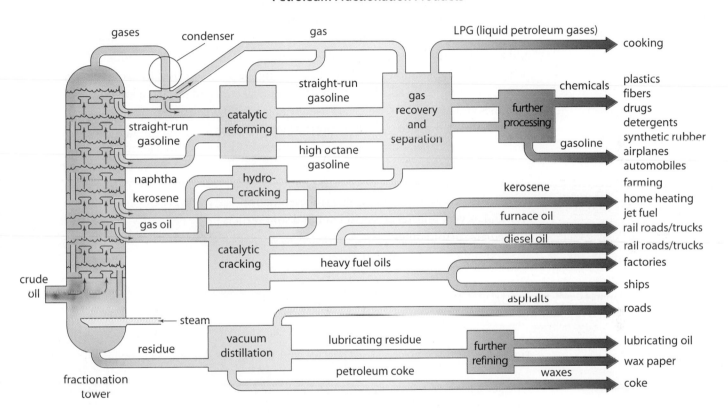

Figure 14.23 Notice the terms *catalytic reforming, hydrocracking, and catalytic cracking*. Each of these is a different process, but they all change the structure of some of the hydrocarbons into more useful products.

The heating of any hydrocarbon under pressure in the absence of air results in the breaking of carbon–carbon bonds—a process called **cracking**. A cracking operation requires one or more thick-walled reaction vessels and a complex system of furnaces and heat exchangers. For many of the processes, special catalysts are needed. After cracking has occurred, the products are fed through another fractional distillation column to separate the various products. In **steam cracking**, steam at more than 800 °C and pressures slightly above atmospheric normal is combined with hydrocarbon "feedstocks." A feedstock consisting of short-chain alkanes and naphtha will produce short-chain alkenes for use as petrochemicals. (See Figure 14.24.) Using heavier hydrocarbons as a feedstock produces a range of short-chain alkenes and aromatic hydrocarbons.

$$CH_3CH_2CH_2CH_2CH_2CH_3 \xrightarrow{\text{steam, pressure}} CH_3CH_2CH_3 + CH_3CH=CH_2$$

hexane   propane   propene

$$2CH_3CH_2CH_3 \xrightarrow{\text{steam, pressure}} CH_4 + CH_2=CH_2 + CH_3CH=CH_2 + H_2$$

propane   methane   ethene   propene   hydrogen

Figure 14.24 These chemical equations summarize the cracking of hexane and propane. The products, ethene and propene, are industrially valuable, but only small amounts are found in crude oil.

Catalytic cracking is a low-pressure process that involves passing a powdered catalyst through heavy hydrocarbon fractions at temperatures around 600 °C. The catalyst acts to speed up the cracking process and favours the production of diesel oils, gasoline, and kerosene. Alkenes, such as propene and butene, are secondary by-products in this process. The addition of hydrogen gas during the process, called **hydrocracking**, transforms long-chain alkanes into shorter-chain alkanes and results in low-grade gasolines and heating oils, which are usually upgraded by reforming.

Reforming uses heat, very high pressure and catalysts to convert straight-chain alkenes into branched-chain alkanes, cyclic alkanes, and aromatics. Figure 14.25 gives an example of this process. Most of the branched-chain alkanes and aromatic compounds formed are used to upgrade gasoline for high-performance engines. Aromatics, such as benzene, are used to form styrene and eventually polystyrene plastics and foams. It would be very unusual if you did not find some article made of polystyrene in your immediate surroundings.

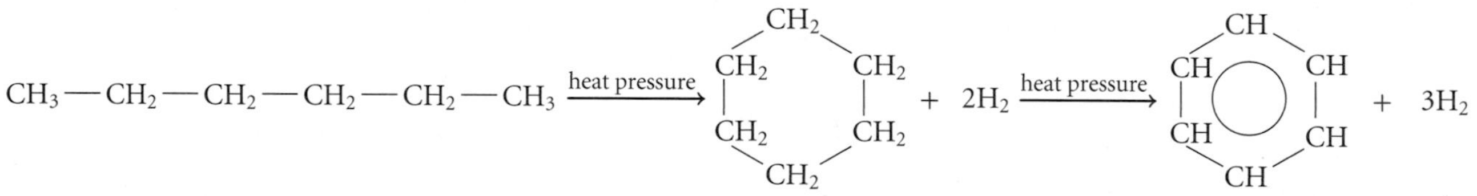

Figure 14.25 The formation of one possible aromatic from an aliphatic in reforming.

One key feature of reforming is the production of hydrogen gas. This by-product can be used within the refining process for hydrocracking or to treat and remove sulfur and nitrogen contaminants.

One special case of reforming is the process of **alkylation**. The short-chain alkenes produced in catalytic cracking are chemically joined under controlled heat and pressure in the presence of an acid catalyst, such as sulfuric acid or hydrofluoric acid. The resulting molecules are used to upgrade gasoline for high-performance engines and aviation fuel. Figure 14.26 shows the formation of 2,2,4-trimethylpentane, the compound that is called octane in "high octane" gasoline.

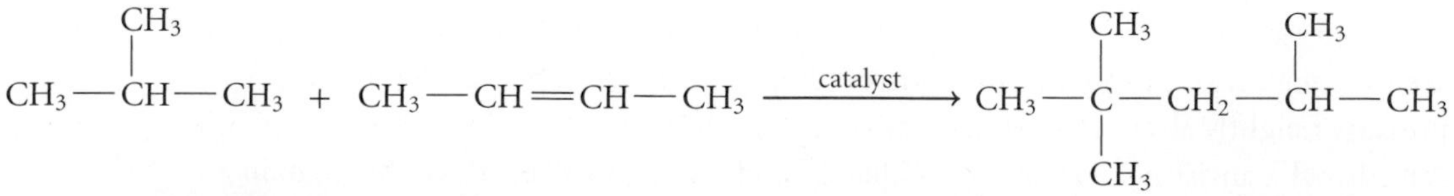

Figure 14.26 The product of this reaction, 2,2,4-trimethylpentane fires and burns in engines at the right speed to prevent engine "knock." The burning of other fuels is compared to this compound to determine the "octane rating."

Separate an Organic Mixture

Fractional distillation is one of the most critical steps in separating and purifying mixtures of organic compounds. The principle behind fractional distillation is simple in theory—two compounds do not have exactly the same boiling point. In this investigation, you will make use of this fact as you design an experiment to separate a mixture of compounds into its individual components.

Question

How can you separate a mixture of corn oil, vinegar (ethanoic acid and water), and cyclohexane?

Safety Precautions

- Organic solvents are flammable. Extinguish all flames in the laboratory area. Use a hot plate. (⚠)

- Organic solvents are toxic. Avoid inhaling or ingesting them. (T)

- Iodine is a severe skin irritant; avoid contact with skin and mucous membranes. Iodine will stain skin and clothing. Use a wood splint or microspatula to obtain and deliver two small crystals.

- Your instructor will introduce you to the distillation apparatus that is available to you in advance of performing this investigation.

- There is no need to exceed 130 °C in this experiment

Materials

- corn oil
- ethanoic acid
- commercial grade cyclohexane
- iodine crystals, $I_2(s)$
- distillation apparatus (varies)
- boiling chips
- cobalt chloride paper
- litmus paper
- wood splint or microspatula
- 5 test tubes or 50 mL beakers
- test-tube rack
- reference tables and access to library resources or the Internet

Experimental Plan

1. A mixture of corn oil, vinegar, and cyclohexane has been prepared for you. Using reference tables or the Internet, determine the boiling points for each component of your mixture.

2. Design a procedure that would allow you to separate those components. In your procedure, be sure to specify the steps that you are to follow.

3. Have your instructor check and approve your procedure before you start.

Analysis

1. Check the effectiveness of your separation by testing each distillate for the presence of the various compounds. Cyclohexane turns magenta with the addition of iodine. You can test for water by using cobalt chloride paper. Litmus paper tests for the presence of an acid.

2. Based on your tests, were the components completely separated?

3. How did you determine when to switch collection containers?

4. How could you improve the efficiency of your procedure?

Conclusion

5. How effective was your distillation procedure?

Finally, the residual, unprocessed crude oil that remains at the bottom of the fractional distillation is processed by vacuum distillation. Heated to more than 900 °C within a vacuum distillation tower, these heavy hydrocarbons crack to form heavy oil, gasoline, cycloalkanes, waxes, and a solid carbon residue called **coke**. The heavy oil fractions that remain are useful as lubricating oils but often contain aromatic hydrocarbon impurities and waxes. In **solvent extraction**, mixing solvents with the heavy oil dissolves those impurities, leaving a purified product. The waxes are recovered later by cooling. The coke is collected in special units, mechanically extracted, usually by high-pressure water jets, and sold for industrial purposes and power generation.

Section 14.4 Summary

- Approximately 5 percent of the petroleum pumped from Alberta's oilfields is used to synthesize petrochemicals such as the plastics that are made from ethene and propene.
- The first step in refining crude oil is fractional distillation. This process is based on the difference in boiling points of the hydrocarbons of differing chain lengths.
- The fractions are then further purified and used for applications from gasoline to asphalt and lubricating oils.
- Short-chain alkenes are in demand because they are used as raw material for synthesizing plastics and other polymers.
- Only trace amounts of these short-chain alkenes are found in crude oil so some of the oil is converted into these compounds by cracking and reforming.
- There are a variety of ways to "crack" crude oil but most of them depend on high temperatures and high pressures in the absence of oxygen.
- Reforming is a process that uses heat, high temperatures, and catalysts to convert straight-chain hydrocarbons into branched chains, cyclic, and aromatic hydrocarbons.

SECTION 14.4 REVIEW

1. What is petroleum?

2. What is a petrochemical?

3. What physical property allows various fractions of crude oil to be separated in a fractionating tower?

4. Describe the process by which crude oil is separated into fractions. Include a diagram in your answer.

5. Society's demand for petrochemicals has increased dramatically over the past century. Describe two techniques that can be used to transform petroleum fractions into more useful substances.

6. What is the difference between catalytic cracking and reforming? How are the two processes similar?

7. In separating two hydrocarbons by distillation, how do you know when to switch your condenser tube from one collection flask to another? Using your knowledge of heat and thermodynamics, explain why.

8. What are the risks to the environment posed in the extraction of bitumen from the oil sands deposits in Northern Alberta?

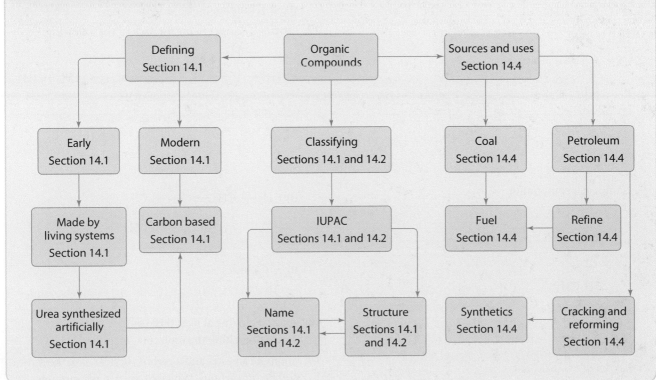

Chapter 14

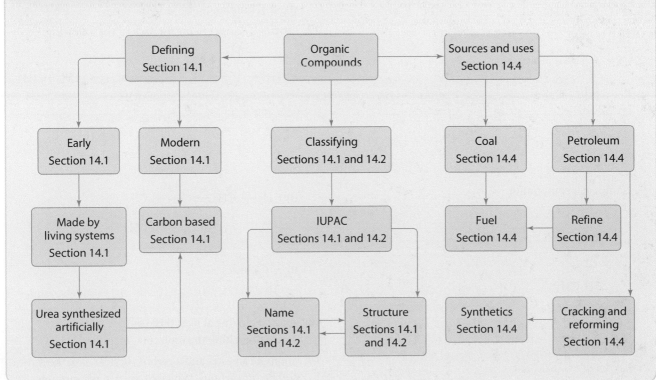

SUMMARY

Chemists originally believed that compounds based on the carbon atom could only be synthesized in the tissues of living organisms, hence the name organic compounds. By the mid 1800s, a few chemists had synthesized some simple organic compounds in the laboratory and the old concepts had to be abandoned. With the exception of carbonates, cyanides, carbides, and the oxides of carbon, all compounds containing carbon are considered organic compounds.

The chemical nature of the carbon atom and its ability to form four covalent bonds with atoms of other elements as well as other carbon atoms, allows it to form millions of different compounds. Organic compounds are often separated into two major classes: hydrocarbons and hydrocarbon derivatives. These two classes are then subdivided into many different sub-classes or families.

Hydrocarbons consist of only carbon and hydrogen atoms yet they have a wide variety of chemical and physical properties. Most fuels are hydrocarbons. Hydrocarbons are also used to make many important chemicals including plastics and pharmaceuticals. Hydrocarbons may contain single bonds (alkanes), double bonds (alkenes), and triple bonds (alkynes). They can have straight chains, branched chains, or closed rings. Aromatic hydrocarbons contain benzene rings, C_6H_6. Each individual hydrocarbon compound has a unique name that specifically describes its

structure. Names include a root that indicates the number of carbons atoms in a main chain, a suffix that indicated the type of hydrocarbon, and a prefix that described the substituent groups on the main chain.

Hydrocarbon derivatives contain functional groups that confer unique chemical and physical properties on the compounds. Alcohols contain a hydroxyl group, —OH. Alkyl halides contain at least one type of halogen atom: —F, —Cl, —Br, or —I. Carboxylic acids contain a carboxyl group, —COOH. Esters contain an ester bond and have the general formula RCOOR′. The names of hydrocarbon derivatives also include a root, suffix, and prefix. The suffix indicates whether the compound is an alcohol, carboxylic acid, or an ester. The prefix indicates whether a compound is an alkyl halide.

The raw material for most synthetic organic compounds is crude oil. Crude oil is first separated into fractions containing a certain range of lengths of carbon chains, by the process called fractional distillation. Each fraction is then further purified for use in a variety of applications such as gasoline, kerosene, fuel oil, diesel oil, asphalt, and waxes. Long-chain hydrocarbons can be cracked and reformed to generate more short-chain compounds that can then be used in the petrochemical industry.

Concept Organizer   Structures and Sources of Organic Compounds

Understanding Concepts

1. What is the basic difference between organic and inorganic compounds?

2. What was the concept of "vitalism?" Why was this concept eventually rejected?

3. What are the key properties of the carbon atom that allow it to form such diverse compounds?

4. Compare organic compounds and inorganic compounds, based on the following criteria:
   • the presence of carbon
   • the variety of compounds
   • the relative size and mass of molecules

5. Briefly compare alkanes, alkenes, and alkynes. (Give both similarities and differences.)

6. a) What is an aromatic hydrocarbon?
   b) How does it differ from an aliphatic hydrocarbon?

7. Write the general formula for
   a) an alkane
   b) an alkene
   c) an alkyne
   d) a cycloalkane
   e) a cycloalkene

8. Classify each hydrocarbon as aliphatic or aromatic:
   a) $CH_3 - CH_2 - CH_2 - CH_3$

   b) 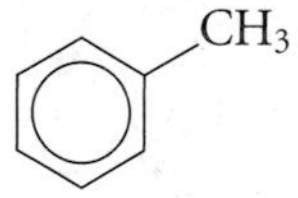
   c) 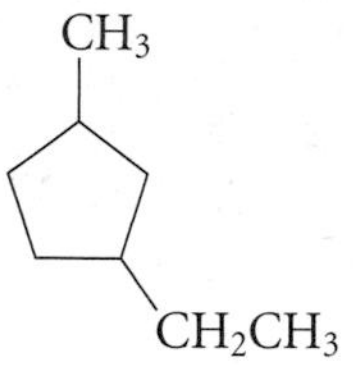
   d)
   e)

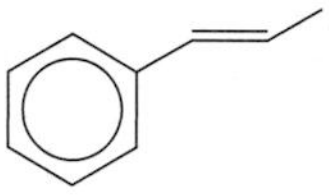

9. Name each compound:

   a) 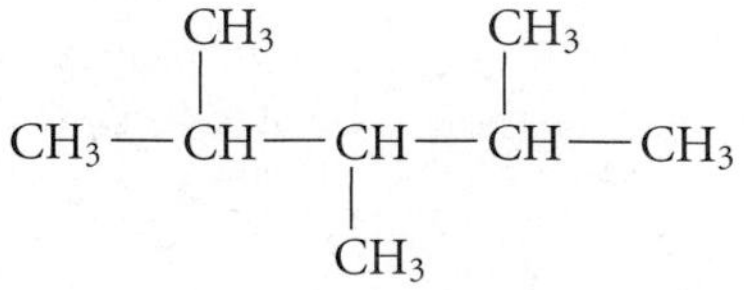 $CH_3 - CH - CH - CH - CH_3$

   b)

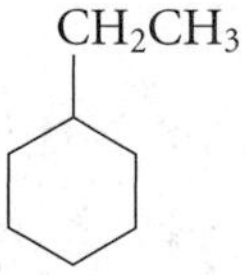

c)

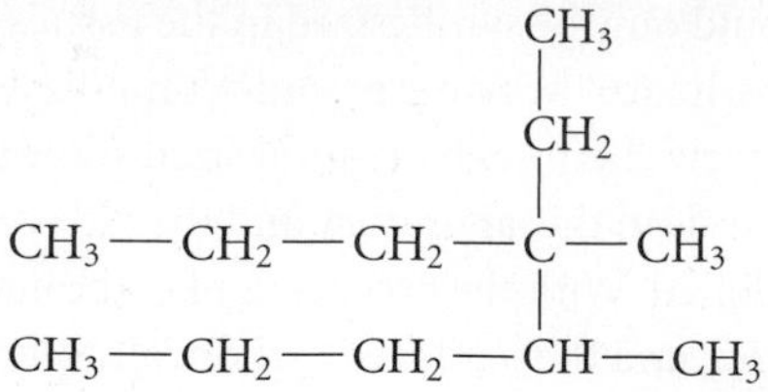

d)

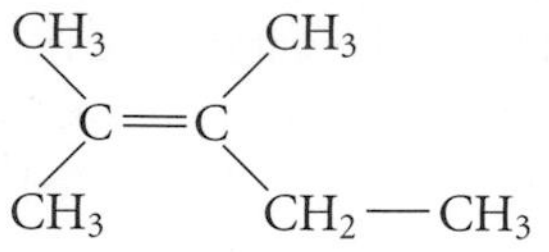

e)

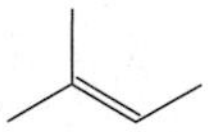

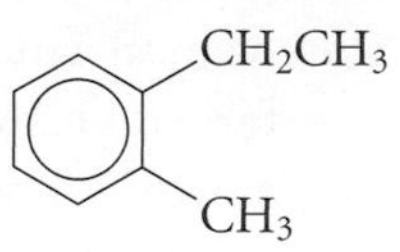

f)

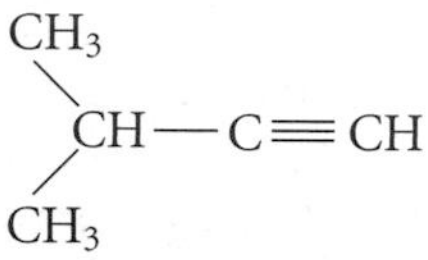

10. Could 1-methylcyclobut-2-ene be the correct name of a compound? To answer this question, draw the condensed structural formula for the compound as written. Then write the correct name.

11. For each of the following, give the correct line structural formula:
    a) heptan-3-ol
    b) 2-iodobutane
    c) 3-propyloctanoic acid
    d) ethyl butanoate

12. Provide a correct name for each of the following:

    a) 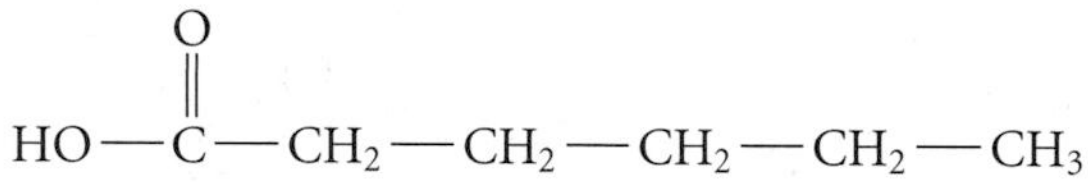 $HO - \overset{O}{\overset{\|}{C}} - CH_2 - CH_2 - CH_2 - CH_2 - CH_3$

    b)

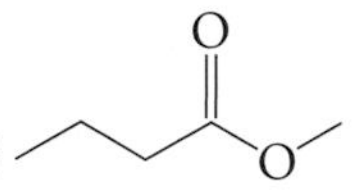

    c)

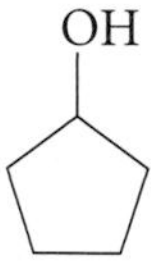

    d) 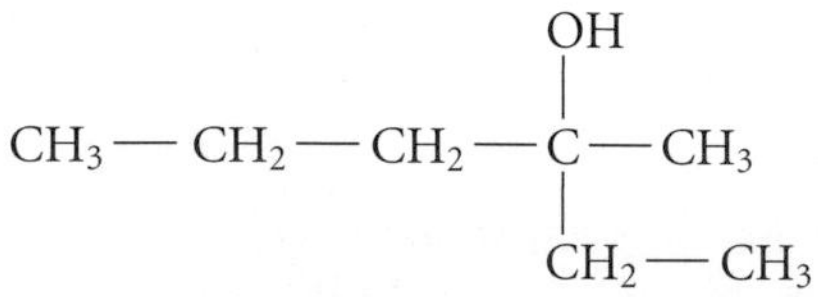 $CH_3 - CH_2 - CH_2 - \overset{OH}{\underset{CH_2 - CH_3}{\overset{|}{C}}} - CH_3$

    e)

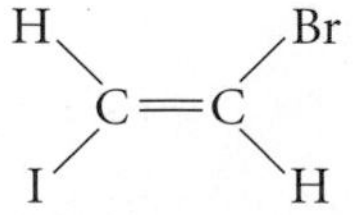

13. Draw and name at least three of the eight possible isomers that have the molecular formula $C_5H_{12}O$.

14. Compare an ester and a carboxylic acid with the same number of carbons. Which of the two compounds would have a higher solubility in water? Explain why.

15. Describe how the boiling point changes as the chain length of an aliphatic compound increases. Explain why this happens.

16. In a fractionating tower, the gaseous fractions are removed from the top of the tower and the solid residues from the bottom. Explain why the fractions separate in this way.

Applying Concepts

17. Benzene is a non-polar solvent. Is an alcohol or an alkane more soluble in benzene? Explain your answer.

18. Consider the compounds $CH_3CH_2CH_2OH$, CH_3CH_2COOH, and CH_3COOCH_3. Which compound is likely to have the highest boiling point? Explain your reasoning.

19. Someone has left two colourless liquids, each in an unlabelled beaker, on the lab bench. You know that one liquid is an alkane, and one is an alkene. Describe a simple test that you can use to determine which liquid is which.

20. a) Describe the bonding in benzene.
   b) Is cyclohexa-1,3,5-triene an acceptable IUPAC name for benzene? Why or why not?

21. Suppose that you were given a beaker labelled C_6H_{12} that contains a liquid. Design a procedure that would let you determine whether the substance was an alkene, a cycloalkane, or an aromatic hydrocarbon.

22. Suppose that you are working with four unknown compounds in a chemistry laboratory. Your teacher tells you that these compounds are ethane, ethanol, ethyl ethanoate, and ethanoic acid.
   a) Use the following table of physical properties to identify each unknown compound.
   b) Draw a complete structural formula for each compound.

Making Connections

23. The National Energy Board has estimated that Canada's original petroleum resources included 4.3×10^{11} m³ (430 billion cubic metres) of oil and bitumen and 1.7×10^{14} m³ (170 trillion cubic metres) of natural gas. Canada's petroleum-producing companies have used only a small fraction of these resources. Why, then, do you think that many people are so concerned about exhausting them?

24. In this chapter, you read about many uses or applications for alkanes, alkenes, alkynes, alcohols, alkyl halides, carboxylic acids, and esters. Write down each of the names of the classes of hydrocarbons or hydrocarbon derivatives. Below each name, list as many ways as you can that you use or encounter these organic compounds in your everyday living. Read labels on foods, medications, cleaning solutions, and other common household items to find more examples of these classes of organic compounds.

25. You are changing a long fluorescent bulb in your recreation room. You find that the fluorescent starter has leaked a very thick tar-like substance onto the plastic covering for the light. You look around your home to find something that you might use to clean the plastic cover. You find some rubbing alcohol (propan-2-ol), some vinegar (5% ethanoic acid), and some vegetable oil (esters of medium length (14 to 18 carbons) carboxylic acids with propane-1,2,3-triol). Which do you think would be the most likely product to dissolve the tar-like substance? Explain why.

Data Table for Question 22

Compound	Solubility in water	Hydrogen bonding	Boiling point	Odour	Molecular polarity
A	Not soluble	None	−89 °C	Odourless	Non-polar
B	Soluble	Accepts hydrogen bonds from water but cannot form hydrogen bonds between its own molecules	77 °C	Sweet	Polar
C	Infinitely soluble	Very strong	78 °C	Sharp, antiseptic smell	Very polar
D	Infinitely soluble	Extremely strong	118 °C	Sharp, vinegar smell	Very polar

CHAPTER 15

Reactions of Organic Compounds

Chapter Concepts

15.1 Types of Organic Reactions

- Combustion, addition, elimination, substitution, and esterification are five types of organic reactions.

- Complete balanced equations can be predicted and written for combustion, addition, elimination, substitution, and esterification reactions.

- Energy is released from fossil fuels through combustion reactions.

15.2 Polymers and the Petrochemical Industry

- Two types of polymerization reactions, addition and condensation, produce synthetic and natural polymers.

- Organic reactions are used to produce economically important compounds from fossil fuels.

TAXOL™, shown on the right, is an organic compound found in the bark of the Pacific yew tree. It has been approved for use in treating ovarian cancer. Unfortunately, the bark of a large yew tree yields only enough TAXOL™ for a single treatment—and cancer patients require repeated treatments over a long period. A solution to this problem has been found, however. The needles and twigs of the European yew contain an organic compound that is similar to TAXOL™. Using just a few organic reactions, scientists can transform this compound into TAXOL™. Recently, researchers with the Canadian Forest Service have discovered another source of TAXOL™ that has the potential to provide large amounts of the drug, along with other related compounds that could be used to synthesize a variety of other drugs. The plant is another species of yew called the Canadian yew or ground hemlock. One day, these shrub-like plants might be cultivated for the production of pharmaceuticals.

How do chemists use organic reactions to transform one organic compound into another? In this chapter, you will learn about some common reactions of organic compounds. You will also learn more about petrochemicals and their applications.

Comparing the Reactivity of Alkanes and Alkenes

What is the difference between saturated fats and unsaturated fats? How can you identify these compounds by using an organic reaction?

When the permanganate ion comes in contact with unsaturated compounds, a reaction occurs and the solution changes colour. When the permanganate ion comes in contact with saturated compounds, no reaction occurs.

Safety Precautions

- $KMnO_4(aq)$ will stain your skin or clothing. If you accidentally spill $KMnO_4(aq)$ on your skin, wash immediately with copious amounts of water. Remove the stain with a solution of sodium bisulfite.

Materials

- samples of vegetable oils, such as margarine, corn oil, and coconut oil
- samples of animal fats, such as butter and lard
- 5.0 mmol/L $KMnO_4(aq)$
- warm-water bath
- hot plate
- medicine droppers (one for each sample)
- test tubes
- test-tube rack
- stoppers

Procedure

1. Read the entire procedure and design a table to record your observations.

2. Melt solids, such as butter, in a warm-water bath (40 °C – 50 °C), and then test them as liquids. Using a different medicine dropper for each substance, place about two full medicine droppers of each test substance into a separate test tube.

3. Use a clean medicine dropper to add one full medicine dropper of potassium permanganate solution to each substance. Seal each tube with a clean rubber stopper and shake the test tube to thoroughly mix the reactants. Record your observations. Dispose of the reactants and products as directed by your teacher.

Analysis

1. List the physical properties of the samples that caused a change in the colour of the potassium permanganate solution upon mixing.

2. List the physical properties of the samples that did not cause a change in the colour of the potassium permanganate solution upon mixing.

3. What physical property of the samples appears to be related to their ability to cause a change in the colour of the potassium permanganate?

4. Based on this observation, infer a relationship between the chemical structure and a physical property of the samples that you analyzed.

TAXOL™

Types of Organic Reactions

Section Outcomes

In this section, you will:

- **define** and give **examples** of combustion, addition, elimination, substitution, and esterification reactions
- **predict** products, and write and interpret balanced equations for combustion, addition, elimination, substitution, and esterification reactions
- **carry out** an investigation to synthesize esters, and describe procedures for the safe handling of materials used in the laboratory
- **relate** the organic reaction types you have learned to reactions for producing energy from fossil fuels
- **investigate** sources of greenhouse gases and the issue of climate change
- **describe** how the science and technology of organic chemistry have developed to meet the needs of society and expand human capabilities
- **assess** some positive and negative consequences for humans and the environment, of society's use of organic compounds
- **work collaboratively in a team,** communicating information and ideas about organic chemistry

Key Terms

complete combustion reaction
addition reactions
elimination reaction
substitution reaction
esterification reaction
condensation reaction

Not only is your body made of mostly organic compounds, but the food you eat, the medicines you take, the clothing you wear, and the plastic implements you use are also all made of organic compounds. In the chapter opener, you saw that scientists can use organic reactions to prepare organic compounds, such as medicines, to help meet society's needs. In this section, you will be introduced to five types of organic reactions: combustion, addition, elimination, substitution, and esterification. In the next section, you will discover more about how these reactions are used to produce consumer products.

Combustion Reactions

The photograph in Figure 15.1A shows a person obtaining energy by eating an apple. Figure 15.1B shows the combustion of natural gas in a Bunsen burner flame. How are these two processes related? Both are examples of the same type of organic reaction.

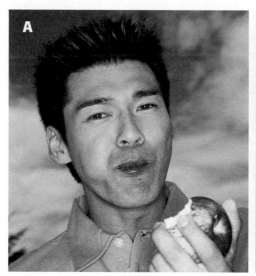 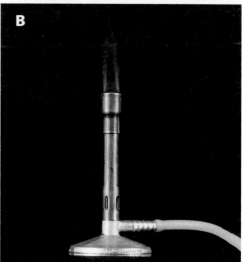

Figure 15.1 Cellular respiration, the process your body uses to obtain energy from food, is a series of organic reactions that produce carbon dioxide, water, and energy. These are the same products that result from the burning, or *combustion*, of natural gas.

Recall that in a combustion reaction, a compound reacts with oxygen to produce the oxides of the elements that make up the compound. The combustion of organic compounds is defined more specifically. In a **complete combustion reaction**, a hydrocarbon reacts with oxygen to produce carbon dioxide, water vapour, and energy. No matter how complicated the structure, if excess oxygen is present, *all hydrocarbons will burn completely to produce carbon dioxide and water vapour.*

$$\text{hydrocarbon} + O_2(g) \rightarrow CO_2(g) + H_2O(g) + \text{energy}$$

Although the term "complete combustion reaction" is usually used in reference to hydrocarbons, other organic compounds can undergo complete combustion reactions. For example, ethanol is added to gasoline to improve combustion in an engine. (See the Thought Lab on page 589 for more information on this topic.) The complete combustion of ethanol is shown below.

$$C_2H_5OH(\ell) + 3O_2(g) \rightarrow 2CO_2(g) + 3H_2O(g)$$

To refresh your memory of writing and balancing combustion reactions, answer the following questions.

Problem Tip

Recall that when you are balancing combustion reactions, you sometimes have an odd number of oxygen atoms in the products. There is no integer that you can place in front of O_2 to give an odd number of oxygen atoms in the reactants. To solve the problem, multiply all other coefficients by 2 and then balance the oxygen atoms.

• • •

1. Write a balanced equation to show the complete combustion of
   a) methane, $CH_4(g)$ **b)** but-2-ene, $C_4H_8(g)$

2. Write a balanced chemical equation to show the complete combustion of propane, $C_3H_8(g)$, in a camping stove.

3. Write a balanced equation for the complete combustion of methanol in a fondue burner.

4. Write and balance an equation to show the complete combustion of butanoic acid.

5. Although cellular respiration in the body is a complex process, the overall reaction can be reduced to the equivalent of a combustion reaction. Write and balance the chemical equation for the complete combustion of glucose, $C_6H_{12}O_6(s)$.

6. Energy is produced from the combustion of hydrocarbons. What is the origin of this energy?

• • •

Thought Lab | **15.1** | **Fossil Fuels and Climate Change**

Target Skills

Understanding that science and technology have both intended and unintended consequences

Assessing the effects of burning fossil fuels on climate change

Natural gas, coal, and petroleum are fossil fuels, created deep in Earth's crust by the compression and heating of ancient organic matter. The combustion of *fossil fuels* to provide energy for cooking, transportation, and warmth is an essential part of your life. Our society's vast consumption of fossil fuels has an environmental cost, however. Both the production and the combustion of fossil fuels release millions of tonnes of gases into the atmosphere. What effect could these gases have on the global climate?

Procedure

Use library and electronic research tools to find answers for the following questions.

1. a) The *greenhouse effect* is a natural system caused by water vapour and other gases in the atmosphere. It protects Earth from extreme temperatures. Draw a diagram illustrating this effect.

   b) Water vapour is a naturally occurring *greenhouse gas*. Define the term greenhouse gas.

2. a) Methane, carbon dioxide, and dinitrogen monoxide (also called nitrous oxide) are present in the atmosphere in increasing amounts as a result of human activities. Why is the increase in the concentration of these gases in the atmosphere a problem?

   b) What are the main sources of these gases?

Analysis

1. What effect do you expect the burning of fossil fuels to have on the global climate? Explain your answer.

2. What other human actions might be affecting the global climate?

3. Suggest three ways to reduce human production of methane, carbon dioxide, and dinitrogen monoxide (commonly called nitrous oxide). Be as specific as possible.

Addition, Elimination, and Substitution Reactions

As you have just seen, the combustion of organic compounds produces energy. Combustion reactions do not yield any useful organic products, however. Organic chemists use many different reactions to synthesize specific organic products. Most of these reactions can be classified into three categories: addition, elimination, and substitution reactions. Each type of reaction transforms certain organic compounds into other organic compounds.

Addition Reactions

Alkenes and alkynes characteristically undergo **addition reactions** in which atoms are added to a double or triple bond. To recognize an addition reaction, check whether the carbon atoms in the product(s) are bonded to *more* atoms than are the carbon atoms in the organic reactant.

$$\underset{/}{\overset{\backslash}{}}C=C\overset{/}{\underset{\backslash}{}} \quad + \quad Y-Z \quad \rightarrow \quad \overset{YZ}{\underset{}{-\overset{|}{C}-\overset{|}{C}-}}$$

In a typical addition reaction, a small molecule adds to an alkene or alkyne. The small molecule is usually one of the following: $H_2O(\ell)$ (when water splits in a reaction, it is often written, HOH), H_2, HX, or X_2 (where X=F, Br, Cl, or I). The reaction results in one major product or two products that are isomers of each other. (Recall that isomers have the same molecular formula but different structures.) Figure 15.2 shows some specific examples for an alkene.

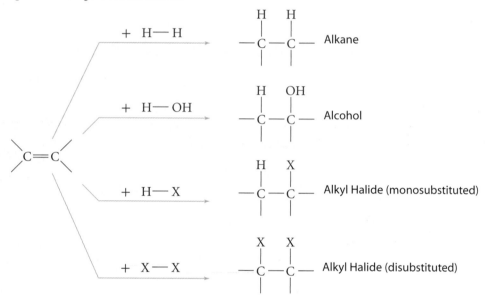

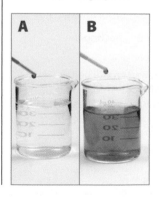

Figure 15.2 In each addition reaction shown here, the electrons in the second bond are rearranged and form bonds with the two additional atoms.

The following equation is an example of an addition reaction in which chlorine gas reacts with but-2-ene.

$$H-\overset{H}{\underset{H}{\overset{|}{\underset{|}{C}}}}-\overset{H}{\underset{H}{\overset{|}{\underset{|}{C}}}}=\overset{H}{\underset{H}{\overset{|}{\underset{|}{C}}}}-\overset{H}{\underset{H}{\overset{|}{\underset{|}{C}}}}-H \quad + \quad Cl-Cl \quad \rightarrow \quad H-\overset{H}{\underset{H}{\overset{|}{\underset{|}{C}}}}-\overset{Cl}{\underset{H}{\overset{|}{\underset{|}{C}}}}-\overset{Cl}{\underset{H}{\overset{|}{\underset{|}{C}}}}-\overset{H}{\underset{H}{\overset{|}{\underset{|}{C}}}}-H$$

but-2-ene   chlorine   2,3-dichlorobutane

In an alkyne, a maximum of four bonding electrons can be rearranged to form bonds with four new atoms, leaving a single bond between the two carbon atoms. The equation below shows what happens when excess chlorine is added to ethyne (acetylene).

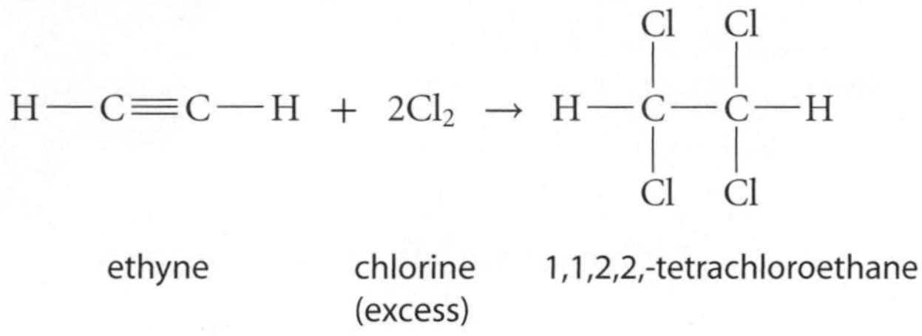

$$H-C\equiv C-H \ + \ 2Cl_2 \ \rightarrow \ H-\overset{\overset{\displaystyle Cl}{|}}{\underset{\underset{\displaystyle Cl}{|}}{C}}-\overset{\overset{\displaystyle Cl}{|}}{\underset{\underset{\displaystyle Cl}{|}}{C}}-H$$

ethyne   chlorine   1,1,2,2,-tetrachloroethane
                (excess)

If the amount of chlorine is limited, an alkene is formed.

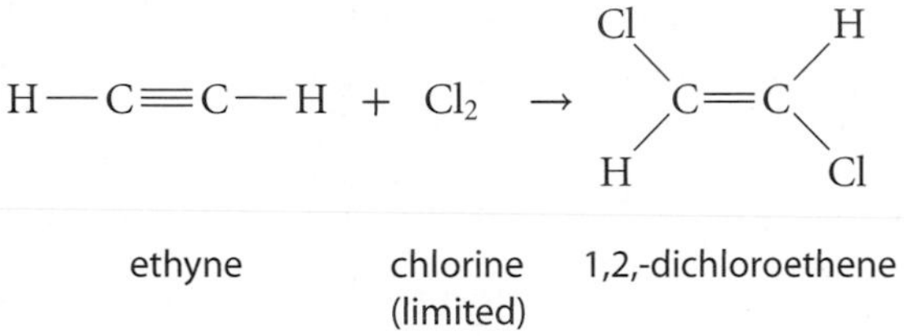

$$H-C\equiv C-H \ + \ Cl_2 \ \rightarrow \ \underset{H}{\overset{Cl}{\diagdown}}C=C\underset{Cl}{\overset{H}{\diagup}}$$

ethyne   chlorine   1,2,-dichloroethene
                (limited)

When alkenes and alkynes that contain more than two carbon atoms react with a small molecule, more than one product might form. For example, if hydrogen chloride, $HCl(g)$, is added to pent-2-ene, two isomers will form as shown here.

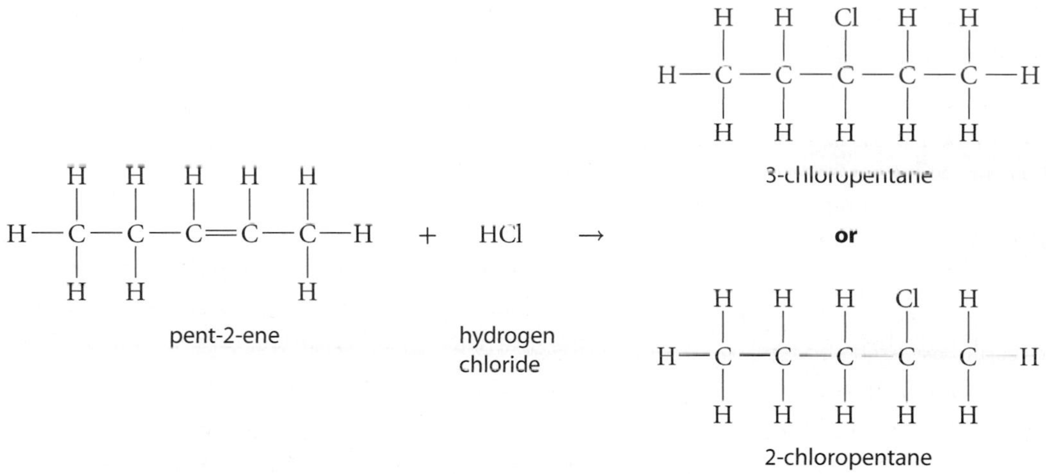

pent-2-ene   hydrogen chloride   →

3-chloropentane

or

2-chloropentane

One of the isomers will usually be more abundant than the other in the products of the reaction. In some cases, such as the reaction between propene and hydrogen chloride shown below, only one isomer will be produced. The Chemistry File in the margin explains how to predict which isomer will predominate.

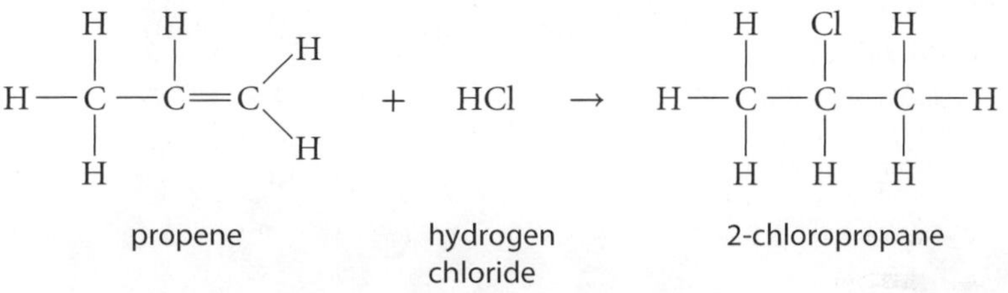

propene   hydrogen chloride   →   2-chloropropane

Chemistry File

FYI

When a small molecule, such as H—Cl(g) or H—OH(ℓ) (H$_2$O(ℓ)), is added to an asymmetric alkene, such as propene (H$_3$C—CH = CH$_2$), the products can contain more than one isomer. You can predict which isomer will be more abundant by using *Markovnikov's rule*. According to this rule, the hydrogen atom of the small molecule will attach to the carbon of the double bond that is already bonded to the most hydrogen atoms. In the equation shown in the text, the major product is 2-chloropropane.

Trans Fats in the Diet

The next time you reach for your favourite snack, have a look at the nutrition label. Does this food contain *trans fats* or partially hydrogenated oils? If so, this tasty treat could be putting you at risk for heart disease and other disorders.

All Fats Are Not Created Equal

Butter, margarine, olive oil, and hydrogenated soybean oil all contain fats. Why is olive oil liquid at room temperature while these other products are solid? Their state depends on the intermolecular forces between the molecules. Fats consist of three *fatty acids* bonded to glycerol. Fatty acids are long-chain carboxylic acids. Unsaturated fatty acids normally have a "cis" configuration around their double bonds, which gives them a bent or kinked structure shown in the space-filling models below. The kinks in the hydrocarbon chains prevent the chains from lining up close to one another thus limiting the forces of attraction between the chains. As a result, unsaturated fats tend to be liquid at room temperature. Saturated fatty acids have straight chains and thus can pack close together. The larger number of London (dispersion) forces make the fats more solid at room temperature.

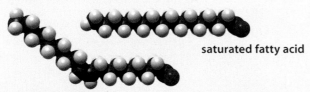

saturated fatty acid

unsaturated fatty acid

Unsaturated fats are readily oxidized and tend to become rancid more rapidly than do saturated fats. To extend the shelf life of foods containing unsaturated fats or to keep them from melting at room temperature, unsaturated fats can be hydrogenated to reduce the number of double bonds. However, under the conditions of hydrogenation, the normal "cis" configuration is changed to the "trans" configuration for those double bonds that do not become saturated. The hydrocarbon chains of trans fatty acids are straight, as shown in the diagram below, and behave similarly to saturated fatty acids because the hydrocarbon chains fit closely together.

A Growing Health Concern

Fats are not only a source of energy but are also a major part of cell membranes. Fats are also precursors for some hormones. Health studies show, however, that trans fats do more harm than good. Most enzymes cannot break down trans fats completely and they can become part of cell membranes. With their unusual trans configuration, trans fats reduce the ability of membranes to function properly.

Research also shows that trans fats affect the body's ability to manage cholesterol. Since cholesterol is insoluble in water, the body uses special complexes to transport fats and cholesterol in the blood. The two types of complexes are called low-density lipoproteins (LDL) and high-density lipoprotein (HDL). High levels of LDL contribute to the build-up of plaque in the arteries. In contrast, HDLs do not contribute to the formation of plaque. Trans fats are thought to raise levels of LDL ("bad cholesterol") and lower levels of HDL ("good cholesterol"), increasing the risk of heart disease.

• • •

1. If fats were fully hydrogenated, there would be no trans fats. Why would food manufacturers choose to only partially hydrogenate fats?

2. Trans fats were used in foods for about 80 years before consumer groups became concerned about their adverse health effects. Why do you think it took this long for people to make the link between the consumption of trans fats and development of heart disease?

3. Do trans fats occur naturally in foods? Are all trans fats, such as conjugated linoleic acid, unhealthy? Research these topics on the Internet. **ICT**

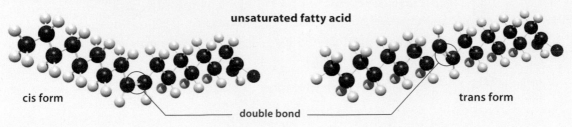

unsaturated fatty acid

cis form   **trans form**

double bond

Elimination Reactions

In an **elimination reaction**, atoms are removed from an organic molecule and a double bond forms between the two carbon atoms from which the atoms were removed. You can envision this type of reaction as the reverse of an addition reaction. One reactant usually loses two atoms, and two products are formed. A double bond is formed in the organic product. To recognize an elimination reaction, determine whether the carbon atoms in the organic product are bonded to *fewer* atoms than were the carbon atoms in the organic reactant.

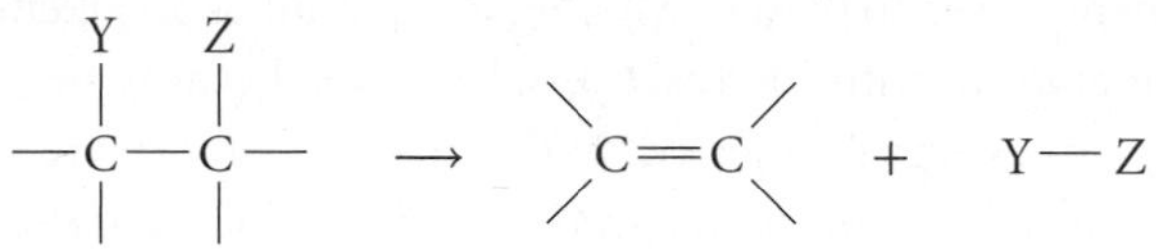

$$\begin{array}{c} Y \quad Z \\ | \quad | \\ -C-C- \\ | \quad | \end{array} \longrightarrow \begin{array}{c} \diagdown \qquad \diagup \\ C=C \\ \diagup \qquad \diagdown \end{array} + \quad Y-Z$$

Alcohols can undergo elimination reactions to produce alkenes when they are heated in the presence of a strong acid, such as sulfuric acid ($H_2SO_4(aq)$). The strong acid acts as a catalyst to speed up the reaction. A small, stable molecule, such as $H_2O(\ell)$ is formed as a second product.

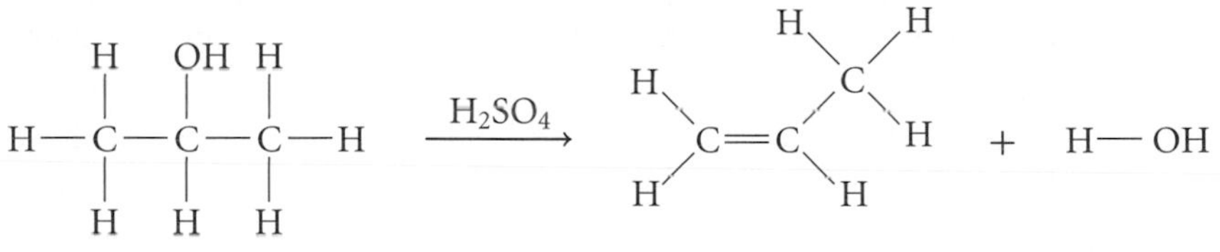

$$\begin{array}{c} H \quad OH \ H \\ | \quad | \quad | \\ H-C-C-C-H \\ | \quad | \quad | \\ H \quad H \quad H \end{array} \xrightarrow{H_2SO_4} \quad C=C \quad + \quad H-OH$$

propan-2-ol propene water

Alkyl halides can undergo elimination reactions to produce alkenes when they are heated in the presence of a strong base, such as sodium ethoxide, $NaOCH_2CH_3$, in ethanol.

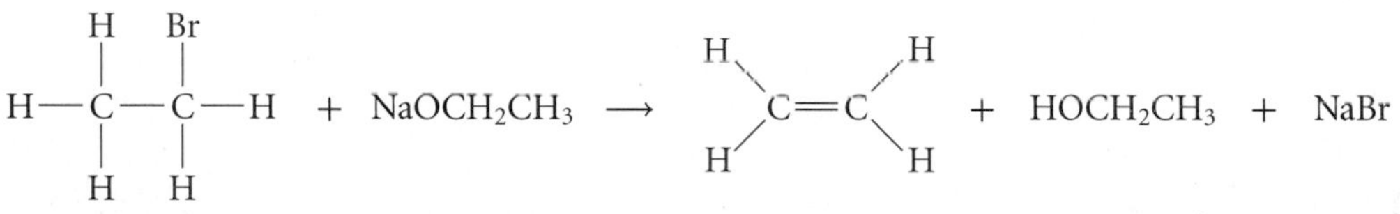

$$\begin{array}{c} H \quad Br \\ | \quad | \\ H-C-C-H \\ | \quad | \\ H \quad H \end{array} + NaOCH_2CH_3 \longrightarrow \quad C=C \quad + HOCH_2CH_3 + NaBr$$

bromoethane sodium ethoxide (strong base) ethene ethanol sodium bromide

Substitution Reactions

In a **substitution reaction**, a hydrogen atom or a functional group is replaced by a different functional group. To recognize this type of reaction, look for the following two features:

- two compounds react to form two different compounds
- carbon atoms bonded to the *same* number of atoms in the product as in the reactant

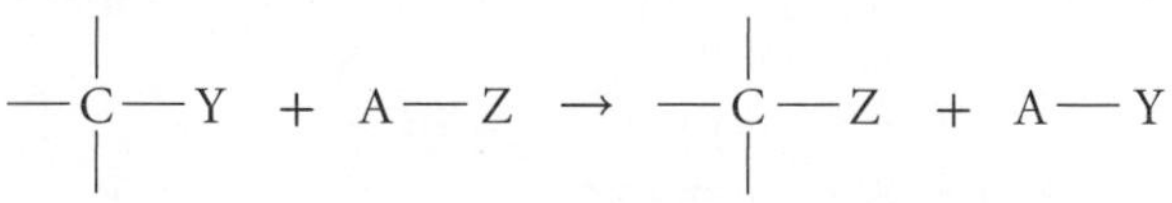

$$\begin{array}{c} | \\ -C-Y \\ | \end{array} + A-Z \rightarrow \begin{array}{c} | \\ -C-Z \\ | \end{array} + A-Y$$

Alcohols and alkyl halides commonly undergo substitution reactions. When an alcohol reacts with an acid that contains a halogen, such as hydrogen chloride or hydrogen bromide, the halogen atom is substituted for the hydroxyl group of the alcohol, as shown below. An alkyl halide is produced.

$$CH_3-CH_2-OH \ + \ HCl \ \rightarrow \ CH_3-CH_2-Cl \ + \ H-OH$$

ethanol hydrogen chloride chloroethane water

Chemistry File

FYI
When an asymmetric molecule undergoes an elimination reaction, more than one isomer can be present among the products. In the example shown below, a hydrogen atom can be removed from either the first carbon in the chain or the third carbon in the chain to form the double bond. As a general rule, the hydrogen atom is most likely to be removed from the carbon atom with the most carbon–carbon bonds. In this example, therefore, a hydrogen atom will be removed from the third carbon to form but-2-ene as the major product.

$CH_3-CH(Br)-CH_2CH_3 \rightarrow$
$CH_3-CH=CH-CH_3$
(major product) +
$CH_2=CH-CH_2-CH_3$
(minor product)

For the purposes of this course, you can assume that an alkyl halide reacting with a hydroxide ion produces an alcohol.

You can also assume that heating an alkyl halide with sodium ethoxide, $NaOCH_2CH_3$, in an ethanol solvent produces an alkene.

An alkyl halide can undergo a substitution reaction with a hydroxide ion to produce an alcohol, as shown below.

$$CH_3{-}CH_2{-}Cl \;+\; OH^- \;\rightarrow\; CH_3{-}CH_2{-}OH \;+\; Cl^-$$

chloroethane | hydroxide ion | ethanol | chloride ion

Recall that alkyl halides can also undergo *elimination reactions* in the presence of a base. Chemists use the reaction conditions, such as the type of base and the type of solvent, as shown on page 593 to predict whether an alcohol or an alkene will form.

Alkanes also undergo substitution reactions. However, because they are relatively unreactive, considerable energy is required. In the presence of ultraviolet light, alkanes react with chlorine or bromine to produce alkyl halides. For example, if enough chlorine is present in a mixture of chlorine and methane, four possible organic products can form:

$$4CH_4 + 10Cl_2 \xrightarrow{UV} CH_3Cl + CH_2Cl_2 + CHCl_3 + CCl_4 + 10HCl$$

methane | chlorine | chloromethane (methyl chloride) | dichloromethane (methylene chloride) | trichloromethane (chloroform) | tetrachloromethane (carbon tetrachloride) | hydrogen chloride

Substitution reactions of alkanes are not very useful in the laboratory because they usually result in a mixture of products. Chemists generally use other reactions to obtain specific alkyl halides.

Like alkanes, most aromatic hydrocarbons are fairly stable. They will undergo substitution reactions with chlorine and bromine but only in the presence of a catalyst. In the equation representing the reaction between benzene and bromine shown below, iron(III) bromide is acting as the catalyst:

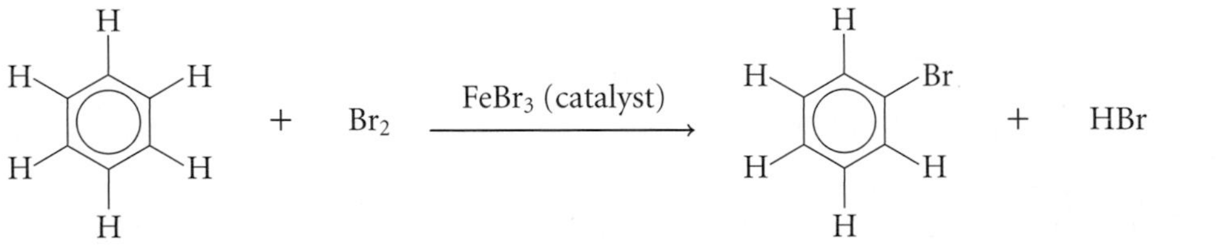

benzene | bromine | bromobenzene | hydrogen bromide

Esterification Reactions

Recall from Chapter 14 that esters are formed by a reaction between two other organic compounds. In a typical **esterification reaction,** an alcohol combines with a carboxylic acid to produce an ester and water, as shown below. The symbols R and R′ in the equation represent different hydrocarbon chains.

$$R{-}\overset{\overset{\large O}{\|}}{C}{-}OH \;+\; HO{-}R' \;\xrightleftharpoons{H_2SO_4}\; R{-}\overset{\overset{\large O}{\|}}{C}{-}O{-}R' \;+\; HO{-}H$$

carboxylic acid | alcohol | ester | water

Esterification is one type of condensation reaction. A **condensation reaction** is a reaction in which two molecules combine to form a larger molecule, producing a small, stable molecule (usually water) as a second product. As shown in Figure 15.3, esterification reactions can be used to produce useful consumer products.

Many of the flavours and aromas of fruits and spices are due to the presence of esters. Through esterification reactions, chemists have learned to duplicate natural esters. Synthesized esters are used to give artificial flavour to juices, candy, and many foods. Figure 15.4 shows the compound responsible for the flavour of cherries. What esterification reaction produces this compound?

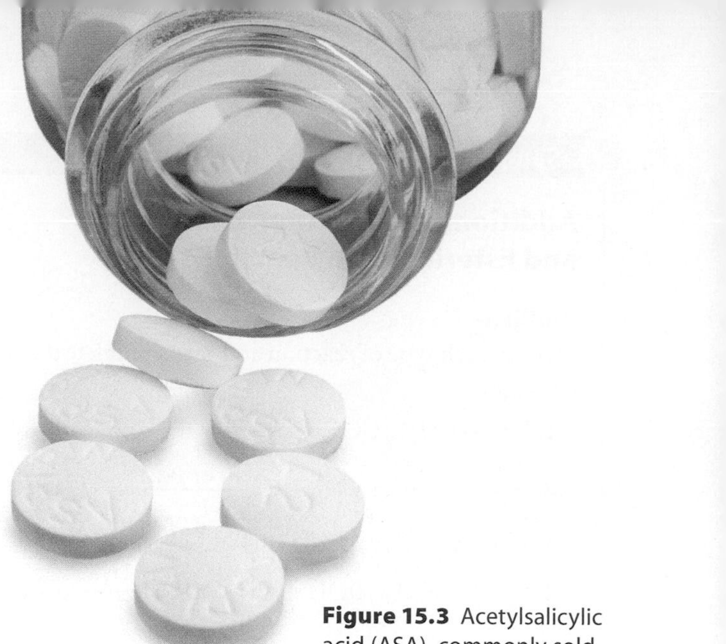

Figure 15.3 Acetylsalicylic acid (ASA), commonly sold as Aspirin™, can be made from salicylic acid by using an esterification reaction. Salicylic acid was first obtained from a natural compound (salicin) found in willow bark. Long before Aspirin™ was produced, Aboriginal peoples in Canada used extracts from willow bark as a remedy for pain and fever.

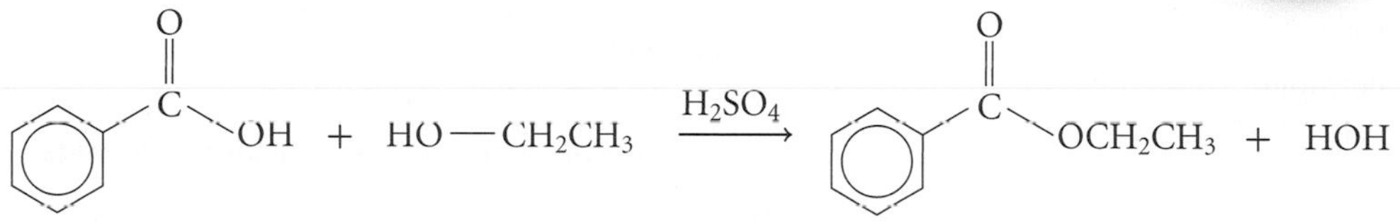

benzoic acid + ethanol $\xrightarrow{H_2SO_4}$ ethyl benzoate + water

What reaction forms pentyl propanoate, a fruity-smelling compound found in apricots?

$$CH_3CH_2-\overset{\overset{\displaystyle O}{\|}}{C}-OH + HO-(CH_2)_4CH_3 \longrightarrow CH_3CH_2-\overset{\overset{\displaystyle O}{\|}}{C}-O(CH_2)_4CH_3 + HOH$$

propanoic acid + pentan-1-ol → pentyl propanoate + water

In Investigation 15.A, you will have the opportunity to make your own esters. But first, work through the Sample and Practice Problems.

Figure 15.4 The organic compound responsible for the flavour of cherries is an ester named ethyl benzoate.

Addition, Substitution, Elimination, and Esterification Reactions

Problem

Identify each type of reaction and then complete the equation.

1. $HO-CH_2CH_2CH_3 \xrightarrow[\Delta]{H_2SO_4}$

2. $H_2C{=}CHCH_2CH_3 + H_2 \xrightarrow{Pt \text{ or } Pd}$

3. $CH_3CH(CH_3)CH_2CH_3 + Br_2 \xrightarrow{UV \text{ light}}$

4. $CH_3CH_2CH_2COOH + CH_3CH_2OH \xrightarrow{H_2SO_4}$

Solution

1. The alcohol is heated in the presence of a strong acid. This reaction is an elimination reaction. An alkene and a small second product are formed in elimination reactions:

$$HO-CH_2CH_2CH_3 \xrightarrow[\Delta]{H_2SO_4} CH_2{=}CHCH_3 + HOH$$
propan-1-ol   **propene**   **water**

2. A small molecule that can react with an alkene in such a way that two parts of the small molecule become bonded to two adjacent carbons in the organic molecule. This reaction is an addition reaction. The addition reaction gives a single product, as follows:

$$H_2C{=}CHCH_2CH_3 + H_2 \xrightarrow{Pt \text{ or } Pd} CH_3CH_2CH_2CH_3$$
but-1-ene   **hydrogen**   **butane**

3. A small molecule that can react with an alkane in the presence of ultraviolet light in such a way that one part of the small molecule is bonded to a carbon of the organic molecule. This is a substitution reaction. Depending on how much bromine is present, the reaction may continue until all the hydrogen atoms have been substituted. Furthermore, the first bromine atom could replace any one of the hydrogen atoms. Only one of the many products that are possible for this reaction is given here:

$$CH_3CH(CH_3)CH_2CH_3 + Br_2 \xrightarrow{UV \text{ light}}$$
2-methylbutane   **bromine**

$$CH_3CH\,Br\,(CH_3)CH_2CH_3 \quad + \quad HBr$$
2-bromo-2-methylbutane   **hydrogen bromide**

4. A carboxylic acid reacts with an alcohol. This is an esterification reaction. An ester and water will be formed as products. The equation is completed as follows:

$$CH_3CH_2CH_2COOH + CH_3CH_2OH \xrightarrow{H_2SO_4}$$
butanoic acid   **ethanol**

$$CH_3CH_2CH_2COOCH_2CH_3 + HOH$$
ethyl butanoate   **water**

Practice Problems

1. Identify each reaction as an addition, a substitution, an elimination, or an esterification reaction:

a) $CH_3-CH{=}CH-\underset{\underset{CH_3}{|}}{CH}-CH_3 + Cl_2 \rightarrow CH_3-\underset{\underset{Cl}{|}}{CH}-\underset{\underset{Cl}{|}}{CH}-\underset{\underset{CH_3}{|}}{CH}-CH_3$

b) $CH_3-CH_2-CH_2-OH + CH_3-\overset{\overset{O}{\|}}{C}-OH \xrightarrow{H_2SO_4} CH_3-\overset{\overset{O}{\|}}{C}-O-CH_2-CH_2-CH_3 + HOH$

c) $CH_3-CH_2-CH_2-CH_3 + Br_2 \xrightarrow{UV \text{ light}} CH_3-CH_2-\underset{\underset{Br}{|}}{CH}-CH_3 + HBr$

d) 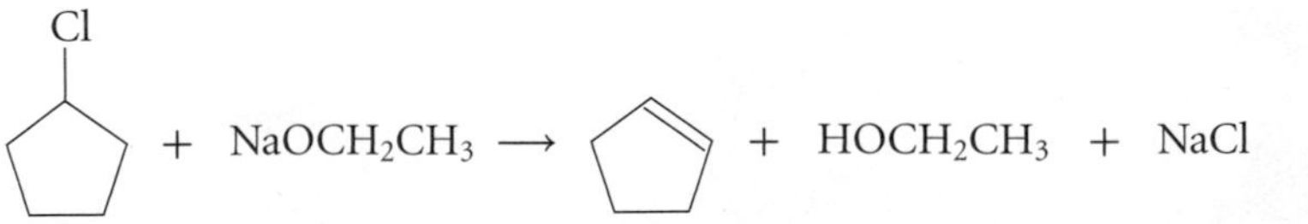 $+ \ NaOCH_2CH_3 \longrightarrow$ $+ \ HOCH_2CH_3 \ + \ NaCl$

2. Complete the equations given on the next page. Draw condensed structural formulas for the missing organic compounds. Name all reactants and products:

a)
$$CH_3 - \underset{\underset{OH}{|}}{CH} - CH_2 - CH_2 - CH_3 \xrightarrow[\triangle]{H_2SO_4}$$

b)
$$\longrightarrow \quad \underset{\underset{H}{|}}{\overset{CH_3-CH_2}{\diagdown}} C = C \overset{\diagup CH_3}{\underset{\diagdown CH_3}{}} + HCl$$

c)
$$\bigcirc + Cl_2 \xrightarrow{FeBr_3}$$

d)
$$H - C \equiv C - CH_3 + \underset{\text{(excess)}}{Br_2} \longrightarrow$$

e)
$$\longrightarrow \quad HC \overset{\overset{O}{\|}}{-} OCH_3 + HOH$$

f)
$$CH_3 - CH_2 - \underset{\underset{CH_3}{|}}{CH} - CH_2 - OH + HBr \longrightarrow$$

3. Identify each type of reaction in the previous question.

4. Write a balanced equation to show how you would form propan-2-ol from an alkene.

5. What series of reactions would you carry out to produce butyl methanoate from 1-chlorobutane? What carboxylic acid would you use? Write a balanced equation for each step.

6. How could you convert 3-chloro-3-methylpentane into 3-methylpentan-3-ol? Write a balanced equation.

INVESTIGATION 15.A

Target Skills

Performing an investigation of esterification reactions by producing an ester

Describing procedures for safe handling, storage, and disposal of materials used in the laboratory with reference to WHMIS labelling

Preparing Esters

You now know that chemists use esterification reactions to synthesize artificial esters. In this investigation, you will synthesize three esters. Make careful observations and see if you can recognize the aromas of these esters.

Question

What observable properties do esters have?

Safety Precautions

- Be sure that there is no open flame in the laboratory. Organic compounds are very flammable. Use a hot plate, not a Bunsen burner.

- Use a fume hood for all steps involving acids. Carry out all procedures in a well-ventilated area.

- Sulfuric acid, ethanoic acid, and butanoic acid are all extremely corrosive. Wear goggles, an apron, and gloves.

- Treat the acids with extreme care. If you spill any acid on your skin, immediately wash it with plenty of cold water and notify your teacher. If you spill the acids on the lab bench or floor, inform your teacher right away.

- Avoid touching the hot plate and the hot-water bath when it has been heated.

- Wash your hands when you have completed the investigation.

Materials

- ice
- distilled water
- ethanoic acid
- ethanol
- propan-1-ol
- butanoic acid
- 6 mol/L sulfuric acid
- 250 mL beakers (2)
- 50 mL small beakers (2)
- 100 mL beaker
- 10 mL graduated pipettes (2)
- watch glass
- hot plate
- thermometer (alcohol or digital)
- retort stand
- 2 clamps
- 4 plastic micropipettes
- medicine dropper
- stopper or paper towel

Procedure

1. Label three pipettes as shown. Use the appropriate pipette for each of the three reactions.
   – ethanoic acid + ethanol
   – ethanoic acid + propan-1-ol
   – butanoic acid + ethanol

2. Prepare your equipment as follows:

   a) Be sure that all the glassware is clean, dry, and free of chips or cracks.

   b) Prepare a hot-water bath. Heat about 125 mL of tap water in a 250 mL beaker on the hot plate to 60 °C. Adjust the hot plate so the temperature remains between 50 °C and 60 °C. Avoid touching the hot plate or beaker.

   c) Prepare a cold-water bath. Place about 125 mL of a mixture of water and ice chips in the second 250 mL beaker. The temperature of the cold-water bath will remain around 0 °C.

   d) Place about 5 mL of distilled water in a 50 mL beaker. You will use this in Procedure step 9.

   e) Set up the retort stand beside the hot-water bath. Use one clamp to hold the thermometer in the hot-water bath. You will use the other clamp to steady the micropipettes when you place them in the hot-water bath.

   f) Cut off the bulb of the unlabelled micropipette, halfway along the wide part of the bulb. You will use this bulb as a cap to prevent vapours from escaping during the reactions.

3. **Note:** In this step, do not inhale any alcohol vapour directly. Use the graduated pipette to measure 1.0 mL of ethanol into the 50 mL beaker. As you do so, you will get a whiff of the odour of the alcohol. Record your observations.

4. **Note:** in this step, do not inhale any acid directly. Use the graduated pipette to add 1.0 mL of ethanoic acid to the ethanol. As you do so, you will get a whiff of the odour of the acid. Record your observations.

5. Your teacher will carefully add four drops of sulfuric acid to the alcohol/acid mixture.

6. Suction the mixture into the appropriately labelled micropipette. Invert the micropipette. Place it, bulb down, in the hot-water bath. (See the diagram.) Place the cap over the tip of the pipette. Use a clamp to hold the pipette in place.

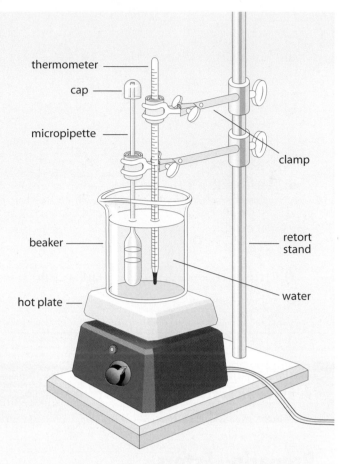

thermometer

cap

micropipette

clamp

beaker

retort stand

hot plate

water

7. Leave the pipette in the hot water for about 10 min to 15 min. Use the thermometer to monitor the temperature of the hot water. The temperature should stay between 50 °C and 60 °C.

8. After 10 to 15 min in the hot-water bath, place the pipette in the cold-water bath. Allow it to cool for about 5 min.

9. Carefully squeeze a few drops of the product onto a watch glass. Mix it with a few drops of distilled water. To smell the odour of the compound, use your hand to wave the aroma toward your nose. Do not inhale. Record your observations of the aroma.

10. Repeat Procedure steps 2 through 8 for the other two reactions. (Butanoic acid has an offensive odour. Do not attempt to smell it.) **Note:** While the first reaction mixture is being heated, you may want to prepare the other mixtures and put them in the hot-water bath. Working carefully, you should be able to stabilize more than one micropipette with the clamp. If you choose to do this, make sure that your materials are clearly labelled. Also, remember to keep a record of the time at which each micropipette is placed in the hot-water bath.

11. Dispose of all materials as your teacher directs. Clean all glassware thoroughly with soap and water.

Analysis

1. Make a table to organize your data. What physical property did you observe?

2. How do you know that a new product was formed in each reaction? Explain.

Conclusion

3. Describe the odour of each product that was formed. Compare the odours with familiar odours (such as the odours of plants, flowers, fruits, and animals) to help you describe them.

Application

4. Research the organic compounds that are responsible for the smell and taste of oranges, pineapples, pears, oil of wintergreen, and apples. Find and record the chemical structure of each compound.

As you have discovered, chemists use organic reactions to synthesize products for society's use. In Thought Lab 15.2, you will consider ways that science and technology of organic compounds have been influenced by society's needs. You will also think about ways that society and the environment have been affected by organic compounds.

Thought Lab 15.2 Problem Solving with Organic Compounds

Target Skills

Assessing the impact of CFCs on the ozone layer

Describing the processes involved in producing gasoline and adjusting the octane rating

Octane-Enhancing Compounds Reduce Engine Knocking

Automobile fuels are graded using an *octane rating*, or *octane numbers*, which measure the combustibility of a fuel. A high octane number means that a fuel requires a higher activation energy (higher temperature or higher pressure or both) to ignite. Racing cars with high-compression engines usually run on pure methanol, which has an octane number of 120.

Gasoline with too low an octane number can cause "knocking" in the engine of a car. This occurs when the fuel ignites too soon and burns in an uncontrolled manner. Knocking lowers fuel efficiency, and it can damage the engine.

As early as 1925, two of the first automobile engineers became aware of the need to improve the octane number of fuels. Charles Kettering advocated the use of a newly developed compound called tetra-ethyl lead, $Pb(C_2H_5)_4$. This compound acts as a catalyst to increase the effi-

ciency of the hydrocarbon combustion reaction. Henry Ford believed that ethanol, another catalyst, should be used instead of tetra-ethyl lead. Ethanol could be produced easily from locally grown crops. As we now know, ethanol is also much better for the environment.

Tetra-ethyl lead became the chosen fuel additive. Over many decades, lead emissions from car exhausts accumulated in many compartments in the environment, such as urban ponds and water systems. Many waterfowl that live in urban areas experience lead poisoning. Lead is also dangerous to human health.

Leaded fuels are now banned across Canada. In unleaded gasoline, simple organic compounds are added instead of lead compounds. These octane-enhancing compounds include methyl-t-butyl ether, t-butyl alcohol, methanol, and ethanol. Like lead catalysts, these compounds

help to reduce engine knocking. In addition, burning ethanol and methanol produces fewer pollutants than burning hydrocarbon fuels, which contain contaminants. Since they can be made from crops, these alcohols are a renewable resource.

2-methoxy-2-methyl propane
(methyl tert-butyl ether)

2-methylpropan-2-ol
(tert-butyl alcohol)

Replacing CFCs—At What Cost?

At the beginning of the twentieth century, refrigeration was a relatively new technology. Early refrigerators depended on the use of toxic gases, such as ammonia and methyl chloride. Unfortunately, these gases sometimes leaked from refrigerators, leading to fatal accidents. In 1928, a new, "miracle" compound was developed to replace these toxic gases. Dichlorodifluoromethane, commonly known as Freon®, was a safe, non-toxic alternative. Freon® and other chlorofluorocarbon compounds, commonly referred to as CFCs, were also used for numerous other products and applications. They were largely responsible for the development of many conveniences, such as air-conditioning, that we now take for granted.

Today we know that CFCs break up when they reach the ozone layer, releasing chlorine atoms according to the reaction:

$$CF_2Cl_2(g) \xrightarrow{\text{UV photon}} CF_2Cl(g) + Cl(g)$$

The chlorine atoms have an unpaired electron making them extremely reactive. They react with ozone according to the following reaction.

$$O_3(g) + Cl(g) \longrightarrow ClO(g) + O_2(g)$$

Finally, the chlorine monoxide reacts with atomic oxygen as show below.

$$ClO(g) + O(g) \longrightarrow Cl(g) + O_2(g)$$

As you can see, after a chlorine atom breaks down ozone to form oxygen, the chlorine atom is regenerated and can continue to destroy more ozone molecules. The chlorine atoms are unchanged after the series of reactions and can therefore be considered to be acting as catalysts. Studies in the past ten years have shown dramatic drops in ozone concentrations at specific locations, especially over the South Pole. Since ozone protects Earth from the Sun's ultraviolet radiation, this decrease in ozone has led to increases in skin cancer, as well as damage to plants and animals. In addition, CFCs are potent greenhouse gases and contribute to global warming. Through the Montreal Protocol, and later "Earth Summit" gatherings, many countries—including Canada—have banned CFC production and use.

Substitutes for CFCs are available, but none provides a completely satisfactory alternative. Hydrofluorocarbons (HFCs) are organic compounds that behave like CFCs but do not harm the ozone layer. For example, 1,1,1,2-tetrafluoroethane and 1,1-difluoroethane are HFCs that can be used to replace CFCs in refrigerators and air conditioners. Unfortunately, HFCs are also greenhouse gases.

Simple hydrocarbons can also be used as CFC substitutes. Hydrocarbons, such as propane, 2-methylpropane (common name: isobutane), and butane, are efficient aerosol propellants. These hydrocarbons are stable and inexpensive, but there are extremely flammable. They are also greenhouse gases.

isobutane   propane   butane

1,1,1,2-tetrafluoroethane   1,1-difluoroethane

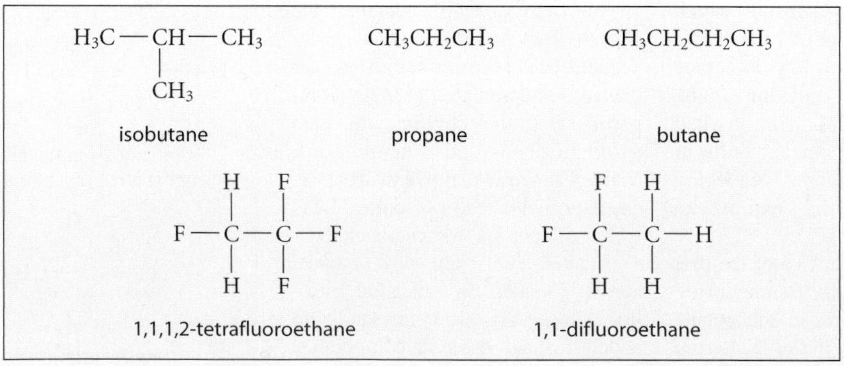

Procedure

After reading the articles, brainstorm with your group to come up with two other issues involving organic compounds, about which you have heard or read. You could consider such issues as food additives, the recycling of plastics, or pesticide pollution in the environment. For each situation, answer the following questions.

1. What was the original human or societal need for which the organic compounds were developed and used to meet?

2. How did organic compounds meet this need?

3. What further problems (if any) were introduced by the use of those organic compounds?

4. Have these additional problems been resolved? If so, how have they been resolved?

5. Do you foresee any new problems arising? Explain your answer.

Analysis

1. List four benefits society receives from its wide use of organic compounds.

2. Describe four unintended results that have occurred as a result of society's use of organic compounds.

3. In your opinion, is it worth finding organic compounds that will solve problems if the compounds may create more problems? Explain your answer. (You may want to debate this issue with your group or members of the class as a whole.)

Section 15.1 Summary

- A complete combustion reaction is one in which a hydrocarbon reacts with an excess of oxygen and produces carbon dioxide and water.
- An addition reaction is one in which atoms from a small molecule react with a double or triple bond in an organic molecule and become part of the molecule.
- When water reacts with a double bond, the product is an alcohol.
- When hydrogen reacts with a double bond, the product is an alkane.
- When a hydrogen halide reacts with a double bond, the product is an alkyl halide.
- When a halogen reacts with a double bond, the product is a disubstituted alkyl halide.
- An elimination reaction is one in which atoms or groups of atoms are removed from an organic molecule and a double bond is formed in the molecule.
- In a strong acid, water will be removed from an alcohol.
- In the presence of a strong base, an alkyl halide will lose a hydrogen atom and the halogen atom.
- A substitution reaction in one in which an atom or small group replaces another atom or small group on the organic molecule.
- In the presence of a hydrogen halide, the halogen atom will replace the hydroxyl group to form an alkyl halide.
- In the presence of a base, the hydroxide ion of the base will replace the halogen atom of an alkyl halide to form an alcohol.
- In the presence of a halogen molecule and ultraviolet light, a halogen atom can replace a hydrogen atom on an alkane to form an alkyl halide.
- In an esterification reaction, an alcohol reacts with a carboxylic acid to form an ester and a water molecule.

1. Identify each type of reaction:

a) 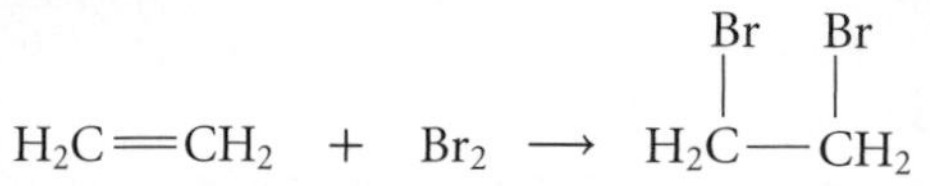 H_2C=CH_2 + Br_2 $\longrightarrow$ $H_2\overset{\overset{\displaystyle Br}{|}}{C}$—$\overset{\overset{\displaystyle Br}{|}}{C}H_2$

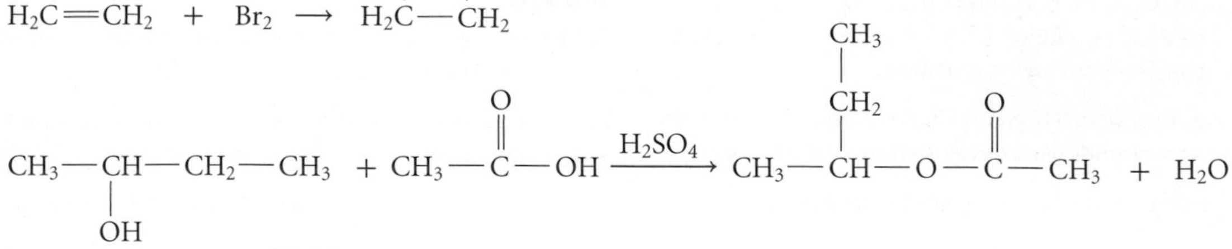

b) CH_3—$\overset{\underset{\displaystyle OH}{|}}{C}H$—$CH_2$—$CH_3$ + CH_3—$\overset{\overset{\displaystyle O}{\|}}{C}$—$OH$ $\xrightarrow{H_2SO_4}$ CH_3—$\overset{\underset{\displaystyle CH_2}{\overset{\overset{\displaystyle CH_3}{|}}{|}}}{C}H$—$O$—$\overset{\overset{\displaystyle O}{\|}}{C}$—$CH_3$ + H_2O

c) CH_3—$\overset{\underset{\displaystyle OH}{|}}{C}H$—$CH_3$ $\xrightarrow{H_2SO_4}$ CH_2=CH—CH_3 + H_2O

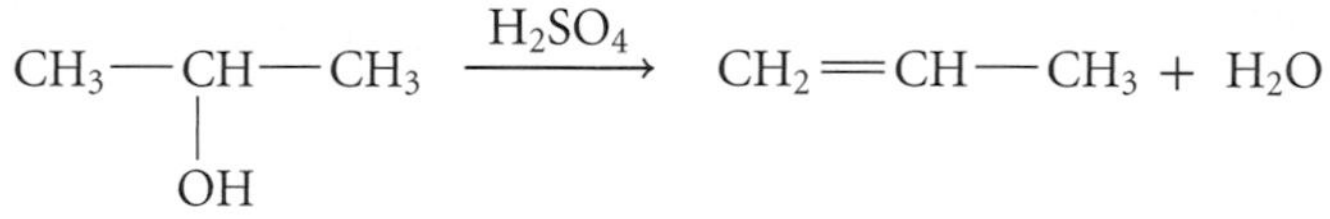

d) 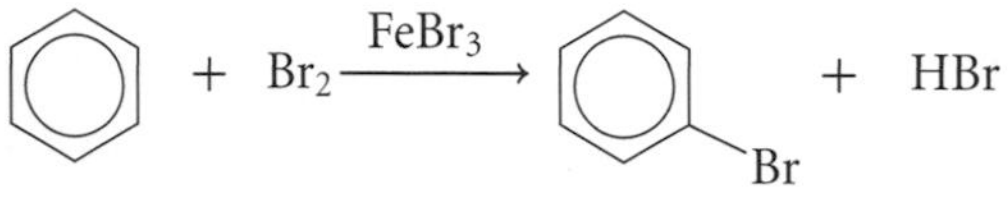 ⬡ + Br_2 $\xrightarrow{FeBr_3}$ ⬡—Br + HBr

2. Write complete chemical equations for each of the following partial equations. Identify the type of reaction and name each reactant and product, where the name is not given:

a) CH_3—CH=CH—CH_3 + H_2O $\longrightarrow$

b) ? + ? $\xrightarrow{?}$ methyl pentanoate + water

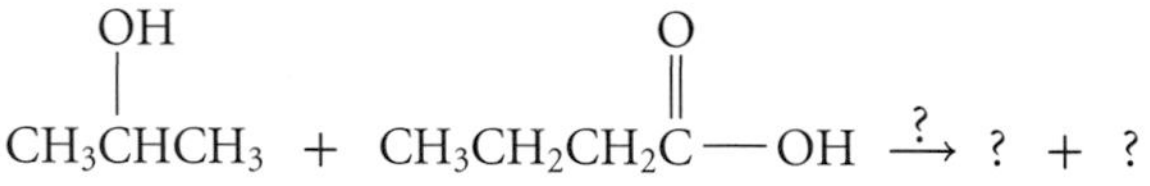

c) 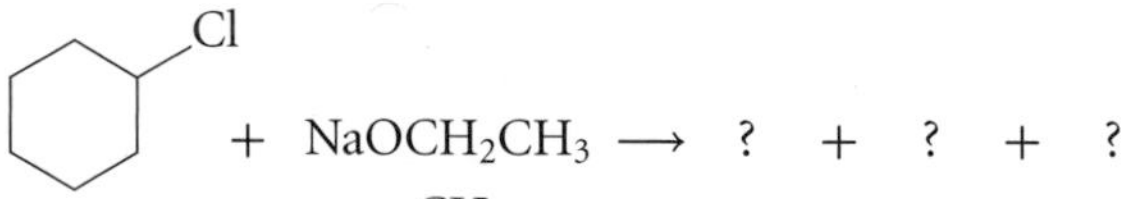 ⬡$\overset{Cl}{}$ + $NaOCH_2CH_3$ $\longrightarrow$? + ? + ?

d) 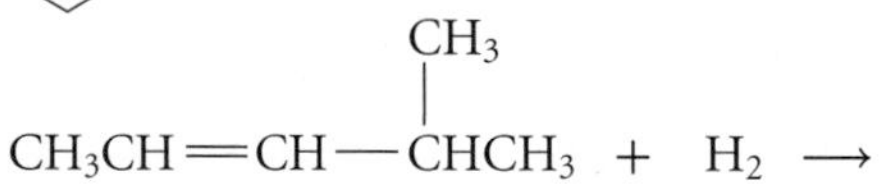 CH_3CH=CH—$\overset{\overset{\overset{\displaystyle CH_3}{|}}{}}{C}HCH_3$ + H_2 $\longrightarrow$

e) $CH_3CH_2CH_2CH_2CH_2$—I + OH^- $\longrightarrow$

f) 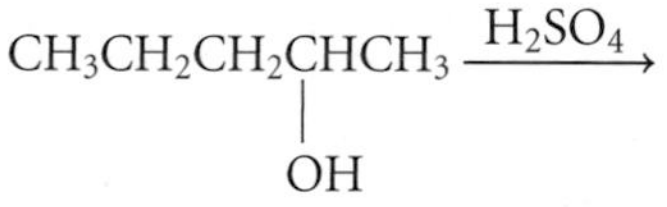 $CH_3CH_2CH_2\overset{\underset{\displaystyle OH}{|}}{C}HCH_3$ $\xrightarrow{H_2SO_4}$

g) $CH_3\overset{\overset{\overset{\displaystyle OH}{|}}{}}{C}HCH_3$ + $CH_3CH_2CH_2\overset{\overset{\displaystyle O}{\|}}{C}$—$OH$ $\xrightarrow{?}$? + ?

h) CH_3—$\overset{\underset{\displaystyle CH_3}{|}}{C}$=$CH_2$ + O_2 $\xrightarrow{spark}$

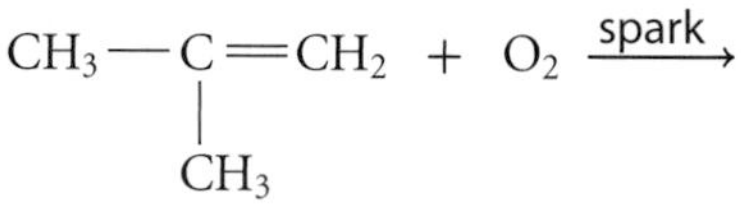

3. Identify the type of organic reaction that would accomplish each of the following changes. Give one example of each type of reaction:
a) alkyl halide $\rightarrow$ alkene   **d)** alkane $\rightarrow$ carbon dioxide
b) alkene $\rightarrow$ alcohol   **e)** alcohol $\rightarrow$ alkyl halide
c) alcohol + carboxylic acid $\rightarrow$ ester

4. What kind of reaction occurred in the Launch Lab? How do you know?

5. How are human activities affecting the global climate? Explain your answer.

6. Hydrofluorocarbons (HFCs) can be used as replacements for CFCs in refrigeration units because, unlike CFCs, they do not damage the ozone layer. Why are HFCs not being widely used in air conditioners or refrigerators?

Polymers and the Petrochemical Industry

Section Outcomes

In this section, you will:

- **define**, **draw**, and **give examples** of monomers and polymers in living and non-living systems
- **build** models to illustrate polymers and polymerization
- **perform** an experiment to make a polymer product
- **relate** the organic reaction types you have learned to reactions for producing important compounds from fossil fuels
- **analyze** a process for producing polymers

Key Terms

polymer
monomer
plastics
addition polymerization
condensation
    polymerization
petrochemicals

From grocery bags and kitchen utensils to furniture and computer equipment, plastics are an integral part of our homes, schools, and workplaces. Plastics belong to a group of organic compounds that you have not yet encountered: polymers. These useful organic compounds have a much larger molecular size than the compounds you have already studied in this unit.

Polymer Chemistry

A **polymer** is a very long molecule that is made by linking together many smaller molecules called **monomers**. To picture a polymer, imagine taking a handful of paper clips and joining them into a long chain. Each paper clip represents a monomer. The long chain of paper clips represents the polymer. Some polymers contain only one type of monomer, as is the case in the paper clip example. Polymers can also be made from a combination of two or more different monomers. Figure 15.5 shows an example of joined monomers in a polymer structure.

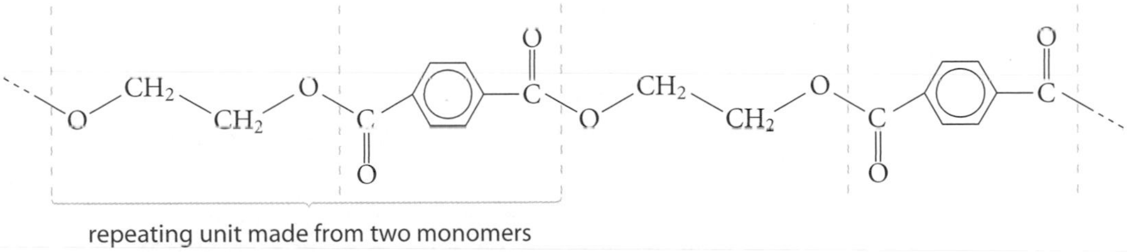

repeating unit made from two monomers

Figure 15.5 Polyethylene terephthalate (PET) is a plastic that is used to make soft drink bottles. A polymer, such as this one, can be several thousand carbons long.

Polymers that can be heated and moulded into specific shapes and forms are commonly known as **plastics**. All plastics are synthetic, or artificial, polymers. Adhesives, chewing gum, and polystyrene are also made from synthetic polymers.

The name of a synthetic polymer is usually written with the prefix poly- (meaning "many") before the name of the monomer. The common name of the monomer is sometimes used instead of the proper IUPAC name. For example, the common name for ethene is ethylene. Polyethene is a polymer that is made from ethene and is used to make plastic bags. Its common name is polyethylene. Similarly, the polymer that is made from chloroethene (common name: vinyl chloride) is named polyvinylchloride (PVC). The polymer that is made from propene (common name: propylene) is called polypropylene instead of polypropene.

As you will discover later in this section, natural polymers can be found in most living systems. Natural polymers have been used for thousands of years to make cotton, linen, and wool for clothing. Some synthetic polymers can also be used to make fabrics, as shown in Figure 15.6. For example, rayon, nylon, and polyester fabrics are made from synthetic polymers.

Chemistry File

Try This
Use paper clips to make a model of a polymer. What can this model tell you about the flexibility and strength of polymers? How is this model an accurate depiction of monomers and polymers, and how is it inaccurate? Design and assemble your own model to represent a polymer.

Figure 15.6 Both synthetic and natural polymers are used to make clothing.

Making Synthetic Polymers: Addition and Condensation Polymerization

Synthetic polymers are extremely useful and valuable. Many polymers and their manufacturing processes have been patented as corporate technology. Polymers are formed by two of the reactions you have already learned: addition reactions and condensation reactions.

Addition polymerization is a reaction in which alkene monomers are joined through multiple addition reactions to form a polymer. Addition reactions are characterized by a reduction in the number of double bonds found on the polymer, as addition reduces a double bond to a single bond. Figure 15.7 illustrates the addition polymerization of ethene to form polyethene. Table 15.1 gives the names, structures, and uses of some common addition polymers.

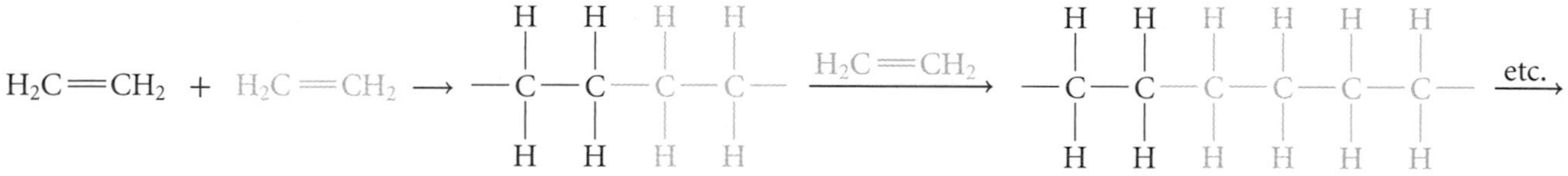

$$H_2C=CH_2 + H_2C=CH_2 \longrightarrow -\overset{\overset{\displaystyle H}{|}}{\underset{\underset{\displaystyle H}{|}}{C}}-\overset{\overset{\displaystyle H}{|}}{\underset{\underset{\displaystyle H}{|}}{C}}-\overset{\overset{\displaystyle H}{|}}{\underset{\underset{\displaystyle H}{|}}{C}}-\overset{\overset{\displaystyle H}{|}}{\underset{\underset{\displaystyle H}{|}}{C}}- \quad \xrightarrow{H_2C=CH_2} \quad -\overset{\overset{\displaystyle H}{|}}{\underset{\underset{\displaystyle H}{|}}{C}}-\overset{\overset{\displaystyle H}{|}}{\underset{\underset{\displaystyle H}{|}}{C}}-\overset{\overset{\displaystyle H}{|}}{\underset{\underset{\displaystyle H}{|}}{C}}-\overset{\overset{\displaystyle H}{|}}{\underset{\underset{\displaystyle H}{|}}{C}}-\overset{\overset{\displaystyle H}{|}}{\underset{\underset{\displaystyle H}{|}}{C}}-\overset{\overset{\displaystyle H}{|}}{\underset{\underset{\displaystyle H}{|}}{C}}- \quad \xrightarrow{\text{etc.}}$$

Figure 15.7 The formation of polyethene from ethene

Table 15.1 Examples of Addition Polymers

Name	Structure of monomer	Structure of polymer	Uses
polystyrene	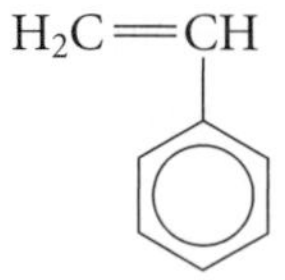 $H_2C=CH$ styrene	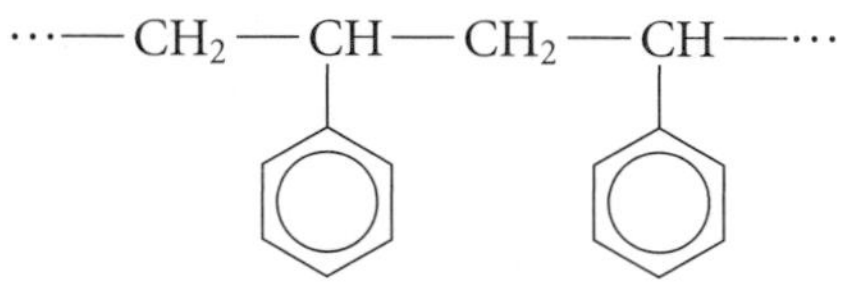 $\cdots—CH_2—CH—CH_2—CH—\cdots$ polystyrene	• styrene and Styrofoam™ cups • insulation • packaging
polyvinylchloride (PVC, vinyl)	Cl $\|$ $H_2C=CH$ vinyl chloride	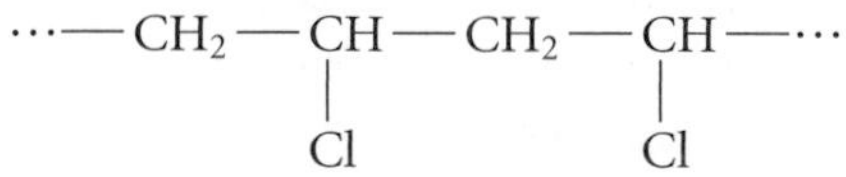 $\cdots—CH_2—CH—CH_2—CH—\cdots$ $\qquad\quad Cl\qquad\quad Cl$ polyvinylchloride	• construction materials • sewage pipes • medical equipment

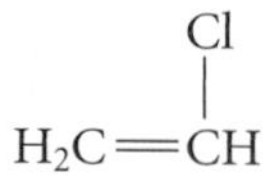

In **condensation polymerization**, monomers are combined through multiple condensation reactions to form a polymer. A second smaller product, usually water, is also produced in the reaction. For condensation polymerization to occur, each monomer must have two functional groups (usually one at each end of the molecule).

In the previous section, you learned that esterification is one type of condensation reaction. The formation of a polymer by esterification reactions between two monomers is shown in Figure 15.8. Note that one of the monomers, terephthalic acid, has a carboxyl group on each end of the molecule. The second monomer, ethane-1,2-diol, also called ethylene glycol, has an alcohol group on each end. Ester linkages can thus form on both ends of each monomer, forming a long polymer chain.

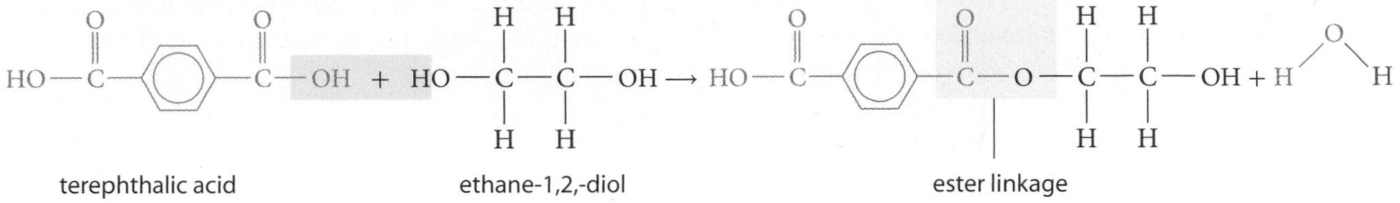

terephthalic acid   ethane-1,2,-diol   ester linkage

Figure 15.8 Water is released in this condensation polymerization reaction. Continued formation of ester linkages between monomers leads to the polymer plastic polyethylene terephthalate (PET). PET is a polyester that is often used to make soft-drink and water bottles.

A second type of condensation reaction that is used to form a polymer involves the formation of an amide linkage, as shown in Figure 15.9. This type of condensation reaction is important in living organisms as the basis for the formation of proteins.

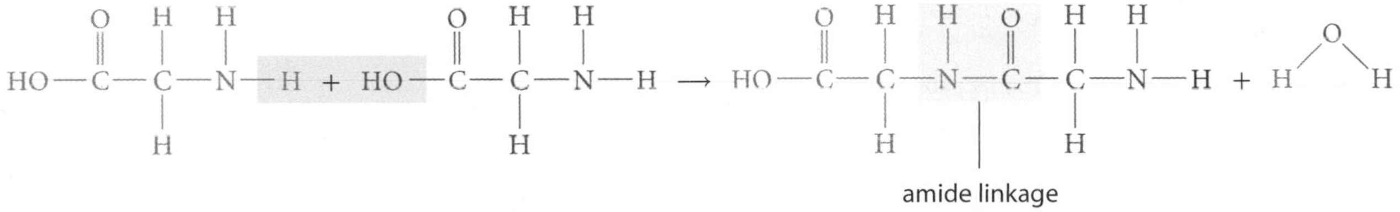

amide linkage

Figure 15.9 The monomers in this reaction have an amino group ($-NH_2$) at one end of the molecule and a carboxyl group ($-COOH$) at the other end. Therefore, they are called amino acids. The amino and carboxyl groups react as shown and form amide bonds. Water is released with the formation of the amide linkage. In proteins, the amide linkage is called a peptide bond.

To determine whether condensation polymerization has occurred, look for the formation of ester or amide linkages in the product, along with a second smaller product, usually water.

Table 15.2 gives the names, structures, and uses of two condensation polymers. Notice that Dacron™ (a trade name for polyethylene terephthalate fibres) contains ester linkages between monomers. Condensation polymers that contain ester linkages are called *polyesters*. Nylon-6 contains amide linkages between monomers. Condensation polymers that contain amide linkages are called *polyamides* or *nylons*.

Table 15.2 Examples of Condensation Polymers

Name	Structure			Uses
Dacron™ (a polyester)				• synthetic fibres used to make fabric for clothing and surgery
Nylon-6 (a polyamide)				• tires • synthetic fibres used to make rope and articles of clothing, such as stockings

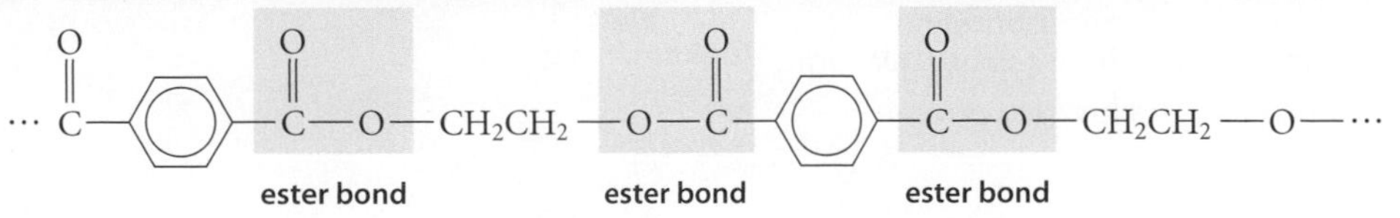

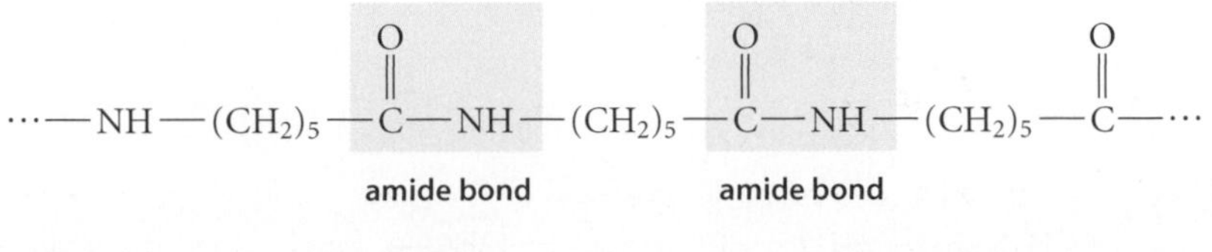

The following Sample Problem shows how to classify a polymerization reaction.

Classifying a Polymerization Reaction

Problem

Tetraflouroethene polymerizes to form the slippery polymer that is commonly known as Teflon™. Teflon™ is used as a non-stick coating in frying pans, among its other uses. Classify the following polymerization reaction to make Teflon™, and name the product. (The letter n indicates that many monomers are involved in the reaction.)

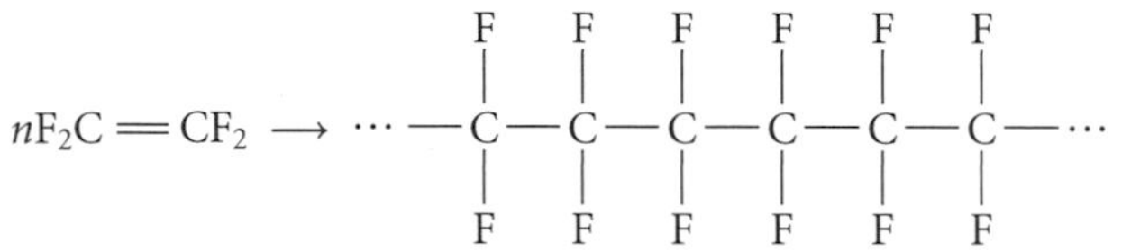

Solution

The monomer reactant of this polymerization reaction contains a double bond. The product polymer has no double bond, so an addition polymerization reaction must have occurred. Since the monomer's name is tetrafluoroethene, the polymer's name is polytetrafluoroethene.

$$n F_2C = CF_2 \longrightarrow \ \cdots -\overset{\overset{\displaystyle F}{|}}{\underset{\underset{\displaystyle F}{|}}{C}}-\overset{\overset{\displaystyle F}{|}}{\underset{\underset{\displaystyle F}{|}}{C}}-\overset{\overset{\displaystyle F}{|}}{\underset{\underset{\displaystyle F}{|}}{C}}-\overset{\overset{\displaystyle F}{|}}{\underset{\underset{\displaystyle F}{|}}{C}}-\overset{\overset{\displaystyle F}{|}}{\underset{\underset{\displaystyle F}{|}}{C}}-\overset{\overset{\displaystyle F}{|}}{\underset{\underset{\displaystyle F}{|}}{C}}- \cdots$$

Practice Problems

7. A monomer called methylmethacrylate polymerizes to form an addition polymer that is used to make bowling balls. What is the name of this polymer?

8. Classify each of the following polymerization reactions as either an addition or a condensation polymerization reaction:

a) 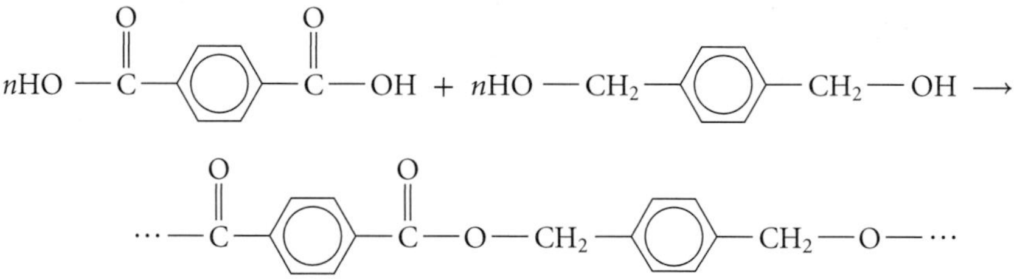 $n\text{HO}-\overset{\overset{\displaystyle O}{\|}}{C}-\bigcirc-\overset{\overset{\displaystyle O}{\|}}{C}-\text{OH} \ + \ n\text{HO}-\text{CH}_2-\bigcirc-\text{CH}_2-\text{OH} \longrightarrow$

$\cdots-\overset{\overset{\displaystyle O}{\|}}{C}-\bigcirc-\overset{\overset{\displaystyle O}{\|}}{C}-\text{O}-\text{CH}_2-\bigcirc-\text{CH}_2-\text{O}-\cdots$

b) 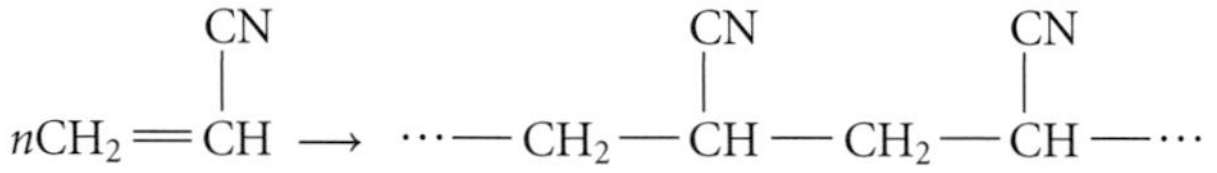 $n\text{CH}_2=\overset{\overset{\displaystyle CN}{|}}{\text{CH}} \longrightarrow \cdots-\text{CH}_2-\overset{\overset{\displaystyle CN}{|}}{\text{CH}}-\text{CH}_2-\overset{\overset{\displaystyle CN}{|}}{\text{CH}}-\cdots$

c) $n\text{HO}-\text{CH}_2-\overset{\overset{\displaystyle O}{\|}}{C}-\text{OH} \longrightarrow \cdots-\text{O}-\text{CH}_2-\overset{\overset{\displaystyle O}{\|}}{C}-\text{O}-\text{CH}_2-\overset{\overset{\displaystyle O}{\|}}{C}-\cdots$

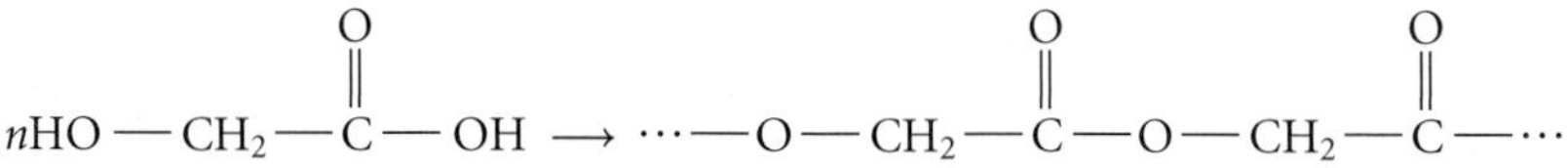

9. Draw the product of each polymerization reaction. Include at least two linkages for each product.

a) 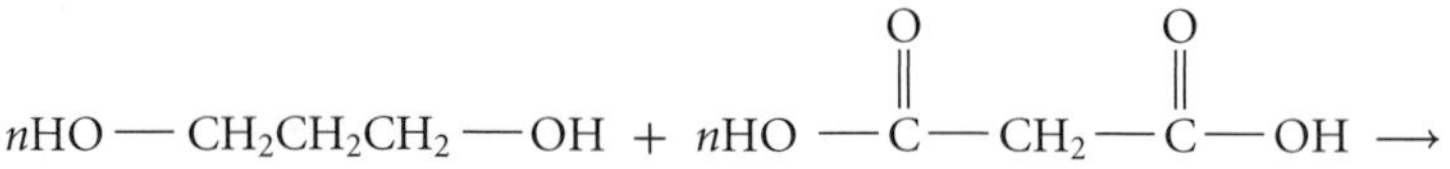 $n\text{HO}-\text{CH}_2\text{CH}_2\text{CH}_2-\text{OH} \ + \ n\text{HO}-\overset{\overset{\displaystyle O}{\|}}{C}-\text{CH}_2-\overset{\overset{\displaystyle O}{\|}}{C}-\text{OH} \longrightarrow$

b) $n\text{H}_2\text{C}=\overset{\overset{\displaystyle CH_3}{|}}{\text{CH}} \longrightarrow$

c) $n\text{H}_2\text{NCH}_2-\bigcirc-\text{CH}_2\text{NH}_2 \ + \ n\text{HO}-\overset{\overset{\displaystyle O}{\|}}{C}(\text{CH}_2)_6\overset{\overset{\displaystyle O}{\|}}{C}-\text{OH} \longrightarrow$

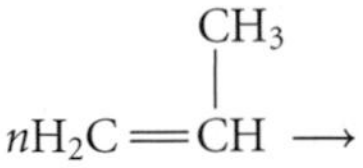

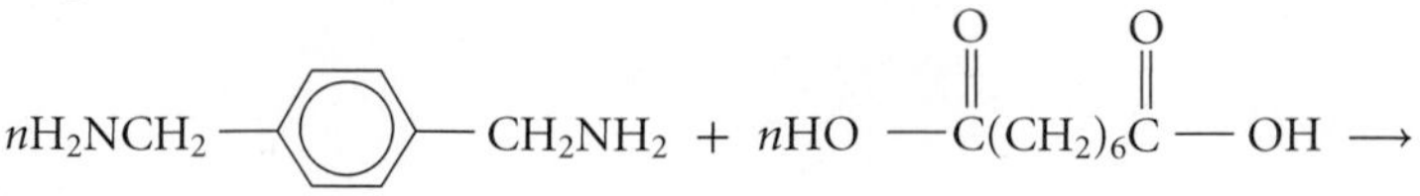

10. Classify each polymer as an addition polymer or a condensation polymer. Then classify each condensation polymer as either a polyester or a polyamide:

a) 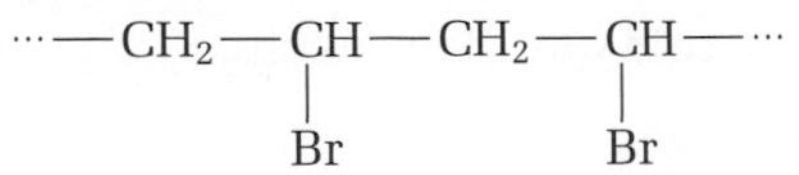 $\cdots\!-\!CH_2\!-\!CH\!-\!CH_2\!-\!CH\!-\!\cdots$
with Br under the first CH and Br under the second CH

b) 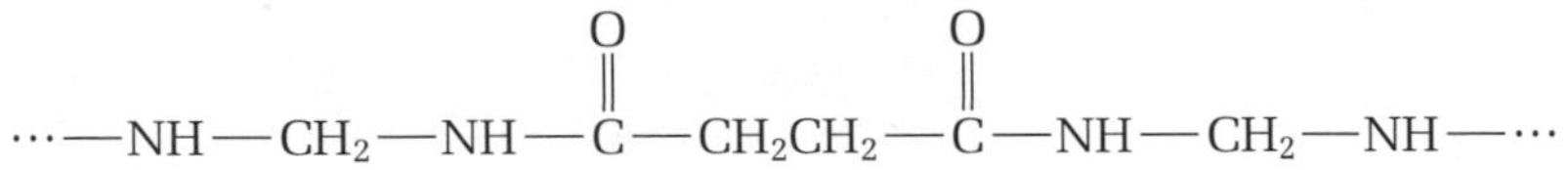 $\cdots\!-\!NH\!-\!CH_2\!-\!NH\!-\!\overset{O}{\overset{\|}{C}}\!-\!CH_2CH_2\!-\!\overset{O}{\overset{\|}{C}}\!-\!NH\!-\!CH_2\!-\!NH\!-\!\cdots$

c) $\cdots\!-\!O\!-\!CH_2CH_2\!-\!\overset{O}{\overset{\|}{C}}\!-\!O\!-\!CH_2CH_2\!-\!\overset{O}{\overset{\|}{C}}\!-\!O\!-\!CH_2CH_2\!-\!\overset{O}{\overset{\|}{C}}\!-\!\cdots$

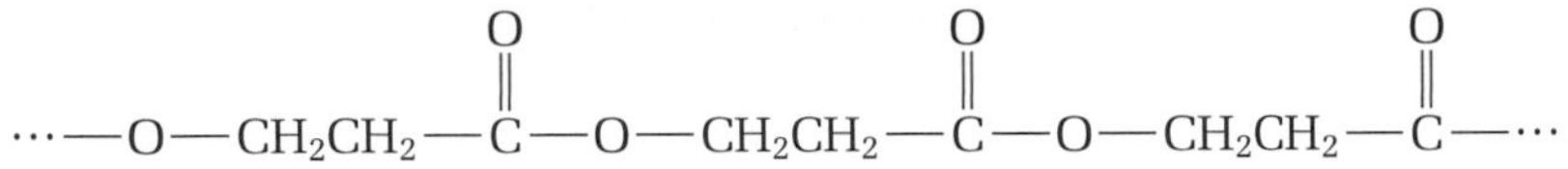

d)

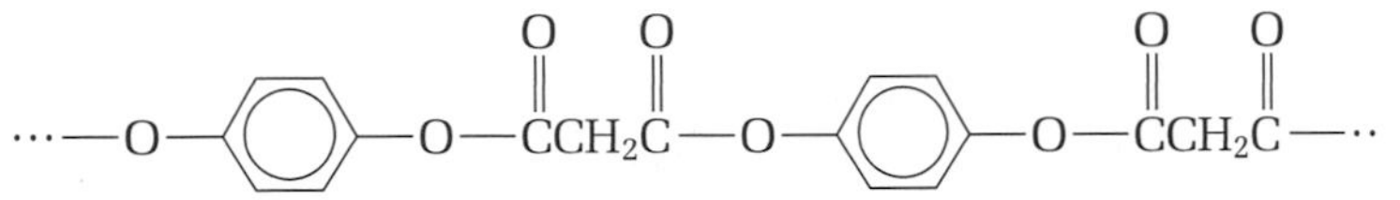

11. Draw the structure of the repeating unit for each polymer in the previous question. Then draw the structure of the monomer(s) used to prepare each polymer.

12. How could you convert 1-bromoethane into polyethene? Write an equation for each step.

Target Skills

Building models depicting the structures of organic compounds

Performing an experiment to synthesize a polymer

Analyzing a process for producing polymers

Modelling and Making Polymers

Polyvinyl alcohol (PVA) is used to make a special type of plastic bag. In the first part of this investigation, you will build a structural model of a short strand of PVA. In the second part of this investigation, you will use PVA to make a different polymer product, "Slime." You will prepare "Slime" by *cross-linking* long strands of polyvinyl alcohol using borax, sodium tetraborate decahydrate, $Na_2B_4O_7 \cdot 10H_2O(aq)$. Cross-linking means that bonds form from one strand to another at several points along the polymer strand.

Questions

- What can you learn about the properties of PVA from building a model of it? What polymerization reaction appears to be responsible for the formation of PVA? What is the structure of this polymer?

- How are the polymers PVA and "Slime" similar? How are they different? How might these compounds be used?

Prediction

Use the model of PVA you build in Part 1 to predict whether PVA forms from an addition reaction or a condensation reaction. Use your observations from Part 1 to predict whether a piece of PVA will dissolve easily in water.

Safety Precautions

- Wear an apron, safety glasses, and gloves while completing Part 2 of this investigation.
- Wash your hands thoroughly after this investigation.

Materials

Part 1

- molecular model kit (1 per group)

Part 2

- pieces of polyvinyl alcohol bags (totalling about 20 cm²)
- 10 mL of very hot water
- food colouring
- 5 mL of 4% borax solution
- kettle or hot plate
- 10 mL graduated cylinder
- 50 mL beaker
- stirring rod

Procedure

Part 1

1. Working in a group of four, use your molecular model kits to build four ethenol monomers, as illustrated below:

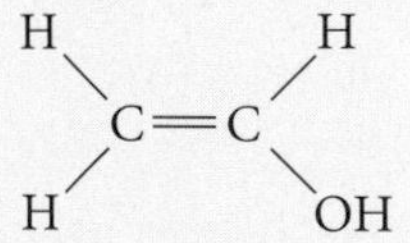

$$H_2C=CH-OH$$

ethenol
(vinyl alcohol)

2. Examine the models you have built.

   a) Decide whether they will react in an addition polymerization or a condensation polymerization reaction to form polyvinyl alcohol.

   b) Predict and draw the structure of PVA.

3. Use the four monomers to build a short strand of PVA with four repeating units.

4. Examine the polymer model.

   a) What intermolecular force(s) might operate between two strands of this polymer?

   b) Use your knowledge of the force(s) to predict whether this polymer would dissolve in water or not.

Part 2

1. Work with a partner. Before starting, examine the pieces of the polyvinyl alcohol bag. Record your observations.

2. Place 10 mL of near-boiling water into a 50 mL beaker.
   Caution Be careful to avoid burning yourself.

3. Add the pieces of the polyvinyl alcohol bag to the hot water. Stir and poke the mixture using a stirring rod until the compound has dissolved.

4. Add a few drops of food colouring to the mixture, and stir again.

5. Add 5 mL of the borax solution and stir.

6. When it has cooled so it can be comfortably handled, examine the "Slime" you have produced. Record your observations.

7. Manipulate the "Slime" sample. For example, roll it into a ball and drop it on the lab bench. Slowly pull it apart between your hands. Pull it apart quickly. Let the ball of "Slime" sit on the bench while you clean up. Record your observations.

Analysis

1. Does PVA appear to be formed by an addition reaction or a condensation reaction? Give reasons to support your answer.

2. Explain why PVA dissolves in water, even though it is a very large molecule.

3. What happened when you manipulated the "Slime" in various ways? Suggest an explanation for what you observed.

Conclusions

4. How do you think a bag made of polyvinyl alcohol might be useful (a) in a hospital? (b) as an adhesive? (c) in the cosmetics industry?

5. How could you tell that changes to the polymer occurred when you added the borax solution to the dissolved solution of polyvinyl alcohol? Explain your observations.

6. Compare the properties of the polyvinyl alcohol polymer and the "Slime" cross-linked polymer you observed in this investigation. Were there any similarities? How were they different?

Application

7. Although polyvinyl alcohol appears to form by a simple polymerization reaction of vinyl alcohol monomers, it does not. Vinyl alcohol is unstable and rearranges to form ethanal. Draw the structure of ethanal. Do library or electronic research to find out how polyvinyl alcohol is synthesized.

Extension

8. Construct a four-monomer strand of a condensation polymer of your choice. Next, take the model apart and use it to build four monomers. What additional atoms are required from the model kit to build the monomers? What does this tell you about a condensation reaction?

Organic Reactions and the Petrochemical Industry in Alberta

Industries use the properties of organic compounds to manufacture consumer products. What is the source of the organic compounds that industries use to make polymers and other products? Most of these compounds start as petroleum, a fossil fuel that contains a mixture of hydrocarbons, such as alkanes and alkenes. As you learned in Section 14.4, petroleum is separated into its various hydrocarbon components through fractional distillation. It is then processed further using cracking and reforming techniques in factories, such as the one in Figure 15.10.

Various organic reactions are used to convert the hydrocarbon components obtained from petroleum into **petrochemicals**, the organic compounds required by industry for the manufacture of plastics and other materials. For example, ethene, commonly known as ethylene, is an important petrochemical in Alberta's petrochemical industry. Ethene is produced on a large scale using the cracking process. During this process, the ethane provided from petroleum refiners is heated to 800 °C and undergoes the following catalytic cracking reaction to produce ethene:

$$C_2H_6(g) \xrightarrow{\text{Pt}} CH_2 = CH_2(g) + H_2(g)$$

As you learned in the previous section, the presence of the double bond makes ethene much more reactive than ethane. Through organic reactions, the petrochemical industry uses ethene to synthesize useful materials, such as ethylene glycol (used in antifreeze), polyethylene (used to make containers and food wrap), and polyvinyl chloride (PVC, used to make items from adhesives to auto parts).

Figure 15.10 The cracking process takes place in petrochemical plants, such as this one, located at Joffre, Alberta. Owned by NOVA Chemicals and DOW Chemical, this plant is the largest single ethane-based cracker in the world.

The Manufacture of PVC

The manufacture of PVC is a good example of the way organic reactions are used in the petrochemical industry. PVC is a type of plastic that is used in almost every industry. For example, the construction industry uses PVC for vinyl windows and doors, wall coverings, siding, PVC pipe, flooring, and fencing.

In the manufacture of PVC, ethene is first reacted with chlorine in an addition reaction to produce 1,2-dichloroethane, as follows:

$$C_2H_4(g) + Cl_2(g) \rightarrow C_2H_4Cl_2(g)$$

The 1,2-dichloroethane product is then cracked at high temperatures to produce chloroethene (vinyl chloride) and hydrogen chloride by the elimination reaction shown below:

$$C_2H_4Cl_2(g) \rightarrow C_2H_3Cl(g) + HCl(g)$$

The by-product of the cracking reaction, $HCl(g)$, is reacted further with oxygen and more ethene to produce even more 1,2-dichloroethane to be cracked:

$$2C_2H_4(g) + 4\,HCl(g) + O_2(g) \rightarrow 2C_2H_4Cl_2(g) + 2H_2O(g)$$

The vinyl chloride is then reacted in an addition polymerization reaction to produce the polymer PVC:

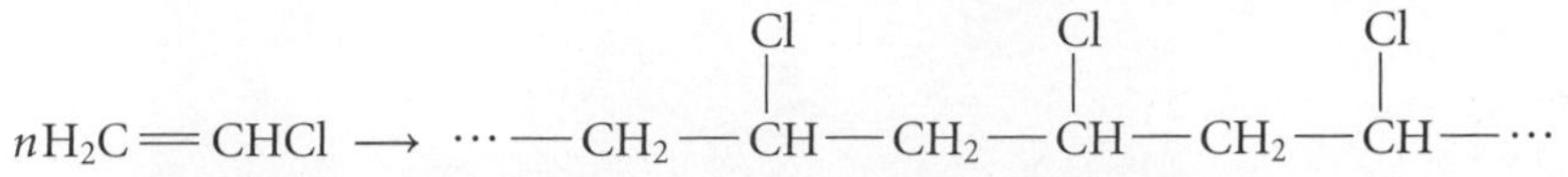

$$n\,H_2C = CHCl \rightarrow \cdots - CH_2 - \overset{\overset{\displaystyle Cl}{|}}{CH} - CH_2 - \overset{\overset{\displaystyle Cl}{|}}{CH} - CH_2 - \overset{\overset{\displaystyle Cl}{|}}{CH} - \cdots$$

chloroethene
(vinyl chloride)

polyvinyl chloride

The polymer molecules that make up most synthetic plastics are too large and complex to be degraded by biological processes. Although chemical processes will eventually degrade plastics back into carbon, hydrogen, and oxygen this may take many hundreds to thousands of years. Degradable plastics are designed in such a way as to accelerate the rate at which this happens. Biodegradable plastics are made from natural polymers, such as starch, that may either dissolve in water, or be changed chemically in water and then be broken down into smaller components by micro-organisms within the environment. Photodegradable plastics include "promoter" complexes that interact with high-frequency light, such as ultraviolet light, causing the polymer to break into smaller fragments. Two common promoter complexes are the carbonyl functional group (carbon double bonded to oxygen) and metallic salts. The small fragments of biodegradable plastics that result are then further degraded by micro-organisms.

Risks of the Polymer Industry

In the 1970s, workers at an American plastic manufacturing plant began to experience serious illnesses. Several workers died of liver cancer before the problem was traced to its source: prolonged exposure to vinyl chloride, a powerful carcinogen. Government regulations now restrict workers' exposure to vinyl chloride. Trace amounts of this dangerous chemical are still present, however, as pollution in the environment. Most vinyl chloride emissions are from gas emissions or waste water contaminants from manufacturing plants producing PVC plastics.

The manufacture and disposal of PVC creates another serious problem. Dioxins, a class of chlorinated aromatic hydrocarbons, such as the one shown in Figure 15.11, are classed as highly toxic chemicals. Dioxins are produced as an unwanted by-product during the manufacture and burning of PVC. Government regulations and voluntary industry efforts have significantly reduced the amount of dioxins being produced by the petrochemical industry. In fact, the largest human source of dioxins emissions now comes from garbage that people burn in their backyards.

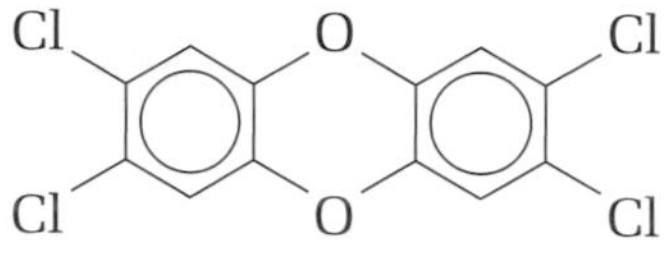

Figure 15.11 The dioxin TCDD, or 2,3,7,8-tetrachlorodibenzo-p-dioxin, has been shown to be extremely toxic in animal studies. As well, it is suspected of causing reduced fertility and birth defects in humans.

Many synthetic polymers, including most plastics, do not degrade in the environment. What, then, can be done with plastic and other polymer waste? As mentioned above, burning plastic waste releases harmful compounds, such as dioxins, into the environment. One solution may be the development and use of *degradable plastics*. These are polymers that break down over time when exposed to environmental conditions, such as light and bacteria.

You are probably already well aware of one solution to the problem of polymer waste: recycling. Plastics make up approximately 7% by mass of the garbage we generate. Because plastics are strong and resilient, they can be collected, processed, and recycled into a variety of useful products, including clothing, bags, bottles, synthetic lumber, and furniture. For example, polyethylene terephthalate (PET) from pop bottles can be recycled to produce fleece fabric. This fabric is used by many clothing companies to make jackets and sweatshirts, as shown in Figure 15.12.

Figure 15.12 To produce this fleece fabric, plastic pop bottles are cleaned and chopped into small flakes. These flakes are melted down and extruded as fibre, which is then knitted or woven into fabric.

Although it has many benefits, recycling does have significant costs for transporting, handling, sorting, cleaning, processing, and storing. Along with recycling, it is essential that society learns to *reduce* its use of polymers and to develop ways of *re-using* polymer products. Together with recycling, these directives are known as the "three Rs": reduce, re-use, and recycle.

Natural Polymers

Natural polymers are found in almost every living system. For example, the natural polymer *cellulose* provides most of the structure of plants. The monomer of cellulose is glucose, a sugar. Wood, paper, cotton, and flax are all composed of cellulose fibres. Figure 15.13 shows part of a cellulose polymer. Figure 15.14 gives a close-up look at cellulose fibres.

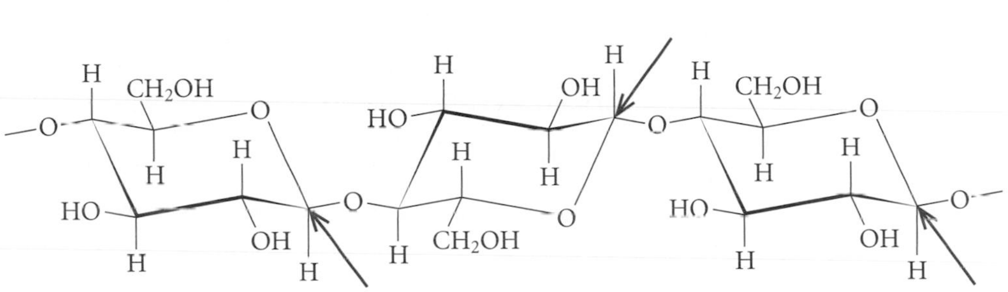

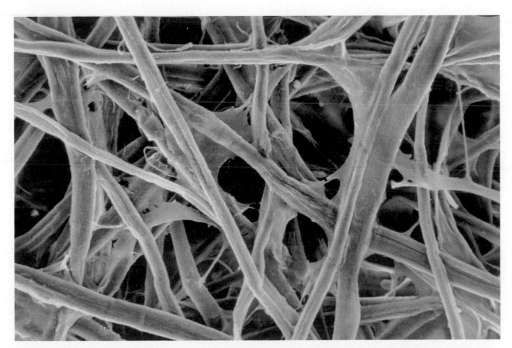

cellulose

Figure 15.13 Cellulose is the main structural fibre in plants and makes up the fibre in your diet. Red arrows point to beta linkages.

Figure 15.14 Cellulose forms fibres that can be seen through a scanning electron microscope.

Starch, the energy storage unit in plants, shown in Figure 15.15, is also a polymer comprising repeating glucose monomers. Humans can digest starch, but they cannot digest cellulose. What is the difference between these polymers? The orientation of individual glucose monomers relative to each other differs in the two polymers. In cellulose it is called a beta linkage, and in starch it is called an alpha linkage. The difference between the two linkages is difficult to visualize in a two-dimensional sketch,

Chemistry File

Web Link
Polymer recycling is becoming increasingly widespread across Canada. Use the Internet and other sources of information to find answers to the following questions:

- Which specific polymers and polymer products can be recycled?
- What products are produced from recycled polymers?
- How does polymer recycling benefit society?
- Are there facilities for collecting and recycling polymers in or near your community? If so, how can you promote these facilities? If not, how could they be set up for the benefit of your community?

www.albertachemistry.ca
WWW

but chemists have agreed on a convention for drawing the alpha and beta linkages. Examine the orientation of the bonds between the carbon atoms and oxygen atoms that are indicated by the arrows in the diagram below. In the alpha linkage, the bond from the carbon atom to the oxygen atom is drawn downward. In the beta linkage, the bond is drawn upward. Enzymes can tell the difference between the three-dimensional shapes of the bonds. Enzymes in your body will recognize and break down a starch molecule. However, animals have no enzymes that recognize the beta linkages and thus cellulose passes through your system.

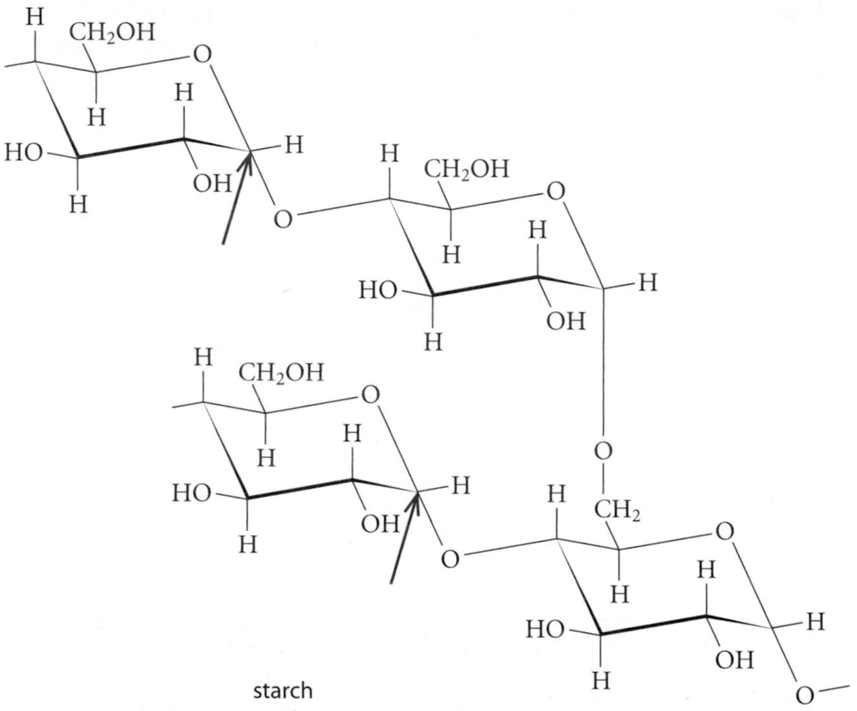

Figure 15.15 The bonds between glucose monomers in starch can be broken by enzymes in the digestive systems of animals. Arrows point to alpha linkages.

starch

Starch also differs from cellulose in that it can have branched chains. Straight-chain starch is called amylose and branched-chain starch is called amylopectin. After about every 24 to 30 glucose monomers in amylopectin, there is a branch like the one in Figure 15.15. *Glycogen*, a third glucose polymer, is the energy storage unit in animals. Glycogen has an alpha linkage, as does starch, and therefore glycogen may also be digested by humans. Glycogen differs from starch in that it has many more branches. Glycogen branches at about every 8 to 12 glucose units. This extensive branching creates many free ends that allow enzymes to act on the glycogen polymer at many points simultaneously, thus increasing the rate at which glucose can be released from storage. Polymers comprising sugar monomers (also called *saccharides*) are called *polysaccharides*.

A *protein* is a natural polymer that is composed of monomers called *amino acids*. Proteins carry out many important functions in your body, such as speeding up chemical reactions (enzymes), transporting oxygen in your blood (hemoglobin), and regulating your body responses (hormones). Figure 15.16 shows the general structure of an amino acid. Amino acids are joined with amide linkages during condensation polymerization to form proteins.

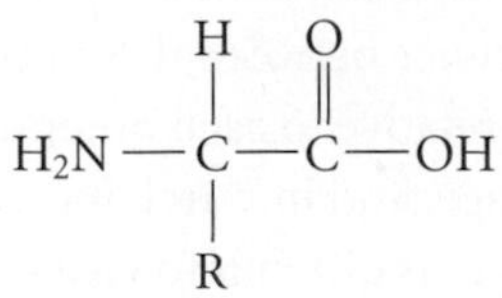

Figure 15.16 The letter R represents a side group. There are 20 common amino acids found in proteins, each with a different side group. For example, the *R* group in serine is —CH_2OH.

DNA (short for 2-deoxyribonucleic acid) is a biological molecule found in cell nuclei that codes for the amino acid sequence in all proteins of an organism and ultimately controls cellular development and function. Each strand of DNA is a polymer composed of repeating units called *nucleotides*. As seen in Figure 15.17, nucleotides have three parts: a sugar (labelled S), a phosphate group (P), and a cyclic organic molecule containing nitrogen known as a nitrogenous base. Through condensation polymerization, the sugar of one nucleotide is linked to the phosphate group of the next to form a strand of DNA.

Section 15.2 Summary

- A polymer is a long molecule consisting of repeating units called monomers.

- Most polymers are formed by addition reactions or condensation reactions.

- The monomers for addition polymerization usually contain double bonds. Each monomer adds to the double bond of the next monomer.

- In condensation polymerization, a hydrogen atom is removed from one monomer and a hydroxyl group is removed from the next. The hydrogen atom and hydroxyl group form a water molecule and a bond forms between the two monomers.

- A polyester is the result of esterification reactions between monomers.

- A polyamide forms when an amino group ($-NH_2$) is at the one end of the monomer and a carboxylic acid is on the other end. The amino group loses a hydrogen atom and the carboxylic acid loses a hydroxyl group forming water. The bond formed between the monomers is called an amide bond.

- In some cases, the monomer unit of a synthetic polymer is toxic, but after polymerization, the polymer is not toxic. Safety precautions must be taken to prevent exposure to the monomer.

- During the synthesis or disposal (by burning) of some polymers, toxic by-products are formed.

- Cellulose, starch, and glycogen are natural polymers consisting of glucose monomers.

- Proteins are natural polyamides. In a protein, however, the amide linkage is called a peptide bond.

- DNA is a natural polymer consisting of nucleotide monomers. Nucleotides consist of a sugar, a phosphate group, and an organic molecule containing cyclic groups that include nitrogen atoms (nitrogenous bases).

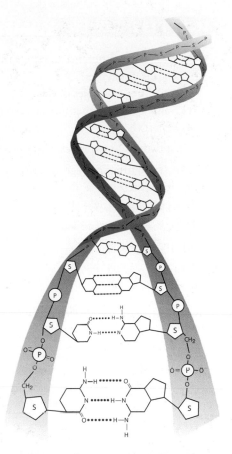

Figure 15.17 Four different bases are found in DNA. The sequence of these bases is the code that determines the sequence of amino acids in proteins. Hydrogen bonds link complementary nitrogenous bases in two strands of DNA to form a double helix.

1. a) What is the difference between a synthetic polymer and a natural polymer?
  b) Give three examples of each.
  c) Name the monomer for each type of polymer that you listed in part (b).

2. What two reaction processes are used to form synthetic polymers?

3. Which two functional groups react to form
  a) a polyester?
  b) a polyamide (nylon)?

4. Draw the product of each polymerization reaction, and classify the reaction.

  a) $n\text{H}_2\text{C}=\text{CH}_2 \longrightarrow$

  b) 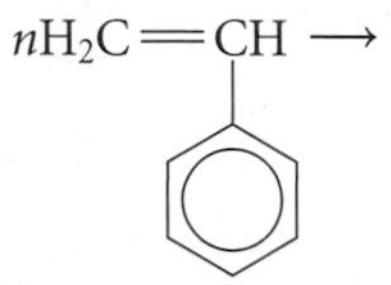 $n\text{H}_2\text{C}=\text{CH} \longrightarrow$

  c) $n\text{HO}-\overset{\overset{\text{O}}{\|}}{\text{C}}-(\text{CH}_2)_3-\overset{\overset{\text{O}}{\|}}{\text{C}}-\text{OH} \ + \ n\text{H}_2\text{N}-\underset{\underset{\text{CH}_3}{|}}{\text{CH}}-\text{NH}_2 \longrightarrow$

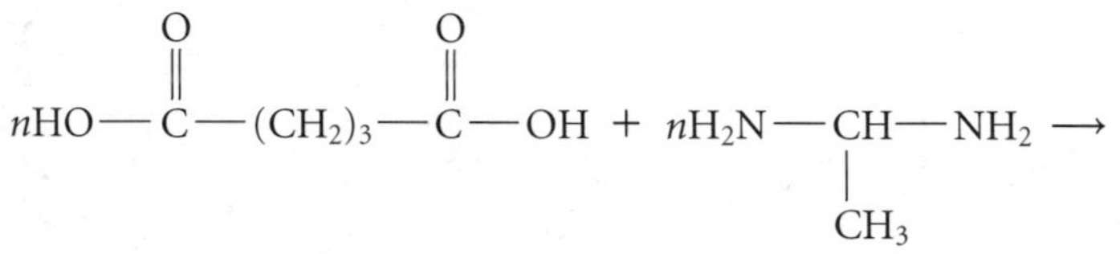

  d) 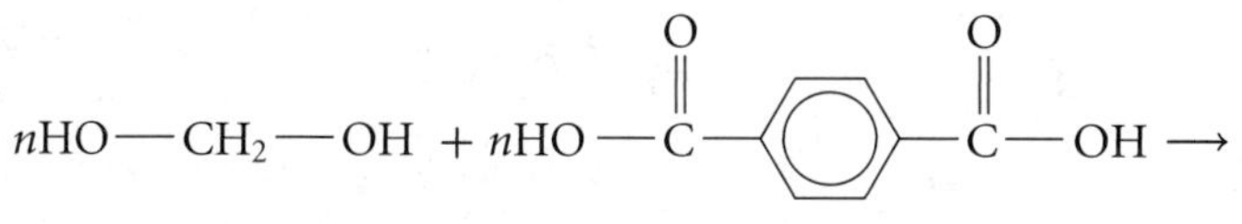 $n\text{HO}-\text{CH}_2-\text{OH} \ + \ n\text{HO}-\overset{\overset{\text{O}}{\|}}{\text{C}}-\!\!\bigcirc\!\!-\overset{\overset{\text{O}}{\|}}{\text{C}}-\text{OH} \longrightarrow$

5. Draw the monomers in each polymerization reaction:

  a) 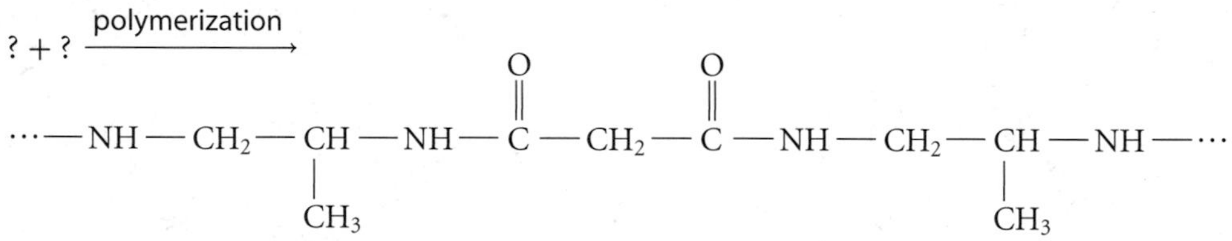 $? + ? \xrightarrow{\text{polymerization}}$

  $\cdots-\text{NH}-\text{CH}_2-\underset{\underset{\text{CH}_3}{|}}{\text{CH}}-\text{NH}-\overset{\overset{\text{O}}{\|}}{\text{C}}-\text{CH}_2-\overset{\overset{\text{O}}{\|}}{\text{C}}-\text{NH}-\text{CH}_2-\underset{\underset{\text{CH}_3}{|}}{\text{CH}}-\text{NH}-\cdots$

  b) 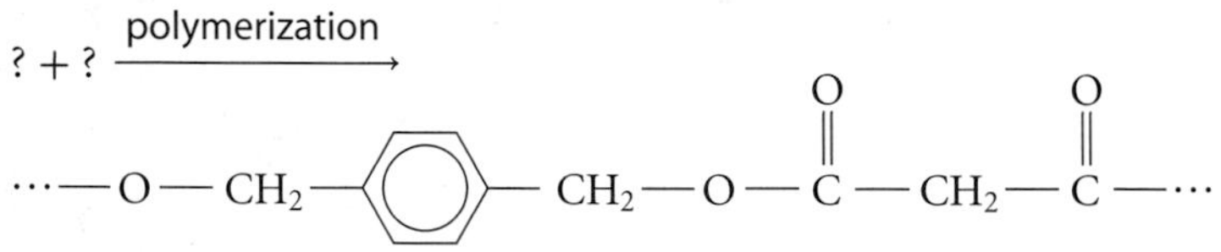 $? + ? \xrightarrow{\text{polymerization}}$

  $\cdots-\text{O}-\text{CH}_2-\!\!\bigcirc\!\!-\text{CH}_2-\text{O}-\overset{\overset{\text{O}}{\|}}{\text{C}}-\text{CH}_2-\overset{\overset{\text{O}}{\|}}{\text{C}}-\cdots$

6. a) What problems are caused by society's use of synthetic polymers?
  b) What benefits do we obtain from polymers?
  c) In your opinion, do the benefits outweigh the risks?
  d) Write a short paragraph to answer these questions.

7. How does the petrochemical industry use organic reactions to produce important compounds from fossil fuels? Include specific examples of reactions.

Combustion of hydrocarbons is used to generate electrical energy, energy for transportation, cooking, and many other purposes. Chemists commonly use reactions such as addition, elimination, and substitution to convert organic compounds derived from petrochemicals to other useful compounds. Alkenes can be converted into alcohols, alkanes, and alkyl halides by using addition reactions. By using elimination reactions, you can convert alcohols or alkyl halides into alkenes. You can control the reaction by using the correct catalysts and conditions.

An acid containing a halogen will react with an alcohol to yield an alkyl halide. You can almost reverse the reaction by treating an alkyl halide with a base to get an alcohol. With the input of energy in the form of ultraviolet light, you can stimulate a substitution reaction between alkanes and halogens. The products will be a mixture of alkyl halides. You can even substitute halogens for the hydrogen atoms on benzene by using a catalyst.

By mixing carboxylic acids with alcohols in the presence of an acid you can produce esters. Natural esters are synthesized by many fruits and they give the fruits their unique aromas. Esterification reactions are a type of condensation reaction. Condensation reactions, including esterification reactions, are important reactions in the formation of polymers. For example, polyesters such as Dacron™, are formed by reacting molecules with an alcohol group on both ends with other molecules that have a carboxyl group on both ends. Another common type of polymer that is formed by using a condensation reaction is a polyamide. One type of monomer that is used to make polyamides has an amino group ($-NH_2$) on one end and a carboxyl group on the other end of the molecule. Nylon is a very common polyamide. Nylon is used in tires and as fibres used to make clothing, stockings, and rope.

Many common polymers are synthesized by using addition reactions. Most of these addition reactions involve the addition to double bonds. Polystyrene and polyvinyl chloride are two very common addition polymers. Many plastics are the product of addition polymerization. Nearly all of these synthetic polymers are derived from petrochemicals.

During the synthesis and disposal of plastics made of petrochemicals, some toxic compounds are formed. The polymers that have provided society with so many convenient products have become a threat to the environment and to the health of the population. It is now important to reduce, re-use, and recycle the plastics that we use in so many ways.

Three of the major classes of biomolecules in your body and in all living organisms are condensation polymers. Carbohydrates such as starch, glycogen, and cellulose are polymers of glucose. Proteins are natural polyamides. DNA is a polymer of monomer units called nucleotides.

Concept Organizer | Organic Reactions

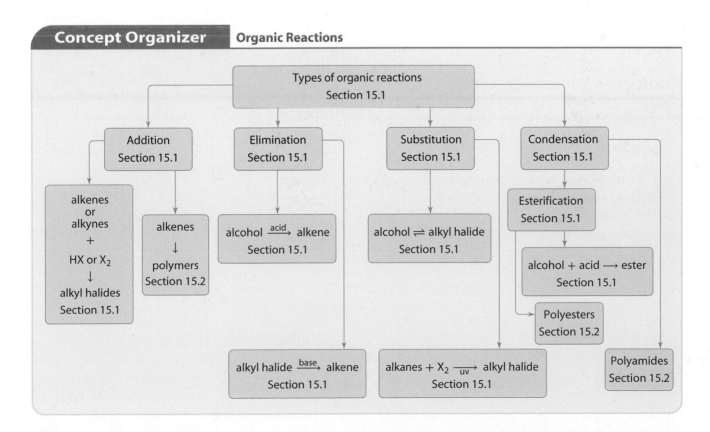

Understanding Concepts

1. Describe each type of organic reaction and provide a reaction equation as an example:
   a) substitution
   b) addition
   c) elimination
   d) esterification
   e) combustion

2. Use your own words to define the following terms:
   a) polymer
   b) monomer
   c) plastic

3. Identify each reaction below as a combustion, substitution, addition, elimination, or esterification reaction:
   a) $CH_3CH_2CH_2OH \xrightarrow{acid} CH_3CH=CH_2 + H_2O$
   b) $(CH_3)_2CHCH=CHCH_3 + 9O_2 \rightarrow 6CO_2 + 6H_2O$
   c) $CH_3COOH + CH_3OH \xrightarrow{acid} CH_3COOCH_3 + H_2O$
   d) $CH_3CH(CH_3)CH=CH_2 + HBr \rightarrow$
   $CH_3CH(CH_3)CH(Br)CH_3$
   e) $CH_3CH_2CH_2CH_3 + Cl_2 \xrightarrow{uv}$
   $CH_3CHClCH_2CH_3 + HCl$

4. Why does "polyethene" end with the suffix -ene if it does not contain any double bonds?

5. Draw and name the product(s) of each reaction that is started below. (**Hint:** Do not forget to include such products as H_2O, HBr, and isomers.)
   a) $CH_3CH=CHCH_3 + Br_2 \rightarrow$
   b) $HO-CH_2CH_2CH_2CH_2CH_3 + HBr \rightarrow$
   c) $CH_3CH_2\overset{\overset{\displaystyle O}{\|}}{C}OH + HOCH_3 \xrightarrow{H_2SO_4}$
   d) $HO-CH_2CH_2CH_3 \xrightarrow[\Delta]{H_2SO_4}$

6. Draw the product(s) of each reaction started below. (**Hint:** Do not forget to include such products as H_2O, HBr, and isomers.)
   a) $CH_3CH_2-\overset{\overset{\displaystyle O}{\|}}{C}-OH + CH_3CH_2CH_2OH \xrightarrow{acid}$
   b) $CH_3CH_2C\equiv CH + Cl_2 \rightarrow (i) + Cl_2 \rightarrow (ii)$
   c) 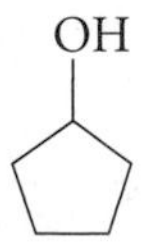 $\xrightarrow[\Delta]{H_2SO_4}$
   d) 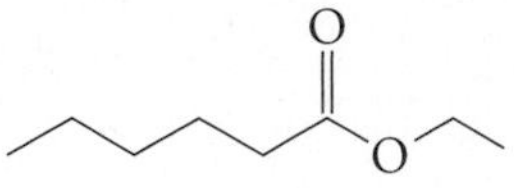 $+ H_2O \xrightarrow[\Delta]{H_2SO_4}$

7. Draw and name the reactant(s) in each reaction:
   a) $? + ? \rightarrow CH_3CH_2\overset{\overset{\displaystyle O}{\|}}{C}OCH_2CH_2CH_2CH_3 + HOH$
   b) $? \xrightarrow[\Delta]{H_2SO_4} \square + H_2O$

Applying Concepts

8. What is the difference between a polymer and any other large molecule?

9. Why are some natural molecules, such as proteins and starches, classified as polymers?

10. Describe each type of polymerization, and give an example with a chemical equation:
    a) addition polymerization
    b) condensation polymerization

11. A short section of the polymer *polypropene* is shown below:
    $$\cdots - CH_2 - CH(CH_3) - CH_2 - CH(CH_3) - \cdots$$
    a) Draw and name the monomer that is used to make this polymer.
    b) What type of polymer is polypropene?
    c) What is the common name of this polymer?

12. What is the difference between a nylon and a polyester? How are they similar?

13. Is a protein an example of a polyester or a polyamide? Explain your answer.

14. Compare and contrast cellulose and starch. Include the important functions they serve.

15. What is the difference between a protein and an amino acid?

16. Draw the product(s) of each polymerization reaction that is started below. (**Hint:** Don't forget to include water as a secondary product where necessary.)
    a) $nHO-\overset{\overset{\displaystyle O}{\|}}{C}-(CH_2)_4-OH \rightarrow$
    b) $nCH_2=CH \rightarrow$
    with $\overset{|}{CH}$ bearing H_3C and CH_3
    c) $nHO(CH_2)_7OH + nHO\overset{\overset{\displaystyle O}{\|}}{C}CH_2\overset{\overset{\displaystyle O}{\|}}{C}OH \rightarrow$

17. Draw the monomer(s) in each polymerization reaction:

a) ? ⟶ ⋯— CH_2 — CH_2 — CH_2 — CH_2 —⋯

b) ? ⟶

$$⋯-\overset{\overset{\displaystyle O}{\|}}{C}-CH_2-\underset{\underset{\displaystyle CH_3}{|}}{CH}-O-\overset{\overset{\displaystyle O}{\|}}{C}-CH_2-\underset{\underset{\displaystyle CH_3}{|}}{CH}-O-⋯$$

c) ? ⟶ ⋯— CH_2 — $\underset{\underset{\displaystyle CH_2CH_3}{|}}{CH}$ — CH_2 — $\underset{\underset{\displaystyle CH_2CH_3}{|}}{CH}$ —⋯

d) ? + ? ⟶

$$⋯-NH-CH_2-\underset{\underset{\displaystyle CH_3}{|}}{CH}-NH-\overset{\overset{\displaystyle O}{\|}}{C}-CH_2-$$

$$\overset{\overset{\displaystyle O}{\|}}{C}-NH-CH_2-\underset{\underset{\displaystyle CH_3}{|}}{CH}-NH-⋯$$

e) ? + ? ⟶

$$⋯-O-CH_2-\bigcirc\!\!\!\!\bigcirc-CH_2-O-\overset{\overset{\displaystyle O}{\|}}{C}-CH_2-\overset{\overset{\displaystyle O}{\|}}{C}-⋯$$

Making Connections

18. Use structural formulas and equations to explain why the addition of hydrogen chloride to but-1-ene may yield two products, but the addition of hydrogen chloride to but-2-ene yields only one product.

19. Bromine water (Br_2(aq)) can be used to test for double bonds.
   a) Design a procedure using bromine water to test fats and oils for the presence of double bonds.
   b) Look up the WHMIS guidelines for bromine. What safety precautions will you include in your procedure?
   c) When bromine water is added to but-2-ene, the colour of the bromine solution changes. What type of reaction has occurred?
   d) Draw the structure for the product of the reaction in part (c).

20. Use a flowchart to describe how PVC is produced from fossil fuel and any other necessary reactant.

21. How has organic chemistry helped to solve the problems caused by the use of the following substances?
   a) leaded gasoline
   b) used plastic, such as PET

22. In this chapter, you carried out research using the Internet and other sources to find out about organic compounds. Not all sources contain accurate information, and many sources have a bias toward one side of an issue. How did you attempt to distinguish between reliable sources and unreliable sources? **ICT**

23. What are some of the properties of plastics that make them so useful? Use your own observation of plastic products to answer this question.

24. Reducing our consumption of polymer products and re-using polymer products help to minimize polymer waste. Come up with suggestions for your school or community on ways to *reduce* and *re-use* polymer products.

25. Biofuel for cars and other engines can be made from biological materials, such as plants and animal fats. Biodiesel and ethanol are the two most common biological fuels.
   a) What factors do you think might be holding back the sale of biofuels on the Canadian market?
   b) What benefits do you think society could obtain from using biofuel to help replace hydrocarbon-based fuel?
   c) What negative consequences might the use of biofuels have on the Canadian economy?
   d) What is your opinion on the development and use of biofuels?

26. Do you think Canada should cut back on the manufacture and use of synthetic polymers? Write a brief paragraph that explains your point of view. Include at least three good reasons to back up your opinion.

27. Muscle-building athletes sometimes drink beverages that contain amino acids. Why might they think that this type of drink could help to build muscles?

28. Use a molecular model kit to build an example of a glucose monomer. Join your monomer to your classmates' monomers to build a short strand of a starch or cellulose molecule.

Career Focus: Ask a Chemical Safety Expert

When Dr. Margaret-Ann Armour joined the safety committee for the Department of Chemistry at the University of Alberta, she couldn't have predicted the results. Armour, an organic chemist, and her colleagues wrote a guidebook for high schools based on their findings. Since then, Armour has written other safety manuals, and continues to advise scientists about chemical safety. In 2002, she received a Governor General's Award in Commemoration of the Persons Case.

Margaret-Ann Armour

Q What made you interested in studying hazardous chemicals?

A When I was on the safety committee in the chemistry department, students used to ask, "What do I do with this left-over chemical?" It made us realize that small quantities of hazardous waste are very difficult to dispose of. When we couldn't find specific procedures for changing the waste into nontoxic, non-environmentally hazardous materials, we decided to get some funding and test some procedures.

Q What makes organic compounds especially challenging to deal with?

A Many chlorinated organic compounds tend to remain in the environment. So, in some ways, they are more hazardous than inorganic substances that are highly toxic. We are not absolutely sure what the long-term exposure to some of the chlorinated organics may do to health.

Q What research are you doing on pesticides?

A We've been collaborating with our colleagues in human ecology to study the effects on the clothing of people who spray pesticides. They're usually wearing caps, and the caps don't get washed very often. And indeed these people sometimes get a rash on the forehead because the cap has absorbed some of the pesticide sprayed and is irritating the skin. We wanted to know if you actually get all of the pesticide out when you wash protective clothing. In some cases, you don't, and people can get a dermal irritation as a result.

Q Does using pesticides cause cancer?

A One can never say that a chemical is not a cancer risk because it is so difficult to identify what people have been exposed to. The development of cancer usually involves exposure over a long period of time. People should reduce their exposure to pesticides as much as possible.

Q Should we be using pesticides in our communities and on farms?

A Pesticides have increased the production of cereal crops, fruits, and vegetables and thus helped to provide food for the increasing population. However, we should be very cautious that we don't *over* use them. They are a health hazard especially to the people who apply them.

People are recognizing that there are certain pesticides that we should *not* be using, because they stay in the soil for a very long time. These pesticides, often chlorinated organics, don't break down.

Q Are the pesticides commonly used on lawns chlorinated organics?

A They are: 2,4-D is a chlorinated organic, but it's quite different in the sense that it breaks down in about seven days in the soil, and so it's not considered the hazard that some of the other chlorinated organics are, which don't break down as readily. Also, 2,4-D breaks down quite readily in the human system, so it's not one that tends to build up in the body as much. However, when you're using 2,4-D, be careful not to breathe the fumes or get it on yourself. Also keep the dog off the lawn for a day or two.

Q Why is it so important to you to encourage youth to pursue their interest in science?

A One very pragmatic reason is that so many careers depend on some kind of technological background. In the future, we're going to be short of people for high-tech jobs. The other reason is that so many of the decisions that we make on a day-to-day basis are better made if we have some scientific knowledge and background.

I'm particularly interested in getting young women interested in science because I believe that the kinds of questions that we women are socialized to ask are just a little different than the ones that young men are socialized to ask. Take Alice Hamilton, whom we think of as the mother of environmental science and occupational health and safety. She was the one who realized at the turn of the century that exposure to environmental contaminants were making people sick and not poor hygiene practices, as Chicago's city fathers thought.

Other Careers Related to Environmental Health and Safety

Environmental Occupational Health and Safety Specialist Experts with good understanding of environmental regulations help ensure safety at the workplace. Specialists in this field monitor and raise awareness about environmental hazards in the workplace and help employers develop policies to make the workplace safer. Junior specialists require a Bachelor's degree in an appropriate science or technical diploma plus work experience. Related occupations include an industrial hygienist, as well as a hazardous materials specialists, who recommend procedures and inspect worksites to ensure safe handling of hazardous materials.

Environmental Technologist Junior technologists conduct surveys and collect data about working environments. More experienced technologists review inspection results, follow-up on violations of environmental regulations, and develop research and training programs. Entry-level technologists require either a specialized Bachelor of Science degree, or an accredited two-year technical diploma in environmental technology, engineering, or applied science technology.

ISO 14000 Consultant ISO 14000 (International Organization for Standardization) is a set of standards for an *environmental management system*, a system that a company or institution can use to minimize its negative impact on the environment. ISO 14000 consultants help companies or institutions to comply with ISO 14000 regulations. The job involves working with people, office work, and site visits. Consultants need a university degree or college diploma in law, occupational health and safety, environmental management, or other field related to environmental science.

Environmental Epidemiologist Health specialists examine statistics in order to make connections between environmental hazards and illnesses. Environmental epidemiologists may also diagnose health conditions, and suggest ways of dealing with illness caused by environmental exposure, such as exposure to a hazardous chemical. A graduate degree in a health-related science is required to work in this field.

Toxicologist Scientists detect and study toxic substances, and treat the conditions they produce. Day-to-day activities can include collecting environmental samples, analyzing them for toxins, conducting laboratory research, and suggesting safety procedures for workers. Most toxicologists have graduate degrees. Some specialize in occupational toxicology.

Remediation Specialist College or university graduates of science and engineering design and implement ways to restore polluted soil and water. Bioremediation experts, in particular, specialize in using bacteria or plants to break down or bind pollutants in soil or water. Junior specialists conduct field and laboratory work, while senior specialists tend to manage and supervise projects.

Go Further...

1. How is it helpful to have a basic knowledge of chemistry and biology in order to make a decision about whether or not, or how to use a particular pesticide?

2. Suppose scientists were to develop a novel organic compound for use in fire-resistant plastic housings for computer and television components. Should the compound be tested to see if it is carcinogenic before manufacturers starting using it? What are some possible benefits and drawbacks in allowing the novel compound to be used right away?

3. What kinds of questions might an occupational health and safety specialist ask about a work environment where unusually high numbers of employees reported feeling unwell?

www.albertachemistry.ca
WWW

Understanding Concepts

1. List and discuss the properties of the carbon atom that make it uniquely suited for forming the basis of organic compounds.

2. State the historic definition of organic compounds. Include the concept of a "vital force" in your definition.

3. What was the basis for abandoning the concept of the "vital force?"

4. State the modern definition of organic compounds.

5. Examine the structural formula shown here. Draw the following formulas for this compound.

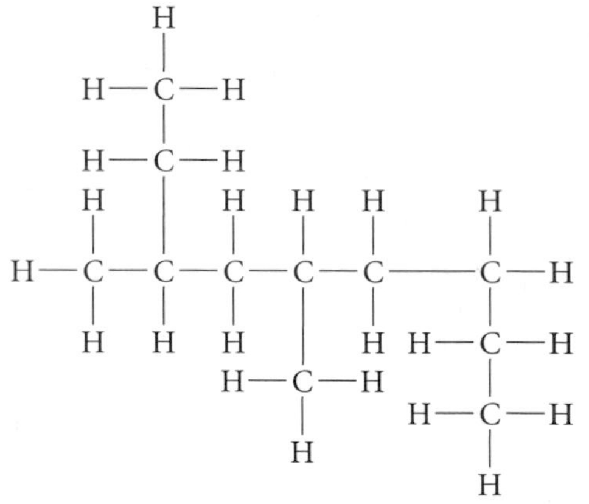

a) condensed structural formula
b) line structural formula
c) empirical molecular formula
d) expanded molecular formula

6. Write the general name of the class of compounds that follow and then write the IUPAC name for each compound.

   a) 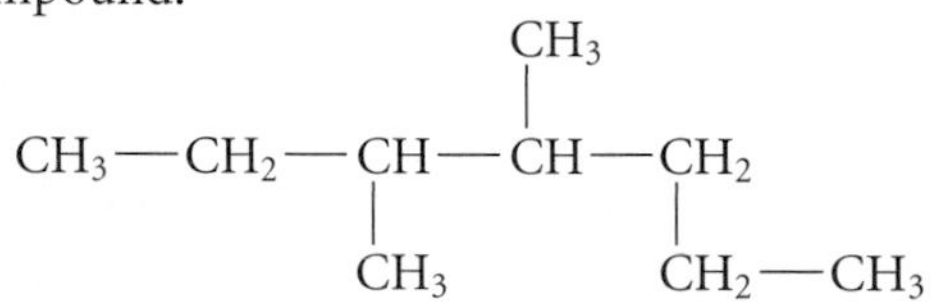 CH_3—CH_2—CH—CH—CH_2
   (with CH_3 below CH, and CH_2—CH_3 below the last)

   b)

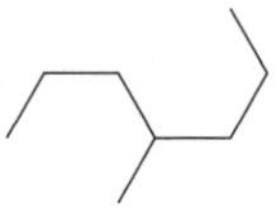

   c)

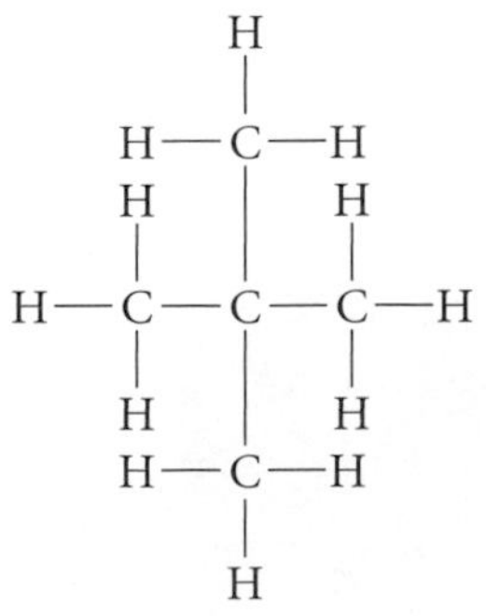

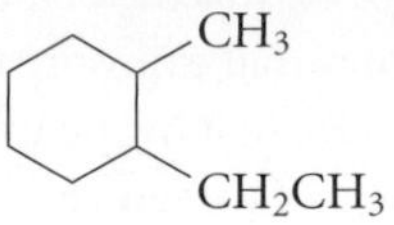

7. Write the IUPAC name for each compound.

   a) 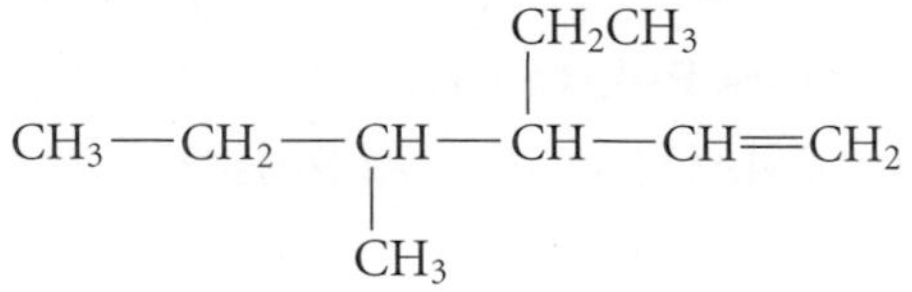 CH_3—CH_2—CH—CH—CH=CH_2
   (with CH_2CH_3 above and CH_3 below)

   b)

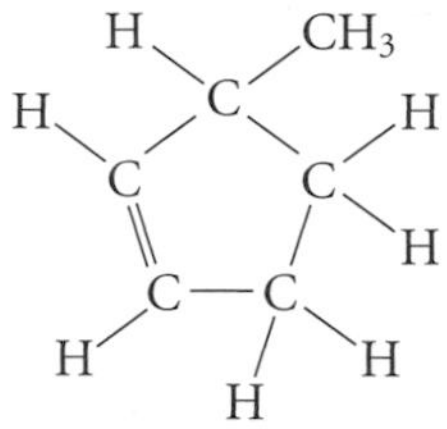

   c)

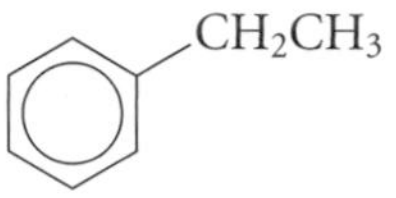

   d) 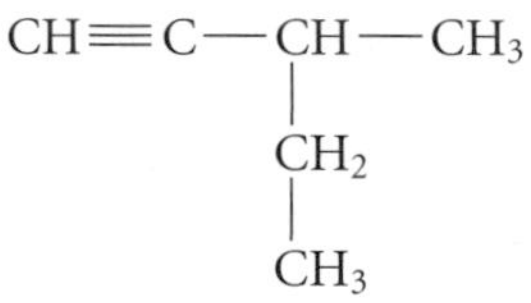 CH≡C—CH—CH_3
   (with CH_2 and CH_3 below)

8. Classify each of the compounds in question 7.

9. Explain why structure A is a better representation of benzene that structure B.

   a)

   b)

10. Draw condensed structural formulas for each of the following compounds.
    a) 2,4-dimethylpentane
    b) 3-ethyloct-1-ene
    c) 3-ethyl-2-methylhexane
    d) 2,3-dimethylcycloheptene
    e) 2,4-dimethylcyclohexene
    f) 3-ethylpent-1-yne
    g) propylbenzene
    h) non-3-yne

11. Write the IUPAC name of each of the following compounds and name the functional group on each compound.

a)

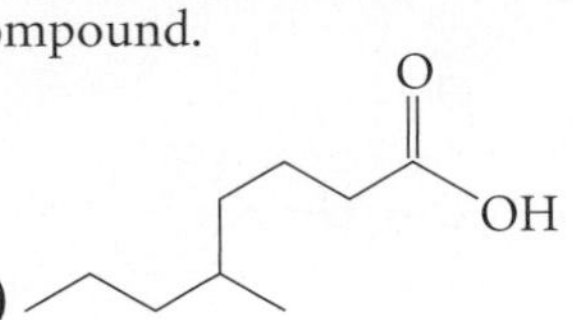

b)

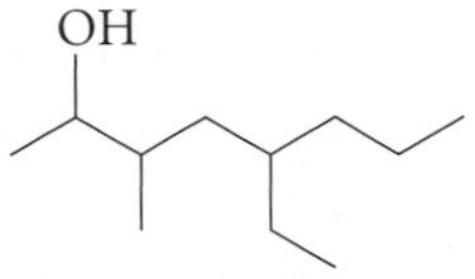

c) $CH_3CH_2CHClCH_2CH_3$

d) 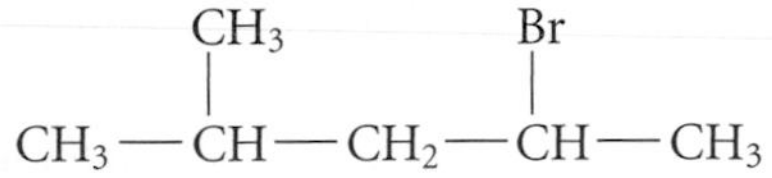 $CH_3-\overset{\overset{\displaystyle CH_3}{|}}{CH}-CH_2-\overset{\overset{\displaystyle Br}{|}}{CH}-CH_3$

e) $CH_3CHOHCH_2CH_3$

f) 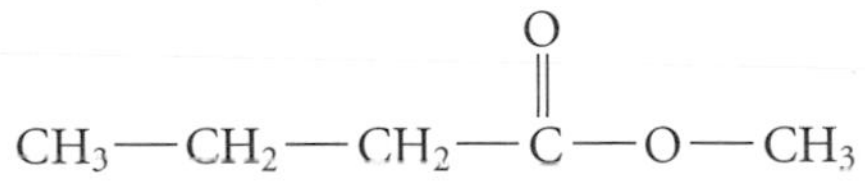 $CH_3-CH_2-CH_2-\overset{\overset{\displaystyle O}{||}}{C}-O-CH_3$

g) 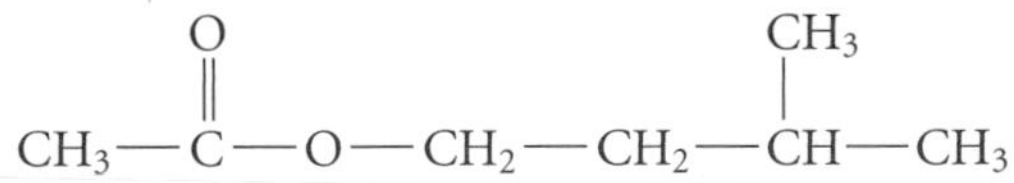 $CH_3-\overset{\overset{\displaystyle O}{||}}{C}-O-CH_2-CH_2-\overset{\overset{\displaystyle CH_3}{|}}{CH}-CH_3$

h) 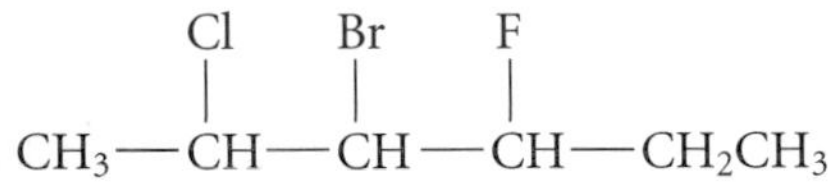 $CH_3-\overset{\overset{\displaystyle Cl}{|}}{CH}-\overset{\overset{\displaystyle Br}{|}}{CH}-\overset{\overset{\displaystyle F}{|}}{CH}-CH_2CH_3$

12. Draw condensed structural formulas for each of the following compounds.

a) octan-2-ol
b) 1,2-dichlorobutane
c) propyl propanoate
d) 2-methylpropanoic acid
e) 2,2,4,4-tetrachlorohexane
f) 3,4-dimethylnonan-2-ol

13. Draw condensed structural formulas of three isomers that have the empirical formula, C_7H_{16}. Name each isomer.

14. Briefly explain the meaning of fractional distillation as it applies to the petroleum industry.

15. Complete the following chemical equations by giving all possible products.

a) $CH_3CH_2CH=CHCH_3 + I_2 \rightarrow$
b) $CH_3CH_2CH_2OH + HCl \rightarrow$
c) propan-1-ol + methanoic acid $\xrightarrow{acid}$
d) heptan-1-ol $\xrightarrow{H_2SO_4}$
e) $CH_2=CHCH_2CH_2OH + 2HCl \rightarrow$
f) $CH_3CH(CH_3)CH_2Br + NaOCH_2C_4H \rightarrow$

16. Draw and name the missing reactant(s) in each of the following reactions.

a) ? + HBr → $CH_2=CHBr$
b) ? + ? → butyl heptanoate
c) ? + ? → $CH_3CH_2CH_3$

17. Complete the following equations by drawing the products. **Hint:** Do not forget to include the second product, such as H_2O or HBr, for a substitution reaction.

a) 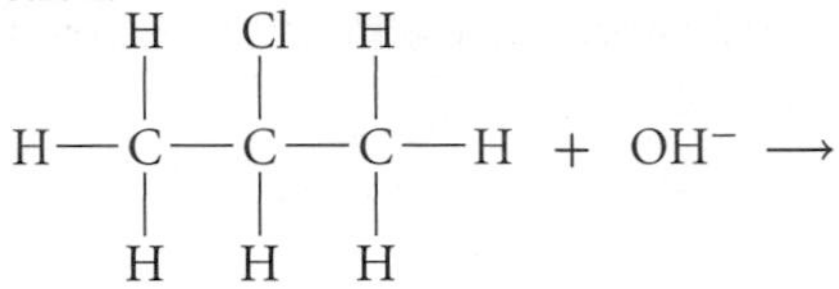 $H-\overset{\overset{\displaystyle H}{|}}{\underset{\underset{\displaystyle H}{|}}{C}}-\overset{\overset{\displaystyle Cl}{|}}{\underset{\underset{\displaystyle H}{|}}{C}}-\overset{\overset{\displaystyle H}{|}}{\underset{\underset{\displaystyle H}{|}}{C}}-H + OH^- \rightarrow$

b) OH + HO $\xrightarrow{acid}$

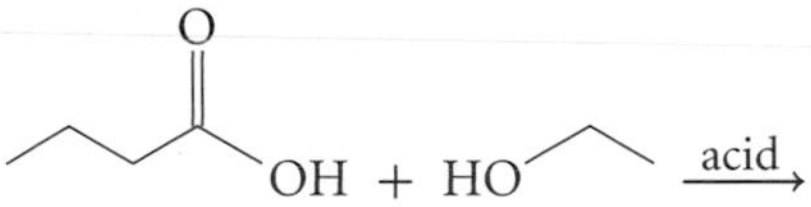

c) $\xrightarrow[\Delta]{H_2SO_4}$

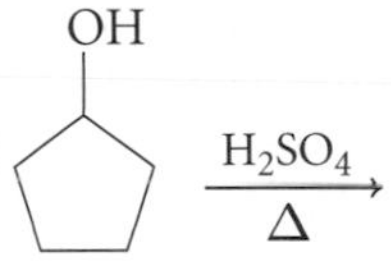

d) $HOCH_2CH_2CH_2CH_2CH_3 + HBr \rightarrow$

18. Draw and name the monomers that were needed to form the following polymers.

a)

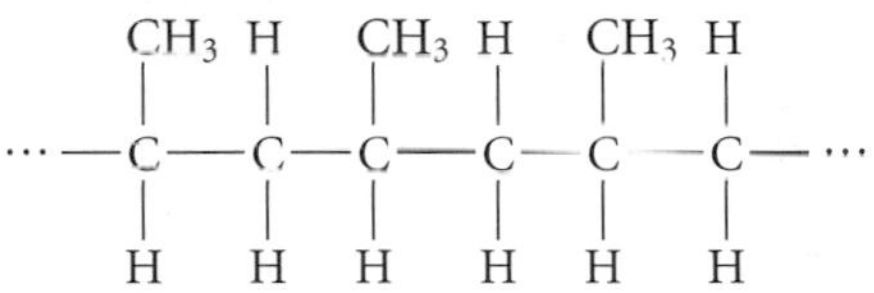

b)

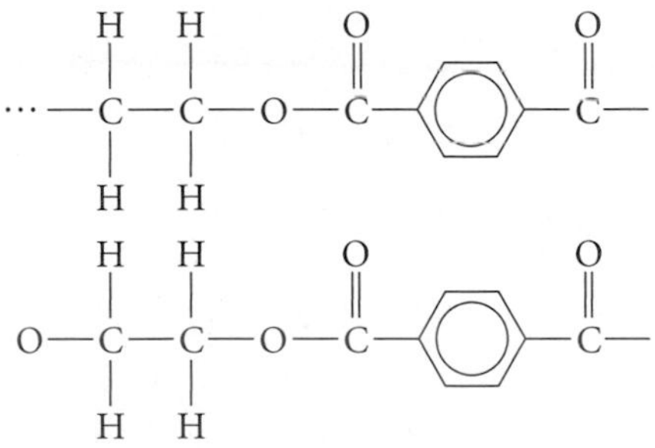

19. For the polymers in question 17, state the type of polymerization that occurred.

20. Identify each equation as representing an addition or a condensation polymerization.

a)

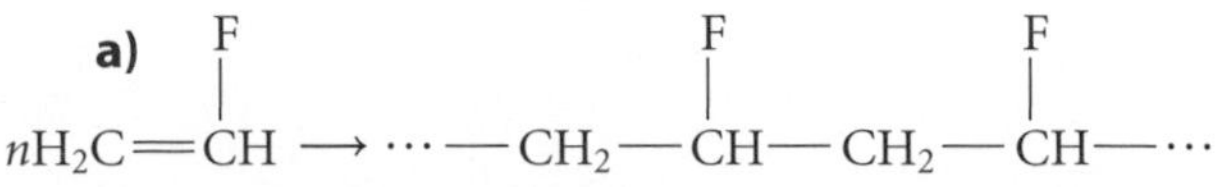

$nH_2C=CH \rightarrow \cdots-CH_2-CH-CH_2-CH-\cdots$

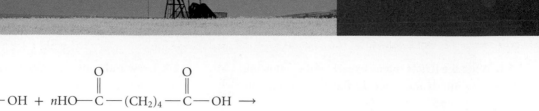

b) 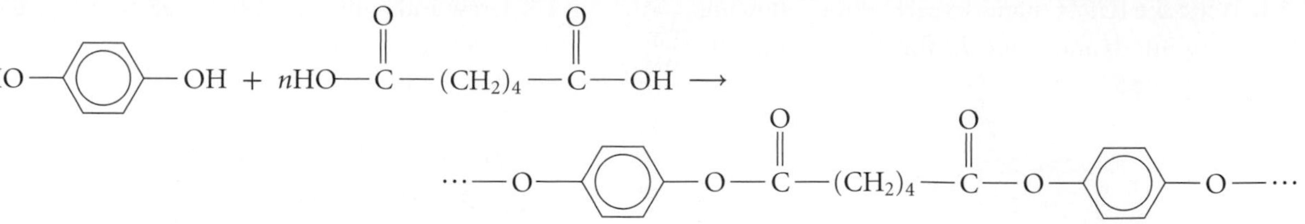 nHO—⬡—OH + nHO—C(=O)—$(CH_2)_4$—C(=O)—OH →

$\cdots$—O—⬡—O—C(=O)—$(CH_2)_4$—C(=O)—O—⬡—O—$\cdots$

Applying Concepts

21. A non-cyclic alkane has 5 carbon atoms. How many hydrogen atoms does it have?

22. A non-cyclic alkyne has 12 hydrogen atoms. How many carbon atoms does it have?

23. You are given two unlabelled beakers. One contains pentan-1-ol and the other contains pentanoic acid. Describe the materials you would need in order to experimentally distinguish between the two liquids.

24. Each of the following names contain an error. Draw the structures of the compounds as they are written. Then correctly rename the compound you have drawn.
 a) 2-propyl-4-pentane
 b) 2-ethylheptan-6-ol
 c) 2-propylpropanoic acid

25. Within the following compounds, there are two pairs of structural isomers. Identify these pairs. Define structural isomer.

 a) 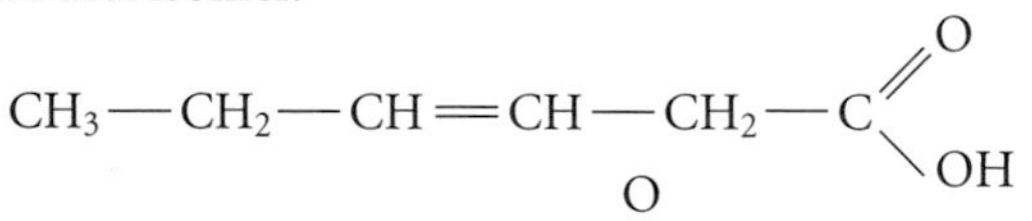 CH_3—CH_2—CH=CH—CH_2—C(=O)OH

 b) CH_2=C=CH—CH_2—C(=O)—CH_3

 c) CH_3—CH—CH—CH_3 with CH_2=C—OH branch and CH_3 branch

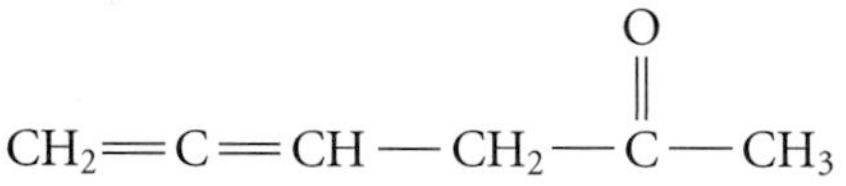

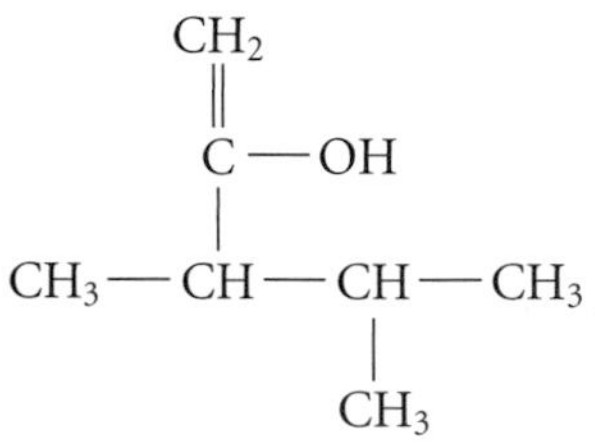

 d) 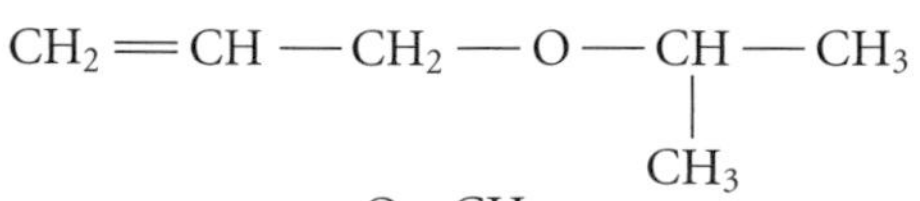 CH_2=CH—CH_2—O—CH—CH_3 with CH_3 branch

 e) 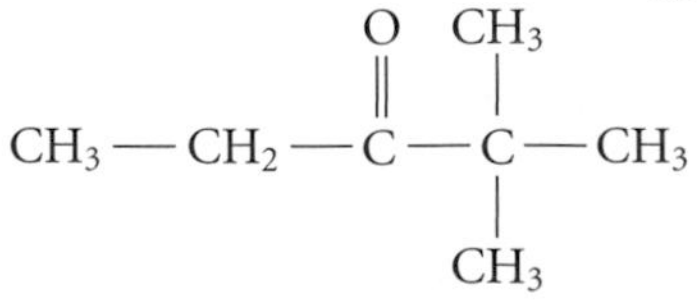 CH_3—CH_2—C(=O)—C(CH_3)—CH_3 with CH_3 branch

 f) 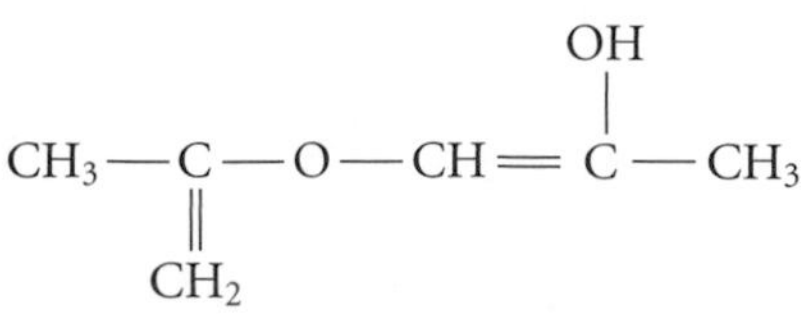 CH_3—C—O—CH=C—CH_3 with OH branch and CH_2 branch

26. You are given three unlabelled test tubes that contain colourless liquids. One test tube contains benzene, another contains ethanol, and the third contains hex-2-ene. Design a procedure that will tell you the contents of each test tube. Describe your expected observations. **Caution** Do not try your procedure in a lab. Benzene is carcinogenic.

27. The following structures cannot exist. By making one change, draw a correct structure. Name your structure.

 a) 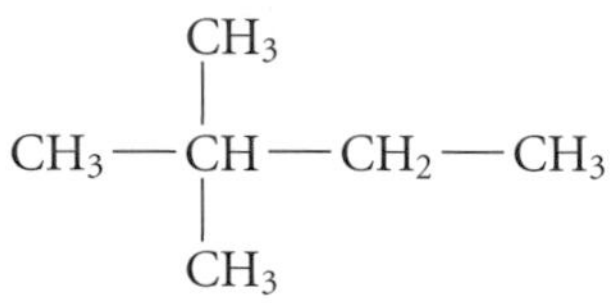 CH_3—CH(CH_3)(CH_3)—CH_2—CH_3

 b) CH_3=CH—CH_2—CH_3

 c) 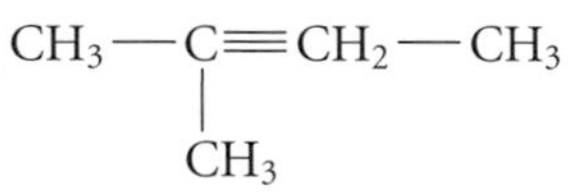 CH_3—C≡CH_2—CH_3 with CH_3 branch

 d) 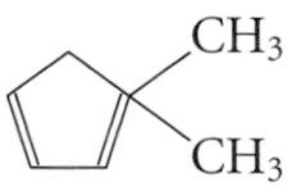 cyclopentene ring with CH_3 and CH_3 branches

28. Explain the difference between addition reactions and substitution reactions.

29. Describe the necessary components of monomers that can be used to form polyesters.

30. You have access to chloroethane, sodium hydroxide, and propanoic acid. Describe the steps you would take to synthesize ethyl propanoate.

31. Design a short procedure to carry out the polymerization of but-2-ene. Your procedure should include:
 • structural diagrams for the monomer and the expected polymer product
 • an appropriate solvent in which but-2-ene dissolves and in which the reaction will be carried out
 • the addition of a catalyst and heat to start the polymerization
 • safety precautions for dealing with *all* the compounds used, including the solvent and product (use WHMIS information)

Making Connections

32.

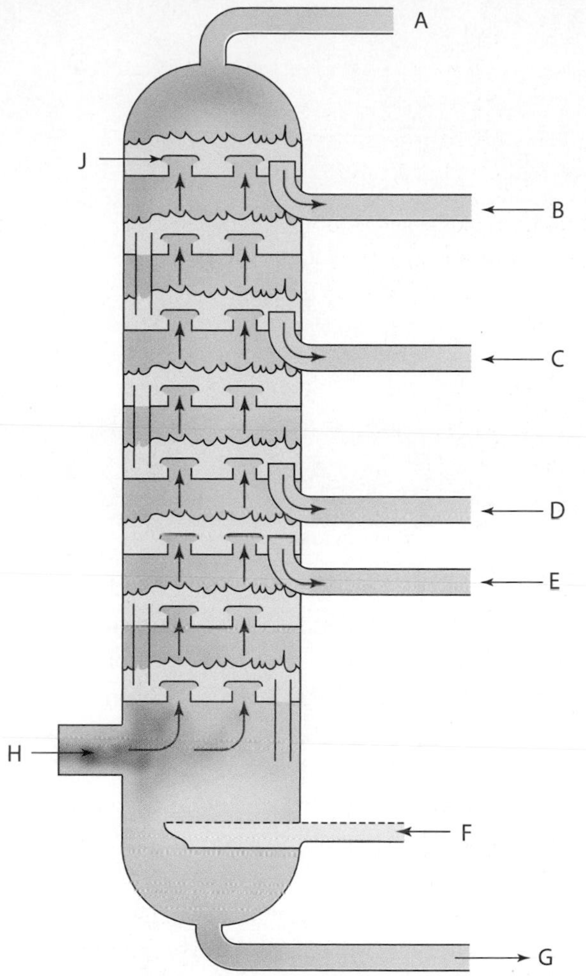

Examine the diagram and answer the following questions.

a) What is the name of the process carried out in this structure?

b) What enters at H?

c) What is the name of the object labelled J? What purpose does it serve?

d) What enters at F? What is its function?

e) What type of material exits at G?

f) In very general terms, what is the nature of the substances that exit at A through E?

g) Explain why some substances rise as high as B before exiting while other substances exit at E.

33. How are organic compounds important to your health and lifestyle? Write a short paragraph that describes any benefits you obtain from an organic compound.

34. Draw a concept map that summarizes the concepts that you have learned about organic chemistry. Include the following topics:
- functional groups
- reactions of organic compounds
- natural and synthetic polymers
- biological molecules and their functions
- petrochemicals

35. Use a flowchart to describe the development and manufacture of PET from the natural compounds from which it is formed. **ICT**

36. Scientists and horticulturalists who manufacture and market pesticides called *pyrethroids* (see figure below) are concerned about public perception. Many people buy the natural pesticide pyrethrin (see below) even though it is more toxic than pyrethroids. Why do you think people make this choice? Do you think this happens for other natural organic products that have synthetic alternatives? Write a brief editorial outlining your opinions and advice to consumers to help them make informed product choices.

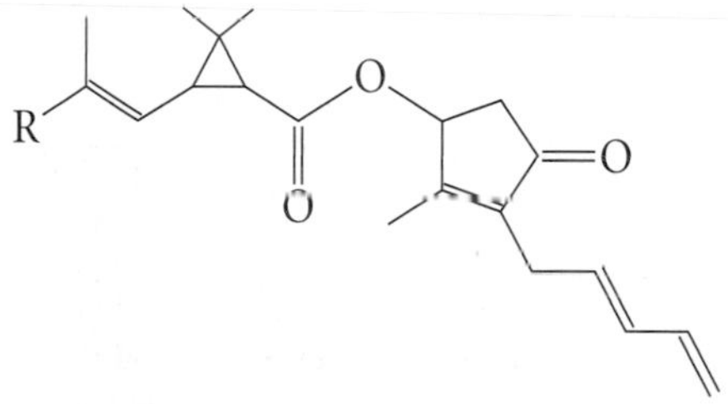

There are two naturally occurring pyrethrins that have the base structure shown here. In pyrethrin I, the R group is a methyl group, $-CH_3$. In pyrethrin II, the R group is an acetyl group, $-COOCH_3$.

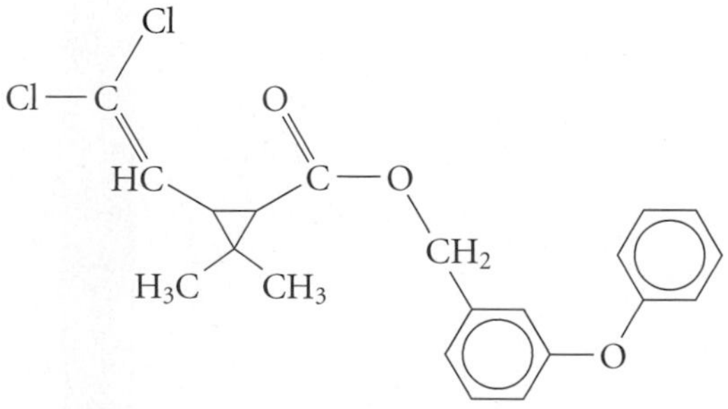

Permethrin, shown here, is one example of the many synthetic pyrethroids that are used as insecticides.

UNIT 8
Chemical Equilibrium Focussing on Acid-Base Systems

General Outcomes

- explain that there is a balance of opposing reactions in chemical equilibrium systems

- determine quantitative relationships in simple equilibrium systems

Unit Contents

Focussing Questions

1. What is happening in a system at equilibrium?

2. How do scientists predict shifts in the equilibrium of a system?

3. How do Brønsted-Lowry acids and bases illustrate equilibrium?

Unit PreQuiz ?

www.albertachemistry.ca

The fertilizer plant in Redwater, Alberta, shown in the photograph, produces 960 000 t of ammonia per year. Some is used directly as fertilizer, and some is used in industrial applications. Much of the ammonia synthesized at the Redwater plant is used on site to produce ammonium sulfate, ammonium nitrate, and urea—all of which are used as fertilizers.

Ammonia is made from hydrogen and nitrogen gases. The atmosphere contains 78 percent nitrogen but only traces of hydrogen. At the Redwater plant, they obtain hydrogen gas by reacting natural gas (mostly methane, $CH_4(g)$) with water in the form of steam to produce hydrogen and carbon dioxide. The process must be completed at a temperature of about 800 °C. Next, air, as a source of nitrogen, is added to the mixture of gases. The carbon dioxide and carbon monoxide gases are removed, and the hydrogen-nitrogen mixture is compressed under extreme pressure and sent to a reactor. There, under high temperatures and in the presence of a catalyst, ammonia is synthesized.

Why is the reaction between natural gas and water carried out at 800 °C? Why is ammonia synthesized under a high pressure and at a high temperature? How do chemists and engineers know what the optimum conditions are for the production of industrial chemicals? In this unit, you will discover that many reactions do not proceed to completion. Instead, at some point, the reaction appears to stop. From that point on, the reaction mixture contains constant amounts of reactants and products. The conditions of the reaction mixture such as temperature and pressure, determine where this point is. Similarly, many acids ionize to a small extent, and then both ionized and unionized forms remain at constant levels in the solution. In this unit, you will learn how to analyze and describe these conditions.

In previous science courses, you might have made your own acid-base indicator using red cabbage, like the student in Figure P8.1 is doing. In this unit, you will learn about the chemistry of indicators and why they change colour with pH as well as many other concepts relating to acids and bases. Before you begin this unit, however, you will benefit from a review of several topics involving acids and bases.

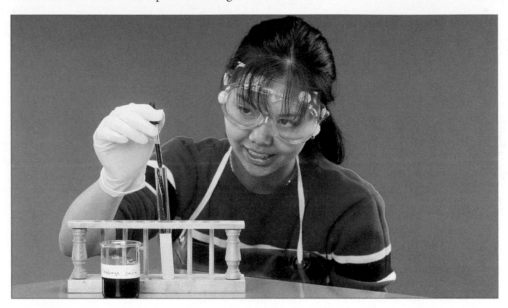

Figure P8.1 This student is testing the acidity of a solution with an acid-base indicator made from red cabbage juice.

Arrhenius Theory of Acids and Bases

The first theory of acids and bases that you learned was the Arrhenius theory. According to this theory, Arrhenius acids are substances that ionize to form hydrogen ions, $H^+(aq)$, in water. Similarly, Arrhenius bases are substances that dissociate in water to produce hydroxide ions, $OH^-(aq)$. This theory did not account for the observations that solutions of some substances, such as ammonia ($NH_3(aq)$), have basic properties. In addition, it was later discovered that hydrogen ions do not exist as such in solutions. Instead, the hydrogen ions react with water to form hydronium ions, $H_3O^+(aq)$. The Arrhenius theory was modified to account for these observations as follows.

An Arrhenius acid is a substance that reacts with water to produce hydronium ions.

An Arrhenius base dissociates in water or reacts with water to produce hydroxide ions.

The following equations illustrate the reactions of modified Arrhenius acids and bases with water.

$$HCl(g) + H_2O(\ell) \rightarrow Cl^-(aq) + H_3O^+(aq)$$

$$NH_3(aq) + H_2O(\ell) \rightleftharpoons NH_4^+(aq) + OH^-(aq)$$

Strong and Weak Acids and Bases

In one flask, you have 0.001 mol/L HCl(aq) and in another flask, you have 6.0 mol/L CH₃COOH(aq) (ethanoic acid also called acetic acid). Which acid of the two is the strong acid? Of these two acids, hydrochloric acid is a strong acid and ethanoic acid is a weak acid. Did you remember that the concentration of an acid does not determine whether it is strong or weak?

A strong acid ionizes completely in water while a weak acid ionizes only to a very small extent. Figure P8.2 illustrates the difference between hydrochloric acid and ethanoic acid.

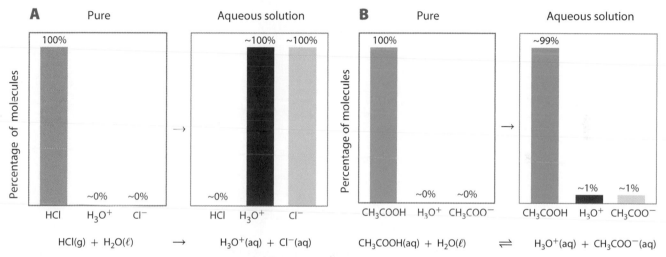

Figure P8.2 (A) HCl gas is a molecular compound. When it is dissolved in water, it reacts with the water to form hydronium and chloride ions. **(B)** Pure ethanoic acid is un-ionized. When dissolved in water, approximately one percent of the ethanoic acid reacts with water to form hydronium ions and ethanoate (CH₃COO⁻) ions.

The double arrow ($\rightleftharpoons$) in the equation under Figure P8.2B indicates that the reaction can proceed in either direction. However, while in solution, most of the ethanoic acid remains in the un-ionized form.

The situation is the same for bases. A strong base dissociates or reacts with water completely while a weak base dissociates or reacts with water, only slightly. One more aspect is important to remember when you are discussing strong and weak bases. Many bases are nearly insoluble in water. However, the solubility of a base does not determine whether it is strong or weak. A base might be nearly insoluble, but if the small fraction that does dissolve dissociates completely, then the base is a strong base.

Acid-Base Titration

In your previous chemistry course, you probably titrated a solution having an unknown concentration of an acid or base with a standardized base or acid. A standardized solution is one for which you know the concentration. You probably used an apparatus similar to the one shown in Figure P8.3.

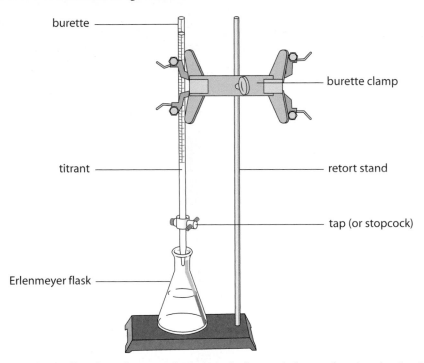

Figure P8.3 By reading the volume on the burette before and after performing the titration, you can determine the volume of titrant that you need to neutralize the acid or base in the Erlenmeyer flask.

In a typical titration experiment, you would put a standardized base (or acid) in the burette and read and record the volume on the scale on the burette. You would then measure the volume of the acid (or base) having an unknown concentration and put it in an Erlenmeyer flask. You would add an acid-base indicator to the flask. An acid-base indicator is a compound such as the red cabbage juice in Figure P8.1, which changes colour over a specific pH range of about 1.5 to 2 pH units. When you are titrating a strong acid with a strong base (or vice versa) you would use an indicator that changes colour near pH 7.0. When everything is in place, you would partially open the stopcock of the burette and very gradually allow the base to drop into the acid solution in the Erlenmeyer flask. You would swirl the flask to mix the solutions.

Suppose, for example, you are titrating an unknown hydrochloric acid solution with a solution of sodium hydroxide for which you know the concentration. The following ionic equation shows the reaction that would be occurring in the flask.

$$Na^+(aq) + OH^-(aq) + H_3O^+(aq) + Cl^-(aq) \rightarrow Na^+(aq) + Cl^-(aq) + 2H_2O(\ell)$$

Notice that there are no acids or bases in the products. They are neutral. Initially, HCl(aq) would be the only reactant. As each drop of NaOH(aq) is added, all of the added hydroxide ions would react with the hydronium ions to form water. When you have added exactly the same number of moles of hydroxide ions as the hydronium ions already present, they would all have formed water. The solution would be neutral or pH 7.0. This point is called the *equivalence point* because the amount of base that you added would be equivalent to the amount of acid present in the unknown solution. In a typical titration, you would add one or two more drops of base. Since all of the acid is gone, the pH would rise significantly and the colour if the indicator would change. This point is called the *endpoint* because the colour tells you that the titration is complete. You would say that you had titrated to the endpoint. At this point, you would read the volume of base remaining in the burette in order to determine how much you had used. A typical set of data and the calculation of the concentration of the unknown acid solution are shown below.

volume of hydrochloric acid of unknown concentration	50.0 mL
concentration of standard sodium hydroxide base	0.750 mol/L
initial burette volume	1.35 mL
final burette volume	14.68 mL

Calculate the volume of NaOH(aq) used.

$$\text{final} - \text{initial} = \text{volume used}$$
$$14.68 \text{ mL} - 1.35 \text{ mL} = 13.33 \text{ mL}$$

Calculate the number of moles of NaOH(aq) used.

$$n = V \times c$$
$$n = \left(0.750 \frac{\text{mol}}{\text{L}}\right)(13.33 \text{ mL})\left(\frac{1 \text{ L}}{1000 \text{ mL}}\right)$$
$$n = 0.009\ 997 \text{ mol}$$

Since 0.009 997 mol of NaOH(aq) were used to neutralize the HCl(aq), there must have been 0.009 997 mol of HCl(aq) in the 50.0 mL sample. Calculate the concentration of the HCl(aq)

$$c = \frac{n}{V}$$
$$c = \left(\frac{0.009\ 997 \text{ mol}}{50.0 \text{ mL}}\right)\left(\frac{1000 \text{ mL}}{\text{L}}\right)$$
$$c = 0.199\ 95 \frac{\text{mol}}{\text{L}}$$
$$c \approx 0.20 \frac{\text{mol}}{\text{L}}$$

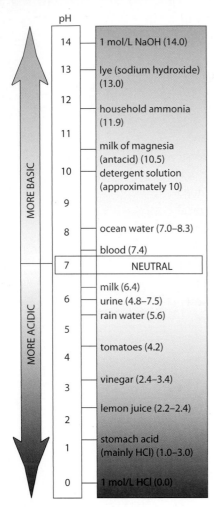

Figure P8.4 Most common solutions have a pH between 0 and 14. This scale shows the pH of some familiar solutions.

Describing Acidic and Basic Solutions

You have probably seen many pH scales similar to the one in Figure P8.4. When you first learned that the pH scale went from 0 to 14 and that neutral was pH 7, did you wonder why chemists chose such a strange scale? Do you recall the answer to that question that you learned in your previous chemistry course? The reasons are based on the definition of pH and the properties of pure water.

The definition of pH is the negative logarithm of the hydronium ion concentration expressed as a molar concentration. It can be expressed mathematically as follows.

$$pH = -\log [H_3O^+(aq)]$$

You might recall that pure water ionizes very slightly. You could write the chemical equation as shown.

$$2H_2O(\ell) \rightleftharpoons H_3O^+(aq) + OH^-(aq)$$

Chemists have determined that the hydronium concentration of pure water is 1.0×10^{-7} mol/L. Therefore the pH of pure water is:

$$pH = -\log [H_3O^+(aq)]$$
$$pH = -\log 1.0 \times 10^{-7}$$
$$pH = -(-7)$$
$$pH = 7$$

Pure water is neutral so pH 7.0 means neutral. Because acids are defined as substances that react with water to produce hydronium ions, acids must have higher concentrations of hydronium ions than pure water. Thus, the hydronium concentration of acids must be greater than 1.0×10^{-7} mol/L. In the sample calculation for the titration described in the previous page, the concentration of hydrochloric acid was 0.20 mol/L. What was its pH?

$$pH = -\log [H_3O^+(aq)]$$
$$pH = -\log 0.20$$
$$pH = -(-0.6989)$$
$$pH = 0.70$$

A pH of less than pH 1.0 is very acidic. Acidic solutions have pH values lower than pH 7.0. Notice that the hydronium ion concentration of the water itself was not considered in this calculation. When you are calculating the pH of acidic solutions, the hydronium concentration is usually so much larger than 1.0×10^{-7} mol/L that you can ignore the hydronium concentration of the water itself.

Although pH is nearly always used to describe the acidity or basicity of a solution, you can define another quantity similar to pH to describe the basicity of a solution. That quantity is the pOH. It is defined as the negative logarithm of the hydroxide ion concentration or $pOH = -\log [OH^-(aq)]$. Essentially all of the mathematical methods used in conjunction with pH can also be applied to pOH. An important relationship to remember is:

$$pH + pOH = 14.0$$

If you have calculated or measured pH or pOH, you can find the other by using this equation.

Do you remember how to calculate the hydronium concentration of a solution for which you know the pH? To stimulate your memory, review the calculation.

$$pH = -\log 1.0 \times 10^{-7}$$
$$pH = -(-7)$$

Notice that the pH is the negative of the exponent of 10 in the calculation. You can think of a logarithm as the exponent of ten that will give you the desired value. Therefore, to find the hydronium ion concentration, you can use the following equation.

$$[H_3O^+(aq)] = 10^{-pH}$$

For example, the hydronium ion concentration of a solution having a pH of 4.70 would be:

$$[H_3O^+(aq)] = 10^{-pH}$$
$$[H_3O^+(aq)] = 10^{-4.70}$$

Using your calculator, you would find, $[H_3O^+(aq)] = 10^{-4.70} = 1.9953 \times 10^{-5}$ mol/L rounded to 2.0×10^{-5} mol/L.

Significant Digits and pH

When looking at the rounded answer to the example above, you might have recalled that the only digits that are significant in a pH value are the digits after the decimal point. Thus, the number of significant digits in pH = 4.70 is two because only the 7 and 0 are counted. You learned that, because the digit before the decimal point represented the exponent of ten, it cannot be considered a significant digit. Did that explanation confuse you at all? If so, the following discussion might help you understand it better. To keep the calculations simpler, use a positive exponent.

Imagine that you are told that 5.379 is the exponent of ten and you are asked to find that number. You could put it directly into your calculator and find the value, but it would not help you understand the number of significant digits in the answer. Therefore, use the rules of exponents to separate the digits before and after the decimal point in the exponent of ten. Remember that when you multiply two powers with the same base and different exponents, you add the exponents. Use that rule backwards to get the following expression. Separate 5.379 to 5 + 0.379.

$$10^{5.379} = 10^5 \times 10^{0.379}$$

Now evaluate each power of ten separately. Use your calculator to evaluate $10^{0.379}$.

$$10^{5.379} = 10^5 \times 2.3933 = 2.3933 \times 10^5 \approx 2.39 \times 10^5$$

Notice that the "5" in 5.379 is the exponent of 10 and only the 0.379 gave you the digits in your answer.

Prerequsite Skills

- **Solving** algebraic equations with one unknown

- **Determining** the molar concentration of a solution and the ions it may contain

CHAPTER
16

Chemical Equilibrium

Chapter Concepts

16.1 Chemical Equilibrium

- When a system is at equilibrium, no further change takes place in the concentration of reactants or products.

16.2 Equilibrium Expressions and Le Châtelier's Principle

- The equilibrium expression for a chemical system can be written from the chemical equation.

- The equilibrium constant can be calculated from the concentration of reactants and products at equilibrium.

- A reaction at equilibrium can be disturbed by changing factors, such as concentration, pressure, volume, or temperature.

16.3 Equilibrium Calculations for Homogeneous Systems

- The equilibrium constant for a reaction can be calculated when equilibrium concentrations are known or when initial concentrations and one equilibrium concentration are known.

- The concentration of each substance in a reaction at equilibrium can be calculated if the equilibrium constant and the concentration of one reactant are known.

16.4 Applications of Equilibrium Systems

- Principles of equilibrium apply to many biological systems.

- Modern chemical processes are the result of extensive experiments and may involve building models that rely on equilibrium concepts.

As the weather turns colder, hardware stores arrange large displays of windshield washer fluid for sale. Motorists use this solution year-round, but the demand for washer fluid increases during winter, when storms bring freezing rain and snow. Most windshield washer fluids are a solution of methanol, water, detergents, and colouring. The dissolved methanol prevents washer fluid from freezing in all but the coldest weather. Because it is also an excellent solvent for some varnishes and lacquers, you may find other products containing methanol in your home. In industry, methanol is mainly used to make other chemicals, which are then turned into various other products we use. Methanol is produced in huge quantities, with about 35 million tonnes being manufactured globally each year. The most economical and widely used raw material for making methanol is methane, which is the largest component of natural gas. With its abundant supply of natural gas, Alberta is a major producer of methanol. The production process depends on the principles of chemical equilibrium. In this chapter, you will extend your knowledge of chemical change and learn how chemists are able to manipulate reaction conditions to maximize the yield obtained from a reaction. In the Launch Lab, you will investigate a reversible change. Reversible changes are the basis of all equilibrium systems.

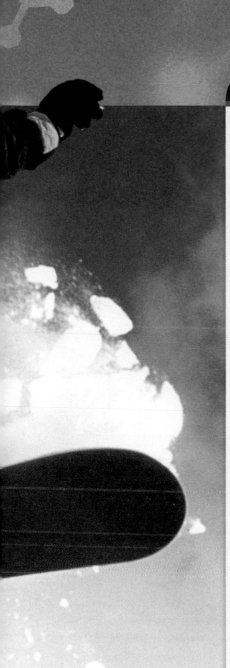

Launch Lab

The Chemical Blues

Reversible chemical reactions are very important in chemistry. In this activity, you will observe an example of a reversible change that involves a change of colour.

Safety Precautions

- Sodium hydroxide is harmful if swallowed or if the fumes are inhaled. It causes irritation to the skin and eyes. If you spill any solution on your skin, immediately rinse the area with plenty of cold water and inform your teacher. If you spill sodium hydroxide solution on the lab bench or floor, immediately inform your teacher.
- Methylene blue will stain clothing and may also cause eye irritation.
- Wash your hands when you have completed the Launch Lab.

Materials

- 500 mL Erlenmeyer flask with stopper
- 8 mol/L NaOH(aq)
- 5% glucose solution (5 g/100 mL)
- methylene blue indicator
- test tube with stopper
- ice bath
- warm-water bath

Procedure

Your teacher will carry out this investigation as a demonstration.

1. Your teacher will have prepared the following solution ahead of time. It will be sitting in a stoppered flask.

   Add fresh solutions of the following to a 500 mL flask:
   - 250 mL glucose solution
   - 7.5 mL 8 mol/L NaOH
   - 1 to 2 drops of methylene blue

2. Record the appearance of the solution in the flask. As a class, discuss the nature of each of the components of the solution.

3. The teacher will give the flask a few vigorous shakes.

4. Observe and record the appearance of the solution in the flask. Wait patiently.

5. Discuss any possible explanation for the change in the colour of the solution.

6. Wait until the appearance of the solution in the flask changes colour again.

7. Propose possible reasons for the changes in the colour of the solution. The list of materials might give you some hints.

8. Suggest ways to test your proposals.

9. Your teacher will test as many of the proposals as possible.

10. As a class, try to agree on an explanation for the observed changes in the colour of the solution.

Analysis

1. What principles were demonstrated by the activity?

2. What unanswered questions were generated by the activity?

Chemical Equilibrium

Key Terms

constant macroscopic
properties
dynamic equilibrium
homogeneous equilibrium
heterogeneous equilibrium

The amount of product formed as the result of a chemical reaction is very important. Previously, when you used stoichiometry to calculate the mass of a product, you assumed that the only factor limiting the mass of a product was the presence of a limiting reactant. In those calculations, you assumed the reaction went to completion because, at the end of the reaction, none of the limiting reactant remained. However, many reactions do not proceed to completion. In these cases, the reaction is finished when there is a mixture of reactants and products with constant properties, such as concentration or temperature.

Figure 16.1 The cars crossing this bridge are modelling equilibrium conditions.

In systems at equilibrium, forward and reverse changes are taking place at the same rate. In Figure 16.1, the flow of traffic across a bridge is a model of forward and reverse change. Notice that, if eight cars go across the bridge from Reactant City to Product Town and eight cars go from Product Town to Reactant City, the number of cars in each location has not changed. There might be 1000 cars in Reactant City and 500 cars in Product Town. As long as the number of cars coming off the bridge is the same as the number that go on to the bridge, however, there are still 1000 cars in Reactant City and 500 cars in Product Town. The system is at equilibrium.

The vehicles crossing in one direction represent molecules combining in the forward reaction, and the vehicles crossing in the opposite direction represent molecules combining in the reverse reaction. At equilibrium, there will be no further change in the concentrations of reactants and products. The concentrations of reactants and products are usually not equal, just as the number of cars in the towns are not equal, but because their concentrations are no longer changing, the mixture will have **constant macroscopic properties**. Examples of constant macroscopic properties include observable properties, such as temperature, colour, and pH. However, just as individual vehicles are crossing the bridge, a great deal of change is taking place at the molecular level. When an equilibrium system is changing at the molecular level, but its macroscopic properties remain constant, it is described as a **dynamic equilibrium**.

Modelling Equilibrium

In this investigation, you will model what happens when forward and reverse reactions take place, and you will take measurements to gain quantitative insight into equilibrium systems. Then you will observe the effect caused by introducing a change to the equilibrium.

Question

How can you model a chemical equilibrium by transferring water into and out of graduated cylinders?

Materials

- water coloured with food dye
- 2 graduated cylinders (25 mL)
- 2 glass tubes of different diameters (e.g., 10 mm and 6 mm)
- 2 labels or a grease pencil

Procedure

1. Copy the following table into your notebook to record your observations.

Transfer number	Volume of water in "reaction" cylinder (mL)	Volume of water in "product" cylinder (mL)
0	25.0	0.0
1		
2		
3		

2. Label one graduated cylinder "reactant." Label the other "product."

3. Fill the reactant graduated cylinder with coloured water up to the 25.0 mL mark. Leave the product graduated cylinder empty.

4. With your partner, transfer water simultaneously from one cylinder to the other as follows: Lower the larger diameter glass tube into the reactant graduated cylinder, keeping the top of the tube open. When the tube touches the bottom of the cylinder, cover the upper end with a finger so that the water will remain in the tube when you raise it out of the cylinder. Lift the tube that now contains water and transfer the water to the product cylinder

by holding the tube over the product cylinder and releasing the water by removing your finger from the tube. *At the same time* that you are transferring liquid into the product cylinder, your partner must use the smaller diameter tube to transfer water from the product cylinder into the reactant cylinder. Of course, for the first transfer, there will be no water in the product cylinder.

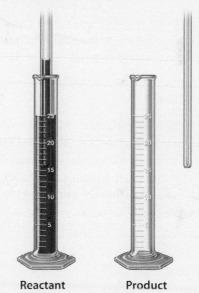

Reactant   **Product**

5. Remove the glass tubes. Record the volume of water in each graduated cylinder to the nearest 0.1 mL.

6. Repeat Procedure steps 4 and 5 until you are sure there is no change in water volume in either graduated cylinder. To be certain that no change has occurred, you will need to complete four transfers that result in approximately the same volume readings.

7. Add approximately 5 mL of water to the graduated cylinder labelled reactant. Record the volume in each cylinder, then repeat Procedure steps 4 and 5 until your results show four transfers that result in no further change in volume.

Analysis

1. Plot a graph of the data you collected. "Transfer number" should be recorded on the *x*-axis and "volume of liquid" on the *y*-axis. Use different symbols or colours to distinguish between the data for the reactant volume and the data for the product volume. Draw the best smooth curves you can through each set of data.

2. In this activity, the volume of liquid represents the concentration of a reactant or product. Each transfer of water represents an amount of reactant that was converted into product or product that was converted back into reactant. How can you use your graph to compare the rate of the forward reaction with the rate of the reverse reaction? What happens to these rates as the reaction proceeds? In your answers, be sure to refer to the shapes of the curves you have drawn.

3. At the point where the two curves cross, is the rate of the forward reaction equal to the rate of the reverse reaction? Explain.

4. How can you recognize when the system is at equilibrium?

5. Were the liquid volumes (that is, the concentrations of reactants and products) in the two graduated cylinders equal at the first equilibrium? Were they equal at the second equilibrium? Make a general statement about the concentrations of reactants and products at equilibrium.

6. In a chemical reaction, what corresponds to the addition of more liquid to the reactant cylinder? How did the final volume of water in the product cylinder change as a result of adding more liquid to the reactant graduated cylinder?

7. Determine the ratio $\dfrac{\text{volume of product}}{\text{volume of reaction}}$ at the end of the first equilibrium and again at the end of the second equilibrium. Within experimental error, were these two ratios the same or different?

8. In this activity, what do you think determines the relative volumes of water in each graduated cylinder? Briefly outline an experiment you could perform to test your answer. In a real chemical reaction, what factors might affect the relative concentrations of reactants and products at equilibrium?

9. Is your experiment a model of an open or a closed system? Explain.

Conclusion

10. Describe what you learned about chemical equilibriums from completing this activity.

Q1 What is the difference between macroscopic and microscopic properties?

Q2 What must occur to create an equilibrium?

Q3 What is happening on the microscopic level during a dynamic equilibrium?

Conditions That Apply to All Equilibrium Systems

Figure 16.2 The equilibrium between carbon dioxide and water is disturbed when the bottle is opened. Eventually a new equilibrium will be reached.

The fundamental requirement for equilibrium is that opposing changes must occur at the same rate. Both physical and chemical changes can reach equilibrium. Three examples of physical changes that reach equilibrium are as follows:

1. a solid in contact with a solution containing this solid, such as sugar crystals in a saturated aqueous sugar solution;

2. a gas above a liquid, such as a closed bottle or can that contains a carbonated drink (see Figure 16.2); and

3. the vapour above a pure solid, such as mothballs in a closed drawer.

Two general chemical processes that reach equilibrium are as follows:

1. A reaction with reactants and products in the same phase, such as a reaction between two gases to produce a gaseous product. In this unit, you will focus on reactions in which the reactants and products are in the same phase. The equilibrium they reach is called **homogeneous equilibrium**.

2. A reaction in which reactants and products are in different phases, such as an aqueous solution of ions in which the ions combine to produce a slightly soluble solid that forms a precipitate. The equilibrium they reach is called **heterogeneous equilibrium**.

What do equilibrium systems like these have in common? What conditions are necessary for equilibrium to become established? As outlined in the box below, four conditions apply to all equilibrium systems.

The Four Conditions That Apply to All Equilibrium Systems

1. Equilibrium is achieved in a reversible process when the rates of opposing changes are equal. A double arrow, $\rightleftharpoons$, indicates reversible changes. For example,

$$H_2O(\ell) \rightleftharpoons H_2O(g)$$
$$H_2(g) + Cl_2(g) \rightleftharpoons 2HCl(g)$$

2. The observable (macroscopic) properties of a system such as temperature, pressure, concentration, and pH are constant at equilibrium.

   You can summarize the first two equilibrium conditions by stating that *equilibrium involves dynamic change at the molecular level but no change at the macroscopic level.*

3. Equilibrium can only be reached in a "closed" system that is maintained at a constant temperature (Figure 16.3). Ideally, a system should be isolated, but an isolated system is difficult to achieve.

4. Equilibrium can be approached from either direction.

A common example of a closed system is carbon dioxide gas in equilibrium with dissolved carbon dioxide in a soft drink at constant temperature. The system remains at equilibrium as long as the container is unopened. Note also that small changes to the components of a system are sometimes negligible. As a result, equilibrium principles can sometimes be applied in situations where the system is not physically closed. For example, consider the equilibrium of a solid in a saturated aqueous solution, such as $CaO(s) + H_2O(\ell) \rightleftharpoons Ca^{2+}(aq) + 2OH^-(aq)$. You can neglect the small amount of water that vaporizes from the open beaker during the course of an experiment.

Chemistry File

FYI

The rate of a reaction is defined as the change in concentration of a reactant or product per unit time. Chemists use square brackets to indicate concentration and the symbol Δ to signify change. For the reversible reaction $A \rightleftharpoons B$, the rate of the forward reaction is

$$\text{Rate} = \frac{[A] \text{ at time } t_2 - [A] \text{ at time } t_1}{t_2 - t_1}$$
$$= \frac{\Delta[A]}{\Delta t}$$

For convenience, rate is always defined as a positive quantity. Since the concentration of a reactant always decreases with time, ΔA is negative. For this reason, a negative sign is inserted to make the rate positive. In the forward direction, the rate of reaction is given by $\text{Rate} = -\frac{\Delta[A]}{\Delta t}$. A sign is not needed for the reverse reaction rate, because the concentration of product is increasing. Therefore, in the reverse direction, the rate of the reaction is given by $\text{Rate} = \frac{\Delta[B]}{\Delta t}$. At equilibrium, the rate of the forward reaction is equal to the rate of the reverse reaction; therefore, $-\frac{\Delta[A]}{\Delta t} = \frac{\Delta[B]}{\Delta t}$.

Figure 16.3 Inside a sealed terrarium, liquid water and water vapour are at equilibrium as long as the temperature is held constant.

An example illustrating the fourth condition can be found in the equilibrium between hydrogen and chlorine gases and hydrogen chloride gas. At equilibrium, the proportions of $H_2(g)$, $Cl_2(g)$, and $HCl(g)$ are the same regardless of whether you started with $H_2(g)$ and $Cl_2(g)$ or $HCl(g)$.

Section 16.1 Summary

- At equilibrium, the macroscopic properties of a system remain constant.
- Macroscopic properties are observable properties such as temperature, pH, and colour.
- When a chemical reaction has reached equilibrium, the concentrations of the reactants and products do not change. However, forward and reverse reactions continue to occur but they occur at the same rate.
- When an equilibrium system is changing at the molecular level, it is called a dynamic equilibrium.
- You can model an equilibrium by transferring water in small tubes between two identical cylinders.
- When the reactants and products of a reaction are in the same phase and the system reaches equilibrium, it is called a homogeneous equilibrium.
- When the reactants and products of a reaction are in different phases and the system reaches equilibrium, it is called a heterogeneous equilibrium.
- An equilibrium can only be achieved in a closed system at a constant temperature or in an isolated system.

SECTION 16.1 REVIEW

1. Describe two physical and two chemical processes that are examples of reversible changes that are not at equilibrium.

2. Refer back to Investigation 16.A. Was each of the four conditions that apply to all equilibrium systems modelled by the activity? Identify ways the model could be changed to improve your understanding of equilibrium.

3. Describe the changes that take place on a molecular level as a reversible reaction approaches equilibrium. Explain why the equilibrium that is reached is dynamic.

4. A sealed carbonated drink bottle contains a liquid with a space above it. The space contains carbon dioxide at a pressure of about 400 kPa.
   a) What changes are taking place at the molecular level?
   b) Which macroscopic properties are constant?

5. Slush is a prominent feature of Canadian winters. Under what conditions do ice and water form an equilibrium mixture?

6. A chemist places a mixture of $NO(g)$ and $Br_2(g)$ in a closed flask. After a time, the flask contains a mixture of $NO(g)$, $Br_2(g)$ and $NOBr(g)$.
   a) Explain this observation.
   b) Is this a homogeneous equilibrium or a heterogeneous equilibrium? Give your reasoning.

7. Copy the following equations into your notebook, and label each system as a homogeneous equilibrium or a heterogeneous equilibrium:
   a) $PCl_5(g) \rightleftharpoons PCl_3(g) + Cl_2(g)$
   b) $CaCO_3(s) \rightleftharpoons CaO(s) + CO_2(g)$
   c) $HF(aq) + H_2O(\ell) \rightleftharpoons H_3O^+(aq) + F^-(aq)$
   d) $N_2O_3(g) \rightleftharpoons NO_2(g) + NO(g)$
   e) $C(s) + CO_2(g) \rightleftharpoons 2CO(g)$

8. What is the meaning of the double arrow, $\rightleftharpoons$, in the equations in question 7?

9. Is the following statement true or false? "In a chemical system at equilibrium, the concentration of reactants is equal to the concentration of products." Support your answer using data from Investigation 16.A.

10. a) When a damp towel is hung on a clothesline in fair weather, it will dry. The same towel left in a gym bag will remain damp. Use equilibrium principles to explain the difference.
    b) Is a running clothes dryer an open system or a closed system? Explain.

11. Is the following statement true or false? "If a chemical system at equilibrium in a closed container is heated, the system will remain at equilibrium." Explain.

Equilibrium Expressions and Le Châtelier's Principle

In Section 16.1, you learned the meaning of equilibrium and the conditions required for a chemical equilibrium to exist. When chemists first observe a phenomenon such as equilibrium, they look for the most fundamental laws that govern the process. If possible, they look for a formula that will describe the process and allow them to make quantitative predictions. Chemists were studying chemical reactions in equilibrium in the mid-1800s when two chemists discovered the law of chemical equilibrium.

The Law of Chemical Equilibrium

In 1864, two Norwegian chemists, Cato Guldberg (1836–1902) and Peter Waage (1833–1900) (See Figure 16.4) collected data on chemical reactions in equilibrium. One of the reactions that they studied was the reaction between ethanoic acid (also called acetic acid), $CH_3COOH(\ell)$, and ethanol, $CH_3CH_2OH(\ell)$, that forms ethyl ethanoate, $CH_3COOCH_2CH_3(\ell)$. The reaction is shown below.

$$CH_3COOH(\ell) + CH_3CH_2OH(\ell) \rightleftharpoons CH_3COOCH_2CH_3(\ell)$$

They would start with varying amounts of ethanol and ethanoic acid and let it form ethyl ethanoate and wait until the reaction reached equilibrium and then measure the concentration of each reactant and product. Then they would start with just ethyl ethanoate and let it form ethanoic acid and ethanol and, again, measure the concentration of each compound at equilibrium. They looked for a pattern in the concentrations of the reactants and products at equilibrium.

Guldberg and Waage also reviewed the results of experiments conducted by other chemists. One reaction that provides visible results is the equilibrium between dinitrogen tetroxide, $N_2O_4(g)$ and nitrogen dioxide, $NO_2(g)$.

Dinitrogen tetroxide, $N_2O_4(\ell)$, boils at 21 °C to form a colourless gas. Dinitrogen tetroxide gas decomposes into $NO_2(g)$, which is dark brown. The coloured haze sometimes seen in heavily polluted areas may be due to the colour of $NO_2(g)$. By observing the intensity of the brown colour in the mixture, the progress of the decomposition reaction can be followed.

Suppose a small quantity of $N_2O_4(g)$ is placed in a sealed flask at 100 °C, and the initial concentration of $NO_2(g)$ is zero. Initially, the flask will be almost colourless. The initial rate of the forward reaction is relatively large, while the initial rate of the reverse reaction is zero because there is no product.

Figure 16.4 Cato Guldberg and Peter Waage discovered the law of chemical equilibrium. In addition to their shared chemical interests, they were brothers-in-law.

The initial conditions correspond to the first exchange in Investigation 16.A when the reactant cylinder was full and the product cylinder was empty. As the reaction proceeds, the rate of the forward reaction decreases because the concentration of $N_2O_4(g)$ decreases, while the rate of the reverse reaction increases because the concentration of $NO_2(g)$ increases (see Figure 16.5). The colour of the gaseous mixture will darken until, at equilibrium, the colour intensity stays constant. At equilibrium, the rate of the forward reaction equals the rate of the reverse reaction, so there will be no further change in the relative amounts of $N_2O_4(g)$ and $NO_2(g)$:

$$N_2O_4(g) \rightleftharpoons 2NO_2(g)$$

colourless   brown

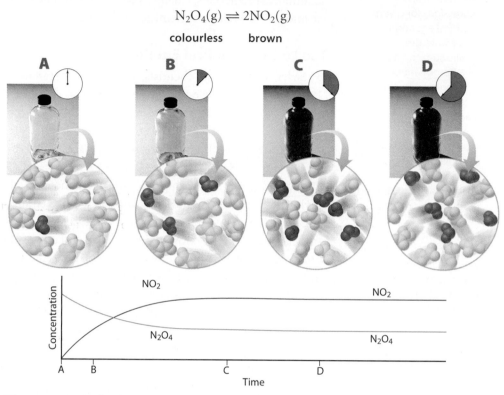

Figure 16.5 Initially, the rate of the forward reaction decreases and the rate of the reverse reaction increases. At equilibrium, the rate of the forward reaction is equal to the rate of the reverse reaction, and the macroscopic properties of the system are constant. Changes at the molecular level take place at equal rates.

Guldberg and Waage analyzed the results of many different experiments and tested a variety of mathematical relationships until they discovered the relationship that always gave consistent results. Their conclusion is summarized in words in the box below.

Law of Chemical Equilibrium

In a chemical system at equilibrium, there is a constant ratio between the concentrations of the products and the concentrations of the reactants.

The constant ratio in the law of chemical equilibrium depends only on the stoichiometry of the equilibrium. The mathematical equation that chemists use today is not identical to that of Guldberg and Waage. By convention, today's chemists always write the product concentrations in the numerator and the reactant concentrations in the denominator. Each concentration term is raised to the power of the coefficient of the term in the chemical equation and the terms are multiplied. You can write a general equilibrium reaction as shown.

$$aA + bB \rightleftharpoons cC + dD$$

Based on this equation, the mathematical equation for the constant ratio discovered by Guldberg and Waage is shown in the following box. The constant, K_c, is called the **equilibrium constant**.

Equilibrium Law Expression

$$K_c = \frac{[C]^c[D]^d}{[A]^a[B]^b}$$

Where [A], [B], [C], and [D] represent the concentrations of the reactants and products after the reaction has reached equilibrium and the concentrations no longer change. The exponents, a, b, c, and d, are the stoichiometric coefficients from the equation.

The following Sample and Practice Problems will help you learn how to write equilibrium law expressions.

Sample Problem

Writing Equilibrium Expressions for Homogeneous Chemical Reactions

Problem

Sulfuric acid is created through the contact process, which involves the catalytic oxidation of sulfur dioxide:

$$2SO_2(g) + O_2(g) \rightleftharpoons 2SO_3(g)$$

Write the equilibrium expression.

What Is Required?

You need to find an expression for K_c.

What Is Given?

You know the chemical equation.

Plan Your Strategy

The expression for K_c is a fraction. The concentration of the product is in the numerator and the concentrations of the reactants are in the denominator. Each concentration term must be raised to the power of its coefficient in the balanced equation.

Act on Your Strategy

$$K_c = \frac{[SO_3]^2}{[SO_2]^2[O_2]}$$

Check Your Solution

The square brackets indicate concentration terms. The product is in the numerator, and each term is raised to the power of its coefficient in the chemical equation. The usual practice in chemistry of not writing a coefficient or power of 1 is followed.

Practice Problems

Write the equilibrium expressions for each of the following homogeneous equilibrium equations.

1. The reaction at 200 °C between ethanol and ethanoic acid to form ethyl ethanoate and water:
$$CH_3CH_2OH(g) + CH_3COOH(g) \rightleftharpoons CH_3COOCH_2CH_3(g) + H_2O(g)$$

2. The reaction between nitrogen gas and oxygen gas at high temperatures:
$$N_2(g) + O_2(g) \rightleftharpoons 2NO(g)$$

3. The reaction between hydrogen gas and oxygen gas to form water vapour:
$$2H_2(g) + O_2(g) \rightleftharpoons 2H_2O(g)$$

4. The reduction-oxidation equilibrium of ferric and iodine ions in aqueous solution:
$$2Fe^{3+}(aq) + 2I^-(aq) \rightleftharpoons 2Fe^{2+}(aq) + I_2(aq)$$

5. The oxidation of ammonia (one of the reactions in the manufacture of nitric acid):
$$4NH_3(g) + 5O_2(g) \rightleftharpoons 4NO(g) + 6H_2O(g)$$

Chemistry File

FYI

At equilibrium, the rate of the forward reaction is equal to the rate of the reverse reaction. On a graph showing concentration against time, the slope of the curve gives the rate of reaction. At equilibrium, the slope of a graph of concentration against time is zero.

Applying the Equilibrium Law Expression

To learn how to use the equilibrium law expression, consider again, the equilibrium reaction between $N_2O_4(g)$ and $NO_2(g)$. Table 16.1 lists data from four experiments on the system.

Table 16.1 Investigating the Decomposition of $N_2O_4(g)$

Experiment	Initial $[N_2O_4(g)]$	Initial $[NO_2(g)]$	Equilibrium $[N_2O_4(g)]$	Equilibrium $[NO_2(g)]$	Ratio: $\frac{[NO_2]^2}{[N_2O_4]}$
1	0.1000	0.0000	0.0491	0.1018	0.211
2	0.0000	0.1000	0.0185	0.0627	0.212
3	0.0500	0.0500	0.0332	0.0837	0.211
4	0.0750	0.0250	0.0411	0.0930	0.210

Notice that the four experiments involved different initial concentrations of reactants and products. In experiment 1, there was initially only $N_2O_4(g)$ in the flask. In experiment 2, there was initially only $NO_2(g)$ in the flask. The last column shows that although the equilibrium concentrations are different in the four experiments, the ratio of concentration terms, $\frac{[NO_2]^2}{[N_2O_4]}$, is constant, within experimental error. Hundreds of experiments on the $N_2O_4(g)$ - $NO_2(g)$ system at 100 °C show the same result. Equilibrium is achieved regardless of the initial concentrations of the reactants and the products. The equilibrium system and results of experiments on it can be summarized as follows:

$$N_2O_4(g) \rightleftharpoons 2NO_2(g)$$

$$K_c = \frac{[NO_2]^2}{[N_2O_4]} = 0.211 \text{ at } 100\ °C.$$

The subscript "c" reminds us that the equilibrium constant is expressed in terms of molar concentration. (There are other equilibrium constants, each with a different subscript, as you will see later.) Develop your skills in using the equilibrium law expression by examining the Sample Problems and completing the Practice Problems.

Sample Problem

Calculating an Equilibrium Constant

Problem

A mixture of nitrogen and chlorine gases was kept at a constant temperature in a 5.0 L reaction flask:

$$N_2(g) + 3Cl_2(g) \rightleftharpoons 2NCl_3(g)$$

When the equilibrium mixture was analyzed, it was found to contain 0.0070 mol $N_2(g)$, 0.0022 mol $Cl_2(g)$, and 0.95 mol $NCl_3(g)$. Calculate the equilibrium constant for this reaction.

What Is Required?

You need to calculate the value of K_c.

What Is Given?

You know the chemical equation and the amount of each substance at equilibrium.

Plan Your Strategy

Step 1 Calculate the molar concentration of each compound at equilibrium.

Step 2 Write the equilibrium expression. Then substitute the equilibrium molar concentrations into the expression.

Act on Your Strategy

The reaction takes place in a 5.0 L flask. Therefore, the equilibrium molar concentrations are

$$[N_2(g)] = \frac{0.0070 \text{ mol}}{5.0 \text{ L}} = 1.4 \times 10^{-3} \text{ mol/L}$$

$$[Cl_2(g)] = \frac{0.0022 \text{ mol}}{5.0 \text{ L}} = 4.4 \times 10^{-4} \text{ mol/L}$$

$$[NCl_3(g)] = \frac{0.95 \text{ mol}}{5.0 \text{ L}} = 1.9 \times 10^{-1} \text{ mol/L}$$

$$K_c = \frac{[NCl_3]^2}{[N_2][Cl_2]^3}$$

$$= \frac{(1.9 \times 10^{-1})^2}{(1.4 \times 10^{-3}) \times (4.4 \times 10^{-4})^3}$$

$$= 3.0 \times 10^{11}$$

Check Your Solution

The equilibrium expression has the product terms in the numerator and the reactant terms in the denominator.

Math Tip

Notice that units are not included when using or calculating the value of K_c. This is the usual practice because the units for K_c are not consistent from one reaction to another. The units of K_c depend on the coefficients in the chemical equation.

Sample Problem

Calculating an Equilibrium Concentration

Problem

Nitrogen and hydrogen gases were mixed in a closed 3500 mL flask and reacted to form ammonia. When the reaction had reached equilibrium, the mixture was sampled and found to contain 0.25 mol of ammonia gas, $NH_3(g)$ and 0.080 mol of hydrogen gas, $H_2(g)$. The equilibrium constant for this reaction under the conditions used in the experiment is known to be $K_c = 5.81 \times 10^5$. What amount of nitrogen gas, $N_2(g)$ was present at equilibrium?

What Is Required?

You need to calculate the amount of nitrogen gas, $N_2(g)$, at equilibrium.

What Is Given?

You know the chemical equation, the equilibrium constant, and the concentrations of ammonia and hydrogen at equilibrium.

$K_c = 5.81 \times 10^5$

$n(H_2) = 0.080$ mol

$n(NH_3) = 0.25$ mol

You know the volume of the container.

$V = 3500$ mL $= 3.5$ L

Plan Your Strategy

Step 1 Calculate the molar concentrations of each compound at equilibrium.

Step 2 Write the equilibrium expression for the equation and solve for the concentration of nitrogen, $[N_2(g)]$. Substitute in known values and calculate $[N_2(g)]$.

Step 3 Use the formula for concentration, $c = \frac{n}{V}$, and solve for n. Substitute in the known values and solve for the numerical value of $n(N_2)$.

Act on Your Strategy

$$N_2(g) + 3H_2(g) \rightleftharpoons 2NH_3(g)$$

$$[NH_3] = \frac{0.25 \text{ mol}}{3.5 \text{ L}}$$

$$[NH_3] = 0.07143 \frac{\text{mol}}{\text{L}}$$

$$[H_2] = \frac{0.080 \text{ mol}}{3.5 \text{ L}}$$

$$[H_2] = 0.02286 \frac{\text{mol}}{\text{L}}$$

$$K_c = \frac{[NH_3]^2}{[H_2]^3[N_2]}$$

$$[H_2]^3[N_2]K_c = \frac{[NH_3]^2}{[H_2]^3[N_2]}\left([H_2]^3[N_2]\right)$$

$$\frac{[H_2]^3[N_2]K_c}{[H_2]^3K_c} = \frac{[NH_3]^2}{[H_2]^3K_c}$$

$$[N_2] = \frac{[NH_3]^2}{[H_2]^3K_c}$$

$$[N_2] = \frac{\left(0.07143 \frac{\text{mol}}{\text{L}}\right)^2}{\left(0.02286 \frac{\text{mol}}{\text{L}}\right)^3(5.81 \times 10^5)}$$

$$[N_2] = 7.3512 \times 10^{-4} \frac{\text{mol}}{\text{L}}$$

$$[N_2] \approx 7.4 \times 10^{-4} \frac{\text{mol}}{\text{L}}$$

$$c = \frac{n}{V}$$

$$cV = \frac{n}{V}V$$

$$n = cV$$

$$n = (7.3512 \times 10^{-4} \frac{\text{mol}}{\text{L}} \text{ N}_2)(3.5 \text{ L})$$

$$n = 2.5729 \times 10^{-3} \text{ mol N}_2$$

$$n \approx 2.6 \times 10^{-3} \text{ mol N}_2$$

There were 2.6×10^{-3} moles of nitrogen in the flask at equilibrium.

Check Your Solution

The equilibrium expression was written correctly. The resulting values are reasonable.

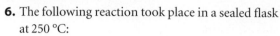

6. The following reaction took place in a sealed flask at 250 °C:

$$PCl_5(g) \rightleftharpoons PCl_3(g) + Cl_2(g)$$

At equilibrium, the gases in the flask had the following concentrations:

$[PCl_5(g)] = 1.2 \times 10^{-2}$ mol/L,
$[PCl_3(g)] = 1.5 \times 10^{-2}$ mol/L, and
$[Cl_2(g)] = 1.5 \times 10^{-2}$ mol/L.

Calculate the value of the equilibrium constant at 250 °C.

7. Iodine and bromine react to form iodine monobromide, $IBr(g)$:

$$I_2(g) + Br_2(g) \rightleftharpoons 2IBr(g)$$

At 150 °C, an equilibrium mixture in a 2.0 L flask was found to contain 0.024 mol iodine, 0.050 mol bromine, and 0.38 mol iodine monobromide. What is the magnitude of K_c for the reaction at 150 °C?

8. At high temperatures, carbon dioxide gas decomposes into carbon monoxide and oxygen gas. The concentration of gases at equilibrium were measured and found to be $[CO_2(g)] = 1.2$ mol/L, $[CO(g)] = 0.35$ mol/L, and $[O_2(g)] = 0.15$ mol/L. Determine K_c at the temperature of the reaction.

9. Hydrogen sulfide is a pungent, poisonous gas. At 1400 K, an equilibrium mixture was found to contain 0.013 mol/L hydrogen, 0.18 mol/L hydrogen sulfide, and an undetermined amount of sulfur in the form of $S_2(g)$. The reaction is

$$2H_2S(g) \rightleftharpoons 2H_2(g) + S_2(g)$$

If the value of K_c at 1400 K is 2.4×10^{-4}, what concentration of $S_2(g)$ is present at equilibrium?

10. Methane, ethyne, and hydrogen form the following equilibrium mixture:

$$2CH_4(g) \rightleftharpoons C_2H_2(g) + 3H_2(g)$$

While studying this reaction, a chemist analyzed a 4.0 L sealed flask containing an equilibrium mixture of the gases at 1700 °C and found 0.64 mol of $C_2H_2(g)$ and 0.92 mol of $H_2(g)$. Given that $K_c = 0.15$ for the reaction at 1700 °C, what concentration should the chemist expect for $CH_4(g)$ at equilibrium?

The Equilibrium Law and Heterogeneous Reactions

All of the examples and problems that you have studied thus far have involved homogeneous equilibriums. Heterogeneous equilibriums often involve a situation that can be handled more easily by modifying the definition of the equilibrium constant. These situations include a pure solid or a pure liquid. The concentration of a pure substance does not change. For example, consider the decomposition of liquid water into hydrogen and oxygen gases shown here.

$$2H_2O(\ell) \rightleftharpoons 2H_2(g) + O_2(g)$$

Initially, you would assume that the equilibrium constant would be written as shown below. In the equation below, a prime ($'$) is added to the K_c to indicate that this is not the final equilibrium constant that you will use.

$$K'_c = \frac{[H_2]^2[O_2]}{[H_2O]^2}$$

The concentration of pure water in the denominator of the expression never changes. It is a constant. (The concentration of pure water under standard conditions is about 55.5 mol/L.) It would be much more convenient to have one constant in an expression rather than two. So, by convention, both sides of the expression are multiplied by the denominator to determine the commonly used constant, K_c.

$$[H_2O]^2 K'_c = \frac{[H_2]^2[O_2]}{\cancel{[H_2O]^2}} \cancel{[H_2O]^2}$$
$$K_c = [H_2O]^2 K'_c = [H_2]^2[O_2]$$
$$K_c = [H_2]^2[O_2]$$

The Development of Equilibrium Theories

The concept of chemical equilibrium is one of the cornerstones of modern chemistry. How was this knowledge arrived at? What steps occurred that allowed modern chemists to understand chemical equilibrium as a scientific concept? It all began with curiosity. Scientists look at everyday phenomena, and seek to understand why those events happen. In fact, the science of chemistry is based on observations of everyday transformations, such as a colour change when two substances are mixed or the burning of wood.

Claude Louis Berthollet (1748–1822)

Chemical Equilibrium

The understanding of chemical equilibrium began with a discovery by a French chemist, C.L. Berthollet (1748–1822). Berthollet taught chemistry to Napoleon, and it was during his travels with him to Egypt in 1798 that he visited Lake Natron, a salt lake located on the edge of the desert. Berthollet noticed that deposits of sodium carbonate, $Na_2CO_3(s)$, surrounded the salt lake. He reasoned that the upper layer of the soil contained sodium chloride that had reacted with the calcium carbonate of nearby limestone. Because he was familiar with the reverse reaction,

$$Na_2CO_3(s) + CaCl_2(s) \rightarrow CaCO_3(s) + 2NaCl(s)$$

he realized that the natural conditions he was seeing could only be the result of the opposite reaction occurring. At the time, scientists believed that a reaction could proceed in one direction only. Consequently, Berthollet's hypothesis was not accepted until almost 50 years later.

At that time, two Norwegian scientists, Cato Guldberg (1836–1902) and Peter Waage (1833–1900), carried out a series of experiments that showed that chemical reactions were made up of forward and reverse reactions in competition with each other. When the rates of the forward and reverse reactions are identical, the chemical reaction is at equilibrium. They went on to summarize their observations in the *law of chemical equilibrium* that states "*In a chemical system, there is a constant ratio between the concentrations of the products and the concentrations of the reactants.*" This law is the foundation of our current understanding of equilibrium.

• • •

1. Berthollet was familiar with the following equation:
   $$Na_2CO_3(s) + CaCl_2(s) \rightarrow CaCO_3(s) + 2NaCl(s)$$
   How should this equation be changed in order to incorporate Berthollet's discovery?

2. Think of an everyday event that you regularly observe but don't truly understand, such as the buildup of different types of ice crystals in your freezer, or the changing colour of leaves in the autumn. What steps would you follow to find out why this event occurs? Assume that there is no previous research on this subject.

3. New scientific understanding is usually based on previous scientific understanding. In some cases, new hypotheses conflict with the accepted understanding of an event. Review the scientific method and explain why science works this way. Why is it important for researchers to publish the results of their experiments?

• • •

Sodium carbonate buildup around a salt lake like this one inspired Berthollet to hypothesize about reverse chemical reactions.

In general, in reactions in which a pure solid or a pure liquid are involved, their constant concentrations are not included in the equation but become part of the equilibrium constant. Note that the concentrations of pure gases can change because they are very compressible and their volumes can be changed. Likewise, the concentrations of dissolved solids change during a reaction.

Qualitatively Interpreting the Equilibrium Constant

The value of the equilibrium constant in a chemical reaction at a specific temperature tells you about the extent of the reaction at that temperature. The extent of the reaction refers to the relative concentration of reactants compared with the concentration of products at equilibrium. As you know, the equilibrium expression is always written with the product terms divided by the reactant terms. Therefore, a large value of K_c means that the concentration of products is larger than the concentration of reactants at equilibrium. When referring to a reaction with a large value of K_c, chemists often say that the position of equilibrium lies to the right or that it favours the products. Similarly, if K_c is small, the concentration of reactants is larger than the concentration of products at equilibrium. Chemists say that the position of equilibrium lies to the left or that it favours the reactants. Thus, the following general statements are true:

- When $K > 1$, products are favoured. The equilibrium lies to the right. Reactions in which K is greater than 10^{10} are usually regarded as proceeding to completion.
- When $K \approx 1$, there are approximately equal concentrations of reactants and products at equilibrium.
- When $K < 1$, reactants are favoured. The equilibrium lies to the left. Reactions in which K is smaller than 10^{-10} are usually regarded as not taking place at all.

Notice that the subscript "c" has been left off K in these general statements. This reflects the fact that there are other equilibrium constants to which these statements apply, not just the general form of the equilibrium constant.

Le Châtelier's Principle

A system at equilibrium must be closed. At equilibrium, the amounts of reactants and products remain the same, the volume and pressure of the system is constant, and the temperature does not change. What happens to an equilibrium system if one or more of these factors are changed? This question has practical importance, because many manufacturing processes are continuous. Products are removed and more reactants are added without stopping the process. For example, consider the Haber-Bosch process to manufacture ammonia:

$$N_2(g) + 3H_2(g) \rightleftharpoons 2NH_3(g) \qquad K_c = 0.40 \text{ at } 500\ °C$$

Ammonia can be removed from the mixture of gases by cooling. When the temperature is decreased, ammonia liquefies before nitrogen or hydrogen does. What happens to an equilibrium mixture if some ammonia is removed? Initially, the system will no longer be at equilibrium. However, because dynamic change is always taking place at the molecular level, equilibrium will be re-established. The value of the equilibrium constant, K_c, will be unchanged, provided the temperature of the system is returned to its previous value. By looking at the equilibrium expression,

$$K_c = \frac{[NH_3]^2}{[N_2][H_2]^3}$$

you can see that the $[NH_3(g)]$ must increase to return K_c to its previous value. The system will respond to the removal of $NH_3(g)$ by forming more ammonia. Similarly, if nitrogen, hydrogen, or both were added, the system would re-establish equilibrium by shifting to the right to form more ammonia. The concentrations of reactants and products will be different, but the ratio of these concentrations as calculated in the equilibrium expression will be the same as before.

A French chemist, Henri Le Châtelier (Figure 16.6), experimented with various chemical equilibrium systems. In 1888, Le Châtelier summarized his work on changes to equilibrium systems in a general statement called **Le Châtelier's principle**. It may be stated as follows:

Figure 16.6 Henri Le Châtelier (1850–1936) was a chemist who specialized in mining engineering. He studied the reactions involved in heating ores to obtain metals. His results led to important advances in the understanding of equilibrium systems.

> **Le Châtelier's Principle**
> A dynamic equilibrium tends to respond so as to relieve the effect of any change in the conditions that affect the equilibrium.

Le Châtelier's principle qualitatively predicts the way that an equilibrium system responds to change. If products are removed from an equilibrium system, more products must be formed to relieve that change. If reactants are added to an equilibrium system, more products will form to relieve that change.

To visualize how this works, consider, again, the equilibrium between $N_2O_4(g)$ and $NO_2(g)$. Figure 16.5 showed you how an equilibrium was established. Now imagine that you inject more $N_2O_4(g)$ into the system. Figure 16.7 shows you what would happen. Notice that the concentration of $N_2O_4(g)$ instantly rises with the injection of more of the gas. Then the concentration immediately begins to drop because the rate of the forward reaction increases in response to an increase in the amount of reactant. As more product accumulates, the rate of the backward reaction begins to increase. The system is responding to "relieve the change." The change was an increase in reactant. The system begins to lower the concentration of reactant and then increase the concentration of product. Soon, a new equilibrium is established with a new concentration of reactant and product. The ratio of the new concentrations of reactants and products, however, is the same as the original ratio because the equilibrium constant has not changed.

For another example of the response of a system to a disturbance in the established equilibrium, examine your data from Investigation 16.A (page 635). After equilibrium had been established, you added water to the reactant cylinder. What change would be predicted by Le Châtelier's principle? Does the graph of your results show the change you expect to see? Le Châtelier's principle also predicts what will happen when other changes are made to an equilibrium system. For example, you can use Le Châtelier's principle to predict the effects of changing the volume of a cylinder containing a mixture of gases or changing the temperature of an equilibrium system.

The Effects of Volume and Pressure Changes on Equilibrium

When the volume of a mixture of gases is decreased, the pressure of the gases must increase in agreement with Boyle's law. Initially, the system will no longer be at equilibrium (Figure 16.8). Le Châtelier's principle predicts a shift in the reaction to relieve this change, so the shift must reduce the pressure of the gases. Molecules striking the walls of a container cause gas pressure, so a reduction in gas pressure at constant temperature must mean fewer gas molecules. Consider the following reaction:

$$2SO_3(g) \rightleftharpoons 2SO_2(g) + O_2(g)$$

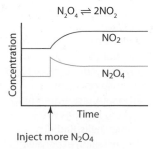

Figure 16.7 If you disturb the equilibrium between $N_2O_4(g)$ and $NO_2(g)$, the system will respond by making an opposing change that will bring the system back to equilibrium.

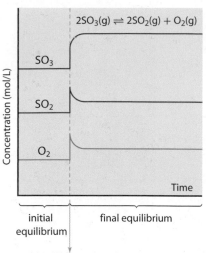

Concentration (mol/L)

$2SO_3(g) \rightleftharpoons 2SO_2(g) + O_2(g)$

SO₃

SO₂

O₂

Time

initial equilibrium | final equilibrium

sudden decrease in volume

Figure 16.8 When the volume is decreased, the concentration of each gas in the mixture increases. Initially, the system is no longer at equilibrium. The system responds to re-establish equilibrium. When equilibrium is re-established, the concentrations of the gases are not the same as before the volume decreased.

There are two moles of gas molecules on the left of the equation, and a total of three moles of gas molecules on the right side. Consequently, if the equilibrium shifts to the left, the pressure of the mixture will decrease. *Reducing the volume of an equilibrium mixture of gases at constant temperature causes a shift in the reaction in the direction of fewer gas molecules.* However, the ratio of the concentrations of products and reactants does not change. The equilibrium constant remains the same. Are you wondering how the ratio of the concentrations can change without changing the equilibrium constant? Imagine that the volume of the system depicted in Figure 16.8 was reduced to half the original volume. The concentration of each component of the system would double. Compare the ratios below. Since the column on the left represents the system at equilibrium, the ratio is equal to the equilibrium constant. The ratio on the right represents the conditions immediately after the volume was reduced to half. Since the system is no longer at equilibrium, the ratio is not equal to the equilibrium constant.

Ratio At Original Equilibrium

$$K_c = \frac{[SO_2]^2[O_2]}{[SO_3]^2}$$

New Ratio

$$\frac{[2[SO_2]]^2[2[O_2]]}{[2[SO_3]]^2} = \frac{2^2[SO_2]^2 2[O_2]}{2^2[SO_3]^2} = 2\frac{[SO_2]^2[O_2]}{[SO_3]^2} = 2K_c$$

The ratio of the doubled concentrations is equal to twice the equilibrium constant. The system must decrease the concentrations of the products and increase the concentrations of the reactants to bring the ratio back to the original equilibrium constant.

You can also increase the pressure on the reactant and product gases by adding an inert or non-reacting gas. Since gases are completely miscible, the gases will mix totally. Thus the same volume is available to the original gases and their concentrations will not change. Consequently, no change will take place in the position of the equilibrium or in the equilibrium constant. There is no change the system could make that would relieve the additional pressure.

What if the reaction equation has the same number of gas molecules on both sides? In this case, a shift in the reaction would cause no change in the pressure of the system. Changing the volume of the container would have no effect on the position of equilibrium.

The Effect of Temperature Changes on the Position of Equilibrium

As you know, the value of the equilibrium constant changes with temperature, because the rates of forward and reverse reactions are affected. Le Châtelier's principle is useful for making qualitative predictions when the temperature of an equilibrium mixture is changed. It can be used to predict the effect of temperature changes on a system when the enthalpy change for the reaction is known. For example, the decomposition of sulfur trioxide is endothermic.

$$2SO_3(g) + 97 \text{ kJ} \rightleftharpoons 2SO_2(g) + O_2(g)$$

As the reaction proceeds from left to right, energy is absorbed by the chemical system and is converted to chemical potential energy. If an equilibrium mixture of these gases is heated, energy is added to the system. Le Châtelier's principle predicts a shift that will relieve the imposed change. Therefore, the shift will tend to remove the added energy. This will happen if the equilibrium shifts to the right, increasing the amount of $SO_2(g)$ and $O_2(g)$ formed. Consequently, the value of K_c will increase. The shift in equilibrium is consistent (as it must be) with the law of conservation of energy.

ChemistryFile

FYI
Chickens have no sweat glands. When the temperature rises, they tend to breathe faster, which lowers the concentration of carbonate ions in their blood. Because eggshells are mostly calcium carbonate, faster-breathing chickens lay eggs with thinner shells. Rather than installing expensive air conditioning, chicken farmers supply carbonated water for them to drink when the temperature reaches "fowl" highs. How does this relate to Le Châtelier's principle?

The effect of temperature on the position of equilibrium can be summarized as described in the box:

> - *Endothermic change*—An increase in temperature shifts the equilibrium to the right, forming more products. The value of K_c increases. A decrease in temperature shifts the equilibrium to the left, forming more reactants. The value of K_c decreases.
> - *Exothermic change*—An increase in temperature shifts the equilibrium to the left, forming more reactants. The value of K_c decreases. A decrease in temperature shifts the equilibrium to the right, forming more products. The value of K_c increases.

A change in temperature is the only change that causes a change in the equilibrium constant.

Sample Problem

Temperature and the Extent of a Reaction

Problem
Carbon monoxide and chlorine react together to form phosgene, $COCl_2(g)$:

$$CO(g) + Cl_2(g) \rightleftharpoons COCl_2(g)$$

At 870 K, the value of K_c for this reaction is 0.20. At 370 K, the value of K_c is 4.6×10^7. Based on only the values of K_c, is the production of phosgene more favourable at higher or lower temperatures?

What Is Required?
You need to choose the temperature that favours the greater concentration of product.

What Is Given?
$K_c = 0.2$ at 870 K
$K_c = 4.6 \times 10^7$ at 370 K

Plan Your Strategy
Phosgene is a product in the chemical equation. A greater concentration of product corresponds to a larger value of K_c.

Act on Your Strategy
The value of K_c is greater at 370 K. The position of equilibrium lies far to the right and favours the manufacture of $COCl_2(g)$ at the lower temperature.

Check Your Solution
The question asked you to choose the conditions that result in a larger concentration of a product. K_c is expressed as a fraction with product terms in the numerator. Therefore, the larger value of K_c must correspond to the larger concentration of product.

Practice Problems

11. For the reaction $H_2(g) + I_2(g) \rightleftharpoons 2HI(g)$, the value of K_c is 25.0 at 1100 K and 8.0×10^2 at room temperature, 300 K. Which temperature favours the decomposition of hydrogen iodide into its component gases?

12. Three reactions and their equilibrium constants are given below:

 I. $N_2(g) + O_2(g) \rightleftharpoons 2NO(g)$ $\qquad K_c = 4.7 \times 10^{-31}$
 II. $2NO(g) + O_2(g) \rightleftharpoons 2NO_2(g)$ $\qquad K_c = 1.8 \times 10^{-6}$
 III. $N_2O_4(g) \rightleftharpoons 2NO_2(g)$ $\qquad K_c = 0.025$

Arrange these reactions in order of their tendency to form products.

13. Identify each of the following reactions as essentially proceeding to completion or not taking place:
 a) $N_2(g) + 3Cl_2(g) \rightleftharpoons 2NCl_3(g)$ $\qquad K_c = 3.0 \times 10^{11}$
 b) $2CH_4(g) \rightleftharpoons C_2H_6(g) + H_2(g)$ $\qquad K_c = 9.5 \times 10^{-13}$
 c) $2NO(g) + 2CO(g) \rightleftharpoons N_2(g) + 2CO_2(g)$
 $\qquad\qquad\qquad\qquad\qquad\qquad K_c = 2.2 \times 10^{59}$

14. Consider the following reaction:

$$H_2(g) + Cl_2(g) \rightleftharpoons 2HCl(g) \quad K_c = 2.4 \times 10^{33} \text{ at } 25\,°C$$

If a quantity of HCl(g) is placed into a reaction vessel, to what extent do you expect the equilibrium mixture will decompose into hydrogen and chlorine?

15. Most metal ions combine with other ions present in solution. For example, in aqueous ammonia, silver (Ag) ions are at equilibrium with different complex ions.

$$[Ag(H_2O)_2]^+(aq) + 2NH_3(aq) \rightleftharpoons [Ag(NH_3)_2]^+(aq) + 2H_2O(\ell)$$

At room temperature, K_c for this reaction is 1×10^7. Which of the two silver complex ions is the more stable? Explain your reasoning.

The Effect of a Catalyst on Equilibrium

The addition of a catalyst speeds up the *rate* of a reaction either by allowing a different reaction mechanism or by providing additional mechanisms. The overall effect is to lower the activation energy, which increases the rate of reaction. However, the activation energy is lowered the same amount for the forward and reverse reactions. The same increase in reaction rate occurs for both reactions. As a result, the addition of a catalyst has no effect on the position of equilibrium. The only effect (although it is an important one) is to decrease the time taken to achieve equilibrium. Table 16.2 summarizes the effects of changing conditions on a system at equilibrium.

Table 16.2 The Effects of Changing Conditions on a System at Equilibrium

Type of reaction	Change to system	Effect on K_c	Direction of change
all reactions	increasing any reactant concentration or decreasing any product concentration	no effect	toward products
	decreasing any reactant concentration or increasing any product concentration	no effect	toward reactants
	using a catalyst	no effect	no change
exothermic	increasing temperature	decreases	toward reactants
	decreasing temperature	increases	toward products
endothermic	increasing temperature	increases	toward products
	decreasing temperature	decreases	toward reactants
equal number of reactant and product gas molecules	changing the volume of the container or adding a non-reacting gas	no effect	no change
more gaseous product molecules than reactant gaseous molecules	decreasing the volume of the container at constant temperature	no effect	toward reactants
	adding a non-reacting gas	no effect	no change
	increasing the volume of the container at constant temperature	no effect	toward products
fewer gaseous product molecules than reactant gaseous molecules	decreasing the volume of the container at constant temperature	no effect	toward products
	adding a non-reacting gas	no effect	no change
	increasing the volume of the container at constant temperature	no effect	toward reactants

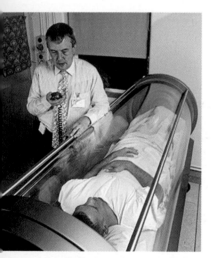

Figure 16.9 Serious burns are complex injuries. One effect of a serious burn can be the destruction of skin tissue and blood supply to the skin. To promote tissue healing in the absence of a normal supply of oxygen, a patient may be placed into a hyperbaric chamber. Inside a hyperbaric chamber, the pressure is increased, changing the equilibrium between gaseous oxygen and oxygen dissolved in tissue fluids. Increased concentration of dissolved oxygen results, promoting tissue growth.

Using Le Châtelier's Principle

Problem

The following equilibrium takes place in a rigid container:

$PCl_5(g) + heat \rightleftharpoons PCl_3(g) + Cl_2(g)$

In which direction does the reaction shift as a result of each of the following changes?

a) adding phosphorus pentachloride gas

b) removing chlorine gas

c) decreasing the temperature

d) increasing the pressure by adding helium gas

e) using a catalyst

What Is Required?

You need to determine whether each change causes the reaction to shift to the left or the right, or whether it has no effect.

What Is Given?

You have the chemical equation. You know the reaction is endothermic.

Plan Your Strategy

Identify the change that was imposed. Then use the chemical equation to determine the shift in the reaction that will minimize the change.

Act on Your Strategy

a) $[PCl_5(g)]$ increases. Therefore, the equilibrium must shift to reduce $[PCl_5(g)]$. The reaction shifts to the right.

b) $[Cl_2(g)]$ is reduced. Therefore, the equilibrium must shift to increase $[Cl_2(g)]$. The reaction shifts to the right.

c) The temperature decreases. Therefore, the equilibrium must shift in the direction in which the reaction is exothermic. From left to right the reaction is endothermic. Thus, the reaction must be exothermic from right to left. The reaction shifts to the left if the temperature decreases.

d) Helium does not react with any of the gases in the mixture. The presence of helium increases the pressure on the gases. However, the concentrations of the gases do not change thus the system is still at equilibrium. Only a change in temperature can change the equilibrium constant.

e) The addition of a catalyst will have no effect on the position of equilibrium. No change will occur.

Check Your Solution

Check the introduced changes. Any change that affects the equilibrium reaction must result in a shift that minimizes the change.

Practice Problems

16. Consider the following reaction:
    $2HI(g) \rightleftharpoons H_2(g) + I_2(g)$ $\Delta H = -52\,kJ$
    In which direction does the reaction shift if the temperature increases? Explain why.

17. In the gaseous equilibrium systems below, the volume of the container is increased, causing a decrease in pressure. In which direction does each reaction shift? Explain why for each case.
    a) $CO_2(g) + H_2(g) \rightleftharpoons CO(g) + H_2O(g)$
    b) $2NO_2(g) \rightleftharpoons N_2O_4(g)$
    c) $2CO_2(g) \rightleftharpoons 2CO(g) + O_2(g)$
    d) $CH_4(g) + 2H_2S(g) \rightleftharpoons CS_2(g) + 4H_2(g)$

18. The following reaction is exothermic from left to right:
    $2NO(g) + 2H_2(g) \rightleftharpoons N_2(g) + 2H_2O(g)$
    In which direction does the reaction shift as a result of each of the following changes?

    a) removing hydrogen gas

    b) increasing the pressure of gases in the reaction vessel by decreasing the volume

    c) increasing the pressure of gases in the reaction vessel by pumping in argon gas while keeping the volume of the vessel constant

    d) increasing the temperature

    e) using a catalyst

19. In question 18, which changes affect the value of K_c? Which changes do not affect the value of K_c?

20. Toluene, $C_7H_8(g)$, is an important organic solvent. It is made industrially from methyl cyclohexane:
    $C_7H_{14}(g) \rightleftharpoons C_7H_8(g) + 3H_2(g)$
    The forward reaction is endothermic. State three different changes to an equilibrium mixture of these reacting gases that would shift the reaction toward greater production of toluene.

Disturbing Equilibrium

In this investigation, you will use Le Châtelier's principle to predict the effect of changing one factor that affects a system at equilibrium. Then you will design a test to check your prediction by assessing a change of colour or the appearance (or disappearance) of a precipitate.

Question

How can Le Châtelier's principle qualitatively predict the effect of a change in a chemical equilibrium?

Part 1 Changes to a Base Equilibrium System

Safety Precautions

- Hydrochloric acid and aqueous ammonia are corrosive to eyes and skin and harmful if swallowed or inhaled. Wash any spills on your skin or clothing with plenty of cool water. Inform your teacher immediately. Also inform your teacher immediately if you spill hydrochloric acid or aqueous ammonia on the lab bench or floor.

- Ammonium chloride is harmful if inhaled or swallowed and causes skin and eye irritation. Avoid contact with eyes and treat spills with copious amounts of cool water. Inform your teacher immediately.

- Phenolphthalein solution may irritate skin, eyes, and mucous membranes. This solution is flammable. Keep away from open flames.

- Wash your hands when you have completed the investigation.

Materials
- 0.01 mol/L NH_3(aq)
- phenolphthalein solution
- NH_4Cl(s)
- 6.0 mol/L HCl(aq)
- 25 mL beaker
- white paper
- 2 test tubes
- test-tube rack
- scoopula

Equilibrium System

$$NH_3(aq) + H_2O(\ell) \rightleftharpoons NH_4^+(aq) + OH^-(aq)$$

You can detect a shift in this equilibrium by using phenolphthalein indicator.

Procedure

1. Pour about 10 mL of NH_3(aq) into a small beaker. Place the beaker on a sheet of white paper. Add two drops of phenolphthalein indicator.

2. Divide the solution equally into two small test tubes. Given the list of materials, design a procedure to test Le Châtelier's principle. Describe how you will shift the equilibrium and predict the colour of the phenolphthalein indicator as a result of the shift. Include guidelines for the safe disposal of all materials.

3. Construct a data table to record your observations.

4. Check your procedure with your teacher. Carry out your procedure and record your observations.

Part 2 Concentration and Temperature Changes

Safety Precautions

- Concentrated hydrochloric acid is corrosive to your eyes, skin, and clothing. Avoid contact with eyes and treat spills with copious amounts of cool water. Inform your teacher immediately. Also inform your teacher if you spill hydrochloric acid on the lab bench or floor.

- Cobalt chloride and silver nitrate solutions are toxic and corrosive. Keep away from eyes and skin. Wash spills on your skin with plenty of cool water. Inform your teacher immediately. Silver nitrate is flammable. Keep away from open flames.

- Ethanol is flammable. Keep ethanol and solutions containing ethanol away from open flames.

Materials
- $CoCl_2$ dissolved in a solution of water and ethanol
- concentrated HCl(aq) in a dropper bottle
- $AgNO_3$(aq) 0.1 mol/L in a dropper bottle
- distilled water in a dropper bottle

- hot water bath
- cold water bath
- 25 mL or 50 mL beaker
- 4 small test tubes
- test-tube rack
- test-tube holder

Equilibrium System

When $CoCl_2(s)$ dissolves in a solution of water and ethanol(aq/et), the salt dissociates and the Co^{2+} ion combines with water to form $Co(H_2O)_6^{2+}$(aq/et). The state (aq/et) indicates a solution of water and ethanol:

$$Heat + Co(H_2O)_6^{2+}(aq/et) + 4Cl^-(aq/et) \rightleftharpoons$$

pink	$CoCl_4^{2-}$(aq/et) + 6H_2O(ℓ/et)
	blue or purple

Procedure

1. Pour 25–30 mL of $CoCl_2$(aq/et) into a small beaker. The solution should be blue or purple. If it is pink, add drops of concentrated HCl(aq) until the solution is blue-purple.

2. Pour about 5 mL of the solution into one of the tubes and put it aside as a control.

3. Given the list of materials, design as many procedures as possible to test Le Châtelier's principle. Describe how you will shift the equilibrium and predict the colour of the solution as a result of the shift. Include guidelines for the safe disposal of all materials.

4. Construct a data table to record your observations.

5. Carry out your procedures and record your observations.

Part 3 Investigating Gaseous Equilibriums

Caution These are not student tests. Due to the poisonous nature of the nitrogen dioxide and dinitrogen tetroxide gases, it is strongly recommended that this part of the investigation be done by analyzing the photographs below and on the following page. However, if your classroom has an efficient fume hood and the apparatus required is available, your teacher may choose to do the demonstrations as described in the Procedure.

Safety Precautions

- Concentrated nitric acid is highly corrosive and a strong oxidizing agent.

- Nitrogen dioxide and dinitrogen tetroxide are poisonous gases.

Materials
- small piece of copper
- concentrated nitric acid ⊞ ☠ ⟁
- boiling water
- ice water
- test tube
- test-tube rack
- one-hole stopper
- glass delivery tube
- short length of rubber tubing
- syringe with a cap or rubber stopper to seal the tip
- $NO_2(g)/N_2O_4(g)$ tubes ☠ ⊞

Equilibrium System

$$Heat + N_2O_4(g) \rightleftharpoons 2NO_2(g)$$
colourless   brown

Procedure

1. Your teacher will use sealed tubes containing a mixture of $N_2O_4(g)$ and $NO_2(g)$. One tube will be placed in boiling water, and a second tube will be placed in ice water. A third tube (if available) will remain at room temperature as a control. Compare and record the colour of the gas mixture at each temperature.

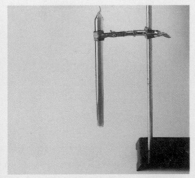

These three tubes contain a mixture of $NO_2(g)$ and $N_2O_4(g)$. The tube on the left is in an ice-water bath. The centre tube is at room temperature. The tube on the right is in boiling water. Given that $NO_2(g)$ is brown, can you explain the shift in equilibrium? Think about Le Châtelier's principle and the enthalpy of the reaction between the two gases.

2. $NO_2(g)$ can be prepared by reacting copper with concentrated nitric acid. The gas is poisonous. The reaction, if your teacher performs it, must take place in a fume hood.

3. By using a one-hole stopper, glass delivery tube, and a short length of rubber tubing, some $NO_2(g)$ can be collected in a syringe. The syringe is then sealed by attaching a cap or by pushing the needle into a rubber stopper.

4. Observe what happens when the syringe plunger is pressed down sharply, changing the volume of the equilibrium mixture. You will observe an immediate change in colour. Then, if the plunger is held in a fixed position the colour will change over a few seconds as the system re-establishes equilibrium. Carefully record these colour changes.

Analysis

1. Compare the predictions you made using Le Châtelier's principle with the observations you made in your tests. Account for any differences.

2. In which tests did you increase the concentration of a reactant or product? Did your observations indicate a shift in equilibrium to form more or less of the reactant or product?

3. In which tests did you decrease the concentration of a reactant or product? Did your observations indicate a shift in equilibrium to form more or less of the reactant or product?

4. In two of the systems you studied, energy changes were indicated:

$$\text{Heat} + Co(H_2O)_6{}^{2+}(aq/et) + 4Cl^-(aq/et) \rightleftharpoons$$
$$CoCl_4{}^{2-}(aq/et) + 6H_2O(\ell/et)$$
$$\text{Heat} + N_2O_4(g) \rightleftharpoons 2NO_2(g)$$

a) Are these systems endothermic or exothermic when read from left to right?

b) When heated, did these systems shift to the left or to the right? In terms of the energy change, was the observed shift in equilibrium toward the endothermic or exothermic side of the reaction?

c) Do you think the value of K_c changed or remained the same when the equilibrium mixture was heated? Explain your answer.

5. Think about the $N_2O_4(g)/NO_2(g)$ equilibrium.

a) How was the total pressure of the mixture affected when the plunger was pushed down?

b) How was the pressure of the mixture affected by the total number of gas molecules in the syringe?

c) Explain the observed shift in the reaction when the plunger was pushed down. In your explanation, refer to Le Châtelier's principle and the total amount of gas in the syringe.

6. Did any groups perform experiments on an equilibrium system that differed from the investigations of the same system performed by your group? If so, briefly describe the experiment and the observations.

Conclusion

7. How did your results compare with your predictions? Discuss with your class and resolve any discrepancies.

Application

8. What would be the effect, if any, on the following equilibrium system if the volume were reduced at constant temperature? Explain.

$$2IBr(g) \rightleftharpoons I_2(g) + Br_2(g)$$

The sealed syringe contains a mixture of $NO_2(g)$ and $N_2O_4(g)$. The photograph on the left shows an equilibrium mixture at atmospheric pressure. The middle photograph shows that the plunger has been pushed down, increasing the pressure. The darker appearance of the gas mixture is caused by two changes. First, the concentration of gases is greater. Second, decreasing the volume heats the gas. Why does heating the mixture of gases result in the colour becoming darker? The photograph on the right shows the result a few seconds after the plunger was forced down. The gas has cooled back to room temperature. The colour of the mixture is less brown. Explain this observation.

Section 16.2 Summary

- According to the law of equilibrium, when a chemical system reaches equilibrium, there is a constant ratio of the products and the reactants in a chemical reaction.

- The equilibrium law expression is $K_c = \dfrac{[C]^c[D]^d}{[A]^a[B]^b}$ for the general reaction, $aA + bB \rightleftharpoons cC + dD$

- In heterogeneous reactions, there is often a pure liquid or a pure solid. Since the concentration of these pure substances does not change, their concentrations are not included in the expression for the equilibrium constant.

- Qualitatively, when $K > 1$, products are favoured. When $K \approx 1$, there are approximately equal concentrations of products and reactants at equilibrium. When $K < 1$, reactants are favoured.

- Le Châtelier's principle states that a dynamic equilibrium tends to respond so as to relieve the effect of any change in the conditions that affect the equilibrium.

- A change in temperature is the only condition that can cause a change in the equilibrium constant.

SECTION 16.2 REVIEW

1. Write equilibrium expressions for each of the following homogeneous reactions:
   a) $SbCl_5(g) \rightleftharpoons SbCl_3(g) + Cl_2(g)$
   b) $2H_2(g) + 2NO(g) \rightleftharpoons N_2(g) + 2H_2O(g)$
   c) $2H_2S(g) + CH_4(g) \rightleftharpoons 4H_2(g) + CS_2(g)$

2. When 1.0 mol of ammonia gas is injected into a 0.50 L flask at a given temperature, the following reaction proceeds to equilibrium:

   $$2NH_3(g) \rightleftharpoons N_2(g) + 3H_2(g)$$

   At equilibrium, the mixture contains 0.30 mol of $H_2(g)$, 0.10 mol of $N_2(g)$, and 0.80 mol of $NH_3(g)$.
   a) Write the equilibrium expression for the reaction.
   b) What is the value of the equilibrium constant?

3. Phosphorus trichloride reacts with chlorine to form phosphorus pentachloride as follows:

   $$PCl_3(g) + Cl_2(g) \rightleftharpoons PCl_5(g)$$

   When 0.75 mol each of $PCl_3(g)$ and $Cl_2(g)$ were placed into an 8.0 L reaction vessel at 500 K, the equilibrium concentration of the mixture was found to contain 0.035 mol/L of $PCl_3(g)$ and $Cl_2(g)$. The value of K_c at 500 K is 49. Calculate the equilibrium concentration of $PCl_5(g)$.

4. Nitrogen gas and oxygen gas are present in large quantities in the atmosphere. At a certain temperature, the value of K_c for the following reaction is 4.2×10^{-8}:

   $$N_2(g) + O_2(g) \rightleftharpoons 2NO(g)$$

   What can you say about the equilibrium concentration of $NO(g)$ at this temperature?

5. At one time, methanol was obtained by heating wood without allowing the wood to burn. The products were collected, and methanol (sometimes called "wood alcohol") was separated by distillation. Today, methanol is manufactured by reacting carbon monoxide with hydrogen gas:

   $$CO(g) + 2H_2(g) \rightleftharpoons CH_3OH(g)$$

   At 210 °C, K_c for this reaction equals 14.5. Is this temperature favourable for the formation of methanol? Explain.

6. In which direction does the reaction shift as a result of the change to each homogeneous equilibrium system?
   a) Adding $Cl_2(g)$: $2Cl_2(g) + O_2(g) \rightleftharpoons 2Cl_2O(g)$
   b) Removing $N_2(g)$: $2NO_2(g) \rightleftharpoons N_2(g) + 2O_2(g)$
   c) Using a catalyst: $CH_4(g) + 2H_2O(g) \rightleftharpoons CO_2(g) + 4H_2(g)$
   d) Decreasing the total volume of the reaction container: $2NO_2(g) \rightleftharpoons N_2O_4(g)$
   e) Increasing the temperature: $CO(g) + 3H_2(g) \rightleftharpoons CH_4(g) + H_2O(g)$ $\Delta H = -230$ kJ

7. For each reversible reaction, determine whether high temperatures or low temperatures favour the forward reaction:
   a) $N_2O_4(g) \rightleftharpoons 2NO_2(g)$ $\Delta H = +59$ kJ
   b) $2ICl(g) \rightleftharpoons I_2(g) + Cl_2(g)$ $\Delta H = -35$ kJ
   c) $2CO_2(g) + 566 \text{ kJ} \rightleftharpoons 2CO(g) + O_2(g)$
   d) $2HF(g) \rightleftharpoons H_2(g) + F_2(g)$ $\Delta H = -536$ kJ

8. In each reaction, the volume of the reaction vessel is decreased. What is the effect (if any) on the equilibrium?
   a) $N_2(g) + O_2(g) \rightleftharpoons 2NO(g)$
   b) $4HCl(g) + O_2(g) \rightleftharpoons 2Cl_2(g) + 2H_2O(g)$
   c) $2H_2S(g) + CH_4(g) \rightleftharpoons 4H_2(g) + CS_2(g)$

SECTION 16.3

Equilibrium Calculations for Homogeneous Systems

Section Outcomes

In this section, you will:

- **determine** equilibrium constants and concentrations for homogeneous systems
- **apply** stoichiometric principles to chemical equations at equilibrium
- **solve** equilibrium problems using an approximation method
- **determine** an equilibrium constant by preparing equilibrium mixtures with different concentrations

In Section 16.2, you learned about the concept of equilibrium and the expression for the equilibrium law constant. You practiced calculating the equilibrium constant from given data about the concentrations of reactants and products at equilibrium. However, it is not always possible to determine the concentration of each individual component of a reaction at equilibrium. Also, you might want to make predictions about the amount of products that would be produced from certain amounts of reactants, given the equilibrium constant. In such cases, you might need to use stoichiometry and the mole ratios of reactants and/or products to determine needed concentrations of other reactants or products in an equilibrium mixture. In this section, you will learn a method for solving this type of problem.

Creating and Using ICE Tables

An ICE table is a convenient way to organize data for an equilibrium experiment and calculation. The letter "I" represents the initial molar concentrations of the reactants and products. The letter "C" represents the amount of change in each reactant and product from initial conditions to equilibrium. The letter "E" represents the molar concentrations of the reactants and products at equilibrium. To set up the table, you write out the equilibrium equation and make a column below each reactant and product. Then label the rows with **I**nitial, **C**hange, and **E**quilibrium. Fill in the table with all of the known values that are available. Use x to represent the value that you do not know. You then fill in the other boxes in the table with expressions that include the known and unknown values. For example, imagine that you wanted to experimentally determine the equilibrium constant for the decomposition of hydrogen iodide at 453 °C. You filled an evacuated 2.0 L flask with 0.200 mol of hydrogen iodide, HI(g) at 453 °C. You let the hydrogen iodide gas decompose into hydrogen gas and iodine gas until it reached equilibrium. You were able to determine that the concentration of hydrogen iodide gas was 0.078 mol/L at equilibrium but you could not determine the concentration of the hydrogen or iodine gases. First, set up your ICE table as shown below. You know the initial concentrations because you put 0.200 mol of HI(g) into a 2.0 L flask so the concentration is 0.200 mol/2.0 L or 0.100 mol/L. Put it into the table. Initially, there was no hydrogen or iodine so put zero for their concentrations. From the stoichiometry of the reaction, the mole ratio of hydrogen gas to iodine gas is 1:1. Therefore, at equilibrium, the concentrations of hydrogen gas and iodine gas will be the same. Let x represent the amount of hydrogen and iodine at equilibrium. Place an x in the boxes for hydrogen gas and iodine gas. Since their concentrations started at zero and ended at x, the change in their concentrations was $+x$. Put those symbols in the table. From the coefficients in the equation, you know that the mole ratios of HI(g) to both $H_2(g)$ and $I_2(g)$ is 2:1. For every mole of hydrogen and iodine that formed, two moles of hydrogen iodide decomposed. Therefore, the change in the hydrogen iodide concentration was $-2x$. Put that in the table. Finally, you know that the concentration of HI(g) at equilibrium is the initial concentration plus the change or $0.100 - 2x$. Put that value into the table.

	2HI(g)	⇌	$H_2(g)$	+	$I_2(g)$
	[HI(g)] mol/L		$[H_2(g)]$ mol/L		$[I_2(g)]$ mol/L
Initial	0.100		0		0
Change	$-2x$		$+x$		$+x$
Equilibrium	$0.100 - 2x$		x		x

You also measured the equilibrium concentration of hydrogen iodide to be 0.078 mol/L. Therefore, $0.100 - 2x = 0.078$ mol/L. Solve for x.

$$0.100\,\frac{mol}{L} - 2x = 0.078\,\frac{mol}{L}$$

$$-2x = 0.078\,\frac{mol}{L} - 0.100\,\frac{mol}{L}$$

$$\frac{-2x}{-2} = \frac{-0.022\,\frac{mol}{L}}{-2}$$

$$x = 0.011\,\frac{mol}{L}$$

Now solve for the equilibrium constant.

$$K_c = \frac{[H_2][I_2]}{[HI]^2}$$

$$= \frac{x \times x}{(0.078)^2}$$

$$= \frac{(0.011)(0.011)}{0.006084}$$

$$= 0.019888$$

$$K_c \approx 0.020$$

Sample Problem

Using Stoichiometry to Calculate K_c

Problem
Hydrogen gas is considered to be a non-polluting, sustainable fuel source. It can be used in internal combustion engines and in fuel cells to produce heat and electricity. Hydrogen gas can be produced in several different ways. The following reaction illustrates one method using water and carbon monoxide:

$$CO(g) + H_2O(g) \rightleftharpoons H_2(g) + CO_2(g)$$

The reaction has been studied at different temperatures to find the optimum conditions. At 700 K, the equilibrium constant is 0.83. Suppose that you start with 1.0 mol of CO(g) and 1.0 mol of $H_2O(g)$ in a 5.0 L container. What amount of each substance will be present in the container when the gases are in equilibrium at 700 K?

What Is Required?
You need to find the amount (in moles) of CO(g), $H_2O(g)$, $H_2(g)$, and $CO_2(g)$ at equilibrium.

What Is Given?
You have the chemical equation. You know the initial amount of each gas, the volume of the container, and the equilibrium constant.

Plan Your Strategy
Step 1 Calculate the initial concentrations.

Step 2 Set up an ICE table. Record the initial concentrations you calculated in step 1 in your ICE table. Let the change in molar concentrations

of the reactants be x. Use the stoichiometry of the chemical equation to write and record expressions for the equilibrium concentrations.

Step 3 Write the equilibrium expression. Substitute the expressions for the equilibrium concentrations into the expression for K_c. Solve the equilibrium expression for x.

Step 4 Calculate the equilibrium concentration of each gas. Then use the volume of the container to find the amount of each gas present.

Act on Your Strategy
Step 1 The initial amount of CO(g) is equal to the initial amount of $H_2O(g)$:

$$[CO(g)] = [H_2O(g)] = \frac{1.0\,mol}{5.0\,L} = 0.20\,mol/L$$

Step 2 Set up an ICE table.

	CO(g) + H₂O(g) ⇌ H₂(g) + CO₂(g)			
	[CO(g)] (mol/L)	[H₂O(g)] (mol/L)	[H₂(g)] (mol/L)	[CO₂(g)] (mol/L)
Initial	0.20	0.20	0	0
Change	$-x$	$-x$	$+x$	$+x$
Equilibrium	$0.20 - x$	$0.20 - x$	x	x

Note that change in concentration for the reactants is negative because they are being used up in the reaction, while change in concentration for the products is positive because products are being created.

Step 3 $K_c = \dfrac{[H_2][CO_2]}{[CO][H_2O]}$

$0.83 = \dfrac{(x)(x)}{(0.20 - x)(0.20 - x)} = \dfrac{(x)^2}{(0.20 - x)^2}$

$0.9110 = \dfrac{(x)}{(0.20 - x)}$

$x = 0.095\,34$

Step 4 The concentrations of the reactants at equilibrium are as follows:

$[H_2(g)] = [CO_2(g)] \approx 0.095\,34$ mol/L

$[CO(g)] = [H_2O(g)] \approx 0.2000 - 0.095\,34 = 0.104\,66$ mol/L

Round to two significant digits:

$[H_2(g)] = [CO_2(g)] \approx 0.095$ mol/L

$[CO(g)] = [H_2O(g)] \approx 0.10$ mol/L

To find the amount of each gas, multiply the concentration of each gas by the volume of the container (5.0 L).

Amount of $H_2(g) = CO_2(g) = 0.48$ mol

Amount of $CO(g) = H_2O(g) = 0.50$ mol

Check Your Solution

The equilibrium expression has product concentrations in the numerator and reactant concentrations in the denominator. The concentration of each chemical present at equilibrium is given in mol/L. Check K_c by substituting these concentrations back into the equation.

$K_c = \dfrac{(0.095\,343)^2}{(0.104\,66)^2} = 0.8298 \approx 0.83$

Math Tip

FYI

In this problem, the right side of the equilibrium expression is a perfect square in step 3. Noticing perfect squares, then taking the square root of both sides makes solving the equation easier. Also, when you use an answer to one part of a problem as data in another calculation, always use the unrounded value. If you round numbers at every step, the error will become quite large. For example, if you had rounded every step in the solution to this Sample Problem to two significant figures, your answer for K_c would have been 0.75, which is an error of about 10%.

Practice Problems

21. At 1100 K, hydrogen and iodine combine to form hydrogen iodide:

$H_2(g) + I_2(g) \rightleftharpoons 2HI(g)$

At equilibrium in a 1.0 L reaction vessel, the mixture of gases contained 0.30 mol of $H_2(g)$, 1.3 mol of $I_2(g)$, and 3.4 mol of HI(g). What is the value of K_c at 1100 K?

22. At 25 °C, the following reaction takes place:

$I_2(g) + Cl_2(g) \rightleftharpoons 2ICl(g)$

K_c for the reaction is 82. If 0.83 mol of iodine gas and 0.83 mol of chlorine gas are placed in a 10 L container at 25 °C, what will the concentrations of the various gases be at equilibrium?

23. A chemist was studying the following reaction at a certain temperature:

$SO_2(g) + NO_2(g) \rightleftharpoons NO(g) + SO_3(g)$

In a 1.0 L container, the chemist added 1.7×10^{-1} mol of sulfur dioxide to 1.1×10^{-1} mol of nitrogen dioxide. At equilibrium, the concentration of $SO_3(g)$ was found to be 0.089 mol/L. What is the value of K_c for the reaction at this temperature?

24. Hydrogen bromide decomposes at 700 K:

$2HBr(g) \rightleftharpoons H_2(g) + Br_2(g) \qquad K_c = 4.2 \times 10^{-9}$

0.090 mol of HBr is placed into a 2.0 L reaction vessel and heated to 700 K. What is the equilibrium concentration of each gas?

Solving Problems with a Small Equilibrium Constant

In many cases, the mathematical equations that you create when you substitute expressions into the formula for the equilibrium constant will include quadratic equations making them more difficult to solve. For example, in the last Sample Problem, the concentrations of carbon monoxide and water were both $(0.20 - x)$ and when you multiplied them together, you obtained $(0.20 - x)^2$. You simplified the equation by taking the square root. However, you often have a situation in which you will obtain an expression such as $(0.20 - x)(0.30 - x)$ and your result will be $0.04 - 0.50x + x^2$ on one side of the equation and x^2 on the other. You will end up with an equation such as $2x^2 + 0.50x - 0.04 = 0$. Do you remember how to solve a quadratic equation that cannot be factored? Fortunately, in many such situations, you can make an approximation that will simplify the equation.

If the equilibrium constant for a reaction is very small, then a very small amount of product will be present at equilibrium. The amount of reactant at equilibrium will be almost the same as the initial amount. For example, assume that the initial amount of a reactant is 0.065 mol/L and the change in the amount of reactant is 0.000032 mol/L. The amount of reactant present at equilibrium would be 0.065 mol/L − 0.000032 mol/L = 0.064968 mol/L. The rule for significant digits when adding or subtracting tells you that the sum or difference cannot have any more decimal places than the data with the fewest number of decimal places. The number 0.065 has three decimal places so you must round the answer to three decimal places. When you round 0.064968 mol/L to three decimal places, the result is 0.065 mol/L. Your result is identical to the initial concentration of reactant. Therefore, when your equilibrium constant is very small, you can make the approximation that the amount of reactant at equilibrium is the same the initial concentration. To decide whether to use approximation, chemists often compare certain quantities in the data. An approximation may be made if the initial concentrations of reactants are at least 1000 times greater than K_c. The following Sample Problem will show you how to use this approximation.

Math Tip

Generally, it is best to let x represent the substance with the smallest coefficient in the chemical equation. This avoids fractional values of x in the equilibrium expression. Fractional values make solving the equation more difficult.

Sample Problem

Using the Approximation Method

Problem

The atmosphere contains large amounts of oxygen and nitrogen. These two gases do not react at ordinary temperatures; however, they do react at high temperatures, such as those produced by a lightning flash or inside a running car engine. In fact, nitrogen oxides from exhaust gases are a serious pollution problem.

An environmental chemist is studying the following equilibrium reaction:

$$N_2(g) + O_2(g) \rightleftharpoons 2NO(g)$$

At the temperature of the exhaust gases from a particular engine, the value of K_c is 4.2×10^{-8}. The chemist puts 0.085 mol of nitrogen and 0.038 mol of oxygen in a rigid 1.0 L cylinder. What is the concentration of nitrogen monoxide in the mixture at equilibrium?

What Is Required?

You need to find the concentration of NO at equilibrium.

What Is Given?

You have the chemical equation. You know the value of K_c and the following initial concentrations:

$[N_2(g)] = 0.085$ mol/L and $[O_2(g)] = 0.038$ mol/L.

Plan Your Strategy

Step 1 Set up an ICE table. Let x represent the change in $[N_2(g)]$ and $[O_2(g)]$. Because there are two

molecules of NO(g) in the product, the change in concentration, x, must be multiplied by two.

Step 2 Write the equilibrium expression. Substitute the equilibrium concentrations into the expression for K_c.

Step 3 If a quadratic equation is required to solve for x, test to see whether you can use the approximation.

Step 4 Solve for x.

Step 5 Calculate $[NO(g)]$ at equilibrium.

Act on Your Strategy

Step 1

	N_2(g)	+	O_2(g)	$\rightleftharpoons$	2NO(g)
	$[N_2(g)]$ (mol/L)		$[O_2(g)]$ (mol/L)		$[NO(g)]$ (mol/L)
Initial	0.085		0.038		0
Change	$-x$		$-x$		$+2x$
Equilibrium	$0.085 - x$		$0.038 - x$		$2x$

Step 2
$$K_c = \frac{[NO]^2}{[N_2][O_2]}$$

$$4.2 \times 10^{-8} = \frac{x^2}{(0.085 - x)(0.038 - x)}$$

$$x^2 + 5.166 \times 10^{-9}x - 1.3566 \times 10^{-10} = 0$$

A quadratic equation is obtained when the equation is expanded.

Step 3 To find out whether you may use an approximation, multiply K_c by 1000 and compare it to the initial concentrations of the reactants.

$$1000 \times 4.2 \times 10^{-8} = 4.2 \times 10^{-5}$$

$$[N_2(g)] = 0.085 \text{ mol/L}$$

$$[O_2(g)] = 0.038 \text{ mol/L}$$

$1000 \times K_c$ is much less than the initial concentration of either reactant. Therefore you may make the approximation below:

$$0.085 - x \cong 0.085$$

$$0.038 - x \cong 0.038$$

Step 4 $4.2 \times 10^{-8} = \dfrac{(2x)^2}{0.085 \times 0.038}$

$$4x^2 = 1.3566 \times 10^{-10}$$

$$x^2 = 3.3915 \times 10^{-11}$$

$$x = \sqrt{3.39 \times 10^{-11}}$$

$$= 5.82 \times 10^{-6}$$

Step 5 $[NO(g)] = 2x$

$$[NO(g)] = 2(5.82 \times 10^{-6})$$

$$= 1.2 \times 10^{-5}$$

The concentration of $NO(g)$ at equilibrium is $1.2 \times 10^{-5} \text{ mol/L}$.

Check Your Solution

Check the equilibrium values.

$$K_c = \frac{(1.2 \times 10^{-5})^2}{0.085 \times 0.038} = 4.5 \times 10^{-8}$$

This is equal to the equilibrium constant, within rounding errors in the calculation.

Practice Problems

25. The following equation represents the equilibrium reaction for the dissociation of phosgene gas:

$$COCl_2(g) \rightleftharpoons CO(g) + Cl_2(g)$$

At 100 °C, the value of K_c for this reaction is 2.2×10^{-8}. The initial concentration of $COCl_2(g)$ in a closed container at 100 °C is 1.5 mol/L. What are the equilibrium concentrations of $CO(g)$ and $Cl_2(g)$?

26. Hydrogen sulfide is a poisonous gas with a characteristic offensive odour. At 1400 °C, the gas decomposes, with K_c equal to 2.4×10^{-4}:

$$2H_2S(g) \rightleftharpoons 2H_2(g) + S_2(g)$$

4.0 mol of H_2S is placed in a 3.0 L container. What is the equilibrium concentration of hydrogen gas at 1400 °C?

27. At a particular temperature, K_c for the decomposition of carbon dioxide gas is 2.0×10^{-6}:

$$2CO_2(g) \rightleftharpoons 2CO(g) + O_2(g)$$

3.0 mol of CO_2 is put in a 5.0 L container. Calculate the equilibrium concentration of each gas.

28. At a certain temperature, the value of K_c for the following reaction is 3.3×10^{-12}:

$$2NCl_3(g) \rightleftharpoons N_2(g) + 3Cl_2(g)$$

A certain amount of nitrogen trichloride is put in a 1.0 L reaction vessel at this temperature. At equilibrium, 4.6×10^{-6} mol of nitrogen gas is present. What amount of $NCl_3(g)$ was put in the reaction vessel?

29. At a certain temperature, the value of K_c for the following reaction is 4.2×10^{-8}:

$$N_2(g) + O_2(g) \rightleftharpoons 2NO(g)$$

0.45 mol of nitrogen gas and 0.26 mol of oxygen gas are put in a 6.0 L reaction vessel. What is the equilibrium concentration of $NO(g)$ at this temperature?

Measuring Equilibrium Concentrations

The equilibrium constant, K_c, is calculated by substituting equilibrium concentrations into the equilibrium expression. Experimentally, this means that a reaction mixture must come to equilibrium, at which point one or more properties are measured. The properties that are measured depend on the reaction. Common examples of properties that may be measured include colour, pH in aqueous solution, and pressure of a gaseous reaction. From these measurements, the concentrations of the reacting substances can be determined. As you know, it is not necessary to measure all the concentrations in the mixture at equilibrium. You can determine equilibrium concentrations if you know the initial concentrations and the concentration of one reactant or product at equilibrium.

Ethyl ethanoate is an important ester. It is used as a solvent in paints, adhesives, and nail polish remover. Ethanoic acid reacts with ethanol to form ethyl ethanoate and water:

$$CH_3COOH(aq) + CH_3CH_2OH(aq) \rightleftharpoons$$
$$CH_3COOCH_2CH_3(aq) + H_2O(aq)$$

This is a homogeneous equilibrium system in which each substance takes part in the reaction and forms a solution. The reaction is catalyzed by hydrochloric acid.

A group of students investigated this reaction by using the following method.

Procedure

1. A known mass of ethanoic acid was placed in a flask. Then some ethanol was measured and added to the flask.

2. A measured volume of hydrochloric acid of known concentration was added to the mixture of ethanoic acid and ethanol.

3. The flask was sealed with a stopper and placed in a water bath to keep the temperature of the mixture constant at 20 °C. The flask was left for a week to allow the mixture to reach equilibrium.

4. After leaving the flask for a week, the volume of the solution was measured. Then the solution was titrated against a freshly prepared standardized solution of sodium hydroxide, using phenolphthalein as an indicator.

Analysis

1. Using the titration data, the total amount of ethanoic acid and hydrochloric acid present at equilibrium was calculated. Because it is a catalyst, the amount of hydrochloric acid remains constant throughout the reaction. By subtracting the amount of hydrochloric acid from the total amount of acid, the amount of ethanoic acid at equilibrium was determined. The following data were obtained from five different experiments.

The Equilibrium Reaction to Form Ethyl Ethanoate at 20 °C

Experiment	Initial CH₃COOH (aq) (mol)	Initial CH₃CH₂OH (aq) (mol)	Equilibrium CH₃COOH (aq) (mol)	Total volume (mL)
1	0.220	0.114	0.125	38.1
2	0.184	0.115	0.0917	40.3
3	0.152	0.121	0.0631	39.4
4	0.214	0.132	0.110	42.6
5	0.233	0.137	0.122	41.5

2. In a spreadsheet program, enter the data from the table and calculate the initial $[CH_3COOH(aq)]$, the initial $[CH_3CH_2OH(aq)]$, and the equilibrium $[CH_3COOH(aq)]$.

3. The ICE table for experiment 1 is partially filled in below. One more digit has been carried to reduce rounding error. Check that your spreadsheet returns the values shown in the ICE table for experiment 1.

4. Add calculations to your spreadsheet program to calculate the terms missing from the ICE table for each of the five experiments.

5. Use your spreadsheet program to calculate the equilibrium constant, K_c, for each of the five experiments.

6. The five determinations of K_c should be the same *within experimental error*. Comment on the results of the experiments.

7. Use your spreadsheet program to investigate other mathematical relationships among the equilibrium concentrations. Does one of these relationships give a more constant value than the relationship you used to calculate K_c?

$$CH_3COOH(aq) + CH_3CH_2OH(aq) \rightleftharpoons CH_3COOCH_2CH_3(aq) + H_2O(aq)$$

	[CH₃COOH(aq)] (mol/L)	[CH₃CH₂OH(aq)] (mol/L)	[CH₃COOCH₂CH₃(aq)] (mol/L)	[H₂O(aq)] (mol/L)
Initial	5.774	2.992	0	0
Change				
Equilibrium	3.281			

When a reaction involves a coloured substance, change in colour intensity can be measured and used to determine the equilibrium constant for the reaction. For example, an aqueous mixture of iron(III) nitrate and potassium thiocyanate reacts to form iron(III) thiocyanate, $Fe(SCN)^{2+}(aq)$. The reactant solutions are nearly colourless. The product solution ranges in colour from orange to blood red, depending on its concentration. The nitrate and potassium ions are spectator ions, so the net ionic equation is as follows:

$$Fe^{3+}(aq) + SCN^-(aq) \rightleftharpoons Fe(SCN)^{2+}(aq)$$
$$\text{nearly colourless} \qquad \text{orange/blood red}$$

Because the reaction involves a colour change, you can determine the concentration of $Fe(SCN)^{2+}(aq)$ by measuring the intensity of the colour. You will find out how to do this in Investigation 16.C. For now, assume that it can be done. From the measurements of colour intensity, you can calculate the equilibrium concentration of $Fe(SCN)^{2+}(aq)$. Then, knowing the concentration of each solution, you can calculate the equilibrium concentration of each ion by using the chemical equation.

Suppose, for instance, that the initial concentration of $Fe^{3+}(aq)$ is 6.4×10^{-3} mol/L and the initial concentration of $SCN^-(aq)$ is 0.0010 mol/L. When the solutions are mixed, the orange/red complex ion forms. By measuring the intensity of its colour, you are able to determine that the concentration of $Fe(SCN)^{2+}(aq)$ is 4.5×10^{-4} mol/L. From the stoichiometry of the equation, each mole of $Fe(SCN)^{2+}(aq)$ forms when equal amounts of $Fe^{3+}(aq)$ and $SCN^-(aq)$ react. So, if there is 4.5×10^{-4} mol/L of $Fe(SCN)^{2+}(aq)$ at equilibrium, then the same amount of both $Fe^{3+}(aq)$ and $SCN^-(aq)$ must have reacted. This value represents the overall change in the concentrations of the reactants. The equilibrium concentration of any reacting species is the sum of its initial concentration and the change that results from the reaction. For instance, the initial concentration of $Fe^{3+}(aq)$ was 6.4×10^{-3} mol/L. The change in the concentration of $Fe^{3+}(aq)$ as a result of the reaction was -4.5×10^{-4} mol/L. The value is negative because $Fe^{3+}(aq)$ was used up in the reaction. Therefore, the concentration of $Fe^{3+}(aq)$ at equilibrium is $(6.4 \times 10^{-3} - 4.5 \times 10^{-4})$ mol/L = 5.95×10^{-3} mol/L, or 6.0×10^{-3} mol/L. You can determine the equilibrium concentration of $SCN^-(aq)$ the same way and complete the table below.

	$Fe^{3+}(aq)$	+	$SCN^-(aq)$	$\rightleftharpoons$	$Fe(SCN)^{2+}(aq)$
	$[Fe^{3+}(aq)]$ (mol/L)		$[SCN^-(aq)]$ (mol/L)		$[Fe(SCN)^{2+}(aq)]$ (mol/L)
Initial	6.4×10^{-3}		1.0×10^{-3}		0
Change	-4.5×10^{-4}		-4.5×10^{-4}		4.5×10^{-4}
Equilibrium	6.0×10^{-3}		5.5×10^{-4}		4.5×10^{-4}

Finally, you can calculate K_c by substituting the equilibrium concentrations into the equilibrium expression. In the following investigation, you will collect experimental data to determine an equilibrium constant for the reaction between iron(III) nitrate and potassium thiocyanate.

INVESTIGATION 16.C

Using Experimental Data to Determine an Equilibrium Constant

The colour intensity of a solution is related to the type of ions present, their concentration, and the depth of the solution (the linear measure of the solution through which you are looking). By adjusting the depth of a solution with an unknown concentration until it has the same intensity as a solution with known concentration, you can determine the concentration of the unknown solution. For example, if the concentration of a solution is lower than the standard, the depth of the solution has to be greater in order to have the same colour intensity. For this reason, the ratio of the concentrations of two solutions with the same colour intensity is in inverse ratio to their depths.

Which solution is the least concentrated? Why is the colour intensity the same when you look vertically through the solutions?

In this investigation, you will examine the homogeneous equilibrium between iron(III) ions and thiocyanate ions, and iron(III) thiocyanate ions, $Fe(SCN)^{2+}(aq)$:

$$Fe^{3+}(aq) + SCN^-(aq) \rightleftharpoons Fe(SCN)^{2+}(aq)$$

You will prepare four equilibrium mixtures with different initial concentrations of $Fe^{3+}(aq)$ and $SCN^-(aq)$. You will calculate the initial concentrations of these reacting ions from the volumes and concentrations of the stock solution used and the *total* volumes of the equilibrium mixtures. Then you will determine the concentration of $Fe(SCN)^{2+}(aq)$ ions in each mixture by comparing the colour intensity of the mixture with the colour intensity of a solution with known concentration. After you find the concentration of $Fe(SCN)^{2+}(aq)$ ions, you will use it to calculate the concentrations of the other two ions at equilibrium. You will substitute the three concentrations for each mixture into the equilibrium expression to determine the equilibrium constant.

Your teacher might choose to do this investigation as a demonstration. Alternatively, your teacher might have stations set up with solutions already prepared and you will make observations and record the data.

Question

What is the value of the equilibrium constant at room temperature for the following reaction?

$$Fe^{3+}(aq) + SCN^-(aq) \rightleftharpoons Fe(SCN)^{2+}(aq)$$

Prediction

Write the equilibrium expression for this reaction.

Safety Precautions

- The $Fe(NO_3)_3(aq)$ solution is acidified with nitric acid. It should be handled with care. Nitric acid is corrosive and will damage eyes, skin, and mucous membranes. Immediately wash any spills on your skin or clothing with plenty of water and inform your teacher. Also inform your teacher if you spill any solutions on the lab bench or floor.

- All glassware used in this experiment must be clean and dry before using it.

- Wash your hands when you have completed the investigation.

Materials

- 30 mL 0.0020 mol/L KSCN (aq)
- 30 mL 0.0020 mol/L $Fe(NO_3)_3(aq)$ (acidified)
- 25 mL 0.200 mol/L $Fe(NO_3)_3(aq)$ (acidified)
- distilled water
- 5 test tubes (18 mm × 150 mm)
- 5 flat-bottom vials
- 3 beakers (100 mL)
- test-tube rack

- labels or grease pencil
- 1 pipette (20.0 mL)
- 3 pipettes (5.0 mL)
- pipette bulb
- stirring rod
- paper towel
- thermometer (alcohol or digital)
- strip of paper
- diffuse light source, such as a light box (used by doctors to look at x-rays)
- medicine dropper

Procedure

1. Copy the following tables into your notebook, and give them titles. You will use the tables to record your measurements and calculations.

Test tube #	Fe(NO₃)₃(aq) (mL)	H₂O(ℓ) (mL)	KSCN(aq) (mL)	Initial [SCN⁻(aq)] (Mol/l)
2	5.0	3.0	2.0	
3	5.0	2.0	3.0	
4	5.0	1.0	4.0	
5	5.0	0	5.0	

Vial #	Depth of solution in vial (mm)	Depth of standard solution (mm)	Depth of standard solution / Depth of solution in vial
2			
3			
4			
5			

2. Label five test tubes and five vials with the numbers 1 through 5. Label three beakers with the names and concentrations of the stock solutions: 2.00×10^{-3} mol/L KSCN(aq), 2.00×10^{-3} mol/L Fe(NO$_3$)$_3$(aq), and 0.200 mol/L Fe(NO$_3$)$_3$(aq). Pour about 30 mL of each stock solution into its labelled beaker. Be sure to distinguish between the different concentrations of the iron(III) nitrate solutions. Make sure you use the correct solution when needed in the investigation. Measure the volume of each solution as carefully as possible to ensure the accuracy of your results.

3. Prepare the standard solution of Fe(SCN)$^{2+}$(aq) in test tube 1. Use the 20 mL pipette to transfer 18.0 mL of 0.200 mol/L Fe(NO$_3$)$_3$(aq) into the test tube. Then use a 5 mL pipette to add 2.0 mL of 2.00×10^{-3} mol/L KSCN(aq). The large excess of Fe^{3+}(aq) is to help ensure that essentially all of the SCN$^-$(aq) will react to form Fe(SCN)$^{2+}$(aq).

4. Use the pipette to transfer 5.0 mL of 2.0×10^{-3} mol/L Fe(NO$_3$)$_3$(aq) into each of the other four test tubes (labelled 2 to 5).

5. Use the pipette to transfer 3.0, 2.0, 1.0, and 0 mL of distilled water into test tubes 2, 3, 4, and 5, respectively.

6. Use the pipette to transfer 2.0, 3.0, 4.0, and 5.0 mL of 2.0×10^{-3} mol/L KSCN(aq) into test tubes 2, 3, 4, and 5, respectively. Each of these test tubes should now contain 10.0 mL of solution. Notice that the first table you prepared (in step 1) shows the volumes of the liquids you added to the test tubes. Use a stirring rod to mix each solution, being careful to rinse the rod with water and then dry it with paper towel before stirring the next solution. Measure and record the temperature of one of the solutions. Assume that all the solutions are at the same temperature.

7. Pour about 5 mL of the standard solution from test tube 1 into vial 1.

8. Pour some of the solution from test tube 2 into vial 2. Look down through vials 1 and 2. Add enough solution to vial 2 to make its colour intensity appear about the same as the colour intensity in vial 1. Use a sheet of white paper as background to make your rough colour intensity comparison.

9. Wrap a sheet of paper around vials 1 and 2 to prevent light from entering the sides of the solutions. Looking down through the vials over a diffuse light source, adjust the volume of the standard solution in vial 1 until the colour intensity in the vials is the same. Use a medicine dropper to remove or add standard solution. Be careful not to add standard solution to vial 2.

10. When the colour intensity is the same in both vials, measure and record the depth of solution in each vial as carefully as possible.

11. Repeat Procedure steps 9 and 10 using vials 3, 4, and 5.

12. Discard the solutions into the container supplied by your teacher. Rinse the test tubes and vials with distilled water, then return all the equipment. Remember to wash your hands when you have finished.

13. Copy Table 1 into your notebook to summarize the results of your calculations.

   a) Calculate the equilibrium concentration of Fe(SCN)$^{2+}$(aq) in the standard solution you prepared in test tube 1. The [Fe(SCN)$^{2+}$(aq)]$_{standard}$ is essentially the same as the starting concentration of SCN$^-$(aq) in tube 1. The large excess of Fe^{3+}(aq) ensured that the reaction of SCN$^-$(aq) was almost complete. However, remember to include the volume of Fe(NO$_3$)$_3$(aq) in the total volume of the solution for your calculation.

   b) Calculate the initial concentration of Fe^{3+}(aq) in test tubes 2 to 5. [Fe^{3+}(aq)]$_i$ is the same in these four test tubes. They all contained the same volume of Fe(NO$_3$)$_3$(aq), and the total final volume was the same. Remember to use the total volume of the solution in your calculation.

   c) Calculate the initial concentration of SCN$^-$(aq) in tubes 2 to 5. [SCN$^-$(aq)]$_i$ is different in each test tube.

   d) Calculate the equilibrium concentration of Fe(SCN)$^{2+}$(aq) in test tubes 2 to 5. Use the following equation:

$$[\text{FeSCN}^{2+}(aq)]_{eq} = \frac{\text{Depth of standard solution}}{\text{Depth of solution in vial}} \times [\text{FeSCN}^{2+}(aq)]_{standard}$$

Table 1

Test tube #	Initial concentration (mol/L)		Equilibrium concentration (mol/L)			Equilibrium constant, K_c
	[Fe^{3+}(aq)]$_i$	[SCN$^-$(aq)]$_i$	[Fe^{3+}(aq)]$_{eq}$	[SCN$^-$(aq)]$_{eq}$	[Fe(SCN)$^{2+}$(aq)]$_{eq}$	
1						
2						

e) Based on the stoichiometry of the reaction, each mole of $Fe(SCN)^{2+}(aq)$ is formed by the reaction of one mole of $Fe^{3+}(aq)$ with one mole of $SCN^-(aq)$. Thus, you can find the equilibrium concentrations of these ions by using the equations below:

$$[Fe^{3+}(aq)]_{eq} = [Fe^{3+}(aq)]_i - [Fe(SCN)^{2+}(aq)]_{eq}$$
$$[SCN^-(aq)]_{eq} = [SCN^-(aq)]_i - [Fe(SCN)^{2+}(aq)]_{eq}$$

f) Calculate four values for the equilibrium constant, K_c, by substituting the equilibrium concentrations into the equilibrium expression. Find the average of your four values for K_c.

Analysis

1. How did the colour intensity of the solutions in test tubes 2 to 5 vary at equilibrium? Explain your observation.

2. How consistent are the four values you calculated for K_c? Suggest reasons that could account for any differences.

3. Suppose the following reaction were the equilibrium reaction:

$$Fe^{3+}(aq) + 2SCN^-(aq) \rightleftharpoons Fe(SCN)^{2+}(aq)$$

a) Would the equilibrium concentration of the product be different from the concentration of the product in the actual reaction? Explain.

b) Would the value of K_c be different from the value you calculated earlier? Explain.

Conclusion

4. Write a short conclusion, summarizing your results for this investigation.

Section 16.3 Summary

- You can often determine an equilibrium constant or, given an equilibrium constant, predict the amount of product you will obtain from a small amount of experimental information by organizing data into an ICE table.

- An ICE table organizes known and unknown values for the molar concentrations of all reactants and products in a reaction according to the amount present initially, the change in the initial amount, and the amount present at equilibrium.

- If the equilibrium constant for a reaction is very small, you can make assumptions that will simplify the mathematical calculations for predicting the resultant concentrations of components of a reaction.

- You can experimentally measure the amount of a reactant or a product of a reaction at equilibrium from the intensity of its colour.

SECTION 16.3 REVIEW

1. Write equilibrium expressions for each reaction:
   a) $2H_2S(g) + CH_4(g) \rightleftharpoons 4H_2(g) + CS_2(g)$
   b) $P_4(g) + 3O_2(g) \rightleftharpoons 2P_2O_3(g)$
   c) $7N_2(g) + 2S_8(g) \rightleftharpoons 2N_3S_4(g) + 4N_2S_2(g)$

2. When 1.0 mol of ammonia gas is injected into a 0.50 L flask, the following reaction proceeds to equilibrium:

$$2NH_3(g) \rightleftharpoons N_2(g) + 3H_2(g)$$

   At equilibrium, 0.30 mol of hydrogen gas is present. Calculate the equilibrium concentrations of nitrogen and ammonia and the value of K_c.

3. At a certain temperature, K_c for the following reaction between sulfur dioxide and nitrogen dioxide is 4.8:

$$SO_2(g) + NO_2(g) \rightleftharpoons NO(g) + SO_3(g)$$

   $SO_2(g)$ and $NO_2(g)$ have an initial concentration of 0.36 mol/L. What amount of $SO_3(g)$ is present in a 5.0 L container at equilibrium?

4. Phosphorus trichloride reacts with chlorine to form phosphorus pentachloride:

$$PCl_3(g) + Cl_2(g) \rightleftharpoons PCl_5(g)$$

   0.75 mol of $PCl_3(g)$ and 0.75 mol of $Cl_2(g)$ are placed in an 8.0 L reaction vessel at 500 K. At equilibrium, the concentration of $PCl_5(g)$ is 0.059 mol/L. What is the value of K_c for the reaction at 500 K?

5. Hydrogen gas has several advantages and disadvantages as a potential fuel. Hydrogen can be obtained by the thermal decomposition of water at high temperatures:

$$2H_2O(g) \rightleftharpoons 2H_2(g) + O_2(g) \quad K_c = 7.3 \times 10^{-18} \text{ at } 1000 \,°C$$

   a) The initial concentration of water in a reaction vessel is 0.055 mol/L. What is the equilibrium concentration of hydrogen gas at 1000 °C?

   b) Comment on the practicality of thermal decomposition of water to obtain $H_2(g)$.

Applications of Equilibrium Systems

Section Outcomes

In this section, you will:

- **identify** physical and chemical equilibrium important in blood gases
- **illustrate** the involvement of prototypes, experiments, and theory in the development of technology
- **analyze** the application of equilibrium principles in an industrial process
- **evaluate** various factors that are important to the construction of a chemical plant

Key Terms

acidosis
alkalosis

Equilibrium in Blood Gases: Scuba Diving

Your body is an active chemical system. It must maintain a delicate balance between acidic and basic properties and the concentration of gases in the blood, a process that involves both physical and chemical equilibrium. The scuba diver shown in Figure 16.10 must understand the importance of several gas laws to safely enjoy diving. One of these laws is related to the physical equilibrium of gases dissolved in the blood.

Figure 16.10 At depth, the diver must breathe air at greater pressure. How does that change the solubility of gases in the blood?

The total pressure exerted on a diver increases by one atmosphere for every 10 m descent below the surface. The diver's lungs would collapse if they were not supplied with air (through the tank and a regulator) at the same pressure as the surrounding water. Because the total pressure of the air the diver breathes has increased, the partial pressure of each of the gases present has increased proportionately. As you know, the solubility of a gas in a liquid increases as the pressure of that gas above the liquid increases. Therefore, as the diver descends, an increasing amount of nitrogen and oxygen dissolves in the blood.

High concentrations of nitrogen gas in the blood impair the conduction of electrical signals along the nerves. The effect, called nitrogen narcosis, begins at depths below about 30 m. For every additional 15 m below the surface, a diver experiences an effect often described as similar to consuming an alcoholic drink. Thus, when 60 m below the surface, a diver would experience the impaired judgment, confusion, and drowsiness of a person who had consumed two alcoholic drinks. At depths of 90 m, nitrogen narcosis can lead to hallucinations and unconsciousness.

As a diver returns to the surface, the solubility of gases in the blood decreases. The effect of suddenly reducing the pressure above a solution containing a dissolved gas was shown in Figure 16.2. A diver returning to the surface must do so slowly enough to allow nitrogen to be breathed out. If the ascent is too rapid, bubbles will form in the blood and nitrogen gas may collect in the joints and other parts of the body. The pain caused by increased gas pressure in the joints is called the "bends." However, nitrogen bubbles in the vessels can cause damage to any of the tissues or organs where the bubbles collect. If a diver remains at a depth of less than 20 m, there is little danger of getting the bends. As divers go deeper, the bends can occur at shorter and shorter times. What about the effect of dissolved oxygen gas on the scuba diver? Oxygen is toxic at increased concentration in the blood caused by deep dives or dives where oxygen-enriched gas is breathed. Toxic levels of oxygen can lead to a seizure.

Carbon Monoxide Poisoning

Carbon monoxide is a poisonous gas that is formed when incomplete combustion of a hydrocarbon occurs. Burning fuel with a limited air supply will form carbon monoxide gas, which is colourless and odourless. Note the similarity between the formula of carbon monoxide, $CO(g)$, and the formula of oxygen molecules, $O_2(g)$. Carbon monoxide is able to combine with hemoglobin, but the equilibrium constant for the reaction with carbon monoxide is about 250 times as great as K_c for the reaction with oxygen:

$$HHb(aq) + CO(aq) \rightleftharpoons H^+(aq) + HbCO^-(aq)$$

Because hemoglobin binds more readily with carbon monoxide than with oxygen, breathing even small concentrations of carbon monoxide can be deadly. Arterial blood pumped to the cells and organs will carry too little oxygen, with potentially lethal effects.

Carbonate Ions in Caves and Corals

When carbon dioxide reacts with rainwater in the atmosphere, it dissolves, forming carbonic acid. When carbonic acid, $H_2CO_3(aq)$, comes into contact with limestone, $CaCO_3(s)$, the limestone dissolves. This reaction is shown below:

$$CaCO_3(s) + H_2CO_3(aq) \rightleftharpoons Ca^{2+}(aq) + 2HCO_3^-(aq)$$

The position of equilibrium lies to the right. Water saturated with calcium ions and hydrogen carbonate ions eventually drips through a crack in the roof of a limestone cave. As droplets form, carbon dioxide leaves the solution because air in the cave has a very low concentration of $CO_2(g)$. This reduces the concentration of carbonic acid in solution and calcium carbonate precipitates, forming a stalactite from the roof of the cave. Stalactites grow at the rate of about 1 cm every 40 years. At the floor of the cave, $CaCO_3(s)$ also builds up as a stalagmite. Sometimes, a stalactite and a stalagmite can join to form a solid column, as in Figure 16.11.

Coral reefs are diverse ecosystems that protect fish and provide natural breakwaters to protect coastlines (Figure 16.12). Reefs are formed by tiny coral polyps, which feed on small particles floating in the water. These polyps also get nutrients and sugars from tiny photosynthetic algae that live inside the coral. A coral colony may consist of thousands of polyps that leave behind a hard, branching structure made of calcium carbonate when they die. The algae cement various corals and shells together with calcium compounds to create the coral reefs.

Figure 16.11 In this cave near Canmore, a stalactite and a stalagmite have joined to form a calcite column.

Figure 16.12 Coral reefs are very sensitive ecosystems.

Reefs grow very slowly and are sensitive to damage from environmental changes. Many factors affect the health of coral, but increasing amounts of carbon dioxide in the atmosphere present a long-term threat. The burning of fossil fuels releases about 25 billion tonnes of $CO_2(g)$ into the atmosphere each year. Increased levels of atmospheric carbon dioxide enhance the greenhouse effect over time and, consequently, result in global warming. Higher global temperatures ultimately raise the temperature of the world's oceans. Even a small increase in seawater temperature causes coral to be stressed and take on a bleached appearance. It is estimated that the oceans absorb about one third of the $CO_2(g)$ released by fossil fuel combustion. This equilibrium, which involves carbon dioxide and carbonic acid, is very important to the chemistry of the oceans and the health of corals around the world.

Carbon dioxide is not very soluble in water, and carbonic acid is a weak acid:

$$CO_2(g) + H_2O(\ell) \rightleftharpoons H_2CO_3(aq)$$
$$H_2CO_3(aq) + H_2O(\ell) \rightleftharpoons H_3O^+(aq) + HCO_3^-(aq)$$
$$HCO_3^-(aq) + H_2O(\ell) \rightleftharpoons H_3O^+(aq) + CO_3^{2-}(aq)$$

The equilibrium in seawater can be simplified to:

$$CO_2(g) + CO_3^{2-}(aq) + H_2O(\ell) \rightleftharpoons 2HCO_3^-(aq)$$

As more $CO_2(g)$ dissolves in the oceans, Le Châtelier's principle predicts a shift to the right and the concentration of carbonate ion, $[CO_3^{2-}(aq)]$, will decrease. The decrease in carbonate ion affects the amount of dissolved calcium carbonate:

$$CaCO_3(s) \rightleftharpoons Ca^{2+}(aq) + CO_3^{2-}(aq)$$

Le Châtelier's principle predicts the reaction will shift to the right when $[CO_3^{2-}(aq)]$ is decreased. When less $CO_3^{2-}(aq)$ is available, corals grow more slowly.

The Alberta Chemical Industry

The chemical industry is an important contributor to Canada's economy (Figure 16.13). It is the third-largest manufacturing industry, behind the automotive and agricultural sectors, with production valued at about $35 billion per year. Canada's chemical industries are concentrated in three provinces: Ontario, Quebec, and Alberta.

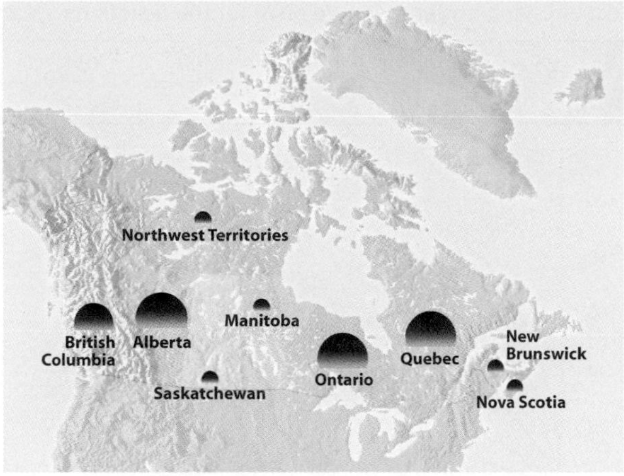

Figure 16.13 This map shows the location of major chemical plants in Canada.

In Alberta, production from the oil sands deposits is estimated to account for 50% of Canada's total crude oil output. The total reserves of oil in the deposits at Athabasca, Peace River, and Cold Lake are estimated to be well in excess of the reserves in Saudi Arabia. Crude oil, natural gas liquids, and refined petroleum are transported through the Enbridge pipeline to Sarnia, Ontario and Chicago, Illinois. Canada is the third-largest producer of natural gas in the world and the second-largest exporter. About 83% of Canada's gas production occurs in Alberta and most of the remainder in British Columbia. Natural gas is mostly methane and ethane. In Alberta, methane is processed into methanol, and ethane is converted into ethylene (ethene). The world's largest ethylene plant, capable of producing 1 million tonnes a year, is located at Joffre. Most of the ethylene produced is sent through the Cochin pipeline to other locations for processing. The pipeline, which is more than 3000 kilometres long, links Fort Saskatchewan with Sarnia and Chicago.

The chemical industry also manufactures acids and bases for various uses, including agriculture. On a global scale, sulfuric acid and ammonia are manufactured in the largest quantities. Canada produces more ammonia than sulfuric acid, and in 2004 alone, our nation produced more than 5 million tonnes of the chemical.

The Haber-Bosch Process for Manufacturing Ammonia

The Haber-Bosch process for manufacturing ammonia was the first application of modern chemical principles to produce a chemical in very large quantities. Before the work of Fritz Haber (1868–1934), sodium nitrate was imported from Chile and made into fertilizers and explosives. Sodium nitrate was a critical resource, and several chemists undertook the challenge of finding a replacement. In 1909, working at a technical college in Germany, Haber produced an equilibrium mixture containing nitrogen, hydrogen, and ammonia:

$$N_2(g) + 3H_2(g) \rightleftharpoons 2NH_3(g) \qquad \Delta H = -92 \text{ kJ/mol}$$

At 25 °C, $K_c = \dfrac{[NH_3]^2}{[N_2][H_2]^3} = 4.1 \times 10^8$. The large equilibrium constant suggests that the reaction proceeds to completion. However, the position of equilibrium is only one of several factors important in chemical reactions. Another important factor in chemical reactions is the *rate* of the reaction. At 25 °C, the rate of reaction between nitrogen and hydrogen is so slow that there is essentially no ammonia formed.

Haber knew that he could increase the rate of reaction by raising the temperature of the reacting gases. However, he also knew the reaction is exothermic, so Le Châtelier's principle would predict a shift to the left in the position of equilibrium at higher temperature. Table 16.3 shows that the value of K_c falls very rapidly as temperature increases.

Chemistry File

FYI

Fritz Haber was the son of a wealthy Jewish chemical merchant in Germany. In 1913, the first ammonia plant was making about 30 tonnes of ammonia per day. Ammonia plays an important role in the manufacture of ammunition, and there is no doubt that Haber's invention helped to prolong World War I. Haber also worked on the use of chlorine as a poisonous gas against British and French troops in France. His wife Clara, also a chemist, bitterly opposed Haber's work on chemical weapons. Haber had converted to Christianity as a young man, but when the Nazis came to power in 1933, he was labelled a Jew. The civil service was purged of Jews, and although his war service would have allowed him to continue in his position, Haber resigned. He left Germany and died in Basel, Switzerland the next year.

Table 16.3 The Effect of Changing Temperature on K_c for the Reaction to Manufacture Ammonia

Temperature (°C)	25	100	200	300	400	500	600
K_c	4.1×10^8	2.3×10^5	4.4×10^2	7.3	0.41	0.05	9.5×10^{-3}

Figure 16.14 Haber performed thousands of experiments to find the best conditions to make ammonia.

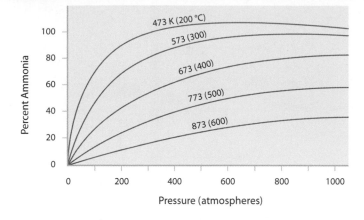

Again, using Le Châtelier's principle, Haber knew that he could shift the position of equilibrium back toward the right by running the reaction at high pressure. Figure 16.14 shows how the yield of ammonia varies with the temperature and pressure of the reaction. Although the yield of ammonia increases with pressure, the cost of running a chemical plant also increases. A plant operating at higher pressure would incur higher costs because more pumps would be required. The reaction vessel and pipes would also need to be thicker to withstand the greater pressure. Because safety risks increase with pressure, repair and maintenance costs also increase. Haber reasoned that he could afford to have a smaller equilibrium constant if the rate of the reaction increased. Also, finding a suitable catalyst would shorten the time taken for the mixture of gases to reach equilibrium. The removal of ammonia would also shift the equilibrium toward the production of more ammonia. Ammonia is removed from the reaction vessel by cooling the mixture of gases. Once the ammonia has been removed, the gases are recycled back to the reaction vessel in a continuous operation. Haber performed more than 6500 experiments, changing the temperature and pressure of the mixture and varying the catalyst, measuring the rate and yield of ammonia in the reaction. Carl Bosch (1874-1940) was an engineer who solved the engineering problems associated with the high pressures used in the process Haber had discovered (Figure 16.15). Haber won the Nobel prize for chemistry in 1918, and Bosch won the prize in 1931.

Figure 16.15 The Haber-Bosch process has been called one of the most important inventions of the twentieth century.

Although ammonia is an inorganic compound, its synthesis depends greatly on a plentiful supply of methane. Modern ammonia plants combine nitrogen from the air with hydrogen derived from methane. Different catalysts are used for the different reactions (see Figure 16.16). There are a number of ammonia producers in Canada, including Agrium, with four plants in Alberta; Simplot in Brandon, Manitoba; and Pacific Ammonia in Kitimat, British Columbia.

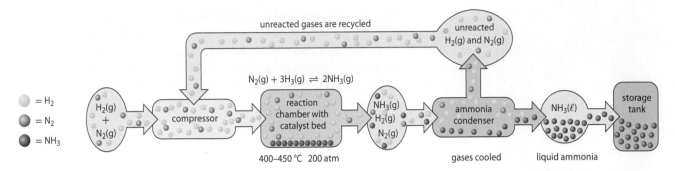

Figure 16.16 The catalyst that is often used with the Haber-Bosch process, shown in the diagram, is a mixture of MgO(s), Al_2O_3(s), and SiO_2(s), with embedded iron crystals.

The Manufacture of Sulfuric Acid

More sulfuric acid is produced than any other chemical in the world (Figure 16.17). Because sulfuric acid is used in so many processes, it is often said that the amount of sulfuric acid manufactured is a good indicator of the general economy.

Sulfuric acid can be made from the waste gases formed during the smelting of ores that contain sulfur. Sulfuric acid can also be made from the sulfur removed from sour oil and gas. Canada is the second-largest producer of sulfur in the world, extracting about 8 million tonnes annually. Sulfur is made into sulfuric acid in the contact process, using air and water as the other raw materials. First, sulfur is burned in air:

$$S(s) + O_2(g) \rightarrow SO_2(g)$$

Then, the sulfur dioxide is oxidized over a vanadium catalyst to form sulfur trioxide:

$$2SO_2(g) + O_2(g) \rightleftharpoons 2SO_3(g) \qquad \Delta H = -197 \text{ kJ/mol}$$

The reaction is exothermic, so Le Châtelier's principle predicts that the forward reaction is favoured at low temperatures. However, as you saw with the Haber-Bosch process to manufacture ammonia, lower temperatures slow the rate of reaction. As a compromise, a temperature of about 450 °C is used. This gives a reasonable rate of reaction and a yield of 97% sulfur trioxide. Le Châtelier's principle predicts greater yield if the pressure is increased. However, the yield is high enough that it does not justify the increased cost of pressurizing the gases more than is necessary to move the gas mixture through the reaction vessel. In the final two steps, the SO_3(g) is dissolved in 98% sulfuric acid, $H_2SO_4(\ell)$, to create pyrosulfuric acid, $H_2S_2O_7(\ell)$. Water is then added to the pyrosulfuric acid to produce sulfuric acid. The reaction is completed as follows:

$$SO_3(g) + H_2SO_4(\ell) \rightarrow H_2S_2O_7(\ell)$$
$$H_2S_2O_7(\ell) + H_2O(\ell) \rightarrow 2\,H_2SO_4(\ell)$$

Notice that sulfuric acid is used as a reactant in the contact process to make more sulfuric acid. However, because two moles of sulfuric acid are produced for each mole used as a reactant, the process results in a net production of sulfuric acid.

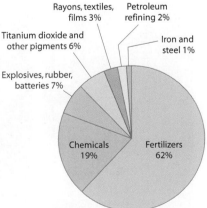

Figure 16.17 The annual world production of sulfuric acid is more than 150 million tonnes.

The Manufacture of Methanol

Methane, obtained from natural gas, can be converted to a mixture of carbon monoxide and hydrogen gas. This mixture, called synthesis gas, or syngas, is the raw material used to make methanol (see Figure 16.18). Syngas is also the source of the hydrogen used in the Haber-Bosch process to manufacture ammonia. Syngas can be produced by reacting methane and steam at moderate pressure (1–2 MPa) and high temperatures (850–1000 °C) over a nickel catalyst. This reaction is called the methane-steam reaction:

$$CH_4(g) + H_2O(g) + heat \rightleftharpoons CO(g) + 3H_2(g)$$

The reaction is endothermic, so the relatively high temperature of the reaction helps to push the equilibrium to the right, as well as increase the rate of reaction. According to Le Châtelier's principle, low pressure would push the equilibrium to the right, but moderate pressure is needed to force the gases through the reaction chamber. The ratio of carbon monoxide and hydrogen can be adjusted by mixing the syngas with steam over an iron catalyst. This reaction is mildly exothermic, and the heat produced can be used for the methane-steam reaction:

$$CO(g) + H_2O(g) \rightleftharpoons CO_2(g) + H_2(g) + heat$$

The syngas is passed over a catalyst at 5–10 MPa pressure at about 250 °C to form methanol:

$$CO(g) + 2H_2(g) \rightleftharpoons CH_3OH(g) + heat$$

In this reaction, higher pressures are used to shift the reaction to the right, according to Le Châtelier's principle. The reaction is very exothermic, and heat must be removed (of course, it can be used for the methane-steam reaction) because if the temperature increases, the position of equilibrium will shift to the left.

The production of methanol consumes two moles of $H_2(g)$ for every mole of $CO(g)$. However, the methane-steam reaction to make syngas results in three moles of $H_2(g)$ for every mole of $CO(g)$. Excess hydrogen can be used to remove sulfur from gasoline and diesel at nearby plants.

In Alberta, methanol is produced at the Celanese plant in Edmonton, which has a capacity of about 850 000 tonnes per year. Methanol can be processed into a number of other chemicals, as Figure 16.19 shows. One of these chemicals, a gasoline additive called MTBE, illustrates how the chemical industry is affected by the price of raw materials and changes in demand for its products.

Figure 16.18 Methanol was used by the ancient Egyptians to prepare an embalming solution.

Uses of Methanol

MTBE

MDF Plywood

Silicone sealants, PET bottles

Paints, Adhesives

Figure 16.19 Methanol is a versatile chemical that can be processed into other chemicals that are used to make products we commonly use. Borden Chemical opened a plant in 2004 in West Edmonton to process methanol into methanal.

MTBE (methyl tertiary butyl ether) is manufactured from methanol and 2-methylpropene:

$$CH_3OH(g) + (CH_3)_2CCH_2(g) \rightleftharpoons (CH_3)_3COCH_3(g)$$

MTBE is one of a number of chemicals called oxygenates that can be added to gasoline. Oxygenates raise the oxygen content of gasoline, which makes the fuel burn more efficiently and reduces atmospheric pollution. In the United States, the Clean Air Act of 1990 resulted in the production of large quantities of MTBE to help reduce pollution from automobiles in large cities. However, MTBE was found to cause cancer in animals, and because the substance dissolves in water, concerns were raised. Leaking storage tanks could introduce MTBE into ground water supplies, and this might affect the health of people who draw their drinking water from the sources contaminated with MTBE. Very little MTBE is used in Canada, and several states, especially California, have now banned its use. Fortunately, other oxygenates, such as ethanol, can be added to gasoline to reduce atmospheric pollution. Although the demand for MTBE has declined significantly, methanol production throughout the world has increased, resulting in a sharp drop in price. In 1999, Methanex closed one of its two plants in Medicine Hat and, a year later, closed its plant in Kitimat, British Columbia.

Chemistry File

FYI
Before modern methods of manufacturing methanol were developed, the chemical was obtained by heating wood in a limited supply of air. Under these conditions, the wood chars and releases a vapour that contains a number of substances, including methanol. Today, methanol is still sometimes called wood alcohol. Methanol is a very poisonous substance, and a fatal dose is about 100–125 mL.

Section 16.4 Summary

- The large pressure on a scuba diver in deep water changes the equilibrium between gaseous nitrogen and dissolved nitrogen in the lining of the lungs and thus in the bloodstream. High concentrations of nitrogen in the blood can cause nitrogen narcosis.

- If a diver rapidly returns to the surface from a deep dive, the change in the equilibrium between dissolved and gaseous nitrogen causes bubbles to form in the blood vessels causing a serious condition called "the bends."

- The equilibrium between free carbon monoxide and that bound to hemoglobin in the blood lies much further in the direction of bound carbon monoxide than does the same equilibrium with oxygen and hemoglobin. The binding of carbon monoxide prevents the binding of oxygen, starving the tissues of the needed oxygen.

- The equilibrium between solid calcium carbonate and dissolved carbonic acid is responsible for the formation of stalagmites and stalactites in caves and also of the coral reef in the ocean.

- An understanding of the effects of pressure and temperature on the position of equilibrium between nitrogen and hydrogen gases and ammonia allowed Haber and Bosch to develop an efficient method for synthesizing ammonia.

- A knowledge of the equilibrium reactions that are important in the production of sulfuric acid and of methanol make it possible to produce these very important commodities in large amounts.

- Alberta is a major producer of both sulfuric acid and methanol.

1. a) How does the solubility of a gas in water change as the pressure of the gas over the solution increases?
   b) How is this relevant to a scuba diver?
   c) What are the effects of increased concentrations of nitrogen and oxygen in the blood?

2. The snails you see in the garden have shells made of calcium carbonate. In deep crevasses on the ocean floor, hydrothermal vents emit very hot, acidic water. Explain why snails living near these vents do not have a shell.

3. The Simplot Manufacturing complex in Brandon, Manitoba is capable of producing 1250 tonnes of ammonia each day. Most of the ammonia produced is made into fertilizer products at the same site. The raw materials for the process are natural gas (from Alberta), air, and water, which are combined in the Haber-Bosch process. The process can be simplified and understood in two steps. The first step is the reaction between methane and steam to form hydrogen and carbon dioxide:

$$CH_4(g) + 2H_2O(g) \rightleftharpoons 4H_2(g) + CO_2(g) \quad \Delta H = +206 \text{ kJ}$$

To increase the yield of hydrogen gas, should the reaction take place
   a) at a high or low pressure?
   b) at a high or low temperature? What effect will the temperature have on the rate of the reaction?
   c) A nickel catalyst is used for the reaction. What effect does the catalyst have on the position of equilibrium?

d) In the second step, nitrogen and hydrogen are combined to form ammonia at a pressure of 200–300 atm and a temperature of about 500 °C. Write the equilibrium expression for the reaction.
e) Why are these pressure and temperature conditions used?
f) What is the chemical equation for the reaction?
g) At 500 °C, the value of K_c for the reaction is about 5.0×10^{-2}. If the reaction chamber contains is 2.0 mol/L $H_2(g)$ and 1.2 mol/L $N_2(g)$ at equilibrium, what is the concentration of ammonia gas at equilibrium?

4. During a thunderstorm, the air near the lightning flash is raised to a temperature high enough for the following reaction to take place:

$$N_2(g) + O_2(g) \rightleftharpoons 2NO(g)$$

A chemist studying this reaction collected the following data.

Is the reaction in the forward direction endothermic or exothermic? Explain.

Temperature (K)	K_c
300	4×10^{-31}
700	5×10^{-13}
1100	4×10^{-8}
1500	1×10^{-5}

Many chemical reactions do not proceed to completion. Instead the reactions reach a condition called equilibrium in which the amounts of reactants and products remain constant. Although chemical reactions continue to occur, the rates of the reactions in the forward and reverse directions are equal. A true equilibrium can only be achieved in an isolated system or in a closed system that is maintained at a constant temperature.

Norwegian chemists, Guldberg and Waage, discovered the law of chemical equilibrium which states that there exists a constant ratio of the concentrations of products and reactants at equilibrium. The law of chemical equilibrium can be expressed mathematically as, $K_c = \dfrac{[C]^c[D]^d}{[A]^a[B]^b}$, where A and B are the reactants and C and D are the products of a generalized chemical reaction. The exponents, a, b, c, and d, are the stoichiometric coefficients of A, B, C, and D respectively. Given experimental data, you can calculate the equilibrium constant of a chemical reaction. Since temperature affects the equilibrium constant, the temperature must be reported along with the equilibrium constant. Given the equilibrium constant, you can predict the equilibrium concentrations of one component of a reaction if you have information about the other components.

If an equilibrium is disturbed by the addition or removal of one of the components of the reaction, a change in temperature or pressure of the system, Le Châtelier's principle allows you to predict the response of the system. Le Châtelier's principle states that an equilibrium responds to a disturbance in a way that will relieve the change. For example, an increase in the pressure on a system of gases will cause the reaction to proceed in the direction of a decrease in the number of moles of all gases in the system. A change in temperature is the only change that actually changes the value of the equilibrium constant. A catalyst has no effect on the value of the equilibrium constant and no effect on the position of equilibrium. The only effect of a catalyst is to bring the system to equilibrium much more quickly.

Building an ICE table is a method of organizing experimental data that helps you to set up a mathematical equation for calculating concentrations of the components of a reaction involving a system going to equilibrium. The mnemonic, ICE, represents initial, change, and equilibrium. To use the method, you must have experimental data regarding the system at equilibrium. If one of the components of the reaction has a colour, you can determine the amount of that component present at equilibrium by measuring the intensity of the colour.

Understanding equilibrium reactions is critical to many practical industrial processes and the understanding of natural and physiological processes.

Concept Organizer · Chemical Equilibrium

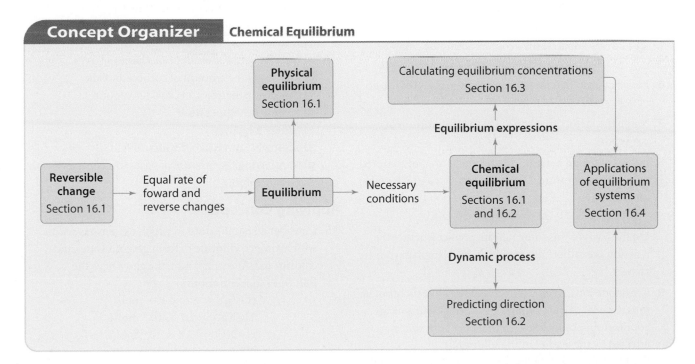

Understanding Concepts

1. Agree or disagree with the following statement, and give reasons to support your answer: "In a sealed jar of water at equilibrium, the quantity of water molecules in the liquid state equals the quantity of molecules in a gaseous state."

2. At equilibrium, there is no overall change in the concentrations of reactants and products. Why then is this state described as dynamic?

3. In a chemical system at equilibrium, are the concentrations of reactants and products in the same ratio as the coefficients in the chemical equation? Explain your answer.

4. Label each system as a homogeneous equilibrium or a heterogeneous equilibrium:
   a) $CaCO_3(s) + 2HF(g) \rightleftharpoons CaF_2(s) + H_2O(g) + CO_2(g)$
   b) $Na_2SO_4(aq) \rightarrow 2Na^+(aq) + SO_4^{2-}(aq)$
   c) $2NaHSO_3(s) \rightleftharpoons Na_2SO_3(s) + H_2O(g) + SO_2(g)$

5. Write the equilibrium expression for each of the following homogeneous reactions:
   a) The reaction between propane and oxygen to form carbon dioxide and water vapour:
   $$C_3H_8(g) + 5O_2(g) \rightleftharpoons 3CO_2(g) + 4H_2O(g)$$
   b) The reaction between hydrazine, $N_2H_4(g)$, which is used as a rocket fuel, and oxygen:
   $$N_2H_4(g) + 2O_2(g) \rightleftharpoons 2NO(g) + 2H_2O(g)$$
   c) The reaction to form a copper-ammonia complex:
   $$Cu^{2+}(aq) + 4NH_3(aq) \rightleftharpoons Cu(NH_3)_4^{2+}(aq)$$

6. For a reaction that goes to completion, is K_c very large or very small? Explain why.

7. Name the factors that can affect the equilibrium of a reaction.

8. In the following reaction, the products are not favoured at room temperature, but are favoured at a much higher temperature.
   $$CO_2(g) + H_2(g) \rightleftharpoons CO(g) + H_2O(g)$$
   On the basis of this information, predict whether the reaction is exothermic or endothermic. Explain your answer.

9. Chromate ions, $CrO_4^{2-}(s)$, are yellow, while dichromate ions, $Cr_2O_7^{2-}(s)$, are orange. In aqueous solution, an equilibrium occurs between these two ions that is affected by the acidity of the solution:
   $2CrO_4^{2-}(aq) + 2H_3O^+(aq) \rightleftharpoons Cr_2O_7^{2-}(aq) + 3H_2O(\ell)$
   Predict and explain the change in colour of an equilibrium mixture after the addition of
   a) dilute acid   b) dilute base

10. The following reaction is at equilibrium:
    $$H_2(g) + Cl_2(g) \rightleftharpoons 2HCl(g) + heat$$
    Which condition will produce a shift to the right: a decrease in volume or a decrease in temperature? Explain why.

11. The following system is at equilibrium:
    $$2CO_2(g) \rightleftharpoons 2CO(g) + O_2(g)$$
    Will an increase in pressure result in a shift to the left or to the right? How do you know?

12. Consider the following reaction:
    $$CO(g) + 3H_2(g) \rightleftharpoons CH_4(g) + H_2O(g)$$
    The volume and temperature are kept constant, but pumping in argon, a non-reacting gas, increases the pressure on the system. Does this affect the equilibrium? If so, how? If not, explain.

13. Jacques Cousteau was a famous underwater explorer. When diving to depths of 100 m, he used a mixture of gases consisting of 98% helium and 2% oxygen. At 100 m below the surface, the pressure of water is 10 atmospheres. Why did Cousteau use this particular mix of $He(g)$ and $O_2(g)$?

14. Hydrogen for the Haber-Bosch process can be obtained by the following endothermic reaction:
    $$CH_4(g) + 2H_2O(g) \rightleftharpoons CO_2(g) + 4H_2(g)$$
    How would the concentration of hydrogen be affected by each of the following changes to an equilibrium mixture of gases in a rigid container?
    a) increasing the amount of methane
    b) increasing the amount of carbon dioxide
    c) decreasing the amount of water vapour
    d) raising the temperature
    e) adding a catalyst
    f) adding helium gas to the mixture
    g) transferring the mixture of gases to a container with greater volume

Applying Concepts

15. Consider an equilibrium in which oxygen gas reacts with hydrogen chloride to form gaseous water and chlorine gas. At equilibrium, the gases have the following concentrations:
    $$[O_2(g)] = 8.6 \times 10^{-2} \text{ mol/L}$$
    $$[HCl(g)] = 2.7 \times 10^{-2} \text{ mol/L}$$
    $$[H_2O(g)] = 7.8 \times 10^{-3} \text{ mol/L}$$
    $$[Cl_2(g)] = 3.6 \times 10^{-3} \text{ mol/L}$$
    a) Write the chemical equation for this reaction.
    b) Calculate the value of the equilibrium constant, K_c.

16. The following results were collected for two experiments involving the reaction, at 600 °C, between gaseous sulfur dioxide and oxygen to form gaseous sulfur trioxide. Show that the value of K_c was the same in both experiments.

Measurements of Initial and Equilibrium Concentrations for the Reaction Between $SO_2(g)$ and $O_2(g)$

Experiment 1	
Initial concentration (mol/L)	Equilibrium concentration (mol/L)
$[SO_2(g)] = 2.00$	$[SO_2(g)] = 1.50$
$[O_2(g)] = 1.50$	$[O_2(g)] = 1.25$
$[SO_3(g)] = 3.00$	$[SO_3(g)] = 3.50$

Experiment 2	
Initial concentration (mol/L)	Equilibrium concentration (mol/L)
$[SO_2(g)] = 0.500$	$[SO_2(g)] = 0.590$
$[O_2(g)] = 0$	$[O_2(g)] = 0.0450$
$[SO_3(g)] = 0.350$	$[SO_3(g)] = 0.260$

17. Write the chemical equation for the reversible reaction that has the following equilibrium expression:

$$K_c = \frac{[NO]^4[H_2O]^6}{[NH_3]^4[O_2]^5}$$

Assume that, at a certain temperature, at equilibrium $[NO(g)]$ and $[NH_3(g)]$ are equal. If $[H_2O(g)] = 2.0$ mol/L and $[O_2(g)] = 3.0$ mol/L at this temperature, what is the value of K_c?

18. At one time, methanol was obtained by heating wood without allowing the wood to burn. The products were collected, and methanol (sometimes called "wood alcohol") was separated by distillation. Today, methanol is manufactured by reacting carbon monoxide with hydrogen gas (obtained from natural gas):
$CO(g) + 2H_2(g) \rightleftharpoons CH_3OH(g)$
At 210 °C, K_c for this reaction equals 14.5. A gaseous mixture at 210 °C contains the following concentrations of gases: $[CO(g)] = 0.25$ mol/L and $[H_2(g)] = 0.15$ mol/L. What is the concentration of methanol in the equilibrium mixture at 210 °C?

19. The oxidation of sulfur dioxide to sulfur trioxide is an important reaction in the process to manufacture sulfuric acid. At 1000 K, the equilibrium constant for the reaction is 3.6×10^{-3}:
$2SO_2(g) + O_2(g) \rightleftharpoons 2SO_3(g)$
If a closed flask originally contains 3.8 mol/L of sulfur dioxide and the same concentration of oxygen, what is the concentration of $SO_3(g)$ at equilibrium when the reaction vessel is maintained at 1000 K?

Solving Problems

20. Equal amounts of hydrogen gas and iodine vapour are heated in a sealed flask.
   a) Sketch a graph to show how $[H_2(g)]$ and $[I_2(g)]$ change over time.
   b) Would you expect a graph of $[I_2(g)]$ and $[HI(g)]$ to appear much different from your first graph? Explain why.

21. Unfortunately, people die every year from carbon monoxide poisoning. Prepare a flyer that could be distributed to households in your neighbourhood describing the dangers of carbon monoxide. Describe specific ways in which the gas forms, and include measures that should be taken to prevent carbon monoxide poisoning.

Making Connections

22. Does a small value for the equilibrium constant of a reaction indicate the reaction proceeds slowly? Justify your answer.

23. Although the pH of blood must be maintained within strict limits, the pH of urine can vary. The sulfur in foods, such as eggs, is oxidized in the body and excreted in urine. Does the presence of sulfide ions in urine tend to increase or decrease the pH? Explain.

24. One of the steps in the Ostwald process for the manufacture of nitric acid involves the oxidation of ammonia:
$4NH_3(g) + 5O_2(g) \rightleftharpoons 4NO(g) + 6H_2O(g) \quad \Delta H = -905$ kJ
   a) State the reaction conditions that favour the production of nitrogen monoxide.
   b) A rhodium/platinum alloy is used as a catalyst. What effect does using a catalyst have on the position of equilibrium?
   c) Explain why the reaction temperature is relatively high, typically about 900 °C.
   d) A relatively low pressure is used, about 7 atmospheres. Suggest an explanation.
   In the next step of the Ostwald process, nitrogen monoxide is mixed with oxygen to form nitrogen dioxide:
$2NO(g) + O_2(g) \rightleftharpoons 2NO_2(g) \quad \Delta H = -115$ kJ
   e) Why are the gases cooled for this reaction? What do you think happens to the heat extracted?
   Finally, the nitrogen dioxide reacts with water to form nitric acid:
$3NO_2(g) + H_2O(\ell) \rightleftharpoons 2HNO_3(aq) + NO(g)$
   f) What is done with the $NO(g)$ that is formed?
   g) Name at least three uses for nitric acid.

CHAPTER 17

Acid-Base Equilibrium Systems

Chapter Concepts

17.1 Acid Deposition

- Acid deposition, which includes acid rain, is caused by natural processes in the atmosphere and by industrial activity.

17.2 Understanding Acids and Bases

- In the Brønsted-Lowry theory, an acid is a substance capable of donating a proton, H^+, and a base is a substance capable of accepting a proton.

17.3 Acid-Base Equilibriums

- The percent ionization of a weak acid or a weak base is related to its equilibrium constant, K_a or K_b.

- The ion product constant for pure water, K_w, is the equilibrium constant for the ionization of water into its ions.

- For a conjugate acid-base pair, $K_a \times K_b = K_w$.

- The constants K_a and K_b can be used to determine pH, pOH, $[H_3O^+(aq)]$, and $[OH^-(aq)]$ of acidic and basic solutions.

- The direction of an acid-base reaction can be predicted by using the values of K_a and K_b for an acid and base reacting together.

17.4 Titration Curves and Buffers

- An acid-base titration curve is a plot of the pH of a solution as a function of the volume of titrant added.

- Buffer solutions resist change in pH when a small amount of an acid or a base is added.

In Alberta, the advance and retreat of ice sheets over millions of years has carved the landscape you see today. In most locations, Alberta is covered by the material left behind when the last ice sheets melted, about 10 000 years ago. Some of this silt, sand, and gravel was transformed by immense pressure into sandstone, shale, and other types of sedimentary rocks. Most of Alberta's fresh drinking water has passed through these sedimentary rocks. Although fresh water resources, especially lakes, are threatened worldwide by acid deposition formed as the result of industrial activity, most natural sources of water in Alberta have not been affected much by acid rain. Alberta's lakes are nearly acid-free because the soils and sandstones prevalent in the province act as natural buffers. In the Launch Lab, you will investigate the buffering activity of water samples. You will use equilibrium concepts from the previous chapter to calculate the pH of solutions of weak acids and bases. Finally, you will investigate acid-base reactions, prepare a buffer solution, and examine how a buffer solution resists change in pH when small amounts of acid or base are added to it.

Buffering Ground Water: A Delicate Balance

Many areas in Alberta have alkaline soils and rocks that neutralize acid deposition. In this lab, you will represent different types of soil and rock by using samples of granite or quartz, and marble. Different types of ground water will be modelled by adding water to these solids. You will compare the impact of acid rain on the models of ground water you prepare.

Safety Precautions

Wash your hands when you have completed the Launch Lab.

Materials
- dilute vinegar solution
- granite or quartz chips
- marble chips
- sample of local soil or rock
- 250 mL beakers (5)
- universal indicator paper (pH paper)
- pH meter (optional)
- 100 mL graduated cylinder
- 50 mL burette
- stirring rod
- retort stand
- burette clamp
- a label or a grease pencil

Procedure

1. Create a table in your notebook to record your observations using the following headings: acid rain added (mL), tap water + acid rain (pH), granite or quartz + water + acid rain (pH), marble + water + acid rain (pH), and local soil or rock + water + acid rain (pH). Leave six rows under the headings to fill in your data.

2. Pour 120 mL of the dilute vinegar solution into a beaker labelled "acid rain." Measure and record the pH of the solution. Set up a retort stand, burette clamp, and burette. Fill the burette with the simulated acid rain solution. Record the initial volume.

3. Measure 100 mL of tap water into a labelled beaker. Measure and record the pH. Add 5.0 mL of acid rain to the tap water. Stir, then measure and record the pH. Repeat until a total of 20.0 mL of acid rain has been added. Put the solution aside.

4. Place a few granite or quartz chips into a labelled beaker, then add 100 mL of tap water. Stir the mixture, then measure and record the pH of the solution. Add 5.0 mL samples of acid rain to the beaker. After each addition, stir and measure the pH. After 20.0 mL of acid rain has been added, put the solution aside.

5. Repeat Procedure step 4 using a few marble chips. Repeat again using local soil or rock.

6. Set aside each labelled beaker until your next class. Stir each solution before you make a final measurement of the pH. Pour the solutions down the sink, being careful not to discard any solid chips. Return the chips to your teacher.

Analysis

1. Use a spreadsheet program to graph the data in your table. Spreadsheet programs allow data to be plotted in different ways. Which type of graph is best for your data? Explain. ICT

2. Rank the water samples from least resistant to change in pH when acid rain was added to most resistant. Does your ranking agree with other students?

3. Compare the pH of each sample after the last addition of acid rain with the measurement made after the solution was left standing. If any measurement changed, suggest a reason for the change.

Acid Deposition

Key Terms

acid rain
acid deposition
liming

In the complete absence of pollution, rain would still be acidic. Carbon dioxide in the atmosphere dissolves slightly in water and reacts with water to form carbonic acid:

$$CO_2(g) + H_2O(\ell) \rightleftharpoons H_2CO_3(aq)$$

$$H_2CO_3(aq) + H_2O(\ell) \rightleftharpoons HCO_3^-(aq) + H_3O^+(aq)$$

Carbonic acid ionizes and reaches equilibrium with hydrogen carbonate ions, $HCO_3^-(aq)$, carbonate ions, $CO_3^{2-}(aq)$, and hydronium ions, $H_3O^+(aq)$. Because of dissolved carbonic acid and other naturally occurring acids in rain, the pH of unpolluted rainwater is about 5.3. **Acid rain** is the term used to describe the lowering of pH of rain to less than pH 5 and is caused by human activities. However, *acid deposition* is a better term, because acid can also precipitate through other means. For instance, dry acidic particles can settle out from the atmosphere, while acidic snow and fog can also cause damage. **Acid deposition** is the total effect of acid falling in rain, in snow, or as fine solid particles. Figure 17.1 shows some of the ways that acid deposition can occur.

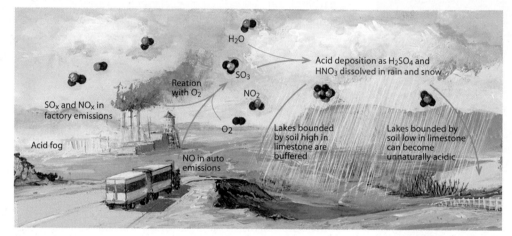

Figure 17.1 Modern industrial societies produce various emissions that contribute to acid deposition.

Acid Deposition: SO_x and NO_x ("Sox" and "Nox")

The most important cause of acid deposition is the presence of dissolved oxides of sulfur ($SO_2(g)$ and $SO_3(g)$) and oxides of nitrogen ($NO(g)$ and $NO_2(g)$) in water droplets in the atmosphere. These gases are sometimes collectively referred to as SO_x and NO_x, often called "sox" and "nox." Metal ores and many fuels, especially coal and some natural gas supplies, contain compounds of sulfur. When these fuels are burned, or metal ores are smelted to purify the metal, sulfur combines with oxygen in the air:

$$S_8(s) + 8O_2(g) \rightarrow 8SO_2(g)$$

Sulfur dioxide reacts with water in the atmosphere, forming sulfurous acid:

$$SO_2(g) + H_2O(\ell) \rightleftharpoons H_2SO_3(aq)$$

Sulfur dioxide can also react with oxygen in the air to form sulfur trioxide:

$$2SO_2(g) + O_2(g) \rightleftharpoons 2SO_3(g)$$

Sulfur trioxide reacts with water to form sulfuric acid:

$$SO_3(g) + H_2O(\ell) \rightleftharpoons H_2SO_4(aq)$$

ChemistryFile

Web Link

Find out more about acid deposition and summarize some points you would like to raise in a letter to your federal member of Parliament. What are the consequences of acid deposition in Canada? What are the principal sources of acid deposition from industry? How is Alberta's air quality managed and monitored?

www.albertachemistry.ca
WWW

Figure 17.2 This part of the plant at Syncrude's facility in Fort McMurray splits large molecules in crude oil into the smaller molecules that are used for fuel. The company is expanding its output of oil and will spend more than $350 million to reduce $SO_2(g)$ emissions.

In Alberta, the pollutants that cause acid rain come mainly from removing sulfur from natural gas and oil (see Figure 17.2), and from burning coal that contains sulfur compounds, such as pyrite, $FeS_2(s)$. Natural gas often contains significant amounts of hydrogen sulfide. Because this poisonous sulfur compound has a characteristic smell, gas containing it is called sour gas. Sour gas must be removed before the natural gas is distributed. This removal process is known as sweetening the gas. A large percentage of the sulfur removed through this method can be recovered through the Claus process. Sulfur dioxide is formed, however, when excess compounds that contain sulfur are burned off in a flare stack. It is also formed when coal is burned at a power station.

Another major source of acid rain comes from dissolved nitrogen oxides. Nitrogen and oxygen combine at the high temperatures generated in the engines of cars, trucks, and in furnaces:

$$N_2(g) + O_2(g) \rightleftharpoons 2NO(g)$$

Then, a slower reaction with oxygen in the air forms nitrogen dioxide:

$$2NO(g) + O_2(g) \rightleftharpoons 2NO_2(g)$$

Nitrogen dioxide reacts with moisture in the atmosphere to form nitrous acid, $HNO_2(aq)$, and nitric acid, $HNO_3(aq)$:

$$2NO_2(g) + H_2O(\ell) \rightleftharpoons HNO_2(aq) + HNO_3(aq)$$

Automobile Exhaust and Le Châtelier's Principle

There is essentially no reaction between nitrogen and oxygen at room temperature, because K_c for the reaction is very small:

$$N_2(g) + O_2(g) \rightleftharpoons 2NO(g) \qquad K_c = 4.7 \times 10^{-31} \text{ at SATP}$$

The reaction is endothermic from left to right. You can use Le Châtelier's principle to predict that this reaction will shift to the right at higher temperatures. At the temperature inside an automobile engine, about 1100 K, the value of K_c is approximately 6×10^{-9}. The exhaust gases cool down too rapidly for nitrogen monoxide to decompose back into $O_2(g)$ and $N_2(g)$. Thus, automobile and diesel engines, as well as furnaces, produce nitrogen monoxide pollution, contributing to acid deposition. Engines operate more efficiently at high temperatures, so it is not practical to reduce the amount of pollution by lowering the temperature at which an engine operates.

A

Granite is a mixture of crystals formed when molten rock cools slowly deep within the ground.

Figure 17.3

B

Chalk, limestone, and marble are different forms of calcium carbonate, $CaCO_3$(s). Most limestone is formed by the accumulation of seashell fragments from ancient seas over millions of years. Marble is formed from limestone by heat and pressure under Earth's crust.

C

Quartz is the common name of silicon dioxide, SiO_2(s). Oxygen and silicon are the two most abundant elements (by mass) in our planet's crust. Amethyst, the birthstone for February, is quartz coloured by the presence of iron oxide.

Acid Deposition and Lakes

Acid deposition can damage living things on land or in the water. Trees and other forms of plant life have been destroyed in some area. However, lakes often are the most seriously affected because they collect rain and run-off from a wide area. Different organisms have different tolerances for acidic water. Snails, for example, cannot tolerate water with a pH below 6.0. Perch and pike die at about pH 5.0 while eel and brook trout survive to about pH 4.5.

It is often not the pH of the water itself that affects aquatic animals but the minerals that are leached from rocks and soil by the acidic water. Toxic metals including aluminium, mercury, and lead, are more soluble in acidic water than in neutral water. Mercury and lead can accumulate in fish with lethal results. Dissolved aluminium interferes with gill function. The blockage of the gills is usually not fatal but can cause stress that results in reduced body size and weight.

The Role of Soil and Rocks as Natural Buffers

Acid deposition does not affect Alberta as much as other parts of Canada because the soil and rock offers natural protection (Figure 17.3). Alkaline soils cover much of Alberta and this neutralizes the effect of acid deposition. In Central and Southern Alberta, the soils contain limestone. Limestone, $CaCO_3$(s), neutralizes the hydronium ions from acid deposition:

$$2H_3O^+(aq) + CaCO_3(s) \rightleftharpoons Ca^{2+}(aq) + CO_2(g) + 3H_2O(\ell)$$

The carbon dioxide formed in this reaction dissolves to form carbonic acid. The result is that the acids in acid rain (such as sulfuric acid) are replaced by carbonic acid. Carbonic acid ionizes much less than the acid it has replaced, so the result is a lower concentration of hydronium ion, H_3O^+(aq), thus a higher pH. Conversely, the Canadian Shield, which covers much of Eastern, Central, and Northern Canada, is made of granite rock. Granite does not contain carbonates to neutralize the effects of acid deposition, so fish and plants in many lakes in this region have been adversely affected by acid deposition.

•••

1 Which rock type is most effective at neutralizing the effects of acid deposition?

2 Your local member of Parliament believes that the government should aim to completely eliminate the effects of acid rain. Suppose you are an aid to this member of Parliament. Draft a memo explaining why this goal is unreasonable.

•••

Protecting Lakes from Acid Shock

In some regions, snow can collect acid deposition throughout the winter. In the spring, when the snow melts, a relatively concentrated surge of acidic water runs into lakes and streams. This severe acid shock causes many fish to die suddenly. When spring acid shock is expected to harm a lake, the lake and surrounding watershed may be treated with limestone. In a process known as **liming,** limestone is added to the water to neutralize acid deposition and buffer lake water against a rapid change in pH. The calcium carbonate in limestone raises the pH of the water, reduces the solubility of toxic metal ions, and supplies calcium ions, an essential plant nutrient. A simple way to lime a lake is to shovel the limestone into the wake of a moving boat or to drop it from a helicopter (see Figure 17.4). To counter spring acid shock, lime can be spread from a snowmobile while the lake is frozen.

In a previous chemistry course, you learned about weak acids such as carbonic acid that is naturally found in rain. You know that weak acids ionize to a very small extent before they reach equilibrium. In Chapter 16, you learned new methods for working with equilibrium reactions. In the next section, you will apply these methods to acids and bases.

Figure 17.4 When a lake lacks natural limestone to buffer the effect of acid rain, liming improves the overall water quality.

Section 17.1 Summary

- Carbon dioxide in the air reacts with rainwater to form carbonic acid which makes natural rainwater slightly acidic.
- Emissions from motor vehicles release nitrogen oxides into the atmosphere. Power and heating plants that burn fossil fuels and ore smelting plants release sulfur oxides into the atmosphere.
- When nitrogen and sulfur oxides react with water in the atmosphere, they form nitrous acid, nitric acid, sulfurous acid, and some sulfuric acid which fall to Earth as acid deposition.
- Acid deposition harms plant and animal life. The pH of lakes and streams can become dangerously low.
- Alberta's predominately limestone geological formations buffer pond and lake water.
- Acidic lakes and ponds can be treated with limestone to reduce the damage caused by acidic shock in the spring when acidic snow melts and flows into the bodies of water.

SECTION 17.1 REVIEW

1. List three major sources of emissions that cause acid deposition. Name the gases associated with each type of emission.

2. Which areas of Canada would you expect to have the highest emissions of the gases causing acid deposition?

3. Why might a lake or stream have a lower pH in the spring than at any other time of the year?

4. Combustion reactions result in much greater emissions of carbon dioxide than of SO_x or NO_x. Explain why there is less concern over the contribution to acid rain from the $CO_2(g)$ emitted compared with the contribution to acid rain from SO_x and NO_x.

5. How does acid rain affect building materials?

6. List three strategies you could use to lower emissions of gases that cause acid deposition.

Understanding Acids and Bases

Section Outcomes

In this section, you will:

- **define** acid-base reactions as proton-transfer reactions
- **write** Brønsted-Lowry equations for acid-base reactions and identify conjugate acid-base pairs
- **identify** amphiprotic compounds and **write** Brønsted-Lowry equations to explain their acid-base reactions
- **recall** the names and formulas of some polyprotic acids and polyprotic bases
- **distinguish** between strong acids and bases, and weak acids and bases

Key Terms

Brønsted-Lowry theory of acids and bases
conjugate acid-base pair
conjugate base
conjugate acid
amphiprotic
polyprotic acid
polyprotic base
strong acid
weak acid
strong base
weak base

The Brønsted-Lowry Theory of Acids and Bases

When you learned about acids and bases before, the modified Arrhenius theory provided a simple way to understand acid-base reactions. But the Arrhenius theory is limited by its assumption that all acid-base reactions occur in water. Because many acid-base reactions take place in solvents other than water, the limitations of the Arrhenius theory of acids and bases were removed by a more general theory proposed independently in 1923 by Johannes Brønsted, a Danish chemist, and Thomas Lowry, an English chemist (see Figure 17.5). The **Brønsted-Lowry theory** recognizes acid-base reactions as a chemical equilibrium having both a forward and a reverse reaction that involve the transfer of a proton.

The Brønsted-Lowry Definitions of Acids and Bases

An acid is a substance that can donate a proton, H^+.
A base is a substance that can accept a proton, H^+.

Figure 17.5 Johannes Brønsted (1879–1947), left, and Thomas Lowry (1874–1936), right. Because Brønsted published many more articles about ions in solution than Lowry, some chemistry resources refer to the "Brønsted theory of acids and bases."

Like an Arrhenius acid, a Brønsted-Lowry acid must contain H in its formula. Not every compound that contains a hydrogen atom is an acid, only those able to provide a proton, H^+. This means that all Arrhenius acids are also Brønsted-Lowry acids. However, many molecules other than $OH^-(aq)$ can be a Brønsted-Lowry base. In addition, water is not the only solvent that can be used in a Brønsted-Lowry acid-base reaction. The only requirement for an acid-base reaction according to the Brønsted-Lowry theory is that one substance must be capable of providing a proton, and another substance must be capable of receiving the same proton. In other words, an acid-base reaction involves the transfer of a proton. Note that the word "proton" refers to the nucleus of a hydrogen atom—an H^+ ion that has been removed from the acid molecule. It does not refer to a proton removed from the nucleus of another atom, such as oxygen or sulfur, which may be present in the acid molecule.

The idea of proton transfer has major implications for understanding the nature of acids and bases. According to the Brønsted-Lowry theory, any substance that behaves as an acid can only do so if another substance behaves as a base at the same time. Similarly, if a substance behaves as a base, another substance must behave as an acid at the same time. For example, consider the reaction between hydrochloric acid and water shown in Figure 17.6. In this reaction, hydrochloric acid is an acid because it donates a proton, H^+, to water. According to the Brønsted-Lowry theory, the water molecule is a base because it accepts the proton. When the water molecule accepts the proton, it becomes a hydronium ion, $H_3O^+(aq)$.

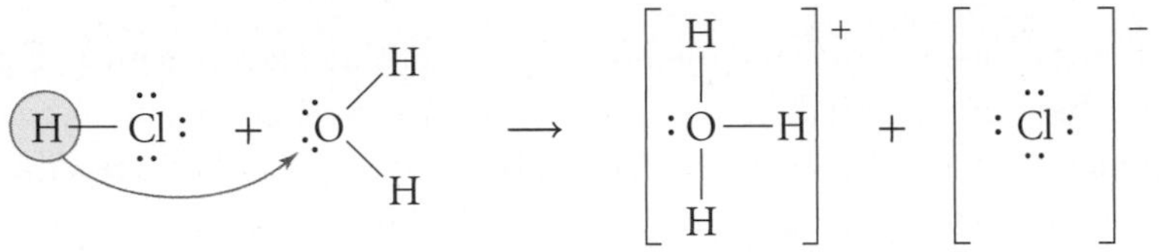

Figure 17.6 The Brønsted-Lowry theory explanation of the reaction between hydrochloric acid and water

Two molecules or ions that differ because of the transfer of a proton are called a **conjugate acid-base pair.** (Conjugate means "linked together.") The **conjugate base** of an acid is the molecule that remains when a proton is removed from the acid. The **conjugate acid** of a base is the molecule formed when the base receives the proton from the acid. An acid-base reaction always contains two conjugate acid-base pairs. In the reaction between hydrochloric acid and water, the chloride ion is the conjugate base of hydrochloric acid. Water acts as a base in this reaction, and the hydronium ion is its conjugate acid.

In the Brønsted-Lowry theory, every acid has a conjugate base, and every base has a conjugate acid. The ionization of ethanoic acid in water is represented in Figure 17.7. Ethanoic acid is a Brønsted-Lowry acid because it provides a proton, H^+, to the water. In receiving the proton, the water molecule is acting as a base in this reaction and becomes a hydronium ion. Notice that this reaction proceeds in both directions, and equilibrium is established. When ethanoic acid reacts with water, only a few molecules react to form ions. The position of equilibrium lies to the left, and the equilibrium constant for the reaction is small. Because the equilibrium lies to the left, the reverse reaction is favoured. In the reverse reaction, the hydronium ion gives up a proton to the ethanoate ion. Thus, in the reverse reaction, the hydronium ion is a Brønsted-Lowry acid and the ethanoate ion is a Brønsted-Lowry base. The acid on the left, $CH_3COOH(aq)$, and the base on the right, $CH_3COO^-(aq)$, differ by one proton, so they are a conjugate acid-base pair. Similarly, $H_2O(\ell)$ and $H_3O^+(aq)$ are a conjugate acid-base pair, because they too differ by one proton.

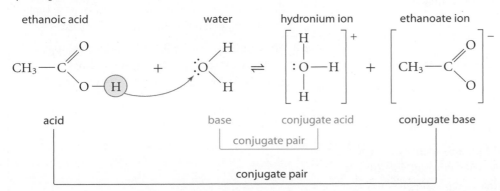

Figure 17.7 Conjugate acid-base pairs in the ionization of ethanoic acid in water.

3 In what way(s) is the Brønsted-Lowry theory more general than the Arrhenius theory of acids and bases?

4 Name two substances that are Brønsted-Lowry bases that would not be considered Arrhenius bases.

5 Must a Brønsted-Lowry acid contain H^+? Must a Brønsted-Lowry base contain $OH^-(aq)$?

Chemists sometimes use a shorthand method of representing reactions involving weak acids and weak bases in general discussions. They use **HA** to represent a weak acid and **B** to represent a weak base. To illustrate these symbols, Figure 17.8 shows reactions of the general weak acid with water and the general weak base with water.

HA	+	H_2O	$\rightleftharpoons$	H_3O^+	+	A^-
acid		water		hydronium ion		conjugate base
B	+	H_2O	$\rightleftharpoons$	OH^-	+	BH^+
base		water		hydroxide ion		conjugate acid

Figure 17.8 In the first reaction, the general acid, HA, donates a proton (H^+) to water and becomes the conjugate base, A^-. In the second reaction, the general base, B, accepts a proton from water and becomes the conjugate acid, BH^+.

If you compare the first reaction in Figure 17.8 with the reaction in Figure 17.7, you will see that the weak acid, HA, represents ethanoic acid and the conjugate base, A^-, represents the ethanoate ion. Also notice that, in the first reaction in Figure 17.8, water is acting as a base when it accepts the proton from HA. The hydronium ion is the conjugate acid. In the second reaction, water is acting as an acid when it donates a proton (H^+) to the base, B. The hydroxide ion is the conjugate base. Water can act as either an acid or a base.

Sample Problem

Conjugate Acid-Base Pairs

Problem

Hydrogen bromide is a gas at room temperature. It is soluble in water, forming hydrobromic acid. Identify the conjugate acid-base pairs in the reaction between HBr(g) and water to form $H_3O^+(aq)$ and $Br^-(aq)$.

What Is Required?

You need to identify the two sets of conjugate acid-base pairs.

What Is Given?

You know that hydrogen bromide forms hydrobromic acid in aqueous solution.

Plan Your Strategy

Step 1 Write a chemical equation.

Step 2 On the left side of the equation, identify the acid as the molecule that donates a proton. Identify the base as the molecule that accepts the proton.

Step 3 Identify the conjugate base on the right side of the equation as the particle with one less proton than the acid on the left side of the equation. Identify the particle on the right that has one proton more than the base on the left as the conjugate acid.

Act on Your Strategy

Step 1 $HBr(g) + H_2O(\ell) \rightarrow H_3O^+(aq) + Br^-(aq)$

Step 2 Hydrogen bromide donates a proton, so it is the Brønsted-Lowry acid in the reaction. Water accepts the proton, so it is the Brønsted-Lowry base.

Step 3 The conjugate base is $Br^-(aq)$, and the conjugate acid is $H_3O^+(aq)$. The conjugate acid-base pairs are $HBr(g)/Br^-(aq)$ and $H_2O(\ell)/H_3O^+(aq)$.

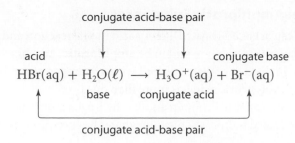

conjugate acid-base pair

acid — conjugate base

$$HBr(aq) + H_2O(\ell) \longrightarrow H_3O^+(aq) + Br^-(aq)$$

base — conjugate acid

conjugate acid-base pair

Check Your Solution

The formulas of the conjugate acid-base pairs differ by one proton, H^+.

Sample Problem

More Conjugate Acid-Base Pairs

Problem

Ammonia is a pungent gas at room temperature. Its main use is in the production of fertilizers and explosives. You may be familiar with the odour of ammonia because it is used in some glass cleaners. Ammonia is very soluble in water, forming a solution with the properties of a base, such as turning red litmus blue. Because the Arrhenius theory defines a base as a substance containing a hydroxide ion that can ionize, aqueous ammonia, $NH_3(aq)$, cannot be an Arrhenius base. Use the Brønsted-Lowry theory to identify the conjugate acid-base pairs in the reaction between aqueous ammonia and water:

$$NH_3(aq) + H_2O(\ell) \rightleftharpoons NH_4^+(aq) + OH^-(aq)$$

What Is Required?

You need to identify the conjugate acid-base pairs.

What Is Given?

The chemical equation is given.

Plan Your Strategy

Step 1 Identify the proton-donor on the left side of the equation as the acid. Identify the proton-acceptor on the left side of the equation as the base.

Step 2 Identify the conjugate acid and base on the right side of the equation by the difference of a single proton from the acid and base on the left.

Act on Your Strategy

The conjugate acid-base pairs are $NH_3(aq)/NH_4^+(aq)$ and $H_2O(\ell)/OH^-(aq)$.

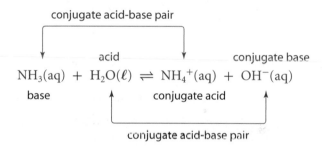

conjugate acid-base pair

acid — conjugate base

$$NH_3(aq) + H_2O(\ell) \rightleftharpoons NH_4^+(aq) + OH^-(aq)$$

base — conjugate acid

conjugate acid-base pair

Check Your Solution

The formulas of the conjugate acid-base pairs differ by one proton, H^+, as expected.

Practice Problems

1. Name and write the formula of the conjugate base of each molecule or ion:
   a) $HF(aq)$
   b) $HCO_3^-(aq)$
   c) $H_2SO_4(aq)$
   d) $N_2H_5^+(aq)$
2. Name and write the formula of the conjugate acid of each molecule or ion:
   a) $NO_3^-(aq)$
   b) $OH^-(aq)$
   c) $H_2O(\ell)$
   d) $HCO_3^-(aq)$
3. When perchloric acid dissolves in water, the following reaction occurs:
   $HClO_4(aq) + H_2O(\ell) \rightarrow H_3O^+(aq) + ClO_4^-(aq)$
   Identify the conjugate acid-base pairs.
4. Identify the conjugate acid-base pairs in the following reactions:
   a) $HS^-(aq) + H_2O(\ell) \rightleftharpoons H_2S(aq) + OH^-(aq)$
   b) $O^{2-}(aq) + H_2O(\ell) \rightleftharpoons 2OH^-(aq)$
   c) $H_2S(aq) + NH_3(aq) \rightleftharpoons NH_4^+(aq) + HS^-(aq)$
   d) $H_2SO_4(aq) + H_2O(\ell) \rightleftharpoons H_3O^+(aq) + HSO_4^-(aq)$

ChemistryFile

FYI

The word "amphiprotic" can be split into two parts, the prefix *amphi-* and the ending *-protic*. The prefix *amphi-* comes from a Greek word meaning "on both sides." The ending *-protic* refers to a proton. Thus, an amphiprotic substance is literally on both sides of a proton transfer. In one reaction, it is capable of donating a proton; and in another reaction, it is capable of accepting a proton. Can you think of another word beginning with the prefix *amphi-*?

Both an Acid and a Base: Amphiprotic Substances

Substances, such as water, that can act as a Brønsted-Lowry acid in one reaction and as a Brønsted-Lowry base in a different reaction are said to be **amphiprotic**. Amphiprotic substances can be molecules, as in the case of water, or ions with a hydrogen atom that can ionize, such as $HSO_4^-(aq)$. A **polyprotic acid** has more than one proton that can ionize. After the first ionization, the ion must contain another hydrogen atom that can also ionize. For example, carbonic acid is diprotic, meaning that it has two protons that can ionize:

$$H_2CO_3(aq) + H_2O(\ell) \rightleftharpoons HCO_3^-(aq) + H_3O^+(aq)$$
$$HCO_3^-(aq) + H_2O(\ell) \rightleftharpoons CO_3^{2-}(aq) + H_3O^+(aq)$$

The hydrogen carbonate ion, $HCO_3^-(aq)$, is an acid when it reacts with water. In the presence of hydronium ions, $HCO_3^-(aq)$ accepts a proton and acts as a base:

$$HCO_3^-(aq) + H_3O^+(aq) \rightleftharpoons H_2CO_3(aq) + H_2O(\ell)$$

Thus, the hydrogen carbonate ion is amphiprotic. The carbonate ion, $CO_3^{2-}(aq)$, is a **polyprotic base**. A polyprotic base is a substance capable of accepting more than one proton. Another common polyprotic base is the sulfite ion:

$$SO_3^{2-}(aq) + H_2O(\ell) \rightleftharpoons HSO_3^-(aq) + OH^-(aq)$$
$$HSO_3^-(aq) + H_2O(\ell) \rightleftharpoons H_2SO_3(aq) + OH^-(aq)$$

Sample Problem

An Amphiprotic Ion

Problem

The hydrogen sulfite ion, $HSO_3^-(aq)$, is amphiprotic. Write a chemical equation to show how the ion reacts with water first as an acid and then as a base.

What Is Required?

You need to write two chemical equations. In the first equation, $HSO_3^-(aq)$ must act as an acid. In the second equation, the ion must act as a base.

What Is Given?

The formula of the hydrogen sulfite ion, $HSO_3^-(aq)$, is given.

Plan Your Strategy

Step 1 As an acid, $HSO_3^-(aq)$ must donate a proton to water.

Step 2 As a base, $HSO_3^-(aq)$ must accept a proton from water.

Act on Your Strategy

Step 1 $HSO_3^-(aq) + H_2O(\ell) \rightleftharpoons SO_3^{2-}(aq) + H_3O^+(aq)$

Step 2 $HSO_3^-(aq) + H_2O(\ell) \rightleftharpoons H_2SO_3(aq) + OH^-(aq)$

Check Your Solution

When $HSO_3^-(aq)$ acts as an acid, the conjugate base, $SO_3^{2-}(aq)$, has one less proton. When $HSO_3^-(aq)$ acts as a base, the conjugate acid, $H_2SO_3(aq)$, has one more proton.

Practice Problems

5. Hydrogen sulfide is a gas at room temperature. It has a very unpleasant smell and is not very soluble in water, forming a dilute solution of hydrosulfuric acid, $H_2S(aq)$.
   a) Hydrosulfuric acid is diprotic. Write two chemical equations to show the ionization of $H_2S(aq)$ in water.
   b) The hydrogen sulfide ion is amphiprotic. Write a chemical equation to show $HS^-(aq)$ acting as a base in water.

6. a) Write a chemical equation to show the hydrogen carbonate ion, $HCO_3^-(aq)$, acting as an acid in the presence of $OH^-(aq)$.
   b) Write a chemical equation to show $HCO_3^-(aq)$ acting as a base in the presence of $HF(aq)$.

7. Sodium dihydrogen phosphate, $NaH_2PO_4(s)$, is soluble in water. Write chemical equations to show that $H_2PO_4^-(aq)$ is amphiprotic. Label each equation as an acid or a base reaction, as appropriate.

Predicting the Direction of Reaction for an Acid-Base Reaction

All acid-base reactions involve two conjugate acid-base pairs. On the left side of the equation, you can label the acid and base. On the right side of the equation, you can identify the conjugate base and conjugate acid. How does the strength of each species determine the direction of the reaction?

Recall the reaction of hydrochloric acid with water:

$$HCl(aq) + H_2O(\ell) \rightarrow H_3O^+(aq) + Cl^-(aq)$$

acid	base	conjugate acid	conjugate base

You know this reaction goes completely to the right, because hydrochloric acid is a strong acid. Since $HCl(aq)$ must lose its proton more easily than $H_3O^+(aq)$ does, $HCl(aq)$ is a stronger acid than $H_3O^+(aq)$. You could also consider the reaction in terms of the bases, $H_2O(\ell)$ and $Cl^-(aq)$. Water gains a proton more readily than $Cl^-(aq)$ does; therefore, water must be a stronger base than $Cl^-(aq)$. The following generalization can help you predict the direction of an acid-base reaction: *An acid-base reaction usually proceeds to the right if the stronger acid and stronger base are on the left side of the equation and products are favoured.* This makes sense because a stronger acid more easily donates a proton, and a stronger base more easily accepts the proton.

Over many years, chemists have performed countless experiments involving acids and bases. The data from these experiments have enabled chemists to rank acids and bases according to their strengths in relation to one another, as shown in Table 17.1. You can use this table to predict the direction in which an acid-base reaction will proceed.

Table 17.1 Relative Strengths of Selected Acids and Bases

Acid	Base
HCl	Cl^-
H_2SO_4	HSO_4^-
HNO_3	NO_3^-
H_3O^+	H_2O
H_2SO_3	HSO_3^-
HSO_4^-	SO_4^{2-}
H_3PO_4	$H_2PO_4^-$
HF	F^-
CH_3COOH	CH_3COO^-
H_2CO_3	HCO_3^-
H_2S	HS^-
HSO_3^-	SO_3^{2-}
$H_2PO_4^-$	HPO_4^{2-}
NH_4^+	NH_3
HCO_3^-	CO_3^{2-}
HPO_4^{2-}	PO_4^{2-}
H_2O	OH^-

Increasing Acid Strength ↑

Increasing Base Strength ↓

Sample Problem

Predicting the Direction of an Acid-Base Reaction

Problem

Predict the direction in which the following reaction will proceed. Briefly explain your answer.

$$SO_4^{2-}(aq) + CH_3COOH(aq) \rightleftharpoons$$
$$HSO_4^-(aq) + CH_3COO^-(aq)$$

What Is Required?

You need to determine whether reactants or products are favoured at equilibrium.

What Is Given?

You have the chemical equation and Table 17.1.

Plan Your Strategy

Step 1 Identify the species acting as an acid on each side of the equation.

Step 2 Use Table 17.1 to find the stronger acid. The reaction proceeds from the stronger acid toward the weaker acid.

Act on Your Strategy

Step 1 On the left side of the equation, $CH_3COOH(aq)$ is the acid. On the right side of the equation, $HSO_4^-(aq)$ acts as an acid.

Step 2 In Table 17.1, $HSO_4^-(aq)$ is above $CH_3COOH(aq)$; therefore, $HSO_4^-(aq)$ is the stronger acid. The stronger acid more easily gives up a proton, so the equilibrium is to the left. Reactants are favoured in this reaction.

$$SO_4^{2-}(aq) + \underset{\textbf{weaker acid}}{CH_3COOH(aq)} \rightleftharpoons \underset{\textbf{stronger acid}}{HSO_4^-(aq)} + CH_3COO^-(aq)$$

Check Your Solution

Comparing the relative strengths of $HSO_4^-(aq)$ and $CH_3COOH(aq)$, ethanoic acid is weaker. The reaction proceeds from the stronger acid toward the weaker acid.

8. Predict the direction for the following reactions. State whether reactants or products are favoured, and give reasons to support your decision:

a) $NH_4^+(aq) + H_2PO_4^-(aq) \rightleftharpoons$
$$NH_3(aq) + H_3PO_4(aq)$$

b) $H_2O(\ell) + HS^-(aq) \rightleftharpoons OH^-(aq) + H_2S(aq)$

c) $HF(aq) + SO_4^{2-}(aq) \rightleftharpoons F^-(aq) + HSO_4^-(aq)$

9. In which direction will the following reactions proceed? In each case, explain your decision:

a) $HPO_4^{2-}(aq) + NH_4^+(aq) \rightleftharpoons$
$$H_2PO_4^-(aq) + NH_3(aq)$$

b) $H_2SO_4(aq) + H_2O(\ell) \rightarrow HSO_4^-(aq) + H_3O^+(aq)$

c) $H_2S(aq) + NH_3(aq) \rightleftharpoons HS^-(aq) + NH_4^+(aq)$

10. Write equilibrium equations for each of the following reactions. State whether the reactants or the products are favoured. (Consult Table 17.1 or the table in the Appendix G.)

a) aqueous carbonic acid is combined with ammonia

b) sodium hydrogen sulfite is dissolved in water (**Hint:** the sodium hydrogen sulfite dissociates completely into sodium ions and hydrogen sulfite ions. What happens to the hydrogen sulfite ion in water?)

c) hydrofluoric acid is mixed with potassium nitrate (Consider the hint in part b.)

d) all possible reactions between water and phosphoric acid

Section 17.2 Summary

- Brønsted and Lowry developed a theory of acids and bases that was more general than that of Arrhenius.

- A Brønsted-Lowry acid is a proton donor.

- A Brønsted-Lowry base is a proton acceptor.

- According to the Brønsted-Lowry theory, an acid must react with a base and vice versa.

- When an Brønsted-Lowry acid donates a proton, it becomes the conjugate base of the acid.

- When an Brønsted-Lowry base accepts a proton, it becomes the conjugate acid of the base.

- A compound that can either donate or accept a proton is called an amphiprotic substance.

- Water in an amphiprotic substance.

- An acid-base reaction proceeds to the right if the acid on the left is stronger than the acid on the right.

SECTION 17.2 REVIEW

1. At one time, sodium ethanoate was used as a fungicide. It is a good electrolyte and in water the ethanoate ion reacts as follows:

$CH_3COO^-(aq) + H_2O(\ell) \rightleftharpoons$
$$CH_3COOH(aq) + OH^-(aq)$$

Identify the conjugate acid-base pairs.

2. Predict the direction for the reaction between dihydrogen phosphate ions and aqueous ammonia:

$H_2PO_4^-(aq) + NH_3(aq) \rightleftharpoons NH_4^+(aq) + HPO_4^{2-}(aq)$

State whether reactants or products are favoured, and give reasons to support your decision.

3. Boric acid, $H_3BO_3(s)$, is a household chemical that is deadly to all insects. Boric acid is a weak, triprotic acid. Write three equations that show that boric acid is triprotic in water.

4. Identify the conjugate acid-base pairs in the following reactions:

a) $H_2PO_4^-(aq) + CO_3^{2-}(aq) \rightleftharpoons$
$$HPO_4^{2-}(aq) + HCO_3^-(aq)$$

b) $HCOOH(aq) + CN^-(aq) \rightleftharpoons$
$$HCOO^-(aq) + HCN(aq)$$

c) $H_2PO_4^-(aq) + OH^-(aq) \rightleftharpoons HPO_4^{2-}(aq) + H_2O(\ell)$

Acid-Base Equilibriums

Section Outcomes

In this section, you will:

- **calculate** acid ionization constants, K_a, and base ionization constants, K_b, given the concentration of a weak acid or base and the pH of the solution
- **solve** acid-base equilibrium problems using an approximation method
- **measure** an acid ionization constant by preparing a solution of a weak acid and measuring the pH of the solution
- **calculate** the ionization constant for a conjugate base or a conjugate acid, given K_a or K_b, respectively
- **predict** whether products or reactants are favoured in an acid-base reaction

Key Terms

acid ionization constant, K_a
percent ionization
base ionization constant, K_b
ion product constant for water, K_w

Many common foods (such as citrus fruits), pharmaceuticals (such as Aspirin™ and some sedatives), and some vitamins (such as niacin and vitamin B3) are weak acids. In Section 17.2, you learned that you can write equilibrium equations for weak acids and you can predict whether reactants or products are favoured. The equilibrium reactions for weak acids and bases are essentially the same as any other equilibrium reaction and can be treated mathematically in the same way. The equilibrium constant is denoted as K_a instead of K_c and is called the **acid ionization constant**.

Acid Ionization Constant

By using the general formula HA to represent any monoprotic acid, the equilibrium of a weak monoprotic acid in aqueous solution can be written as follows:

$$HA(aq) + H_2O(\ell) \rightleftharpoons H_3O^+(aq) + A^-(aq)$$

This is a heterogeneous equilibrium containing a pure liquid, water. Remember, that pure liquids (and pure solids) have constant concentrations. In dilute solutions, the concentration of water is almost constant. For these reasons the concentration of water is, in effect, included in the equilibrium constant and is not written in the equilibrium expression. The equilibrium expression for this reaction is

$$K_a = \frac{[H_3O^+][A^-]}{[HA]}$$

This expression is called the acid ionization expression or K_a expression. In a weak acid, the number of acid molecules that ionize is small, so the position of equilibrium lies far to the left. The concentration of acid is larger than the concentration of the other ions in solution. For this reason, values of K_a for weak acids are always small numbers. The smaller the value of K_a, the weaker is the acid. Very weak acids have K_a values that are less than 1×10^{-15}. Appendix G lists the acid ionization constants for selected acids at 25 °C. You will notice that the K_a value of the hydronium ion is relatively large compared with other weak acids; however, the hydronium ion is still a weak acid. Strong acids have very large K_a values that are much greater than 1.

Sample Problem

Acid Ionization Expressions

Problem

Sulfurous acid, $H_2SO_3(aq)$ is a weak diprotic acid. Write an acid ionization expression for each of the ionizations that take place in an aqueous solution of sulfurous acid. Include the value of K_a.

Solution

$$H_2SO_3(aq) + H_2O(\ell) \rightleftharpoons HSO_3^-(aq) + H_3O^+(aq)$$

$$K_{a1} = \frac{[HSO_3^-][H_3O^+]}{[H_2SO_3]} = 1.4 \times 10^{-2}$$

$$HSO_3^-(aq) + H_2O(\ell) \rightleftharpoons SO_3^{2-}(aq) + H_3O^+(aq)$$

$$K_{a2} = \frac{[SO_3^{2-}][H_3O^+]}{[HSO_3^-]} = 6.3 \times 10^{-8}$$

Chemistry File

FYI

Niacin is found in many foods, including Indian corn. The niacin in Indian corn, however, cannot be absorbed in the intestinal tract. In regions where this food is a major part of the diet, niacin deficiency can occur. The addition of calcium oxide or wood ash to cooking water forms a basic solution that allows niacin to be absorbed from the corn. This is the first step in the preparation of tortillas.

11. Write the K_a expression for each acid ionizing in an aqueous solution. Include the value of K_a from Appendix G:

    a) nitrous acid, $HNO_2(aq)$

    b) benzoic acid, $C_6H_5COOH(aq)$

    c) citric acid, $H_3C_6H_5O_7(aq)$

    d) hydrogen citrate ion, $HC_6H_5O_7{}^{2-}(aq)$

12. Carbonic acid, $H_2CO_3(aq)$, is a weak diprotic acid. Write an acid ionization expression for each of the ionizations that take place in an aqueous solution of carbonic acid.

13. Phenol, $C_6H_6O(aq)$, is a weak monoprotic acid used as a disinfectant. The K_a expression is $K_a = \dfrac{[C_6H_5O^-][H_3O^+]}{[C_6H_6O]}$. Write the ionization reaction for phenol in an aqueous solution.

Math Tip

In the ICE table, $[H_2O(\ell)]$ is left blank because it is not included in the expression for K_a. The $[H_3O^+(aq)]$ in pure water is 1.0×10^{-7} at 25 °C. This is the initial $[H_3O^+(aq)]$. However, compared with the equilibrium concentration of hydronium ion, 1.0×10^{-7} is not significant in the problems you will solve. To show this, write ~0 ("almost zero") in the ICE table for the initial $[H_3O^+(aq)]$.

Acid Ionization Constant and Percent Ionization

In the previous chapter, you calculated K_c values given initial concentrations and the concentration of one substance at equilibrium. Using the same approach, you can calculate a value for the K_a of a weak acid when its initial concentration and the pH of the aqueous solution are known. This is because the pH of the solution can be used to calculate the hydronium ion concentration, which is one of the equilibrium concentrations. As you know, only a few molecules of a weak acid ionize. The **percent ionization** of a weak acid is the fraction of molecules that ionize compared with the initial concentration of the acid, expressed as a percentage. The percent ionization depends on the value of K_a for the acid, as well as the initial concentration of the weak acid. The following Sample Problem shows how to calculate the K_a of a weak acid and its percent ionization.

Sample Problem

Determining K_a and Percent Ionization

Problem

Propanoic acid, $CH_3CH_2COOH(aq)$, is a weak monoprotic acid that is used to inhibit mould formation in bread. A student prepared a 0.10 mol/L solution of propanoic acid and found the pH was 2.96. What is the acid ionization constant for propanoic acid? What percentage of its molecules were ionized in the solution?

What Is Required?

You need to find K_a and the percent ionization for propanoic acid.

What Is Given?

You have the following data:

Initial $[C_2H_5COOH(aq)] = 0.10$ mol/L

$\qquad\qquad\quad$ pH $= 2.96$

Plan Your Strategy

Step 1 Write the equation for the ionization equilibrium of propanoic acid in water. Then set up an ICE table.

Step 2 Write the expression for the acid ionization constant.

Step 3 Calculate $[H_3O^+(aq)]$ using $[H_3O^+(aq)] = 10^{-pH}$. Note that $[C_2H_5COO^-] = [H_3O^+]$

Step 4 Use the stoichiometry of the equation and $[H_3O^+(aq)]$ to complete the ICE table. Calculate K_a.

Step 5 Calculate the percent ionization by expressing the fraction of molecules that ionize out of 100.

Act on Your Strategy

Step 1 Use the equation for the ionization equilibrium of propanoic acid in water to set up an ICE table.

	$C_2H_5COOH(aq)$	$+$	$H_2O(\ell)$	$\rightleftharpoons$	$C_2H_5COO^-(aq)$	$+$	$H_3O^+(aq)$
	$[C_2H_5COOH(aq)]$ (mol/L)		$[H_2O(\ell)]$ (mol/L)		$[C_2H_5COO^-(aq)]$ (mol/L)		$[H_3O^+(aq)]$ (mol/L)
Initial	0.10				0		~ 0
Change	-1.1×10^{-3}				$+1.1 \times 10^{-3}$		$+1.1 \times 10^{-3}$
Equilibrium	$0.10 - 1.1 \times 10^{-3}$				1.1×10^{-3}		1.1×10^{-3}

Step 2 Remember, pure water is not included in the expression for K_a.

$$K_a = \frac{[C_2H_5COO^-][H_3O^+]}{[C_2H_5COOH]}$$

Step 3 Calculate the value of $[H_3O^+(aq)]$.

$[H_3O^+(aq)] = 10^{-pH}$

$[H_3O^+(aq)] = 10^{-2.96}$

$[H_3O^+] = 1.1 \times 10^{-3}$ mol/L

Step 4 Recall that $[C_2H_5COO^-] = [H_3O^+]$

$$K_a = \frac{(1.1 \times 10^{-3})^2}{0.10 - (1.1 \times 10^{-3})}$$

$= 1.2 \times 10^{-5}$

Step 5 Percent ionization $= \dfrac{1.1 \times 10^{-3} \, \frac{mol}{L}}{0.10 \, \frac{mol}{L}} \times 100\%$

$= 1.1\%$

Check Your Solution

The value for K_a and the percent ionization are reasonable for a weak acid.

Math Tip

As you know, powers of ten do not count in the number of significant digits in a measurement. For this reason, the part of a logarithm that depends on the power of ten does not count in the number of significant digits in the logarithm. For example, the number 3.57×10^4 has three significant digits. The logarithm of 3.57×10^4 is 4.5527. The numeral 4 to the left of the decimal represents the exponent of 10, so it does not count in the number of significant digits. The numerals to the right of the decimal determine the 3.57, so they count as significant digits. (**Hint:** Find the logarithm of 3.57×10^8.) Therefore, the logarithm of 3.57×10^4 with the correct number of significant digits is 4.553. If the concentration of hydronium ion is 1.1×10^{-5} mol/L, there are two significant digits. The pH is 4.96, correct to two significant digits.

Practice Problems

14. In low doses, barbiturates act as sedatives. Barbiturates are made from barbituric acid, a weak monoprotic acid that was first prepared by the German chemist Adolph von Baeyer in 1864. The formula of barbituric acid is $HC_4H_3N_2O_3(s)$. A chemist prepares a 0.10 mol/L solution of barbituric acid. The chemist measured the pH of the solution and recorded the value as 2.50. What is the acid ionization constant for barbituric acid? What percentage of its molecules were ionized?

15. The word "butter" comes from the Greek butyros. Butanoic acid, $HC_4H_7O_2(aq)$, (common name butyric acid) gives rancid butter its distinctive odour. If the pH of a 1.00×10^{-2} mol/L solution of butanoic acid is 3.41, calculate the acid ionization constant for butyric acid. What percentage of butyric acid molecules in this solution is ionized?

16. Wild almonds taste bitter (and are dangerous to eat!) because they contain hydrocyanic acid, $HCN(aq)$. When a chemist prepared a 0.75 mol/L solution of $HCN(aq)$, the pH was found to be 4.67. What is the acid ionization constant, K_a?

17. Many sunscreen lotions contain salts of para-aminobenzoic acid (PABA). The structure of PABA is shown here: A saturated solution of PABA was prepared by dissolving 4.7 g in a 1.0 L solution. The pH of the solution was found to be 3.19. Calculate the acid ionization constant, K_a, for PABA.

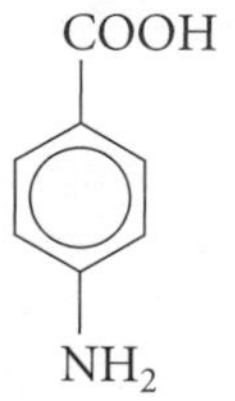

18. Aspirin™ (acetylsalicylic acid) is a monoprotic acid with molar mass of 180. An aqueous solution containing 3.3 g/L was found to have pH = 2.62. What percentage of acetylsalicylic acid molecules were ionized in the solution?

You can determine the value of K_a for a particular acid by measuring the pH of a solution of the acid and determining the concentration of the acid. In your previous chemistry course, you learned how to perform a titration experiment to determine the molar concentration of a solution of an acid or a base. You reviewed indicators in the Unit 8 Preparation. In the following investigation, you will use your titration skills and knowledge of acid equilibrium to determine the value of K_a for a weak acid.

INVESTIGATION | 17.A

Determining K_a for Ethanoic Acid

To find the concentration of an acid in a titration experiment, you can use a pipette to place a known volume of the acid into an Erlenmeyer flask and then add a few drops of an indicator to the flask. Next, you can use a burette to add a basic solution with known concentration to the Erlenmeyer flask until the indicator changes colour. In this investigation, you will be given a sample of ethanoic acid with an unknown concentration. You will measure the pH of the solution using pH paper or a pH meter if one is available. Then you will perform a titration experiment to find the molar concentration of the ethanoic acid solution. Using this data, you will calculate K_a for ethanoic acid and find the percentage of ethanoic acid molecules that ionized in the solution.

Question

What is the acid ionization constant of ethanoic acid? What percentage of its molecules ionize in an aqueous solution?

Prediction

Predict the value of K_a and the percent ionization of $CH_3COOH(aq)$.

Safety Precautions

- Sodium hydroxide is toxic and is harmful if swallowed or inhaled. Both ethanoic acid and sodium hydroxide are corrosive. Immediately wash any spills on your skin or clothing with plenty of cool water and inform your teacher. Also inform your teacher immediately if you spill sodium hydroxide or ethanoic acid on the lab bench or floor.

- Phenolphthalein solution may irritate skin, eyes, and mucous membranes. This solution is flammable. Keep away from open flames.

- Wash your hands when you have completed the investigation.

Materials

- unknown concentration ethanoic acid solution, $CH_3COOH(aq)$
- known concentration sodium hydroxide solution, $NaOH(aq)$
- dropper bottle containing phenolphthalein
- distilled water
- 10 mL pipette
- labels

- 100 mL beakers (2)
- pH meter or pH paper
- 250 mL beaker for waste solutions
- burette and burette clamp
- retort stand
- meniscus reader
- funnel
- pipette bulb or pipette pump
- 150 mL Erlenmeyer flask
- sheet of white paper

Procedure

1. Your teacher will give you the concentration of $NaOH(aq)$. Record this concentration in your notebook, as well as the volume of the pipette. Design a table to record your titration data.

2. Label a clean, dry 100 mL beaker for each liquid. Obtain about 40 mL of ethanoic acid and approximately 70 mL of $NaOH(aq)$.

3. Measure the pH of the ethanoic acid solution using pH paper, or a pH meter if one is available. Record this value.

4. Rinse a clean burette with about 10 mL of $NaOH(aq)$. Discard the rinse into the 250 mL beaker. Then set up a retort stand, burette clamp, meniscus reader, and funnel. Fill the burette with $NaOH(aq)$. Make sure that the solution fills the tube below the burette tap and contains no air bubbles. Remove the funnel.

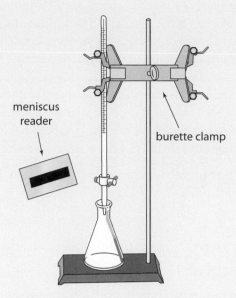

meniscus reader

burette clamp

5. Obtain a clean 10 mL pipette and a suction bulb or pipette pump. Rinse the pipette with a few mL of $CH_3COOH(aq)$ and discard the rinse. Pipette 10.00 mL of $CH_3COOH(aq)$ into the Erlenmeyer flask. Add two or three drops of phenolphthalein indicator. Place a sheet of white paper under the flask.

6. Perform the titration. The endpoint is a faint pink colour that remains after swirling the contents of the Erlenmeyer flask for at least ten seconds. Measure the volume of base required to reach the endpoint. Repeat the titration as time permits until you have at least two sets of data that agree with each other within 2%.

7. Discard waste liquids into the beaker you have been using for this purpose. Give the beaker containing waste liquids from your experiments to your teacher for safe disposal.

8. Rinse the pipette and burette with distilled water. Leave the burette tap open and store the burette upside down.

Analysis

1. Write the chemical equation for the neutralization reaction you performed.

2. Calculate the molar concentration of the ethanoic acid, $[CH_3COOH(aq)]$. Use the ratio in which the acid and base react, determined from the chemical equation.

3. Calculate $[H_3O^+(aq)]$ using your measurement of the pH of the ethanoic acid solution.

4. Write the expression for K_a, the ionization constant of ethanoic acid in water.

5. Set up an ICE table and substitute equilibrium concentrations into your expression for K_a. Calculate the value of K_a and the percentage of ethanoic acid molecules that ionized in solution.

Conclusion

6. Calculate the percentage difference between your value for K_a of ethanoic acid and the accepted value. State two sources of error that might account for any differences.

7. Compare your value for K_a and the percentage of ethanoic acid molecules that ionized with the values determined by others in your class. Discuss the results.

Application

8. Do the values you calculated for $[H_3O^+(aq)]$ and $[CH_3COOH(aq)]$ demonstrate that ethanoic acid is a weak acid? Explain.

The Ion Product Constant for Water

The most common solvent for acids and bases is water. Pure water contains a few ions produced by the ionization of water molecules:

$$2H_2O(\ell) \rightleftharpoons H_3O^+(aq) + OH^-(aq)$$

At 25 °C, only about two water molecules in a billion ionize. This is why pure water is such a poor conductor of electricity. In pure water at 25 °C, the concentration of hydronium and hydroxide ions is 1.0×10^{-7} mol/L. The concentrations must be equal, because the ionization of water produces an equal number of hydronium and hydroxide ions. The equilibrium constant for the ionization of water is given the symbol K_w.

$$K_w = [H_3O^+][OH^-]$$

K_w is called the **ion product constant for water**. At 25 °C, $K_w = 1.0 \times 10^{-14}$. As in the expression for K_a, the concentration of water is not included in the equilibrium expression for K_w. When an acid is added to pure water, the acid molecules ionize and increase the concentration of $H_3O^+(aq)$. The increased $[H_3O^+(aq)]$ pushes the equilibrium reaction between water molecules to the left, in accordance with Le Châtelier's principle. Therefore, when an acid is added to water, $[H_3O^+(aq)]$ increases, and $[OH^-(aq)]$ decreases. Similarly, when a base is added to pure water the increased $[OH^-(aq)]$ pushes the equilibrium reaction between water molecules to the left, and $[H_3O^+(aq)]$ decreases. The value of K_w stays at 1.0×10^{-14}.

• • •

6 How can the ion product for water stay at 1.0×10^{-14} (at 25 °C) if you add an acid to the water?

7 Why is the ionization of water usually not significant when you calculate the pH of a solution of an acid or a base?

8 Consider the ionization of a weak acid in water, such as ethanoic acid:

$CH_3COOH(aq) + H_2O(\ell) \rightleftharpoons CH_3COO^-(aq) + H_3O^+(aq)$

Use Le Châtelier's principle to predict how the percent ionization of the acid will be affected by diluting the acid.

• • •

The Relationship between pH and pOH

The ion product constant for water can be used to calculate $[OH^-(aq)]$ when the pH of a solution has been measured.

$K_w = 1.0 \times 10^{-14} = [H_3O^+][OH^-]$. Taking logarithms of both sides, $-14.0 = \log[H_3O^+] + \log[OH^-]$. Reversing each sign, $14.0 = -\log[H_3O^+] - \log[OH^-]$. Because $pH = -\log[H_3O^+]$ and $pOH = -\log[OH^-]$, the relationship simplifies to $14.0 = pH + pOH$ (at 25 °C).

The Base Ionization Constant K_b

The equilibrium of weak bases uses concepts similar to the equilibrium of weak acids. Because the lone pair of electrons on a nitrogen atom can bond with H^+ from water, many nitrogen-containing compounds are Brønsted-Lowry bases. Ammonia and many amines are common examples of weak bases containing nitrogen. Many plant-based compounds, such as caffeine in coffee and piperidine in black pepper, are also weak bases.

A weak base, represented by B, reacts with water to form an equilibrium solution of ions:

$$B(aq) + H_2O(\ell) \rightleftharpoons HB^+(aq) + OH^-(aq)$$

The equilibrium constant for this general reaction is given the symbol K_b.

$$K_b = \frac{[HB^+][OH^-]}{[B]}$$

K_b is called the **base ionization constant**. Note that the expression for K_b does not include pure water, for the same reasons water was not included in the K_a expression for an acid.

The conjugate base of a weak acid can also act as a weak base. For example, the fluoride ion is a weak base:

$$F^-(aq) + H_2O(\ell) \rightleftharpoons HF(aq) + OH^-(aq)$$

$$K_b = \frac{[HF][OH^-]}{[F^-]}$$

You can calculate K_b for a weak base in aqueous solution if the initial concentration of the base and the concentration of $OH^-(aq)$ are known. Chemists almost always measure the pH of a solution, and the $[OH^-(aq)]$ can be calculated from this measurement by using the equilibrium constant for the ionization of water.

Sample Problem

Calculating K_b for a Weak Base

Problem

One of the uses for aniline, $C_6H_5NH_2(\ell)$, is in the manufacture of dyes. Aniline is soluble in water and acts as a weak base. When a solution containing 5.0 g/L of aniline was prepared, the pH was found to be 8.68. Calculate the base ionization constant for aniline.

What Is Required?

You need to find K_b for aniline.

What Is Given?

You have the following data:
The formula for aniline is $C_6H_5NH_2(\ell)$
The solution contains 5.0 g/L $C_6H_5NH_2(aq)$
pH = 8.68

Plan Your Strategy

Step 1 Calculate the molar concentration of the solution using the molar mass of aniline and the mass of aniline dissolved in one litre of solution.

Step 2 Calculate the hydroxide ion concentration using the following:
$$pH + pOH = 14.0$$
$$[OH^-(aq)] = 10^{-pOH}$$

Step 3 Write the equation for the ionization equilibrium of aniline in water. Then set up an ICE table.

Step 4 Write the expression for the base ionization constant. Substitute equilibrium terms into the expression and calculate K_b.

Act on Your Strategy

Step 1 Calculate the molar concentration of the solution.
Molar mass of aniline, $C_6H_5NH_2(\ell)$, = 93.12 g/mol
$$[C_6H_5NH_2(aq)] = \frac{5.0\,\frac{g}{L}}{93.12\,\frac{g}{mol}}$$
$$= 0.0537\,\frac{mol}{L}$$

Step 2 Calculate $[OH^-(aq)]$.
$$pOH = 14.0 - 8.68$$
$$= 5.32$$
$$[OH^-(aq)] = 10^{-5.32}$$
$$= 4.79 \times 10^{-6}$$

Step 3 Use the equation for the ionization equilibrium of aniline in water to set up an ICE table.

Step 4 Write the expression for K_b. Substitute equilibrium terms into the expression.
$$K_b = \frac{[C_6H_5NH_3^+][OH^-]}{[C_6H_5NH_2]}$$
$$= \frac{(4.79 \times 10^{-6})(4.79 \times 10^{-6})}{0.0537}$$
$$= 4.3 \times 10^{-10}$$

	$C_6H_5NH_2(aq)$	$+$	$H_2O(\ell)$	$\rightleftharpoons$	$C_6H_5NH_3^+(aq)$	$+$	$OH^-(aq)$
	$[C_6H_5NH_2(aq)]$ (mol/L)		$[H_2O(\ell)]$ (mol/L)		$[C_6H_5NH_3^+(aq)]$ (mol/L)		$[OH^-(aq)]$ (mol/L)
Initial	0.0537				0		~0
Change	-4.79×10^{-6}				$+4.79 \times 10^{-6}$		$+4.79 \times 10^{-6}$
Equilibrium	$(0.0537 - 4.79 \times 10^{-6})$				$+4.79 \times 10^{-6}$		$+4.79 \times 10^{-6}$

Math Tip

Remember, in the ICE table $[H_2O(\ell)]$ is left blank. Compared with the equilibrium concentration of hydroxide ion, 1.0×10^{-7} is not significant in the problems you will solve. To show this, write ~0 ("almost zero") in the ICE table for the initial $[OH^-(aq)]$.

Check Your Solution

The value for K_b is reasonable for a weak base. The answer has the correct number of significant digits (two).

19. Write the chemical equation for each base ionizing in an aqueous solution:
   a) ammonia, $NH_3(aq)$
   b) trimethylamine, $(CH_3)_3N(aq)$
   c) hydrogen sulfite ion, $HSO_3^-(aq)$
   d) carbonate ion, $CO_3^{2-}(aq)$

20. Write the K_b expression for each base in the previous question.

21. When a 0.25 mol/L aqueous solution of methylamine was prepared, the pH of the solution was found to be 10.04. What percentage of methylamine molecules ionized in the solution?

22. Codeine, $C_{18}H_{21}NO_3(s)$, is added to some cough medicines. When a 0.020 mol/L aqueous solution of codeine was prepared, the pH of the solution was found to be 10.26. Calculate K_b for codeine.

23. A Material Safety Data Sheet (MSDS) describes pyridine, $C_5H_5N(\ell)$, as a clear liquid with a putrid odour. A 16 g/L solution of pyridine has pH = 9.23. Use these data to calculate K_b for pyridine.

Calculating the pH of a Solution of a Weak Acid or a Weak Base

Below are a few general steps to guide you when you need to solve an acid or a base equilibrium problem. You recall that in calculations involving the equilibrium constant, if the concentrations of reactants were at least 1000 times greater than K_c, you could make the approximation that the concentrations of the reactants at equilibrium were the same as the initial concentrations. The situation is the same when using acid and base ionization constants. The steps below give you a method for determining whether to use the approximation method. This rule is sometimes called the *rule of a thousand*. Read the steps below and then study the Sample Problem to find out how to apply the rule.

- Write the chemical equation. Use the chemical equation to set up an ICE table for the reacting substances whenever initial acid or base concentrations are known. Enter into the table any values that are given in the problem.

- For problems in this textbook, you can assume that the concentrations of $H_3O^+(aq)$ and $OH^-(aq)$ in pure water are negligible compared with the concentrations of these ions when a weak acid or a weak base is dissolved in water.

- For problems that give the initial concentration of the acid, HA, compare the initial concentration of the acid with the acid ionization constant.

- If $[HA] > 1000 \times K_a$, the change in the initial concentration is not significant compared with the initial [HA]. Therefore, [HA] at equilibrium can be considered to be the same as it was initially.

- If $[HA] < 1000 \times K_a$, the change in the initial concentration may be significant compared with the initial [HA]. You cannot use the approximation method.

- For a base B, compare [B] to $1000 \times K_b$ to determine whether the change in the initial concentration is or is not significant compared with the initial [B].

Calculating the pH of a Weak Acid

Problem
Methanoic acid, $HCOOH(\ell)$, is present in the sting of certain ants. What is the pH of a 0.25 mol/L solution of methanoic acid?

What Is Required?
You need to calculate the pH of the solution.

What Is Given?
You know the concentration of methanoic acid:
$[HCOOH(aq)] = 0.25$ mol/L

The acid ionization constant for methanoic acid is listed in Appendix G: $K_a = 1.8 \times 10^{-4}$.

Plan Your Strategy
Step 1 Write the equation for the ionization equilibrium of methanoic acid in water. Then set up an ICE table.

Step 2 Write the equation for the acid ionization constant. Substitute equilibrium terms into the equation.

Step 3 If you get a quadratic equation in Step 2, test to find out if you may use an approximation.

Step 4 Solve the equation for x.

Step 5 Calculate the pH by using the equation
$pH = -\log [H_3O^+(aq)]$

Act on Your Strategy
Step 1

	HCOOH(aq)	+	H₂O(ℓ)	⇌	HCOO⁻(aq)	+	H₃O⁺(aq)
	$[HCOOH(aq)]$ (mol/L)		$[H_2O(\ell)]$ (mol/L)		$[HCOO^-(aq)]$ (mol/L)		$[H_3O^+(aq)]$ (mol/L)
Initial	0.25				0		~0
Change	$-x$				$+x$		$+x$
Equilibrium	$0.25 - x$				x		x

Step 2
$$K_a = \frac{[HCOO^-][H_3O^+]}{[HCOOH]}$$

$$1.8 \times 10^{-4} = \frac{x^2}{(0.25 - x)}$$

$$x^2 + 1.8 \times 10^{-4}x - 4.5 \times 10^{-5} = 0$$

Step 3 The substitutions resulted in a quadratic equation so compare $1000 \times K_a$ to the initial concentration of methanoic acid.
$$1000 \times 1.8 \times 10^{-4} = 0.18$$
$$[HCOOH] = 0.25$$

$0.25 > 0.18$ so you may use the approximation that $[HCOOH]$ at equilibrium is the same as its initial concentration of $0.25 - x \cong 0.25$.

Step 4 $1.8 \times 10^{-4} = \dfrac{x^2}{0.25}$
$$x^2 = 4.5 \times 10^{-5}$$
$$x = 6.71 \times 10^{-3}$$
$$x = 6.71 \times 10^{-3} \frac{mol}{L}$$
$$= [H_3O^+(aq)]$$

Step 5 $pH = -\log 6.71 \times 10^{-3}$
$$= 2.17$$

The pH of a solution of 0.25 mol/L methanoic acid is 2.17.

Check Your Solution
The pH indicates an acidic solution, as expected. Data that were given in the problem have two significant digits, and the pH has two digits following the decimal place.

24. Calculate the pH of a sample of vinegar that contains 0.83 mol/L ethanoic acid. What is the percent ionization of the ethanoic acid?

25. A solution of hydrofluoric acid has a molar concentration of 1.00 mol/L. What is the pH of the solution? What percentage of its molecules ionize?

26. Hypochlorous acid, $HOCl(aq)$, is used to make bleach. A chemist finds that 0.027% of hypochlorous acid molecules are ionized in a 0.40 mol/L solution. What is the value of K_a for the acid?

27. Hexanoic acid, commonly known as caproic acid, $C_5H_{11}COOH(s)$, occurs naturally in coconut and palm oil. It is a weak monoprotic acid, with $K_a = 1.3 \times 10^{-5}$. A certain aqueous solution of hexanoic acid has pH of 2.94. How much acid was dissolved to make 100 mL of this solution?

Calculating the pH of a Weak Base

Problem

The characteristic bitter taste of tonic water is due to the addition of quinine. Quinine, $C_{20}H_{24}N_2O_2(s)$, is a naturally occurring, white crystalline compound. It is also used to treat malaria. The base ionization constant, K_b, for quinine is 3.3×10^{-6}. What is the hydroxide ion concentration and pH of a 3.6×10^{-3} mol/L solution of quinine?

What Is Required?

You need to find $[OH(aq)]$ and pH.

What Is Given?

$K_b = 3.3 \times 10^{-6}$
Concentration of quinine $= 3.6 \times 10^{-3}$ mol/L

Plan Your Strategy

Step 1 Let Q represent the formula of quinine. Write the equation for quinine acting as a base in water. Then set up an ICE table.

Step 2 Write the expression for the base ionization constant. Substitute equilibrium terms into the equation.

Step 3 If the you get a quadratic equation in Step 2, test to find out if you may use an approximation.

Step 4 Solve the equation for x.

Step 5 Calculate the pH by using the equations
$$pOH = -\log [OH^-(aq)]$$
$$pH = 14.00 - pOH$$

Act on Your Strategy

Step 1

	Q(aq)	+	H₂O(ℓ)	⇌	HQ⁺(aq)	+	OH⁻(aq)
	[Q(aq)] (mol/L)		[H₂O(ℓ)] (mol/L)		[HQ⁺(aq)] (mol/L)		[OH⁻(aq)] (mol/L)
Initial	3.6×10^{-3}				0		~0
Change	$-x$				$+x$		$+x$
Equilibrium	$(3.6 \times 10^{-3} - x)$				x		x

Step 2
$$K_b = \frac{[HQ^+][OH^-]}{[Q]}$$

$$3.3 \times 10^{-6} = \frac{(x)(x)}{3.6 \times 10^{-3} - x}$$

$$x^2 + 3.3 \times 10^{-6} x - 1.188 \times 10^{-8} = 0$$

Step 3 The substitutions resulted in a quadratic equation so compare $1000 \times K_b$ with $[Q]$ to see if you may use an approximation.
$$1000 \times 3.3 \times 10^{-6} = 3.3 \times 10^{-3}$$
$$[Q] = 3.6 \times 10^{-3}$$

$3.6 \times 10^{-3} > 3.3 \times 10^{-3}$ so you can use the approximation that $[Q]$ at equilibrium is the same as its initial concentration, or
$$3.6 \times 10^{-3} - x \cong 3.6 \times 10^{-3}$$

Step 4 $3.3 \times 10^{-6} = \dfrac{x^2}{3.6 \times 10^{-3}}$

$$x^2 = 1.188 \times 10^{-8}$$
$$x = 1.09 \times 10^{-4}$$
$$x \approx 1.1 \times 10^{-4} \frac{mol}{L}$$
$$= [OH^-(aq)]$$

Step 5 $pOH = -\log 1.09 \times 10^{-4}$
$$= 3.96$$
$$pH = 14.00 - pOH$$
$$pH = 10.04$$

Check Your Solution

The pH of the solution is greater than 7, as expected for a basic solution. The answer has the correct number of significant digits (two).

Math Tip

If an acid or a base has a complex molecular formula, you can represent the formula using a shortened notation, such as Q for quinine. Remember that a Brønsted-Lowry acid always provides a proton to water, and a Brønsted-Lowry base always accepts a proton from water.

Quinine helps to prevent malaria, but it has a bitter taste that is characteristic of a base. This unpalatable taste made quinine an unpopular medicine among the British who colonized India. Tonic water contains dissolved quinine, and the gin and tonic cocktail was invented to help make taking quinine easier.

TONIC WATER

CONTAINS QUININE   2 LITRES

28. An aqueous solution of household ammonia, $NH_3(aq)$, has a molar concentration of 0.105 mol/L. Calculate the pH of the solution.
$$K_b (NH_3) = 1.8 \times 10^{-5}$$

29. Hydrazine, $N_2H_4(\ell)$, has been used as a rocket fuel. The concentration of an aqueous solution of hydrazine is 5.9×10^{-2} mol/L. Calculate the pH of the solution.
$$K_b (N_2H_4) = 1.3 \times 10^{-6}$$

30. Morphine, $C_{17}H_{19}NO_3(s)$, is a naturally occurring base that is used to control pain. A 4.5×10^{-3} mol/L solution has a pH of 9.93. Calculate K_b for morphine.

31. Methylamine, $CH_3NH_2(g)$, is a fishy-smelling gas at room temperature. It is used in the manufacture of several prescription drugs, including methamphetamine. Calculate $[OH^-(aq)]$ and pOH of a 1.5 mol/L aqueous solution of methylamine.
$$K_b (CH_3NH_2) = 4.6 \times 10^{-4}$$

32. At room temperature, trimethylamine, $(CH_3)_3N(g)$, is a gas with a strong ammonia-like odour. Calculate $[OH^-(aq)]$ and the percentage of trimethylamine molecules that react with water in a 0.22 mol/L aqueous solution.
$$K_b ((CH_3)_3N) = 6.3 \times 10^{-5}$$

33. An aqueous solution of ammonia has a pH of 10.85. What is the concentration of the solution?

The Relationship Between Conjugate Acid-Base Pairs

Recall that Table 17.1 on page 689 shows an inverse relationship between the strength of an acid and its conjugate base. Strong acids have weaker conjugate bases, and weak acids have stronger conjugate bases. Note that this relationship is relative to the strength of the acid. For instance, the conjugate base of a weak acid is stronger than the weak acid because K_a for the weak acid is less than K_b for the conjugate base. But this does not mean that the conjugate base is necessarily a "strong base". In fact, relative to these "strong bases," the conjugate base may be considered to be a "weak base," but it is still stronger than the weak acid of its conjugate acid-base pair. There is also an important relationship between K_a and K_b for a conjugate acid-base pair. This relationship can be seen using the following specific example.

Ethanoic acid ionizes in water:

$$CH_3COOH(aq) + H_2O(\ell) \rightleftharpoons H_3O^+(aq) + CH_3COO^-(aq)$$

The acid ionization constant for ethanoic acid is:

$$K_a = \frac{[CH_3COO^-][H_3O^+]}{[CH_3COOH]}$$

The conjugate base of ethanoic acid is $CH_3COO^-(aq)$. The ethanoate ion acts as a base in water as follows:

$$CH_3COO^-(aq) + H_2O(\ell) \rightleftharpoons CH_3COOH(aq) + OH^-(aq)$$

The base ionization constant for the ethanoate ion is

$$K_b = \frac{[CH_3COOH][OH^-]}{[CH_3COO^-]}$$

The product $K_a \times K_b$ is

$$K_a \times K_b = \frac{[\cancel{CH_3COO^-}][H_3O^+]}{[\cancel{CH_3COOH}]} \times \frac{[\cancel{CH_3COOH}][OH^-]}{[\cancel{CH_3COO^-}]} = [H_3O^+][OH^-] = K_w$$

$$K_w = K_a \times K_b \qquad K_a = \frac{K_w}{K_b} \qquad K_b = \frac{K_w}{K_a}$$

Equilibrium Constants for Conjugate Acid-Base Pairs

Problem

a) Calculate K_b for the conjugate base of benzoic acid, $C_6H_5COOH(aq)$.

b) Calculate K_a for the conjugate acid of ethylamine, $C_2H_5NH_2(aq)$.

Solution

a) The conjugate base of benzoic acid is the benzoate ion, $C_6H_5COO^-(aq)$.

From Table 1 in Appendix G, K_a for $C_6H_5COOH(aq)$ $= 6.3 \times 10^{-5}$.

$$K_b = \frac{K_w}{K_a}$$

$$= \frac{1.0 \times 10^{-14}}{6.3 \times 10^{-5}}$$

$$= 1.6 \times 10^{-10}$$

K_b for the benzoate ion, $C_6H_5COO^-(aq)$, is 1.6×10^{-10}.

b) The conjugate acid of ethylamine is $C_2H_5NH_3^+(aq)$.

From Table 2 in Appendix G, K_b for $C_2H_5NH_2(aq) = 4.7 \times 10^{-4}$.

$$K_a = \frac{K_w}{K_b}$$

$$= \frac{1.0 \times 10^{-14}}{4.7 \times 10^{-4}}$$

$$= 2.1 \times 10^{-11}$$

K_a for $C_2H_5NH_3^+(aq)$ is 2.1×10^{-11}.

Math Tip

The relationship $K_w = K_a \times K_b$ applies *only* to a *conjugate acid-base pair*, such as $CH_3COOH(aq)/CH_3COO^-(aq)$ or $NH_3(aq)/NH_4^+(aq)$. When you use the equation to calculate K_a or K_b, make sure the chemical formula of the acid and base you are using differ by just one proton.

Predicting the Direction of an Acid-Base Reaction

Problem

Predict the direction for the reaction between ammonium ions and carbonate ions in aqueous solution:

$$NH_4^+(aq) + CO_3^{2-}(aq) \rightleftharpoons NH_3(aq) + HCO_3^-(aq)$$

State whether reactants or products are favoured, and give reasons to support your decision.

K_a for $HCO_3^-(aq) = 4.7 \times 10^{-11}$.
K_b for $NH_3(aq) = 1.8 \times 10^{-5}$.

What Is Required?

You need to determine whether reactants or products are favoured at equilibrium.

What Is Given?

You know the chemical equation.
K_a for $HCO_3^-(aq) = 4.7 \times 10^{-11}$.
K_b for $NH_3(aq) = 1.8 \times 10^{-5}$.

Plan Your Strategy

The reaction will proceed in the direction of the weakest acid.

Step 1 Identify the acid on the left side of the equation, and the conjugate acid on the right side of the equation.

Step 2 Calculate K_a for the conjugate acid using

$$K_a = \frac{K_w}{K_b}$$

Step 3 Compare the two K_a values. The weaker acid has the smaller K_a value. The reaction proceeds toward the weaker acid.

Act on Your Strategy

Step 1 $NH_4^+(aq)$ is the acid on the left side of the equation. On the right side of the equation, $HCO_3^-(aq)$ acts as the acid.

Step 2 $NH_4^+(aq)$ is the conjugate acid of the base, $NH_3(aq)$. K_b for ammonia is 1.8×10^{-5}. Calculate K_a for $NH_4^+(aq)$.

$$K_a = \frac{K_w}{K_b} = \frac{1.0 \times 10^{-14}}{1.8 \times 10^{-5}} = 5.6 \times 10^{-10}$$

Step 3 Comparing K_a for $HCO_3^-(aq) = 4.7 \times 10^{-11}$ with K_a for $NH_4^+(aq) = 5.6 \times 10^{-10}$, the hydrogen carbonate ion is a weaker acid. Therefore, the reaction proceeds to the right. Products are favoured in the reaction.

Check Your Solution

Both acids in the reaction are expected to be weak. The calculation of K_a for the ammonium ion looks reasonable. A number with a larger negative exponent is smaller than another number with a smaller negative exponent.

Practice Problems

34. For each of the following acids, write the name and formula of its conjugate base. Use Appendix G to calculate K_b for the conjugate base:
   a) hydrocyanic acid, $HCN(aq)$
   b) nitrous acid, $HNO_2(aq)$

35. Write the formula for the conjugate acid of methylamine, $CH_3NH_2(aq)$, and pyridine, $C_5H_5N(aq)$. Then use the data below to decide which of these conjugate acids is stronger. Explain your choice.
$$K_{b(CH_3NH_2)} = 4.6 \times 10^{-4}$$
$$K_{b(C_5H_5N)} = 1.7 \times 10^{-9}$$

36. Which is the stronger Brønsted-Lowry base, $CN^-(aq)$ or $F^-(aq)$? Give a reason for your choice.

37. Hydrosulfuric acid is a weak diprotic acid:
$$H_2S(aq) + H_2O(\ell) \rightleftharpoons HS^-(aq) + H_3O^+(aq)$$
$$K_{a1} = 8.9 \times 10^{-8}$$
$$HS^-(aq) + H_2O(\ell) \rightleftharpoons S^{2-}(aq) + H_3O^+(aq)$$
$$K_{a2} = 1.0 \times 10^{-19}$$
To calculate K_b for the sulfide ion, should you use K_{a1} or K_{a2} for hydrosulfuric acid? Explain. Calculate K_b for $S^{2-}(aq)$.

38. Predict the direction for the reaction between dihydrogen phosphate ions and aqueous ammonia:
$$H_2PO_4^-(aq) + NH_3(aq) \rightleftharpoons NH_4^+(aq) + HPO_4^{2-}(aq)$$
State whether reactants or products are favoured, and give reasons to support your decision.
K_a for $H_2PO_4^-(aq) = 6.2 \times 10^{-8}$.
K_b for $NH_3(aq) = 1.8 \times 10^{-5}$.

Section 17.3 Summary

- The equilibrium for the ionization of a weak acid can be treated in a similar way as the equilibrium for any reaction.
- The equilibrium constant for the ionization of an acid is called the acid ionization constant and is given the symbol, K_a.
- The equilibrium expression for an acid is $K_a = \dfrac{[H_3O^+][A^-]}{[HA]}$, where HA represents any weak acid and A^- is its conjugate base.
- You can use the same strategy when solving problems involving K_a that you used when solving problems involving K_c.
- The ion product constant for water is symbolized K_w and is expressed as $K_w = [H_3O^+][OH^-]$.
- At 25 °C, $K_w = 1.0 \times 10^{-14}$.
- The base ionization constant is: $K_b = \dfrac{[HB^+][OH^-]}{[B]}$.
- At 25 °C, $pH + pOH = 14$.
- The rule of 1000 allows you to determine whether you can use the approximation that the equilibrium concentration of a weak acid (or base) is approximately equal to the initial concentration.
- $K_w = K_a \times K_b$ and $K_a = \dfrac{K_w}{K_b}$ and $K_b = \dfrac{K_w}{K_a}$

Chemistry File

FYI
In principle, every anion is capable of acting as a Brønsted-Lowry base. However, the anion formed from a strong acid is extremely weak. (Remember the inverse relationship between an acid and its conjugate base, summarized in the equation for a conjugate acid-base pair: $K_a \times K_b = K_w$). For example, the chloride ion $Cl^-(aq)$ is not able to remove a proton from water. Because the conjugate bases of strong acids are so weak, they are sometimes called ineffective bases.

1. Sodium hydrogen phosphate, $Na_2HPO_4(s)$ is soluble in water.

a) Show that $HPO_4^{2-}(aq)$ is amphiprotic.

b) Write the acid ionization constant expression, K_a.

c) Write the base ionization constant expression, K_b.

d) K_a for $HPO_4^{2-}(aq)$ is 6.2×10^{-8}. K_b for $HPO_4^{2-}(aq)$ is 1.6×10^{-7}. Will an aqueous solution of sodium hydrogen phosphate be acidic or basic? Explain your answer.

2. Which of the following would you expect to be the stronger Brønsted-Lowry acid? Briefly justify each choice:

a) $H_3O^+(aq)$ or $H_2CO_3(aq)$

b) $HSO_4^-(aq)$ or $H_3PO_4(aq)$

c) $H_2AsO_4^-(aq)$ or $HAsO_4^{2-}(aq)$

3. Which of the following would you expect to be the stronger Brønsted-Lowry base? Briefly justify each choice:

a) $H_2O(\ell)$ or $SO_3^{2-}(aq)$

b) $F^-(aq)$ or $Cl^-(aq)$

c) $OH^-(aq)$ or $O^{2-}(aq)$

4. For each of the following acids, write the formula of its conjugate base and calculate K_b.

a) ascorbic acid, $C_6H_8O_6(aq)$ ($K_a = 9.1 \times 10^{-5}$)

b) ethanoic acid, $CH_3COOH(aq)$ ($K_a = 1.8 \times 10^{-5}$)

c) methanoic acid, $HCOOH(aq)$ ($K_a = 1.8 \times 10^{-4}$)

5. Use the data in Appendix G to decide which is the stronger acid, $CH_3NH_3^+(aq)$ or $C_2H_5NH_3^+(aq)$. Explain your choice.

6. For each of the following acid-base reactions, predict whether reactants or products are favoured. Explain your answers:

a) $NH_4^+(aq) + H_2O(\ell) \rightleftharpoons NH_3(aq) + H_3O^+(aq)$

b) $HSO_4^-(aq) + H_3O^+(aq) \rightleftharpoons H_2SO_4(aq) + H_2O(\ell)$

c) $HSO_3^-(aq) + HS^-(aq) \rightleftharpoons H_2S(aq) + SO_3^{2-}(aq)$

d) $CO_3^{2-}(aq) + HF(aq) \rightleftharpoons HCO_3^-(aq) + F^-(aq)$

Titration Curves and Buffers

In your previous chemistry course, you titrated strong acids and bases to determine their concentration. In doing so, you used an indicator to determine the endpoint. The endpoint of a titration occurs when the indicator changes colour and you stop adding acid or base. There are many acid-base indicators, and together they cover the whole range of possible pH values. Figure 17.9 shows the colours and endpoints of various indicators. The colour change of an indicator should be as dramatic as possible. For example, if the equivalence point in a titration is at about pH 8, most chemists would prefer to use phenolphthalein rather than phenol red. Phenol red changes colour from yellow to red, and it is harder to distinguish the orange colour at the endpoint than the colourless to pink change of phenolphthalein.

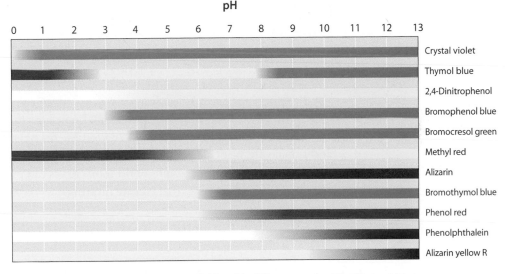

Figure 17.9 Many indicators are available with different endpoints. Thymol blue and alizarin are diprotic. They change colour over two different ranges.

Indicators Are Weak Acids

What makes the indicator change colour? Indicators are, themselves, weak acids. In the acid form, the indicator is one colour and the conjugate base is another colour. You can write a general equilibrium equation for indicators, as follows, where HIn represents the acid form of the indicator and In⁻ represents the conjugate base.

$$HIn(aq) + H_2O(\ell) \rightarrow H_3O^+(aq) + In^-(aq)$$
$$\text{colour 1} \qquad\qquad\qquad \text{colour 2}$$

The structures of indicators can be somewhat complex. For example, Figure 17.10 shows the structures of phenolphthalein in the acid and conjugate base forms.

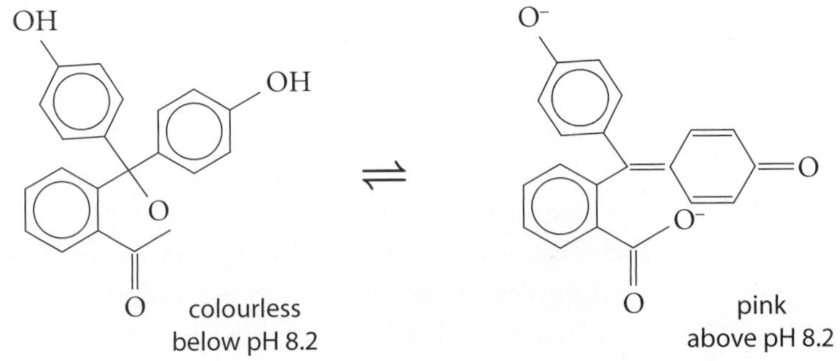

Figure 17.10 When the pH of a solution of phenolphthalein increases, the molecule loses two protons and changes significantly in structure.

colourless
below pH 8.2

pink
above pH 8.2

You have probably noticed that different indicators change colour at a wide range of pH values. The range of the colour change depends on the strength of the acid. As you know, the K_a of a weak acid indicates the relative strength of the acid. Table 17.2 lists the range of the colour change and the K_a of some of the indicators that are illustrated in Figure 17.9. Notice the relationship between the exponent of K_a and the pH range of the colour change.

Table 17.2 pH Range and K_a or Some Indicators

Indicator	pH Range of colour change	K_a
thymol blue	1.2–2.8 8.0–9.6	2.2×10^{-2} 6.3×10^{-10}
bromocresol green	3.8–5.4	1.3×10^{-5}
methyl red	4.8–6.0	1.0×10^{-5}
bromothymol blue	6.0–7.6	5.0×10^{-8}
phenol red	6.6–8.0	1.0×10^{-8}
phenolphthalein	8.2–10.0	3.2×10^{-10}
alizarin yellow R	10.1–12.0	6.9×10^{-13}

When you titrate to an endpoint, you do not know how the pH of the solution changes while you are adding base (or acid). To determine the pH of the solution with each additional volume of added base, you need to use a pH meter. When you do so, you can plot a titration curve. The graphs of titration curves can provide information about the nature of the acid (or base) that you are titrating.

The Titration Curve for a Strong Acid with a Strong Base

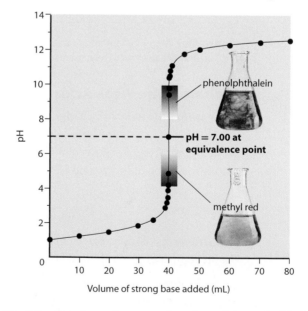

Figure 17.11 The curve for a strong acid–strong base titration, such as hydrochloric acid with sodium hydroxide. The range over which an indicator changes colour must lie on the portion of the titration curve where the pH changes rapidly.

A graph of the pH of the titration mixture against the volume of added acid (or base) is called a **titration curve**. Figure 17.11 shows the titration curve for a strong acid, such as HCl(aq), with a strong base, such as NaOH(aq). The titration of a strong acid with a strong base forms a solution with pH = 7 at equivalence. Any indicator that changes colour in the approximate range from pH 4 to pH 10 could be used for a strong acid–strong base titration. Examples of good indicators for this type of titration include methyl red, phenolphthalein, and bromothymol blue. Many chemists prefer phenolphthalein because the colour change from colourless (in acidic solution) to pink (in basic solution) is easy to see.

How is the titration curve different if the acid or base is weak? Weak acids and weak bases contain ions that react with water in equilibrium reactions. In these cases, the shape of the titration curve will be different, and the pH at equivalence will not be 7.

Titration Curve for a Weak Acid with a Strong Base

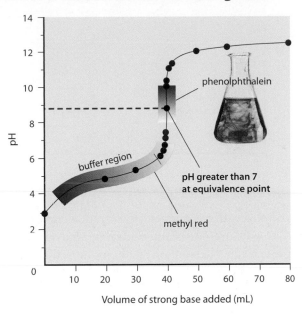

Figure 17.12 The curve for a weak acid–strong base titration, such as ethanoic acid with sodium hydroxide. In this case, equivalence occurs when 40 mL of sodium hydroxide has been added. Note that the middle of the buffer region is at the half-titration point, when half (20 mL) of the NaOH(aq) has been added.

When a weak acid (such as ethanoic acid) is titrated with a strong base (such as sodium hydroxide) the reaction forms an aqueous solution of a salt:

$$CH_3COOH(aq) + NaOH(aq) \rightarrow NaCH_3COO(aq) + H_2O(\ell)$$

Sodium ethanoate ionizes in aqueous solution as follows:

$$NaCH_3COO(aq) \rightarrow CH_3COO^-(aq) + Na^+(aq)$$

Remember that the cation generated by the ionization of a strong base (in this case, $Na^+(aq)$) is a very weak acid and does not react with water. However, the anion (in this case, $CH_3COO^-(aq)$) is the conjugate base of a weak acid. These ions combine with water molecules and act as a base as follows:

$$CH_3COO^-(aq) + H_2O(\ell) \rightleftharpoons CH_3COOH(aq) + OH^-(aq)$$

The added hydroxide ions will make the solution basic. Therefore, the titration of a weak acid with a strong base forms a solution with pH greater than 7 at equivalence. Phenolphthalein has an endpoint range that includes the pH at equivalence. Thus, phenolphthalein is a good indicator for this titration. Methyl red has an endpoint range that does not include the pH at equivalence. It changes colour before the endpoint of the titration and should not be used as an indicator for this titration.

If you compare Figure 17.12 with Figure 17.11, you will notice another difference. In Figure 17.12, there is a gently rising portion of the curve labelled "buffer region." At this stage of the titration, adding a small amount of base has very little effect on the pH of the solution. You will see in Figure 17.13, that the titration curve of a weak base with a strong acid also includes a buffer region.

Titration Curve for a Weak Base with a Strong Acid

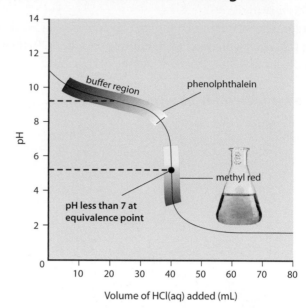

Figure 17.13 The curve for a weak base-strong acid titration, such as aqueous ammonia with hydrochloric acid.

When a weak base is titrated with a strong acid, the titration curve can be understood using the same concepts as before. For example, titrating aqueous ammonia with hydrochloric acid forms an aqueous solution of ammonium chloride at equivalence:

$$HCl(aq) + NH_3(aq) \rightleftharpoons NH_4Cl(aq) + H_2O(\ell)$$

The ammonium chloride then ionizes completely as follows:

$$NH_4Cl(aq) \rightarrow NH_4^+(aq) + Cl^-(aq)$$

The anion from a strong acid (in this case $Cl^-(aq)$) does not react with water. The cation from a weak base will react with water to change the equilibrium concentration of ions:

$$NH_4^+(aq) + H_2O(\ell) \rightarrow NH_3(aq) + H_3O^+(aq)$$

The cation from a weak base will increase the concentration of hydronium ions. Therefore, the titration of a weak base with a strong acid will form a solution with pH less than 7 at equivalence. In this titration, the endpoint range of methyl red includes the pH at equivalence. Methyl red is a good indicator for this reaction, but phenolphthalein is not. Again, there is a buffer region where the addition of a small amount of acid has very little effect on the pH of the solution.

Titrating a Weak Acid with a Weak Base

The salt formed by the reaction between a weak acid and a weak base will contain an ion that is a weak acid and another ion that is a weak base. For this reason, the pH at equivalence will be close to 7, but it may be slightly acidic or slightly basic, depending on the particular ions present. For example, when nitrous acid ($HNO_2(aq)$, $K_a = 7.2 \times 10^{-4}$) is titrated against aqueous ammonia ($NH_3(aq)$, $K_b = 1.8 \times 10^{-5}$), a solution of ammonium nitrite, $NH_4NO_2(aq)$ is formed. The ammonium ion acts as an acid ($K_a = 5.6 \times 10^{-10}$), and the nitrite ion acts as a base ($K_b = 1.4 \times 10^{-11}$). The solution will be slightly acidic because K_a for $NH_4^+(aq)$ is slightly greater than K_b for $NO_2(aq)$. The titration curve for a weak acid with a weak base changes slowly near equivalence. The lack of a steep change near equivalence means it is difficult to judge the endpoint of the indicator. For this reason, chemists avoid titrating a weak acid with a weak base by using an indicator. If the titration must be done, a pH meter is used.

A student used a pH meter to collect data for the titration of ethanoic acid having an unknown concentration with a 0.150 mol/L solution of sodium hydroxide. The data table from the experiment is shown in the table below.

The Titration of Ethanoic Acid with Sodium Hydroxide
Volume of Ethanoic acid = 25.00 mL
$[NaOH(aq)] = 0.150 \dfrac{mol}{L}$

Volume of NaOH(aq) added (mL)	pH
None	2.83
2.00	3.84
4.11	4.20
7.98	4.64
11.95	5.03
14.08	5.27
16.05	5.61
17.00	5.90
17.21	6.00
17.39	6.09
17.62	6.23
17.99	6./4
18.18	8.80
18.39	10.92
18.80	12.24
20.00	12.56
22.03	12.69

Procedure

1. Enter the data from the table into a spreadsheet program. Use the program to plot the results, with pH on the vertical axis and volume of base added on the horizontal axis. Make sure you enter labels for each axis, and provide a suitable title for your graph. Print your graph.

2. On your graph, shade the buffer region. Identify the pH range of an indicator suitable for this titration. Name two indicators that would have endpoints suitable for this titration.

3. The titration curve should show a steep change in pH near the equivalence point. Choose a point halfway along the portion of rapid change on the graph. Label this the equivalence point.

Analysis

1. From your graph, find the pH and volume of base added at equivalence.

2. Calculate the concentration of the ethanoic acid solution.

3. Explain why the pH at equivalence is not 7, the pH of neutral water.

4. What ions and molecules are in solution well before the equivalence point, such as at 10 mL of base added? How does this compare with the solution well after the equivalence point, such as at 20 mL of base added?

Titration Curves for Polyprotic Acids and Polyprotic Bases

Sulfurous acid is a weak diprotic acid ($K_{a1} = 1.4 \times 10^{-2}$; $K_{a2} = 6.3 \times 10^{-8}$). Figure 17.14 shows the titration curve for sulfurous acid with a strong base. As you might expect, there are two equivalence points. These points correspond to the complete reaction of the base with each mole of protons removed from the acid.

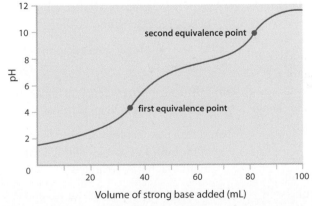

Figure 17.14 The curve obtained by titrating a weak diprotic acid with a strong base. Notice the two equivalence points—one for each proton that ionizes from the acid.

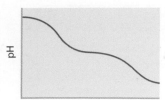

Figure 17.15 The curve obtained by titrating a weak diprotic base with a strong acid.

The general shape of a titration curve for a diprotic base is shown in Figure 17.15. Again, there are two equivalence points. Each equivalence point occurs with the addition of a proton to the base.

● ● ●

9 **a)** Distinguish between the equivalence point and the endpoint for a titration.

b) When choosing an indicator, should the pH at equivalence coincide exactly with the pH at the endpoint of the titration? Explain.

10 Acid-base reactions are often called neutralization reactions. Although "neutral" water has a pH of 7, not every acid-base reaction results in a solution with a pH of this value. Explain.

11 Use Figure 17.9 to estimate the pH of a solution in which bromocresol green is blue and methyl red is orange.

12 Suggest an indicator that could be used to titrate sodium hydroxide with nitric acid. Explain your choice.

● ● ●

The titration curves that represent the titration of a weak acid with a strong base or of a weak base with a strong acid, you observed a region that was labelled, "buffer region." In this region, the addition of a strong base or acid changed the pH of the solution very gradually. The presence of the weak acid or weak base caused the solution to resist a change in pH over that region. Such a solution is called a buffer solution. In the following investigation, you will prepare a buffer solution and observe some of its properties.

INVESTIGATION 17.B

Target Skills

Designing a buffering system

Preparing a buffer to investigate the relative abilities of a buffer and a control to resist a pH change when a small amount of strong acid or strong base is added

Preparing a Buffer and Investigating Its Properties

In this investigation, you will first prepare a buffer solution. Then you will compare how the buffer resists a change in pH when an acid or a base is added with how water resists the same changes.

Question

How can you prepare a buffer solution? How much does the pH of a buffer change when a small amount of a strong acid or strong base is added? How much strong acid or base must be added to a buffer solution to change its pH by one unit?

Prediction

a) Calculate the volume of 0.20 mol/L NaOH(aq) to make the concentration of OH⁻ ions equal to half the concentration in 40 mL of 0.20 mol/L of ethanoic acid.

b) The centre of the buffer region occurs at the half-titration point between a weak acid and a strong base. What volume of 0.20 mol/L NaOH(aq) is required to prepare a buffer solution with 40.0 mL of 0.20 mol/L ethanoic acid?

c) Make a reasonable guess as to the pH of the buffer solution.

d) Calculate the pH when 1 mL of 0.20 mol/L NaOH(aq) is added to 20 mL of water.

e) Make a reasonable guess as to how the pH of 20 mL of the buffer solution is affected when 1 mL of 0.20 mol/L NaOH(aq) is added.

f) Repeat your predictions for (d) and (e), substituting 0.20 mol/L hydrochloric acid for the sodium hydroxide.

Safety Precautions

- Sodium hydroxide and hydrochloric acid are corrosive to eyes and skin and harmful if swallowed or inhaled. Wash any spills on your skin or clothing with plenty of cool water. Inform your teacher immediately. Also inform your teacher immediately if you spill hydrochloric acid or sodium hydroxide on the lab bench or floor.

- Dispose of all materials as instructed by your teacher

- Wash your hands when you have completed the investigation.

Materials

- distilled water
- 0.20 mol/L ethanoic acid (acetic acid), $CH_3COOH(aq)$
- 0.20 mol/L sodium hydroxide, $NaOH(aq)$
- 0.20 mol/L $HCl(aq)$
- 50 mL graduated cylinder
- 50 mL beakers (4)
- 100 mL beaker
- universal indicator paper (pH paper)
- pH meter (optional)
- clean straw
- stirring rod
- burettes

Procedure

Part 1 Acid Breath

1. Measure 20 mL of distilled water and pour it into a 50 mL beaker.

2. Use universal indicator paper or a pH meter (if available) to measure the pH of the water. Record your results.

3. Use a clean straw to blow into the water for about two minutes. Then record the pH of the solution again.

Part 2 Preparing the Buffer

1. Rinse the graduated cylinder with a few mL of 0.20 mol/L ethanoic acid. Discard the rinse as directed by your teacher. Measure 40 mL of 0.20 mol/L ethanoic acid into a 100 mL beaker.

2. Rinse the graduated cylinder several times with tap water. Then rinse the graduated cylinder using a few mL of 0.20 mol/L NaOH(aq). Discard the rinse as directed by your teacher.

3. Have your teacher check the volume of 0.20 mol/L NaOH(aq) you calculated in (a), under Predictions. After receiving approval, add it to the contents of the beaker from Procedure step 1 to make the buffer solution.

4. Divide the buffer solution from Procedure step 3 into three equal portions, using the graduated cylinder and three 50 mL beakers.

Part 3 The Control

1. Copy the following table into your notebook. Give the table a suitable title.

Volume NaOH(aq) added (mL)	pH of water + added NaOH(aq)	pH of buffer + NaOH(aq)	pH of buffer + HCl(aq)
0.0			
1.0			

2. Measure 20 mL of distilled water and pour it into a 50 mL beaker.

3. Record the pH of the water in the second column of the table (for 0.0 mL of added NaOH), measured with universal indicator paper or with a pH meter (optional).

4. Go to one of the burettes set up by your teacher and add 1.0 mL of NaOH(aq). Stir thoroughly, then measure and record the pH of the solution in the second column.

5. Repeat Procedure step 4 until the pH of the solution is at least one unit greater than the initial pH.

Part 4 Adding Base to the Buffer

1. Obtain one of the beakers containing buffer solution from Part 2. Measure the pH of the solution and record the value in the third column of the table (for 0.0 mL of NaOH added).

2. Add 1.0 mL of NaOH(aq) to the buffer. Stir thoroughly, then measure and record the pH of the solution in the third column of the table.

3. Repeat Procedure step 2 until the pH of the solution is at least one unit more than the initial pH.

Part 5 Adding Acid to the Buffer

1. Using a second beaker containing buffer solution, measure the pH of the solution and record the value in the last column of the table (for 0.0 mL HCl added).

2. Add 1.0 mL of HCl(aq) to the buffer. Stir thoroughly, then measure and record the pH of the solution in the last column of the table.

3. Repeat Procedure step 2 until the pH of the solution is at least one unit less than the initial pH.

Part 6 Controlling Acid Breath

1. Obtain the last portion of buffer solution. Use a clean straw to blow into the solution for about two minutes. Then record the pH of the solution.

Analysis

1. The pH of distilled water may not be 7.0. Give reasons why this might be the case.

2. How did the pH of the water change when you blew air into it? Explain by using a chemical equation.

3. What was the effect on pH of blowing air into the buffer solution?

4. Compare the calculated pH when 1 mL of 0.20 mol/L NaOH(aq) was added to 20 mL of water with the value you measured. If different, account for the difference.

5. Compare your estimate of the pH of the buffer solution with the value you measured. If it is outside the range you estimated, explain the difference.

6. Use a spreadsheet program to graph the data in your table. Compare the effect on pH of adding base to water with the effect of adding the same amount of base to the buffer. Compare the effect on pH of adding acid to the buffer solution with the effect of adding the same amount of base to the buffer solution.

Conclusion

7. Write a statement about the effect of adding small amounts of either an acid or a base to a buffer solution.

Extension

8. How could you prepare a buffer solution that is basic? If time permits, obtain the reactants from your teacher and test your prediction.

Buffer Solutions

A solution that contains a weak acid–conjugate base mixture, or a weak base–conjugate acid mixture, is called a **buffer solution**. A buffer solution resists changes in pH when a moderate amount of an acid or a base is added (see Figure 17.16). For example, adding 10 mL of 1.0 mol/L hydrochloric acid to 1 L of water changes the pH from 7 to about 3, a difference of 4 units. Adding the same amount of acid to 1 L of buffered solution might change the pH by only 0.1 units.

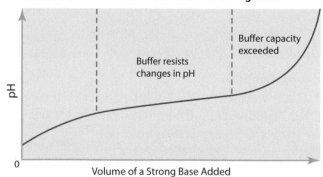

Figure 17.16 Adding a moderate amount of an acid or a base to a buffer solution causes little change to pH.

At the centre of the buffer region, there are equal amounts of weak acid and conjugate base present in the solution. Therefore, when you prepare a buffer, you want to ensure that this condition exists. In Investigation 17.B, you added enough sodium hydroxide to neutralize half of the ethanoic acid. The reaction of the sodium hydroxide with the ethanoic acid can be written:

$$CH_3COOH(aq) + NaOH(aq) \rightarrow NaCH_3COO(aq) + H_2O(\ell)$$

Since you added half as much sodium hydroxide as ethanoic acid, the concentration of ethanoic acid, $CH_3COOH(aq)$, was equal to the concentration of sodium ethanoate, $NaCH_3COO(aq)$. The sodium ethanoate will ionize completely as shown:

$$NaCH_3COO(aq) \rightarrow Na^+(aq) + CH_3COO^-(aq)$$

The ethanoate ion is the conjugate base of ethanoic acid, thus you prepared a buffer with a pH in the centre of the buffer region for ethanoic acid.

Adding a strong base to a weak acid is one way to prepare a buffer. You could also add equal amounts of the weak acid (such as ethanoic acid) and its conjugate base in the form of an ionic compound (such as sodium ethanoate).

When you measured the pH of the buffer that you made in Investigation 17.B, you probably discovered that it was close to pH 4.8. To decide which weak acid to use to make a buffer, you must choose the weak acid that will have equal amounts of weak acid and conjugate base at the pH of your choice for the experiment.

How does a buffer solution resist changes in pH when an acid or a base is added? Consider a buffer solution that is made using ethanoic acid and sodium ethanoate. Ethanoic acid is weak, so most of its molecules are not ionized and $[CH_3COOH(aq)]$ is high. Sodium ethanoate is soluble and a good electrolyte, so $[CH_3COO^-(aq)]$ is also high. Adding an acid or a base has little effect because the added $H_3O^+(aq)$ or $OH^-(aq)$ are removed by one of the components in the buffer solution. The equilibrium of the reactions between the ions in solution shifts, as predicted by Le Châtelier's principle and described below:

- *Adding an acid to the buffer*: ethanoate ions react with hydronium ions added to the solution:

$$CH_3COO^-(aq) + H_3O^+(aq) \rightleftharpoons CH_3COOH(aq) + H_2O(\ell)$$

The position of equilibrium shifts to the right. Here hydronium ions are removed by ethanoate ions from the sodium ethanoate component.

- *Adding a base to the buffer*: hydroxide ions react with hydronium.

$$H_3O^+(aq) + OH^-(aq) \rightleftharpoons 2H_2O(\ell)$$

The hydroxide ions are removed but they removed some hydronium ions as well. The loss of hydronium ions pulls the ethanoic acid/ethanoate ion equilibrium to the right thus replacing the hydronium ions.

$$CH_3COOH(aq) + H_2O(\ell) \rightleftharpoons CH_3COO^-(aq) + H_3O^+(aq)$$

The position of this water equilibrium shifts to the right, removing hydroxide ions. Buffer solutions have two important characteristics. One is the pH of the solution. The other is its **buffer capacity**: the amount of acid or base that can be added before significant change occurs to the pH. The buffer capacity depends on the amount of acid–conjugate base (or base–conjugate acid) in the buffer solution. When the ratio of the concentration of buffer components is close to 1, the buffer capacity has reached its maximum.

Chemistry File

Web Link
Computer animations can help you review acids and bases, and to understand what happens when an acid or a base is added to a buffer solution.

**www.albertachemistry.ca
WWW**

Chemistry File

FYI
A type of dye called an acid dye bonds with negative sites on certain fibres. Acid dyes contain a functional group, such as $-SO_3H$ or $-COOH$, attached to a larger molecule that can be represented as [dye]. The dye ionizes in water, giving dye fragments with a negative charge:
[dye]$-COOH(aq) + H_2O(\ell) \rightarrow$ [dye]$-COO^-(aq) + H_3O^+(aq)$
The colour of the dye changes with pH, because the colour depends on the amount of the dye that ionizes. The polymer chains in wool fibres contain positive $-NH_3^+$ groups, and negatively charged dye fragments bond to these positive sites.

Aboriginal Canadians were skilled in using plants and bark extracts to dye linen and wool.

Blood, Sweat, and Buffers: The Control of pH

For a person to remain healthy, blood must maintain a very narrow range of pH. The pH of blood is about 7.4. If it drops to 7.0, or rises above 7.5, life-threatening problems develop. What types of substances and processes help maintain blood at a constant pH?

Figure 17.17 Exercising adds $CO_2(g)$ and $H_3O^+(aq)$ to the blood. The lungs, kidneys, and buffers in the blood prevent the pH from dropping too low.

During exercise, the body's metabolism increases. The muscles use oxygen, which is used in the metabolism of glucose, releasing $CO_2(g)$ and $H_2O(\ell)$ into the blood stream (see Figure 17.17). During strenuous exercise, oxygen demand can exceed the supply from the lungs. When this happens, other biochemical (anaerobic) processes begin. These processes release lactic acid into the blood stream. To maintain its pH within a narrow range, blood contains a number of buffer systems. The most important buffer system in the blood depends on the equilibrium between hydrogen carbonate ions and carbonate ions. Dissolved carbon dioxide reacts with water to form carbonic acid, $H_2CO_3(aq)$, which ionizes in water to form hydrogen carbonate ions, $HCO_3^-(aq)$. The reaction occurs in the blood as follows:

$$CO_2(aq) + 2H_2O(\ell) \rightleftharpoons H_2CO_3 + H_2O(\ell) \rightleftharpoons H_3O^+(aq) + HCO_3^-(aq)$$

When metabolic changes (as the result of strenuous exercise, for example) add hydronium ions to the blood, the excess $H_3O^+(aq)$ ions are removed by combining with $HCO_3^-(aq)$:

$$H_3O^+(aq) + HCO_3^-(aq) \rightleftharpoons CO_2(aq) + 2H_2O(\ell)$$

If a person hyperventilates, more carbon dioxide is exhaled and their blood pH will increase. This is called respiratory alkalosis. Excess hydroxide ions in the blood are removed by reaction with the hydrogen carbonate ions:

$$OH^-(aq) + HCO_3^-(aq) \rightleftharpoons CO_3^{2-}(aq) + H_2O(\ell)$$

In addition to buffers in the blood, the kidneys remove excess $HCO_3^-(aq)$ from the body, while the lungs help to remove $CO_2(aq)$ from the blood.

Buffers and Oxygen Transport

Oxygen gas is not very soluble in water. At body temperature, 37 °C, the solubility of $O_2(g)$ is about 3×10^{-3} g/L. Your blood can carry large amounts of oxygen because it is bound to hemoglobin and is not in solution. A red blood cell contains about 200 million hemoglobin molecules. Figure 17.18 shows the structure of the hemoglobin molecule. It consists of four protein strands, and each hemoglobin molecule is capable of carrying four oxygen molecules.

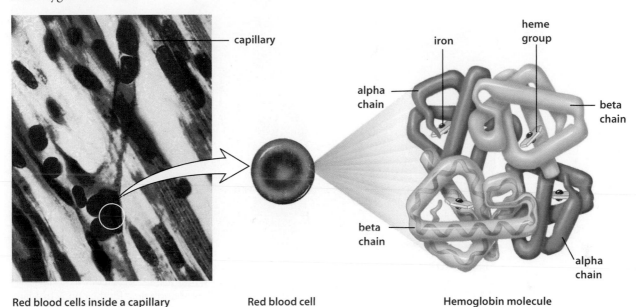

Red blood cells inside a capillary   Red blood cell   Hemoglobin molecule

Figure 17.18 The hemoglobin molecule binds with oxygen gas to transport $O_2(g)$ to cells throughout the body.

Hemoglobin can be represented as HHb(aq) and the oxygenated form, oxyhemoglobin, by $HbO_2^-(aq)$:

$$HHb(aq) + O_2(g) + H_2O(\ell) \rightleftharpoons H_3O^+(aq) + HbO_2^-(aq)$$

Because $H_3O^+(aq)$ is part of the equilibrium equation, the ability of hemoglobin to combine with oxygen depends critically on the pH of the blood. The pH of blood has to remain close to 7.4. If the blood pH becomes more acidic, the condition is called **acidosis**. Acidosis may be caused by a number of conditions, including diabetes, severe diarrhea/dehydration, and drug overdose. If the blood pH increases above 7.4, the condition is called **alkalosis**. Alkalosis can be caused by hyperventilation, prolonged vomiting, or kidney disease.

During the condition of acidosis, the presence of excess $H_3O^+(aq)$, will shift the equilibrium to the left and oxygen will be released from the hemoglobin. Thus, the hemoglobin is not capable of carrying enough oxygen. During the condition of alkalosis, the concentration of $H_3O^+(aq)$ is decreased and the equation is shifted to the right. This might appear to be good because the hemoglobin can carry a lot of oxygen. However, in order for the oxygen to be used by the tissues, the hemoglobin must release it. If oxygen is bound too tightly to hemoglobin, less will be released in the tissues where it is needed.

When the amount of oxygen in the air is reduced as it is at high altitudes (see Figure 17.19), the equilibrium will shift to the left and the hemoglobin will not carry enough oxygen to the tissues. After a period of time at high altitudes, the body releases stored red blood cells from the spleen. Soon thereafter, the body makes more red blood cells and hemoglobin so a sufficient amount of oxygen can reach the tissues.

Elite athletes, training for an endurance event, sometimes live and train at high altitude before their competition. The increased level of hemoglobin may improve their performance.

Figure 17.19 Mountain climbers may suffer from altitude sickness (hypoxia) at high altitudes, where the concentration of oxygen is less than that to which they are accustomed. This is especially dangerous when climbers ascend too quickly and do not allow enough time for their body to adapt.

Oxygen is transported to tissues via the bloodstream, where it is used in the breakdown of glucose to generate energy for cellular metabolism. Carbon dioxide is formed as a product of these reactions. The carbon dioxide dissolves in the blood and reacts with water to form carbonic acid, which ionizes into hydrogen carbonate ions:

$$H_2O(\ell) + CO_2(g) \rightleftharpoons H_2CO_3(aq)$$

$$H_2CO_3(aq) + H_2O(\ell) \rightleftharpoons HCO_3^-(aq) + H_3O^+(aq)$$

In the lungs, these reactions are shifted to the left by the protons released when oxygen combines with hemoglobin molecules. Because the equilibrium is shifted to the left, you breathe out carbon dioxide gas (Figure 17.20). The carbonic acid and hydrogen carbonate ions form the principal buffer controlling blood pH.

Figure 17.20 When a person has consumed alcohol, it enters the bloodstream. In the lungs, an equilibrium is reached between alcohol dissolved in the blood and alcohol present in the breath. A roadside Breathalyser™ test measures the concentration of alcohol vapour present in the breath.

Buffers in the Blood

A buffer is a mixture of dissolved substances that tends to keep the pH of a solution constant when modest additions of acid or base are made. During exercise, muscles use up oxygen, $O_2(g)$, as they convert the chemical energy in glucose to mechanical energy. Oxygen is transported through the bloodstream via the hemoglobin in the red blood cells. Hemoglobin binds with oxygen, transporting it to cells throughout the body. Carbon dioxide, $CO_2(g)$, and hydrogen, H^+, are produced during the breakdown of glucose and are removed from the muscles by the blood. The production and removal of $CO_2(g)$ and H^+, together with the use and transport of $O_2(g)$, cause chemical changes in the blood. These chemical changes, unless offset by other physiological functions, cause the pH of the blood to drop.

How Buffers Work

In all multicellular organisms, the fluids within and surrounding the cells have a characteristic and nearly constant pH of 7.4. If the blood becomes too acidic, buffers take up hydrogen atoms, decreasing the H^+ concentration, thus increasing the pH. If the blood is not acidic enough, the reverse happens; buffers lower pH by releasing H^+ atoms into the blood.

There are three types of buffers. One is the phosphate buffer system that operates in the internal fluid of all cells. This buffer system consists of dihydrogen phosphate ions, $H_2PO_4^-(aq)$, as the proton provider (acid) and hydrogen phosphate ions, $HPO_4^{2-}(aq)$, as the proton acceptor (base). These two ions are in equilibrium:

$$H_2PO_4^-(aq) \rightleftharpoons H^+ + HPO_4^{2-}(aq)$$

If additional hydrogen ions enter the cellular fluid, they are consumed in the reaction with $HPO_4^{2-}(aq)$, and the equilibrium shifts to the left. If additional hydroxide ions enter the cellular fluid, they react with $H_2PO_4^-(aq)$, producing $HPO_4^{2-}(aq)$, and the equilibrium shifts to the right.

A second type of buffer is carbonic acid, $H_2CO_3(aq)$, which is formed when water and carbon dioxide react in blood plasma. In this buffer, the carbonic acid, $H_2CO_3(aq)$, is the proton provider (acid) and hydrogen carbonate ion, $HCO_3^-(aq)$, is the proton acceptor (base). The reaction occurs in the blood as follows:

$$CO_2(aq) + H_2O(\ell) \rightleftharpoons H_2CO_3(aq) \rightleftharpoons H^+ + HCO_3^-(aq)$$

Proteins are the third type of buffer. The blood contains many proteins in addition to hemoglobin. Proteins contain basic and acidic groups that help to maintain blood pH.

Disruptions in the Equilibrium

The body can tolerate only a small change in blood pH (pH 7.4 ±0.4). A decrease in blood pH is called acidosis; an increase is called alkalosis. *Respiratory acidosis* occurs when blood pH falls because of decreased respiration caused by asthma, pneumonia, emphysema, or smoke inhalation. The concentration of dissolved carbon dioxide in the blood increases, making the blood too acidic. *Metabolic acidosis* is the decrease in blood pH that results when excessive amounts of acidic substances enter the blood. This can happen because of diabetes or through prolonged physical exertion. The normal response to both kinds of acidosis is increased breathing to reduce the amount of dissolved carbon dioxide in the blood; hence, we breathe more heavily after strenuous exercise. *Respiratory alkalosis* results from excessive breathing that produces an increase in blood pH. Hyperventilation causes too much carbon dioxide to be removed from the blood, decreasing the carbonic acid concentration and raising the blood pH. *Metabolic alkalosis* is an increase in blood pH resulting from alkaline materials entering the blood by ingestion or overuse of diuretics. The body usually responds to alkalosis by slowing breathing, possibly through fainting.

•••

1. Research the roles of the kidneys and lungs in maintaining blood pH.

2. All enzymes—biological catalysts—have an optimum pH at which enzyme activity is greatest. Explain how a change in blood pH might affect chemical reactions in body fluids.

3. Find out the causes, symptoms, and treatments for ketoacidosis.

Section 17.4 Summary

- Indicators change colour over a small pH range. They can be used to show the endpoint of a titration.
- Indicators are weak acids.
- To observe the pH throughout a titration of a weak acid or a weak base, you need to use a pH meter.
- When titrating a weak acid or a weak base with a strong base or a strong acid, the titration curve shows a region in which the pH changes slowly. This is called the buffer region.
- A buffer solution resists a change in pH on addition of a little strong base or strong acid.
- You choose a weak acid for a buffer depending on the pH range over which you want the buffer to control the pH.
- The most important buffer system in the blood stream is the equilibrium among carbon dioxide and water, carbonic acid, and hydrogen carbonate ion and water.
- The ability of hemoglobin in the red blood cells to carry oxygen to the tissues is sensitive to pH. If the pH varies away from pH 7.4, the hemoglobin will not be able to carry enough oxygen and release it to the tissues where it is needed.

SECTION 17.4 REVIEW

1. Write a chemical equation for each neutralization reaction:
   a) $KOH(aq)$ with $HNO_3(aq)$
   b) $H_3PO_4(aq)$ with $NaOH(aq)$
   c) $HBr(aq)$ with $Ca(OH)_2(aq)$
   d) $NH_3(aq)$ with $HClO_4(aq)$

2. For each reaction in question 1, predict whether the mixture at equivalence is acidic, neutral, or basic. Explain.

3. Congo red is an indicator that changes colour in the pH range of 3.0 to 5.0. Suppose that you are titrating nitric acid with an aqueous sodium hydroxide solution added from a burette. Is Congo red a suitable indicator? If so, explain why. If not, will the endpoint occur before or after the equivalence point? Explain your answer.

4. If you were titrating a weak acid with a strong base, which of these indicators might be suitable: bromophenol blue, bromothymol blue, or phenolphthalein? (Refer to Figure 17.9)

5. Interpret the following titration curve:

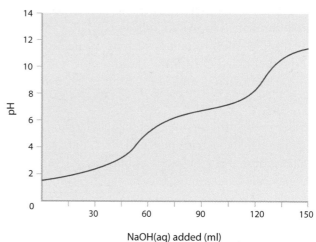

NaOH(aq) added (ml)

6. Describe how a buffer solution differs from an aqueous acidic or basic solution.

7. Explain why an aqueous mixture of $NaCl(aq)$ and $HCl(aq)$ does not act as a buffer, but an aqueous mixture of $NH_3(aq)$ and $NH_4Cl(aq)$ does.

8. Explain the function and importance of buffers in the blood.

When the engines of motor vehicles or industrial plants burn fossil fuels for energy, the trace amounts of sulfur in the fuels and the nitrogen in the air react with oxygen under these high temperatures. The sulfur and nitrogen oxides ("sox" and "nox") are released into the atmosphere. These gases react with moisture in the air to form acids that then fall to Earth as acid deposition. This acid harms plants and causes lakes and streams to become acidic. Many fish are either killed or weakened by the acidic water. Scientists and engineers are working on methods to reduce the release of these SO_x and NO_x gases and are having some success. However, more efforts are needed to clean up the environment. Understanding the chemistry of acids and bases is necessary for finding ways to reduce their harmful effects.

In the early 1900s, chemists Brønsted and Lowry independently developed a new theory of acids and bases, which accounted for some properties that the Arrhenius theory could not explain. The Brønsted-Lowry definition of an acid is a compound that is a proton donor. A base is a proton acceptor. An acid must donate a proton to a base. They cannot function independently. When an acid donates a proton it becomes the conjugate base of that acid. Likewise, when a base accepts a proton it becomes the conjugate acid of that base. Some compounds, such as water, can either donate or accept a proton. These compounds are called amphiprotic substances.

Chemists have tested a large number of weak acids and bases and have determined the order of their strength and weakness. They have used the concept of an equilibrium constant for the ionization of acids and called it the acid ionization constant, K_a. The mathematical expression is: $K_a = \dfrac{[H_3O^+][A^-]}{[HA]}$. As the acid ionization constant becomes smaller, the acid becomes weaker. Strong acids have K_as greater than one. You can determine the value of K_a for a given acid by measuring the pH of a solution of known concentration. Although it is not used as much as the K_a, you can define the base dissociation constant as $K_b = \dfrac{[HB^+][OH^-]}{[B]}$. The dissociation constant for water is called the ion product constant: $K_w = [H_3O^+][OH^-]$. In pure water at 25 °C, $K_w = 10^{-14}$. You can show that $K_w = K_a \times K_b$.

When you use a pH meter to measure the pH of a solution of a weak acid or base and plot the pH versus the volume of strong base or strong acid added, you have produced a titration curve. The titration curve for a weak acid or weak base shows a region in which the pH changes only slightly with the addition of strong acid or base. This region is called the buffer region. A buffer is a solution that resists a change in pH. Weak acids act as buffers. The fluids in your body have buffers that maintain the pH of your blood very near pH 7.4. This is the optimum pH for the transport of gases through your body, as well as other chemical reactions that take place in the blood.

Concept Organizer   Acid-Base Equilibrium Systems

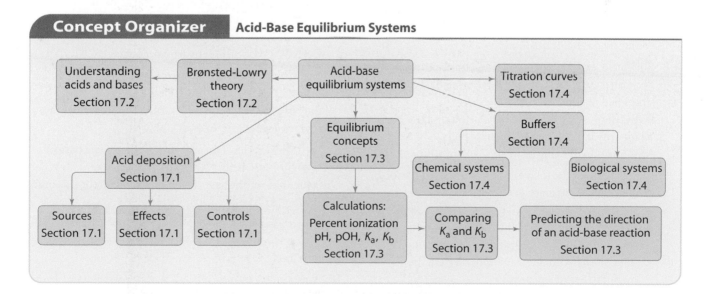

Understanding Concepts

1. Unpolluted rainwater has a pH of 5.60. If a sample of acid rain has a pH of 4.52, how much more hydronium ion is present in acid rain than in unpolluted rain?

2. Developing countries, such as China, tend to emit relatively more SO_x than NO_x than do developed countries, such as Canada. Suggest a reason for the difference. How is this situation likely to change in the future?

3. A liquid shampoo has a hydroxide ion concentration of 6.8×10^{-5} mol/L at 25 °C.
   a) Is the shampoo acidic, basic, or neutral?
   b) Calculate the hydronium ion concentration.
   c) What are the pH and the pOH of the shampoo?

4. The pH of urine can be an indicator of medical problems if it is outside the normal range of values. The pH of a urine sample was measured to be 5.53 at 25 °C. Calculate pOH, $[H_3O^+(aq)]$, and $[OH^-(aq)]$ for the sample.

5. Phenol, $C_6H_6O(s)$, is used as a disinfectant. An aqueous solution of phenol was found to have a pH of 4.72. Is phenol acidic, neutral, or basic? Calculate the $[H_3O^+(aq)]$, $[OH^-(aq)]$, and pOH of the solution.

6. When a patient's blood was tested, the hydroxide concentration in their blood was found to be 2.34×10^{-7} mol/L. Is the patient healthy or suffering from acidosis or alkalosis?

7. Para-aminobenzoic acid (PABA) is a weak monoprotic acid used in some sunscreen lotions. Its formula is $C_6H_4NH_2COOH(s)$. What is the formula of the conjugate base of PABA?

8. Boric acid, $H_3BO_3(aq)$, is used as a mild antiseptic in eyewash solutions. In aqueous solution, the following simplified reaction takes place:
   $H_3BO_3(aq) + H_2O(\ell) \rightleftharpoons H_2BO_3^-(aq) + H_3O^+(aq)$
   a) Identify the conjugate acid-base pairs.
   b) Is boric acid strong or weak? Explain your answer.

9. Classify each of the following compounds as a strong acid, weak acid, strong base, or weak base:
   a) butyric acid, $CH_3CH_2CH_2COOH(aq)$, responsible for the odour of rancid butter
   b) hydroiodic acid, $HI(aq)$, added to some cough syrups
   c) potassium hydroxide, $KOH(aq)$, used in the manufacture of soft soaps
   d) red iron(III) oxide, $Fe_2O_3(s)$, used as a colouring pigment in paints

10. In each pair of bases, which is the stronger base?
    a) $HSO_4^-(aq)$ or $SO_4^{2-}(aq)$
    b) $S^{2-}(aq)$ or $HS^-(aq)$
    c) $HPO_4^{2-}(aq)$ or $H_2PO_4^-(aq)$
    d) $HCO_3^-(aq)$ or $CO_3^{2-}(aq)$

11. Sodium methanoate, $NaHCOO(aq)$, and methanoic acid, $HCOOH(aq)$, can be used to make a buffer solution. Explain how this combination resists changes in pH when small amounts of acid or base are added.

Applying Concepts

12. Lactic acid, $CH_3CHOHCOOH(aq)$ ($K_a = 1.38 \times 10^{-4}$), is a monoprotic acid produced by muscle activity. It is also produced in milk by the action of bacteria. What is the pH of a 0.12 mol/L solution of lactic acid?

13. Salicylic acid is a solid used in the manufacture of Aspirin™. A student prepared a saturated solution of salicylic acid and measured the pH of the solution. The student then carefully evaporated 100 mL of solution and collected the solid. If the pH of the solution was 2.43, and 0.22 g was collected after evaporating 100 mL of solution, what is the acid ionization constant for salicylic acid?

salicylic acid

14. Use the table of K_a values in Appendix G to list the conjugate bases of the following acids in order of increasing base strength: methanoic acid, $HCOOH(aq)$; hydrofluoric acid, $HF(aq)$; benzoic acid, $C_6H_5COOH(aq)$; and hydrocyanic acid, $HCN(aq)$.

15. At normal body temperature of 37 °C, the value of K_w for water is 2.5×10^{-14}. Calculate the $[H_3O^+(aq)]$ and the $[OH^-(aq)]$ at this temperature. Is pure water at this temperature acidic, neutral, or basic?

16. Write the chemical formula for the conjugate base of hypobromous acid, $HOBr(aq)$ ($K_a = 2.1 \times 10^{-9}$). What is the value of the base ionization constant for this ion?

17. The K_b values for ammonia, $NH_3(aq)$, and trimethylamine, $(CH_3)_3N(aq)$, are 1.8×10^{-5} and 6.5×10^{-5}, respectively. Which is the stronger acid, $NH_4^+(aq)$ or $(CH_3)_3NH^+(aq)$?

18. Ammonium nitrate, $NH_4NO_3(s)$, is used as a fertilizer. If the run-off from a farm contains an aqueous solution of 0.10 mol/L $NH_4NO_3(aq)$, what will be the pH of the water? Only the ammonium ion affects the pH of the water.

19. The active ingredient in household bleach is a 5% solution of sodium hypochlorite, $NaOCl(aq)$. If 5.0 g of $NaOCl(s)$ is dissolved to make 100 mL of bleach, what is the pH of the solution? Only the $OCl^-(aq)$ ion affects the pH of the solution.

20. Sodium hydrogen phosphate, $Na_2HPO_4(s)$, is soluble in water. The hydrogen phosphate ion is amphiprotic. For the $HPO_4^{2-}(aq)$ ion:
   a) write the acid ionization constant expression, K_a.
   b) write the base ionization constant expression, K_b.
   c) Will an aqueous solution of sodium hydrogen phosphate be acidic or basic? Explain your answer. (K_a for $HPO_4^{2-}(aq)$ is 6.2×10^{-8}; K_b for $HPO_4^{2-}(aq)$ is 1.6×10^{-7}.)

Solving Problems

21. The greatest reserves of fossil fuel in the world are in the form of coal. The main component of coal is carbon. There are different types of coal, distinguished by the percentage of carbon present, which ranges from about 30% to just over 80%. Sulfur present in coal can lead to emissions of sulfur dioxide that are damaging to the environment. Using the Internet, research the different types of coal, and find out about the coal used in some of the electrical generating plants in Alberta. How is sulfur dioxide removed from the gases released to the atmosphere? **ICT**

22. Using the Internet, research the use of hypochlorous acid in the management of swimming pools and write a report on your findings. **ICT**

23. Use Appendix G to find the acid ionization constants for hydrosulfuric acid and sulfurous acid.
   a) Write equations for the base ionization constants of $HS^-(aq)$ and $HSO_3^-(aq)$.
   b) Calculate the value for the base ionization constant of each ion.
   c) Which is the stronger base, $HS^-(aq)$ or $HSO_3^-(aq)$? Explain.

24. Consider the following acid-base reactions:
$HBrO_2(aq) + CH_3COO^-(aq) \rightleftharpoons$
$$CH_3COOH(aq) + BrO_2^-(aq)$$
$H_2S(aq) + OH^-(aq) \rightleftharpoons HS^-(aq) + H_2O(\ell)$
$HS^-(aq) + CH_3COOH(aq) \rightleftharpoons$
$$H_2S(aq) + CH_3COO^-(aq)$$
If each equilibrium lies to the right, arrange the following acids in order of increasing strength: $HBrO_2(aq)$, $CH_3COOH(aq)$, $H_2S(aq)$, and $H_2O(\ell)$.

25. a) Sketch the titration curves you would expect if you titrated the following:

- a strong acid with a strong base
- a strong acid with a weak base
- a weak acid with a strong base

   b) Congo red changes colour over a pH range of 3.0 to 5.0. For which of the above titrations would Congo red be a good indicator to use?

26. A buffer can be made using an aqueous solution of boric acid and sodium borate. Although boric acid is a component of the buffer, the pH of the solution is 9.2. Explain why the buffer is basic.

Making Connections

27. On several occasions during the past few years, you have studied the environmental issue of acid rain. Now that you have further developed your understanding of acids and bases in this chapter, reflect on your earlier understanding of the issue.
   a) List two facts that you now understand in a more comprehensive way. Explain the difference between your previous and your current understanding in each case.
   b) Identify three questions that could be assigned as a research project on acid deposition. The emphasis of the research must be on how an understanding of chemistry can help to clarify questions and possible solutions involved in the issue. Develop a rubric that could be used to assess this research project.

28. Benzoic acid is a weak, monoprotic acid with a K_a of 6.3×10^{-5}. Calculate the pH and the percent ionization of the following solutions of benzoic acid:
   a) 1.0 mol/L
   b) 0.10 mol/L
   c) 0.01 mol/L

benzoic acid

Use Le Châtelier's principle to explain the trend in percent ionization of the acid as the solution becomes more dilute.

29. Gallic acid is the common name for 3,4,5-trihydroxybenzoic acid.
   a) Draw the structure of gallic acid.
   b) K_a for gallic acid is 3.9×10^{-5}. Calculate K_b for the conjugate base of gallic acid. Then write the formula of the ion.

Career Focus: Ask a Dairy Farmer

Alberta Dairy Farmer
Richard Lavoie

For Richard Lavoie, the well-being of his dairy herd is a top priority. He and his brothers own Enterprise Lavoie (1999) Ltd., a large dairy farm in St. Isadore, Alberta. The farm is a growing success, and in recognition of its contribution to the local and national economies, the operation was named one of the 2005 Lauriers de la PME (laureates of small and medium enterprises) in rural development. Since good nutrition keeps the cows healthy and their milk of high quality, feeding the herd is an important part of the business. For example, when cows do not eat enough fibre, they spend less time chewing their cud and produce less saliva. Saliva contains bicarbonate ions, which buffer acids in the cow's rumen (one of its stomachs). Without these biological buffers, the pH in the rumen drops, resulting in rumen acidosis. This acidity inhibits the growth of desirable rumen bacteria.

$$HCO_3^-(aq) \ + \ H_3O^+(aq) \ \rightleftharpoons \ H_2CO_3(aq) \ + \ H_2O(\ell)$$

bicarbonate ion   **hydronium ion**   **carbonic acid**   **water**

Q   When did you start dairy farming?

A   I have been a dairy farmer since 1974. My family has been farming in St. Isadore since 1953. Enterprise Lavoie Ltd. was established in 1999. It is still a family farm and family owned.

Q   What are your responsibilities at the farm?

A   I am the CEO and manager of Enterprise Lavoie. This means that I manage the daily operations. I especially enjoy working with the animals and people, as well as organizing the farm on a day-to-day basis. Altogether, we have about 950 head of cattle. We milk close to 400 of them. I then sell the raw milk to Alberta Milk.

Q   Why is good nutrition important for your cows?

A   My cows produce so much milk—about 10 000 kg/yr. On average a cow can produce 35–40 kg of milk a day. To produce this much milk, the cows need to eat a large amount of food, so it is critical that the nutrition is balanced according to the needs of the animal. That is why I am working with a specialist in dairy cow nutrition. Because nutrients in a cow's diet need to be balanced, there is a lot of chemistry involved. You have to have vitamin E, you have to have copper sulfate, you have to have zinc—and everything in the right amount.

Q   Why is rumen acidosis a problem?

A   It is a problem because the appetite of the cows decreases as a result, and they no longer eat as much as

they should. For this reason, we supplement the cows' diet with calcium and other supplements to make sure they do not get acidosis. Rumen acidosis also reduces the milk fat slightly; however, the main problem is the effect on the cows' health. This is the important thing. They lose weight faster and they do not produce as much milk as they should. Rumen acidosis is a stress on the cows.

Q   What do you do to prevent rumen acidosis?

A   A few days after the cows have calved, I usually drench (administer vital nutrients directly to the cow's rumen during a period when nutrient and energy levels are deficient) the cows with propylene glycol. It has a very high sugar content that provides energy for the cows. I also drench them with a special formula so that all the right bacteria build up in the stomach. Because of the cows' increased food intake and stomach acidity, quite a metabolic change occurs in the stomach at this time. Anything I can do to help the right stomach bacteria repopulate is a plus for the animal.

Q   Is it true that if the cows do not eat enough fibre, they will not produce enough saliva to buffer the acids in the rumen?

A   That is right. I make sure that the grain (low fibre) to forage (high fibre) ratio is 40:60. I have to make sure that I get a feed analysis with the fibre content. Then my nutritionist works out the balance, or ratio, for the particular forage that I purchased.

Other Careers Related to Buffers, Acids, and Bases

Food Scientist/Food Science Technologist Scientists and technologists in the food industry often work in food-processing plants. Their duties range from testing the quality and safety of food, to maximizing the efficiency of microbial fermentation reactions used to make products, such as yogurt, beer, and wine. A bachelor's degree in chemistry, biochemistry, or microbiology is required to work as a food scientist. Food science technologists generally have two-year technical diplomas.

Gastroenterologist Medical degree graduates can obtain further practical training in order to specialize in health and diseases of the gastrointestinal (digestive) system. Common health problems that gastroenterologists deal with are reflux and heartburn, which are caused by the backflow of stomach acid into the upper digestive tract, and stomach ulcers, which occur where the inner stomach lining has degraded. Antacid tablets, such as calcium carbonate, can be used to buffer stomach acids and treat the symptoms of these conditions. Gastroenterologists can also prescribe antibiotics and perform internal examinations to check for cancer or other diseases.

Nephrology Technologist Also called dialysis technicians, these registered health professionals monitor and operate dialysis machines. People with kidney failure rely on dialysis to remove toxic wastes from the blood or to add sodium bicarbonate to buffer the blood if it is too acidic. To work in a hospital or clinic, a nephrology technologist should have a college diploma in biomedical engineering, or instrumentation technology, plus first-aid and CPR training. Technologists work closely with registered nurses, who have specialized training in dialysis, and may work as hospital dialysis nurses or nurse educators, who instruct patients on the use of home dialysis equipment. The Canadian Association of Nephrology Nurses and Technologists runs a certification program for workers in this field.

Nutritionist Nutritionists may specialize in human or veterinary nutrition. A bachelor of science degree in nutrition and food science is required to work as a nutritional educator or consultant, or to make recommendations about nutritional labels. In Alberta, nutritionists (also called dietitians) must register with the College of Dietitians of Alberta. Animal nutritionists have undergraduate training in agriculture or nutritional sciences or graduate training in a specialized field, such as ruminant nutrition. Animal nutritionists may work for feed producers, for research institutions, or with farmers to help manage livestock.

Perfusionist During open-heart surgery, a heart-lung machine is used to add oxygen to and remove carbon dioxide from the blood. Perfusionists operate heart-lung machines. This task is critical because the blood gases and blood pH must remain within narrow parameters for the patient's health. Although this is a highly specialized field, employment opportunities are good. Perfusionists are health professionals with post-graduate training in perfusion and certification from the Canadian Society of Clinical Perfusion. In Alberta, perfusionists are physicians.

Go Further...

1. Cow saliva contains phosphates, which, in addition to bicarbonate ions, help to buffer acids in the rumen. In your notebook, write the equation for the reversible reaction that occurs when hydrogen phosphate, HPO_4^{2-}(aq), buffers hydrogen ions. (**Hint:** Water will be one of the products of this reaction.)

2. The main buffer in mammalian blood is carbonic acid, H_2CO_3(aq). Carbon dioxide in the blood reacts with water to form carbonic acid. Predict what would happen to the blood pH if an excessive amount of carbon dioxide were to build up in the blood because someone was not breathing properly.

3. Suggest a chemical supplement that a farmer could give a cow that had rumen acidosis. Write a chemical equation for the reaction that would take place in the rumen with the addition of this supplement.

www.albertachemistry.ca
WWW

Understanding Concepts

1. In many industrial reactions, a constant flow of chemicals enters the reaction vessel, and a constant flow of products leaves it. Is a system like this at equilibrium? Explain.

2. Classify each of the following reactions as either a homogeneous equilibrium or a heterogeneous equilibrium:
   a) $H_2(g) + O_2(g) \rightleftharpoons H_2O(\ell)$
   b) $2SO_2(g) + O_2(g) \rightleftharpoons 2SO_3(g)$
   c) $H_2(g) + CO_2(g) \rightleftharpoons H_2O(g) + CO(g)$
   d) $2NaHCO_3(s) \rightleftharpoons Na_2CO_3(s) + CO_2(g) + H_2O(g)$
   e) $C_6H_{12}(\ell) + HBr(aq) \rightleftharpoons C_6H_{13}Br(aq)$

3. The reactions below are involved in the formation of stalactites and stalagmites from calcium carbonate. Identify each reaction as a homogeneous equilibrium or a heterogeneous equilibrium:
   a) $CO_2(g) + H_2O(\ell) \rightleftharpoons H_2CO_3 (aq)$
   b) $H_2CO_3(aq) + H_2O(\ell) \rightleftharpoons HCO_3^-(aq) + H_3O^+(aq)$
   c) $Ca^{2+}(aq) + 2HCO_3^-(aq) \rightleftharpoons$
   $\qquad CaCO_3(s) + CO_2(g) + H_2O(\ell)$
   d) $CaCO_3(s) + H_2CO_3(aq) \rightleftharpoons Ca^{2+}(aq) + 2HCO_3^-(aq)$

4. Write the equilibrium constant expression for each of the following reactions:
   a) $CH_4(g) + H_2O(g) \rightleftharpoons 3H_2(g) + CO(g)$
   b) $2N_2O(g) + O_2(g) \rightleftharpoons 4NO(g)$
   c) $2PCl_3(g) + O_2(g) \rightleftharpoons 2POCl_3(g)$
   d) $2NO(g) + 2H_2(g) \rightleftharpoons N_2(g) + 2H_2O(g)$
   e) $C_7H_{14}(g) \rightleftharpoons C_7H_8(g) + 3H_2(g)$

   Use the following chemical equations for questions 5 and 6:
   a) $H_2(g) + I_2(g) \rightleftharpoons 2HI(g) \qquad \Delta_rH = +53 \text{ kJ}$
   b) $2NO_2(g) \rightleftharpoons 2NO(g) + O_2(g) \qquad \Delta_rH = +116.2 \text{ kJ}$
   c) $N_2(g) + 3H_2(g) \rightleftharpoons 2NH_3(g) + \text{heat}$
   d) $PCl_3(g) + Cl_2(g) \rightleftharpoons PCl_5(g) + \text{heat}$
   e) $NO_2(g) + NO(g) \rightleftharpoons N_2O(g) + O_2(g)$
   $\qquad \Delta_rH° = -43 \text{ kJ}$

5. Which reaction(s) will shift to the right to re-establish equilibrium if the volume of the container is decreased?

6. Which reaction(s) will shift to the left to re-establish equilibrium if the temperature of the system is decreased?

7. Consider the following equilibrium reaction:
   $2SO_2(g) + O_2(g) \rightleftharpoons 2SO_3(g) \qquad \Delta_rH = -197.8 \text{ kJ}$
   Indicate if reactants or products are favoured or if no change occurs when:
   a) the temperature is increased
   b) helium gas is added at constant volume

   c) helium gas is added at constant pressure

8. Explain the reasoning you used to answer question 7 b).

9. For the following reaction at equilibrium, predict the shift, if any, that will occur as a result of each identified change:
   $4NH_3(g) + 5O_2(g) \rightleftharpoons 4NO(g) + 6H_2O(g) + \text{heat}$
   a) removing $O_2(g)$
   b) adding $NO(g)$
   c) adding $H_2O(g)$
   d) adding a catalyst
   e) decreasing the volume of the container
   f) cooling the container

10. For the following equilibrium reaction,
    $N_2(g) + O_2(g) \rightleftharpoons 2NO(g)$
    $K_c = 1.0 \times 10^{-15}$ at 25 °C and $K_c = 0.05$ at 2200 °C: Based on this information, is the reaction exothermic or endothermic? Briefly explain your answer.

11. Decide whether each of the following statements is true or false. For any false statement(s), rewrite each statement so that it is true.
    a) The equilibrium constant does not depend on temperature.
    b) A fast reaction is one that has a large equilibrium constant.
    c) At equilibrium, the concentration of reactants is equal to the concentration of products.
    d) The equilibrium constant is a ratio of reactant concentrations to product concentrations.

12. In the industrial manufacture of ammonia, why is a compromise made between the rate of reaction and the position of equilibrium for the reaction?
    $N_2(g) + 3H_2(g) \rightleftharpoons 2NH_3(g) + \text{heat}$

13. What is syngas? How can the ratio of the main gases present in syngas be adjusted to manufacture different chemicals?

14. Which gases are common pollutants responsible for acid deposition?

15. What are the major sources of pollution that affect Alberta?

16. Name and write the formula of the conjugate acid of each molecule or ion:
    a) $HSO_4^-(aq)$
    b) $NH_3(aq)$
    c) $H_2O(\ell)$
    d) $HPO_4^{2-}(aq)$

17. Name and write the formula of the conjugate base of each molecule or ion:
  a) $HSO_4^-(aq)$
  b) $HNO_2(aq)$
  c) $H_2SO_3(aq)$
  d) $OH^-(aq)$

18. When chlorous acid dissolves in water, the following reaction occurs:
$$HClO_2(aq) + H_2O(\ell) \rightarrow H_3O^+(aq) + ClO_2^-(aq)$$
Identify the conjugate acid-base pairs.

19. Identify the conjugate acid-base pairs in the following reaction:
$$HBO_3^{2-}(aq) + HSIO_3^-(aq) \rightleftharpoons SIO_3^{2-}(aq) + H_2BO_3^-(aq)$$

20. Write the expression for K_a for each of the following in aqueous solution:
  a) $HCN(aq)$
  b) $H_2PO_4^-(aq)$
  c) $HCOOH(aq)$ (methanoic acid)
  d) $HOOCCOO^-(aq)$ (hydrogen oxalate ion)

21. Write the expression for K_b for each of the following in aqueous solution:
  a) $F^-(aq)$
  b) $HC_6H_6O_6^-(aq)$ (ascorbate ion)
  c) $N_2H_4(aq)$ (hydrazine)
  d) $PH_3(aq)$ (phosphine)

22. Arrange the following in order of increasing acid strength: $H_3PO_4(aq)$, $HI(aq)$, $CH_3COOH(aq)$, and $HF(aq)$.

23. Sort the conjugate bases for the acids in question 22 in order of increasing strength.

24. Which of these mixtures is a buffer solution?
  a) Excess $HCl(aq)$ + $NaOH(aq)$
  b) Excess $HCl(aq)$ + $NH_3(aq)$
  c) $HCl(aq)$ + excess $NH_3(aq)$

25. Buffers are solutions that contain a conjugate acid-base pair. Write the formula of a conjugate acid-base pair that
  a) is a buffer
  b) is not a buffer

Applying Concepts

26. The equilibrium constant for the following reaction is 2.8×10^2 at 727 °C:
$$2SO_2(g) + O_2(g) \rightleftharpoons 2SO_3(g)$$
Is the percentage of sulfur dioxide that reacts with oxygen at this temperature relatively large or relatively small? How does the percentage of $SO_2(g)$ that reacts compare with the percentage of $O_2(g)$ that reacts?

27. A mixture is prepared for the following equilibrium:
$$I_2(aq) + I^-(aq) \rightleftharpoons I_3^-(aq)$$
The initial concentrations of $I_2(aq)$ and $I^-(aq)$ are both 0.002 mol/L. When equilibrium is established, $[I_2(aq)] = 5.0 \times 10^{-4}$ mol/L. What is K_c for this equilibrium?

28. The reaction between hydrogen and carbon dioxide was being studied:
$$H_2(g) + CO_2(g) \rightleftharpoons H_2O(g) + CO(g)$$
When 3.60 mol of $H_2(g)$ and 5.20 mol of $CO_2(g)$ were placed in a 2.00 L container, it was found that 75.0 percent of the hydrogen had reacted at equilibrium. What is the equilibrium constant?

29. Nitrogen gas is relatively unreactive at room temperature but combines with oxygen at higher temperatures as follows:
$$N_2(g) + O_2(g) \rightleftharpoons 2NO(g)$$
When equilibrium is attained at 1800 °C, 5.3 percent of $N_2(g)$ will react with $O_2(g)$. What is the value of K_c at this temperature?

30. Nitrogen dioxide and dinitrogen tetroxide form an equilibrium mixture:
$$2NO_2(g) \rightleftharpoons N_2O_4(g)$$
At 25 °C, 0.100 mol of $NO_2(g)$ was placed in a 4.00 L container and equilibrium was established. At equilibrium, 74 percent of the nitrogen dioxide had reacted. What is the equilibrium constant for this reaction at 25 °C?

31. Hypochlorous acid, $HOCl(aq)$, is a weak acid ($K_a = 4.0 \times 10^{-8}$) used in the chlorination of drinking water. It is made by dissolving chlorine gas in water:
$$Cl_2(g) + 2H_2O(\ell) \rightleftharpoons H_3O^+(aq) + Cl^-(aq) + HOCl(aq)$$
  a) Calculate the pH and percent ionization of a 0.065 mol/L solution of hypochlorous acid.
  b) What is the conjugate base of hypochlorous acid? What is its value of K_b?

32. A weak acid with a molar concentration of 0.25 mol/L is 3.0 percent dissociated. Calculate $[H_3O^+(aq)]$, $[OH^-(aq)]$, pH, pOH, and K_a for this acid.

33. The two acid ionization constants for selenous acid, $H_2SeO_3(aq)$, are $K_{a1} = 3.5 \times 10^{-3}$ and $K_{a2} = 5 \times 10^{-8}$. What is the base ionization constant, K_b, for $HSeO_3^-(aq)$?

34. Cyanide ion, $CN^-(aq)$, reacts with $Fe^{3+}(aq)$ to form the blue dye that is used in blueprint paper. Hydrocyanic acid, $HCN(aq)$, is a weak acid, with $K_a = 6.2 \times 10^{-10}$.
  a) Calculate K_b for the conjugate base of $HCN(aq)$.
  b) Calculate the pH of 0.12 mol/L $HCN(aq)$.

35. Calculate the percent ionization of 0.25 mol/L benzoic acid, $C_6H_5COOH(aq)$, $K_a = 6.3 \times 10^{-4}$.

36. $BH(aq)$ is a weak base with a K_b of 1.55×10^{-4}. What is the pH of 0.16 mol/L $BH(aq)$?

37. When hydrocyanic acid reacts with a solution containing hydrogen sulfide ions, it is found that reactants are favoured:
$$HCN(aq) + HS^-(aq) \rightleftharpoons CN^-(aq) + H_2S(aq)$$
Based on this information, which is the stronger acid, $HCN(aq)$ or $H_2S(aq)$? Explain your choice.

38. Determine the relative strengths of the acids and bases in each of the following reactions. Then decide whether the equilibrium mixture favours reactants or products:
a) $H_2S(aq) + CH_3COO^-(aq) \rightleftharpoons HS^-(aq) + CH_3COOH(aq)$
b) $NH_4^+(aq) + CO_3^{2-}(aq) \rightleftharpoons NH_3(aq) + HCO_3^-(aq)$
c) $SO_4^{2-}(aq) + HCN(aq) \rightleftharpoons HSO_4^-(aq) + CN^-(aq)$
d) $CN^-(aq) + H_2CO_3(aq) \rightleftharpoons HCN(aq) + HCO_3^-(aq)$

39. Use the information in the box to answer the question that follow.

A student was given a sample of a clear liquid and told that it contained a monoprotic acid at a concentration of 0.100 mol/L. The student was asked to determine whether the sample contained a strong or a weak acid. After some thinking, the student identified a number of designs that might work to identify the acid as strong or weak.

1. Measure the pH of the solution.
2. Titrate a sample of the acid with 0.100 mol/L $NaOH(aq)$, measuring pH after each addition of base, plotting a pH curve, and identifying the pH at the equivalence point.
3. Observe what occurs when a small sample of $Zn(s)$ is placed in 25.0 mL of the acid.
4. Measure the volume(s) of gas(es) produced in the electrolysis of the acidic solution using graphite electrodes.
5. Measure the temperature change that occurs in a calorimeter when equivalent volumes of the acidic solution and a 0.100 mol/L $NaOH(aq)$ solution are mixed.
6. Measure the final pH of a solution made by mixing 25.0 mL of the acidic solution with 25.0 mL of 0.10 mol/L $NaOH(aq)$.
7. Measure the voltage generated in a voltaic cell made with a graphite electrode in the acidic solution, a functioning salt bridge, and an iron electrode in 0.100 $Fe(NO_3)_2(aq)$.

8. Compare the rate of reaction of the acidic solution with a known mass of $NaHCO_3(s)$ with the same volume of a 0.100 mol/L $HCl(aq)$ solution.

Assume that the student has material commonly available in a high school laboratory.
a) Evaluate each design for its potential to allow the student to solve the problem.
b) Select one design that should work and create a materials list for the design.
c) Prepare an observation table for design #2.
d) Identify the important fixed or controlled variables in design #8.
e) If the acidic sample is strong, what hypothesis could be made in design #1?
f) If the acid sample is strong, predict the balanced net ionic reaction that occurs in method #3.
g) If the acidic sample is strong and hydrogen ion is the oxidizing agent, predict the voltage that will be observed using method #7.

Solving Problems

40. Coal that contains sulfur may contribute to acid deposition when the coal is burned, for example, at an electrical generating plant. Use print resources to find the percentage of sulfur present in coal burned at an Alberta plant. Then determine how much coal is burned in a year and estimate the mass of sulfur dioxide the plant produces annually. Write a report outlining the controls used at the plant to reduce damaging emissions.

41. a) List all of the types of acids that contribute to acid deposition.
b) Write the chemical reactions that produce these acids in the atmosphere.
c) List the sources of all of the gases that contribute to acid deposition.
d) List the effects of acid deposition on the environment.
e) Use your lists to create a concept organizer for acid deposition.

42. What is an amphiprotic substance? Write the formula of an amphiprotic substance, and use chemical equations to explain your answer.

43. Use equations to show how a buffer system, such as $HNO_2(aq)$ and $NO_2^-(aq)$, reacts with $H_3O^+(aq)$ and $OH^-(aq)$.

44. Explain the difference between the endpoint and the equivalence point of a titration. Do you always reach

the equivalence point first? Explain.
- **a)** Explain how you would select an indicator to use in a titration involving a strong acid and a weak base.
- **b)** Would you select a different indicator if you were titrating a weak acid and a strong base? Explain why or why not.

45. Use sketches to compare titration curves for the following:
- **a)** nitric acid with potassium hydroxide
- **b)** ethanoic acid with potassium hydroxide
- **c)** nitric acid with aqueous ammonia

46. How does the titration curve for a polyprotic acid compare with the titration curve you drew in question 44 a)?

47. A student was given a clear solution containing an unknown concentration of an acid or base. The student measured the pH and found it to be about pH 3. The student decided to titrate the solution with a strong base and draw a titration curve. The results are shown in the diagram. Examine the curve and use it to answer the following questions.

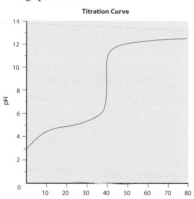

Titration Curve

- **a)** Did the solution contain a strong acid, weak acid, strong base, or weak base? Explain your reasoning.
- **b)** Approximately what was the pH at the equivalence point?
- **c)** Did the solution act as a buffer? If so, what was the pH range of the buffering region?
- **d)** If you were going to titrate a sample of this solution to an endpoint, what indicator would you choose to use? Explain your reasoning.
- **e)** Which of the following compounds is most likely the solute in the solution? Explain your reasoning.
  - **i)** HCl(aq)
  - **ii)** CH$_3$COOH(aq)
  - **iii)** NaHCO$_3$(aq)
  - **iv)** NH$_3$(aq)
  - **v)** KOH(aq)

Making Connections

49. Imagine that you have collected a sample of rainwater in your community. The pH of your sample is 4.52. Unpolluted rainwater has a pH of about 5.6. Suggest at least two possible factors that could be responsible for the pH you measured.

50. Bones and teeth consist mostly of a compound called hydroxyapatite, Ca$_{10}$(PO$_4$)$_6$(OH)$_2$(s). This compound contains PO$_4$$^{3-}$ and OH$^-$ ions.
- **a)** From the formula of hydroxyapatite, do you expect it will be an acid or a base? Explain.
- **b)** Foods that contain sucrose form lactic acid in the mouth, causing the pH to drop. As a result, eating candy promotes a reaction between hydroxyapatite and H$_3$O$^+$(aq). Balance the following reaction:
$$Ca_{10}(PO_4)_6(OH)_2(s) + H_3O^+(aq) \rightleftharpoons$$
$$CaHPO_4(s) + H_2O(\ell) + Ca^{2+}(aq)$$
- **c)** At lower pH values, the CaHPO$_4$(s) also reacts with hydronium ions:
$$CaHPO_4(s) + H_3O^+(aq) \rightleftharpoons$$
$$Ca^{2+}(aq) + H_2PO_4^-(aq) + H_2O(\ell)$$
Dentists and toothpaste manufacturers warn that eating candy promotes tooth decay. What evidence is there to support this advice?

51. Putrescine is the common name for butane-1,4-diamine, NH$_2$CH$_2$CH$_2$CH$_2$CH$_2$NH$_2$(aq). Putrescine gives rotting animal cells their distinctive odour. Calculate K_b for a 0.10 mol/L aqueous solution of putrescine, in which [OH$^-$(aq)] = 2.1 × 10^{-3}.

52. Vanadium and its compounds are highly toxic to humans. Vanadic acid, H$_3$VO$_4$(aq), is triprotic.
- **a)** Write chemical equations to show that vanadic acid is triprotic.
- **b)** Show that H$_2$VO$_4^-$(aq) can act either as an acid or as a base.
- **c)** Which is the stronger acid: H$_3$VO$_4$(aq) or H$_2$VO$_4^-$(aq)? Explain your answer.

53. Ammonium hydrogen carbonate, NH$_4$HCO$_3$(s), is used in baked foods as a leavening agent. In aqueous solution, both NH$_4^+$(aq) and HCO$_3^-$(aq) react with water.
$$NH_4^+(aq) + H_2O(\ell) \rightleftharpoons NH_3(aq) + H_3O^+(aq)$$
$$HCO_3^-(aq) + H_2O(\ell) \rightleftharpoons H_2CO_3(aq) + OH^-(aq)$$
Use the data in Appendix G to help you decide whether an aqueous solution of NH$_4$HCO$_3$(aq) is acidic, basic, or neutral. Explain your answer.

Answers to Practice Problems

Unit 1 The Diversity of Matter and Chemical Bonding

Chapter 1 Chemical Bonding

Page 25

1. 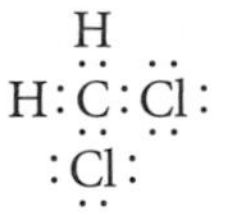 H
H:C:Cl:
:Cl:

2. 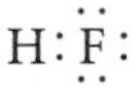 H:F:

3. 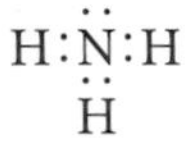 H:N:H
H

4.  :S::C::S:

5.  H:C::N:

6. 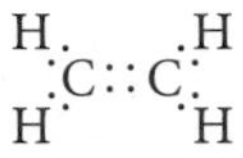 H   H
C::C
H   H

Page 27

7. 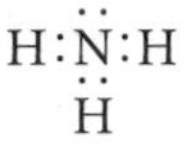 H:N:H
H

8. 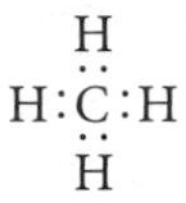   H
H:C:H
H

9. 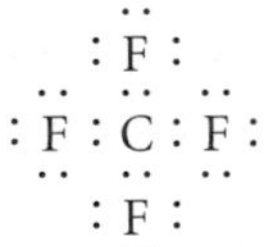   :F:
:F:C:F:
:F:

10. 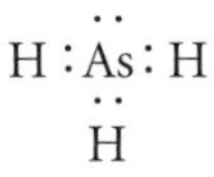 H:As:H
H

11.  H:S:H

12.  H:O:O:H

13.  :Cl:N::O:

14. 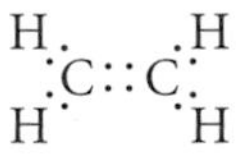 H   H
C::C
H   H

15. 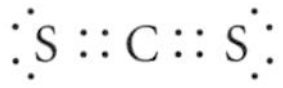 :S::C::S:

16.  H:O:Cl:

17.  H:C::C:H

Page 29

18. 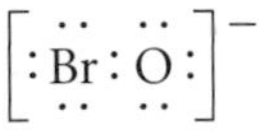 $\left[:Br:O: \right]^{-}$

19.  $\left[:N::O: \right]^{+}$

20. 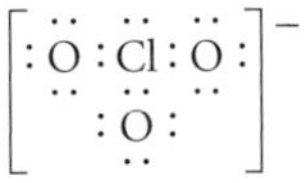 $\left[:O:Cl:O: \atop :O: \right]^{-}$

21.  $\left[:O:S:O: \atop :O: \right]^{2-}$

22.    H:Cl:
H:C:C:F:
H:Cl:

23. a) 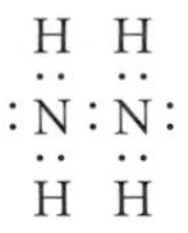 H H
:N:N:
H H

b)  :F:N::N:F:

24. 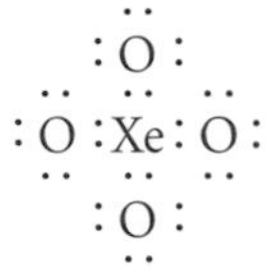   :O:
:O:Xe:O:
:O:

Chapter 2 Diversity of Matter

Page 56

1. linear   H:C:::N

2. tetrahedral (pyramidal)
:F:P:F:
:F:

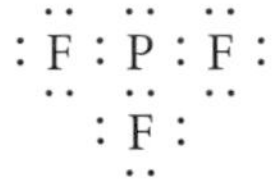

3. trigonal planar
:O::S:O: ⟷ :O::S:O: ⟷ :O:S::O:
:O: .O.   :O:

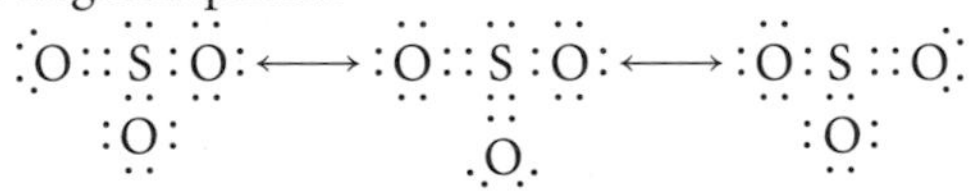

4. trigonal planar
:Cl: :Cl:
C
::
.O.

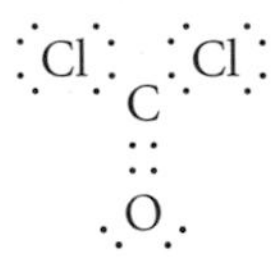

5. pyramidal
:Cl:As:Cl:
:Cl:

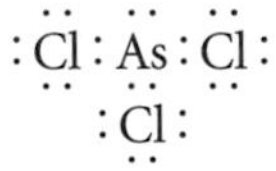

6. bent
:I:S:I:

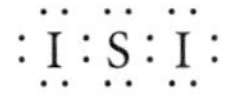

7. tetrahedral
H
H:C:F:
H

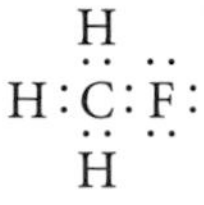

8. tetrahedral
H
:F:C:F:
H

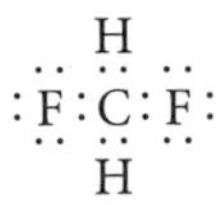

9. linear
:S::C::S:

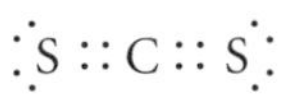

10. trigonal planar
.O.
H   H
H:C:C:C:H
H   H

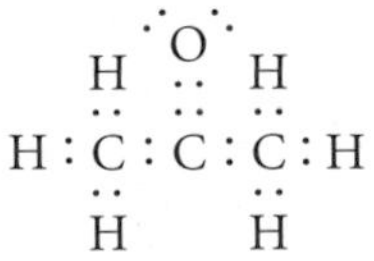

Page 59

11. Following the steps for drawing a Lewis structure results in a central carbon atom surrounded by four electron groupings, all four of which are bonding pairs. The predicted shape of this molecule is tetrahedral. If all four bonds are the same, the polarity of the bonds will give a zero resultant when added together as vectors. In the case of CH_3F, the polarity of the C-H bond is different from the polarity of the C-F bond. Therefore, when these polarities are added as vectors, there will be a resultant polarity. CH_3F is a polar molecule.

$$\overset{H}{\underset{H}{H:C:F:}}$$

12. Following the steps for drawing a Lewis structure results in a central carbon atom surrounded by 3 electron groupings, all three of which are bonding pairs. The predicted shape of this molecule is trigonal planar (refer to Table 1). If all three bonds are the same, the polarity of the bonds will give a zero resultant when added together as vectors. In the case of CH_2O, the polarity of the C-H bond is different from the polarity of the C=O bond. Therefore, when these polarities are added as vectors, there will be a resultant polarity. CH_2O is a polar molecule.

$$\overset{H}{\underset{H}{:C::O:}}$$

13. Following the steps for drawing a Lewis structure results in a central arsenic atom surrounded by four electron groupings, three bonding pairs, and one lone pair. The predicted shape of this molecule is pyramidal (refer to Practice Problem 24 for $AsCl_3$). For any pyramidal shape, regardless of the polarities of the bonds, when the polarities are added as vectors, there will be a resultant polarity. AsI_3 is a polar molecule.

$$\underset{I}{\overset{As}{I \quad I}}$$

14. Following the steps for drawing a Lewis structure results in the structure shown below. Each oxygen atom has two bonding pairs (BP) and two lone pairs (LP), and shares another BP. The BP-LP repulsions from one central oxygen atom exactly balance the equivalent repulsions from the other central oxygen atom. Each H–O–O bond will be bent. Each O–H bond will be polar. Depending on the orientation of the bends with each other, the vectors might or might not cancel. (Experimental observation indicates that the shape causes the molecule to be polar.)

$$\underset{H \qquad H}{O — O}$$

15. Following the steps for drawing a Lewis structure results in a central carbon atom surrounded by four electron groupings, all four of which are bonding pairs. The predicted shape of this molecule is tetrahedral (similar to Practice Problem 8). If all four bonds are the same, the polarity of the bonds will give a zero resultant when added together as vectors. In the case of $CCl_2F_2(g)$, the polarity of the C-Cl bond is different from the polarity of the C-F bond. Therefore, when these polarities are added as vectors, there will be a resultant polarity. CCl_2F_2 is a polar molecule.

$$\underset{:Cl:}{\overset{:Cl:}{:F:C:F:}}$$

16. Following the steps for drawing a Lewis structure of these two molecules results in a central nitrogen atom surrounded by four electron groupings, three bonding pairs and one lone pair. The predicted shape of both molecules is pyramidal. For any pyramidal shape, regardless of the polarities of the bonds, when the polarities are added as vectors, there will be a resultant polarity. Since the N-F bond has a greater polarity than the N-Cl bond, the resultant polarity of the NF_3 molecule will be greater than for the resultant polarity of the NCl_3 molecule.

$$\underset{F}{\overset{N}{F \quad F}} \qquad \underset{Cl}{\overset{N}{Cl \quad Cl}}$$

Unit 2 Forms of Matter: Gases

Unit 2 Preparation

Page 95

1. 2.55×10^3 g	**2.** 1.1 m^2
3. 1.7 kN	**4.** 1.42×10^7 cm^3
5. 380 g	**6.** 39.95 g/mol

Chapter 3 Properties of Gases

Page 110

1. 24 L	**2.** 203 kPa
3. 2.25 L	**4.** 2.0 L
5. 1.03 bar	**6.** 0.97 atm

7. a) 300.4 K   b) 311.0 K
   c) 395.6 K   d) 248 K
   e) 233 K

8. a) 100.0 °C   b) 2 °C
   c) −100 °C   d) −249.6 °C
   e) 600 °C

9. 2.9×10^2 mL or 0.29 L

10. The final volume of the balloon was 3.2 L, so the balloon did not burst in the hot car.

Page 120

11. 308 K

12. The Kelvin temperature increases by 25%, or by a factor of 1.25, and is proportional to the increase in volume. This is not true of the Celsius temperature. For example, if the Celsius room temperature was 22.0 °C, the temperature of the water bath would be 95.6 °C.

13. 606 °C

14. −214 °C

Chapter 4 Exploring Gas Laws

Page 130

1. 586 mL   2. 92 mL
3. 14 mL   4. 1.2 atm
5. 84 °C   6. 56.9 °C

Page 133

7. 250 mL   8. 5.0 mL
9. 700 mL

Page 141

10. 118 kPa   11. $V = 1.4 \times 10^2$ L
12. 1.01×10^{-3} mol   13. 2.73 L
14. 3.0×10^5 L   15. 166 °C
16. 38.0 g/mol

Page 142

17. 1.78 g/L   18. 0.17 g/L
19. 97.6 kPa

Unit 3 Matter as Solutions, Acids, and Bases

Chapter 5 Solutions

Page 186

1. a) 20.7% (m/m)   b) 16.61% (m/m)
   c) 15.1%

2. 35.5% (m/m)   3. 85 g

4. 6.35 g   5. 15 g

Page 188

6. a) 0.33 ppm   b) 3.3×10^2 ppb
   c) 3.3×10^{-5}% (m/m)

7. 0.0700% m/m 8. 0.040 g

9. 1.4 ppm to 4.0 ppm   10. 0.0032 g

Page 191

11. a) 1.7 mol/L   b) 2.41 mol/L
12. a) 8.05×10^{-1} mol/L   b) 1.8×10^{-1} mol/L
13. 7.35×10^{-4} mol/L   14. 5.6×10^{-6} mol/L
15. 1.05×10^{-2} mol/L   16. 3.1 mol/L

Page 193

17. 0.29 mol/L   18. 3.2×10^{-7} mol/L
19. 2×10^{-6} mol/L   20. 3.45×10^{-7} mol/L
21. 2.4×10^{-2} mol/L   22. 1.00×10^{-2} mol/L

Page 194

23. a) 1.46 g   b) 135 g
24. a) 0.16 g   b) 1.5×10^2 g
25. 7.45 g   26. 4.5×10^2 g

Page 195

27. a) 0.034 L   b) 0.052 L
28. a) 0.0854 L   b) 3.1 L
29. 50 L   30. 1.0 L

Page 198

31. a) 0.040 L   b) 0.128 L
32. a) 0.300 mol/L   b) 0.0900 mol/L
    c) 0.00720 mol/L
33. 0.021 mol/L

Chapter 6 Acids and Bases

Page 230

1. a) 2.57 **b)** -0.716
  c) 11.082 **d)** 4.01
  e) 3.792

2. 2.30, acidic **3.** 3.54, acidic

4. 9.181 **5.** 2.201

6. 2.319

Page 237

7. a) 2.21 **b)** -0.597
  c) 10.3 **d)** 2.62
  e) 0.279 **f)** 1.94

8. 5.70, basic **9.** 3.60, basic

10. 2.460 **11.** 10.02

12. 3.564

Page 241

13. a) 1×10^{-4} mol/L, acidic
  b) 1.7×10^{-9} mol/L, basic
  c) 3.78×10^{-6} mol/L, basic
  d) 4.5×10^{-12} mol/L, acidic

14. 1.6×10^{-5} mol/L **15.** 6.2×10^{-5} mol/L

16. 1.5×10^{-3} mol/L **17.** 1.3 g

18. 1.2×10^{-2} g

Unit 4 Quantitative Relationships in Chemical Changes

Unit 4 Preparation

Page 258

1. a) $CH_4(g) + 2O_2(g) \rightarrow CO_2(g) + 2H_2O(g)$, this is a hydrocarbon combustion reaction
  b) $P_4(s) + 4I_2(s) \rightarrow 2P_2I_4(s)$, this is a formation reaction
  c) $Cl_2(g) + 2CsBr(aq) \rightarrow Br_2(\ell) + 2CsCl(aq)$, this is a single replacement reaction
  d) $3Ba(ClO_3)_2(aq) + 2Na_3PO_4(aq) \rightarrow Ba_3(PO_4)_2(s) + 6NaClO_3(aq)$, this is a double replacement reaction
  e) $2Li_3N(s) \rightarrow 6Li(s) + N_2(g)$, this is a decomposition reaction.
  f) $C_6H_{12}O_6(aq) \rightarrow 2C_2H_5OH(aq) + 2CO_2(g)$, this reaction may be classed as a decomposition

2. a) $2C_4H_{10}(g) + 13O_2(g) \rightarrow 10H_2O(g) + 8CO_2(g)$
  b) $Zn(s) + Pb(NO_3)_2 \rightarrow Pb(s) + Zn(NO_3)_2(aq)$
  c) $8Mg(s) + S_8(s) \rightarrow 8MgS(s)$

d) $Sr(OH)_2(aq) + H_2SO_4(aq) \rightarrow SrSO_4(aq) + 2H_2O(\ell)$
  e) $2LiN_3(s) \rightarrow 2Li(s) + 3N_2(g)$

Chapter 7 Stoichiometry

Page 264

1. a) $Cl_2(g) + 2Br^-(aq) \rightarrow Br_2(\ell) + 2Cl^-(aq)$
  b) $Cu(s) + 2Ag^+(aq) \rightarrow Cu^{2+}(aq) + 2Ag(s)$
  c) $2Al(s) + 3Cu^{2+}(aq) \rightarrow 3Cu(s) + 2Al^{3+}(aq)$
  d) $Zn(s) + Pb^{2+}(aq) \rightarrow Zn^{2+}(aq) + Pb(s)$
  e) $H_2(g) + 2Na^+(aq) \rightarrow 2Na(s) + 2H^+(aq)$

2. a) $3Ba^{2+}(aq) + 2PO_4^{3-}(aq) \rightarrow Ba_3(PO_4)_2(s)$
  b) $Sr^{2+}(aq) + SO_4^{2-}(aq) \rightarrow SrSO_4(s)$
  c) $2Al^{3+}(aq) + 3Cr_2O_7^{2-}(aq) \rightarrow Al_2(Cr_2O_7)_3(s)$
  d) $H^+(aq) + OH^-(aq) \rightarrow H_2O(\ell)$
  e) $Mg^{2+}(aq) + 2OH^-(aq) \rightarrow Mg(OH)_2(s)$

Page 274

3. a) 2 mol $AgNO_3(aq)$:1 mol $Na_2CrO_4(aq)$: 1 mol $Ag_2CrO_4(s)$:2 mol $NaNO_3(aq)$
  b) 2 mol $AgNO_3(aq)$:1 mol $Ag_2CrO_4(s)$
  c) 0.25 mol

4. a) 2 mol $NH_3(g)$:1 mol $CO_2(g)$:1 mol $NH_2CONH_2(s)$:1 mol $H_2O(g)$
  b) 2.60 mol
  c) 6.0 mol

5. a) 2 mol $NH_3(g)$:1 mol $H_2SO_4(aq)$: 1 mol $(NH_4)_2SO_4(s)$
  b) 40 000 mol or 40 kmol
  c) 1.64 mol

6. a) 1 mol $Xe(g)$:2 mol $F_2(g)$:1 $XeF_4(s)$
  b) 4.70 mol
  c) 0.0244 mol or 24.4 mmol

7. a) 2 mol $N_2(g)$:1 mol $O_2(g)$:2 mol $N_2O(g)$; 1 mol $N_2(g)$:2 mol $O_2(g)$:2 mol $NO_2(g)$
  b) 0.0468 mol
  c) 0.187 mol or 187 mmol

Page 278

8. 1.3×10^5 g

9. 5.84×10^5 g or 5.84×10^2 kg

10. a) $Cu(s) + 2AgNO_3(aq) \rightarrow 2Ag(s) + Cu(NO_3)_2(aq)$
  b) 3.39 g

11. 112 g **12.** 8.06 g

13. 20.1 g **14.** 40.3 g

15. 417 g

16. 18 mL   **17.** 12.9 mL

18. a) $2 NaI(aq) + Pb(NO_3)_2(aq)$
$\rightarrow 2NaNO_3(aq) + PbI_2(s)$
   b) 40.0 mL   **c)** 1.15 g

19. a) 0.33 L   **b)** 0.13 L
   c) 0.46 L

Page 286

20. a) 1 mol O_2(g):2 mol H_2O(g)
   b) 1 mol H_2(g):1 mol H_2O(g)
   c) 20 L

21. 24.0 L   **22.** 30 L

23. a) $2A(g) + B(g) \rightarrow C(g)$
   b) A_4B_2(g)

Page 287

24. a) 67 L   **b)** 0.016 mol
   c) 4.8×10^{-4} mol

25. 0.787 L   **26.** 5.7×10^6 L

27. 2.73×10^{-3} g

Chapter 8 Applications of Stoichiometry

Page 299

1. $Fe(NO_3)_2$(aq)   **2.** 1.9 g

3. 22 g   **4.** 7.7 g

5. The zinc would react completely.

6. NH_3(g) is the limiting reagent.

Page 303

7. $Mg(NO_3)_2$(aq) is the limiting reactant.

8. 4 g

9. There is sufficient Na_2S(aq) to precipitate all of the Hg^{2+}(aq).

10. 15 g

11. a) $CuSO_4(aq) + Sr(NO_3)_2(aq)$
$\rightarrow Cu(NO_3)_2(aq) + SrSO_4(s)$
   b) The filtrate will be red.
   c) 1.47 g
   d) 97.4%

12. a) $2AgNO_3(aq) + Na_2SiO_3(aq)$
$\rightarrow Ag_2SiO_3(s) + 2NaNO_3(aq)$
   b) $AgNO_3$(aq) is the limiting reagent.
   c) 4.9 g
   d) Yes, all of the Ag^+(aq) are removed because the $AgNO_3$(aq) is the limiting reagent.

Page 308

13. a) 11.4 g   **b)** 79%

14. 60%

15. Trial 1: 95.5%; Trial 2: 94.3%; Trial 3: 97.2%;
Average: 95.7%

16. 95.2%   **17.** 26.7%

18. a) 2.88 g   **b)** 97.3%

19. a) 15.0 g   **b)** 10.5 g
   c) 12.8 g

Page 315

20. 1.1 mol/L   **21.** 0.210 mol/L

22. 0.19 mol/L

23. Trial 1: 0.26 mol/L; Trial 2: 0.26 mol/L; Trial 3: 0.26 mol/L; Average: 0.26 mol/L

24. a) 3×10^1 mL
   b) By performing this calculation before the titration, you know approximately the volume of titrant that will be used. The NaOH(aq) can be added quickly up to a few mL of the end point, then proceed to titrate slowly towards the endpoint. Therefore, time is saved in performing the titration.

25. 0.524 mol/L

Unit 5 Thermochemical Changes

Unit 5 Preparation

Page 337

1. 6.1×10^3 J or 6.1 kJ

2. The second beaker absorbs twice as much energy as the first since it has twice the mass.

3. −214 kJ (or 214 kJ are released)

4. 0.7899 J/g•°C. The substance is likely granite.

Chapter 9 Energy and Chemical Reactions

Page 349

1. −7.17 kJ

2. a) −405.60 kJ/mol O_2(g)
   b) −648.96 kJ/mol CO_2(g)
   c) −540.80 kJ/mol $H_2O(\ell)$

3. a) -2.553×10^3 kJ
   b) -6.04×10^4 kJ
   c) -1.044×10^6 kJ

4. -50.0 kJ

5. 1.03 kg

6. a) -227 kJ/mol $NH_3(g)$
b) -181 kJ/mol $O_2(g)$
c) -227 kJ/mol $NO(g)$
d) -151 kJ/mol $H_2O(g)$

Page 355

7. -49.8 kJ/mol $HCl(aq)$

8. a) -4.1×10^2 kJ/mol
b) All the thermal energy released by the reaction is absorbed by the solution; the solution has the same heat capacity and density as water; the reaction proceeds to completion.

9. -2.9×10^2 kJ/mol $Na(s)$

10. $29.7\,°C$

11. $HBr(aq) + KOH(aq) \rightarrow KBr(aq) + H_2O(\ell)$
$\Delta_rH = -57$ kJ

12. -41.9 kJ/mol $NaOH(aq)$ or -41.9 kJ/mol $HCl(aq)$

Page 363

13. $616\,°C$ (It will be lower than predicted; thermal energy will be absorbed by the surroundings.)

14. -3.9×10^3 kJ/mol $C_8H_{18}(\ell)$

15. 2.17 g $C_3H_8(g)$

16. -20.3 kJ/g

17. -25 kJ/g of fuel

Chapter 10 Theories of Energy and Chemical Changes

Page 374

1. -44.2 kJ **2.** -509 kJ

3. -121.0 kJ **4.** 30 kJ

5. 52.6 kJ/mol **6.** -128.8 kJ/mol

Page 382

7. -136.4 kJ **8.** -882.6 kJ

9. -77.6 kJ

10. a) -637.9 kJ/mol **b)** 2.49×10^3 kJ

11. -102.7 kJ/mol **12.** -224.2 kJ/mol

Page 387

13. 18% **14.** 69%

15. 0.88 g **16.** 38.7%

Chapter 11 Activation Energy and Catalysts

Page 409

1. The reaction is endothermic.

2. $E_{a(fwd)} = 99$ kJ **3.** $\Delta H = -89$ kJ

4. $E_{a(fwd)} = 42$ kJ; $E_{a(rev)} = 67$ kJ; $\Delta H = -25$ kJ. This reaction is exothermic.

5. $E_{a(rev)} = 411$ kJ

Unit 6 Electrochemical Changes

Chapter 12 Oxidation-Reduction Reactions

Page 448

1. a) $5Ag(s) + MnO_4^-(aq) + 8H^+(aq) \rightarrow 5Ag^+(aq) + Mn^{2+}(aq) + 4H_2O(\ell)$; Since the MnO_4^- gains electrons, it is reduced and it is the oxidizing agent. Since the Ag loses electrons, it is oxidized and it is the reducing agent.
b) $Hg(\ell) + 4Cl^-(aq) + 4H^+(aq) + 2NO_3^-(aq) \rightarrow HgCl_4^{2-}(s) + 2NO_2(g) + 2H_2O(\ell)$; Since the NO_3^- is reduced it is the oxidizing agent. Since the Hg is oxidized it is the reducing agent.
c) $AsH_3(g) + 4H_2O(\ell) + 4Zn^{2+}(aq)$
$\rightarrow H_3AsO_4(aq) + 8H^+(aq) + 4Zn(s)$; Since the $Zn^{2+}(aq)$ gains electrons, it is reduced and it is the oxidizing agent. Since the $AsH_3(g)$ loses electrons, it is oxidized and it is the reducing agent.
d) $I_2(s) + 5OCl^- + H_2O(\ell) \rightarrow 2IO_3^-(aq) + 2H^+(aq) + 5Cl^-(aq)$; Since $I_2(aq)$ is oxidized, it is a reducing agent. Since $OCl^-(aq)$ is reduced, it is an oxidizing agent.

2. a) $I^-(aq) + 6OH^-(aq) + 6MnO_4^-(aq)$
$\rightarrow 6MnO_4^{2-}(aq) + IO_3^-(aq) + 3H_2O(\ell)$; Since the $MnO_4^-(aq)$ is reduced, it is the oxidizing agent. Since the $I^-(aq)$ is oxidized it is the reducing agent.
b) $3H_2O_2(aq) + 2OH^-(aq) + 2ClO_2(aq) \rightarrow 2ClO^-(aq) + 3O_2(g) + 4H_2O(\ell)$; Since the $ClO_2(aq)$ is reduced, it is the oxidizing agent. Since the $H_2O_2(aq)$ is oxidized it is the reducing agent.
c) $2CrO_2^-(aq) + 6ClO^-(aq) + 2H_2O(\ell) \rightarrow 3Cl_2(g) + 4OH^-(aq) + 2CrO_4^{2-}(aq)$; Since the $ClO^-(aq)$ is reduced, it is the oxidizing agent. Since the $CrO_2^-(aq)$ is oxidized it is the reducing agent.
d) $4Al(s) + OH^-(aq) + 3NO^-(aq) + 4H_2O(\ell) \rightarrow 3NH_3(g) + 4AlO_2^-(aq)$; Since the $NO^-(aq)$ is reduced, it is the oxidizing agent. Since the $Al(s)$ is oxidized it is the reducing agent.

3. $2PbSO_4(aq) + 2H_2O(\ell) \rightarrow Pb(s) + PbO_2(aq) + 2H_2SO_4(aq)$

4. $3NO_2(g) + H_2O(\ell) \rightarrow NO(g) + 2HNO_3(aq)$

5. $Cl_2(g) + 2OH^-(aq)$
$\rightarrow Cl^-(aq) + ClO^-(aq) + H_2O(\ell)$

6. $3I_3^-(aq) + 3H_2O(\ell) \rightarrow 8I^-(aq) + IO_3^- + 6H^+(aq)$

Page 457

7. a) $+3$ **b)** 0
   c) $+6$ **d)** $+5$
   e) 0 **f)** $+2$

8. a) $H = +1, S = +4, O = -2$
   b) $H = +1, O = -2$
   c) $H = +1, P = +5, O = -2$

9. a) $O = +2$ **b)** $O = -1$

10. a) $Al = +3, H = +1, C = +4, O = -2$
    b) $N = -3, H = +1, P = +5, O = -2$
    c) $K = +1, H = +1, I = +7, O = -2$

Page 461

11. a) This is a redox reaction.
   b) There is no change in the oxidation number of any element. This is not a redox reaction.
   c) This is a redox reaction.
   d) This is a redox reaction. Since N in $NO_2(g)$ undergoes an increase and a decrease in oxidation number, this is a disproportionation reaction.

12. a) $H_2O_2(aq)$ undergoes reduction. It is an oxidizing agent. $Fe(OH)_2(s)$ undergoes oxidation. It is a reducing agent.
   b) This is not a redox reaction. There is no oxidizing agent or reducing agent.
   c) $O_2(g)$ undergoes reduction. It is the oxidizing agent. $C_2H_6(g)$ undergoes oxidation. It is the reducing agent
   d) $NO_2(g)$ undergoes both reduction and oxidation. It is both an oxidizing agent and a reducing agent.

13. The oxidation number of Br decreases from 0 in $Br_2(\ell)$ to -1 in $2Br^-(aq)$. This is reduction. $Br_2(\ell)$ is an oxidizing agent. The oxidation number of Cl increases from $+3$ in $ClO_2^-(aq)$ to $+4$ in $ClO_2(aq)$. This is oxidation. $ClO_2^-(aq)$ is a reducing agent.

14. The oxidation number of the nickel decreases from $+2$ in the sulfide to 0 in its elemental form. This is a reduction. A reducing agent must be used to achieve this change. This is a redox reaction. The same analysis can be used with CuS and Cu.

The oxidation number of the copper decreases from $+1$ to 0.

Page 464

15. $CS_2(g) + 3O_2(g) \rightarrow CO_2(g) + 2SO_2(g)$

16. a) $B_2O_3(aq) + 6Mg(s) \rightarrow 3MgO(s) + Mg_3B_2(aq)$
    b) $8H_2S(g) + 8H_2O_2(aq) \rightarrow S_8(s) + 16H_2O(\ell)$

17. a) $Cr_2O_7^{2-}(aq) + 6Fe^{2+}(aq) + 14H^+(aq)$
$\rightarrow 2Cr^{3+}(aq) + 6Fe^{3+}(aq) + 7H_2O(\ell)$
    b) $I_2(g) + 10NO_3^-(aq) + 8H^+(aq)$
$\rightarrow 2IO_3^-(aq) + 10NO_2(g) + 4H_2O(\ell)$
    c) $2PbSO_4(aq) + 2H_2O(\ell) \rightarrow Pb(s) + PbO_2(aq) + 2SO_4^{2-}(aq) + 4H^+(aq)$

18. a) $3Cl^-(aq) + 2CrO_4^{2-}(aq) + H_2O(\ell)$
$\rightarrow 3ClO^-(aq) + 2CrO_2^-(aq) + 2OH^-(aq)$
    b) $3Ni(s) + 2MnO_4^-(aq) + H_2O(\ell) \rightarrow 3NiO(s) + 2MnO_2(s) + 2OH^-(aq)$
    c) $I^-(aq) + 6Ce^{4+}(aq) + 6OH^-(aq)$
$\rightarrow IO_3^-(aq) + 6Ce^{3+}(aq) + 3H_2O(\ell)$

Page 469

19. $0.067\,74$ g; 5.276% **20.** $0.138\,7\%$

21. $0.011\,0$ mol/L   **22.** $0.174\,3$ mol/L

Chapter 13 Cells and Batteries

Page 487

1. $+0.29$ V   **2.** $+3.17$ V

3. $+0.14$ V   **4.** 0.96 V

Page 508

5. a) 0.14 V
   b) The anode is the Cd(s) electrode. The cathode is the Ni(s) electrode.
   c) Oxidation: $Cd(s) \rightarrow Cd^{2+}(aq) + 2e^-$
Reduction: $Ni^{2+}(aq) + 2e^- \rightarrow Ni(s)$
Overall cell reaction: $Cd(s) + Ni^{2+}(aq) \rightarrow Ni(s) + Cd^{2+}(aq)$

6. -0.14 V   **7.** Co(s) and $Cl_2(g)$

8. Ag(s) and $O_2(g)$

Page 516

9. 2.97 g   **10.** 8.90 min

11. 1.85 A

12. a) 1.09 g increase
    b) Since the overall balanced equation can be determined from the given information, the mole ratio is sufficient information to calculate the mass change.

Unit 7 Chemical Changes of Organic Compounds

Chapter 14 Structure and Physical Properties of Organic Compounds

Page 549

1. a) 2-methylbutane
b) 2,2-dimethylpropane
c) 3-ethyl-2,5-dimethylheptane
d) 2,2,4,4-tetramethylhexane
e) 2,2,4-trimethyl-4-propylheptane

2. a) 2,2,3-trimethylbutane
b) 5-ethyl-3-methylnonane
c) 4,5-diethyl-3-methylnonane

3. a) heptane
b) 2,3-dimethylpentane
c) 4-ethyl-2,3-dimethylhexane.

Page 550

4. a)

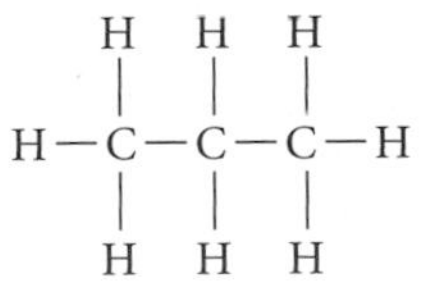

b)

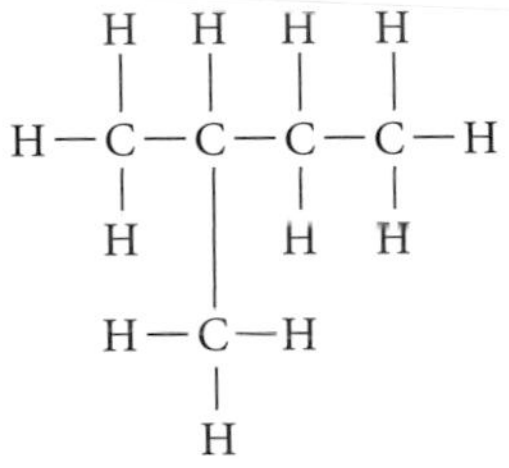

5. a)

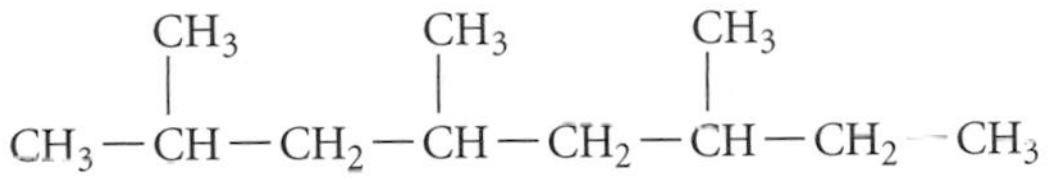

$$CH_3-CH-CH_2-CH-CH_2-CH-CH_2\!-\!CH_3$$
with CH_3 branches

b)

$$CH_3-CH_2-CH-CH-CH_2-CH_2-CH_3$$
with CH_3 and CH_2-CH_3 branches

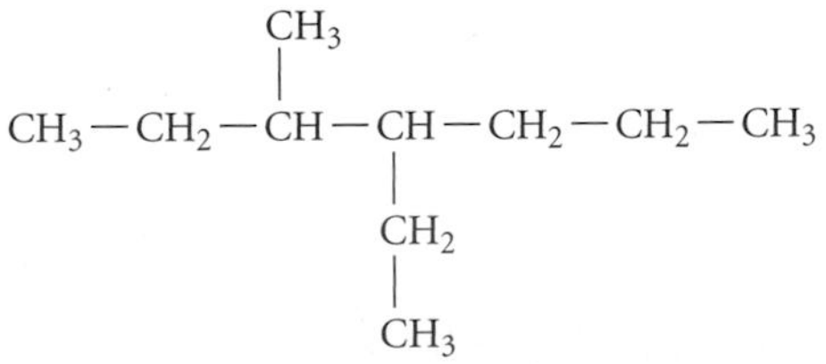

6. a) $CH_3CH(CH_3)CH_2CH(CH_3)CH_2CH(CH_3)CH_2CH_3$
b) $CH_3CH_2CH(CH_3)CH(C_2H_5)CH_2CH_2CH_3$

7. a) 3-methylhexane **b)** 4-methylheptane
c) 2-methylpentane

8. a)

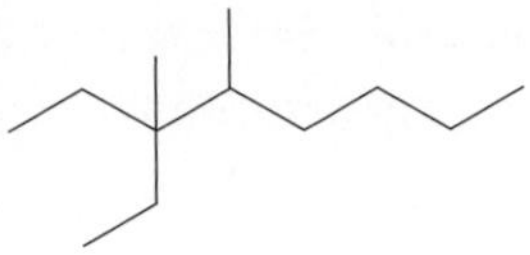

b)

c)

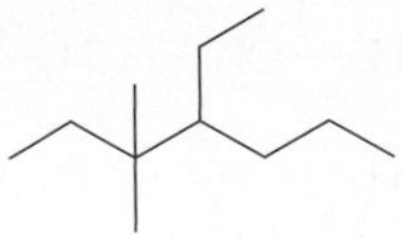

d)

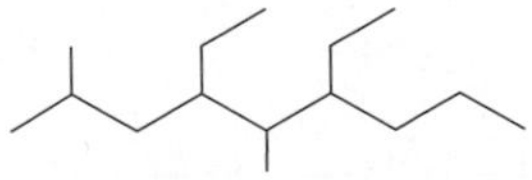

9. a) The longest continuous chain has 6 carbons, not 5, with two methyl groups (CH_3-) on carbons C2 and C4. The correct name is 2,4-dimethylhexane.
b) The longest continuous chain of 6 carbons should be numbered from right to left. The correct name is 2,3-dimethylhexane.
c) The longest continuous chain has 6 carbons, not 5, with an ethyl group (C_2H_5-) and a methyl group (CH_3-) on carbon C3. The correct name is 3-ethyl-3-methylhexane.

Page 554

10. a) hex-3-ene **b)** 3-propylhept-2-ene

11. 4-ethyl-2,3-dimethyloct-4-ene

12. 4,4-dimethylpent-2-ene

13. 3-ethyl-5-methyl-3-propylhex-1-ene

Page 555

14. a)

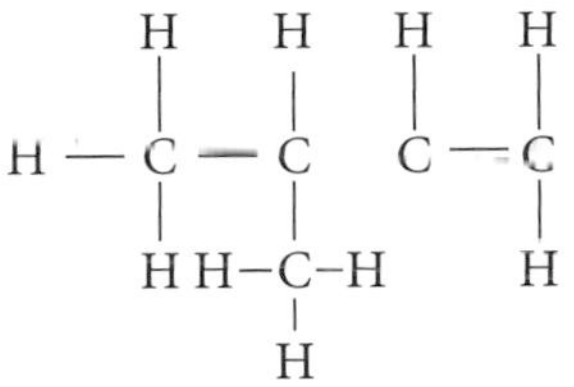

b)

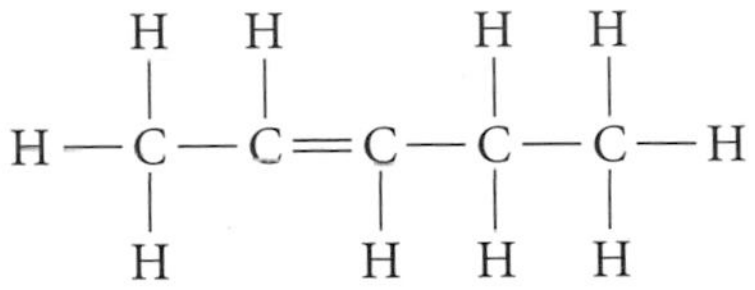

c)

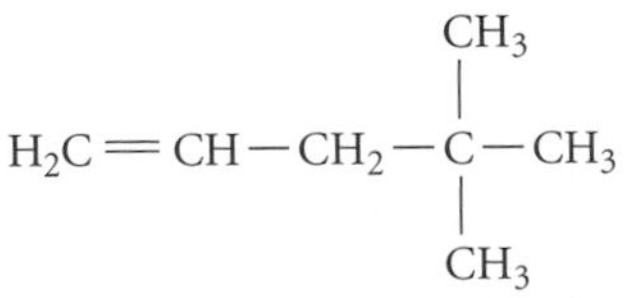

$$H_2C=CH-CH_2-\underset{\underset{CH_3}{|}}{\overset{\overset{CH_3}{|}}{C}}-CH_3$$

15. a)

$$CH_3-CH=C-CH-CH-CH-CH_2-CH_3$$
with CH_3, CH_3, CH_3 and CH_2-CH_3 branches

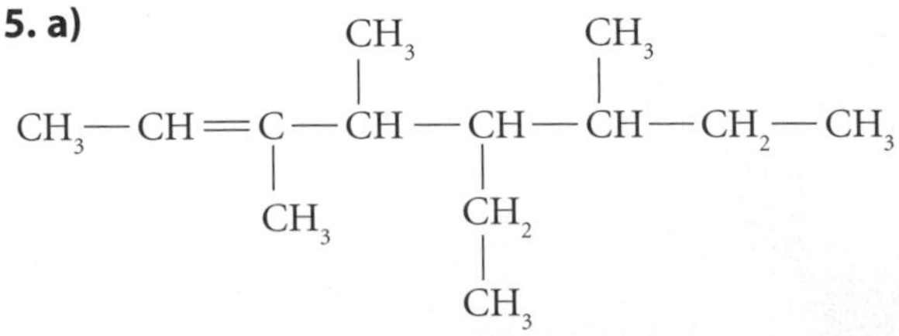

b)

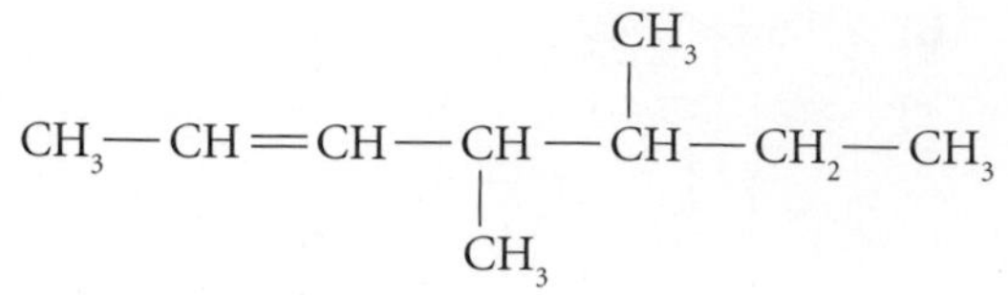

$$CH_3-CH=CH-CH-\underset{\underset{CH_3}{|}}{CH}-\underset{\underset{}{}}{CH}-CH_2-CH_3$$

CH₃ above, CH₃ below

$$CH_3-CH=CH-\overset{\overset{\displaystyle CH_3}{|}}{CH}-\overset{\overset{}{}}{CH}-CH_2-CH_3$$

c)

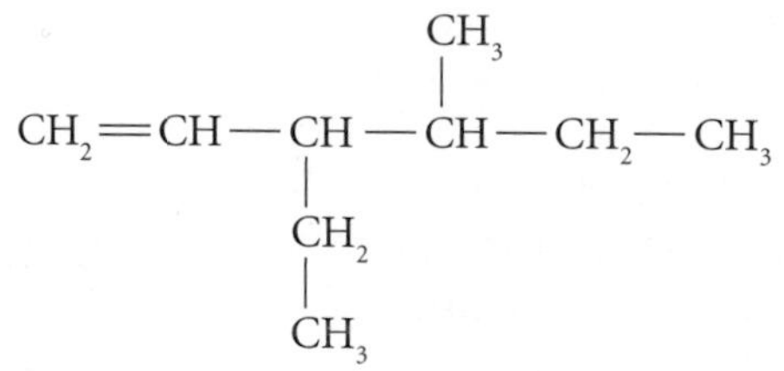

$$CH_2=CH-CH-\overset{\overset{\displaystyle CH_3}{|}}{CH}-CH_2-CH_3$$
with CH₂—CH₃ branch below

d)

$$CH_3-\overset{\underset{\displaystyle CH_3}{|}}{CH}-CH=CH-\overset{\underset{\displaystyle CH_3}{|}}{CH}-CH_2-\overset{\underset{\displaystyle CH_3}{|}}{CH}-CH_3$$

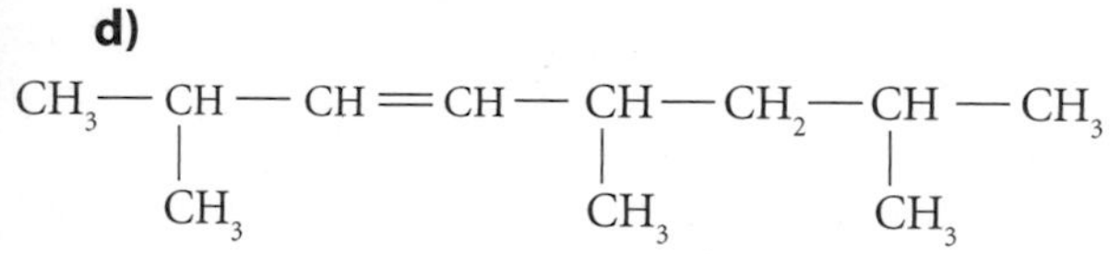

Page 556

16. a) 4,4-dimethylpent-2-yne
b) 3-ethyl-5-methyl-3-propylhex-1-yne

17. a) $CH_3-C\equiv C-CH_2-CH_3$

b)

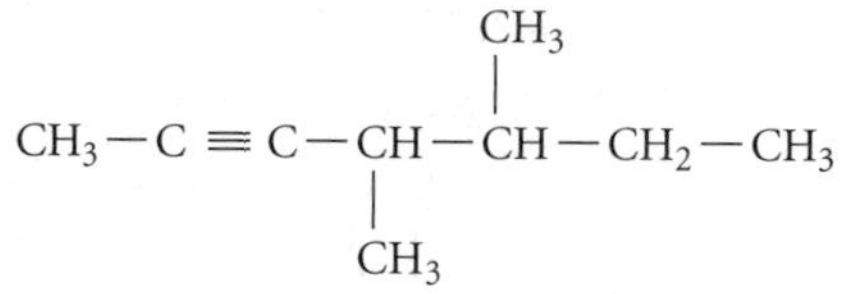

$$CH_3-C\equiv C-CH-\overset{\overset{\displaystyle CH_3}{|}}{CH}-CH_2-CH_3$$
with CH₃ below first CH

c)

$$HC\equiv C-CH-\overset{\overset{\displaystyle CH_3}{|}}{CH}-CH_2-CH_3$$
with CH₂—CH₃ below first CH

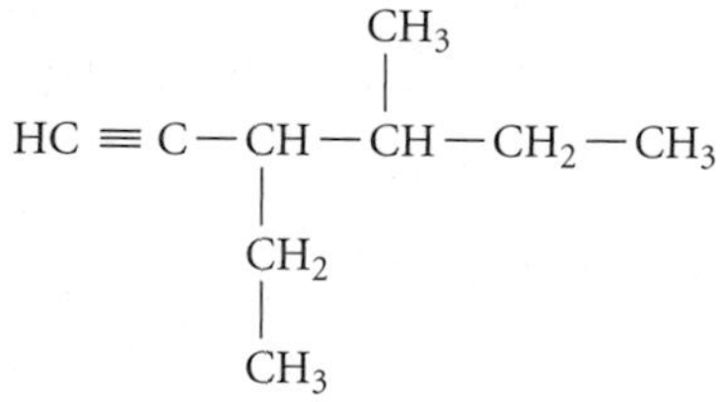

d)

$$CH_3-\overset{\underset{\displaystyle CH_3}{|}}{CH}-C\equiv C-\overset{\underset{\displaystyle CH_3}{|}}{CH}-CH_2-CH-CH_3$$
with CH₃ above last CH

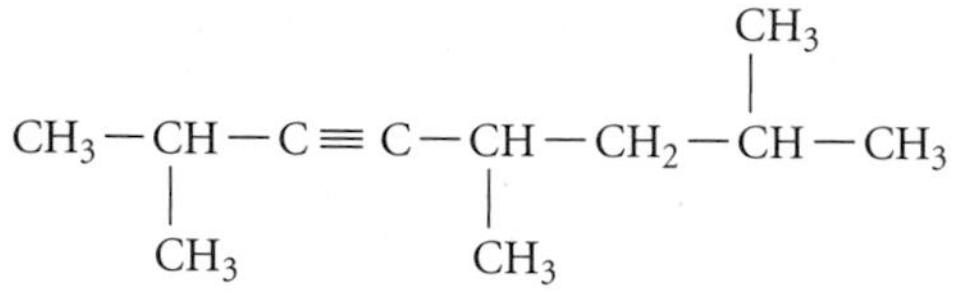

Page 559

18. a) methylcyclobutane
b) 1-ethyl-3-methylcyclopentane

19. a)

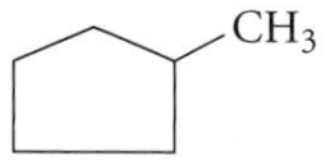

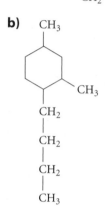

b)

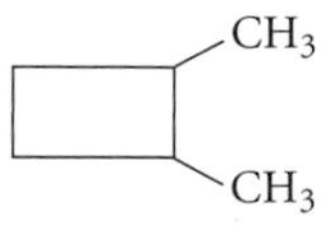

20. a) 4-methylcyclooctene
b) 3-methyl-5-propylcyclopentene

21. a)

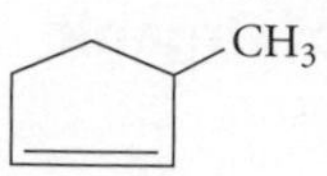

b)

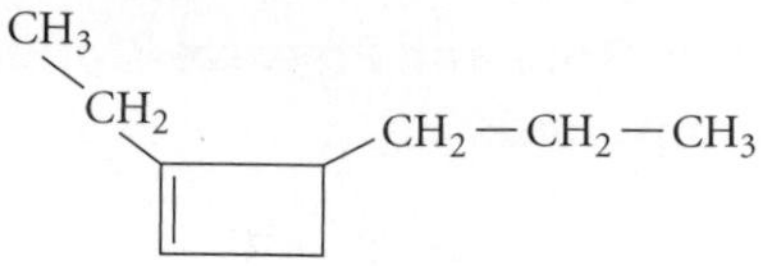

22. a) 5-ethyl-3,4-dimethylcyclononene
b) 3,5-diethylcyclohexene

23. a)

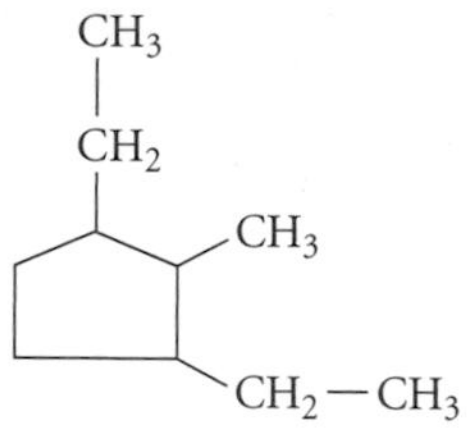

b)

Page 562

24. 1,3,5-trimethylbenzene

25. a)

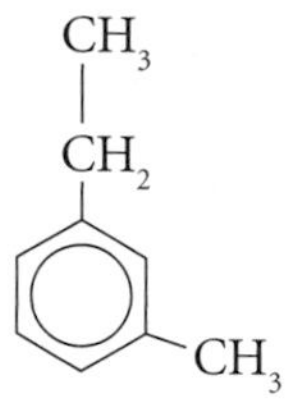

b)

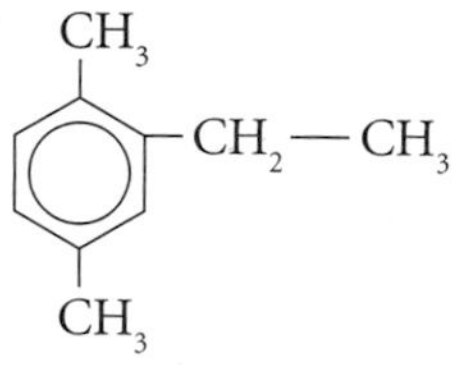

26.

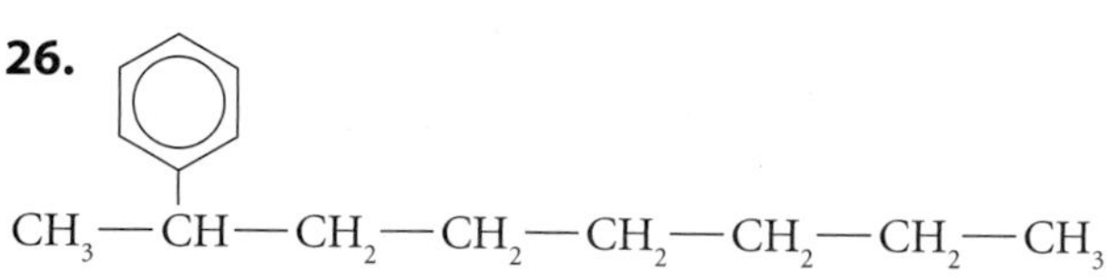

$$CH_3-CH-CH_2-CH_2-CH_2-CH_2-CH_2-CH_3$$

27. 2,3-diphenylnonane

Page 567

28. a) propan-1-ol
b) butan-2-ol
c) cyclobutanol
d) pentane-2,3-diol
e) 2,4-dimethylheptan-1-ol

29. a) CH_3-OH
b) 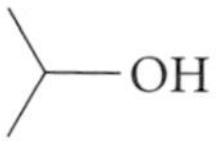 OH

c)

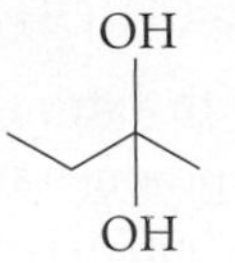

d)

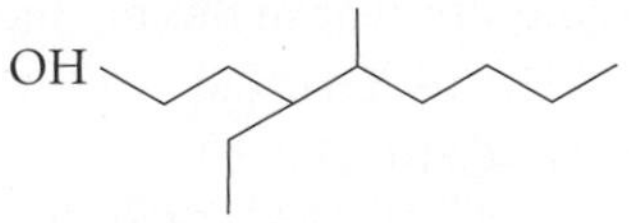

e)

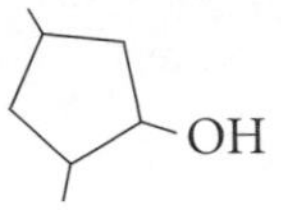

30. a) pentane-1,3-diol
b) 3,4-diethyldecan-1-ol
c) 3-methylpentan-2-ol

Page 569

31. a) 2-bromopropane
b) 2,3-dichloro-4-fluorohexane
c) 2,2-dibromo-4-methylpentane

32. a) $CH_3—CH_2—Br$

b)

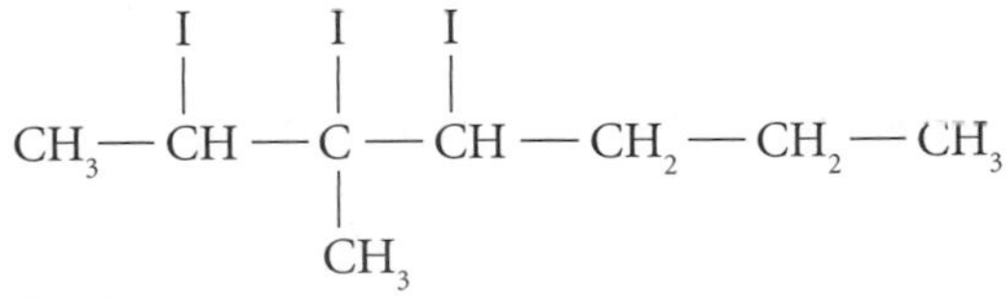

$$CH_3—\underset{\underset{I}{|}}{CH}—\underset{\underset{CH_3}{|}}{\overset{\overset{I}{|}}{C}}—\underset{\underset{I}{|}}{CH}—CH_2—CH_2—CH_3$$

Page 570

33. a) propanoic acid
b) 2,2-dimethylbutanoic acid
c) 2-ethyl-4,5-dimethylhexanoic acid

34. a)

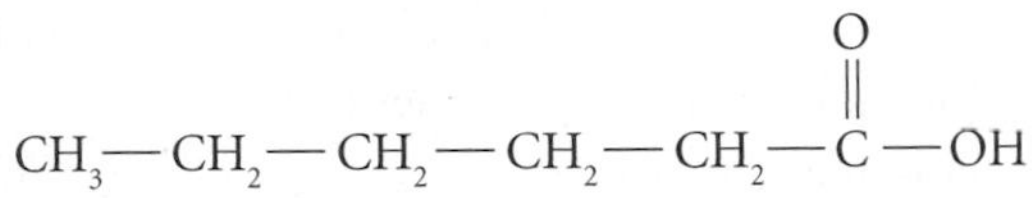

$$CH_3—CH_2—CH_2—CH_2—CH_2—\overset{\overset{O}{\|}}{C}—OH$$

b)

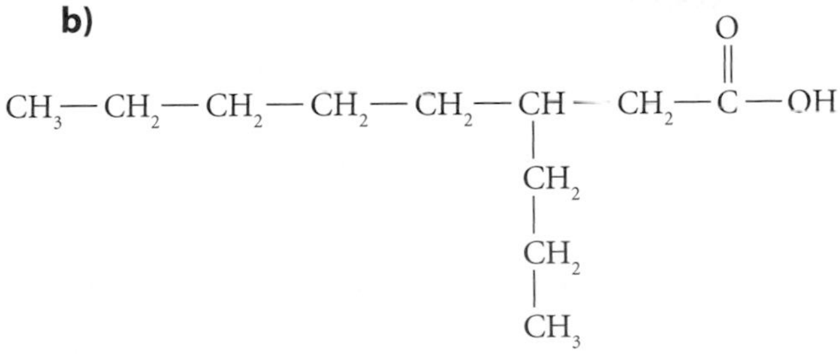

c)

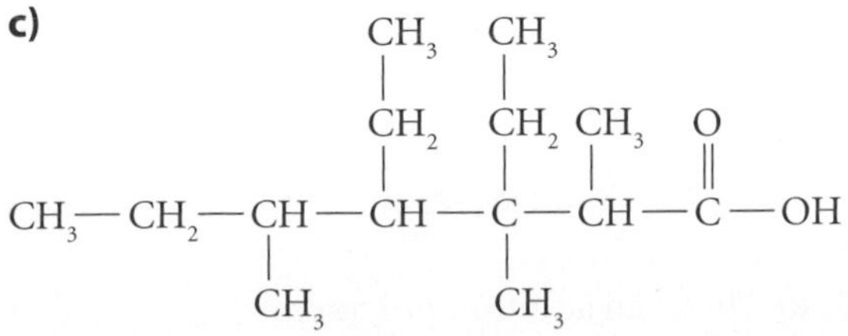

35. a)

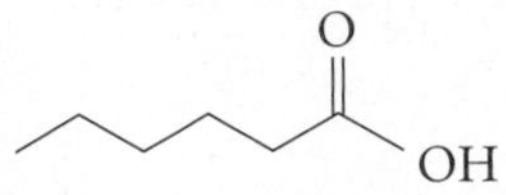

b)

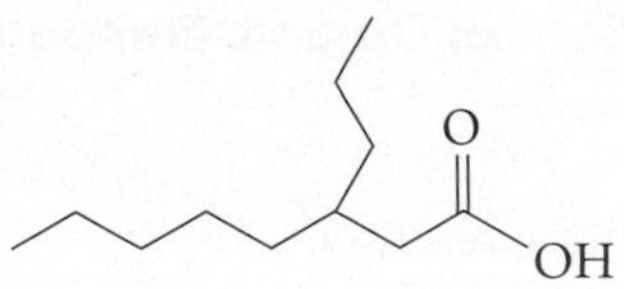

c)

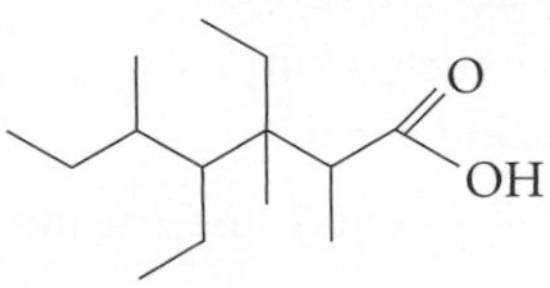

36. butanoic acid

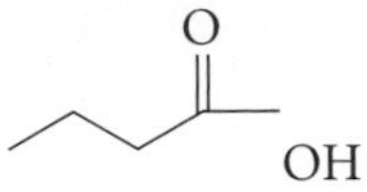

2-methylpropanoic acid

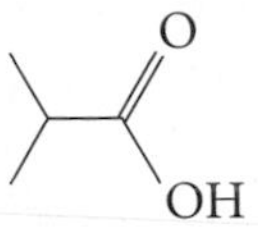

Page 572

37. a) ethyl methanoate
b) methyl butanoate
c) pentyl pentanoate

38. a)

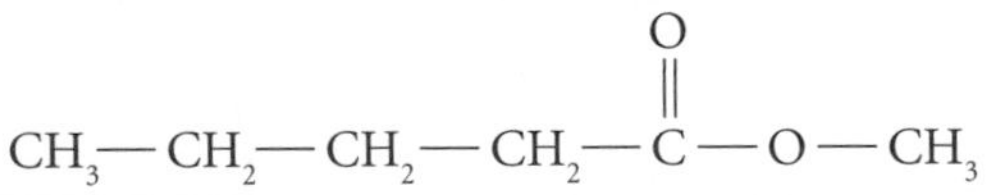

$$CH_3—CH_2—CH_2—CH_2—\overset{\overset{O}{\|}}{C}—O—CH_3$$

b)

$$\underset{\underset{O}{\|}}{HC}—O—CH_2—CH_2—CH_2—CH_2—CH_2—CH_2—CH_3$$

c) $H_3C—\underset{\underset{O}{\|}}{C}—O—CH_2—CH_2—CH_2—CH_3$

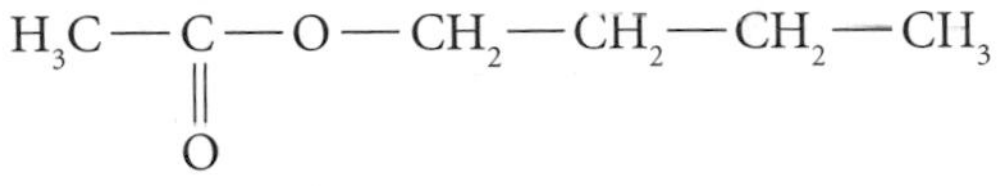

d)

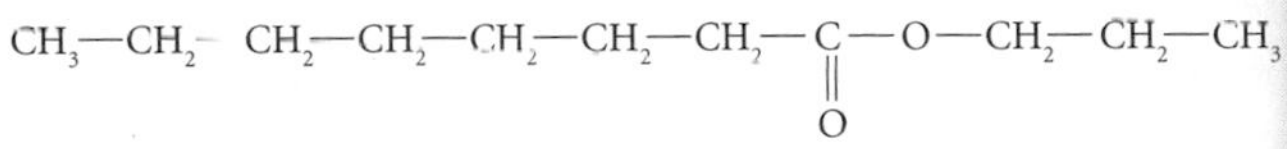

$$CH_3—CH_2—CH_2—CH_2—CH_2—CH_2—CH_2—\underset{\underset{O}{\|}}{C}—O—CH_2—CH_2—CH_3$$

e)

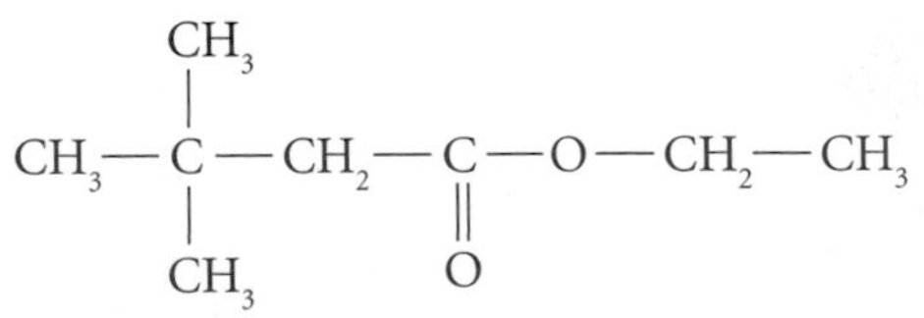

$$CH_3—\underset{\underset{CH_3}{|}}{\overset{\overset{CH_3}{|}}{C}}—CH_2—\underset{\underset{O}{\|}}{C}—O—CH_2—CH_3$$

f)

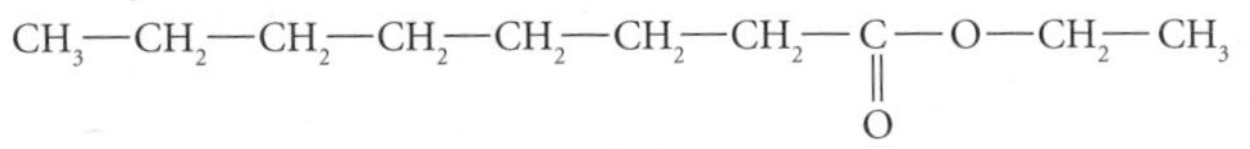

$$CH_3—CH_2—CH_2—CH_2—CH_2—CH_2—CH_2—\underset{\underset{O}{\|}}{C}—O—CH_2—CH_3$$

g) 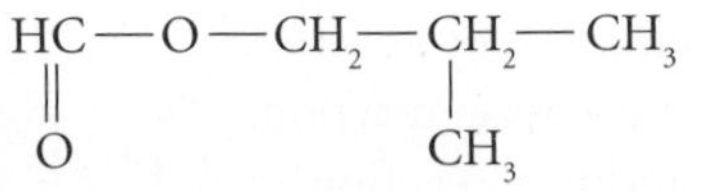 $\underset{\underset{O}{\|}}{HC}—O—CH_2—\underset{\underset{CH_3}{|}}{CH}—CH_3$

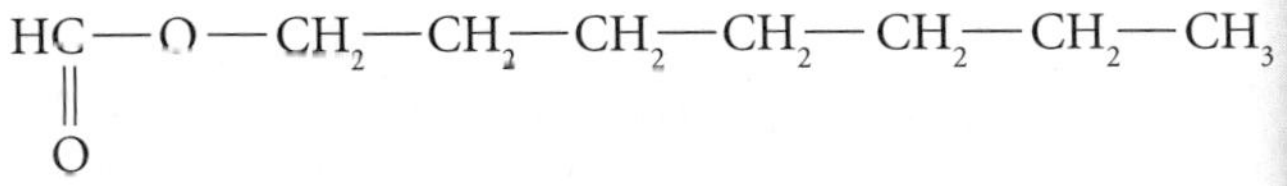

Page 596

1. a) This is an addition reaction.

b) This is an esterification reaction.

c) This is a substitution reaction.

d) This is an elimination reaction.

2. a) pentan-2-ol $\xrightarrow{H_2SO_4}$ pent-1-ene + water

$$CH_3-\overset{\overset{\displaystyle OH}{|}}{CH}-CH_2-CH_2-CH_3 \xrightarrow{H_2SO_4}$$

$$CH_2 = CH-CH_2-CH_2-CH_3 + H_2O$$

b) 2-chloro-2-methylpentane

$\rightarrow$ 2-methylpent-2-ene + hydrogen chloride

$$CH_3-CH_2-CH_2-\overset{\overset{\displaystyle Cl}{|}}{\underset{\underset{\displaystyle CH_3}{|}}{C}}-CH_3 \rightarrow \overset{CH_3-CH_2}{\underset{H}{}}C=C\overset{CH_3}{\underset{CH_3}{}} + HCl$$

c) benzene + chlorine gas $\xrightarrow{FeBr_3}$ chlorobenzene + hydrogen chloride

$$\bigcirc + Cl_2 \xrightarrow{FeBr_3} \bigcirc -Cl + HCl$$

d) propyne + bromine vapour (excess) $\rightarrow$ 1,1,2,2-tetrabromopropane

$$H-C \equiv C-CH_3 + 2Br_2(g) \rightarrow H-\overset{\overset{\displaystyle Br}{|}}{\underset{\underset{\displaystyle Br}{|}}{C}}-\overset{\overset{\displaystyle Br}{|}}{\underset{\underset{\displaystyle Br}{|}}{C}}-CH_3$$

e) methanol + methanoic acid

$\xrightarrow{H_2SO_4}$ methyl methanoate + water

$$H-\overset{\overset{\displaystyle O}{||}}{C}-OH + CH_3-OH \xrightarrow{H_2SO_4} H-\overset{\overset{\displaystyle O}{||}}{C}-O-CH_3 + HOH$$

f) 2-methylbutan-1-ol + hydrogen bromide $\rightarrow$ 1-bromo-2-methylbutane + water

$$CH_3-CH_2-\overset{\overset{\displaystyle}{}}{\underset{\underset{\displaystyle CH_3}{|}}{CH}}-CH_2-OH + HBr \rightarrow$$

$$CH_3-CH_2-\underset{\underset{\displaystyle CH_3}{|}}{CH}-CH_2Br$$

$$+ HOH$$

3. a) This is an elimination reaction.

b) This is an elimination reaction.

c) This is a substitution reaction.

d) This is an addition reaction.

e) This is an esterification reaction.

f) This is a substitution reaction.

4. $CH_2 = CH-CH_3 + HOH \rightarrow CH_3-CH(OH)-CH_3$

5. Step 1: An alkyl halide will react with a strong base $OH^-(aq)$ at room temperature to produce an alcohol, in a substitution reaction. butan-1-ol can be made from 1-chlorobutane in this manner.

$CH_3-CH_2-CH_2-CH_2Cl + OH^-(aq)$

$\rightarrow CH_3-CH_2-CH_2-CH_2OH + Cl^-$

Step 2: Methanoic acid will react with the butan-1-ol to form butyl methanoate and water.

$$CH_3-CH_2-CH_2-CH_2OH + H\overset{\overset{\displaystyle O}{||}}{C}-OH \rightarrow$$

$$H\overset{\overset{\displaystyle O}{||}}{C}-O-CH_2CH_2CH_2CH_3 + H_2O$$

6. This reaction can be carried out in one step with a substitution reaction in the presence of a strong base.

3-chloro-3-methylpentane + $OH^-(aq) \rightarrow$
3-methyl-pentan-3-ol + $Cl^-(aq)$

$$CH_3-CH_2-\overset{\overset{\displaystyle CH_3}{|}}{\underset{\underset{\displaystyle Cl}{|}}{C}}-CH_2-CH_3 + OH^-(aq) \rightarrow$$

$$CH_3-CH_2-\overset{\overset{\displaystyle CH_3}{|}}{\underset{\underset{\displaystyle OH}{|}}{C}}-CH_2-CH_3 + Cl^-(aq)$$

Page 606

7. The name of the polymer is based on the name of the monomer. The polymer is polymethylmethacrylate (commonly known as PMMA).

8. a) This is a condensation polymerization.

b) This is an addition polymerization.

c) This is a condensation polymerization.

9. a)

$$\cdots-O-(CH_2)_3-O-\overset{\overset{\displaystyle O}{||}}{C}-CH_2-\overset{\overset{\displaystyle O}{||}}{C}-O-(CH_2)_3-O-\cdots$$

b)

$$\cdots-CH_2-\underset{\underset{\displaystyle CH_3}{|}}{CH}-CH_2-\underset{\underset{\displaystyle CH_3}{|}}{CH}-CH_2-\underset{\underset{\displaystyle CH_3}{|}}{CH}-\cdots$$

c)

$$\cdots-NHCH_2-\bigcirc-CH_2NH-\overset{\overset{\displaystyle O}{||}}{C}(CH_2)_6\overset{\overset{\displaystyle O}{||}}{C}- NHCH_2-\bigcirc-CH_2NH-\cdots$$

10. a) This is an addition polymer.

b) This is a condensation polymer and is a polyamide.

c) This is a condensation polymer and is a polyester.

d) This is a condensation polymer and is a polyester.

11. a) repeating unit:

$$\cdots - CH_2 - CH - \cdots$$
$$\quad\quad\quad\quad | $$
$$\quad\quad\quad\quad Br$$

monomer:

$$CH_2 = CH$$
$$\quad\quad\quad | $$
$$\quad\quad\quad Br$$

b) repeating unit:

$$\cdots - NH - CH_2 - NH - \overset{\displaystyle O}{\overset{\|}{C}} - CH_2CH_2 - \overset{\displaystyle O}{\overset{\|}{C}} - \cdots$$

monomers:

$$H_2N - CH_2 - NH_2 \quad\quad HO - \overset{\displaystyle O}{\overset{\|}{C}} - CH_2CH_2 - \overset{\displaystyle O}{\overset{\|}{C}} - OH$$

c) repeating unit:

$$\cdots - O - CH_2CH_2 - \overset{\displaystyle O}{\overset{\|}{C}} - \cdots$$

monomer:

$$HO - CH_2CH_2 - \overset{\displaystyle O}{\overset{\|}{C}} - OH$$

d) repeating unit:

$$\cdots - O - \langle \bigcirc \rangle - O - \overset{\displaystyle O}{\overset{\|}{C}} CH_2 \overset{\displaystyle O}{\overset{\|}{C}} - \cdots$$

monomers:

$$HO - \langle \bigcirc \rangle - OH \quad\quad HO - \overset{\displaystyle O}{\overset{\|}{C}} CH_2 \overset{\displaystyle O}{\overset{\|}{C}} - OH$$

12. Step 1: Production of ethene, C_2H_4. An alkyl halide can undergo elimination to produce an alkene when heated in the presence of a strong base, such as sodium ethoxide ($NaOCH_2CH_3(aq)$) mixed with an ethanol solvent

$$CH_3-CH_2Br + NaOCH_2CH_3$$
$$\rightarrow CH_2 = CH_2 + HOCH_2CH_3 + NaBr$$

Step 2: Polymerization

$$n\, CH_2 = CH_2 \rightarrow \cdots - \underset{\underset{H}{|}}{\overset{\overset{H}{|}}{C}} - \underset{\underset{H}{|}}{\overset{\overset{H}{|}}{C}} - \underset{\underset{H}{|}}{\overset{\overset{H}{|}}{C}} - \underset{\underset{H}{|}}{\overset{\overset{H}{|}}{C}} - \underset{\underset{H}{|}}{\overset{\overset{H}{|}}{C}} - \cdots$$

Unit 8 Chemical Equilibrium Focussing on Acid-Base Systems

Chapter 16 Chemical Equilibrium

Page 641

1. $K_c = \dfrac{[CH_3CH\,OOCH_2CH_3][H_2O]}{[CH_3CH_2OH][CH_3COOH]}$

2. $K_c = \dfrac{[NO]^2}{[N_2][O_2]}$

3. $K_c = \dfrac{[H_2O]^2}{[H_2]^2[O_2]}$

4. $K_c = \dfrac{[Fe^{2+}]^2[I_2]}{[Fe^{3+}]^2[I^-]^2}$

5. $K_c = \dfrac{[NO]^4[H_2O]^6}{[NH_3]^4[O_2]^5}$

Page 644

6. $K_c = 1.9 \times 10^{-2}$ **7.** $K_c = 1.2 \times 10^2$

8. $K_c = 0.013$ **9.** $[S_2(g)] = 0.046$ mol/L

10. $[CH_4(g)] = 0.11$ mol/L

Page 649

11. 1100 K **12.** III, II, I

13. a) completion **b)** no reaction

c) completion

14. essentially no dissociation

15. $[Ag(NH_3)_2]^+(aq)$ is more stable

Page 651

16. left

17. a) no change **b)** left

c) right **d)** right

18. a) left **b)** right

c) no change **d)** left

e) no change

19. Only **d)** affects the value of K_c.

20. Adding $C_7H_{14}(g)$; removing $C_7H_8(g)$; removing $H_2(g)$; decreasing the pressure by increasing the volume of the container; and/or increasing the temperature.

Page 658

21. $K_c = 30$

22. $[I_2(g)] = [Cl_2(g)] = 0.015$ mol/L; $[ICl(g)] = 0.14$ mol/L

23. $K_c = 4.7$

24. $[HBr(g)] = 0.045$ mol/L; $[H_2(g)] = [Br_2(g)] = 2.9 \times 10^{-6}$ mol/L

Page 660

25. $[CO(g)] = [Cl_2(g)] = 1.8 \times 10^{-4}$ mol/L

26. $[H_2(g)] = 9.4 \times 10^{-2}$ mol/L

27. $[CO_2(g)] = 0.60$ mol/L; $[CO(g)] = 1.1 \times 10^{-2}$ mol/L; $[O_2(g)] = 5.6 \times 10^{-3}$ mol/L

28. initial $[NCl_3(g)] = 6.0 \times 10^{-5}$ mol/L

29. $[NO(g)] = 1.2 \times 10^{-5}$ mol/L

Chapter 17 Acid-Base Equilibrium Systems

Page 687

1. a) $F^-(aq)$
   b) $CO_3^{2-}(aq)$
   c) $HSO_4^-(aq)$
   d) $N_2H_4(aq)$

2. a) $HNO_3(aq)$
   b) $H_2O(\ell)$
   c) $H_3O^+(aq)$
   d) $H_2CO_3(aq)$

3. $HClO_4(aq)/ClO_4^-(aq)$ and $H_2O(\ell)/H_3O^+(aq)$

4. a) $HS^-(aq)/H_2S(aq)$ and $H_2O(\ell)/OH^-(aq)$
   b) $O^{2-}(aq)/OH^-(aq)$ and $H_2O(\ell)/OH^-(aq)$
   c) $H_2S(aq)/HS^-(aq)$ and $NH_3(aq)/NH_4^+(aq)$
   d) $H_2SO_4(aq)/HSO_4^-(aq)$ and $H_2O(\ell)/H_3O^+(aq)$

Page 688

5. a) $H_2S(aq) + H_2O(\ell) \rightleftharpoons HS^-(aq) + H_3O^+(aq)$;
   $HS^-(aq) + H_2O(\ell) \rightleftharpoons S^{2-}(aq) + H_3O^+(aq)$
   b) $HS^-(aq) + H_2O(\ell) \rightleftharpoons H_2S(aq) + OH^-(aq)$

6. a) $HCO_3^-(aq) + OH^-(aq) \rightleftharpoons CO_3^{2-}(aq) + H_2O(\ell)$
   b) $HCO_3^-(aq) + HF(aq) \rightleftharpoons H_2CO_3(aq) + F^-(aq)$

7. Acid: $H_2PO_4^-(aq) + H_2O(\ell) \rightleftharpoons HPO_4^{2-}(aq) + H_3O^+(aq)$

   Base: $H_2PO_4^-(aq) + H_2O(\ell) \rightleftharpoons H_3PO_4(aq) + OH^-(aq)$

Page 690

8. a) left, favouring reactants
   b) left, favouring reactants
   c) left, favouring reactants

9. a) left, favouring reactants
   b) right, favouring products
   c) right, favouring products

10. a) products favoured
   $H_2CO_3(aq) + NH_3(aq) \rightleftharpoons NH_4^+(aq) + HCO_3^-(aq)$
   b) reactants favoured
   $Na^+(aq) + HSO_3^-(aq) + H_2O(\ell) \rightleftharpoons Na^+(aq) + SO_3^-(aq) + H_3O^+(aq)$
   c) reactants favoured
   $K^+(aq) + NO_3^-(aq) + HF(aq) \rightleftharpoons K^+(aq) + HNO_3(aq) + F^-(aq)$
   d) reactants favoured
   $H_3PO_4(aq) + H_2O(\ell) \rightleftharpoons H_2PO_4^-(aq) + H_3O^+(aq)$

$H_2PO_4^-(aq) + H_2O(\ell) \rightleftharpoons HPO_4^{2-}(aq) + H_3O^+(aq)$
$HPO_4^{2-}(aq) + H_2O(\ell) \rightleftharpoons PO_4^{3-}(aq) + H_3O^+(aq)$

Page 692

11. a) $K_a = \dfrac{[NO_2^-][H_3O^+]}{[HNO_2]} = 5.6 \times 10^{-4}$
   b) $K_a = \dfrac{[C_6H_5COO^-][H_3O^+]}{[C_6H_5COOH]} = 6.3 \times 10^{-5}$
   c) $K_a = \dfrac{[H_2C_6H_5O_7^-][H_3O^+]}{[H_3C_6H_5O_7]} = 7.4 \times 10^{-4}$
   d) $K_a = \dfrac{[C_6H_5O_7^{3-}][H_3O^+]}{[HC_6H_5O_7^{2-}]} = 1.7 \times 10^{-5}$

12. $K_a = \dfrac{[HCO_3^-][H_3O^+]}{[H_2CO_3]}$ $K_{a1} = \dfrac{[CO_3^{-2}][H_3O^+]}{[HCO_3^-]}$

13. $C_6H_6O(aq) + H_2O(\ell) \rightleftharpoons C_6H_5O^-(aq) + H_3O^+(aq)$

Page 693

14. $K_a = 1.0 \times 10^{-4}$; 3.2% ionization
15. $K_a = 1.6 \times 10^{-5}$; 3.9% ionization
16. $K_a = 6.1 \times 10^{-10}$
17. $K_a = 1.2 \times 10^{-5}$
18. 13% ionization

Page 698

19. a) $NH_3(aq) + H_2O(\ell) \rightleftharpoons NH_4^+(aq) + OH^-(aq)$
   b) $(CH_3)_3N(aq) + H_2O(\ell) \rightleftharpoons (CH_3)_3NH^+(aq) + OH^-(aq)$
   c) $HSO_3^-(aq) + H_2O(\ell) \rightleftharpoons H_2SO_3(aq) + OH^-(aq)$
   d) $CO_3^{2-}(aq) + H_2O(\ell) \rightleftharpoons HCO_3^-(aq) + OH^-(aq)$

20. a) $K_b = \dfrac{[NH_4^+][OH^-]}{[NH_3]}$
   b) $K_b = \dfrac{[(CH_3)_3NH^+][OH^-]}{[(CH_3)_3N]}$
   c) $K_b = \dfrac{[H_2SO_3][OH^-]}{[HSO_3^-]}$
   d) $K_b = \dfrac{[HCO_3][OH^-]}{[CO_3^{2-}]}$

21. 0.044%   **22.** $K_b = 1.7 \times 10^{-6}$
23. $K_b = 1.4 \times 10^{-9}$

Page 699

24. pH = 2.41; 0.47% ionization
25. pH = 1.60; 2.5% ionization
26. $K_a = 2.9 \times 10^{-8}$ **27.** 1.18 g

28. pH = 11.14   **29.** pH = 10.44

30. $K_b = 1.6 \times 10^{-6}$

31. $[OH^-(aq)] = 2.6 \times 10^{-2}$ mol/L; pOH = 1.58

32. $[OH^-(aq)] = 3.7 \times 10^{-3}$ mol/L; 1.7% ionization

33. $[NH_3(aq)] = 2.8 \times 10^{-2}$ mol/L

34. a) cyanide ion, $CN^-(aq)$ $K_b = 1.6 \times 10^{-5}$
b) nitrite ion, $NO_2^-(aq)$, $K_b = 1.8 \times 10^{-11}$

35. $CH_3NH_3^+(aq)$ and $C_5H_5NH^+(aq)$ (stronger acid)

36. $CN^-(aq)$ is the stronger base

37. Use K_{a2}; $K_b = 1.0 \times 10^5$

38. $NH_4^+(aq)$ is the weaker acid. The reaction proceeds to the right and products are favoured.

Answers to Supplemental Practice Problems

Unit 1 The Diversity of Matter and Chemical Bonding

Chapter 1 Chemical Bonding

1. a) $Sn(BrO_4)_4$ **b)** UO_2
c) $Zn(H_2PO_4)_2$ **d)** Li_2CrO_4
e) $In(OH)_3$

2. a) radium phosphate
b) mercury(II) sulfide
c) ammonium sulfite
d) lead(IV) fluoride
e) cobalt(III) sulfide

3. a) $TiBr_4$
b) either lead(IV) oxide, PbO_2 or lead(II) oxide, PbO
c) either dinitrogen tetroxide, N_2O_4, or nitrogen dioxide, NO_2
d) $CaHPO_4$
e) mercury(II) oxide

4. a) 1+   **b)** 4+
c) 3+   **d)** 3+
e) 2+

5. a) T_2O **b)** T_2SO_4
c) $TBrO_2$ **d)** TCN

6. a) I—I   **b)**

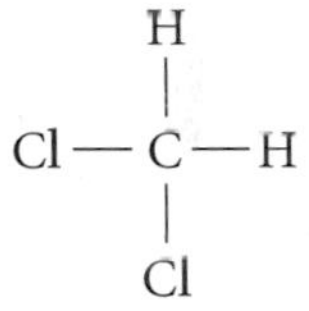

c)

d)

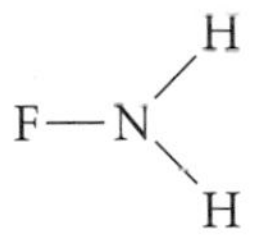

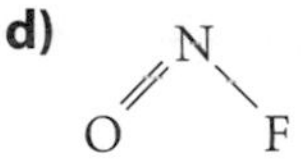

7. $\left[Mg \right]^{2+}$ $\left[:\ddot{F}: \right]^{1-}$ $:\ddot{N}e:$ $\left[:\ddot{O}: \right]^{2-}$ $\left[:\ddot{N}: \right]^{3-}$ $\left[Na \right]^+$

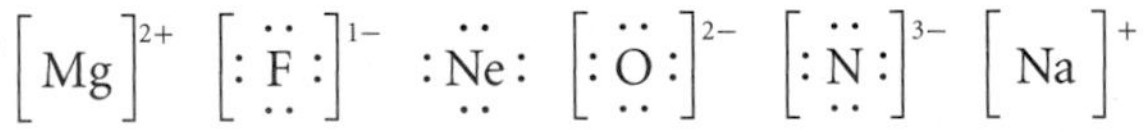

8. a) $\left[Ba \right]^{2+}$ $\left[:\ddot{S}: \right]^{2-}$ **b)** $3\left[Mg \right]^{2+}$ $2\left[:\ddot{N}: \right]^{3-}$

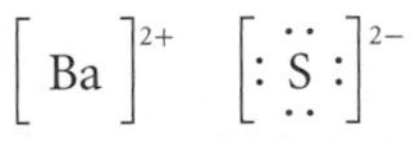

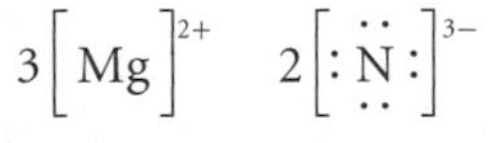

c) $\left[Li \right]^{1+}$ $\left[:\ddot{I}: \right]^{1-}$ **d)** $\left[Ba \right]^{2+}$ $\left[:\ddot{O}: \right]^{2-}$

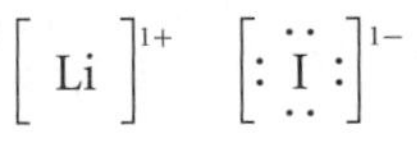

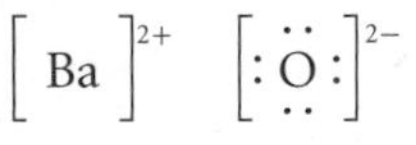

e) $2\left[Li \right]^{1+}$ $\left[:\ddot{O}: \right]^{2-}$ **f)** $\left[Al \right]^{3+}$ $3\left[:\ddot{F}: \right]^{1-}$

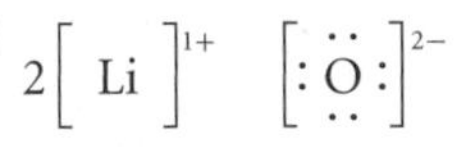

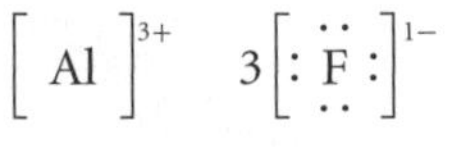

9. a) polar covalent
b) unequally shared pair of electrons lying closer to oxygen

10. Cl H>Cl F>Al P>H S>S C

11. a) Both atoms have electrons only in the first energy level. The hydrogen atom is larger in diameter since it has a smaller nuclear charge to attract the electron in the 1st energy level. Helium has twice the nuclear charge to hold the two electrons in the 1st energy level.

b) The S atom is larger since its electrons are found in 3 energy levels. Oxygen has two energy levels.

12. carbon has zero lone pair electrons in CO_2; sulfur has 1 lone pair in SO_2

13. Two fluorine atoms attain a noble gas configuration by sharing one pair of electrons. Two lithium atoms both must lose electrons to attain a noble gas configuration.

Chapter 2 Diversity of Matter

14. a) Na^+ has the smaller diameter since the electrons are in 2 energy levels compared to K^+ that has 3 energy levels.

b) Cl^- has the smaller diameter since there is a greater nuclear attraction per electron than in Cl^{2-}. (Note: Cl^{2-} does not exist)

15. $CH_3Cl(\ell)$ has the lower boiling point because it is less polar and has weaker dipole-dipole attractions between molecules. Fewer electrons in this molecule leads to smaller dispersion forces.

16. Cl_2O bent; ClO_2^- bent

17. a) CH_3F tetrahedral; NH_2F pyramidal

b) CH_3F dipole-dipole, dispersion; NH_2F dipole-dipole, dispersion, hydrogen bonding

c) NH_2F, more polar and has hydrogen bonding

18. a)

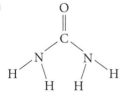

b) dispersion forces, dipole-dipole attraction and hydrogen bonding

c) same shape; dispersion forces, dipole-dipole attraction but no hydrogen bonding.

19. a) $S_8(\ell)$ dispersion, $Hg(\ell)$ metallic bonding, $O_2(\ell)$, dispersion.

b) $O_2(\ell)$, $Hg(\ell)$, $S_8(\ell)$ (Boiling points: $Hg = 357\,°C$; $S_8 = 445\,°C$)

20. a) $\ddot{\ddot{F}} : \overset{..}{\underset{..}{Sb}} : \ddot{\ddot{F}}$ $: \ddot{\ddot{F}} :$

b) $: \overset{..}{\underset{..}{O}} :$ $\ddot{\ddot{O}} : \overset{..}{\underset{..}{S}} : \ddot{\ddot{F}} :$ $: \ddot{\ddot{F}} :$

c) $: \overset{..}{\underset{..}{Cl}} \, \overset{..}{\underset{..}{N}} \, \overset{..}{\underset{..}{O}} \cdot$

21. a) pyramidal   **b)** tetrahedral
c) bent

22. polar since the shape is pyramidal

23. For the sodium halides, the difference in the size of the metal ion and halide ion becomes greater from Cl to Br to I. This increases the distance between the ions and weakens the attraction between the ions and the strength of the ionic bonding. As the molar mass of the hydrogen halides increase, the number of electrons increase and the dispersion forces increase.

24. $BeCl_4^{2-}$ tetrahedral; $SbCl_4^+$ tetrahedral

25. phosphate, PO_4^{3-}, tetrahedral; phosphite, PO_3^{3-}, pyramidal

Unit 2 Forms of Matter: Gases

Chapter 3 Properties of Gases

26. Add more gas at constant temperature and pressure; reduce the pressure around the balloon at constant temperature and amount of gas; increase the temperature at constant pressure and amount of gas.

27. No because 1 °C is equivalent to 1 K.

28. The volume would likely increase by a factor less than expected ideally because some of the kinetic energy would be used to overcome the intermolecular attractions leading to inelastic collisions.

29. 5.00 L   **30.** $P_2 = \frac{1}{3}P_1$
31. $T_2 = -73.05\,°C$ **32.** 1.04×10^4 mm
33. $P_2 = \frac{1}{4}P_1$ **34.** 128 kPa
35. 510 L   **36.** 172 kPa

Chapter 4 Exploring Gas Laws

37. 138 °C   **38.** 2.66 g/L

39. Molecules at the same temperature have the same average kinetic energy. Since $E_k = \frac{1}{2}mv^2$, a more massive molecule moves at lower velocity and therefore will collide less often than a lighter molecule. The total pressure exerted by the two gases will be the same.

40. From the ideal gas equation, density $= \frac{PM}{RT}$. Density is directly proportional to the molar mass at constant temperature and pressure. Since the average molar mass of air is lower when water vapour is present, moist air will have a lower density than dry air.

41. 4.76 kPa **42.** 29.4 mL

43. 4.00 g/mol **44.** 43.64 L

45. b) and e)

46. volume, number of moles, number of molecules, average kinetic energy

47. $\frac{3}{2}$ mol **48.** 46.7 °C

49. 17.1 kPa **50.** 611 L

51. 7.99 L

Unit 3 Matter as Solutions, Acids, and Bases

Chapter 5 Solutions

52. Incorrect, 44.4% m/m

53. No since the volume of the solution is not known.

54. 0.00564 g

55. Excess solute crystallizes from the solution and the concentration is constant.

56. 0.15 mol/L Na^+

57. 2.9×10^{-5} % m/m

58. 8.39×10^{-3} mol/L

59. 0.125 mol/L

60. 0.625 L

61. 0.800 mol/L

62. 0.49 g

63. 0.0170 % m/m

64. Add 238 g $NaNO_3$ to water. Add more water, until the total volume is 2.00 L.

65. 0.348 mol/L

Chapter 6 Acids and Bases

66. a) 12.76 **b)** 11.176
 c) 0.308 **d)** 2.40

67. a) 5.4×10^{-3} mol/L **b)** 6.11×10^{-5} mol/L
 c) 1×10^{-8} mol/L

68. 9.30

69. $[H_3O^+(aq)] = 6.7 \times 10^{-4}$ mol/L;
 $[OH^-(aq)] = 1.5 \times 10^{-11}$ mol/L

70. a) 2.5×10^{-9} mol/L; **b)** 7.2×10^{-1} mol/L

71. 1.40×10^{-3} g **72.** 6.872×10^{-6} mol/L

73. a) No since in both solutions the HCl(aq) is 100% ionized
 b) 0.01 mol/L is less concentrated having fewer $H_3O^+(aq)$ ions to react with Mg(s).

74. b), c), d) **75.** 2.30

76. a) $BaO(s) + H_2O(\ell) \rightarrow Ba(OH)_2(aq)$
 $Ba(OH)_2(aq) \rightarrow Ba^{2+}(aq) + 2OH^-(aq)$; base

b) $SO_3(g) + H_2O(\ell) \rightleftharpoons H_2SO_4(aq)$
 $H_2SO_4(aq) + H_2O(\ell) \rightarrow H_3O^+(aq) + HSO_4^-(aq)$; acid

c) $K_2CO_3(s) \rightarrow 2K^+(aq) + CO_3^{2-}(aq)$
 $CO_3^{2-}(aq) + H_2O(\ell) \rightleftharpoons HCO_3^-(aq) + OH^-(aq)$; base

d) $P_4O_{10}(s) + 6H_2O(\ell) \rightarrow 4H_3PO_4(aq)$
 $H_3PO_4(aq) + H_2O(\ell) \rightleftharpoons H_3O^+(aq) + H_2PO_4^-(aq)$; acid

77. a) 100 times **b)** 100 times

78. 2.3

Unit 4 Quantitative Relationships in Chemical Changes

Chapter 7 Stoichiometry

79. a) total ionic equation:
 $Pb^{2+}(aq) + 2NO_3^-(aq) + 2NH_4^+(aq) + CO_3^{2-}(aq) \rightarrow PbCO_3(s) + 2NH_4^+(aq) + 2NO_3^-(aq)$
 Net ionic equation: $Pb^{2+}(aq) + CO_3^{2-}(aq) \rightarrow PbCO_3(s)$
 Spectator ions: $NH_4^+(aq) + NO_3^-(aq)$

b) $Hg^{2+}(aq) + 2NO_3^-(aq) + CH_3COO^-(aq) + Na^+(aq) \rightarrow$ no reaction

c) total ionic equation:
 $3Ba^{2+}(aq) + 6Cl^-(aq) + 2Al^{3+}(aq) + 3SO_4^{2-}(aq) \rightarrow 3BaSO_4(s) + 2Al^{3+}(aq) + 6Cl^-(aq)$
 net ionic equation: $Ba^{2+}(aq) + SO_4^{2-}(aq) \rightarrow BaSO_4(s)$
 spectator ions: $Al^{3+}(aq) + Cl^-(aq)$

d) total ionic equation:
 $2Fe^{3+}(aq) + 6NO_3^-(aq) + 6Na^+(aq) + 3S^{2-}(aq) \rightarrow Fe_2S_3(s) + 6Na^+(aq) + 6NO_3^-(aq)$
 net ionic equation: $2Fe^{3+}(aq) + 3S^{2-}(aq) \rightarrow Fe_2S_3(s)$
 spectator ions: $Na^+(aq) + NO_3^-(aq)$

80. 137.7 mL **81.** 26.2g

82. 36.4 g **83.** 4.606×10^5 L

84. 36.4 g **85.** 150.0 L

86. 37.9 L

87. a) $H_2S(g) + 2AgNO_3(aq) \rightarrow 2HNO_3(aq) + Ag_2S(s)$
 b) double replacement
 c) 2.726 g

88. 0.820 mol/L **89.** 0.1515 g

90. 0.964 L **91.** 0.135 mol/L

Chapter 8 Applications of Stoichiometry

92. a) 56.7 g   **b)** 90.3%

93. 14.7 L   **94.** 2.3 g

95. 40.0 mL   **96.** $P_4(s)$

97. 36.6%   **98.** 72.4%

99. 600 mL

100. 90.7%   **101.** yes

102. 1.76 L   **103.** 40.0%

Unit 5 Thermochemical Changes

Chapter 9 Energy and Chemical Reactions

104. $-4.08 \times 10^4 \, \text{kJ}$

105. 0.134 L

106. a) $C_7H_8(\ell) + 9O_2(g) \longrightarrow 7CO_2(g) + 4H_2O(\ell) + 3904 \, \text{kJ}$
     b) 23.6 g

107. 73.5%

108. a) 5.48 kJ   **b)** -5.48 kJ
     c) 0.996 kJ/g

109. 89.8 °C

110. 23.9 kJ   **111.** 1.30×10^3 kJ/mol

112. a) 25.7 kJ/mol   **b)** decrease
     c) 16.8 °C

113. 1.38×10^4 kJ

114. a) 200.3 kJ   **b)** 864.6 kJ/mol
     c) $CH_3COOH(\ell) + 2O_2(g) \longrightarrow 2CO_2(g) + 2H_2O(\ell) + 864.6 \, \text{kJ}$

115. 71.1 °C

Chapter 10 Theories of Energy and Chemical Changes

116. 74.3%   **117.** -521 kJ

118. -96.7 kJ   **119.** $+131.3$ kJ

120. a) $C_2H_4(g) + 3O_2(g) \longrightarrow 2CO_2(g) + 2H_2O(\ell) + 337 \, \text{kJ}$
     b) 24.6 kg

121. 114.1 kJ/mol

122. The H-Cl bond is stronger than the H-I bond

123. a) $C_3H_7COOH(\ell) + 5O_2(g) \longrightarrow 4CO_2(g) + 4H_2O(\ell) + 2184 \, \text{kJ}$
     b) -533 kJ/mol
     c) -366.2 kJ

124. 6.65 kJ

125. $Br_2(g) \longrightarrow Br_2(\ell) + 29.99 \, \text{kJ/mol}$

Chapter 11 Activation Energy and Catalysts

126. a) 360 kJ
     b) No effect because E_a depends upon the bond energy of the reactants

127. No. In a) frequency of collisions is the only factor. In b) orientation of the reactant molecules is an additional factor. Reaction a) will be affected more by temperature change.

128. a) 420.0 kJ   **b)** 404.5 kJ
     c) -400.0 kJ

129. c) There is less attraction between particles in the gas phase and there will a greater frequency of collision of these particles.

130. a) No. E_a depends on the bond energy of the reactant molecules
     b) Same for both. E_a depends on the bond energy of the reactant molecules. Kindling ignites faster because it has larger surface area compared with a larger piece of wood.

131. a) yes   **b)** no

132. The enzyme inhibitor may block or distort the active site.

133. An exothermic reaction because the heat given off would continue to supply the necessary E_a.

134. a) and e)

135. a) E **b)** -100kJ
     c) 250 kJ   **d)** C
     e) difference in $E_{a(\text{fwd})}$ uncatalyzed vs catalyzed

Unit 6 Electrochemical Changes

Chapter 12 Oxidation-Reduction Reactions

136. c) and d)

137. a) $\overset{0}{S_8}(s) + 8\overset{+1\ +4\ -2}{Na_2SO_3}(aq) \longrightarrow 8\overset{+1\ +2\ -2}{Na_2S_2O_3}(aq)$
     b) $Na_2SO_3(aq)$ is reduced
     c) $S_8(s)$ is the reducing agent

138. a) -2 in $HS^-(aq)$, 0 in $S_4(\ell)$, $+4$ in $SO_3{}^{2-}(aq)$, $+2$ in $S_2O_3{}^{2-}(aq)$, $+5/2$ in $S_4O_6{}^{2-}$
     b) $+3$ in $B_4O_7{}^{2-}(aq)$, $+5$ in $BO_3{}^-$, $+3$ in $BO_2{}^-(aq)$, $+3$ in $B_2H_6(g)$, $+3$ in $B_2O_3(s)$

139. $N_2O_5(g)$, $N_2O_4(g)$, $NO_2(g)$, $N_2O_3(g)$, $NO(g)$, $N_2O(g)$

140. $SF_4(g)$ is the oxidizing agent; $BCl_3(g)$ is the reducing agent.

141. In, Ga, Mn, Np

142. a) $3Ti^{3+}(aq) + RuCl_5^{2-}(aq) + 6OH^-(aq) \rightarrow$
$Ru(s) + 3TiO^{2+}(aq) + 5Cl^-(aq) + 3H_2O(\ell)$

b) $2ClO_2(g) + 2OH^-(aq) \rightarrow ClO_2^-(aq) +$
$ClO_3^-(aq) + H_2O(\ell)$

143. a) $5ClO_3^-(aq) + 3I_2(s) + 3H_2O(\ell)$
$\rightarrow 6IO_3^-(aq) + 5Cl^-(aq) + 6H^+(aq)$

b) $Cu(s) + SO_4^{2-}(aq) + 4H^+(aq) \rightarrow Cu^{2+}(aq) +$
$SO_2(g) + 2H_2O(\ell)$

c) $8Al(s) + 3NO_3^-(aq) + 5OH^-(aq) +$
$18H_2O(\ell) \rightarrow 8Al(OH)_4^-(aq) + 3NH_3(g)$

d) $5C_2H_4(g) + 12MnO_4^-(aq) + 36H^+(aq) \rightarrow$
$10CO_2(g) + 12Mn^{2+}(aq) + 28H_2O(\ell)$

e) $2NO_2(g) + 2OH^-(aq) \rightarrow NO_2^-(aq) +$
$NO_3^-(aq) + H_2O(\ell)$

f) $CrO_4^- + 2HSnO_2^-(aq) \rightarrow CrO_2^-(aq) +$
$2HSnO_3^-(aq)$

144. a) $3P_4(s) + 20NO_3^-(aq) + 20H^+(aq) + 8H_2O(\ell)$
$\rightarrow 12H_3PO_4(aq) + 20NO(g)$

b) $MnO_2(s) + NO_2^-(aq) + 2H^+(aq) \rightarrow NO_3^-(aq)$
$+ Mn^{2+}(aq) + H_2O(\ell)$

c) $TeO_3^{2-}(aq) + 2N_2O_4(g) + 2OH^-(aq) \rightarrow Te(s)$
$+ 4NO_3^-(aq) + H_2O(\ell)$

d) $4MnO_4^-(aq) + 3N_2H_4(g) \rightarrow 3N_2(g) +$
$4MnO_2(s) + 4H_2O(\ell) + 4OH^-(aq)$

e) $S_2O_3^{2-}(aq) + 4OCl^- + 2OH^-(aq) \rightarrow$
$2SO_4^{2-}(aq) + 4Cl^-(aq) + H_2O(\ell)$

f) $Br_2(\ell) + SO_2(g) + 2H_2O(\ell) \rightarrow 2Br^-(aq) +$
$SO_4^{2-}(aq) + 4H^+(aq)$

g) $PbO_2(s) + 4Cl^-(aq) + 4H^+(aq) \rightarrow PbCl_2(aq)$
$+ Cl_2(g) + 2H_2O(\ell)$

145. a) $5CH_3OH(aq) + 4MnO_4^-(aq) + 12H^+(aq) \rightarrow$
$5HCOOH(aq) + 4Mn^{2+}(aq) + 11H_2O(\ell)$

b) 20.0 g

146. $Mn(OH)_2(s) + 4MnO_4^-(aq) + 6OH^-(aq) \rightarrow$
$5MnO_4^{2-}(aq) + 4H_2O(\ell)$
128.1 mL of $MnO_4^-(aq)$

Chapter 13 Cells and Batteries

147. a) Pt   **b)** Al to Pt
c) Ce^{4+} **d)** decrease

148. $Fe(s)|Fe^{2+}(aq)\|Ag^+(aq)|Ag(s)$

149. -2.50 V

150. a) $Ni(s) + Pb^{2+}(aq) \rightarrow Ni^{2+}(aq) + Pb(s)$
c) $Sn(s) + 2Fe^{3+}(aq) \rightarrow Sn^{2+}(aq) + 2Fe^{2+}(aq)$
e) $Pb(s) + Sn^{4+}(aq) \rightarrow Pb^{2+}(aq) + Sn^{2+}(aq)$

151. a) $Fe(s) + Sn^{2+}(aq) \rightarrow Fe^{2+}(aq) + Sn(s)$
b) Sn
c) 0.31 V
d) 0.706 g

152. $H_2(g)$ and $I_2(s)$

153. Using the two equations, $E^o_{cell} = +0.90$ V.
Therefore $S_2O_3^{2-}(aq)$ can exist in acid solution.

154. The half-cell reduction potential, E^o, for
$MnO_4^-(aq)$ is $+1.51$ V and for $Cr_2O_7^{2-}(aq)$ is
1.23 V. $MnO_4^-(aq)$ is more easily reduced and is
the stronger oxidizing agent.

155. $E^o_{cell} = +0.15$ V. Reaction will occur.

156. 1.45×10^3 C

157. 3.56 g

158. a) anode: $2Cl^-(\ell) \rightarrow Cl_2(g) + 2e^-$
cathode: $Al^{3+}(\ell) + 3e^- \rightarrow Al(\ell)$

b) No. Standard reduction potentials are
measured for ions in aqueous solution at a
concentration of 1.00 mol/L

Unit 7 Chemical Changes of Organic Compounds

Chapter 14 Structure and Physical Properties of Organic Compounds

159. a) 2-methylpentane
b) 2-methyloctane
c) methylcyclohexane
d) 3,7,7-trimethyldecane

160. a)

$$CH_3-\underset{\underset{\displaystyle CH_3}{|}}{CH}-CH_2-CH_2-\underset{\underset{\displaystyle CH_3}{|}}{CH}-CH_3$$

b)

$$CH_3-\underset{\underset{\displaystyle CH_3}{|}}{CH}-CH_2-CH_2-\underset{\underset{\displaystyle CH_3}{|}}{CH}-CH_2-CH_3$$

c)

$$CH_3-\underset{\underset{\displaystyle CH_3}{|}}{CH}-CH-CH_2-CH_3$$

d)

$$CH_3-\underset{\underset{\displaystyle CH_3}{|}}{\overset{\overset{\displaystyle CH_3}{|}}{C}}-\underset{\underset{\displaystyle CH_3}{|}}{CH}-\underset{\underset{\displaystyle CH_2}{|}}{CH}-CH_2-CH_2-CH_2-CH_3$$

e)

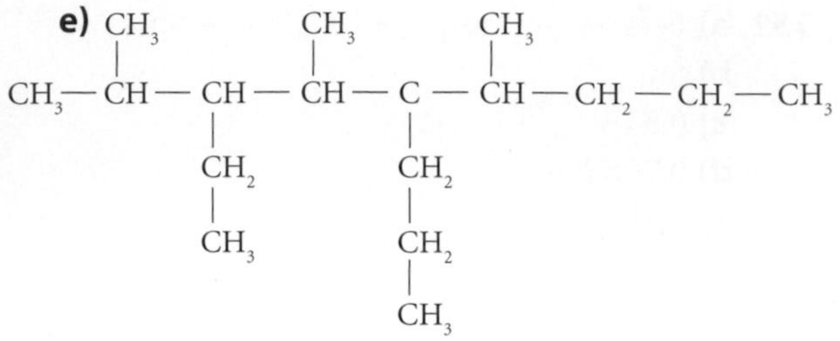

$$CH_3-CH(CH_3)-CH(CH_2CH_3)-CH(CH_3)-C(CH_2CH_2CH_3)(CH_3)-CH(CH_3)-CH_2-CH_2-CH_3$$

161. a) 3,4,4,5-tetramethyloctane

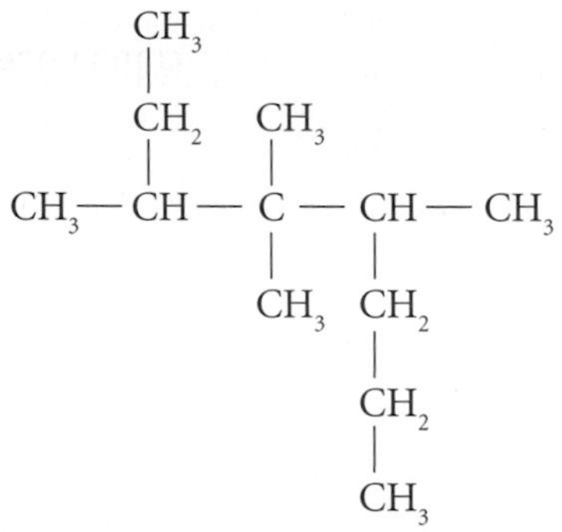

(structure)

b) 2,3-dimethylheptane

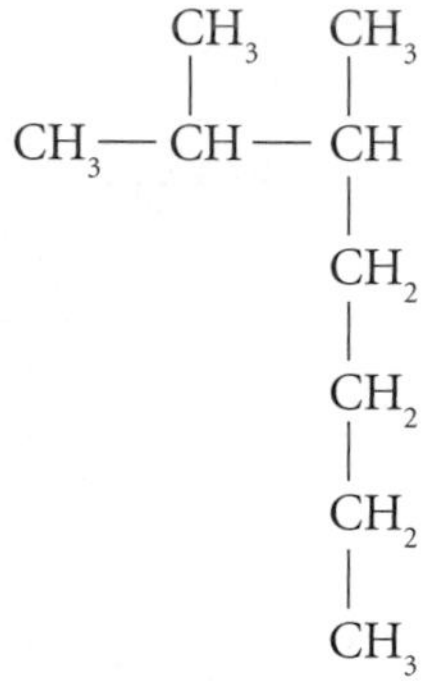

(structure)

c) heptane

(structure)

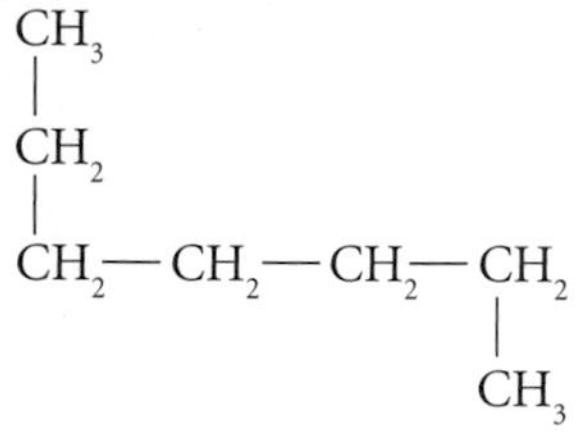

162. a) propene
b) pent-2-yne
c) 4-methylcyclohexene
d) oct-2-ene
e) 4-methylpent-1-yne
f) 3-ethylcyclohexene

163. a) 2,4-dimethylheptane
b) 1,2-dimethylcyclohexane
c) 3-ethyl-4-methylhexane
d) 3-methylcyclobutene

164. a)

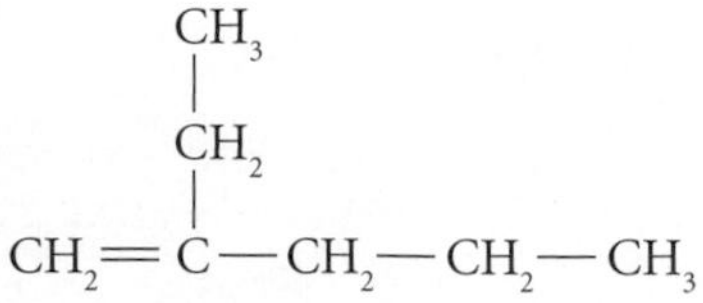

(structure)
$$CH_2=C(-CH_2-CH_3)-CH_2-CH_2-CH_3$$

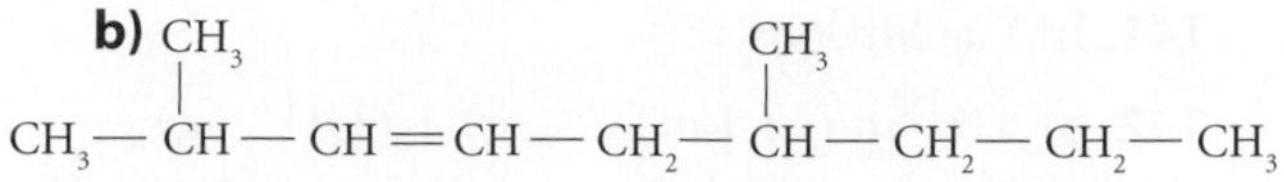

b)

$$CH_3-CH(CH_3)-CH=CH-CH_2-CH(CH_3)-CH_2-CH_2-CH_3$$

c)

$$CH_3-CH(CH_3)-C\equiv C-CH(CH_3)-CH_3$$

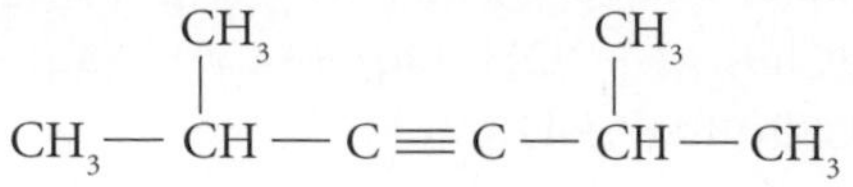

d)

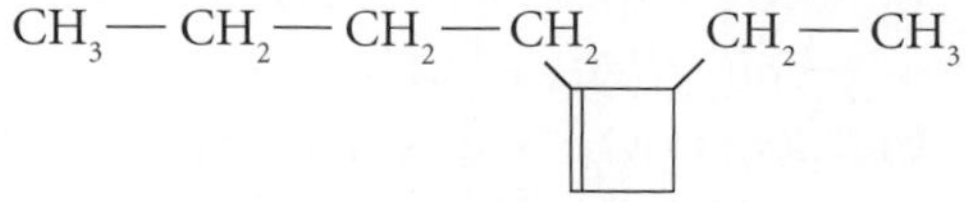

$$CH_3-CH_2-CH_2-CH_2 \quad CH_2-CH_3$$ (cyclobutene structure)

e)

$$CH_3-CH_2-CH_2-CH_2 \quad CH_2-CH_2-CH_3$$ (cycloheptene structure)

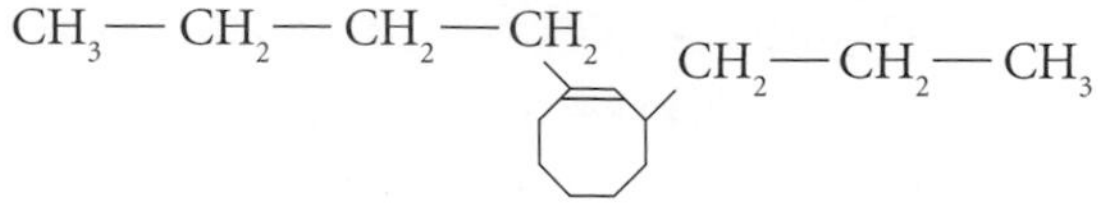

165. a) 4-ethyl-3-methylcyclohexene
b) butylcyclopentane
c) 1-butyl-3-ethyl-4-methyl-2-propylcyclohexane
d) 4-ethyl-5-methylhept-2-ene
e) 3,5-dimethylcyclohexene

166. a) propylbenzene
b) 1-ethyl-3-methyl-5-propylbenzene
c) 1,2-diethylcyclohexane

167. a) 4-methylpentan-2-ol
b) 2,4-dimethylhexan-1-ol
c) cyclopentanol

168. a) $FC\equiv CCHF–CH(CH_3)–CH_3$
b) $CH_3–CHCl–CHCl–CHCl–CH_3$
c)

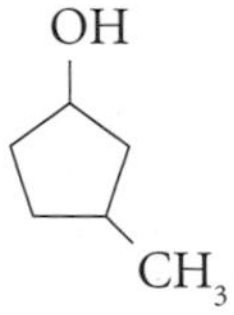

(cyclopentane with OH and CH₃ substituents)

d) $CH_2=CH–CH_2–CH_2–CH_2–CH_2–C(CH_3)_2–CH_3$

169. a) methanoic acid
b) 2-methylpropanoic acid
c) butyl methanoate
d) 2-methylbutyl ethanoate

Chapter 15: Reactions of Organic Compounds

170. a) elimination   **b)** substitution
      c) elimination   **d)** elimination

171. a) 2,3-dibromopentane

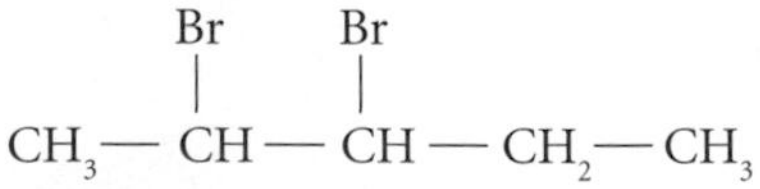

$$CH_3-CH(Br)-CH(Br)-CH_2-CH_3$$

b) methylcyclopentane

(cyclopentane with CH₃ substituent)

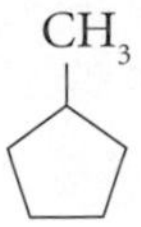

c) 1,2-dichloroheptane

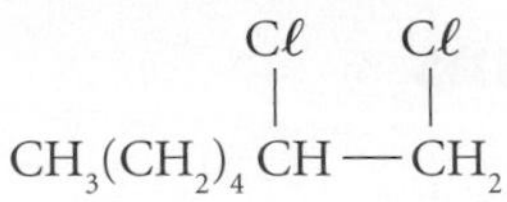

$$CH_3(CH_2)_4 \underset{|}{C}\!\!\overset{C\ell}{H} - \underset{|}{C}\!\!\overset{C\ell}{H_2}$$

d) 1,2-dimethylchlorocyclobutane

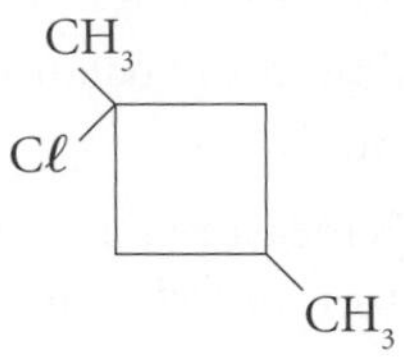

172. a) benzene + chlorine

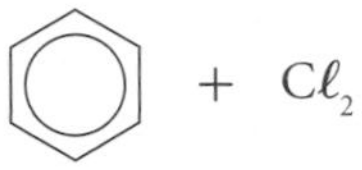

$+\ C\ell_2$

b) cyclohexene + chlorine

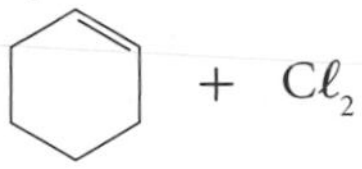

$+\ C\ell_2$

c) cyclopentene + water

$+\ H-OH$

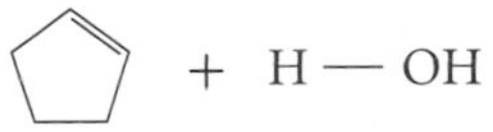

d) cyclobutene + hydrogen bromide

$+\ HBr$

173. a)

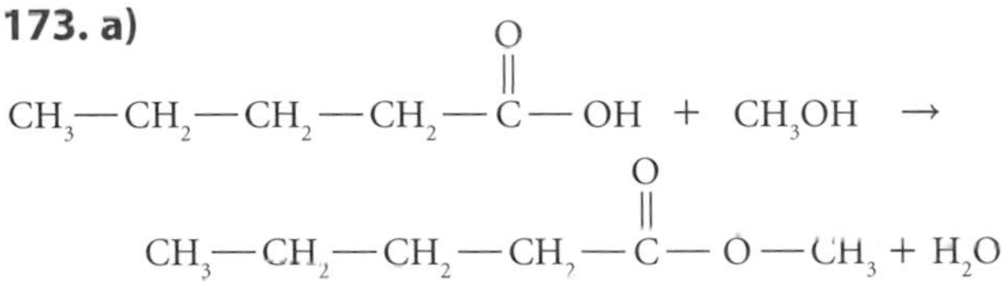

$$CH_3-CH_2-CH_2-CH_2-\overset{O}{\underset{\|}{C}}-OH\ +\ CH_3OH\ \rightarrow$$
$$CH_3-CH_2-CH_2-CH_2-\overset{O}{\underset{\|}{C}}-O-CH_3 + H_2O$$

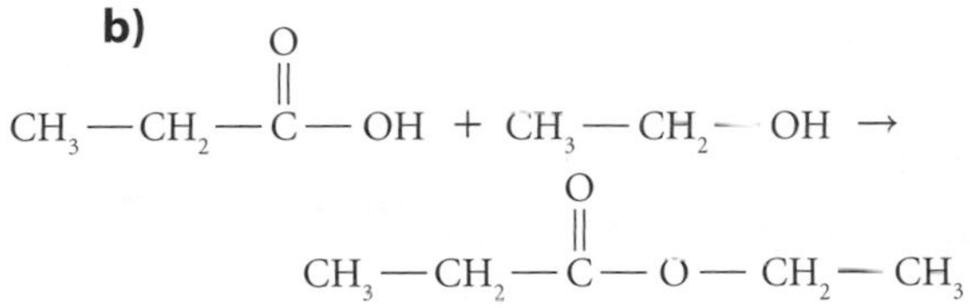

b)

$$CH_3-CH_2-\overset{O}{\underset{\|}{C}}-OH\ +\ CH_3-CH_2-OH\ \rightarrow$$
$$CH_3-CH_2-\overset{O}{\underset{\|}{C}}-O-CH_2-CH_3$$

c)

$$H_2O + CH_3-CH_2-CH_2-\overset{O}{\underset{\|}{C}}-O-CH_2-CH_2-CH_2-CH_3\ \rightarrow$$
$$CH_3-CH_2-CH_2-\overset{O}{\underset{\|}{C}}-OH + CH_3-CH_2-CH_2-CH_2OH$$

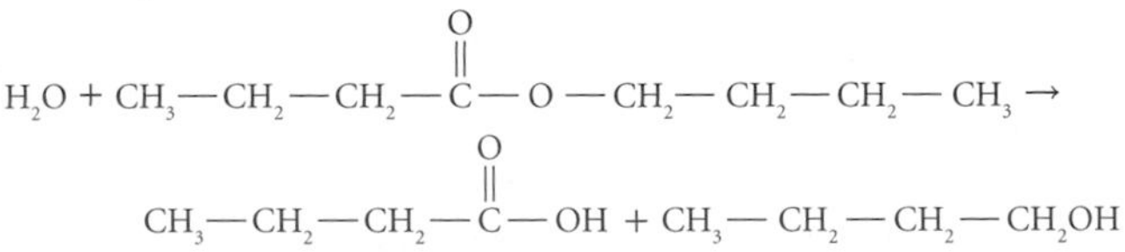

d)

$$CH_3-\underset{\underset{OH}{|}}{C}H-CH_3 + CH_3-CH_2-CH_2-\overset{O}{\underset{\|}{C}}-OH\ \rightarrow$$
$$CH_3-CH_2-CH_2-\overset{O}{\underset{\|}{C}}-O-\underset{\underset{CH_3}{|}}{C}H-CH_3$$

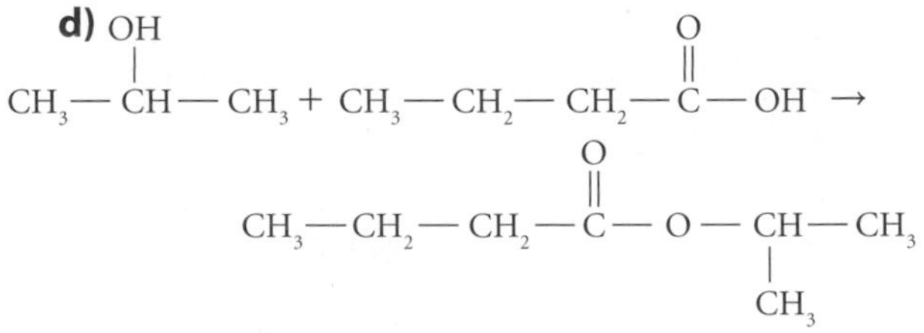

174. a) $BrHC=CBr-CH_2-CH_2F$
1,2-dibromo-4-fluorobut-1-ene

b) $CH_3-CHCl-CCl(CH_3)-CH_3$
2,3-dichloro-3-methylbutane (one of many possible dichloro substituted products)

c) $CH_3-C\equiv C-CH_2-CH_3$ pent-2-yne

175. a) addition   **b)** elimination
c) substitution   **d)** elimination

176. a) substitution; 2-chloro-3-methylbutane + water
b) substitution; 2-methylpropan-1-ol + I⁻
c) substitution; bromocyclohexane + water
d) elimination ; 3,4-dimethylcyclopentene
e) elimination; 4-methylpent-2-ene + Cl⁻ + H₂O

177. a) alkene   **b)** alkene
c) ester   **d)** alkyl halide
e) alkyl halide

178. a) $\ldots CH_2-CH-CH_2-CH-CH_2-CH\ldots$

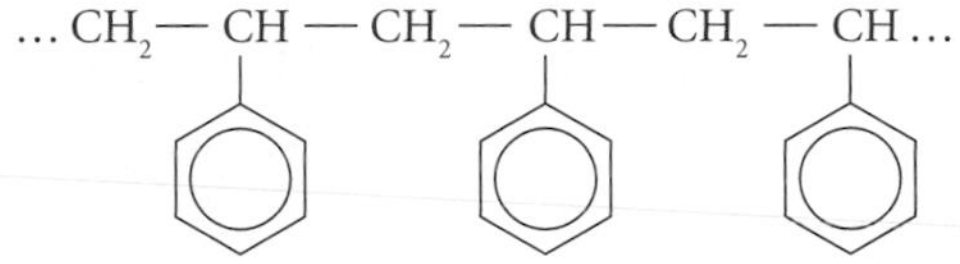

b) addition

179. a) $H_2N-CH_2-CH_2-NH_2\ +$

$$HO-\overset{O}{\underset{\|}{C}}-(CH_2)_3-\overset{O}{\underset{\|}{C}}-OH$$

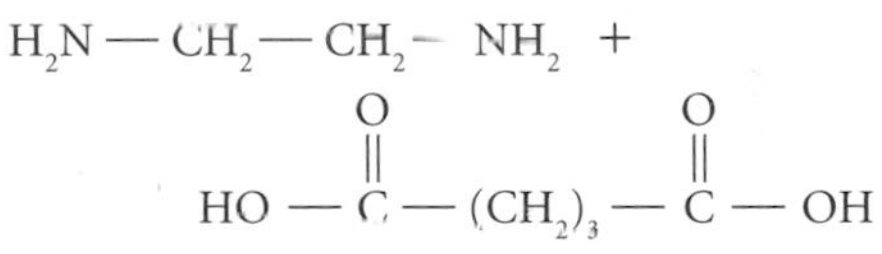

b) condensation

Unit 8: Chemical Equilibrium Focussing on Acid-Base Systems

Chapter 16: Chemical Equilibrium

180. 22.7 g

181. a) $\dfrac{[NO_2(g)]^2}{[N_2(g)][O_2(g)]^2}$

b) 1.94×10^{-3}

c) no change

d) decrease; equilibrium shifts to the right

e) increase; since the volume is smaller, the concentration increases (even though there is a shift to the right)

182. a) $K_c = \dfrac{[P_4O_{10}(g)]}{[P_4(g)][O_2(g)]^5}$

b) $K_c = [HF(g)]$

c) $K_c = \dfrac{1}{[NH_3(g)][HCl(g)]}$

183. a) decrease   **b)** no change
c) no change   **d)** no change
e) increase   **f)** decrease

184. 0.958

185. CO(g) = 0.993 mol; O₂(g) = 0.996 mol

186. $[SO_2Cl_2(g)] = 0.305$ mol/L; $[Cl_2(g)] = 1.19$ mol/L; $[SO_2(g)] = 0.665$ mol/L

187. 63.8 g

188. a) 0.22 g   **b)** no effect

189. left

190. a) left   **b)** right
 c) right   **d)** left

191. $[Fe^{3+}(aq)] = [SCN^-(aq)] = 0.037$ mol/L; $[FeSCN^{2+}(aq)] = 1.46$ mol/L

Chapter 17: Acid-Base Equilibrium Systems

192. a) $HF(aq)/ F^-(aq)$ and $NH_4^+(aq)/ NH_3(aq)$
 b) $Fe(H_2O)_6^{3+}(aq)/Fe(H_2O)_5(OH)^{2+}(aq)$ and $H_3O^+(aq)/ H_2O(\ell)$
 c) $NH_4^+(aq)/NH_3(aq)$ and $HCN(aq)/ CN^-(aq)$
 d) $H_2O(\ell)/OH^-(aq)$ and $(CH_3)_3NH^+(aq)/ (CH_3)_3N(aq)$

193. 2.07   **194.** 4.025

195. a) basic   **b)** 8.372

196. 2.7×10^{-5} mol/L; 0.10 mol/L

197. a) $N_3^-(aq)$ **b)** 3.57×10^{-10}
 c) 8.409

198. a) acid: $HPO_4^{2-}(aq) + H_2O(\ell) \rightleftharpoons H_3O^+(aq) + PO_4^{3-}(aq)$
 base: $HPO_4^{2-}(aq) + H_2O(\ell) \rightleftharpoons H_2PO_4^-(aq) + OH^-(aq)$
 b) basic. Since $K_b > K_a$ $HPO_4^{2-}(aq)$ has a greater tendency to gain a proton from water than to donate a proton to water.
 c) $H_2PO_4^-(aq)$ **d)** 6.2×10^{-8}

199. 8.17×10^{-3} mol/L

200. 1.8 g   **201.** 0.028%

202. $[Cl^-(aq)] = 0.1$ mol/L; $[H_3O^+(aq)] = 0.1$ mol/L (essentially all from HCl); $[CH_3COO^-(aq)] = 1.8 \times 10^{-5}$ mol/L

203. 11.276

204. a) 2.71
 b) equilibrium shifts left, decreasing $[H_3O^+(aq)]$ and pH increases

205. a) $HCl(aq)$
 b) 0.10 mol/L HCl, pH = 1.00; 0.20 mol/L HCl, pH = 0.70
 0.10 mol/L HCN, pH = 5.10; 0.20 mol/L HCN, pH = 4.95

Answers to Appendix D Practice Problems

Precision, Error, and Accuracy

1. The results are precise, but not accurate.

2. The improperly zeroed triple-beam balance resulted in systematic error.

3. A student carries out mass measurements in a room with fluctuating temperature and pressure.

4. A student carries out a titration, but consistently fails to rinse the burette with titrant after rinsing with water. Therefore the acid in the burette is always more dilute than its recorded concentration.

5. a) -0.05 s (assume last digit is estimated)
 b) The uncertainty is likely not sufficient, since human reflexes introduce error as well.

Significant Digits

1. a) 101.45 g   **b)** 2.5 mm
 c) 1.70 L   **d)** 3.07 mL
 e) 35.2 cm^2 **f)** 8.0 g/cm^3

Scientific Notation

1. a) 9.34×10^{-4} **b)** 7.983×10^9
 c) 8.2057×10^{-10} **d)** 4.96×10^8
 e) 6×10^{-4} **f)** 3.0972×10^{-4}

2. a) 9.2×10^2 **b)** 9.02×10^2
 c) 1.0053×10^1 **d)** 1.0×10^{14}
 e) 1.53×10^{-2}

Logarithms

1. a) 0   **b)** 0.69897
 c) 1   **d)** 1.69897
 e) 2   **f)** 2.69897
 g) 4.69897   **h)** 5

2. a) 1   **b)** 10
 c) 0.1   **d)** 100
 e) 0.01   **f)** 1000
 g) 0.001

3. a) $10^{-1} \times 100 = 10$
 b) $10^{-2} \times 10^4 = 100$
 c) $10^{-3} \times 10^6 = 1000$
 d) 3162   **e)** 0.0003162
 f) 3162
 g) The numbers 3.5 and -3.5 are 7 units apart, and their antilogarithms are 7 places of magnitude different.

4. a) $1.88 + 1.74$
 b) 3.62
 c) 4180

Unit Analysis

1. 1.10 g   **2.** 55 g/L
3. 3.7 kg   **4.** 93.6 g

Supplemental Practice Problems

Unit 1 The Diversity of Matter and Chemical Bonding

Chapter 1 Chemical Bonding

1. Write the formula for the following compounds.
   a) tin(IV) perbromate
   b) uranium dioxide
   c) zinc dihydrogen phosphate
   d) lithium chromate
   e) indium hydroxide

2. Name each of the following compounds:
   a) $Ra_3(PO_4)_2$
   b) HgS
   c) $(NH_4)_2SO_3$
   d) PbF_4
   e) Co_2S_3

3. Correct the error in each pairing shown below.
   a) titanium(IV) bromide, Ti_2Br_4
   b) lead(II) oxide, PbO_2
   c) nitrogen dioxide, N_2O_4
   d) calcium hydrogen phosphate, CaH_2PO_4
   e) mercury oxide, HgO

4. For each of the following ionic compounds, indicate the electric charge on the metal ion.
   a) Rb_3N
   b) ZrO_2
   c) AsF_3
   d) Cr_2O_3
   e) SrI_2

5. Atoms of a metallic element are known to lose 1 electron when an ionic bond is formed with a non-metal. Use the symbol T for the metal atom and write the formula of the following compounds of the element.
   a) oxide
   b) sulfate
   c) bromite
   d) cyanide

6. Draw a structural formula for the following molecular compounds.
   a) I_2
   b) CCl_2H_2
   c) NFH_2
   d) NOF

7. Draw electron dot diagrams to represent ions or atoms that have the same electron population as Al^{3+}.

8. Draw electron dot diagrams to represent the following ionic compounds.
   a) BaS
   b) Mg_3N_2
   c) LiI
   d) BaO
   e) Li_2O
   f) AlF_3

9. The electonegativities of oxygen and sulfur are respectively 3.4 and 2.6.
   a) What type of bond will form between these two atoms?
   b) What does this indicate about the electrons between the two atoms that are bonded?

10. Arrange the following bonds in order of decreasing polarity.
    a) Al P
    b) H S
    c) Cl F
    d) S C
    e) Cl H

11. Predict which atom in the following pairs will have the larger diameter. Explain your answers.
    a) H or He
    b) O or S

12. Compare the number of lone pair of electrons that are found around the central atom in molecules of CO_2 and SO_2.

13. Explain why two atoms of a group 17 element such as fluorine can combine but two atoms of a group 1 element such as Li do not combine.

Chapter 2 Diversity of Matter

14. Predict which ion in each of the following pairs has the smaller diameter. Explain your answers.
    a) K^+ or Na^+
    b) Cl^- or Cl^{2-}

15. Predict which substance will have the lower boiling point, $CHCl_3(\ell)$ or $CH_3Cl(\ell)$. Give a reason for your answer.

16. Use VSEPR theory to compare the molecular shape of Cl_2O and ClO_2^-.

17. a) Use VSEPR theory to predict the shapes of molecules of CH_3F and NH_2F.
    b) Compare the intermolecular bonding between molecules of each compound.
    c) Predict which compound will have the higher boiling point.

18. a) Draw a structural formula for a molecule of urea, $H_2NC(O)NH_2$.
    b) What type of intermolecular bonding is expected in a sample of urea?
    c) In what way, if any, would the shape and type of intermolecular bonding change if the molecule was $H_3CC(O)CH_3$?

19. a) Predict the type of intermolecular bonding that will be found in samples of the following elements: $S_8(\ell)$, $Hg(\ell)$, $O_2(\ell)$.
    b) List these elements in order of increasing boiling point.

20. Draw Lewis structures for the following molecules
    a) SbF_3
    b) SO_2F_2
    c) ClNO

21. Use VSEPR theory to determine the shape of each molecule in question 20.

22. The following information is known about a compound: a central atom from group 15 is bonded to three identical atoms with bonds that are classed as polar covalent. Is the molecule polar or non-polar. Give a reason for your answer.

23. Explain why the melting points of the sodium halides (NaCl, NaBr, NaI) decrease as the molar mass increases whereas the melting points of the hydrogen halides (HCl, HBr, HI) increase as the molar mass increases.

24. Use VSEPR theory to determine the molecular shape of $BeCl_4^{2-}$ and $SbCl_4^+$.

25. Use VSEPR theory to compare the shape of the phosphate ion with the phospite ion.

Unit 2 Forms of Matter

Chapter 3 Properties of Gases

26. A balloon contains 1.00 L of an ideal gas. List ways that the volume can be increased, changing only one variable at a time. Indicate what factors are held constant for each change.

27. Does the volume of a fixed amount of gas at constant pressure change differently if the temperature increases by 10 °C compared with 10K degrees? Give a reason for your answer.

28. Charles's law applies to ideal gases. At high pressures the molecules are close together and there are significant intermolecular attractions between molecules. Will an increase in temperature result in less, more or the same volume change expected ideally? Explain your reasoning.

29. The gas in a large balloon occupies 25.00 L at ambient pressure of 100 kPa. If the temperature remains constant, what will be the volume of the balloon at 500 kPa?

30. If the volume of a fixed amount of gas is tripled at constant temperature, by what factor will the pressure change?

31. To what temperature must an ideal gas originally at 27.00 °C be cooled to reduce its volume by one third?

32. A Torricelli type barometer is constructed using water instead of mercury. The density of mercury is 13.6 times the density of water. If the barometric pressure was measured to be 765 mm Hg, what will be the barometric pressure in mm H_2O?

33. By what factor must the pressure change if it is to have the same effect on a fixed amount of gas in a rigid container as quadrupling the Kelvin temperature?

34. A balloon is filled with 5.00 L of helium and sealed at 24.0 °C on a day when the barometric pressure is 102.4 kPa. The next day the temperature is again 24.0 °C, but the volume of the balloon is now changed to 4.00 L. What is the pressure on the balloon on the second day?

35. What volume of gas can be released at 98.0 kPa pressure from a cylinder in which 200.0 L of gas is stored at 250 kPa pressure if there is no change in temperature?

36. An aerosol can having a fixed volume of 0.400 L is heated from 20.00 °C to 86.40 °C. If the pressure inside the can was originally 140 kPa, what will be the pressure after heating the can?

Chapter 4 Exploring Gas Laws

37. A 5.00 L sample of gas at 18.00 °C and standard pressure is heated until the volume becomes 6.50 L and the pressure changes to 110.0 kPa. To what temperature was the gas heated?

38. What is the density of chlorine gas at 15.00 °C and 90.0 kPa pressure?

39. At constant volume and temperature, 1 mole of any gas will exert the same pressure. Why do heavier molecules not exert more pressure in a container than lighter molecules?

40. When compared at the same temperature and pressure, which is less dense, moist air or dry air?

41. A gas sample is made up of 5.00 g each of oxygen, hydrogen and carbon dioxide. The total pressure exerted by the mixture of gases is 115.0 kPa. What is the partial pressure of the carbon dioxide gas?

42. 30.00 mL of hydrogen is collected over water at 20.0 °C and 98.6 kPa pressure. The vapour pressure of water at 20.0 °C is 2.34 kPa. What will be the volume of the dry gas at SATP?

43. What is the molar mass of a gas if data is obtained that shows 0.1967 mg of the gas occupy 1.861 mL at 60.00 kPa and 0 °C?

44. What volume will 80.62 g of carbon dioxide occupy at 38.00 °C and 108.6 kPa pressure?

45. A sample of gas in a closed container of constant volume is heated until the Kelvin temperature doubles. Which of the following properties will double?
a) density
b) average kinetic energy of the molecules
c) number of molecules
d) average speed of the molecules
e) pressure

46. Two containers of gas are at the same temperature and pressure. One contains 8.0 g of oxygen gas and the other 0.50 g of hydrogen gas. What is the same regarding these two gas samples?

47. A container holds one mole of neon gas at a certain temperature and pressure. A second, identical container holds nitrogen gas at three times the pressure and twice the Kelvin temperature. How many moles of nitrogen are in the second container?

48. The volume of an automobile tire does not change appreciably when the car is driven. Before starting on a trip, a tire contains air at 220 kPa and 20.0 °C. After driving for an hour, the tire and the air in it become warmer and the pressure increases to 240 kPa. What is the temperature of the air inside the tire?

49. The total pressure exerted by a mixture of gases is 150.0 kPa. In the mixture, 75.0% of the pressure is due to $N_2(g)$ and 12.0% is due to $O_2(g)$. Some water vapour is present that accounts for 2.4 kPa of pressure. The remaining gas is an unknown amount of $CO_2(g)$. Calculate the pressure exerted by the $CO_2(g)$.

50. A cylinder having a volume of 50.0 L contains 100 g of helium at 14.0 °C. What volume will the helium occupy if it released at 30.0 °C and 103.2 kPa?

51. At its boiling point of −183 °C, liquid oxygen is 1.14 times as dense as water (density = 1.00 g/mL). 10.0 mL of liquid oxygen is warmed from its boiling point to STP. What volume of oxygen gas will be present?

Unit 3 Matter as Solutions, Acids, and Bases

Chapter 5 Solutions

52. A saturated solution of sodium nitrate at 10 °C contains 80.0 g $NaNO_3(s)$ in 100 g H_2O. A student reports that this is an 80.0% m/m solution. Comment on the accuracy of this statement.

53. Using the information given in question #1, can the concentration of the solution be determined in moles/L?

54. What mass of $NaOH(s)$ is present in 112 mL of 0.00126 mol/L NaOH(aq)?

55. A saturated solution of salt is left at room temperature to evaporate to dryness. Assuming that the temperature does not change, how will the concentration of solute change during the evaporation process?

56. Blood serum contains 3.4 g Na^+(aq)/L. What is the molar concentration of Na^+(aq)?

57. What is the mass percent concentration of nicotine in the body of a 70 kg person who smokes a pack of cigarettes (20 cigarettes) in one day? Assume that there is 1.0 mg of nicotine per cigarette, and that all the nicotine is absorbed into the person's body.

58. A fertilizer contains 3.46 g $NH_4NO_3(s)$ and 2.02 g $(NH_4)_3PO_4(s)$ in 10.0 L of solution. What is the molar concentration of NH_4^+(aq)?

59. What is the final concentration of Na^+(aq) in a solution made by mixing 50.0 mL 0.200 mol/L NaCl(aq) with 150.0 mL 0.100 mol/L NaCl(aq)?

60. What volume of 4.00×10^{-2} mol/L calcium nitrate solution will contain 5.00×10^{-2} mol of $NO_3^-(aq)$ ions?

61. What is the concentration of a sulfuric acid solution after 80.0 mL of 4.00 mol/L $H_2SO_4(aq)$ is diluted with water to 400.0 mL?

62. What mass of $Ca^{2+}(aq)$ has been ingested into your body when you drink 5.0 L of bottled water labeled as having "98 ppm Ca"? 1.0 L of water has a mass of 1.00 kg.

63. A dilute solution of KCl(aq) has a concentration of 0.002 28 mol/L. What is the concentration of this solution expressed as mass percent concentration? (for dilute solutions, assume that the density of the solution is 1g/mL)

64. Describe how to prepare 2.00 L of 1.4 mol/L $NaNO_3(aq)$.

65. A concentrated solution of hydrochloric acid is 37.2% HCl(aq) by mass and has a density of 1.137 g/mL. 15.0 mL of this acid is diluted to 500.0 mL with water. What is the concentration of the diluted solution in mol/L?

Chapter 6 Acids and Bases

66. Calculate the pH of each of the following solutions.
a) 5.0 g LiOH(s) in 3.62 L of solution
b) 0.0642 g $Ba(OH)_2$(s) in 500 mL of solution
c) 0.492 mol/L HCl(aq)
d) 0.0040 mol/L HNO_3(aq)

67. What is the molar concentration of $OH^-(aq)$ in each of the following solutions?
a) 1.00 g $Ca(OH)_2$(s) in 5.0 L of solution
b) 100 mL of solution of pOH = 4.214
c) a solution having a pH = 6.0

68. What is the pH of 1.0×10^{-5} mol/L $Ca(OH)_2$(aq)?

69. 2.0×10^{-4} mol of HCl(g) is dissolved in water to make 300 mL of solution. What are the molar concentrations of $H_3O^+(aq)$ and $OH^-(aq)$?

70. Calculate the $[H_3O^+(aq)]$ given the following:
a) pH = 8.60
b) pOH = 13.86

71. What mass of KOH(s) must be dissolved in 250.0 mL of solution to make the pH = 10.000?

72. What is the $[H_3O^+(aq)]$ in acid solution of pH = 5.1629?

73. a) Is 0.10 mol/L aqueous hydrogen chloride a stronger acid solution than a 0.010 mol/L solution? Give a reason for your answer?
b) Why does 0.010 mol/L HCl(aq) react more slowly with Mg(s)?

74. Consider equal volumes of 0.10 mol/L HCl(aq) of pH = 1.00 and of 0.10 mol/L CH_3COOH(aq) of pH = 2.37. Which of the following statements are true?
a) HCl(aq) is more concentrated than CH_3COOH(aq)
b) CH_3COOH(aq) is less ionized than HCl(aq)
c) HCl(aq) is a stronger acid than CH_3COOH(aq)
d) HCl(aq) will react more quickly with Fe(s)
e) There are more H_3O^+(aq) ions in the sample of CH_3COOH(aq)

75. 500 mL of HCl(aq) of pH = 2.00 is mixed with NaOH(s) until half of the acid has been neutralized. Assuming that there is no change in total volume, what is the pH of the resulting solution?

76. Classify the following as either an Arrhenius acid or base. Write equations to illustrate your answer.
a) BaO(s)
b) SO_3(g)
c) K_2CO_3(s)
d) P_4O_{10}(s)

77. a) How much more acidic is a solution of pH = 3 compared to pH = 5?
b) How much more acidic is a solution of pH = 3.5 compared to pH = 5.5?

78. What is the pH of the solution that results from mixing equal volumes of HCl(aq) of pH = 2.0 and pH = 4.0?

Unit 4 Quantitative Relationships in Chemical Changes

Chapter 7 Stoichiometry

79. For each pair of aqueous solutions that are combined, write a total ionic equation. If a net reaction occurs, write the net ionic equation and list the spectator ions. Refer to Table P3.1 page 161 for solubility data.
a) $Pb(NO_3)_2$(aq) + $(NH_4)_2CO_3$(aq)
b) $Hg(NO_3)_2$(aq) + CH_3COONa(aq)
c) $BaCl_2$(aq) + $Al_2(SO_4)_3$(aq)
d) $Fe(NO_3)_3$(aq) + Na_2S(aq)

80. Traces of bleach, sodium hypochlorite, can be removed from a sample by a reaction with 3% hydrogen peroxide solution. What volume of $H_2O_2(\ell)$ is needed to react completely with 9.045 g of sodium hypochlorite? Assume that the density of $H_2O_2(\ell)$ is 1.000 g/mL.

$$NaClO(aq) + H_2O_2(\ell) \rightarrow O_2(g) + H_2O(\ell) + NaCl(aq)$$

81. What mass of antimony(III) sulfide is required to react completely with 28.4 g of potassium chlorate?

$$Sb_2S_3(s) + 3KClO_3(s) \rightarrow Sb_2O_3(s) + 3KCl(s) + 3SO_2(g) + heat$$

82. What volume of 0.500 mol/L $H_2SO_4(aq)$ will react completely with 45.0 g of magnesium metal?

$$Mg(s) + H_2SO_4(aq) \rightarrow MgSO_4(aq) + H_2(g)$$

83. A simplified equation for the combustion of wood is shown below.

$$C_6H_{10}O_5(s) + 6O_2(g) \rightarrow 6CO_2(g) + 5H_2O(g) + energy$$

What volume of $CO_2(g)$ at 35.00 °C and 102.9 kPa pressure will be added to the atmosphere when a 500.0 kg of wood burn according to this equation?

84. What mass of iron can be replaced from 1.55 L of 0.4208 mol/L $Fe(NO_3)_3(aq)$ by reacting with an excess of magnesium metal?

85. What volume of $NO(g)$ can be produced when 450.0 L of $NO_2(g)$ react at 22.15 °C and 500 kPa pressure according to the reaction $3NO_2(g) + H_2O(\ell) \rightarrow 2HNO_3(aq) + NO(g)$?

86. A disposable butane lighter contains 5.00 mL of liquid butane ($C_4H_{10}(\ell)$). The density of liquid butane is 0.579 g/mL. What volume of air that is 21.2% oxygen by volume will be required to have all of the butane in a lighter undergo complete combustion to $CO_2(g)$ and $H_2O(g)$ at SATP?

87. An excess of $H_2S(g)$ is bubbled through 100.0 mL of 0.2200 mol/L $AgNO_3(aq)$
   a) Write the balanced equation for the reaction that occurs.
   b) Classify this type of reaction.
   c) What mass of precipitate is produced?

88. What was the concentration of a silver nitrate solution if 4.74 g of $Ag(s)$ is produced when 53.62 mL of $AgNO_3(aq)$ react as shown below?

$$2AgNO_3(aq) + H_2O_2(aq) \rightarrow 2Ag(s) + 2HNO_3(aq) + O_2(g)$$

89. 57.26 mL of hydrogen gas at 30.00 °C and 102.00 kPa pressure is produced when a sample of zinc reacts with dilute sulfuric acid. What mass of zinc reacted?

$$Zn(s) + H_2SO_4(aq) \rightarrow ZnSO_4(aq) + H_2(g)$$

90. What volume of $H_2(g)$ at 50.0 °C and 105.8 kPa pressure will react completely with 315.0 mL of 0.241 mol/L $FeCl_3(aq)$?

$$2FeCl_3(aq) + H_2(g) \rightarrow 2FeCl_2(aq) + 2HCl(aq)$$

91. What is the concentration of $HNO_3(aq)$ if 50.8 mL of 0.0800 mol/L $NaOH(aq)$ neutralize 30.0 mL of the acid?

Chapter 8 Applications of Stoichiometry

92. A reaction that can be used to launch a small rocket is

$$3Al(s) + 3NH_4ClO_4(s) \rightarrow Al_2O_3(s) + AlCl_3(s) + 3NO(g) + 6H_2O(g)$$

   a) What is the theoretical yield of $Al_2O_3(s)$ when 45.0 g of $Al(s)$ react completely with excess $NH_4ClO_4(s)$?
   b) Determine the percentage yield if 51.2 g of $Al_2O_3(s)$ is produced.

93. Pure liquid hydrogen peroxide has a density of 1.46 g/mL. What volume of oxygen gas measured at 40.0 °C and 98.42 kPa pressure is produced when 30.0 mL of $H_2O_2(\ell)$ decomposes if the reaction has an 86.5% yield?

$$2H_2O_2(\ell) \rightarrow 2H_2O(\ell) + O_2(g)$$

94. What mass of $PbI_2(s)$ is precipitated when 50 mL of 0.200 mol/L KI is mixed with 65.0 mL of 0.140 mol/L $Pb(NO_3)_2(aq)$?

95. What is the theoretical yield of SO_3 when 40.0 mL of $SO_2(g)$ and 40.0 mL of O_2 react? All gases are at the same temperature and pressure.

96. The following reaction occurs when a "strike-anywhere" match burns.

$$P_4(s) + 3S(s) + 2KClO_3(s) + 5O_2 \rightarrow P_4O_{10}(s) + 3SO_2(g) + 2KCl(s) + heat$$

Determine the limiting reagent when 0.600 g of $P_4(s)$, 1.58 g S(s) and 1.187 g $KClO_3(s)$ react with sufficient oxygen gas.

97. 40.0 mL of $KOH(aq)$ of pH = 13.400 is titrated with 24.5 mL of 0.15 mol/L $HCl(aq)$. What percentage of the $KOH(aq)$ has been neutralized at this point in the titration?

98. In the reaction shown below, 100.0 mL of 0.200 mol/L $HBrO_3(aq)$ is reacted with excess $SO_2(g)$.

Calculate the percentage yield if 3.55 g of $H_2SO_4(aq)$ is produced.

$$2HBrO_3(aq) + 5SO_2(g) + 4H_2O(\ell)$$
$$\rightarrow Br_2(g) + 5H_2SO_4(aq)$$

99. Assuming a percentage yield of 80.0 %, what volume of oxygen is required to use up 160.0 mL of $C_2H_4(g)$ if all gases are measured at 50.0 °C and standard pressure?

$$C_2H_4(g) + 3O_2(g) \rightarrow 2CO_2(g) + 2H_2O(g)$$

100. What is the percentage purity of a sample of sodium azide, $NaN_3(s)$, if 5.00 L of $N_2(g)$ is released at 25.00 °C and 103.71 kPa pressure when a 10.00 g sample of this solid decomposes as shown in the equation below?

$$2NaN_3(s) \rightarrow 2Na(s) + 3N_2(g)$$

101. Determine if all of the $Pb^{2+}(aq)$ in 50.0 mL of 0.350 mol/L $Pb(NO_3)_2(aq)$ is precipitated when mixed with 70.0 mL of 0.500 mol/L $NaBr(aq)$.

102. What volume of acetylene, $C_2H_2(g)$, can be produced at 20.0 °C and 102.4 kPa pressure when 5.00 g of 95.0% pure $CaC_2(s)$ reacts with water?

$$CaC_2(s) + 2H_2O(\ell) \rightarrow Ca(OH)_2(aq) + C_2H_2(g)$$

103. What percentage of a 2.50 g sample of pure $CaCO_3(s)$ will react with 50.0 mL of 0.400 mol/L $HCl(aq)$?

$$CaCO_3(s) + 2HCl(aq)$$
$$\rightarrow CaCl_2(aq) + CO_2(g) + H_2O(\ell)$$

Unit 5 Thermochemical Changes

Chapter 9 Energy and Chemical Reactions

104. What is the enthalpy change for the complete combustion of 1.00 L of octane, $C_8H_{18}(\ell)$. The density of octane is 0.918 g/mL.

105. What volume of methane, $CH_4(g)$, measured at 35.1 °C and 99.86 kPa, must be burned to provide enough heat energy to warm 100.0 mL of water by 10.0 °C? Assume no loss of heat to the surroundings and the density of water is 1.00 g/mL.

106. The heat of combustion, Δ_cH, for methylbenzene $(C_7H_8(\ell))$ has been measured as −3904 kJ/mol.

a) Write the thermochemical equation for this combustion.

b) What mass of methylbenzene must be burned in an open system to release 1.00 MJ of heat energy?

107. An impure sample of zinc has a mass of 7.35 g. The sample reacts with 150.0 g of dilute hydrochloric acid solution inside a calorimeter.

The calorimeter has a mass of 520.57 g and a specific heat capacity of 0.400 J/g °C.

$$Zn(s) + 2HCl(aq)$$
$$\rightarrow ZnCl_2(aq) + H_2(g) + 153.9 \text{ kJ}$$

When the reaction occurs, the temperature of the solution rises from 14.5 °C to 29.7 °C. What is the percentage purity of the sample? Assume that the specific heat capacity of hydrochloric acid solution is 4.19 J/g °C and that all of the zinc in the impure sample reacts.

108. A solid is dissolved in water to determine its heat of solution. The following data was recorded:

mass of solid = 5.50 g

mass of water in calorimeter = 120.0 g

initial temperature of water = 21.7 °C

final temperature of solution = 32.6 °C

a) Calculate the heat gained by the surroundings.

b) Calculate the heat lost by the system.

c) Calculate the heat of solution, $\Delta_{soln}H$ per g of solid.

109. A 100.0 g sample of food is placed in a bomb calorimeter calibrated at 7.23 kJ/°C. When the food is burned, the calorimeter gains 512 kJ of heat. If the initial temperature of the calorimeter was 19.0 °C, what is the final temperature of the calorimeter and its contents?

110. For the reaction $C(s) + PbO(s) \rightarrow Pb(s) + CO(g)$, Δ_rH is +106.9 kJ/mol. How much heat is needed to convert 50.0 g of $PbO(s)$ to $Pb(s)$?

111. A 0.160 g sample of ethanol is burned and the heat released is absorbed by 75.42 g of water. The temperature of the water rises from 26.9 °C to 41.2 °C. Use this information to calculate the heat of combustion of ethanol, Δ_cH. Assume all the heat was absorbed by the water.

112. A sample of NH_4NO_3 having a mass of 30.5 g is dissolved in water in an insulated cup to make 500.0 mL of solution.

$$NH_4NO_3(s) + 25.7 \text{ kJ} \rightarrow NH_4^+(aq) + NO_3^-(aq)$$

a) What is the heat of solution, $\Delta_{soln}H$?

b) Will the temperature increase or decrease during the dissolving process?

c) If the initial temperature of the water is 21.5 °C, determine the final temperature after dissolving is complete. Assume no loss of heat to the surroundings, the density of the solution is equal to 1.00 g/mL and the specific heat capacity of the solution is 4.19 J/g°C.

113. 500 g each of $NH_3(g)$ and $CH_4(g)$ is burned in an excess of oxygen to produce hydrogen cyanide.
$$2NH_3(g) + 3O_2(g) + 2CH_4(g)$$
$$\rightarrow 2HCN(g) + 6H_2O(g) + 939 \text{ kJ}$$
How much heat is given off in this reaction?

114. A calorimeter and its contents have a heat capacity of 18.14 kJ/°C. A 13.91 g sample of ethanoic acid, $CH_3COOH(\ell)$, is burned in an excess of oxygen in this calorimeter. The temperature was observed to change from 20.00 °C to 31.04 °C.

   a) How much heat was gained by the calorimeter?
   b) Calculate the heat of combustion, $\Delta_c H$ for ethanoic acid.
   c) Write the thermochemical equation for this combustion reaction.

115. 1.0 L of 6.00 mol/L of sodium hydroxide solution is prepared by adding $NaOH(s)$ to water initially at 22.00 °C. Given that the heat of solution for NaOH is −41.86 kJ/mol, what will be the final temperature when dissolving is complete? Assume that the density of the solution is 1.22 g/mL, the specific heat capacity of the solution is 4.19 J/g°C and that no heat is lost to the surroundings.

Chapter 10 Theories of Energy and Chemical Changes

116. 8.025 g of methane, $CH_4(g)$, is burned to heat water in a beaker. The methane undergoes complete combustion as shown in the following equation.
$$CH_4(g) + 2O_2(g)$$
$$\rightarrow CO_2(g) + 2H_2O(\ell) + 802.5 \text{ kJ}$$
Use the data that follows to determine the efficiency of this heating.
mass of water = 1.642 kg
mass of beaker = 0.3413 kg
specific heat capacity of water = 4.19 kJ/kg°C
specific heat capacity of glass (beaker) = 0.811 kJ/kg°C
initial temperature of water in beaker = 5.57 °C
final temperature of water in beaker = 47.21 °C

117. Calculate the enthalpy of formation of manganese(IV) oxide, based on the following information:
$$4Al(s) + 3MnO_2(s)$$
$$\rightarrow 3Mn(s) + 2Al_2O_3(s) + 1790 \text{ kJ}$$
$$2Al(s) + \frac{3}{2}O_2(g) \rightarrow Al_2O_3(s) + 1676 \text{ kJ}$$

118. Use the information listed below to determine ΔH for the reaction
$$Ca^{2+}(aq) + 2OH^-(aq) + CO_2(g)$$
$$\rightarrow CaCO_3(s) + H_2O(\ell)$$
$$CaO(s) + H_2O(\ell) \rightarrow Ca(OH)_2(s)$$
$$\Delta H = -65.2 \text{ kJ/mol}$$
$$CaCO_3(s) \rightarrow CaO(s) + CO_2(g)$$
$$\Delta H = +178.1 \text{ kJ/mol}$$
$$Ca(OH)_2(s) \rightarrow Ca^{2+}(aq) + 2OH^-(aq)$$
$$\Delta H = -16.2 \text{ kJ/mol}$$

119. Use the enthalpies of combustion for the burning of $CO(g)$, $H_2(g)$ and $C(s)$ to determine $\Delta_r H$ for the reaction $C(s) + H_2O(g) \rightarrow H_2(g) + CO(g)$.
$$CO(g) + \frac{1}{2}O_2(g) \rightarrow CO_2(g)$$
$$\Delta_c H^\circ = -283.0 \text{ kJ/mol}$$
$$H_2(g) + \frac{1}{2}O_2(g) \rightarrow H_2O(g)$$
$$\Delta_c H^\circ = -241.8 \text{ kJ/mol}$$
$$C(s) + O_2(g) \rightarrow CO_2(g)$$
$$\Delta_c H^\circ = -393.5 \text{ kJ/mol}$$

120. The enthalpy of combustion for ethene, $C_2H_4(g)$, is −337 kJ/mol
   a) Write the balanced thermochemical equation for the complete combustion of ethene in air.
   b) The heat that is produced from burning 1.00 kg of ethene warms a quantity of water from 15.0 °C to 85.0 °C. What is the mass of the water if the heat transfer is 60.0% efficient?

121. In a commercial process to prepare tungsten, hydrogen is used to reduce tungsten trioxide. The heat of formation of $WO_3(s)$ is −839.5 kJ/mol. Calculate the heat of reaction per mole of tungsten for this process.
$$3H_2(g) + WO_3(s) \rightarrow W(s) + 3H_2O(g)$$

122. What information can be concluded about the relative bond strengths in molecules of HI and HCl from the thermochemical equations given below?
$$\frac{1}{2}H_2(g) + \frac{1}{2}I_2(g) + 26.5 \text{ kJ} \rightarrow HI(g)$$
$$\frac{1}{2}H_2(g) + \frac{1}{2}Cl_2(g) \rightarrow HCl(g) + 92.3 \text{ kJ}$$

123. The heat of combustion for butanoic acid, $C_3H_7COOH(\ell)$, is −2184 kJ/mol.
   a) Write the thermochemical equation for the combustion of butanoic acid.
   b) Use the information in the Chemistry Data Booklet to calculate the heat of formation of butanoic acid.
   c) A 15.00 g sample of butanoic acid that is 98.50% pure undergoes complete combustion in a bomb calorimeter. How much heat is given off?

124. Ethyne can be prepared from the reaction of calcium carbide and water.

$$CaC_2(s) + 2H_2O(\ell) \rightarrow C_2H_2(g) + Ca(OH)_2(s)$$
$$\Delta H_r = -128.0 \text{ kJ}$$

If 3.91 g of $CaC_2(s)$ that is 85.2% pure is added to water, how much heat is given off?

125. Experimental evidence shows that 187.7 J of energy is required to vaporize 1.000 g of $Br_2(\ell)$. Write the thermochemical equation for the condensation of $Br_2(g)$.

Chapter 11 Activation Energy and Catalysts

126. For the reaction $CO(g) + NO_2(g)$ $\rightarrow CO_2(g) + NO(g)$ that occurs at 330 °C, $E_{a(fwd)} = 134$ kJ and $\Delta_r H = -226$ kJ/mol
  a) What is $E_{a(rev)}$ for this reaction?
  b) What will be the effect on $E_{a(fwd)}$ if the temperature is increased to 500 °C? Give a reason for your answer.

127. Will an increase in temperature have an equal effect on the rate of the reactions shown below? Explain your reasoning.
  a) $Cl \cdot + Cl \cdot \rightarrow Cl_2$
  b) $I^- + CH_3 - Br \rightarrow CH_3 - I + Br^-$

128. A reaction is exothermic in the forward direction with $E_{a(fwd)} = 20.0$ kJ and $\Delta_r H = -400.0$ kJ/mol. A catalyst is found for the reaction and $E_{a(fwd)}$ becomes 4.50 kJ.
  a) What is $E_{a(rev)}$ for the uncatalyzed reaction?
  b) What is $E_{a(rev)}$ for the catalyzed reaction?
  c) What is $\Delta_r H°$ for the catalyzed reaction?

129. Use collision theory to predict which reaction is most likely to be affected by an increase in temperature of 10 °C.
  a) $A(s) + B(s) \rightarrow AB(s)$
  b) $A(aq) + B(aq) \rightarrow AB(s)$
  c) $A(g) + B(g) \rightarrow AB(g)$

130. a) Is the activation energy dependant upon the average kinetic energy of the reacting particles?
  b) Wood chopped into small pieces of kindling ignites more quickly than a larger piece of wood. How do the activation energies compare for the burning of both samples of wood?

131. a) For a reaction that is exothermic in the forward direction, can the activation energy, $E_{a(fwd)}$, ever be less than the enthalpy change?
  b) For a reaction that is endothermic in the forward direction, can the activation energy $E_{a(fwd)}$, ever be less than the enthalpy change?

132. Enzymes generally make reactions proceed much faster than they would in the absence of an enzyme. However, there are other molecules called enzyme inhibitors that can reduce the speed of an enzyme catalyzed reaction. Suggest a mechanism that would explain how an inhibitor could slow a reaction when the enzyme is present.

133. Is a spark more likely to activate an exothermic reaction or an endothermic reaction? Give a reason for your answer.

134. When a catalyst is used for a certain exothermic reaction, the rate of the reaction doubles. Which of the following statements are true?
  a) The $E_{a(fwd)}$ for the catalyzed reaction is less than $E_{a(fwd)}$ for the uncatalyzed reaction.
  b) The $\Delta_r H$ for the catalyzed reaction is less than for the uncatalyzed reaction.
  c) The catalyst lowers the activation energy for the forward reaction only.
  d) If more of the catalyst was used, the rate of reaction would increase even more.
  e) The catalyzed reaction goes through more transition states than the uncatalyzed reaction.

135. Examine the potential energy diagram shown below and answer the questions that follow.

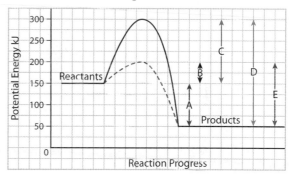

  a) Which letter represents $E_{a(rev)}$ for the catalyzed reaction?
  b) What is $\Delta_r H$?
  c) What is $E_{a(rev)}$ for the uncatalyzed reaction?
  d) Which letter represents $E_{a(fwd)}$ for the uncatalyzed reaction?
  e) What does the quantity $[C-B]$ represent?

Unit 6 Electrochemical Changes

Chapter 12 Oxidation-Reduction Reactions

136. Identify the redox-reactions.
  a) $CCl_4(g) + HF(g) \rightarrow CFCl_3(g) + HCl(g)$
  b) $Al_2O_3(s) + 3H_2SO_4(aq) \rightarrow Al_2(SO_4)_3(aq) + 3H_2O(\ell)$

c) $CH_4(g) + 2O_2(g) \rightarrow CO_2(g) + 2H_2O(g)$
d) $P_4(s) + 3OH^-(aq) + 3H_2O(\ell) \rightarrow PH_3(s) +$
$3H_2PO_2^-$

137. Consider the following reaction:
$S_8(s) + 8Na_2SO_3(aq) \rightarrow 8Na_2S_2O_3(aq)$
a) Assign oxidation numbers to all the elements.
b) Identify the reactant that undergoes reduction.
c) Identify the reducing agent.

138. a) Assign the oxidation number to sulfur in each
of the following: $HS^-(aq)$, $S_4(\ell)$, $SO_3^{2-}(aq)$,
$S_2O_3^{2-}(aq)$, $S_4O_6^{2-}(aq)$
b) Assign the oxidation number to boron in each
of the following: $B_4O_7^{2-}(aq)$, $BO_3^-(aq)$,
$BO_2^-(aq)$, $B_2H_6(g)$, $B_2O_3(s)$

139. List the following oxides of nitrogen in order of
decreasing oxidation number of nitrogen: $NO_2(g)$,
$N_2O_5(g)$, $NO(g)$, $N_2O_3(g)$, $N_2O(g)$, $N_2O_4(g)$

140. Identify the oxidizing agent and the reducing
agent in the following reaction:
$3SF_4(g) + 4BCl_3(g) \rightarrow 4BF_3(g) + 3SCl_2(g) +$
$3Cl_2(g)$

141. The metals $Ga(s)$, $In(s)$, $Mn(s)$, and $Np(s)$ and
their salts react as shown in the equations that
follow.
$3Mn^{2+}(aq) + 2Np(s) \rightarrow 3Mn(s) + 2Np^{3+}(aq)$
$In^{3+}(aq) + Ga(s) \rightarrow In(s) + Ga^{3+}(aq)$
$Mn^{2+}(aq) + Ga(s) \rightarrow$ no reaction
Use this information to list the reducing agents
from weakest to strongest.

142. Balance the following equations that occur in
basic solution using the oxidation number
method.
a) $Ti^{3+}(aq) + RuCl_5^{2-}(aq) \rightarrow Ru(s) + TiO^{2+}(aq)$
$+ Cl^-(aq)$
b) $ClO_2(g) \rightarrow ClO_2^-(aq) + ClO_3^-(aq)$

143. Balance the following redox reactions using the
half reaction method.
a) $ClO_3^-(aq) + I_2(s) \rightarrow IO_3^-(aq) + Cl^-(aq)$
(acidic)
b) $Cu(s) + SO_4^{2-}(aq) \rightarrow Cu^{2+}(aq) + SO_2(g)$
(acidic)
c) $Al(s) + NO_3^-(aq) \rightarrow Al(OH)_4^-(aq) + NH_3(g)$
(basic)
d) $C_2H_4(g) + MnO_4^-(aq) \rightarrow CO_2(g) + Mn^{2+}(aq)$
(acidic)
e) $NO_2(g) \rightarrow NO_2^-(aq) + NO_3^-(aq)$ (basic)
f) $CrO_4^- + HSnO_2^-(aq) \rightarrow CrO_2^-(aq) +$
$HSnO_3^-(aq)$ (basic)

144. Balance the following redox reactions:
a) $P_4(s) + NO_3^-(aq) \rightarrow H_3PO_4(aq) + NO(g)$
(acidic)
b) $MnO_2(s) + NO_2^-(aq) \rightarrow NO_3^-(aq) +$
$Mn^{2+}(aq)$ (acidic)
c) $TeO_3^{2-}(aq) + N_2O_4(g) \rightarrow Te(s) + NO_3^-(aq)$
(basic)
d) $MnO_4^-(aq) + N_2H_4(g) \rightarrow N_2(g) + MnO_2(s)$
(basic)
e) $S_2O_3^{2-}(aq) + OCl^- \rightarrow SO_4^{2-}(aq) + Cl^-(aq)$
(basic)
f) $Br_2(\ell) + SO_2(g) \rightarrow Br^-(aq) + SO_4^{2-}(aq)$
(acidic)
g) $PbO_2(s) + Cl^-(aq) \rightarrow PbCl_2(aq) + Cl_2(g)$
(acidic)

145. The reaction shown below occurs in acid solution.
Balance this equation and calculate the mass of
methanol that will react completely with 3.33 L of
0.150 mol/L $MnO_4^-(aq)$.
$CH_3OH(aq) + MnO_4^-(aq) \rightarrow HCOOH(aq) +$
$Mn^{2+}(aq)$.

146. What volume of 0.3160 mol/L $MnO_4^-(aq)$ will
react completely with a 1.000 g sample that is
90.00% $Mn(OH)_2(s)$? The unbalanced reaction
shown below takes place in basic solution.
$Mn(OH)_2(s) + MnO_4^-(aq) \rightarrow MnO_4^{2-}(aq)$

Chapter 13 Cells and Batteries

147. Consider the electrochemical cell represented as
$Al(s) | Al^{3+}(aq) \| Ce^{4+}(aq) | Ce^{3+}(aq) | Pt$.
a) At which electrode will reduction occur?
b) In which direction will electrons flow?
c) What is the oxidizing agent?
d) Will the aluminium electrode increase or
decrease in mass?

148. Given the list of metals and aqueous solutions of
their salts, determine which pair will give the
greatest cell voltage in a voltaic cell.
$Ni(s) | Ni^{2+}(aq)$
$Cu(s) | Cu^{2+}(aq)$
$Fe(s) | Fe^{2+}(aq)$
$Ag(s) | Ag^+(aq)$

149. The cell voltage for the voltaic cell
$No | No^{3+}(aq) \| Cu^{2+}(aq) | Cu(s)$ is 2.84 V.
Refer to the reduction half-cell potential for the
$Cu^{2+}(aq) | Cu(s)$ in the Chemistry Data Booklet
and calculate reduction potential, E°, for the
$No | No^{3+}$ half-cell reaction.

150. Determine if the following single-replacement reactions can occur spontaneously. Write a balanced equation for the reactions that occur spontaneously.
a) $Ni(s) + Pb^{2+}(aq)$
b) $Fe(s) + Cr^{2+}(aq)$
c) $Sn(s) + Fe^{3+}(aq)$
d) $Sn(s) + Fe^{2+}(aq)$
e) $Pb(s) + Sn^{4+}(aq)$

151. A voltaic cell is set up using tin in a 1.0 mol/L $Sn^{2+}(aq)$ solution and iron in a 1.0 mol/L $Fe^{2+}(aq)$ solution.

a) Write the balanced equation for the overall reaction that occurs in this cell.
b) Which electrode is positive?
c) What is the expected cell voltage?
d) Calculate the change in mass at the anode when the cathode undergoes a change in mass of 1.50 g

152. Predict the products expected from passing a direct current through 1.0 mol/L NaI. Use the non-standard potentials for the half-reactions involving water.

153. Given the half-reactions shown below, determine if the thiosulfate ion, $S_2O_3^{2-}(aq)$, can exist in an acidic solution under standard conditions.
$S_2O_3^{2-}(aq) + H_2O(\ell)$
$\rightarrow 2SO_2(g) + 2H^+(aq) + 4e^- \qquad E^o = 0.40\ V$
$2S(s) + H_2O(\ell) \rightarrow S_2O_3^{2-}(aq) + 6H^+(aq) + 4e^-$
$E^o = -0.50\ V$

154. How can E^o values be used to decide which is the stronger oxidizing agent, $MnO_4^-(aq)$ or $Cr_2O_7^{2-}(aq)$.

155. Use half-cell reduction potentials to determine if $MnO_2(s)$ can oxidize $Br^-(aq)$ to $Br_2(\ell)$ in acid solution under standard conditions.

156. A voltaic cell contains 50.0 mL of 0.150 mol/L $CuSO_4(aq)$. If the $Cu^{2+}(aq)$ react completely, what quantity of electric charge is generated?

157. A current of 3.0 A flows for 1.0 h during an electrolysis of $CuSO_4(aq)$. What mass of Cu(s) is deposited at the cathode?

158. To recover aluminium metal, $Al_2O_3(s)$ is first converted to $AlCl_3(\ell)$. An electrolysis of $AlCl_3(\ell)$ is carried out using inert carbon electrodes.
a) What is the half-cell reaction that occurs at the anode and cathode?

b) Can the standard reduction potentials be used to calculate the external voltage needed for this process? Give a reason for your answer.

Unit 7 Chemical Changes of Organic Compounds

Chapter 14 Structure and Physical Properties of Organic Compounds

159. Name each of the following hydrocarbons.
a) 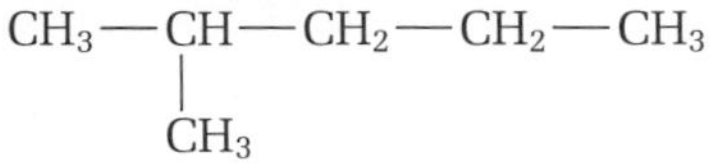 $CH_3 - CH - CH_2 - CH_2 - CH_3$
$\qquad\qquad |$
$\qquad\quad CH_3$

b) 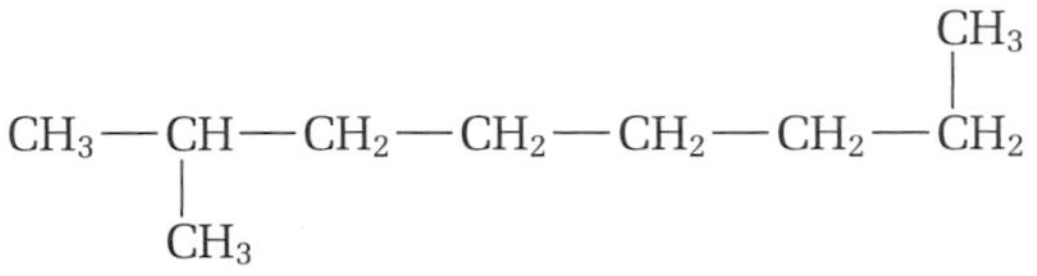 $CH_3 - CH - CH_2 - CH_2 - CH_2 - CH_2 - CH_2$
$\qquad\qquad |\qquad\qquad\qquad\qquad\qquad\qquad\qquad CH_3$
$\qquad\quad CH_3$

c) 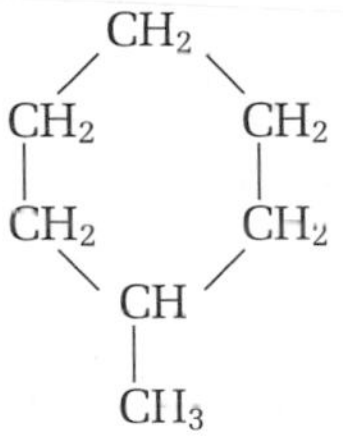
ring of CH_2 groups with CH and CH_3

d) 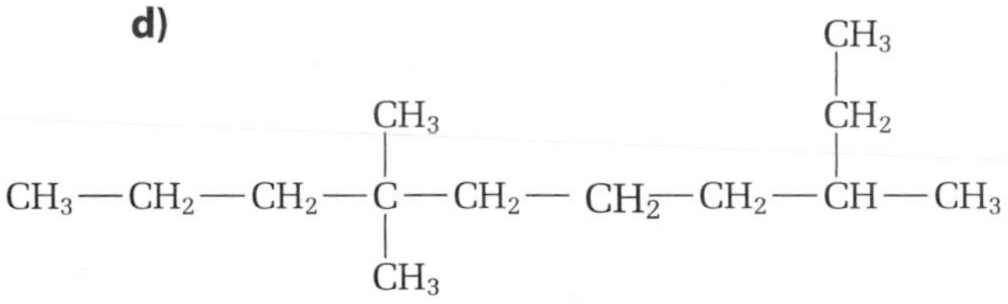 $CH_3 - CH_2 - CH_2 - C - CH_2 - CH_2 - CH_2 - CH - CH_3$
with CH_3 substituents

e) 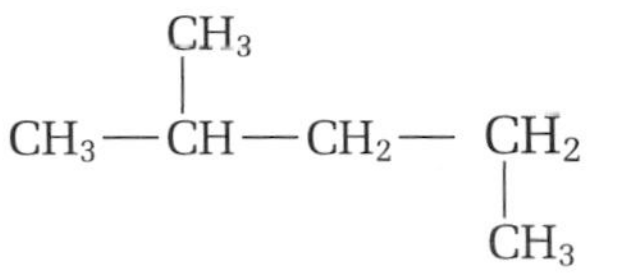 $CH_3 - CH - CH_2 - CH_2$
$\qquad\qquad |\qquad\qquad\qquad |$
$\qquad\quad CH_3\qquad\qquad CH_3$

160. Draw a condensed structural formula for each of the following compounds.
a) 2,5-dimethylhexane
b) 2,5-dimethylheptane
c) 2,3,-dimethylpentane
d) 2,2,3-trimethyl-4-propyloctane
e) 3-ethyl-2,4,6-trimethyl-5-propylnonane

161. Use each *incorrect* name to draw the corresponding hydrocarbon. Examine your drawing and rename the hydrocarbon correctly.
a) 2-ethyl-3,3-dimethyl-4-propylpentane
b) 2,3-dimethyl-3-butylpropane
c) 1-ethyl-4-methylbutane

162. Name the following compounds.

a) $CH_2\!=\!CH\!-\!CH_3$

b) 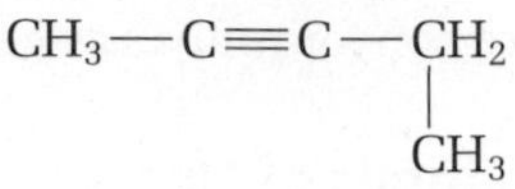 $CH_3\!-\!C\!\equiv\!C\!-\!CH_2$
         $|$
         CH_3

c)
```
          CH₂
        /      \
    CH          CH₂
    ‖            |
    CH          CH
        \      /
         CH₂    CH₃
```

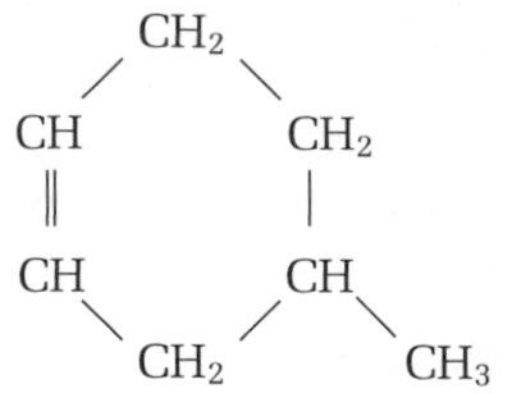

d)
$CH_3\!-\!CH\!=\!CH\!-\!CH_2\!-\!CH_2\!-\!CH_2\!-\!CH_2\!-\!CH_3$

e) 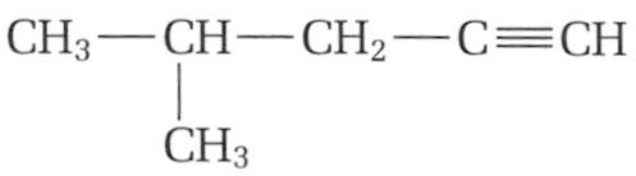 $CH_3\!-\!CH\!-\!CH_2\!-\!C\!\equiv\!CH$
         $|$
         CH_3

f) 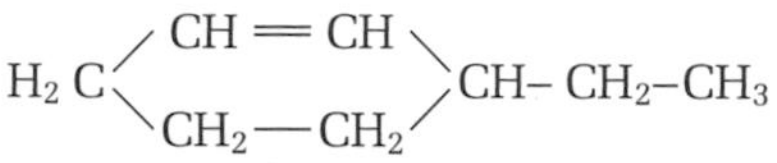
```
           CH = CH
         /         \
   H₂ C            CH– CH₂–CH₃
         \         /
           CH₂ — CH₂
```

163. Examine the structural formula and the corresponding name for each compound shown below. Correct the error by renaming the compound.

a) 4-ethyl-2-methylpentane
```
          CH₃
          |
          CH₂
          |
CH₃—CH—CH₂—CH—CH₂—CH₃
     |
     CH₃
```

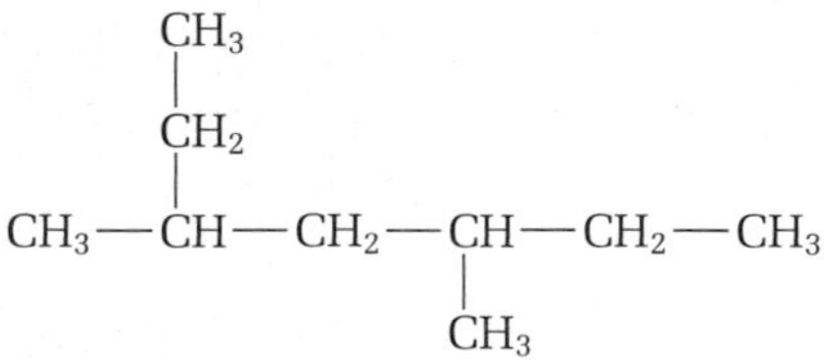

b) 4,5-dimethylhexane
```
              CH₂
            /      \
       CH₂          CH₂
       |            |
       CH           CH₂
      /             |
   CH₃      CH
            |
            CH₃
```

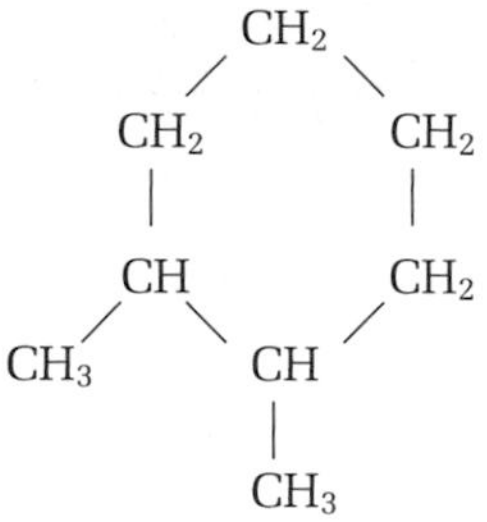

c) 2-methyl-3-ethylpentane
```
          CH₃
          |
          CH₂
          |
CH₃—CH—CH—CH₂—CH₃
          |
          CH₂
          |
          CH₃
```

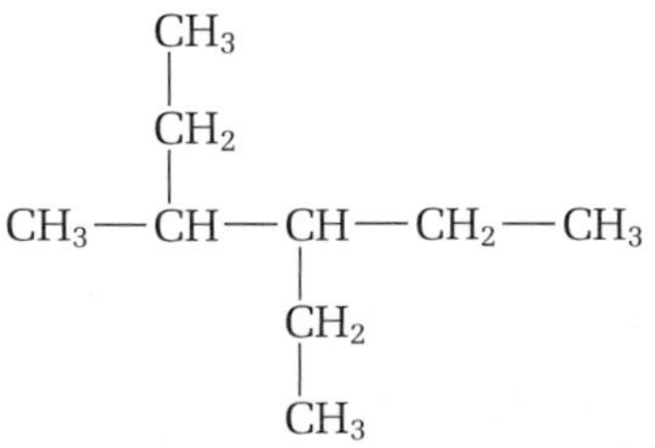

d) 1-methyl-3-cyclobutene
```
   CH = CH
   |
   CH₂—CH—CH₃
```

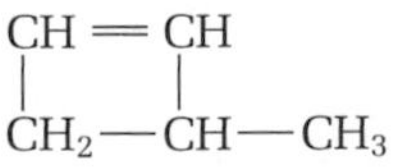

164. Draw a condensed structural formula for each of the following compounds.

a) 2-ethylpent-1-ene

b) 2,6-dimethylnon-3-ene

c) 2,5-dimethylhex-3-yne

d) 2-butyl-3-ethylcyclobutene

e) 1-butyl-3-propylcyclooctene

165. Name each compound.

a) 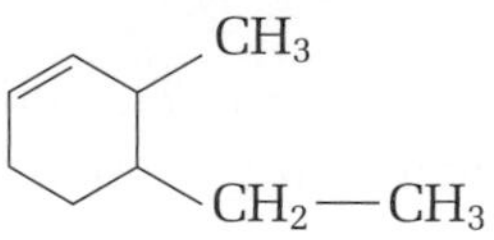
```
              CH₃
        _____/
       /     \
      |       CH₂—CH₃
```

b) 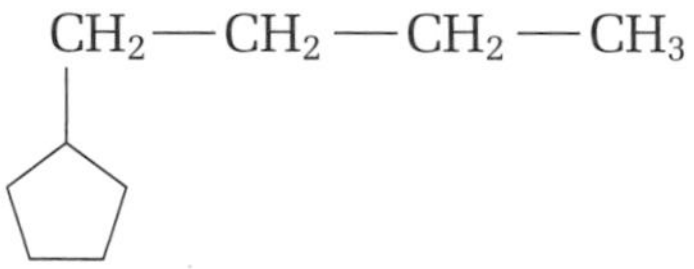 $CH_2\!-\!CH_2\!-\!CH_2\!-\!CH_3$
```
      |
     /\
    /  \
```

c) 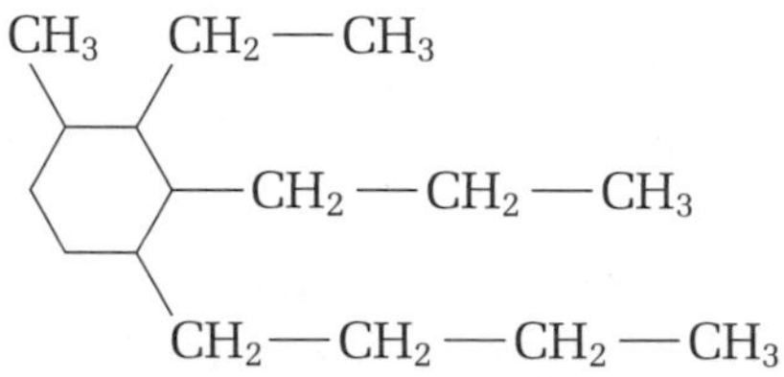 CH_3 $CH_2\!-\!CH_3$
```
       \    /
        ____
       /    \—CH₂—CH₂—CH₃
       \    /
        ‾‾‾‾
      CH₂—CH₂—CH₂—CH₃
```

d) 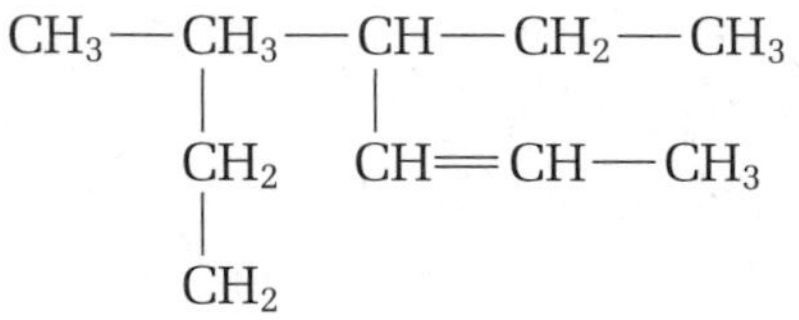 $CH_3\!-\!CH_3\!-\!CH\!-\!CH_2\!-\!CH_3$
```
              |         |
              CH₂   CH=CH—CH₃
              |
              CH₂
```

e) 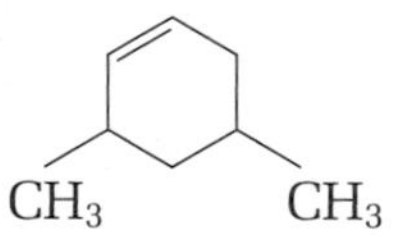
```
       ____
      /    \
     |      |
      \    /
   CH₃      CH₃
```

166. Name the following compounds.

a) 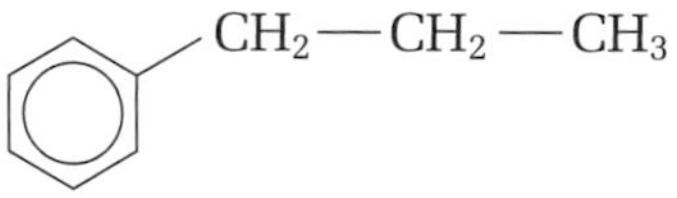
```
   ___    CH₂—CH₂—CH₃
  /   \__/
  \___/
```

b) 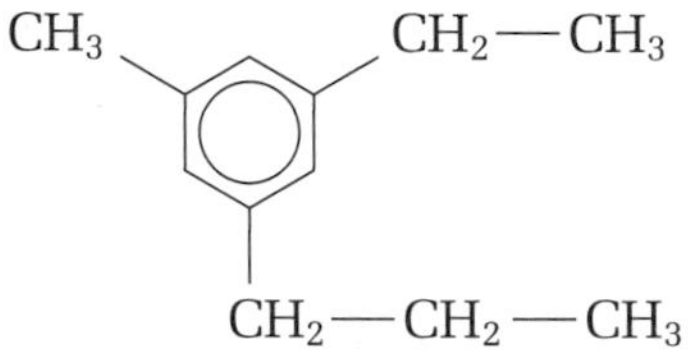 CH_3 $CH_2\!-\!CH_3$
```
       \          /
        \___    _/
        /   \__/
        \___/
         |
      CH₂—CH₂—CH₃
```

c) 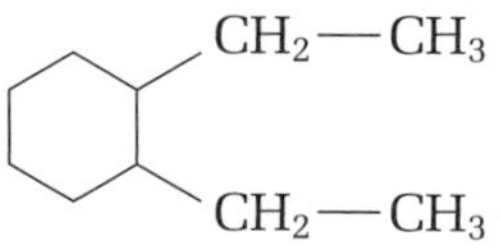 $CH_2\!-\!CH_3$
```
    ____   /
   /    \_/
   \____/
        \
      CH₂—CH₃
```

167. Name the compounds.

a) 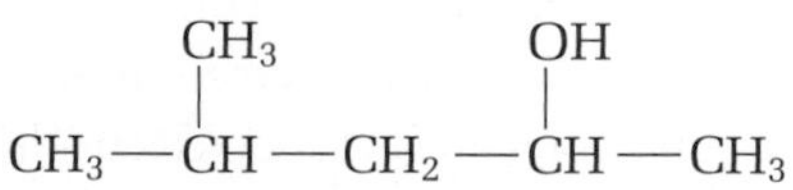
```
          CH₃              OH
          |               |
CH₃—CH—CH₂—CH—CH₃
```

b) 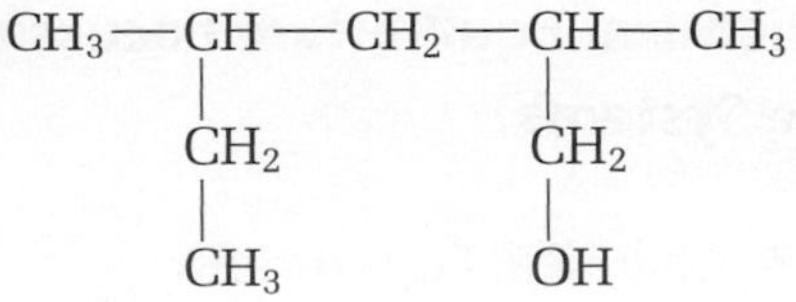 $CH_3-CH-CH_2-CH-CH_3$
$\quad\quad\quad\;\; | \quad\quad\quad\quad |$
$\quad\quad\quad CH_2 \quad\quad\quad CH_2$
$\quad\quad\quad\;\; | \quad\quad\quad\quad |$
$\quad\quad\quad CH_3 \quad\quad\quad OH$

c) 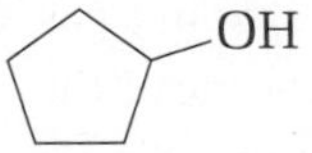 (cyclopentane ring with) OH

168. Draw a condensed structural formula for each of the following compounds.
   a) 1,3-difluoro-4-methylpent-1-yne
   b) 2,3,4-trichloropentane
   c) 3-methylcyclopentanol
   d) 7,7-dimethyloct-1-ene

169. Name the compounds.
   a) 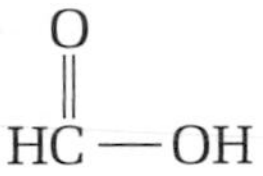
   $$\overset{\displaystyle O}{\underset{\displaystyle \|}{}}$$
   $HC-OH$

   b) 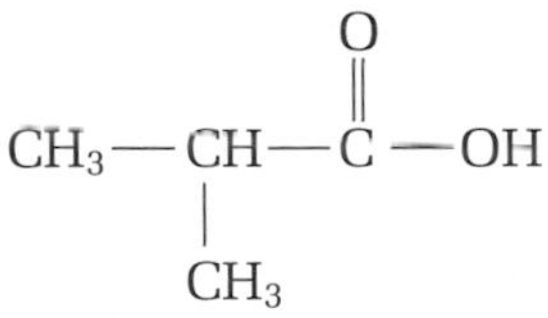
   $$CH_3-CH-\overset{O}{\overset{\|}{C}}-OH$$
   $$\quad\quad\;\; | $$
   $$\quad\quad CH_3$$

   c)
   $$CH_3-CH_2-CH_2-CH_2-O-\overset{O}{\overset{\|}{CH}}$$

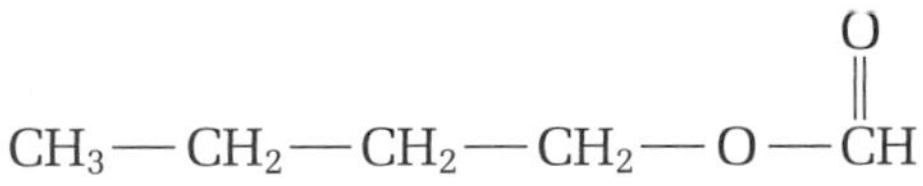

   d) 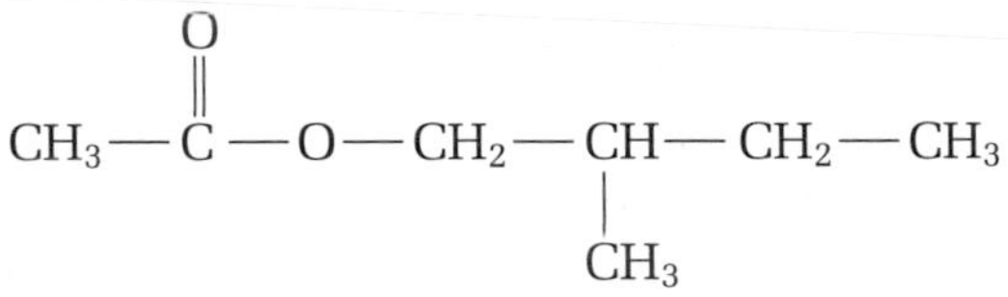
   $$CH_3-\overset{O}{\overset{\|}{C}}-O-CH_2-CH-CH_2-CH_3$$
   $$\quad\quad\quad\quad\quad\quad\quad\;\; | $$
   $$\quad\quad\quad\quad\quad\quad\quad CH_3$$

Chapter 15 Reactions of Organic Compounds

170. Identify the type of reaction.
   a) $CH_3CH_2OH \rightarrow CH_2 = CH_2 + H_2O$
   b) $CH_3CH_2CH_2CH_2OH + HBr$
   $\quad \rightarrow CH_3CH_2CH_2CH_2Br + HOH$
   c) $CH_3CH_2CH(OH)CH_3$
   $\quad \rightarrow CH_3CH_2CH = CH_2 + H_2O$
   d) $CH_3CH_2CH(Br)CH_3 + OH^-$
   $\quad \rightarrow CH_3CH_2CH = CH_2 + H_2O + Br^-$

171. Draw and name the products of each reaction.
   a) $CH_3-CH=CH-CH_2-CH_3 + Br_2 \rightarrow$
   b) 1-methylcyclopentene $+ H_2 \rightarrow$
   c) $CH_3(CH_2)_4CH = CH_2 + Cl_2 \rightarrow$
   d) 1,3-dimethylcyclobutene $+ HCl \rightarrow$

172. Draw and name the reactants for the following reactions.
   a) 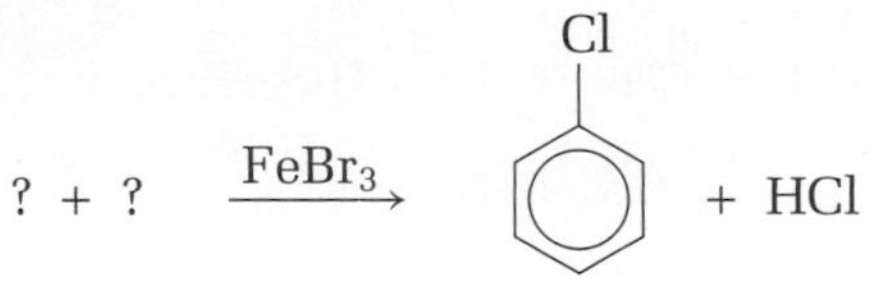
   $? + ? \xrightarrow{FeBr_3}$ (chlorobenzene) $+ HCl$

   b) 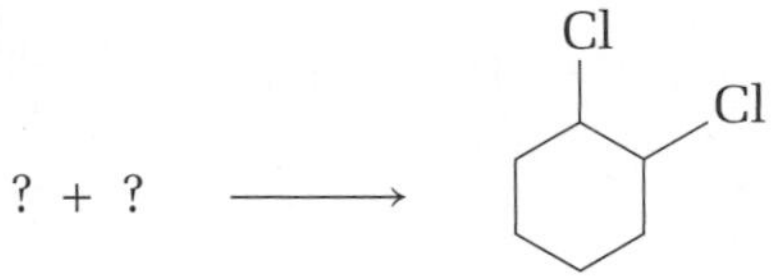
   $? + ? \longrightarrow$ (cyclohexane with Cl, Cl)

   c) 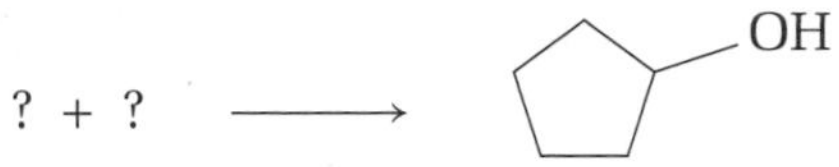
   $? + ? \longrightarrow$ (cyclopentane with OH)

   d) 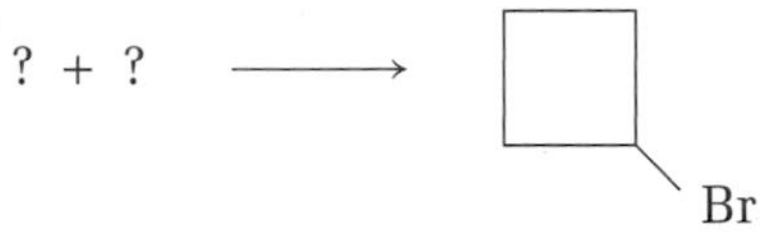
   $? + ? \longrightarrow$ (cyclobutane with Br)

173. Write reactions for the following.
   a) synthesis of methyl pentanoate
   b) esterification of ethanol and propanoic acid
   c) water with butyl butanoate
   d) esterification of propan-2-ol and butanoic acid

174. Draw and name the missing compound for each reaction shown below.
   a) $HC\equiv C-CH_2-CH_2F + Br_2 \rightarrow ?$
   b) $CH_3-CH_2-CH(CH_3)-CH_3 + 2Cl_2 \xrightarrow{UV}$
   $\quad ? + 2HCl$
   c) $? + HBr \rightarrow CH_3-CBr=CH-CH_2-CH_3$

175. Identify each reaction as an addition, substitution, or elimination reaction.
   a) 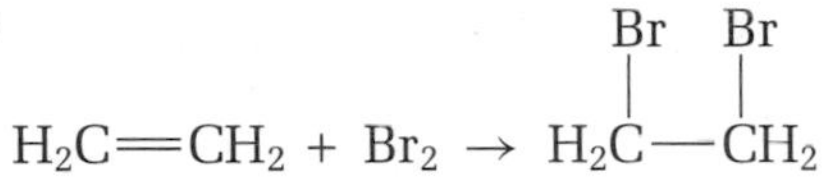
   $$H_2C=CH_2 + Br_2 \rightarrow H_2\overset{Br}{\overset{|}{C}}-\overset{Br}{\overset{|}{C}}H_2$$
   b) 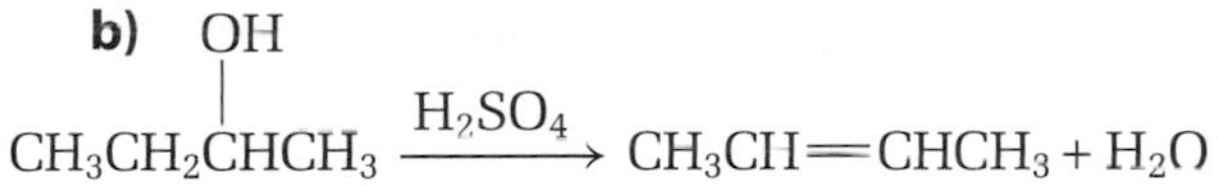
   $$CH_3CH_2\overset{OH}{\overset{|}{C}}HCH_3 \xrightarrow{H_2SO_4} CH_3CH=CHCH_3 + H_2O$$
   c) $CH_3CH_2CH_2Br + H_2NCH_2CH_3 \rightarrow$
   $\quad CH_3CH_2CH_2NHCH_2CH_3 + HBr$
   d) 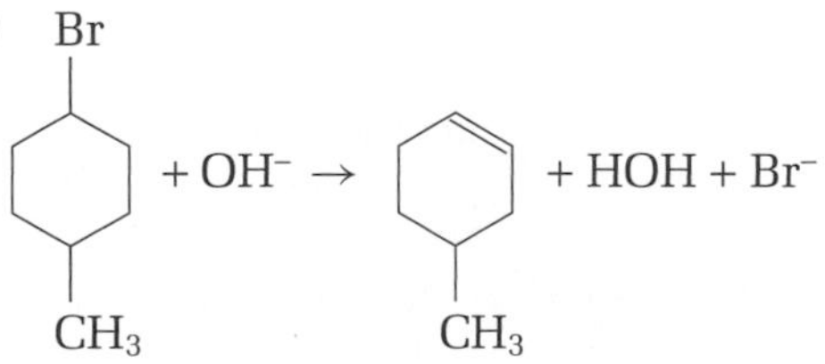
   (cyclohexane with Br and CH$_3$) $+ OH^- \rightarrow$ (cyclohexene with CH$_3$) $+ HOH + Br^-$

176. Predict what type of reaction will occur.

a)

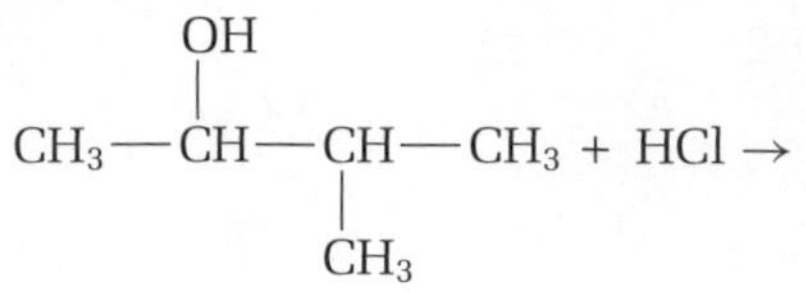

$$CH_3-\underset{\underset{\displaystyle OH}{|}}{CH}-\underset{\underset{\displaystyle CH_3}{|}}{CH}-CH_3 + HCl \rightarrow$$

b)

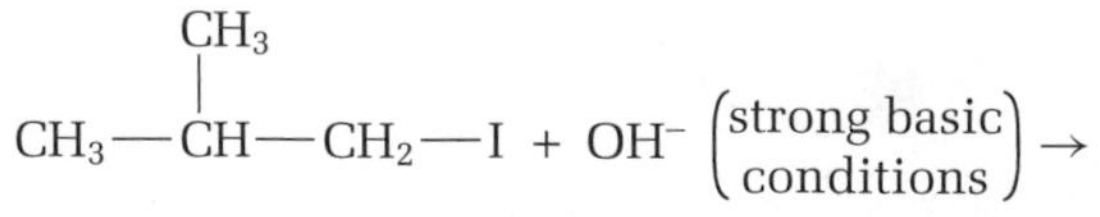

$$CH_3-\underset{\underset{\displaystyle CH_3}{|}}{CH}-CH_2-I + OH^- \left(\begin{array}{c}\text{strong basic}\\\text{conditions}\end{array}\right) \rightarrow$$

c) OH

(cyclohexanol structure) + HBr →

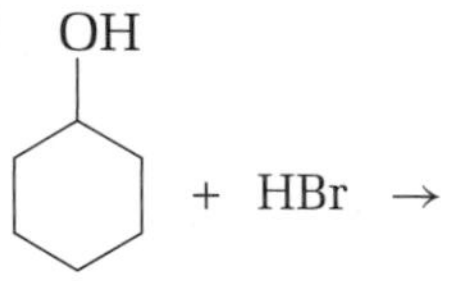

d)

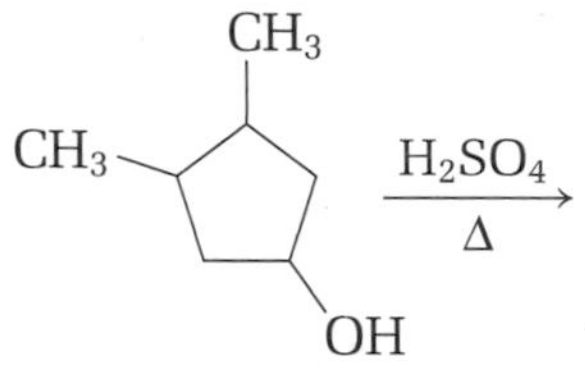

(cyclopentane with CH$_3$ and OH substituents) $\xrightarrow[\Delta]{H_2SO_4}$

e)

$$CH_3-\underset{\underset{\displaystyle CH_3}{|}}{CH}-\underset{\overset{\displaystyle CH_2CH_3}{|}}{CH}-Cl + OH^- \left(\begin{array}{c}\text{high temperature}\\\text{in alcohol solution}\end{array}\right) \rightarrow$$

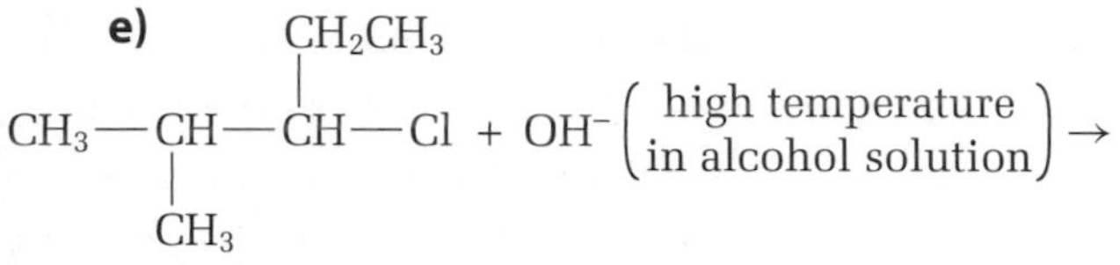

177. What type of compound forms in the following reactions?
   a) An alkane undergoes elimination in basic solution.
   b) An alcohol is heated in the presence of H_2SO_4
   c) Methanoic acid reacts with methanol
   d) pent-2-yne reacts with $2Br_2$
   e) 3-methylbutan-2-ol reacts with HCl

178. a) Sketch three units of the polymer made from the structure shown here.

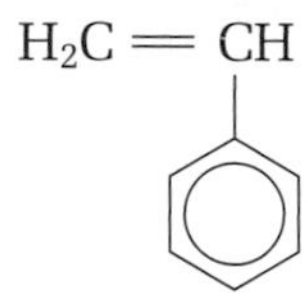

$$H_2C = CH$$

(with phenyl group attached)

   b) What type of polymerization reaction has occurred?

179. a) Draw the monomer units that were used to make the polymer shown here.

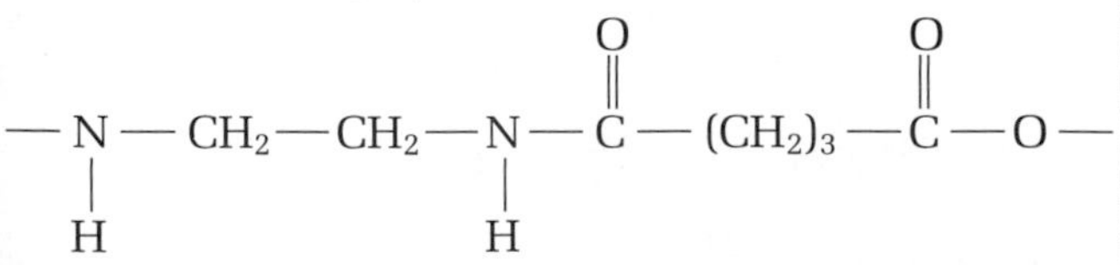

$$-\underset{\underset{\displaystyle H}{|}}{N}-CH_2-CH_2-\underset{\underset{\displaystyle H}{|}}{N}-\overset{\overset{\displaystyle O}{\|}}{C}-(CH_2)_3-\overset{\overset{\displaystyle O}{\|}}{C}-O-$$

   b) What type of polymerization reaction has occurred?

Unit 8 Chemical Equilibrium Focussing on Acid-Base Systems

Chapter 16 Chemical Equilibrium

180. A 23.0 g sample of $I_2(g)$ is sealed in a gas bottle having a volume of 500 mL. Some of the molecular $I_2(g)$ dissociates into iodine atoms and after a short time the following equilibrium is established.
$I_2(g) \rightleftharpoons 2I(g)$
For this system, $K_c = 3.80 \times 10^{-5}$. What mass of $I_2(g)$ will be in the bottle when equilibrium is established?

181. A flask having a volume of 1.75 L contains 4.00 mol of $N_2(g)$ and 6.00 mol of $O_2(g)$ in equilibrium with 0.400 mol of $NO_2(g)$ at a certain temperature.
   a) Write the equilibrium expression for this system.
   b) What is the value of the equilibrium constant?
   c) If more $O_2(g)$ is added to the system at the same temperature, what will happen to the value of the equilibrium constant?
   d) If the volume of the flask is reduced, will the amount of $N_2(g)$ increase, decrease or remain the same? Give a reason for your answer.
   e) If the volume of the system is reduced, will the concentration of $N_2(g)$ increase, decrease or remain the same? Give a reason for your answer.

182. Write the equilibrium expression for each of the following reactions.
   a) $P_4(g) + 5O_2(g) \rightleftharpoons P_4O_{10}(g)$
   b) $NaF(s) + H_2SO_4(\ell) \rightleftharpoons NaHSO_4(s) + HF(g)$
   c) $NH_3(g) + HCl(g) \rightleftharpoons NH_4Cl(s)$

183. For the equilibrium system $CO_2(g) + H_2(g) \rightarrow CO(g) + H_2O(g) + heat$, what is the effect upon the concentration of $CO_2(g)$ for each change to the system?
   a) the temperature is lowered
   b) a catalyst is used
   c) an inert gas is added at constant pressure
   d) an inert gas is added at constant volume
   e) more $CO_2(g)$ is added
   f) more $H_2(g)$ is added

184. 12.0 mol of $N_2O(g)$ is placed in a 10.0 L reaction vessel and the following equilibrium is established.
$2N_2O(g) \rightleftharpoons 2N_2(g) + O_2(g)$
At equilibrium, 4.60 mol of $N_2O(g)$ is present. What is the value of the equilibrium constant?

185. At a certain temperature, the following equilibrium mixture is formed.
$$2CO(g) + O_2(g) \rightleftharpoons 2CO_2(g)$$
The value of the equilibrium constant at this temperature is 1.00×10^{-4}. Determine the number of moles of $CO(g)$ and $O_2(g)$ present in the mixture at equilibrium if initially 1.00 mol of $CO(g)$ and 1.00 mol of $O_2(g)$ are placed in a 2.00 L reaction vessel for the reaction to proceed.

186. The following reaction took place in a 5.00 L reaction chamber at 100 °C.
$$SO_2Cl_2 \rightleftharpoons Cl_2(g) + SO_2(g)$$
Initially 2.65 mol of $Cl_2(g)$ and 4.85 mol of $SO_2Cl_2(g)$ were placed in the chamber. Calculate the concentration of each component at equilibrium. K_c at this temperature is 2.60.

187. The ester ethyl ethanoate can be produced in the following equilibrium mixture.
$$CH_3COOH(\ell) + C_2H_5OH(\ell) \rightleftharpoons H_2O(\ell) + CH_3COOC_2H_5(\ell)$$
At a certain temperature, K_c for this equilibrium is 6.86. If the initial concentration of the acid and alcohol are 1.00 mol/L, what mass of ester will be present at equilibrium in a 1.00 L container?

188. For the equilibrium $H_2(g) + S(s) \rightleftharpoons H_2S(g)$ at 370 K, the equilibrium constant is 6.8×10^{-2}.
a) 0.500 mol of $S(s)$ and 0.100 mol of $H_2(g)$ are allowed to reach equilibrium in a 1.00 L reaction vessel. What mass of $H_2S(g)$ will be present at equilibrium?
b) If initially 0.200 mol of $S(s)$ was used to establish the equilibrium, how would this affect the final mass of $H_2S(g)$?

189. An equilibrium is established for the following reaction.
$$2HI(g) \rightleftharpoons H_2(g) + I_2(g)$$
The equilibrium concentrations of the components are $[HI(g)] = 1.00$ mol/L, $[H_2(g)] = 0.16$ mol/L, and $[I_2(g)] = 0.12$ mol/L. Determine the direction in which the reaction must shift for equilibrium to be established when the components have the following concentrations: $[HI(g)] = 0.75$ mol/L, $[H_2(g)] = 0.14$ mol/L, and $[I_2(g)] = 0.10$ mol/L.

190. For the equilibrium system shown below, in which direction will the equilibrium shift for each change?
$$CH_3OH(g) + heat \rightleftharpoons CO(g) + 2H_2(g)$$
a) $CH_3OH(g)$ is removed from the system.
b) the system remains at constant pressure while an inert gas is added.
c) the temperature is raised.
d) $CO(g)$ is added to the system.

191. For the equilibrium $FeSCN^{2+}(aq) \rightleftharpoons Fe^{3+}(aq) + SCN^-(aq)$ at a certain temperature, $K_c = 9.10 \times 10^{-4}$. What are the equilibrium concentrations of each species if the initial concentration of $FeSCN^{2+}(aq)$ is 1.5 mol/L?

Chapter 17 Acid-Base Equilibrium Systems

192. For each reaction shown below, identify the conjugate acid-base pairs.
a) $HF(aq) + NH_3(aq) \rightleftharpoons NH_4^+(aq) + F^-(aq)$
b) $Fe(H_2O)_6^{3+}(aq) + H_2O(\ell) \rightleftharpoons Fe(H_2O)_5(OH)^{2+}(aq) + H_3O^+(aq)$
c) $NH_4^+(aq) + CN^-(aq) \rightleftharpoons HCN(aq) + NH_3(aq)$
d) $(CH_3)_3N(aq) + H_2O(\ell) \rightleftharpoons (CH_3)_3NH^+(aq) + OH^-(aq)$

193. A flask contains 400 mL of 0.400 mol/L methanoic acid, $HCOOH(aq)$. The K_a for this acid is 1.80×10^{-4}. What is the pH of this solution?

194. Hydrosulfuric acid, $H_2S(aq)$, has a $K_{a1} = 8.90 \times 10^{-8}$ and $K_{a2} = 1.00 \times 10^{-14}$. What is the pH of 0.100 mol/L $H_2S(aq)$?

195. a) Is an aqueous solution of sodium ethanoate, $NaC_2H_3O_2(aq)$, acidic or basic?
b) Given that the K_a for ethanoic acid, $HC_2H_3O_2$, is 1.80×10^{-5}, what is the pH of 0.0100 mol/L $NaC_2H_3O_2(aq)$?

196. 0.25 mol of ethanoic acid, CH_3COOH, and 0.10 mol sodium hydroxide, $NaOH$, are mixed with enough water to make 1.0 L of solution. What is the resulting molar concentration of $H_3O^+(aq)$ and $CH_3COO^-(aq)$?

197. a) What is the conjugate base of hydrazoic acid, $HN_3(aq)$?
b) The K_a for $HN_3(aq)$ is 2.80×10^{-5}. What is K_b for its conjugate base?
c) What is the pH of a solution that has 0.600 g of sodium azide, NaN_3, in 500 mL of solution?

198. For the anion $HPO_4^{2-}(aq)$, $K_a = 4.8 \times 10^{-13}$ and $K_b = 1.6 \times 10^{-7}$.
a) Write equations to show $HPO_4^{2-}(aq)$ acting as an acid and as a base.
b) Is an aqueous solution of the sodium salt of $HPO_4^{2-}(aq)$ acidic or basic? Give a reason for your answer.

c) What is the conjugate acid of $HPO_4^{2-}(aq)$?

d) Use the above information to calculate the K_a for the conjugate acid of $HPO_4^{2-}(aq)$.

199. K_b for $CN^-(aq)$ is 1.61×10^{-5}. What is the molar concentration of a solution of potassium cyanide that has a pH = 10.550.

200. K_a for $NH_4^+(aq)$ is 5.6×10^{-10}. What mass of $NH_4Cl(s)$ must be dissolved in 500 mL of solution for the pH to be 5.21?

201. What is the percent ionization of 1.8×10^{-2} mol/L pyridine, C_5H_5N, given that K_b is 1.4×10^{-9}?

202. Equal volumes of 0.2 mol/L ethanoic acid, $CH_3COOH(aq)$, and 0.2 mol/L HCl are mixed. What is the molar concentration of each ion in the resulting solution?

203. What is the pH of a solution containing 3.406 g NH_3 in 1.00 L of solution?

204. Dichloroethanoic acid, $HC_2HCl_2O_2$, has a K_a of 5.0×10^{-2}.

a) What is the pH of a 0.00205 mol/L solution of this acid?

b) A small amount of sodium dichloroethanoate, $NaC_2HCl_2O_2(s)$, is added to the solution. How will this affect the pH of the solution? Give a reason for your answer.

205. a) Predict which acid will have a greater change in pH when the concentration is doubled, HCl or HCN.

b) Determine if your prediction is correct by comparing the pH of 0.10 mol/L and 0.20 mol/L solutions of HCl and HCN. K_a for HCN is 6.2×10^{-10}.

Alphabetical List of Elements

Element	Symbol	Atomic Number
Actinium	Ac	89
Aluminium	Al	13
Americium	Am	95
Antimony	Sb	51
Argon	Ar	18
Arsenic	As	33
Astatine	At	85
Barium	Ba	56
Berkelium	Bk	97
Beryllium	Be	4
Bismuth	Bi	83
Bohrium	Bh	107
Boron	B	5
Bromine	Br	35
Cadmium	Cd	48
Calcium	Ca	20
Californium	Cf	98
Carbon	C	6
Cerium	Ce	58
Cesium	Cs	55
Chlorine	Cl	17
Chromium	Cr	24
Cobalt	Co	27
Copper	Cu	29
Curium	Cm	96
Dubnium	Db	105
Dysprosium	Dy	66
Einsteinium	Es	99
Erbium	Er	68
Europium	Eu	63
Fermium	Fm	100
Fluorine	F	9
Francium	Fr	87
Gadolinium	Gd	64
Gallium	Ga	31
Germanium	Ge	32
Gold	Au	79
Hafnium	Hf	72
Hassium	Hs	108
Helium	He	2
Holmium	Ho	67
Hydrogen	H	1
Indium	In	49
Iodine	I	53
Iridium	Ir	77
Iron	Fe	26
Krypton	Kr	36
Lanthanum	La	57
Lawrencium	Lr	103
Lead	Pb	82
Lithium	Li	3
Lutetium	Lu	71
Magnesium	Mg	12
Manganese	Mn	25
Meitnerium	Mt	109
Mendelevium	Md	101
Mercury	Hg	80
Molybdenum	Mo	42

Element	Symbol	Atomic Number
Neodymium	Nd	60
Neon	Ne	10
Neptunium	Np	93
Nickel	Ni	28
Niobium	Nb	41
Nitrogen	N	7
Nobelium	No	102
Osmium	Os	76
Oxygen	O	8
Palladium	Pd	46
Phosphorus	P	15
Platinum	Pt	78
Plutonium	Pu	94
Polonium	Po	84
Potassium	K	19
Praseodymium	Pr	59
Promethium	Pm	61
Protactinium	Pa	91
Radium	Ra	88
Radon	Rn	86
Rhenium	Re	75
Rhodium	Rh	45
Rubidium	Rb	37
Ruthenium	Ru	44
Rutherfordium	Rf	104
Samarium	Sm	62
Scandium	Sc	21
Seaborgium	Sg	106
Selenium	Se	34
Silicon	Si	14
Silver	Ag	47
Sodium	Na	11
Strontium	Sr	38
Sulfur	S	16
Tantalum	Ta	73
Technetium	Tc	43
Tellurium	Te	52
Terbium	Tb	65
Thallium	Tl	81
Thorium	Th	90
Thulium	Tm	69
Tin	Sn	50
Titanium	Ti	22
Tungsten	W	74
Ununbium	Uub	112
Ununhexium	Uuh	116
Ununnilium	Uun	110**
Ununquadium	Uuq	114
Unununium	Uuu	111
Uranium	U	92
Vanadium	V	23
Xenon	Xe	54
Ytterbium	Yb	70
Yttrium	Y	39
Zinc	Zn	30
Zirconium	Zr	40

**The names and symbols for elements 110 through 118 have not yet been chosen

1	2	3	4	5	6	7	8	9

Table of Common Polyatomic Ions

Name	Formula	Name	Formula	Name	Formula
acetate (ethanoate)	CH_3COO^-	chromate	$CrO_4{}^{2-}$	phosphate	$PO_4{}^{3-}$
ammonium	$NH_4{}^+$	dichromate	$Cr_2O_7{}^{2-}$	hydrogen phosphate	$HPO_4{}^{2-}$
benzoate	$C_6H_5COO^-$	cyanide	CN^-	dihydrogen phosphate	$H_2PO_4{}^-$
borate	$BO_3{}^{3-}$	hydroxide	OH^-	silicate	$SiO_3{}^{2-}$
carbide	$C_2{}^{2-}$	iodate	$IO_3{}^-$	sulfate	$SO_4{}^{2-}$
carbonate	$CO_3{}^{2-}$	nitrate	$NO_3{}^-$	hydrogen sulfate	$HSO_4{}^-$
hydrogen carbonate (bicarbonate)	$HCO_3{}^-$	nitrite	$NO_2{}^-$	sulfite	$SO_3{}^{2-}$
perchlorate	$ClO_4{}^-$	oxalate	$OOCCOO^{2-}$	hydrogen sulfite	$HSO_3{}^-$
chlorate	$ClO_3{}^-$	hydrogen oxalate	$HOOCCOO^-$	hydrogen sulfide	HS^-
chlorite	$ClO_2{}^-$	permanganate	$MnO_4{}^-$	thiocyanate	SCN^-
hypochlorite	ClO^- or OCl^-	peroxide	$O_2{}^{2-}$	thiosulfate	$S_2O_3{}^{2-}$
		persulfide	$S_2{}^{2-}$		

Periodic Table

1 1.01 — 1+,1− — 2.2 — **H** hydrogen

3 6.94, 1+, 1.0, **Li** lithium	**4** 9.01, 2+, 1.6, **Be** beryllium
11 22.99, 1+, 0.9, **Na** sodium	**12** 24.31, 2+, 1.3, **Mg** magnesium

Group 3	Group 4	Group 5	Group 6	Group 7	Group 8	Group 9
19 39.10, 1+, 0.8, **K** potassium	**20** 40.08, 2+, 1.0, **Ca** calcium	**21** 44.96, 3+, 1.4, **Sc** scandium	**22** 47.87, 4+,3+, 1.5, **Ti** titanium	**23** 50.94, 5+,4+, 1.6, **V** vanadium	**24** 52.00, 3+,2+, 1.7, **Cr** chromium	**25** 54.94, 2+,4+, 1.6, **Mn** manganese
						26 55.85, 3+,2+, 1.8, **Fe** iron
						27 58.93, 2+,3+, 1.9, **Co** cobalt

37 85.47, 1+, 0.8, **Rb** rubidium	**38** 87.62, 2+, 1.0, **Sr** strontium	**39** 88.91, 3+, 1.2, **Y** yttrium	**40** 91.22, 4+, 1.3, **Zr** zirconium	**41** 92.91, 5+,3+, 1.6, **Nb** niobium	**42** 95.94, 6+, 2.2, **Mo** molybdenum	**43** (98), 7+, 2.1, **Tc** technetium
						44 101.07, 3+,4+, 2.2, **Ru** ruthenium
						45 102.91, 3+, 2.3, **Rh** rhodium

55 132.91, 1+, 0.8, **Cs** cesium	**56** 137.33, 2+, 0.9, **Ba** barium	**57** 138.91, 3+, 1.1, **La** lanthanum	**72** 178.49, 4+, 1.3, **Hf** hafnium	**73** 180.95, 5+, 1.5, **Ta** tantalum	**74** 183.84, 6+, 1.7, **W** tungsten	**75** 186.21, 7+, 1.9, **Re** rhenium
						76 190.23, 4+, 2.2, **Os** osmium
						77 192.22, 4+, 2.2, **Ir** iridium

87 (223), 1+, 0.7, **Fr** francium	**88** (226), 2+, 0.9, **Ra** radium	**89** (227), 3+, 1.1, **Ac** actinium	**104** (261), **Rf** rutherfordium	**105** (262), **Db** dubnium	**106** (266), **Sg** seaborgium	**107** (264), **Bh** bohrium
						108 (277), **Hs** hassium
						109 (268), **Mt** meitnerium

lanthanides and actinides series begin

58 140.12, 3+, 1.1, **Ce** cerium	**59** 140.91, 3+, 1.1, **Pr** praseodymium	**60** 144.24, 3+, 1.1, **Nd** neodymium	**61** (145), 3+, —, **Pm** promethium	**62** 150.36, 3+,2+, 1.2, **Sm** samarium
90 232.04, 4+, 1.3, **Th** thorium	**91** 231.04, 5+,4+, 1.5, **Pa** protactinium	**92** 238.03, 6+,4+, 1.7, **U** uranium	**93** (237), 5+, 1.3, **Np** neptunium	**94** (244), 4+,6+, 1.3, **Pu** plutonium

References

Lide, D.R. 2001. *CRC Handbook of Chemistry and Physics.* 86th ed. Boca Raton: CRC Press.

Dean, John A. 1999. *Lange's Handbook of Chemistry.* 70th ed. New York: McGraw-Hill, Inc.

IUPAC *commision on atomic weights and isotopic abundances.* 2002. http://www.chem.qmw.ac.uk/iupac/AtWt/index.html.

10	11	12	13	14	15	16	17	18

Legend for Elements

	Metallic Solids		Gases
	Non-metallic solids		Liquids

Note: The legend denotes the physical state of the elements at exactly 101.325 kPa and 298.15 K.

Key

Atomic number → **26** 55.85 ← Atomic molar mass (g/mol)*
3+, 2+ ← Most stable ion charges
Electronegativity → 1.8
Symbol → **Fe**
Name → iron

* Based on $^{12}_{6}C$
( ) Indicates mass of the most stable isotope

2	4.00
He	—
helium	

5	10.81	6	12.01	7	14.01	8	16.00	9	19.00	10	20.18
2.0		2.6	—	3.0	—	3.4	—	4.0	—	—	—
B		**C**		**N**		**O**		**F**		**Ne**	
boron		carbon		nitrogen		oxygen		fluorine		neon	

13	26.98	14	28.09	15	30.97	16	32.07	17	35.45	18	39.95
1.6	3+	1.9	—	2.2	—	2.6	—	3.2	—	—	—
Al		**Si**		**P**		**S**		**Cl**		**Ar**	
aluminium		silicon		phosphorus		sulfur		chlorine		argon	

28	58.69	29	63.55	30	65.41	31	69.72	32	72.64	33	74.92	34	78.96	35	79.90	36	83.80
1.9	2+, 3+	1.9	2+, 1+	1.7	2+	1.8	3+	2.0	4+	2.2		2.6		3.0		—	—
Ni		**Cu**		**Zn**		**Ga**		**Ge**		**As**		**Se**		**Br**		**Kr**	
nickel		copper		zinc		gallium		germanium		arsenic		selenium		bromine		krypton	

46	106.42	47	107.87	48	112.41	49	114.82	50	118.71	51	121.76	52	127.60	53	126.90	54	131.29
2.2	2+, 4+	1.9	1+	1.7	2+	1.8	3+	2.0	4+, 2+	2.1	3+, 5+	2.1	—	2.7	—	2.6	—
Pd		**Ag**		**Cd**		**In**		**Sn**		**Sb**		**Te**		**I**		**Xe**	
palladium		silver		cadmium		indium		tin		antimony		tellurium		iodine		xenon	

78	195.08	79	196.97	80	200.59	81	204.38	82	207.2*	83	208.98	84	(209)	85	(210)	86	(222)
2.2	4+, 2+	2.4	3+, 1+	1.9	2+, 1+	1.8	1+, 3+	1.8	2+, 4+	1.9	3+, 5+	2.0	2+, 4+	2.2	—	—	—
Pt		**Au**		**Hg**		**Tl**		**Pb**		**Bi**		**Po**		**At**		**Rn**	
platinum		gold		mercury		thallium		lead		bismuth		polonium		astatine		radon	

110	(271)	111	(272)
Ds		**Rg**	
darmstadtium		roentgenium	

* The isotopic mix of naturally occurring lead is more variable than other elements preventing precision to greater than tenths of a gram per mole.

63	151.96	64	157.25	65	158.93	66	162.50	67	164.93	68	167.26	69	168.93	70	173.04	71	174.97
—	3+, 2+	1.2	3+	—	3+	1.2	3+	1.2	3+	1.2	3+	1.3	3+	—	3+, 2+	1.0	2+
Eu		**Gd**		**Tb**		**Dy**		**Ho**		**Er**		**Tm**		**Yb**		**Lu**	
europium		gadolinium		terbium		dysprosium		holmium		erbium		thulium		ytterbium		lutetium	

95	(243)	96	(247)	97	(247)	98	(251)	99	(252)	100	(257)	101	(258)	102	(259)	103	(262)
—	3+, 4+	—	3+	—	3+, 4+	—	3+	—	3+	—	3+	—	2+, 3+	—	2+, 3+	—	3+
Am		**Cm**		**Bk**		**Cf**		**Es**		**Fm**		**Md**		**No**		**Lr**	
americium		curium		berkelium		californium		einsteinium		fermium		mendelevium		nobelium		lawrencium	

APPENDIX D

Math and Chemistry

Precision, Error, and Accuracy

A major component of the scientific inquiry process is the comparison of experimental results with predicted or accepted theoretical values. In conducting experiments, realize that all measurements have a maximum degree of certainty, beyond which there is uncertainty. The uncertainty, often referred to as "error," is not a result of a mistake, but rather, it is caused by the limitations of the equipment or the experimenter. The best scientist, using all possible care, could not measure the height of a doorway to a fraction of a millimetre accuracy using a metre stick. The uncertainty introduced through measurement must be communicated using specific vocabulary.

Experimental results can be characterized by both their accuracy and their precision.

Precision describes the exactness and repeatabilty of a value or set of values. A set of data could be grouped very tightly, demonstrating good precision, but not necessarily be accurate. The darts in illustration (A) missed the bull's-eye and yet are tightly grouped, demonstrating precision without accuracy.

Differentiating between accuracy and precision

Accuracy describes the degree to which the result of an experiment or calculation approximates the true value. The darts in illustration (B) missed the bull's-eye in different directions, but are all relatively the same distance away from the centre. The darts demonstrate three throws that share approximately the same accuracy, with limited precision.

The darts in illustration (C) demonstrate accuracy and precision.

Random Error

- Random error results from fluctuations in measurements. For example, it is not possible to push the stem of a stop watch exactly the same every time and your reflexes vary. No measurement is perfect.
- Repeating trials will reduce but never eliminate the effects of random error.

- Random error is unbiased.
- Random error affects precision, and, usually, accuracy.

Systematic Error

- Systematic error results from consistent bias in observation.
- Repeating trials will not reduce systematic error.
- Three sources of systematic error are natural error, instrument-calibration error, and personal error.
- Systematic error affects accuracy.

Error Analysis

Error exists in every measured or experimentally obtained value. The error could deal with extremely tiny values, such as wavelengths of light, or with large values, such as the distances between stars. A practical way to illustrate the error is to compare it to the specific data as a percentage.

Relative Uncertainty

Relative uncertainty calculations are used to determine the error introduced by the natural limitations of the equipment used to collect the data. For instance, measuring the width of your textbook will have a certain degree of error due to the quality of the equipment used. This error, called "estimated uncertainty," has been deemed by the scientific community to be half of the smallest division of the measuring device. A metre stick with only centimetres marked would have an error of ± 0.5 cm. A ruler that includes millimetre divisions would have a smaller error of ± 0.5 mm (0.05 cm or ten-fold decrease in error). The measure should be recorded showing the estimated uncertainty, such as 22.0 ± 0.5 cm. Use the relative uncertainty equation to convert the estimated uncertainty into a percentage of the actual measured value.

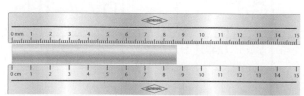

Estimated uncertainty is accepted to be half of the smallest visible division. In this case, the estimated uncertainty is ± 0.5 mm for the top ruler and ± 0.5 cm for the bottom ruler.

$$\text{relative uncertainty} = \frac{\text{estimated uncertainty}}{\text{actual measurement}} \times 100\%$$

Example:

Converting the error represented by
22.0 ±0.5 cm to a percentage

$$\text{relative uncertainty} = \frac{0.5 \text{ cm}}{22.00 \text{ cm}} \times 100\%$$

relative uncertainty = 2%

Percent Deviation

In conducting experiments, it frequently is unreasonable to expect that accepted theoretical values can be verified, because of the limitations of available equipment. In such cases, percent deviation calculations are made. For instance, the standard value for acceleration due to gravity on Earth is 9.81 m/s² toward the centre of Earth in a vacuum. Conducting a crude experiment to verify this value might yield a value of 9.6 m/s². This result deviates from the accepted standard value. It is not necessarily due to error. The deviation, as with most high school experiments, might be due to physical differences in the actual lab (for example, the experiment might not have been conducted in a vacuum). Therefore, deviation is not necessarily due to error, but could be the result of experimental conditions that should be explained as part of the error analysis. Use the percent deviation equation to determine how close the experimental results are to the accepted or theoretical value.

percent deviation =

$$\left| \frac{\text{experimental value} - \text{theoretical value}}{\text{theoretical value}} \right| \times 100\%$$

Example:

$$\text{percent deviation} = \frac{\left| 9.6 \frac{m}{s^2} - 9.8 \frac{m}{s^2} \right|}{9.8 \frac{m}{s^2}} \times 100\%$$

percent deviation = 2%

Percent Difference

Experimental inquiry does not always involve an attempt at verifying a theoretical value. For instance, measurements made in determining the width of your textbook do not have a theoretical value based on a scientific theory. You still might want to know, however, how precise your measurements were. Suppose you measured the width 100 times and found that the smallest width measurement was 21.6 cm, the largest was 22.4 cm, and the average measurement of all 100 trials was 22.0 cm. The error contained in your ability to measure the width of the textbook can be estimated using the percent difference equation.

percent difference =

$$\frac{\text{maximum difference in measurements}}{\text{average measurement}} \times 100\%$$

Example:

$$\text{percent difference} = \frac{(22.4 \text{ cm} - 21.6 \text{ cm})}{22.0 \text{ cm}} \times 100\%$$

percent difference = 4%

Practice Problems

1. In Sèvres, France, a platinum–iridium cylinder is kept in a vacuum under lock and key. It is the standard kilogram with mass 1.0000 kg. Imagine you were granted the opportunity to experiment with this special mass, and obtained the following data: 1.32 kg, 1.33 kg, and 1.31 kg. Describe your results in terms of precision and accuracy.

2. You found that an improperly zeroed triple-beam balance affected the results obtained in question 1. If you used this balance for each measure, what type of error did it introduce?

3. Describe a fictitious experiment with obvious random error.

4. Describe a fictitious experiment with obvious systematic error.

5. (a) Using common scientific practice, find the estimated uncertainty of a stopwatch that displays up to a hundredth of a second.

   (b) If you were to use the stopwatch in part (a) to time repeated events that lasted less than 2.0 s, could you argue that the estimated uncertainty from part (a) is not sufficient? Explain.

Significant Digits

All measurements involve uncertainty. One source of this uncertainty is the measuring device itself. Another source is your ability to perceive and interpret a reading. In fact, you cannot measure anything with complete certainty. The last (farthest right) digit in any measurement is always an estimate.

The digits that you record when you measure something are called *significant digits*. Significant digits include the digits that you are certain about, and a final, uncertain digit that you estimate. Follow the rules below to identify the number of significant digits in a measurement.

Rules for Determining Significant Digits

Rule 1 All non-zero numbers are significant.
- 7.886 has four significant digits.
- 19.4 has three significant digits.
- 527.266 992 has nine significant digits.

Rule 2 All zeros that are located between two non-zero numbers are significant.
- 408 has three significant digits.
- 25 074 has five significant digits.

Rule 3 Zeros that are located to the left of a measurement are not significant.
- 0.0907 has three significant digits: the 9, the third 0 to the right, and the 7.

Rule 4 Zeros that are located to the right of a measurement with no decimal point are usually considered ambiguous. However, for purposes of practice problems and exams in Alberta, consider them significant.
- 22 700 has have five significant digits.

When you take measurements and use them to calculate other quantities, you must be careful to keep track of which digits in your calculations and results are significant. Why? Your results should not imply more certainty than your measured quantities justify. This is especially important when you use a calculator. Calculators usually report results with far more digits than your data warrant. Always remember that calculators do not make decisions about certainty. You do. Follow the rules given below to report significant digits in a calculated answer.

Rules for Reporting Significant Digits in Calculations

Rule 1 Multiplying and Dividing
The value with the fewest number of significant digits, going into a calculation, determines the number of significant digits that you should report in your answer.

Rule 2 Adding and Subtracting
The value with the fewest number of decimal places, going into a calculation, determines the number of decimal places that you should report in your answer.

Rule 3 Rounding
To get the appropriate number of significant digits (rule 1) or decimal places (rule 2), you may need to round your answer.
- If your answer ends in a number that is greater than 5, increase the preceding digit by 1. For example, 2.346 can be rounded to 2.35.

- If your answer ends with a number that is less than 5, leave the preceding number unchanged. For example, 5.73 can be rounded to 5.7.
- If your answer ends with 5, increase the preceding number by 1 if it is odd. Leave the preceding number unchanged if it is even. For example, 18.35 can be rounded to 18.4, but 18.25 is rounded to 18.2.

Sample Problem

Using Significant Digits

Problem
Suppose that you measure the masses of four objects as 12.5 g, 145.67 g, 79.0 g, and 38.438 g. What is the total mass?

What Is Required?
You need to calculate the total mass of the objects.

What Is Given?
You know the mass of each object.

Plan Your Strategy
- Add the masses together, aligning them at the decimal point.
- Underline the estimated (farthest right) digit in each value. This is a technique you can use to help you keep track of the number of estimated digits in your final answer.
- In the question, two values have the fewest decimal places: 12.5 and 79.0. You need to round your answer so that it has only one decimal place.

Act on Your Strategy
$$
\begin{array}{r}
12.\underline{5} \\
145.6\underline{7} \\
79.\underline{0} \\
+\ 38.43\underline{8} \\
\hline
275.\underline{608}
\end{array}
$$

Total mass = 275.608 g
Therefore, the total mass of the objects is 275.6 g.

Check Your Solution
- Your answer is in grams. This is a unit of mass.

- Your answer has one decimal place. This is the same as the values in the question with the fewest decimal places.

Significant Digits

1. Express each answer using the correct number of significant digits.

 a) 55.671 g + 45.78 g

 b) 1.9 mm + 0.62 mm

 c) 87.9478 L − 86.25 L

 d) 0.350 mL + 1.70 mL + 1.019 mL

 e) 5.841 cm × 6.03 cm

 f) $\dfrac{17.51 \text{ g}}{2.2 \text{ cm}^3}$

Scientific Notation

One mole of water, H_2O, contains
602 214 199 000 000 000 000 000 molecules.
Each molecule has a mass of
0.000 000 000 000 000 000 000 029 9 g. As you can see, it would be very awkward to calculate the mass of one mole of water using these values. To simplify large numbers (and clarify the number of significant digits), when reporting them and doing calculations, you can use scientific notation.

Step 1 Move the decimal point so that only one non-zero digit is in front of the decimal point. (Note that this number is now between 1.0 and 9.99999999.) Count the number of places that the decimal point moves to the left or to the right.

Step 2 Multiply the value by a power of 10. Use the number of places that the decimal point moved as the exponent for the power of 10. If the decimal point moved to the right, exponent is negative. If the decimal point moved to the left, the exponent is positive.

6.02 000 000 000 000 000 000 000.

23 21   18   15   12   9   6   3

6.02×10^{23}

Figure D.1 The decimal point moves to the left.

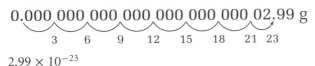

0.000 000 000 000 000 000 000 02.99 g

  3   6   9   12   15   18   21 23

2.99×10^{-23}

Figure D.2 The decimal point moves to the right.

Figure D.3 shows you how to calculate the mass of one mole of water using a scientific calculator. When you enter an exponent on a scientific calculator, you do not have to enter (× 10).

Keystrokes	Display
6 . 0 2 EXP 2 3	6.02 23
× 2 . 9 9 EXP 2 3 ±	299 −23
=	17998

Round to three significant digits and express in scientific notation: 1.80×10^1 g/mol

Figure D.3 On some scientific calculators, the EXP key is labelled EE. Key in negative exponents by entering the exponent, then striking the ± key.

Rules for Scientific Notation

Rule 1 To multiply two numbers in scientific notation, add the exponents.

$$(7.32 \times 10^{-3}) \times (8.91 \times 10^{-2})$$
$$= (7.32 \times 8.91) \times (10^{-3} \times 10^{-2})$$
$$= 65.2212 \times 10^{-5}$$
$$\rightarrow 6.52 \times 10^{-4}$$

Rule 2 To divide two numbers in scientific notation, subtract the exponents.

$$(1.842 \times 10^6 \text{ g}) \div (1.0787 \times 10^2 \text{ g/mol})$$
$$= (1.842 \div 1.0787) \times 10^{(6-2)}$$
$$= 1.707611 \times 10^4 \text{ g}$$
$$\rightarrow 1.708 \times 10^4 \text{ g}$$

Rule 3 To add or subtract numbers in scientific notation, first convert the numbers so they have the same exponent. Each number should have the same exponent as the number with the greatest power of 10. Once the numbers are all expressed to the same power of 10, the power of 10 is neither added nor subtracted in the calculation.

$$(3.42 \times 10^6 \text{ cm}) + (8.53 \times 10^3 \text{ cm})$$
$$= (3.42 \times 10^6 \text{ cm}) + (0.00853 \times 10^6 \text{ cm})$$
$$= 3.42853 \times 10^6 \text{ cm}$$
$$\rightarrow 3.43 \times 10^6 \text{ cm}$$

$$(9.93 \times 10^1 \text{ L}) - (7.86 \times 10^{-1} \text{ L})$$
$$= (9.93 \times 10^1 \text{ L}) - (7.86 \times 10^{-1} \text{ L})$$
$$= (9.93 \times 10^1 \text{ L}) - (0.0786 \times 10^1 \text{ L})$$
$$= 9.8514 \times 10^1 \text{ L}$$
$$\rightarrow 9.85 \times 10^1 \text{ L}$$

Practice problems are given on the following page.

Scientific Notation

1. Convert each value into correct scientific notation.
   a) 0.000 934
   b) 7 983 000 000
   c) 0.000 000 000 820 57
   d) 496×10^6
   e) $0.000\ 06 \times 10^1$
   f) $309\ 72 \times 10^{-8}$

2. Add, subtract, multiply, or divide. Round off your answer, and express it in scientific notation to the correct number of significant digits.
   a) $(3.21 \times 10^{-3}) + (9.2 \times 10^2)$
   b) $(8.1 \times 10^3) + (9.21 \times 10^2)$
   c) $(1.010\ 1 \times 10^1) - (4.823 \times 10^{-2})$
   d) $(1.209 \times 10^6) \times (8.4 \times 10^7)$
   e) $(4.89 \times 10^{-4}) \div (3.20 \times 10^{-2})$

Logarithms

Logarithms are a convenient method for communicating large and small numbers. The *logarithm*, or "log," of a number is the value of the exponent that 10 would have to be raised to, in order to equal this number. Every positive number has a logarithm. Numbers that are greater than 1 have a positive logarithm. Numbers that are between 0 and 1 have a negative logarithm. Table D1 gives some examples of the logarithm values of numbers.

Table D.1 Some Numbers and Their Logarithms

Number	Scientific notation	As a power of 10	Logarithm
1 000 000	1×10^6	10^6	6
7 895 900	7.859×10^5	$10^{5.8954}$	5.8954
1	1×10^0	10^0	0
0.000 001	1×10^{-6}	10^{-6}	−6
0.004 276	4.276×10^{-3}	$10^{-2.3690}$	−2.3690

Logarithms are especially useful for expressing values that span a range of powers of 10. The Richter scale for earthquakes, the decibel scale for sound, and the pH scale for acids and bases all use logarithmic scales.

Logarithms and pH

The pH of an acid solution is defined as $-\log[H_3O^+]$. (The square brackets mean "concentration.") For example, suppose that the hydronium ion concentration in a solution is 0.0001 mol/L (10^{-4} mol/L). The pH is $-\log(0.0001)$. To calculate this, enter 0.0001 into your calculator. Then press the [LOG] key. Press the [±] key. The answer in the display is 4. Therefore, the pH of the solution is 4.

There are logarithms for all numbers, not just whole multiples of 10. What is the pH of a solution if $[H_3O^+] = 0.004\ 76$ mol/L? Enter 0.00476. Press the [LOG] key and then the [±] key. The answer is 2.322. This result has three significant digits—the same number of significant digits as the concentration.

> For logarithmic values, only the digits to the right of the decimal point count as significant digits. The digit to the left of the decimal point fixes the location of the decimal point of the original value. For example:
> $[H_3O^+] = 0.0476$ mol/L; the pH = 1.322
> $[H_3O^+] = 0.00476$ mol/L; the pH = 2.322
> $[H_3O^+] = 0.000476$ mol/L; the pH = 3.322
>
> Notice only the decimal place in the $[H_3O^+]$ changes; therefore, only the first digit in the pH changes.

What if you want to find $[H_3O^+]$ from the pH? You would need to find 10^{-pH}. For example, what is $[H_3O^+]$ if the pH is 5.78? Enter 5.78, and press the [±] key. Then use the $[10^x]$ function. The answer is $10^{-5.78}$. Therefore, $[H_3O^+]$ is 1.7×10^{-6} mol/L.

Remember that the pH scale is a negative log scale. Thus, a decrease in pH from pH 7 to pH 4 is an increase of 10^3, or 1000, in the acidity of a solution. An increase from pH 3 to pH 6 is a decrease of 10^3, or 1000, in acidity.

Logarithms

1. Calculate the logarithm of each number. Note the trend in your answers.

   a) 1   **c)** 10   **e)** 100   **g)** 50 000

   b) 5   **d)** 50   **f)** 500   **h)** 100 000

2. Calculate the antilogarithm of each number.

   a) 0   **c)** −1   **e)** −2   **g)** −3

   b) 1   **d)** 2   **f)** 3

3. **a)** How are your answers for question 2, parts b) and c), related?

   b) How are your answers for question 2, parts d) and e), related?

   c) How are your answers for question 2, parts f) and g), related?

   d) Calculate the antilogarithm of 3.5.

   e) Calculate the antilogarithm of −3.5.

   f) Take the reciprocal of your answer for part d).

   g) How are your answers for parts e) and f) related?

4. **a)** Calculate log 76 and log 55.

   b) Add your answers for part a).

   c) Find the antilogarithm of your answer for part b).

The Unit Analysis Method of Problem Solving

The unit analysis method of problem solving is extremely versatile. You can use it to convert between units or to solve some formula problems. If you forget a formula, you may still be able to solve the problem using unit analysis.

The unit analysis method involves analyzing the units and setting up conversion factors. You match and arrange the units so that they divide out to give the desired unit in the answer. Then you multiply and divide the numbers that correspond to the units.

Steps for Solving Problems Using Unit Analysis

Step 1 Determine which data you have and which conversion factors you need to use. (A conversion factor is usually a ratio of two numbers with units, such as 1000 g/1 kg. You multiply the given data by the conversion factor to get the desired units for the answer.) It is often convenient to use the following three categories to set up your solution: Have, Need, and Conversion factor.

Step 2 Arrange the data and conversion factors so that you can cross out the undesired units. Decide whether you need any additional conversion factors to get the desired units for the answer.

Step 3 Multiply all the numbers on the top of the ratio. Then multiply all the numbers on the bottom of the ratio. Divide the top result by the bottom result.

Step 4 Check that you have cancelled the units correctly. Also check that the answer seems reasonable, and that the significant digits are correct.

Remember that counting numbers for exact quantities are considered to have infinite significant digits. For example, if you have 3 apples, the number is exact, and has an infinite number of significant digits. Conversion factors for unit analysis are a form of counting or record keeping. Therefore, you do not need to consider the number of significant digits in conversion factors, such as 1000 mL/1 L, when deciding on the number of significant digits in the answer.

Active ASA

Problem

In the past, pharmacists measured the active ingredients in many medications in a unit called grains (gr). A grain is equal to 64.8 mg. If one headache tablet contains 5.0 gr of active acetylsalicylic acid (ASA), how many grams of ASA are in two tablets?

What Is Required?

You need to find the mass in grams of ASA in two tablets.

What Is Given?

There are 5.0 gr of ASA in one tablet. A conversion factor for grains to milligrams is given.

Plan Your Strategy

Multiply the given quantity by conversion factors until all the unwanted units cancel out and only the desired units remain.

Have	Need	Conversion factors
5.0 gr	? g	64.8 mg/1 gr and 1 g/1000 mg

Act on Your Strategy

$$\frac{5.0 \,\cancel{gr}}{1 \,\cancel{tablet}} \times \frac{64.8 \,\cancel{mg}}{1 \,\cancel{gr}} \times \frac{1 \,g}{1000 \,\cancel{mg}} \times 2 \,\cancel{tablets}$$

$$= \frac{5.0 \times 64.8 \times 1 \times 2 \,g}{1000}$$

$$= 0.648 \,g$$

$$= 0.65 \,g$$

There are 0.65 g of active ASA in two headache tablets.

Check Your Solution

There are two significant digits in the answer. This is the least number of significant digits in the given data.

Notice how conversion factors are multiplied until all the unwanted units are cancelled out, leaving only the desired unit in the answer.

The next sample problem will show you how to solve a stoichiometric problem.

Stoichiometry and Unit Analysis

Problem

What mass of oxygen, O_2, can be obtained by the decomposition of 5.0 g of potassium chlorate, $KClO_3$? The balanced equation is given below.

$$2KClO_3 \rightarrow 2KCl + 3O_2$$

What Is Required?

You need to calculate the mass of oxygen, in grams, that is produced by the decomposition of 5.0 g of potassium chlorate.

What Is Given?

You know the mass of potassium chlorate that decomposes.

$$\text{Mass} = 5.0 \,g$$

From the balanced equation, you can obtain the molar ratio of the reactant and the product.

$$\frac{3 \text{ mol } O_2}{2 \text{ mol } KClO_3}$$

Plan Your Strategy

Calculate the molar masses of potassium chlorate and oxygen. Use the molar mass of potassium chlorate to find the number of moles in the sample. Use the molar ratio to find the number of moles of oxygen produced. Use the molar mass of oxygen to convert this value to grams.

Act on Your Strategy

The molar mass of potassium chlorate is

$$\begin{array}{l} 1 \times M_K = 39.10 \\ 1 \times M_{Cl} = 35.45 \\ \underline{3 \times M_O = 48.00} \\ 122.55 \,g/mol \end{array}$$

The molar mass of oxygen is

$$2 \times M_O = 32.00 \,g/mol$$

Find the number of moles of potassium chlorate.

$$\text{mol } KClO_3 = 5.0 \,\cancel{g} \times \left(\frac{1 \text{ mol}}{122.55 \,\cancel{g}\, KClO_3}\right)$$

$$= 0.0408 \text{ mol}$$

Find the number of moles of oxygen produced.

$$\frac{\text{mol } O_2}{0.0408 \text{ mol } KClO_3} = \frac{3 \text{ mol } O_2}{2 \text{ mol } KClO_3}$$

$$\text{mol } O_2 = 0.0408 \,\cancel{\text{mol } KClO_3} \times \frac{3 \text{ mol } O_2}{2 \,\cancel{\text{mol } KClO_3}}$$

$$= 0.0612 \text{ mol}$$

Convert this value to grams.

$$\text{mass } O_2 = 0.0612 \,\cancel{\text{mol}} \times \frac{32.00 \,g}{1 \,\cancel{\text{mol}}\, O_2}$$

$$= 1.96 \,g$$

$$= 2.0 \,g$$

Therefore, 2.0 g of oxygen are produced by the decomposition of 5.0 g of potassium chlorate. As you become more familiar with this type of question, you will be able to complete more than one step at once. Below, you can see how the conversion factors we used in each step above can be combined. Set these conversion ratios so that the units cancel out correctly.

$$\text{mass } O_2 = 5.0 \text{ g KClO}_3 \times \left(\frac{1 \text{ mol}}{122.6 \text{ g KClO}_3}\right) \times$$
$$\left(\frac{3 \text{ mol } O_2}{2 \text{ mol KClO}_3}\right) \times \left(\frac{32.0 \text{ g}}{1 \text{ mol } O_2}\right)$$
$$= 1.96 \text{ g}$$
$$= 2.0 \text{ g}$$

Check Your Solution

The oxygen makes up only part of the potassium chlorate. Thus, we would expect less than 5.0 g of oxygen, as was calculated.

The smallest number of significant digits in the question is two. Thus, the answer must also have two significant digits.

Practice Problems

Unit Analysis

Use the unit analysis method to solve each problem.

1. The molar mass of copper(II) chloride, $CuCl_2(s)$, is 134.45 g/mol. What is the mass, in grams, of 8.19×10^{-3} mol of this compound?

2. To make a salt solution, 0.82 mol of $CaCl_2$ are dissolved in 1650 mL of water. What is the concentration, in g/L, of the solution?

3. The density of solid sulfur is 2.07 g/cm³. What is the mass, in kg, of a 1.8 dm³ sample?

4. How many grams of dissolved sodium bromide are in 689 mL of a 1.32 mol/L solution?

Tips for Writing the Diploma Exam Written Response Questions

The Diploma Exam Preparation feature at *www. albertabiology.ca* shows examples of the different types of questions that require a written response, although the tips here will assist you in answering any type of question.

Your answers will be assessed on the basis of how well you communicate both your understanding of the information presented and your understanding of the applicable science. Evaluators will be looking for examples of your understanding of scientific principles and techniques. For more information about the Diploma Exams, visit the Alberta Education Diploma Exam web site at **http://www.education.gov.ab.ca/k%5F12/testing/diploma/**

Key Diploma Exam Skills

In order to successfully answer the questions, you must be able to:

1. Read critically and identify
   - key words, phrases, and data that deliver useful information
   - distractor information and data that can be ignored because it does not have any bearing on the answer to the question
   - if the question is an open-response style that requires a unified response, or if it is a closed-response style that requires a more analytical approach
   - precisely what the question is asking
   - pay close attention to the process words (see list below)
   - pay close attention to the directing words (see list opposite) to determine how you should answer the question. The directing words are always highlighted in boldface type.
   - the scientific concept(s) that you should include in your answer
   - any formula(s) that you will have to use
   - the information in your Data Booklet that you will need

2. Interpret and Analyze
   - process words and directing words
   - information including the key words, phrases, and data presented in the information box
   - information that is presented in charts, tables, and graphs

3. Communicate
   - conclusions by making a formal statement
   - results in the form of charts, graphs, or diagrams

   - ideas or answers to questions in the form of complete sentences, paragraphs, or short essays

4. If you are asked to perform an experiment, you must write the experimental design as follows:
   - state the problem or question to be answered
   - formulate a hypothesis or make a prediction
   - identify the manipulated and responding variable(s) if required
   - provide a method for controlling variables
   - identify the required materials clearly
   - describe any applicable safety procedures
   - list the steps in the experimental procedure
   - provide a sketch(es) of the apparatus to help make the set-up clear
   - provide a method for collecting and recording pertinent data—include the units for the data being collected

Guidelines

The Diploma Exam feature includes guidelines that direct you in the skills and knowledge you need to use to answer the questions successfully. This type of information will *not* appear on the exam.

Assess the type of question and the guidelines that are provided to help you answer it. Use the information provided here to develop your own technique for answering each type of question.

Alberta Science Process and Directing Words

The following "process" and "directing" words have specific meanings when they are used in the Diploma Exam questions. Study this list so you will understand exactly what you are being asked to do in the questions. The success of your answer depends not only on your interpretation of the information in the question but also on your understanding of the wording of the questions.

Process Words

Hypothesis: A single proposition intended as a possible explanation for an observed phenomenon; e.g., a possible cause for a specific effect

Conclusion: A proposition that summarizes the extent to which a hypothesis and/or a theory has been supported or contradicted by the evidence

Experiment: A set of manipulations and/or specific observations of nature that allow the testing of hypotheses and/or generalizations

Variables: Conditions that can change in an experiment. Variables in experiments are categorized as:

- *manipulated variables* (independent variables): conditions that were deliberately changed by the experimenter
- *controlled variables* (fixed or restrained variables): conditions that could have changed but did not, because of the intervention of the experimenter
- *responding variables* (dependent variables): conditions that changed in response to the change in the manipulated variables

Directing Words

Discuss — The word "discuss" **will not** be used as a directing word on math and science diploma examinations because it is not used consistently to mean a single activity.

The following words are specific in meaning.

Algebraically — Using mathematical procedures that involve letters or symbols to represent numbers

Analyze — To make a mathematical, chemical, or methodical examination of parts to determine the nature, proportion, function, interrelationship, etc. of the whole

Compare — Examine the character or qualities of two things by providing characteristics of both that point out their *similarities* and *differences*

Conclude — State a logical end based on reasoning and/or evidence

Contrast/ Distinguish — Point out the *differences* between two things that have similar or comparable natures

Criticize — Point out the *demerits* of an item or issue

Define — Provide the essential qualities or meaning of a word or concept; make distinct and clear by marking out the limits

Describe — Give a written account or represent the characteristics of something by a figure, model, or picture

Design/Plan — Construct a plan; i.e, a detailed sequence of actions for a specific purpose

Determine — Find a solution, to a specified degree of accuracy, to a problem by showing appropriate formulas, procedures, and calculations

Enumerate — Specify one by one or list in concise form and according to some order

Evaluate — Give the significance or worth of something by identifying the good and bad points or the advantages and disadvantages

Explain — Make clear what is not immediately obvious or entirely known; give the cause of or reason for; make known in detail

Graphically — Using a drawing that is produced electronically or by hand and that shows a relation between certain sets of numbers

How — Show in what manner or way, with what meaning

Hypothesize — Form a tentative proposition intended as a possible explanation for an observed phenomenon; i.e., a possible cause for a specific effect. The proposition should be testable logically and/or empirically

Identify — Recognize and select as having the characteristics of something

Illustrate — Make clear by giving an example. The form of the example must be specified in the question; i.e., word description, sketch, or diagram

Infer — Form a generalization from sample data; arrive at a conclusion by reasoning from evidence

Interpret — Tell the meaning of something; present information in a new form that adds meaning to the original data

Justify/ Show How — Show reasons for or give facts that support a position

Model — Find a model (in mathematics, a model of a situation is a pattern that is supposed to represent or set a standard for a real situation) that does a good job of representing a situation

Outline — Give, in an organized fashion, the essential parts of something. The form of the outline must be specified in the question; eg., list, flow chart, concept map

Predict — Tell in advance on the basis of empirical evidence and/or logic

Prove — Establish the truth of validity of a statement for the general case by giving factual evidence or logical argument

Relate — Show logical or causal connection between things

Sketch — Provide a drawing that represents the key features of an object or graph

Solve — Give a solution for a problem; i.e., explanation in words and/or numbers

Summarize — Give a brief account of the main points

Trace — Give a step-by-step description of the development

Verify — Establish, by substitution for a particular case or by geometric comparison, the truth of a statement

Why — Show the cause, reason, or purpose

Designing and Reporting Scientific Investigations

Scientific knowledge is built on the results of countless experiments conducted by many investigators. Complete details of investigations are shared with colleagues in order to ensure that results can be replicated. Methods must be refined and investigations repeated. Then, when findings have been confirmed by others, a scientific question is considered to have been answered. This scientific process cannot be successful unless every aspect of an investigation is documented and published. Most important are accurate reports of the results and the analysis of those results.

The types of scientific questions that can be asked are varied. Each type of question requires a slightly different model of investigation in order to find an answer or achieve a goal. During your current chemistry studies, you will be most likely to:

- conduct an inquiry into a specific question. For example, testing a procedure such as titrating an acid with a base, or carrying out a chemical analysis of a substance.
- attempt to solve a problem. You may be asked to try to improve the energy efficiency of a heating device, Or perhaps to improve the treatment of waste water.

- analyze an issue and make a recommendation. For example, how can the province's natural resources be used without creating excessive waste and damage to the environment?

A review of **scientific inquiries** shows that they share certain elements, including:
- a clear purpose;
- a specific, purpose-related hypothesis to be tested or question related to the purpose. This element may be broken down into a number of sub-questions that will help you analyze the results of the experiment and gather data that can be used to answer the main question or test the hypothesis.

- a prediction of the outcome based on known scientific laws and theories;
- a written plan for conducting the investigation. This plan must include a brief description of diagnostic tests and the variables and controls that will be used. Outline how results will be observed and measured, and how they will be analyzed.
- a full list of all apparatus and materials required, including the measuring and recording instruments to be used;
- a detailed set of instructions on how to proceed with the investigation. This must include safety precautions and clean up and waste disposal procedures when the lab work is done.
- A report on the success or failure of the investigation is also necessary, in order to share the results with colleagues.

When you document your design for a scientific inquiry, use the following headings: Purpose, Question, Hypothesis, Prediction, Experimental Plan, Materials, Procedure (including Safety Precautions), Data and Observations, Analysis, Conclusion, and Application (optional).

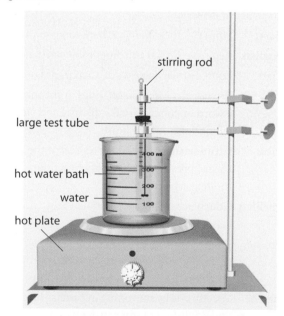

Verify that the design tests the prediction or hypothesis by writing an explanation of how it does so. Ensure that the procedure is written in a logical, step-by-step sequence, and that it is complete and clear enough that some else could carry it out. Create diagrams if necessary. Ask someone else to read the procedure through and check their understanding. Consider the safety aspects of all steps and document precautions. Double check that you listed all materials, including such specifications as size and quantity.

Ensure that both the dependent and independent variables are clearly identified and defined. The independent variable must be controlled and measured accurately. Specify how the results are to be measured (including units) and recorded. For example, will sketches be made or photos taken to record qualitative observations? Will a data table showing the measurement of variables over time be required to track the results of quantitative observations?

Plan for repeated trials, and explain how to collect data so it can be compared. You should also plan to estimate the margin of error in results.

Other types of investigations follow the same principles, adjusted to suit the investigation. A **problem-solving investigation** usually requires you to design a process or a technology to solve a problem. In designing this type of investigation, you must:

- define the problem;
- write the design specifications for whatever is being considered to solve the problem; and
- write the plans, physically construct the technology, and then evaluate it in terms of whether or not it solves the problem.

General headings for this type of investigation are: Problem, Design Specifications, Plan and Construction, Evaluation, and Communicating.

You may also be called upon to conduct an investigation in order to aid **decision-making** on an issue. In this case, you need to:

- define the issue;
- choose the relevant data and information, and gather it; and
- organize your findings so you can form your opinion and make recommendations.

Headings for this design are: Issue, Gathering Data and Information, Organizing Findings, and Opinions and Recommendations.

After the investigation is complete, you should also take some time to analyze other factors related to the design of the activity.

- Did the investigation validate or refute the scientific principles you used to make your prediction?
- Did the investigation yield unexpected results that create the need for a new set of inquiries? For example, did an analysis of various energy efficiencies raise questions about a particular technology that was so much more efficient (or inefficient)?
- Were the materials and apparatus the best for the task? Why or why not? Is there other technology available that would have been better?

Reporting the Results of an Investigation

In order for an investigation to be useful, the results must be communicated clearly. A lab report is the best format for the information. Use the following guidelines to create a neat and legible lab report that can be circulated for comment and used by others to repeat the investigation and test the results. Remember that failures must be documented as carefully as successes, and the all reports should be circulated. The report of a "failed" investigation will save others from wasting their time on the same experiments. This will allow more rapid progress on the study, or it may give some insight into a new area for investigation.

Introduction

- Begin the report with a title that clearly states the independent variable and the dependent variable, but not the outcome.
- Give the names of all participants, the date(s) of the investigation, and location of the investigation.
- Summarize the investigation concisely. Include a statement of the problem, the hypothesis, the procedures, the main results, and the conclusions. Limit this summary to 250 words or less.

Statement of the Problem

- Summarize the background of the problem.
- Cite any relevant scientific principles or literature.
- Restate the hypothesis, this time predicting the influence of the independent variable on the dependent variable.

Experimental Design

- List the steps for controlling and measuring variables through repeated trials.

Data Collection and Display

- Set out the data in clearly organized and labelled tables.
- Graph data accurately. Clearly show labels, and use the correct scale and units.

Data Analysis

- Include all data, and ensure that you can defend your analysis.

Conclusion

- Relate your conclusion to the hypothesis.
- Use the data to make extrapolations, and justify those extrapolations.
- Make recommendations for application of your conclusion or further study of the question.

Chemistry Data Tables

Table G.1 Fundamental Physical Constants (to six significant digits)

acceleration due to gravity (g)	$9.806\ 65\ \text{m/s}^2$
Avogadro constant (N_a)	$6.022\ 14 \times 10^{23}/\text{mol}$
charge on one mole of electrons (Faraday constant)	$96\ 485.3\ \text{C/mol}$
mass of electron (m_s)	$9.109\ 38 \times 10^{-31}\ \text{kg}$
mass of neutron (m_n)	$1.674\ 93 \times 10^{-27}\ \text{kg}$
mass of proton (m_p)	$1.672\ 62 \times 10^{-27}\ \text{kg}$
universal gas constant (R)	$8.314\ 47\ \dfrac{\text{kPa·L}}{\text{mol·K}}$
molar volume of gas at STP	$22.414\ 0\ \text{L/mol}$
speed of light in vacuo (c)	$2.997\ 92 \times 10^8\ \text{m/s}$
unified atomic mass (u)	$1.660\ 54 \times 10^{-27}\ \text{kg}$

Table G.2 Common SI Prefixes

tera (T)	10^{12}
giga (G)	10^{9}
mega (M)	10^{6}
kilo (k)	10^{3}
deci (d)	10^{-1}
centi (c)	10^{-2}
milli (m)	10^{-3}
micro (m)	10^{-6}
nano (n)	10^{-9}
pico (p)	10^{-12}

Table G.3 Conversion Factors

Quantity	Relationships between units
length	$1\ \text{m} = 10^{-3}\ \text{km}$ $= 10^3\ \text{mm}$ $= 10^2\ \text{cm}$
	$1\ \text{pm} = 10^{-12}\ \text{m}$
mass	$1\ \text{kg} = 10^3\ \text{g}$ $= 10^{-3}\ \text{t}$
	$1\ \text{u} = 1.66 \times 10^{-27}\ \text{kg}$
temperature	$0\ \text{K} = -273.15\ ^\circ\text{C}$
	Kelvin $\quad T = t + 273.15$ Celsius $\quad t = T - 273.15$
	mp of $H_2O = 273.15\ \text{K}$ (0°C) bp of $H_2O = 373.15\ \text{K}$ (100°C)
volume	$1\ \text{L} = 1\ \text{dm}^3$ $= 10^{-3}\ \text{m}^3$ $= 10^3\ \text{mL}$
	$1\ \text{mL} = 1\ \text{cm}^3$
pressure	$101\ 325\ \text{Pa} = 101.325\ \text{kPa}$ $= 760\ \text{mm Hg}$ $= 1\ \text{atm}$
density	$1\ \text{kg/m}^3 = 10^3\ \text{g/m}^3$ $= 10^{-3}\ \text{g/mL}$ $= 1\ \text{g/L}$
energy	$1\ \text{J} = 6.24 \times 10^{18}\ \text{eV}$

Table G.4 Alphabetical Listing of Common Polyatomic Ions

Most common ion		Common related ions	
ethanoate	CH_3COO^-		
ammonium	NH_4^+		
arsenate	AsO_4^{3-}	arsenite	AsO_3^{3-}
benzoate	$C_6H_5COO^-$		
borate	BO_3^{3-}	tetraborate	$B_4O_7^{2-}$
bromate	BrO_3^{3-}		
carbonate	CO_3^{2-}	hydrogen carbonate	HCO_3^-
chlorate	ClO_3^-	perchlorate chlorite hypochlorite	ClO_4^- ClO_2^- ClO^-
chromate	CrO_4^{2-}	dichromate	$Cr_2O_7^{2-}$
cyanide	CN^-	cyanate thiocyanate	OCN^- SCN^-
glutamate	$C_5H_8NO_4^-$		
hydroxide	OH^-	peroxide	O_2^{2-}
iodate	IO_3^-	iodide	I^-
nitrate	NO_3^-	nitrite	NO_2^-
oxalate	$OOCCOO^{2-}$		
permanganate	MnO_4^-		
phosphate	PO_4^{3-}	phosphite tripolyphosphate	PO_3^{3-} $P_3O_{10}^{5-}$
silicate	SiO_3^{2-}	orthosilicate	SiO_4^{4-}
stearate	$C_{17}H_{35}COO^-$		
sulfate	SO_4^{2-}	hydrogen sulfate sulfite hydrogen sulfite thiosulfate	HSO_4^- SO_3^{2-} HSO_3^- $S_2O_3^{2-}$
sulfide	S^{2-}	hydrogen sulfide	HS^-

Table G.5 Standard Molar Enthalpies of Formation

Substance	Δ_fH° (kJ/mol)	Substance	Δ_fH° (kJ/mol)	Substance	Δ_fH° (kJ/mol)
$Al_2O_3(s)$	−1675.7	$HBr(g)$	−36.3	$NH_3(g)$	−45.9
$CaCO_3(s)$	−1207.6	$HCl(g)$	−92.3	$N_2H_4(\ell)$	+50.6
$CaCl_2(s)$	−795.4	$HF(g)$	−273.3	$NH_4Cl(s)$	−314.4
$Ca(OH)_2(s)$	−985.2	$HCN(g)$	+135.1	$NH_4NO_3(s)$	−365.6
$CCl_4(\ell)$	−128.2	$H_2O(\ell)$	−285.8	$NO(g)$	+91.3
$CCl_4(g)$	−95.7	$H_2O(g)$	−241.8	$NO_2(g)$	+33.2
$CHCl_3(\ell)$	−134.1	$H_2O_2(\ell)$	−187.8	$N_2O(g)$	+81.6
$CH_4(g)$	−74.6	$HNO_3(\ell)$	−174.1	$N_2O_4(g)$	+11.1
$C_2H_2(g)$	+227.4	$H_3PO_4(s)$	−1284.4	$PH_3(g)$	+5.4
$C_2H_4(g)$	+52.4	$H_2S(g)$	−20.6	$PCl_3(g)$	−287.0
$C_2H_6(g)$	−84.0	$H_2SO_4(\ell)$	−814.0	$P_4O_6(s)$	−2144.3
$C_3H_8(g)$	−103.8	$FeO(s)$	−272.0	$P_4O_{10}(s)$	−2984.0
$C_6H_6(\ell)$	+49.1	$Fe_2O_3(s)$	−824.2	$KBr(s)$	−393.8
$CH_3OH(\ell)$	−239.2	$Fe_3O_4(s)$	−1118.4	$KCl(s)$	−436.5
$C_2H_5OH(\ell)$	−277.6	$FeCl_2(s)$	−341.8	$KClO_3(s)$	−397.7
$CH_3COOH(\ell)$	−484.3	$FeCl_3(s)$	−399.5	$KOH(s)$	−424.6
$CO(g)$	−110.5	$FeS_2(s)$	−178.2	$Ag_2CO_3(s)$	−505.8
$CO_2(g)$	−393.5	$PbCl_2(s)$	−359.4	$AgCl(s)$	−127.0
$COCl_2(g)$	−219.1	$MgCl_2(s)$	−641.3	$AgNO_3(s)$	−124.4
$CS_2(\ell)$	+89.0	$MgO(s)$	−601.6	$Ag_2S(s)$	−32.6
$CS_2(g)$	+116.7	$Mg(OH)_2(s)$	−924.5	$SF_6(g)$	−1220.5
$CrCl_3(s)$	−556.5	$HgS(s)$	−58.2	$SO_2(g)$	−296.8
$Cu(NO_3)_2(s)$	−302.9	$NaCl(s)$	−411.2	$SO_3(g)$	−395.7
$CuO(s)$	−157.3	$NaOH(s)$	−425.6	$SnCl_2(s)$	−325.1
$CuCl(s)$	−137.2	$Na_2CO_3(s)$	−1130.7	$SnCl_4(\ell)$	−511.3
$CuCl_2(s)$	−220.1				

Note: The enthalpy of formation of an element in its standard state is defined as zero.

Table G.6 Solubility of Some Common Ionic Compounds in Water at 298.15 K

Ion	$H^+, Na^+, NH_4^+,$ $NO_3^-, ClO_3^-,$ ClO_4^-, CH_3COO^-	F^-	$Cl^-, I^-,$ Br^-	SO_4^{2-}	$CO_3^{2-},$ $PO_4^{3-},$ SO_3^{2-}	$IO_3^-,$ $C_2O_4^{2-}$	S^{2-}	OH^-
Solubility greater than or equal to 0.1 mol/L (**very soluble**)	most Except: $RbClO_4$, $CsClO_4$, $AgCH_3COO$, $Hg_2(CH_3COO)_2$	most	most	most	$H^+, Na^+,$ K^+, NH_4^+ Except: Li_2CO_3	$H^+, Na^+, K^+,$ $NH_4^+, Li^+,$ Ni^{2+}, Zn^{2+} Except: $Co(IO_3)_2,$ $Fe_2(C_2O_4)_3$	$H^+, Na^+,$ $K^+, NH_4^+,$ $Li^+, Mg^{2+},$ Ca^{2+}	$H^+, Na^+,$ $K^+, NH_4^+,$ $Li^+, Sr^{2+},$ Ca^{2+}, Ba^{2+}
Solubility less than 0.1 mol/L (**slightly soluble**)	none	$Li^+, Mg^{2+},$ $Ca^{2+}, Sr^{2+},$ $Ba^{2+}, Fe^{2+},$ Hg_2^{2+}, Pb^{2+}	$Cu^+, Ag^+,$ $Hg_2^{2+}, Hg^{2+},$ PbI_2	$Ca^{2+}, Sr^{2+},$ $Ba^{2+}, Hg_2^{2+},$ Pb^{2+}, Ag^+	most	most	most	most

Note: The solubility table is only a guideline that is established using the K_{sp} values. A concentration of 0.1 mol/L corresponds to approximately 10 g/L to 30 g/L depending on molecular mass.

Table G.7 Standard Reduction Potentials

Reduction half reaction	$E°$(V)
$F_2(g) + 2e^- \rightleftharpoons 2F^-(aq)$	2.87
$Co^{3+}(aq) + e^- \rightleftharpoons Co^{2+}(aq)$	1.92
$H_2O_2(aq) + 2H^+(aq) + 2e^- \rightleftharpoons 2H_2O(\ell)$	1.78
$Ce^{4+}(aq) + e^- \rightleftharpoons Ce^{3+}(aq)$	1.72
$PbO_2(s) + 4H^+(aq) + SO_4^{2-}(aq) + 2e^- \rightleftharpoons PbSO_4(s) + H_2O(\ell)$	1.69
$MnO_4^-(aq) + 8H^+(aq) + 5e^- \rightleftharpoons Mn^{2+}(aq) + 4H_2O(\ell)$	1.51
$Au^{3+}(aq) + 3e^- \rightleftharpoons Au(s)$	1.50
$PbO_2(s) + 4H^+(aq) + 2e^- \rightleftharpoons Pb^{2+}(aq) + 2H_2O(\ell)$	1.46
$Cl_2(g) + 2e^- \rightleftharpoons 2Cl^-(aq)$	1.36
$Cr_2O_7^{2-}(aq) + 14H^+(aq) + 6e^- \rightleftharpoons 2Cr^{3+}(aq) + 7H_2O(\ell)$	1.23
$O_2(g) + 4H^+(aq) + 4e^- \rightleftharpoons 2H_2O(\ell)$	1.223
$MnO_2(s) + 4H^+(aq) + 2e^- \rightleftharpoons Mn^{2+}(aq) + 2H_2O(\ell)$	1.22
$IO_3^-(aq) + 6H^+(aq) + 6e^- \rightleftharpoons I^-(aq) + 3H_2O(\ell)$	1.09
$Br_2(\ell) + 2e^- \rightleftharpoons 2Br^-(aq)$	1.07
$AuCl_4^-(aq) + 3e^- \rightleftharpoons Au(s) + 4Cl^-(aq)$	1.00
$NO_3^-(aq) + 4H^+(aq) + 3e^- \rightleftharpoons NO(g) + 2H_2O(\ell)$	0.96
$2Hg^{2+}(aq) + 2e^- \rightleftharpoons Hg_2^{2+}(aq)$	0.92
$Ag^+(aq) + e^- \rightleftharpoons Ag(s)$	0.80
$Hg_2^{2+}(aq) + 2e^- \rightleftharpoons 2Hg(\ell)$	0.80
$Fe^{3+}(aq) + e^- \rightleftharpoons Fe^{2+}(aq)$	0.77
$O_2(g) + 2H^+(aq) + 2e^- \rightleftharpoons H_2O_2(aq)$	0.70
$I_2(s) + 2e^- \rightleftharpoons 2I^-(aq)$	0.54
$Cu^+(aq) + e^- \rightleftharpoons Cu(s)$	0.52
$O_2(g) + 2H_2O(\ell) + 4e^- \rightleftharpoons 4OH^-(aq)$	0.40
$Cu^{2+}(aq) + 2e^- \rightleftharpoons Cu(s)$	0.34
$AgCl(s) + e^- \rightleftharpoons Ag(s) + Cl^-(aq)$	0.22
$4H^+(aq) + SO_4^{2-}(aq) + 2e^- \rightleftharpoons H_2SO_3(aq) + H_2O(\ell)$	0.17
$Cu^{2+}(aq) + e^- \rightleftharpoons Cu^+(aq)$	0.15
$2H^+(aq) + 2e^- \rightleftharpoons H_2(g)$	0.000
$Fe^{3+}(aq) + 3e^- \rightleftharpoons Fe(s)$	−0.04
$Pb^{2+}(aq) + 2e^- \rightleftharpoons Pb(s)$	−0.13
$Sn^{2+}(aq) + 2e^- \rightleftharpoons Sn(s)$	−0.14
$Ni^{2+}(aq) + 2e^- \rightleftharpoons Ni(s)$	−0.26
$Cd^{2+}(aq) + 2e^- \rightleftharpoons Cd(s)$	−0.40
$Cr^{3+}(aq) + e^- \rightleftharpoons Cr^{2+}(aq)$	−0.41
$Fe^{2+}(aq) + 2e^- \rightleftharpoons Fe(s)$	−0.45
$Cr^{3+}(aq) + 3e^- \rightleftharpoons Cr(s)$	−0.74
$Zn^{2+}(aq) + 2e^- \rightleftharpoons Zn(s)$	−0.76
$2H_2O(\ell) + 2e^- \rightleftharpoons H_2(g) + 2OH^-(aq)$	−0.83
$Al^{3+}(aq) + 3e^- \rightleftharpoons Al(s)$	−1.66
$Mg^{2+}(aq) + 2e^- \rightleftharpoons Mg(s)$	−2.37
$La^{3+}(aq) + 3e^- \rightleftharpoons La(s)$	−2.38
$Na^+(aq) + e^- \rightleftharpoons Na(s)$	−2.71
$Ca^{2+}(aq) + 2e^- \rightleftharpoons Ca(s)$	−2.87
$Ba^{2+}(aq) + 2e^- \rightleftharpoons Ba(s)$	−2.91
$K^+(aq) + e^- \rightleftharpoons K(s)$	−2.93
$Li^+(aq) + e^- \rightleftharpoons Li(s)$	−3.04

Table G.8 Colours of Some Common Aqueous Ions

Ionic Species	Solution Concentration	
	1.0 mol/L	0.010 mol/L
chromate	yellow	pale yellow
chromium(III)	blue-green	green
chromium(II)	dark blue	pale blue
cobalt(II)	red	pink
copper(I)	blue-green	pale blue-green
copper(II)	blue	pale blue
dichromate	orange	pale orange
iron(II)	lime green	colourless
iron(III)	orange-yellow	pale yellow
manganese(II)	pale pink	colourless
nickel(II)	blue-green	pale blue-green
permanganate	deep purple	purple-pink

Table G.9 The Flame Colour of Selected Metallic Ions

Ion	Symbol	Colour
lithium	Li^+	red
sodium	Na^+	yellow
potassium	K^+	violet
cesium	Cs^+	violet
calcium	Ca^{2+}	red
strontium	Sr^{2+}	red
barium	Ba^{2+}	yellowish-green
copper	Cu^{2+}	bluish-green
boron	B^{2+}	green
lead	Pb^{2+}	bluish-white

Table G.10 End-point Indicators

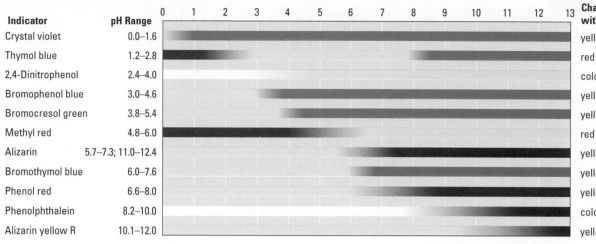

Indicator	pH Range	pH 0–13 scale	Change of Colour with Increasing pH
Crystal violet	0.0–1.6		yellow to blue
Thymol blue	1.2–2.8		red to yellow
2,4-Dinitrophenol	2.4–4.0		colourless to yellow
Bromophenol blue	3.0–4.6		yellow to blue
Bromocresol green	3.8–5.4		yellow to blue
Methyl red	4.8–6.0		red to yellow
Alizarin	5.7–7.3; 11.0–12.4		yellow to red; red to violet
Bromothymol blue	6.0–7.6		yellow to blue
Phenol red	6.6–8.0		yellow to red
Phenolphthalein	8.2–10.0		colourless to pink
Alizarin yellow R	10.1–12.0		yellow to red

Table G.11 Relative Strengths of Acids and Bases (concentration = 0.10 mol/L) at 25°C

Acid Name	Acid Formula	Formula of Conjugate Base	K_a
perchloric acid	$H(aq)ClO_4(aq)$	$ClO_4^-(aq)$	very large
hydroiodic acid	$HI(aq)$	$I^-(aq)$	very large
hydrobromic acid	$HBr(aq)$	$Br^-(aq)$	very large
hydrochloric acid	$HCl(aq)$	$Cl^-(aq)$	very large
sulfuric acid	$H_2SO_4(aq)$	$HSO_4^-(aq)$	very large
nitric acid	$HNO_3(aq)$	$NO_3^-(aq)$	very large
hydronium ion	$H_3O^+(aq)$	$H_2O(\ell)$	1
oxalic acid	$HOOCCOOH(aq)$	$HOOCCOO^-(aq)$	5.6×10^{-2}
sulfurous acid ($SO_2 + H_2O$)	$H_2SO_3(aq)$	$HSO_3^-(aq)$	1.4×10^{-2}
hydrogen sulfate ion	$HSO_4^-(aq)$	$SO_4^{2-}(aq)$	1.0×10^{-2}
phosphoric acid	$H_3PO_4(aq)$	$H_2PO_4^-(aq)$	6.9×10^{-3}
nitrous acid	$HNO_2(aq)$	$NO_2^-(aq)$	5.6×10^{-4}
citric acid	$H_3C_6H_5O_7(aq)$	$H_2C_6H_5O_7^-(aq)$	7.4×10^{-4}
hydrofluoric acid	$HF(aq)$	$F^-(aq)$	6.3×10^{-4}
methanoic acid	$HCOOH(aq)$	$HCOO^-(aq)$	1.8×10^{-4}
hydrogen oxalate ion	$HOOCCOO^-(aq)$	$OOCCOO^{2-}(aq)$	1.5×10^{-4}
ascorbic acid	$C_6H_8O_6(aq)$	$C_6H_7O_6^-(aq)$	9.1×10^{-5}
benzoic acid	$C_6H_5COOH(aq)$	$C_6H_5COO^-(aq)$	6.3×10^{-5}
ethanoic (acetic) acid	$CH_3COOH(aq)$	$CH_3COO^-(aq)$	1.8×10^{-5}
dihydrogen citrate ion	$H_2C_6H_5O_7^-(aq)$	$HC_6H_5O_7^{2-}(aq)$	1.7×10^{-5}
carbonic acid ($CO_2 + H_2O$)	$H_2CHO_3(aq)$	$HCO_3^-(aq)$	4.5×10^{-7}
hydrogen citrate ion	$HC_6H_5O_7^{2-}(aq)$	$C_6H_5O_7^{3-}(aq)$	4.0×10^{-7}
hydrosulfuric acid	$H_2S(aq)$	$HS^-(aq)$	8.9×10^{-8}
hydrogen sulfite ion	$HSO_3^-(aq)$	$SO_3^{2-}(aq)$	6.3×10^{-8}
dihydrogen phosphate ion	$H_2PO_4^-(aq)$	$HPO_4^{2-}(aq)$	6.2×10^{-8}
hypochlorous acid	$HOCl(aq)$	$OCl^-(aq)$	4.0×10^{-8}
hydrocyanic acid	$HCN(aq)$	$CN^-(aq)$	6.2×10^{-10}
ammonium ion	$NH_4^+(aq)$	$NH_3(aq)$	5.6×10^{-10}
hydrogen carbonate ion	$HCO_3^-(aq)$	$CO_3^{2-}(aq)$	4.7×10^{-11}
hydrogen ascorbate ion	$C_6H_7O_6^-(aq)$	$C_6H_6O_6^{2-}(aq)$	2.0×10^{-12}
hydrogen phosphate ion	$HPO_4^{2-}(aq)$	$PO_4^{3-}(aq)$	4.8×10^{-13}
water (55.5 mol/L)	$H_2O(\ell)$	$HO^-(aq)$	1.0×10^{-14}

Increasing Acid Strength →

Increasing Base Strength →

How to Titrate

Figure H.1 Draw more solution than you need into the pipette. It is easier to reduce this volume than it is to bring more solution into the pipette.

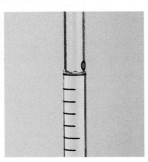

Figure H.2 The bottom of the meniscus must be exactly in line with the etched mark on the pipette.

Rinsing the Pipette

As you learned in Chapter 5, a volumetric pipette is calibrated to accurately deliver a certain volume of solution. A graduated pipette can deliver different measured volumes of solution. A pipette should be cleaned and then rinsed with the sample solution whose volume you are measuring.

1. Pour a sample of standard solution into a clean dry beaker and ensure that its volume is about two to three times that of the pipette.
2. Place the pipette tip in a beaker of distilled water. Squeeze the suction bulb. Maintain your grip on the bulb while placing it on the end of the pipette and hold it there firmly. (If the suction bulb has valves, your teacher will show you how to use them.)
3. Relax your grip on the bulb and let it draw up a small volume of distilled water.
4. Remove the bulb and discard the distilled water by letting it drain out the end of the pipette.
5. Rinse the pipette by drawing up several millilitres of the standard solution from the beaker. Rotate and rock the pipette to coat the inner surface with the solution. Discard this rinsing solution in the appropriate container. Rinse the pipette twice in this way. The pipette is now ready to fill with the standard solution.

Filling the Pipette

6. Place the pipette tip below the surface of the solution you are collecting.
7. Hold the suction bulb loosely on the end of the glass stem. Use the bulb to draw liquid up just past the volume mark (see Figure H.1).
8. As quickly and smoothly as you can, remove the bulb and place your index finger over the end of the glass stem. The solution level must still be above the volume mark.
9. Slowly rotate the pipette stem back and forth under your finger to let some of the solution drain out.
10. When the bottom of the meniscus is in line with the volume mark, as in Figure H.2, press down on the stem so that the solution stops draining out of the pipette.
11. Touch the tip of the pipette to the inside wall of the beaker to remove any clinging drop of solution. The measured volume is now ready to be delivered to an Erlenmeyer flask.

Transferring the Solution

12. Place the tip of the pipette against the inside glass wall of the flask. Let the solution drain slowly by removing your finger from the stem.
13. After the solution drains, wait several seconds, then touch the tip of the pipette to the inside wall of the flask to remove any drop on the end (see Figure H.3). Note that a small quantity of liquid remains in the tip of the pipette. The pipette was calibrated to retain this small volume. Do not try to remove it.

Adding the Indicator

14. Add two or three drops of the appropriate indicator to the solution in the Erlenmeyer flask. Do not use too much indicator. Using more than a few drops does not make a colour change easier to see. Also, many indicators are weak acids or bases. Adding too much indicator can change the volume of titrant needed for neutralization, thereby affecting the accuracy of your titration. You are now ready to prepare the apparatus for a titration.

Rinsing the Burette

A burette is used to accurately measure the volume of liquid during a titration. It is a graduated glass tube with a tap (also called a *stopcock*) at one end.

15. To rinse the burette, close the tap and fill it with about 10 ml of distilled water from a wash bottle.

16. Hold the burette at an angle and roll it back and forth so that the water comes in contact with all the inner surfaces.

17. Hold the burette over the sink, open the tap, and let the water drain out. As you do this, check the tap for leaks. Make sure that the tap turns smoothly.

18. Rinse the burette with 5–10 mL of the solution that will be measured. As you discard this solution into a waste beaker, remember to rinse the lower portion of the burette. Rinse the burette twice, discarding the liquid each time.

Filling the Burette

19. Assemble a retort stand and a burette clamp to hold the burette. Place a clean dry funnel in the top of the burette. Place a waste beaker under the burette.

20. Close the tap. Add solution to the burette until the liquid level is just above the zero mark. Remove the funnel. Carefully open the tap and drain some liquid into the waste beaker. Stop when the meniscus is at or below the zero mark.

21. Touch the tip of the burette against the inside wall of the beaker to remove any clinging drop of solution. Check that the portion of the burette below the tap is filled with solution and does not contain air bubbles.

22. Record the initial burette volume in your notebook. (The scale on a burette is read top to bottom.) Do not try to get an initial reading of 0.00 mL.

23. Replace the waste beaker with the Erlenmeyer flask containing the sample you prepared earlier. Make sure that the tip of the burette is just below the lip of the flask. Place a sheet of white paper under the flask to help highlight the colour change that is the endpoint of the titration. You are now ready to titrate the sample.

Titrating

24. Carefully add titrant to the Erlenmeyer flask until you reach the endpoint, as shown in Figure H.4. When you notice the indicator changing colour as the solution enters the flask from the burette, you are near the endpoint. Begin to add solution drop by drop from the burette. Mix the solution thoroughly after each drop.

25. When you reach the endpoint, read the final volume. When you read the burette volume, always make sure that your eye is level with the bottom of the meniscus. You may want to use a meniscus reader. A meniscus reader is a small white card with a thick black line on it. Hold the card behind the burette, with the black line just under the meniscus, as shown in Figure H.5. Record the volume of solution added from the burette to the nearest 0.05 mL.

Figure H.3 The last drop of solution on the tip of the pipette is delivered to the sample by touching the side of the Erlenmeyer flask as shown. A small amount of solution will remain in the pipette.

Figure H.4 You may need to practise in order to operate the tap and swirl the flask at the same time.

Titration Tip

Use one hand to operate the tap and control the flow of titrant. Use your other hand to swirl the contents of the Erlenmeyer flask. As you get close to the endpoint, you may want to use both hands to control the tap, adding one drop and then swirling after each drop.

Figure H.5 A meniscus reader helps you read the volume of solution in the in the burette.

How to Prepare and Dilute a Standard Solution

What do the effectiveness of a medicine, the safety of a chemical reaction, the cost of an industrial process, and the taste of a soft drink have in common? They all depend on solutions that are made carefully with known concentrations. A solution with a known concentration is called a **standard solution**. There are two ways to prepare an aqueous solution with known concentration. You can make a solution by dissolving a measured mass of pure solute in a certain volume of solution. Alternatively, you can dilute a standard solution.

Preparing a Standard Solution

To prepare a solution of known concentration from a solute and a solvent, you need an electronic balance, a beaker, a scoopula, and a volumetric flask. A **volumetric flask** is a pear-shaped glass container with a flat base and a long neck. Volumetric flasks are used to make up standard solutions. They are available in a variety of sizes. Each size is calibrated to measure a fixed volume of solution to ± 0.1 mL at a particular temperature, usually 20 °C.

To prepare an aqueous solution of known concentration, follow these steps:

1. Measure the required mass of solute.

2. Dissolve the solute in some distilled water in a beaker. Use about 50% to 60% of the desired final volume of the solution. For example, if you are making 100.0 mL of solution, use about 50 mL to 60 mL of distilled water to dissolve the solute. Ensure the solute is completely dissolved.

3. Rinse your volumetric flask with fresh distilled water several times. Discard the water.

4. Transfer your solution from the beaker to the volumetric flask using a funnel.

5. Rinse the beaker and transfer the rinsing to the flask. Rinse the funnel and allow the rinsing to flow into the flask, as shown in Figure I.1.

6. Carefully fill the volumetric flask just to the line with distilled water, as shown in Figure I.2.

7. Place a cap on the flask and invert several times to mix completely, as shown in Figure I.3.

Figure I.1 After adding the solution to the flask, rinse all apparatus (beaker, funnel, stirring rod, etc.) into the flask as well.

Figure I.2 Add distilled water until it nearly reaches the line on the volumetric flask. Then add water very slowly, drop by drop, until the bottom of the meniscus touches the line.

Figure I.3 Ensure the cap fits well before you invert the flask to mix the solution. Keep your thumb on the cap.

If you were performing an experiment in which significant digits and errors were important, you would record the volume of a solution in a 500 mL volumetric flask as 500.0 mL ±0.1 mL. The volume of a solution in a 100 mL volumetric flask would be 100.0 mL ±0.1 mL.

Standard solutions are not usually stored in volumetric flasks. Instead, they are transferred to another bottle or flask that has a secure cap. Often, the storage flask is made of dark glass to keep light-sensitive substances from degrading.

Diluting a Solution

You can use a standard solution to make a less-concentrated solution by adding additional solvent. For example, if you wanted to prepare 100.0 mL of 0.100 mol/L solution, you could dilute a 1.00 mol/L solution. Since the standard solution is 10 times as concentrated as the solution you want to prepare, you would dilute the standard solution by a factor of 10. The volume of the standard solution to be diluted is precisely measured using a **pipette**. A pipette is a thin glass tube that is used to measure a precise volume of solution. As you can see in Figure I.4, there are two varieties of pipettes: graduated and volumetric.

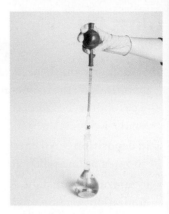

Figure I.5 Squeeze some air from the pipette bulb before lightly placing it over the pipette to draw up the solution. This technique is also used to rinse the pipette.

Figure I.4 Pipette A is a graduated pipette: It is calibrated to measure a number of volumes, like a graduated cylinder. Pipette B is a volumetric pipette: It is calibrated to measure one volume only, like a volumetric flask. A pipette bulb is used with both types.

To dilute a solution, follow these steps:

1. Rinse the pipette with the standard solution. Discard the rinse.

2. Draw the solution into the pipette as shown in Figure I.5.

3. Remove the bulb and quickly place your index finger over the top of the pipette.

4. By slightly removing your index finger, release the solution from the pipette until the bottom of the meniscus is on the line etched into the glass. You may need to practise to get the hang of this technique.

5. Release the standard solution into a clean volumetric flask. *Do not* force the last drop from the pipette using the bulb. Instead, touch the pipette to the side of the flask as shown in Figure I.6. A small volume of standard solution will remain in the pipette. Pipettes are calibrated in the factory based on how much solution they *deliver*, not on how much solution they *contain*.

6. Fill the volumetric flask to the etched line with solvent, stopper the flask, and invert several times to mix.

Figure I.6 Drain the contents of the pipette into the volumetric flask. Do not force out the last drop in the pipette, just touch the tip to the side of the flask.

If pipettes are not available, you can use a graduated cylinder to measure the volume of standard solution to be diluted. A graduated cylinder is less precise than a pipette, however. The precision of the glassware depends on the magnitude of the smallest gradation. A 10 mL pipette is generally calibrated to ±0.05 mL, whereas most common 10 mL graduated cylinders are calibrated to ±0.1 mL.

Glossary

absolute zero the theoretically the lowest possible temperature; the temperature at which the volume of a gas approaches zero; equivalent to −273.15 °C, or 0 on the Kelvin scale (3.3)

acid deposition total effect of acid falling in rain, in snow, or as fine solid particles (17.1)

acid ionization constant, K_a equilibrium constant for ionization reaction of acids (17.3)

acid rain term used to describe the lowering of pH of rain to less than pH 5, caused by human activities (17.1)

acid-base indicator a weak acid that has different colours in its acid and base forms (6.3)

acidosis condition in which the blood and tissues are more acidic than is normal (17.4)

activated complex chemical species temporarily formed by the reactant molecules as a result of a collision before they form the product (11.1)

activation energy, E_a the minimum amount of energy (collision energy) required to initiate a chemical reaction (11.1)

active site the region of any enzyme at which catalysis occurs, is the site of binding for the reactant molecules (substrate) (11.2)

addition polymerization reaction in which alkene monomers are joined through multiple addition reactions to form a polymer (15.2)

addition reaction reaction in which atoms are added to a carbon-carbon double or triple bond (15.1)

alcohol hydrocarbon derivative that contains an −OH, or hydroxyl, functional group (14.3)

aliphatic hydrocarbon compound containing only carbon and hydrogen in which carbon atoms form chains and/or non-aromatic rings (14.2)

alkali metal element in Group 1 of the periodic table; is highly reactive and reacts explosively with water (Unit 1 Prep.)

alkaline battery battery in which ammonium chloride and zinc chloride used in a dry cell are replaced by strongly alkaline potassium hydroxide, KOH (13.2)

alkaline earth metal element in Group 2 of the periodic table; is not as reactive as the alkali metals but is nevertheless quite reactive(Unit 1 Prep.)

alkaline fuel cell fuel cell developed for NASA's Apollo program and still used in space shuttles; the electrolyte is aqueous potassium hydroxide; compare *direct methanol fuel cell (DMFC)*, *phosphoric acid fuel cell*, and *proton exchange membrane (PEM) fuel cell* (13.2)

alkalosis condition in which the blood and tissues are more alkaline than is normal (above pH 7.4); compare *acidosis* (17.4)

alkane an aliphatic hydrocarbon molecule in which the carbon atoms are joined by single covalent bonds (also considered saturated hydrocarbons because they contain no double or triple bonds) (14.2)

alkene an aliphatic hydrocarbon molecule that contains one or more carbon-carbon double bonds (14.2)

alkyl group a side group that is based on an alkane (14.2)

alkyl halide hydrocarbon derivitive that contains at least one halogen atom (14.3)

alkylation type of reforming in which short-chain alkenes produced in catalytic cracking are chemically joined under controlled heat and pressure in the presence of an acid catalyst, such as sulfuric acid or hydrofluoric acid; the resulting molecules are used to upgrade gasoline for high-performance engines and aviation fuel (14.4)

alkyne an aliphatic hydrocarbon molecule that contains one or more carbon-carbon triple bonds (14.2)

allotrope one of two or more crystalline or molecular forms of an element (14.1)

amide organic functional group characterized by a carbonyl group (C = O) linked to a nitrogen atom (N) (15.2)

amino acid monomer of protein (15.2)

amount concentration unit of concentration expressed as the number of moles of solute dissolved in one litre of solution; also known as *molar concentration* and *molarity* (5.3)

amphiprotic describes substances, such as water, that can act as a Brønsted-Lowry acid in one reaction and as a Brønsted-Lowry base in a different reaction (17.2)

anode the electrode at which oxidation occurs; compare *cathode* (13.1)

antilog the number for which a given logarithm stands; for example, where log x equals y, then x is the antilogarithm of y. (6.3)

aqueous solution any solution for which water is the solvent (5.1)

aromatic having a strong odour (14.2)

aromatic compound organic compound based on the aromatic benzene ring (14.2)

aromatic hydrocarbon compound containing only carbon and hydrogen and based on the aromatic benzene ring (14.2)

Arrhenius acid substance that ionizes to form hydrogen ions, $H^+(aq)$, in aqueous solutions (6.1)

Arrhenius base substance that dissociates to produce hydroxide ions, $OH^-(aq)$, in aqueous solutions (6.1)

Arrhenius theory of acids and bases theory explaining the nature of acids and bases in terms of their structure and the ions produced when they dissolve in water; see *Arrhenius acid* and *Arrhenius base*; compare *modified Arrhenius theory of acids and bases* (6.1)

atomic molar mass mass of one mole of atoms of the element; a weighted average of the molar masses of the naturally occurring isotopes of the element (Unit 1 Prep.)

atomic number number of protons in the nucleus, symbolized by Z (Unit 1 Prep.)

Avogadro's law gas law stating that equal volumes of all ideal gases at the same temperature and pressure contain the same number of molecules (4.1)

B

base ionization constant, K_b equilibrium constant for a dissociation reaction or reaction with water for bases (17.3)

battery a set of voltaic cells connected in series, so the negative electrode of one cell is connected to the positive electrode of the next; the voltage is the sum of the voltages of the individual cells (13.2)

bent angular molecular shape that results when a central atom has one or two lone pairs and two electron groupings around it (2.1)

benzene a cyclic, aromatic hydrocarbon, C_6H_6, in which all six carbon-carbon bonds are intermediate in length between a single and double bond; delocalized electrons are shared by all six carbon atoms (14.2)

binary ionic compound ionic compound that consists of two different elements, one metal and one non-metal, grouped together in a lattice structure (Unit 1 Prep.)

bioaccumulation build-up of substances to higher concentrations in the tissues of living organisms compared to the non-living environment (water or soil) (5.3)

bioluminescence production of light by a living organism through a chemical reaction which converts chemical energy to light energy (12.3)

biomagnifications accumulation of a compound in body tissues of predators such that the concentration increases in organisms higher in the food chain (5.3)

blast furnace furnace in which smelting occurs (12.2)

boiling point temperature at which a liquid becomes a gas (2.3)

bomb calorimeter device that measures heat released during a combustion reaction at a constant volume (9.1, 9.2)

bond angle angle formed by two bonds around a central atom (2.1)

bond dipole bond in which the electrons spend more time around one atom (more electronegative) than the other atom creating a partial positive charge at one end of the bond and a partial negative charge at the other end (1.2)

bonding pair electron pair shared by two atoms in a molecule (1.1)

Boyle's law gas law stating that the volume of a fixed amount of gas at a constant temperature is inversely proportional to the applied (external) pressure on the gas: $P_1V_1 = P_2V_2$ (3.2)

Brønsted-Lowry theory of acids and bases theory defines an acid as a substance that can donate a proton (H^+ ion), and a base as a substance that can accept a proton (17.2)

buffer capacity amount of acid or base that can be added to a solution before considerable change occurs to the pH of the solution (17.4)

buffer solution solution that resists changes in pH when a moderate amount of acid or base is added; contains a weak acid–conjugate base mixture (17.4)

button battery a battery much smaller than an alkaline; commonly used in watches, hearing aids, pacemakers, and some calculators and cameras (13.2)

C

calorimeter device used to measure the heat released or absorbed during a chemical or physical process occurring within it (9.2)

calorimetry technological process of measuring the heat released or absorbed during a chemical or physical process (9.2)

carboxylic acid hydrocarbon derivative that contains a –COOH, or carboxyl, functional group (14.3)

carboxyl group name of the functional group for carboxylic acid (–COOH) (14.3)

catalyst a substance that increases the rate of a chemical reaction without being consumed by the reaction (11.2)

catalytic cracking low-pressure process that involves passing a powdered catalyst through heavy hydrocarbon fractions at temperatures around 600°C; catalyst speeds up the cracking process and favours the production of diesel oils, gasoline, and kerosene (14.4)

cathode the electrode at which reduction occurs; compare *anode* (13.1)

cathodic protection method of preventing the rusting of an iron object, using a more reactive metal that acts as the sacrificial anode; unlike in galvanizing, the metal used in cathodic protection does not completely cover the iron (13.2)

cell notation a shorthand method for representing voltaic cells (13.1)

cell potential, E_{cell} electrical potential difference between the electrodes of an electrochemical cell when no current flows (13.1)

cellulose natural polymer of glucose molecules that provides most of the structure of plants (15.2)

charge in the process of smelting, term used for the mixture containing pulverized iron ore—usually hematite, $Fe_2O_3(s)$—limestone, $CaCO_3(s)$, and coke, $C(s)$, which is poured into the top of the blast furnace; also refers to electric charge (12.2)

Charles's law gas law stating that the volume of a fixed amount of gas at a constant pressure is directly proportional to the absolute (Kelvin) temperature of the gas: $\frac{V_1}{T_1} = \frac{V_2}{T_2}$, where T is the Kelvin temperature (3.3)

chlor-alkali process industrial method that electrolyzes concentrated aqueous sodium chloride and produces chlorine gas, hydrogen gas, and aqueous sodium hydroxide (13.3)

closed system system which allows energy to enter or leave but not matter; compare *isolated system* and *open system* (Unit 5 Prep.)

coke fuel made from coal by heating it in a closed oven until the gases have been removed; burns with much heat and little smoke, and is used in furnaces, for melting metal, etc.; solid waste left after the purification of petroleum (14.4)

collision theory model that explains reaction rate as being the result of particles (atoms, molecules, or ions) colliding with a certain minimum energy (11.1)

competing reaction secondary reaction that occurs when the same two chemicals can react to give different products (8.2)

complete balanced equation equation for a chemical reaction showing all the reactants and products as if they were intact compounds that had not dissociated in solution (7.1)

complete combustion reaction process by which hydrocarbon reacts with oxygen to produce carbon dioxide, water vapour, and energy (Unit 4 Prep.; 7.1, 15.1)

concentrated contains a large ratio of solute to solvent; compare *dilute* (5.3)

concentration quantity of solute per quantity of solvent in a solution (5.3)

condensation polymerization reaction in which monomers are combined through multiple condensation reactions to form a polymer (15.2)

condensation reaction reaction in which two molecules combine to form a larger molecule, producing a small, stable molecule, usually water, as a second product (15.1)

conjugate acid of a base, is the molecule formed when the base receives a proton from an acid (17.2)

conjugate acid-base pair two molecules or ions that differ because of the transfer of a proton (17.2)

conjugate base of an acid, is the molecule formed when a proton is removed from the acid (17.2)

corrosion the natural redox process that results in unwanted oxidation of a metal; e.g., rusting (13.2)

covalent bond electrostatic attraction between the nuclei of two adjacent atoms (usually non-metal) and a pair of shared bonding electrons (Unit 1 Prep., 1.1)

cracking process that uses heat under pressure, in the absence of air, to break the carbon-carbon bonds in large hydrocarbon molecules to form smaller molecules (14.4)

cross-linking in a polymer, the formation of bonds from one strand to another at several points along the polymer strand (15.2)

cyclic hydrocarbon an aliphatic hydrocarbon chain that forms a ring (but not a benzene ring); can be a cycloalkane, cycloalkene, or cycloalkyne; is non-polar and has physical properties similar to its straight chain counterpart; compare *aromatic hydrocarbon* (14.2)

D

Dalton's law of partial pressures gas law stating that, in a mixture of unreacting gases, the total pressure is the sum of the partial pressures of the individual gases: $P_{total} = P_1 + P_2 + ...$ (4.2)

Daniell cell voltaic cell in which one half cell consists of a zinc electrode in zinc sulfate solution and the other half cell consists of a copper electrode in copper(II) sulfate solution (13.1)

decomposition reaction class of chemical reaction in which a compound breaks down into two or more smaller compounds or elements (Unit 4 Prep., 7.2)

degradable plastic polymer that breaks down over time when exposed to environmental conditions such as light and bacteria (15.2)

delocalized refers to electrons that are free to move from one atom to the next (1.1)

delocalized electron in a molecule, electron not associated with a single atom but are shared by three or more atoms; e.g., as found in benzene rings (14.2)

dilute contains a small ratio of solute to solvent; compare *concentrated* (5.3)

dipole a molecule in which the electrons spend more time around one nucleus than another nucleus creating a partial positive charge in one part of the molecule and a partial negative charge in another part of the molecule (2.2)

dipole-dipole attraction intermolecular force between oppositely charged ends of two polar molecules (molecules with dipoles) (2.2)

direct methanol fuel cell (DMFC) fuel cell currently under development to use methanol rather than hydrogen; compare *alkaline fuel cell*, *phosphoric acid fuel cell*, and *proton exchange membrane (PEM) fuel cell* (13.2)

dispersion force a weak intermolecular force of attraction that is present between all molecules due to temporary dipoles; also known as *London force* or *London (dispersion) force* (2.2)

disproportionation reaction reaction in which some atoms of an element undergo oxidation and other atoms of the same element undergo reduction (12.2)

dissociate to break apart to form separate ions (as when ionic substances such as NaOH break apart); the process is not called ionization because the substances are already made of ions (6.1)

dissociation equation chemical equation representing the process that occurs when compounds dissociate (5.1)

double bond covalent bond that consists of two bonding pairs; two atoms sharing four electrons (1.1)

double-replacement reaction chemical reaction in which cations of two ionic compounds exchange places, resulting in the formation of two new compounds (Unit 4 Prep., 7.1)

dry cell voltaic cell with the electrolyte contained in a thick paste (13.2)

dynamic equilibrium describes the state of an equilibrium system when it is changing at the molecular level, but its macroscopic properties remain constant (16.1)

E

efficiency ratio of useful energy produced (energy output) to energy used in its production (energy input), expressed as a percentage (10.2)

elastic collision collision in which molecules exchange kinetic energy with one another but the total kinetic energy of all of the molecules remains constant (3.1)

electric current net movement of electrical charge; often the flow of electrons in a circuit (13.1)

electrical potential difference difference in the electrical potential energy of a unit charge between two points; for an electrochemical cell, between the anode and the cathode, measured in volts (V) (13.1)

electrochemical cell a system that incorporates a redox reaction to produce or use electrical energy; examples are *voltaic cell*, *electrolytic cell* (13.1)

electrochemistry study of the processes involved in converting chemical energy to electrical energy, and in converting electrical energy to chemical energy (13.1)

electrode the part of an electrochemical cell that conducts the electric current between the cell and the surroundings (13.1)

electrolysis the nonspontaneous lysing (splitting) of a substance, often to its component elements, by supplying electrical energy (13.3)

electrolyte solute that conducts an electric current in an aqueous solution (5.1, 13.1)

electrolytic cell electrochemical cell that uses an external source of electric energy to drive a non-spontaneous chemical (redox) reaction; converts electrical energy into chemical energy (13.3)

electron dot diagram model that consists of the chemical symbol for the element with dots representing the valence electrons (electrons in the lower, filled energy levels are not shown in these diagrams) (Unit 1 Prep., 1.1)

electron pair two electrons in the same energy level of an atom that interact in a unique way that allows them to be closer together, rather than repelling each other as two negative electrons would normally do; are less likely to participate in bond formation than are the unpaired electrons (1.1)

electronegativity (EN) relative measure of an atom's ability to attract the shared electrons in a chemical bond (1.2)

electronegativity difference difference in electronegativities between the atoms in a bond; symbol is *EN* (1.2)

elimination reaction reaction in which atoms are removed from an organic molecule to form a double bond (15.1)

enamel shiny, hard, and very unreactive type of glass that can be melted onto a metal surface (13.2)

end point the point in a titration at which the indicator changes colour (Unit 8 Prep., 8.3)

endothermic refers to a process which requires energy; occurs with an absorption of heat from the surroundings and therefore an increase in the enthalpy of the system (5.1)

endothermic reaction reaction that results in a net absorption of energy (9.1)

energy input energy used (10.2)

energy output energy produced (10.2)

enthalpy, *H* thermodynamic quantity that is a property of a system and is closely related to the energy of the system (9.1)

enthalpy change, ΔH the difference between the potential energy of the reactants and the potential energy of the products in a process; under conditions of constant pressure, is equal to the heat, *Q*, absorbed or released by a system (9.1)

enthalpy of fusion amount of energy required to melt 1 mole of a solid substance; also known as *heat of fusion* (2.3)

enthalpy of reaction, $\Delta_r H$ the difference between the enthalpies of the products and the enthalpies of the reactants (9.1)

enthalpy of vaporization amount of energy required for one mole of a liquid to become a gas; also known as *heat of vaporization* (2.3)

enzyme biological catalyst (usually a protein) (11.2)

equilibrium in chemistry, when a process and the reverse of the process take place at the same rate in a closed system; e.g., equal rates of dissolving and crystallization (5.2, 16.1)

equilibrium constant, K_c ratio of the concentrations of the products divided by the concentrations of the reactants, with all concentration terms raised to the power of the coefficients in the chemical equation and the terms multiplied; see also *equilibrium law expression* (16.2)

equilibrium law expression expression to determine the equilibrium constant, represented as

$$K_c = \frac{[C]^c[D]^d}{[A]^a[B]^b}, \text{ where } [A], [B], [C], \text{ and } [D]$$

represent the concentrations of the reactants and products after the reaction has reached equilibrium and the concentrations no longer change, and the exponents a, b, c, and d are the stoichiometric coefficients from the complete balanced equation: $aA + bB \rightleftharpoons cC + dD$ (16.2)

equivalence point the point in a titration at which the number of moles of the unknown solution is stoichiometrically equal to the number of moles of the standard solution, (8.3, Unit 8 Prep.)

ester hydrocarbon derivative that contains the functional group RCOOR′, the symbol R representing any hydrocarbon or just a hydrogen atom, and R′ representing a hydrocarbon which cannot simply be a hydrogen atom (14.3)

esterification reaction reaction of a carboxylic acid with an alcohol to form an ester and water; a specific type of condensation reaction (15.1)

excess reactant reactant that remains after a chemical reaction is over (8.1)

exothermic refers to a process which releases energy; occurs with a release of heat to the surroundings and therefore a decrease in the enthalpy of the system (5.1)

exothermic reaction reaction that results in a net release of energy (9.1)

experimental yield quantity of product actually obtained from a reaction; also known as *actual yield* (8.2)

external circuit in a voltaic cell, an electric circuit outside the reaction vessel (13.1)

extraction process by which a pure compound is obtained from a mixture; e.g., a metal from an ore (13.4)

Faraday's law law stating that the amount of a substance produced or consumed in an electrolysis reaction is directly proportional to the quantity of charge that flows through the circuit (13.4)

fatty acid long-chain carboxylic acid (15.1)

first law of thermodynamics law stating that energy can be converted from one form to another, but cannot be created or destroyed; can be represented as $E_{system} = -E_{surroundings}$ (9.1)

flame test qualitative procedure for identifying the presence of metal ions, by which a granule of a compound or a drop of its solution is placed in a flame to observe a characteristic colour (7.1)

formation reaction class of chemical reaction in which two or more reactants combine to form one new product (Unit 4 Prep., 7.1)

formula unit smallest ratio of ions in a crystal such as NaCl (2.1)

fraction refers to each group of hydrocarbon compounds of petroleum, with its own range of boiling points and relative densities (14.4)

fractional distillation process that uses the specific boiling points of substances to separate (refine) a mixture into its components; used in oil refineries to separate petroleum into its hydrocarbon components (14.4)

fuel cell battery that produces electric energy while chemical reactants are supplied continuously from an external source (13.2)

fullerene class of spherical allotropes of carbon (2.1)

functional group special arrangement of atoms that is mainly responsible for the chemical and physical behaviour of a molecule (14.3)

galvanizing process in which iron is covered with a protective layer of zinc (13.2)

gas stoichiometry determination of the quantity of one reactant or product if the quantity of another reactant or product is known; uses volumes, temperatures, and pressures of gaseous reactants or products to determine the quantities of other reactants or products (7.2)

global warming gradual increase in the average global temperature due to the increase in greenhouse gases in the atmosphere (10.3)

glycogen natural polymer of glucose molecules that serves as the energy storage unit in animals (15.2)

gravimetric stoichiometry stoichiometric analysis involving mass (7.2)

greenhouse gas gas that traps heat in Earth's atmosphere and prevents the heat from escaping into outer space (10.3)

group in chemistry, refers to the columns in a periodic table (Unit 1 Prep.)

half-reaction reaction equation that explicitly shows the electrons involved in either the oxidation half or the reduction half of a redox reaction (12.2)

halogen any element in Group 17 of the periodic table; e.g., fluorine (F), chlorine (Cl), bromine (Br), iodine (I) (Unit 1 Prep., 14.3)

heat capacity, C the amount of energy required to raise the temperature of a substance by one degree Celsius; measured in J/°C or kJ/°C; compare *specific heat capacity* (9.2)

heat content of a fuel, is the amount of energy released per kilogram of a fuel (10.3)

Hess's law law stating that the enthalpy change of a physical or chemical process depends only on the initial and final states; the enthalpy change of the overall process is the sum of the enthalpy changes of its individual steps (10.1)

heterogeneous equilibrium equilibrium in which the reactants and products in the chemical system are not all in the same phase (16.1)

heterogeneous mixture mixture in which the composition of components varies throughout the mixture; compare *homogeneous mixture* (5.1)

homogeneous equilibrium equilibrium in which all reactants and products in the chemical system are in the same phase (16.1)

homogeneous mixture mixture with a uniform composition of solvent and one or more solutes; examples are tap water, filtered air, and iced tea; also known as a solution; compare *heterogeneous mixture* (5.1)

homologous series series of molecules in which each member differs from the next by an additional specific structural unit such as —CH_2— (14.2)

hydrocarbon compound that contains only carbon atoms and hydrogen atoms (14.2)

hydrocarbon combustion see *complete hydrocarbon combustion*

hydrocracking process by which hydrogen gas is added during the cracking process to transform long-chain alkanes into shorter-chain alkanes, resulting in low-grade gasolines and heating oils, which are usually upgraded by reforming (14.4)

hydrogen bond electrostatic attraction between the exposed proton (or nucleus) of a hydrogen atom that is covalently bonded to a highly electronegative atom and the partial negative charge of a highly electronegative atom (usually oxygen or nitrogen) on an adjacent molecule; considered a special type of dipole-dipole attraction (2.2)

hydronium ion a proton covalently bonded to a water molecule; written as $H_3O^+(aq)$ (6.1)

hydroxide ion name of the ion containing hydrogen and oxygen (OH^-) (14.3)

hydroxyl group name of the functional group for alcohols (—OH) (14.3)

ideal gas hypothetical gas with particles that have mass but no volume or attractive forces between them (3.1)

ideal gas law equation that expresses the relationship among volume, pressure, temperature, and amount (moles) of an ideal gas: $PV = nRT$ (4.2)

inert electrode electrode made from a material that is neither a reactant nor a product of the cell reaction; can carry a current and provide a surface on which redox reactions can occur (13.1)

inorganic compound compound that does not contain carbon, or that contains carbon but with no carbon-carbon or carbon-hydrogen bonds; e.g., carbonates ($CO_3{}^{2-}$), cyanides ($CN-$), carbides (C^{2-}), and oxides of carbon (CO_2, CO) (14.1)

insoluble does not dissolve in a given solvent to any great extent; also known as *slightly soluble* (5.1)

intermolecular forces forces that are exerted *between* molecules to influence the physical properties of compounds (2.2)

intramolecular forces forces that are exerted within a molecule or polyatomic ion (2.2)

ion product constant for water, K_w equilibrium constant for water; equals 1×10^{-14} at 25 °C (17.3)

ionic bond electrostatic attraction between oppositely charged ions (Unit 1 Prep., 1.1)

ionic compound positively and negatively charged ions combined in a ratio that results in a net zero charge; compare *molecular compound* (Unit 3 Prep.)

ionic equation equation showing all the high-solubility ionic compounds dissociated into ions; is a more accurate representation of a reaction in solution than is a complete balanced equation; also known as *total ionic equation* (7.1, 12.1)

ionize the process in which a molecular substance, often an acid, dissolves in water and separates into ions (6.1)

isolated system system which allows neither matter nor energy to enter or leave; compare *closed system* and *open system* (9.1)

isomer one of two or more compounds with the same molecular formula but different properties as a result of different arrangements of atoms; see also *structural isomer* (14.3)

isotope atoms of the same element that have different numbers of neutrons (Unit 1 Prep.)

kinetic energy (E_k) the energy of motion (in chemistry, this usually means the energy of the motion of particles or thermal energy); compare *potential energy* (9.1)

L

law of chemical equilibrium law stating that in a chemical system at equilibrium, there is a constant ratio between the concentrations of the products and the concentrations of the reactants (16.2)

law of combining volumes law stating that when gases react, the volumes of the gaseous reactants and products, measured at the same conditions of temperature and pressure, are always in whole-number ratios (4.1)

law of conservation of energy law stating that the total quantity of energy in the universe is constant, expressed as $\Delta E_{universe} = 0$ (9.1)

law of conservation of mass law stating that matter can be neither created nor destroyed; in any chemical eaction, the mass of the products is always equal to the mass of the reactants (7.2)

Le Châtelier's principle a principle stating that if a system in a state of equilibrium is disturbed, it will undergo a change that shifts its equilibrium position in a direction that reduces the effect of the disturbance (16.2)

Lewis structure electron dot diagram of a molecule (1.1)

liming process by which limestone is added to a body of water to neutralize acid deposition and buffer water against a rapid change in pH (17.1)

limiting reactant reactant that is completely consumed during a chemical reaction, limiting the amount of product produced (8.1)

linear in a straight line (2.1)

London force a weak intermolecular force of attraction that is present between all molecules due to temporary dipoles; also known as *dispersion force* or *London (disperson) force* (2.2)

lone pair two electrons in an atom's valence energy level that are paired but are not involved in covalent bonding within a molecule (1.1)

macroscopic referring to properties of gases: can be directly observed using the senses or a measuring instrument (3.1)

main group elements belonging to Groups 1, 2, and 13 through 18; for most of the main group elements, the number of valence electrons can be predicted from position of the element on the periodic table (1.1)

mass number sum of the number of protons (Z) and neutrons (N) in an element, symbolized as A (A = Z + N) (Unit 1 Prep.)

mass percent fraction by mass expressed as a percentage; a concentration term expressed as the mass in grams of solute dissolved per 100 g of solution; also known as *percent m/m* and *percent by mass* (5.3)

Maxwell-Boltzmann distribution curve resulting from plotting the number of molecules against the speed of the molecules at a given temperature (11.1)

mechanical loss small amount of product lost during transfer from glassware to filter paper in the lab (8.2)

melting point temperature at which the change from a solid to a liquid occurs (2.3)

metallic bonding idealized type of bonding based on the attraction between metal ions and their delocalized valence electrons (those free to move from one atom to the next); all the atoms share all the valence electrons (1.1)

miscible describes substances that are able to mix completely with each other (3.1)

modified Arrhenius acid substance that reacts with water to produce $H_3O^+(aq)$ in aqueous solutions (6.1)

modified Arrhenius base substance that dissociates or reacts with water to produce $OH^-(aq)$ in aqueous solutions (6.1)

molar concentration unit of concentration expressed as the number of moles of solute dissolved in one litre of solution; also known as *amount concentration* and *molarity* (5.3)

molar enthalpy change in enthalpy when one mole of a substance undergoes a process (9.1)

molar enthalpy of combustion, $\Delta_c H$ enthalpy change associated with the complete combustion of one mole of a given substance (9.1)

molar volume amount of space occupied by 1 mole of a substance; equal to 22.4 L for an ideal gas at standard temperature and pressure (STP) and 24.8 L for an ideal gas at SATP (4.1)

molarity unit of concentration expressed as the number of moles of solute dissolved in one litre of solution; also known as *amount concentration* and *molar concentration* (5.3)

mole ratio ratio between the molar amounts of any two elements, ions, or compounds in a complete balanced equation (7.2)

molecular compound consists mainly of non-metal elements covalently bonded to one another; compare *ionic compounds* (1.1, Unit 3 Prep.)

molecular formula formula that gives the actual number of atoms of each element in a molecule (1.1)

monomer a small molecule, linked covalently to others of the same or similar type to form a polymer (15.2)

monoprotic acid acid having only a single hydrogen atom per molecule that ionizes in water; e.g., hydrochloric acid (6.2)

monoprotic base base that dissociates or reacts with water in one step, forming hydroxide ions (6.2)

net ionic equation representation of a chemical reaction in a solution that shows only the ions involved in the chemical change; the spectator ions are not included (7.1, 12.1)

network solid solid in which dozens or even millions of atoms, often of the same element, are bonded covalently into continuous two-dimensional or three-dimensional arrays (2.1)

neutral solution solution that is neither acidic nor basic (6.1)

neutralization reaction reaction between an acid and a base resulting in the formation of water and a salt (6.2)

noble gas non-metal element in Group 18 of the period table; is almost completely unreactive (Unit 1 Prep.)

nomenclature naming of chemical compounds (1.1)

non-electrolyte solute that does not conduct an electric current in an aqueous solution (5.1)

non-polar covalent bonds bonds in which electrons are equally shared between two atoms whose difference in electronegativity is zero (1.2)

non-renewable energy source resource (such as coal, oil, or natural gas) that is gone once used up, because it takes millions of years to form and is used at a much faster rate than it can be replenished (10.3)

nucleotide monomer of deoxyribonucleic acid (DNA) (15.2)

octet rule rule stating that when bonds form, the atoms gain, lose, or share electrons in such a way as to create a filled outer energy level containing eight electrons (an octet); the octet rule should not be considered a law of nature as bonds can form that result in one atom having more than eight electrons in its outer energy level (Unit 1 Prep., 1.1)

open system system which both matter and energy are allowed to enter or leave; compare *closed system* and *isolated system* (9.1)

organic compound carbon-containing compound in which carbon atoms are nearly always bonded to each other and to hydrogen, and often to other elements (usually oxygen, nitrogen, sulfur, and phosphorus) (14.1)

overpotential increase in potential difference beyond the calculated value for the cell potential (13.3)

oxidation loss of electrons, accompanied by an increase in oxidation number (12.1, 12.3)

oxidation number an assigned number that is the same as the charge that an atom in a compound would have if electrons were not shared but were completely held by the atom with the greatest electronegativity; does not represent actual charges on atoms, but is a way to describe some properties of atoms in a compound (12.3)

oxidation number method method for balancing redox reactions without using half reactions (12.3)

oxidation-reduction reaction also known as redox reaction; reaction in which electrons are gained by one atom or ion (oxidizing agent) and lost by another (reducing agent); reaction in which the oxidation numbers of at least two atoms change (12.1)

oxidizing agent substance that accepts electrons in a reaction and therefore oxidizes another reactant (12.1)

P

parent alkane hydrocarbon chain that is the basis for a more complex hydrocarbon derivative molecule; used as the root for naming the complex molecule (14.3)

partial pressure the portion of the total pressure contributed by one gas in a mixture of gases (4.2)

parts per billion mass of a solute present in a billion units of mass of a the solution (5.3)

parts per million mass of a solute present in a million units of mass of a the solution (5.3)

percent by mass fraction by mass expressed as a percentage; concentration expressed as the mass in grams of solute dissolved per 100 g of solution; also known as *mass percent* and *percent m/m* (5.3)

percent ionization in a weak acid, is the fraction of molecules that are ionized compared with the initial concentration of the acid, expressed as a percentage (17.3)

percent *m/m* fraction by mass expressed as a percentage; concentration expressed as the mass in grams of solute dissolved per 100 g of solution; also known as *mass percent* and *percent by mass* (5.3)

percentage yield experimental (actual) yield of a reaction, expressed as a percent of the predicted (theoretical) yield (8.2)

period in chemistry, refers to the rows in a period table (Unit 1 Prep.)

petrochemical product derived from petroleum; a basic hydrocarbon, such as ethene and propene, that is converted into plastics and other synthetic materials (14.4, 15.2)

petroleum fossilized remains of plankton transformed geologically under extremely high temperatures and pressures into a complex mixture of solid, liquid, and gaseous hydrocarbons (14.4)

pH the negative logarithm (base-ten) of the molar concentration of hydronium ions in a solution (6.3)

pH titration curve plot of the pH of the reaction mixture versus the volume of titrant added (8.3)

phenyl group term used for a benzene ring that forms a substituent group on a hydrocarbon chain (14.2)

phosphoric acid fuel cell fuel cell in which the oxidation and reduction half-reactions are the same as those for the proton exchange membrane (PEM) fuel cell, but the electrolyte is phosphoric acid; runs at slightly higher temperatures than the PEM fuel cells but can achieve slightly higher efficiencies and can tolerate fuels that are not as pure as those for the PEM fuel cells; compare *alkaline fuel cell* and *direct methanol fuel cell (DMFC)* (13.2)

pig iron term for the crude molten iron that is poured into long narrow moulds to form bars, then transported to other locations, where most of it is converted into steel, cast iron, wrought iron (12.2)

plastics synthetic polymers that can be heated and moulded into specific shapes and forms (15.2)

pOH the negative logarithm (base-ten) of the molar concentration of hydroxide ions in a solution (6.3)

point mass a mass that takes up no space (has no volume) (3.1)

polar covalent bond bond in which electrons are unequally shared between two atoms, thus forming a bond dipole (1.2)

polyamide condensation polymer that contains amide linkages; e.g., nylon (15.2)

polyatomic ion ion consisting of several atoms that are covalently bonded together and are charged (Unit 1 Prep., 1.1)

polyester condensation polymer that contains ester linkages (15.2)

polymer a large long-chain molecule with repeating units of small molecules called monomers (15.2)

polyprotic acid acid that has two or more hydrogen atoms that ionize can in water; e.g., sulfuric acid (6.2, 17.2)

polyprotic base a base that reacts with water in two or more steps, forming hydroxide ions; e.g., calcium hydroxide (6.2, 17.2)

polysaccharide polymer made up of sugar monomers (saccharides) (15.2)

porous barrier barrier in electrochemical cell that prevents ions in one half cell from coming into direct contact with the electrode in the other half cell, but allows ions to gradually diffuse through (13.1)

potential energy (E_p) energy that is stored (in chemistry, usually energy stored in chemical bonds); compare *kinetic energy* (9.1)

potential energy diagram in thermochemistry, graphical representation of the change in enthalpy during a chemical or physical change (9.1)

precipitate solid product of a precipitation reaction (7.1)

precipitation reaction a double replacement reaction in solution that produces a solid product (7.1)

predicted yield maximum quantity of product that could be produced in a chemical reaction; also known as *theoretical yield* (8.2)

prefix the part of the IUPAC name of any organic compound that gives the positions and names of any branches from the main chain (14.2)

primary battery disposable battery (13.1)

protein natural polymer composed of monomers called amino acids; carries out many important functions in the body (15.2)

proton exchange membrane (PEM) fuel cell type of fuel cell in which the electrolyte is a membrane that will allow protons to pass through but not electrons; first used by NASA on space flights for their Gemini program; also under development by Ballard Power Systems for use in cars and buses; compare *alkaline fuel cell, direct methanol fuel cell (DMFC),* and *phosphoric acid fuel cell* (13.2)

pure substance any material with a definite chemical composition; cannot be separated into other substances by any mechanical process (5.1)

pyramidal molecular shape that results when a central atom has one lone pair and three single bonds (2.1)

qualitative analysis determination by experiment *whether* a certain substance is present in a sample; compare *quantitative analysis* (7.1)

quantitative analysis determination by experiment *how much of* a certain substance is present in a sample; compare *qualititative analysis* (7.1)

R

reaction rate speed at which a reaction occurs, or the change in the amount of reactants consumed or products generated over time (11.1)

redox reaction also known as oxidation-reduction reaction; reaction in which electrons are gained by one atom or ion (oxidizing agent) and lost by another (reducing agent); reaction in which the oxidation numbers of at least two atoms change (12.1, 12.3)

reducing agent substance that donates electrons in a redox reaction and therefore reduces another reactant (12.1)

reduction gain of electrons, accompanied by a decrease in oxidation number (12.1, 12.3)

refining in the making of steel, the removal of impurities from pig iron involving oxidation reactions; (12.2, 13.4)

reforming technique that uses heat, pressure, and catalysts to convert naturally occurring straight-chain hydrocarbons into a more useful form (such as branched-chain alkanes, cyclic alkanes, and aromatics); usually involves the removal of hydrogen atoms (14.4)

renewable energy source resource that exists in infinite supply (such as solar energy) or that can be replaced within a reasonable amount of time (such as wood) (10.3)

root the part of the IUPAC name of any organic compound that denotes the number of carbon atoms in the longest continuous chain of carbon atoms for alkanes or the longest continuous chain that includes the functional group (14.2)

rule of a thousand rule stating if the quotient of the initial concentration of the acid divided by K_a is greater than 1000, then the equilibrium concentration of the acid can be considered to be the same as the initial concentration (17.3)

S

saccharide sugar monomer (15.2)

sacrificial anode anode made of material that is more easily oxidized than iron; is destroyed to protect the iron (13.2)

salt ionic compound formed in a neutralization reaction; composed of an anion from an acid and a cation from a base (6.2)

salt bridge laboratory device used to provide an electrical connection between the oxidation and reduction half-cells of an electrochemical cell (13.1)

saturated hydrocarbon hydrocarbon that contains only single bonds, and no double or triple bonds, i.e., each carbon atom is bonded to the maximum possible number of atoms (either hydrogen or carbon atoms) (14.2)

saturated solution containing the maximum amount of solute that can be dissolved in a solvent at a given temperature; compare *unsaturated* (5.2)

secondary battery rechargeable battery (13.2)

side group atom or group of atoms substituted in place of a hydrogen atom on the parent chain of a hydrocarbon; also known as *substituent group* (14.2)

silane silicon analogue to the simplest type of organic carbon compounds; will bond to organic carbon molecules and bond readily to inorganic substances, such as glass and metals (14.1)

simple calorimeter calorimeter made with basic components: water, a thermometer, nested polystyrene cups, and a plastic lid (9.2)

single bond bond that consists of one electron pair (1.1)

single-replacement reaction a class of chemical reaction in which one element in a compound is replaced by another element (Unit 4 Prep., 7.1)

slightly soluble does not dissolve in a given solvent to any great extent; also known as *insoluble* (5.1)

smelting process involving the chemical reduction of iron ions to iron atoms with carbon as the reducing agent (12.2)

solubility the mass of solute that will dissolve in a given quantity of solvent at a specific temperature (5.2)

soluble able to be dissolved in a given solvent (5.1)

solute substance that is dissolved in a solvent (5.1)

solution mixture with a uniform composition of solvent and one or more solutes; also known as a *homogeneous mixture* (5.1)

solution stoichiometry determination of the quantity of one reactant or product in solution when the quantity (concentration) of another reactant or product is known (7.2)

solvation the dissolving process for aqueous solutions; also known as *hydration* (5.1)

solvent the substance present in the largest quantity in a solution (5.1)

solvent extraction in purifying heavy oil fractions of crude oil, mixing solvents with the heavy oil fractions to dissolve the impurities such as aromatic hydrocarbon and waxes and remove them with the solvent (14.4)

specific heat capacity, c the amount of energy required to raise the temperature of one gram of a substance by one degree Celsius; measured in J/g · °C (Unit 5 Prep.)

spectator ion dissolved ion that is in the reactants before a reaction begins and in the products once a reaction is complete (7.1) (12.1)

stainless steel an alloy of iron that contains at least 10% chromium, by mass, in addition to small quantities of carbon and occasionally metals such as nickel; is much more resistant to corrosion than is pure iron (13.2)

standard ambient temperature and pressure (SATP) condition defined as temperature of 25 °C (298.15 K) and pressure of 100 kPa (4.1)

standard atmospheric pressure atmospheric pressure in dry air at 0 °C at sea level (3.2)

standard cell potential, $E°_{cell}$ potential difference between the electrodes of a voltaic cell when the concentrations of the salt solutions are 1.0 mol/L, the atmospheric pressure is 1.01325×10^5 Pa (1.0 atm), the electrodes are pure metals, the temperature is 25 °C, and there is no current flowing in the cell (13.1)

standard enthalpy of reaction, $\Delta_r H°$ change in enthalpy of a system during a chemical reaction that occurs at standard atmospheric temperature and pressure (SATP: 25 °C and 100 kPa) and all solutions have a concentration of 1.0 mol/L; the degree symbol denotes SATP (9.1)

standard molar enthalpy of formation, $\Delta_f H°$ change in enthalpy when one mole of a compound is formed directly from its elements in their most stable state at standard ambient temperature and pressure (SATP: 25 °C and 100 kPa) and all solutions have a 1.0 mol/L concentration (10.1)

standard reduction potential potential difference between a given half cell and a hydrogen half cell, written as a reduction reaction, when all solutes are present in 1 mol/L concentrations at SATP and hydrogen gas is at 1 atm of pressure (13.1)

standard solution solution of known concentration (5.4)

standard temperature and pressure (STP) condition defined as temperature of 0 °C (273.15 K) and pressure of 1 atm (101.325 kPa) (4.1)

standardizing establishing the concentration of a solution by analyzing its reaction with a substance of known concentration (a standard) (8.3)

starch natural polymer of glucose molecules that serves as an energy storage unit in plants (15.2)

steam cracking process by which steam at more than 800 °C and pressures slightly above atmospheric normal are combined with hydrocarbon "feedstocks" to produce petrochemicals (14.4)

stoichiometric amount reactants present in amounts that correspond exactly to the mole ratio determined by a balanced equation for the reaction (8.1)

stoichiometric coefficient number placed in front of the formula of a reactant or product in a balanced chemical equation to indicate the mole ratio of the reactants and products in a reaction (8.1)

stoichiometry study of the relative quantities of reactants and products in chemical reactions; provides a method of predicting the quantity of a reactant or product in a chemical reaction based on the quantity of another reactant or product in the same reaction (7.2)

strong acid acid that ionizes nearly 100% when in solution (6.2, 17.2)

strong base base that dissociates nearly 100% when in solution (6.2, 17.2)

structural formula diagrammatic representation that shows every individual atom in the molecule, their relative placement, and the bonds between them (1.2, 14.2)

structural isomer compound that has the same molecular formula as another but a different structural formula (14.3)

substituent group atom or group of atoms substituted in place of a hydrogen atom on the parent chain of an organic compound; commonly referred to as a *side group* (14.2)

substitution reaction reaction in which a hydrogen atom or functional group is replaced by a different atom or functional group (15.1)

substrate reactant that binds to the active site in an enzyme-catalyzed reaction (11.2)

suffix the part of the IUPAC name of any organic compound that indicates the series to which the molecule belongs; sometimes includes position number of functional group (14.2)

supersaturated containing more dissolved solute than can be dissolved in a solvent at a given temperature (5.2)

surroundings the rest of the universe outside the system (9.1)

system any specific part of the universe that is of interest (9.1)

tetrahedral contains four atoms bonded to a central atom with no lone pairs; the atoms are positioned at the four corners of an imaginary tetrahedron, and the angles between the bonds are approximately 109.5° (2.1)

theoretical yield maximum quantity of product in a chemical reaction; also known as *predicted yield* (8.2)

thermal equilibrium state achieved when all components in a system have the same final temperature (9.2)

thermal stability ability of a substance to resist decomposition when heated (10.1)

thermally excited an electron in an atom that has been raised to a higher energy level by absorbing thermal energy (7.1)

thermochemical equation complete balanced equation that indicates the amount of heat that is absorbed or released (i.e., the change in enthalpy) by a reaction that the equation represents (9.1)

thermochemistry study of the energy changes involved in physical and chemical processes (9.1)

thermodynamics study of energy transfer and energy changes (9.1)

titrant the solution that is being added and its volume measured during a titration (8.3)

titration laboratory process to determine the concentration of a compound in solution by quantitatively observing its reaction with a solution of known concentration (8.3)

titration curve graph of the pH of the titration mixture against the volume of added acid (or base) (17.4)

total ionic equation equation showing all the high-solubility ionic compounds dissociated into ions; also known as *ionic equation* (12.1)

trans fatty acid also known as trans fat; type of unsaturated fat in which the double bonds have a trans configuration; not produced by living systems (15.1)

transition state the condition of the reactants and products of a reaction in the activated complex (11.1)

trigonal planar molecular shape in which three bonding groups surround a central atom; the three bonded atoms are all in the same plane as the central atom, at the corners of an invisible triangle (2.1)

triple bond covalent bond that consists of three bonding pairs of electrons (1.1)

unit cell smallest set of ions in a crystal for which the pattern is repeated over and over (2.1)

universal gas constant (R) proportionality constant that relates the pressure on and the volume of an ideal gas to its amount and temperature; $R = 8.314\ 47$ kPa·L/mol·K (4.2)

universal indicator liquid or paper which contains a number of pH indicators; indicators have different characteristic colours over the range of pH values, therefore providing a more accurate pH estimate than does one indicator alone (6.3)

unpaired electron in an atom, an electron that is not paired with another (1.1)

unsaturated hydrocarbon hydrocarbon that contains carbon-carbon double or triple bonds, whose carbon atoms can potentially bond to additional atoms (14.2)

unsaturated solution containing less dissolved solute than can be dissolved in a solvent at a given temperature; (5.2)

valence electron electron that occupies the outermost energy level of an atom (Unit 1 Prep., 1.1)

Valence-Shell Electron-Pair Repulsion theory also known as VSEPR theory: the theory allowing the prediction of the three-dimensional shape of a molecule around a central atom (2.1)

viscosity measure of the resistance of a liquid to flow (3.1)

voltage unit of measurement for electrical potential difference (13.1)

voltaic cell electrochemical cell that uses redox reactions, to transform chemical potential energy into electrical energy (13.1)

VSEPR theory Valence-Shell Electron-Pair Repulsion theory: the theory allowing the prediction of the three-dimensional shape of a molecule around a central atom (2.1)

weak acid acid that ionizes only slightly when in solution (6.2, 17.2)

weak base base that dissociates only slightly when in solution (6.2, 17.2)

Index

Boldface numbers refer to boldface terms in the text.
Italic numbers refers to italicized terms in the text.
CF refers to Chemistry File.
LL refers to Launch Lab and is boldface.
TL refers to Thought Lab and is boldface.
SP refers to Sample Problems.
inv refers to investigations and is boldface.
tab refers to tables and is boldface.
dia refers to diagrams.
fig refers to figures.

Photo Credits

254 (center right), ©Ian Crysler; 260–261 (spread), ©Roy Ooms/Masterfile; 261 (bottom left), ©GRANT HEILMAN PHOTOGRAPHY, INC; 262 (center left), ©Dave Buston/Getty Images; 263 (top left), McGraw-Hill Higher Education/ Stephen Frisch, photographer; 263 (bottom right), ©Daryl Benson/Masterfile; 266 (top center), ©P. Freytag/zefa/Corbis; 270 (bottom right), ©Ian Crysler; 275 (bottom right), ©NASA/Roger Ressmeyer/CORBIS; 279 (bottom left), ©John Edwards/ Getty Images; 281 (top left), ©Ian Crysler; 282 (bottom right), ©Ian Crysler; 287 (bottom right), ©Tek Image/SPL/PUBLIPHOTO; 289 (bottom right), ©E.R. Degginger/Colour-Pic, Inc.; 291 (top center), ©Roy Ooms/Masterfile; 292 (top center), ©Roy Ooms/ Masterfile; 293 (top left), ©1990 Robert Mathena, Fundamental Photographs, NYC; 294–295 (spread), ©AP/Wide World Photos; 295 (bottom left), ©NASA/JPL/ZUMA/Corbis; 297 (bottom right), ©Suzanne & Nick Geary/Getty Images; 304 (bottom left), ©Btand X/Fotosearch; 305 (center right), ©Ariel Skelley/CORBIS; 306 (bottom left), ©Tom Pantages; 313 (top right), ©Tom Pantages; 313 (bottom right), ©Ian Crysler; 320 (top right), ©Ottmar Bierwagen/Spectrum Photofile; 323 (top center), ©AP/Wide World Photos; 324 (top center), ©AP/Wide World Photos; 328 (top center), ©D. Hurst/Alamy; 329 (top right), ©PHOTOTAKE Inc./Alamy; 329 (center left), ©Michael Newman/PhotoEdit, Inc.; 330 (bottom right), ©Tim Davis/CORBIS; 330 (top center), ©D. Hurst/Alamy; 330 (bottom left), ©Rosal/SPL/PUBLIPHOTO; 331 (center right), ©Sygma/Corbis; 332–333 (spread), ©Sebastien Starr/Getty Images; 336 (top left), ©Paul A. Souders/CORBIS; 338–339 (spread), ©www.pamdoylephoto.com; 339 (bottom left), ©dino&tino; 341 (center right), ©Ian Crysler; 342 (center left), ©Angela Wyant/ Getty Images; 344 (bottom), ©Bettmann/CORBIS; 346 (top left), ©Royalty-Free/Corbis; 346 (center left), ©Michael Mahovlich/ Masterfile; 351 (center), ©Romilly Lockyer/Getty Images; 351 (center right), ©Gazimal/Iconica/Getty Images; 352 (bottom left), ©Stockdisc/Getty Images; 360 (top), ©Stephen Frisch Photography; 361 (top right), ©Mark Tomalty/Masterfile; 362 (center right), ©W. Perry Conway/CORBIS; 364 (bottom), ©Nathan Bilow/Getty Images; 364 (bottom), ©Justin Pumfrey/Getty Images; 365 (top center), ©www.pamdoylephoto.com; 366 (center right), ©Hemera Technologies/Alamy; 366 (top center), ©www.pamdoylephoto.com; 367 (bottom right), ©World Perspectives/Getty Images; 368–369 (spread), ©MC DONALD, JOE/ Animals Animals/Earth Scenes; 369 (bottom left), ©Gunter Marx/Alamy; 375 (bottom right), ©Ian Crysler; 378 (center left), ©Artbase Inc.; 378 (center right),   ©Thom Lang/CORBIS; 388 (bottom left), ©Stuart Dee; 388 (bottom center), ©R. W. Harwood, photographersdirect.com; 389 (top right), ©Stuart Dee; 393 (bottom), ©Mike Grandmaison; 394 (top), ©iStockphoto; 395 (center), ©CP/Edmonton Sun (Walter Tychnowicz); 396 (center left), ©Lester Lefkowitz/Corbis; 399 (top center), ©MC DONALD, JOE/Animals Animals/Earth Scenes; 400 (top center), ©MC DONALD, JOE/Animals Animals/ Earth Scenes; 402–403 (spread), ©John Foster/Masterfile; 403 (center left), ©Vicky Kasala/Getty Images; 404 (top center), ©SHAUN BEST/Reuters/Corbis; 411 (center), ©E.R. Degginger/Color-Pic, Inc.; 411 (center right), ©Image Source/Alamy; 411 (center), ©Stockbyte/Getty Images; 411 (center right), ©John Millar/Getty Images; 412 (bottom left), From Chemistry: The Molecular Nature of Matter and Change, ©2000, The McGraw-Hill Companies, Inc; 412 (bottom right), From Chemistry: The Molecular Nature of Matter and Change, ©2000, The McGraw-Hill Companies, Inc; 412 (bottom left), From Chemistry: The Molecular Nature of Matter and Change, ©2000, The McGraw-Hill Companies, Inc; 412 (bottom right), From Chemistry: The Molecular Nature of Matter and Change, ©2000, The McGraw-Hill Companies, Inc; 413 (bottom right), ©James L. Amos/ Corbis; 414 (top right), ©AC/Peter Arnold, Inc; 416 (center), ©Dan Dalton/Getty Images; 416 (center right), ©Rachel Epstein/ PhotoEdit; 416 (bottom left), ©David Young-Wolff/PhotoEdit; 419 (top center), ©John Foster/Masterfile; 420 (top center), ©John Foster/Masterfile; 423 (bottom), ©PAUL RAPSON/SCIENCE PHOTO LIBRARY; 424 (top center), ©Sebastien Starr/ Getty Images; 425 (center left), ©Stuart Dee; 426 (top center), ©Sebastien Starr/Getty Images; 428–429 (spread), ©Artbase Inc.; 432–433 (spread), ©Dale Wilson/Masterfile; 433 (top right), ©dino&tino; 433 (bottom left), ©Tanya Constantine/Getty Images; 433 (center left), ©Photo 24/Getty Images; 434 (top), ©Charles O'Rear/CORBIS; 434 (bottom left), ©Charles D. Winters/Photo Researchers, Inc; 436 (top left), ©1987 Richard Megna, Fundamental Photographs, NYC; 436 (top right), ©1987 Richard Megna, Fundamental Photographs, NYC; 441 (center), ©Brownie Harris/CORBIS; 443 (center), ©Ian Crysler; 452 (center left), ©Museum of London, UK,/The Bridgeman Art Library; 453 (bottom left), ©Adam Woolfitt/CORBIS; 453 (bottom right), ©Gabor Geissler/Getty Images; 455 (top center), ©Sylvia Cordaly Photo Library Ltd/Alamy; 467 (center right), ©Chris Howes/ Wild Places Photography/Alamy; 468 (center), ©Ian Crysler; 473 (top center), ©Dale Wilson/Maskerfile; 474 (top center), ©Dale Wilson/Masterfile; 476–477 (spread), ©ThinkStock/SuperStock; 477 (top right), ©Ian Crysler; 477 (center left), ©dino&tino; 478 (bottom center), ©Ian Crysler; 480 (top left), McGraw-Hill Higher education/Stephen Frisch, photographer; 491 (bottom center), ©StudiOhio; 492 (center left), McGraw-Hill Companies, Inc/Pat Watson, photographer; 493 (bottom right), ©Mike Blake/ Reuters/CORBIS; 494 (top center), ©NASA; 496 (bottom center), ©YOSHIKAZU TSUNO/Getty Images; 497 (bottom), ©AP images/Robert Nichols; 499 (bottom right), ©RC Hall Photography; 504 (center), ©Stephen Frisch; 511 (top right), ©Charles D.Winters/Photo Researchers, Inc.; 512 (top left), ©Myrleen Ferguson Cate/Photo Edit; 514 (top center), © Bettmann/ CORBIS; 519 (top right), ©Artbase Inc.; 519 (bottom center), ©Chris R. Sharp/Photo Resarchers, Inc.; 521 (top center), ©ThinkStock/SuperStock; 522 (top center), ©ThinkStock/SuperStock; 525 (bottom left), ©Lara Solt/Dallas Morning News/Corbis; 526 (top center), ©Artbase Inc.; 528 (top center), ©Artbase Inc.; 530–531 (spread), ©Alec Pytlowany/ Masterfile; 536–537 (spread), ©Wolfgang Kaehler/CORBIS; 537 (center left), ©George & Judy Manna/Photo Researchers, Inc; 538 (bottom), ©SPL/PUBLIPHOTO; 542 (center left), ©Tek Image/SPL/PUBLIPHOTO; 544 (center), ©Rob Melnychuk/Getty Images;

551 (bottom right), ©Felicia Martinez/ Photoedit, 556 (top left), ©Geostock/Getty Images; 557 (bottom right), ©Royalty-free/ CORBIS; 569 (center left), ©Fred Lyons/Cole Group/Getty Images; 571 (center right), ©Laurie Rubin/Masterfile; 575 (center), ©Mike Dobel/Masterfile; 576–577 (spread), ©Dan Lamont/Corbis; 577 (bottom right), ©Greg Smith/Corbis; 577 (center right), ©Dan Lamont/Corbis; 578 (center left), ©Courtesy of Rhine Ruhr Pty. Ltd; 583 (top center), ©Wolfgang Kaehler/CORBIS; 584 (top center), ©Wolfgang Kaehler/CORBIS; 588 (center lef), ©Noel Hendrickson/Masterfile; 588 (center right), ©Yoav Levy/ PhototakeUSA.com; 592 (top right), ©SuperStock, Inc./SuperStock; 595 (top right), ©Royalty-Free/Corbis; 595 (bottom left), ©Marie-Louise Avery/Alamy; 604 (top center), ©Neal Preston/CORBIS; 604 (top right), ©Jacqui Hurst/CORBIS; 609 (top right), ©NOVA Chemicals Corporation; 610 (bottom), ©Grambo/Firstlight; 611 (bottom center), ©Dennis Kunkel/Phototake; 618 (top left), ©Thomas Fricke; 619 (bottom right), ©Lester Lefkowitz/CORBIS; 620 (top center), Alec Pytlowany/Masterfile; 622 (top center), Alec Pytlowany/Masterfile; 624–625 (spread), ©Courtesy of Agrium Inc.; 632–633 (top center), ©Zoom Agence / Getty Images; 637 (bottom right), ©Will Crocker/Getty Images; 639 (bottom right), ©Riffarth, H; 645 (top right), ©Stefano Bianchetti/CORBIS; 645 (bottom left), ©B. Edmaier/SPL/PUBLIPHOTO; 647 (center right), ©SPL/PUBLIPHOTO; 650 (center left), ©J. King-Holmes/SPL/PUBLIPHOTO; 653 (bottom left), ©Ian Crysler; 653 (bottom center), ©Ian Crysler; 653 (bottom right), ©Ian Crysler; 654 (bottom left), ©Ian Crysler; 654 (bottom center), ©Ian Crysler; 654 (bottom right), ©Ian Crysler; 655 (top right), ©Tom Brownold; 663 (top left), ©Ian Crysler; 663(top left), ©Ian Crysler; 666 (center), ©Jeff Hunter/Getty Images; 667 (center right), ©Fritz Polking/Peter Arnold, Inc.; 668 (top right), ©Digital Vision/Getty Images; 670 (bottom left), ©Emilio Segre Visual/ SPL/PUBLIPHOTO; 670 (center), ©Corbis; 670 (center right), ©Archiv zur Geschichte der Max-Planck-Gesellschaft, Berlin-Dahlem; 672 (top left), ©LANDMANN PATRICK/CORBIS SYGMA; 672 (bottom), ©Jeff Greenberg/Photoedit; 672 (bottom right), ©Rob Brimson/Getty Images; 672 (bottom left), ©Gary Garovac/Masterfile; 672 (bottom), ©Westend61/Alamy; 675 (top center), ©Zoom Agence/Getty Images; 676 (top center), ©Zoom Agence/Getty Images; 678–679 (top center), ©Royalty-Free/CORBIS; 681 (top left), ©CP PHOTO/Jeff McIntosh; 682 (top left), ©Visuals Unlimited; 682 (top center), ©Wally Eberhart/Visuals Unlimited; 682 (top right), ©Andrew J. Martinez/Photo Researchers, Inc; 683 (center right), ©MARK EDWARDS/Peter Arnold, Inc.; 684 (center), ©SCIENCE PHOTO LIBRARY; 684 (center right), ©Edgar Fahs Smith Collection, University of Pennsylvania Library; 691 (bottom right), ©Felicia Martinez/PhotoEdit, Inc; 700 (bottom right), ©dino&tino; 713 (bottom right), ©Glenbow Museum; 714 (center), ©Dennis MacDonald/PhotoEdit Inc.; 716 (top center), ©Phillip & Karen Smith/Getty Images; 716 (bottom center), ©Jim Varney/Photo Researchers, Inc.; 719 (top center), ©Royalty-Free/CORBIS; 720 (top center), ©Royalty-Free/CORBIS; 723 (bottom right), ©Sherman Hines/Masterfile; 724 (top right), Courtesy of Agrium Inc.; 726 (top center), Courtesy of Agrium Inc.; 778 (center left), ©Troy and Mary Parlee/Alamy

Illustration Credits

57 (center left) Theresa Sakno, 76 (top, bottom) Theresa Sakno, 77 (top, bottom) Theresa Sakno, 100 (center) Jane Whitney, 101 (bottom right) Paulette Dennis, 102 (center, bottom left) Jane Whitney, 103 (top right) Jun Park, 106 (bottom right) Jun Park, 108 (top right) Theresa Sakno, 112 (center left, center right) Jun Park, 116 (center left) Colin McGill, 131 (center left, center right) Dave Mckay, 131 (center right) Dave McKay, 143 (bottom) Jun Park, 146 (center right, bottom right) Rob Schuster, 150 (bottom right) Jun Park, 157 (top left) Jane Whitney, 157 (top right) Dave Mckay, 169 (center) Theresa Sakno, 179 (top) Theresa Sakno, 180 (center right) Theresa Sakno, 202 (center left) Theresa Sakno, 284 (top) Jun Park, 296 (bottom left) Dave Mckay, 317 (center left) Theresa Sakno, 320 (bottom) Jun Park, 340 (center) Michael Rothman, 352 (top left) Theresa Sakno, 371 (bottom) John Harvey, 391 (center) Dave McKay, 395 (bottom) Dave McKay, 414 (bottom left) Dave McKay, 465 (top right) Dave McKay, 477 (bottom left) Jane Whitney, 479 (bottom) Theresa Sakno, 480 (bottom) Theresa Sakno, 481 (center) Theresa Sakno, 484 (top)Theresa Sakno, 485 (bottom) Theresa Sakno, 494 (bottom) Dave McKay, 495 (bottom) Dave McKay, 538 (bottom right) Theresa Sakno, 592 (center left, bottom) Dave Mazierski, 634 (center) Jane Whitney, 635 (center right) John Harvey, 680 (center) Michael Rothman, 778 (center right) Theresa Sakno